GIS and Environmental Modeling:
Progress and Research Issues

GIS and Environmental Modeling:
Progress and Research Issues

Michael F. Goodchild, Louis T. Steyaert, Bradley O. Parks,
Carol Johnston, David Maidment, Michael Crane, and Sandi Glendinning, Editors

GIS World Books
155 East Boardwalk Drive, Suite 250
Fort Collins, CO 80525
USA

Publisher: H. Dennison Parker
Vice President, Publishing: Nora Sherwood
Chief Operating Officer: Derry Eynon
Editor, GIS World, Inc.: Donald F. Hemenway Jr.
Managing Editor, Books: Terri Bates Eyden
Production Manager: Christine Thompson
Design and Composition: Wade L. Smith
Printer: Edwards Brothers, Inc.

ISBN 1-882610-11-3

Library of Congress Cataloging-in-Publication Data
GIS and environmental modeling : progress and research issues / Michael F. Goodchild . . . [et al.], editors.
p. cm.
Includes bibliographical references.
ISBN 1-882610-11-3
1. Geographic information systems. 2. Atmospheric sciences–Mathematical models. I. Goodchild, Michael F.
G70.2.G57416 1996
910′ .285—dc20 95-46477
CIP

This book is printed on acid-free paper.

DEDICATION

This book is dedicated to the memory of
Ian D. Moore
(1951–1993)

Ian Moore was a native Australian. He received his B.S. and M.S. degrees in civil engineering from Monash University in Melbourne; his Ph.D. in agricultural engineering from the University of Minnesota. He divided his professional career between the United States and Australia, as a faculty member of the universities of Kentucky and Minnesota, and as a scientist for the Commonwealth Scientific and Industrial Research Organization (CSIRO), Canberra, Australia. At the time of his death, he was Professor and Jack Beale Chair of Water Resources at the Center for Resource and Environmental Studies of the Australian National University in Canberra.

Ian's contribution to science lay in his ability to mathematically characterize the shape of the landscape from a hydrologic viewpoint. He devised new topographic indices that illuminate the spatial patterns of soil wetness, erosion, and related hydrologic processes. Ian had a broad vision of the link between hydrology and ecology, such as the relationship between soil moisture and vegetative growth. He was truly an "environmental scientist"—his knowledge transcended traditional disciplinary boundaries.

Ian was a plenary speaker at the First International Conference on GIS and Environmental Modeling held in Boulder, Colorado, in September 1991, and he was also to have been a plenary speaker at the second conference in Breckenridge, Colorado, until he was suddenly struck with cancer a few weeks before the conference began. He died on September 28, 1993, in Canberra on the morning he was to have delivered his paper at the conference. He is survived by his wife, Laura; his daughter, Natalie; and his son, Nicholas.

The editors dedicate this book to Ian Moore, in celebration of his life and his accomplishments. He was an outstanding environmental scientist and a fine man, whom we feel privileged to have known, as a colleague and as a friend.

CONTENTS

PLATES

Plate 3-1. Visualization of uncertainty in the position of the 350-m contour for the State College, Pennsylvania, 7.5-minute quadrangle. Shown are the 95% confidence limits on the contour position, based on the standard 7-m root mean square error in ground surface elevation and an assumed Gaussian distribution. (Source: U.S. Geological Survey 30-m DEM.)

Plate 16-1. Estimated standard error surface for the annual mean rainfall plotted in Figure 16-4.

Plate 16-2. Logarithm of rainfall variance versus logarithm of annual mean rainfall.

Plate 18-1. Montana digital elevation grid, 1-km horizontal resolution, 6-m vertical resolution, with markers indicating the location of weather stations used in this analysis.

Plate 18-2. Spatial distribution of β_1 parameters for (a) T_{max}, (b) T_{min}, and (c) *Prcp* regressions, averaged over the entire year at each grid cell.

Plate 18-3. Spatial distribution of 1990 annual average for (a) T_{max}, (b) T_{min}, and (c) annual total *Prcp.*

Plate 25-1. Present and future canonical trend surface maps of the main environmental gradients for the chaparral and yellow pine forest Munz community types (RSRU/UCSB).

Plate 32-1. Map output at selected time steps from a 500-year LANDIS simulation of a 10,000-ha landscape at 60-m cell size (approximately 37,500 cells): (a) initial state, (b) 50 years, (c) 300 years, and (d) cumulative disturbance over the 500-year model run.

Plate 33-1. Output from MOSAIC displayed by GRASS. Four roles at simulation years 0 and 200. Increasing intensity of color in a cell corresponds to increasing values of the proportion of that role in the cell.

Plate 41-1a. Orogenic precipitation index image computed as altitude from a DEM divided by distance from the Continental Divide (west of the site). Minimum value is 0.32 (dark blue); maximum value is 1.97 (red).

Plate 41-1b. Growing degree days image computed from daily temperature grids derived using environmental lapse rates and meteorological station data. Minimum value is 6.26 (dark blue); maximum value is 317.77 (red).

Plate 41-1c. Topoclimatic slope-aspect index image. Minimum value is 0.00 (dark blue); maximum value is 148.11 (red).

Plate 41-1d. Insolation index image. Minimum value is 10.39 (dark blue); maximum value is 25.79 (red).

Plate 41-1e. Snow probability image derived from Landsat MSS data archive. Minimum value is 0.00 (dark blue); maximum value is 1.00 (red).

Plate 41-2. Landsat MSS band 7 images of the Niwot Ridge LTER acquired on (a) June 19, 1986; (b) June 22, 1987. Landsat MSS images are used to monitor year-to-year and seasonal variations in the spatial extent of snow cover. Photographs of the Saddle snowfield taken during (c) June 1986 and (d) June 1987 field campaigns show the effect of a decrease in snowfall on plant growth.

Plate 42-1. Conceptual model of predator production in a spatially explicit framework. The water mass is divided by columns and rows into a grid of cells, depicted here in the vertical and horizontal dimension. Spatial heterogeneity is indicated by shading of the cells. Cell i has volume V_i. Prey density, prey size, and water temperature are used as inputs to a foraging model (f) and a growth model (g) to produce the growth-rate potential for the predator in that volume of water. The integration of growth-rate potential over all water volumes produces an estimate of system production potential.

Plate 42-2. Prey biomass density (upper five panels) and striped bass growth-rate potential (lower five panels) taken across five geometrically separated transects (approximately 25–45 km between transects) stretching from Still Pond in the north (top panel) to Smith Point in the south (bottom panel). Data were collected at night during (a) April 1991 and (b) August 1991. Maximum bottom depth is about 34 m. Cell size is 0.5 m in the vertical dimension and 30 m in the horizontal dimension.

Plate 48-1. Example of three conditional simulations (CS) and a kriged (KRIG) distribution of hydraulic conductivity on the Mahantango Creek watershed using conditioning data given in Table 48-1.

Plate 48-2. Distribution of recharge flux associated with (a) the 25th and (b) the 75th quartiles of travel time to the water table.

Plate 51-1. GEO-WAMS screen showing three-dimensional segmentation editing utility with a segmentation scheme for the Buffalo River depicted in the edit windows.

Plate 51-2. Snapshot of GEO-WAMS model output animation module display of dissolved oxygen in the lower waters of the Buffalo River.

Plate 55-1. (a) Mean January minimum temperature, spatially interpolated with adiabatic adjustment from surface station means. (b) Sand content (%) of the top 0–50-cm soil layer for the dominant soil type (first modal soil). Grid interval is 0.5° latitude by 0.5° longitude.

Plate 74-1. CLIMEX: Comparison of thematic maps.

Plate 74-2. ISC: Abatement strategy optimization module.

Plate 74-3. XGW: Three-dimensional display of groundwater contamination.

Plate 74-4. XENVIS: Visualization of risk contours.

PREFACE

The interdisciplinary chapters in this book summarize recent progress and identify key research issues concerning the integration of geographic information systems (GIS) and environmental modeling. Our goal in assembling this book was to promote cross-disciplinary communication among environmental modelers, remote-sensing scientists, and GIS specialists who have mutual interests and common goals. This volume discusses current applications of GIS in environmental modeling, recent accomplishments, advanced modeling and analysis, impediments to GIS and model integration, and future research directions.

The impetus for publishing this book stemmed from our desire to answer the basic question, "What is the potential role of GIS technology in contemporary environmental simulation modeling?" In a simplistic sense, this amounts to determining the appropriate links and interactions between two pieces of computer software code and their associated databases. However, a deeper understanding of both GIS and the environmental modeling process is essential for further progress. On one hand, GIS is a sophisticated emerging technology for processing, analyzing, and visualizing the growing amounts of digital spatial data. However, only recently have users attempted to move GIS beyond a traditional approach that deals mainly with a two-dimensional, static representation of the world.

In contrast, environmental models deal with the numerical simulation of three-dimensional, time-dependent processes with roots in atmospheric, hydrologic, geologic, soil, biologic, ecologic, and other natural sciences. For example, contemporary simulation modeling emphasizes cross-disciplinary approaches in which atmospheric, hydrologic, and ecologic models can be linked across various space and time scales to investigate, understand, parameterize, and predict interactions between the biosphere and other Earth systems. Such modeling features the need for scaling up from plots to regions, or scaling down from the globe to river basins, ecosystems, and watersheds. Remote-sensing technology is critical for understanding and monitoring complex land processes and for building necessary multiresolution, multitemporal land cover characteristic data sets. Increasingly, these advanced modeling approaches are being used to support decision making related to land and water resource management, air-quality analysis, ecosystem vulnerability assessment, and environmental risk studies.

This book is divided into three parts: Environmental Databases and Mapping, Environmental Modeling Linked to GIS, and Building Environmental Models with GIS. The works within each of these sections help to broaden our understanding of potential links between GIS and environmental models. For example, the chapters in Part I examine the role and use of GIS as a tool to build and tailor spatial data for environmental models. Part II describes various GIS applications used in environmental modeling for managing spatial data, integrating diverse data types, conducting spatial analysis, and visualizing model results. The chapters in Part III focus on advanced research issues, including embedding models directly within a GIS. Each section begins with two overview chapters, followed by a series of chapters that provide more case-specific discussions.

Author contributions are based on presentations originally made at the Second International Conference/Workshop on Integrating Geographic Information Systems and Environmental Modeling at Breckenridge, Colorado, September 26–30, 1993. This conference was organized by the National Center for Geographic Information and Analysis (NCGIA) at the University of California, Santa Barbara, with assistance from the U.S. Geological Survey and the U.S. Environmental Protection Agency. The publication of this book by GIS World, Inc., complements Environmental Modeling with GIS, published in 1993 by the Oxford University Press following the first international conference (Boulder, Colorado, September 1991). Plans are already underway for a third conference in Santa Fe, New Mexico, in January 1996.

The generous contributions of the agencies, institutions, and organizations that provided conference support are sincerely appreciated. Conference sponsors include:

U.S. Army Corps of Engineers

U.S. Department of Agriculture/Soil Conservation Service

U.S. Department of Commerce/NOAA

U.S. Department of the Interior/Geological Survey

U.S. Department of the Interior/Minerals Management Service

U.S. Department of the Interior/Fish and Wildlife Service

U.S. Department of Energy/Oak Ridge National Laboratory

U.S. Environmental Protection Agency

NASA

ESRI

Consortium for International Earth Science Information Network (CIESIN)

Ogden Environmental and Energy Services Co., Inc.

The many contributions of the conference steering committee are gratefully acknowledged. Their efforts in helping to design and conduct the conference are sincerely appreciated. A list of conference steering committee members is provided at the end of this Preface.

The staff of the NCGIA provided exceptional support in preparing for and administering the conference and in preparing this manuscript. The editors would especially like to commend the superior contributions of Judith Parker and Karen Kline.

The NCGIA is supported by the National Science Foundation, through cooperative agreement #SBR 88-10917 with the University of California, Santa Barbara. References to hardware and software products in this book do not imply endorsement by the authors, editors, or sponsoring organizations. The names of many of these products are registered trademarks.

CONFERENCE STEERING COMMITTEE MEMBERS*

Steering Committee
Lawrence Band, University of Toronto
Francis Bretherton, University of Wisconsin
Ingrid Burke, Colorado State University
Peter Burrough, University of Utrecht
Thomas Carroll, National Weather Service/NOAA
Josef Cihlar, Canada Center for Remote Sensing
David Clark, National Geophysical Data Center/NOAA
Russell Congalton, University of New Hampshire
Robert Coulson, Texas A&M University
*Michael Crane, U.S. Geological Survey
Evan Englund, U.S. EPA
Kurt Fedra, IIASA, Austria
Christopher Field, Carnegie Institute
Andrew Frank, Technical University Vienna
Janet Franklin, San Diego State University
*Sandi Glendinning, UC, Santa Barbara
*Michael Goodchild, UC, Santa Barbara
Robert Gurney, University of Reading
Ann Henderson-Sellers, Macquarie University
Mason Hewitt III, U.S. EPA
Carolyn Hunsaker, Oak Ridge National Laboratory
Michael Hutchinson, Australian National University
Chris Johannsen, Purdue University
*Carol Johnston, University of Minnesota
Timothy Kittel, UCAR
Sachio Kubo, Keio University
David Lam, Canada Center for Inland Waters
George Leavesley, U.S. Geological Survey
Richard Liston, U.S. Department of Agriculture
Thomas Loveland, U.S. Geological Survey
*David Maidment, University of Texas
Timothy McGrath, U.S. Geological Survey
James Merchant, University of Nebraska
William Michener, University of South Carolina
Ian Moore, Australian National University
Ramakrishna Nemani, University of Montana
Joan Novak, U.S.EPA
*Bradley Parks, University of Colorado, CIRES
Donna Peuquet, Pennsylvania State University
Roger Pielke, Colorado State University
Jean Poitevin, Environment Canada
Rene Reitsma, University of Colorado, CADSWES
David Rejeski, U.S. EPA
David Schimel, UCAR
William Schlesinger, Duke University
Hank Shugart, University of Virginia
*Louis Steyaert, U.S. Geological Survey
John Townshend, University of Maryland
Kenneth Turgeon, Minerals Management Service
Monica Turner, Oak Ridge National Laboratory
Jan van Wagtendonk, Yosemite National Park
Denis White, Oregon State University
Paul Whitehead, Institute of Hydrology, United Kingdom
Jeffrey Wright, Purdue University
*coorganizer

THE EDITORS

Michael F. Goodchild

Professor of Geography at the University of California, Santa Barbara, and Director, National Center for Geographic Information and Analysis. His research interests center on geographic data models and the accuracy of geographic databases.

Department of Geography
University of California
Santa Barbara, California 93106-4060

Louis T. Steyaert

Remote-sensing scientist in the Science and Applications Branch, EROS Data Center of the U.S. Geological Survey. He has worked full time on global change research since 1988, conducting land surface characterization research with space-based remote sensing and geographic information systems technologies as related to land data needs of environmental models. Dr. Steyaert also coordinated the test and evaluation of USGS land data by atmospheric, hydrologic, and ecologic modelers. He currently is an investigator in the NASA-led Boreal Ecosystem-Atmosphere Study in central Canada, while also coordinating the land surface characterization research plan for the GEWEX Continental-Scale International Project in the Mississippi River basin. Since 1991, he also has been involved in cooperative research with atmospheric modelers at Colorado State University concerning the use of mesoscale modeling to investigate the role of historical land cover change as a forcing factor in regional climate change.

U.S. Geological Survey
EROS Data Center
Science and Applications Branch
c/o NASA Goddard Space Flight Center
Greenbelt, Maryland 20771

Bradley O. Parks

Environmental scientist with the University of Colorado, Boulder, Cooperative Institute for Research and Environmental Sciences (CIRES). His research interests in natural resources management and environmental assessment and planning focus on: improving analytic modeling and decision-making methods for whole-watershed management, coastal ecosystem assessment, and sustainable land-use practice.

Cooperative Institute for Research in Environmental Sciences (CIRES)
University of Colorado
Boulder, Colorado 80301

Carol A. Johnston

Senior Research Associate at the Natural Resources Research Institute of the University of Minnesota, Duluth. Carol conducts research on landscape ecology, wetlands, and the application of GIS and GPS to ecological problems.

Natural Resources Research Institute
University of Minnesota
Duluth, Minnesota 55811

David R. Maidment

Professor of Civil Engineering at the University of Texas at Austin. His research involves applying GIS to hydrologic modeling. David is also editor of *Journal of Hydrology* and editor in chief of *Handbook of Hydrology.*

Department of Civil Engineering
University of Texas
Austin, Texas 78712

Michael P. Crane

Director of the Central Region Geographic Information Systems Laboratory, a facility that provides survey scientists with access to a variety of spatial/analytic tools, training, data, and technical assistance necessary to accomplish research and applications projects. His research interests are graphic communication and changes in land cover and use.

U.S. Geological Survey
PO Box 25046, MS-516
Denver Federal Center
Denver, Colorado 80225

Sandi Glendinning

Visitor and program coordinator at the National Center for Geographic Information and Analysis in Santa Barbara, California. Sandi served with distinction on the organizing committees for both International Conferences/Workshops on Integrating GIS and Environmental Modeling.

NCGIA
3510 Phelps Hall
University of California
Santa Barbara, California 93106-4060

Introduction Part 1

Environmental Databases and Mapping

Environmental models require data for calibration, verification, and specification of boundary conditions. The lack of suitable data has been one of the most serious impediments to the development and use of environmental models. Part I of this volume deals with a wide range of data-related issues, including the availability of data, the design of environmental databases, methods for interpolation and resampling, the problem of accuracy and related issues of error propagation in modeling, and the sensitivity of model results to data models and data quality.

The section begins with a short essay by Francis Bretherton on the implications of global environmental change. As we face monumental problems in coping with anticipated population growth and natural resource use, geographic information systems (GIS) will be essential in our efforts to understand what is happening to planet Earth and in developing appropriate policy. Roberta Miller picks up the policy theme in Chapter 2, examining the relationship between policy development and data supply, particularly in meeting the need for information as it relates to global environmental change. She warns of the potential for information misuse and argues for openness in data ownership as a reliable safeguard.

In Chapter 3, on data requirements of environmental modeling, Michael Goodchild reviews the accuracy problem and the limitations imposed by the approximate nature of most geographic data. Recent efforts to build a national framework of geographic data for the United States promise a coordinated and open basis for environmental data; however, the actual availability of national and international frameworks falls far short of popular expectations. Chapter 4, by Louis Steyaert, provides an extensive review of geographic data resources, focusing particularly on the importance of detailed land characterization in supplying boundary conditions for models of atmospheric, biogeochemical, and ecological processes.

The next four chapters elaborate on the accuracy issue. Peter Burrough et al. look at the role of GIS in environmental modeling and review work done at the University of Utrecht on modeling error and its propagation through GIS-based models. In Chapter 6, Richard Aspinall and Diane Pearson examine the relationship between accuracy and scale and their impact on ecological models: they believe that model predictions should be accompanied by estimates of the uncertainty that results from known uncertainties in model inputs. Brian Lees discusses the rapidly expanding paradigm of machine learning, typified by neural network models, and deals with the difficult problems analysts face in trying to design suitable samples both for calibration of models and for validation and testing. Steve Carver and his colleagues examine the issue of accuracy within the context of field sampling and data validation in Chapter 8.

Chapters 9 through 12 discuss soils data and associated issues. Dennis Lytle, Norman Bliss, and Sharon Waltman describe the U.S. Department of Agriculture's STATSGO database and illustrate its use in providing input to land surface model processes. In contrast, in Chapter 10 Paul Gessler et al. describe the Australian approach to estimate relevant soil properties by taking advantage of relationships between soils and land forms as well as recent advances in automated methods of terrain analysis. S. D. DeGloria and R. J. Wagenet focus on the use of simulation models to represent the spatiotemporal effects of various processes, on land use and soils, and techniques for visualizing those effects. Stephen Ventura and his colleagues examine the data structures inherited from mapping practice that are commonly used to represent spatial variation in soils. They also propose novel alternatives capable of capturing the inherent three-dimensional variability of soil systems.

The next three chapters describe various large database projects. In Chapter 13, James Frew presents an overview of the Sequoia 2000 project—an effort to build new computing environments for global change research—and the major research challenges it has tackled. Robert Lozar discusses the global data sets assembled and widely disseminated on CD-ROM by the U.S. Army Corps of Engineers Construction Engineering Research Laboratory. His chapter includes a simple tutorial designed to introduce students to the power of data and GRASS-GIS. Next, John Kineman et al. review the Global Ecosystems Database Project and some technical challenges associated with integrating data from widely disparate sources.

Environmental modeling requires accurate representation of the spatial variation of key data fields; often these must be built by interpolation from very limited sets of observations. In Chapter 16, Mike Hutchinson discusses the problems of interpolating rainfall fields from point samples, incorporating knowledge of topographic effects. Christopher Daly and George Taylor describe the application of the PRISM model in a comparable effort to estimate the precipitation field of Oregon using complex terrain information. Steven Running and Peter Thornton describe a model for generating fields of maximum and minimum air temperature as well as precipitation.

A very different approach is taken by Roger Pielke and his colleagues in Chapter 19: They use terrain and landscape pattern information to disaggregate the predictions of general circulation models from coarse grids. Hoyt Walker and John Leone examine the effects of uncertainty in terrain representation on a model of mesoscale atmospheric circulation and its prediction of nocturnal drainage wind. Scott Robeson and Cort Willmott compare a variety of methods for interpolating fields of climatological variables over the sphere and demonstrate the somewhat severe impacts of the observation network configuration on estimated parameters such as air temperature averages. Chapter 22, by John Corbett, continues the theme of climate field interpolation, again, using data on terrain and demonstrating significant information loss that occurs in the traditional classification approach of the cartographer.

In Chapter 23, Bernard Engel demonstrates the importance of the underlying GIS data model on predictions from environmental process models. GIS provides a convenient tool kit for exploring these effects and for comparing options such as hydrologic response units, grid cells, and TIN (triangulated irregular network). The issue of data model choice also underlies the work of Ralph Dubayah and Paul Rich on modeling solar radiation.

The section ends with a chapter by Dean Fairbanks et al. who analyze relationships between natural vegetation and climatic variables in California, allowing subsequent prediction of the effects of climatic change. Their analysis is based on grid cells rather than polygons, allowing the database to approximate continuous spatial change. Like its two preceding chapters, it serves to emphasize the importance of the underlying data representation model on this type of environmental forecasting.

As a whole, Part I illustrates the wide range of data issues raised by the combination of GIS and environmental modeling. The number of chapters allotted each topic may bear little relationship to its importance or to the likelihood of significant progress being made within the research community. Perhaps other issues will arise as environmental modelers make greater use of GIS and spatial data. However, taken together, the chapters present quite an accurate cross section of current data-related issues. Many of these issues are raised again in Parts II and III, but these early chapters provide the opportunity to examine and discuss them in detail. Clearly, significant progress has been made in harnessing the power of digital geographic data processing to address issues that for years have hindered analysis and modeling in a geographic context.

1

Why Bother?

Francis Bretherton

WHAT IS THE CARRYING CAPACITY OF THIS PLANET?

Let us pause for a moment to consider what the world might be like fifty years from now. The lead question is prompted by the expectations of demographers that by then, the human population will have at least doubled and that the goal of almost every nation is to increase its standard of living, which, in practical terms, translates to increasing per capita resource use. Of course, we do not know what the carrying capacity really is, but a simple calculation should provoke some reflection.

A fundamental property of our environment is the conversion of solar energy into biological energy through photosynthesis, and one reference point might be the fraction of global primary production we humans currently eat. A daily intake of 2,500 kcal per person (mostly carbohydrate) corresponds to about 200 g carbon, which, after multiplying by a global population of five billion, gives an annual uptake of 0.4 GtC/y, which is less than 1% of the land net primary production, that is, the chemical energy available to plants for growth. This fraction seems comfortably small, but an entirely different impression comes from a study by Vitousek et al. (1986) that considered the fraction of net primary production that is co-opted, that is, falls on cultivated fields, pastures, or other land that is substantially devoted to human use. They estimated this fraction to be as large as 40%. The factor of forty above direct consumption reflects the inefficiencies of conversion of primary biological energy into grains and animal products, which are our preferred foodstuffs, together with the proclivity of the human species to dominate the other inhabitants of this planet. Even so, it does not seem outrageous to respond, "So what, that means 40% for us and 60% for nature."

However, now double the world population. If other things were equal, the equation would become, "80% for us and 20% for nature." This would clearly be a state of acute crisis in which nature could scarcely continue to survive in its present form. The system is already showing signs of strain, and a further reduction of living space by a factor of three must have profound consequences, both for the other species on this planet and for ourselves. Of course this is a very simplistic calculation, but it serves to illustrate what has to happen if such consequences are not to follow. From an ecologist's perspective, the human species seems destined for a population crash. If we are to avoid it, it will be through unusual survival skills that cannot be taken for granted.

Of course, other things would not be equal. We can expect some help from technology, particularly genetic engineering. In addition, the number 40% is clearly somewhat soft, and some increased efficiency in land use is surely available. On the other hand, the part so far co-opted is the easiest, and efficient cultivation of marginal land will not be as straightforward. In addition, an increased standard of living has generally so far included eating a higher percentage of protein and animal products as opposed to direct consumption of cereals, which increases the requirement for primary production per capita. Furthermore, trends in climate, soil erosion, acid rain, deforestation, water pollution, and so on, must call into question our ability to sustain even the present population without drastic changes, so we should at least be seriously examining our options.

COPING SCENARIOS

Thoughtful individuals with expertise in several of the critical areas have indeed been examining technological scenarios which might address the twin challenges of increasing population and per capita resource use. Focusing on how we might be coping fifty years from now, most agronomists foresee greatly enhanced fertilizer use—probably ten times the present. This does not seem unreasonable, seeing that fifty years ago use was only 10% of what it is today, and the energy requirements for manufacturing it, though substantial, are not overwhelming. However, on a global scale, the natural fixation of nitrogen is presently approximately the same as that from fertilizer, and experience shows that fertilizer does not stay confined to the fields on which it is applied; therefore, one implication of this scenario is that the natural environment will have to adjust to an entirely new situation in which nutrients are no longer the major limitation on growth. In the energy arena, constraints from climate change require a focus on renewables, with significant new contributions from wind and solar power, and especially from plantation-grown biomass converted to methanol or hydrogen. Though ambitious and requiring fundamental institutional changes, such a scenario may well be practical, but it also has major implications for land use. Freshwater resources are critical in many regions of the world, and runoff is the dominant source. Except possibly for climate change this resource is fixed, and we will need much more efficient allocation and end use, in addition to reversing the trend toward polluting the supplies we have.

Each of these examples illustrates a crucial need for efficient science-driven natural resource management and provokes a fundamental question: Are our information systems and models good enough?

SOME FUNDAMENTAL ISSUES

Other critical issues that will not be discussed here are:

- Can social structures evolve for collaboration in the common interest?
- When and how will populations stabilize?
- How do we redistribute wealth where it is critically needed?
- What if science and technology cannot deliver in time?

Individually, none of these is necessarily insuperable, and history shows that human societies have a remarkable capacity for innovation and adaptation. However, this analysis points to a major debate that still has not been effectively joined by all the groups of experts and people around the world who need to be involved: What are the critical limitations imposed on modern society by our natural environment and the finiteness of this planet?

PERSPECTIVES ON POLICY

Homo sapiens has evolved in an environment that is at the same time both challenging and dangerous, and it should not be surprising that we each come equipped with mechanisms for coping, which are deep seated and not necessarily consistent with each other. Our curiosity drives us to explore our environment and understand how it functions. We have individually and collectively benefited greatly from such mastery. At the same time we are afraid of the unknown, an emotion that serves to keep us relatively out of danger. When faced with the prospects of global environmental change and all the uncertainties associated with it, we each react accordingly.

Table 1-1 presents a highly simplified characterization of different approaches to issues of global change, associated with the different ways individuals tend to deal with uncertainty in their everyday lives. The perspective is a personal one, with the icons inspired by Steve Rayner. On the left of each row is a schematic world view, in which the circle represents a ball and the curve a container. No one world view is right or wrong, just different. The balancing between them is essentially a political decision. The role of scientists is to provide reliable information for each group without prejudging the outcome.

In the first row, the world is seen as stable when perturbed, and the viewer naturally takes risks and concentrates on adapting to a new situation. Such people tend to be entrepreneurs and see global change as a challenge, perhaps even as an opportunity to make money. They tend to focus on regional economic analyses, in particular, what are the most sensitive sectors and what is their resilience to any kind of change. What this group most wants out of natural scientists are plausible, self-consistent scenarios of possible change, not actual predictions. In addition, as with all groups in Table 1-1, they need reliable monitoring of what changes are actually occurring.

For the second row in the table, the world is seen as unstable, and the viewer instinctively wishes to preserve the status quo. All change is bad, and too rapid change is potentially disastrous. Faced with the prospect of climate change, the instinct is to stop it through emission controls, or at least to slow it down. Besides reliable monitoring, the need of this group is for analyses of cost-effective options for mitigation.

In the third row, the vessel is so rough that the future is totally unpredictable, and the natural reaction is to deny that problems exist.

In the fourth row (to which I personally probably belong), the world is stable for small perturbations but with real, misunderstood

Table 1-1. Perspectives of global change.

World View	Premise	Agenda	Analyses	Science Implications
	Change inevitable	Adaptation Insurance	Regional impacts	Regional scenarios Impact models Monitoring
	Change bad, possibly disastrous	Mitigation Emission controls	Relative costs	Reliable monitoring Technological options
	Unpredictable	Denial	Ignore issue	Hard proof Practical options
	Limits to growth	Sustainable development	Human - environmental interactions	Social - natural science interactive models
	Collaboration essential	Develop insitutional framework	Communication Education Action groups	Scientists as communicators, advocates

limits. The agenda is to recognize those limits before we transgress them. Though no one quite knows how to define sustainable development, the goal could be paraphrased as humankind both having its cake and eating it too. It requires both adaptation and mitigation, so this banner also provides a refuge for members of both the first and third groups, temporarily concealing their fundamental differences. However, the scientific requirements are extremely ambitious, implying a quantitative understanding of the two-way interactions between human societies and the natural environment.

The fifth row is the option found in all good multiple-choice examinations: none of the above. It is for those who see the issue as how the peoples of the world will learn to collaborate with each other about things over which they have to collaborate, that is, "What is the new world order?" From this perspective, the actions taken under rows one through four are just social experiments. The key is the development of appropriate institutions, in the broadest sense of that term, including how to develop understanding among diverse groups.

The point here is that each world view tends to be associated with a different agenda, with quite different expectations from science. It is arguable that in the United States, November 1992 marked a transition from a political climate dominated by rows one and three to a climate more concerned with rows two and four. Such transitions will undoubtedly occur again, and, if it is to survive long enough to be effective, the basic research program in global change has to be sensitive to the needs of each group.

THE U.S. GLOBAL CHANGE RESEARCH PROGRAM (USGCRP)

The USGCRP was a presidential initiative in the budget for fiscal year 1990, prepared under the Reagan administration as a focused, interdisciplinary, long-term program of basic research "to establish the scientific basis for national and international policy making related to natural and human-induced changes in the global earth system" (Committee on Earth Sciences 1989). From the beginning, it included an embryo program on human interactions, but the primary emphasis has remained on the natural sciences, particularly issues related to global warming. The total annual budget is now around $1.5 billion. This effort is embedded in a combination of the International Geosphere-Biosphere Program, the World Climate Research Program, and the Human Dimensions Program, coordinated by various international agencies. Under the Clinton administration, a closer link to active policy concerns is being sought, but it is still unclear to what extent the basic science will be affected.

Table 1-2 presents a personal summary of the outputs to which the USGCRP seems currently to be committed, phrased from the point of view of potential users of the information, as opposed to that of the scientists formulating the fundamental questions. When stated in this way, almost every output has a human-dimensions component, though closer examination shows that the expertise required varies substantially within the list. In addition, the interface to policy issues is markedly diverse, reflecting the various agendas described in Table 1-1. Of particular interest to users of geographic information systems (GIS) may be outputs 2, 3, 8, and 9. The basic questions raised by the scenario with which this discussion began are all under 9, not because it is the least important, but because the state of our understanding is such that measurable progress seems furthest away. However, research proceeds in bite-sized pieces—the cumulative result of many instances of individual dedication and creativity. Perhaps this quick overview will help identify where each piece fits in and stimulate a fresh look past the day-to-day frustrations of making progress. It seems that we do indeed have good reason to bother.

Table 1-2. Products of USGCRP as seen from a policy perspective.

Each product results from ongoing activities, with anticipated periodic upgrades and improvements. Items are listed in approximately the order in which significant improvements can be expected, given their present state.
1. Detection, causes, and impacts of significant changes in the stratospheric ozone layer.
2. Regularly scheduled, ongoing predictions one year ahead of interannual climate fluctuations associated with El Nino, together with regional impact and forecast utilization studies.
3. Plausible scenarios for regional climate, ecosystem, and habitat change, in a form suitable for various impact models.
4. Estimates of the relative global warming potential of various gases and aerosols, including interactions and the indirect effects of other chemical species.
5. Ability to determine regional sources and sinks for atmospheric carbon dioxide, for application as part of the monitoring system for a greenhouse gas-emissions reduction agreement.
6. Reduction in uncertainty of effects on climate of clouds, ocean heat storage, and land-surface processes and, hence, in the range of predictions of the rate of global warming over the next century.
7. Predictions of anthropogenic interdecadal changes in regional climate, in the context of statistics for the natural, unpredictable, interannual, and interdecadal variability.
8. Detection beyond reasonable doubt of greenhouse gas-induced global warming and documentation of other significant changes in the global environment.
9. Understanding of the major interactions of human societies with the global environment, enabling quantitative analyses of existing and anticipated patterns of change.

References

Committee on Earth Sciences. 1989. *Our Changing Planet: A U.S. Strategy for Global Change Research.* Washington, D.C.: Executive Office of the President.

Vitousek, P. M., P. R. Ehrlich, A. H. Ehrlich, and P. A. Matson. 1986. Human appropriation of the products of photosynthesis. *BioScience* 36:368–373.

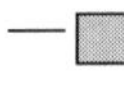

Francis Bretherton
Space Science and Engineering Center
University of Wisconsin
Madison, Wisconsin
E-mail: *fbretherton@ssec.wisc.edu*

2

Information Technology for Public Policy

Roberta Balstad Miller

This chapter examines some of the implications of the new information era for public policy. Its purpose is not to predict the shape or the nature of society or policy in the future. Rather, it will discuss: 1) the technological innovations that are making the new information era possible, 2) ongoing technical and scientific problems that inhibit our ability to make full use of information innovations in public policy, and 3) the dangers and opportunities the information era poses for public policy.

INTRODUCTION

Electronic technologies and communications are propelling us into a new era in human history. The use of the computer in word processing and information retrieval has been compared, particularly in its social and intellectual implications, to the invention of printing in the fifteenth century.[1] The advent of global communications networks and our rapidly expanding capacity to store, manipulate, model, and process data promise to have an even greater impact on human society and politics.[2] We are, in essence, creating a new context within which the lives of men and women, and the fortunes of nations and empires, will be played out—a context unlike anything we have previously experienced.

WHAT IS POSSIBLE NOW?

Electronically, it is possible to communicate in real time with people around the world. Computers can store and manipulate immense quantities of information. This information can be arrayed in a geographic information system (GIS) that simultaneously incorporates data on economic and social attributes and on physical conditions above and below the surface of the earth. Modeling can be done using GIS, and data can be shared through archives without walls—virtual data archives where the data are stored and updated in locations around the world, even as they are made accessible as part of the same global electronic data system.

Much of the new information-sharing technology is being developed within the scientific community for research on global environmental change. In part, this is serendipitous. The expansion of information technologies occurred at about the same time as political and scientific interest in global environmental change intensified. In part, however, this development is also quite logical. Global change research requires vast quantities of disparate data on a global scale. Twenty, even ten, years ago, much of the research currently underway could not have been undertaken because we lacked the capacity to either acquire, manipulate, or analyze such large databases.

One example of the new data-sharing technologies is CIESIN, the Consortium for International Earth Science Information Network. CIESIN was created by the U.S. Congress to provide data on human interactions in global environmental change. Established during a period when the administration was reluctant to invest in the infrastructure necessary to understand environmental change, CIESIN was placed within the Mission to Planet Earth program at the U.S. National Aeronautics and Space Administration (NASA). With its focus on socioeconomic data and information, CIESIN has been very much a foreign element in an agency devoted primarily to space technologies and secondarily to Earth science. Nonetheless, it is a credit to NASA leadership that they took what was an unanticipated mandate from Congress and made CIESIN a part of the larger Earth-Observing System Data and Information System (EOSDIS).

EOSDIS now consists of nine data and information centers, eight of which are charged with providing specific types of natural science data for understanding global change, and one, CIESIN's SEDAC (the Socioeconomic Data and Applications Center), charged with providing socioeconomic data on the anthropogenic causes and impacts of global change. These data are intended for use in research, in policy making, and in education.

CIESIN obtains data for SEDAC through the Information Cooperative (Info Co-op), an international network of data-collecting institutions and archives that have agreed to catalog and share their data electronically through CIESIN. The Info Co-op plays the role within SEDAC that the satellite Earth-Observing System plays within the other EOSDIS centers. In a very real sense, it is the platform for another kind of Earth-observing system, one which is closer to earth, observing and recording the economic and policy climate, the ecology of transportation and settlement, and the dynamics of anthropogenic emissions and pollution. As shown in Figure 2-1, members of the Info Co-op range from United Nations agencies and regional statistical organizations to social science data archives and data-collecting organizations with an interest in the environment.

Given the rapidly advancing technology for data collection, analysis, and sharing—and the deepening scientific understanding of global environmental change—Dr. John Gibbons, Director of the White House Office of Science and Technology Policy (OSTP), has

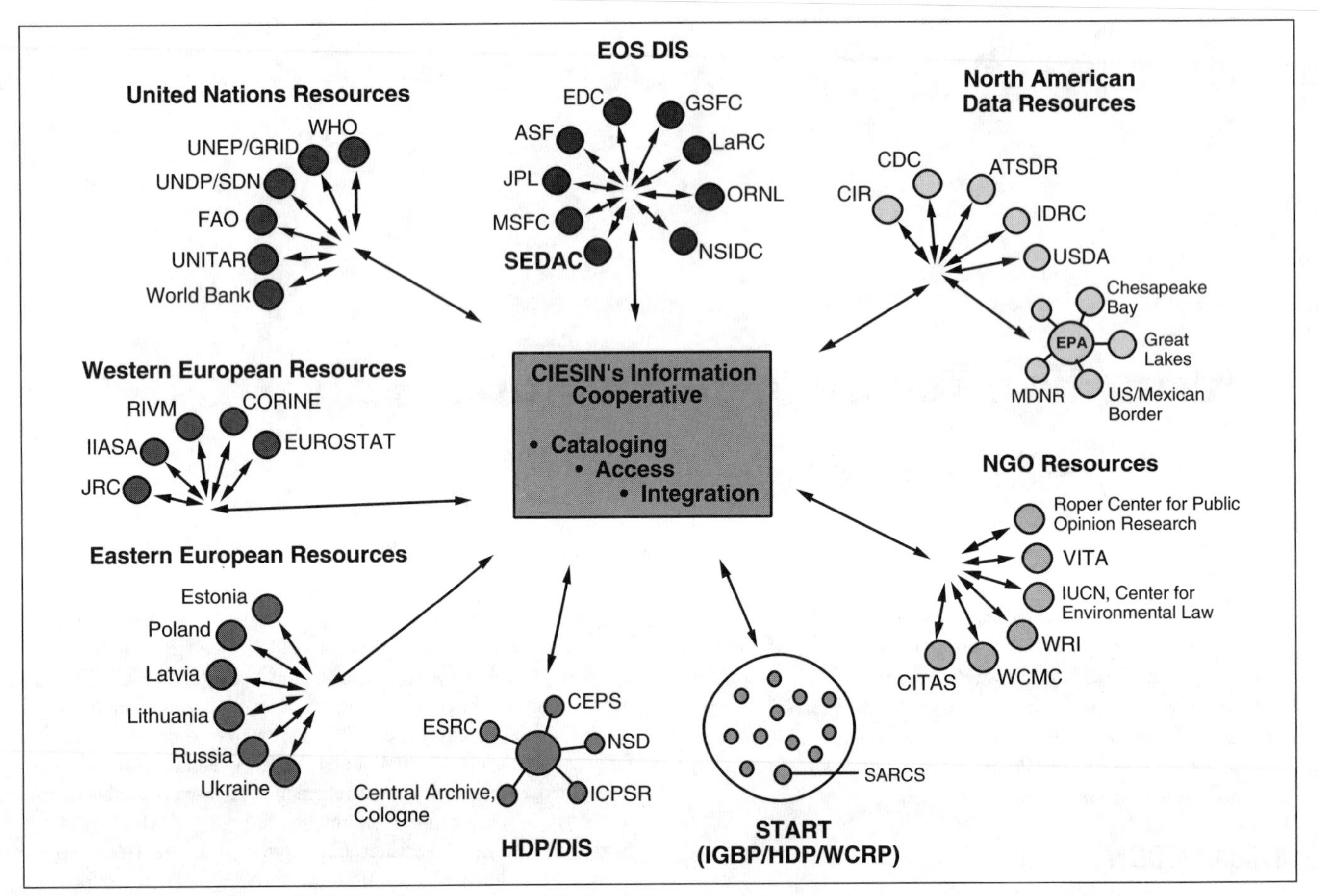

Figure 2-1. Information cooperative development plan.

made public policy utility a priority for the U.S. Global Change Research Program. Implicitly using an argument that can be traced back to Plato, who envisioned a world where wise and informed men rationally discussed alternative policies and only then determined affairs of state, Gibbons wishes to see U.S. global change research and data programs directed toward policy needs. In effect, this will require that *the capacities of the information era be made available for use in environmental policy* so that they can be instrumental in creating a world in which those who are responsible for policy and governance have ready access to the data and information they need to make reasonable and just decisions.

This is the basic argument. Unfortunately, it is something of an unrealized Platonic ideal itself. In the real world, policy makers almost never have the luxury of disinterested contemplation in decision making. Political decisions are an amalgam of long- and short-term priorities, the influence of constituents, and the need for political compromise. In the political arena, the impact of data and information is often eclipsed by the imperatives of the political process.

Clearly, the goal of serving the national interest by providing a sounder informational base for public policy is one that everyone in the United States could support. Just as clearly, however, we need to go beyond the rhetoric of "meeting policy needs" to identifying what those needs are. For example, policy analysts provide data and information to those who are in positions of policy responsibility. These policy analysts, working for elected officials in universities and in public nongovernmental capacities, are the real links between decision makers and new sources of data and information.[3]

TECHNICAL AND SCIENTIFIC PROBLEMS

Before policy analysts will be able to take full advantage of new information-rich technologies in environmental policy and assessment, a number of scientific and technical problems must be addressed, for there are significant technical problems in developing policy-useful data systems. Among them are: (1) the need to create merged data sets (i.e., data sets which encompass both socioeconomic and physical/biological data); (2) the need to develop both time-series databases and baseline data; and (3) the need to expand data access and electronic capability in developing countries.

The Need for Merged Data Sets

Our capacity to create extensive, sophisticated databases has developed within the traditional parameters of distinct scientific disciplines or fields of inquiry. That is, scientific databases are in most cases intended to answer specific research questions. These questions arise from both disciplinary research concerns and, increasingly, from multidisciplinary concerns in evolving research fields such as global environmental change. Because of these disciplinary origins, there is a significant gap between databases constructed for use in the social sciences and those developed for research in the natural sciences. This divide persists, despite the fact that there is widespread recognition that understanding of the ways that human actions affect global change is dependent upon the creation of merged, georeferenced, time-series databases that contain both socioeconomic data and physical data that reflect the interaction of human and physical forces over time. Moreover, before significant progress can be made in developing merged databases, scientists must develop a matrix of carefully focused research questions.

The raw material for merged databases is already available. Socioeconomic data have been collected by governments, churches, private organizations, and others for hundreds of years, and social scientists are extremely sophisticated in assessing and interpreting these data. But when social and economic statistics are combined with physical data, serious problems of scale and distribution arise. For example, socioeconomic data are often available only at the national level. That is, a single figure, whether aggregate or representative, is usually given for, say, a country's population or its income or its energy consumption. If there is finer resolution to the data, it is provided for smaller political units rather than for regular geographical units or on a grid scale as physical data are arrayed. Most socioeconomic data are collected and intended for use within political systems or are obtained from probability samples of the national population which cannot be disaggregated geographically without doing violence to sampling assumptions or violating privacy and confidentiality restrictions.[4] Thus, merging these data with georeferenced physical data in a GIS—and then analyzing them—is not only extremely difficult, but it may also be misleading.

The Need for Time-Series and Baseline Data

A second issue is the need for time-series and baseline data. Time-series data are those which extend similar measures over a period of time, and baseline data are those which establish a measure at one point in time for use in comparisons with future and past measurements. To be able to measure—as opposed to merely describe—human interactions in global change, scientists and policy makers need both baseline data and time-series data over large areas of the globe. Baseline data will be used to create a standard from which future change and deviation can be measured and which can be used in future environmental monitoring; time-series data will provide information on the pace and trajectory of change, both in the past and in years to come.

In modeling, time-series data are essential. Process, change, and causation cannot be modeled using cross-sectional data. Yet many extant databases consist of data collected at a single point in time. Time-series data can be constructed based on these data, but this is an expensive process that must be informed by scientific and technical understanding of the data. Much of the data collected in the past are already lost or irretrievable. In addition, a great deal of potentially valuable data were never collected. This puts a premium on preserving and rescuing the data we now have and using them to construct retrospective time series.

Expanding Data Access in Developing Countries

Still another problem is related to the need to work closely with scientists and statisticians in developing countries. The days when scientific research could be conducted entirely in one country—or in a small group of leading scientific countries—are over. Global change research and the construction of databases for this research require the active participation of scientists from many countries. The diffusion of scientific talent worldwide and the need to amortize the costs of equipment and instrumentation, education, and data collection all demand the involvement of scientists in both developed and developing countries in global research projects.

An immediate problem that hinders this global collaboration is that scientists in many countries do not have easy access to the Internet. For example, the entire EOSDIS system of data archives, the U.S. government's Global Change Data and Information System (GCDIS), and CIESIN's data catalog system and the Info Co-op all depend on the Internet. Within the Info Co-op, the Internet is critical, both for obtaining data from developing countries and for sharing data with scientists in those countries. What is needed to permit cross-national collaboration in research on global environmental change is immediate and equal access to Internet-based information.

Unlike their colleagues in the developing countries, researchers in the United States and the more developed regions of the world have pervasive, high-speed access to the Internet for both communications and research. In most cases, this access is subsidized at the national level. In developing countries, even where there is Internet availability, researchers generally must pay for network access each time they use it. In many cases, such access is not even interactive, but is limited to electronic mail (e-mail) or simple "store and forward" solutions. An additional barrier in less-developed countries is that Internet access is as much as 1,000 times slower than it is elsewhere.

Because e-mail has become the de facto electronic lowest common denominator, CIESIN is seeking to provide "catalog interoperability" for researchers in developing countries using Internet mail. This will effectively make it possible for everyone with the capacity to send and receive e-mail to search a worldwide directory of information resources and databases relevant to global and environmental change issues. As a result of this access mechanism, it will be possible for scientists around the world to access the same pool of scientific data, regardless of the speed or nature of their Internet access. The next step will be to provide access to data resources for those who are in countries without Internet access.

THE DANGERS AND OPPORTUNITIES THE INFORMATION ERA POSES FOR PUBLIC POLICY

The history of scientific influence on public policy is replete with examples of the ways that scientific ideas and technological innovations have improved the general welfare. This is particularly evident in those fields of science that deal with human and institutional behavior. One of the best examples is the extension of economic democracy in Sweden in the 1930s and 1940s based on social research and the arguments developed by social scientists such as Gunnar and Alva Myrdal. Another is the contribution of social science research to the intellectual rationale for racial equality in the United States, including the integration of the armed forces in the 1940s and the decisive arguments of *Brown vs. Board of Education* (1954), the Supreme Court case which ruled that separate but equal education was not equal. There is also a rich tradition of social science contribution to administration and governance in less frequently recognized ways, including the development of tools such as cost-benefit analysis, national income accounts, program evaluation, economic indicators, social surveys, and the measurement of unemployment and poverty, to give but a few examples.

Yet the public influence of science and new technologies has not always been benign. Science has been used to give a patina of credibility to policies that are morally and ethically reprehensible. Examples include the complicity of anthropologists and psychologists in the sterilization of hundreds of thousands of Germans in the Nazi period through the work of the Kaiser-Wilhelm Institute in Berlin and the use of new computer technologies by the Nazi government to register, identify, and deport the victims of its racial policies.[5] In another part of the world during this same period, social science concepts were used to give a scientific underpinning to the legalization of segregation and exclusionary racial policies in South Africa.[6] What can be learned from these disparate examples is that scientific research and technological innovations take on a life of their own and cannot be controlled by those who develop them, and that if science and scientific information are held to be the instruments of a select and powerful few, they can be used as instruments to dominate the many. This is most significant in fields that directly relate to human behavior and to social and political policy and institutions. It is also most dangerous in periods of rapid political change, where science and technology may either be used as the instruments of political change or may provide the rationale for such change.

Given what is known about past uses of science and technological innovation, it would be naive to assume that the new communications and information technologies and the rapidly expanding scope of environmental data would be exempt from uses that might subvert, rather than improve, the general welfare. Clearly, providing a means of communicating in real time across great distances, and permitting open access to vast quantities of data could be used to influence or even control environmental policy. It is entirely possible that this control could be used to place the welfare of one political group, one set of interests, or one nation ahead of others.

On the other hand, the new data and information technologies could instead be major tools for the expansion of democracy. Certainly access to data and information resources could expand the capacity of scientists and policy analysts to understand the environment and the ability of policy makers to respond wisely to environmental data. In addition, it could provide governments with a means of measuring both the social impacts of environmental changes and the extent of compliance with environmental treaties. But access to data should not be restricted to scientists or policy makers. If environmental data and information are accessible only to a few, they could be used to promote policies at odds with the broader public interest. Data for policy makers must also be accessible to the public so that these data can be used for democratic oversight. In essence, if new types of data resources are to be a component of public policy, it should be a matter of principle that those data should be made usable and available to all.[7]

Clearly, the public use of new information technologies will not be cost-free. It will require user-friendly software that can be accessed by individuals who do not have the scientific training required for complex information systems. The commitment to making data for policy available to voters requires additional commitments: (1) to education in statistical methods and the analysis of data and environmental information in secondary schools, and (2) to the continual improvement of software that permits interactive use by nonscientists.

The issue of data access has international ramifications as well. Environmental policy will be advanced if those who are responsible for policy in the United States and other countries are working from the same basic data and the same understanding of the scale, nature, and extent of environmental change. Similarly, solutions to global environmental problems will also be advanced. Disagreements will, of course, be inevitable, but they will be more easily resolved if discussion proceeds from a common base of knowledge and understanding. For example, both CIESIN's Information Cooperative and CIESIN's data catalog system will be internationally distributed and accessible. Equally important for their credibility for research and policy, they are not controlled by the U.S. government, but are being built as multinational data archives by representatives of contributing data centers—regardless of their country of origin.

If the new data and information technologies are to serve democratic purposes, then the infrastructure that makes this possible must be maintained. This is not an insignificant task. If scientific data are seen as a public resource, then data must be made available free of charge. But we must also find a way to maintain the system that does not impose controls over the data nor compromise its credibility. Environmental data and information systems can play a role in policy in the United States and other countries, but they must have a mixed base of institutional support. They should not be maintained by a single agency or a single government, lest they be controlled by it. Moreover, the support base should include private as well as government organizations. In addition to ensuring that no single set of interests can capture the data system, a diverse base of support for cooperative, publicly accessible environmental information systems will also mean that the cost can be amortized across a number of sponsors and substantially reduced for each one.

The new information age that we have entered has widespread implications. Both policy makers and the public have the capacity to know more about the world than earlier generations thought of asking. We still face technical problems in obtaining and sharing data, but we are moving swiftly to solve these problems. Yet in the face of rapid advances in electronic data and information systems, we must not forget that scientific and technical innovations can be used for ill as well as for good. To ensure that data systems become an instrument of democracy rather than a tool for subverting it, there must be many sources of data, open and multiple forms of access to data, and user-friendly data systems. If they are to be freely and widely accessible, these data systems cannot be owned by one set of interests. They must be supported by a diverse constituency, and, in the final analysis, they must belong to the people.

Notes

1. See, for example, Michael Heim, *Electric Language: A Philosophical Study of Word Processing* (Yale University Press, 1987).
2. There are any number of sources that examine implications of the new information age. A good starting point for establishing the background of the technology revolution is O. B. Hardison, Jr., *Disappearing Through the Skylight* (Viking, 1989).
3. To examine this problem, CIESIN has established a Commission on Global Environmental Change Information Policy in conjunction with the Kennedy School at Harvard University. This commission is examining the data and information needs of policy analysts and will make recommendations to CIESIN on data and information it should acquire and make available for policy uses.
4. There is some research on this topic underway through the Human Dimensions of Global Environmental Change Program's work toward a Global Omnibus Environmental Survey (GOES).
5. See, for example, Sybil Milton, "Re-Examining Scholarship on the Holocaust," *The Chronicle of Higher Education* (April 21, 1993); Michael Burleigh, *Germany Turns Eastward: A Study of Ostforschung in the Third Reich* (Cambridge University Press, 1988); and Michael Burleigh and Wolfgang Wippermann, *The Racial State: Germany, 1933–1945* (Cambridge University Press, 1991).
6. Roberta Balstad Miller, "Science and Society in the Early Career of H. F. Verwoerd," *Journal of Southern African Studies*, 19(4), 1993.
7. Sten Johanson, "Mot en Teori for Social Rapportering," Institutet for Social Forskning (Stockholm, Institutet för Social Forskning, 1979).

Roberta Balstad Miller is president of the Consortium for International Earth Science Information Network (CIESIN). Prior to joining CIESIN, Dr. Miller was the director of the Division of Social and Economic Science at the National Science Foundation. During the 1991–1992 academic year, she was senior associate member of St. Antony's College, University of Oxford. She has published on global environmental change, science policy, and the history of the social sciences and has served as chairperson of the Committee on Science, Engineering, and Public Policy of the American Association for the Advancement of Science. She is currently vice-president of the International Social Science Council and vice-chairperson of its Human Dimensions of Global Environmental Change Program.

3

The Spatial Data Infrastructure of Environmental Modeling

Michael F. Goodchild

Widespread use of spatial data in environmental modeling and associated policy formulation and decision making raises a series of research issues of fundamental significance. The problem of uncertainty in spatial data is used as an illustration. Spatial data are used in environmental modeling first as a framework and second in the form of measured or sampled fields. The status of framework data is reviewed within the context of the U.S. National Spatial Data Infrastructure (NSDI). The chapter concludes by arguing for an extension of NSDI to the international and global levels.

INTRODUCTION

Within the general topic of integrating GIS and environmental modeling, the organizers of the second conference felt that three distinct themes stood out: (1) issues of spatial data, including availability, access, common formats, resampling, and accuracy; (2) issues of modeling, including the development and structuring of models; and (3) issues of systems, including the design of GIS, data models, GIS functionality, and user interfaces. The abstracts submitted for the conference seemed to fall into these three themes in some degree of balance, as did the general subject matter of the conference.

The purpose of this chapter is to provide an introduction to some of the issues addressed in subsequent chapters on spatial data problems. Individual chapters cover a range of research issues, from accuracy and access to spatial interpolation and land-surface characterization. This chapter focuses briefly on one of those topics, the problem of spatial data quality, and then looks at the status of spatial data internationally and at some of the spatial data policies that affect environmental modeling with GIS.

THE ACCURACY PROBLEM

As with any assemblage of scientific measurements, spatial data are subject to errors of measurement that depend on the accuracy of the measuring instrument and propagate into the results of modeling. Two further characteristics of spatial data also contribute to the accuracy problem: (1) the process of spatial averaging or generalization that is present in much spatial data of limited spatial resolution; and (2) the use of data that embeds the subjective judgment of the observer. For example, measurements obtained from Earth-imaging satellites, such as AVHRR, are averaged over areas as large as 1 km^2. When the land surface is characterized by assigning a class to each AVHRR pixel instead of a measured spectral response, the resulting class captures a comparatively sophisticated analysis of spatiotemporal variation, but it is probable that two analysts would produce somewhat different classifications, and consequently the pattern of classes is not entirely reproducible from one analyst to another. This lack of reproducibility of land surface characterization may be a substantial source of uncertainty in environmental modeling.

In recent years significant advances have been made in modeling uncertainty in spatial data. Error models have been developed by Fisher (1991, 1992), Goodchild et al. (1992), Haining and Arbia (1993), Mark and Csillag (1989), and others, and the field of geostatistics also provides a comprehensive theoretical framework for discussion of spatial data uncertainty (Isaaks and Srivastava 1989). But while suitable models exist, along with techniques for error analysis by propagating uncertainty through GIS processes into confidence limits on results, the implications of uncertainty are often difficult to understand, especially for users not familiar with the concepts of spatial statistics (Cressie 1993). Beard et al. (1991) have proposed that visual techniques be developed to communicate information on uncertainty, and that data quality be an essential component of any spatial data display. Recent research on the visualization of spatial data quality is described by Buttenfield (1993).

A simple example is shown in Plate 3-1. In this standard U.S. Geological Survey 30-m digital elevation model (DEM), sample points form a square grid with a spacing of 30 m on the ground, and were converted into a contour map using a standard contouring algorithm. The DEM on which this illustration is based has a stated 7-m root mean square error (RMSE), indicating that the root mean square difference between true and observed elevation at each sample point is 7 m. For the purposes of this discussion, we assume that the distribution of elevation errors is Gaussian about the true value, with a standard deviation equal to the RMSE (see Hunter and Goodchild 1995 for a more detailed discussion). Plate 3-1 shows the probability that the elevation at each point is in reality greater than 350 m, in other words, that the point lies above the true 350-m contour.

Although the example and the illustration of uncertainty appear straightforward, closer examination shows that this is in fact far from the truth. Consider what happens when this landscape is flooded to 350 m, perhaps by construction of a reservoir. Had the elevations been lower, the consequences of sea level rise might have provided an alternate motivation.

Suppose first that the 7-m RMSE refers not to the uncertain elevation of the ground surface, but to the uncertain elevation of the floodwater relative to an error-free ground surface. For example, the elevation of the reservoir surface on any given day might be sampled from a Gaussian distribution with a mean of 350 m and a standard deviation of 7 m. Water surface elevations of 347 m or 355 m are comparatively likely, but 371 m lies three standard deviations from the mean, and is therefore highly unlikely. From the statistics of the Gaussian distribution we know that the risk of flood in areas mapped at 359 m (1.29 standard deviations above the mean) is approximately 10%, and the appropriate areas with this risk of flood are clearly identified in the visualization in Plate 3-1.

Now suppose that the elevation of the water surface is exactly 350 m, and that the uncertainty exists in the ground elevation. Although the same Gaussian distribution models the sampling of elevation, it is likely that some areas are characterized by positive errors (observed elevation greater than true elevation) and some by negative errors (observed elevation less than true elevation). In the extreme, the error at each sample point might be statistically independent. It seems more likely that errors are statistically correlated, such that a positive error at one sample point implies similar positive errors at neighboring sample points, but the correlation is likely less than perfect.

In these circumstances, the interpretation of Plate 3-1 changes. Although the visualization still shows correctly the probability that any given point is above 350 m, the probability that an area is above 350 m depends on the number of points lying in the area, and on the spatial autocorrelation of errors. If errors are statistically independent, an area of n sample points now has a probability of p^n of being above 350 m, if p is the (constant) probability associated with each of the points in the area. In a visualization such as Plate 3-1, where sample point locations are not shown, it is virtually impossible to make this interpretation successfully. The risk that a patch containing ten sample points will be *entirely* flooded lies somewhere between .10 and $.10^{10}$, if each contained sample point has a risk of flood of 10%, depending on the structure of spatial dependence of errors.

Unfortunately virtually nothing is known about the spatial dependence of errors in digital elevation models, or in any other type of spatial data. While this may be unimportant in many applications, the complex processing of spatial data that increasingly occurs in GIS and environmental modeling makes this issue hard to ignore. For example, the ability to estimate slope (Burrough 1986) depends directly on spatial dependence in errors between neighboring sample points. If slope is calculated from adjacent pairs of elevation estimates, by taking the difference in elevation and dividing by 30 m, a simple error analysis shows that the standard deviation of the uncertainty in slope estimates is given by:

$$\sigma^2_{slope} = 2\sigma^2_e \frac{(1 - r)}{h^2} \quad (3\text{-}1)$$

where σ_e is the standard deviation of elevation estimates (7 m), h is the spacing, and r is the correlation between elevation errors at adjacent points. If the correlation is zero (adjacent estimates are independent), then the uncertainty in slope estimates is no less than 33%, far too high for slope calculations to have practical value. On the other hand, if errors at adjacent points are perfectly correlated (r=1) then slope estimates have no uncertainty. In practice, r is very close to 1 and slope estimates are reliable enough to be useful. But since r varies over the map, there may be some areas where slope errors are notably higher.

In summary, while our ability to model and propagate uncertainty has improved markedly in the past few years, lack of knowledge of fundamental parameters of spatial structure of errors makes it impossible to answer even the most simple question about propagation, and makes it very easy to misinterpret simple visualizations of uncertainty such as Plate 3-1. Englund (1993) provides another illustration of this problem in the geostatistical context of GIS analysis using kriging.

THE NATURE OF SPATIAL DATA

To the outsider, a focus on spatial data might seem somewhat surprising or nonintuitive. Why not, for example, focus on meteorological data or ecological data? Essentially, the answer lies in experience that shows that spatial data have certain common characteristics and working with them raises a series of common or generic issues that are significant to environmental modelers, whether they be atmospheric scientists, ecologists, or integrated modelers. Accuracy, and the modeling of error, is merely one of an increasingly important set of data-related issues that must be addressed in any extensive application of spatial data.

Spatial data, or more precisely geospatial or geographic data, are defined as data that contain explicit geographic reference, and thus are tied to locations on the surface of the earth. The accuracy of the locational tie varies by many orders of magnitude, from the cm needed for analysis of crustal movement to the 100 m that is adequate for many environmental modeling purposes. For global climate modeling, or for international comparisons of UV exposure, a positional accuracy of tens or even hundreds of km may be adequate.

Within the domain established by this broad definition of spatial data, two major types are of importance to environmental modeling. They are termed here positional *framework* data sets, and sampled or measured *field* data sets. The defining characteristics of each type, and major issues associated with them, are discussed in the next two sections.

Framework Data

Framework data sets serve to establish positions of activities, samples, features, and observations with respect to the *geoid*, a mathematical function approximating the shape of the earth and anchored by the axis of rotation and the Greenwich meridian. To measure geographic position directly, one must rely on astronomical observation coupled with accurate timekeeping, or more recently on the global positioning system (GPS), which provides civilian users with a positional accuracy of 40 m for 50% of observations and 100 m for 98%. Differential GPS is capable of cm accuracy, but only with respect to a fixed point whose position with respect to the geoid must have been established by some other means.

For most practical purposes, position is established not by direct measurement but by indirect measurement with respect to some fixed point of known location. Clearly this fixed point must be well defined, a property more likely to be found in cultural than in natural features. Table 3-1 shows the sources of well-defined features and associated positions commonly used in the United States to measure position indirectly for various levels of positional accuracy.

The framework function of spatial data is to provide Earth locations of well-defined features, allowing other spatial data of various

Table 3-1: Sources of U.S. framework data for establishing geographic position.

Positional accuracy	Commonly used source of well-defined features
1cm	Geodetic control network
1m	Fixed monuments
10m	1:24,000 topographic mapping
100m	1:250,000 topographic mapping, GPS
1km	1:2,000,000 topographic mapping, Landsat
10km	AVHRR

types to be registered to the earth frame. Today, framework data sets are increasingly available in digital form on the Internet, allowing a fast, cheap, and reliable means whereby environmental scientists can build spatial data sets on a common framework. For many purposes, the most important framework data sets contain the locations of political or local government boundaries or centerlines of streets. For scientific modeling, however, it is important that the framework contain features of environmental significance, such as hydrographic unit boundaries, or the locations of weather stations.

In summary, framework data sets allow one to find oneself, or the locations of sample points or observations, with respect to well-defined points, and hence the geoid. Thus they allow data to be registered to the geoid. They provide suitable sources of control or registration points for Earth imagery, or for transformations of coordinates or the merger of data sets. Finally, framework data sets provide a reasonably well-defined positional accuracy that can be matched to the needs of any application.

Sampled or Measured Fields

The second major class of spatial data includes sample measurements of continuous variation of some parameter over a two-dimensional surface, or in three dimensions. Fields are often time-dependent, in which case sampling in time is also important. Fields provide the boundary conditions for many environmental models, as in the use of a field of topographic elevation in mesoscale atmospheric modeling. They provide the means to scale up or down, and to aggregate or disaggregate. Many fields are sampled remotely, either from the air or from space, using such instruments as AVHRR, Landsat, SPOT, airborne cameras, or airborne video. In other cases fields are sampled at points on the ground, for such parameters as surface temperature, elevation, soil moisture content, wind direction, or ground slope. In many cases the raw fields are processed, interpreted, and classified to produce fields of soil, land use, or land cover class.

In all such cases the focus of sampling or measurement is the more or less well-defined value of some variable which is conceived as having a unique value at every point on the two-dimensional surface or in the three-dimensional space. The variable is often a scalar but sometimes a vector, such as a wind field. It is represented digitally using a variety of methods (raster, triangular mesh [TIN], digitized contours, polygons [Goodchild 1993]), depending on the context and the nature of the data. Fields are registered to the geoid using a framework data set, which provides a source of control points and allows the sampling interval or spatial resolution of the measured field to be matched to the positional accuracy of the framework.

Technical problems abound in this process. For example, an Earth image can be registered to the geoid using only control points that are clearly visible in the image. This provides an effective upper limit of about 10 m on positional accuracy, and much worse in unpopulated areas with no recognizable well-defined features, unless special targets are placed on the landscape.

Framework Data Sets: A Paradox

Framework data sets have very little inherent scientific value to an environmental modeler. It is difficult, for example, to use the content of topographic maps as a data source for geomorphological studies of land forms (see Mark 1983 for an excellent example of the problems this can cause). Almost all spatial data of significance in environmental modeling fall into the category of sampled or measured fields. Yet framework data sets are essential to the effective use of fields. Without them, for example, it would be impossible to find field areas, to register data sets to each other, to compare Earth images from different time periods, or to map the nation's biological resources. Each of these processes requires the existence of framework data sets at appropriate levels of positional accuracy.

SUPPLYING THE NEED FOR FRAMEWORK DATA

Framework data sets are produced for numerous purposes, including a vast array of civilian and military activities in which science, and particularly environmental science, is a comparatively minor player.

Traditionally, the functions of framework data sets have been provided by paper maps, which have acted as readily available sources of position at approximately known levels of positional accuracy. The production of paper topographic maps has been a central government function in most countries, along with the maintenance of the primary geodetic control network that provides positional accuracies in the submeter range. However, the transition to digital dissemination of framework data has coincided with increasing pressure on central governments to recover costs. Traditional production of framework data sets is enormously costly and labor intensive, and it is difficult to build a broad base of popular support for this kind of expenditure. Moreover, science is traditionally cash poor and unable to contribute substantially to the cost of framework data set production. Another factor affecting the availability of framework data sets for science is national security, since accurate positioning of well-defined features is regarded in many countries as a valuable military resource.

Estes and Mooneyhan (1994) explore what they see as a series of popular misconceptions about mapping that ultimately affect the supply of framework data for environmental science. Although the American public and many other well-informed people in the world community believe that the world is well mapped, in reality much of the framework does not exist, is not available because of concerns for national security, or is old. This is a particularly worrisome problem for those fields of basic importance to science such as soils, land-surface characteristics, hydrography, or climate, where fields are either not observable from space, or require spatial resolutions that are not currently available from space, or are too expensive to create from Earth images.

For example, in 1993, 1:24,000 topographic map data, which potentially provide 10-m positional registration, were available in digital form for only a fraction of the conterminous United States (the transportation and hydrography layers were each available for 12% of the 54,145 quadrangles [Estes and Mooneyhan 1994]). Mapping at 1:100,000 or better (50-m positional accuracy) existed in paper or digital form for only 59% of the earth's land surface and was available to environmental scientists for some fraction of that percentage. Digital elevation data were available at a 30-m sampling interval for only 50%

of the conterminous United States. In summary, there are serious gaps in the framework both for the United States and for the globe as a whole, and there seems little prospect that the situation will improve dramatically in the near future. New technologies may provide an abundant and economical supply of digital elevation data and digital orthophotography, but the identification of cultural features and features invisible from space will remain costly and labor-intensive.

The National Spatial Data Infrastructure

Recently, an umbrella concept has emerged to embrace discussions of the national supply of framework data in the United States. The National Spatial Data Infrastructure (NSDI) is defined by the National Research Council as "the means to assemble geographic information that describes the arrangement and attributes of features and phenomena on the earth" (National Research Council 1993). The concept of NSDI clearly mirrors many of the objectives of the National Information Infrastructure (NII), or "information superhighway," in the specific context of spatial data. NSDI is conceived as including the human resource issues of education and training, the development of standards for data exchange, and improvement in Internet access mechanisms through the development of digital catalogs and metadata. At the core of NSDI, however, are the digital framework data sets that allow users to tie their own information and observations to the earth frame, and thus to integrate them geographically with other information.

Within the federal government, responsibility for the development of NSDI rests with the Federal Geographic Data Committee, an interagency group staffed by the U.S. Geological Survey and with representation from all major spatial data agencies. The creation of NSDI was mandated by an executive order signed by the president on April 11, 1994.

Although the precise contents of the NSDI framework data sets are still subject to intense debate, it is likely that they will include some combination of the following:

- The national geodetic control network of precisely defined point monuments, maintained by the National Geodetic Survey;
- Digital line graphs (DLG) created from the 1:24,000 topographic map series of the U.S. Geological Survey (in June 1993 approximately 11% of these maps of the conterminous United States were available in DLG form);
- Digital orthophoto quads (DOQ), corrected panchromatic images with 1-m pixel size and approximately 6-m positional accuracy, each covering one quarter of the area of a 1:24,000 quadrangle (although only a small fraction of the conterminous United States is currently covered by DOQ, there are plans to produce them rapidly through a series of interagency and federal-state partnerships);
- Street centerline data, digital representations of the centerlines of roads, highways, and streets, with a positional accuracy of 10 m or poorer (the currently available TIGER data, produced by the U.S. Geological Survey and the Bureau of the Census, is being upgraded through a combination of public and private sector efforts); and
- Digital elevation models (DEM), digital representations of terrain based on a 30-m horizontal sampling interval (roughly 50% of the conterminous United States available in June 1993).

In the terminology of this chapter, DEM are measured fields, but they are frequently used as high-accuracy sources of such framework features as watershed boundaries.

Although the federal government is now committed to establishing NSDI, its potential costs are enormous. With modern technology it is possible to produce a DOQ for as little as $2,000, with a DEM as a by-product. But with 200,000 DOQ needed to cover the conterminous United States, the cost of a single complete coverage is at least $400 million. One advantage of the relatively low cost of DOQ is that they can be updated frequently, but the cost of update is no less than the cost of initial coverage. Constant efforts are being made by the mapping agencies to find funding for programs like DOQ coverage, but they face the harsh political reality that mapping is not the most effective way to buy votes. To return to an earlier point, the taxpayer believes, by and large, that the United States is well mapped and that mapping is a simple and cheap process. After all, it has been decades since the last blank parts of Africa and Antarctica were filled in, and centuries since "terra incognita" appeared on world maps.

Toward a Global Infrastructure

While recent debate over NSDI suggests a widely recognized need for a coordinated approach to framework data in the United States, it seems appropriate at this stage to ask about parallel efforts at a global or international level. The Digital Chart of the World (DCW) provides a framework data set with a positional accuracy of 500 m, based on United States 1:1,000,000 mapping. Global DEM data is available with similar levels of resolution. Although there are no truly international format standards for spatial data to promote data exchange, many national and commercial standards are widely recognized, and NATO has developed standards in the interests of common defense. There are many excellent efforts to establish spatial data clearinghouses, including the U.N. Environment Program's GRID, the efforts of CIESIN, and the World Conservation Monitoring Center's spatial data resources.

These efforts notwithstanding, remote sensing remains virtually the only reliable source of truly global coverage. Dimensions of spatial variation that are not visible from space, including political boundaries, place names, and other cultural features, continue to pose enormous problems to environmental scientists needing reliable sources of framework data, and little progress has been made in utilizing remote sensing to detect social, demographic, industrial, or economic variation. The problems of building international spatial databases are illustrated by CIESIN's CITAS (China in Time and Space) program, which aims to build a spatiotemporal database of Chinese demography. Although the Chinese population is arguably the world's best-counted, with accurate records extending back hundreds of years, on average one change per month is made in the internal boundaries of Chinese counties, making any longitudinal perspective virtually unattainable without complex spatial data handling technology; variation and duplication of place names merely adds to the difficulties.

Clearly, the problems being addressed nationally by NSDI need to be tackled also at the global level. Earth systems science requires a reliable source of framework data so that observations can be positioned with respect to the earth frame and integrated with other data for modeling and policy formulation. Emerging environmental programs like GLOBE cannot succeed without a sound framework, and associated programs to address the needs of education and training, standards, and protection of copyright, national security, and intellectual property.

CONCLUSION

The first two of the following six concluding points echo Estes and Mooneyhan (1994):

1. We need to dispel the myths that exist among the public, politicians, policy makers, key agencies, and our administra-

tors and colleagues in the scientific community that the maps and spatial data needed to support environmental modeling and related science exist, that mapping is easy, and that there is no science left to do in mapping.

2. The availability of spatial data is often adversely affected by concerns over national security and sovereignty. We need to find a better balance between these concerns and the scientific need to understand Earth as a system.
3. We must develop and maintain the framework data sets that are needed to register scientific measurements, models, and predictions to the surface of the earth with accuracy appropriate to each application.
4. Besides the framework data sets themselves, the objectives of NSDI include the development of the means to assemble, process, store, analyze, and disseminate spatial data, and the human resource base that will make this feasible.
5. We need to implement the objectives of NSDI at the international and global levels and to find appropriate mechanisms for doing so.
6. Many of the objectives of NSDI and its proposed global equivalent raise questions about research issues such as data accuracy that will have to be solved before these programs can be fully implemented. We need a concerted effort to address the outstanding research issues of spatial data, many of which are identified in other chapters in this volume.

References

Beard, M. K., B. P. Buttenfield, and S. Clapham. 1991. *Visualizing the Quality of Spatial Information: Scientific Report of the Specialist Meeting.* Technical Report 91-26. Santa Barbara: National Center for Geographic Information and Analysis.

Burrough, P. A. 1986. *Principles of Geographical Information Systems for Land Resources Assessment.* Oxford: Clarendon.

Buttenfield, B. P., ed. 1993. Mapping data quality: A collection of essays. *Cartographica* 30(2–3):v-viii, 1–46.

Cressie, N. A. C. 1993. *Statistics for Spatial Data.* New York: Wiley.

Englund, E. 1993. Spatial simulation: Environmental applications. In *Environmental Modeling with GIS,* edited by M. F. Goodchild, B. O. Parks, and L. T. Steyaert, 432–46. New York: Oxford University Press.

Estes, J. E., and D. W. Mooneyhan. 1994. Of maps and myths. *Photogrammetric Engineering and Remote Sensing* 60(5):517–24.

Fisher, P. F. 1991. Modeling soil map-unit inclusions by Monte Carlo simulation. *International Journal of Geographical Information Systems* 5:193–208.

———. 1992. First experiments in viewshed uncertainty—simulating fuzzy viewsheds. *Photogrammetric Engineering and Remote Sensing* 58:345–52.

Goodchild, M. F. 1993. Data models and data quality: Problems and prospects. In *Environmental Modeling with GIS,* edited by M. F. Goodchild, B. O. Parks, and L. T. Steyaert, 94–104. New York: Oxford University Press.

Goodchild, M. F., G. Q. Sun, and S. Yang. 1992. Development and test of an error model for categorical data. *International Journal of Geographical Information Systems* 6:87–104.

Haining, R. P., and G. Arbia. 1993. Error propagation through map operations. *Technometrics* 35:293–305.

Hunter, G. J., and M. F. Goodchild. 1995. Dealing with uncertainty in spatial databases: A simple case study. *Photogrammetric Engineering and Remote Sensing* 61(5):529–37.

Isaaks, E. H., and R. M. Srivastava. 1989. *Applied Geostatistics.* New York: Oxford University Press.

Mark, D. M. 1983. Relations between field-surveyed channel networks and map-based geomorphometric measures, Inez, Kentucky. *Annals, Association of American Geographers* 73(3):358–72.

Mark, D. M., and F. Csillag. 1989. The nature of boundaries on "area-class" maps. *Cartographica* 26(1):65–78.

National Research Council. 1993. *Toward a Coordinated Spatial Data Infrastructure for the Nation.* Washington, D.C.: National Academy Press.

Michael F. Goodchild is Professor of Geography at the University of California, Santa Barbara, California. He holds a B.A. in physics from Cambridge University and a Ph.D. in geography from McMaster University. His research interests include accuracy of spatial databases, geographic data models, and geographical analysis. He was editor of *Geographical Analysis,* and has coedited three books on GIS including *Geographical Information Systems: Principles and Applications* and *Environmental Modeling with GIS.*

National Center for Geographic Information and Analysis
University of California
Santa Barbara, California 93106-4060
E-mail: *good@ncgia.ucsb.edu*

4

Status of Land Data for Environmental Modeling and Challenges for Geographic Information Systems in Land Characterization

Louis T. Steyaert

Ongoing research to develop a new generation of environmental simulation models has significantly increased the demand for land surface data and the corresponding need for the expanded use of remote sensing and geographic information systems (GIS) technologies for database development. The expanded requirements for land data are clearly illustrated by land-atmosphere interactions research, a component of Earth system modeling concepts developed in the mid-1980s. Multiresolution land surface data on topography, soils, and land cover characteristics are needed in nested grid model simulations. These simulations may involve the scaling of atmospheric, hydrologic, and ecologic processes from local to regional and global scales, and conversely, from larger to smaller scales. To meet these data needs, database developers are working with the simulation modelers to define land data requirements, promote interagency coordination and data transfer standards, address spatial data accuracy and error propagation issues, publish reference guides, establish online metadata searches, and develop new databases, especially by using remote-sensing technology for land cover characterization. New regional databases in the conterminous United States include the 1:250,000-scale State Soil Geographic Database, consistently processed 1-km advanced very high-resolution radiometer (AVHRR) time-series image composites, and a prototype 1-km AVHRR seasonal land cover characteristics database. Historic data sets now consolidated for continental to global studies include a global ecosystems database, global GRASS databases and the Digital Chart of the World. New global databases scheduled for release in 1994 include ten-day time-series image composites of 1-km AVHRR for global land areas and a comprehensive 1° x 1°-gridded database for global change research modelers. Enhanced GIS mapping tools are needed to conduct multiscale land characterization research and to tailor land surface data sets for use in environmental simulation models.

INTRODUCTION

Since the mid-1980s, ongoing research to develop a new generation of environmental simulation models has significantly increased the demand for multiresolution land data. The purpose of these computer-based, numerical models is to simulate three-dimensional, time-dependent environmental processes in a realistic manner, in part, as the basis for understanding the natural environment, making model predictions, and assessing impacts. The cross-disciplinary research approach emphasizes coupled-systems modeling in which nested atmospheric models are linked with various hydrologic and ecologic models. Although originally conceived as part of multiscale land-atmosphere interactions modeling for global climate change research programs, this coupled-systems modeling approach is increasingly contributing to the understanding of terrestrial ecosystem dynamics, management of soil and water resources, and assessment of environmental risk from contaminants.

Because the physical and biophysical properties of the land surface exert fundamental controls on land surface processes, these environmental simulation models require a wide variety of land surface data. Basic topographic, soils, and land cover characteristics data sets at local, regional, and global scales are an essential starting point. In addition, these three land surface data sets are increasingly integrated with various operational near-surface hydrometeorological data sets as part of model development and assessment activities. Tailoring these environmental databases for individual model specifications, developing derivative data products, and integrating and analyzing these types of land surface databases are some of the opportunities for developing advanced geographic information systems (GIS) tools.

This chapter reviews the state of land surface data development for environmental simulation modeling and suggests some challenges for GIS technology in land characterization research and environmental mapping. It focuses on the availability of multiresolution topographic, soils, and land cover characteristics data to develop and test land-atmosphere interactions models. This chapter emphasizes regional- to global-scale modeling typically based on 1-km resolution data. Several land data sets in progress are noted, including various hydrometeorological land data sets. The following sections describe the models and their land data needs, summarize recent advances in land database development for these types of models, and suggest opportunities and challenges for GIS technology to facilitate land characterization research and land database development. Several chapters in this volume are cited as examples illustrating some of these new databases and GIS tools.

Note: Any use of trade, product, or firm names is for descriptive purposes only and does not imply endorsement by the U.S. Government.

LAND DATA NEEDS OF ENVIRONMENTAL SIMULATION MODELS

Global change research programs have defined an advanced coupled-systems approach for the integrated modeling of the Earth system (Bretherton 1985; Earth System Sciences Committee [ESSC] 1986, 1988; National Research Council [NRC] 1990; and International Geosphere-Biosphere Programme [IGBP] 1990). Land-atmosphere interactions, also termed biosphere-atmosphere exchange processes, form a major subsystem of the earth system and deal with understanding and modeling of exchanges between the terrestrial ecosystems and the atmosphere. One goal is to model multiscale processes and their interactions at spatial scales that range from local to global, and in temporal scales that range from seconds to millennia.

Land-atmosphere interactions modeling provides a framework for discussing the expanded role of land data in contemporary environmental simulation modeling. This coupled-systems modeling approach illustrates the types of models and the preliminary steps to link these models across time and space scales, hence the requirement for multiresolution land data. The modeling approach also illustrates the wide variety of land surface characteristics that are needed for the simulations, the role of these land characteristics in controlling basic land surface processes, and the importance of remote-sensing technology in process-oriented field experiments as one key to database development.

Overall descriptions of the land-atmosphere modeling approach are provided in the context of Earth system modeling by the NRC (1990), Trenberth (1992), and the National Aeronautics and Space Administration (NASA) (NASA 1993a). The modeling builds on previous research; for example, see Pielke (1984) for a discussion of mesoscale meteorological modeling. Farmer and Rycroft (1991) discuss computer simulation modeling in the environmental sciences and provide examples of the application of atmospheric and hydrologic models for air and water quality assessment. Goodchild et al. (1993) also review land-atmosphere interactions modeling, as well as other types of environmental models, but in the context of integrating GIS technology with the models.

The NRC (1990) categorized land-atmosphere interactions models according to the characteristic time-step interval of each numerical iteration within the simulation: (1) short time-step intervals of seconds to hours for water and energy exchange processes in the soil-plant-atmosphere system; (2) intermediate time-step intervals of days to seasons for biologic and ecosystem dynamics processes (nutrient cycling, growth and development, biomass production) and some soil biogeochemical processes; and (3) annual time steps for longer-term processes (decades to centuries), such as biogeochemical cycling (soil development) and ecological processes.

Although somewhat arbitrary, this categorization does illustrate three major classes of models representing short, medium, and longer time-scale processes. Contemporary research on atmospheric, hydrologic, and ecologic modeling is focused on improving each type of model at its respective time scale, while also making the first steps to link the models across time and space scales. Land surface parameterizations, one major class of models representing short time scales, are used in global-scale atmospheric general circulation models (GCM) and mesoscale meteorological models to account for water, energy, and mass transfer between the land surface and the atmosphere. Examples include the biosphere-atmosphere-transfer-scheme (BATS) (Dickinson et al. 1986), the simple biosphere model (SiB) (Sellers et al. 1986) and the land-ecosystem-atmosphere-feedback (LEAF) model (Lee et al. 1993). Examples of distributed parameter watershed models and other types of hydrologic models operating at intermediate time scales include Leavesley and Stannard (1990), Gao et al. (1993), Engel et al. (1993), and Smith et al. (1993). Examples of ecological models operating at various time scales include ecosystem dynamics (daily and annual time steps: Running and Coughlan 1988), biogeochemical cycles (monthly time steps: Parton et al. 1987; Schimel et al. 1991), and ecological succession models (annual time steps: Botkin et al. 1972; Shugart 1984). A more detailed discussion on each type of model is provided by Steyaert (1993).

The coupling of atmospheric and hydrologic models across scales is a key research topic, especially for water resource assessment (i.e., see Leavesley et al. 1992; Hay et al. 1993). This coupled-systems modeling approach also demonstrates the need for multiresolution land data for models operating at different grid cell sizes and computational domains. The modeling illustrates the wide variety of land surface properties needed in the simulations. For example, Lakhtakia et al. (this volume) describe the initial steps for coupling atmospheric, hydrologic, and assessment models for water resource assessment and potential climatic impact studies within the Susquehanna River basin. In the research project, a set of nested mesoscale models are linked to an atmospheric GCM to scale atmospheric processes from the global to the watershed levels. Typically, the nested mesoscale models are based on nested grids of 64, 16, and 4 km. Land surface processes are accounted for in each atmospheric model by using the BATS land surface parameterization for biosphere-atmosphere coupling. As described by Smith et al. (1993), a soil hydrology model is used to estimate soil moisture for initializing BATS. The mesoscale models provide the atmospheric forcing data to various types of watershed and impact assessment models, typically operating at daily to seasonal time scales for different scale watersheds.

The atmospheric and hydrologic models described in the preceding example require a wide variety of multiresolution land characteristics data on terrain, soil, and land cover. Some of the specific requirements for land data to develop, initialize, and validate land-atmosphere interactions models are described by the NRC (1990), IGBP (1990, 1992), World Climate Research Programme/International GEWEX Project Office (1993), and NASA (1993b). For example, topographic data are needed to derive estimates of slope, aspect, stream networks, hydrologic response units, and many other watershed characteristics important to land processes. Essential soils data include class, texture, available water capacity, soil depth, rooting depth, albedos, emissivities, porosity, pH, and thermal conductivity.

The following four types of general land cover characteristics data are needed: (1) land use, (2) community, composition, and structure, (3) disturbance history, and (4) biophysical properties. Community, composition, and structure include data on vegetation class, percentage composition, secondary ground cover, stand age, tree density, percent crown closure, height, patch size, ratio of area vegetation cover to area bare soil, percent sunlit canopy, percent shadow, and percent sunlit background. Disturbance history includes data on fires, storm damage, clearing, grazing, pests and disease, and successional stage. A wide range of biophysical characteristics are required including temporal data on albedos, leaf area index, seasonal phenology, biomass (foliar, total, etc.), canopy conductance, leaf architecture, leaf optical properties, and fraction of absorbed photosynthetically active radiation.

There are several important land data issues in land-atmosphere research. The role of land surface characteristics in controlling fundamental processes is not entirely understood; therefore, sensitivity analysis studies are conducted. The development of parameterizations to account for subgrid processes is a key research topic. This includes the aggregation of heterogeneous landscape patches that have fundamentally different biophysical characteristics. Most types of land surface characteristics data have not been extensively measured and mapped. Even the basic data sets on terrain, soils, and land cover are limited, especially at global scales. For example, the NRC (1990) noted the need for consistent global land use change and soils data. The inconsistencies among existing global land cover databases are clearly illustrated by Townshend et al. (1991).

Many of these land data issues are addressed in process-oriented field experiments such as the International Satellite Land Surface Climatology Project (ISLSCP) (NASA 1993b) and the Global Energy and Water Cycle Experiment (GEWEX) Continental International Project (GCIP) (WCRP/IGPO 1993). Such field experiments are designed to understand basic processes (for example, water and energy exchange, mass transfer, trace gas fluxes, photosynthetic and other biological activity, carbon budget, atmospheric convection, and mesoscale circulations) and their multiscale interactions.

The role of these field experiments in process research, remote-sensing algorithm development, and land-atmosphere model improvement is defined by NASA (1993b). Process-level research and multistage observations based on gound systems, airborne instruments, and satellites contribute to advanced remote-sensing algorithms for developing land cover characteristics data sets and making regional extrapolations. Research is determining how land surface properties help to control land surface processes such as the carbon budget or water and energy exchange. Increasingly, such field experiments are addressing the parameterization and modeling of land surface processes within a complex terrain and heterogeneous landscape environment.

Remote sensing helps in understanding processes and in developing data sets to initialize and validate models. Remote-sensing algorithms are needed to develop land cover characteristics data for models and to make regional extrapolations based on process-level research results. Remote sensing provides information concerning the spatial distribution and the temporal variability of dynamic land cover characteristics such as albedo, aerodynamic roughness, canopy conductance, and other biophysical attributes that are needed by the models. Because of the land data development and analysis requirements, the integration of remote-sensing technology and environmental simulation models is increasingly complemented by the use of GIS.

STATUS OF LAND DATA FOR ENVIRONMENTAL MODELING

Sources on the availability of land data for environmental simulation modeling have significantly improved in recent years. Environmental research programs, especially global change, and the growing use of GIS for environmental decisions, such as those involving air and water quality management, are contributing to enhanced data and information resources. The producers of land data are working more closely with modelers to define requirements, promote interagency coordination and data transfer standards, address spatial data accuracy and error propagation issues, publish reference guides, establish online metadata searches, develop new land databases, and integrate technologies such as remote sensing and GIS with the models. The following subsections describe some of these activities and review the status of some important land data sets now available for environmental simulation modeling.

Finding Information on Land Databases

General information on land data quality, accuracy, standards, availability, proposed management and information systems, and reference guides are available from published reference materials. The number of facilities for computer-based searches is growing. Recent publications addressing the general state, quality, availability, utility, and information system requirements of land data for global change include Mounsey and Tomlinson (1988), Townshend (1991), Clark et al. (1991), and Glaeser and Ruttenberg (1992). Estes and Mooneyhan (1994) provide a detailed summary on the worldwide availability of cartographic data sets. Burrough (this volume) and Aspinall and Pearson (this volume) address land data quality and error analysis issues. The Committee on Earth and Environmental Sciences (1992) outlines the management program plan for the U.S. Global Change Data and Information System (GCDIS), and a GCDIS implementation plan is being reviewed. These supplement the references cited in the preceding section.

Several published reference guides summarize the available land data sets. For example, many georeferenced databases of interest to environmental modelers are documented by the Federal Geographic Data Committee (FGDC) (FGDC 1992) in a manual that describes federal geographic data products, including maps, digital data, aerial photographs and multispectral images, Earth science, and other geographically referenced data sets. This FGDC manual is part of a larger effort to develop a strategic plan for the National Spatial Data Infrastructure (NSDI) (FGDC 1993; NRC 1993). Some major components of the proposed NSDI would include strengthening federal spatial data coordination efforts, forming or strengthening state coordinating mechanisms, cultivating partnerships with the private sector, and ensuring effective means for finding and sharing spatial data.

As another example, Kelmelis and Rowland (1994) provide extensive summaries of data sources and descriptive information concerning data availability for research on land use and land cover change. Other examples include guides concerning selected national environmental statistics in the U.S. Government (EPA 1992), a selective guide to climatic resources (NOAA 1988), special data sets concerning carbon dioxide research (ORNL 1993), and core data sets for the long-term ecological research network (Michener et al. 1990).

Many data sets are now published on CD-ROM. The Special Interest Group on CD-ROM Applications and Technology (SIGCAT) has published a CD-ROM compendium (USGPO 1993). This compendium provides information on the source data, the vendor or supplier, technical details, and the status of a wide variety of CD-ROMs containing data sets of interest to environmental modelers. Although data sets published by federal agencies are emphasized, information is also provided on many private sector vendors that sell federal data sets. The types of CD-ROM data listed in the SIGCAT compendium include scientific abstracts, data directories, map indexes, a wide variety of satellite data, digital orthophotoquad data, water quality data, streamflow data, climatic data, geographic names, geologic data, oceanographic data, numerical model gridded data, upper atmospheric data, and radar and sonar data sets.

Environmental modelers also have access to growing resources for computerized searches for data, metadata, and other types of information concerning land data, especially through client/server systems on the Internet. Typically, the client/server architecture provides direct access to open systems such as Wide Area Information Servers (WAIS) and World Wide Web (WWW). Abbott (1994) lists more than 500 WAIS-accessible databases containing tens of thousands of text-searchable files and documents. Gopher, another resource for searches on the Internet, provides a user-friendly interface between library catalogs, campus information services, and government databases (Abbott 1994). These full-text, keyword search utilities are complemented by the

WWW, a hypermedia system that provides text, graphics, sound, and animation. The National Center for Supercomputer Applications has written a WWW client called MOSAIC. MOSAIC is available by anonymous ftp (*ftp.ncsa.uiuc.edu*) for X-Windows, PC Microsoft Windows, and Macintosh environments. MOSAIC permits direct access to databases through WWW.

These facilities for open-system searches on the Internet complement the more traditional structured systems for metadata searches that provide information on available data sets and supplementary information on data centers, projects, observation systems and instruments, and sensors. Some of the online metadata search systems include the Global Change Master Directory (GCMD) of GCDIS at NASA, the NOAA Environmental Services Data Directory (NOAADIR) of the National Oceanic and Atmospheric Administration (NOAA), and the Global Land Information System (GLIS) of the U.S. Geological Survey (USGS). The Consortium for International Earth Science Information Network (CIESIN) has developed the Information Gateway to provide access to CIESIN's information guides, applications, and services. Beier (1992) lists global change data sets based on excerpts from the NASA Master Directory (MD), now part of the GCMD. Kelmelis and Rowland (1994) provide significant background information on the NASA MD and other online metadata information sources. The NASA MD/GCMD will soon be an X-client server. The GCMD, NOAADIR, GLIS, and other metadata information systems are linked within the GCDIS.

In a related activity, the FGDC is developing a prototype geospatial data clearinghouse that will propose metadata standards and search capabilities within the federal, state, regional, and local governments.

Status of Key Land Surface Data Sets

As described in the preceding sections, some of the key land data sets required by environmental simulation models include multiresolution data on topography, soils, and land cover. Increasingly, environmental simulation models also require daily and seasonal information on the dynamic biophysical characteristics of land cover conditions. The status of these data sets is summarized in terms of data at local and regional scales for the conterminous United States and also at the global scale. The focus is on land data sets that are now available for modeling; selected data sets now in progress are noted. Some brief remarks are included on the growing availability of land surface meteorological and hydrological observations, Doppler rainfall estimates, and mesoscale model outputs. These data sets provide relevant information concerning the dynamic nature of landscape characteristics. Such data are needed for integration with more traditional data sets for environmental mapping, monitoring, and modeling. Hydrometeorological data are increasingly needed for remote-sensing applications; for example, atmospheric correction and vegetation seasonality studies.

Hydrographic, transportation, and other support databases. Ancillary spatial data on political boundaries, transportation networks, streams and water bodies, hydrographic boundaries, and cultural features contribute to the development and application of environmental simulation models at local to global scales. These data facilitate ground-truth study and interpretation of remote-sensing data, analysis of results from model simulations, investigation of possible human factors in simulation modeling, and assessment of the socioeconomic impacts of environmental processes. The overall status of these types of spatial data is summarized by Goodchild (this volume).

Several vector databases are available on CD-ROM for use in environmental simulation modeling. The USGS has produced a CD-ROM for the 1:2,000,000-scale digital line graph (DLG) data extracted from the *National Atlas of the United States for all 50 States*. Data types include political boundaries, administrative boundaries, streams, water bodies, hypsography (Continental Divide only), railroads, and cultural features. The USGS has placed 1:100,000-scale DLG hydrography and transportation on CD-ROMs for the United States (minus Alaska).

The USGS has also developed an ancillary data set for the conterminous United States and Alaska to help interpret 1-km advanced very-high-resolution radiometer (AVHRR) data sets (see *Land cover characteristics data*). These raster data sets are coregistered to the AVHRR data. Some of the ancillary data sets on the U.S. Companion CD-ROM include USGS hydrologic units, DLG state and county boundaries, U.S. Department of Agriculture Soil Conservation Service (SCS) land resource areas, Environmental Protection Agency (EPA) ecoregions, and NOAA climatic regions.

The Digital Chart of the World (DCW), now available on a set of four CD-ROMs, is a 1:1,000,000-scale base map of the world developed by the Defense Mapping Agency (DMA) with the cooperation of Australia, Canada, and the United Kingdom (DMA 1992). The primary source of the database is the DMA Operational Navigation Chart series, which is the largest scale unclassified map series with consistent, continuous global coverage and essential base map features. The DCW is organized into seventeen thematic layers, including political boundaries, ocean coastlines, cities, transportation networks, drainage, land cover, and elevation contours. The database also includes a global index of more than 100,000 place names. For some regions of the world, DCW data such as elevation contours must be used with caution and may require additional processing and checks for accuracy and consistency.

Digital elevation data. Multiresolution digital elevation data are essential to environmental simulation modeling. Topography influences atmospheric, hydrologic, and ecologic processes. For example, microclimate, local wind circulations, precipitation-runoff processes, and vegetation composition are strongly related to elevation, slope, and aspect. Terrain correlates with soils. One challenge for modelers, including the spatial analysis community, is to understand these interdependencies. In addition to researching watershed processes, hydrologists use digital elevation models (DEM) to generate stream networks and watershed boundaries; for example, based on algorithms developed by Band (1986) and Jenson and Domingue (1988).

The use of DEM in environmental models depends on the type of model and the scale of analysis. For example, atmospheric models generally require DEM at a cell size that is about one-quarter the size of the model grid cell. The grid cell sizes for atmospheric GCMs can go down to about 100 km, and mesoscale model runs based on a 1-km grid cell are not uncommon. In contrast, hydrologic models may require DEM for spatial scales that include small watersheds, entire river basins and continents as part of the global hydrologic modeling. These scales correspond to DEM of a few meters, 100 meters, and a kilometer in resolution.

Hydrologists have several types of DEM available for regional research, and work is proceeding to develop enhanced global-scale data sets. Hydrologic modeling of small watersheds within the conterminous United States can use USGS DEM data in 7.5-minute units that are now available for many regions of the country, especially the mountainous western states. River basin scale studies can use USGS 1°-DEM data that have a 3-arc-second horizontal resolution. The EROS Data Center of the USGS has also developed a

DEM with a 0.5-km resolution for the conterminous United States and Alaska. Previously, the NOAA National Geophysical Data Center developed a 30-arc-second DEM for the United States.

However, the availability of consistent DEM data for the globe is currently limited to the ETOP05 data set (nominally 10-km resolution). Work is in progress at the EROS Data Center to use quality control in developing a consistent global DEM at an approximate resolution of 1 km, based on digital elevation contours within the DCW. As continental and global hydrologic modeling techniques mature, consistent global DEM at higher resolutions than 1 km will be required (NASA 1993b).

Digital soils data. Environmental simulation models need multiscale soils data to model soil moisture variability, water and trace gas fluxes, nutrient uptake, biogeochemical cycles, ecological succession, regional analysis, and many other processes that depend on soil characteristics. The SCS is developing three key soil-geographic databases, including the Soil Survey Geographic Database (SSURGO), the State Soil Geographic Database (STATSGO), and the National Soil Geographic Database (NATSGO) (Reybold and TeSelle 1986). Regional modeling within the conterminous United States will benefit from the 1:250,000-scale STATSGO data that are being placed on CD-ROM by the SCS. The status of the STATSGO, SSURGO, and NATSGO databases is described by Lytle et al. (this volume). A consistent global soils database at a 1:1,000,000 scale is needed by environmental simulation modelers (NASA 1993b).

Land cover characteristics data. Detailed multiresolution data on land cover and its biophysical attributes are essential to the entire range of land-atmosphere models. For example, key land surface characteristics such as albedo, surface roughness, and leaf stomatal conductance act in concert to control fundamental land processes such as the radiative, momentum, and heat balance fluxes, respectively. Other important biophysical properties that are related to vegetation structure and spatial distribution include measures of the fractional vegetation cover, leaf area index, leaf optical properties, and leaf angle distributions, as described by Dickinson et al. (1986) and Sellers et al. (1986).

Because such detailed land cover attributes are typically available only from limited field observations and laboratory measurements, the land cover classification takes on added importance (IGBP 1992). One commonly used approach is to regionally extrapolate or infer such land cover characteristics based on the land cover classification for each model grid cell. In some cases, remote-sensing technology is used to derive vegetation characteristics such as leaf area index, a key variable in many models. The use of daily, 1-km AVHRR data as the basis for land cover characterization is a significant step toward consistently developed regional and global land cover databases.

The USGS EROS Data Center has compiled 1-km resolution AVHRR time-series data sets at selective times during the past five years for the conterminous United States, Alaska, Mexico, and Eurasia (Eidenshink 1992; Loveland and Scholz 1993). Data sets now available on CD-ROM include 1990–1994 for the conterminous United States and 1991–1994 for Alaska. The data are georeferenced, calibrated, and composited on the maximum value normalized difference vegetation index (NDVI) "greenness" image compositing technique. The time-series images are composited biweekly for the conterminous United States and every fifteen days for Alaska. These conterminous United States and Alaska data sets consist of ten channels of composited information, including channels 1–5 (calibrated), NDVI, satellite zenith, solar zenith, relative solar and satellite azimuth, and date of pixel observation.

The EROS Data Center, with support from the University of Nebraska-Lincoln, has developed a prototype land cover characteristics database for the conterminous United States (Loveland et al. 1991). The experimental database was developed as part of the USGS global change research program as a prototype for a potential global land cover database. Development focused on two aspects: (1) defining a conceptual format for a global land cover database that meets the broad needs of modelers, and (2) testing classification methods that could be applied globally.

Details of the methods used to produce the AVHRR land cover characteristics database are provided by Loveland et al. (1991) and Brown et al. (1993). In brief, 1990 maximum value composite 1-km multitemporal AVHRR NDVI data were classified using an unsupervised classification technique. Postclassification refinement using ancillary data—including elevation, selected climatic normals, and ecoregions—was used to resolve cases in which classes represented disparate land cover types. This resulted in 159 seasonal land cover classes, each representing unique combinations of land cover mosaics, seasonal properties (onset and peak of greenness and length of green period), and relative levels of primary production. Detailed attributes describing vegetation and land cover, seasonality, spectral properties, elevation, climate, ecoregions, and soil characteristics were developed from ancillary data. In addition, translation tables that link the 159 classes to commonly used land cover classification systems, such as BATS, SiB, LEAF, and the USGS Anderson Level II system, were developed so the database could be tailored to suit specific user needs. The prototype database is now available on CD-ROM.

These AVHRR-based products complement other available land cover data and new digital data for land characterization. The USGS digital land cover and land-use data at 1:250,000 and 1:100,000 scales are available for the United States. Historical Landsat data are now available for selected regions worldwide through GLIS. The program to develop high-resolution digital orthophotoquads based on aerial photography is expanding. Three new sources for high-resolution radar data include the European Space Agency's Remote Sensing Satellite (ERS-1), the Japanese Space Agency's JERS-1 data, and the planned Canadian remote-sensing satellite, RADARSAT.

Two recent programs based on Landsat data, the North American Landscape Characterization (NALC), and the Multi-Resolution Land Characteristics (MRLC) Monitoring System, will contribute to land cover data resources. The EPA is sponsoring the NALC project, which is designed to use the twenty-plus years of historical Landsat multispectral scanner (MSS) data for characterizing landscape features and conducting change detection studies. The goal of NALC is to produce standardized data sets for the majority of North America. land cover change products for the 1970s, 1980s, and 1990s will be based on Landsat MSS data forming a set of three-date, georeferenced, MSS triplicates. The NALC activity is one of NASA's Pathfinder projects developed as part of the U.S. Global Change Research Program to be a "pathfinder" for advanced Earth Observation System technologies. The USGS EROS Data Center is providing support in acquiring and processing data, managing the MSS triplicate data archive, and producing and disseminating data sets for the United States and countries to the south.

In the second effort, several federal agencies are collaborating to develop a regional land cover characteristics database for the United

States using Landsat Thematic Mapper (TM) data. Participating agencies and programs include the EPA's Environmental Monitoring and Assessment Program, the U.S. Fish and Wildlife Service Gap Analysis Project, the USGS National Water Quality Assessment Program, and the NOAA Coastwatch-Change Analysis Program. Wall-to-wall Landsat TM data will be used in combination with various ancillary data sets to develop a regional land cover characteristics database that will permit the flexibility to tailor land cover characteristics for individualized project needs. This is analogous to the flexible database concept demonstrated with the 1-km AVHRR land cover characteristics database previously described. The integrated use of AVHRR and Landsat TM is the foundation for MRLC. The USGS EROS Data Center is providing technical support on MRLC database development.

Various agencies and international organizations are addressing the need for global scale land cover data. For example, recent data set development activities for continental to global studies include a Global Ecosystems Database developed by NOAA and EPA (see Kineman et al., this volume). The database was produced by NOAA's NGDC and includes data on land use, ecosystems, wetlands, satellite-derived vegetation indexes, climate, topography, and soils. Another major contribution is the Global GRASS Databases developed by the Army Corp of Engineers and Rutgers University (Lozar, this volume). There are now four global GRASS data sets on CD-ROM (Global GRASS 1–4). The NOAA and GRASS CD-ROMs bring together many key global land data sets such as the land cover and soils data described by Henderson-Sellers et al. (1986).

These global databases have been supplemented by an international effort since April 1992 to collect and prepare consistently processed, daily 1-km AVHRR data sets for global land areas; this work is being done under the auspices of the U.S. Global Change Research Program, the IGBP, and other national and international agencies. The proposed plan for this international activity was established by IGBP (1992). Following processing by the USGS EROS Data Center, ten-day time-series image composites of consistently processed 1-km AVHRR data will be available for widespread distribution.

The IGBP Data and Information System's Land Cover Working Group (LCWG) is encouraging and coordinating the development of a global land cover database derived from these ten-day AVHRR image composites. In addition to the USGS, participating organizations include the Committee on Earth Observing Satellites, NASA, NOAA, the European Space Agency, the Commission of the European Communities, and the Commonwealth Scientific and Industrial Research Organization of Australia. Over the next several years the LCWG will develop a global land cover classification at 1-km resolution, estimate by direct parameterization key land cover variables (i.e., albedo and LAI), map functional vegetation and land cover classes, and validate the classifications.

As proposed by NASA (1993b), a global 1°-land cover database was made available in 1994 as part of the ISLSCP CD-ROM. This data set was developed from an 8-km database derived from AVHRR global area coverage (GAC) data. In addition to providing land cover type, the vegetation database includes estimates of phenology, LAI, fraction of absorbed photosynthetically active radiation, and other biophysical properties. Additional types of data on this CD-ROM are described in the following section.

Hydrometeorological land surface data. There is a broad class of hydrometeorological land surface data sets of interest to the GIS community. Examples of such hydrometeorological data at the land surface include precipitation, temperatures, wind, humidity, cloud conditions, solar insolation, photosynthetically active radiation, reflected solar radiation, downward atmospheric infrared radiation, outgoing terrestrial infrared radiation, sensible heat flux, latent heat flux, ground heat flux, soil moisture, infiltration, and surface runoff. These operational land surface data sets are integral to the development, testing, application, and validation of environmental models and remote-sensing algorithms (NASA, 1993b; WCRP/IGPO 1993). The modeling research increasingly focuses on the integration of these hydrometeorological data with land surface data sets defined in the preceding sections. As noted in the following section, these operational, near-surface hydrometeorological, and satellite-derived data sets provide many challenges for GIS.

The sources of these types of hydrometeorological land surface data include: (1) networks of meteorological and hydrological observing stations maintained by NOAA, USGS, the United States Department of Agriculture, etc.; (2) meteorological satellites; (3) NOAA's evolving Doppler radar network; and (4) meteorological forecast models. For example, the NOAA National Climatic Data Center has recently placed several climatic databases on CD-ROM. These include: (1) National Climate Information Disc, Volume 1, containing monthly time series of temperature, precipitation, and drought data and graphics (1895–1989) for 344 climate divisions in the conterminous United States; (2) SAMSON CD Set, containing hourly Solar and Meteorological Surface Observational Network data on three CD-ROMs for 237 NOAA weather stations for the period 1961–1990; (3) Radiosonde Data for North America; and (4) Global Daily Summary (GDS) containing a 10,000-station set of daily precipitation, daily maximum/minimum temperature, and present weather for the period 1977–1991.

The USGS Water Resources Division operates stations to collect water resources data as part of the Water-Data Program. Data collected through the program include streamflow (discharge) and height (stage), reservoir and lake stage and storage, groundwater levels, well and spring discharge, and the quality of surface and groundwater. The data are archived in the USGS Water Data Storage and Retrieval System (WATSTORE). The WATSTORE database includes files of station header indexes, daily values, peak flow, water quality, unit values for water parameters measured more frequently than on a daily basis, groundwater site inventory, and water use. The data are accessible through the National Water Data Exchange, which consists of federal, state, local government, academic, and private organizations that collect, store, and use water data. The USGS National Water Information System II (NWIS II) is being implemented to upgrade data access. The WRD district offices and NWIS II are good starting points for locating water data (1-800-H2O-9000). The USGS has recently released a CD-ROM entitled USGS Hydroclimatic Data Network that contains streamflow data (monthly and annual summaries) for more than 1,600 stations with at least a twenty-year continuous record during the period of 1874–1988.

As part of NASA's Pathfinder project, several satellite data sets are now being developed and will contribute various types of land surface data. The North American Land Characterization pathfinder project to produce Landsat MSS triplicates is one example involving NASA, EPA, and the USGS. There are also NASA and NOAA pathfinder data projects for AVHRR GAC data, geostationary operational environmental satellite (GOES) data, and the special sensor microwave/imager (SSM/I) data. Various types of land surface gridded and image data sets will be derived from the visible, infrared, and microwave radiometric data available from these satellite data sets. For example, hourly estimates of solar insolation and photosynthetically active radiation are derived from GOES data. Passive microwave data available from the SSM/I sensor have been used in research on snow cover and soil moisture estimates. Although not a pathfinder project, the NOAA National Operational Hydrologic Remote Sensing Center has placed airborne-derived snow water equivalent data and AVHRR- and GOES-derived snow-cover estimates on CD-ROM for 1990–1992.

Although still in the implementation and developmental stages, NOAA's planned operational NEXRAD WSR-88D Doppler radar network will provide a large set of analysis products (Klazura and Imy 1993). One product will be hourly digital precipitation estimates on an approximate 4-km × 4-km grid for most of the conterminous United States. Precipitation, which is highly variable in space and time and not adequately measured by the current surface station network, is one of the most important meteorological parameters that determine land processes. Designed for operational use by hydrologic forecast centers, archive data sets will benefit climatologic, hydrologic, and ecologic research. These data sets will significantly enhance the existing surface station network.

Meteorological forecast models are increasingly viewed as a source of near-surface meteorological or "atmospheric forcing" data of interest to the land surface process modeling, remote sensing, and assessment communities. Model forecast data and observations (i.e., precipitation and cloud reports) are combined in a postanalysis process called four-dimensional data assimilation (4DDA) to produce enhanced, consistently gridded data sets of meteorological fields. These 4DDA outputs of near-surface information on precipitation, wind, temperature, pressure, humidity, and surface radiation parameters are of interest to land process modelers, to land and water resource managers, and increasingly to remote-sensing researchers; for example, to make atmospheric corrections to satellite data. These types of assimilated data sets are produced by global and mesoscale forecast models. For example, the NOAA National Meteorological Center is implementing the NOAA Eta mesoscale model for North America, while simultaneously conducting land processes research in conjunction with the GCIP global change project (Mitchell 1994). The model outputs will be a 30-km horizontal grid spacing every six hours with plans for a 15-km spacing by 1997. As described by Mitchell, the NOAA Eta model and the NOAA Doppler radar precipitation products are essential to GCIP research on improving the understanding and modeling of atmospheric and land surface hydrologic and energy cycles over large regions. The use of these improved 4DDA data sets in land process models and remote-sensing algorithms will enhance the operational assessment of how climate and weather affect land and water resources.

These types of hydrometeorological land data sets are integral to ISLSCP field experiments (NASA 1993b) and to the GEWEX/GCIP project for the Mississippi River basin (WCRP/IGPO 1993). The NASA ISLSCP global data CD-ROM includes many of these types of data sets. In addition to the vegetation data previously described, the ISLSCP CD-ROM includes 1°-data sets on near-surface meteorology based on 4DDA products from operational GCMs plus global precipitation, radiation fluxes, and other data sets.

CHALLENGES FOR GIS IN LAND SURFACE CHARACTERIZATION AND ENVIRONMENTAL MAPPING

The preceding section summarizes a wide variety of land data now available for developing, testing, validating, and using environmental simulation models, specifically land-atmosphere interactions models. These land surface data for environmental modeling present opportunities and challenges for using GIS as a tool to support land surface characterization research and environmental mapping. These challenges are consistent with environmental simulation modelers' use of GIS to build spatial databases, to maintain spatial data, including results of model simulations, to provide tools for exploratory spatial analysis, and to display and visualize model results (Steyaert and Goodchild 1994).

However, there is tremendous overlap between the GIS database development and GIS analysis roles. Land characterization research is one key to successful database development. A close working relationship between the GIS and remote-sensing specialists and modelers is desirable. For example, these land data sets provide the first opportunity to investigate and understand complex interrelationships among topography, land cover, and soils from the scale of individual landscapes to the conterminous United States. Moreover, multisource data describing the temporal behavior of land cover characteristics (i.e., vegetation seasonality from NDVI) and associated hydrometeorological parameters are becoming available for the first time. Understanding the interrelationships between dynamic land characteristics and basic land data on land cover, soil, and topography is a key science objective. Such understanding contributes to building better databases and models. This overlap of the use of GIS for data development and for land characterization research is illustrated in the following discussion of the role of GIS in data accuracy, data development, and analysis.

Data Accuracy and Error Analysis

GIS can help resolve data accuracy and error propagation problems. These land data sets are derived from field surveys, surface observation networks, various types of satellites, and simulation models such as meteorological forecast models. In many cases, the quality and accuracy of these data are quite variable. Estimates of some data types are frequently available from multiple sources. Burrough (this volume) and Aspinall and Pearson (this volume) point out some of the pitfalls in the intercomparison and overlay of spatial data with different levels of accuracy. The propagation of error through the analysis is of real concern. In many models, spatial data such as vegetation class and soil properties are aggregated to grid cells on the basis of percentage composition or predominant class. Mismatches can occur. All of these examples are of interest to modelers and database developers working with complex land cover characteristics and soils properties. Better GIS tools are needed to track, flag, and visualize minimum accuracy levels or inconsistencies.

GIS and the Flexible Land Database Approach

Environmental models require many types of land data, including dynamic land cover characteristics that must be derived and tailored from existing data according to variable grid cell and polygon size requirements. Although most models operate on grid cells within some specified computational domain, the use of polygons as a basis for model analysis is growing, especially in those models dealing with precipitation-runoff and evapotranspiration processes in watersheds, or modeling based on an object-oriented approach. Using coupled models, such as nested mesoscale models linked to an atmospheric GCM and providing data to various watershed and ecologic models, requires multiresolution land surface data for the entire set of land-atmosphere interactions models. The modeling process requires multiresolution tailored data for developing, initializing (boundary conditions), testing, validating, and intercomparing simulation models. There are many different types of modeling applications, for example those involving air, water, and land resource assessment.

A flexible land database approach for multiple user requirements can help meet the tailored land data needs of individual modelers, even though many different types of applications are represented (Lauer 1986; Loveland et al. 1991). In this approach, the land database is populated with sufficiently detailed land cover characteristics, DEM soils data, and other land data to permit the flexible tailoring of data for a wide variety of applications. This concept was incorporated by Loveland et al. (1991) and Brown et al. (1993) in developing the AVHRR land cover characteristics database described in the preceding section. As described by Steyaert et al. (in press), this concept of a flexible database with GIS tools for tailoring land data was recently demonstrated as part of the test and evaluation of AVHRR land cover characteristics data by land-atmosphere interactions modelers. GIS tools were used to tailor land data to meet the specific modeling regional domain, grid cell size, land cover classification, and method of spatial data aggregation for each grid cell (i.e., predominant class) of various atmospheric, hydrologic, and ecologic modelers.

Environmental modelers have adopted two approaches for meeting tailored database needs. For example, Lakhtakia et al. (this volume) used GIS functionality to meet database needs. However, some modelers have incorporated "homegrown" GIS functionality directly into their modeling system. Lee and Pielke (this volume) describe the development of a land data processing module that is incorporated into the regional atmospheric modeling system.

Existing GIS is an integral part of the flexible database concept for tailoring land data to meet individualized applications requirements. However, the requirements for land data are becoming more complex. Remote sensing technology is providing the foundation for a growing number of multitemporal measurements of dynamic, biophysical land cover characteristics. The need for developing derivative products from remotely sensed data is expanding. Advanced GIS tools for land characterization research are needed to meet these needs.

Advanced Data Analysis

Several GIS tools for enhanced land characterization research are suggested to help understand multiscale interrelationships in land data and to develop improved land data products. These tools involve algorithms for interpolating and extrapolating weather data in complex terrain, portable color notebook PC with GIS and handheld GPS capabilities in the field, and capabilities to better handle hierarchical, temporal, and four-dimensional data structures.

Land surface characterization research. Land surface characterization represents both opportunities and challenges for GIS as a tool to analyze complex data interrelationships while focused on developing improved tailored data sets. For the first time, relatively detailed land data sets appropriate for continental- to global-scale analysis are being developed to meet the growing demands of environmental models. The STATSGO soils data and 1-km AVHRR land cover characteristics databases permit detailed studies of terrain, soil and land cover interrelationships throughout the conterminous United States. Equally important, the diurnal and seasonal dynamics of the landscape can be studied with various satellite-derived data sets (GOES, AVHRR, SSM/I) and available hydrometeorological data. As Landsat-derived products become more uniformly available across the country, multiscale landscape characterization becomes possible. These types of studies are of interest to many scientists and modelers. Knowledge gained from these studies helps build better models and databases. GIS have a large role.

Several chapters in this volume illustrate how GIS can contribute to land characterization research and suggest some new tools for data analysis. For example, four chapters discuss the use of GIS to analyze soil landscapes (Lytle et al., Gessler et al., DeGloria and Wagenet, and Ventura et al.) They illustrate the use of GIS in soils research. Mackey et al. describe the use of spatial interpolation tools for ecosystem analysis in the boreal forest.

GIS can play an expanded role in the spatial interpolation and extrapolation of environmental data. For example, modelers have developed various algorithms designed to interpolate and extrapolate weather data in complex terrain. Because terrain interacts with atmospheric flow patterns, it is an important determinant of local weather and microclimate conditions. In fact, terrain is a major factor determining precipitation and potential solar insolation patterns that help to define microclimate, vegetation, and soils interrelationships. GIS tools are needed to extrapolate and interpolate surface weather data and to estimate potential solar radiation. Such tools will help scientists understand interrelationships between microclimate and land cover.

Several chapters in this volume describe approaches for interpolating and extrapolating available surface weather data. Hutchinson and Gessler describe the use of thin plate smoothing splines to interpolate precipitation according to elevation and aspect within a DEM. Daly and Taylor describe the PRISM model, which smooths the terrain for an effective aspect to the prevailing rain-producing wind patterns. Thorton and Running describe the MTCLIM algorithm for extrapolating daily solar radiation, maximum and minimum temperatures, and precipitation from limited surface weather station data. Pielke et al. describe the potential use of mesoscale models as a basis for simulating seasonal weather patterns in complex terrain. Dubayah and Rich discuss an algorithm to estimate potential solar radiation in terrain; the potential solar radiation estimate includes contributions of both solar direct and diffuse components, but assumes a clear, transparent atmosphere.

As discussed by Carver et al. (this volume), a portable GIS with GPS capabilities is needed to collect data in the field. However, taking this suggestion further, such a system is an essential tool for scientific research in the field for land characterization research, not just a means to validate data sets. A color notebook PC with GIS and GPS used to collect and interactively analyze data from automobiles and aircraft is also feasible to help understand the landscape structure. As remote-sensing algorithms are increasingly used to regionally extrapolate vegetation parameters such as biomass, the role for GIS in the field will expand.

Hierarchical GIS. The coupled systems modeling approach uses nested grid domains for different types of models, and multiresolution data are needed to meet requirements for local, regional, continental, and global-scale data. One of the major research themes is the "scaling up" of processes from local to global scales and the "scaling down" from global back to local scales. The chapter by Lakhtakia et al. (this volume) illustrates this point. Much of the research is focused on understanding the multiscale behavior of such processes as evapotranspiration, net primary production, and precipitation runoff. GIS can do much to support this research in terms of building multiresolution databases.

However, some spatial research issues on hierarchical scaling and generalization involve both raster and vector data structures. More research is needed on the scaling of topographic, soils, and land cover data at scales ranging from the local to the global level. This scaling involves generalizing data structures, as well as aggregating the thematic attributes in the case of soils and land cover data. More work needs to be done on generalizing stream networks and hydrologic boundaries from the watershed and basin to continental scales.

As pointed out by Engel (this volume), one of the main hydrologic research themes is the concept of the hydrologic response unit (HRU) for modeling precipitation-runoff processes (Leavesley and Stannard 1990). These HRUs are a function of the topography, land cover, and soils. Enhanced GIS tools are needed to support research on the development of HRUs and their hierarchical data structures.

Temporal GIS. Traditionally, analysts have treated the element of time in topographic, soils, land cover, and land-use data as a relatively static concept, with typically a one-time change detection. Current GIS are generally well suited for conducting change detection studies on these types of static data sets. However, biophysical land cover characteristics and the associated hydrometeorological data are highly variable, with time scales ranging from hours to seasons. Expanded temporal GIS tools are needed to incorporate the time dimension of these data. Current GIS are simply not well suited to efficiently analyze multitemporal images and gridded data sets.

Four-dimensional GIS. New GIS tools are needed for enhanced three-dimensional and temporal analysis of landscapes, watersheds, large regions, continents, and the globe. Concepts of a four-dimensional GIS need to be explored. In the preceding sections, several examples of multidimensional data were cited, including operational data sets on surface meteorological and hydrological station reports, Doppler radar rainfall estimates for most of the conterminous United States, near-surface meteorological data based on mesoscale modeling, and four-dimensional data assimilations, as well as routine satellite data from a wide variety of sensors. Better tools are needed for diagnostic analysis of these time-space data fields. The ability to simultaneously analyze these types of data sets within the land surface layer (lowest 100 m of atmosphere), at the land surface, and within the soil moisture zone with terrain coordinates, is essential. The ability to run and interactively analyze offline simulation models using these data is desirable.

User-Friendly GIS

Perhaps the largest single impediment to the expanded use of GIS by the modeling community, and in some cases, by the remote-sensing community, is the user-friendliness issue. For many modelers, GIS appear overly complicated. More user-friendly interfaces are needed to get more modelers involved with basic GIS capabilities.

CONCLUSION

The expanded requirements for land data are clearly illustrated by land-atmosphere interactions research, a component of Earth system modeling concepts developed in the mid-1980s. Several new multiresolution land surface databases on topography, soils, land cover characteristics, and hydrometeorological elements are now available for use in nested grid model simulations. GIS technology has a large role to play in improving land characterization research and database development. Understanding the multiscale and multitemporal interrelationships of land characteristics from the landscape to the globe is a challenge for GIS. Such land characterization research is one key to successful database development. Suggested research topics for enhancing GIS include spatial interpolation and extrapolation algorithms and new tools for temporal, hierarchical, and four-dimensional GIS analysis.

Acknowledgments

Research for this paper was conducted in conjunction with the global change research program of the U.S. Geological Survey.

References

Abbott, T. 1994. *Internet World's ON INTERNET 94, An International Guide to Electronic Journals, Newsletters, Texts, Discussion Lists, and Other Resources on the Internet.* Westport: Mecklermedia.

Band, L. E. 1986. Topographic partitioning of watersheds with digital elevation models. *Water Resources Research* 22:15–24.

Beier, J. 1992. *Global Change Data Sets: Excerpts from the Master Directory.* World Data Center "A" for Rockets and Satellites, National Space Science Data Center, Goddard Space Flight Center, National Aeronautics and Space Administration, Greenbelt, Maryland.

Botkin, D. B., J. F. Janak, and J. R. Wallis. 1972. Rationale, limitations, and assumptions of a northeastern forest growth simulator. *IBM J. Res. Dev.* 16:101–16.

Bretherton, F. P. 1985. Earth system science and remote sensing. *Proceedings of the IEEE* 73:1,118–27.

Brown, J. F., T. R. Loveland, J. W. Merchant, B. C. Reed, and D. O. Ohlen. 1993. Using multisource data in global land cover characterization: Concepts, requirements, and methods. *Photogrammetric Engineering and Remote Sensing* 59:977–87.

Clark, D. M., D. A. Hastings, and J. J. Kineman. 1991. Global databases and their implications for GIS. In *Geographical Information Systems: Principles and Applications,* vol. 2, edited by D. J. Maguire, M. F. Goodchild, and D. W. Rhind, 217–31. London: Longman Scientific and Technical.

Committee on Earth and Environmental Sciences. 1992. *The U.S. Global Change Data and Information Management Program Plan.* Federal Coordinating Council for Science, Engineering, and Technology, Office of Science and Technology Policy, Washington, D.C.

Defense Mapping Agency (DMA). 1992. *VPFVIEW 1.0 User's Manual for the Digital Chart of the World for Use with the Disk Operating System (DOS).* Washington, D.C.: Defense Mapping Agency.

Dickinson, R. E., A. Henderson-Sellers, P. J. Kennedy, and M. F. Wilson. 1986. *Biosphere-Atmosphere Transfer Scheme (BATS) for the NCAR Community Climate Model.* NCAR Technical Note NCAR/TN-275+STR. Boulder, Colorado.

Earth System Sciences Committee (ESSC). 1986. *Earth System Science Overview: A Program for Global Change.* National Aeronautics and Space Administration, Washington, D.C.

———. 1988. *Earth System Science: A Closer View.* Washington, D.C.: National Aeronautics and Space Administration.

Eidenshink, J. C. 1992. The 1990 conterminous U.S. AVHRR data set. *Photogrammetric Engineering and Remote Sensing* 58:809–13.

Engel, B. A., R. Srinivasan, and C. Rewerts. 1993. A spatial decision support system for modeling and managing agricultural nonpoint source pollution. In *Environmental Modeling with GIS,* edited by M. F. Goodchild, B. O. Parks, and L. T. Steyaert, 231–37. New York: Oxford University Press.

Environmental Protection Agency (EPA). 1992. *A Guide to Selected National Environmental Statistics in the U.S. Government.* Office of Policy, Planning, and Evaluation, Environmental Protection Agency, Washington, D.C.

Estes, J. E., and D. W. Mooneyhan. 1994. Of maps and myths. *Photogrammetric Engineering and Remote Sensing* 60(5):517–24.

Farmer, D. G., and M. J. Rycroft. 1991. *Computer Modeling in the Environmental Sciences*. New York: Clarendon Press.

Federal Geographic Data Committee (FGDC). 1992. *Manual of Federal Geographic Data Products*. Falls Church: Vigyan.

———. 1993. *A Draft Strategic Plan for the National Spatial Data Infrastructure: Building the Foundation of an Information-Based Society*. FGDC Secretariat, U.S. Geological Survey, Reston, Virginia.

Gao, X., S. Sorooshian, and D. Goodrich. 1993. Linkage of a GIS and a distributed rainfall-runoff model. In *Environmental Modeling with GIS*, edited by M. F. Goodchild, B. O. Parks, and L. T. Steyaert, 182–87. New York: Oxford University Press.

Glaeser, P. S., and S. Ruttenberg. 1992. Data for global change and other Earth science papers. *Proceedings of 13th CODATA International Conference, CODATA Bulletin*. CODATA Secretariat, Paris.

Goodchild, M. F., B. O. Parks, and L. T. Steyaert. 1993. *Environmental Modeling with GIS*. New York: Oxford University Press.

Hay, L. E., W. A. Battaglin, R. S. Parker, and G. H. Leavesley. 1993. Modeling the effects of climate change on water resources in the Gunnison River Basin, Colorado. In *Environmental Modeling with GIS*, edited by M. F. Goodchild, B. O. Parks, and L. T. Steyaert, 173–81. New York: Oxford University Press.

Henderson-Sellers, A., M. F. Wilson, G. Thomas, and R. E. Dickinson. 1986. *Current Global Land Surface Data Sets for Use in Climate-Related Studies*. NCAR Technical Note NCAR/TN272+STR, Boulder, Colorado.

IGBP Secretariat. 1990. *Geosphere-Biosphere Program: A Study of Global Change: The Initial Core Projects*. IGBP Global Change Report No. 12. Stockholm, Sweden.

———. 1992. *Improved Global Data for Land Applications: A Proposal for a New High-Resolution Data Set*, edited by J. R. G. Townshend. IGBP Global Change Report No. 20. Stockholm, Sweden.

Jenson, S. K., and J. O. Domingue. 1988. Extracting topographic structure from digital elevation model data for geographic information system analysis. *Photogrammetric Engineering and Remote Sensing* 54:1,593–1,600.

Kelmelis, J., and F. Rowland. 1994. Data useful for land-use/cover change analysis. In *Changes in Land Use and Land Cover: A Global Perspective*, edited by W. Meyer and B. L. Turner. New York: Cambridge University Press.

Klazura, G., and D. A. Imy. 1993. A description of the initial set of analysis products available from the NEXRAD WSR-88D system. *Bulletin of the American Meteorological Society* 74:1,293–1,311.

Lauer, D. T. 1986. Applications of Landsat data and the database approach. *Photogrammetric Engineering and Remote Sensing* 52:1,193–99.

Leavesley, G. H., M. D. Branson, and L. E. Hay. 1992. Using coupled atmospheric and hydrologic models to investigate the effects of climate change in mountainous regions. *Managing Water Resources During Global Change*, 691–700. AWRA 28th Annual Conference and Symposium, Reno, Nevada.

Leavesley, G. H., and L. G. Stannard. 1990. Application of remotely sensed data in a distributed parameter watershed model. *Proceedings of Workshop on Applications of Remote Sensing in Hydrology*. National Hydrologic Research Center, Environment Canada.

Lee, T. J., R. A. Pielke, T. G. F. Kittel, and J. F. Weaver. 1993. Atmospheric modeling and its spatial representation of land surface characteristics. In *Environmental Modeling with GIS*, edited by M. F. Goodchild, B. O. Parks, and L. T. Steyaert, 108–122. New York: Oxford University Press.

Loveland, T. R., J. W. Merchant, D. O. Ohlen, and J. F. Brown. 1991. Development of a land cover characteristics database for the conterminous United States. *Photogrammetric Engineering and Remote Sensing* 57:1,453–63.

Loveland, T. R., and D. K. Scholz. 1993. Global data set development and data distribution activities at the U.S. Geological Survey's EROS Data Center. *ASPRS Technical Papers, Remote Sensing: Looking to the Future with an Eye on the Past*, vol. 2. American Society for Photogrammetry and Remote Sensing and American Congress on Surveying and Mapping, Bethesda, Maryland.

Michener, W. K., A. B. Miller, and R. Nottrott. 1990. *Long-Term Ecological Research Network: Core Data Set Catalogue*. Belle W. Baruch Institute for Marine Biology and Coastal Research, University of South Carolina, Columbia, South Carolina.

Mitchell, K. E. 1994. GCIP initiatives in operational mesoscale modeling and data assimilation at NMC. *Preprint Volume of the AMS Fifth Conference on Global Change Studies* 1:23–28. Nashville, Tennessee.

Mounsey, H., and R. F. Tomlinson. 1988. *Building Databases for Global Science*. London: Taylor & Francis.

National Aeronautics and Space Administration (NASA). 1993a. *Modeling the Earth System in the Mission to Planet Earth Era*, prepared by S. Unninayer and K. H. Bergman. Washington, D.C.: National Aeronautics and Space Administration.

———. 1993b. *Remote Sensing of the Land Surface for Studies of Global Change: Models-Algorithms-Experiments*, edited by P. J. Sellers. International Satellite Land Surface Climatology Project (ISLSCP). Workshop Report, NASA/Goddard Space Flight Center, Greenbelt, Maryland.

National Oceanic and Atmospheric Administration (NOAA). 1988. *Selective Guide to Climatic Data Sources*. U.S. Department of Commerce, NOAA National Environmental Satellite, Data, and Information Service, National Climate Data Center, Asheville, North Carolina.

National Research Council (NRC). 1990. *Research Strategies for the U.S. Global Change Research Program*. Committee on Global Change, U.S. National Committee for the IGBP. Washington, DC: National Academy Press.

———. 1993. *Toward a Coordinated Spatial Data Infrastructure for the Nation*. Mapping Sciences Committee, Board on Earth Science and Resources, Commission on Geosciences, Environment, and Resources. Washington, D.C.: National Academy Press.

Oak Ridge National Laboratory (ORNL). 1993. *Carbon Dioxide Information Analysis Center Catalog, Numeric Data Packages and Computer Model Packages*. Carbon Dioxide Information Analysis Center, Oak Ridge National Laboratory, Oak Ridge, Tennessee.

Parton, W. J., D. S. Schimel, C. V. Cole, and D. S. Ojima. 1987. Analysis of factors controlling soil organic levels in Great Plains grasslands. *Soil Sci. Soc. Amer.* 51: 1,173–79.

Pielke, R. A. 1984. *Mesoscale Meteorological Modeling*. San Diego: Academic Press.

Reybold, W. U., and G. W. TeSelle. 1986. Soil geographic databases. *J. of Soil and Water Conservation* 44:28–29.

Running, S. W., and J. C. Coughlan. 1988. A general model of forest ecosystem processes for regional applications, I: Hydrologic balance, canopy gas exchange, and primary production processes. *Ecological Modeling* 42:125–54.

Schimel, D. S., T. G. F. Kittel, and W. J. Parton. 1991. Terrestrial biogeochemical cycles, global interactions with the atmosphere and hydrology. *Tellus* 43:188–203.

Sellers, P. J., Y. Mintz, Y. C. Sud, and A. Dalcher. 1986. A simple biosphere model (SiB) for use within general circulation models. *Journal of Atmospheric Science* 43: 505–31.

Shugart, H. H. 1984. *A Theory of Forest Dynamics: The Ecological Implications of Forest Succession Models.* New York: Springer-Verlag.

Smith, C. B., M. N. Lakhtakia, W. J. Capehart, and T. N. Carlson. 1993. Initialization of soil-water content for regional-scale atmospheric prediction models. *Conference on Hydroclimatology,* 24–27. Seventy-third AMS Annual Meeting, Anaheim, California.

Steyaert, L. T. 1993. A perspective on the state of environmental simulation modeling. In *Environmental Modeling with GIS,* edited by M. F. Goodchild, B. O. Parks, and L. T. Steyaert, 16–30. New York: Oxford University Press.

Steyaert, L. T., and M. F. Goodchild. 1994. Integrating GIS and environmental simulation models, a status review. In *Environmental Information Management and Analysis, Ecosystem to Global Scales,* edited by W. Michener, J. Brunt, and S. Stafford, 333–56. London: Taylor & Francis.

Steyaert, L. T., T. R. Loveland, J. F. Brown, and B. C. Reed. In press. Integration of environmental simulation models with satellite remote sensing and geographic information systems technologies, case studies. *Proceedings of Pecora 12 Symposium.* American Society for Photogrammetric Engineering and Remote Sensing, Bethesda, Maryland.

Townshend, J. R. G. 1991. Environmental databases and GIS. In *Geographical Information Systems: Principles and Applications,* vol. 2, edited by D. J. Maguire, M. F. Goodchild, and D. W. Rhind, 201–16. London: Longman Scientific and Technical.

Townshend, J. R. G., C. O. Justice, W. Li, C. Gurney, and J. McManus. 1991. Global land cover classification by remote sensing, present capabilities and future possibilities. *Remote Sensing of Environment* 35:243–55.

Trenberth, K. E., ed. 1992. *Climate System Modeling.* Cambridge: Cambridge University Press.

U.S. Government Printing Office (USGPO). 1993. *SIGCAT CD-ROM Compendium, Special Interest Group on CD-ROM Applications and Technology (SIGCAT).* Washington, D.C.: USGPO.

World Climate Research Programme-International GEWEX Project Office (WCRP-IGPO). 1993. *Implementation Plan for the GEWEX Continental-Scale International Project (GCIP),* Volume I, *Data Collection and Operational Model Upgrade,* compiled by J. A. Leese. IGPO Publication Series No. 6, Washington, D.C.

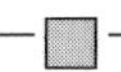

Louis T. Steyaert is a remote-sensing scientist in the Science and Applications Branch, EROS Data Center of the U.S. Geological Survey. Since 1988, he has worked full time on global change research, conducting land surface characterization research with space-based remote sensing and GIS technologies. He coordinated the test and evaluation of USGS land data by atmospheric, hydrologic, and ecologic modelers. Dr. Steyaert currently is an investigator in the NASA-led Boreal Ecosystem-Atmosphere Study in central Canada. Prior to his arrival at USGS in 1987, he was a research meteorologist for the National Oceanic and Atmospheric Administration where he helped develop an operational drought-monitoring and disaster early-warning program for food security assessment in developing countries. He completed his Ph.D. in atmospheric science in 1977 at the University of Missouri–Columbia.

U.S. Geological Survey
EROS Data Center
Science and Applications Branch
c/o NASA Goddard Space Flight Center
Code 923
Greenbelt, Maryland 20771
Phone: (301) 286-2111
Fax: (301) 286-0239
E-mail: *steyaert@ltpmail.gsfc.nasa.gov*

5

Spatial Data Quality and Error Analysis Issues:
GIS Functions and Environmental Modeling

P. A. Burrough, R. van Rijn, and M. Rikken

The main issues in linking mathematical models of environmental processes to GIS are discussed in terms of hardware and software options, representations of time and space, and classes of space-time models. The advantages and disadvantages of computer modeling are presented, and the factors affecting the quality of model predictions are discussed. Utrecht University software tools used for following error propagation in GIS modeling are presented: their use is illustrated by a cost-benefit study of predicting heavy-metal concentrations on the floodplain of the river Maas in the Netherlands. The case study demonstrates the value of integrating error propagation tools in GIS and illustrates the need to present GIS users with good advice on how to choose the best combination of model and data types to achieve results with a required level of quality.

ENVIRONMENTAL MODELING AND GIS

Increasingly, geographic information systems (GIS) is being used for the inventory, analysis, modeling, and management of the natural environment (Goodchild et al. 1993; Pflug and Harbaugh 1992). In this paper, the term "geographic information systems" is used to indicate a set of hardware and software tools for storing, retrieving, analyzing, and displaying spatial data, rather than the whole complex of organization, personnel, and databases that some define as GIS; the accent of this paper is more on the way these tools are used than on the organizations that use them. It is axiomatic that for the analysis and modeling of environmental phenomena, basic GIS tools must be accompanied by reliable data and sensible analytical models to enable the analysis of trends or to anticipate the possible results of planning decisions. By using GIS as a trainee pilot would use a flight simulator, managers, planners, and decision makers can examine a range of scenarios and explore the possible consequences before expensive, irrevocable mistakes have been made. The value of GIS for environmental analysis is determined by the costs and the accuracy of these predictions, which are dependent on the functionality of the GIS tools, the level of understanding in the models, and the quality of data available.

Issues in Modeling

Model building, data inventory, and the building of GIS are three separate kinds of activities that until relatively recently have had little to do with each other. Only recently have people found it expedient to move from simple, or complex lumped models of a phenomenon (such as runoff or erosion), to distributed models that attempt to describe the transport of material over and through whole landscapes (de Roo et al. 1989). Although modelers may have spent much effort on working out the detailed physics and chemistry of processes such as the movement and degradation of pesticides in soil, they are rarely specialists in describing how landscape properties critical for their process actually vary, and they often have little understanding of the scale problems involved. Conventional resource surveyors, on the other hand, have frequently recorded the properties of the landscape without any idea of a specific use for the data in mind, and also without recourse to the recognition of gradual change, either in space or time.

Linking nonspatial lumped models to static, soil, or land-use units in a choropleth map is relatively easy in most commercial GIS, when the main problem is that of bringing all data to the same geometric base, but it is a simple procedure that fails to appreciate the intrinsic aspects of how processes actually operate in a four-dimensional landscape. An absolutely critical aspect of linking models and data in GIS is that there are very few opportunities for determining whether the results are reliable or not. There is a tendency among modelers to assume that complex physical models are superior to empirical models, but there is little systematic work to demonstrate that this is indeed so. Indeed, the situation is probably little better now than a few years ago when Morgan (1986) commented that "despite the desirability of physically based models, the present state of model development is such that a simple empirical model is often more successful in predicting soil erosion than a complex physically based one which is difficult to operate and has been only partially evaluated." His comments may apply to more areas than just the modeling of soil erosion, and often the only criterion of "quality" in GIS-based modeling is the cartographic display of the results.

Linking Models and GIS

There are two options for linking environmental models with GIS. The first is to run the model outside the GIS, using the latter as a source of data and a means of displaying the results. The second is to integrate the model in the GIS by writing it using standard analysis functions, such as those provided in the command language by SQL or cartographic algebra. The first option has advantages when the process being modeled is complex, or has been modeled by "experts," or when processing on dedicated hardware or by special programming languages is essential. Against this there is a need for converting data formats and for special knowledge in running the model. The second option has advantages in that the process can be modeled using generic tools on a single integrated database, but computation may not be optimal; there can be difficulties with iteration, and it may be difficult to write the model in terms of the standard GIS functionality that is available.

Representations of Space: Exact Objects or Discretized Continua?

Most GIS and databases are based on the fundamental reductionist assumption that the world can be described in terms of sets of basic entities—points, lines, polygons, pixels, voxels—that carry sets of exact valued attributes (Robinove 1986). This view is limited and excludes the proper treatment of a wide range of continuous and stochastic phenomena (Burrough 1992a), such as the spatial and temporal variation of properties of land form, geology, soil, biological systems, water, and atmosphere. Many models of environmental processes implicitly assume continuous spatial variation, as do interpolation techniques such as kriging. For some processes, the variation of spatial patterns over time is also essential. In practice, continuous variation is discretized; the size of the regular or irregular units used for approximating continuous variation determines the levels of spatial and temporal detail that can be resolved.

Classes of Models

We can group mathematical models of environmental processes according to two kinds of criteria. The first is the way in which the model is an algorithmic compression of reality; the second describes how the model operates in space time.

Algorithmic compression. Models contain approximations of how the world works. The simpler the process, the easier it is to formulate it in simple mathematical terms:

> Without the development of algorithmic compressions of data, all science would be replaced by mindless stamp collecting—the indiscriminant accumulation of every available fact. Science is predicated upon the belief that the universe is algorithmically compressible . . . a belief that there is an abbreviated representation of the logic behind the universe's properties that can be written down in finite form by human beings (Barrow 1991).

For the sake of argument, we can group models into four classes of algorithmic compression: (1) rule-based (logical models); (2) empirical (regression models); (3) deterministic physical (everything known); and (4) stochastic physical (only probabilities known). Further, many empirical and physical models of processes are linear because linear models are easy to handle computationally and they behave predictably when feedback loops are included. In contrast, some nonlinear models with feedback loops can be numerically unstable yielding chaotic, and therefore unpredictable results (Stewart 1991).

Space-time extents. Models can transform data locally, within a given neighborhood, or globally over a distance in space or time. Local transformations refer to the derivation of new attributes for single entities or locations from other attributes of the same entity or location; neighborhood transformations refer to the derivation of new attributes for a single entity or location from the attributes and entities within a localized domain that surrounds it (e.g., the computation of slope or aspect for a cell in a digital elevation model [DEM] from altitude data at surrounding cells). Global transformations compute new attributes for entities and locations as a result of a process that extends throughout a large part of the database. Examples are the computation of the hydrological output of a drainage basin or the variation of air quality in a city street as a result of temperature effects, wind speeds, and traffic flows.

Table 5-1. A typology of models.

Kind of model	Local	Spatial extent neighborhood	Global
Rule-based (Including geometric rules)	2c,d,t_0	2c,d,t_0	2c,d,t_0
Empirical	2c,d,t_0,t_1	2c,d,t_0	1,2c,d,t_0
Process (Deterministic)	1,2c,d,t_1	1c,d,t_1	1c,d,t_1
Process (Stochastic)	1c,d,t_1	1c,d,t_1	1c,d,t_1

1: Model external to GIS.
2: Model integrated in GIS (cartographic algebra).
c: Discretized spatial/temporal variation.
d: Defined spatial entities.
t_0: Time-independent models.
t_1: Time-dependent models.

Advantages and Disadvantages of Environmental Modeling

Varcoe (1990) has summarized the advantages and disadvantages of using crop simulation models, and the following is an extension of his points to computerized environmental modeling in general.

Advantages. Many variables and complex interactions can be accommodated, limited only by the degree of understanding and the size of research effort able to be invested. Data storage is standardized and computer simulation provides a single organized approach. Environmental modeling demands interdisciplinary cooperation during development, and the simulation/modeling language can be a means of communicating ideas. Simulation of results is a rapid and cheap method of investigation, particularly useful when time frames or finances do not allow for data collection. It is possible to run simulations for periods equivalent to several years within minutes or seconds. It is not necessary to collect data which may often have been collected before. Simulation is often claimed to be scale-neutral in concept although this is not usually the case, given real-world limitations on the resolution of environmental data collection. The success or otherwise of simulations can identify areas where research is necessary

Further advantages are that models can be continuously updated and modified to reflect improvements in understanding. The control of key variables and organizational structures gives the investigator a greater degree of control than would be possible in the real world. The modeling process identifies particularly important inputs to each modeled system. Simulation permits the extrapolation of experimental results to sites with different environmental conditions so that in yield modeling, for example, production assessments can be made for crops not previously grown in a region or different crops can be compared. The simulation of processes such as crop growth, runoff, level of acid deposition, etc., provides a quantifiable method of classifying land. Models can be extended to incorporate economic and/or social constraints. Finally, the advantage of using models with a GIS is that the data can reflect as wide a range of spatial and temporal conditions as the database permits, thereby enabling considerable refinements (de Roo et al. 1989).

Disadvantages of modeling. One of the main dangers of computer modeling is that unskilled users may uncritically accept the results and assume that complex models perform adequately. Even experts may accept simulated results without adequate validation. The mechanistic use of simulation models by nonspecialists is a potential danger to rational decision making.

Other disadvantages are that simulation does not replace field or experimental work but can increase the efficiency of the investigative process. Very careful validation is required to use models both inside and outside the areas for which they were developed. There are few guidelines for carrying out acceptable validation, and for some models validation is very difficult or almost impossible. Many relationships in environmental data are only empirical, and therefore are area- or event-specific, as complete (physical) mechanistic models have yet to be developed. The use of computer models requires data to be available and standardized in technical format, in method of recording and classifying, and in spatial and temporal resolution. Many models of environmental processes are complex and may require data at levels of spatial and temporal resolution that are too costly to collect. In addition, the initial development of complex models is labor intensive and it may be necessary to learn a simulation language or to understand the principles of multiple regression.

In addition, overemphasis on modeling may reduce field experience so that significant site-specific factors can be ignored. Even though simulations may predict correct results in an area, this does not guarantee that the model describes a process correctly. Cynical observers may also suggest that models are often used to confirm what is already suspected.

FACTORS AFFECTING THE QUALITY OF MODEL PREDICTIONS

With GIS, investigators can apply models to spatially and temporally differentiated data from large areas in order to achieve new insights or to impose some form of control. Increasingly the ability to model environmental processes is not limited by computer technology but by human understanding, encapsulated in mathematical models and the availability of sufficient reliable data to drive them. So, the added value of using GIS for environmental modeling can be expressed as:

$$\textit{information} = \textit{conceptual models} + \textit{data} \quad (5\text{-}1)$$

This is only satisfactory if the results are not misleading, which is to say that they have been achieved with a certain minimum level of reliability. Just as in any other area of science or industrial production system, the output of a process, whether in the form of useful hypotheses, automobiles, planes, household furniture, or compiled information, needs to meet minimum standards of quality in order to be satisfactory (Burrough 1992b; Moore and Rowland 1990).

If (5-1) expresses the link between information, models, and data, then the quality of that information is determined by:

$$\textit{quality}[\text{information}] = f(\text{quality}[\text{model}], \text{quality}[\text{data}]) \quad (5\text{-}2)$$

Alternatively, the error in the information is given by:

$$\textit{error}[\text{information}] = f(\text{error}[\text{model}], \text{error}[\text{data}]) \quad (5\text{-}3)$$

Model Errors

In addition to the factors given in Table 5-1, which relate to the conceptual basis of the model, model predictions can be affected by uncertainty and errors in the values of coefficients and boundary conditions. The residual error in a regression model describes the ultimate limit of precision that the particular algorithmic compression can reach. Rule-based models and deterministic models are often assumed to have no need of a residual error term. Complex models may include several submodels for which the residual errors are unknown.

Data Quality and Errors

Many factors can affect the quality of spatial and temporal data supplied to a model by a GIS (Burrough 1993). These include:

1. Provenance and timeliness.
2. Method of measurement, recording, and analysis technique.
3. Assumptions about the kind of spatial/temporal variation—discrete or continuous.
4. Spatial and temporal resolution—support and scale.
5. Spatial and temporal variability.
6. Number or density of observations.
7. Method used to extrapolate data from point observations to the regular or irregular space-time lattice used for modeling.
8. Data representation in the computer (integer/real/double precision)—methods for data conversion between different kinds of spatial representation, observation technique, or data model.

Some scientists believe that the limitations to successful modeling are caused more by data availability and quality than by lack of scientific insight (de Roo et al. 1992).

TOOLS FOR EVALUATING THE QUALITY OF GIS MODELING

A complete enumeration of all combinations of the factors affecting model quality and data quality given above provides nearly 2,000 separate possibilities. Certainly, some of these will be redundant, or will occur rarely, but there is still an enormous variety. Clearly, without experience, guidance, or both, it is impossible to decide on the best combination of model and data that will achieve a prediction that has a given quality or error for a given cost. In practice, people use "recommended" models uncritically for want of better alternatives.

Recently Burrough (1992b) argued for an "intelligent GIS" that would assist users to select the best combination of model and data to suit their needs. The intelligent GIS would request the user to specify the acceptable level of error in the model predictions and would then automatically test to see if results of such a quality were feasible. If not, the system should advise the user on how best to proceed. Although, to the best of my knowledge, no such system actually exists as a fully integrated tool, at the University of Utrecht we have created the main components of such a system as a series of prototypes (Heuvelink 1993).

Error Propagation Studies in GIS Modeling: The Components

The following components are required:

1. A suitable GIS with provision for mathematical modeling.
2. Suitable methods for obtaining error estimates.
3. Theory and tools for computing error propagation.

In principle, any GIS will serve. We use our own PC-RASTER because of its unified raster-data structure which handles integers or single- and double-precision real numbers, and its ability to carry out most kinds of logical and algebraical arithmetic on gridded data using the syntax and grammar of the C programming language (van Deursen and Wesseling 1991). Standard statistical packages can serve for determining regression models, coefficients and their errors, and residual errors; error surfaces of interpolated data are obtained by geostatistical interpolation (univariate or multivariate point or block kriging)—or conditional simulation. The error propagation tool called ADAM (Heuvelink 1993) uses statistical theory of error propagation (Taylor 1982) and model approximation techniques to set up control files for running models and calculating errors which can be executed by the GIS arithmetical functions. ADAM requires that the model and data be declared formally in a control file. It then analyzes the model, recommends an error propagation strategy (Taylor series approximation, Rosenbleuth's method, or Monte Carlo conditional simulation) and compiles a control file which computes the results using the GIS commands.

These tools can be used to analyze how errors are propagated through models, and different modeling procedures can be compared in terms of costs and prediction errors, as the following example illustrates. Other recent examples of error propagation in environmental modeling with GIS are given by Heuvelink (1993), Heuvelink et al. (1989), Heuvelink and Burrough (1993), de Roo et al. (1992), and Wesseling and Heuvelink (1993).

A Practical Example: Modeling Heavy-Metal Pollution on Floodplains

When a river loaded with heavy-metal polluted sediment overflows its floodplain, the soils become contaminated. The degree of pollution depends on the frequency and duration of flooding and on the distance from the river. Variation of heavy-metal concentrations in the soil is expected to be continuous in line with the deposition process. Because many factors affect the way sediment is transported and deposited, the exact concentration of heavy metal at any given site can only be specified within a tolerance limit. The aim of this example is to demonstrate the value of error propagation studies by comparing two kinds of prediction methods. The first, empirical regression modeling, predicts heavy-metal (zinc) concentrations as a function of distance from the river and the relative elevation of the site; the second is a data-driven approach using kriging. The question to be decided is which method gives the best results for different investments in data (numbers of analyzed samples of soil).

This study was carried out on the floodplain of the river Maas near the town of Stein in South-Limburg, the Netherlands. The area is about 5 km^2: it is almost totally inundated when the discharge of the Maas exceeds 1,500–2,000 m^3/s. Land use is pasture and crops. One hundred fifty-five topsoil samples were collected at sites located by stratified random sampling using information on elevation relative to the river and geomorphology. Bulk soil samples were collected for the 0–10-cm layer at ten places within a circle of radius 10 m (effectively a 20-m × 20-m square; Rikken and van Rijn 1993). Previous research (Leenaers et al. 1989) showed that heavy-metal concentrations in floodplain soils in the Limburg area is related to relative elevation on the floodplain, and also to distance from the riverbank. Both these properties can be determined cheaply from 1:10,000 elevation maps. This study uses the following data: easting and northing (accurate to about 1 m), zinc content (laboratory determination on sample in ppm), elevation with respect to the river (accuracy ± 0.7 m) and the distance in meters of the sample from the nearest point on the riverbank (accuracy ±5 m). Preliminary analysis showed that zinc content and distance to the river were strongly log-normally distributed, so these attributes were transformed to natural logarithms. All statistical analysis proceeded with the log-normally-transformed zinc and distance data.

Correlation analysis carried out on all samples showed that there were strong negative correlations between *ln* (zinc) and *ln* (distance) [-0.768], between *ln* (zinc) and relative elevation [-0.624], and a positive correlation between *ln* (distance) and relative elevation [0.481]. Before proceeding further, the total data set was split randomly into two parts: 102 samples were used for comparing prediction methods and the remaining 53 were kept apart for validation. Statistical comparisons of the two sub-data sets suggested that they were not significantly different samples of the study area. The following mapping methods were used:

(a) linear regression of *ln* (zinc) on *ln* (distance to river) and relative elevation

$$ln(zinc) = B_0 + B_1.ln\ (distance) + B_2.elevation + \varepsilon \tag{5-4}$$

(b) ordinary kriging of *ln* (zinc) levels

Both methods were applied to the whole data set of 102 samples: two sets of 51 samples, two sets of 27 samples, and two sets of 14 samples. Error propagation and kriging standard errors relative to the same areal units were used as criteria of success. All predictions were made for a 20-m grid and were validated independently by the 53 test samples. The costs of field and laboratory work were used to rank the results. Figure 5-1 presents some examples of the resulting maps and the resulting standard error surfaces.

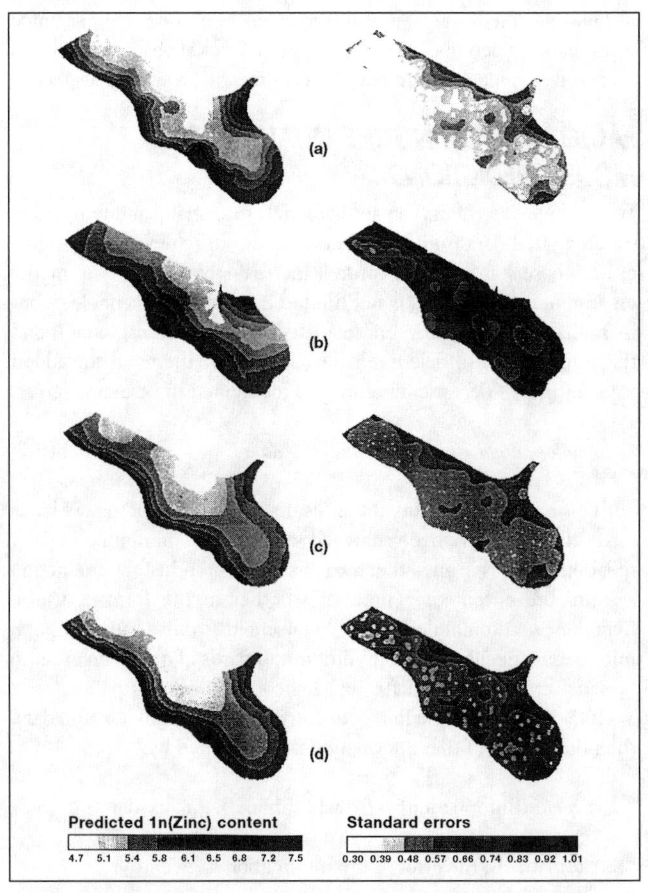

Figure 5-1. Variation of map quality with method and number of samples: A, B—kriging; C, D—regression; A, C—102 sites; B, D—27 sites.

Table 5-2 shows the control file for the linear regression model. The model coefficients, their standard errors and correlations were obtained by standard regression for every subdata set respectively, and the input maps and errors of *ln* (distance) and relative elevation were obtained by ordinary kriging from the 102 sites. ADAM recommended that this model could be approximated by a second-order Taylor series.

For mapping the *ln* (zinc) directly, variograms were computed for each subset of the data and used for interpolating from the same data set. This was not possible with 14 data as consistent variograms could not be obtained: variograms based on 27 data were also poorly defined.

The mean standard error per cell as given by error propagation or kriging is an indication of the quality that the method expects to achieve; the root mean square difference between predicted data and the 53 validation samples is an estimate of actual achievement. Costs of collecting samples in the field and laboratory analyses were determined from commercial sources. Figure 5-2 shows the mean prediction standard error (PSE) given by each method and data set and the root mean square error (RMSE) of the same maps as determined by the fifty-three validation samples. Figure 5-2 shows that the differences between predicted error and validated error are small. The average prediction errors are larger than the validation errors because they include contributions from sites with relatively large errors at the edge of the study area. Because these sites are built up they were not sampled for the validation data set.

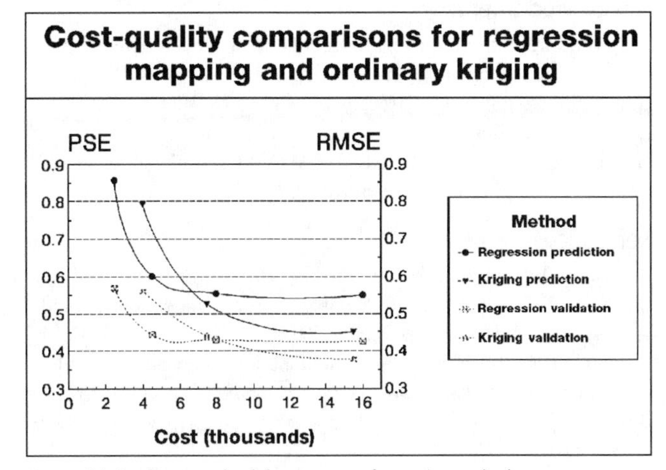

Figure 5-2. Prediction and validation errors of mapping methods versus cost.

Table 5-2. Control file for error propagation with regression modeling.

```
/*********************************************************************/
/* predict ln zinc content linear regression on lnDM & RA - 102 sites */
/*********************************************************************
/* model parameters */
co1 : random (normal, 9.973, 0.299);
co2 : random (normal, -0.333, 0.033);
co3 : random (normal, -0.291, 0.041);
/* correlations between model coefficients */
r0(co1, co2) : -0.014;
r0(co1, co3) : -0.860;
r0(co2, co3) : -0.487;
/* residual noise term of the regression */
residual : random (normal, 0, 0.365);
r0(residual, $) : 0;
/* data characteristics - input data and error surfaces */
/* ln dist Mass m */
lndm : random(normal, "lndm102.est", "lndm102.sd");
/* relative elevation m */
ra : random( normal, "ra102.est" , "ra102.sd");
/* correlations between data */
r0(lndm, ra) : 0.4870;
r0(lndm, $) : 0;
r0(ra, $) : 0;
/* output maps from model */
calc("lznm102.est",                  */ model results
 "lznm102.sd",                       */ model error
 "zmod102.sd" : co1 & co2 &          */ contribution model
co3 & residual,
 "da102dm.sd" : lndm,                */ contribution distance
 "da102ra.sd" : ra                   */ contribution elevation
) : co1 + co2*lndm + co3*ra + residual;.
```

The results show clearly that when only few samples can be afforded the empirical model yields better results than interpolation. When more data can be purchased, however, then kriging interpolation is superior. Analysis of the error contributions suggests that in this case there is little point in refining the model further, and improvements can be more easily obtained by resolving spatial variations.

CONCLUSION

Environmental modeling with GIS is currently a haphazard cottage industry in which the method of coupling between the model and the GIS, the coding of the model, and the implementation are all dependent on the specific GIS, the data resolution and format, and the whims of the investigator. Under these circumstances it is no wonder that there have been few serious attempts to evaluate the propagation of errors in environmental modeling with GIS, and the results are usually judged on the cartographic excellence of the final map. If the results are found to be unsatisfactory, then model builders will tend to want to develop more complex models and data collectors will want to collect more data; there is no integrated approach.

This chapter argues for the proper integration of models and data in GIS so that users can make rational decisions about the best way to achieve model predictions that match required levels of quality for given purposes and given costs. This integration includes the standardization of data formats and data resolution together with the provision of standard methods for estimating errors and for following error propagation. Work at the University of Utrecht has demonstrated that error propagation tools can be integrated with GIS models to give reliable estimates of the quality of results obtained by different methods.

There is a strong need, however, to obtain detailed empirical understanding of how errors propagate through the large number of possible combinations of model types, data types, data sources, and kinds of error, and to make this available to users in an easily accessible form. Both the error propagation tools and the skills needed to use them need to be developed and broadcast widely in the coming years.

Acknowledgments

The work presented in this paper could not have been done without major contributions from Willem van Deursen, Cees Wesseling, Gerard Heuvelink, Ad de Roo, Edzer Pebesma, Victor Jetten, and Lodewijk Hazelhoff. Software for environmental modeling with error propagation (PC-RASTER and ADAM) is available for a modest fee upon application to Cees Wesseling.

References

Barrow, J. D. 1991. *Theories of Everything.* Vintage.

Burrough, P. A. 1992a. Are GIS data structures too simpleminded? *Computers and Geosciences* 18:395–400.

Burrough, P. A. 1992b. Development of intelligent geographical information systems. *Int. J. Geographical Information Systems* 6:1–15.

Burrough, P. A. 1993. Soil variability—revisited. *Soils and Fertilizers* 56:529–62.

de Roo, A., L. Hazelhoff, and P. A. Burrough. 1989. Soil erosion modeling using ANSWERS and geographical information systems. *Earth Surface Processes & Landforms* 14:517–32.

de Roo, A., L. Hazelhoff, and G. B. M. Heuvelink. 1992. Estimating the effects of spatial variability of infiltration on the output of a distributed runoff and soil erosion model using Monte Carlo methods. *Hydrological Processes* 6:127–43.

Goodchild, M. F., B. O. Parks, and L. T. Steyaert, eds. 1993. *Environmental Modeling with GIS.* New York: Oxford University Press.

Heuvelink, G. B. M. 1993. "Error Propagation in Quantitative Spatial Modeling: Applications in Geographical Information Systems." Ph.D. diss., University of Utrecht.

Heuvelink, G. B. M., and P. A. Burrough. 1993. Error propagation in cartographic modeling using Boolean logic and continuous classification. *Int. J. Geographical Information Systems* 7:231–46.

Heuvelink, G. B. M., P. A. Burrough, and A. Stein. 1989. Propagation of errors in spatial modeling with GIS. *Int. J. Geographical Information Systems* 3:303–22.

Leenaers, H., P. A. Burrough, and J. P. Okx. 1989. Efficient mapping of heavy-metal pollution on floodplains by co-kriging from elevation data. In *Three-Dimensional Applications in Geographic Information Systems,* edited by J. Raper, 37–50. London: Taylor and Francis.

Moore, R. D., and J. D. Rowland. 1990. Evaluation of model performance when the observed data are subject to error. *Physical Geography* 11:379–92.

Morgan, R. P. C. 1986. *Soil Erosion and Conservation.* Longman.

Pflug, R., and J. W. Harbaugh, eds. 1992. *Computer Graphics in Geology: Three-dimensional Computer Graphics in Modeling Geologic Structures and Simulating Geologic Processes.* Lecture Notes in Earth Sciences 41. Springer-Verlag.

Rikken M. G. J., and R. P. G. van Rijn. 1993. "Soil Pollution with Heavy Metals: An Inquiry into Spatial Variation, Cost of Mapping and the Risk-Evaluation of Copper, Cadmium, Lead and Zinc in the Floodplains of the Meuse West of Stein, the Netherlands." Master's thesis, Department of Geography, University of Utrecht.

Robinove, C. J. 1986. Principles of logic and the use of digital geographic information systems. U.S. Geol. Survey Circ. 977.

Stewart, I. 1989. *Does God Play Dice? The Mathematics of Chaos.* Oxford: Blackwell.

Taylor, J. R. 1982. *An Introduction to Error Analysis.* Oxford: OUP.

van Deursen, W. P. A., and C. G. Wesseling. 1991. *The PC-RASTER Package.* Department of Physical Geography, University of Utrecht.

Varcoe, V. J. 1990. A note on the computer simulation of crop growth in agricultural land evaluation. *Soil Use and Management* 6:157–60.

Wesseling, C. G., and G. B. M. Heuvelink. 1993. Manipulating quantitative attribute accuracy in vector GIS. In *Proceedings EGIS '93,* edited by J. Harts, H. F. L. Ottens, and H. J. Scholten, 675–84. Utrecht: EGIS Foundation.

Peter A. Burrough is Professor of Physical Geography (Land Resources Assessment and Geographical Information Systems) at the University of Utrecht in the Netherlands where he lectures in geostatistics, spatial analysis, multivariate analysis, and GIS for all aspects of environmental and landscape studies. He is chairman of the GISLA research program (Geographical Information Systems for Landscape Analysis), which covers the use of geostatistics and error propagation studies in spatial modeling with geographical information systems.

R. van Rijn and *M. Rikken* are M.S. students at Utrecht.

Department of Physical Geography
Institute of Geographical Sciences
Rijksuniversiteit Utrecht
PO Box 80.115
3508 TC Utrecht
The Netherlands
Phone: 31 30 53 2749
Fax: 31 30 54 0604
E-mail: *iaapab1@cc.ruu.nl*

6

Data Quality and Spatial Analysis:
Analytical Use of GIS for Ecological Modeling

Richard J. Aspinall and Diane M. Pearson

This chapter discusses scale, spatial dependence, and data quality in spatial data for use in environmental modeling and use of spatial analysis to model the distribution of a species. Relationships between distribution data for a species and environmental data thought to influence the distribution are investigated through simulation of distribution as a stochastic process; spatial output has the resolution of the environmental data and a series of ecological relationships are also developed. Difference in resolution and other data-quality issues are an integral component of the approach, effects of uncertainty in data and assumptions made by the modeling process being tested throughout the analysis and output being accompanied by estimates of uncertainty. The analysis is structured to use environmental data from a hierarchical database. Hypotheses describing ecological relationships for the species are established in relation to the information content of the environmental data; important spatial scales and environmental relationships for the species are identified.

INTRODUCTION

This chapter considers some issues associated with spatial data that impact on the quality of output from environmental models in GIS. Relatively few environmental models are based on analytical methods that consider the quality and spatial nature of the input data although scale, spatial dependence, and data quality in environmental data have important influences. We discuss these issues and then outline a spatial analysis method that can be used to investigate their importance. The approach is used to model the distribution of animal species in Scotland by generating spatial and ecological hypotheses about distribution: the hypotheses suggest why species are where they are. The method deals explicitly with the quality of spatial data and provides a way of investigating the importance of scale by coupling the approach with a structured database of spatial data to analyze an ecological question. The spatial analysis approach, database structure and ecological problem are all based in hierarchy theory (Allen and Starr 1982; O'Neill et al. 1986; O'Neill 1989; Urban et al. 1987) allowing relationships between distribution and environmental variables to be investigated for different source scale or spatial grain environmental data to characterize the spatial scales and environmental variability to which species respond. The approach has generic application in many fields of environmental modeling by providing an analytical tool for structured interpretation of data from different geographic scales.

DATA QUALITY

The quality of data is fundamentally important to the reliability of output generated by GIS, particularly when complex analytical functions are applied in environmental models and output is to be used for decision support. Three aspects of spatial data are considered: scale, spatial dependence, and data quality.

Scale and Spatial Resolution

Scale is the ratio of length on a map to length on the ground; the basis for taking measurements from maps. It also influences the size of the phenomena that the map can represent, although this can be modified through cartographic processes depending on the purpose of the map. For example, on a map that is to communicate location of point features, these features may be enhanced for emphasis; other maps may be primarily for data recording and scale/size relationships will be more strictly applied. Tobler (1988) has shown a simple rule that describes the interaction of scale, resolution, and detectable size for features: the detectable size is the map scale divided by 1,000, the resolution is half of this amount. Thus a 1:25,000 scale map has a detectable size for features of 25 m and a resolution of 12.5 m.

Allen and Starr (1982) define the scale of a structure by "the time and space constants whereby it receives and transmits information." This is a useful definition of scale in the context of spatial data in GIS databases as the source scales of data can be viewed as a series of filters that indicate the nature of the information contained in the different data sets. Treating scale as a filter offers opportunity for comparison of data to investigate interactions of scaling, generalization, and enhancement in geographic data. Consider the data shown in Figure 6-1. The four levels show 64×64, 32×32, 16×16, and 8×8 pixel resolution maps for the same geographic area, equivalent to doubling the scale between levels. The same data are shown at each level and the influence of resolution and hierarchical position is clearly seen. The pattern in the coarse resolution (8×8) level appears as noise in lower level spatial scales (fine grain, high resolution), while the fine-resolution pattern in the 64×64 map is smoothed in the coarser resolution levels (large grain, low spatial resolution). This is the basis of hierarchical understanding of data and explanation: patterns mapped in spatial scales which are too fine (fine grain, high resolution) appear as

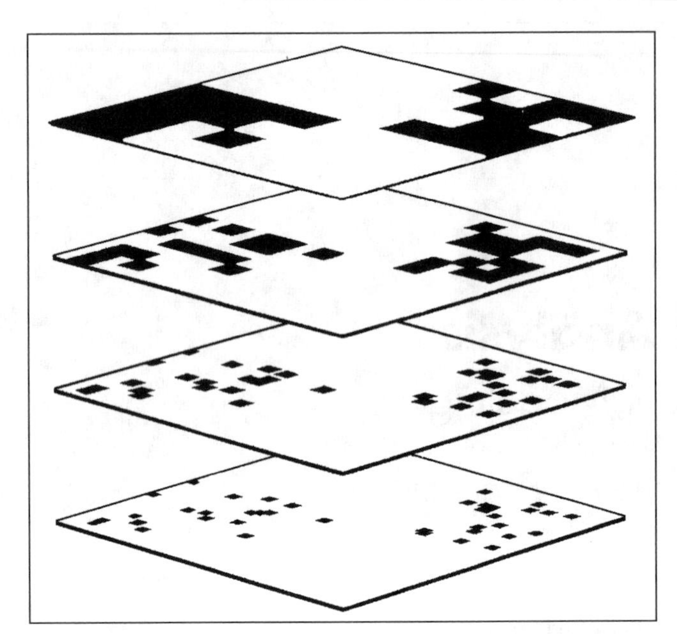

Figure 6-1. Hierarchy in spatial resolution and representation of distribution.

noise; scales that are too coarse (large grain, low spatial resolution) appear as constants (Shugart et al. 1991).

Ecological phenomena (and questions) can also be organized according to a range of hierarchies, including scale in space and time (Delcourt and Delcourt 1988). For example, Walker and Walker (1991) have used GIS for the north slope in Alaska to investigate questions related to energy development and climate change. The GIS is organized with an hierarchical database based on spatial and temporal scaling of data and natural disturbance phenomena (Delcourt et al. 1983; Delcourt and Delcourt 1988) and are used to analyze a range of phenomena at a variety of scales. Important technical and methodological issues which emerge include the question of scaling and the relationship between data sources and scale and topic of investigation.

Spatial Dependence and Modeling

Spatial dependence is the tendency for objects near each other in geographic space to have similar values. It can be measured using an approach based on a weights matrix that looks at similarity between spatially adjacent values. The geographic distances over which spatial dependence is apparent are described using semivarigrams. Spatial dependence in data is important for statistical analysis of spatial data (Cressie 1991) and has an unknown influence in other forms of mathematical analysis such as modeling.

Spatial dependence is not only a property of geographic data sets but also describes a spatial process that influences operation of environmental processes represented in models. In linking GIS and environmental modeling, these two effects may need to be separated to distinguish spatial patterns in output from models that are due (1) to operation of spatial processes in models, and (2) from consequences of display of aspatial process models using data exhibiting spatial dependence. Representing spatial processes is one of the challenges for environmental modelling (Hunsaker et al. 1993).

Data Quality and Error Propagation

The issue of data quality in GIS has received extensive treatment and review (Goodchild 1989; Goodchild and Gopal 1989; Thapa and Bossler 1992). The National Committee for Digital Cartographic Data Standards identified six components of digital cartographic data: lineage, positional accuracy, attribute accuracy, logical consistency, completeness, and temporal accuracy. Although these components often can be described, few GIS provide methods for routine use of the information they provide. Lanter and Veregin (1992) present a method for propagation of error in a layer-based GIS, and Monte Carlo simulation has also found application in modeling error propagation in GIS (Openshaw 1989). To use these methods, data quality must be described using formats that can be incorporated into GIS processing. Data quality is, however, only one aspect of error propagation in modeling as model parameters and the mathematical operations of the model itself contribute to error in output. The description of data quality in a format that can be used for analysis of error propagation and the issue of management of error in modeling remain important areas for research.

MODELING ACCURACY

These issues identify four major needs in relation to accuracy in linking models with GIS: (1) description of uncertainty in spatial data in a format that can be used in both GIS and environmental models; (2) management of these properties to enable models to be applied geographically while considering the quality of input, functional performance of the model, and quality of output; (3) analysis and management of error propagation in models; and (4) provision of tools for analysis and identification of influences of scale in description, data, process models, and output.

Spatial Questions in Ecological Modeling

Issues of space, scale, and modeling in ecology have been reviewed recently (Levin 1992; Hunsaker et al. 1993). Levin argued that the problem of pattern and scale is the central problem in ecology, providing a unifying concept and linking basic and applied ecology. Interaction of spatial and temporal scales with each other and with phenomena at different levels of ecological organization provides a framework for analysis and synthesis of many ecological problems and provides a framework within which methodological and technical and analytical tools can be identified and developed. These issues are generic and apply to all subdisciplines of environmental modeling, particularly in the context of linking models with GIS.

Linkage between environmental modeling and GIS attempts to address these issues by gaining understanding of process, representing that understanding in models, and displaying the results of the process model as a pattern through GIS. Since few process models are inherently spatial, use of (spatially neutral) process models linked to a GIS seldom reveals the influence and importance of scale on pattern and process. The ability to display results of a model in space and time using a GIS does not mean that spatial effects are represented in models (Hunsaker et al. 1993), rather the GIS displays results of independent effects that appear spatial only because of spatial dependence in the data sets.

The spatial aspects of interaction between pattern and scale and between processes that operate at different spatial and temporal scales can be more rigorously and fruitfully addressed by using exploratory and experimental spatial analyses that directly investigate the role of scale and space in environmental patterns and processes. This provides a basic research topic for use of GIS in environmental modeling, demanding novel, generic methodologies for modeling and analysis with spatial data that allow theoretical understanding of spatial processes and interaction to be developed for incorporation into links between GIS and environmental models. This will complement development of models that describe process relationships in environmental subjects.

Spatial Analysis

Data. The approach is developed for modeling species distributions across Scotland. Data for species distribution are mapped as series of Atlases with 10-km grid cell resolution. Environmental data used are climate, soils, and land cover from a range of scales. Climate data are produced as monthly means of mean, minimum and maximum temperature, and total rainfall from thirty-year mean data calculated from meteorological station records for 1951–1980; these data are mapped with a spatial resolution of 1 km using a combination of trend surface analysis and point kriging with a 1:250,000 scale digital terrain model to assist in interpolation between meteorological stations (Smith et al. 1989; Aspinall and Miller 1990). Data quality for the climate data are estimated from the variability of monthly mean climate records around the thirty-year mean values and the standard deviation estimates associated with the geostatistical analysis. Soils data are from a 1:250,000 soils map produced using standard field-based soil survey techniques. The map has been digitized into raster format with 100-m pixel resolution. Land cover data are from a 1:25,000 scale map of land cover and have a 50-m pixel resolution. These data are from air photo interpretation of 1:24,000 scale air photographs, class identification being validated with field survey. The sequence of processes involved in generating the land cover data set have been described, and a comprehensive set of data-quality descriptions have been developed and integrated into GIS (Aspinall and Pearson 1993). The data-quality description for the land cover data set includes class interpretation accuracy, absolute and relative accuracy of map and feature position, assessment of class heterogeneity, and arc specific accuracy for class boundary position (Aspinall et al. 1993; Aspinall and Pearson 1993).

Spatial Analysis. The approach to spatial analysis here is based on exploiting the hierarchy in the scale and resolution of the spatial data. Differences in resolution between data sets offer analytical opportunities (Openshaw 1989) and are the basis for investigating influences of scale and spatial relationships. The use of data sets with different resolution and source scale means that error analysis is an important component of the analytical method. The analytical method we use is based on Bayesian probability statistics and has been developed as a process for inductive exploratory spatial data analysis in GIS (Aspinall 1992). One of the advantages of this approach is that it can be implemented in almost any GIS.

Analysis uses the difference in resolution between dependent (species distribution) and independent (climate, soils, and land cover) data sets; the independent data are of finer spatial resolution or more detailed geographic scale. The coarser resolution units of the dependent data allow multiple (stochastic) realizations of distribution to be generated at the finer spatial resolution of the independent data using the coarse resolution units as a sampling frame for generating the realizations. The method searches for patterns contained in the fine resolution data that match the pattern observed at the coarser resolution. The spatial analysis becomes, therefore, a form of spectral analysis that searches for pattern across scales and spatial resolutions. Realizations of patterns are compared with a set of realizations for a random distribution within the study area, allowing the distribution of the phenomena to be compared with a reference distribution generated through a stochastic spatial process. This is similar to the approach adopted by Openshaw et al. (1987) in their point-pattern-analysis modeling procedure. An important difference is that our approach analyzes data across geographic scales and thereby provides a tool for exploration of importance of scale in description and explanation of pattern. Multiple realizations allow uncertain estimates to be generated for use in analysis of error propagation.

Error Analysis

Error analysis is integral to derivation, interpretation, and use of the results. Two error analyses are performed. First, an iterative procedure in the model generation process is used to test the assumption that the species is distributed in all the finer scale units within each of the coarser spatial units. Second, Monte Carlo simulation is used to determine sensitivity of the model to uncertainties associated with input data sets and model parameters (conditional probabilities) propagated through the analysis (Openshaw 1989; Aspinall 1992).

The assumption of a ubiquitous distribution within the finer units within each coarse unit is tested by generating conditional probabilities in a second run of the model, but allocating realizations at the finer scale to either presence or random classes based on the results of the first model. Reallocation of the random class is not carried out. The rationale for this is that a species need not be everywhere throughout each of the map units within which it is recorded but only somewhere in it. The model is updated based on its first estimate. This indicates geographic areas in which the model is sensitive to assumptions made in changing scale.

Outputs and Applications

Results of analysis represent hypotheses concerning ecological relationships at the range of spatial scales of the input data. Maps visualize this information and the associated error sensitivities. These outputs together describe and locate the geographic areas in which a species might be expected to occur on the basis of the environmental data and also define ecological relationships between species and environmental conditions (represented at a given scale). The approach makes effective use of biogeographic data and environmental data sets in GIS, these data being an important resource whose use and interpretation can provide valuable insights into ecological relationships. Comparing results of analyzing the species distribution against the different environmental data shows the extent to which these data can explain the distribution. For some species, either climate or habitat (land cover and soils) is equally able to explain the distribution; for other species, the intersection of climate and habitat is important—climate providing general limit to distribution and habitat providing pattern within this climatic range. This approach is one way of gaining such insights, and the outputs generated provide useful information for a range of applications including assessing impact of climate change and automating generation of testable hypotheses about process-based species-environment relationships.

More generally, the approach has potential for hypothesis generation from spatial data, and as a tool for inductive generation of synthetic spatial data (fine-resolution spatial data from coarse-resolution inputs) with known error tolerances. It also generates output that has defined relationships established for a range of spatial environmental data to address the importance of scale and data quality in explanation of spatial phenomena.

CONCLUSION

The purpose of spatial analysis in ecological research is to support scientific investigation of ecological phenomena and to provide tools that allow ecologists to manage and analyze interactions between scale in space and time and ecological organization (Levin

1992). The approach here is part of a GIS toolbox for spatial analysis and ecological modeling. This allows methods to be developed through particular applications as a generic set of spatial analysis methods with applications beyond those in which they are developed. The use of these tools is to address issues of scale, data quality, error propagation, and interaction of pattern and process across levels of spatial, temporal, and ecological organization. They also have wide application in research and environmental management. The tools allow questions concerning processes operating at different spatial scales to be addressed and offer potential for gaining new insights into pattern and process relationships through detailed analysis of extensive databases held and managed in GIS.

Use of the approach in an interactive computing environment enhances the value of the approach for modeling and investigating the importance of scale by allowing the modeler to interact with the model to test hypotheses as well as generate them. The output model can be integrated into decision-support and management systems, the information content of the modeled ecological relationships being of value for understanding species ecology and biology and of significance for conservation management and predicting impacts of environmental change.

Acknowledgments

Funding for the projects reported herein was provided by the Scottish Office Agriculture and Fisheries Department.

References

Allen, T. F. H., and T. B. Starr. 1982. *Hierarchy: Perspectives for Ecological Complexity.* Chicago: The University of Chicago Press.

Aspinall, R. J. 1992. An inductive modeling procedure based on Bayes theorem for analysis of pattern in spatial data. *International Journal of Geographical Information Systems* 6:105–21.

Aspinall, R. J., and D. R. Miller. 1990. Mixing climate change models with remotely sensed data using raster-based GIS. In *Remote Sensing and Global Change: Proceedings of the 16th Annual Conference of the Remote Sensing Society,* edited by M. G. Coulson, 1–11.

Aspinall, R. J., D. R. Miller, and A. Richman. 1993. Data quality and error analysis in GIS: Measurement and use of metadata describing uncertainty in spatial data. In *Proceedings of ARC/INFO Users Conference,* vol. 2, 279–90. Palm Springs.

Aspinall, R. J., and D. M. Pearson. 1993. Error analysis using ARC/INFO and ORACLE: Integrating data-quality measures for categorical maps into GIS processing. In *Proceedings ESRI UK Users Conference.* Nottingham.

Cressie, N. 1991. *Statistics for Spatial Data.* New York: John Wiley and Sons.

Delcourt, H. R., and P. A. Delcourt. 1988. Quaternary landscape ecology: Relevant scales in space and time. *Landscape Ecology* 2:23–44.

Delcourt, H. R., P. A. Delcourt, and T. A. Webb III. 1983. Dynamic plant ecology: The spectrum of vegetation change in space and time. *Quaternary Science Reviews* 1:153–75.

Goodchild, M. F. 1989. Modeling error in objects and fields. In *Accuracy of Spatial Databases,* edited by M. F. Goodchild and S. Gopal, 107–13. London: Taylor and Francis.

Goodchild, M. F., and S. Gopal, eds. 1989. *Accuracy of Spatial Databases.* London: Taylor and Francis.

Hunsaker, C. T., R. T. Nisbet, D. Lam, J. A. Browder, W. L. Baker, M. G. Turner, and D. Botkin. 1993. Spatial models of ecological systems and processes: The role of GIS. In *Geographic Information Systems and Environmental Modeling,* edited by M. F. Goodchild, B. O. Parks, and L. T. Steyaert. Oxford: Oxford University Press.

Lanter, D. P., and H. Veregin. 1992. A research paradigm for propagating error in layer-based GIS. *Photogrammetric Engineering and Remote Sensing* 6:825–33.

Levin, S. A. 1992. The problem of pattern and scale in ecology. *Ecology* 73(6):1,943–67.

O'Neill, R. V. 1989. Hierarchy theory and global change. In *Scales and Global Change: SCOPE 35,* edited by T. Rosswall, R. G. Woodmansee, and P. G. Risser, 29–45. New York: John Wiley and Sons.

O'Neill, R. V., D. L. DeAngelis, J. B. Waide, and T. F. H. Allen. 1986. *A Hierarchical Concept of Ecosystems.* Princeton: Princeton University Press.

Openshaw, S. 1989. Learning to live with errors in spatial databases. In *Accuracy of Spatial Databases,* edited by M. F. Goodchild and S. Gopal, 263–76. London: Taylor and Francis.

Openshaw, S., M. Charlton, C. Wymer, and A. Craft. 1987. A Mark I geographical analysis machine for the automated analysis of point data sets. *International Journal of Geographical Information Systems* 1:335–58.

Shugart, H. H., G. B. Bonan, D. L. Urban, W. K. Lauenroth, W. J. Parton, and G. M. Hornberger. 1991. Computer models and long-term ecological research. In *Long-term Ecological Research: An International Perspective,* edited by P. G. Risser, 211–39. New York: John Wiley and Sons.

Smith, J. M., D. R. Miller, and J. G. Morrice. 1989. An evaluation of a low-resolution DTM for use with satellite imagery for environmental mapping and analysis. In *Remote Sensing for Operational Applications: Proceedings of the 15th Annual Conference of the Remote Sensing Society,* 393–98. Bristol: Remote Sensing Society.

Thapa, K., and J. Bossler. 1992. Accuracy of spatial data used in geographic information systems. *Photogrammetric Engineering and Remote Sensing* 6:835–41.

Tobler, W. R. 1988. Resolution, resampling and all that. In *Building Databases for Global Science,* edited by H. Mounsey and R. F. Tomlinson, 129–37. London: Taylor and Francis.

Urban, D., R. V. O'Neill, and H. H. Shugart. 1987. Landscape ecology. *Bioscience* 37:119–27.

Walker, D. A., and M. D. Walker. 1991. History and pattern of disturbance in Alaskan arctic terrestrial ecosystems: A hierarchical approach to analyzing landscape change. *Journal of Applied Ecology* 28:244–76.

Richard J. Aspinall and *Diane M. Pearson*
GIS and Remote Sensing Group
Macaulay Land Use Research Institute
Craigiebuckler, Aberdeen AB9 2QJ
Scotland
E-mail: *mi019@uk.ac.sari.mluri*

7

Sampling Stratgies for Machine Learning Using GIS

Brian Lees

Machine-learning algorithms offer a way to produce cost-effective, practical GIS-based ecological models. One major limitation of these systems is that many environmental data sets used for modeling were initially collected for other purposes. It is unusual to have sufficient preexisting data to provide a learning sample for machine-learning algorithms. If a barely adequate sample is partitioned to provide learning and test samples, the result is unlikely to be either representative or independent. Comparison of existing data with a multistage sampling model, using a statistic such as Moran's I, allows the identification of areas to be sampled and the cost-effective transformation of preexisting data into a representative sample. It is better to build learning, validation, and test samples independently than to split an existing body of samples and hope that the result remains representative.

INTRODUCTION

GIS-based ecological modeling using machine-learning techniques is being rapidly accepted as a conventional tool. Unfortunately, this approach is moving from research to production environments before its limitations are widely understood.

The enthusiastic application of these techniques to "real-world" problems is understandable. Too much emphasis is given in the literature to the development of increasingly complex models that use increasingly sophisticated mathematical techniques and computer technology. In his review of environmental modeling, carried out for UNEP, Asit Biswas (Biswas et al. 1990) noted that only one of the 750 papers on modeling published in Water Resources Research (1965–1985) was ever implemented. Many neglected the complexity of real-world environmental problems. They made assumptions that simplified the problems for analytical purposes, but which reduced the validity of the result. We can all recognize the solution-in-search-of-a-problem syndrome. Both GIS and remote sensing already suffer from this. Modeling suffers similarly. There is a real opening for cost-effective, practical models. Machine-learning research has led to spin-offs of problem-solving-type systems that can help us meet this need.

MACHINE LEARNING

Over the last ten years, machine learning has undergone dramatic changes. The original goal of computer scientists to develop intelligent machines that can learn or adapt to a simulation of human-learning processes depends upon an understanding of human-learning processes. Neural modeling and evolutionary processes have brought new perspectives of how a system may learn.

The three main types of systems currently are: artificial neural networks, inductive reasoning/decision trees, and genetic algorithms. How do they differ from current approaches to environmental modeling? Traditional GIS-based modeling, using climate and landscape ecology principles, assumes that there has been no disturbance. Model output must be constrained by digitizing forest boundaries, for example, and "cookie cutting" the model predictions out where relevant. This still leaves the disturbance caused by fire, hurricane, selective logging, or whatever, as a major cause for error within the surviving model polygons.

The typical study area may have areas of partially disturbed forest, rural areas, and possibly urban areas, and rural–urban fringe areas; each of these requires a different approach. Defining these areas can be done using a number of methods: inspection, using thematic land-use maps, and so on. Automatic spatial stratification using remote sensing is an attractive approach. If one considers that the spatial extension of good field data (quadrat data) using environmental parameters should be valid within polygons of uniform disturbance in forest areas, then one can make use of the sensitivity of remote sensing to patterns of disturbance to define these polygons.

Combined analysis of GIS and remote sensing is harder than it sounds. In early analyses of remotely sensed and ancillary data, cluster analyses were performed on data sets combining spectral and nonspectral data such as slope, aspect, or geology. The operational space in which these cluster analyses were carried out was essentially nonsensical and, like principal component analyses, not reproducible from data set to data set. It assumed that the data distributions and character were similar and normally distributed, which was not the case. Rather, there are three or four quite separate operational data spaces, which are multidimensional, coherent, and sensible. Environmental data space is defined by environmental variables such as temperature, moisture availability, and nutrient status. In more complex analyses, variables such as temperature of the coolest month, rainfall in the driest quarter, etc., can be used. Spectral space is a data space defined by discrete slices of the electromagnetic spectrum. Geographic space is conventionally defined using latitude, longitude, and elevation (Figure 7-1).

Some motion in these spaces can correlate with an apparently equivalent vector in another data space. For example, movement parallel to the elevation axis in geographic space often has an equivalent

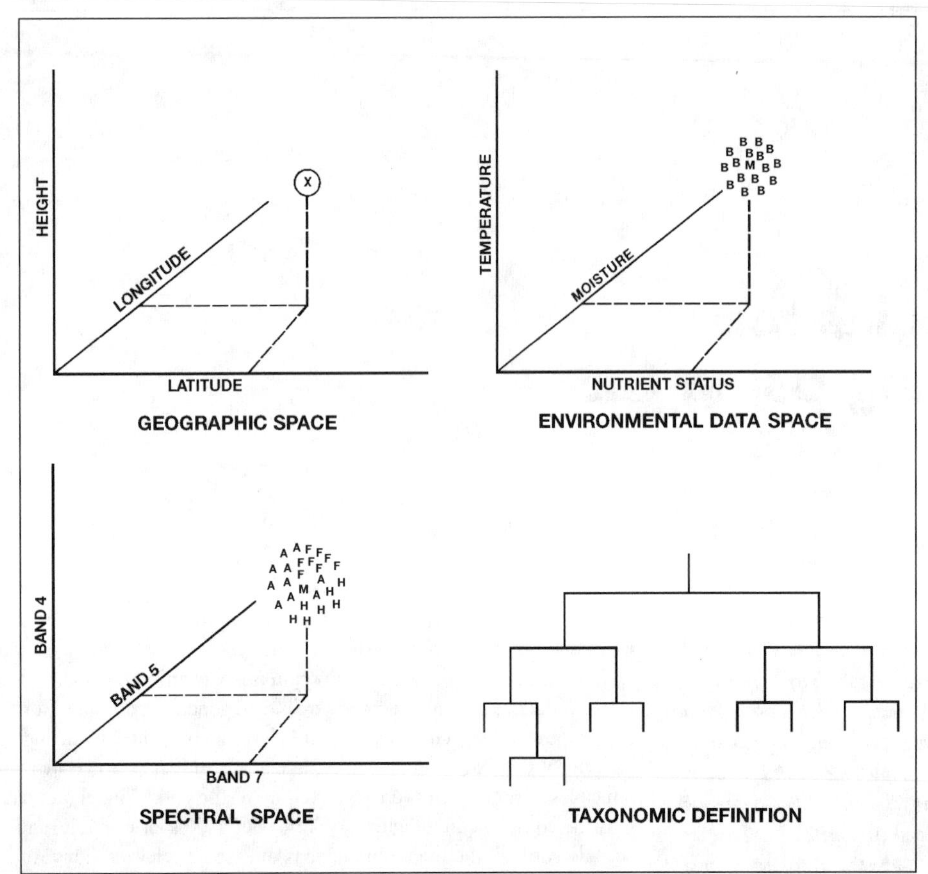

Figure 7-1. Operational data spaces.

translation along the temperature and moisture axes of environmental data space. Similarly, movement along the (soil) moisture axis of environmental data space is often matched by a vector in spectral space.

Other movements, such as movement along the nutrient status axis of environmental data space may be matched by jumps across geographic space. A point in environmental data space can represent extensive, disjunct areas in geographic space, and discrete, remote volumes in spectral space. At the smallest scales, a point in environmental data space can be occupied by a climax vegetation association in one biogeographic region—and a quite different assemblage of plants in another biogeographic region. The same point can be occupied by disturbed vegetation associations at various stages of succession or by any range of agriculture, or even by a city.

There is no reason for a discrete point in environmental data space to be represented by an equivalent single point in spectral space. It is also quite possible for plants occupying the same environmental domain, and having similar spectral properties, to be quite remote taxonomically. A great deal of confusion exists because of fundamentally incorrect assumptions about correlation between these data spaces. A more rational approach to analysis is to see the analysis as a series of steps, each of which can be carried out in the appropriate data space, and to use an analytical approach that makes no assumptions about data distribution. The character and recursive nature of most machine-learning algorithms are ideally suited to this.

Decision Trees

Inductive learning/decision trees are a sequential means to analyze data. They work by recursive partitioning with the partitioning being driven by some optimization procedure. Classification and characterization are accomplished through a series of splits (decisions) based on values in the various fields (Breiman et al. 1984; Liepins et al. 1990; Julien 1992; Lees and Ritman 1991).

Building and using a decision tree can be separated into three phases: (1) a learning set is used with a heuristic to obtain a decision tree; (2) a test set is passed down through the tree and the rules simplified; and (3) the tree is used to classify any new examples. The tree structure can also be used as an efficient data-selection algorithm.

Specific implementations differ in terms of the type of data the trees are designed for (categorical integer, real, or mixed) and the criteria that the trees are designed to satisfy (maximum homogeneity at the leaf nodes, least-squares predictors, least-absolute deviation). Other differences include the splitting criteria used (minimum misclassification cost, information theoretical measures, Gini, or two-ing), whether the splits are binary or not, stopping rules, and pruning rules.

One major limitation of these learning systems is that they are good at interpolating within the envelope of the learning sample but very poor at extrapolating beyond it. The rules derived inductively are plausible hypotheses but are limited by the available data. It is essential to have some domain knowledge to guide the selection of rules. This background knowledge will provide guidance for selecting or rejecting inductive alternate rules.

Inductive learning programs are not very noise tolerant. Sampling errors or mislabeling can provide problems, but as long as there is a large enough sample with a high signal-to-noise ratio, statistical methods can be used to control the situation.

A large learning sample is also necessary so that the level of confidence at each splitting rule remains at an acceptable level. In general terms, the lower limit is where $n = 30$. This requirement means that the number of splits is directly limited by the size of the learning sample. This is an important restriction. In environmental analyses using decision-tree rule induction, we typically cascade through a series of deterministic variables. The order in which these are addressed is surprisingly constant and usually takes the form:

- Climate
- Geology
- Topography (geometric)
- Remote sensing
- Topographic (hydrological)

Under these broad headings are very specific data sets; for example, the climatic indices, in their fullest form (Mackey 1991), comprise:

Radiation regime (kj/m²/day)

1. Annual mean monthly daily radiation.
2. Monthly mean daily radiation of minimum month.
3. Monthly mean daily radiation of maximum month.
4. Range.

5. Coefficient of variation.
6. Monthly mean daily radiation of wettest quarter.
7. Monthly mean daily radiation of driest quarter.

Thermal regime (degrees Celsius)

8. Annual mean temperature.
9. Mean daily maximum temperature of hottest month.
10. Mean daily minimum temperature of coldest month.
11. Range (9–10).
12. Coefficient of variation.
13. Mean temperature of wettest quarter.
14. Mean temperature of driest quarter.

Moisture regime (mm of rainfall)

15. Total annual precipitation.
16. Wettest month precipitation.
17. Driest month precipitation.
18. Range (16–17).
19. Coefficient of variation.
20. Wettest quarter precipitation.
21. Driest quarter precipitation.

Clearly, few such variables can be included in decision-tree-type analyses with large learning samples. One solution is to limit analyses to small areas where regional climate variations are not important and to use elevation as an analog for climate variation within the area. The problem of learning sample size with decision-tree systems is likely to be a major constraint on their future application.

Neural Networks

Such a limitation is not present in the application of neural-network algorithms to similar problems (Fitzgerald and Lees 1991). In the most common neural-network family used for such problems—Backpropagation (BPN)—the learning sample is passed through the network many times until the network stabilizes. This continued reuse of the learning sample means that the size of the learning sample can be much smaller than that necessary for decision-tree building. Indeed, Liang et al. (1992) showed that fewer training cases provide fewer degrees of freedom and therefore require fewer hidden nodes to cover all cases. Keeping the number of hidden nodes low is important in BPN to force the system to generalize. This reduces the "grandmother node" effect where individual hidden nodes become too specialized. A network with a large number of hidden nodes may perform well reclassifying the learning sample, but would perform less well when faced with new cases.

Neural networks provide an interesting alternative to decision trees but lack one of the most useful attributes of decision-tree analysis: the ability to simply examine the internal linkages of the classifier. It is possible to examine the internal weightings of a neural network, but only with great difficulty. They do provide a useful alternate inductive reasoning system that will take its place in our growing armory of nonparametric statistical tools. Like decision trees, they demand accurate and precise learning samples to give reasonable results.

SAMPLE CHARACTERISTICS

Sample Size

One of the most significant and persistent problems is the quality of data available to these systems. A large proportion of many environmental data sets used for modeling were initially collected for other purposes. There are well-understood problems associated with this that relate to the scale, currency, and accuracy of the data. Only in very unusual situations is there sufficient preexisting data to provide an adequate size of learning samples for machine-learning algorithms.

The minimum size of learning sample for neural networks should approximate two bits of data for each weight. Neural networks are considerably more efficient users of training data than decision trees. Learning samples for decision-tree analyses should typically be an order of magnitude larger than those necessary for neural networks. The sample size is determined by the minimum splitting size, the type of splitting, and the number of variables in the analysis. In geomorphological modeling, large samples are easily generated, but this is a serious limitation on their operational utility for vegetation and habitat work.

Representativeness

Given the particular sensitivity of machine-learning algorithms to unrepresentative learning samples, if a barely adequate sample is partitioned to provide learning and test samples, the result is unlikely to be satisfactory. The learning, validation (to optimize in Backpropagation), and test samples must be both independent and representative.

The requirement for representativeness means that, in a multidimensional data space, the sample envelope should match the environmental data envelope and that the distribution of sample points within that envelope should match that of the environmental data. This is not easy to achieve by simple random sampling, unless the sample size is large, as these distributions tend not to be normal. Given the cost of quadrat data, such large sample sizes are rarely practicable in vegetation or habitat studies.

SAMPLING STRATEGIES

While nonparametric techniques such as neural networks, decision trees, or genetic algorithms are particularly attractive tools for dealing with data sets with disparate characteristics and distributions because they make no assumptions about the data, they are particularly sensitive to biased learning samples. Use of GIS to structure samples is one way of minimizing these problems.

The most cost-effective way to generate suitable sample sets is to use a GIS to structure a multistage sampling strategy (Foreman 1991). Multistage sampling, rather than stratified random sampling, allows each of the different operational data spaces to be sampled in turn.

Multistage sampling also allows control of spatial autocorrelation. Using a statistic such as Moran's I to examine the important independent variables, the most homogeneous variable can be selected to form the upper stage of the sample hierarchy. Climate or mapped geology (rather than real geology) tend to be common selections. It is important that highly correlated variables are not included in these types of analyses. For example, modeled climate variables (which are largely derivations of elevation) and elevation are inappropriate in combination.

Within the first-stage attribute classes, the next most homogeneous variable can be selected to form the second stage. It is important for the sampling to meet the criterion of independence for subsampling in a given primary sampling unit to be carried out independently of subsampling in any other primary sampling unit. Sampling for the learning sample, validation sample, and test sample should be invariant. This means that every time a primary sampling unit is included in the first-stage sample, the same subsampling design must be used (Sarndal et al. 1992).

Stratification beyond a third stage is normally not necessary. If there is a choice between equally suitable variables, they should be chosen so that there is movement between the operational data spaces. Below the final stage, random sampling within a specified area of 5 × 5 cells prevents overspecification. In some studies, particularly where GIS and remote sensing are integrated in the analysis, there may be a temptation to select all of these cells on the basis that local variability in reflectance values may provide useful information about, say, vegetation structure (Walker et al. 1986; Jupp et al. 1986). This is dangerous, and local variability in reflectance values can be more usefully introduced as a separate attribute data set.

It is possible to use a GIS to examine the representativeness of the existing sample by comparing it with the global characteristics of the area under consideration. The use of bivariate plots and histograms allows a rapid identification of deficiencies in the sample. Coupling this with conventional spatial statistics allows the identification of areas to be sampled and the cost-effective transformation of preexisting data into a representative sample. It is important that the sample envelope be examined in each of the operational data spaces for representativeness.

Realistically, the total number of units selected will depend on time and cost restrictions. In general, the sample size should be the largest possible for given costs. The larger the sample size, the lower the range of estimates from all possible samples and the smaller the variance and standard error. It is much more effective to build learning, validation, and test samples independently than to split an existing body of samples and hope that the results remain representative.

Samples for use with decision-tree analysis are constrained by the minimum splitting size that is chosen. A learning sample with fewer samples of a class than the minimum splitting size will almost certainly result in a decision tree incapable of predicting that class. In such a situation it would be prudent to predict probability surfaces, or use neural networks, rather than to force the collection of a very large and expensive sample to meet the requirements of probability proportional-to-size sampling driven by histogram matching.

CONCLUSION

Integration of GIS and other data requires nonparametric procedures that make few, if any, assumptions about the structure and characteristics of the data. In parametric statistics the assumptions about data distribution mean that comparatively small samples can be used. Where such assumptions cannot be made, larger samples are required. Decision trees, in particular, make significant demands on sample size even where there are only minimal numbers of deterministic variables. Moving to neural networks as an alternative approach, with many of the attractive characteristics of decision trees, means that the simplicity of interpretation of decision trees is lost. Hopefully we can look forward to a simpler approach to examining the internal weightings of neural networks in the future. In the meantime, the characteristics of both these useful methods of integrating data mean that we must impose an increasingly strict discipline on our field data collection. We must also consider the increasing requirements for larger samples, something that will, of course, add to costs.

References

Biswas, A. K., T. N. Khoshoo, and A. Khosla. 1990. *Environmental Modelling for Developing Countries*. London: Tycooly.

Breiman, L. H., R. Friedman, A. Olshen, and C. J. Stone. 1984. *Classification and Regression Trees*. Belmont: Wadsworth International.

Fitzgerald, R. W., and B. G. Lees. 1991. The application of neural networks to the floristic classification of remote sensing and GIS data in complex terrain. In *Proceedings of the XVII Congress Int. Soc. Photogrammetry and Remote Sensing*. Washington, D.C.

Foreman, E. K. 1991. *Survey Sampling Principles*. New York: Marcel Dekker.

Julien, B. 1992. Experience with four probability-based induction methods. *AI Applications* 6:51–56.

Jupp, D. L. B., J. Walker, and L. K. Penridge. 1986. Interpretation of vegetation structure in Landsat MSS imagery: A case study in disturbed semiarid eucalypt woodlands, part 2: Model-based analysis. *Journal of Environmental Management* 23:35–37 .

Lees B. G., and K. Ritman. 1991. Decision-tree and rule-induction approach to integration of remotely sensed and GIS data in mapping vegetation in disturbed or hilly environments. *Environmental Management* 5:828–31.

Liang, T., H. Moskowitz, and Y. Yih. 1992. Integrating neural networks and semi-Markov processes for automated knowledge acquisition. *Decision Sciences* 23:1,297–313.

Liepins, G., R. Goeltz, and R. Rush. 1990. Machine-learning techniques for natural resource data analysis. *AI Applications* 4:9–18.

Mackey, B. G. 1991. "The Spatial Extension of Vegetation Site Data: A Case Study in the Rain Forests of the Wet Tropics of Queensland, Australia." Ph.D. diss., Australian National University, Canberra.

Sarndal, C., B. Swensson, and J. Wretman. 1991. *Model-Assisted Survey Sampling*. New York: Springer-Verlag.

Walker, J., D. L. B. Jupp, L. K. Penridge, and G. Tian. 1986. Interpretation of vegetation structure in Landsat MSS imagery: A case study in disturbed semiarid eucalypt woodlands, part 1: Field data analysis. *Journal of Environmental Management* 3:19–35.

Brian Lees started his professional career as a flight navigator in the RAF. After service in the Middle East and Germany, he migrated to Australia where he earned his Senior Commercial Pilot's Licence. He worked with a number of air survey and exploration companies before completing a Ph.D. at the University of Sydney in 1984 on the sediment dynamics of shallow continental shelves. He was appointed to a position at the Australian National University in 1985.

Department of Geography
Australian National University
Canberra, ACT 0200
Australia
E-mail: *brian.lees@anu.edu.au*

8

Evaluating Field-Based GIS for Environmental Characterization, Modeling, and Decision Support

Steve Carver, Ian Heywood, Sarah Cornelius, and David Sear

This chapter examines the potential of using GIS in the field for environmental characterization and modeling in isolated areas. Observations are based on experiences gained during two Anglo–Russian expeditions to the Altai Mountains of Siberia aimed at evaluating proposals for a new national park in the Katunsky Ridge area of the Belukha Massif. The use of GIS together with GPS, EDM surveying equipment, and portable data loggers for primary data collection and verification/update of existing data is described, and the use of field-based systems for on-the-spot modeling and decision support is evaluated. Despite disadvantages associated with taking sensitive equipment into a harsh environment, adopting an integrated approach to data collection and database creation holds greater attractions of interactive feedback from field surveys and ground truthing. This is enhanced by the provision of more accurate data derived from field surveys and local knowledge than would be possible from studies based on maps and remotely sensed imagery alone. Conclusions from the study are expanded in terms of GIS decision support for relief planning and response management where up-to-date field information may be important.

INTRODUCTION

This chapter evaluates the use of portable geographical information systems (GIS) for field-based mapping, modeling, and management, which has proven to be valuable for studies aimed at environmental characterization, modeling, and decision support. However, as a technique, it has received little attention by either the academic or commercial research community. From the literature (for example, Burrough 1986; Maguire et al. 1991; Haines-Young et al. 1993) it is apparent that more time on GIS-based environmental modeling research is spent in the laboratory than in the field. This is understandable since a wider range of environmental data sets are now available that cover most locations at a variety of scales than there were before, and this in some way seems to negate the necessity to collect data in the field. Problems of data error (Carver 1991), including data quality, coordinate accuracy and precision, attribute uncertainty, etc., give rise to problems of confidence in the data which are often opaque to the remote user, particularly if the user has not visited the study area. This is likely to become a major problem area in the future as GIS become more widely used. This is compounded by the fact that most users of environmental data sets will not have been involved in the data collection process and may therefore be largely unaware of the limitations of the data. This inevitably leads to problems where such data are used for environmental modeling, and ultimately for assisting in policy formulation.

In approaching some of the problems in GIS-based environmental modeling outlined above, this chapter will do the following:

1. Outline a methodology for field-based GIS for environmental modeling, characterization, and decision support.
2. Highlight the advantages and disadvantages of this approach with reference to fieldwork in remote environments.
3. Review the above in the context of an ongoing environmental research program in the Altai Mountains of Siberia.
4. Provide guidelines for the use of field-based GIS in similar projects.
5. Illustrate how such an approach might be used to assist in other areas such as disaster-relief planning where up-to-date field information may be important.

BACKGROUND

It has already been pointed out that the integration of environmental modeling and GIS can be deficient in that:

- confidence in the data is lacking where existing digital data sets are used;
- modeling can be divorced from field knowledge and local input; and
- the modeler is unable to verify model predictions through direct field observations.

The research presented here attempts to rectify these basic problems through the development and testing of a field-based GIS approach to environmental characterization, modeling, and decision support during the GeoAltai 1992 and 1993 expeditions.

The GeoAltai Research Program

GeoAltai is a joint Anglo–Russian program of research based in the Altai Mountains of Siberia near the Russian/Kazakstan/Mongolian borders. The initial aim of the research is to provide baseline environmental information for the Katunsky Ridge area of the Belukha Massif (4506 m) and use this as the basis for landscape assessment, simple modeling, and environmental zoning of the area in order to evaluate proposals for a new national park (Heywood et al. 1992). The location of the study area is shown in Figure 8-1. The information provided by this survey includes spatial databases on the biophysical characteristics (geology, soils, vegetation, hydrology, etc.) of two contrasting watersheds and the anthropogenic impacts (tourism, forestry,

Figure 8-1. Location of study area, Altai Mountains, Siberia.

grazing, pollution, etc.) which impinge on the area as a whole. Further research will be based on using the above information and models as the basis for a decision support system for helping define the boundaries of the park and, ultimately, in managing the park.

GeoAltai embraces the use of GIS throughout the entire research program, but it is its use in the field where the greatest advantages are to be gained. The whole ethos of this approach is based on the interactive use of GIS in the field, not only for primary data collection, but also for ground truthing and updating existing databases, visualization, design of sampling strategies, integrating various data sources (field observations, digital map data, remotely sensed imagery, video, etc.), and as a medium for generating ideas and focusing research effort (Brokes et al. 1992).

METHODOLOGY

Technologically, there is little new in using GIS with global positioning systems (GPS) and total station surveying equipment in the field for data collection. This has been done before, though perhaps not on the scale and level described here. What is new is the "total" approach taken to the whole process, from initial desk-based data collection, through field studies and modeling, to the development of a spatial decision support system (SDSS). Of particular interest is the integrated approach to data collection and modeling in the field, bringing together various techniques, data sources, and types in order to create a fully integrated database and modeling environment during the fieldwork period.

Methodological Outline

The methodology developed here can be divided into three distinct stages. These are outlined below and illustrated in Figure 8-2.

1. Desk-based collection and processing of existing data sources on the area of interest.
2. Field-based data collection, verification, update, and modeling.
3. Development of an SDSS.

Stage One: Prefieldwork Data Collection/Processing. The first stage, involving the desk-based collection and processing of existing data sources, is carried out prior to the fieldwork period. This provides the field party with a baseline data set covering the area of interest from which to work once in the field. This normally would include information digitized from paper maps and aerial photographs as well as from existing digital data sets such as relevant satellite imagery and environmental databases. Some potentially useful generic data sources are listed in Table 8-1. The resulting database may be used to define an initial sampling strategy, though this may be altered once in the field.

Stage Two: Field-Based Data Collection, Verification, Update, and Modeling. The second stage, consisting of field-based data collection, verification, update, and modeling is crucial to the whole approach proposed here. It involves the use of GPS to fix locations of field observations to within 1–5-m accuracy (using differential and averaging techniques) and EDM surveying equipment for more accurate surveys such as fixing survey control points and river/valley cross sections, etc. Observations located in this manner can then be used to map new environmental information and to verify/update existing data while in the field and on a day-to-day basis. New information (both positional and attributional) is added or used to update the GIS database in the usual way. Visualization of GIS databases created in this manner is an important

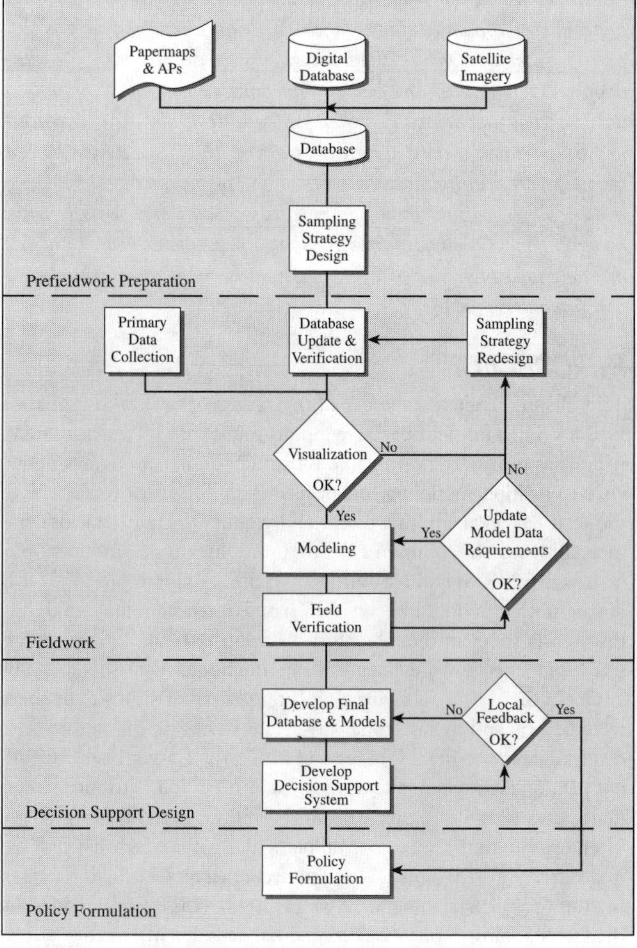

Figure 8-2. Conceptual diagram of field-based GIS methodology.

Table 8-1. Generic data sources.

Type	Typical sources and scale/resolution
Analog maps	USAF Tactical Pilotage Charts @ 1:1,000,000 Local maps (scale, quality, and availability depending on study area)
Aerial photographs	Local agencies (scale, quality, and availability depending on study area) ... or take your own
Digital vector data	Digital Chart of the World (DCW) @ 1:1,000,000
Digital raster data	Global Resources Information Database (GRID) @ 1-km grid
Satellite imagery	NOAHH imagery @ 80-m pixel Landsat TM imagery @ 25-m pixel

ability of a field-based GIS system. The user can see the data on the screen, interrogate it, and make a hard copy for use in planning the next day's fieldwork. Areas where the quality of existing data is uncertain (e.g., vegetation classification using satellite imagery) or boundaries where positional error may be important (e.g., boundaries between soil units) can be highlighted and positively targeted for sampling/investigation in this way. The ability to redesign sampling strategies in the field as new information is collected, old data is updated, and areas of uncertainty are cleared up is an important advantage of this field-based approach. In effect, this interactive approach allows a positive feedback mechanism of data collection, database update, visualization, sampling strategy design, data collection, and so on, to operate.

Perhaps the greatest advantage of this approach is the ability to run any GIS-based environmental model developed using desk-based databases in the field and to verify their results through the data collection program. Through an examination of the differences between model predictions and empirical evidence collected in the field (from spot samples or monitoring devices), inadequacies in the model can be seen. Depending on the process being modeled, it may then be possible to review the workings of the model and alter it in such a way that its outputs fit reality more closely. This may entail the simple adjustment of model parameters or constants, the collection of further field data, the rewriting of parts of the model, or even the addition of new processes within the model structure itself. Updated models can then be rerun, the new results compared with the empirical data, the model altered again if necessary and rerun, etc., as is often done in the computing laboratory. Again, a positive feedback mechanism operates with the advantages associated with the ability to update databases and review processes affecting the model on an interactive basis while in the field. It is noted that this approach is not entirely suitable for all environmental models. For those models where the processes operating are very slow (e.g., weathering or climate change), then it will be impossible to measure process rates during the duration of the fieldwork period. However, it may be possible to improve model performance on the basis of better field information on relevant variables/parameters and processes affecting the one being modeled.

Stage Three: Development of Spatial Decision Support Systems. The third and final stage of the proposed methodology is the use of field databases and improved environmental models to develop SDSS for use in strategic decision making. Through working with local decision makers and researchers, either in the field or at relevant institutions afterward, it is possible to design better systems. The ability to visualize problems through the medium of GIS databases is a useful tool for focusing attention and generating ideas. Again, a mechanism of positive feedback operates with local decision makers commenting on the prototype SDSS, the results of which may then be built into the SDSS and presented for further comment. By working in this interactive manner, the SDSS envisaged by the research team and by the decision makers should eventually converge. At this point, the SDSS may be considered for use in policy formulation (though perhaps only on a trial basis).

Technical Specification

Following is a brief technical specification of the equipment and data used on the recent GeoAltai 1993 expedition. Figure 8-3 shows the basic setup in diagram form.

Equipment used. Three portable notebook-type PCs (386/486 models) were used to run IDRISI GIS software in the field. These were also used to run a variety of supporting software including statistical, spreadsheet, graphics, and word processing packages. Two Magellan NAV 5000 PRO™ handheld GPS receivers were used in differential mode to obtain positional fixes accurate to within 3–5 m. Used singly, these GPS receivers gave positional fixes accurate to within 25 m (single fix) or 12 m (average). An electronic distance meter (EDM)

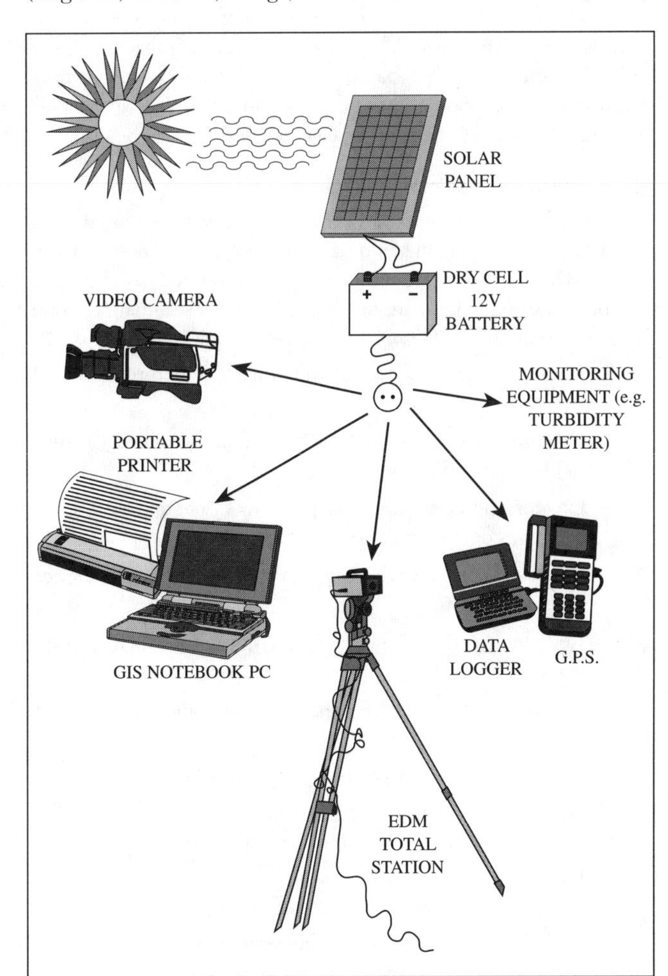

Figure 8-3. Technical setup.

"total station" theodolite was used for more accurate survey work. Handheld data loggers were used in conjunction with the GPS and EDM to store data in the field before input into the GIS database. A combination of video and still photography was used to record images in the field for later analysis and integration into multimedia presentations. Aerial video coverage from a helicopter used to bring personnel and equipment into the field proved very useful in the absence of large-scale aerial photographs (aerial stills were taken for postfieldwork processing). A variety of scientific instrumentation (e.g., turbidity meters, stage recorders, meteorological stations, etc.) was used across a range of studies (e.g., geomorphological, hydrological, glaciological, etc.), and the data were included in the GIS database where appropriate.

Data used. A variety of existing digital databases were tapped to provide as comprehensive a coverage of the study area as possible prior to the fieldwork period. These included both raster and vector data sources. Global databases such as Digital Chart of the World (DCW) and the Global Resource Information Database (GRID) were used to provide a regional perspective to the study area. More detailed information on the study area was provided from digitized 1:50,000 scale maps and Russian (KOSMOS) satellite imagery (hard copy only). All other data in the final GIS database were collected and input in the field.

DISCUSSION

The following discussion of the methodology outlines the advantages and disadvantages of the field-based GIS approach and suggests not only how GIS-based environmental modeling can be improved, but also how it may be applied in other areas such as disaster-relief planning where up-to-date information may be useful.

Evaluation of Field-Based GIS

In evaluating the field-based GIS methodology for environmental characterization, modeling, and decision support in remote environments as outlined above, a number of advantages and disadvantages can be identified. These are best considered in a summarized form. The suggested advantages of the field-based approach are:

1. Interactive development of sampling strategy through visualization and feedback.
2. On-the-spot environmental modeling and feedback through field-based verification.
3. Greater appreciation of the problem in context and of the processes operating.
4. Improved confidence in the data and greater user awareness of its limitations.
5. Reduction in the number of field visits and associated project costs.
6. Ability to integrate local ideas and knowledge at the start of the analyses.
7. Increased ability to convince local decision makers of the approach adopted.

The suggested disadvantages of the field-based approach are:

1. Logistical problems associated with power source(s) and transport/protection of equipment in the field.
2. Lack of technical backup facilities/services.
3. Problems of data availability, sensitivity, and security in some countries/regions.
4. Education and training of the field team in GIS concepts and techniques.
5. Long fieldwork preparation lead times.
6. It is very hard work.

The GeoAltai Experience

The real experience gained in the field during the GeoAltai expedition serves to underline the advantages and disadvantages of the field-based GIS outlined above, although item 7 in the list of advantages remains to be fully tested/proven through practical experience. For developing sampling strategies and visualization, the field-based use of the GIS proved extremely useful. While the GIS environmental modeling actually done in the field was very simple, it also proved advantageous in that comparisons between field knowledge and the GIS database demonstrated which analyses were actually sensible in terms of GIS modeling and, through interactive verification procedures, whether the results gained were meaningful.

The bulk of the problems experienced in the field were related to data availability and technical hardware/software problems. As with satellite imagery, use of Landsat MSS data proved infeasible as the study area fell into a black spot between orbits and the Russian KOSMOS data were only made available to us in hardcopy form. High-quality Russian maps and aerial photographs are hard to come by as many are still considered military secrets. In fact, none of the state-produced large-scale maps (1:25000, 1:50000, and 1:100000) have any longitude/latitude or projection information. And while they do have grids they are not referenced with coordinates, nor is there any indication as to their orientation or origin. This had to be worked out in the field using the GPS and fixed geographical features to transform digitized data (still in digitizer coordinates) to latitude/longitude. Six days of cold weather and no sun meant that the batteries ran down, leaving us without power for three days, while programming bug fixes in a wet and mosquito-infested tent was not pleasant.

Looking at the list of the advantages/disadvantages of field-based GIS, it must be concluded that the benefits far outweigh the costs. This is particularly true when we consider (1) the advantages gained through improvements to environmental models through interactive field verification procedures and the greater confidence in the data gained through direct involvement in the data collection process, (2) the positive feedback mechanisms operating, and (3) the input of local knowledge and experience. None of the disadvantages listed and experienced by GeoAltai are insurmountable, although it is extremely frustrating when equipment malfunctions or bureaucratic/logistical problems occur in the field. Built-in redundancy to field equipment and proper planning/organization should prevent or at least soften the impact of these problems.

Improvements to GIS-Based Environmental Modeling

The beneficial opportunities offered by the interactive, field-based testing and modification of GIS-based environmental models should, by now, be obvious. However, they may be summarized as follows:

1. Ability to run the models and verify outputs through empirical observation in the field.
2. Opportunity to update model variables/parameters from empirical field measurements.
3. Ability to make modifications to the model in the field via positive feedback from 1 and 2 above.

4. Greater confidence in model outputs as a result of greater confidence in the input data and the model itself.

The disadvantages of this approach to integrating environmental modeling with GIS are the same as the general disadvantages stated above for the field-based methodology as a whole. However, the widespread adoption of such an approach must inevitably depend largely on GIS users and their preference for working either in harsh and remote field conditions or in the comfort of their offices or laboratories.

Other Application Areas

A number of other application areas beyond the particular field of environmental modeling may benefit from the adoption of a field-based GIS approach. These may include disaster-relief planning, emergency planning, traffic control, park management, forestry, etc. Taking the example of disaster-relief planning, up-to-date information may be required and could be collected by workers operating in the field. Important information—for example, in the case of a major earthquake—could be input/updated by teams working in the field with portable GIS/GPS systems, which could include road blockages, location of landslides, broken services (water, gas, electricity, telephone, etc.), building collapses, etc. In such an example, a field-based GIS could be extremely useful in assisting on-the-spot decision support. In another example, that of famine relief, the time frame of operations is somewhat longer, though no less urgent. Here, field-based GIS and environmental modeling may be used more in a traditional planning capacity, not only in ensuring that short-term aid is managed properly (efficiently distributed and allocated, etc.), but also to ensure that long-term measures aimed at promoting development and preventing environmental degradation are planned and managed properly.

CONCLUSION

This chapter outlined a methodology for using GIS and GIS-based environmental modeling in the field for environmental characterization and decision support. The proposed methodology was illustrated with reference to its use during a program of environmental research in the Altai Mountains of Siberia. The discussion focused on the advantages and disadvantages of such an approach and how such an approach may be of benefit in integrating environmental modeling and GIS. We maintain that the advantages far outweigh the disadvantages, although from personal experience, it is extremely hard work. To reiterate, the suggested advantages include: (1) the improvement of environmental models through interactive field verification procedures and greater confidence in the data gained through direct involvement in the data collection process, (2) the operation of positive feedback mechanisms, and (3) the input of local knowledge and experience.

References

Brokes, P., A. Buecek, I. Downey, I. Heywood, E. Pauknerova, and J. P. Petch. 1992. Remote sensing and GIS in management of protected areas: Pilot survey of Zdarske Vrchy, Czechoslovakia. In *Science and Management of Protected Areas*, edited by J. H. M. Wilson et al. Amsterdam: Elsevier Science Publishers.

Burrough, P. A. 1986. *Principles of Geographical Information Systems for Land Resource Assessment.* Oxford: Oxford Science Publications.

Carver, S. 1991. Error modeling in GIS: Who cares? In *The Association for Geographic Information Yearbook 1991*, edited by J. Cadoux-Hudson and D. I. Heywood. London: Taylor and Francis/Miles Arnold.

Haines-Young, R., D. R. Green, and S. Cousins. 1993. *Landscape Ecology and Spatial Information Systems.* London: Taylor and Francis.

Heywood, D. I., S. Carver, S. Cornelius, and N. Mikhailov. 1992. The development of a GIS for landscape management in the Altai Mountains of Siberia. In *Proceedings: EuroCarto X.* Oxford: Trinity College.

Maguire, D. J., M. F. Goodchild, and D. W. Rhind, eds. 1991. *Geographical Information Systems: Principles and Applications.* London: Longman.

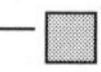

Steve Carver
School of Geography
University of Leeds
Leeds LS2 9JT
England
Phone: 44 113 2333318
Fax: 44 113 2333308
E-mail: *steve@geography.leeds.ac.uk*

Ian Heywood
Department of Environmental and Geographical Sciences
Manchester Metropolitan University
Manchester M1 5GD
England
Phone: 44 161 287 1574
Fax: 44 161 247 6318
E-mail: *i.heywood@mmu.ac.uk*

Sarah Cornelius
Department of Environmental and Geographical Sciences
Manchester Metropolitan University
Manchester M1 5GD
England
Phone: 44 161 2471578
Fax: 44 161 2476318
E-mail: *s.cornelius@mmu.ac.uk*

David Sear
Department of Geography
Southampton University
Southampton SO9 5NH
England
Phone: 44 1703 594614
Fax: 44 1703 593729
E-mail: *d.sear@soton.ac.uk*

9

Interpreting the State Soil Geographic Database (STATSGO)

Dennis J. Lytle, Norman B. Bliss, and Sharon W. Waltman

Soil databases provide detailed information on the physical and chemical properties of soils and the data can be interpreted for use in environmental simulation models. In the United States, the State Soil Geographic Database (STATSGO) is used to make digital maps of soil organic carbon, available water capacity, soil depth, and other properties. The maps are a basis for spatially extrapolating the results of field studies of environmental processes. Simulation models provide insight into the dynamic aspects of the processes. Techniques for generalizing the STATSGO are being used to develop a national soil geographic database and prototypes for improved global data sets.

INTRODUCTION

Understanding the movement of water in soil is critical for understanding many environmental processes, including mass and energy transfers during evapotranspiration, plant growth, and pollutant transport. Simulation models of water in the environment require information on the behavior of water in the soil. Both watershed models and atmospheric general circulation models need information on infiltration, soil water storage, and evapotranspiration.

Water movement in ecological and agricultural ecosystems is influenced by precipitation patterns, soil properties, vegetation cover, and topography. Characteristics of the soil that influence water movement can be extracted from soil geographic databases. By linking the soil data to simulation models, processes at smaller or larger scales in space and time can be modeled.

This chapter presents examples in which soil data have been used in conjunction with simulation models. Soil maps and databases produced by the National Cooperative Soil Survey (NCSS), a partnership between the U.S. Department of Agriculture (USDA), states, and universities, were analyzed to provide interpretations that are suitable for watershed models and atmospheric general circulation models, such as soil rooting depth and available water capacity.

Development of STATSGO

In 1983, the NCSS conference recognized the nation's need for small-scale digital soil maps. Participants at the NCSS conference recommended the establishment of a nationally consistent general soil geographic database. This recommendation was made to overcome the deficiencies in the existing general soil maps of individual states. These state general soil maps often did not join accurately with maps of adjacent states and varied in scale, level of detail, age, and concept.

Development of STATSGO began in 1984 and was completed in 1993. STATSGO was intended to replace the existing, inconsistent, hard-copy general soil map record with a more consistent digital record. This digital record consists of generalized soil map unit delineations linked to an attribute database, which contains information related to map unit acreage, proportionate extent of the component soils within each map unit, and the chemical and physical properties of the component soils.

STATSGO maps were manually compiled using U.S. Geological Survey (USGS) 1:250,000-scale quadrangles as base maps. The 1:250,000 scale is similar to the scale used for traditional NCSS general soil maps, which is useful for broad planning. The scale provides a resolution of soil mapping that is compatible with existing USGS databases for terrain, land use and land cover, political boundaries, and federally owned land. U.S. Fish and Wildlife Service National Gap Analysis Program data and Landsat multispectral scanner (MSS) data also share a comparable scale and resolution.

The STATSGO specifications identify two basic rules—a cartographic rule and a map unit composition rule. The cartographic rule requires a minimum-size delineation of about 625 ha (1,544 acres—an area about 1 × 1 cm on the 1:250,000-scale base map). A recommended density of 100–200 soil map delineations (polygons) per 1° × 2° 1:250,000-scale quadrangle is specified, with exceptions to 400 polygons. For reference, there are 128, 7.5-minute quadrangle units in each of these quadrangles. STATSGO map units are delineated based on available references such as county soil survey maps (which are available for about 90% of the conterminous United States), county general soil maps, and other ancillary data such as Landsat MSS data and vegetation, terrain, climate, and geology maps. These specifications provide a nationally consistent cartographic representation of data that forms a relatively continuous coverage for the United States.

The map units are formed by grouping phases of soil series from larger-scale maps. The generalization rule for the aggregation of attributes allows up to twenty-one different component soils to be identified for each STATSGO map unit. The composition of STATSGO map unit delineations is usually determined by measuring transects on county level soil survey maps. The STATSGO map unit components are soil series phases, and their percent composition represents the estimated areal proportion of each within STATSGO map units. The composition for a STATSGO map unit is generalized to represent the statewide extent of that map unit and not the extent of any single STATSGO map unit delineation. These specifications provide a nationally consistent representation of STATSGO attribute data.

The STATSGO is designed for multicounty, state, and regional environmental resource planning and management. Based on this design, states use regionally unique criteria to group soils into relatively homogeneous map units that will respond similarly to use and management. These are the same criteria that have been used to develop county level soil survey maps.

STATSGO Data in a Geographic Information System

Users are able to store, retrieve, analyze, and display STATSGO soil data as well as connect soil data with other spatially referenced resource and demographic data using geographic information system (GIS) technology. The digital map is related to attribute tables by the GIS software. The attribute data for the STATSGO are contained in a set of relational tables. The relational tables named MAPUNIT, COMP (component), and LAYER have attributes for progressively more detailed soil areas and volumes. The mapped delineations (polygons) are related to a MAPUNIT table. For each map unit, there may be multiple components. Data on the COMP table include slope, soil texture, flooding regime, depth to water table, depth to rock or hardpan, subsidence, hydrologic group, and capability class. A generalized soil profile is described by records in the LAYER table, including attributes for soil texture, particle size distribution, bulk density, pH, organic matter, available water capacity, permeability, and selected chemical measurements (Soil Survey Staff 1993b; Lytle 1993).

METHODS OF INTERPRETATION

Techniques were developed for using a GIS to manage and interpret the component attributes associated with the STATSGO. These interpretation techniques focus on some form of aggregation that generalizes the more detailed component attribute values either to a single value or to probability values that are then mapped using a GIS according to the composition of individual map units. Two techniques are used to present original STATSGO attributes (such as available water capacity) as well as the results of environmental simulation models, which may have used either STATSGO attribute data or more specific soil pedon data.

The first technique involves preparing multiple probability maps to portray a given interpretation. There is one map for each interpretation category, and the map shows the percentage of the area having STATSGO components that match the criterion. This technique relies on an aggregation of component ratings using areal weighting values, which are given by the component percent values for each STATSGO map unit. The second technique results in a single aggregated value reported for each STATSGO map unit; these values are mapped using legend classes of user-determined value ranges.

Multiple Probability Maps

The earliest efforts at interpreting the STATSGO displayed data using multiple probability maps. The first STATSGO interpretations were prepared by soil geographic database developers within USDA-Soil Conservation Service (SCS) and USGS (Bliss and Reybold 1989). They used an early draft of the STATSGO for Fairfax County, Virginia, to develop an interpretation technique that rated each map unit component for a given set of criteria. In a simple example, they used the following criterion: the high range of slope class greater than or equal to 7% (the SLOPEH attribute data element found in the component table). Using the component percent values (COMPPCT in the component table), the component ratings for this criterion were then accumulated into an overall percentage for each individual map unit. The legend for the resulting map was titled "percent of map unit meeting criteria." The legend classes were ranges of 0–20%, 21–40%, 41–60%, 61–80%, and 81–100%, plus any unrated categories, such as water. The researchers also explored more complex queries involving criteria such as available water capacity less than or equal to 0.14 inches of water per inch of soil calculated for the top 30 inches of soil. The results of this query were mapped using the same "percent of map unit meeting criteria" legend.

A similar approach was used in the STATSGO/GRASS Interface developed by USDA-SCS (Minzenmayer 1992). This software is an interface between the U.S. Army Corps of Engineers Geographic Resource Analysis Support System (GRASS) GIS package and STATSGO spatial and attribute data. This menu-driven interface, which does not require a commercial database management system, was developed to meet SCS NCSS needs. It allows the user to conduct routine or custom soils queries of the STATSGO. For example, if a user were to prepare a STATSGO-based pesticide leaching interpretation showing low, moderate, and high ratings, three maps would be created, one for each rating class. Each map uses the percent of map unit meeting criteria legend.

STATSGO Linked to the Agricultural Census

Monds et al. (1991) combined the STATSGO with agricultural census data to produce a map showing areas that would be most vulnerable to groundwater contamination from agricultural chemicals and animal wastes within the Pennsylvania portion of the Chesapeake watershed. Researchers used preliminary Pennsylvania STATSGO data to set high leaching potential criteria for soils, which included map units that contained greater than 70% Hapludalfs, greater than 50% moderate soil pesticide leaching potential, or greater than 35% high soil pesticide leaching potential. The query result was mapped as a single rating value, called "high leaching potential," for each STATSGO map unit meeting the criteria. The map has a single aggregated legend category. This map was overlaid with another derived from county-based agricultural census data that identified lands with a high percentage of chemically treated cropland and a high rate of manure production. A single map resulted with the following legend classes: soils with high leaching potential; high percentage of chemically treated land; high rate of manure production; and highest potential for groundwater contamination (those areas where all classes overlapped).

Parameters for Hydrologic Models

Soil information can be extracted from the STATSGO and reformatted to match the requirements for the portions of climate models that simulate land surface processes, including hydrologic processes. To illustrate this process, analyses were performed for available water capacity and depth of soil for the states of Colorado and Minnesota (Soil Survey Staff 1993a, 1993b).

These analyses were used to understand the total capacity of the soil to store water. They may become more valuable when used in conjunction with other measures, such as soil texture or permeability.

Available water capacity (AWC). AWC is the amount of water available to plants; specifically, it is the difference between the water stored in the soil at field capacity and the water stored at the permanent wilting point. Field capacity is defined as the amount of water retained if a saturated soil is allowed to drain freely (for several days). The permanent wilting point is the water content of the soil at which plants wilt and fail to recover their turgidity when placed in a dark humid atmosphere (Soil Conservation Society 1982). To standardize the measurement of these parameters, soil scientists measure AWC in terms of the water retention difference:

> Water retention difference (WRD) is the volume fraction for water in the whole soil that is held between 15-bar water retention and an upper limit, that is, usually the ⅓- or the ⅒-bar water retention. . . . The ⅓- and 15-bar gravimetric water contents are converted to a whole soil volume basis by multiplying by the bulk density of the whole soil. The WRD is a derived value and is reported as centimeters of water per centimeter of depth of soil (cm cm^{-1}). The WRD is an estimate of the available water capacity (AWC) (Soil Survey Staff 1992a).

In the STATSGO, the available water capacity attribute on the layer table is defined as a range. The "available water capacity—low" (AWCL) attribute is defined as the "minimum value for the range of available water capacity for the soil layer or horizon, expressed as inches/inch" (Soil Survey Staff 1992b). The maximum end of the range is defined by the "available water capacity—high" attribute (AWCH).

Because the units for AWC form a "dimensionless ratio," they may be interpreted as centimeters of water per centimeter of soil, inches of water per inch of soil, or cubic meters of water per cubic meter of soil.

Figure 9-1 is a map of the AWC for Minnesota. The map was formed by calculating the midpoint between AWCL and AWCH for each layer as an estimate of AWC (cm cm^{-1}, interpreted as m^3m^{-3}). This was multiplied by the thickness of the layer (m) to compute the AWC for the layer (m^3m^{-2}). These values were summed over the layers in the profile to give a measure of the AWC for the component (m^3m^{-2}). These values were multiplied by the component area (m^2) and summed to give a measure of the AWC of the map unit (m^3). The total AWC for the state was calculated by summing over all map units. To form a legend for coloring the map,

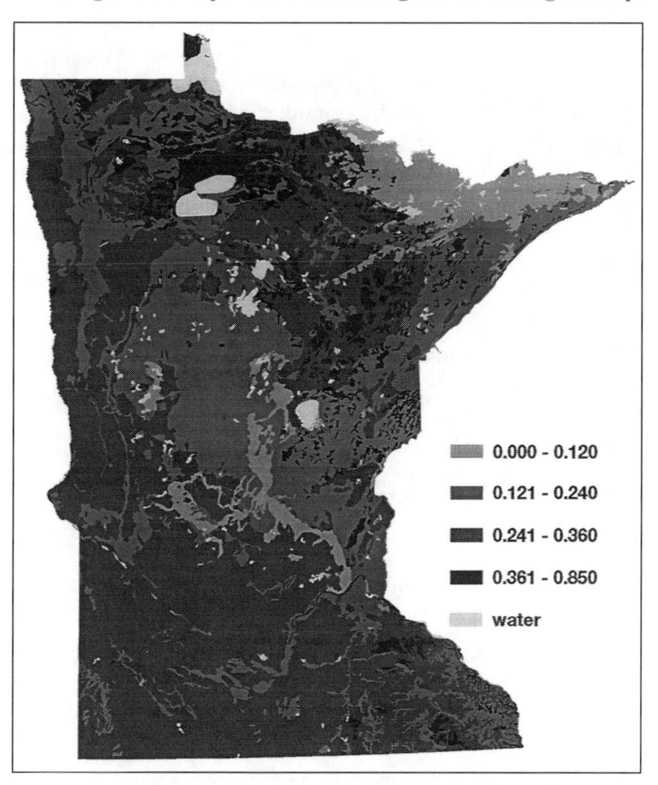

Figure 9-1. Available water capacity (m^3m^{-2}) for Minnesota from data in the State Soil Geographic database (STATSGO).

the total AWC was divided by the area of analyzed soil (m^2) to calculate an average AWC for each map unit (m^3m^{-2}).

Rooting depth. The data for the rooting depth calculation are on the component and layer tables. Four data items in the component table were used in the rooting depth calculation: depth to bedrock low and high (ROCKDEPL, ROCKDEPH), and depth to hardpan low and high (PANDEPL, PANDEPH). The midpoint values for these depth measures were calculated. The texture data on the layer table were evaluated, and if a layer was classified as weathered bedrock or unweathered bedrock, the layer depth low (LAYDEPL) measure was saved as an intermediate value. The minimum of this intermediate value and the two midpoint values were used as the limiting soil depth (m) for the component. The limiting soil depth was multiplied by the component area (m^2) to calculate the volume (m^3) of soil in the map unit available for roots. This was divided by the area of the map unit (m^2) to calculate an average depth (m) of soil within the map unit. The depths were classified into legend categories, and a summary map was produced.

The average depth calculation is appropriate if the purposes of modeling do not require detailed distributions of soil depth. If a detailed hydrological study is required, it is possible to compute the distribution of soils of various depths, and a series of maps could be produced showing percentage of the land area in each depth category.

Parameters for Biogeochemical Models

Bliss et al. (1993) used the STATSGO to calculate soil organic carbon values for thirty-six eastern and midwestern states. A single soil organic map was produced using areal extent weighting values (COMPPCT in the COMP table) to aggregate the component carbon values into a single map unit value. For each state, plots were produced that have data quality inset maps. These maps illustrate areas where data gaps exist, nonrated components that are present within the map units, and the varying soil depths used in the calculation.

Link to Soil Laboratory Data

Grossman et al. (1992) linked soil survey laboratory characterization data to Iowa STATSGO components through taxonomic classification (Soil Survey Staff 1975) and field knowledge of the soils for Pottawattamie County. A pedon-based quantity of organic carbon for the surface meter of soil was computed and assigned to the appropriate STATSGO component. A map unit average was aggregated using an areal weighting technique (COMPPCT attribute in the COMP table). A single map was produced with legend classes of discrete values.

Link to Hydrologic Simulation

Landre (1991) used the New York STATSGO to assess the susceptibility of soils and landscapes of the Genesee River basin to pesticide leaching. The STATSGO identified all possible component soils found within the basin. Published soil characterization pedon data were then assigned to gently sloping component soils (SLOPEH < 8%) using soil classification and field knowledge. These data became the parameters for a soil leaching simulation model called LEACHM. The LEACHM results included a flux value for atrazine (at a specified depth within the soil profile under a corn crop after a one-year period) for each gently sloping component of the STATSGO map units. The component-specific LEACHM results were aggregated for each map unit using the areal weighting technique (COMPPCT in the COMP table) for the modeled components. These aggregated map unit values were presented on maps for areas with agricultural land use.

Wagenet et al. (1993) followed a similar approach to using the STATSGO with the LEACHM simulation model, except that the parameters relied on the STATSGO rather than on pedon data and the study area included seven northeastern states. The LEACHM results were presented in multiple probability maps illustrating high, moderate, and low atrazine leaching potential in agricultural land, which used a percent of map unit meeting criteria legend.

Additional Formats for Results

Modelers may need to have the data formatted according to a specified data structure, such as a 10 × 10-km grid. The results of a STATSGO analysis can be formatted in this way. The proportion of each map unit polygon within a grid cell is calculated, and the areas of the contributing components within each legend category are accumulated to produce estimates of the proportions of the legend categories in each grid cell. This accounting can be done for detailed distributions, or weighted averages can be calculated for quantitative measures.

CONCLUSION

Soil is an integral part of ecosystems, and soil water relations are an important part of hydrologic and atmospheric models. State and national soil geographic databases are now available that can provide data for these models over large areas of the earth's surface. Techniques have been developed to analyze the data and format them to match the modeler's requirements. Future efforts in soil database development will develop a hierarchy of soil databases, so that there is a link from field samples of soil pedons to global soil maps.

Acknowledgments

Work performed under USGS contract 1434-92-C-40004.

References

Bliss, N. B., and W. U. Reybold. 1989. Small-scale digital soil maps for interpreting natural resources. *Journal of Soil and Water Conservation* 44:30–34.

Bliss, N. B., S. W. Waltman, and G. W. Petersen. 1993. Preparing a soil carbon inventory for the United States using geographic information systems. *Advances in Soil Science.*

Grossman, R. B., E. C. Benham, J. R. Fortner, S. W. Waltman, J. M. Kimble, and C. E. Branham. 1992. A demonstration of the use of soil survey information to obtain areal estimates of organic carbon. In *Proceedings Resource Technology 92, 4th International Symposium on Advanced Technology in Natural Resource Management,* 457–65. Washington, D.C.

Landre, P. T. 1991. "Integrating Geographic Information Systems and a Simulation Model for Assessing Groundwater Vulnerability to Pesticide Contamination in the Genesee River Watershed." Master's thesis, Cornell University, Ithaca, New York.

Lytle, D. J. 1993. Digital soils databases for the United States. In *Environmental Modeling with GIS,* edited by M. F. Goodchild, B. O. Parks, and L. T. Steyaert, 386–91. New York: Oxford University Press.

Minzenmayer, F. E. 1992. *STATSGO/GRASS Interface User's Guide Version 1.2.* Internal Publication, U.S. Department of Agriculture, Soil Conservation Service, Lincoln, Nebraska.

Monds, D., G. Osborn, and H. C. Smith. 1991. Chesapeake Bay water-quality maps—Pennsylvania example. Unpublished maps prepared for the USDA-SCS Chesapeake Bay Council Representatives to the Chesapeake Bay Program. U.S. Department of Agriculture, Soil Conservation Service, Chester, Pennsylvania.

Soil Conservation Society of America (SCSA). 1982. *Resource Conservation Glossary.* Soil Conservation Society of America, Ankeny, Iowa.

Soil Survey Staff. 1975. *Soil Taxonomy: A Basic System of Soil Classification for Making and Interpreting Soil Surveys.* Agriculture Handbook 436, U.S. Department of Agriculture, Soil Conservation Service.

———. 1992a. *Soil Survey Laboratory Methods Manual.* Soil Survey Investigations Report No. 42, Version 2.0, National Soil Survey Center, U.S. Department of Agriculture, Soil Conservation Service, Lincoln, Nebraska.

———. 1992b. *State Soil Geographic Database (STATSGO) Data Users Guide.* Miscellaneous Publication 1492, U.S. Department of Agriculture, Soil Conservation Service.

———. 1993a. *Colorado State Soil Geographic Database (STATSGO).* U.S. Department of Agriculture, Soil Conservation Service, Lakewood, Colorado.

———. 1993b. *Minnesota State Soil Geographic Database (STATSGO).* U.S. Department of Agriculture, Soil Conservation Service, St. Paul, Minnesota.

Wagenet, R. J., J. Hutson, S. D. DeGloria, and R. G. Perritt. 1993. *Mapping Groundwater Contamination Potential Using Integrated Simulation Modeling and GIS.* Final Report—CSRS Agreement #89-COOP-1-471B. Washington, D.C.: U.S. Department of Agriculture.

Dennis J. Lytle and *Sharon W. Waltman* are responsible for developing national soil geographic databases for the Soil Conservation Service.
USDA-SCS
National Soil Survey Center
Federal Building, Room 152
100 Centennial Mall North
Lincoln, Nebraska 58508-3866
E-mail: *sgis@calmit.unl.edu*

Norman B. Bliss specializes in the application of national and global soil databases for hydrological, ecological, and agricultural studies.
Hughes STX Corporation
EROS Data Center
Sioux Falls, South Dakota 57198
E-mail: *bliss@dgl.cr.usgs.gov*

STATSGO data are available from the National Cartography and GIS Center, U.S. Department of Agriculture, Soil Conservation Service, PO Box 6567, Fort Worth, Texas 76115; phone (817) 334-5559, fax (817) 334-5290.

10

Soil-Landscape Modeling in Southeastern Australia

P. E. Gessler, I. D. Moore, N. J. McKenzie, and P. J. Ryan

Explicit and quantitative soil-landscape models are required for environmental modeling and management. Advances in the spatial representation of hydrological and geomorphological processes using terrain analysis techniques are integrated with the development of a sampling and soil-landscape model building strategy. Preliminary results are encouraging and show useful relationships between terrain and soil attributes. These techniques may provide a more appropriate methodology for spatial prediction and for understanding soil-landscape processes.

INTRODUCTION

Environmental models require spatial representation of the soil landscape due to its role as a modifier of material and energy fluxes. Ideally, spatial predictions of soil layers, specific soil properties and, eventually, soil-landscape processes are needed at a scale appropriate for environmental management (Moore et al., *Soil*, 1993). The challenge is to develop explicit, quantitative, and spatially realistic models of the soil-landscape continuum useful for a variety of purposes beyond taxonomic classification (McSweeney et al. 1993). A promising development is the potential for correlating soil properties with simple-to-measure terrain and environmental attributes that have physical meaning (Moore et al., *Soil*, 1993; McKenzie and Austin 1993). The underlying hypothesis of this work is that catenary soil-landscape development occurs in response to the way water moves through and over the landscape—water movement is in turn controlled by the geometry of the land surface.

There has been a trend in recent work (Bouma 1989; Gessler et al. 1989; Baize and Girard 1992) toward using soil layers rather than soil profiles or pedons as the basic object for study. Soil layers may have a pedogenic (soil horizon) or geomorphic (stratigraphic unit) origin. Regardless of origin, they form a logical building block for spatial modeling and interpretation of how sequences of layers behave. The soil layers at any point are a result of integrated pedogeomorphic and hydrological processes (Butler 1964; Simonson 1959). As such, a description of the arrangement, dimension, and nature of the soil layers at points in the landscape may be used as a link or pointer to the spatial distribution of processes and vice versa.

However, soil-landscape processes operate across a range of spatial and temporal scales (Allen and Starr 1982; Kachanowski 1988) and it is clear that imprinting, truncation, and synergisms occur (Malanson et al. 1990; Allison 1991). These complexities cause soil properties to exhibit different and complex scales of variation (Butler 1964; Beckett and Webster 1971; Burrough 1993). Thus, our expectations for deciphering the relationship between pattern and process should vary in different physiographic domains. This reinforces the need to develop environmental correlations using exploratory data analysis followed by explicit definition based on physically interpretable statistical models (McKenzie and Austin 1993). McKenzie and Austin propose explicit model development using generalized linear models but do not provide a spatial implementation. Moore et al. (*Soil*, 1993) demonstrate a spatial implementation for soil-property prediction, but only for the soil surface of a small toposequence study plot with a dense grid of sample points.

Australia contains vast areas with scant land resource information. The resources for collecting basic data sets to understand environmental function and management are limited (McKenzie 1991). This chapter presents initial results on the testing of a method for developing explicit soil-landscape models using advanced spatial analysis, field sampling, exploratory data analysis, and statistical modeling techniques. The primary aim is a more rational and efficient soil sampling strategy to develop robust statistical models for the spatial prediction of soil properties. These models may then be used to parameterize models for environmental management.

TWO-DIMENSIONAL SPATIAL CHARACTERIZATION OF PROCESSES: TERRAIN ANALYSIS

Moore et al. (1991) review terrain analysis and its application in the earth sciences. Primary and secondary (or compound) topographic attributes are recognized, and a table summarizing the significance of these attributes for characterizing the spatial distribution of landscape processes is presented. Many of the attributes have potential use as spatial predictors of soil properties. Primary attributes are directly calculated from elevation data and include areal measures such as specific catchment area and point measures, including the first and second derivatives such as slope, aspect, plan and profile curvature. Secondary attributes involve combinations of the primary attributes that quantify the contextual nature of points or characterize the spatial variability of specific processes occurring in the landscape or both. Methods of computation are presented by Moore et al. (1991; *Soil*, 1993).

Digital topographic attributes are scale dependent and if these effects are not considered, computed attributes may be meaningless or the processes of interest may be masked (Moore et al. 1991; *Modeling*, 1993). Moore et al. (*Modeling*, 1993) report critical differences in the computation methods of primary and secondary topographic attributes and, for example, advise against the use of the D8 method of flow-direction computation. This method does not allow flow dispersion and produces unrealistic flow patterns. This significantly influences the computation of flow accumulation that is critical to the computation of many spatial hydrological and soil-landscape attributes such as catchment and dispersal areas. Differences in environmental attribute correlations and model development will occur due to physiographic setting, scale of analysis, computation methods, and other factors (data structure, quality, and error). It is essential for a modeling framework to have explicit definition of decisions relating to the particular combination of methods applied (McSweeney et al. 1993).

STUDY REGION

The study region is the Wagga Wagga 1:100,000 topographic map sheet located on the western slopes of the Great Dividing Range in southeastern Australia (147°E,35°S; 147°E,35°30'S; 147°30E,35°30'S; 147°30'E,35°S). This region was chosen because it incorporates several physiographic domains typical of the Murray-Darling River basin. The major physiographic domains have been delineated and soil-landscape models will be developed in each. However, this chapter focuses on initial methodology development in a 100-km² pilot study area (centered on 147°27'E, 35°24'S) dominated by gently rolling erosional land forms on Ordovician metasediments.

SOIL-LANDSCAPE MODEL DEVELOPMENT

Two methods have recently been proposed for development of explicit and quantitative soil-landscape models (McKenzie and Austin 1993; McSweeney et al. 1993). Both methods are similar in approach and require a definition of purpose, scale of application, and stratification of the sampling domain. This work is aimed at developing a spatial model of soil layer patterns within a particular soil-landscape system. This is viewed as critical to eventual spatial prediction of specific soil properties and soil-landscape processes. The scale of application is the hillslope within small catchments, and it corresponds with the size of management unit in the study area. The soil layer is used as the basic object of study, and the catchment is the boundary of the system, due to its significance for spatially related hydrological and erosional processes.

Stratification is an important issue, particularly in the polygenetic landscapes common to Australia. However, its quality depends on the availability of prior information such as soil, geology, vegetation, land form, and stratigraphy maps. At the onset of this work, a 1:100,000 geology map (Raymond 1992) was generated and initial stratification was performed using these data. Additional data layers (soils, land form, stratigraphy, vegetation, climate) are being generated as part of a collaborative project and subsequent work will look more specifically at stratification using these integrated data. The focus here is on the methods of explicit soil-landscape model development.

Digital contours (10-m contour interval), stream lines, and spot heights registered to the Australian Map Grid (AMG-UTM) were obtained from the New South Wales Land Information Center in digital form. A baseline 20-m × 20-m grid DEM for the 100-km² study area was developed using the program ANUDEM (Hutchinson 1989). This involved interpolation to a 5-m grid resolution and resampling to the 20-m level. Scaling parameters, factual, and error properties of this surface are reported elsewhere (Moore et al., *Modeling*, 1993). Seventeen catchments were delineated using an automatic catchment delineation algorithm and a full range of primary and secondary topographic attributes were generated for each catchment using the methods of Moore et al. (*Soil*, 1993; *Modeling*, 1993). Flow or area accumulation (i.e., specific catchment area) was calculated using the FRho8 flow-dispersion algorithm (Moore et al., *Modeling*, 1993) and a 100-cell channel initiation threshold. An algorithm was developed for the creation of graphical displays of the probability density function and listing of the summary and Moran statistics for each attribute on a catchment basis using the S language (Statistical Sciences 1992). This enables the rapid characterization of a catchment and quantitative comparison of the overall differences between catchments or specific zones within a catchment. A contiguous five catchment subarea (20,868 cells or 834.72 ha) was selected for soil-landscape model development because it encompassed a range of the topographic variability (including aspect) characteristic of the physiographic domain as a whole.

Development of a Rational and Quantitative Sampling Strategy

An iterative sampling strategy using four criteria was used to select field sample sites: (1) the sampling plan must reflect the provisional predictive pedologic model by sampling evenly along the predictive variable(s) in attribute space; (2) strict randomization is necessary to ensure an unbiased sample; (3) inefficiencies due to spatial dependence in soil properties must be minimized; and (4) locational error between the digital model and the real world must be minimized.

When soil surveyors begin work in unmapped territory, they often begin with an implied model and begin testing hypotheses with sample points. This provisional predictive pedologic model (McKenzie and Austin 1993) evolves as points are sampled. But much of this information about continuous soil-landscape variation is lost or subsumed when map unit lines are drawn. The most common provisional model is the *catena* (Latin = a chain) soil-landscape model (Milne 1935) that implies a concordance of soil pattern with land form as one traverses from hilltop to valley bottom along toposequences. The compound topographic index (CTI), often referred to as the steady state wetness index, is a quantification of catenary landscape position. It is defined as:

$$\mathrm{CTI} = \ln(A_S / \tan\beta) \qquad (10\text{-}1)$$

where A_S is specific catchment area (area per unit width orthogonal to the flow direction) and β is slope angle. Moore et al. (*Soil*, 1993) show that the CTI is correlated with several soil properties such as silt percentage (r=0.61), organic matter content (r=0.57), phosphorus (r=0.53), and *A* horizon depth (r=0.55) in the soil surface of a small toposequence. The CTI is used here as an explicit and quantitative provisional predictive pedologic model. To develop a robust statistical model for testing hypothesized correlations, it is sensible to sample evenly in CTI attribute space. Thus, the CTI was divided evenly into five, twenty percentile classes (Figure 10-1a). The goal of this work is to develop a soil-landscape model applicable to the broader Ordovician metasediment physiographic domain. Therefore, the percentile break points were computed using all the grid cells falling on this bedrock type in the 100-km² study area. Figure 10-1b shows a spatial display of the percentile classes for the study catchments. The percentile classes also provide convenient patches that can be used for randomization to meet the second sampling criterion.

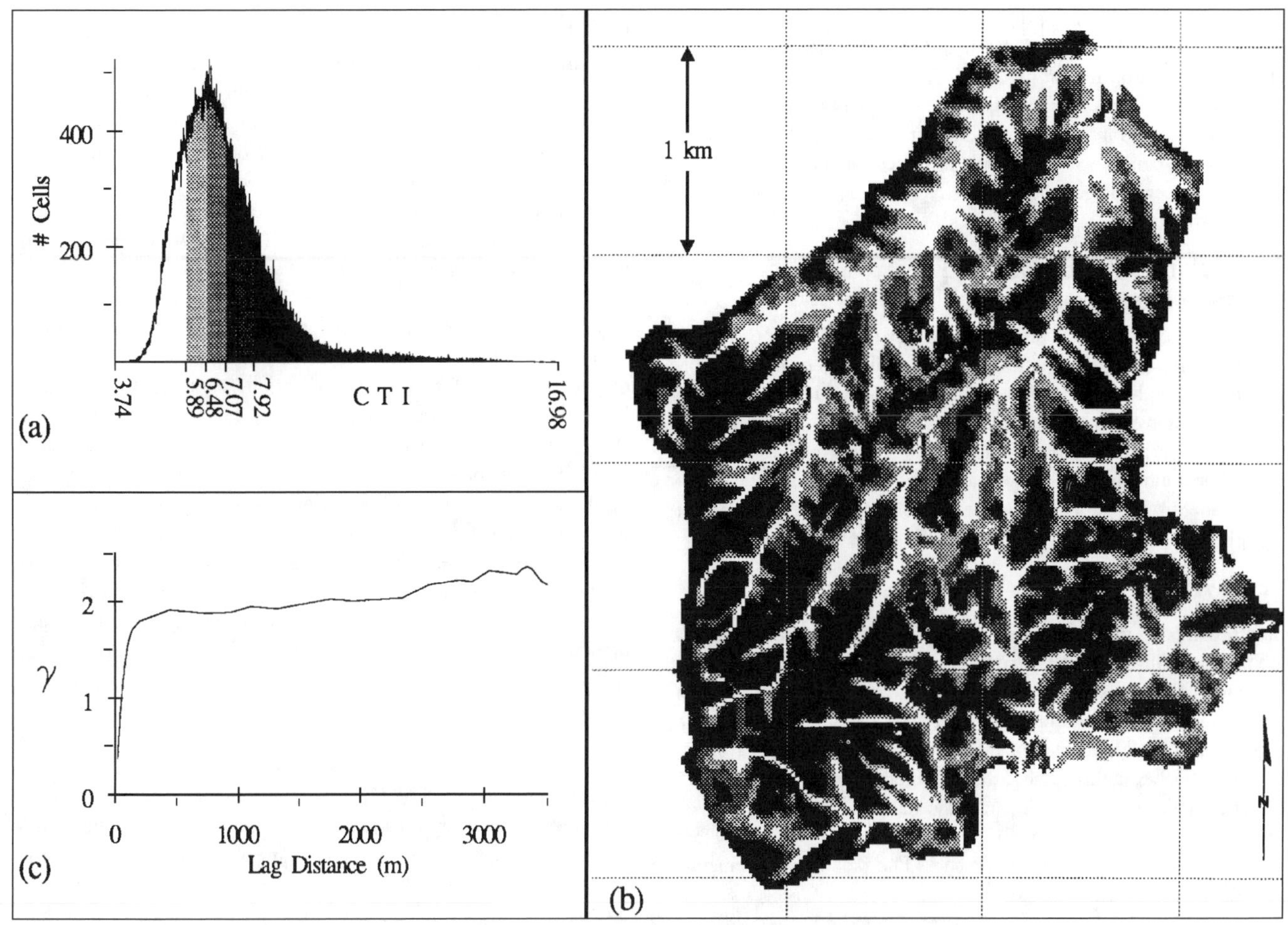

Figure 10-1. (a) Twenty percentile histogram of CTI, (b) spatial display of CTI for study catchments, and (c) CTI variogram.

Soil properties show varying degrees of spatial dependence (McBratney and Webster 1981), which reduces the efficiency of random sampling. Spacing sample sites using information about the spatial-dependence structure increases the information content of samples. We have no *a priori* information on the spatial-dependence structure of the soil properties of interest. Instead, we postulate that the spatial-dependence structure of the CTI relates in a general way to the spatial-dependence structure of the soil properties of interest. Moran's I coefficient (Goodchild 1986), which characterizes the overall strength of spatial dependence, is 0.70 for the CTI cells on the Ordovician metasediments in the 100-km^2 study area. This indicates strong spatial dependence in the CTI. The variogram (Webster and Oliver 1990) is a common method of quantifying the spatial-dependence structure of a regionalized variable (Matheron 1971). Figure 10-1c shows the computed variogram for the CTI cells on the Ordovician metasediments. This variogram shows a range (distance within which spatial dependence occurs) of approximately 500 m. This suggests that statistical independence can best be maintained by spacing samples 500 m or more apart. An assumption is that the spatial dependence is stationary across the landscape. Subsequent sampling may be useful at nested scales within this distance to develop a useful understanding of short-range variation for individual properties. Short-range variation was not our primary interest and will not be covered in this chapter.

Accurate location of field sample points as allocated using the GIS is critical to the development of robust empirical models. To minimize locational errors, samples are located only in attribute patches with a minimum size of 3 × 3 grid cells (0.36 ha). This is accomplished by passing a 3 × 3 filter through each quantile class to eliminate smaller patches.

Sample Site Allocation and Data Collection

Sites were allocated in two batches of thirty samples. Six samples were distributed in each CTI percentile class according to the following iterative scheme. The filtered patches for each class were numbered from 1–*n* (total number of patches for percentile class). A random number generator was used to produce a random number vector of length *n*. Sites were selected sequentially from randomly selected patches and Australian Map Grid coordinates produced for each site. Sites within 500 m of previously selected sites were discarded, and the next random patch selected until six sites were allocated for each class. Each site was located in the field using a global positioning satellite (GPS) receiver. The slope, aspect, elevation, and specific catchment area attributes for each site were output from the GIS and used in the field to refine site placement and ensure consistency. At each site a 71-mm diameter core was taken to a maximum depth of 2.3 m. The cores were described according to McDonald et al. (1990).

Diagnostic morphologic properties that characterize the soil layers were used for model development. These properties were: *A* horizon depth; *E* horizon presence/absence; *E* horizon depth, mottle presence/absence, depth to mottles; *A* horizon clay percentage; and *B* horizon clay percentage and solum depth (*A*+*E*+*B* horizon depths). *A* horizon depth, solum depth, and the probability of encountering an *E* horizon are used to demonstrate the results. The *A* horizon depth is a general guide to nutrient status of soils in the study area and also an indicator of surface stability to erosional and depositional processes. An *E* horizon is indicative of downward or lateral percolation and leaching processes and periods of water logging. This has an impact on biological productivity and trafficability. Solum depth provides an indication of the available water capacity and this also exerts a major control on biological productivity.

Exploratory Data Analysis and Statistical Model Development

A simple matrix of scatter plots was developed to identify pattern or structure within the data and to provide an indication of attribute correlations. An exhaustive search technique (Statistical Sciences 1992) was also employed to assist in identifying a parsimonious subset of explanatory variables. Statistical modeling was performed using generalized linear models (McCullagh and Nelder 1989). Diagnostic methods for identifying outliers, influential observations, and violations of model assumptions were used routinely (Statistical Sciences 1992).

RESULTS

Statistical models that predict soil properties using topographic attributes are presented in Table 10-1. Two types of generalized linear model are presented. The first is a multiple regression with an identity link function and normal errors; it is equivalent to a classical least squares multiple regression. The second type of model is used for predicting a binary response variable, in this instance the probability of encountering an *E* or bleached horizon in the upper part of the soil profile. This generalized linear model uses a logistic link function and binomial errors and is often referred to as a logistic regression model. The proportion of variation accounted for by a logistic model cannot be expressed using a statistic analogous to R^2. Model adequacy is assessed in terms of the prediction errors and the reduction in deviance which is distributed approximately like χ^2 (McCullagh and Nelder 1989).

Table 10-1. Regression equations for prediction of soil properties (standard errors are shown in parentheses)*.

Least squares multiple regression models

A depth =	0.92 (14.1)	+ 5.67 plancrv (1.4)	+ 4.88 CTI (1.9)	$R^2 = 0.63$
Solum depth =	-57.95 (39.4)	+ 12.83 plancrv (3.9)	+ 21.46 CTI (5.2)	$R^2 = 0.68$

Logistic regression model

ln(p/(1-p)) = 2.52 + 1.68 umplancrv

rearranging gives:

p = exp(2.52 + 1.68 umplancrv) / (1 + exp(2.52 + 1.68 umplancrv))

Analysis of deviance

Model	Deviance	Resid. Deviance	Df	Pr(Chi)
Null		69.31	49	
umplancrv	29.43	39.88	48	<0.001

* CTI = compound topographic index
plancrv = plan curvature
umplancrv = upslope mean plan curvature
p = probability that an *E* horizon is present

CONCLUSION

The ubiquitous and substantial short-range variation of soil attributes places a fundamental limit on the quality of spatial prediction. Webster (1977) concluded that the variation accounted for by a typical general-purpose survey would range from about ½ the total variance for physical properties to less than ⅒ for some chemical properties. This provides an informal measure for judging the success of a statistical model. The results in Table 10-1 are encouraging because of the comparatively large R^2 values.

As expected, CTI is a useful predictor because it combines contextual and site information via the upslope catchment area and slope respectively. Plan curvature was not expected to have a strong predictive power because it does not include contextual information. However, it was significant in predicting *A* horizon and solum depth along with CTI. This suggests that local scale pedogenic as well as hill-slope scale processes are influencing soil profile development. Upslope mean plan curvature provided the best logistic model fit for the probability of an *E* horizon occurrence. This indicates that the overall upslope convergent and divergent flow processes may control *E* horizon development. The next best logistic fit was provided by CTI, which in part measures some of the same types of landscape processes as upslope mean plan curvature. Figure 10-2 displays the spatial extension of the logistic model for *E* horizon presence/absence (a) and solum depth (b) for the study catchments.

We began with a provisional pedogenic model where CTI was hypothesized to be a strong controlling variable and designed our sampling plan accordingly. The field data supported this assertion and provided evidence of other useful explanatory variables. The identification of plan curvature and upslope mean plan curvature as useful predictors demonstrates a key feature of our methodology. Models are proposed and then tested. During the testing phase, new hypotheses of landscape processes controlling soil distribution are formulated, and these may be tested to further improve our capacity for spatial prediction. In a conventional survey, this process is undertaken in the surveyor's mind as he or she traverses a region and develops a mental, and sometimes verbal, model for spatial prediction.

We developed a quantitative and statistical analog to the conventional method that is explicit, consistent, and repeatable; evidence is not confused with interpretation, and models can be communicated in an objective way. At present, a large body of knowledge is trapped within the minds of soil surveyors and eventually will be lost. Our procedure meets with Hewitt's (1993) demands for a scientific rather than subjective procedure for developing explicit and quantitative soil-landscape models for spatial prediction. These methods provide a basis for understanding soil-landscape processes and may be integrated with other spatial interpolation techniques such as kriging and splines (Hutchinson and Gessler 1994). Information about scale (Moore et al., *Modeling*, 1993) and error (Burrough, this volume) must also be explicitly incorporated.

Acknowledgments

The authors thank Linda Ashton for her overall assistance and John Hutka for help with fieldwork. This study was funded in part by grant #NRMS-M218 from the Murray-Darling Basin Commission and by the Water Research Foundation of Australia.

References

Allen, F., and T. B. Starr. 1982. *Hierarchy: Perspectives for Ecological Complexity*. Chicago: The University of Chicago Press.

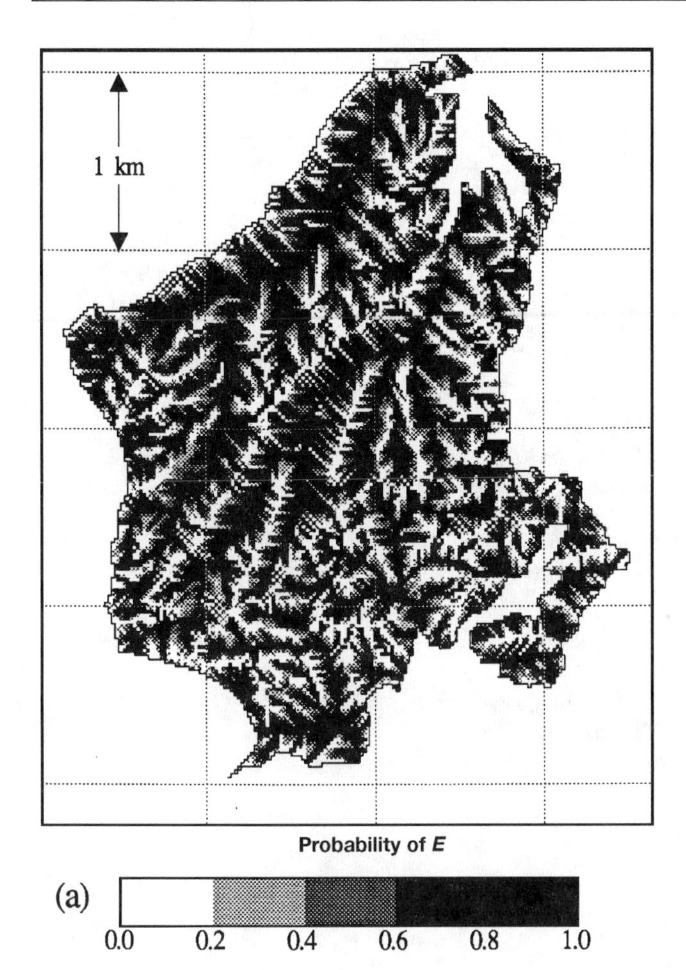

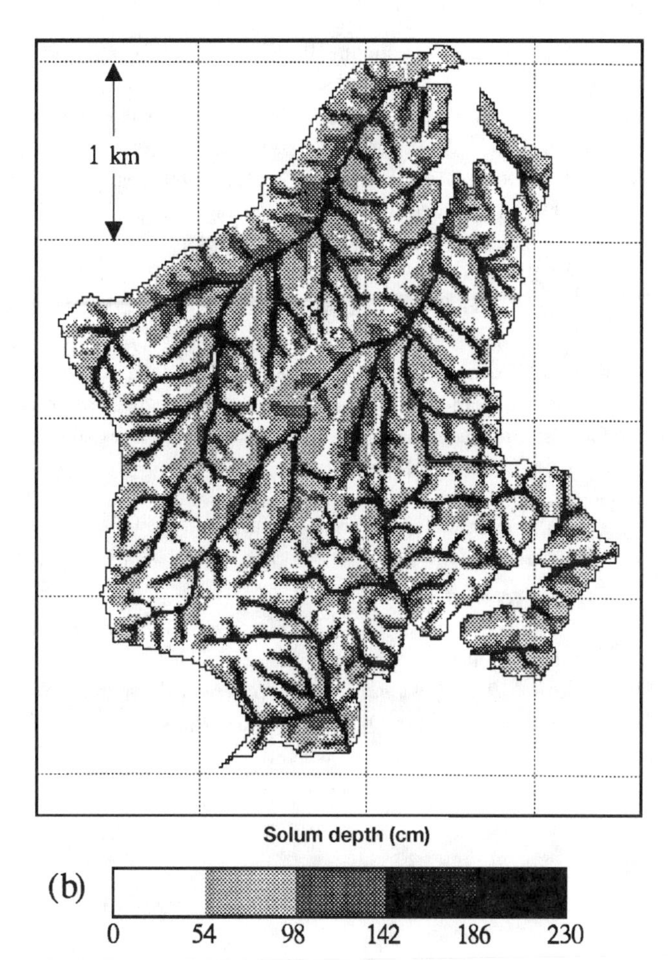

Figure 10-2. (a) Probability of *E* horizon and (b) predicted solum depth.

Allison, R. J. 1991. Slopes and slope processes. *Progress in Physical Geography* 15(4):423–37.

Baize, D., and M. C. Girard. 1992. Pedological reference base: Main soils of Europe. Referentiel pedologique. *Principaux sols d'Europe*. Versailles: INRA.

Beckett, P. H. T., and R. Webster. 1971. Soil variability: A review. *Soils and Fertilizers* (5):529-62.

Bouma, J. 1989. Land qualities in space and time. In *Proceedings of International Society of Soil Science*, 3–13. Wageningen: Pudoc.

Burrough, P. A. 1993. Soil variability: A late 20th-century view. *Soils and Fertilizers* 56(5):529–62.

Butler, B. E. 1964. Can pedology be rationalized: A review of the general study of soils. Presidential address, publication no. 3. Australian Soc. Soil Sci., CSIRO Div. Soils, Canberra.

Gessler, P. E., K. McSweeney, R. Kiefer, and L. Morrison. 1989. Analysis of contemporary and historical soil/vegetation/land-use patterns in southwest Wisconsin utilizing GIS and remote-sensing technologies. In *Tech. Papers 1989 ASPRS/ACSM Ann. Convention*, vol. 4, 85–92. Falls Church, Virginia.

Goodchild, M. F. 1986. Spatial autocorrelation. *CATMOG—Concepts and Techniques in Modern Geography*. Norwich: Geo Abstracts.

Hewitt, A. E. 1993. Predictive modeling in soil survey. *Soils and Fertilizers* 56(3):305–14.

Hutchinson, M. F. 1989. A new procedure for gridding elevation and stream line data with automatic removal of spurious pits. *J. Hydrol.* 106:211–32.

Hutchinson, M. F., and P. E. Gessler. 1994. Splines: More than just a smooth interpolator. *Geoderma* 62:45–67.

Kachanowski, R. G. 1988. Processes in soils—from pedon to landscape. In *Scales and Global Change*, edited by T. Rosswall et al., 153–77. New York: John Wiley and Sons.

Malanson, G. P., D. R. Butler, and S. J. Walsh. 1990. Chaos theory in physical geography. *Phys. Geog.* 11:293–304.

Matheron, G. 1971. The theory of regionalized variables and its applications. Les Cahiers du centre de morphologie mathematique de Fontainbleu. Ecole Nationale Superieur des Mines de Paris.

McBratney, A. B., and R. Webster. 1981. The design of optimal sampling schemes for local estimation and mapping of regionalized variables: Program and examples. *Computers & Geosciences* 7:335–65.

McCullagh, P., and J. A. Nelder. 1989. Generalized linear models. In *Monographs on Statistics and Applied Probability*, No. 37. London: Chapman and Hall.

McDonald, R. C., R. F. Isbell, J. G. Speight, J. Walker, and M. S. Hopkins. 1990. *Australian Soil and Land Survey Field Handbook*, 2nd ed. Sydney: Inkata Press.

McKenzie, N. J. 1991. *A Strategy for Coordinating Soil Survey and Land Evaluation in Australia*. CSIRO Division of Soils, Divisional Report 114.

McKenzie, N. J., and M. P. Austin. 1993. A quantitative Australian approach to medium- and small-scale surveys based on soil stratigraphy and environmental correlation. *Geoderma* 57:329–55.

McSweeney, K. M., P. E. Gessler, B. Slater, R. D. Hammer, J. Bell, and G. W. Peterson. 1993. Toward a new framework for modeling the soil-landscape continuum. In *Factors of Soil Formation: A Fiftieth Anniversary Perspective*. SSSA Special Publication 34.

Milne, G. 1935. Some suggested units of classification and mapping particularly for east African soils. *Soil Res.* 4:3.

Moore, I. D., R. B. Grayson, and A. R. Ladson. 1991. Digital terrain modeling: A review of hydrological, geomorphological, and biological applications. *Hydrological Processes* 5:3–30.

Moore, I. D., P. E. Gessler, G. A. Nielsen, and G. A. Petersen. 1993. Soil attribute prediction using terrain analysis. *Soil Sci. Soc. Am. J.* 57:443–52.

Moore, I. D., A. Lewis, and J. C. Gallant. 1993. Terrain attributes: Estimation methods and scale effects. In *Modelling Change in Environmental Systems*, edited by A. J. Jakeman, B. Beck, and M. McAleer. London: Wiley.

Raymond, O. L. 1992. *Geology of the Wagga Wagga Sheet*. Sheet 8327. Australian Geological Survey Organization.

Simonson, R. W. 1959. Outline of a generalized theory of soil genesis. *Soil Sci. Soc. of Am. Proc.* 23:152–56.

Statistical Sciences. 1992. *S-PLUS Programmer's Manual*. Seattle: Statistical Science.

Webster, R. 1977. *Quantitative and Numerical Methods in Soil Classification and Survey*. Oxford: Clarendon Press.

Webster, R., and M. A. Oliver. 1990. *Statistical Methods in Soil and Land Resource Survey*. Oxford: Oxford University Press.

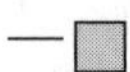

Paul Gessler is a native of Wisconsin and has a B.S. in soil science and a M.S. in environmental monitoring—a remote sensing and GIS graduate program—University of Wisconsin, Madison, Wisconsin. He is currently a research scientist/GIS specialist with the Commonwealth Scientific and Industrial Research Organization (CSIRO) Division of Soils in Canberra, Australia. He is also pursuing a Ph.D. at the Center for Resource and Environmental Studies (CRES) at the Australian National University.

P. E. Gessler and *N. J. McKenzie*
CSIRO Division of Soils
GPO Box 639
Canberra, ACT 2601
Australia
Phone: 61 6 246 5955
Fax: 61 6 246 5965
E-mail: *paulg@cbr.soils.csiro.au*

I. D. Moore (deceased)

P. J. Ryan
CSIRO Division of Forestry
Canberra, Australia

11

Modeling and Visualizing Soil Behavior at Multiple Scales

S. D. DeGloria and R. J. Wagenet

Understanding soil behavior and characterizing soil quality are required to develop effective environmental management strategies, implement sustainable farming systems, and minimize human-induced soil degradation. Our goal is to place in a conceptual framework the mechanisms by which we define soil information requirements for integrated modeling and visualizing of soil behavior. Data requirements are a function of information needs, type of simulation model, scale of application, quality of data, and experiences of soil users. Recommendations to address data requirements include: (1) provide improved field characterization of physical, chemical, and biological properties that support a recognized concept of soil quality in soil resource inventories; (2) improve understanding of soil processes and the influence of environmental factors on soil quality at discrete spatial and temporal scales; (3) develop methods for field verification of simulation model predictions assuming valid application of models and use of inventory data; and (4) characterize and communicate to decision makers the uncertainty of model predictions and the appropriate use of maps constructed to provide a visualization of model predictions.

INTRODUCTION

In agronomic systems soil behavior can be defined as the response of soil material to interacting environmental processes and cultural practices. Soil quality is a measure of the soil's productive and regenerative capacity and is directly influenced by the behavior of soil under environmental stress. Understanding soil behavior and characterizing soil quality at field and landscape scale are required to develop effective environmental management strategies, implement sustainable farming systems, and minimize natural and human-induced soil degradation.

The importance of soil quality in developing sustainable agricultural systems is reviewed by Papendick and Parr (1992). Soil data required to characterize soil quality have physical, chemical, and biological components that are derived from field observations and measurements, simulation models, and knowledge and experiences of soil users, experts, and specialists (Bouma 1993). Compilation, integration, spatial analysis, and visualization of these data are best accomplished using the information technologies embodied in geographic information systems (GIS) and simulation models.

Integration of dynamic simulation models and GIS and technology serves to represent our current state of knowledge regarding soil behavior and soil quality at multiple scales of space, time, and complexity. This integration also provides the framework by which we advance our knowledge and understanding of soil systems under variable environmental management regimes. Our goal is to place in a conceptual framework the mechanisms by which we define soil information requirements for modeling and visualizing soil behavior. Specific objectives are to: (1) review recent advances in the inventory of soil resources and the simulation of soil processes at multiple scales; (2) describe representative examples and experiences in integrating simulation models and environmental databases at landscape scale using GIS technology, and (3) make recommendations to improve the integration of soil information and process models for characterizing soil behavior and quality in support of environmental policy and decision making.

SOIL INFORMATION FRAMEWORK

A conceptual framework for modeling and visualizing soil behavior in support of environmental policy and decision making is comprised of four components: specification, computation, understanding, and policy (Figure 11-1). Emphasis is placed here on soil resource inventories and soil process models, their interrelationships, and landscape-scale applications for meeting the data and information needs of environmental management and planning programs.

Soil Resource Inventory

Soil resource inventories generate relevant environmental data used in most soil process models at field and landscape scale. Three major soil survey databases used in simulation modeling activities are described by Reybold and TeSelle (1989). For most landscape-scale modeling applications, the State Soil Survey Geographic Database (STATSGO) is used (Soil Conservation Service 1991). The soil survey database of choice is a function of information needs. A coherent expression of information needs establishes the degree of complexity, application scale, and boundary conditions of both resource inventories and simulation models.

Major limitations of soil resource inventories and associated databases relate to the nature and quality of numerical data describing soil physical and chemical properties. These data are summarized for typifying pedons, and without quantitative information on spatial distribution of properties within map units. Traditionally, numeric

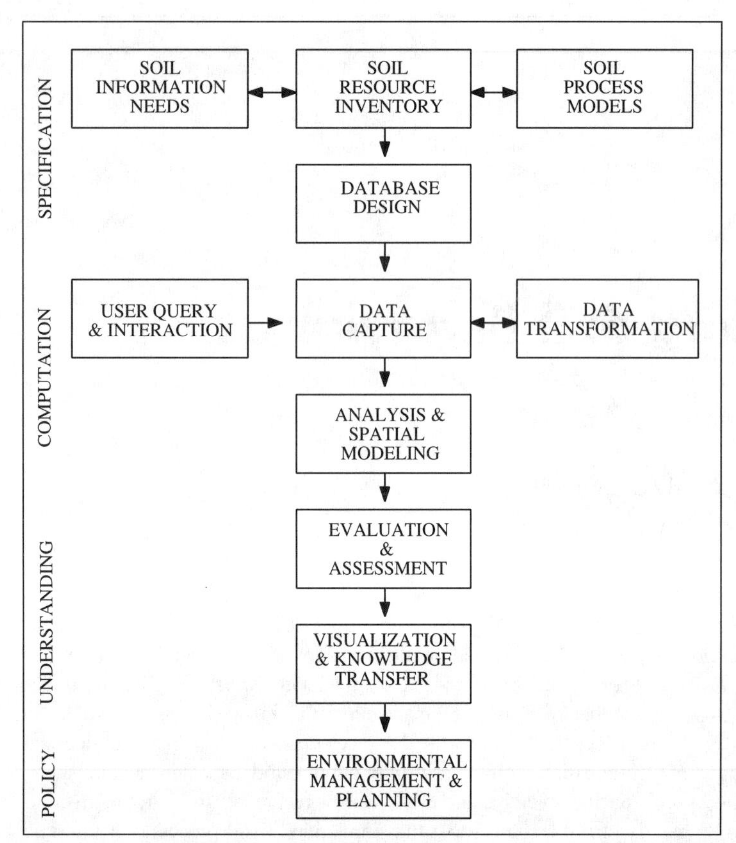

Figure 11-1. Conceptual framework for modeling and visualizing soil behavior in support of environmental policy and decision making.

values of soil properties are extrapolated to similar landscape positions from the location of the typifying pedon sampled for laboratory characterization. In addition, probability distribution functions of these variables are derived from the experiences of others when needed for some forms of simulation modeling. Recent advances in our ability to characterize field soils and land forms through the development and integration of landscape models, sampling designs, characterization of spatial variability, and soil map unit refinements for enhancing soil interpretations are presented by Mausbach and Wilding (1991).

Several investigators are developing methods to improve the spatial characterization of soil properties to overcome these limitations in current soil survey procedures. Such methods include landscape models which incorporate digital elevation models and the *catena* concept to estimate various soil map unit attributes (Bell et al. 1992; Moore et al. 1993), geostatistical models to produce maps of selected soil attributes and associated variance structure from point observations and measurements of soil properties and indicators of soil quality (Bregt et al. 1991, 1992; Smith et al. 1993; Yost et al. 1993), and expert systems to map soil-landscape features using local knowledge and environmental databases (Skidmore et al. 1991). A comprehensive treatment of statistical methods used in soil survey and land evaluation is presented by Webster and Oliver (1990).

Soil Process Models

Simulation models are now used to improve our understanding of soil dynamics that vary with respect to intensity of computation, complexity of design, and prediction and scale of application. Generally, soil process models are either qualitative, in which understanding of a soil process is characterized in conceptual or descriptive terms, or quantitative, in which the process is characterized using mathematical or statistical algorithms. Definition of soil process models and related data requirements are described by Addiscott (1993), Addiscott and Wagenet (1985), and Wagenet et al. (*Quantitative*, 1993).

Quantitative models are defined as either *deterministic*, in which a unique prediction is based on a unique set of input variables, or *stochastic*, in which a statistical uncertainty is associated with model parameters or predictions. Deterministic simulation models are further divided into two types: *mechanistic* or *functional*. Mechanistic deterministic simulation models incorporate fundamental mechanisms of soil processes to the degree they are understood by the model developer. Functional deterministic simulation models use simplified, empirical representations of fundamental soil processes while striving to maintain a high degree of sensitivity in predicting complex interacting soil processes.

The advantages of functional simulation models include reduced computation time, minimized data requirements for parameterization and verification, and simplified and expanded scales of application. In most cases, mechanistic and functional deterministic models of soil processes require estimation of some model parameters from soil variables tabulated in soil inventory databases. This estimation is achieved using transfer functions, such as empirically derived regression equations or decision trees, as described by Bouma and van Lanen (1987).

INTEGRATION OF PROCESS MODELS AND GIS

Once soil information needs have been defined, the integration of soil process models and environmental databases is achieved by first correlating the nature of information needs, type of model, scale of application, quality of data, and experience of soil users, experts, and specialists. Considerable computer processing and networking is required to compile, sort, and transform inventory data to meet input requirements of the given model (Burrough 1989).

Landscape-scale simulation modeling, which relies on soil survey databases, is exemplified by the recent work of Burke et al. (1991), Hamlett et al. (1992), and Wagenet et al. (*Mapping*, 1993). In these cases, as in others, STATSGO-level soil survey data and related environmental information were used to drive both simulation and descriptive models. A representative example of applying a functional deterministic model to simulate pesticide transport at landscape scale is shown in Figure 11-2. This application effectively expanded the spatial scale of a related modeling application using the mechanistic version of the LEACHM solute transport model with GIS-derived environmental data (Petach et al. 1991). The landscape-scale application presented in Figure 11-2 was successful in developing an effective methodology for modifying a computationally and data-intensive mechanistic simulation model for predicting pesticide transport at field scale (Wagenet and Hutson 1989) to a practical, computationally efficient, field- and landscape-scale functional deterministic simulation model (Hutson 1993; Hutson and Wagenet 1993).

The significant results from this landscape-scale integration of GIS and simulation modeling indicated that (1) modifying a mechanistic solute transport model permitted the valid application of the functional form of the model at variable spatial scales using available soil and environmental data; (2) improved estimation of soil and climate variables such as soil organic carbon and distribution, intensity, and timing of precipitation events is needed to improve model pre-

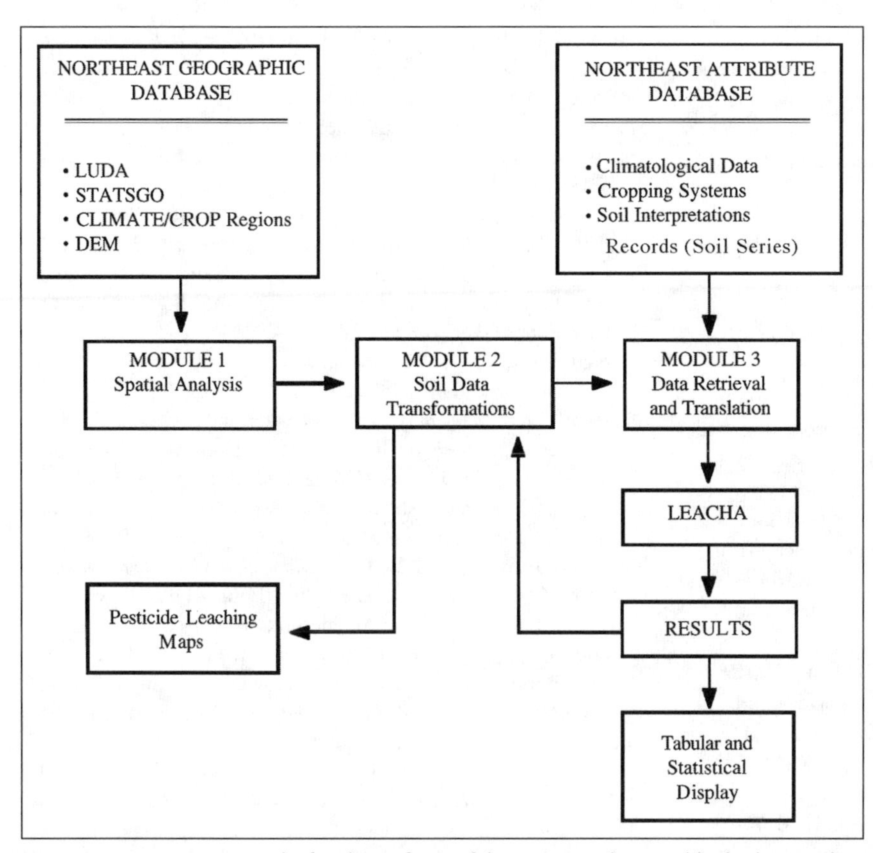

Figure 11-2. Representative example of applying a functional deterministic simulation model at landscape scale.

dictions at multiple scales; and (3) proper and effective uses of model predictions and resulting map products for policy development need to be explored.

The most significant result of these recent landscape-scale simulations of soil processes is the need to characterize and relate to decision makers both the *intrinsic* and *information* uncertainty of our predictions and the proper interpretation and use of maps that provide a visualization of model output. Intrinsic uncertainty results from spatial and temporal variability of soil properties which are difficult to characterize in most modeling applications given the current state of soil resource inventories and databases. Knowledge from laboratory and field-based experiments is used to characterize parameter variability at larger scales. Information uncertainty results from measurement and computation errors, model misspecification, and field sampling. Several investigators have assessed uncertainty in field- and landscape-level simulation modeling activities, some of which incorporate the use of GIS technology (Carsel 1988; Loague et al. 1989, 1990).

FUNCTIONALITY REQUIREMENTS

To improve our understanding and characterization of soil behavior and quality under variable environmental management regimes, advances in GIS technology will need to address: (1) functional computer software interfaces between simulation model libraries and environmental databases, (2) artificial and interactive modification of landscape patterns in concert with simultaneous simulation modeling, and (3) development of interactive linkages with scientific visualization systems.

The functional interfaces should allow either the scientist or decision maker to interactively select the appropriate model and environmental data as a function of application scale, information need, and data quality. The interface needs to minimize uncertainty in the modeling process by accessing soil survey databases that incorporate some measures of spatial and temporal variability of soil physical, chemical, and biological properties.

The ability to modify landscape patterns as a function of land-use policy with interactive simulations of soil behavior would permit the scientist and policy analyst to assess sensitivity of modified landscape patterns on soil processes and environmental quality. Modifications to the landscape would be based on land-use planning scenarios such as the creation of stream buffers or erosion control practices to improve water quality or implementation of sustainable farming systems to enhance soil productivity and quality. If designed as an iterative process, model predictions of soil processes as a function of pattern can be used to "design" landscapes that maximize productive capacity of the land, maintain biological diversity, and minimize environmental degradation.

The realm of decision and policy making requires an effective medium by which we can convey the complexity of soil processes under variable environmental management regimes or landscape patterns. Technology of scientific visualization provides one mechanism to improve our understanding of soil behavior and convey current knowledge to decision makers (DeGloria 1993). The functionality of GIS can be improved by facilitating the interactive, real-time spatial visualization of simulation model predictions. Complex processes such as soil water dynamics, pedogenesis, sediment and nutrient transport, and nutrient cycling can be modeled and visualized at variable spatial and temporal scales with the primary purpose of improving both understanding of soil systems and land-use decision making.

CONCLUSION

Resource inventories should incorporate estimates and associated spatial and temporal variability of physical, chemical, and biological properties of soils that support a recognized concept of soil quality. Development of spatial and temporal indicators of soil quality from this set of soil characteristics should be possible using descriptive or empirical relationships where necessary. Inventories should incorporate the use of summary statistics that minimize the use of soil property ranges and should include measures of central tendency and other statistical measures more appropriate to simulation modeling, such as probability distribution functions derived from field studies.

Simulation model developers and GIS practitioners need enhanced understanding of soil processes and the influence of environmental factors upon soil behavior and quality at multiple scales. The need to vary simulation model type and complexity based on scale of application and nature of environmental data used in model parameterization and output visualization must be recognized.

Methodologies for the field verification of simulation model predictions need to be developed. Field procedures are needed to confirm model predictions within a defined level of uncertainty. The ability to field-verify model predictions is necessary for integration of simulation models and GIS within the context of the environmental decision-making process. Standards for verifying simulated model predictions of soil dynamics and associated map products need to be established with respect to field sampling regimes, measurements, and mechanisms for knowledge transfer to decision makers.

References

Addiscott, T. M. 1993. Simulation modeling and soil behavior. *Geoderma* 60:15–40.

Addiscott, T. M., and R. J. Wagenet. 1985. Concepts of solute leaching in soils: A review of modeling approaches. *J. Soil Science* 36:11–24.

Bell, J. C., R. L. Cunningham, and M. W. Havens. 1992. Calibration and validation of a soil-landscape model for predicting soil drainage class. *Soil Sci. Soc. Am. J.* 56:1,860–66.

Bouma, J. 1993. Soil behavior under field conditions: Differences in perception and their effects on research. *Geoderma* 60:1–14.

Bouma, J., and H. A. J. van Lanen. 1987. Transfer functions and threshold values: From soil characteristics to land qualities. In *Proc. Workshop on Quantified Land Evaluation*, edited by K. J. Beek, et. al, 106–10. The Netherlands: Enschede.

Bregt, A. K., A. B. McBratney, and M. C. S. Wopereis. 1991. Construction of isolinear maps of soil attributes with empirical confidence limits. *Soil Sci. Soc. Am. J.* 55:14–19.

Bregt, A. K., H. J. Gesink, and Alkasuma. 1992. Mapping the conditional probability of soil variables. *Geoderma* 53:15–29.

Burke, I. C., T. G. F. Kittel, W. K. Lauenroth, P. Snook, C. M. Yonker, and W. J. Parton. 1991. Regional analysis of the central Great Plains. *Bioscience* 41:685–92.

Burrough, P. A. 1989. Matching spatial databases and quantitative models in land resource assessment. *Soil Use and Management* 5:3–8.

Carsel, R. F., R. S. Parrish, R. L. Jones, J. L. Hansen, and R. L. Lamb. 1988. Characterizing the uncertainty of pesticide leaching in agricultural soils. *J. Contaminant Hydrology* 2:111–24.

DeGloria, S. D. 1993. Visualizing soil behavior. *Geoderma* 60:41–55.

Hamlett, J. M., D. A. Miller, R. L. Day, G. W. Peterson, G. M. Baumer, and J. Russo. 1992. Statewide GIS-based ranking of watersheds for agricultural pollution prevention. *J. Soil & Water Conservation* 47:399–404.

Hutson, J. L. 1993. Applying one-dimensional deterministic chemical fate models on a regional scale. *Geoderma* 60:201–12.

Hutson, J. L., and R. J. Wagenet. 1993. A pragmatic field-scale approach for modeling pesticides. *J. Environ. Qual.* 22:494–99.

Loague, K. M., R. E. Green, T. W. Giambelluca, T. C. Liang, and R. S. Yost. 1990. Impact of uncertainty in soil, climatic, and chemical information in a pesticide leaching assessment. *J. Contaminant Hydrology* 5:171–94.

Loague, K. M., R. S. Yost, R. E. Green, and T. C. Liang. 1989. Uncertainty in pesticide leaching assessment in Hawaii. *J. Contaminant Hydrology* 4:139–61.

Mausbach, M. J., and L. P. Wilding, eds. 1991. *Spatial Variabilities of Soils and Landforms.* Special publication no. 28, Soil Science Society of America, Madison, Wisconsin.

Moore, I. D., P. E. Gessler, G. A. Nielsen, and G. A. Peterson. 1993. Soil attribute prediction using terrain analysis. *Soil Sci. Soc. Am. J.* 57:443–52.

Papendick, R. I., and J. F. Parr, eds. 1992. Soil quality—The key to a sustainable agriculture. *Am. J. Alternative Agriculture* 7:2–3.

Petach, M. C., R. J. Wagenet, and S. D. DeGloria. 1991. Regional water flow and pesticide leaching using simulations with spatially distributed data. *Geoderma* 48:245–69.

Reybold, W. U., and G. W. TeSelle. 1989. Soil geographic databases. *J. Soil and Water Cons.* 44:28–29.

Skidmore, A. K., P. J. Ryan, W. Dawes, D. Short, and E. O'Loughlin. 1991. Use of an expert system to map forest soils from a geographical information system. *Int. J. Geographical Information Systems* 5:431–45.

Smith, J. L., J. J. Halvorson, and R. I. Papendick. 1993. Using multiple-variable kriging for evaluating soil quality. *Soil Sci. Soc. Am. J.* 57:743–49.

Soil Conservation Service. 1991. *State Soil Geographic Database (STATSGO) Data Users Guide.* Misc. Publ. #1492. Washington, D.C.: U.S. Department of Agriculture, Soil Conservation Service.

Wagenet, R. J., J. Bouma, and J. L. Hutson. 1993. Modeling water and chemical fluxes as driving forces of pedogenesis. In *Quantitative Modeling of Soil Forming Processes Symp.* Am. Soc. Agronomy, Madison, Wisconsin.

Wagenet, R. J., R. B. Bryant, S. D. DeGloria, and R. G. Perritt. 1993. *Mapping Groundwater Contamination Potential Using Integrated Simulation Modeling and GIS.* Final Rept. Coop. State Res. Ser. Washington, D.C.: U.S. Department of Agriculture.

Wagenet, R. J., and J. L. Hutson. 1989. *LEACHM: Leaching Estimation and Chemistry Model: A Process-Based Model of Water and Solute Movement, Transformations, Plant Uptake, and Chemical Reactions in the Unsaturated Zone.* Continuum #2. Water Resources Institute, Cornell University, Ithaca, New York.

Webster, R., and M. A. Oliver. 1990. *Statistical Methods in Soil and Land Resource Survey.* Oxford: Oxford University Press.

Yost, R., K. Loague, and R. Green. 1993. Reducing variance in soil organic carbon estimates: Soil classification and geostatistical approaches. *Geoderma* 57:247–62.

S. D. DeGloria's research interests focus on the development and application of advanced resource inventory methods that incorporate remote sensing and geographic information technologies. Research activity is directed at quantifying and visualizing the type, spatial distribution, and quality of environmental resources. Current activity is focused on regional-scale characterization and mapping of biodiversity using multispectral and multitemporal imagery, soil geographic and climatic databases, and digital terrain models.

Department of Soil, Crop, and Atmospheric Sciences
Emerson Hall
Cornell University
Ithaca, New York 14853

12

Data Structures for Representation of Soil Stratigraphy

Stephen J. Ventura, Barbara J. Irvin,
Brian K. Slater, and Kevin McSweeney

Traditional soil surveys do not provide information about the three-dimensional structure, characteristics, and variability of soils. This is primarily a deficiency in the data structure—two-dimensional polygons—used to record and convey soils information. Hybrid geographic information system (GIS) data structures are proposed to represent the shape and properties of layers of soils and near-surface geology. Two interconvertible data structures are used, the need for which arises from differing criteria about the acquisition, analysis, modeling, and display of soils data. The first data structure is based on horizons or layers of soil. It uses a vector representation of stratigraphic boundaries—triangulated irregular networks (TINs)—which are derived from digital elevation models (DEMs) of surface configuration and from boundary position determined from subsurface sampling or interpolation between sample points. The second data structure is used to represent soil volumes in the form of volume cells referenced to boundaries—"Z-relative voxels"—which are suitable both for storing the results of spatial interpolation of soil properties and for providing data for finite element-based models. These data structures provide a basis for storing all the original observations without loss of data. Sampling and aggregation algorithms can be used to convert between these data structures.

DESCRIBING SOIL VOLUMES

Soil scientists have long recognized the limitations of traditional soil surveys for representing complex three-dimensional properties of soils (McSweeney et al. 1994; Slater et al. 1994). Soil mapping units and associated keys only partially capture horizontal variability of soils and do very little to convey variability of soil strata and the variance structure of different attributes. This is primarily a deficiency in the data structure—two-dimensional polygons—used to record and convey soils information.

Traditional representations of soils are inadequate for many applications. For landscape-scale analyses, such as soil-forming processes or groundwater movement, data in the third and sometimes fourth dimensions (time) are needed. These data include information about the three-dimensional spatial structure and properties of soil horizons, and about the location and characteristics of boundaries between horizons. Other users of soil surveys may want access to original measurements for analyses or interpretation not anticipated by the initial mappers. Original data, such as the location of and observations along transects, are rarely available to subsequent users of soil surveys. Finally, educators need ways to depict soils that convey both the themes of soil properties and processes and their variability. A typical soil survey document conveys only a limited piece of soil scientists' much broader knowledge about soils and landscapes. Soil scientists need understandable, affordable tools for analyzing and portraying the three-dimensional structure, characteristics, and variability of soils.

Mausbach and Wilding (1991) presented several state-of-the-practice articles reviewing methods for measuring and conveying information about soils. Many of these articles assumed that two-dimensional mapping units would continue to be the basic data structure. Authors proposed various techniques for enhancing the information content of polygons, such as more explicit information about polygon purity (Brubaker and Hallmark 1991; Nordt et al. 1991), transect location and analyses (Upchurch and Edmonds 1991), more precise definition of mapping units (Burrough 1991), and correlation with landforms and landscape positions (Hall and Olson 1991). Burrough, Upchurch, and Edmonds suggested geostatistical techniques such as kriging to better characterize soil properties, though the analyses were done in two dimensions. In their review of GIS techniques, Hammer et al. (1991) described some of the GIS tools and corresponding data structures that may be applicable to characterization of soils. Again, these applications were focused on soil surface shape or map unit composition, not on the three-dimensional structure of soils.

Scientists working to understand soil landscape patterns and processes have developed both qualitative and quantitative terms for the description of land surface configuration. Quantitative geomorphometric techniques have been facilitated by the development of DEMs, regularly spaced matrices of spot elevations, and corresponding GIS tools for determining surface characteristics such as slope, aspect, curvature, landscape position, and so forth (Zevenbergen and Thorne 1987; Dikau 1989; Skidmore 1990). However, these techniques are used to describe surfaces, not volumes—"2½-D" in the jargon of GIS experts. The relations between landforms, landscape position, and soil characteristics such as horizonation are broadly understood, but little has been done to quantitatively describe these relations.

Three-Dimensional Representations

Three-dimensional data structures have been described in the GIS literature. Raper (1989) presented a number of articles that described various approaches to representation of surfaces and volumes. In general terms, surfaces are represented as DEMs or triangulated irregular networks (TINs) and visualized by fishnet, wirenet, shaded relief, and terrain drape depictions. Tools for management and display of surface data are available in commercial GIS software packages.

The representation of volumes has taken two courses, which are essentially the extensions of raster and vector planar data structures. The volume cell, or "voxel," is the three-dimensional equivalent of a raster cell or pixel. The voxel comes from a regular tessellation of space wherein matrix position within a data file corresponds to a location in three-dimensional space as defined by orthogonal (*X*, *Y*, *Z*) axes. The value in the file represents a single characteristic of that position in space, for example, a soil texture class or map unit identifier.

A voxel approach has advantages and limitations similar to those of raster data. Because location is implicit in file position, many types of spatial analyses are easy to program, including providing data for finite element or finite difference models. Raster and voxel structures tend to be difficult to link with database management systems, do not explicitly encode topological relations, and have large data volumes when fully expanded. Though schemes such as run-length or octree encoding (Mark and Cebrian 1986; Samet 1990) can be used to improve data storage and access, this approach is known to be inefficient, particularly for multiple attributes of single locations and for large areas of relatively homogeneous characteristics. If attributes are not completely correlated, a separate file must be maintained for each attribute or else data are lost in aggregation or classification. Fixed cell size means that the resolution of data must be defined before data are incorporated, potentially biasing some types of analyses and affecting the visual quality of relatively large-scale displays.

Vector-based approaches to volume representation correspond either to the graphic vector data structure of CAD (computer-aided drafting) systems or the topological vectors of GIS. In both CAD and GIS, spatial features are represented as points, lines, or areas. In GIS, these geographic primitives are also described in terms of their topology, the spatial relations of adjacency and connectivity. Vector data are intuitively understandable to users of hard-copy maps. Precision of representation is limited only by data measurement and automation techniques. However, vector-based approaches are algorithmically complex and are not immediately suitable to many of the statistical and landscape process tools currently in use in pedology.

The three-dimensional extension of CAD data structures is generally referred to as solids modeling. Several approaches can be used to describe the configuration of a three-dimensional surface, including constructive solid geometry, boundary representation, and localized mathematical shape descriptors (Bak and Mill 1989; Jones and Wright 1991). These data structures are used in a variety of engineering applications, primarily for design and display. Like CAD, they represent the shape of an object, not its characteristics, so the analysis of processes and spatial relations is quite limited.

Hazelton et al. (1990) and Pigot (1991) described the extension of topological vectors. Nodes, edges, and facets (equivalent to points, lines, and bounded areas) and their explicit spatial relations can be used to describe the shape and characteristics of complex volumes. Though theoretically simple and complete, such data structures and their analysis are algorithmically very complex. For subsurface applications, it is difficult to sample with sufficient density to adequately delineate volumes with the potential (and implied) precision this approach provides. For these reasons, there is no commercially available software supporting this approach.

Geologists exploring mineral and petroleum resources have used both solids modeling and voxel approaches to describe phenomena—geological strata—with similarities to soil scientists' realm of interest. Both are interested in information about the attributes of volumes, not just shape and location. To characterize large areas that cannot be directly observed, both must use sampling and interpolation techniques such as kriging, which evolved out of the geological exploration industry. Some very sophisticated geologic software has become available commercially, using several of the three-dimensional data structures described above for analysis and visualization of subsurface data (Fried and Leonard 1990).

Differences also exist between the geological and soil science applications that have implications on the selection of appropriate data structures. In general, it is possible to make a major simplifying assumption about soils—recognizable units generally lie one on top of another in simple strata. Soil scientists can derive much more information from the configuration of the earth's surface than geologists. Typically, a substantial degree of correlation exists between soil properties and slope, aspect, curvature, and landscape position. These correlations are directly and continuously observable. Soil scientists typically deal with more attributes than do geologists; the attributes often are not highly covariant, boundaries between attribute classes may be indistinct, and complex mixtures occur. This increases the importance of facile database management system links and continuous or fuzzy classification systems. In practical terms, mineral or petroleum geologists deal with small areas of high monetary value while soil scientists work with large areas of relatively small value. This results in intensive versus extensive data requirements and is perhaps an explanation of why (quite expensive) hardware and software have been developed to support the needs of geologists.

Requirements for Describing Soil Volumes

To discard traditional assumptions of "mapping units" and horizons, data structures that support representation of soils as continuously and differently variable on multiple properties in all spatial dimensions must be developed. (A data structure should be related intuitively to the way soils are organized to be understandable.) If the assumption is made that soils are simple layers, then the data structure must accommodate information about the position and shape of layer boundaries as well as the content of each stratum. Data structures to represent continuously variable soils data should support a variety of queries, modeling techniques, and visualization methods. It should be possible to portray three-dimensional soil boundaries and characteristics on two-dimensional media such as monitors and paper plots.

Since we cannot exhume an entire volume of soil to study it, we will never have complete knowledge across extensive areas; however, a data structure should not preclude that possibility. It should maintain precise information where characteristics are directly measured, should support the interpolation of properties between sample sites, and should provide information about the reliability or uncertainty of estimates between sample points. The data structure must also accommodate a variety of data sources and data-generating methods as well as incorporate new data. Some types of spatial analysis such as kriging can provide information about the reliability of estimates by location. This information can be used to densify sampling for better characterization of properties or variability. Consequently, the data structure must accommodate additional measurements—extensible in terms of spatial resolution or attribute values.

Field survey of soils is an expensive and time-consuming process, and we cannot always anticipate all the subsequent analyses of collected data. By using techniques that minimize the aggregation or generalization of original measurements, we maintain greater flexibility in subsequent analyses. For example, if data were stored as voxels, subsequent users would be constrained to use the selected cell size or subject the data to error-prone resampling techniques. Similarly, a data structure may impose limitations on the types of analysis that can be performed. For example, it is difficult to use a vector isoline (contour) approach for slope or aspect calculations. Therefore, data structures that are readily converted to other forms with minimal information loss are desirable—converted to whatever format is most appropriate for a given type of analysis or rendition (Scarlatos and Pavlidis 1993).

Several new technologies have emerged that have or will have a substantial impact on the conduct of soil surveys and the resulting information products. These include positioning technologies such as global positioning systems (GPS), "soft-copy" photogrammetric techniques producing digital orthophotography and high-resolution DEMs, and geostatistical techniques for interpolation of sparse data. GIS will become the workhorse for integration and analysis of multiple sources of spatially referenced data, as well as for the storage and portrayal of these data. For any of these technologies to be incorporated in the tool bags of pedologists involved in day-to-day production of soil surveys, they will have to be accessible and their cost will have to be commensurate with the value of the survey. Software with three-dimensional capabilities is available for specific applications such as engineering design or geologic exploration. However, these packages are too narrow in their application, too expensive for soil survey, or both. For these reasons, data structures should be built from commercial GIS software. Though none specifically support three-dimensional data, several have the flexibility to extend their inherent structures through additional programming and application development.

Criteria for Data Structures

Our immediate requirements for a three-dimensional data structure have arisen out of a project to interpret Earth-surface processes in a portion of the Driftless area of southwestern Wisconsin. Though the landscapes of this area have generally developed over millions of years (as differentiated from much younger soil landscapes present in areas of Wisconsin covered by Pleistocene glaciers), it is an area that has experienced considerable landscape instability during the Quaternary. A variety of processes have shaped the soils of this area, including mass wasting, *in situ* weathering, loess deposition, and anthropogenically accelerated erosion. The resulting soil patterns generally correspond to landforms but have considerable variability within and between topographically defined landscape units. We intend to identify and define soil stratigraphy by dense, georeferenced sampling and detailed soil profile characterization at nodes of a 50-m grid in conjunction with a DEM (10-m horizontal and 0.3-m vertical resolution) and geomorphometric, geostatistical, and fuzzy classification techniques (Slater et al. 1994).

We have specified several criteria for appropriate data structures to represent soil stratigraphy, based on requirements from our study site and projected improvements in soil surveying in general. The spatial data structure should:

- describe the properties and variability of soils in three dimensions;
- be readily understandable and easy to manage;
- provide a basis for developing graphic representations;
- be easily converted to other structures or formats as appropriate for particular types of analysis;
- be readily linked to a database management system for management of multiple attributes of geographic objects or locations;
- incorporate data from a variety of sources, and incorporate new or more spatially dense data without restructuring the database or losing information; and
- be built from existing, readily available, geoprocessing tools.

DATA STRUCTURES FOR SOIL/LANDSCAPE ANALYSIS

Given the above requirements, no single data structure is suitable. We propose a system in which data are kept in the form most appropriate for their generation or use, with methods for interconversion as needed.

A TIN Structure for Representation of Soil Strata Boundaries

We will use multiple "stacked" TINs to represent boundaries between layers of soils. Nodes of the TINs will be used to maintain sample point attributes. Additional nodes will be added as needed to accommodate new data from sampling, modeling, or interpolation. TIN facets will be used to approximate the shape of the boundary and maintain attributes of the boundary and volumes above and/or below the boundary.

An initial soil surface TIN will be derived from a DEM, based on elevations of sample points and significant terrain features such as points along ridges, drainages, and slope breaks. The relation between the TIN and the DEM from which it is derived are shown for two areas between sample points (Figure 12-1a). For one area, the difference between the TIN surface generated only from elevations at sample points and other cells of the DEM exceed a tolerance, and so another node is added to the TIN to achieve better conformance between the surfaces (Figure 12-1b). The TIN can be densified until residual differences between surfaces are less than the vertical accuracy of the DEM.

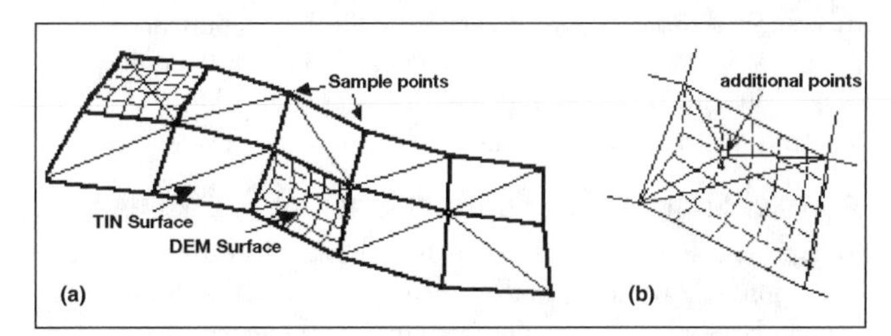

Figure 12-1. (a) Relationship between TIN and DEM surfaces; (b) Densification of TIN.

The three-dimensional location of subsequent TIN layers will be developed from depth-of-change observations at sample points (Figure 12-2). The TIN-defined boundaries do not necessarily correspond to the limits of horizons as defined by pedologists, though they could be the same. Instead, they merely represent the depth at which conceptually significant differences in soil characteristics are

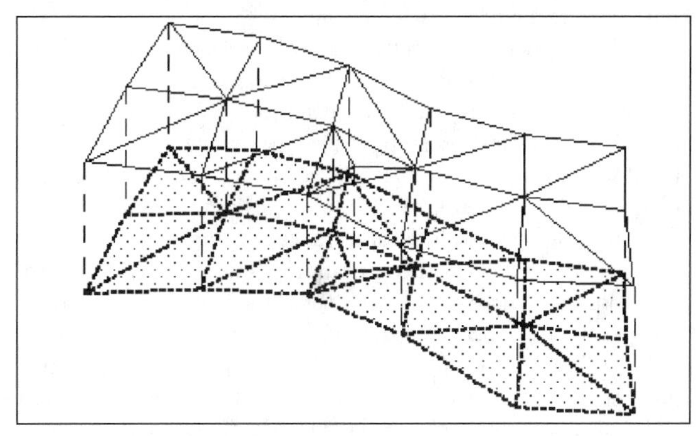

Figure 12-2. Multiple TIN surfaces.

observed at sample points, whether or not this was considered a horizon break. These boundaries may be based on categories of a single attribute (e.g., percent clay) or on classification of multiple attributes by interpretation, statistical clustering, fuzzy classification, or other techniques. If depth-of-change data at TIN nodes are generated by sampling results of a technique that can generate a continuous surface such as kriging, these TINs can also be densified to achieve better conformance between the model surface and the TIN surface, with tolerances again based on the reliability of the model surface. The new nodes could carry the attributes of the model-generated data, eliminating the need for large data files of model results.

As explicit geographic objects (polygons), it will be possible to maintain multiple attributes of TIN facets. If the layers correspond to pedological horizons, attributes could include descriptions of horizon characteristics such as shape (e.g., straight, wavy) and thickness of transition zone. The facets can also maintain data about volumes above and/or below their position. These attributes could simply be data derived from nodes. For categorical data, the attributes would be the class at all three nodes. This would allow queries to show homogenous areas, areas of transition, and areas with particular combinations of attributes. For continuous data, techniques will be needed to spatially interpolate attribute values across the facet as needed for voxelation or modeling.

Currently available GIS software can be used to depict multiple TIN surfaces. Packages provide support for various depictions, including single and double lines (wirenets and fishnets) showing the surface configuration from an oblique perspective, and terrain drapes and shaded relief depictions showing one or more attributes conforming to the surface. Though none specifically support multiple horizontally registered surfaces "stacked" in the same graphic, this can be achieved with cartographic editing tools provided by the packages. Slicing vertically through multiple TINs will require additional software programming for data sampling, using techniques similar to those described in the next section.

A Voxel Structure for Representation of Soil Volumes

To represent soil volumes, we will use volume cells with the *Z*-value (elevation) relative to a surface—the ground surface or a stratigraphic boundary—rather than referenced to an absolute datum as is typical of most cell approaches. Cell data will be maintained as a series of two-dimensional raster arrays, each series representing a depth increment below the surface. To determine absolute position of a cell or to determine its horizontal neighbors, it will be necessary to know its depth increment and to reference an array of surface elevations (with the same horizontal resolution and axes). For our study, we will use voxels that are horizontally 10 × 10 m and 0.05-m thick. For ease of computation in adjacency determinations, voxel elevations can be constrained to the nearest 0.05-m multiple.

We have chosen to use a surface-relative voxel for two reasons. First, it allows us to use existing raster GIS software for most of the data manipulation and display. Most importantly, it readily supports the spatial analyses of interest—splining and kriging. Ideally, we would like to model soil variability in all three dimensions simultaneously, possibly using three-dimensional kriging. However, existing tools do not adequately handle data with variance structures very different in one dimension versus the other two, as is the case with soil strata. Models of semivariograms cannot account for major discontinuities in horizons, for example, when distinct pedogenic processes are involved.

We will separately interpolate attributes in vertical and horizontal dimensions to populate voxel data files. For example, we have used an "equal area" splining technique (Ponce-Hernandez et al. 1986) to generate data down sample points (Figure 12-3). Attributes are measured in the field and samples are taken for laboratory analysis at selected depths along a sample core, representing what appear to be horizons or at least distinct units of soils. A spline fit to these depth units creates a continuous description of the attribute(s). If warranted, connectivity constraints can be relaxed to account for discontinuities in attributes. The spline function can be sampled at whatever depth increments are needed to populate the voxels at that sample point.

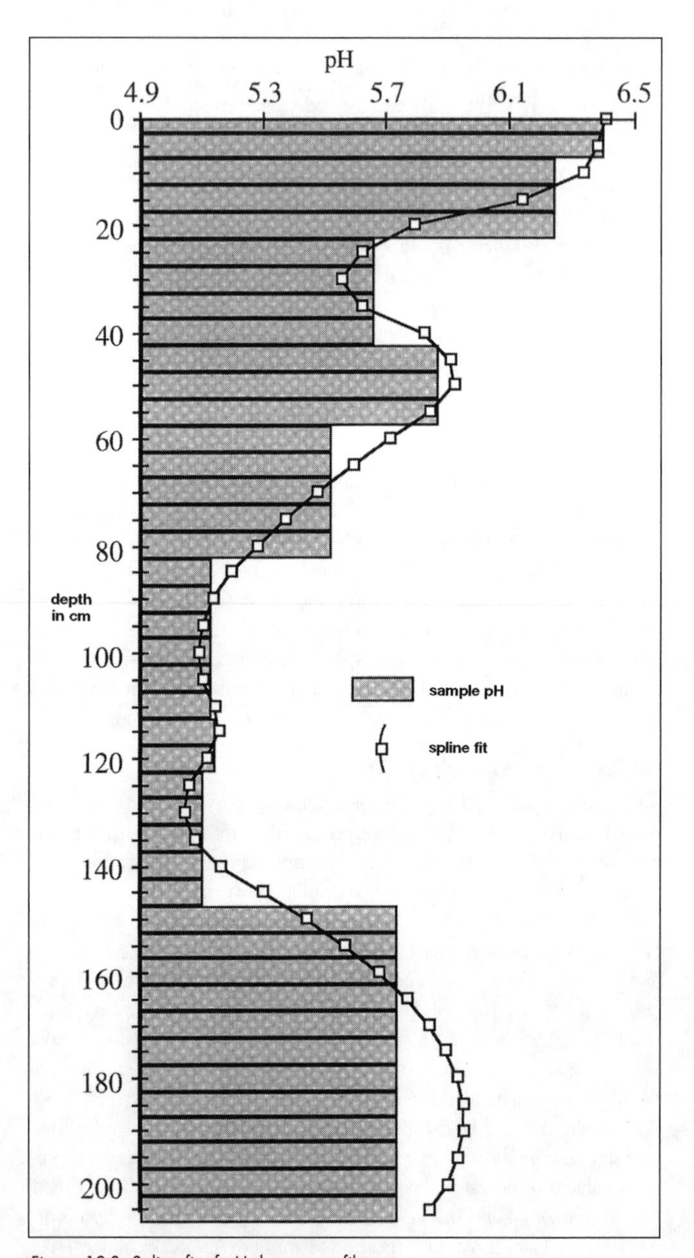

Figure 12-3. Spline fit of pH down a profile.

Two-dimensional kriging or cokriging (with DEM derivatives such as slope, curvature, landscape position, etc.) can be used to interpolate horizontally between sample points for any given depth. Though these points do not have the same absolute elevation, they are treated as a simple relative plane, on the assumption that pedogenic processes will be similar at a particular depth. If subsurface

TINs have been created based on soil stratigraphy, the voxels can also be generated relative to these surfaces, creating a greater likelihood that nearby pixels will have undergone similar pedogenesis. In this case, they are no longer of equal depth from the surface, which will account for variation in depth and thickness of horizons.

When data interpolated horizontally and vertically are combined, a complete voxel data file can be created. The file actually consists of multiple two-dimensional raster arrays. If these data are shown in plain view, the horizontal distribution of attributes at any given surface-relative depth can be shown. A profile slice can be depicted by positioning a line of surface cells based on their absolute elevation and filling in subsequent cells below by sampling raster layers. Figure 12-4 is an oblique depiction of such profiles down and across a slope. It was generated with a graphic arts program; commercial GIS do not have cartographic tools for such a rendering. Planar profiles in two dimensions (X, Z or Y, Z) can be generated with existing GIS tools.

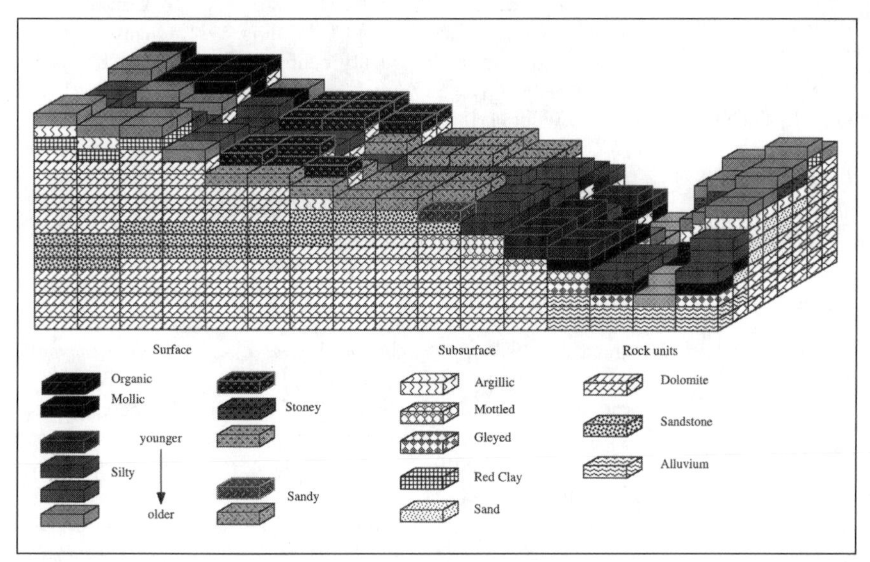

Figure 12-4. Voxel representation of a soil landscape (Slater et al. 1994).

Converting between TINs and Voxels

In addition to traditional uses of soil survey data, we anticipate that information about the structure and variability of soils in all three spatial dimensions will provide support for new kinds of investigations and modeling. In the near term, we will investigate the relationships between soil formation and landscape processes. Because these efforts will require data in various forms, spatial extents, and resolutions, interconversion between the data structures will be necessary.

The conversion from a TIN-based representation of soils (wherein characteristics of volumes are represented as "above" or "below" attributes of TIN facets) to a surface-relative voxel representation will be necessary for finite element-based modeling and some kinds of visualization. The conversion process is akin to rasterization of two-dimensional vector data, with an additional step of determining whether the sampling point has crossed another surface. The origin, X and Y axes, and resolution for a regular grid are defined. The first raster "slice" is then simply a point-in-polygon attribute determination for all the defined grid points within the TIN-vector space. The process is repeated after a depth increment is subtracted from each point's elevation on the first surface, with an additional determination to make sure it is still above the second surface (Figure 12-5). In other words, for every grid point, it is necessary to determine the absolute elevation of each X, Y position on both the surface of interest and the next surface below by linear interpolation across the TIN facet, and then determine if the depth increment has taken the sample point below the next surface. If all voxels' positions are relative to the (initial or ground) surface, then cells are attributed by whatever layer they are immediately below. If multiple surfaces are used to generate voxels, then a voxel falling below a second surface would be attributed as "below next layer" in the array relative to the first surface; attributes of the volume it entered would be accounted for in the voxelation of the next stratum (relative to the second surface).

The polygonization of cell data could take several routes, depending on required products. Within a raster layer, standard vectorization methods could be used to generate (planar) polygons of equal or similar cells; these could be displayed in 2 or in 2½ dimensions (draped across the surface to which the voxels are related). By scanning down raster columns and determining the absolute depth at which a transition occurs (e.g., change in soil texture), it will be possible to generate subsurface stratigraphic boundaries. The result of determining the depth of transition for all voxel columns would be a subsurface DEM, which could be sampled and/or filtered, and then converted to a subsurface TIN for display and analysis. This would be particularly appropriate when boundaries are not field-determined or vary depending on the attribute(s) chosen to define strata.

CONCLUSION

This discussion is based on what we believe is needed to support our study of soil-landscape processes in a small intensively sampled study area in southwest Wisconsin. The operational feasibility of the concepts will be tested in this arena before their use in soil survey and landscape analyses in general are assessed. As such, we run the risk of developing methods that are only applicable to a narrow domain. However, extrapolation of concepts from 2 and 2½-dimensional GIS approaches suggests at least some gen-

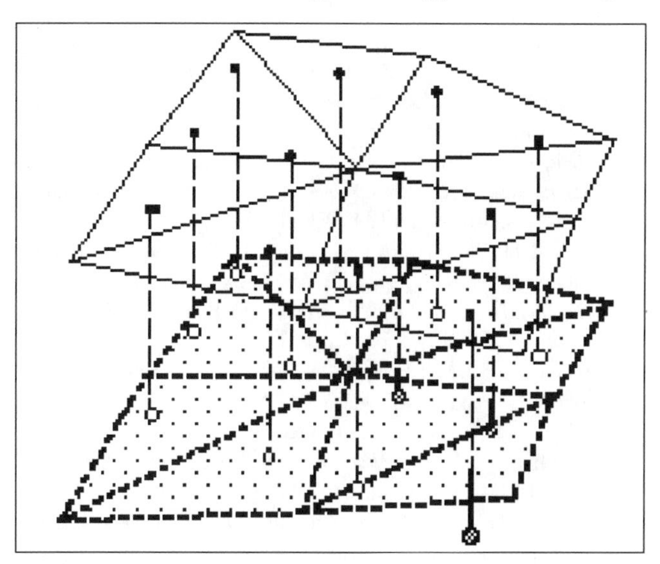

Figure 12-5. TIN-to-voxel conversion. Points that fall below the second surface will be assigned second horizon attribute.

eral areas of optimism and concern. The use of TINs for stratigraphic boundaries and surface relative voxels should meet most of our previously specified criteria for a three-dimensional data structure because:

1. Graphic renditions of both TINs and voxels can be readily generated, the basic concepts should be understandable. The concepts of soil strata and continuous variability of multiple attributes are increasingly recognized, and these data structures provide a direct translation of the concepts. Whether the data are easy to manage depends on the cleverness of our programming.
2. TINs can be densified or generalized as needed and sampled at whatever resolution is appropriate when generating cells.
3. The TIN structure (as vector polygons) provides for a linkage of geographic objects with database management systems for management of multiple attributes and loading of attributes into voxel cells.
4. The voxel structure provides a very easy mechanism to load finite element models. The only aspect that will require some care is the case when voxels are relative to subsurface TIN layers.
5. We will be using primarily ARC/INFO version 6.1, which has both raster and vector capabilities and a macro language for programming. We should be able to do most of the data management with this software and other existing tools.

If the proposed data structures prove to be viable means in which to store and analyze three-dimensional soil and landscape structure and variability, we believe they should have a substantial impact on the way soil surveys are conducted and the way in which these data are managed and conveyed.

Acknowledgments

This work was supported in part by a grant from the National Science Foundation (SES-0210093).

References

Bak, P. R. G., and A. J. B. Mill. 1989. Three-dimensional representation in a geoscientific resource management system for the minerals industry. In *Three Dimensional Applications in Geographic Information Systems*, edited by J. Raper, 155–82. London: Taylor and Francis.

Brubaker, S. C., and C. T. Hallmark. 1991. A comparison of statistical methods for evaluating map unit composition. In *Spatial Variability of Soils and Landforms*, edited by M. J. Mausbach and L. P. Wilding, 73–88. Madison: SSSA.

Burrough, P. A. 1991. Sampling designs for quantifying map unit composition. In *Spatial Variability of Soils and Landforms*, edited by M. J. Mausbach and L. P. Wilding, 89–125. Madison: SSSA.

Dikau, R. 1989. The application of a digital relief model to landform analysis in geomorphology. In *Three Dimensional Applications in Geographic Information Systems*, edited by J. Raper, 51–77. London: Taylor and Francis.

Fried, C. C., and J. E. Leonard. 1990. Petroleum three-dimensional models come in many flavors. *GEOBYTE* 5(1):27–30.

Hall, G. F., and C. G. Olson. 1991. Predicting variability of soils from landscape models. In *Spatial Variability of Soils and Landforms*, edited by M. J. Mausbach and L. P. Wilding, 9–24. Madison: SSSA.

Hammer, R. D., J. H. Astroth, Jr., and G. S. Henderson. 1991. Geographic information systems for soil survey and land-use planning. In *Spatial Variability of Soils and Landforms*, edited by M. J. Mausbach and L. P. Wilding, 243–70. Madison: SSSA.

Hazelton, N. W. J., F. J. Leahy, and I. P. Williamson. 1990. On the design of temporally referenced three-dimensional GIS: Development of four-dimensional GIS. In *GIS/LIS '90 Proceedings*, 357–72. Anaheim.

Jones, N. L., and S. G. Wright. 1991. Solid modeling for site representation in geotechnical engineering. *Geotechnical Engineering Congress*, 1,021–31. ASCE.

Mark, D. M., and J. A. Cebrian. 1986. Octrees: A useful data structure for the processing of topographic and subsurface data. *ACSM/ASPRS Annual Convention Technical Papers*, 104–13. Washington, D.C.

Mausbach, M. J., and L. P. Wilding, eds. 1991. *Spatial Variability of Soils and Landforms*. SSSA Special Publication 28. Madison: SSSA.

McSweeney, K., P. E. Gessler, B. K. Slater, R. D. Hammer, J. Bell, and G. W. Petersen. 1994. Toward a new framework for modeling the soil-landscape continuum. In *Factors of Soil Formation: A Fiftieth Anniversary Retrospective*, edited by R. Amundson, et al. SSSA Special Publication 34. Madison: ASA, CSSA, and SSSA.

Nordt, L. C., J. S. Jacob, and L. P. Wilding. 1991. Quantifying map unit composition for quality control in soil survey. In *Spatial Variability of Soils and Landforms*, edited by M. J. Mausbach and L. P. Wilding, 183–98. Madison: SSSA.

Pigot, S. 1991. Topological models for three-dimensional spatial information systems. In *Proceedings, AutoCarto 10*, 368–92. Baltimore.

Ponce-Hernandez, R., F. H. C. Marriott, and P. H. T. Beckett. 1986. An improved method for reconstructing a soil profile from analyses of a small number of samples. *Journal of Soil Science* 37:455–67.

Raper, J. F., ed. 1989. *Three Dimensional Applications in Geographic Information Systems*. London: Taylor and Francis.

Samet, H. 1990. *Applications of Spatial Data Structures: Computer Graphics, Image Processing, and GIS*. Reading: Addison-Wesley.

Scarlatos, L. L., and T. Pavlidis. 1993. Techniques for merging raster and vector features with three-dimensional terrain models in real time. ACSM/ASPRS *Annual Convention Technical Papers*, 372–81. Washington, D.C.

Skidmore, A. K. 1990. Terrain position as mapped from a gridded digital elevation model. *International Journal of Geographical Information Systems* 4(1):33–49.

Slater, B. K., K. McSweeney, A. B. McBratney, S. J. Ventura, and B. J. Irvin. 1994. A spatial framework for integrating soil-landscape and pedogenic models. In *Quantitative Modeling of Soil Forming Processes*. Madison: SSSA.

Upchurch, D. R., and W. J. Edmonds. 1991. Statistical procedures for specific objectives. In *Spatial Variability of Soils and Landforms*, edited by M. J. Mausbach and L. P. Wilding, 49–72. Madison: SSSA.

Zevenbergen, L. W., and C. R. Thorne. 1987. Quantitative analysis of land surface topography. *Earth Surface Processes and Landforms* 12:47–56.

Stephen J. Ventura is an assistant professor of Environmental Studies and Soil Science at the University of Wisconsin, Madison, Wisconsin. His research interests include GIS technology transfer and the use of GIS with environmental and resource management models. He has worked on problems such as nonpoint source pollution, groundwater contamination, vegetation mapping, habitat evaluation, and invasion of exotic species.

Institute for Environmental Studies and Department of Soil Science
1525 Observatory Drive
University of Wisconsin-Madison
Madison, Wisconsin 53706-1299
E-mail: *sventura@macc.wisc.ed*

13

The Sequoia 2000 Project

James Frew

The Sequoia 2000 Project is a large-scale collaboration between the Digital Equipment Corporation, the University of California, and several industrial partners and government agencies, for the purpose of developing new computing environments for global change research. The primary focus of the project is to develop solutions for massive data storage, data access, data analysis and visualization, and wide area networking. Some of these solutions are embodied in Bigfoot, a computing environment the project is constructing at the Berkeley campus. Bigfoot comprises 10 terabytes of tertiary storage supporting a variety of access methods, managed by an extended relational database management system supporting Earth science data types and operations. Bigfoot is linked to a complex of private high-speed local and wide area networks, running both standard and experimental protocols. In addition to its daily use by Sequoia 2000 investigators, this distributed computing environment supports research projects in visualization, integrating databases with GCMs and GIS, and full-text retrieval.

WHAT IS THE SEQUOIA 2000 PROJECT?

Earth science—particularly studies of global change and other aspects of the whole Earth system—is facing a data management crisis, precipitated by the drastically increasing data volumes from new sensor systems (Gershon and Dozier 1993). Simultaneously, computer science has been devoting increasing attention to data management issues, both in terms of how to organize massive data stores and how to most effectively deliver massive amounts of data to ever-faster processing units (Katz 1991). Thus, there is a growing convergence between the data management needs of Earth scientists and the research agendas of computer scientists.

The Sequoia 2000 Project exploits this convergence by uniting UC Earth scientists and computer scientists in an effort to build a large-scale data management environment tailored for the demands of interdisciplinary global change research.

WHO IS THE SEQUOIA 2000 PROJECT?

The Sequoia 2000 Project is organized around a partnership between the Digital Equipment Corporation and the University of California. DEC is Sequoia 2000's primary sponsor and considers Sequoia 2000 its "flagship" external research project, the successor to MIT's project Athena. Other industrial sponsors of Sequoia 2000 contribute products or direct financial support and provide invaluable feedback on the viability of the technologies being developed by Sequoia 2000.

Five UC campuses are involved in Sequoia 2000. Most of the computer science and engineering activities are concentrated at UC Berkeley, with significant contributions from UC San Diego and the San Diego Supercomputer Center. Earth scientists participating in Sequoia 2000 are affiliated with the Los Angeles, Santa Barbara, and Davis campuses, and with the Scripps Institution of Oceanography.

Several government agencies sponsor Sequoia 2000. Most of them face data management challenges that could possibly be mitigated by early adoption of some Sequoia 2000 technologies. Others are integrating their own data management technologies into the Sequoia 2000 computing environment.

FIVE COMPUTATIONAL CHALLENGES

The following broad research areas were identified in which Sequoia 2000 computer scientists could make substantial research contributions, and in which Sequoia 2000 Earth scientists had the most urgent needs:

- Storage
- Retrieval
- Analysis
- Visualization
- Remote access

Data Storage

A typical local computing environment for a Sequoia 2000 Earth scientist is a local area network of workstations sharing perhaps up to 20 gigabytes[1] of on-line disk storage. Compare this to the sizes of some of the data sets these scientists would like to process, as shown in Table 13-1.

Only the smallest of these data sets fits entirely on-line in our typical non-Sequoia environment; the remainder live on some off-line media (tapes, CD-ROMs, etc.) Access to these very large data sets requires, at a minimum, human intervention to retrieve and mount a particular volume. If the media are sequential (e.g., tape) but the data are to be accessed randomly, then the data will also have to be copied to disk. For very large off-line data sets, the sheer number of volumes can vastly complicate access to a particular piece, and effectively thwart any attempt to process the entire data set.

Therefore, the first computational challenge to Sequoia 2000 is to deliver a storage system that permits terabytes[2] of data to be stored and retrieved without human intervention or excessive copying.

Data Retrieval

In addition to the problem of physically storing terabytes, earth science data management is hindered by the logical data access models of cur-

Table 13-1.

Gigabytes	Data set
4	Biweekly composite normalized difference vegetation index (NDVI) from NOAA Advanced Very High Resolution Radiometer (AVHRR); one year; continental United States (Eidenshink 1992)
23	UCLA hybrid coupled ocean-atmosphere model (HCM) output; one simulated year; global (Weibel et al. 1993)
813	The complete Coastal Zone Color Scanner (CZCS) data set (Feldman 1989)
3000	AVHRR global area coverage (GAC); level 1B; 1981–present (EOS Pathfinder data set) (Wiscombe 1992)

rent computing environments; i.e., the way in which a particular portion of a data set is specified for retrieval. Current access models tend to be inordinately coupled to the physical structure of the data; we often speak of satellite image "tapes" or calibration "files" even when the data are no longer stored on tapes or in individual files. The only metadata such an access model provides are the names of the files and volumes (usually severely limited in length and drawn from a restricted character set) and whatever structure (usually hierarchical) is imposed by the filesystem (if the data are on-line.) For example, a series of satellite images might be stored in files whose names denote their spectral coverage, and whose directory hierarchy encodes their location; but this would not assist a search for data for a particular date and time.

This problem is compounded when heterogeneous data types are to be retrieved, because their organization into volumes and files may be quite different. For example, meteorological station data are usually in tabular form and, compared to satellite imagery, they are geographically sparse. A single file might contain data corresponding to hundreds of satellite images, thus presenting two disjointed access models to an investigator wishing to retrieve coincident (in space or time) data from both data sets.

The second computational challenge to Sequoia 2000 is thus to manage the terabyte data store, and to maintain associated metadata, in a way that permits the retrieval of arbitrary portions of multiple, heterogeneous data sets according to some common set of attributes, without regard to the underlying storage structure.

Data Analysis

Earth science data analysis in current computing environments typically involves the successive application of several barely compatible tools, each of which uses a different data format, different units, etc. Users must spend a great deal of time both converting data as they flow from one tool to the next and keeping track of which sequence of tools was applied to which data sets. The overhead of this bookkeeping tends to stifle the interactive nature of an analysis. Sequoia 2000's third computational challenge is to provide a data management environment that handles as much as possible of the overhead of a multistep analysis procedure by:

- allowing a multistep analysis to be specified graphically as a flowchart;
- "gluing" the individual tools together with transparent data conversions; and
- maintaining lineage (audit trails) for all modified data sets.

Data Visualization

Visualization tools are moving from a desirable to a critical part of the Earth scientist's analytical tool kit, yet current visualization tools have two distinct shortcomings. First, they are very large, slow, cumbersome software systems. Some of this is due to visualization being an inherently complex activity, but much is due to the visualization packages having to devote a great deal of effort to data management. The complexity of current visualization software makes it difficult to use effectively—many Earth scientists employ specially trained programmers to operate their visualization software.

A second shortcoming of current visualization tools is that they are usually output only—the screen, film recorder, or whatever is viewed as a data sink. This makes these tools suitable for preparing publication graphics, but less useful for the kind of feedback processing necessary to refine, or even direct, an analysis sequence. Imagine that a visualization tool applied to the output of a general circulation model (GCM) shows boiling water surface temperatures over North America—it might be desirable to use the visualization tool to interactively probe the executing GCM in an attempt to discover which parameters were causing the (presumed) error.

Our fourth computational challenge is to build visualization tools that overcome these limitations by:

- using, rather than duplicating, the data management features of the Sequoia 2000 computing environment;
- simplifying their interfaces so nonprogrammers can use them effectively; and
- allowing tight coupling to other Sequoia 2000 analysis tools.

Remote Data Access

Earth scientists routinely collaborate with remote colleagues. Analysis of remote data must become similarly routine. Huge-capacity data storage systems are expensive, physically massive, and mechanically complex, and they require more administrative attention than workstation storage. Very large data sets will always be restricted to a relatively few storage sites, and the majority of scientists must access them remotely. Thus, a high-bandwidth wide area network (WAN) becomes an essential component of a science data management system. The Sequoia 2000 Project is an instance of this problem: most of the data storage is located at the Berkeley campus, while the earth scientists are located at other campuses. The final computational challenge to Sequoia 2000 is to build an intraproject WAN fast and reliable enough to support use of the Berkeley storage systems by scientists anywhere in the project.

THE SEQUOIA 2000 COMPUTING ENVIRONMENT

The Sequoia 2000 Project is currently building a data management environment in response the challenges outlined above. While not complete, the overall architecture of this environment is in place and comprises the following layers:

- Applications
- Network
- Database management system
- Hierarchical storage

These layers will be presented bottom-up, since each layer builds on the one beneath it. The bottom two layers (storage and database) are implemented as an integrated system at Berkeley called "Bigfoot." The bottom three layers (Bigfoot plus the network) constitute the computing infrastructure of the project.

Hierarchical Storage

Current computing environments all have at least a two-level storage hierarchy, with primary memory (RAM) at the top and magnetic disk underneath. To store very large data sets, Sequoia 2000 adds a third layer (i.e., "tertiary storage") to the bottom of this hierarchy, comprising various persistent storage media (tapes, optical disks, etc.) configured as multivolume "jukeboxes" with robotic media manipulation. These robotic jukeboxes are called "robo-line" storage (Katz 1991) because, while all their contents are not simultaneously on-line, any volume may be brought on-line subject only to the latency of the robotics (e.g., the time for a robot arm to fetch and load a tape), as opposed to the much greater and more variable latency of a human exchanging tapes in a tape drive.

As shown in Table 13-2, Bigfoot currently supports about 40 gigabytes of magnetic disk storage and 10 terabytes of tertiary storage on various devices.

Table 13-2.

Device	Capacity (Tbytes)
Hewlett-Packard magneto-optical disk jukebox	0.1
Sony WORM optical disk jukebox	0.36
Exabyte 8-mm tape jukebox	0.58
Metrum VHS tape jukebox	9

The Hewlett-Packard and Metrum jukeboxes are virtual file systems: they are controlled by commercial software packages that make each device appear to Bigfoot as if it were a single enormous disk file system. The Exabyte jukebox is a virtual tape drive: it is controlled by locally developed software that makes it appear to Bigfoot as a single enormous tape from which files may be requested by name. The Sony jukebox is managed directly by the POSTGRES database management system, described below. Providing a variety of access methods for tertiary storage allows us to evaluate which method is most appropriate for which application.

Database Management System

In addition to traditional file-based data access, Sequoia 2000 supports a database management system (DBMS). In this access model, data set owners associate metadata with their data sets, and the DBMS uses these metadata to perform attribute-based retrieval of portions of the data sets. Since the DBMS has access to the data as well as the metadata, it can also perform content-based retrieval. Moreover, a retrieve query to the DBMS may reference multiple data sets. If portions of more than one data set are returned, a new data structure is automatically created to hold them. The benefits of this approach may be summarized as:

1. The DBMS retrieves only the data specied, not merely the files or volume3s that contain the data, and users are freed from having to remember arbitrary file and volume names and how they map into (portions of) data sets.
2. The DBMS subsets the data sets before they are delivered to the user, which may dramatically reduce the user's workload.
3. The DBMS can access multiple data sets as easily as a single one.

Bigfoot runs the POSTGRES extended relational database management system developed at Berkeley. In addition to the usual relational DBMS capabilities, POSTGRES provides some major additional functionality of particular importance to the management of Earth science data:

- *Large objects:* POSTGRES supports binary large objects of unlimited size, and a file systemlike mechanism to access their contents. This mechanism is called "Inversion" because it implements a file system on top of a database, including simulating UNIX-style file names and directories (Olson 1993), the inverse of the usual relationship between a database and a file system.
- *User-specified types:* In addition to the usual scalar types supported by any DBMS (characters, integers, text strings, floating-point numbers, etc.), POSTGRES allows users to define their own types and the functions that operate on them. These types may be implemented as large objects, if necessary. Multidimensional arrays, polygons, and physical quantities such as temperature are some of the types added to POSTGRES for Sequoia 2000.
- *User-specified functions:* Users may define functions that operate on either built-in or user-specified types. These functions may be written in the POSTQUEL query language, in which case they are stored in POSTGRES; or they may be written in C and compiled, in which case they are dynamically loaded when they are referenced.

Collectively, these POSTGRES extensions support a new model of scientific computing in which the DBMS is treated less as a data server and more as a procedure server—instead of asking the DBMS for data, one asks it for results.

Network

All components of the Sequoia 2000 Project are geographically separated to some extent. The Bigfoot system itself spans several buildings on the Berkeley campus, and most of its users are at least 300 miles away. This has driven the project to organize itself around a hierarchy of dedicated local and wide area networks. The Sequoia 2000 WAN is currently a set of dedicated DS3 (45 megabits/sec) connections linking the participating campuses. Each campus gateways the WAN onto a local FDDI (100 megabits/sec) fiber optic ring, to which the scientists' workstations are connected. Project-developed driver and routing software plays a major role in the operation of these networks.

All Sequoia 2000's components are networked: the DBMS, the file systems and tertiary storage devices, and the display systems. Standard protocols like X and NFS are used where applicable. The Sequoia 2000 networking strategy is to make all services available to all users. The network should not be the transfer bottleneck; remote users should receive the same level of service as local users.

In addition to providing a guaranteed level of end-to-end bandwidth for daily use, the Sequoia 2000 network is being used as a testbed for a new suite of protocols called RTIP (Ferrari 1990), developed at UC Berkeley. The RTIP protocols provide guaranteed delivery, whereby a client program may reserve a fixed portion of network bandwidth. This is important for continuous media applications (sound, video), where the delay between successive transfers is critical to the application's performance. Sequoia 2000 researchers are using RTIP, together with prototype video compression hardware, to develop a desktop teleconferencing system that can be installed in any Sequoia 2000 investigator workstation.

APPLICATIONS

This section briefly describes a few of the many applications that have been or are being built by Sequoia 2000 scientists, using the data management infrastructure outlined above.

Tioga

Tioga is an attempt to build an integrated programming and visualization system within and on top of POSTGRES (Stonebraker et al. 1992). Tioga has three major components:

- *Recipes:* Tioga allows users to construct programs or recipes from POSTGRES functions and to store those recipes in POSTGRES.
- *Graphical programming and query language:* Tioga includes a graphical programming environment analogous to AVS, in which functions are depicted as boxes and data flows by directed lines. Programs constructed graphically are stored as

recipes. A skeletal program, with some or all boxes replaced by regular expressions, may be used to query POSTGRES and retrieve any recipes whose structure matches skeleton.

- *Smart renderer:* Tioga will send renderable forms of recipe outputs over a network connection to a "smart" renderer being constructed at the San Diego Supercomputer Center (Kochevar 1993). In addition to the usual mechanisms for visualizing geometry and images, the smart renderer will incorporate a knowledge base allowing it to suggest appropriate representations for specific types of objects (e.g., use of a thermometer as an icon for temperature).

The Big Lift

The Big Lift[3] is a project to connect the UCLA GCM directly to POSTGRES; i.e., the output from the GCM will be deposited directly into the DBMS without being saved in intermediate files. This involves modifying both the GCM, replacing its output routines, and POSTGRES, to allow it to accept data at the rate at which the GCM generates it.

There are two expected benefits from The Big Lift: (1) the ability to browse the intermediate output of the GCM while it is running in order to detect errors and adjust parameters or restart the model (made possible by the ability of POSTGRES to subset the voluminous output of the GCM); and (2) the ability to visually control the GCM's execution. This is long term and depends on Tioga; however, if successful, it will radically change the way GCMs are used.

Lassen

Lassen is a full-text document retrieval system built into and on top of POSTGRES. Text pages are stored as large objects, with weighted keyword indices constructed automatically by POSTGRES as the text is entered. Lassen includes a separate natural language query tool that interfaces with POSTGRES, allowing text retrieval in a manner familiar to users of automated library catalogs.

Lassen will be used to access the growing on-line collection of text on Bigfoot. The UC Berkeley Computer Science Division is currently scanning all its technical reports into Bigfoot, saving the text as images, and using OCR to extract keywords. We expect that eventually all printed material associated with Sequoia 2000 will be accessible via Lassen.

POST-GRASS

With the cooperation of the U.S. Army Construction Engineering Research Laboratory (the originators of the GRASS-GIS), the Sequoia 2000 Project is working on subsuming the functionality of GRASS into POSTGRES. This effort is proceeding incrementally. The first phase, already complete, has replaced the input/output subsystem of GRASS with calls to POSTGRES, with the relevant GRASS external data structures replaced by a custom POSTGRES schema. This GRASS-on-top-of-POSTGRES has been dubbed "POST-GRASS." The second phase of POST-GRASS, currently in progress, will migrate specific GRASS commands into POSTGRES as internal functions. The GRASS user interface will then be modified to issue POSTQUEL instead of UNIX commands. In the third phase, Tioga will replace the GRASS user interface, and POSTGRES will be able to provide equivalent functionality to GRASS, but with access to all the data in Bigfoot.

References

Eidenshink, J. 1992. The 1990 conterminous United States AVHRR data set. *Photogrammetric Engineering and Remote Sensing* 58(6):809–13.

Feldman, G. 1989. Ocean color: Availability of the global data set. *Eos* 70(23):634–40.

Ferrari, D. 1990. Client requirements for real-time communication services. *IEEE Communications Magazine* 28:11.

Gershon, N., and J. Dozier. 1993. The difficulty with data. *BYTE* 18(4):143–48.

Katz, R. 1991. High-performance network and channel-based storage. *Sequoia 2000 Technical Report* 91/2. Berkeley: University of California.

Katz, R., T. Anderson, J. Ousterhout, and D. Patterson. 1991. Roboline storage: Low latency, high-capacity storage systems over geographically distributed networks. *Sequoia 2000 Technical Report* 91/3. Berkeley: University of California.

Kochevar, P. 1993. A visualization architecture for the Sequoia 2000 Project. *Sequoia 2000 Technical Report.* San Diego: San Diego Supercomputer Center.

Olson, M. 1993. The design and implementation of the Inversion File System. *USENIX Associates Winter 1993 Conference Proceedings*. San Diego.

Stonebraker, M. J. Chen, N. Nathan, and C. Paxson. 1992. Tioga: Providing data management support for scientific visualization applications. *Sequoia 2000 Technical Report* 92/20. Berkeley: University of California.

Stonebraker, M., and G. Kemnitz. 1991. The POSTGRES next-generation database management system. *Communications of the ACM* 34(10):78–92.

Weibel, W., J. D. Neelin, and H. H. Syu. 1993. An end-to-end processing scenario for hybrid coupled ocean–atmosphere model data under the Sequoia 2000 project. *Sequoia 2000 Technical Report.* Los Angeles: University of California.

Wiscombe, W. 1992. Personal communication. NASA Goddard Space Flight Center, Greenbelt, Maryland.

Notes

1. 1 gigabyte = 2^{30} (approximately 1 billion) bytes.
2. 1 terabyte = 2^{40} (approximately 1 trillion) bytes.
3. The name "Big Lift" is borrowed from the large pumping plant in the California state water system that pumps water from northern California over the Tehachapi Mountains to southern California.

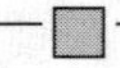

James Frew is Associate Director of the Sequoia 2000 Project. He is currently employed as a specialist by the Center for Remote Sensing and Environmental Optics at the University of California, Santa Barbara. He received his Ph.D. in geography from UCSB and is the principal architect of the widely used Image Processing Workbench (IPW) software package. His research interests include remote sensing, software design, digital spatial data libraries, and Earth science data management.
Sequoia 2000 Project
Center for Remote Sensing and Environmental Optics
University of California
Santa Barbara, California 94720
E-mail: *frew@crseo.ucsb.edu*

Sequoia 2000 Project documentation, including a series of technical reports, is available from the World Wide Web: http://s2k-ftp.cs.berkeley.edu:8000 or by anonymous FTP from s2k-ftp.cs.berkeley.edu. Paper copies may be requested from:

Claire Mosher
Sequoia 2000 Project
617 Soda Hall, University of California
Berkeley, California 94720
Phone: (510) 642-4662
E-mail: *claire@postgres.berkeley.edu.*

Source code for most Sequoia 2000 Project-developed software, including the POSTGRES database management system, is also available electronically.

14

Global Data and Tutorial to Model Climatic Change and Environmental Sensitivity

Robert Lozar

The Construction Engineering Research Laboratory (CERL) has made large portions of its internal global data available to the public as the GLOBAL GRASS CD-ROMs I, II, and III. The CD-ROMs are accompanied with a tutorial, utilizing simplified climatic change analysis examples that can be used in an intro to GIS techniques course. The data represent the availability of over 2.5 gigabytes of global data in a single standard GIS format for use by scientists and educators. A portion of this chapter is a tutorial on learning and using GRASS to answer questions regarding climatic change. Because the GLOBAL GRASS CD-ROM data are all in the same simple format, GIS as well as GRASS users have immediate access to large amounts of environmentally significant global data. These sets of data are chosen for wide distribution because these concerns have been identified as the most useful data for a variety of purposes. The tutorial is intended for global environmental analysis and can be used in high schools as well as grade schools. Since the GLOBAL GRASS CD-ROMs are inexpensive and the GRASS-GIS is public domain, this configuration of global data, powerful GIS, and tutorial empowers a wide audience to become involved in global analysis issues. When the exercises in the tutorial are completed, the user will be able to explore the characteristics of the earth with confidence and with the latest technology.

BACKGROUND

The idea that one can compare the relative sensitivity of an area or the relative environmental risk of carrying out an action which may affect very large areas anywhere on Earth is surprisingly new. This is due in part to that fact that data to execute global analysis were not available—the technology was not available, nor was the need for global analysis well defined. But things have changed: data from many sources recently have been translated into a common format, GIS technology has matured, and the need has arisen to question the affects of global climatic change.

The overall program to study climatic change is called the Mission to Planet Earth (MTPE) initiative. MTPE is an international cooperative effort to understand Earth processes. What is being developed in support of this work will have a major influence on the direction of GIS technology, remote sensing techniques, computer manipulation capabilities, and data analysis and extraction. The purpose of this chapter and tutorial is to teach the GRASS community how GRASS and GIS tools can be applied to climatic change questions. GIS technology is only beginning to be applied to questions of climatic change.

Another purpose in writing this chapter is to let scientists within the climatic change community know that the application of GIS technology to these questions is feasible and in many cases straightforward. The tutorial concepts can be taught as exercises in an initial GIS course.

It makes sense to apply GRASS capabilities to global change analysis. Scientists have been developing increasingly sophisticated Global Climatic Models (GCMs) for several years. Though GCMs deal with spatial information, they are of a statistical heritage. Dealing with spatial relationships within a statistical framework is inherently labor intensive, requires programming talent, and is time consuming. They are usually very complex programs written for a single purpose. In contrast, GIS are usually designed as a series of modules which are versatile enough to be put together repetitively and used for a wide variety of purposes. This flexibility implies that it may take some effort to apply the GIS tools to specific problems. However, the flexibility allows for greater longevity with less effort and a potentially wider audience—thus the tutorial.

GIS can be important tools for manipulation and use in scientific and technological information analysis. They are logical tools by which many of the environmental analyses resulting from the Mission to Planet Earth can be carried out. But many researchers in this area are using their traditional tools, primarily statistical mathematical modeling. This is adequate for those few who have a natural ability to extract from the numbers and equations the significance they hold. However, the ability to manipulate data in a spatial context and then visualize the results with a graphical interface provides for a clearer understanding of the significance of the data. This makes access to tools for scientific research available to a greater number of people. Advances in visualization research will enhance this.

GRASS is a public domain software set. Many governmental agencies (including those dealing with climatic change research) find it a desirable and powerful tool in supporting their spatial analysis requirements. GRASS provides for the integration of site-collected data and remotely sensed data. A large section of GRASS is dedicated to remote sensing processing techniques. Remotely sensed data are often handled only by image processing techniques, but GRASS has the ability to integrate image processing techniques with a collection of GIS tools, allowing for greater versatility and information extraction. Many universities and government agencies use GRASS because the source code is available. Therefore they enthusiastically develop additional GRASS capabilities for their own research. These new tools are integrated into the next release of GRASS so that its depth and breadth are continually being expanded. The adoption of GRASS will be particularly useful to those working in the Mission to Planet Earth initiative.

DATA

When CERL began the global work in 1989, we intended to use "the standard" global digital data set. It quickly became evident that no standard set existed. Thus, CERL began developing that set by ordering the best existing digital data and translating it into GRASS format. In addition, an extensive search of hard-copy library sources fed into a high-priority digitizing program. Both were integrated into a single GRASS data set. One of the strengths of GRASS is that it can integrate data of differing resolutions.

The data now available on GLOBAL GRASS CD-ROMs represent a portion of the global data set developed by CERL. The original purpose of gathering the data was to provide the in-house capability to evaluate environmental sensitivity anywhere in the world. This work is a part of CERL's ongoing research into GIS development, global analysis, climatic change, and visualization techniques.

It became apparent that these data had value to other scientists and educators. Though a good deal of data existed, very little of it covered the entire globe, and that which did exist was often in different formats so that one data set could not be compared or combined with others. For purposes of global analysis, a data set in a single consistent format is required for evaluation of environmental sensitivity, climatic change, Earth resources analysis, and economic and political demography. The data are significantly refined from those developed originally for internal CERL usage. GLOBAL GRASS I CD-ROM made these data publicly available (March 1992) in a single format for use in a standard GIS. (See Table 14-1 for a complete list). It included elevation, vegetation types, green leaf production, certain threatened species, national boundaries, soils characteristics, and marine productivity, among others. GLOBAL GRASS II (March 1993) has three primary topics: global climatic characteristics, global hydrology (including individually named basins), and human population distribution. (Table 14-2.) GLOBAL GRASS III (March 1993) is a set of fifty-three temporal satellite sensing layers of the entire Earth for the complete year of 1988 for chlorophyll production (normalized difference vegetation index [NDVI]). GLOBAL GRASS III data enable individuals to animate global vegetation production on their computers. GLOBAL GRASS IV and GLOBAL GRASS V were issued in April 1994. They include many more themes, including terrain and ocean characters. GLOBAL GRASS V has highly detailed raster and vector data, plus over six-million named sites (in about 700 categories). There are over 200 digital maps included on CD-ROMs I, II, and III, providing the user with more than 2.5 gigabytes of global coverage data. (For users of other GIS, all of the data sets on the GLOBAL GRASS CD-ROMs are in the exact format, so importing 150 different maps is no more difficult than importing one.)

As CERL's global research program produces additional data layers and as other groups contribute data of global interest, additional CD-ROMs will be made available to the community of GRASS users (and GIS users at large). An unlimited number of analyses become possible with this data. The tutorial details some global climatic change analyses for GRASS users.

The availability of these data are the result of CERL's global data development and global analysis program.[1] To perform quality control on the CERL internal data and distribute the actual CD-ROMs, CERL and Cook College Remote Sensing Center of Rutgers University in New Brunswick, New Jersey, implemented an agreement allowing the Remote Sensing Center to provide public distribution and support for these data layers and to make the tutorial available.[2]

Table 14-1. GRASS raster layers available on GLOBAL GRASS I CD-ROM.

• Surface albedo for January season	• Marine conservation projects
• Surface albedo for April season	• General range of marine otters
• Surface albedo for July season	• Ocean productivity in mg/m^3 (czcs data)
• Surface albedo for October season	• Phytoplankton productivity
• Aspect map of the world (direction of slope)	• Soil map of the world
• Continents (with nearby islands)	• World soil slope groups
• Continental shelf	• Soil texture groups
• Coral reef distribution	• Soils for general circulation climate models
• Primary cover/vegetation types	• Reliability of soils data
• Reliability of land cover data	• Shaded relief map of the world
• Secondary cover/vegetation types	• World topographic elevation ranges
• Intensity of agricultural cultivation	• Vegetation production for January
• Major world ecosystem complexes	• Vegetation production for February
• Highest biomass productivity	• Vegetation production for March
• Lowest biomass productivity	• Vegetation production for April
• Medium biomass productivity	• Vegetation production for May
• Fisheries productivity by major areas	• Vegetation production for June
• Major world fisheries areas	• Vegetation production for July
• Greenturtle distribution and characteristics	• Vegetation production for August
• Land (elevation greater than 0)	• Vegetation production for September
• Nations—world political boundaries	• Vegetation production for October
• Ocean biogeographic zones based on surface temperature	• Vegetation production for November
	• Vegetation production for December
• Biological ocean zones based on bathymetry	• Natural vegetation (preagriculture)
• Ocean floor biomass production	• Abundance of zooplankton

THE GRASS GLOBAL CLIMATIC CHANGE TUTORIAL

Global analysis has been done as if the globe consisted of a set of homogeneous nations, with one piece of information for the entire world, or as the result of image processing of

Table 14-2 GLOBAL GRASS II CD-ROM population, hydrology, and climatology.

• Natural color version of the earth with shaded topography	• Precipitation for March
• Major watershed basins of the world (in five files)	• Precipitation for April
• Named rivers based on the major basins	• Precipitation for May
• Major well-defined ridge lines	• Precipitation for June
• Bathymetry (under sea level elevation information)	• Precipitation for July
• Percentage urbanized corrected to equatorial area comparison	• Precipitation for August
• Population density distribution	• Precipitation for September
• Soil elements	• Precipitation for October
• Soil formative groups characteristics	• Precipitation for November
• Percent cloud cover for January	• Precipitation for December
• Percent cloud cover for February	• Temperature for January
• Percent cloud cover for March	• Temperature for February
• Percent cloud cover for April	• Temperature for March
• Percent cloud cover for May	• Temperature for April
• Percent cloud cover for June	• Temperature for May
• Percent cloud cover for July	• Temperature for June
• Percent cloud cover for August	• Temperature for July
• Percent cloud cover for September	• Temperature for August
• Percent cloud cover for October	• Temperature for September
• Percent cloud cover for November	• Temperature for October
• Percent cloud cover for December	• Temperature for November
• Precipitation for January	• Temperature for December
• Precipitation for February	• Shaded relief map (more detailed, replaces the first on GLOBAL GRASS I)

satellite data. Little GIS global analysis has occurred based on a fully integrated data set. That which has occurred has been largely done by scientists with extensive training and access to sophisticated hardware configurations.

To expand the pool of expertise, examples of global analyses which can be done using the data available from the GLOBAL GRASS CD-ROMs were developed to show the basic concepts associated with:

- global analysis
- GIS techniques
- use of GRASS tools
- use of the publicly available global data set, GLOBAL GRASS

When a student completes these exercises, he or she will have an understanding as to how GIS techniques can be applied to problems of global climatic analysis. Each of the exercises illustrates a problem and the procedure to generate the results. This allows educators at the university level the ability to deal with global questions as part of their schoolwork. The tutorial can also be used by high school and grade school students, inspiring them to take up roles in global climatic change research and environmental analysis and management. The ability to inspire young minds and therefore influence the next generation in global environmental care inherently resides within this combination of technologies.

Following are the tutorial analyses. For each, the analysis technique is stated, a short background outlined, the problem posed, the procedure described, and the output results given. (SUN systems often places a period after the layer name on the CD-ROM. This is the convention used in this tutorial. All GRASS commands look like: "grasscommand options."

(The author acknowledges and encourages the development of user-friendly interfaces, however, these vary greatly. The procedures presented here should be applicable to all GRASS systems.)

Analysis: Possible Sea Level Change

Background: A great deal of freshwater is locked up in the ice caps and glaciers. If there is global warming, ice will melt, sea level will rise, and low-lying areas will flood.

Question: What areas are likely to be inundated?

Results:

1. Map of distribution of sea level change.
2. Identification of nations with greatest amount of area flooded.
3. What kind of ecosystems are most likely effected by this change?

Assumptions:

1. That there will be no relative warping in the elevations of the continents.

GRASS programs used:

1. *r.mask*
2. *r.report*
3. *g.copy*

Usage:

1. Decide on a reasonable rise in sea level. The data layer to use is "topo" with elevations between 0–20 m (categories 1–4) above current sea level. Even in recent times, the change in sea level has been as much as 500 feet in 10,000 years (Thompson 1990). Thus, 20 m is only a small amount compared to what is possible. To carry this analysis out in GRASS, use the command *r.mask* to create a mask that limits the analysis only to those areas of interest (elevations of 0–19 m). Type: "r.mask." *R.mask* is a user-interactive GRASS tool, so you will get a menu. Choose option two: "make a new mask." Request use of the data layer called "topo." A new menu will appear, based on the category information in the file you have requested ("land"). You tell the system which areas should *not* be masked (that is, where the analysis *should* be carried out) by changing the zeros in the last column to one (or any other positive integer). Press "enter" to move down to the "next category" line. Make sure you enter "end" in the next category line; otherwise you will be entering "Xs" to make this mask all day long. When done, the screen should look like Table 14-3.

When you have exited the *r.mask* program, only those locations of 0–19-m elevation will be analyzed. Now run *r.report* to see how much area is affected by a change in sea level which would flood areas up to 20 m above sea level. You will see which nations are affected. First

Table 14-3. Menu page for *r.mask*.

Identify Those Categories to Be Included in the Mask		
Old category name	**Category number**	
Less than 0 Meters Elevation	0	0__
Elevation Range from 0 Meters to 4 Meters	1	1__
Elevation Range from 5 Meters to 9 Meters	2	1__
Elevation Range from 10 Meters to 14 Meters	3	1__
Elevation Range from 15 Meters to 19 Meters	4	1__
Elevation Range from 20 Meters to 24 Meters	5	0__
Elevation Range from 25 Meters to 29 Meters	6	0__
Elevation Range from 30 Meters to 34 Meters	7	0__
Elevation Range from 35 Meters to 39 Meters	8	0__
Elevation Range from 40 Meters to 44 Meters	9	0__
Next category: end__ (of 210)		
AFTER COMPLETING ALL ANSWERS, HIT <ESC> TO CONTINUE (OR <Ctrl-C> TO CANCEL)		

use the interactive version of the command: *r.report*. Answer the question as to which layer to use with "nations."

Since we only want information on this layer, just press "enter" when asked for another layer. On the next menu, you will be asked for the units of measure for the report. Place an "X" next to the square miles, cell count, and percent cover. After you make your submission, the system will generate the appropriate report. It will look similar to Table 14-4.

We can determine what kind of ecosystems are most likely affected by this change. We use the same procedure with the *ecosys.* layer. This time, use the command line version of *r.report* rather than the interactive version. If you need to know the required format and options for any command, you can always find out by typing: "some.command.of.interest help." Do this for *r.report*: "r.report help."

To generate the report on the ecosystems impacted by this sea level rise (i.e., the area showing through the current MASK), the command is: "r.report map=ecosys. units=miles,cell_counts,percent_cover pl=55 output=p.sea.change.by.ecosys."

A look at the created file—"more p.sea.change.by.ecosys"—shows that the most highly affected ecosystems (in terms of percent) are as indicated in Table 14-5.

Table 14-4. Nations most affected by sea level rise (in terms of percent of land area affected, in order of decreasing impact).

Category description	Square miles	Percent cover
United States	9.8243e+04	15.02
Australia	9.4521e+04	14.45
Russia	5.8250e+04	8.90
Mexico	3.3511e+04	5.12
Canada	2.8298e+04	4.33
Indonesia	2.4628e+04	3.76
Iraq	1.5443e+04	2.36
Italy	1.1511e+04	1.76

It is clear that vegetation of a warm/hot character will carry the bulk of the lowlands inundation impacts.

Save the mask you created for future usage. When a mask is created, a new temporary file is made in GRASS called "MASK." To save this for future use (and at the same time, remove the mask from masking a portion of the world) simply change the name of this file to a new name. This will also prevent the file from being overwritten next time a mask is created (which would also be called "MASK"). To do this, use the *g.copy* command. The command line version is: "g.copy rast=MASK,sea.level.rise."

CONCLUSION

The existence of the three GLOBAL GRASS CD-ROMs opens a new era in the public availability of global environmental data for application to questions of climatic change and environmental sensitivity and risk evaluation. Because the data are all in the same simple format, GIS and GRASS users have immediate access to about 2.5 gigabytes of environmentally significant global data. These data were chosen for wide distribution because the developers have had project experience in dealing with global analysis questions and have identified these as the most useful sets for a variety of purposes. In this tutorial, analysis examples have been developed based on a knowledge of the needed areas of research. These procedures have been simplified so that they can be taught in an initial course on GIS techniques. Unlike some other data sets, the emphasis here has not been to indiscriminately blaze all available data (particularly massive amounts of remotely sensed data) onto a CD-ROM. Instead, those data types that have the greatest versatility for the widest range of global analysis purposes are included. Since the data are publicly available and the GRASS-GIS is public domain, the configuration is inexpensive and easy to set up. It is expected that the general concept will be integrated into high school, grade school, and college level curricula within the next few years. This configuration of global data, powerful GIS, and usage tutorial empowers the widest audience to become involved in global analysis questions.

References

Aho, A. V., B. W. Kerninghan, and P. J. Weinberger. 1988. *The AWK Programming Language*. Addison-Wesley.

Balbach, H., and R. Lozar. 1990. *Global Commons Environmental Review*. Environmental Division, Construction Engineering Research Laboratory, Champaign, Illinois.

EOS Program and Project Offices, NASA. 1990. EOS and Pre-EOS Reference Handbooks. Greenbelt: Goddard Space Flight Center.

Lozar, R. 1992a. Geographical information system (GIS) technology in global environmental evaluation—An overview. In *SPACE '92: Proceedings of the Third International Conference, American Society of Civil Engineers*, New York, New York, 2,103–127.

Lozar, R. 1992b. Global climatic change management by watershed basin units. *Technical Papers of the American Society of Photogrammetry and Remote Sensing* vol 4, 150–59. Bethesda.

Lozar, R., and D. Artis. 1992. The Army Corps of Engineer's (ACE) interaction with the Mission to Planet Earth Initiative. In *SPACE '92: Proceedings of the Third International Conference, American Society of Civil Engineers*, New York, New York, 2,094–103.

Thompson, Ida. 1990. *Field Guide to North American Fossils*. New York: Alfred A. Knopf.

U.S. Army Corps of Engineers. 1992. Mission to Planet Earth Task Force Report, Directorate of Research and Development, (CERD-ZA). Washington, D.C.

Table 14.5 Major world ecosystem complexes most affected by sea level rise (in terms of percent, in order of decreasing impact).

Category number	Category description	Square miles	Cell count	Percent cover
31	Nonwoods-warm/hot farms/towns	1.0816e+05	4642	16.53
41	Nonwoods-main warm/hot scrub and grassland	6.5994e+04	2537	10.09
58	Interrupted woods-trop/temp wds, fields, grass, scrub	3.5637e+04	1339	5.45
45	Wetland/coastal-major warm/hot mangrove/tropical swamp forest	3.2715e+04	1181	5.00
56	Interrupted woods-2nd grow trop/subtrop, humid/temp/boreal forest	3.2615e+04	1253	4.99
32	Major woods-trop/subtrop dry forest and woodland	3.1420e+04	1070	4.80
37	Nonwoods-other irrigated dryland	2.8209e+04	1081	4.31

Notes

1. CERL Point of Contact for global analysis work is Robert Lozar: (217) 373-6736, fax (217) 373-7222, e-mail: *lozar@zorro.cecer.army.mil.*
2. Copies of the CD-ROMs and the full tutorial are available from: GLOBAL GRASS CD-ROM, Department of Environmental Resources, Cook College, Rutgers University, PO Box 231, New Brunswick, New Jersey 08903, phone: (908) 932-9631, fax (908) 932-8644.

Robert Lozar
Environment Sustainment Laboratory
Spatial Analysis Team
U.S. Corps of Engineers
Champaign, Illinois 61826
E-mail: *lozar@zorro.cecer.army.mil*

15

Global Ecosystems Database Project:
An Experiment in Data Integration for Global Change

John J. Kineman, Donald L. Phillips, and Mark A. Ohrenschall

The Global Ecosystems Database Project (GEDP) is developing an integrated global database to support global change characterization and modeling. One CD-ROM of integrated data with supporting documents was published in 1992 (NOAA-EPA1992), and another is being developed. This chapter describes current data integration work and discusses future directions. The current integration effort adds regional databases and model outputs along with additional global data. A research effort is planned to evaluate effectiveness of this approach in supporting modeling and research.

INTRODUCTION

The Global Ecosystems Database Project (GEDP) is an interagency project between the National Geophysical Data Center (NGDC) of the U.S. National Oceanic and Atmospheric Administration (NOAA), and the Environmental Research Laboratory-Corvallis (ERL-C) of the U.S. Environmental Protection Agency (EPA). It is a part of NGDC's Global Change Database Program (GCDP), whose goal is to provide modern, global, and continental scale data (for the entire Earth's surface) needed by the global change research community.

The project began in 1991 as a five-year effort to build a reviewed database with an integrated approach to analysis. The philosophy and conceptual design of the project was described in more detail in the first NCGIA conference on GIS and modeling (Kineman 1993b). This chapter will focus on current data integration activities and plans for the future. In the effort of building an integrated database, the GEDP is also making advances in integrating database structure and function, and defining how this integrated approach connects with characterization and modeling for global change. Results of this work include the distribution of CD-ROMs containing successive improvements of the database and supporting software, database documentation manuals, and a user's guide with reviews of the database and its effectiveness in supporting global change research (NOAA-EPA 1992).

The goal of the GEDP is to develop a research database for *global environmental and ecological characterization* within the U.S. Global Change Program. Ecosystem characterization has been defined as:

> a study to obtain and synthesize available environmental data and to provide an analysis of the functional relationships between the different components of an ecosystem and the dynamics of that system...it is simply a structured approach to combining information from physical, chemical, biological, and socioeconomic sciences into an understandable description of an ecosystem (Watson 1978).

Characterization is descriptive, goal-directed, and organized by conceptual models of the ecosystem, which provide the scientific context, experimental designs, and information priorities needed to focus the effort. To apply this concept to the global change scientific program, which is a goal-directed multidisciplinary ecosystem study, characterization efforts must be developed as a complementary link to process-oriented research and modeling efforts (i.e., our present understanding of the earth system).

The GEDP thus represents a modern approach to characterization that relies on digital information (both remotely and directly observed) and computer analysis technology represented in the rapidly developing field of analytical GIS. It is presently linked with ongoing modeling research at the EPA's Environmental Research Laboratory in Corvallis, Oregon, and links are being formed with other modeling groups. This research linkage determines the short- and long-term priorities for database development, as well as the functional requirements for analytical tools. An extensive peer-review effort ensures scientific quality and relevance to broadly based research efforts.

INTEGRATION METHODS

The overall database structure and functional design was developed from initial experiments in a pilot project for the International Geosphere–Biosphere Program, in collaboration with the IDRISI Project of Clark University and UNEP/GRID. This basic approach was then refined in the first years of the current project and is described in the user's guide that was published with the first public CD-ROM release (Kineman 1992).

The technical objective of data integration in the GEDP is to provide the ability to compare or combine data sets geographically and temporally within a common GIS framework, with reference to appropriate metadata, documentation, and scientific publications. Part of this objective is to represent data as accurately as possible, but also to support raster-based GIS analysis related to environmental modeling. To ensure geographic comparability in the raster structure, a standard for edge-matched "nested" grids has been established for the project database. This allows only integer multiples of grids that can thus be compared or combined without resampling by the user. Allowable grids are five, ten, thirty, and sixty minutes and may eventually be extended above and below this range. In practice, this involves correcting registration and coding errors, regridding when necessary, restructuring data into a geographic object convention, performing quality tests of the data, and producing improved documentation. Full lineage referencing is provided back to the source data, and examples of source data are included in the final database when needed to allow users to duplicate and test processing methods. A report of data processing and quality studies conducted by the project is provided for each data set, compiled in a published documentation manual for each database release.

The most common problem occurring in the majority of data sets contributed to the project is geographic registration. The example of elevation data that follows illustrates a worse-case scenario, where data are internally misregistered in the same grid without lineage information. The more typical cases involve incompatible grids, simple and correctable registration offsets, projection uncertainties, and digitizing errors. Many of these problems are actual processing or documentation mistakes rather than errors in the original measurements. Aside from simple registration and projection fixes, unit and numerical type conversions, and reproduction of legend information, there are often more complex problems, due to the way data are structured, requiring special processing. Vector- and raster-based GIS techniques have been developed in the project to test and correct registration errors using Micro World Data Bank II (MWDBII) (Pospeschil 1992) as a default standard. By processing other data sets to reveal coastal features and using overlay functions to compare them with MWDBII, rigorous tests of registration and projection accuracy are performed.

Determining the resolution of data sets is another important issue affecting comparability and decisions about rasterization. The following examples demonstrate that there is a difference between resolution and appropriate grid or cell size for raster representation. Sampling densities for vector data sets are determined by average point spacings and can be indicative of the maximum resolution possible. It is often the case, however, that finer cell sizes than the average vector sampling resolution are required for proper raster representation. For example, a general rule of thumb in remote sensing is that feature resolution is usually no better than 2–3 pixels. The same concept seems to apply to rasterizing vector polygons where we have found that rasters of about ½–⅓ the average vector sampling resolution are needed to avoid losing important relative information, and often higher detail, at region boundaries.

DISCUSSION

Critique of the 1992 Database

The contents of *GED Version 1.0: Disc A*, released in June 1992 and updated in September 1992, are shown in Table 15-1. The data processing and quality control for those data sets are described in documentation that accompanies the CD-ROM (Kineman and Ohrenschall 1992).

Table 15-1. Contents of *GED Version 1.0: Disc A*.

Chapter	Title (resolution)
GED Global Geographic (lat./long.) Raster Data-Sets	
A01	NGDC Monthly Generalized Global Vegetation Index from NOAA-9 (April 1985-December 1988) (ten minutes)
A02	EDC-NESDIS Monthly Experimental Calibrated Global Vegetation Index from NOAA-9 and 11 (April 1985-December 1990) (ten minutes)
A03	Leemans and Cramer IIASA Mean Monthly Values of Temperature, Precipitation, and Cloudiness on a Global Grid (thirty minutes)
A04	Legates and Willmott Average Monthly Surface Air Temperature and Precipitation (regridded) (thirty minutes)
A05	Olson World Ecosystems (mixed thirty minutes/ten minutes)
A06	Leemans Holdridge Life Zone Classifications (thirty minutes)
A07	Matthews Vegetation, Land Use, and Seasonal Albedo (one deg.)
A08	Lerner, Matthews, and Fung Methane Emissions from Animals (one deg.)
A09	Matthews and Fung Global Distribution, Characteristics, and Methane Emissions of Natural Wetlands (one deg.)
A10	Wilson and Henderson-Sellers Global Land Cover and Soils Data for GCMs (one deg.)
A11	Staub and Rosensweig Zobler Soil Type, Soil Texture, Surface Slope, and Other Properties (one deg.)
A12	Webb, Rosenzweig, and Levine Global Soil Particle Size Properties (one deg.)
A13	FNOC Elevation, Terrain, and Surface Characteristics (ten minutes)
GED Global Geographic (lat./long.) Vector Data Sets	
A14	Pospeschil Micro World Data Bank II (one minute)
Experimental Source Data (non-GED structure)	
A15X	Edwards Global Gridded Elevation and Bathymetry (five minutes)
A16X	UNEP/GRID Gridded FAO/UNESCO Soil Units (two minutes)

Table 15-2. Contents of *GED Version 1.0: Disc A*; abbreviations.

EDC =	U.S. Geological Survey EROS Data Center
FAO =	United Nations Food and Agriculture Organization
FNOC =	U.S. Navy Fleet Numerical Oceanographic Center
GCM =	General Circulation Model
GED =	NOAA-EPA Global Ecosystems Database
GRID =	Global Resource Information Database
NESDIS =	NOAA National Environmental Satellite, Data, and Information Service
NGDC =	NOAA National Geophysical Data Center
NOAA =	U.S. National Oceanic and Atmospheric Administration
UNEP =	United Nations Environment Program
UNESCO =	United Nations Education, Scientific, and Cultural Organization

No data set in the project can be considered error free, and it is often not feasible to completely correct contributed data sets. The best that can be done in many cases is to improve documentation through quality testing and comparison with other data. It is also important to distinguish between uncertainty or error in the data, and suitability of data sets for modeling support, including the effectiveness of their integration. In this latter category, data processing issues as well as quality control are critically important.

The 1992 database has been in public use and evaluation for over a year. During this time we have learned more about the data sets. As expected, the greatest problems are with three data sets originally labeled as "experimental." These cases are discussed in the following paragraphs, and lessons are summarized in the Conclusion.

1. The composite elevation data developed by Edwards (A15X), and also distributed by NGDC as "ETOPO5," was placed in the experimental category because its grid origin (centroid aligned with standard parallels and meridians) is out of phase with the GED convention (edge aligned with standard parallels and meridians), and because known registration errors between its land and bathymetry portions exceed the geographic standards of the project. On subsequent investigation it was determined that these internal errors were larger than thought and could not be corrected without accurate lineage information for each pixel. A new successor to this data set will be included in the 1993 CD-ROM release of the Ecosystems Database, but is itself only a precursor to other products in development at NGDC.
2. The FAO/UNESCO Soils data set (A16X) has been revised by FAO. Distribution of the earlier version was discouraged by 1992; however, because many of the other soils data sets in the GED were derived from the original FAO data set, or in some way used it, and because considerable scientific work has been based on it, we felt that scientists should have access to the original version for validation purposes. Because of the FAO "embargo," we were only able to provide a gridded version of the soil units (not the complete data set), which was obtained from UNEP/GRID before FAO's restriction. Because of the sensitive nature of this issue, and the fact that the UNEP/GRID version was itself a derived product, it was included in the experimental category on the 1992 CD-ROM with suitable caveats. The new version from FAO, in the meantime, has been considerably delayed.
3. Extensive evaluation had indicated that the Experimental Monthly Global Vegetation Index (GVI) data set, with prelaunch calibrations (A02), contains processing errors that make it difficult to use in GIS and other applications involving spatial/temporal mapping. There is a uniform 1-pixel registration shift at several times in the time series, and erroneous values occur between 1–100 as a result of averaging flag values embedded in the original biweekly version with true data values. Also the attempted calibration did not improve the uniformity of the time series, which shows a significant discontinuity at the satellite transition and retains the sensor drift problems of earlier data sets.

Less serious problems have been noted in some of the other data sets. These will be reported in future addenda and updates to the documentation, but some notes are highlighted here:

4. The Generalized Global Vegetation Index (A01) time series is apparently not exclusively composed of NOAA-9 data as originally intended. The last two months of the forty-five-month time series are from NOAA-11 data, thus incorporating different sensor calibration and drift characteristics, which could bias studies of those characteristics (one of the intended uses of this data set). One should thus exclude the last two months if attempting to correct for systematic trends from sensor drift.
5. The Olson World Ecosystems (A05), which contains updates since the original version of the Olson data published through CDIAC (Olson et al. 1985), also has incorporated errors in coastal areas due to use of the FNOC terrain data set (A13). The effect is restricted to a few areas and, in an overall evaluation, is probably outweighed by other improvements and the fact that significant new information was also incorporated.
6. It has been noted that the Micro World Data Bank II data set has some discontinuities in its line work and that some of the features appear in the wrong file. Other than these problems, however, the data set has proven its usefulness, at resolutions above one minute, for testing registration of other data sets. Methods have also been worked out to rigorously determine the best overall registration fit between this data set and others. It has thus become the registration standard for the project. Nevertheless, since the data set was produced from finer-scale data by truncation to one minute of arc (lat./long.) rather than rounding, it should be used at resolutions no finer than two minutes without correcting the bias in registration that is an artifact of truncation. Other higher-resolution sources are now being investigated, such as the World Vector Shoreline and Digital Chart of the World, from the U.S. Defense Mapping Agency, and other full-resolution versions of World Data Bank II.

1993 Data Integration

Additional data derived from satellite observations and other sources are being processed for the next CD-ROM release. Also with this next release, the project begins to focus on regional databases and associated numerical outputs from static and dynamic models, including various general circulation models (GCMs). The new product will be distributed to approximately one hundred reviewers in 1993, after which it will be publicly available from NGDC on a data exchange basis or at the cost of publication, as are the earlier products. It is interesting to note some of the typical problems encountered in integration and quality testing of these data. The following provides some highlights of data-quality issues encountered in this year's integration effort.

There are a growing number of gridded data sets generated from vector data sets (digitized lines and polygons), which themselves were created by digitizing by hand from a paper map. Often physical distortions of the paper map and digitizing error accumulate to the point that accuracy is far below that of the original data, while the digital data are actually represented with much greater precision. This was the case with a digitized version of Kuchler's vegetation classes for the United States, which clearly contained features approaching two 2 km in size, but also exhibited spatial location errors up to 8 km. In this case a grid representation of 8 km (five minutes in the Geographic projection) was chosen as the appropriate precision (although we will also include the vector data for users to experiment with).

To partially fill the void for elevation data, an improved five-minute composite data set, called "EL5," and an accompanying bathymetry data set called "BAT5," have been produced. This was done by overlaying available five-minute source data for Japan, Europe, North America, Brazil, and Australia. In areas for which no five-minute data could be obtained, ten-minute data from the U.S. Navy Fleet Numeric Oceanographic Center were replicated to the five-minute grid. The five-minute bathymetry data were interpolated to match the standard GED grid alignment. Separate maps give the data source for each pixel, allowing one to estimate quality in each region. No attempt was made to correct discontinuities between data sets that might occur at their boundaries.

Another challenge, and an interesting spatial data issue, was presented by general circulation model (GCM) data, also intended for release in the 1993 database. Each of the major GCMs is calculated on a different grid, none of which match the allowable grid sizes in the GED. Since the project seeks to provide comparison capabilities, an obvious solution would be to regrid. However, GCM modelers are very sensitive about distributing these data on anything but the original grids for fear of misrepresenting the spatially compartmented nature of the model. In essence these data are very similar to site data, with multiple "observations" at preestablished and spatially important locations (in this case a regular lattice of points). The solution was to treat the data set as one would handle station observations, and provide it in tabular form, referencing data values to their original grids as point locations, which are stored in vector GIS form. This preserves the original grid while providing the user with convenient tools for interpolating to any desired grid for comparison or modeling purposes. An important issue in this choice was if the GCMs represent "finite differences" or "finite elements." The former are values that can be treated as samples in a theoretical surface and interpolated by assuming an appropriate spatial model, while the latter are unique to a defined spatial unit. GCMs tend to be mixed but are generally used as finite differences.

So far, there has not been a data set in the project that could not be represented in the GIS, or "geographic object," structure. The most challenging data are tabular data that often do not contain adequate referencing to resolve spatial ambiguities. The result is that spatial maps derived from these tables can be multivalued. For example, a Forest Practices data set is being processed, which contains multiple forestry practices and project data. These data are referenced by country and Bailey's Ecoregion, but not by stand. Extracting mappable information results in choropleth maps with often multiple policies and/or practices. This makes it impractical to convert the entire data set into digital map form because one would have to anticipate the range of user searches that result in mappable output. Alternatively, we can provide the means for users to map the results of searches, which can then be used with the GIS database.

Finally, to end this discussion with a near horror story with a happy ending, our attempts to use various digitized versions of Bailey's Ecoregions of the World have been particularly illustrative. The projection of the original map was unknown, and the digital version produced by the USDA Forest Service was distributed with corresponding caveats. This prompted several attempts to discover the original projection by inquiry and by trial and error, both of which failed. In the course of this investigation, however, we located one version that had supposedly been "corrected" to an equal area projection and was subsequently being used in a major modeling project. We found that a mistake had been made in the correction that resulted in spatial errors on the order of several hundred km; however, this was not noticed by either the data processing group or the modelers! We have since determined that Bailey's map was based on a Russian map, and we have now received information for approximating the projection (which apparently has no mathematical form) from the (former) U.S.S.R. Geodetic and Cartographic Institute in Moscow. Meanwhile, we located a correctly reprojected version (approximated by rubber sheeting) from the World Conservation Monitoring Center in England and have successfully integrated that into the GED. In testing registration and projection accuracy, using GIS techniques developed by the project, we were able to determine that much of the data may be of mixed resolution—some owing to its Russian origins at a 1:80,000,000 scale and some from later work at 1:30,000,000. Nevertheless, the data set seems usable at about 0.5°–1° resolution and is important to modelers at this scale. Also, given the Russian lineage in the Bailey data, there may be correspondences with another higher-resolution vegetation class (and phytomass) data set produced by Dmitry Varlyguin (from the Academy of Sciences in Moscow) from historical Russian data sources, thus allowing further testing and comparison (by users).

Plans

The Ecosystems Project, now halfway into its five-year plan, is at a very interesting point. It has long been recognized that a project to support global change based solely on data integration would soon lose its relevance due to lack of firm priorities. Having achieved some measure of success in the initial effort of integration, based on relatively generic requirements for modeling, the project is now incorporating regional databases and outputs associated with specific ecosystem modeling projects conducted elsewhere. Also, since inception of this project various research groups have become established and are just now at the point of needing an integrated database. We are thus seeking close ties with such users, to better evaluate the effectiveness of the integration concept and to glean clearer priorities for future data processing. The remaining years of the project will see an increased emphasis on such evaluations and research support. The project has been reorganized to accommodate this change by defining three roughly equal components to the work. These are (1) CD-ROM publication and distribution, (2) continued data integration, and (3) research and review. The research and review component will continue the peer-review process and begin testing the effect of uncertainty in the multiple versions of different variables in several modeling projects, with which close ties have been formed.

CONCLUSION

Case (1), regarding topographic data, illustrates the need to produce quality or lineage data along with the primary data, especially for composite data sets where the nature and quality of source data will vary geographically. Had this been done in the original data set, its correction might have been possible. "Lineage tracking" and "error tracking" have been the subject of considerable concern for the future development of GIS, yet development will continue to be slow without the necessary metadata and ancillary quality/lineage data sets to support the needed capabilities. In the 1992 database, only one data set (A10, Wilson and Henderson-Sellers) was provided with separate quality layers. A quality layer was developed by the project for the Olson data (A05) to distinguish between thirty-minute data and ten-minute updates. Quality layers for the Monthly Generalized GVI might be produced from cloud data, thus allowing users to flag or compensate for unreliable data in chronically cloudy regions.

Case (2), involving the FAO soils data, illustrates an important principle, that older versions of data are necessary for validating other scientific publications that may be based on them. If an older data set can be "withdrawn" from circulation, the process of scientific validation is hindered. For this reason, the GED seeks to establish a precedent of "published" data, meaning that the data sets themselves are placed into a permanent public form for future generations to refer to. Political sensitivities and the wishes of the principal investigators must be respected at the time of publication, for ethical reasons and to preserve good relations with data suppliers. Nevertheless, for scientific reasons, the publication process must also be respected. This case fell into a gray area between the two concerns because prior public "availability" does not necessarily imply "publication." The issue is still not fully resolved, but it seems clear that the data community needs to advance in the direction of more rigid publication conventions.

Case (3), regarding the Experimental Calibrated GVI, illustrates a major principle of this project, that multiple versions of derived variables should be encouraged and that no single expert view should limit creative development. The data set was originally included at the urging of the IGBP Working Group on Land Cover Change, based on the then-current feelings of GVI experts regarding the appropriateness of processing methods and the strong recommendation that this data set *replace* the "Monthly Generalized GVI" (A01), which was produced specifically for earlier versions of the integrated database. While in no way disparaging expert advice, the lesson here is that it should be used only to indicate the most reasonable starting point at the time, and that educated feelings cannot substitute for free and open experimentation and empirical testing. The Monthly Generalized GVI data set has resulted in considerable positive feedback over four years since the IGBP recommendation, and has been shown to be a useful derivative of the original NOAA data for geographic and temporal studies. Gaston et al. (1994) have successfully used the data set for land cover classification where other versions have failed; Eastman and Fulk (1993a, 1993b) have successfully used the data set (again where others failed) for time-series analysis using principal components analysis. Over four years later, a potentially improved version was designed (by S. Goward, University of Maryland). Even now, because of the uniquely useful character of the generalized version, consideration is being given to continuing it as a NOAA product.

In general, we have learned that the widest possible testing of data sets within various contexts should be supported through data integration projects that are strongly linked with research and quality testing. While many researchers have the individual capability for performing data-quality testing, there is often little time (or funding) for such work, which can be time consuming and costly. The added problem that individual data sets are not usually distributed in a geographically comparable form means that very little validation and cross checking gets done; even less is reported since it is usually done in the context of other research. The scientific establishment tends not to consider descriptive studies involving issues of accuracy as equal in importance to environmental science, indeed some claim it is not science at all.

Another important lesson concerns the potential for propagation of error in combining data sets to "improve" the database. This, and loss of comparative information when data sets are combined, implies that updates and improvements to the integrated database should incorporate independent data sources. Each of the data sets in the GED are known to contain errors and are provided for intercomparison and evaluation as much as for careful and informed use. If they are used to update each other, the result is to destroy the value of intercomparison for quality purposes and to homogenize the database. For this reason, hybrid data sets are considered less valuable than more independent compilations, unless they incorporate significantly new information.

In each case, as seen in the example of GCM data and other tabular data sets, but also in complicated grid data such as the Webb soil profiles (A12), and nonstandard grids such as all the GVI data (A01, A02, and a new data set being processed), ETOPO5 (A15X), and Legates and Willmott climate data (A04), the proper method of representing a data set spatially must be determined to achieve integration while avoiding the loss of information to the greatest extent possible. If regridding is required, which necessarily implies a choice of methods, each with different effects, source examples should be provided (along with information about obtaining the original grids) so that users can assess these losses with regard to their intended use. An alternative solution of representing nonstandard grids as point data, as is currently being employed for GCM data sets, works well where the grid size is large, the original grids have special meaning, and the point values represent "finite differences."

The question of how far to go with integration has been raised often with regard to this project. Some feel that only original versions of data should be distributed and that integration should not include regridding, but only quality studies. The experience of this project strongly suggests the opposite view, however there is a need to ensure that integration methods keep the results as close to the original data as possible and support the greatest number of uses. This, we believe, is the value of maintaining a GIS context, which represents an analytical environment with evolving conventions for a wide interdisciplinary community. The need for regridding, and other carefully designed restructuring of certain data sets, is to achieve geographic comparability between data sets, which is considered to be a benefit both to the user and to the original PI. The reasons for this criterion are:

1. There are many interpolation and aggregation methods to choose from, and the appropriate method is specific to the design of the data. Yet few users will consult the original PI. Also, most systems have a limited set of "default" methods, the details of which are hidden from the user, or worse yet, they may even be performed automatically.
2. If, as a result of our distribution, we refer a large number of users directly to the original investigators for advice on integration methods, this could have negative effects on future relations. The same may result if people don't consult the PI, but publish erroneous conclusions based on improper resampling.
3. By involving the PI during production of the database in the issue of geographic registration and comparability, we estab-

lish rapport, obtain the "best" method according to the original investigator, and encourage them to correct the problem in the original version, at least for future data sets.

4. The alternative of geographically incompatible data does not tend to reveal the quality issues that geographic integration does. When resampling is needed for geographic comparison, it is thus also necessary for many of the quality checks (Q/C) and intercomparisons done by the project. Without an integration requirement, much of the quality and documentation work would be lost.
5. If the geographic comparison and Q/C are done separately from the database distribution effort, then separate versions of the data must be used than are actually published on the CD-ROM (i.e., resampled only for Q/C purposes). This makes results less applicable, especially since some Q/C results may depend on integration method.
6. Users who might question our choice of interpolation or aggregation method can access "source" examples on the CD-ROM to assess quality and error issues. If a different method is wanted, full references to the source data are provided. Thus researchers have an immediately usable working version, an enhanced ability to learn about and assess integration issues, and the ability to redo any part of the integration process by requesting the source version.

Having established an integrated global change database on CD-ROM, the Global Ecosystems Database Project is now investigating more deeply its existing and potential links with modeling. An important interim conclusion of the project is that data integration efforts such as these are essential not only for general use, but also for testing and improving the quality of the empirical database. Without such efforts there may be virtually no validation of basic boundary conditions and input data for models in the rush to investigate more "scientific" issues. This is not a criticism of the research and modeling groups, whose priorities are not primarily on improving data, but rather a criticism of the current state of the art where descriptive studies and data-quality investigations are not well formalized or funded. In this regard, the concept of "ecological characterization" that is being developed in this project (Kineman 1993a) can be an important way of focusing this and similar efforts within an overall descriptive science agenda. This agenda must continue to develop conventions on methods and publication standards and produce a clearer understanding of how integrated "characterization" databases link with and support other research.

References

Eastman, J. R., and M. Fulk. 1993a. Long sequence time-series evaluation using standardized principal components. *Photogrammetric Engineering* 59(8):1,307–12.

Eastman, J. R., and M. Fulk. 1993b. Time-series analysis of remotely sensed data using standardized principal components. In *Proceedings, of the 25th International Symposium*, vol. 1, *Remote Sensing and Global Environmental Change.* Graz, Austria.

Gaston, G., T. Vinson, P. Jackson, and T. Kolchugina. 1994. Identification of carbon quantifiable regions in the former Soviet Union using unsupervised classification of AVHRR Global Vegetation Index. *Int. J. of Remote Sensing* 15(16):3,199–221.

Kineman, J. J. 1992. *Global Ecosystems Database Version 1.0: User's Guide.* Key to Geophysical Records Documentation No. 26. Boulder: USDOC/NOAA National Geophysical Data Center.

Kineman, J. J. 1993a. Global environmental characterization: Interim results of the NOAA-EPA Global Ecosystems Database Project. *Proceedings of the NSF International Symposium on Environmental Information Management and Analysis: Ecosystem to Global Scales, Albuquerque, New Mexico.* New York: Taylor and Francis.

Kineman, J. J., and M. Ohrenschall. 1992. *Global Ecosystems Database Version 1.0: Disc-A, Documentation Manual.* Key to Geophysical Records Documentation No. 27. Boulder: USDOC/NOAA National Geophysical Data Center.

———. 1993b. What is a scientific database? Design considerations for global characterization in the NOAA-EPA Global Ecosystems Database Project. In *Environmental Modeling with GIS,* edited by M. F. Goodchild, B. O. Parks, and L. T. Steyaert, 372–78. New York: Oxford University Press.

NOAA-EPA Global Ecosystems Database Project. 1992. *Global Ecosystems Database Version 1.0: User's Guide.* Documentation Manual, Reprints, and Digital Data on CD-ROM. Boulder: USDOC/NOAA National Geophysical Data Center.

Olson, J. S., J. A. Watts, and L. J. Allison. 1985. *Major World Ecosystem Complexes Ranked by Carbon in Live Vegetation: A Database.* NDP-017. Oak Ridge: Carbon Dioxide Information Center, Oak Ridge National Laboratory.

Pospeschil, F. 1992. Micro World Databank II (MWDB-II): Coastlines, country boundaries, islands, lakes, and rivers. Digital vector data at one-minute resolution. In *Global Ecosystems Database Version 1.0: Disc A.* Boulder: NOAA National Geophysical Data Center.

Watson, J. F. 1978. Ecological characterization of the coastal ecosystems of the United States and its territories. In *Proceedings: Energy/Environment '78,* 47–53. Los Angeles: Society of Petroleum Industry Biologists. Also, subsequent publications of the Coastal Ecosystems Project. Washington, D.C.: Coastal Ecosystems Project, Office of Biological Services, Fish and Wildlife Service, U.S. Department of the Interior.

John J. Kineman
NOAA National Geophysical Data Center
Boulder, Colorado
Donald L. Phillips
USEPA Environmental Research Laboratory
Corvallis, Oregon
Mark A. Ohrenschall
University of Colorado
Cooperative Institute for Research in the Environmental Sciences
Boulder, Colorado

16

Thin Plate Spline Interpolation of Mean Rainfall:

Getting the Temporal Statistics Correct

M. F. Hutchinson

A statistical model of the spatial variation of rainfall means, based on incomplete observations at a set of point locations, is presented in this chapter. The model permits the smooth interpolation from these data of mean rainfall for a specified standard period using thin plate spline functions of position and elevation. The incorporation of a spatially varying dependence on elevation makes the dominant contribution to the accuracy of the interpolated surfaces. Incorporating a dependence on aspect makes only a marginal further improvement. The error structure of the statistical model has two components which allow for strongly spatially correlated departures of observed means from the standard period mean and for deficiencies in the representation of the standard period mean rainfall by a thin plate spline. Simplified versions of the model that use minimal summary data at each location are also presented. The accuracy of these models is similar to that obtained using more complete statistical models.

INTRODUCTION

Climate means for months, seasons, and years resolve much of the spatial variability of climate and of dependent biological activity (Hutchinson et al. 1992). Effective determination of the spatial distribution of mean rainfall and other mean climate variables is also a necessary first step toward the development of stochastic models of the weather for more refined assessments of the impacts of climate. Of the standard climate variables, rainfall displays the greatest variability in time and space and so makes the greatest demands on both temporal and spatial analysis techniques. Methods for the interpolation of mean rainfall for a standard period from ground-based point data have ranged from simple trend surface analyses (Edwards 1972; Hughes 1982) and local bivariate interpolation techniques based on Thiessen polygons and Delaunay triangulations, to more sophisticated statistical methods which can incorporate various dependencies on topography. These include geostatistical methods (Hevesi et al. 1992; Phillips et al. 1992) and thin plate splines (Hutchinson and Bischof 1983), all of which have incorporated dependencies of rainfall on elevation.

It is also important to take proper account of the large year-to-year variability in rainfall when interpolating rainfall for a standard period. An appropriate statistical model of the spatial distribution of observed rainfall means is presented that allows for departures of observed rainfall means from standard period means due to missing records and for deficiencies in the representation of mean rainfall as a smooth function of position and elevation. The first component of the error is due to the relatively large year-to-year variation of rainfall. The broad-scale processes which give rise to this variation imply strong correlations in observed means with overlapping periods of record. This component can therefore exhibit strong dependencies between different data locations, depending on the number of years of common record. The second component is due essentially to local effects below the spatial resolution of the data network. These can be assumed to be independent between different data locations.

The nondiagonal error covariance structure can either be accommodated directly or the observed short-period rainfall means can first be standardized to more accurate estimates of the standard-period means by performing linear regressions with nearby stations and then smoothly interpolating using a spline with a diagonal error covariance structure. Both methods permit the use of rainfall means recorded over all periods, no matter how short. Since sites measuring rainfall for relatively few years often make up the majority of sites in a network, this can significantly improve the density of data coverage and consequently improve the accuracy of the interpolated rainfall fields. The main drawback of these methods is the additional data required. Annual mean rainfall data for a region in southeastern Australia are used to show that the error structure can be replaced by simpler uncorrelated error structures which give rise to interpolated surfaces of only slightly inferior accuracy. For the simplest of these models, only the rainfall means and their lengths of record are required.

Thin plate smoothing splines permit the incorporation of varying degrees of topographic dependence, particularly when they are extended to include linear parametric submodels. Such functions are called *partial* splines. The chapter concludes by examining several spline and partial spline models which incorporate varying degrees of dependence on topography, including both aspect and elevation.

PARTIAL THIN PLATE SPLINES

Partial thin plate splines have been described in detail by Wahba (1990). They include thin plate splines as a special case. A summary of the basic methodology, with climate interpolation problems principally in mind, can be found in Hutchinson (1991), while more recent comparisons with geostatistical (kriging) methods are presented in Hutchinson and Gessler (1994) and Hutchinson (1993a). A useful variant of splines has been presented by Mitasova and Mitas (1993). The partial thin plate spline observational model is that there are n data values z_i at positions x_i given by:

$$z_i = f(x_i) + \sum_{j=1}^{p} \beta_j \psi_j(x_i) + \varepsilon_i \qquad (i = 1,\ldots,n;\ j = 1,\ldots,p) \quad (16\text{-}1)$$

where f is an unknown smooth function to be estimated, the ψj are a set of p known functions and the β_j are a set of unknown parameters which have also to be estimated. The x_i commonly represent coordinates in two- or three-dimensional euclidean space. The ε_i are zero mean random errors with covariance structure given by:

$$E(\varepsilon\varepsilon^T) = V\sigma^2 \quad (16\text{-}2)$$

where V is a positive definite n × n matrix and σ^2 may be known or unknown. The errors ε_i are uncorrelated if V is diagonal and correlated otherwise. The function f and the parameters β_j are estimated by minimizing:

$$(z - g)^T V^{-1}(z - g) + \rho J_m(f) \quad (16\text{-}3)$$

where $z=(z_1,\ldots,z_n)^T$ and $g=(g_1,\ldots,g_n)^T$ with

$$g_i = g(x_i) = f(x_i) + \sum_{j=1}^{p} \beta_j \psi_j(x_i). \quad (16\text{-}4)$$

$J_m(f)$ is a measure of the roughness of the spline function f, defined in terms of m, the order derivatives of f, and ρ is a positive number called the smoothing parameter. The solution to this minimization problem can be solved explicitly, with the estimate of f having an expansion in terms of a scalar function of distance from each data position. The form of the scalar function depends on the dimension of the x_i and the order of derivative m defining the roughness penalty (Wahba 1990; Hutchinson and Gessler 1994). The solution reduces to an ordinary thin plate smoothing spline when there is no parametric submodel (i.e., when ρ=0).

The smoothing parameter ρ determines a trade-off between data infidelity and surface roughness. It is usually calculated by minimizing the generalized cross validation (GCV). This is a measure of the predictive error of the fitted surface which is calculated by removing each data point in turn and summing, with appropriate weighting, the square of the discrepancy of each omitted data point from a surface fitted to all the other data points. It is possible to calculate the GCV implicitly, and hence efficiently. Intuitively, the GCV is a good measure of the predictive power of the fitted surface, as has been verified both theoretically and in applications to real and simulated data (Wahba 1990; Hutchinson and Gessler 1994).

Computational techniques for optimizing ρ and solving for f and β_j have been optimized for speed and storage space (Hutchinson 1984; Bates et al. 1987). The number of data points is limited by the requirement to perform a tridiagonalization of a matrix with dimension slightly less than n. If there are more than a few hundred data points, the above minimization problem can be solved using a thin plate spline function f defined in terms of a restricted set of data positions or knots. The computations can then be arranged so that only matrices with order equal to the number of knots have to be stored and decomposed (Hutchinson 1984). This permits data sets with up to a few thousand points to be easily processed on quite modest computers. Implementations of these techniques in FORTRAN have been described by Hutchinson (1984) and Bates et al. (1987). (A commercial package called ANUSPLIN is available from the author.)

A SERIALLY INCOMPLETE ANNUAL MEAN RAINFALL DATA SET

Annual rainfall data for various subsets of the thiry-four years between 1955–1988 were available at 195 locations for the region of southeastern Australia shown in Figure 16-1. Topographic relief across the entire region plotted in Figure 16-1 was over 1,600 m. As shown in Figure 16-2, relatively few stations had complete or even near-complete records. To assess the accuracy of the various interpo-

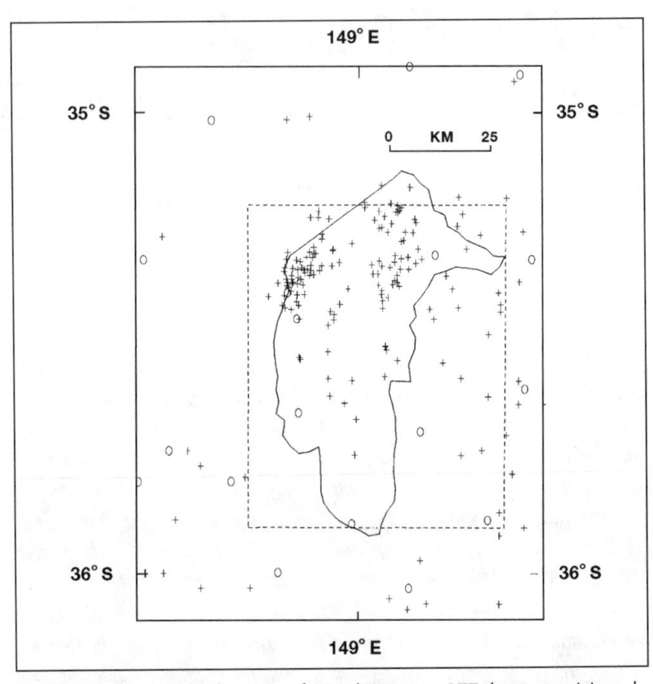

Figure 16-1. Geographic locations of 195 data points: 177 data points (+) used for surface fitting and 18 data points (0) withheld for validation. The inner solid line delimits the Australian Capital Territory. The dashed line delimits the area plotted in Figures 16-4 and 16-5.

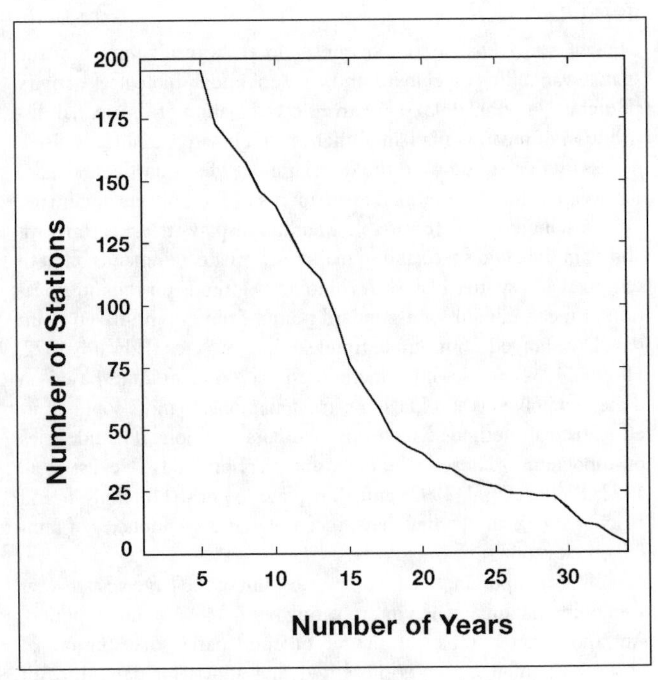

Figure 16-2. Number of stations for each length of record.

lation strategies, eighteen stations were withheld from the analysis, as also shown in Figure 16-1. These stations were selected, using the SELNOT program in the ANUSPLIN package, to approximately equi-sample the three-dimensional space spanned by the data. The selection procedure imposed a vigorous test on the proposed interpolation methods, since the withheld data points, most of which had more than thirty years of record, would normally be seen as the most desirable to be included in the data set used to fit the surfaces. Some of the withheld stations were fairly remote from the other data points, especially in terms of elevation.

TWO STATISTICAL PARADIGMS FOR INTERPOLATING MEAN RAINFALL

Nondiagonal Error Covariance Model

A relatively complete statistical model for the observed rainfall means at each location is given by equations (16-1) and (16-2) with the error structure of the ε_i decomposed into two components by setting $V = R + S$ where R accounts for departures of the observed mean from the thirty-four year standard period mean and S accounts for deficiencies in the spatial model of mean rainfall given by the partial thin plate spline function g. It can be shown that the elements of R are given by:

$$r_{ij} = \left(\frac{n_{ij}}{n_i n_j} - \frac{1}{34} \right) \rho_{ij} \sigma_i \sigma_j \qquad (16\text{-}5)$$

where n_i and n_j are the numbers of years of record at locations i and j, n_{ij} is the number of years in common, ρ_{ij} is the correlation between annual rainfall at each location, and σ_i^2 and σ_j^2 are the variances of annual rainfall at each location. It is reasonable to assume that the deficiencies of the fitted spline model, due to measurement error and local effects below the spatial resolution of the data network, are independent from one location to another, and moreover that the magnitude of this error is proportional to the annual rainfall variance at each location. Thus S is assumed to be a diagonal matrix with diagonal elements given by:

$$s_{ii} = \sigma_i^2 / m \qquad (16\text{-}6)$$

where m is an unknown constant factor.

The variance σ_i^2 at location i can be estimated directly from the annual rainfall data. However, in order to ensure that the matrix R was positive definite, a parametric form for the interstation correlation was estimated by fitting to the observed correlations an exponential function of station separation d_{ij} in km given by:

$$\rho_{ij} = \exp(-d_{ij} / 400). \qquad (16\text{-}7)$$

This model of interstation correlation is displayed in Figure 16-3.

A thin plate spline function of longitude and latitude in degrees and elevation in km was fitted using this nondiagonal error covariance model. The value of the unknown scale factor m was determined by adjusting its value until the value of σ_i^2 in equation (16-2) was estimated to be 1.0 by the minimum GCV model, making the relative error structure given by $V = R + S$ absolute. This procedure needs to be examined further, but the result presented in Table 16-1 appears to be reasonable.

The fitted value of m was 80, indicating that the root mean square of the departures of the unknown smooth function g from the actual standard period mean annual rainfall was 26.6 mm, about 3% of the network rainfall mean. The root mean square of the standard errors of the fitted spline estimate of g at the eighteen withheld data points was estimated by the method indicated in Hutchinson (1993a) and Hutchinson and Gessler (1994) to be 45.2 mm. The estimated root mean square validation error given in Table 16-1 was then calculated by allowing for these two errors and for the standard errors of the standardized means at the eighteen withheld data points, given by 16.5 mm. The estimated root mean square validation error was thus given by:

$$(28.2^2 + 45.2^2 + 16.5^2)^{1/2} = 55.0\ mm \qquad (16\text{-}8)$$

which agrees closely with the actual root mean square residual of 60.3 mm. It is in fact less than the root mean square error of the original data means given by 65.9 mm.

A three-dimensional perspective view of the interpolated annual rainfall overlying a $\frac{1}{400}$°-DEM is shown in Plate 16-1 and a similar view of the estimated standard errors is given in Plate 16-2. Both figures clearly show a strong dependence on elevation. The standard errors were greatest in those areas which were remote from the data points and had high rainfall. The mean rainfall over the total area shown in Figure 16-1 was determined, using the fitted spline and a $\frac{1}{400}$°-DEM, to be 907 mm with an estimated standard error of 38 mm. This was somewhat larger than the data network mean of 870 mm.

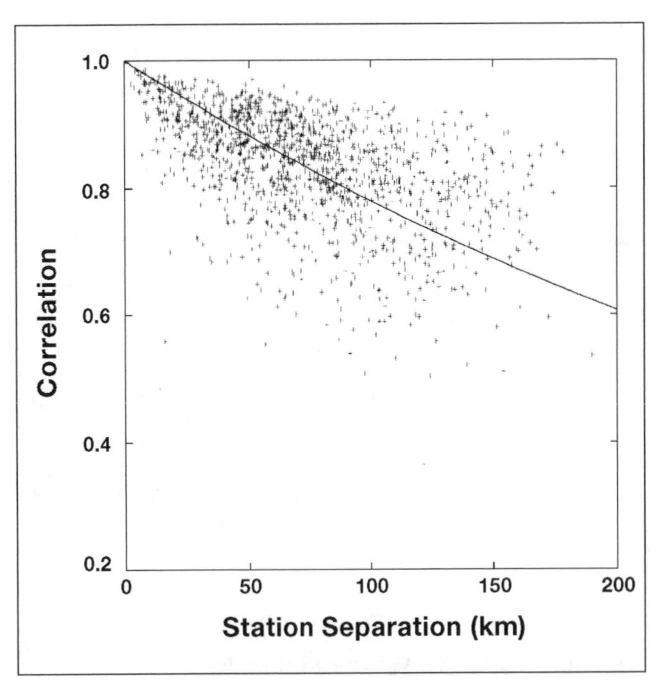

Figure 16-3. Interstation correlations versus separation in km.

Table 16-1. Estimated and actual root mean square residuals from trivariate splines using nondiagonal and prestandardized observational models.

Method	Estimated RMS residual of withheld data (mm)	RMS residual of withheld data (mm)
Nondiagonal error model	55.0	60.5
Prestandardized data	48.7	62.1

Diagonal Prestandardized Model

An alternative to the above model makes use of the strong interstation correlations to first standardize the short period means to standard period means using linear regressions. As for the nondiagonal procedure, this requires complete knowledge about the actual years of record at each location, but does avoid having to estimate a positive definite error structure for the interstation correlations. As for the nondiagonal model, it can be assumed that the variances of the errors in the unknown function *g*, which now describes the standardized data means, are uncorrelated and proportional to the data variance at each location.

The prestandardized values were smoothly interpolated by a minimum GCV thin plate spline function of longitude, latitude, and elevation with this diagonal relative error structure. The actual root mean square residual in Table 16-1 agreed closely with the residual for the nondiagonal procedure and was in reasonable agreement with the estimated value.

PRACTICAL APPROXIMATE METHODS

The statistically precise models investigated above were replaced by progressively simpler, less precise models. These can be applied when only brief summary data are available at each location.

Weighting by Local Variance Estimates

The actual data means were first weighted by a diagonal error structure given by:

$$v_{ii} = \sigma_i^2 / n_i \qquad (16\text{-}9)$$

where σ_i^2 is the local variance estimated from the data at location *i* and n_i is the number of years of record. This is the weighting used by Hutchinson and Bischof (1983). It recognizes that stations with higher variances or shorter periods of record should be subjected to more data smoothing than stations with smaller variances or longer periods of record. It ignores the strong correlations between means at locations with overlapping periods of record. Nevertheless, the root mean square residual of the withheld data in Table 16-2 is quite comparable with the estimated value and only slightly inferior to the values obtained with the statistically complete models.

Table 16-2. Estimated and actual root mean square residuals from trivariate splines using four approximate error covariance structures.

Method	Estimated RMS residual of withheld data (mm)	RMS residual of withheld data (mm)
Local variance estimate	77.2	72.0
Regression estimate	79.4	74.2
Length of record	75.8	75.4
Uniform weighting	68.6	76.8

Weighting by Local Variances When Variances Are Unknown

In some cases variances are not easily obtained. In the next step of the simplification, actual observed variances were replaced by estimates obtained by a simple regression model of the logarithm of the variance versus the logarithm of the mean. A graph of this model is given in Figure 16-4. Though the estimates of the variances were not particularly precise, Table 16-2 shows that the model performed virtually as well as the model which uses actual variances. This method can also be applied to monthly mean data since reasonably linear relationships are known to hold between the logarithms of monthly variances and the logarithms of monthly means.

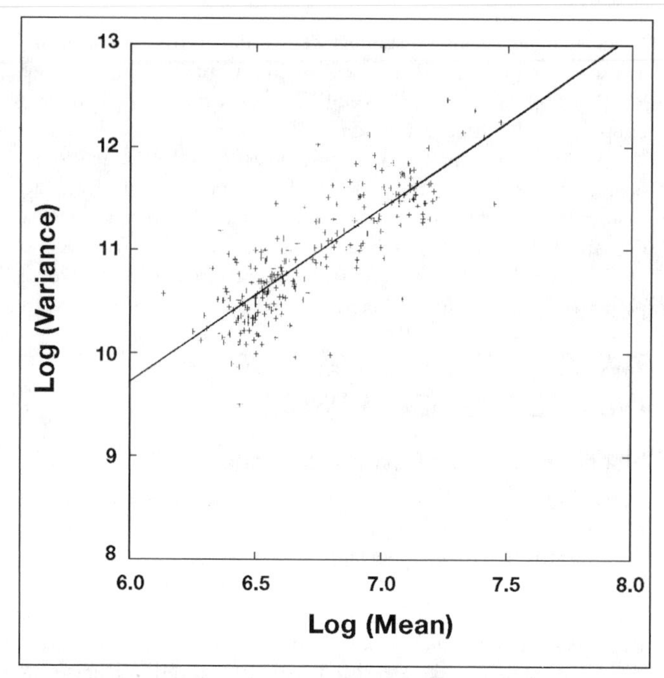

Figure 16-4. Logarithm of rainfall variance versus logarithm of annual mean rainfall.

Weighting by Length of Record

If variances do not vary greatly across a data network, it may be reasonable to weight the observed rainfall means simply by setting:

$$v_{ii} = 1 / n_i . \qquad (16\text{-}10)$$

Experience in fitting monthly rainfall means across large areas, for which there can be large spatial variation in rainfall means and variances, suggests that this is not always a viable option. However, it does enjoy the practical advantage of giving the same weighting for each month when interpolating monthly mean rainfall. This can permit significant savings in computation. For the small region considered here, the root mean square residual in Table 16-2 was similar to that obtained with the preceding models.

Uniform Weighting

Uniform, diagonal weighting is often the default option for interpolation methods. This is not generally recommended for interpolating mean rainfall. However, again because the region considered here is relatively small, the root mean square residual in Table 16-2 is only slightly larger than the values obtained with the preceding models.

INCORPORATING ADDITIONAL TOPOGRAPHIC EFFECTS

It is generally acknowledged that there are significant effects on rainfall related to the aspect. It is also acknowledged that the effects of topography on received rainfall are spatially coarser than the finest descriptions of topography. Both of these effects have been examined using local regression methods in the PRISM model (Daly et al. 1994). Alternative approaches based on partial thin plate splines are examined here. Using the prestandardized data described above, partial thin plate splines, with increasingly complex dependencies on topography, were fitted as follows:

- bivariate thin plate spline function of longitude and latitude only;
- trivariate *partial* thin plate spline incorporating a bivariate thin plate spline function of longitude and latitude and a constant linear dependence on elevation obtained from a 1⁄40°-DEM (Hutchinson 1993b);
- trivariate thin plate spline function of longitude, latitude, and elevation obtained from a 1⁄40°-DEM; and
- quintvariate *partial* thin plate spline incorporating a trivariate thin plate spline function of longitude, latitude and elevation derived from a 1⁄40°-DEM and a linear dependence on the two horizontal components of the unit normal vector on the same 1⁄40°-DEM.

The results are presented in Table 16-3. The bivariate thin plate spline and the trivariate partial spline were plainly inferior to the splines which incorporated a *spatially varying* dependence on elevation, both in terms of actual and estimated root mean square residuals. The trivariate thin plate spline function, which used elevations from the 1⁄40°-DEM, performed similarly to the trivariate spline function derived from actual station elevations (see Table 16-1). The quintvariate partial spline performed only marginally better than the trivariate thin plate spline. The fitted coefficients of the eastern and northern components of the unit normal vector were 113 and 27 mm/m, with estimated standard errors of 55 and 80 mm/m respectively (see Hutchinson 1993a). This implied that the rainfall means were perturbed positively from a relatively simple function of elevation for areas facing the neighboring east coast of Australia and perturbed negatively for areas facing away from the coast.

Table 16-3. Estimated and actual root mean square residuals from four spline models using varying degrees of dependence on a 1⁄40°-DEM.

Method	Estimated RMS residual of withheld data (mm)	RMS residual of withheld data (mm)
Bivariate spline	86.9	179.5
Trivariate partial spline	87.5	139.9
Trivariate spline	41.6	57.2
Quintvariate spline	40.2	56.4

CONCLUSION

Partial thin plate splines have been shown to be a flexible tool for interpolating mean rainfall for a standard period. They can allow for the correlated error structure inherent in rainfall means obtained from incomplete records and can provide accurate estimates of the error of the fitted surfaces which take account of the larger variability in higher rainfall regions. The method which first prestandardizes observed means to long-term estimates has the same data requirements as the method which uses the nondiagonal error covariance structure, but is otherwise simpler to use, since it avoids having to estimate and process a full, positive definite covariance matrix. Once the data have been standardized they can be processed by methods which permit only diagonal weighting.

The simpler noncorrelated error structure posed by Hutchinson and Bischof (1983) has been shown to produce only a modest reduction in accuracy. The practical advantage of this structure is that it depends only on simple summary data at each location and so further reduces computational requirements. It also gives an indication of the robustness of thin plate smoothing splines when error variances are not accurately specified.

In keeping with the results of Chua and Bras and Phillips et al. (1992), the incorporation of a constant linear dependence on elevation produced a more accurate result than having no elevation dependence. However, errors were much more significantly reduced by using a trivariate thin plate spline which incorporated a spatially varying dependence on elevation. Previous attempts to incorporate more complex dependencies on elevation using cokriging have not led to significant improvements over methods with constant linear dependence on elevation (Chua and Bras 1982; Phillips et al. 1992). The large estimated and actual validation residuals obtained with surfaces which did not incorporate a spatially varying dependence on elevation clearly indicate the unreliability of these methods, especially in areas which are remote from the data points.

Trivariate thin plate splines naturally lend themselves to application across large heterogeneous areas (Hutchinson 1991, 1993b). They also produce a spatially continuous interpolated rainfall field. This may be contrasted with the PRISM method (Daly et al. 1994) for which *post hoc* filtering and other adjustments need to be made to produce an acceptable interpolated field. Both the incorporation of aspect and the use of elevations obtained from a broad-scale DEM have led to only marginal improvements over the trivariate thin plate spline function of position and actual station elevation. The appropriate scale of such elevation and aspect effects clearly demand further investigation.

An important practical advantage of using actual station elevations when fitting thin plate splines is that access to a DEM is not required, neither when fitting the surface nor when calculating fitted values at particular points. A DEM is only required when producing images such as those in Plates 16-1 and 16-2 and in calculating mean areal rainfall. The fitted rainfall surface can be stored compactly as a set of surface coefficients describing the fitted function of longitude, latitude, and elevation. Files of these coefficients can be used to construct a readily accessible climate surface database. Such databases have already been constructed for whole continents, including Australia and Africa, and used for a variety of environmental analyses.

References

Bates, D., M. Lindstrom, G. Wahba, B. Yandell. 1987. GCVPACK—routines for generalized cross validation. *Commun. Statist. B-Simul. Comput.* 16:263–97.

Daly, C., R. P. Neilson, and D. L. Phillips. 1994. A statistical-topographic model for mapping climatological precipitation over mountainous terrain. *J. Appl. Meteorol.* 33(2):140–58.

Edwards, K. A. 1972. Estimating areal rainfall by fitting surfaces to irregularly spaced data. In *Proceedings of the International Symposium on the Distribution of Precipitation in Mountainous Areas 2,* 565–87. World Meteorological Organization.

Hevesi, J. A., J. D. Istok, and A. L. Flint. 1992. Precipitation estimation in mountainous terrain using multivariate geostatistics, part I: Structural analysis. *J. Appl. Meteorol.* 31:661–76.

Hughes, D. A. 1982. The relationship between mean annual rainfall and physiographic variables applied to the coastal region of southern Africa. *South African Geographic Journal* 64:41–50.

Hutchinson, M. F. 1984. Some surface fitting and contouring programs for noisy data. CSIRO Division Mathematics and Statistics Consulting Report ACT 84:6.

———. 1991. Continent-wide data assimilation using thin plate smoothing splines. In *Data Assimilation Systems*, edited by J. D. Jasper, 104–13. BMRC Report No. 27. Melbourne: Bureau of Meteorology.

———. 1993a. Thin plate splines and kriging. *Interface '93*.

———. 1993b. Development of a continent-wide DEM with applications to terrain and climate analysis. In *Environmental Modeling with GIS*, edited by M. F. Goodchild, B. O. Parks, and L. T. Steyaert, 392–99. New York: Oxford University Press.

Hutchinson, M. F., and R. J. Bischof. 1983. A new method for estimating mean seasonal and annual rainfall for the Hunter Valley, New South Wales. *Austral. Meteorol. Mag.* 31:179–84.

Hutchinson, M. F., and P. E. Gessler. 1994. Splines—more than just a smooth interpolator. *Geoderma* 62:45–67.

Hutchinson, M. F., H. A. Nix, and J. P. McMahon. 1992. Climate constraints on cropping systems. In *Ecosystems of the World: Field Crop Ecosystems*, edited by C. J. Pearson, 37–58. Amsterdam: Elsevier.

Mitasova, H., and L. Mitas. 1993. Interpolation by regularized spline with tension: Theory and implementation. *Math. Geol.* 25:641–55.

Phillips, D. L., J. Dolph, and D. Marks. 1992. A comparison of geostatistical procedures for spatial analysis of precipitation in mountainous terrain. *Agric. Forest Meteorol.* 58:119–41.

Wahba, G. 1990. Spline models for observational data. *CBMS-NSF Regional Conference Series in Applied Mathematics*. Philadelphia: Society for Industrial and Applied Mathematics.

M.F. Hutchinson
Center for Resource and Environmental Studies
Australian National University
G.P.O. Box 4
Canberra ACT 2601
Australia
E-mail: *hutch@cres.anu.edu.au*

17

Development of a New Oregon Precipitation Map Using the PRISM Model

Christopher Daly and George H. Taylor

Significant progress in our ability to distribute point monthly and annual precipitation data to a regular grid in complex terrain has recently been achieved through the development of PRISM (Precipitation Elevation Regressions on Independent Slopes Model). PRISM is well suited to regions with mountainous terrain because it incorporates a conceptual framework that addresses the spatial scale and pattern of orographic precipitation. In a model comparison, PRISM exhibited superior performance to various methods of kriging in the Willamette River basin, Oregon, and has been applied to the entire United States with excellent results. PRISM was used to develop a new official isohyetal analysis for Oregon—the first update since 1964. The PRISM-generated map equals or exceeds the accuracy and detail of the hand-drawn 1964 analysis but required only a small fraction of the time and resources.

INTRODUCTION

Estimates of the amount and spatial distribution of monthly and annual precipitation are critical inputs to a variety of ecological, agricultural, and hydrological models. These include models of vegetation, water supply, water quality, drought severity, fire risk, crop production, and others. The demand for precipitation fields on a regular grid and in digital form is growing dramatically as models become increasingly linked to geographic information systems (GIS) that spatially represent and manipulate model output.

There are many methods of interpolating precipitation from monitoring stations to grid points. Some provide estimates of acceptable accuracy in flat terrain, but none have been able to adequately explain the extreme, complex variations in precipitation that occur in mountainous regions. Inadequacies in these methods typically must be overcome by adding numerous estimated "pseudostations" to the data set and tediously modifying the resulting output by hand. Even then, there is no provision for easily updating the precipitation maps with new data or developing maps for other years or months.

Significant progress in this area has recently been achieved through the development of PRISM. PRISM is an analytical model that uses point data and a digital elevation model (DEM) to generate gridded estimates of monthly and annual precipitation. PRISM is well suited to regions with mountainous terrain, because it incorporates a conceptual framework that addresses the spatial scale and pattern of orographic precipitation.

OVERVIEW OF PRISM

The primary effect of orography on a given mountain slope face is to cause precipitation to increase with elevation. Orographic effects operate at a hierarchy of spatial scales. For example, a large mountain barrier can enhance precipitation over a broad area windward of the crest through forced uplift of moist airflow. Imbedded within this major effect are minor, smaller-scale precipitation perturbations caused, for example, by vertical air motion associated with river valleys carved into the barrier. Orographic effects caused by the large-scale (>10 km) terrain features typically explain most of the spatial variation in climatological precipitation in mountainous terrain (see Burns 1953; Henry 1919; Schermerhorn 1967; Spreen 1947). In addition, the density of most routine precipitation networks is sufficient to resolve only the larger-scale orographic effects at best. Therefore, it is not surprising that relationships between observed precipitation and elevation are generally strengthened when the elevation of each data point is given in terms of its height on a smoothed terrain (e.g., at 10-km resolution), which we might term its effective "orographic" elevation.

The relationship between precipitation and orographic elevation varies from one slope face to another, depending on location and orientation. Thus, a mountainous landscape can be thought of as a mosaic of smoothed topographic faces, or "facets," each experiencing a different orographic regime. Each topographic facet is a contiguous area over which the slope orientation is reasonably constant. Topographic facets are best delineated by using a DEM at a resolution that closely matches the smallest orographic scale supported by the data, thereby reducing the number of facets delineated at terrain scales too small to be resolved by the data.

In operation, PRISM (1) estimates the "orographic" elevation of each precipitation station using a smoothed DEM; (2) assigns each DEM grid cell to a topographic facet by assessing slope orientation, then estimates precipitation at each DEM cell by using a windowing technique to develop a precipitation/DEM-elevation regression function from nearby rainfall stations on the cell's topographic facet; and (3) predicts precipitation at the cell's DEM elevation with this regression function. Whenever possible, PRISM calculates a prediction interval for the estimate, which is an approximation of the uncertainty involved.

EVALUATION OF PRISM

PRISM has been compared to kriging, detrended kriging, and cokriging in the Willamette River basin, Oregon. In a jackknife cross-validation exercise, PRISM exhibited lower overall bias and mean absolute error (Phillips et al. 1992; Daly et al. 1994). PRISM was also applied to northern Oregon and to the entire western United States. Detrended kriging and cokriging could not be used in these regions because there was no overall relationship between elevation and precipitation. PRISM's cross-validation bias and absolute error in northern Oregon increased a small to moderate amount compared to those in the Willamette River basin; errors in the western United States showed little further increase. PRISM has recently been applied to the entire United States in three separate runs (western, central, and eastern) with excellent results, even in regions where orographic processes do not dominate precipitation patterns.

By using many precipitation/DEM-elevation relationships developed within local windows and on individual topographic facets rather than a single domain-wide relationship, PRISM continually adjusts its frame of reference to accommodate local and regional changes in orographic regime.

DEVELOPMENT OF A NEW OREGON PRECIPITATION MAP

Development of isohyetal analyses is a lengthy and difficult process, particularly in areas with significant terrain features. In the western United States, the problem of complex terrain is compounded by lack of precipitation data in many areas.

The latest isohyetal analysis for the state of Oregon was published in 1964. In recent years, several attempts have been made to procure funding for a new analysis, but the labor-intensive nature of typical isohyetal analyses has made such an effort prohibitively expensive.

Development of the updated analysis began with collection of precipitation data from Oregon locations and nearby stations within contiguous states (Washington, Idaho, Nevada, and California). This included two primary data sets: National Climatic Data Center (NCDC) cooperative stations and Soil Conservation Service (SCS) SNOTEL stations. The period 1961–1990 was used for the analysis. Mean monthly values for stations which had at least twenty-seven years of data were used without modification. Data sets for those with at least fifteen but less than twenty-seven years of data were modified by applying a least squares fit to data from a representative nearby station, yielding a thirty-year average.

PRISM was run for each monthly data set as well as for the annual averages. Output consisted of precipitation values for each grid cell on an 5-minute latitude/longitude DEM. The gridded output was then imported into a GIS for plotting and map creation. Data were made available to interested users in both hard-copy (map) and digital (numerical matrix) format. Data have been successfully imported into a variety of GIS programs and other spatial-distribution software, including ARC/INFO, GRASS, IDRISI, and Spyglass Transform.

Based on suggestions from prospective users of the product, it was decided to offer three different versions of the map. The smallest is an 11 × 17-inch color shaded polygon map of the state (1:2,000,000 scale). A black-and-white contour map, with 5-inch resolution, was also produced (scale 1:1,000,000). Finally, a large 36 × 50-inch color map, showing 5-inch contours, lakes and rivers, and urban areas was printed; scale was 1:600,000. Figure 17-1 is a reproduction of a small section of the 1:1,000,000 scale map.

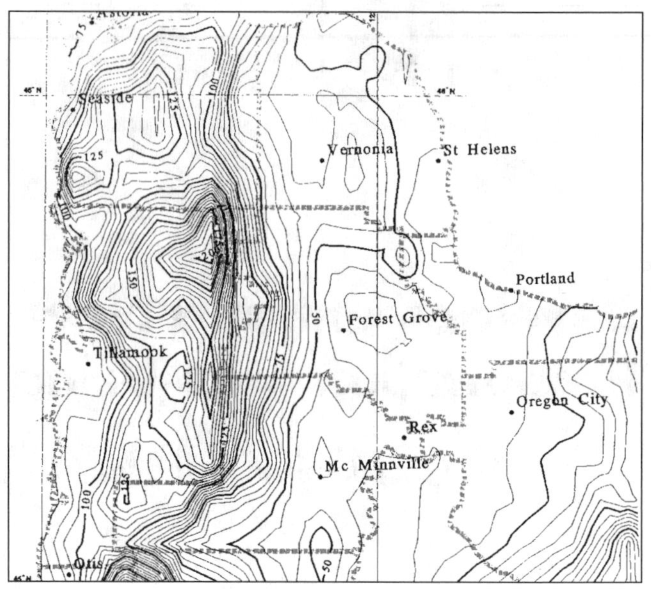

Figure 17-1. Reproduction of a section of a new isohyetal map for Oregon.

CONCLUSION

The new isohyetal analysis for Oregon is thought to be of quality and accuracy comparable to that of traditional labor-intensive isohyetal methods at a fraction of the cost. The PRISM analysis incorporates the latest station data with easy updating capabilities and uses a reasonably objective method that is reproducible and GIS compatible.

Currently PRISM is being studied for use in predicting distribution of other meteorological parameters such as temperature, snowfall, evapotranspiration, and degree-day totals. Other current research topics include the following:

1. Investigation of optimal grid resolution for precipitation modeling.
2. Application of the model to short-term extreme events, including updating of precipitation frequency–duration information.
3. Use in streamflow forecasts.
4. Use in conjunction with water balance and ecological models to provide predictions of the effects of future climate-change scenarios on native vegetation.
5. Comparison of distribution of precipitation during different climatic regimes (e.g., ENSO events) with those during normal years.
6. Research into rainfall and snowfall undercatch, especially for high-elevation stations.
7. Studying the optimal placement of sensors for temperature and precipitation measurement.

References

Burns, J. I. 1953. Small-scale topographic effects on precipitation distribution in San Dimas Experimental Forest. *Transactions of the American Geophysical Union* 34:761–68.

Daly, C., R. P. Neilson, and D. L. Phillips. 1994. A statistical-topographic model for mapping climatological precipitation over mountainous terrain. *Journal of Applied Meteorology* 33(2):140–58.

Henry, A. J. 1919. Increase of precipitation with altitude. *Monthly Weather Review* 47:33–41.

Phillips, D. L., J. Dolph, and D. Marks. 1992. A comparison of geostatistical procedures for spatial analysis of precipitation in mountainous terrain. *Agricultural and Forest Meteorology* 58:19–141.

Schermerhorn, V. P. 1967. Relations between topography and annual precipitation in western Oregon and Washington. *Water Resources Research* 3:707–11.

Spreen, W. C. 1947. A determination on the effect of topography upon precipitation. *Transactions of the American Geophysical Union* 28:285–90.

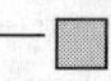

Christopher Daly
Oregon State University
U.S. EPA Environmental Research Laboratory
200 SW 35th St.
Corvallis, Oregon 97333
E-mail: *chris@heart.cor.epa.gov*

George H. Taylor
Oregon Climate Service
326 Strand Ag Hall
Oregon State University
Corvallis, Oregon 97333
E-mail: *oregon@ats.orst.edu*

18

Generating Daily Surfaces of Temperature and Precipitation over Complex Topography

Steven W. Running and Peter E. Thornton

We present the logical and algorithmic framework of a numerical model which generates daily interpolated surfaces of maximum air temperature, minimum air temperature, and precipitation over a gridded terrain, and we demonstrate its application in the state of Montana for a one-year-period. The model generates daily estimates of the relationship between each meteorological variable and elevation for each grid cell in the terrain model and uses these estimates to generate interpolated surfaces of temperature and precipitation based on daily observations from a network of recording stations. Interpolations are based on a vertically exaggerated proximal polygon algorithm, in combination with a Gaussian-weighted spatial convolution kernel. Elevation relationships are estimated with a least squares weighted linear regression model that varies in space and in time by subsetting both the spatial and temporal domains. Standard lapse-rate regressions are used to model the relationship between maximum and minimum temperature and elevation. Precipitation relationships are based on a normalized difference algorithm. Predictions of precipitation occurrence employ a local event frequency statistic. We present some of the principal results of simulations across the state of Montana on a 1-km resolution grid for 1990. Cross-validation was used to generate daily predictions of maximum and minimum temperature and precipitation. Predicted annual averages were compared to observed annual averages, resulting in mean absolute errors for daily prediction of maximum and minimum temperatures of 0.69° and 0.98°C/day, respectively. Mean absolute errors for predicted precipitation were 0.03 cm/day (11.83 cm/year, or 20.0% measured as a proportion of total annual precipitation).

INTRODUCTION

Meteorological variables are the principal drivers of plant ecosystem processes, and mechanistic numerical simulations of these processes depend on accurate inputs of meteorological information. Concurrent simulations of ecosystem processes such as evaporation, transpiration, photosynthesis, respiration, decomposition, and nutrient cycling require meteorological inputs such as air temperature, humidity, precipitation, and incident solar radiation. In this chapter we consider the generation of surfaces of meteorological inputs for terrestrial ecosystem simulations at regional (subcontinental) spatial scales and at daily temporal resolution, for one year or longer.

Spatial and temporal resolution of the modeling domain for these sorts of simulations are dictated in large part by the practicalities of data acquisition and computational loads. Our goal here is to provide meteorological inputs to a regional ecosystem process model which employs satellite-derived estimates of leaf area index as a principal ecosystem descriptor (Running and Coughlan 1988), and so the availability of useful satellite data products places a constraint on both the spatial and temporal resolution of the simulations. At regional spatial scales the NOAA/AVHRR satellite instrument, with a spatial resolution of about 1 km and daily global coverage, has been shown to be useful in the estimation of leaf area and the parameterization of terrestrial ecosystem models (Running and Nemani 1988; Running et al. 1989; Goward et al. 1991; Tucker et al. 1985). Other input requirements to this regional ecosystem model are elevation and land-cover class, and high-quality digital elevation and seasonal land-cover data sets are available at the same resolution as the AVHRR data. Based on these considerations we chose 1 km as a minimum spatial resolution for these simulations, and we retained the Lambert Azimuthal Equal Area projection of the satellite and topographic databases.

The temporal resolution of these applications is a compromise between a time step, which is short enough to provide accurate predictions of ecosystem processes, particularly hydrologic balances and canopy gas exchange, and long enough to allow multiple-year simulations of carbon and nitrogen dynamics over many grid cells at an acceptable computational cost (Running 1984, 1992). Also critical to the choice of time step is the native temporal resolution of raw sources of meteorological data. Archives of National Weather Service station and Soil Conservation Service SNOTEL station observations provide point information with a spatial coverage appropriate to these simulations. These archives record observations daily, which proves to be both an adequate and efficient time step for our plant ecosystem process models (Running 1984). We therefore have selected a daily time step as the temporal resolution for these simulations.

While extremely useful, the currently available archives do not meet all the meteorological input needs of regional ecosystem modeling: daily maximum temperature, minimum temperature, and precipitation are required and are commonly recorded, but daily incident solar radiation and humidity are also critical inputs to these models, and archived observations of these variables are rare. In addition, none of the currently available databases provide terrain-sensitive gridded data sets required by the applications considered here.

Over the past ten years we have developed a climate simulator that is sensitive to the influences of terrain and that meets the need for accurate, high-spatial-resolution meteorological data on a daily time step (MT-CLIM) (Running et al. 1987). The original MT-CLIM model was a point-to-point extrapolator. Input consisted of daily observations of maximum and minimum temperatures and precipitation from one or, at most, two points of observation, as well as a table of physical characteristics describing the observation point(s) and the point to which these observations were to be extrapolated. Other inputs included a parameter list defining lapse rates for temperature and dew point, and base and site isohyets for extrapolation of precipitation amounts. Output consisted of daily predictions of temperatures, precipitation, humidity, vapor pressure deficit, and incident solar radiation for the extrapolated point. The logic employed by MT-CLIM is applicable to areas small enough that one or two base stations can adequately describe the spatial and temporal variation in meteorological variables, and studies have demonstrated its usefulness in simulating meteorological conditions for ecosystem process models on the single watershed scale (Running et al. 1989; Band et al. 1991; White and Running 1994).

We have now developed an extension of the MT-CLIM logic, MT-CLIM-3D, which generates daily surfaces of meteorological variables over large regions, incorporating topographic databases and observations from an unlimited number of arbitrarily spaced weather stations. Daily predictions of maximum and minimum temperature and precipitation for each cell over a gridded surface are generated by an interpolation algorithm, after which the standard MT-CLIM algorithms are employed to generate surfaces of humidity and incoming shortwave radiation. The interpolation logic also provides a spatially and temporally explicit diagnosis of the temperature and precipitation relationships with elevation, eliminating the need for the user to supply those parameters.

The methods described here are unique for their focus on meteorological as opposed to climatological interpolation, as well as for their unified approach to the interpolation of multiple meteorological fields. Recent work by Daly et al. (1994) has shown that similar methods can be used to interpolate fields of monthly or annual climatological precipitation over large and climatically diverse regions, and that those methods provide better estimates of annual climatological precipitation than do common geostatistical methods based on kriging. No recent research has focused on the application of spatial interpolation methods to the generation of daily surfaces of meteorological variables in complex terrain.

The MT-CLIM-3D interpolation logic is presented here, along with examples of output from a one-year simulation over the state of Montana. We focus on the generation of interpolated fields of daily maximum and minimum temperature and daily precipitation, although in practice this system is used to generate surfaces of humidity and incoming shortwave radiation as well. We use cross-validation to generate estimates of the errors associated with the interpolated surfaces of temperature and precipitation. Other recent work has demonstrated the validity of the radiation and humidity routines at a regional scale (Glassy and Running 1994), and we retain the functionality of those routines unchanged.

MODEL DESCRIPTION

Input Requirements

Our model requires, as input, a topographic database of elevation, which also defines the spatial domain of the simulation. The simulations described here use a gridded digital elevation data set for the state of Montana (an area of 377,884 km^2) at a horizontal resolution of 1 km and a vertical resolution of approximately 6 m obtained on CD-ROM from the EROS Data Center. The interpolation logic requires a meteorological database with daily observations of maximum temperature (T_{max}), minimum temperature (T_{min}), and precipitation (*Prcp*) at each station in an arbitrarily spaced network. The stations need not lie within the bounds of the simulations' spatial domain. It is common for records of observations at a station to have occasional missing values, and these gaps must be filled prior to use by linear interpolation or a similar procedure. Stations with excessive missing data are dropped from the analysis. For the simulations described here, any station with more than three consecutive missing values in any field or with more than twenty-five missing values for a given field for more than one year was dropped from the station database. Starting with all stations in Montana that are archived by the National Climatic Data Center (NCDC) as well as all the Soil Conservation Service SNOTEL stations in the state, the above criteria resulted in the inclusion of 167 NCDC and forty-eight SNOTEL stations, for a total of 215 stations (Plate 18-1). Elevation, latitude, and longitude are also required for each station. Although the model is presented here in its application to a uniformly gridded spatial domain, this logic is applicable to arbitrarily partitioned surfaces with only minor modifications.

Interpolation Logic

The interpolation logic employed by MT-CLIM-3D is based on the generation, for each grid cell in the model domain, of a weighted list of stations that contributes to the prediction of meteorological fields at that cell. The weights associated with the stations in the list are determined by a spatial filtering process that gives greater weights to stations that are near the point of prediction, and that also modifies the weights based on the topographically weighted density of stations near the point of prediction. The spatial filtering consists of two steps: (1) the spatial domain is partitioned into regions (proximal polygons) that associate each unit of area within the domain (each cell in the grid) with a single station based on a nearness algorithm; and (2) a truncated Gaussian filter kernel is convolved with the proximal polygon surface, generating both a list of stations that are to be included in the interpolation for a point and a list of weights associated with those stations. These two steps are detailed below.

Proximal polygon formation: The first step in the interpolation process is the generation of a set of proximal polygons (variously known as Voronoi or Dirichlet polygons) which partition the spatial domain such that: given a set S of n points (the stations, in this case), each member s_i of S is associated with a polygon which is the locus of points (x,y,z) in the spatial domain that are nearer to s_i than to any other member of S (after Preparata and Shamos 1985). The simplest criteri for nearness on a topographic surface is:

$$D_{pi} = \sqrt{(x_p - x_i)^2 + (y_p - y_i)^2 + (z_p - z_i)^2} \tag{18-1}$$

where D_{pi} is the distance from an arbitrary point (x_p,y_p,z_p) to a point (x_i,y_i,z_i) which is a member s_i of the set S.

Neglecting the contribution from the z-component of D_{pi}, these criteria are the foundation of the often-reviewed Thiessen method for the spatial distribution of precipitation (Thiessen 1911; Tabios and Salas 1985). The average density of stations in the Montana database is about one station per 1,500 km^2, resulting in contributions to D_{pi} that are much greater for the horizontal dimensions than for the vertical dimension, and proximal polygon surfaces generated using (18-1) are indistinguishable from Thiessen polygon surfaces for the same area. We hypothesize that distances in the vertical are climatologically more influential than distances in the horizontal, and that therefore some degree of exaggeration of vertical distances for the purposes of the calculation of D_{pi} will reduce interpolation error. Reformulating (18-1) with a new parameter, VEXAG, a unitless vertical distance multiplier, gives:

$$D_{pi} = \sqrt{(x_p - x_i)^2 + (y_p - y_i)^2 + VEXAG(z_p - z_i)^2} \quad (18\text{-}2)$$

where symbols are as in (18-1).

Cross-validation errors were used as a measure of the influence of VEXAG on the interpolation logic. (Cross-validation is a jackknifing technique which, in general, consists of the imposed ignorance of a portion of the available data, the generation of predictions for that portion, and the comparison of those predictions with the actual suppressed data.) Here we are testing a very simple hypothesis: that the inclusion of VEXAG leads to an increased similarity between a station and the points in its proximal polygon, and so the prediction algorithm employed is equally simple. One station at a time was removed from the database, and predictions of annual average values at that point were made based on the observed annual averages at the nearest neighbor station (where closeness was defined as in (18-2) over a range of VEXAG. In order to eliminate the gross effects of elevation differences between the predictor station and the suppressed station we employed corrections for temperature and precipitation relationships with elevation that were based on the annual averages for all stations. An optimal value for VEXAG was determined as the value which minimized cross-validation error.

Spatial filtering—kernel convolution: As the second step in the interpolation process, we defined for each point in the spatial domain, a weighted list of stations which contribute to predictions at that point. Our approach to this problem employs a spatial filtering kernel of circular extent and fixed diameter that is convolved with the proximal polygon surface to determine which stations are to be included in the predictions for a given cell and to assign weights to each included station. The kernel is defined on a regular grid of the same resolution as the model spatial domain, and the value, V_{ij}, associated with each cell within the kernel is defined as:

$$V_{ij} = 0 \qquad for: \; D_{ij} > R_k$$
$$V_{ij} = e^{\frac{-D_{ij}^2 \alpha}{R_k^2}} - e^{\alpha} \qquad for: \; D_{ij} \le R_k \quad (18\text{-}3)$$

where R_k is the radius of the circular kernel, D_{ij} is the distance from the center of cell (ij) to the center of the kernel grid, and α is a shape parameter.

These equations define a truncated Gaussian-weighting function within the circular kernel, with the greatest weight located at the center of the kernel and the weight decreasing radially outward until, at a distance R_k from the center of the kernel grid, the weight is 0. Convolution of this kernel with the proximal polygon surface generates a weighted list of stations for each cell in the spatial domain, as follows: With the kernel centered on a domain cell in the proximal polygon surface, the station list for that cell consists of all stations which have any part of their proximal polygons within the non-0-weighted region of the kernel. It is therefore not required that the station itself fall within the bounds of the kernel; only some portion of its proximal polygon must do so for the station to be included in the list. The weight assigned to each station is defined as the sum of the kernel weightings for each grid cell occupied by that station's proximal polygon within the kernel bounds. Since it is common at the edges of the domain to have a part of the kernel extending outside the domain, weights derived in this way are summed over all included stations at a point and normalized to that sum, forcing the normalized sum of weights for all stations at a given point to 1.0.

This method simultaneously favors stations that are near the point of prediction and distributes weights in proportion to the local density of stations. This kernel convolution method requires two parameters: (1) the kernel radius, R_k, and (2) the shape parameter for the Gaussian function, α. The magnitude of R_k determines, on average, how many stations will be included in the weighted list; for a given R_k that number will vary spatially, increasing in areas of high station density and decreasing in areas of low density. The shape parameter determines the rate of change of weighting at the kernel edges: larger values generate more smoothing at the edges of the kernel. The weighted list of included stations is generated one time for each grid cell; the same list is used for predictions of all the principal variables at each time step.

Estimation of Elevation Relationships

For each grid cell, maximum and minimum temperature lapse rates are independently estimated for each day in the temporal domain. These estimates are based on a weighted least squares regression of the observed T_{max} and T_{min} data for the stations in a grid cell's weighted list. There is one point in these regressions for each unique pairing of stations in the cells list, and the number of points in each regression is given as:

$$N = \frac{n_{list}^2 - n_{list}}{2} \quad (18\text{-}4)$$

where N is the number of points in the regression, and n_{list} is the number of stations in the weighted list at the point of prediction.

The weight axssociated with each regression point is defined as the product of the two weights associated with the stations in a pair. The independent variable is defined as the difference in elevation between the stations in a pair, and the dependent variable is defined as the difference in the daily value of either T_{max} or T_{min} between the stations in a pair, giving a regression equation of the form:

$$Temp_1 - Temp_2 = \beta_0 + \beta_1(Elev_1 - Elev_2) \quad (18\text{-}5)$$

where $Temp_1$ and $Temp_2$ are the observed values of either T_{max} or T_{min} at two points making up a pair, β_0 and β_1 are the y-intercept and slope, respectively, of the regression line, and $Elev_1$ and $Elev_2$ are the elevations of the two points.

The case for estimation of the relationship of precipitation to elevation is complicated by the sparsity of the observation network and the stochastic nature of precipitation events. Unlike the case for temperatures, where there are always values with which to build a lapse regression, there are frequently days with very few observed precipitation events, making the estimation of the lapse rates associated with those events difficult. We recognize that the spatial scale of variation in the frequency and intensity of precipitation events is much smaller than the average spatial scale of sampling provided by the current observation network, and so our estimations of daily precipitation lapse relationships are designed more to infer the locally averaged behavior than to accurately predict the details of each event. In order to arrive at a more continuous description of precipitation events for the purposes of estimating elevation relationship statistics, a linearly ramped window smoothing algorithm is applied to the observed precipitation data. The width, in days, of this smoothing window is denoted W_{ps}. The smoothed data is used only in the generation of the elevation relationship; predictions of precipitation amounts are based on the unsmoothed data.

We explored both a semilogarithmic and a linear form for the estimation of relationships between precipitation and elevation. For the case of simple predictions based on the regression of annual average precipitation data for all the stations in the state against elevation, the semilogarithmic transformation leads to lower cross-validation prediction errors than the untransformed data. However, due to the spatial and temporal subsampling of this relationship through the process of spatial convolution and the estimation of precipitation-elevation relationships on a daily (albeit smoothed) basis, the linear formulation results in lower cross-validation prediction errors when applied in the context of our interpolation logic. We chose the linear model, resulting in a regression of the form:

$$\frac{Prcp_{S1} - Prcp_{S2}}{\left(\frac{Prcp_{S1} + Prcp_{S2}}{2}\right)} = \beta_0 + \beta_1(Elev_1 - Elev_2) \qquad (18\text{-}6)$$

where $Prcp_{S1}$ and $Prcp_{S2}$ are the smoothed values of daily precipitation at two points making up a pair, β_0 and β_1 are the y-intercept and slope, respectively, of the regression line, and $Elev_1$ and $Elev_2$ are the elevations of the two points.

The left-hand side of (18-6) expresses a normalization of the smoothed daily precipitation difference by the average of the two smoothed precipitation amounts for the pair for that day. Since the smoothed precipitation data is not quantitatively comparable to the unsmoothed data, which is used in the prediction algorithm described below, we require an algorithm which is independent of the magnitude of the dependent variable. By normalizing the independent variable for each point in the regression to the average daily smoothed value for the two stations in a pair, β_0 and β_1 lose their dependence on the absolute precipitation amount. A disadvantage of this formulation is that corrections must be made for negative predictions of precipitation, whereas a semilog transformation always generates positive precipitation predictions.

Prediction of Temperatures and Precipitation

The prediction equation for both T_{max} and T_{min} at a particular location-day takes the form:

$$T_p = \sum_{i=1}^{n_{list}} W_i \left(T_i + \beta_0 + \beta_1(Elev_p - Elev_i)\right) \qquad (18\text{-}7)$$

where T_p is the predicted temperature, i is the index for stations in the station list, n_{list} is the number of stations in the list, W_i is the weight associated with station i, T_i is the observed temperature for station i, β_0 and β_1 are the regression parameters defined above, $Elev_p$ is the elevation of the point of prediction, and $Elev_i$ is the elevation of station i.

The prediction of daily surfaces of precipitation presents a unique challenge. The sampling density afforded by the station network is much too coarse to resolve the detailed spatial patterns of precipitation occurrence. This lack of spatial detail is somewhat offset by the use of a daily time step, since over the course of one day a winter frontal precipitation event will be recorded over a wide range even if the precipitating front is relatively narrow, and warm season convective precipitation events, although of generally shorter duration, will likewise show more spatially coherent patterns when integrated over a day than they would at any instant. In general, a simple prediction of the form of (18-7) is inappropriate for the diagnosis of precipitation, since the weight contributed by stations that receive no precipitation for a day combined with the weight of stations that recorded precipitation tends to result in the overprediction of event frequency and the underprediction of event intensity; we call this the "constant drizzle" problem.

Based on the assumption that over one day there is some level of spatial coherence in the precipitation event frequency, we propose a diagnostic, the precipitation trigger statistic (*PT*), which is used to make a binomial prediction of precipitation occurrence at a point. We define this statistic as the sum of the weights of the stations in a cell's weighted list that record any precipitation for the day. If that sum exceeds a prescribed value (PT_{crit}), a precipitation event is triggered and we continue by generating a prediction of the event intensity, otherwise we predict a dry day. If all stations in a cell's list record precipitation for a day then $PT = 1.0$, if no stations record precipitation then $PT = 0.0$. Values of PT_{crit} close to 1.0 result in relatively few predicted events, whereas values close to 0.0 result in many predicted events. At present we parameterize PT_{crit} once for the entire spatial and temporal domain, based on a cross-validation analysis of predicted versus observed event frequencies, but we are exploring potential spatial and temporal variation in this parameter.

If $PT > PT_{crit}$, a precipitation event intensity prediction is made. While the incorporation of *PT* helps to solve the "constant" side of the constant drizzle problem, another correction must be made for the "drizzle." By incorporating in the daily predictions only those stations which recorded precipitation, we generate event intensities which are more realistic than if all stations are included. The weights for the new list of included stations (the "wet list") are renormalized to sum to 1.0, and the final event intensity prediction takes the form:

$$P_p = \sum_{i=1}^{n_{wetlist}} W_i \, P_i \left(\frac{2 + \beta_0 + \beta_1(Elev_p - Elev_i)}{2 - \beta_0 - \beta_1(Elev_p - Elev_i)}\right) \qquad (18\text{-}8)$$

where P_p is the predicted precipitation amount, i is the variable indexing the stations in the station list that recorded precipitation for the day, $n_{wetlist}$ is the number of such stations, W_i is the renormalized weight associated with station i, P_i is the observed unsmoothed precipitation at station i, β_0 and β_1 are the *y*-intercept and slope of the regression line for the regression defined in (18-5).

The fraction on the right side of (18-8) expresses the conversion from the normalized regression parameters to predictions of actual precipitation amounts. We have found this formulation to be quite stable, with the exception of predictions at the wettest of the high elevation grid cells. In those cases the denominator on the right side of (18-8) can become negative if very low and dry sites are included in the predictions. We find it necessary to correct for this problem by setting an upper limit on the product of $\beta_1\,(Elev_1 – Elev_2)$. This correction was necessary to avoid gross prediction errors for the highest elevation grid cells in the northwestern mountains of the study area (about 0.003% of the study area).

RESULTS

Parameters

Tests of cross-validation error for a simple prediction algorithm over a range of VEXAG resulted in the selection of VEXAG = 50 (unitless multiplier) for the Montana 1990 test case. Based on experience and the station density in this case, we set $R_k = 201$ km and $\alpha = 4.0$. This results in an average of twenty-four stations included in the predictions at each point in the domain, with the actual number of stations included at a point ranging from six to fifty-one. The precipitation smoothing parameter W_{ps} was set to thirty-one days. The precipitation trigger parameter PT_{crit} was set to 0.45, based on tests of observed versus predicted event frequency over a range of the parameter from 0.2–0.8. The maximum value of $\beta_1(Elev_1–Elev_2)$ was denoted $MAX\beta_1$ and was set at 0.7, based on experience and bounded by the value at which the denominator of (18-8) goes to 0.

Elevation Relationships

Averaging the daily diagnosed β_1 parameters from (18-5) and (18-6) for the three primary variables over each grid cell in the domain, we arrived at an annual time series of the spatial average of these parameters (Plate 18-2). Averaging the daily diagnosed β_1 parameters for T_{max}, T_{min}, and $Prcp$ over the year at each point in the domain, we arrived at a spatial distribution of the temporal average of those parameters (Plate 18-3). It is clear from Plates 18-2 and 18-3 that there are significant variations in the slope parameters for all three primary variables in both the temporal and spatial dimensions, and that more rigorous analysis than is presented here has demonstrated that cross-validation errors of annual average values based on daily predictions are decreased significantly by the successive incorporation of this temporal and spatial variation into the prediction algorithms. The algorithms presented here reflect the full incorporation of the spatial and temporal variation.

Predicted Meteorological Fields

Table 18-1 lists three different measures of prediction error for predictions of each of the primary variables. Daily predictions at each station, generated under the cross-validation protocol, were averaged for the year, and the predicted annual averages were compared with the actual annual averages. Standard errors of estimation (SEE) for the regression of predicted against observed averages for each variable over the range of stations are given in the first column of Table 18-1. The errors for precipitation prediction are listed both as annual average errors (cm/day) and as annual total errors (cm/year). Mean absolute errors (MAE), defined as the average of the absolute differences between observed and predicted annual averages for all stations, are given in the second column. Following the example of Daly et al. (1994), the MAE is also reported for precipitation as a percentage of the annual totals. Bias is also listed as a measure of tendencies in the algorithms toward overprediction or underprediction, and, again, precipitation is shown as daily and annual errors as well as a percentage of annual totals.

The daily surfaces for the entire year were combined to generate surfaces of annual average maximum and minimum temperature, and annual total precipitation for 1990 (Figure 18-1).

Table 18-1. Cross-validation error estimates for observed versus predicted variables, averaged over all stations.

	SEE	MAE	Bias
T_{max} (°C/day)	0.90	0.69	+0.012
T_{min} (°C/day)	1.11.	0.98	-0.052
Prcp (cm/day)	0.0470	0.032	-0.005
Prcp (cm/year)	17.15	11.83	-1.97
Prcp (%)	NA	20.0%	-3.7%

SEE = standard error of estimation
MAE = mean absolute error

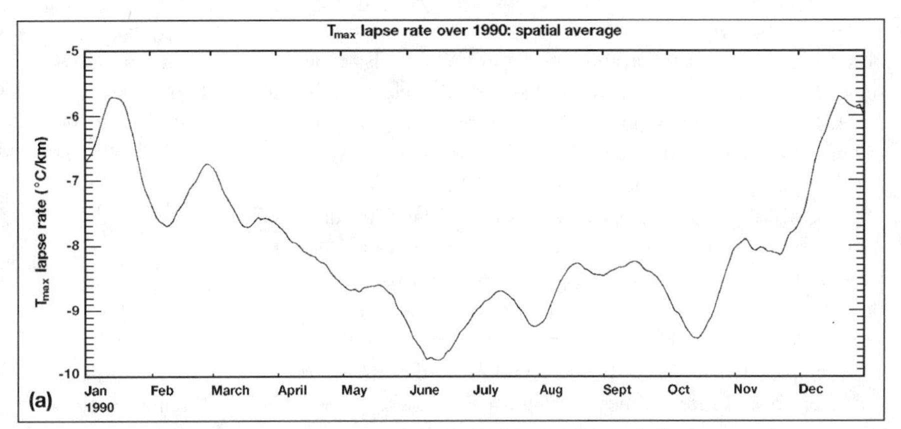

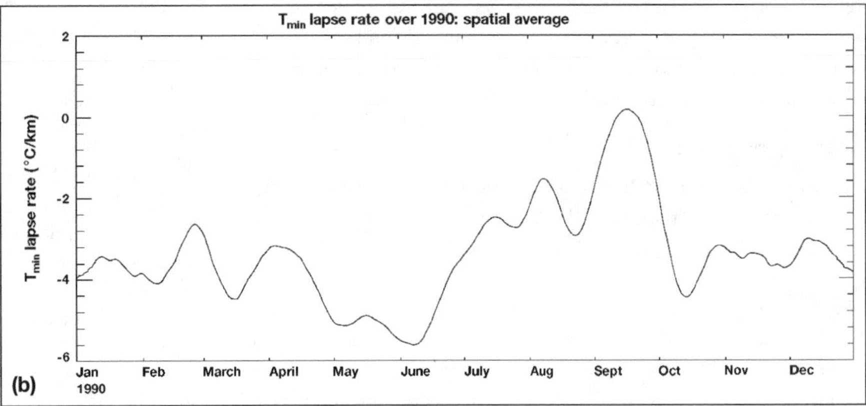

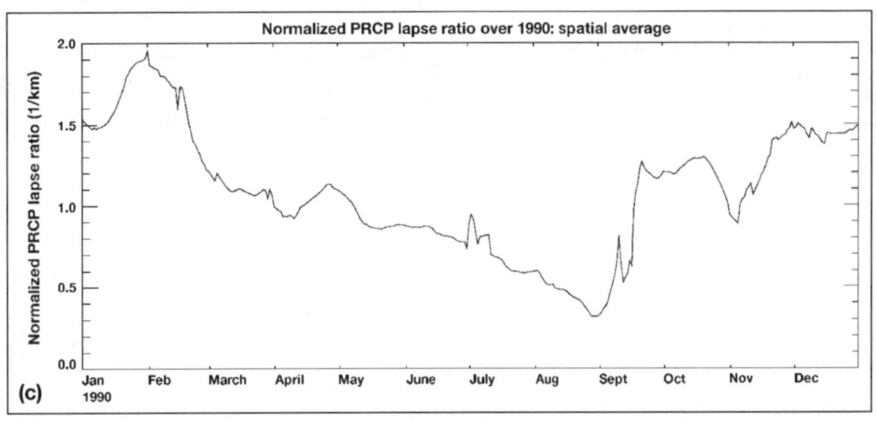

Figure 18-1. (a) Time-series of β_1 parameters for T_{max}, T_{min}, and $Prcp$ regressions, averaged over the entire spatial domain for each day. (b) Spatial distribution of β_1 parameters for T_{max}, T_{min}, and $Prcp$ regressions, averaged over the entire year at each grid cell. (c) Spatial distribution of 1990 annual average T_{max}, T_{min}, and annual total $Prcp$.

CONCLUSION

The methods presented here provide us with our first opportunity to perform ecosystem process simulations at a regional scale with meteorological input surfaces derived from daily observations. The algorithms employed allow these surfaces to be generated rapidly (runs illustrated in Figure 18-1a–c took eight hours running on an IBM RS6000 Model 350 workstation). The list of parameters required by these algorithms is rather short, making the application of these methods to a new model domain relatively painless. We are currently exploring the effects of varying spatial resolution of the topographic input layer on the model behavior: Daly et al. (1994) report that for annual and monthly long-term average precipitation, a smoothed digital elevation map provides better estimates of the effective climatological elevation of an observation point than the actual elevation at the point, and the same may be true for daily observations of precipitation and temperature. We are also exploring the potential for incorporating thermal infrared data collected by the NOAA/AVHRR instrument into these algorithms as a way to restrain the regressions of observed variables against elevation.

The application of the methodology described here is not limited to the specific case of plant ecosystem process modeling. All studies of regional-scale ecological questions and resource management issues are intimately linked to the meteorological variation across and through their spatial and temporal domains. MT-CLIM-3D can serve as a general tool for cases in which daily surfaces of meteorological variables over complex terrain are useful, for example in the assessment of ecosystem disturbance regimes or land-management scenarios. The logical framework presented here can be configured to accommodate a wide range of input densities and spatial resolutions, and can provide both interpolated surfaces of meteorological variables as well as surfaces of the prediction statistics associated with those variables.

Acknowledgments

This study was funded by the U.S. Forest Service, contract #53-0343-3-00027; the Cooperative Park Service, project #1268-0-9001; and NASA, contract #NAS5-31368.

References

Band, L. E., D. L. Peterson, S. W. Running, J. C. Coughlan, R. Lammers, J. Dungan, and R. R. Nemani. 1991. Forest ecosystem processes at the watershed scale: Basis for distributed simulation. *Ecol. Modeling* 56:171–96.

Daly, C., R. P. Nielson, and D. L. Phillips. 1994. A statistical-topographic model for mapping climatological precipitation over mountainous terrain. *J. App. Meteorology* 33:140–58.

Glassy, J. M., and S. W. Running. 1994. Validating diurnal climatology logic of the MT-CLIM model across a climatic gradient in Oregon. *Ecol. App.* 4(2):248-57.

Goward, S. N., B. Markham, D. G. Dye, D. L. Dulaney, and J. Yang. 1991. Normalized difference vegetation index measured from the Advanced Very High Resolution Radiometer. *Rem. Sens. Env.* 35:257–77.

Preparata, F. P., and M. I. Shamos. 1985. *Computational Geometry.* Texts and Monographs in Computer Science, edited by D. Gries. New York: Springer-Verlag.

Running, S. W. 1984. Microclimate control of forest productivity: Analysis by computer simulation of annual photosynthesis/transpiration balance in different environments. *Ag. and For. Meteorology* 32:267–88.

———. 1992. A bottom-up evolution of terrestrial ecosystem modeling theory and ideas toward global vegetation modeling. In *Modeling the Earth System,* edited by D. Ojima. Boulder: UCAR/Office for Interdisciplinary Earth Studies.

Running, S. W., and J. C. Coughlan. 1988. A general model of forest ecosystem processes for regional applications: Hydrologic balance, canopy gas exchange, and primary production processes. *Ecol. Modeling* 42:125–54.

Running, S. W., and R. R. Nemani. 1988. Relating seasonal patterns of the AVHRR vegetation index to simulated photosynthesis and transpiration of forests in different climates. *Rem. Sens. Env.* 24:347–67.

Running, S. W., R. R. Nemani, and R. D. Hungerford. 1987. Extrapolation of synoptic meteorological data in mountainous terrain and its use for simulating forest evapotranspiration and photosynthesis. *Can. J. For. Res.* 17:472–83.

Running, S. W., R. R. Nemani, D. L. Peterson, L. E. Band, D. F. Potts, L. L. Pierce, and M. A. Spanner. 1989. Mapping regional forest evapotranspiration and photosynthesis by coupling satellite data with ecosystem simulation. *Ecology* 70(4):1,090–101.

Tabios, III, G. Q., and J. D. Salas. 1985. A comparative analysis of techniques for spatial interpolation of precipitation. *Wat. Res. Bul.* 21:365–80.

Thiessen, A. H. 1911. Precipitation averages for large areas. *Mon. Wea. Rev.* 39:1,082–84.

Tucker, C. J., J. R. G. Townshend, and T. E. Goff. 1985. African land-cover classification using satellite data. *Science* 277:369–75.

White, J. D., and S. W. Running. 1994. Testing scale dependent assumptions in regional ecosystem simulations. *J. Veg. Sci.* 5(5):687–702.

Steven W. Running and *Peter E. Thornton*
Numerical Terradynamics Simulation Group
School of Forestry
University of Montana
Missoula, Montana 59810

19

Use of Mesoscale Models for Simulation of Seasonal Weather and Climate Change for the Rocky Mountain States

R. A. Pielke, J. Baron, T. Chase, J. Copeland, T. G. F. Kittel, T. J. Lee, R. Walko, and X. Zeng

This chapter discusses the procedure of using mesoscale/regional models to downscale GCM output to local and regional seasonal weather and climate patterns. While the accuracy of this procedure is limited by the skill of the GCM simulations, when used together with detailed characterizations of terrain and landscape patterns, it offers an effective framework to study climate-change scenarios. The need for coupled atmospheric/terrestrial ecosystem models is also discussed in the chapter.

INTRODUCTION

Current general circulation models (GCMs) have inadequate spatial resolution to represent such geographic features as the mountain ranges in the Rocky Mountains (Figure 19-1). Because at least four grid intervals are required to reasonably resolve terrain and atmospheric features, as shown by Pielke (1984) and Derickson (1992), the effective resolution of GCMs is on the order of about half the width of the contiguous United States. However, it is well known that smaller-scale features can significantly influence local climate and weather (Atkinson 1981; Pielke 1984; Pielke and Segal 1986; Avissar and Verstraete 1990; Cotton and Pielke 1992; Doran et al. 1992; Segal and Arritt 1992). The more detailed representations of terrain (Figure 19-2a) and land cover (Figure 19-2b) illustrate the substantial finer-scale variability in surface forcing, which must be represented to properly represent climate at regional and larger scales.

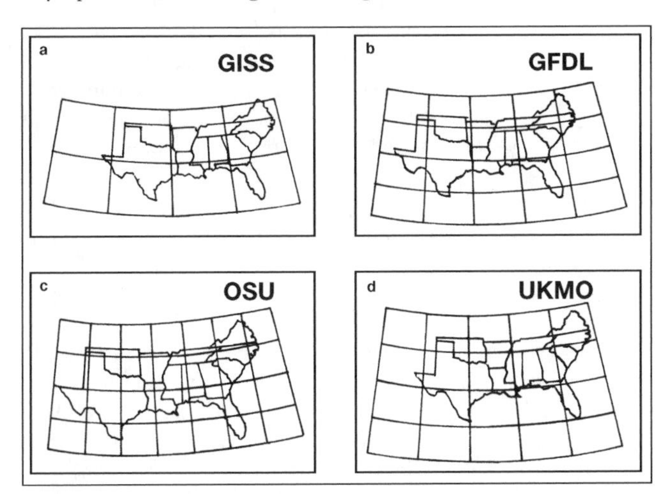

Figure 19-1. Horizontal resolution of four general circulation models (GCMs) across the southern United States (Cooter et al. 1993).

In order to provide a more realistic representation of geographic features, mesoscale models have recently been introduced as tools to downscale larger-scale model output (Avissar and Verstraete 1990).

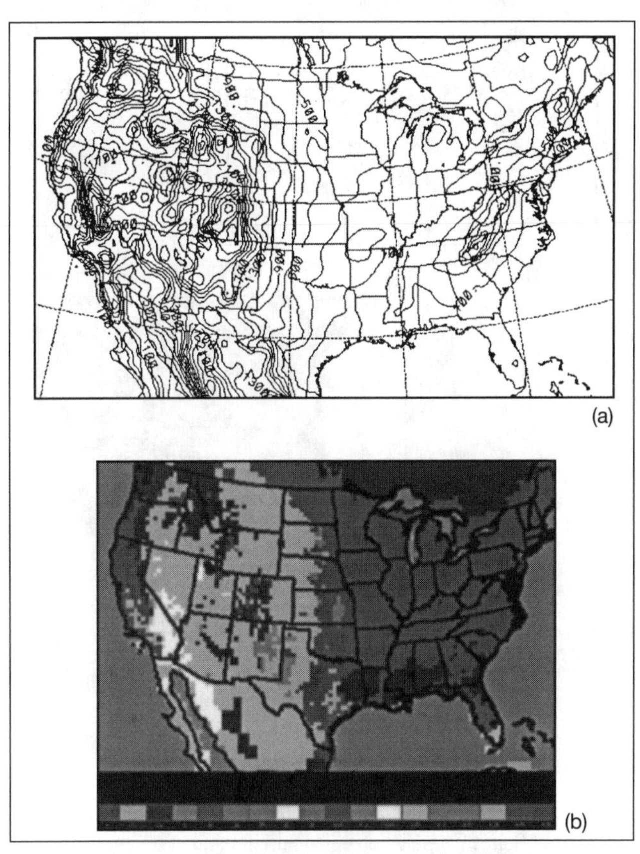

Figure 19-2. (a) Terrain of United States at 40 km Δx and Δy (200-m contour interval), (b) land cover (using the BATS classification; Dickinson et al. 1986) of United States at 40 km Δx and Δy (based on EROS Data Center Land Cover Classification database; Loveland et al. 1991). BATS classes are: (1) crop/mixed farming, (2) short grass (3) evergreen needleleaf tree, (4) deciduous needleleaf tree, (5) deciduous broadleaf tree, (6) evergreen broadleaf tree, (7) tall grass, (8) desert, (9) tundra, (10) irrigated crop, (11) semidesert, (12) ice cap/glacier, (13) bog or marsh, (14) inland water, (15) ocean, (16) evergreen shrub, (17) deciduous shrub, and (18) mixed woodland.

Lateral boundaries are supplied by a GCM simulation, with the higher-resolution mesoscale model able to create smaller-scale features through nonlinear interactions within the atmosphere, and by interactions with terrain features. The value of the mesoscale simulations is limited to the accuracy of the lateral boundary conditions and their relative impact on the solution in the interior of the mesoscale grid domain. However, these tools, when utilized as part of a sensitivity analysis, provide valuable estimates of local and regional climate that potentially could occur as a result of large-scale climate change, and/or due to man-caused or natural changes in the landscape.

EXAMPLES OF MODEL SIMULATIONS

The most extensive examples of long-term mesoscale simulations using input from a larger-scale model include those reported in Giorgi et al. (1993). It was found, for instance, that winter precipitation patterns could be reasonably well predicted, although summertime cumulonimbus rainfall is less accurate. Figure 19-3 from Giorgi et al. 1993 illustrates the type of output and its spatial structure for the western United States where a horizontal grid interval of 60 km was used.

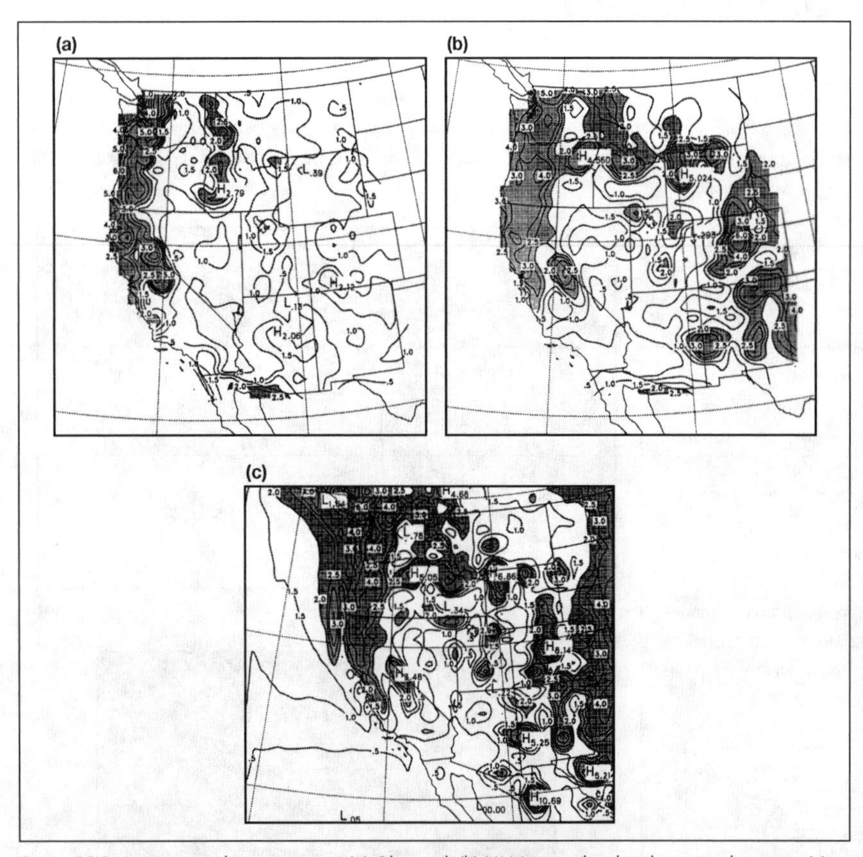

Figure 19-3. Average yearly precipitation: (a) Observed, (b) MM4 interpolated at the station locations, (c) MM4 gridded precipitation. The averages are taken over 1982, 1983, and 1988. Contours are plotted at 0.5, 1.0, 1.5, 2.0, 2.5, 3.0, 4.0, 5.0, and 6.0-mm day^{-1}. Shaded areas indicated precipitation greater than 2-mm day^{-1} (Giorgi et al. 1993).

An important issue yet to be defined, however, is the optimal spatial resolution required to provide adequate input to spatially distributed hydrologic and ecological system models (Band 1993; Band et al. 1993; Parton et al. 1987, 1993). For winter orographically forced precipitation, it is straightforward that the terrain structure must be reasonably represented. Using a Fourier decomposition of terrain structure through the Front Range of Colorado, Young and Pielke (1983) and Young et al. (1984) found that significant variability exists even on scales of hundreds of meters, although the accuracy of the Giorgi et al. (1993) simulations suggest that this smaller-scale variability is smoothed to some extent as sensed by the atmosphere. However, such effects as snow accumulation resulting from drifting and heterogeneous melt patterns will be a function of small-scale terrain forms. Summer precipitation, unfortunately, is not as well homogenized as a result of localized structure of deep cumulus clouds. As shown by Gibson and Vonder Haar (1990), for example, such clouds often develop initially over mountain ridges such as illustrated in Figure 19-2a.

Such clouds and resultant precipitation are also sensitive to the vegetation cover. This effect is well known over flatter terrain as shown by André et al. (1990), Avissar and Segal (1988), Avissar and Pielke (1991), and Pielke et al. (1991). Pielke et al. (*Proceedings*, 1992; *International*, 1993) demonstrated this effect in the Colorado Rockies where, using identical large-scale meteorology, rainfall patterns and amounts were shown to be sensitive to tree-line elevation.

METHODOLOGY

One-Way Interaction

The procedure to complete seasonal meteorological simulations in the absence of dynamic hydrologic and ecological feedbacks is straightforward. The meteorological initial and lateral boundary conditions under current climate can be obtained from the National Meteorological Center (NMC) or the European Center for Medium-range Weather Forecasting (ECMWF) analyzed fields, available from the NCAR data archives (Jenne 1989). Simulations should be completed for several years in order to assure a representative sample of the current interannual range of large-scale conditions. The mesoscale model output must be validated against surface and other observational data, which are not used in the model initialization of lateral boundary conditions, in order to quantify the mesoscale model skill, and level of improvement over that achieved by the larger-scale models.

Land surface information (terrain, elevation, soil, and vegetation characteristics) at the same spatial resolution as the mesoscale model grid structure is also required. While experiments are needed in which the background meteorology changes year to year, identical and actual landscape characteristics for each year should be used to permit the most appropriate comparison with observations. This is needed because vegetation characteristics such as leaf area index, vegetation albedo, etc., will be different at the same time of year in different years. The United States Geological Survey (USGS) is the source for this data (Loveland et al. 1991; Brown et al. 1993).

For climate change experiments, the resolution of the large-scale data should be degraded to that of GCMs in order to evaluate the loss of realism due to the poorer representation of meteorology. Once credible results of the mesoscale model are obtained using GCM-resolution data as input, climate-change scenarios as performed by GCMs can be used to provide initial and lateral boundary conditions for the mesoscale model.

To test the sensitivity of a mesoscale model to the resolution of large-scale initial and lateral boundary conditions, three numerical experiments have been performed using RAMS with three nested grids with 125-km, 25-km, and 6.25-km, horizontal grid intervals, respectively. Initial and lateral boundary conditions for the mesoscale model were provided by the RAMS isentropic analysis package (Pielke et al., *Meteor.*, 1992; Tremback 1990), which combines ECMWF analyses with surface and rawinsonde observations for use as the RAMS initialization.

The control experiment was initialized and laterally forced on a 1° × 1°-grid with large-scale fields smoothed to a 1,000-km wavelength (a standard RAMS initialization) on July 26, 1985. To degrade this analysis and make it consistent with current GCM model resolution, the initial and lateral boundary fields were inserted on a 4°-grid and smoothed to a wavelength of 1,600 km. Finally, an upgraded experiment was performed where observational data were heavily weighted to provide fields on a 1°-grid but including features with wavelengths above 400 km.

Figure 19-4 shows the results of initial and lateral boundary condition resolution on spatially averaged total precipitation over a 312-km × 262-km area for the three experiments. Figure 19-4 suggests that degrading the large-scale forcing has only a minor impact relative to a standard RAMS initialization. There is a larger difference, however, between the upgraded experiment and the control experiment, which raises the possibility of significantly different solutions in the mesoscale model interior with improved detail in initial and boundary conditions.

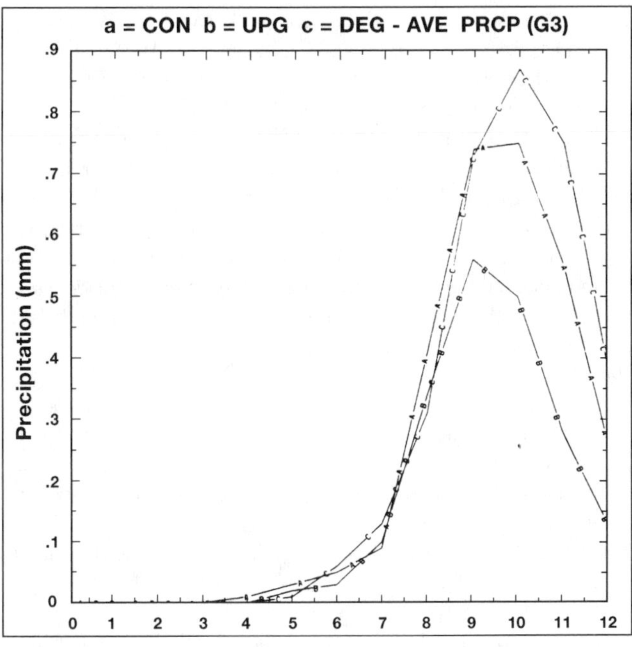

Figure 19-4. Spatially averaged total precipitation over a 312 km × 262-km area centered at 40.4°N and 105.7°W for the (a) control, (b) upgraded, and (c) degraded simulations.

Two-Way Interactive Scenarios

While landscape and hydrologic feedbacks are in a sense implicit in the lateral boundary conditions, results for climate-change scenarios could be artificially constrained by using a current landscape configuration. Such an approach is analogous to specifying ocean surface temperatures in a GCM simulation as opposed to coupled atmosphere–ocean GCM calculations.

Thus to permit interactive atmospheric–land surface modeling, an atmospheric model must be coupled to an ecological/hydrologic modeling system such as RHESSys (Band 1993; Band et al. 1993) or CENTURY (Parton et al. 1987, 1993). Pielke et al. (*Ecological*, 1993) discuss the need for this dynamic feedback capability. Only with this approach can we explore whether, for example, drought conditions would be mitigated or exacerbated as a result of landscape responses to this weather stress.

Simulations of the above two-way interactions are restricted to regional scales at present. Global two-way couplings are still impractical due to a lack of detailed global surface data, a lack of understanding of the coupling processes, and inadequate parameterizations of subgrid-scale interactions such as the nonlinear effects of the lateral redistribution of surface water on regional energy and water vapor fluxes (Famiglietti and Wood 1994). The coupling between atmospheric models and surface models involves two classes of strategies. One class of coupling strategy is that an atmospheric model such as RAMS (Pielke et al., *Meteor.*, 1992) or the Penn State/NCAR Mesoscale Model Version 4 (MM4) (Anthes et al. 1987) is directly coupled to an ecological/hydrological model such as RHESSys (Band 1993; Band et al. 1993) or CENTURY (Parton et al. 1987, 1993), which have been developed by the ecological community. Another approach is that an atmospheric model is directly coupled to a vegetation module such as BATS (Dickinson et al. 1986), SiB (Sellers et al. 1986), or LEAF (Lee et al. 1993), developed in the meteorological community, whereby the vegetation module is then coupled to an ecological/hydrological model. It is unclear at this time which coupling strategy is best.

The feedback from ecological/hydrological processes to the atmosphere involves the aggregation of inhomogeneous surface contributions. Qualitatively, when the horizontal scale of surface inhomogeneities is small, a homogeneous surface can be assumed, at least in terms of the atmospheric response (Pielke et al. 1991). When the horizontal scale is not too large, the mosaic approach (Avissar and Pielke 1989) can be used. When the horizontal scale is larger, nonlinear circulations induced by surface inhomogeneities should be considered. These nonlinear coherent structures may be parameterized based on mesoscale kinetic energy (Avissar and Chen 1993) or on several dimensionless parameters which combine both large-scale and local effects. The latter method is currently being investigated by our group (Dalu et al. 1993).

CONCLUSION

Current GCM grid resolutions are unable to realistically represent local and regional seasonal weather patterns, even if the large-scale atmospheric structure is correctly simulated. Mesoscale models offer a procedure to link GCM output to these smaller-scale weather and climate patterns. Even with higher-resolution mesoscale models, however, the representation will be incomplete unless the land surface and atmosphere can dynamically interact.

Acknowledgments

Department of the Interior under Assistance #ATM-14-08-0001-A0929; the National Park Service Global Change Program under contracts #CA1268-2-9004 and #CA1268-2-9004 CSU 28; the National Science Foundation under grant #ATM-8915265; and the Climate System Modeling Program, and the National Oceanic and Atmospheric Administration under contract #NA36GP0378. Dallas McDonald is thanked for preparing this work.

References

André, J. C., P. Bougeault, and J. P. Goutorbe. 1990. Regional estimates of heat and evaporation fluxes over nonhomogeneous terrain. Examples from the HAPEX-MOBILHY Programme. *Bound.-Layer Meteor.* 50:77–108.

Anthes, R. A., E. Y. Hsie, and Y. H. Kuo. 1987. Description of the Penn State/NCAR Mesoscale Model Version 4 (MM4). NCAR/TN-282+STR.

Atkinson, B. W. 1981. *Mesoscale Atmospheric Circulations*. New York: Academic Press.

Avissar, R., and F. Chen. 1993. Development and analysis of prognostic equations for mesoscale kinetic energy and mesoscale (subgrid-scale) fluxes for large-scale atmospheric models. *J. Atmos. Sci.* 50(22):3,751–74.

Avissar, R., and R. A. Pielke. 1989. A parameterization of heterogeneous land surfaces for atmospheric numerical models and its impact on regional meteorology. *Mon. Wea. Rev.* 117:2,113–36.

———. 1991. The impact of plant stomatal control on mesoscale atmospheric circulations. *Agric. Forest Meteor.* 54:353–72.

Avissar, R., and M. Segal. 1988. Evaluation of vegetation effects on the generation and modification of mesoscale circulations. *J. Atmos. Sci.*

Avissar, R., and M. M. Verstraete. 1990. The representation of continental surface processes in atmospheric models. *Rev. Geophys.* 28:35–52.

Band, L. E. 1993. Effect of land surface representation on forest water and carbon budgets. *J. Hydrology* 150(2-4):749–72.

Band, L. E., P. Patterson, R. Nemani, and S. W. Running. 1993. Forest ecosystem processes at the watershed scale: Incorporating hill slope hydrology. *Agric. For. Meteor.* 63:93–126.

Brown, J. F., T. R. Loveland, J. W. Merchant, B. C. Reed, and D. O. Ohlen. 1993. Using multisource data in global land cover characterization—Concepts, requirements, and methods. *Photo. Eng. Rem. Sens.* 59:977–87.

Cooter, E. J., B. K. Eder, S. K. LeDuc, and L. Truppi. 1993. General circulation model output for forest climate change research and applications. *Southeastern Experiment Station Technical Bulletin.*

Cotton, W. R., and R. A. Pielke. 1992. Human impacts on weather and climate. In *Geophysical Science Series*, vol. 2. Fort Collins: ASTeR Press.

Dalu, G. A., and R. A. Pielke. 1993. Vertical heat fluxes generated by mesoscale atmospheric flow induced by thermal inhomogeneities in the PBL. *J. Atmos. Sci.* 50(6):919–26.

Derickson, Russ. 1992. "Finite Difference Methods in Geophysical Flow Simulations." Ph.D. diss., Department of Civil Engineering, Colorado State University.

Dickinson, R. E., H. Henderson-Sellers, P. J. Kennedy, and M. R. Wilson. 1986. Biosphere Atmosphere Transfer Scheme (BATS) for the NCAR Community Climate Model. NCAR/TN-275+STR.

Doran, J. C., F. J. Barnes, R. L. Coulter, T. L. Crawford, D. D. Baldocchi, L. Balick, D. R. Cook, D. Cooper, R. J. Dobosy, W. A. Dugas, L. Fritschen, R. L. Hart, L. Hipps, J. M. Hubbe, W. Gao, R. Hicks, R. R. Kirkham, K. E. Kunkel, T. J. Martin, T. P. Meyers, W. Porch, J. D. Shannon, W. J. Shaw, E. Swiatek, and C. D. Whiteman. 1992. The Boardman regional flux experiment. *Bull. Amer. Meteor. Soc.* 73:1,785–95.

Famiglietti, J. S., and E. F. Wood. 1994. Multiscale modeling of spatially variable water and energy balance processes. *Water Resour. Res.* 30(11):3,061–78.

Gibson, H. M., and T. H. Vonder Haar. 1990. Cloud and convection frequencies over the southeast United States as related to small-scale geographic features. *Mon. Wea. Rev.* 118:2,215–27.

Giorgi, F., G. T. Bates, and S. J. Nieman. 1993. The multiyear surface climatology of a regional atmospheric model over the western United States. *J. Climate* 6:75–95.

Jenne, R. L. 1989. Data availability at NCAR. SCD UserDoc. Boulder: Scientific Computing Division, National Center for Atmospheric Research.

Lee, T. J., R. A. Pielke, T. G. F. Kittel, and J. F. Weaver. 1993. Atmospheric modeling and its spatial representation of land-surface characteristics. In *Environmental Modeling with GIS*, edited by M. Goodchild, B. Parks, and L. T. Steyaert, 108–122. New York: Oxford University Press.

Loveland, T. R., J. W. Merchant, D. O. Ohlen, and J. F. Brown. 1991. Development of a land cover characteristics database for the conterminous U.S. *Photo. Eng. Rem. Sens.* 57:1,453–63.

Parton, W. J., D. S. Schimel, C. V. Cole, and D. S. Ojima. 1987. Analysis of factors controlling soil organic matter levels in the Great Plains grasslands. *Soil Sci. Amer. J.* 51:1,173–79.

Parton, W. J., D. S. Schimel, D. S. Ojima, and C. V. Cole. 1993. A general model for soil organic matter dynamics: Sensitivity to litter chemistry, texture, and management. *Soil Sci. Amer. J.*

Pielke, R. A. 1984. *Mesoscale Meteorological Modeling*. New York: Academic Press.

Pielke, R. A., J. S. Baron, T. G. F. Kittel, T. J. Lee, T. N. Chase, and J. M. Cram. 1992. Influence of landscape structure on the hydrologic cycle and regional and global climate. In *Proceedings, Managing Water Resources During Global Change, American Water Resources Association (AWRA)*, 283–96. Reno.

Pielke, R. A., W. R. Cotton, R. L. Walko, C. J. Tremback, W. A. Lyons, L. D. Grasso, M. E. Nicholls, M. D. Moran, D. A. Wesley, T. J. Lee, and J. H. Copeland. 1992. A comprehensive meteorological modeling system—RAMS. *Meteor. Atmos. Phys.* 4:69–91.

Pielke, R. A., G. Dalu, J. S. Snook, T. J. Lee, and T. G. F. Kittel. 1991. Nonlinear influence of mesoscale land use on weather and climate. *J. Climate* 4:1,053–69.

Pielke, R. A., T. J. Lee, T. G. F. Kittel, T. N. Chase, J. M. Cram, and J. S. Baron. 1993. Effects of mesoscale vegetation distributions in mountainous terrain on local climate. In *International Conference on Mountain Environments in Changing Climates*, edited by M. Beniston. Routledge Press.

Pielke, R. A., D. S. Schimel, T. J. Lee, T. G. F. Kittel, and X. Zeng. 1993. Atmosphere terrestrial ecosystem interactions: Implications for coupled modeling. *Ecological Modelling* 7:5–18.

Pielke, R. A., and M. Segal. 1986. Mesoscale circulations forced by differential terrain heating. In *Mesoscale Meteorology and Forecasting*, edited by P. Ray, 516–48. AMS.

Segal, M., and R. W. Arritt. 1992. Nonclassical mesoscale circulations caused by surface heat-flux gradients. *Bull. Amer. Meteor. Soc.* 73:1,593–1,604.

Sellers, P. J., Y. Mintz, Y. C. Sud, and A. Dalcher. 1986. A simple biosphere model (SiB) for use within general circulation models. *J. Atmos. Sci.* 43:505–31.

Tremback, C. J. 1990. "Numerical Simulation of a Mesoscale Convective Complex: Model Development and Numerical Results." Ph.D. diss., Department of Atmospheric Science, Colorado State University.

Young, G. S., and R. A. Pielke. 1983. Application of terrain height variance spectra to mesoscale modeling. *J. Atmos. Sci.* 40:2,555–60.

Young, G. S., R. A. Pielke, and R. C. Kessler. 1984. A comparison of the terrain height variance spectra of the Front Range with that of a hypothetical mountain. *J. Atmos. Sci.* 41:1,249–50.

Roger Pielke is a professor at Colorado State University, Department of Atmospheric Science. For the last twenty years, he has concentrated on the study of terrain-induced mesoscale systems, including the development of a three-dimensional mesoscale model of the sea breeze, for which he received the NOAA Distinguished Authorship Award for 1974. Dr. Pielke received the 1984 Abell New Faculty Research and Graduate Program Award and the 1987/1988 Abell Research Faculty Award. He was declared "Researcher of 1993" by the Colorado State University Research Foundation. His books include *Mesoscale Meteorological Modeling*, *The Hurricane*, and *Human Impacts on Weather and Climate*. In addition, he has published over 170 papers in peer-reviewed journals, thirteen book chapters, and has coedited three books.

R. A. Pielke, T. Chase, J. Copeland, T. J. Lee, R. Walko, and *X. Zeng*
Department of Atmospheric Science
Colorado State University
Fort Collins, Colorado 80523
E:mail: *dallas@leuropa.atmos.colostate.edu*

J. Baron and *T. G. F. Kittel*
Natural Resource Ecology Laboratory
Colorado State University
Fort Collins, Colorado 80523

20

The Effect of Elevation Data Representation on Nocturnal Drainage Wind Simulations

Hoyt Walker and John M. Leone, Jr.

A critical requirement for accurately representing surface–atmosphere interactions within an atmospheric model is to realistically characterize the land surface. Data must be extracted from a geographical database and transformed so that it is consistent with the needs of the numerical model. In such a process, there are two major classes of error that must be understood and minimized whenever possible. The first class involves the accuracy, precision, and resolution of the geographical data itself. The second is error introduced by the transformations used to assimilate the data into the atmospheric model. Thus, this research has two coupled objectives: to understand the effects of errors within the geographical database upon the accuracy of the atmospheric model simulation and to design optimal techniques for the transformation of the data.

A series of model experiments has been designed to determine the response of a hydrostatic mesoscale atmospheric model to lower boundary forcing due to variations in the representation of an idealized mountain valley system during nocturnal cooling. Early results indicate that the variations in the model response are substantial when the valley widths are 2Δx, but that the general flow features are well represented when the valley widths are 4Δx. Details of the flow do vary for 4Δx features and these details may be important in certain applications.

INTRODUCTION

Powerful new computing capabilities are permitting researchers to model atmospheric phenomena that are sensitive to, or driven by, land-surface characteristics. In the context of atmospheric modeling, the surface characteristics of interest are the elevation, surface roughness, and sensible and latent heat fluxes (Lee et al. 1993). Appropriate descriptions of these characteristics in the form of model boundary conditions must be developed from geographical databases in a way that balances the need for descriptive detail and accuracy with the spatial resolution and numerical characteristics of a model. Thus, certain features of the geographical data, such as its resolution, accuracy, and the manner in which the data are processed, are especially important to predictive models of the planetary boundary layer. This is due to the nonlinearity of the physics being expressed in such models and to the close coupling of the land surface with boundary layer phenomena under many conditions (Pielke 1984). While issues of model sensitivity to geographic data range from local to global scales, this study restricts itself to the atmospheric mesoscale (20–2,000 km).

For mesoscale atmospheric models, elevation data are an important class of geographic information. The development of nocturnal drainage winds is one example of how terrain affects atmospheric behavior. This phenomenon offers a useful test case for examining the effects of elevation data on mesoscale models (Leone and Lee 1989). Because such winds are driven by inhomogeneities in the temperature field, which are caused by the cooling of the sloping land surface relative to the adjacent air, alterations in the representation of that surface can be expected to have a significant effect on a simulation of the associated flow.

It often happens that the resolution of the elevation data needed in support of an application is not ideal. If the resolution is too low, then other sources of higher-resolution data should be sought out, since the use of overly coarse data in any analysis can result in meaningless conclusions. If the resolution is too high, given the computing resources available, then the data must be adjusted to an appropriate scale. This chapter summarizes the results of numerous simulations of drainage wind development within an idealized multivalley system. A control run, in which the valley system has been highly resolved, is compared against various alternative representations of the valley that correspond to coarser sampling of the terrain surface.

APPROACH

The experiments described herein were performed using a hydrostatic finite element mesoscale atmospheric model. Elevation data were generated to represent valleys incised in a mesa using a software tool developed for this purpose. This program allows the specification of an analytic description of multivalley (or hill) systems. This analytic surface can be sampled at any point; the sample points serve as the basis for generating the model terrain. Details of this process along with descriptions of the simulations are presented in the following subsections.

The SABLE Model

The atmospheric model used in these tests is called SABLE, a hydrostatic mesoscale model developed at the Lawrence Livermore National Laboratory. SABLE solves the hydrostatic, anelastic equations for velocity, potential temperature, and Exner function (a scaled pressure variable) in three dimensions (Zhong et al. 1991). The equations are solved by using a unique blend of numerical techniques. The prognostic equations for the horizontal velocity components and the potential temperature are solved using trilinear, isoparametric finite elements in space combined with a semiimplicit time integration scheme wherein the diffusion terms are integrated implicitly while all other terms are explicit. The linear systems generated by this approach are solved by using the diagonally scaled conjugate gradient method. The diagnostic equations for vertical velocity and Exner function are solved by integrating up or down vertical columns, respectively, via centered finite differences.

Terrain Generation

To support this research, a program, SIMULATERR, was developed to generate terrain surfaces, including hills, hemispheres, or valleys. Cross sections of the hills and valleys are described by Gaussian logit curves described by the equation:

$$\log\left[\frac{h(x)}{1-h(x)}\right] = b_0 + b_1 x + b_2 x^2 = a - (1/2)\frac{(x-u)^2}{t^2} \qquad (20\text{-}1)$$

where h is the relative height or depth of the feature as a function of the distance along the cross-sectional axis, x, and the parameters b_i describe variability in the functional form but have no obvious geometrical interpretation. The parameters, a, u, and t are defined as follows:

$$u = -b_1/(2b_2),$$

$$t = 1/\sqrt{-2b_2},$$

$$h_{\max} = \frac{1}{1+\exp(-b_0 - b_1 x - b_2 x^2)}, \quad 0 \le h_{\max} \le 1, \qquad (20\text{-}2)$$

$$a = \log\left[\frac{h_{\max}}{1-h_{\max}}\right],$$

so as to have a somewhat clearer interpretation than the b_i i.e., u is the x location of the greatest height or depth, h_{max} controls the shape of the feature as well as affecting its width and may be any value between 0 and 1, and t also affects the feature width (ter Braak and Looman 1986). Examples of this set of functions are shown in Figure 20-1. The extremum of the curve, h_{max} is scaled to a height or depth according to various user-selected parameters and its position along the length of the valley. This approach gives substantial control over the shape of the features. It should be noted that h_{max} = 0.5 gives the shape of the standard Gaussian curve.

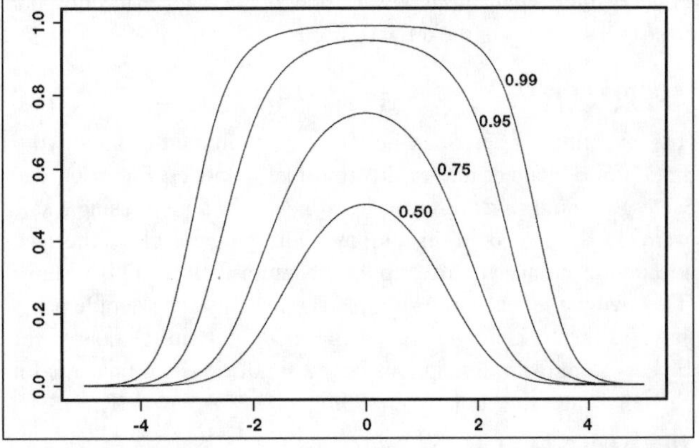

Figure 20-1. Gaussian logit curves with u=0, t=1 for h_{max}=0.5,0.75,0.95,0.99.

There are two classes of valleys in our study: main valleys and side valleys. Side valleys are associated with a main valley or other side valleys in a hierarchical structure. At present, three levels in this hierarchy are supported, although extension to more levels is possible. Each valley is considered in two sections, the valley head and the main valley. The valley geometry of a main valley is controlled by the location of the start of the valley head, the location of the end of the valley head, the height or depth of the end of the valley head, the shape parameter, h_{max}, of the main valley, and the slope of the main valley floor. The side valleys are also controlled by the same first three parameters as the main valleys, but the slope is chosen so that the axis of the side valley intersects the wall of the associated main valley smoothly. Thus, fairly complex geometries can be built within the constraint that each valley traces a straight line.

The system described above provides a means for specifying an analytic value for the surface at all points. This analytic surface can be sampled at each point in a rectangular grid of arbitrary orientation. These sampled values can be the exact value of the surface or they can be weighted averages of the surface values at each point of a filter stencil centered on the sample point. Thus, this tool can be used to investigate the effects of resampling techniques in presenting the analytic surface on a discrete model grid. The sample values are bilinearly interpolated to the model grid of interest.

Control Run

A high-resolution control run was developed to form a basis for comparison with various runs representing lower-resolution terrain. The landform was based on a flat mesa sloping down to a flat plain 400 m below. As seen in Figure 20-2a, three parallel side valleys are associated with the main slope. The depths of the side valley heads are 150, 100, and 50 m above the main valley floor corresponding to relatively steep,

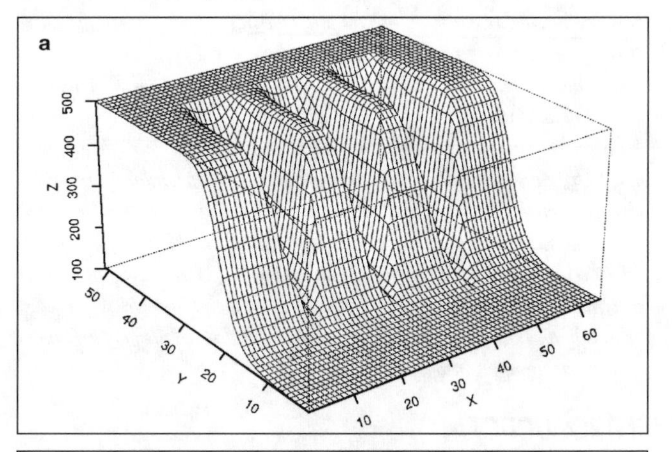

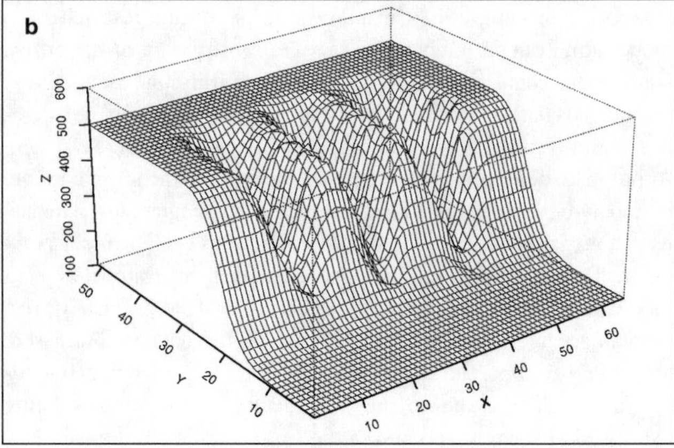

Figure 20-2. Perspective views of terrain for (a) the control run, and (b) the 30°-rotated sampling grid.

moderate and shallow slopes, respectively, along the valley floors. In the horizontal, the finite element grid is composed of regular rectangular grid cells that are 500 m in the cross-side valley direction and 1,000 m in the along-side valley direction. The grid has 64 grid cells in the cross-side valley direction and 53 cells in the along-side valley. Thus, they cover a 32- × 54-km area. The three side valleys are 16 grid cells across with an 8-cell flat area on each side of the model grid. A graded step size was used along the vertical axis with the lowest cell being 15 m in depth. The grading was adjusted so that 15 vertical grid cells extended to an elevation of 2,400 m above the valley floor across the entire grid.

The goal of the simulations described herein is to isolate the effects of terrain representation on a plausible model run, not necessarily to reproduce a true physical situation. Thus, a number of simplifying assumptions were made. For example, the Coriolis parameter was set to 0 to avoid complicating veering motions. The cross-side valley wind component, u, was assumed to be 0 at the appropriate lateral boundaries. At the top boundary, both horizontal wind components, u and v, were set to 0. The lower boundary cooling was specified as a heat flux of -60 W/m^2. The atmosphere was initialized to be slightly stable with a potential temperature lapse rate of 0.002 km. Turbulence was specified via a constant horizontal coefficient, υ_H, of 50.0m^2/s and a constant vertical coefficient, υ_V of 1.0m^2/s. The problems were run for eight hours with a thirty-second time step. These parameters were used in the control and in all comparison runs, thus, the only difference between the runs was the representation of the terrain.

High-Resolution Comparison Runs

The various runs described in this section are all examples of simulated terrain degradation. That is, the model resolution was left unchanged, but the lower boundary was represented by bilinearly interpolating between more widely spaced samples of the actual terrain surface. This approach was taken to avoid convolving the effects of the altered terrain representation with the important numerical effects associated with changing the model step size.

In the first six of nine runs, the resolution was lowered, but only in the cross-valley direction. Thus, the changes mainly affected the valley cross sections. In two of these runs, the 16-cell-wide valleys were represented by sample points that were 8 cells apart, i.e., the valleys represented $2\Delta x$ features on the sampling grid. In run one, the sampling grid was chosen to lie along the axes of the ridges and valleys, while in run two, the sampling grid was shifted 4 cells so that the ridges and valleys were not resolved. In the next four runs, the valleys were represented by points that were 4 grid cells apart. In run three, the sample points were located along the ridge and valley axes. In runs four, five, and six, the sampling grid was shifted 2, 1, and 3 cells from the grid used in run three. Thus, in run three the depth of the valley was represented "correctly." In run four the depth was incorrect but the width was represented quite well, and in runs five and six, there were asymmetries in the representation of the valleys.

In runs seven, eight, and nine the sampling grid was square with a step of 2,000 m but was rotated by 15°, 30°, and 45° with respect to the model grid. As a result, high-frequency periodicities were introduced in the terrain representation (see Figure 20-2b). For the 15°- and 30°-rotations, a small amount of skewness to the ridge and valley axes was also introduced. Figure 20-3 includes sample cross sections for each of these runs with the geometry of the control run superimposed.

Low-Resolution Comparison Runs

In the low-resolution runs, the model grid was also lowered. The sampling grids for each run were the same as in the simulation of low resolution, but the model grid was also changed. For runs one to six, the model grid matched the terrain sampling grid. In runs seven to nine, where the sampling grid was rotated, the model grid was a rectangular grid matching the model grid for low-resolution runs three through seven, i.e., with a grid step of 2 km along the x axis and 1 km along the y axis.

Differences between the control run and comparison runs were determined by visually examining vector plots and contours of down-valley wind speed, horizontal and vertical profiles of down-valley winds through the jet maxima, and jet maximum and down-valley mass flux plots along the length of each valley. The flux was determined by defining a volume that was 16 cells wide on the high-resolution runs and 8 or 4 cells wide in the low-resolution runs for four

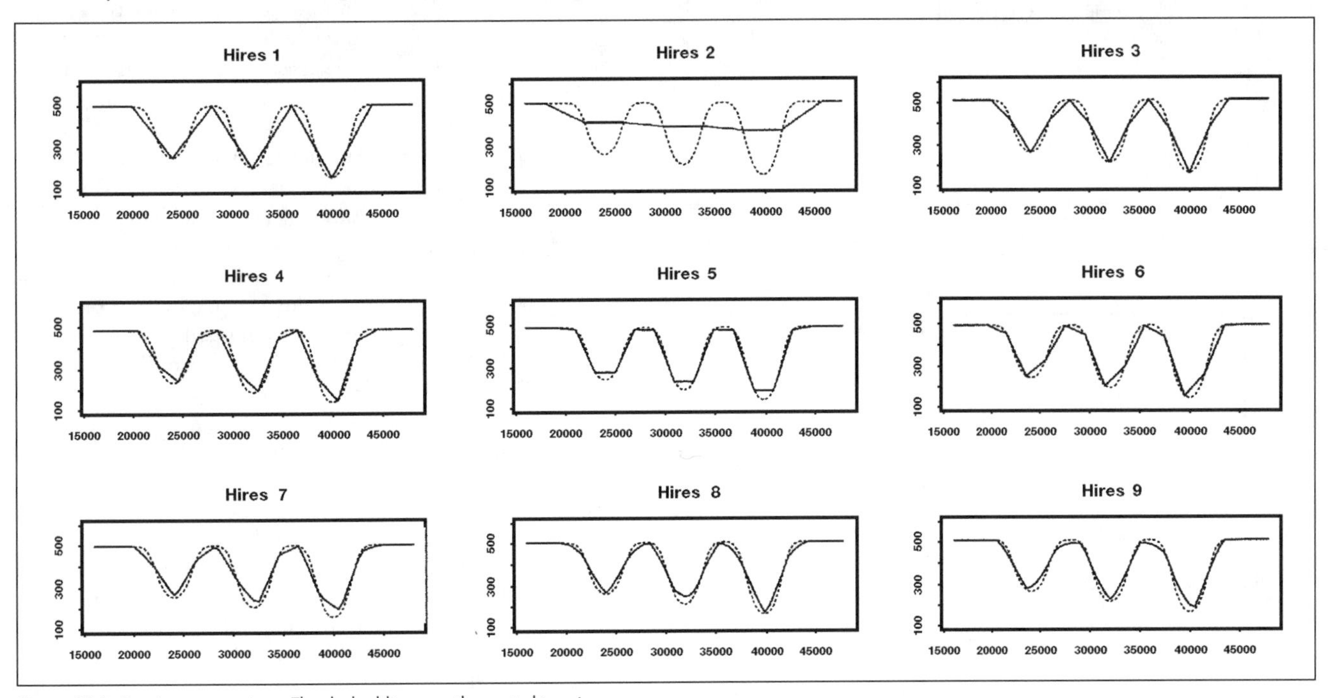

Figure 20-3. Terrain cross sections. The dashed lines are the control terrain.

and two delta features, respectively, centered on each valley axis (i.e., the cells covered the entire width of the valley), 9 cells deep and ranging from the head of each side valley out into the flat area of the main valley. The flux through each plane of this volume normal to the valley axis was computed to provide a diagnostic of the overall jet intensity down the valley; the jet maximum provided a measure of the local jet intensity. Summary statistics of these along-valley diagnostics were also computed.

RESULTS

Surface cooling drove the generation of drainage jets that flowed down each of the valleys onto the plain. The geometry of the jets reflects the shape of the valleys, the rate of cooling, and the time since the beginning of the run. In addition, the location of each valley with respect to the lateral boundaries has an important effect, for example, the jet in the central valley is relatively symmetric, while jets in the left and right valleys are skewed toward the middle of the grid (see Figure 20-4d). This shift is due to strong winds along the valley walls nearest the lateral grid boundaries (see Figure 20-4a). It may reflect the fact that mass is being drawn from a larger flat region adjacent to the lateral boundaries. Some lateral movement of the jet axis over time was noted. This movement first stabilized at the valley head and then began to stabilize down the valleys but did not completely damp in during the eight-hour runs. Also, the jet deepened and the height of the jet maximum rose over the course of the run. Thus, a true steady state was not achieved (the patterns observed at four hours and eight hours are, however, quite similar). In general, the control run shows reasonable behavior and forms the basis for comparison with the low terrain resolution runs.

Runs one and two demonstrate the evanescence in the representation of a $2\Delta x$ feature. In high-resolution run one, the behavior of the control run is reproduced reasonably well as evidenced by the along valley jet maxima for the central valley as shown in Figure 20-5a. Some differences are noted near the head of the valley, reflecting the changed representation, but the overall patterns are quite similar. In contrast, the behavior of the jet maxima in high-resolution run two is completely different, which reflects the complete lack of valley resolution associated with this degradation. In low-resolution run one, the strong $2\Delta x$ forcing, in combination with the proximity of the boundaries in the coarse gridding, caused numerical difficulties and the problem did not run to completion.

It was not surprising that $2\Delta x$ features are not satisfactorily resolved by the model; however, there was an expectation that $4\Delta x$ features would be well represented. In general, this was supported by the simulations, but with an interesting distinction. In high-resolution run three, the lower levels of the valleys were narrowed but the valley depth was maintained. In high-resolution run four, the width was better maintained but the depth was incorrectly resolved (see Figure 20-3). As a result, run three generated a more intense, tightly focused jet that penetrated farther out into the flat main valley than the control, while run four generated a broader, more diffuse jet that propagated a shorter distance than the control (see Figure 20-6). Thus, the change in the terrain representation has noticeable effects on the flow independent of model resolution. In low-resolution run three, the narrow valley again produces a narrow focused jet; however, the jet is only resolved by the wind at one node in each cross-valley plane, and as a result, dissipates rapidly. The skewness of runs five and six and the undulations caused by rotating the sampling grid in runs seven, eight, and nine had little effect on the general flow in the high-resolution runs but did cause a noticeable response in the low-resolution runs as shown in Figure 20-7.

The jet flux was insensitive to the terrain representation, with the exception of run two. The summary statistics confirmed this impression that the general flow is not especially sensitive to the terrain representation, for example, again excluding run two, the maximum absolute deviation of the along valley jet maximum between the control and the degraded runs was typically near 5% but could be as much as 10% in the high-resolution runs and as much as 17% in the low-resolution runs. At different points along the valley, the jet maxima was variously overpredicted and underpredicted, and as a result, the average errors were smaller, ranging from 1%–6% for the high-resolution runs and 3%–12% for the low-resolution runs.

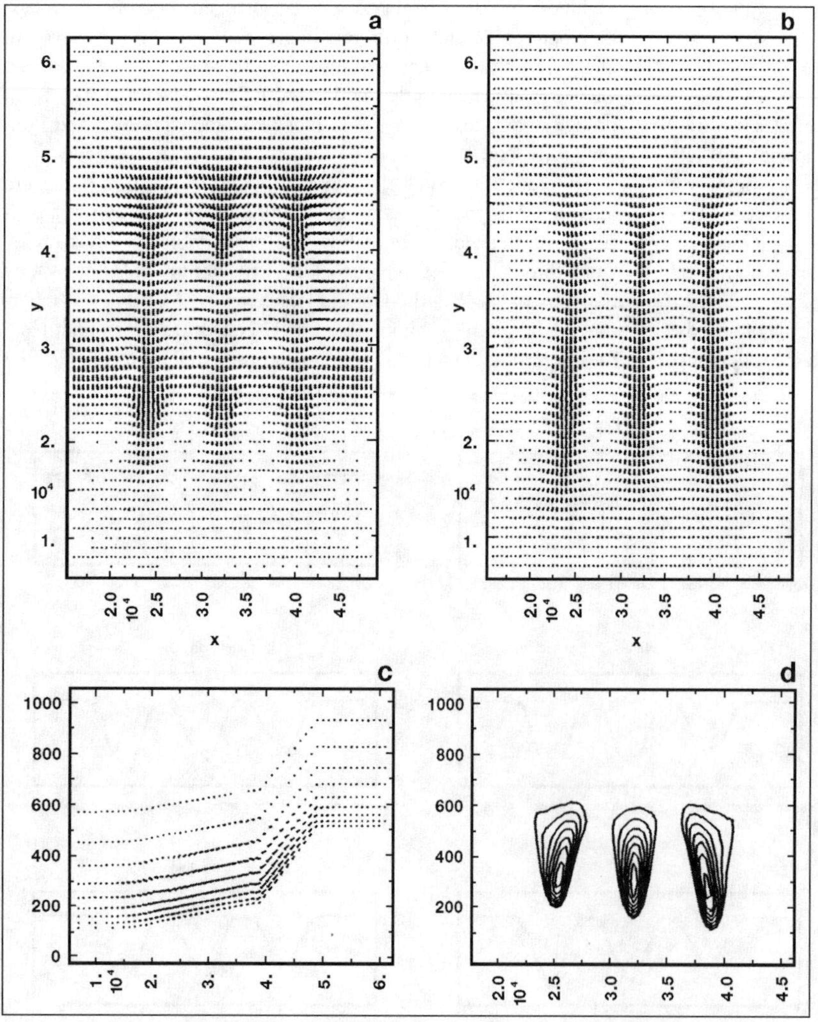

Figure 20-4. The control run at four hours: (a) wind vectors at 15 m above terrain, (b) wind vectors at about 130 m above terrain, (c) wind vectors along the central valley axis, (d) vertical cross sections of isotachs of down-valley wind at *y*=37,000 m.

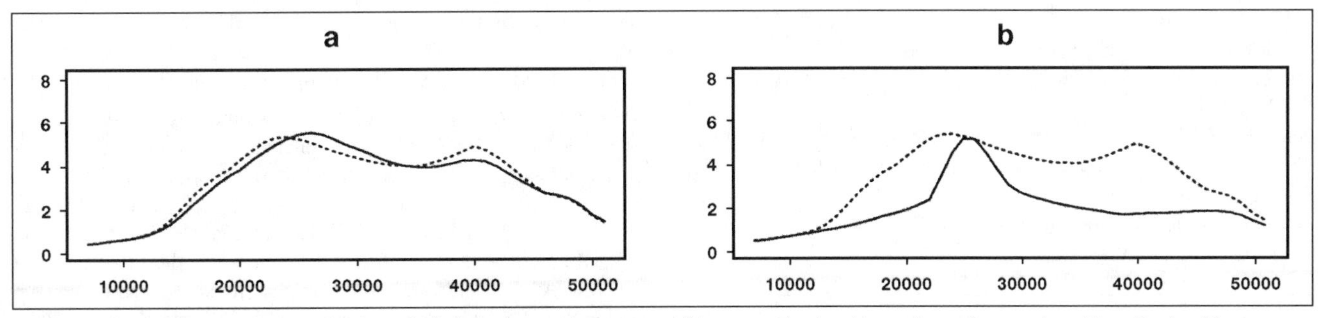

Figure 20-5. Along valley jet maxima at four hours for high resolution: (a) run one and (b) run two. The dotted line indicates the control run. The valley head begins at y=49,000 m along the horizontal axis and opens out onto the plain near y=22,000 m.

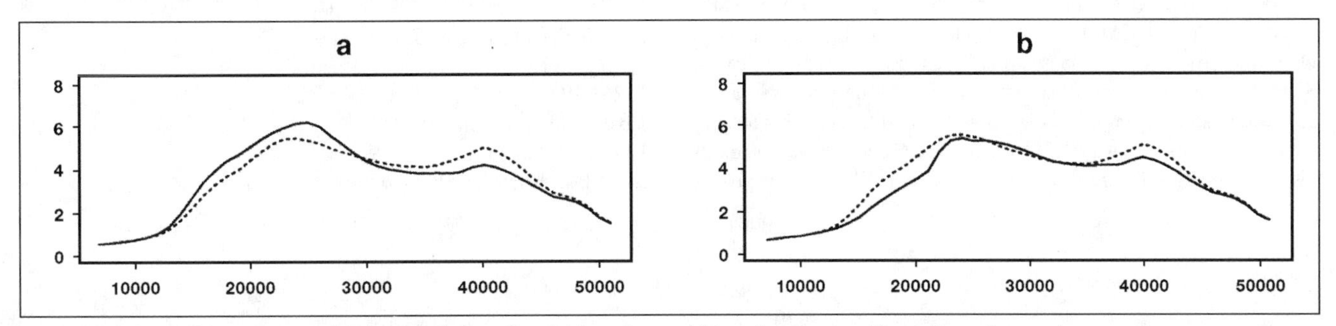

Figure 20-6. Along valley jet maxima at four hours for high resolution: (a) run three and (b) run four. The dotted line indicates the control run.

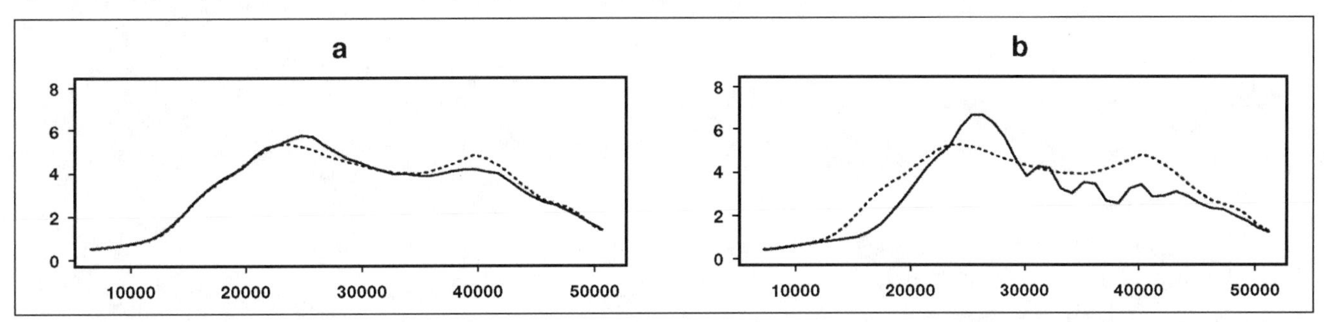

Figure 20-7. Along valley jet maxima at four hours for: (a) high-resolution run eight and (b) low-resolution run eight. The dotted line indicates the control run.

CONCLUSION

The broad conclusion here is that while $2\Delta x$ terrain features will not reliably be represented in model simulations, the general flow produced by $4\Delta x$ features should be adequately resolved. This is not to say that the terrain representation of such features has no effect, just that such effects seems to be restricted to details of the flow. As an illustration, consider a 5% error in wind speed with a 5.5 m/s wind. This leads to a position error for a transported parcel of nearly 1 km every hour, so a run of several hours could result in mislocations of several km. In some applications of mesoscale models, these errors in detail may be significant, such as in modeling the dispersion of a radioactive or toxic material in drainage flow. Other important differences include the distance of penetration of a drainage jet into a plain and the sampling-caused skewness of some of the terrain representations that does translate the jet axis. The mispositioning of the ridges caused by rotations or shifts in the sampling grid could also be important. Such an effect could cause a release location near such a ridge to be incorrectly located in the wrong valley, and thus, a simulation would produce completely erroneous predictions. It is also important to note that these simulations were made with turbulence modeled as a constant, which is known to be strongly smoothing. Thus, we believe that these results represent a lower bound on the model response to terrain-surface representation and that inclusion of a more realistic turbulence model would increase the sensitivity of the atmospheric model behavior.

Continuation of this work will involve analysis of simulations using more complex geometries. The geometries described here are still too simple to generate large differences from different resampling algorithms to lower the resolution. The sensitivity to error will be tested by generating error fields to perturb the terrain representation. This work will also be extended to real elevation data at the sites of drainage wind field experiments.

Acknowledgments

Work for this chapter was performed under the auspices of the U.S. Department of Energy by Lawrence Livermore National Laboratory under Contract #W-7405-Eng-48.

References

Lee, T. J., R. A. Pielke, T. G. F. Kittel, and J. F. Weaver. 1993. Atmospheric modeling and its spatial representation of land-surface characteristics. In *Environmental Modeling with GIS*, edited by M. F. Goodchild, B. O. Parks, and L. T. Steyaert, 108–122. New York: Oxford University Press.

Leone, J. M., and R. Lee. 1989. Numerical simulation of drainage flow in Brush Creek. *Journal of Applied Meteorology* 28:530–42.

Pielke, R. A. 1984. *Mesoscale Meteorological Modeling.* New York: Academic Press.

ter Braak, C. J. F., and C. W. N. Looman. 1986. Weighted averaging, logistic regression, and the Gaussian response model. *Vegetatio* 3–11.

Zhong, S., J. M. Leone, and E. S. Takle. 1991. Interaction of the sea breeze with a river breeze in an area of complex coastal heating. *Boundary-Layer Meteorology* 56:101–39.

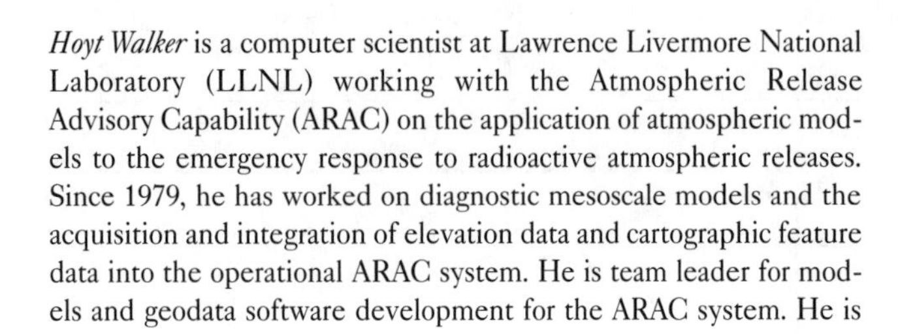

Hoyt Walker is a computer scientist at Lawrence Livermore National Laboratory (LLNL) working with the Atmospheric Release Advisory Capability (ARAC) on the application of atmospheric models to the emergency response to radioactive atmospheric releases. Since 1979, he has worked on diagnostic mesoscale models and the acquisition and integration of elevation data and cartographic feature data into the operational ARAC system. He is team leader for models and geodata software development for the ARAC system. He is also a Ph.D. candidate in geography at the University of California, Santa Barbara. He is studying the relationship between geographic data and predictive mesoscale atmospheric models.

John M. Leone, Jr. has been part of LLNL's finite element fluids group since 1981, investigating the development and application of finite element methods to computational fluid dynamics with a special interest in atmospheric flows. He has been intimately involved in the development of two of LLNL's finite element models: FEM-PBL and SABLE. He is a member of the science team for the Department of Energy's Atmospheric Studies in Complex Terrain (ASCOT) program and has conducted simulations of various ASCOT problems using the ASCOT field data to develop an understanding of the physical processes involved in nocturnal flow and to test and refine the LLNL models.

Hoyt Walker and *John M. Leone, Jr.*
Lawrence Livermore National Laboratory
L-262, PO Box 808
Livermore, California 94550
Fax: (510) 423-4527
E-mail: *hwalker@foggy.llnl.gov*

21

Spherical Spatial Interpolation and Terrestrial Air Temperature Variability

Scott M. Robeson and Cort J. Willmott

Terrestrial air temperature averages are typically obtained from irregularly spaced station data by interpolating to a regular grid and summing the grid-point estimates. Methods of spatial analysis clearly play a role in determining climatological estimates of air temperature and air temperature change. Spatial analysis of interpolation methods and associated errors additionally provides information about the strengths and weaknesses of the station network. Several spherically based interpolation methods—such as inverse distance-weighting, triangulated surface patches, and smoothing by thin plate splines—are evaluated and compared in this chapter. Cross-validation is used to evaluate the interpolation methods and the station network for both air temperature anomalies and actual air temperatures. Cross-validation analysis of air temperature anomalies suggests that errors are nontrivial and, for some years, network bias may be as large as estimates of temperature trends over the last century. Actual air temperature fields may be estimated in several ways, although we have had considerable success by using additional information from a high-resolution climatology.

INTRODUCTION

Air temperature is the most commonly used and perhaps the most comprehensive indicator of global change. Estimating air temperature variability, therefore, is of considerable importance. Although a variety of approaches (e.g., numerical modeling and the analysis of satellite and paleoclimatic data) have been used to analyze air temperature variability, estimates made from observational station networks provide direct measurements over the last few centuries. Estimates of air temperature and air temperature change made from station data, nonetheless, are subject to several types of error. While observational errors have been identified at station locations, errors related to interpolating from sparsely and highly irregularly distributed station networks have not been assessed adequately.

Climatologists have always looked for trends in data; however, concern for human-induced climatic change has led to even closer scrutiny of historical data sources. While much of the recent literature has focused on problems associated with air temperature measurements over the last century (e.g., changing observing practices, urbanization, etc.), perhaps even more uncertainty lies in the uneven and changing spatial sample of climate (Willmott et al. 1991). With poorly conditioned sampling (station) networks, methods of spatial analysis become even more critical to the accurate estimation of global change.

AIR TEMPERATURE VARIABILITY

Resolution and Filtering

Air temperature varies over a wide array of spatial and temporal scales. Much spatial and temporal variability, however, can be eliminated through temporal averaging or filtering processes. Most studies of long-term air temperature change, for instance, have used monthly averages of daily air temperature data (e.g., Jones et al. 1986; Hansen and Lebedeff 1987). When viewed as a rectangular filter, however, the monthly averaging procedure (applied to daily values) is not an optimal way to remove high-frequency information (Hamming 1977). The original data are in the form of daily maxima and minima; but, unfortunately, they are not recoverable for many stations.

Spatial variability is even more problematic since air temperatures are measured at sparse and irregularly spaced locations. Sparse and irregular observation station networks can produce samples with poor resolution and, in turn, spatially aliased air temperature patterns. Unlike regular sampling intervals, the minimum resolvable scale is not well defined for an irregularly spaced sample. Therefore, there is no straightforward choice for grid size. Several studies (e.g., Jones et al. 1986; Yamamoto and Hoshiai 1979) have opted for coarse grids, presumably to match data resolution. In order to: (1) avoid spatial aliasing, (2) produce more accurate spatial integrations, and (3) generate visually realistic fields, however, fine grids probably are a better choice. While aliasing within the historical air temperature network cannot be removed, interpolating from a dense station network to a coarse grid can further alias air temperature signals, distorting space-time patterns (Daley 1991). The original spatial resolution of the data, nonetheless, should be kept in mind when analyses of the gridded fields are performed.

Topographic and Latitudinal Variability

Both topography and latitude exert strong and well-understood influences on the spatial variability of air temperature. To reduce the spatial variability associated with topographic and latitudinal variability, a station mean may be removed from air temperature time series, creating air temperature anomaly series (e.g., Jones et al. 1986; Hansen and Lebedeff 1987). A smoother spatial field is produced while the original temporal variability is maintained at each station. Nearby stations that are located at different elevations consequently become more comparable.

Converting air temperatures to anomalies reduces the effects of elevation and latitude, although useful information is removed with the station mean. Anomalies are most useful for examining climatic change within air temperature records whereas actual air temperatures are needed for other applications (e.g., for verification of climate-modeling studies). Methods for interpolating and analyzing both air temperature anomalies and actual air temperatures are considered below.

SPATIAL INTERPOLATION

Before comparing several interpolants using air temperature data, we present a brief summary of spatial interpolation algorithms. Several extensive reviews of spatial interpolation methods provide additional information (Burrough 1986; Franke 1982; Lam 1983; Shumaker 1976).

Spatial interpolation can be thought of as an estimation problem. Linear combinations of observed values usually are used to estimate values at unsampled locations. Irregularly distributed data are combined or weighted to produce a value at each node of a regular grid. Most procedures, however, treat the spatial dimension as a planar one. Since geographic data are located on the surface of the earth, interpolating on the surface of a sphere—or perhaps an ellipsoid or geoid—is a more consonant approach. Further, when interpolating over large areas of the earth, planar interpolation methods (i.e., interpolation within a cartographic projection) can produce large errors (Willmott et al. 1985). Not only will planar interpolation necessarily result in interpolation errors, but each cartographic projection will produce a different error field from the same data. Using air temperature data, Willmott et al. (1985) showed that, when compared to an equivalent spherical procedure, planar interpolation methods produced errors as large as 10°C in data-sparse regions.

Nearly all interpolation methods may be classified into one of three somewhat distinct categories: (1) distance-weighting (e.g., kriging); (2) tesselation (e.g., triangulation); and (3) functional minimization (e.g., least squares estimation of polynomial surfaces). Distance weighting involves applying a weight to each observation based on its distance from a point where an estimate is desired. How best to determine the weighting function is still a matter of debate (e.g., Bussieres and Hogg 1989; Weber and Englund 1992). Tesselations are a commonly used spatial analysis tool that have wide applicability beyond interpolation (Okabe et al. 1992). A spatial decomposition (often a Delaunay triangulation or the dual Voronoi diagram) is first performed and then a surface is fit over each patch within the decomposition. Tesselation methods have long been used in climatology (e.g., Thiessen 1911). Natural neighbor interpolation provides a nice combination of distance-weighting and tesselation methods (Watson and Philip 1987). Functional minimization is a straightforward application of well-known methods of mathematical analysis. An objective function is established that requires some degree of fidelity between the observations and (usually) a linear combination of basis functions.

While numerous methods of spatial interpolation are available, few spherical interpolants have been developed (although many planar interpolants could be adapted to the sphere). Fortunately, a spherical interpolant from each of the above categories is available. Shepard's (1968) inverse distance weighting algorithm was adapted to the sphere by Willmott et al. (1985). Renka (1984) developed a spherical triangulation routine that applies either a C^0 (continuous and 0 times differentiable, or linear) or C^1 (cubic with constraints) surface to each spherical triangle. Wahba (1981) adapted two-dimensional smoothing splines (thin plate splines) to the sphere.

COMPARISON OF INTERPOLANTS

Three spherical algorithms mentioned above are compared and evaluated using air temperature data. As these three methods represent different approaches to spatial interpolation, comparing and contrasting the estimated temperature fields allows several generalizations to be made.

Interpolation Error

One of the most useful ways to evaluate an estimator is through cross-validation (Efron and Gong 1983). Cross-validation, as implemented here, entails: (1) removing one observation; (2) interpolating the removed value; and (3) repeating (1) and (2) for every observation. For air temperature, the process involves removing one station from the observation station network and, then, using data from surrounding stations, interpolating to the removed station location (to obtain $\hat{T}_i$). Symbolically

$$\hat{T}_i = f(T_1, T_2, ..., T_k, ..., T_n) \; i \neq k, \qquad (21\text{-}1)$$

where f represents a particular spatial interpolation method, and the T_k are air temperatures surrounding the removed station i. By comparing observed and interpolated values at each station, a direct measure of interpolation error can be obtained, allowing the relative merits of each procedure to be assessed quantitatively. By mapping $\hat{T} - T_i$, some spatial features of interpolation error become evident. Data from Jones et al. (1991) for 1890 are used to illustrate interpolation errors for sparse station networks. Interpolation methods also will be compared statistically for all years in the Jones et al. (1991) archive.

Air temperature anomalies: For the station network of 1890 (Figure 21-1), cross-validation error maps for Willmott et al.'s (1985) method (WRP) and Renka's (1984) C^0 approach (RC^0) are much more similar to each other than either is to Wahba's (1981) thin plate splines (TPS) (Figure 21-2). In regions where WRP and RC^0 have large positive errors (e.g., much of Africa), TPS has negative errors. Part of the dissimilarity between the methods is a result of WRP and RC^0 being local interpolants while TPS is a global method. Perhaps even more important than the differences between methods is the magnitude of the cross-validation errors. All methods produce large areas with absolute errors exceeding 0.75°C. Areas with few stations (e.g., Africa, South America, northern Canada) exhibit the largest errors, although some relatively densely sampled areas (e.g., parts of Europe and the United States) have errors exceeding 0.5°C.

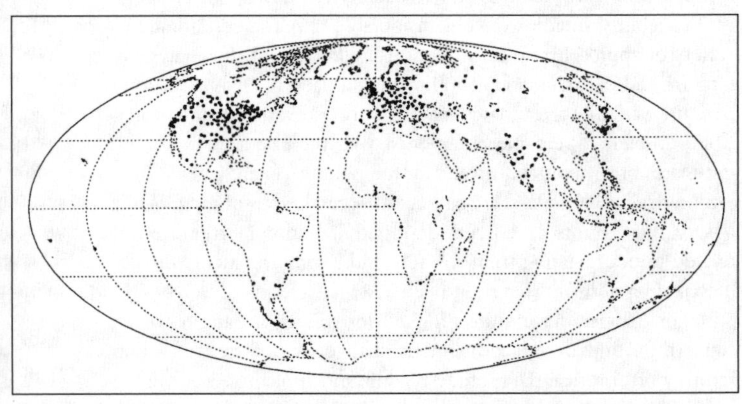

Figure 21-1. Spatial distribution of the 363 air temperature stations available for 1890.

Cross-validation errors at the station locations also can be averaged spatially to estimate trends. Time series of terrestrial mean absolute errors (MAEs) from 1881–1988 were estimated for both WRP and RC^0(Figure 21-3). Cross-validation errors for TPS are available only for a few sparse networks (open circles in Figure 21-3). RC^0 and TPS methods exhibit a

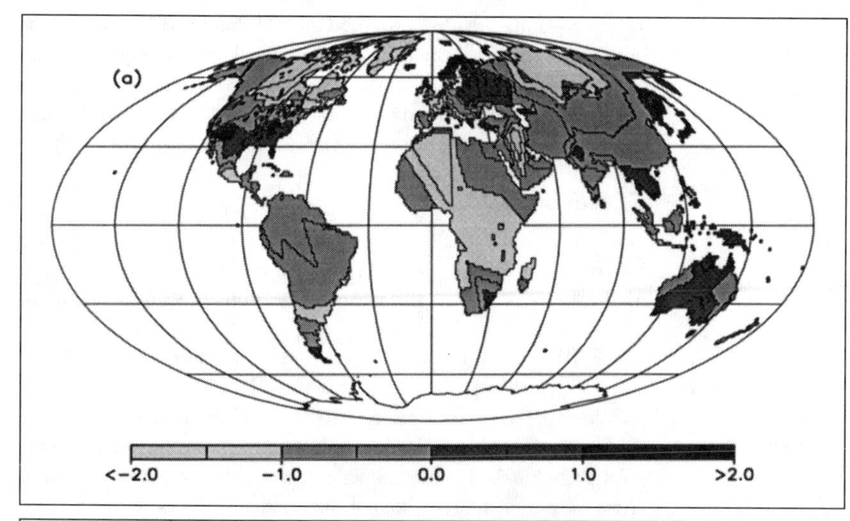

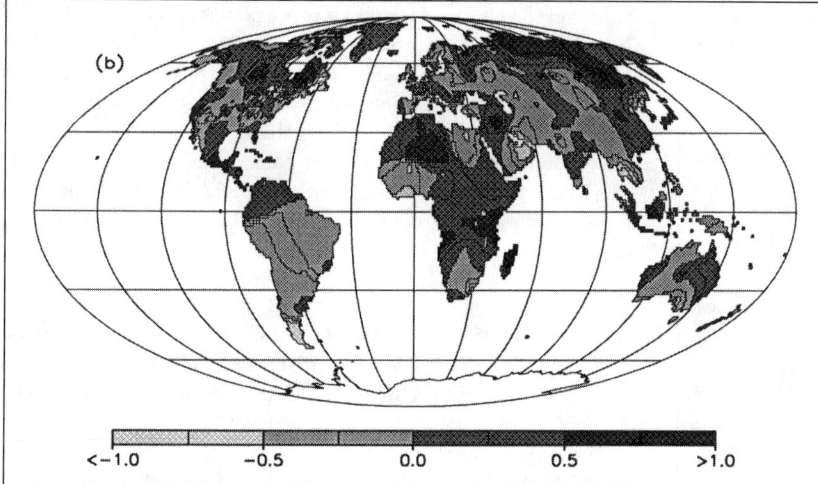

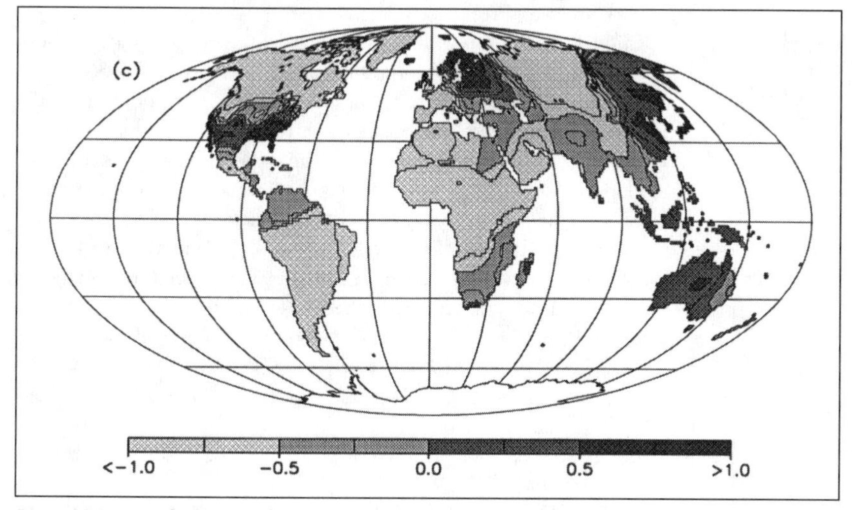

Figure 21-2. Interpolation errors from 1890 cross-validation analysis using (a) RC⁰, (b) WRP, and (c) TPS with air temperature anomaly data.

consistently higher MAE than WRP. When 95% bootstrapped confidence intervals are calculated, however, there often is no statistically significant difference between the cross-validation errors of WRP and RC⁰. Since TPS do not attempt to pass exactly through the data points, direct comparisons between TPS and the interpolations from WRP and RC⁰ are difficult. Renka's second method—the C¹ algorithm—produced even larger cross-validation errors (>1°C for all networks). A modification to the C¹ method that applies arbitrary tension to the triangular surface (Renka 1992, personal communication) also was tested. While variable tension did improve the C¹ results, interpolation errors still were much larger than those of the C⁰ method.

While MAEs of ~0.4°–0.5°C may seem small, anomaly air temperature data exhibit very low variability, particularly when compared to actual air temperature data. Spatial standard deviations for anomaly data (calculated over the entire terrestrial surface) are ~0.5°C. Spatial interpolation of air temperature anomalies, in other words, can produce very large relative errors. It also is disturbing that average interpolation errors of 0.4°–0.5°C are comparable in size to estimated mean air temperature increases over the last century (~0.5°C per 100 years).

Actual air temperatures: Cross-validation time series illustrate that air temperatures are more difficult to interpolate than air temperature anomalies, primarily due to the influence of topography. Mean absolute interpolation errors for actual air temperature range from ~1.3°C for the densest networks to ~1.9°C for the sparsest ones (Figure 21-4). Cross-validation errors also show that WRP seems to perform better than the RC⁰ method for air temperature data, with statistically significant differences in most years.

When using traditional methods of spatial interpolation, the size of cross-validation errors suggests that none of the available station networks represent terrestrial air temperature variability adequately. Interpolation accuracy can be improved, however, by including information from a high-resolution air temperature climatology (Legates and Willmott 1990). We call this climatologically aided interpolation (CAI). It also should be noted that the Legates and Willmott climatology contains an order of magnitude more stations than the best time-series networks.

To incorporate high-resolution climatological information into spatial interpolation for a particular year, long-term station means (from the climatology) are subtracted from annual station temperatures. Deviations from the long-term station means then are interpolated to a regular grid. Once the deviations are gridded, climatological averages at the grid points ($\overline{T}_i$) are added to the interpolated deviations to form the interpolated estimates ($\hat{T}_i$). When simple inverse-distance weighting is used, then the estimated annual air temperature would be

$$\hat{T}_i = \overline{T}_i + \sum_{j=1}^{n} \omega_{ij} T'_j \qquad (21\text{-}2)$$

where $T'_j = T_j - \overline{T}_j$ is the deviation from the climatological mean at station j, ω_{ij} is the weight for station j relative to grid-point i, n is the number of stations near to j, and $\sum_{j=1}^{n} \omega_{ij} = 1$. In this way, CAI considers both the spatially high-resolution climatology as well as current information for a given year at each station.

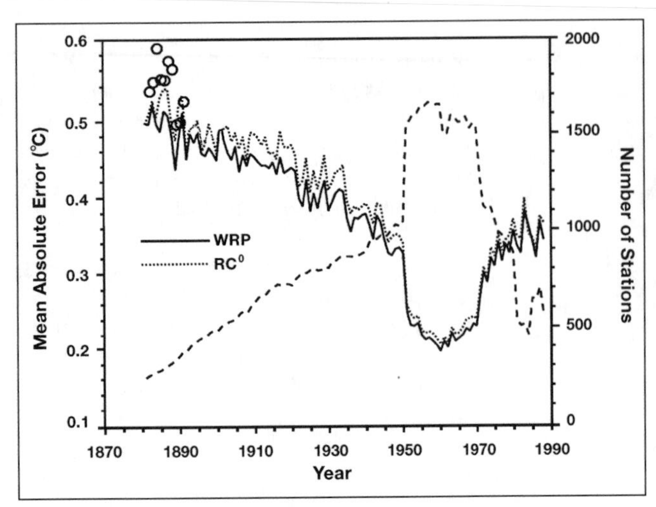

Figure 21-3. Time series of integrated mean absolute error (MAE) from cross-validation analysis using WRP (solid line), RC⁰ (dotted line), and TPS (open circles) with air temperature anomaly data. Dashed line shows the number of stations for each year.

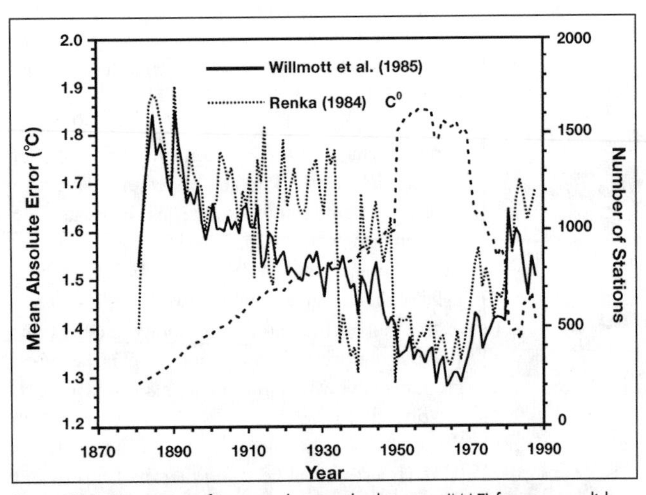

Figure 21-4. Time series of integrated mean absolute error (MAE) from cross-validation analysis using WRP (solid line) and RC⁰ (dotted line) with actual air temperatures. The dashed line gives the number of air temperature stations for each year.

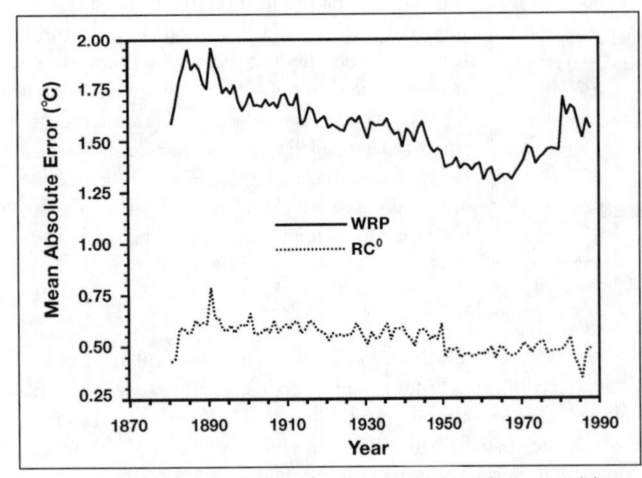

Figure 21-5. Time series of integrated mean absolute error (MAE) from cross-validation analysis using WRP (solid line) and CAI (dotted line) with actual air temperature data.

Cross-validation was performed for CAI using WRP to interpolate both the $\overline{T}_i$ and the deviation field, with substantial improvements over the results using only air temperatures (i.e., not using CAI; Figure 21-5). Interpolation errors are reduced by over two-thirds and approach interpolation errors associated with anomaly data.

CONCLUSION

Since most climatic data are irregularly distributed, values usually are interpolated to a spherical grid for visualization and averaging. Three methods—triangulated surface patches (Renka 1984; C⁰ method), inverse distance-weighting (Willmott et al. 1985), and thin plate splines (Wahba 1981)—were used to interpolate air temperatures and air temperature anomalies. Extensions of Renka's methods and climatologically aided interpolation (CAI) also were employed.

Cross-validation was used to evaluate errors and compare the interpolation methods. For air temperature anomaly data, cross-validation errors were similar for all interpolation methods. Errors, however, were rather large, with average errors for sparse networks being nearly 0.5°C. Denser networks produced lower average errors (~0.2°C). For actual air temperature data, interpolation errors were much larger—on the order of 1.5°C. More accurate results were obtained using CAI, with errors of ~0.6°C. While the average error for CAI is nearly the same as anomaly interpolation error, the relative error (relative to the variance of the field) is much lower for CAI. Regional scale variability induced by interpolation error and network bias can, in some cases, be larger than estimated air temperature increases over the last century.

References

Burrough, P. 1986. *Principles of Geographical Information Systems for Land Resources Assessment.* Oxford: Oxford University Press.

Bussieres, N., and W. Hogg. 1989. The objective analysis of daily rainfall by distance weighting schemes on a mesoscale grid. *Atmosphere Ocean* 27(3):521–41.

Daley, R. 1991. *Atmospheric Data Analysis.* Cambridge: Cambridge University Press.

Efron, B., and G. Gong. 1983. A leisurely look at the bootstrap, the jackknife, and cross-validation. *The American Statistician* 37(1):36–48.

Franke, R. 1982. Scattered data interpolation: Tests of some methods. *Mathematics of Computation* 38:181–200.

Hamming, R. 1977. *Digital Filters.* Englewood Cliffs: Prentice-Hall.

Hansen, J., and S. Lebedeff. 1987. Global trends of measured surface air temperature. *Journal of Geophysical Research* 92(D11):345–413.

Jones, P., S. Raper, R. Bradley, H. Diaz, P. Kelly, and T. M. L. Wigley. 1986. Northern hemisphere surface air temperature variations: 1851–1984. *Journal of Climate and Applied Meteorology* 25:161–79.

Jones, P., S. Raper, S. Cherry, C. Goodess, T. Wigley, B. Santer, and P. Kelly. 1991. *An Updated Global Grid Point Surface Air Temperature Anomaly Data Set: 1851–1990.* Oak Ridge: CDIAC, Oak Ridge National Laboratory.

Lam, N. 1983. Spatial interpolation methods: A review. *American Cartographer* 10:129–49.

Legates, D., and C. Willmott. 1990. Mean seasonal and spatial variability in global surface air temperature. *Theoretical and Applied Climatology* 41:11–21.

Okabe, A., B. Boots, and K. Sugihara. 1992. *Spatial Tesselations.* New York: John Wiley and Sons.

Renka, R. 1984. Interpolation of data on the surface of a sphere. *ACM Transactions on Mathematical Software* 10(4):417–36.

Shepard, D. 1968. A two-dimensional interpolation function for irregularly spaced data. In *Proceedings of the 23rd National Conference, ACM,* 517–23.

Shumaker, L. 1976. Fitting surfaces to scattered data. In *Approximation II*, edited by G. G. Lorentz, et al. New York: Academic Press.

Thiessen, A. 1911. Precipitation averages for large areas. *Monthly Weather Review* 39:1,082–84.

Wahba, G. 1981. Spline interpolation and smoothing on the sphere. *SIAM Journal on Scientific and Statistical Computing* 2(1):5–16.

Watson, D., and G. Philip. 1987. Neighborhood-based interpolation. *Geobyte* 2(2):12–16.

Weber, D., and E. Englund. 1992. Evaluation and comparison of spatial interpolators. *Mathematical Geology* 24(4):381–91.

Willmott, C., S. Robeson, and J. Feddema. 1991. Influence of spatially variable instrument networks on climatic averages. *Geophysical Research Letters* 18(12):2,249–51.

Willmott, C., C. Rowe, and W. Philpot. 1985. Small-scale climate maps: A sensitivity analysis of some common assumptions associated with grid-point interpolation and contouring. *American Cartographer* 12(1):5–16.

Yamomoto, R., and M. Hoshiai. 1979. Recent change of the northern hemisphere mean surface air temperature estimated by optimum interpolation. *Monthly Weather Review* 107:1,239–44.

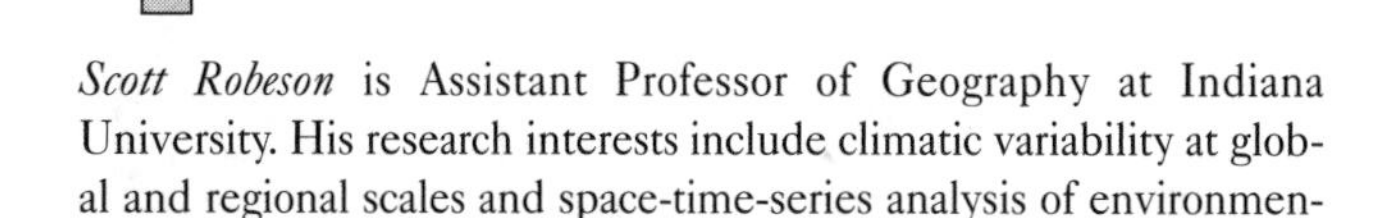

Scott Robeson is Assistant Professor of Geography at Indiana University. His research interests include climatic variability at global and regional scales and space-time-series analysis of environmental processes.
Department of Geography
Indiana University
Bloomington, Indiana 47405
E-mail: *srobeson@indiana.edu*

Cort Willmott is Chair of the Department of Geography at the University of Delaware, Professor of Geography and Marine Studies, and Director of the University's Center for Climatic Research. His research interests include the relationships between land surface processes and climate and the statistical analysis of large-scale climate fields.
Department of Geography
University of Delaware
Newark, Delaware 19716
E-mail: *willmott@brahms.udel.edu*

22

Dynamic Crop Environment Classification Using Interpolated Climate Surfaces

John D. Corbett

Accurate identification and characterization of production zones (and potential zones) is vital to agricultural research. New opportunities exist to greatly improve the mechanisms for characterizing agricultural possibilities. We are no longer limited by historical cartographic restrictions (a static set of boundaries). Past approaches have proven useful, but new methods for accurate spatial interpolation of monthly mean climate permit the development of approaches tailored to specific applications. The accuracy of the interpolated temperature and rainfall surfaces is due to the incorporation of critical dependencies on elevation. The climate surfaces offer robust reproducibility and are built with far more data than traditional cartographic techniques could hope to handle. Subsequent multivariate analyses eliminate the need for a classification method based on fixed discrete temperature, precipitation, and elevation ranges. Most importantly, the classification mechanism targets individual systems, making boundary conditions more precise. Climate surface data are clustered using only data important to the goal of the interpretation. We can thus classify space based on criteria specific to each crop (or ecosystem) and in terms reflecting the complexity of genetic variation within individual crop species.

INTRODUCTION

Accurate identification and characterization of production zones (and potential zones) are vital to agricultural research and development because of their strong effect on the transfer of agrotechnological innovation. Adoption of innovation is dependent on the identification of suitable environmental and human domains. The International Maize and Wheat Improvement Center (CIMMYT, Centro Internacional de Mejoramiento de Maiz y Trigo) has a global mandate which includes the provision of improved maize and wheat germ plasm to the national programs of developing countries. CIMMYT developed the concept of the "megaenvironment" or ME as a means of organizing and focusing its plant breeding programs. An ME is defined as a broad (not necessarily contiguous and frequently transcontinental) area, characterized by similar biotic and abiotic stresses, cropping system requirements, and consumer preferences. Each megaenvironment represents a minimum of 1,000,000 ha, which results in broad production areas where an international center has a comparative advantage. The key to designing crop-specific MEs lies in how they relate to target environments as described by plant breeders.

Initially, agroecological zonation provided a methodology for identifying crop suitability zones, as in the FAO agroecological model (e.g., FAO 1981). However, static, cartographically based, environmental classifications fail to provide sufficiently specific information. A dynamic, spatially oriented approach allows more accurate environmental characterization; integration with critical data on production systems; and the assessment of changing production frontiers (e.g., genotypes) and constraints (ranging from diseases to resource access).

The use of Geographical Information Systems (GIS) greatly enhances CIMMYT's capacity to characterize the diverse maize and wheat production environments throughout the world, provided a consistent resolution of climate data is available. It was therefore necessary to develop a methodology to provide these data for the characterization of the target megaenvironments. Agroclimatic variables, such as temperature, are strongly dependent on elevation. The framework for characterizing crop adaptation environments is based on a data set of climate normals (station or point data), robust algorithms for their spatial interpolation, and the use of a global digital elevation model (DEM). GIS provide the tools to store, manipulate, and process these data.

This chapter describes a dynamic approach toward characterization of environments specific to a crop and crop genotypes. It comprises six parts: (1) the context of agroclimatic characterization, (2) an overview of the dynamic framework, (3) an introduction to climate surface construction, (4) the crop classification technique, (5) applications, and (6) concluding remarks.

The Agroclimatic Context

Over the past decades, environmental characterization of agricultural areas has been the subject of many research efforts (Koppen 1936; Thornthwaite 1948; FAO 1981) that integrated available data and expert opinion and provided powerful interpretations of the resource base for agricultural development. However, recent innovations have demonstrated the limits of traditional "agroecological" models that rely on static zones, cartographic reproduction, and fixed crop environment relationships. Most crops have far more environmental sensitivity than can be handled by such traditional, generalized agroecological zonation strategies.

To illustrate these limitations, the agroecological zones in Mexico which, according to FAO (1981), are suitable for rain-fed maize production are presented in Figure 22-1. Actual rain-fed maize production (1989—dissaggregated to the development district level, a tertiary political unit) has been overlaid with FAO suitability zones. Over 22% (1,340,000 ha) of Mexico's rain-fed maize areas are located in regions the FAO system describes as not suitable, illustrating the inadequacy of the FAO approach. Additionally, the Yucatan Peninsula is described by FAO as having very suitable conditions for maize production, although little maize is grown there due to shallow (lithic phase) soils. The International Center of Tropical Agriculture ([CIAT] Centro Internacional de Agricultural Tropical, Cali, Colombia) found that in Mexico over 50% of beans (*Phaseolus*) are grown in areas "not suitable," according to the FAO system. All of this points to a fundamental difficulty: it is impossible to target genotypes using zones that are not even relevant to the crop in question.

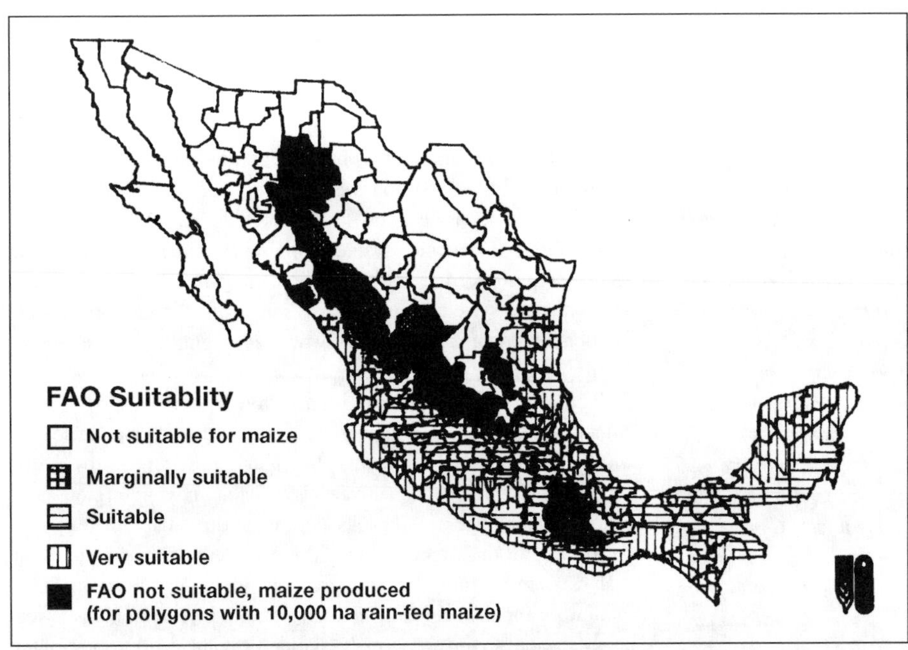

Figure 22-1. FAO agroecological suitability for rain-fed maize and rain-fed maize production by *Distrito Desarollo Rural.*

The major limitations to the traditional, static, agroecological classification, are:

1. A limited and fixed number of climatic parameters. Combinations of variables (from humidity and dewpoint, to extreme events and prediction of future conditions, e.g. warming) and boundary conditions are needed to effectively characterize environmental suitability for each crop and crop genotype.
2. Definition of a single set of zones applicable to all crops.
3. Use of discrete ranges of climatic variables may lead to serious errors. For example, if lowland areas only occur between 0 and 1,000 m elevation, lowland-type environments at slightly higher or lower elevations are, by definition, excluded. Hence, such zoning cannot possibly accurately reflect the potential distribution of individual crops, genotypes, or ecosystems.
4. Lack of risk assessment (timing, frequency, and intensity of abiotic and biotic constraints) excludes a major factor in characterizing cropping environments. For example, farming systems in semiarid areas are characterized by climatic variability rather than climatic normals.
5. Inability to integrate spatial and temporal scales. Static systems are fixed to one scale, while a dynamic approach can take advantage of higher-resolution data and subsequent modeling as data become available.
6. Ignoring the specific requirements of different genotypes makes the static classification overly simple. For example, maize varieties may take 80–300+ days to mature (depending on genotype and location). Although typical agroecological classification systems attempt to take into account the crops' required growing season temperatures and water availability, recognition of the variation within genotypes is either absent or oversimplified. For example, maize is quite sensitive to moisture stress at flowering. One mechanism that improves drought tolerance is reduced anthesis-silking interval, a trait that can be modified by plant breeding. A dynamic system could identify areas where moisture stress at flowering is a common occurrence and where germ plasm improved for this trait could lead to higher yields. The ability to target areas suitable for a particular germ plasm product both facilitates the prioritization of breeding efforts and increases the efficiency or return per dollar invested in agricultural research.

Tropical agronomists may need to look at the length and "depth" of the dry season, but most cereals need about four months of suitable conditions to complete their growth cycle, and off-season climates may or may not be important. Off-season precipitation can play a significant role in agricultural production, and off-season temperatures can affect the expected spectrum of crop challenges (biotic stresses, i.e., diseases, soil nematodes, insects, etc.), but the primary window of opportunity for crop growth is not well reflected by average annual conditions. From this viewpoint, the term agroecological may in fact be an oxymoron if used to predict zones of both natural vegetation and agricultural potential. The use of twelve months of data to determine "natural" conditions is a weak substitute for determining agricultural suitability. A dynamic classification scheme relies on the base data and the process of sifting through those data—it does not depend on a particular classification system. "Agricultural" and "ecological" describe different types of systems that should and can be characterized by conditions critical to their individual performance without compromising either. Thus, to judge the performance of an orchard or forest or to analyze biodiversity, conditions during all twelve months should be taken into account, while analysis of a field crop should focus only on the months important to the cropping cycle.

The Dynamic Framework

There are at least six broad categories of benefits which support adoption of the dynamic framework of environmental characterization:

1. Crop (or ecosystem) specific.
2. Capacity to evaluate germ plasm diversity.
3. Suitability includes assessment of multiple constraints.
4. Risk assessment.

5. Able to integrate various spatial and temporal scales.
6. Reproducible.

The primary advantage to a dynamic approach is the ability to group similar climates using criteria specific to a crop. Genotypic diversity can be accounted for because the climatic classification uses germ-plasm-specific characteristics of adaptation. In addition, crop characteristics can be combined with their constraints (using different sets of variables) to result in a more precise classification. For example, the most susceptible growth stage of a genotype may overlap with the optimum temperature range for a disease. The defining variables for the crop and the disease can be assessed separately, with the final range or suitability reflecting a combination of both analyses. The genotypes' target areas are then defined by both the crop-climate characteristics and the disease's epidemiology. The result is a digital, reproducible surface, targeted to specific cultivars.

Risk assessment adds considerably to the characterization. For example, semiarid areas are not well characterized by the mean climate. Rather, it is the variability which defines production potential. In fact, for some tropical areas (e.g., eastern Kenya), the rainfall season can be thought of as having a switch: the rains either arrive or they do not. The probability of a good rainfall season has a spatial component (for Kenya related to elevation). Thus the targeting of germ plasm (and advice related to the production system) can be greatly enhanced through an evaluation of rainfall variability over space.

In contrast with traditional cartographic approaches, GIS and spatial databases allow exact duplication of resultant zones. This ability to reproduce digital boundaries provides a systematic mechanism for subsequent improvements in the classification. Improved agroclimatic classification is possible with a better understanding of crop environment interactions or the acquisition of additional data.

The climate surfaces (and soils data) become the heart of an agriculturally oriented database, just as they would for any natural resource database. Identification of similar environments can then be tailored toward the specific characteristics of the target crop, genotype, or ecosystem.

Another reason for implementing a GIS-based methodology is the benefit accrued from creating a digital georeferenced database thus allowing integration of agrotechnical data with census and other socioeconomic data. The implications of agricultural policies can also be evaluated within this framework. Finally, this format facilitates the transfer of the results and conclusions, and of the data themselves.

Climate Surface Construction

An essential requirement for implementing this dynamic framework is the precise and accurate construction of georeferenced climatic surfaces. Hutchinson, Nix, and McMahan (1992) effectively argue the case for using mean monthly data to initially describe climatic variation. Mean monthly climate values show significant variation at both the microscale (individual plant) and mesoscale (local topographic inversions). At a still larger mesoscale, temperature and precipitation are affected by elevation. Temperatures decrease by about 6°C per 1,000-m rise in elevation. The effect of elevation on precipitation is less systematic but still significant. These dependencies can be applied when interpolating monthly mean climate values between meteorological stations. The Laplacian method or thin plate smoothing splines (Wahba and Wendelberger 1980, as cited in Hutchinson et al. 1992) are used to incorporate the independent variable elevation in addition to the two usual spatial variables (latitude and longitude). This technique has been used to interpolate monthly mean maximum and minimum temperatures across Australia, to within a standard error of 0.5°C, and monthly mean precipitation to within a standard error of about 10% (Hutchinson et al. 1992). (For additional information on the use of Laplacian splines, see Hutchinson, this volume.)

Temperature has long been recognized as one of the primary climatic factors determining crop growth. Each plant species, and its complementary biotic pests, exhibit characteristic response curves to temperature, which include a lower threshold, a more or less extended optimum range, and an upper threshold. An accurate description of the spatial distribution of the monthly temperature regime is therefore an essential step toward obtaining a reliable, process-based, agroclimatic classification (Hutchinson 1991b).

Thin plate smoothing splines produce a fitted surface. The fitted surfaces represent climate variables (e.g., monthly mean maximum and minimum temperatures) as trivariate spline functions of longitude, latitude, and elevation. This takes advantage of the strong dependence of climate variables, especially temperature, on elevation, but allows the size of this dependence (e.g., the temperature lapse rate) to vary both spatially and temporally (Hutchinson 1991b). Incorporation of the third dependence, in addition to the usual dependence on longitude and latitude, helps overcome the scarcity of data. Standard interpolation packages rarely provide for more than bivariate interpolation (Hutchinson 1991a). Monthly climate surfaces are constructed from period normal data (minimum and maximum temperature, precipitation, evaporation, radiation, and humidity).

The degree of data smoothing imposed by each fitted surface is determined objectively by minimizing the generalized cross-validation (GCV). This minimizes the predictive error of the fitted surface by implicitly withholding each data point in turn in order to validate the fitted surface. The variance of the errors associated with each data value is then estimated by analogy with linear regression (Wahba 1990) from which an estimate of the *true* standard error of the fitted values can be calculated.

Both precipitation and evapotranspiration show significant, if less systematic, dependencies on elevation. For precipitation, the error is strongly related to the density of stations, that is, more stations improve the surface for this spatially highly variable parameter. On complex terrain, the station density is even more important—as is a detailed DEM.

The process of "fitting" the statistical surface serves to demonstrate a vitally important characteristic of the spline technique. For many developing countries, climate data are scarce and of widely varying quality. The results of the spline interpolation provide feedback to the meteorological station database by providing the deviation for each station relative to the fitted surface. Verification of climate data poses many challenges and one tool to assist in that verification is placing a climate station in its spatial context.

Constructed statistical surfaces are not static. As new data are collected or errors identified and corrected, the surface can be recalculated and used to propose modifications to a classification scheme.

Construction of actual monthly climate surfaces (e.g., for risk assessment) requires an extension of the spline technique that is presently being developed.

A Dynamic Crop-Specific Targeting Technique

The key function of a dynamic system is its capacity to provide results specific to the requirements of each classification. Climate surfaces incorporated in a digital database are available for selecting key variables to discriminate specific characteristics of the (agricultural) system. For example, total seasonal precipitation may be important in differentiating rain-fed maize zones, but it may be equally important to characterize the rainfall pattern throughout the growing season, for example, drier conditions during the third month after planting (near flowering). Cluster analysis is one mechanism to group similar areas.

Pollak and Corbett (1993) give an example of the technique in maize adaptation analysis for the area from Mexico to Panama. Monthly climate surfaces were constructed from data covering over 3,000 meteorological stations (many stations only provided precipitation data, while others were complete synoptic stations). The length of records varied, as did the periods of record (for example, the ten-year period for one station may have been from 1970–1980, while for another, 1975–1985). For temperature, the long-term or standard period mean stabilizes after about five years of measurement. This means that variation in five years of monthly averages captures fairly robustly the long-term means (Hutchinson 1991a). For precipitation, the number of years on record were used to weight the influence of a given station on the resultant climate surface calculation. For stations recording long-term monthly means but not the number of years included in the mean, a "standard" period of ten years was assigned for weighting purposes.

Hierarchical cluster analysis using Ward's minimum-variance clustering method (Ward 1963) was used to group similar cells. The algorithm begins by computing a matrix of squared Euclidean distances between every possible pair of grid cells for each variable. Each cell is initially considered a separate cluster. Based on the analysis, clusters with similar characteristics are merged in a stepwise fashion. Ward's minimum-variance method computes distances between clusters, added over the variables. Within-cluster sums of squares, divided by the total sum of squares, give proportions of variance. At each step, the within-cluster sums of squares are minimized by merging the two most similar clusters, until the theoretical last step, when all cells belong to one cluster.

The cluster analysis was stopped at fifty clusters, because a crop adaptation classification based on subjectively identified criteria (the expert system) is best served by evaluating many clusters (Pollak and Corbett 1993), each placed in its appropriate adaptation category based on climate data. In contrast, if the statistical analysis is allowed to proceed to the final targeted number of categories (for example, ten to twelve maize adaptation groups), the methodology loses its sensitivity to boundary conditions.

Figure 22-2 presents the results of a cluster analysis on seven months of monthly climate data interpreted into maize ecologies or maize adaptation zones. These zones represent the first step toward coordinating breeding efforts, since the criteria used to create the zones were derived from maize breeders' perceptions of the differentiation of maize germ plasm across the spectrum of environmental conditions.

The results illustrate one advantage of a clustering technique using agroclimatic and elevation data, namely, that different environments are no longer combined on the basis of an arbitrary and discrete boundary. For example, midaltitude tropical environments may be located below the "boundary" of 1,000-m elevation and lowland tropical maize environments may occur above 1,000 m depending on specific environmental conditions.

Other applications of the climate database: Use of the climate surface database expands as reliable mechanisms relating monthly data to a variety of agricultural issues are identified. CIMMYT's International Maize Testing Program annually distributes seed for evaluation trials

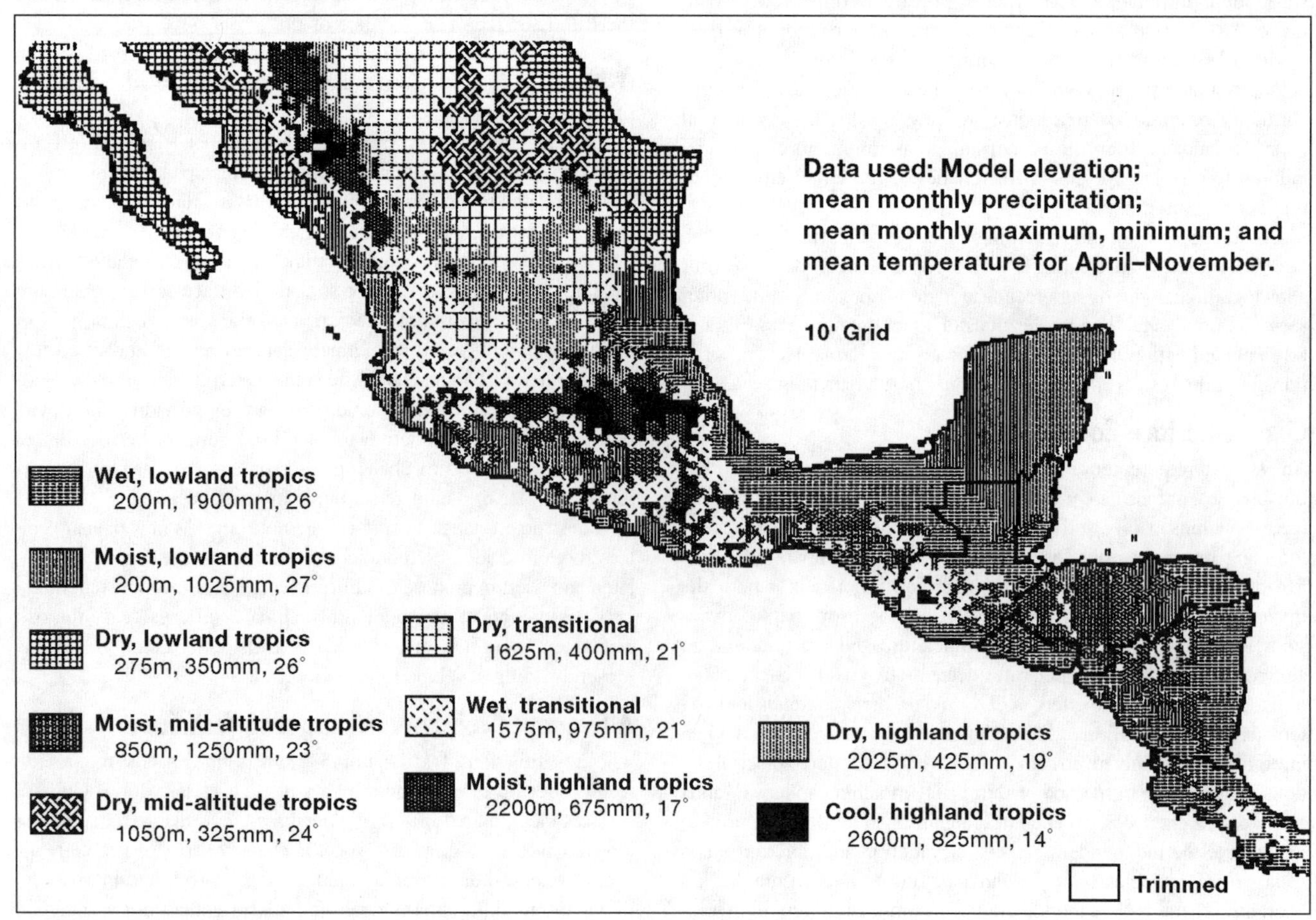

Figure 22-2. Maize ecologies from long-term climate data.

all over the world. Climate surfaces can contribute to the interpretation of the results of international trials through environmental characterization of the test sites. For example, in an iterative process, analysis of international trial results identifies those stations situated in "hot spots" for various biotic stresses (i.e., trial results are screened for the severity of *Exerosilum turcicum*). CIMMYT pathologists have genetic materials with a long history. These "probe" genotypes are included in susceptibility trials planted at various locations to empirically test disease pressure. Empirical models are then constructed (based on climate data) which can be used to extrapolate the results and predict the frequency and intensity of disease pressure elsewhere. The resultant expert diagnosis of a cluster analysis of climate variables, specific to a disease, serves to identify the spatial extent of areas where *E. turcicum* can be expected to be severe. From this information, resistant maize varieties can be better targeted and priority given to improvement for resistance based on demand.

CIMMYT's analysis of twenty-six years of multilocation field trials revealed key characteristics of wheat production environments (DeLacy et al. 1993). Wheat production environments, differentiated on the basis of the wheat germ plasm, can then be identified using environmental characteristics based on the location of the groups of "similar" trial sites. The spatial extent of climatically similar areas can then be assessed, allowing more accurate targeting of wheat germ plasm. For all plant breeders, improved understanding of genotype by environment (G × E) interactions is based on the traditional analysis of field trial results in combination with a biological explanation. A more accurate definition of the environment, as derived from climatic analysis, improves insight of the causal relations underlying G × E interactions.

For a microclimatic analysis, availability and accessibility of daily meteorological data is increasing (globally). Crop simulation models can be used to examine many characteristics of production environments, from land-use planning to optimizing planting dates in response to rainy-season onset to simulating thermal phenological characteristics of particular varieties. The suitability of a specific genotype can be assessed spatially by using these same climate surfaces and interpolated indices suitable for weather generators (for example, we have results using the CERES maize and wheat models, which enable genotype evaluation with respect to spatial variation).

Estimating maize maturity requirements in terms of thermal units demonstrates a relationship to production systems important to tropical agriculture: varietal suitability reflects not only climate and soil conditions but also the needs of the farmer. Many tropical areas with long growing seasons demand early maturing maize so that a second crop can be produced. Production system information has a spatial component that needs to be included in the database. Thus the plant breeder directs his efforts toward improved varieties with full knowledge of the production constraints inherent to the target environment.

Data on actual production system characteristics (typically from census or survey results) can be used in a GIS to further refine target environments. At higher resolutions, soil information is necessary to differentiate production environments and thus target environments for plant breeders. Soil data are often not available in much of the developing world, although small-scale classification systems are available (typically following FAO criteria). Broad geomorphic regions and their characteristics may contribute to the refinement of a classification. Typically, the microscale influences of soil are best handled through computer simulation using actual pedon data (digital databases of georeferenced soil pedons, including their chemical analyses, are being constructed). The results of such simulations can be "extrapolated" based on geomorphic or other terrain and soil information, though such results remain a series of scenarios that are possible within a delineated zone (a table), not mapped as in a summary description.

Finally, global climate surfaces and crop-specific environmental characterization offer a fundamental database for the exploration of the ramifications of global climate change. The potential effects of regional changes in temperature, precipitation, and radiation, as well as increased carbon dioxide concentrations can be tested using a GIS that incorporates potential adaptive responses such as changing to another cultivar, adopting new genotypes, and modifying agronomic practices. Predictions of the types of germ plasm that will be needed in the new climates (and soil environments) are fundamental to enhance the utility of global agricultural research.

CONCLUSION

One of the most powerful characteristics of a dynamic spatial database system is the flexibility afforded individual researchers. Variables can be selected that best differentiate environments based on specific issues. Queries relating environmental characteristics to field trial results strengthen the knowledge base of the entire team involved in improving crop production in the developing world (including plant breeders, physiologists, agronomists, pathologists, systems analysts, economists, policy analysts, administrators, and geographers). Access to these data is facilitated through established computer networks and on-the-shelf software.

A dynamic framework allows evaluation of scenarios not anticipated at the time of database construction. Such spill-over benefits greatly increase the value of database construction using a spatially consistent framework. For example, degradation of the agricultural resource base can be evaluated in terms of: (1) population growth; (2) impact of agricultural technology (e.g., new varieties expanding production area and/or the intensification of agricultural production systems); (3) ramifications of economic policy (e.g., cash crop support versus food security); and (4) dietary preference (e.g., maize versus millet production). The links between spatial data and modeling provide a mechanism to identify target zones which, for much of the developing world, can greatly increase efficiencies in agricultural research.

The capacity to describe long-term climatic normals (and extremes or risk, as needed) is fundamental to formulating an agricultural strategy. The combination of agroclimatic characterization (mesoscale) and computer-simulated crop-environment interactions (microscale) with actual field trial data (for calibration and verification) offers tremendous potential for the accurate assessment of crop suitability as a basis for more efficient use of our land resources.

References

Corbett, J. D., S. C. Chapman, and G. O. Edmeades. 1992. Agronomic interpretation of climate data across space: Using crop models at field and continental scales. Paper presented at the 1992 American Society of Agronomy Annual Meetings, Minneapolis, Minnesota.

DeLacy, I. H., P. N. Fox, J. D. Corbett, J. Crossa, S. Rajaram, R. A. Fischer, and M. van Ginkel. 1993. Long-term association of locations for testing spring bread wheat. *Euphytica* 72(1–2):95–106.

Food and Agricultural Organization of the United Nations. 1981. *Methodology and Results for South and Central America*, vol. 3. Report on the Agro Ecological Zones Project. Rome: FAO.

Hutchinson, M. F. 1990. Climatic analysis in data-sparse regions. In *Climatic Risk in Crop Production: Models and Management for the Semiarid Tropics and Sub-Tropics*, 55–71. Brisbane, Australia.

———. 1991a. The application of thin plate smoothing splines to continent-wide data assimilation. In *Data Assimilation Systems, BMRC Research Report*, no. 27, edited by J. D. Jasper, 104–13. Melbourne: Bureau of Meteorology.

———. 1991b. Estimating the spatial distribution of monthly mean and extreme temperatures across China. *Conference on Agricultural Meteorology*, 319–22. Melbourne: Bureau of Meteorology.

Hutchinson, M. F., H. A. Nix, and J. P. McMahan. 1992. Climate constraints on cropping systems. In *Ecosystems of the World: Field Crop Ecosystems*, edited by C. J. Pearson. Amsterdam: Elsevier.

Kalma, J. I., G. P. Laughlin, A. A. Green, and M. T. O'Brian. 1986. Minimum temperature surveys based on near-surface air temperature measurements and airborne thermal scanner data. *Journal of Climatology* 6:413–30.

Koppen, W., and R. Geiger. 1936. *Handbuch der Klimatologie*. Berlin: Gebr. Borntraeger.

Pollak, L. M., and J. D. Corbett. 1993. Using GIS data sets to classify maize-growing regions in Mexico and Central America. *Agronomy Journal* 85(6):1,133–39.

Thornthwaite, C. W. 1948. An approach to the rational classification of climate. *Geographical Review* 38:55–94.

Wahba, G. 1990. Spline models for observational data. *CBMS-NSF Regional Conference Series in Applied Mathematics*. Philadelphia: Society for Industrial and Applied Mathematics.

Wahba, G., and J. Wendelberger. 1980. Some new mathematical methods for variational objective weather analysis using splines and cross-validation. *Monthly Weather Review* 92:169–76.

Ward, J. H. 1963. Hierarchical grouping to optimize an objective function. *Journal of the American Statistical Association* 58:236–44.

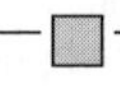

John D. Corbett
International Maize and Wheat Improvement Center
Lisboa 27, Apdo. Postal 6-641
Mexico, D.F. 06600
Mexico
Phone: 52-5-726-9091
E-mail: *j.corbett@cgnet.com*

23

Methodologies for Development of Hydrologic Response Units Based on Terrain, Land Cover, and Soils Data

Bernard A. Engel

Hydrologic/water-quality models use a variety of techniques to represent the areas being simulated. Slight differences in the application of these techniques can result in significantly different estimates of runoff, erosion, and sedimentation. The effects of terrain, land use, and soil-modeling inputs on results of hydrologic/water quality models are examined. The effects of grid cell sizes on model results are also explored. Hydrologic/water quality models that are integrated with GIS provide "laboratories" for such explorations.

INTRODUCTION

Prediction of the hydrologic responses of catchments is required for many purposes. A variety of models are available to predict such responses. The structures commonly used by these models include (Moore et al. 1993):

1. Lumped models that spatially integrate the entire area being modeled.
2. Models based on subdivisions into hydrologic response units (HRUs).
3. Grid-based models.
4. TIN-based models.
5. Contour-based models.
6. Two- and three-dimensional groundwater models.

A hydrologic response unit (HRU) is an irregular but hydrologically similar area. HRUs are usually defined based on land-use properties, soil properties, and topography. The grid-based, TIN-based, and contour-based modeling approaches can be considered as a scaling down and a merging of HRU and routing element structure concepts (Moore et al. 1993).

Lumped models have been the most widely used, largely because of their simplicity. However, inadequacies of lumped models have led to development of other model types as shown above. These other model types have been used primarily as research models until recently when the integration of these models with GIS has made their widespread use feasible. One of the questions associated with the use of these models is how to best define the HRU, grid cell, or TIN.

Wood et al. (1988) introduced the concept of representative elementary area (REA) as a threshold value to quantify runoff generation. They investigated the existence of REA using a modified TOPMODEL. They concluded that at a certain scale, the influence of variability on catchment hydrologic responses is sampled within the area, making it possible to represent the actual catchment response with average values. This area is representative of the continuum for which the mean subcatchment response would stabilize. They also concluded that a watershed's REA is strongly influenced by topography and the length scale of rainfall also influences the size of a watershed's REA. Wood et al. (1990) explored the effects of spatial variability and scale in terms of REA on the quantification and parameterization of a catchment for prediction of flood-frequency analysis. For the watershed investigated, the REA was about 1 km^2.

Many questions remain concerning how to best define an REA, HRU, TIN, or grid cell size for a catchment of interest. The remainder of this chapter explores some of the models that utilize these representation approaches; the influence of terrain, land use, and soil-derived inputs on hydrologic/water quality models that utilize these approaches; and the effect of grid cell sizes on model results.

MODELS THAT USE HRU APPROACHES

HRU-Based Models

HRUs have been most widely used with models that utilize finite element analysis techniques. One of the primary advantages of finite element algorithms over the spatial-time integration of square elements is improved computational efficiency. However, for most applications, computational speed is not the issue it was several years ago.

One of the best-known HRU-based agricultural watershed models is Finite Element Storm Hydrograph Model (FESHM) (Ross et al. 1979; Hession et al. 1987). Wolfe and Neale (1988) and Shanholtz et al. (1990) integrated FESHM with a GIS. FESHM divides a catchment into HRUs determined on the basis of soil types and land use. Spatially distributed rainfall excess and infiltration are computed based on the HRUs. The catchment is also divided into interconnected overland flow elements and channels based on topography and surface flow paths. Area-weighted HRU rainfall excess is then routed from successive elements to the watershed outlet.

Wolfe et al. (1991) describe an object-oriented system for simulating hydrologic processes. Objects are used to represent areas within GIS data layers. The soil data object and the land-use data object are mixed (combined) to obtain hydrologic response units.

Grid-Based Models

Grid-based hydrologic models are perhaps the most common of the more complex hydrologic model types listed above. Areal Nonpoint Source Watershed Environmental Response Simulation (ANSWERS) (Beasley et al. 1982) simulates runoff, erosion, sedimentation, and phosphorus movement from watersheds. A watershed being simulated is divided into a grid of square cells. Runoff, erosion, and sedimentation are computed for each cell and routed. ANSWERS is capable of assessing the effects of land uses, management schemes, and cultural practices on the quality of water leaving a watershed. Its primary applications are watershed planning for erosion and sediment control on complex watersheds and water-quality analysis of sediment-associated chemicals.

The Agricultural Nonpoint Source Pollution model (AGNPS) (Young et al. 1987) has been developed to analyze nonpoint source pollution in agricultural watersheds. Runoff characteristics and transport processes of sediments and nutrients are simulated for each cell and routed. This permits the runoff, erosion, and chemical movement at any grid cell in the watershed to be examined. Thus, AGNPS is capable of identifying sources contributing to a potential problem and prioritizing those locations where remedial measures could be initiated to improve water quality.

SWAT is a continuous spatially distributed watershed model that operates on a daily time step (Arnold et al. 1993). SWAT provides several extensions to the SWRRB model (Arnold et al. 1990). Its primary use is in assisting water resource managers assess water supplies and nonpoint source pollution on watersheds and large basins. SWAT provides considerable flexibility in watershed configuration and discretization, allowing watersheds to be subdivided into cells and/or subwatersheds. SWAT is able to simulate runoff, sediment, nutrient, and pesticide movement through a watershed.

TOPMODEL (TOPography-based MODEL (Beven and Kirby 1979) predicts storm runoff for a catchment using a grid-based approach. TOPMODEL accounts for the effects of catchment topography through a distributed topographic index. It also uses a variable saturated-area concept similar to that used in a simple lumped model. Compared with other distributed parameter models described above, TOPMODEL requires fewer distributed parameter inputs. Numerous researchers describe extensions to the original model (Beven and Wood 1983; Beven 1986, 1987).

Chairat and Delleur (1993) modified TOPMODEL to consider the effects of subsurface drainage systems such as those commonly used for agricultural production. With these modifications, the model performed well for predicting runoff hydrographs from agricultural watersheds in north central Indiana. They also integrated TOPMODEL with the GRASS-GIS to simplify use of the model.

More grid-based hydrologic models have been integrated with GIS than other types of hydrologic models listed above. Yoon et al. (1993) linked a vector-based GIS with AGNPS. Hodge et al. (1988) linked MULTSED (Multiple Watershed Sediment Routing) with the raster-based GRASS-GIS. MULTSED divides a watershed into subbasins for data inputs and simulation purposes. Rewerts and Engel (1991) integrated ANSWERS with a raster-based GIS to simulate runoff, erosion, and sedimentation in watersheds. Srinivasan et al. (1993) and Engel et al. (1993) integrated AGNPS with a raster-based GIS to simulate runoff, erosion, sedimentation, and nutrient movement in agricultural watersheds. Chen et al. (1993) developed a GIS-based transport model to allocate phosphorus yields to fields within a watershed. These integrated systems provide an opportunity to examine the effects of terrain, land use, and soils on the development of HRUs.

INTEGRATED GIS AND HYDROLOGIC MODELS TO DETERMINE HRUs

The derivation of HRUs, selection of TINs, and the selection of grid cell sizes can be a difficult task. For each of these representation structures, the assumption is made that properties of interest (topographic, land use, and soil parameters) are uniform within the structure unit (HRU, TIN, or grid cell). Most GIS tools provide capabilities to assist in selecting such areas. However, these tools can't decide how much "lumping" can be tolerated without adversely affecting model results. In the following sections, integrated GIS and hydrologic modeling systems are used to explore the effects of terrain, land use, and soils on simulated watershed runoff, erosion, and sedimentation. Given this information, one can make better decisions concerning the definition of grid cell size and HRUs.

Effects of Land Use on Hydrologic/Water-Quality Models

Hamed and Engel (1993) explored the effects of land-use GIS layers prepared from four sources on runoff and erosion predictions from an agricultural watershed in north central Indiana using the integrated ANSWERS and GRASS-GIS system (Rewerts and Engel 1991). The four land-use data layers were obtained from a "windshield" survey of the watershed, from interpreted remotely sensed data (Landsat and AVIRIS), from the USGS Land Use Data Analysis GIS layer, and from county crop production statistics. The USGS GIS data layer was modified using county crop production statistics to show specific crops within cropped areas, since the finest level of detail in the USGS data layer categories was crops.

For the watershed simulations with the ANSWERS model, a 1.5-hour-duration, eight-year-return-period rainfall event was selected. An event of this magnitude is often used for design purposes with ANSWERS. Table 23-1 provides a summary of the results obtained.

Table 23-1. ANSWERS results for four land-use maps.

Land use	Runoff (mm)	Peak runoff (mm)	Soil erosion(kg/ha)
Windshield survey	2.80	6.40	192
Interpreted Landsat	3.09	6.88	214
Interpreted AVIRIS	3.48	7.53	285
Modified USGS	2.65	6.04	189

Effects of Topographic Inputs on Hydrologic/Water-Quality Models

Topography plays an important role in the prediction of runoff, erosion, and sedimentation from watersheds. One of the most important parameters is slope. There are four commonly used techniques for estimating slopes from digital elevation maps (DEMs) within GIS (Srinivasan and Engel 1991). These methods are (1) neighborhood, (2) quadratic surface, (3) best fit plane, and (4) maximum slope.

Figure 23-1 shows the distribution of slope values for the Animal Science watershed in north central Indiana obtained using four of the commonly used techniques for computing slopes from DEMs. The watershed is typical of many midwestern watersheds and is approximately 325 ha in size, with slopes that average approximately 1.5%, and soils that are predominantly silty clay loams. Figure 23-2 shows the runoff and sediment yield predicted by ANSWERS at the watershed outlet of this watershed for four events—two, one-hour events with one, two, five, and ten-year-return periods. Results obtained using the topographic inputs derived using the maximum slope algorithm show substantially more runoff and sediment delivery than those obtained from the other methods. Figure 23-3 shows the average erosion and deposition within the watershed as estimated by ANSWERS for the four events. It is interesting to note the differences in average erosion and sedimentation within cells that were obtained. These differences offset one another such that nearly identical sediment yields for the watershed are obtained, as shown in Figure 23-2.

Using the AGNPS model, Srinivasan et al. (1993) explored the effect of these same slope-estimation algorithms on erosion and sediment delivery from watershed Y in Waco County, Texas. Runoff pre-

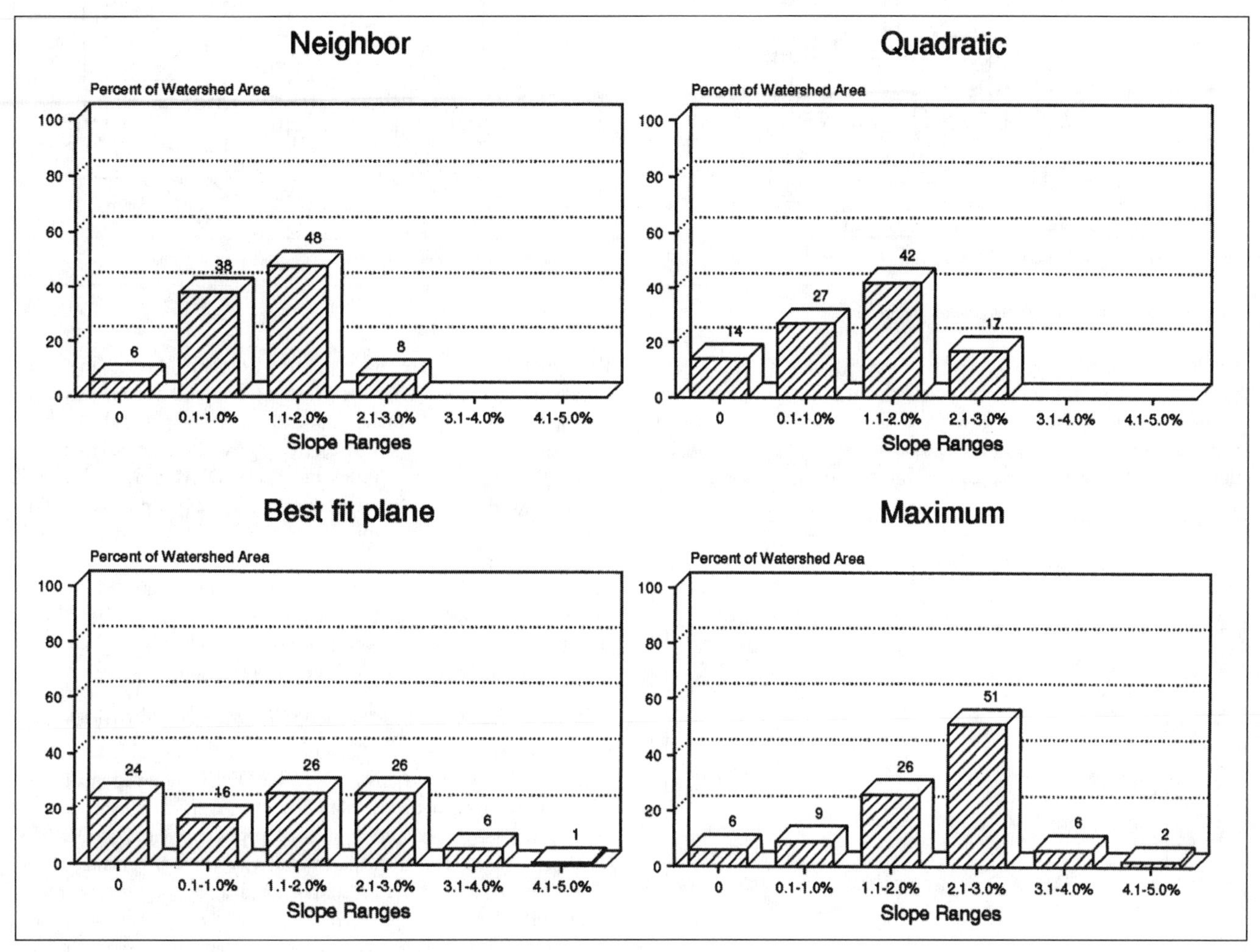

Figure 23-1. Distribution of slope values.

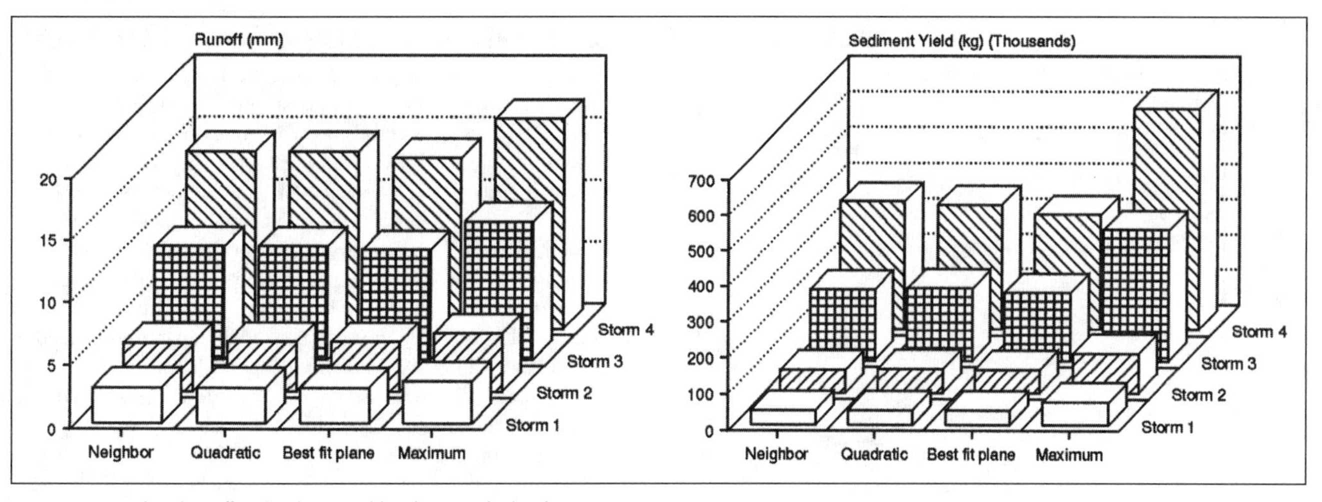

Figure 23-2. Predicted runoff and sediment yield at the watershed outlet.

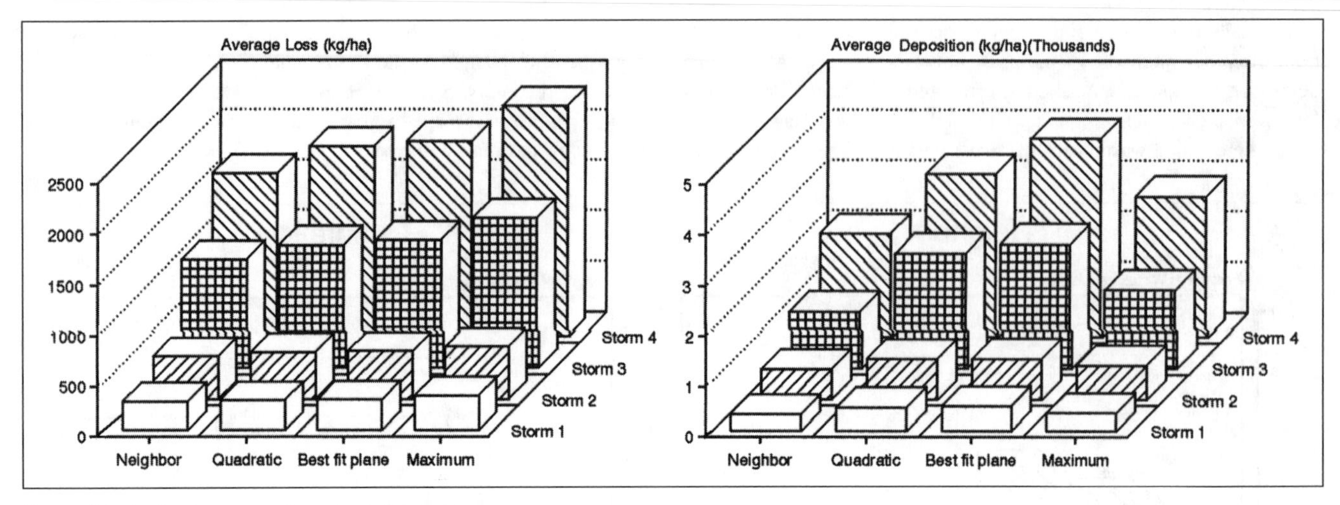

Figure 23-3. Predicted average cell erosion and deposition.

dictions from AGNPS are not sensitive to slope (Srinivasan et al. 1993); however, erosion and sedimentation estimates are quite sensitive. Figure 23-4 shows the average estimated erosion within cells, and Figure 23-5 shows estimated average deposition within cells for ten rainfall events shown in Table 23-2. Simulated sediment delivered to the outlet was compared with observed data for the ten events as shown in Figure 23-6. The best results were obtained using the neighborhood method for estimating slopes (Srinivasan et al. 1993).

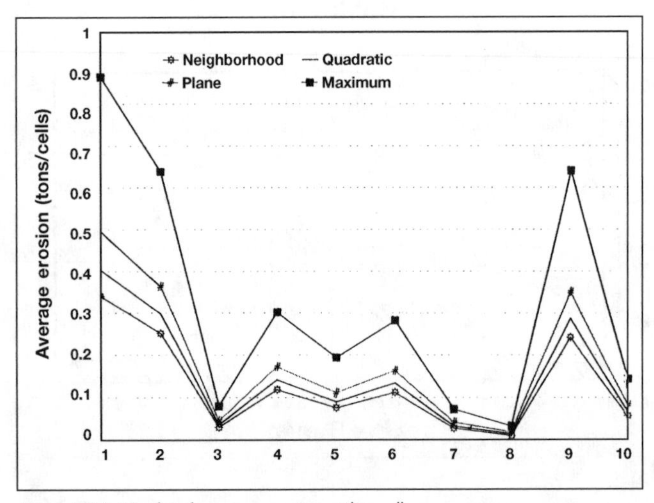

Figure 23-4. Simulated average erosion within cells.

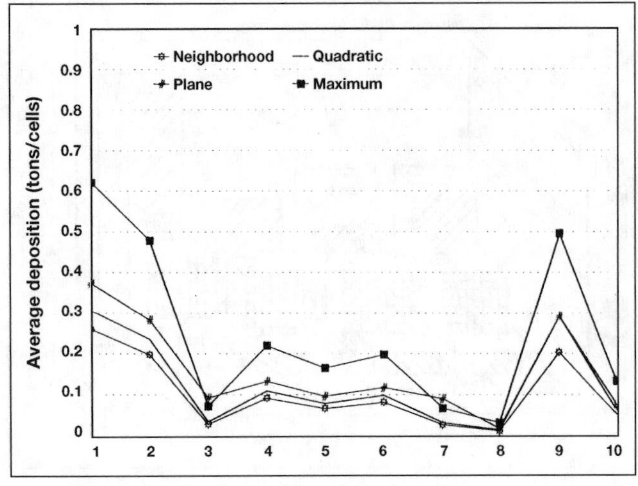

Figure 23-5. Simulated average sedimentation within cells.

Chairat (1993) explored the effects of contour length assumptions on the topographic index used by TOPMODEL. Values of the TOPMODEL diagonal contour length (L2) were varied in simulating runoff from an approximately 325-ha watershed in north central Indiana. Quinn (1990) chose an L2 of the square root of 0.125 dx where dx is the cell resolution. For the watershed explored by Chairat (1993), an L2 value of 0.2 dx gave the best fit to observed data for peak flow-rate prediction, while a value of 0.354 dx yielded the best fit for total runoff prediction.

Effects of Soils on Hydrologic/Water-Quality Models

Chairat (1993) explored the effects of using spatially variable versus uniform hydraulic conductivity within TOPMODEL for an approximately 325-ha watershed in north central Indiana. The average value of hydraulic conductivity for the soils in the watershed was computed using a GIS. The spatial variability of hydraulic conductivity was easily examined since Chairat (1993) had linked TOPMODEL with a GIS. Peak rates of runoff and runoff volumes were predicted to be significantly greater when using spatially variable hydraulic conductivities. However, results obtained using the uniform hydraulic conductivity best matched observed results. Simulated time to peak runoff was identical for both cases examined.

Effects of Grid Size on Hydrologic/Water-Quality Models

The size of HRUs selected can significantly affect the results obtained from the model. The assumption usually made when selecting HRUs and grid cells is that conditions are uniform, or nearly uniform, within the HRU or grid cell. Srinivasan (1992) explored the effect of grid cell size on the AGNPS model for two watersheds: the Animal Sciences watershed in north central Indiana, as described previously, and the Henrietta watershed in Tarrant County, Texas. The Henrietta watershed is approximately 3,275 ha in size, with approximately half of the watershed cropped and half in pasture or rangeland. GIS data for both watersheds was prepared using a 30-m grid cell size. The AGNPS interface with the GRASS-GIS (Srinivasan et al. 1993) was used to explore differences in predicted erosion and sedimentation within these watersheds.

For the Henrietta watershed, cell sizes were varied between 100–400 m, as shown in Table 23-3. For the Animal Sciences watershed, cell sizes were varied between 40–400 m, as shown in

Table 23-2. Observed rainfall, date, antecedent moisture condition (AMC), and energy intensity values for the ten events between 1989–1991 for watershed Y.

Event #	Date	Rainfall (mm)	AMC	Energy intensity (t-m/ha/cm)
1	05/17/89	90.4	I	63.72
2	08/07/89	83.1	I	47.17
3	03/07/90	47.5	I	11.18
4	03/14/90	87.6	I	42.65
5	05/03/90	61.2	I	13.83
6	01/09/91	57.4	II	40.34
7	02/04/91	40.4	I	10.51
8	02/18/91	30.2	I	4.89
9	04/13/91	67.8	I	45.19
10	05/24/91	38.1	I	10.66

Table 23-4. A rainfall of 2.5 inches with a USLE R value of 53 was used for both watersheds. As shown in the tables, differences arose in AGNPS estimates of erosion and sedimentation for the different cell sizes used. Differences would likely be greater for watersheds with more varied topography.

Chairat (1993) explored the effects of grid size (30 m, 60 m, and 90 m) on TOPMODEL results. For this portion of the study, an L2 value of 0.345 dx as recommended by Quinn et al. (1991) was used. Data for the watershed used in the study had been developed at a 30-m resolution within a grid-based GIS. Although differences were not significant, the 30-m resolution provided the best fit between simulated and observed values for both peak flows and total runoff.

CONCLUSION

Hydrologic/water-quality models commonly use one of several structures to represent the catchment of interest. More complex representation techniques have recently become of great interest as a result of the integration of GIS with hydrologic/water-quality models. The complex representation techniques include hydrologic response units (HRUs) which are closely related to the grid-based, TIN-based, and contour-based representation approaches. The delineation of HRUs, TINs, and grids for hydrologic modeling raises several questions. Wood et al. (1988) introduced the concept of representative elementary area (REA) as a threshold value to quantify runoff generation.

Hydrologic/water-quality models that had been integrated with GIS were used to explore the effects of terrain, land use, soils, and grid cell size on runoff, erosion, and sedimentation. The algorithm used for derivation of topographic data from DEMs can have a significant effect on the results obtained with hydrologic/water-quality models. The neighborhood method of slope estimation provided results that gave the best hydrologic/water-quality model results. Land uses for a watershed derived from different sources were used in a hydrologic/water-quality model. Differences in estimated runoff, erosion, and sedimentation were smaller than anticipated. Average and distributed soil properties were used in simulating runoff from a watershed. For the watershed examined, the average soil properties provided a better fit between predicted and observed runoff. The effect of grid cell size on runoff, erosion, and sedimentation was examined for two watersheds. The variation of model results for differing cell sizes was surprisingly large.

The work presented here raises more questions than it answers. At this point, it is difficult to provide guidelines for the selection of HRUs, TINs, and grid cell sizes for hydrologic/water quality modeling. Hydrologic/water-quality models that are integrated with GIS provide tools to develop guidelines for the selection of HRUs, TINs, and grid cell sizes. These integrated systems become "laboratories" for such explorations.

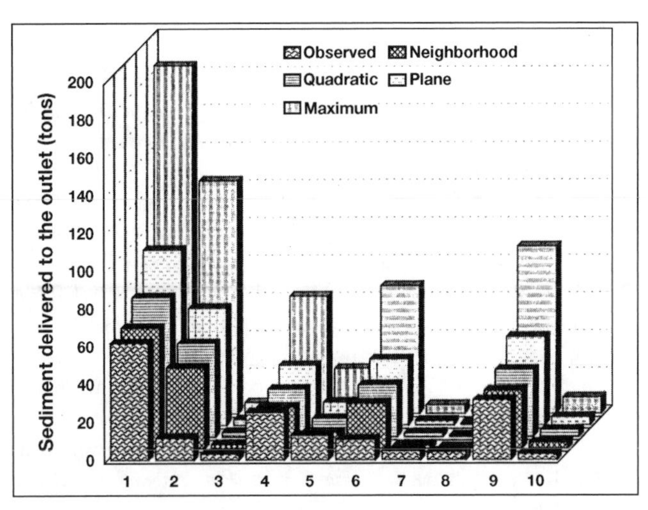

Figure 23-6. Simulated and observed sediment yields.

Table 23-3. Effects of cell resolution on erosion estimates for the Henrietta watershed.

Cell resolution (meters)	Sediment delivered to the outlet (tons)	Average overland erosion (tons/acre)	Average overland deposition (tons/acre)
100	167.3	0.49	0.52
125	170.9	0.48	0.52
200	161.6	0.46	0.46
283	156.2	0.46	0.46
400	309.5	0.42	0.41

Table 23-4. Effects of cell resolution on erosion estimates for the Animal Sciences watershed.

Cell resolution (meters)	Sediment delivered to the outlet (tons)	Average overland erosion (tons/acre)	Average overland deposition (tons/acre)
40	47.6	0.40	0.36
60	46.1	0.39	0.36
80	44.7	0.38	0.35
100	46.0	0.40	0.35
125	49.7	0.30	0.36
200	47.0	0.36	0.32
283	31.4	0.36	0.32
400	38.7	0.31	0.26

References

Arnold, J. G., B. A. Engel, and R. Srinivasan. 1993. Continuous time, grid cell watershed model. *Application of Advanced Information Technologies: Effective Management of Natural Resources.* ASAE Publication 04-93, 267–78.

Arnold, J. G., J. R. Williams, A. D. Nicks, and N. B. Sammons. 1990. *SWRRB: A Basin Scale Simulation Model for Soil and Water Resources Management.* College Station: Texas A&M University Press.

Beasley, D. B., L. F. Huggins, and E. J. Monke. 1982. ANSWERS: A model for watershed planning. *Transactions of ASAE* 23:938–44.

Beven, K. J. 1986. Hillslope runoff processes and flood-frequency characteristics. In *Hillslope Processes,* edited by A. D. Abrahams, 187–202. Winchester: Allen and Unwin.

———. 1987. Toward the use of catchment geomorphology in flood-frequency predictions. *Earth Surface Processes and Landforms* 12:69–82.

Beven, K. J., and M. J. Kirby. 1979. A physically based, variable contributing area model of basin hydrology. *Hydrological Sciences Journal* 24:43–69.

Beven, K. J., and E. F. Wood. 1983. Catchment geomorphology and the dynamics of runoff contributing areas. *Journal of Hydrology* 65:139–58.

Chairat, S. 1993. "Adapting a Physically Based Hydrological Model with a Geographic Information System for Runoff Prediction in a Small Watershed." Ph.D. diss., Civil Engineering, Purdue University, West Lafayette, Indiana.

Chairat, S., and J. W. Delleur. 1993. Integrating a physically based hydrological model with GRASS. *HydroGIS 93, Proceedings of the International Conference on Applications of Geographic Information Systems in Hydrology and Water Resources.*

Chen, Z., D. E. Storm, M. D. Smolen, C. T. Haan, M. S. Gregory, and G. J. Sabbaugh. 1993. Prioritizing nonpoint source loads for phosphorus with a GRASS modeling system. *Proceedings of the Symposium on Geographic Information Systems and Water Resources,* 71–78. AWRA.

Engel, B. A., R. Srinivasan, and C. Rewerts. 1993. A spatial decision support system for modeling and managing agricultural nonpoint source pollution. In *Environmental Modeling with GIS,* edited by M. F. Goodchild, B. O. Parks, and L. T. Steyaert, 231–37. New York: Oxford University Press.

Hamed, K., and B. A. Engel. 1993. Soil Erosion Estimation within the Animal Science Watershed. Class Report for AGEN 526. West Lafayette: Agricultural Engineering Department, Purdue University.

Hession, W. C., V. O. Shanholtz, S. Mostaghimi, and T. A. Dillaha. 1987. *Extensive Evaluation of the Finite Element Storm Hydrograph Model.* ASAE Paper No. 87-2570. St. Joseph: ASAE.

Hodge, W., M. Larson, and W. Goran. 1988. Linking the ARMSED watershed process model with the GRASS geographic information system. *Proceedings of the Modeling Agricultural, Forest, and Rangeland Hydrology Conference,* 501–10. St. Joseph: ASAE

Moore. I. D., A. K. Turner, J. P. Wilson, S. K. Jenson, and L. E. Band. 1993. GIS and land surface-subsurface process modeling. In *Environmental Modeling with GIS,* edited by M. F. Goodchild, B. O. Parks, and L. T. Steyaert, 196–230. New York: Oxford University Press.

Quinn, P. 1990. *Application of the Model TOPMODEL to the Catchment of Booro-Borotou, the Ivory Coast.* Technical Report. Institute of Environmental and Biological Sciences, University of Lancaster.

Quinn, P., K. Beven, P. Chevallier, and O. Planchon. 1991. The prediction of hillslope flow paths for distributed hydrological modeling using digital terrain models. *Hydrological Processes* 5:59–79.

Rewerts, C. C., and B. A. Engel. 1991. *ANSWERS on GRASS: Integrating a Watershed Simulation with a GIS.* ASAE Paper No. 91-2621. St. Joseph: ASAE.

Ross, B. B., D. N. Contractor, and V. O. Shanholtz. 1979. A finite element model of overland and channel flow for assessing the hydrologic impact of land-use change. *Journal of Hydrology* 41:1–30.

Shanholtz, V. O., C. J. Desai, N. Zhang, J. W. Kleene, C. D. Metz, and J. M. Flagg. 1990. *Hydrologic/Water-Quality Modeling in a GIS Environment.* ASAE Paper No. 90-3033. St. Joseph: ASAE.

Srinivasan, R. 1992. "Spatial Decision Support System for Assessing Agricultural Nonpoint Source Pollution Using GIS." Ph.D. diss., Purdue University, West Lafayette, Indiana.

Srinivasan, R., and B. A. Engel. 1991. Effect of slope prediction methods on slope and erosion estimates. *Journal of Applied Engineering in Agriculture* 7:783–99.

Srinivasan, R., B. A. Engel, J. R. Wright, J. G. Lee, and D. D. Jones. 1993. Slope steepness prediction methods effects on topographic attributes and nonpoint source pollution models using GIS. *Transactions of ASAE.*

Wolfe, M. L., and C. Neale. 1988. Input data development for a distributed parameter hydrologic model (FESHM). *Proceedings of the Modeling Agricultural, Forest, and Rangeland Hydrology Conference,* 462–69. St. Joseph: ASAE.

Wolfe, M. L., A. D. Whittaker, and R. Godbole. 1991. Object-oriented simulation of hydrologic process. In *First International Conference/Workshop on Integrating Geographic Information Systems and Environmental Modeling Proceedings.* Santa Barbara: NCGIA.

Wood, E. F., M. Sivapalan, and K. J. Beven. 1990. Similarity and scale in catchment storm response. *Reviews of Geophysics* 28:1–18.

Wood, E. F., M. Sivapalan, K. J. Beven, and L. Band. 1988. Effects of spatial variability and scale with implications to hydrologic modeling. *Journal of Hydrology* 102:29–47.

Yoon, J., G. Padmanabhan, and L. H. Woodbury. 1993. Linking agricultural nonpoint pollution model (AGNPS) to a geographic information system (GIS). *Proceedings of the Symposium on Geographic Information Systems and Water Resources,* 79–87. AWRA.

Young, R., C. Onstad, D. Bosch, and W. Anderson. 1987. AGNPS: A nonpoint source pollution model for evaluating agricultural watersheds. *Journal of Soil and Water Conservation* 44:168–73.

Bernard (Bernie) Engel is an associate professor in the Department of Agricultural Engineering at Purdue University. He teaches courses in soil and water conservation, environmental systems management, agricultural systems engineering, and watershed systems design, including the role of environmental modeling and GIS for such designs. His research interests include the integration of hydrologic and water-quality models with GIS and the application of these integrated systems; the development and application of artificial intelligence techniques to agricultural, natural resources, and environmental problems; and the development of multimedia-based systems to provide natural resources and environmental information.

Agricultural Engineering Department
Purdue University
West Lafayette, Indiana 47907-1146
Phone: (317) 494-1198
Fax: (317) 496-1115
E-mail: *engelb@ecn.purdue.edu*

24

GIS-Based Solar Radiation Modeling

Ralph Dubayah and Paul M. Rich

Incident solar radiation at the earth's surface is the result of a complex interaction of energy between the atmosphere and the surface. Recently much progress has been made toward the creation of accurate, physically based solar radiation formulations that can model this interaction over topographic and other surfaces (such as plant canopies) for a large range of spatial and temporal scales. The ability to implement such models has been facilitated by the development of powerful analysis tools that are part of some GIS environments. In this chapter, we summarize our current work on solar radiation models and their implementation within both GIS and image processing systems. Within this context we focus on several issues including the selection of appropriate physical models, modeling languages, data structures, and model interfaces. An overview of the effects of topography and plant canopies is presented along with a discussion of various options for obtaining the data necessary to drive specific solar radiation formulations. Examples are given from our own work using two models: (1) ATM, a model based within an image processing framework, and (2) SOLARFLUX, a GIS-based model. We consider issues of design including GIS implementation and interface, computational problems, and error propagation.

INTRODUCTION

Topography is a major factor in determining the amount of solar energy incident at a location on the earth's surface. Variability in elevation, slope, slope orientation (aspect), and shadowing, can create strong local gradients in solar radiation that directly and indirectly affect many biophysical processes such as primary production, air and soil heating, and energy and water balances (Geiger 1965; Holland and Steyn 1975; Gates 1980; Kirkpatrick and Nunez 1980; Dubayah 1992). Although it has been recognized that topographic effects are important, until recently little has been done to incorporate them in a quantitative and systematic manner into a modeling environment (Rich and Weiss 1991; Dubayah 1992; Hetrick et al., "GIS," 1993, "Modeling," 1993; Saving et al. 1993). Three factors have limited the progress on topographically based solar radiation models: (1) the complexity of physically based solar radiation formulations for topography; (2) lack of data needed to drive such formulations; and (3) lack of suitable modeling tools.

A GIS, running on a fast, new-generation workstation, can provide the appropriate modeling platform for formulating and running sophisticated solar radiation models (Hetrick et al., "GIS," 1993, "Modeling," 1993). Many of the necessary capabilities are now widely accessible from GIS platforms, including the ability to construct or import digital elevation models (DEM), to integrate diverse databases for input and output, to access viewshed analysis algorithms that permit assessment of sky obstruction and reflectance, and to harness the computational power required for complex calculations.

In this chapter we provide an overview of our research on solar radiation models and their implementation. We first consider topographic effects on direct and diffuse fluxes. We then outline methods for obtaining the data necessary to drive radiation models. Examples of solar radiation modeling using two existing models are presented, one of which (the SOLARFLUX model) is currently implemented within a GIS. We briefly discuss design considerations, data needs, data structures, error propagation, and directions for the future.

MODELING TOPOGRAPHIC AND CANOPY EFFECTS

Detailed descriptions of some topographic solar radiation models can be found in Dozier (1980, 1989), Dubayah et al. (1990), Dubayah (1992), and Hetrick et al. ("GIS," 1993, "Modeling," 1993). Here we briefly summarize the topographic effects most models should consider. There are three sources of illumination on a slope in the solar spectrum: (1) direct irradiance, which includes self-shadowing and shadows cast by nearby terrain; (2) diffuse sky irradiance, where a portion of the overlying hemisphere may be obstructed by nearby terrain; and (3) direct and diffuse irradiance reflected by nearby terrain toward the location of interest.

Direct Irradiance

The direct irradiance is a function of solar zenith angle and the solar flux at the top of the atmosphere (exoatmospheric flux). Zenith angle and exoatmospheric flux vary by date, while transmittance is a function of absorbers and scatterers that can vary greatly over time. Given an optical depth of τ_0, the irradiance is

$$\mu_s S_0 e^{-\tau_0/\cos\theta_0} = [\cos\theta_0 \cos S + \sin\theta_0 \sin S \cos(\phi_0 - A)]\, S_0 e^{-\tau_0/\cos\theta_0} \qquad (24\text{-}1)$$

where S_0 is the exoatmospheric solar flux, θ_0 is the solar zenith angle, ϕ_0 is the solar azimuth, A is the azimuth of the slope, and S is the slope angle. Both S and A are derived from digital elevation data. For clear sky conditions the spatial variability of incoming solar radiation will usually be dominated by (24-1). However, shadowing must also be taken into account.

Diffuse Irradiance

Unlike direct irradiance, exact calculation of the diffuse irradiance on a slope is difficult and almost always involves some degree of approximation. In addition to changing elevation, two factors must be considered: (1) anisotropy in the diffuse irradiance, and (2) the amount of sky visible at a point (its sky view factor).

Anisotropy: In general, diffuse irradiance is not isotropic, that is, it varies depending on sky direction. Experience tells us that for clear sky conditions this is the case, given the familiar observation of a brighter sky near the horizon and near the disk of the sun. However, modeling anisotropy can be complex, especially under partly cloudy conditions. This is further complicated because atmospheric conditions can change rapidly. To simplify the problem we often assume that the diffuse radiation coming from the sky is isotropic.

Sky view factor: At any given location, a portion of the sky may be obstructed by topography, thereby reducing diffuse irradiance from corresponding sky directions. Sky obstruction can result either from "self-shadowing" by the slope itself or from adjacent terrain. A sky view factor V_d can be calculated that gives the ratio of diffuse sky irradiance at a point to that on an unobstructed horizontal surface. In theory, the diffuse flux should be calculated by multiplying the view factor in a particular direction by the amount of diffuse irradiance in that sector of the sky, and integrating over the hemisphere of sky directions. This is computationally complex and storage intensive because it requires the calculation of V_d and diffuse irradiance for each sky sector and for each grid point. If we assume that diffuse irradiance is isotropic, only one view factor is associated with each grid location (as opposed to a factor for each direction). Using an isotropic assumption, the diffuse irradiance is given by

$$V_d \bar{F}\downarrow(\tau_0) \tag{24-2}$$

where $\bar{F}\downarrow(\tau_0)$ is the average diffuse irradiance on a level surface at that elevation, and V_d varies from 1 (unobstructed) to 0 (completely obstructed). Dozier and Frew (1990) provide details on finding V_d using horizon angles. Hetrick et al. ("GIS," 1993) give a highly simplified formulation for diffuse flux on a slope based on Gates (1980). Some GIS environments provide a viewshed capability that delineates for a given point the area that can be seen from that point. The points that make up the border of the viewshed form the horizon for that point. Using this viewshed approach (Hetrick et al., "Modeling," 1993, Rich et al., in press), view factors can be calculated.

Reflected Irradiance

For each point, reflected radiation from surrounding terrain must be estimated. One method of doing this is by calculating an average reflected radiation term and adjusting this by a terrain configuration factor. This configuration factor, C_t, should include both the anisotropy of the radiation and the geometric effects between a particular location and each of the other terrain locations that are mutually visible. The contribution of each of these terrain elements to the configuration factor could be computed, but this is difficult. We can again simplify by assuming (unrealistically) that the radiation reflected off of terrain is isotropic and given that V_d for an infinitely long slope is (1+cosS)/2, approximate C_t by

$$C_t \approx \frac{1+\cos S}{2} - V_d \tag{24-3}$$

The reflected radiation from the surrounding terrain is then

$$C_t \bar{F}\uparrow(\tau_0) = C_t R_0 \bar{F}\downarrow(\tau_0) \tag{24-4}$$

where $\bar{F}\uparrow(\tau_0)$ is the amount of radiation reflected off the surface with an average reflectance of R_0. Hetrick et al. ("GIS," 1993) implement a formulation of Gates (1980) for this as well. As with the sky view factors, there is the possibility that viewshed programs could be exploited to improve the estimate of reflected radiation.

Total Irradiance on a Slope

Given the assumptions above, one physical formulation for the total irradiance on a slope (see Hetrick et al., "GIS" 1993 for another) can now be given as

$$R\downarrow(\text{slope}) = \left[V_d \bar{F}\downarrow(\tau_0) + C_t \bar{F}\uparrow(\tau_0) + \mu_s S_0 e^{-\tau_0/\cos\theta_0} \right] \tag{24-5}$$

where, V_t, C_t, and μ_s are all derived from digital elevation data and all vary spatially. Since τ_0 is a function of pressure, the diffuse radiation will vary spatially with elevation, as will the direct irradiance. Equation (24-5) is implicitly a function of wavelength (i.e., monochromatic). Total irradiance can be found by integrating it with respect to wavelength over the desired spectral interval. A good approximation is to divide the solar spectrum into two broad bands, one mainly scattering and one mainly absorbing, corresponding to the visible and near-infrared, and use (24-5) in each wavelength region.

Canopy Effects and Other Complex Sky Obstruction

Very near the ground, sky obstruction results from local features, in particular plant canopies, nearby terrain, or human-made structures, all of which can present a complex pattern of sky obstruction. Modeling incident solar radiation under circumstances of complex sky obstruction is essentially the same as that already described for locations on a topographic surface. Direct and diffuse components are calculated as the irradiance originating from unobstructed sky directions, integrated over the hemisphere of sky directions (Rich 1989, 1990). However, many problems remain, both in terms of modeling and measurement. Models must account for high temporal variability of sky conditions, anisotropic irradiance distributions, the geometric complexity of plant canopies, and the resulting complex patterns of reflectance (scattering) off of the many canopy surfaces. A comprehensive analysis of sky obstruction would ideally involve detailed three-dimensional reconstruction of canopy architecture combined with viewshed analyses that account for both unobstructed irradiance through canopy openings and scattering. As for terrain models, reflected or scattered components are difficult to measure and model and are commonly ignored because of their relatively small contribution to total irradiance. Rich et al. (1993) suggest that it may be practical to derive digital elevation models of the topographic surface of plant canopies that can be used to provide a first-order estimate of near-ground radiation flux and as input to more complete canopy radiance models.

MODEL DRIVERS

Terrain and Surface Reflectance Data

Terrain data: Topographic radiation models require data about the specific terrain of interest. Specifically, digital elevation and surface reflectance data are needed. Digital elevation data exist for many parts of the world at a variety of grid spacings (Wolf and Wingham 1992). The modeling purpose should determine the grid spacing of the data used (when that option is available). It should be noted that the digital elevation data may not represent terrain, but rather any arbitrary surface, such as buildings and trees. No aspect of the modeling process described here is constrained only to topography.

Methodology for quantifying complex sky obstruction using hemispherical photography is well developed (Rich 1989, 1990), however the technique is limited to locations for which high contrast hemispherical photographs can be obtained. The hemispherical photographs are used as a direct measure of sky directions that are obscured and can be taken in transects or arrays that permit examination of spatial patterns (Galo et al. 1992; Lin et al. 1992; Rich et al. 1993).

Surface reflectance data: Information about the reflectance of the surface is required to compute multiple scattering between the surface and the atmosphere (if using radiative transfer to obtain the diffuse flux), the amount of radiation reflected off of nearby terrain for incoming radiation, and the net solar radiation. The multiple scattering component is usually small for most surfaces other than snow and ice. For simulation purposes, a guess at the average area albedo in the visible and near-infrared is usually sufficient.

Radiation Data

For each location, a radiation formulation such as (24-5) requires an estimate of the direct and diffuse irradiance for a level surface at the corresponding elevation. Obtaining these values can be difficult, especially for the diffuse flux. The source and type of radiative drivers used is perhaps the most important implementation issue because it determines whether the model will produce actual solar radiation fluxes for a given time and location or some type of "potential" radiation. For obtaining actual fluxes, some field-measured data must be available, such as pyranometer data, atmospheric optical data, or atmospheric profile (sounding) data. For potential solar radiation, the state of the atmosphere need not be known and some average or reference conditions are assumed.

Radiative transfer algorithms that describe the flux of energy through the atmosphere can be used to get the direct and diffuse fluxes. One common approximation to the radiative transfer problem is the two-stream method (Meador and Weaver 1980). If we have no information about the atmosphere, other radiative transfer programs, such as LOWTRAN7 (Kneizys et al. 1988), can be used to obtain standard atmospheric optical conditions at particular locations for particular times of year. If radiosonde data are available (providing information on the vertical profiles of temperature, water vapor, and pressure, among others) these data can be used in LOWTRAN7 to get more accurate fluxes.

Where pyranometer data are available, empirical formulations can be used to obtain the diffuse irradiance from global irradiance (Dubayah and van Katwijk 1992). Alternately, the pyranometer data may be used to obtain the optical properties needed to run the two-stream model via inversion, though the inversion is not unique (Dubayah 1991). Semiempirical formulations can be used that combine known optical depths with empirically derived equations for the diffuse flux (e.g., see Gates 1980; Hetrick et al., "GIS," 1993, "Modeling," 1993). For purposes of comparing different locations across a landscape, it is often useful to calculate potential solar radiation under a common set of conditions, for example under clear-sky conditions (Hetrick et al., "GIS," 1993, "Modeling," 1993; Saving et al. 1993).

EXAMPLES: THE ATM AND SOLARFLUX MODELS

In this section we present some examples from our own research using two different models: (1) ATM (Atmospheric and Topographic Model) (Dubayah 1992), which is derived essentially from Dozier (1980, 1989); and (2) SOLARFLUX (Hetrick et al., "GIS," 1993, "Modeling," 1993). ATM is a collection of separate programs, each of which are part of the Image Processing Workbench (IPW) (Frew 1990), which, although raster based, is not explicitly implemented within a GIS. SOLARFLUX has been used effectively in planning, conservation, microclimate, and basic ecology studies (Rich et al. 1992, 1993; Saving et al. 1993; Weiss et al. 1993). Because SOLARFLUX is implemented in the ARC/INFO and GRID GIS platform (Environmental Systems Research Institute) as an ARC Macro Language program (AML), it provides access to a broad range of GIS capabilities.

ATM

One of the main objectives in the development of ATM was to provide inputs for hydrological and snowmelt models in mountainous terrain. Using existing data, ATM can generate detailed topoclimatologies for large river basins. A good example of this is our modeling efforts in the Rio Grande River basin of Colorado (Dubayah and van Katwijk 1992). A mosaic of thirty-nine DEMs at 30-m grid spacing was created covering the upper portion of the basin, above Del Norte and west to the continental divide. A four-year time series of hourly pyranometer measurements of direct and diffuse fluxes was available for four hydrologic years beginning in 1987. The pyranometer data was used with Landsat Thematic Mapper satellite estimates of reflectance and NOAA estimates of snow-covered area to create a four-year monthly climatology of incoming radiation for the entire basin. Figure 24-1 shows a map of net solar radiation for the month of June 1990 for the entire basin. The highest regions at the western end of the basin are still snow covered and hence have low net radiation values because of high surface reflectance. Note that the reflectance features of the surface are barely visible; rather, it is topography that dominates the spatial variability.

Figure 24-1. Map of net solar radiation for June 1990 for the Rio Grande. Range of values on map is from 35 W/m^2–362 W/m^2. The basin is approximately 110-km long.

SOLARFLUX

SOLARFLUX uses input of a topographic surface, specified as a GRID of elevation values, as well as latitude, time interval for calculation, and atmospheric conditions (transmissivity), and provides output of direct radiation flux, duration of direct radiation, sky view factor, hemispherical projections of horizon angles, and diffuse radiation flux for each surface location. Applications of SOLARFLUX have spanned very different temporal and spatial scales. At the landscape level, SOLARFLUX is being used to drive landscape level microclimate-based habitat models for topographically diverse regions (Hetrick et al., "GIS," 1993, "Modeling," 1993; Saving et al. 1993). Potential clear sky solar radiation flux can readily be calculated for any day of the year, for example the winter solstice at the Big Creek reserve in California (Figure 24-2a). A shading index, calculated as the proportional reduction in solar radiation due to topographic shading, permits assessment of the importance of topographic shading for each landscape position (Figure 24-2b). At the scale of individual trees in arid woodlands (Rich et al. 1993), SOLARFLUX has been used to examine microclimate heterogeneity as it affects sites where young trees can become established (Figure 24-3).

DESIGN CONSIDERATIONS

Implementation Issues

Implementation within a GIS: Because of the broad range of applications of solar radiation models, GIS developers should be encouraged to integrate basic solar radiation modeling capabilities as part of their software. Though running SOLARFLUX as an AML has the considerable advantage that it can be customized for a particular application, its performance suffers because many of the routines called from the ARC interpreter take much time to load and run. This can easily be remedied by optimizing and compiling the calculation-intensive steps, such as viewshed analysis, while preserving as much flexibility as possible; for example, permitting the user to specify how many sky sectors are examined. The design challenge is to provide the fundamental set of tools to simulate direct, diffuse, and reflected components of solar radiation without sacrificing the ability to customize the inputs, outputs, and precision of calculation.

Interface with outside programs: A GIS-based radiation model should have the ability to easily interface with existing radiative transfer programs such as LOWTRAN7 for input, and with system models, such as energy or water-balance simulation for output. For example, the ATM model allows the user to specify the same model atmospheres included in LOWTRAN7. It then finds the range of elevations in the DEM, runs LOWTRAN7, and produces a lookup table of diffuse and direct fluxes over the range of elevations. The ability of ATM to interface with these programs is critically linked to its overall design structure. Specifically, the decoupling of elevation from other topographic effects allows for a midstream modeling interaction with radiative transfer programs.

Computational problems: The major computational problems concern the calculation of the sky view and terrain configuration factors and

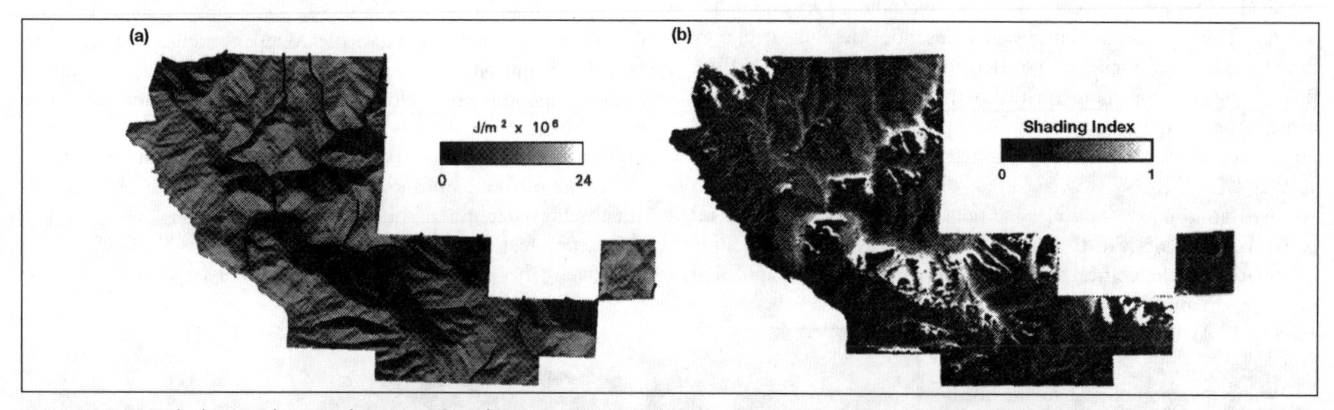

Figure 24-2. (a) Daily direct insolation on the winter solstice for Big Creek reserve (Hetrick et al., "GIS", 1993). (b) Shading index, expressed as proportional decrease in direct insolation due to topographic shading (Hetrick et al., "GIS", 1993).

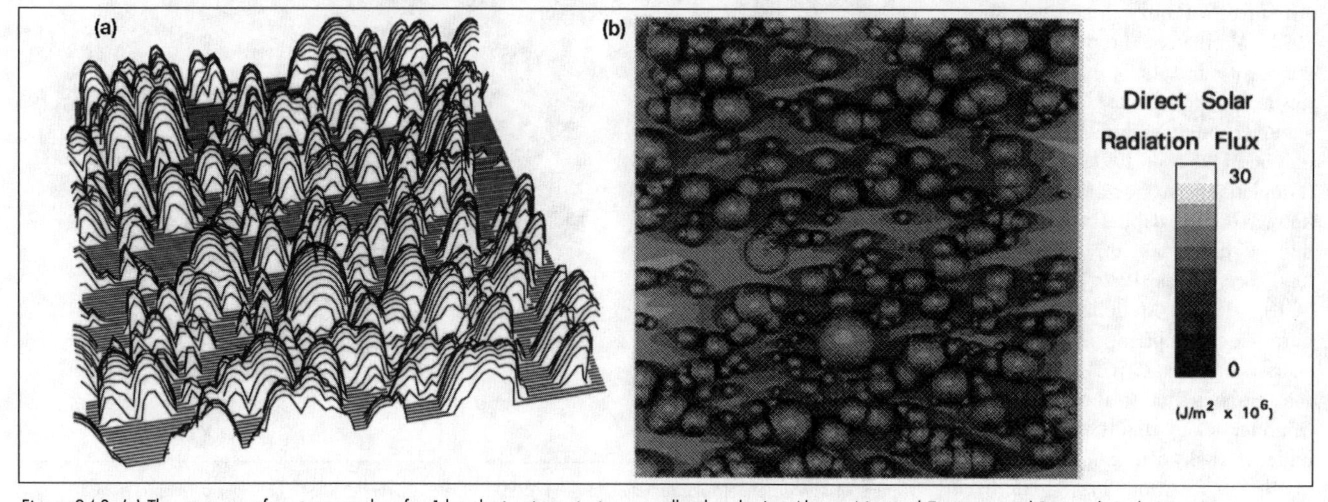

Figure 24-3. (a) The canopy surface topography of a 1-ha plot in pinyon-juniper woodland at the Los Alamos National Environmental Research Park, New Mexico. Surface topography was reconstructed based on maps of individual trees and assuming that crown form could be approximated as the upper half of an ellipsoid, with the height as the major axis and the crown as the minor axis. (b) Simulated daily solar radiation flux on the summer solstice (Rich et. al. 1993).

canopy interactions, although this is not an issue if simple approximations to these are used. If these factors are preprocessed, the majority of the computation is then involved in determining shadowing and terrain reflected flux. If anisotropy is considered, the calculation of reflected flux could be prohibitive. Further research is needed on creating a computationally efficient means for handling anisotropy, especially with an intervening canopy. An approach that uses highly optimized lookup table approaches, such as those employed by Rich (1989) for analysis of hemispherical photographs and incorporated in SOLARFLUX (Rich et al., in press), may be the key to efficient calculations that incorporate anisotropy.

Error Propagation

There are errors associated with every step of solar radiation modeling: (1) associated with the radiative transfer calculations (e.g., the two-stream approximation is no better than 10%–15%); (2) associated with interpolating and extrapolating empirical measurements over a landscape; (3) associated with registration (between reflectance and digital elevation data); and (4) associated with approximations particular to a physical model. The most serious source of error, however, is the poor quality of most digital elevation data. For example, the USGS 30-m DEMs have considerable noise that often produces inaccurate slopes and aspects for any given location. Therefore, not only is the gradient in error, but also view factors and terrain configuration factors. This in turn affects both the direct and diffuse irradiance calculations. If the radiation maps are then used to drive hydrologic or energy balance models, the errors that originate with the DEM are carried very far indeed from their source. For other digital elevation data, such as created from low-flying aircraft, or in the case of plant canopies, reconstructed from hemispherical photographs, the same cautions apply.

FUTURE DIRECTIONS

Given the complex interactions that take place between the atmosphere, topography, and plant canopies, solar radiation models can become highly elaborate. Obtaining increasingly better estimates of actual solar radiation should not be the only goal as the models evolve. The ability to calculate either potential or some type of simulated radiation must be retained. This is especially true in the areas of ecological modeling and global climate-change modeling. As models become more complex they can become more difficult to use, mainly because of the requirement of additional input data. Thus it is important that future models avoid this pitfall by allowing for flexibility with regard to the type of radiation calculated and the input data needed.

There are a variety of extensions that we anticipate in the near future. These include adding anisotropy to the diffuse and reflected terrain calculations and incorporating further canopy effects (see Rich, in press). One important factor, not covered so far, is clouds. Some capability for modeling scattered clouds should be incorporated (see Dubayah et al. 1993). At a local scale, incorporation of ongoing solar radiation measurements can be used to assess either short- or long-term importance of clouds (Rich et al. 1993). Another difficult problem is modeling true three-dimensional surfaces, where there may be more than one height coordinate associated with a given spatial location (as in the case of overlapping plant canopies). Most GIS cannot readily handle true three-dimensional surfaces. Work is needed on the representation of such surfaces and the computation of energy transfer through them.

CONCLUSION

Whether used for hydrologic or ecological modeling, for agriculture or forestry, for conservation or management, or for engineering and design, there is no shortage of applications that require the ability to model solar radiation intercepted by complex topographic surfaces. Much of the theory is now in place, and implementation is progressing rapidly. GIS provide the ideal modeling environment for interface of inputs and outputs of solar radiation models. Models such as SOLARFLUX and ATM serve as prototypes for a future generation of solar radiation models that should be an integral part of any GIS toolbox.

Acknowledgments

This work was supported by NASA grants #NAGW-2928 and #NAG 5-2358; the University of Maryland Laboratory for Global Remote Sensing Studies; the University of Kansas Research Development Fund; the Los Alamos National Laboratory Environmental Research Park; the Kansas Applied Remote Sensing Program; and the Kansas Biological Survey. Additional support was provided by Calcomp, Inc.; Environmental Systems Research Institute, Inc.; and SUN Microsystems, Inc. We thank William Hetrick, Sara Loechel, Shawn Saving, and Stuart Weiss for their various input.

References

Dozier, J. 1980. A clear-sky spectral solar radiation model for snow-covered mountainous terrain. *Water Resources Research* 16:709–18.

———. 1989. Spectral signature of alpine snow cover from the Landsat Thematic Mapper. *Remote Sensing of Environment* 28:9–22.

Dozier, J., and J. Frew. 1990. Rapid calculation of terrain parameters for radiation modeling from digital elevation data. *IEEE Transactions on Geoscience and Remote Sensing* 28:963–69.

Dubayah, R. 1991. Using LOWTRAN7 and field flux measurements in an atmospheric and topographic solar radiation model. In *Proceedings IGARSS '91*. Helsinki, Finland.

———. 1992. Estimating net solar radiation using Landsat Thematic Mapper and digital elevation data. *Water Resources Research* 28:2,469–84.

Dubayah, R., J. Dozier, and F. W. Davis. 1990. Topographic distribution of clear-sky radiation over the Konza Prairie, Kansas. *Water Resources Research* 26:679–90.

Dubayah, R., D. Pross, and S. Goetz. 1993. A comparison of GOES incident solar radiation estimates with a topographic solar radiation model during FIFE. In *Proceedings ASPRS 1993 Annual Conference, New Orleans*. 3:44–53.

Dubayah, R., and V. van Katwijk. 1992. The topographic distribution of annual incoming solar radiation in the Rio Grande River basin. *Geophysical Research Letters* 19:2,231–34.

Frew, J. E. 1990. "The Image Processing Workbench." Ph.D. diss., University of California, Santa Barbara.

Galo, A. T., P. M. Rich, and J. J. Ewel. 1992. Effects of forest edges on the solar radiation regime in a series of reconstructed tropical ecosystems. *American Society for Photogrammetry and Remote Sensing Technical Papers*, 98–108.

Gates, D. M. 1980. *Biophysical Ecology*. New York: Springer-Verlag.

Geiger, R. J. 1965. *The Climate Near the Ground*. Cambridge: Harvard University Press.

Hetrick, W. A., P. M. Rich, F. J. Barnes, and S. B. Weiss. 1993. GIS-based solar radiation flux models. *Proceedings ASPRS 1993 Annual Conference, New Orleans* 3:132–43.

Hetrick, W. A., P. M. Rich, and S. B. Weiss. 1993. Modeling insolation on complex surfaces. *Proceedings of the Thirteenth Annual ESRI User Conference.*

Holland, P. G., and D. G. Steyn. 1975. Vegetational responses to latitudinal variations in slope angle and aspect. *Journal of Biogeography* 2:179–83.

Kirkpatrick, J. B., and M. Nunez. 1980. Vegetation-radiation relationships in mountainous terrain: Eucalypt-dominated vegetation in the Ridson Hills, Tasmania. *Journal of Biogeography* 7:197–208.

Kneizys, F. X., E. P. Shettle, L. W. Abreu, J. H. Chetwynd, G. P. Anderson, W. O. Gallery, J. E. A. Selby, and S. A. Clough. 1988. *Users Guide to LOWTRAN7, Report AFGL-TR-88-0177.* Bedford: Air Force Geophysics Laboratory.

Lin, T., P. M. Rich, D. A. Heisler, and F. J. Barnes. 1992. Influences of canopy geometry on near-ground solar radiation and water balances of pinyon-juniper and ponderosa pine woodlands. In *American Society for Photogrammetry and Remote Sensing Technical Papers,* 285–94.

Meador, W. E., and W. R. Weaver. 1980. Two-stream approximations to radiative transfer in planetary atmospheres: A unified description of existing methods and a new improvement. *Journal of Atmospheric Sciences* 36:630–43.

Rich, P. M. 1989. A manual for analysis of hemispherical canopy photography. Los Alamos National Laboratory Report LA-11733-M.

———. 1990. Characterizing plant canopies with hemispherical photography. *Remote Sensing Reviews* 5:13–29.

Rich, P. M., R. Dubayah, W. A. Hetrick, and S. C. Saving. In press. viewshed analysis for calculation of incident solar radiation: Applications in ecology. *American Society for Photogrammetry and Remote Sensing Technical Papers.*

Rich, P. M., G. S. Hughes, and F. J. Barnes. 1993. Using GIS to reconstruct canopy architecture and model ecological processes in pinyon-juniper woodlands. *Proceedings of the Thirteenth Annual ESRI User Conference.*

Rich, P. M., and S. B. Weiss. 1991. Spatial models of microclimate and habitat suitability: Lessons from threatened species. *Proceedings Eleventh ESRI User Conference, Palm Springs, CA,* 95–99.

Saving, S. C., P. M. Rich, J. T. Smiley, and S. B. Weiss. 1993. GIS-based microclimate models for assessment of habitat quality in natural reserves. *Proceedings ASPRS 1993 Annual Conference, New Orleans,* 3:319–30.

Weiss, S. B., D. D. Murphy, R. R. White, and A. D. Weiss. 1993. Estimation of population size and distribution of a threatened butterfly: GIS applications to stratified sampling. *Proceedings of the Thirteenth Annual ESRI User Conference.*

Wolf, M., and D. Wingham. 1992. The status of the world's public-domain digital topography of the land and ice. *Geophysical Research Letters* 19:2,325–28.

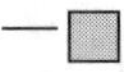

Ralph Dubayah received his M.A. and Ph.D. in geography from the University of California, Santa Barbara. His main research interests are land-surface energy balance, topoclimatology, and remote sensing. He is currently an assistant professor in geography at the University of Maryland.
Geography Department, University of Maryland
College Park, Maryland 20742
E-mail: *ralph_dubayah@umail.umd.edu*

Paul M. Rich is an assistant professor at the University of Kansas after receiving his Ph.D. and M.A. in biology from Harvard University. Dr. Rich has authored over thirty scientific papers and five software packages. His research interests include plant form and function, microclimate and habitat modeling, conservation biology, and use of close-range remote sensing to study ecological systems.
Biological Sciences, University of Kansas
Lawrence, Kansas 66045
E-mail: *prich@oz.kbs.ukans.edu*

25

Sensitivity to Climate Change of Floristic Gradients in Vegetation Communities

Dean Fairbanks, Kenneth McGwire, Kelly Cayocca, Jeffrey LeNay, and John Estes

This research examines the sensitivity of relationships between the distribution of natural vegetation and climate variables for the state of California using an environmental GIS database and vegetation ordination techniques. Chaparral and yellow pine forest vegetation communities are chosen for this study due to their extensive distribution and importance to the California flora. This research is part of a larger effort to assess the potential impacts of climate change on the species diversity and the extent of vegetation communities. Climate variables corresponding to the two most significant ordination axes for the two communities are extracted from the GIS database and modified to reflect the potential changes in temperature and moisture regimes as described in the current literature. Canonical correspondence analysis (CCA) scores for floristic subregions are converted into trend surface maps that show the changing community composition in relation to the modified environmental variables. CCA analysis indicates the pattern of floristic differentiation within each community as related to environmental gradients and provides insight into potential shifts in general species composition in response to predicted climate change. The potential sensitivity of the two communities to changes in corresponding environmental gradients is found to be strong for both chaparral and yellow pine forest. Only the coastal regions for both communities show signs of stability.

INTRODUCTION

The Remote Sensing Research Unit (RSRU) of the University of California, Santa Barbara, is currently investigating the potential of GIS and remote-sensing techniques for modeling and monitoring floristic diversity of large geographic regions. This ongoing research activity is directed toward improving our understanding of the environmental controls on floristic diversity and developing new tools to understand the potential effects of climate change on the diversity of natural ecosystems. A synergistic relationship between GIS-based modeling and remote-sensing-based monitoring is proposed as providing an effective method for regional characterization of floristic diversity.

All organisms are functional components of ecosystems that are knitted together by flows of energy and cycles of materials, which in turn can be thought of as the life-support apparatus of our planet (U.S. Congress Office of Technology Assessment [OTA] 1987; Earth System Sciences Committee 1988). Studies have shown that ecosystems are involved in maintaining the concentrations of CO_2 and other gases that help regulate the earth's climate. Despite the importance of biological diversity, the appropriate methods by which diversity may be understood have yet to be agreed upon (U.S. Congress, 1987; Brown 1988; Soulé and Kohm 1989; Lubchenco et al. 1991; Peters and Lovejoy 1992).

Peters (1992) notes that because vegetation type is a primary determinant of ecosystem type, playing a major role in determining the associated fauna and soil microbiota, it is important that we predict changes in the distributions and compositions of plant associations. The majority of efforts focusing on vegetation response to potential climatic change involve the prediction of species and community range shifts (Kullman 1983; Miller et al. 1987; Davis and Zabinski 1992; Woodward 1992). Species niche modeling is done for specific important taxa (Westman 1991; Averack and Goodchild 1984) and then translated into the spatial distributions of mapped environmental variables. Subsequent modifications to the environmental variables reflecting predicted climatic change for temperature and precipitation are performed, and the new optimal environmental niches for a species are mapped. The typical response to increasing temperature for a species is to shift northward in latitude and/or upslope (Figure 25-1). Paleobiogeographical studies of past plant distributions have shown vegetation zones shifting upslope by as much as 1,000–1,500 m since the last glacial maximum (Flenley 1979; Heusser 1974). MacArthur (1972) provides the theory, based on Hopkins' bioclimatic law, that short climbs in altitude can correspond to a major shift in latitude (e.g., a 3°C cooling of 500 m in elevation equals roughly 250 km in latitude). Westman and Malanson (1992) note that a doubling of CO_2 for California would imply a shift in ecotone boundaries and centroids of distribution of the vegetation types, but not a total displacement of any type from its current geographic range.

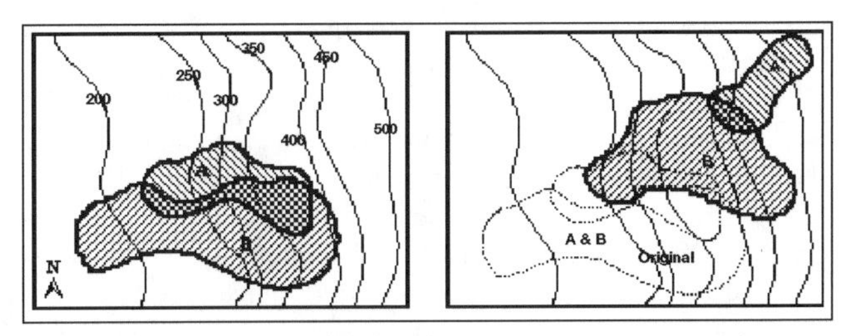

Figure 25-1. In response to climate change, latitudinal and altitudinal shifting occurs at species-specific rates, and the ranges disassociate (Peters 1992).

Typically, emphasis is placed on examining the potential effects of climate change on vegetation distributions at regional (Smith and Tirpak 1989) and global scales (Smith et al. 1991). However, because of the difficulty in quantifying the response of numerous species which are associated with a vegetation community, little direct work has been done to assess the response of regional scale community diversity to potential climatic change. In fact, as one moves toward the scale of quantifying patterns for individual species, the concept of a vegetation community may become ambiguous (Gleason 1926). The methods tested in this study provide a means to quantify within-community floristic variation at a regional scale by integrating traditional techniques from vegetation science with GIS-based analysis. A GIS database for the entire state of California is used to relate maps of vegetation communities, floristic regions, and environmental data. These data sets are coarse regional representations, stored in a 1-km^2 raster format. Data for two specific vegetation communities, chaparral and yellow pine forest (Munz and Keck 1968), are extracted and analyzed using ordination analysis methods. These methods condense information on the presence or absence of numerous individual species into generalized gradients of species turnover. Thus, vegetation communities are examined in terms of the many individuals that make up the "class concept" of a homogeneous community. This quantifies both the relative amount of differentiation between regions and the relationship between these changes in floristic composition and GIS-based environmental variables. This change in species composition, within a community type along climatic gradients or between geographic areas, is referred to as delta diversity (Whittaker 1977).

After quantifying existing relationships between floristic gradients and climate variables for the two communities, data for selected climate variables are modified to reflect climate changes predicted for California by general circulation models (GCMs). By applying ordination regression coefficients to the modified data, trend surfaces are created which display new spatial patterns in the relationship between the environment and species gradients. This research is important for understanding the potential sensitivity of regional vegetation communities to changing environmental conditions. By quantifying the heterogeneity within vegetation communities, potential problems with the assumption of homogenous vegetation units in global ecosystem modeling may be identified. We may also be able to prioritize our monitoring efforts in order to focus on those measures which are most closely related to regional diversity, potentially enhancing long-term environmental decision making.

METHODS

The study areas chosen for this test correspond to all areas of mapped chaparral and yellow pine forest represented as Munz and Keck (1968) community types of California (Figure 25-2). The CALVEG map of California vegetation (Matyas and Parker 1980) created by the U.S. Forest Service has been used to provide an initial map of natural vegetation for the entire state. The CALVEG data set was created from manual interpretation of Landsat MSS film products acquired between 1977 and 1979 and was compiled using a minimum mapping unit of between 1.6 and 3.2 km^2. The CALVEG map was converted to a 1-km resolution raster and spatially registered to the Lambert Azimuthal Equal Area projection selected for this study. The seventy-two individual vegetation classes identified in the CALVEG map were crosswalked to coarser natural vegetation communities as defined by Munz and Keck (1968). Mapped polygons were then visually inspected and edited to correspond more closely with known distributions. Certain label and topology errors for the digitized CALVEG product were also encountered and corrected at this time. It is expected that the accuracy of this map for Munz communities is higher than that of the original CALVEG data set due to specific corrections and the aggregation of classes to a coarser taxonomic grouping.

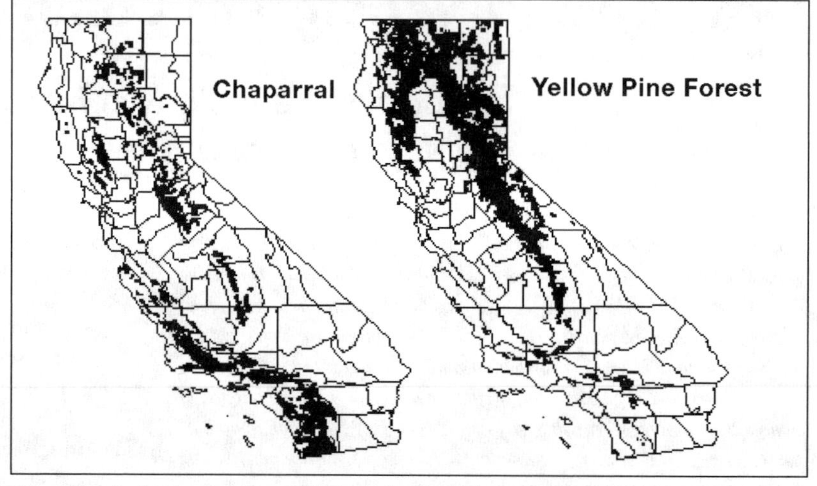

Figure 25-2. Dark polygons represent the mapped extent of the communities; lines indicate the Lum floristic subregions (RSRU/UCSB).

Existing information on species distributions for vascular plants of California is taken from the Lum floristic database (1975). This database was generated from the distribution of plants documented in *A California Flora and Supplement* (Munz and Keck 1968) and was supplemented by county floras, ecological monographs, and individual taxa maps. This database identifies the probable presence or absence of 5,900 species within each of ninety-four regions of the state (Figure 25-2). The membership of species to specific Munz plant community types is also encoded. Generalized floristic variability across the range of a plant community can then be related to climate variables by intersecting these floristic regions with the Munz community type map described previously.

Climate data for the state were obtained from products developed and distributed by ZedX, Inc. These products provide interpolated values for mean precipitation, minimum temperatures, and maximum temperatures on a monthly basis at a 1-km resolution. This data set was created by averaging observations over a thirty-year period. Interpolation of point data to the 1-km raster used digital elevation models (DEM) to account for orographic effects. Climate data files were manipulated to extract factors which may relate to ecologically meaningful limiting factors and to capture the fundamental temporal trends identified by principal components analysis (PCA). Temperature variables extracted from these products include the maximum temperature of all months, minimum temperature of all months, midpoint (nonstatistical average) of the annual maximum and minimum temperatures, and a seasonality index created by differencing the maximum monthly temperatures for January and July. Precipitation variables include mean annual precipitation and a seasonality index created by differencing the monthly means of February and June. Summary statistics (minimum, maximum, average, and standard deviation) were generated to represent the multiple pixel values of climate data occurring within each intersected Munz community and Lum region.

Community ordination analysis is then performed by cross-referencing species presence/absence data for each community with summary climate statistics. Canonical correspondence analysis (ter Braak 1986) is used to ordinate sites along environmentally constrained axes which maximize the variance explained in the species presence/absence data. CCA is designed to detect patterns of variation in species data that can be explained best by observed environmental variables (Jongman et al. 1987). The information resulting from these techniques expresses not only the patterns of variation in species composition but also the main relationships between the species and each of the selected environmental variables. CCA uses an iterative approach which applies reciprocal averaging, a form of matrix normalization, to the matrix of species presence/absence data followed by multiple regression on a user-defined set of environmental variables (Jongman et al. 1987). By iterating between reciprocal averaging and regression, CCA selects the linear combination of environmental variables that maximizes the dispersion of the sample (Lum region) scores and chooses the best weights for the environmental variables. For each of the communities, two environmental variables were chosen based on correlation with floristic variation of the first two CCA axes and a Monte Carlo test for importance (Fairbanks 1993). These variables were subsequently checked against the available literature discussing the geographical trends in each community to confirm their importance and to avoid "data dredging."

The regional maximum of the maximum temperature in any month (axis 1) and regional maximum of the seasonal temperature difference (axis 2) were identified as the two variables explaining the major variations in vegetation for chaparral. This environmental interpretation of the ordination axes on the basis of the correlations is generally consistent with past literature discussing environmental trends for chaparral (Hanes 1988; Keeley and Keeley 1988). The regional minimum of the average yearly precipitation (axis 1) and regional average of the seasonal temperature difference (axis 2) were identified as the two variables explaining the major variations in vegetation for yellow pine forest. The environmental interpretation of the ordination axes for yellow pine based on these correlations is generally consistent with past literature as well (Rundel et al. 1988; Thorne 1988). Axis eigenvalues and correlations for environmental variables are shown in Table 25-1, and the linear relationships of the fits are shown in Figure 25-3. The eigenvalue indicates the importance of the axis and is usually referred to as the percentage of variance accounted for by the axis; however, there is no guarantee that the second axis is uncorrelated with the first. The eigenvalue in this case shows the strength of the floristic variability explained by the

Table 25-1. Ordination results for chosen environmental variables.

	CCA axis	Eigenvalue score	Variable correlation
Chaparral	1	0.41	-0.95
	2	0.18	-0.93
Yellow pine	1	0.25	-0.72
forest	2	0.15	-0.68

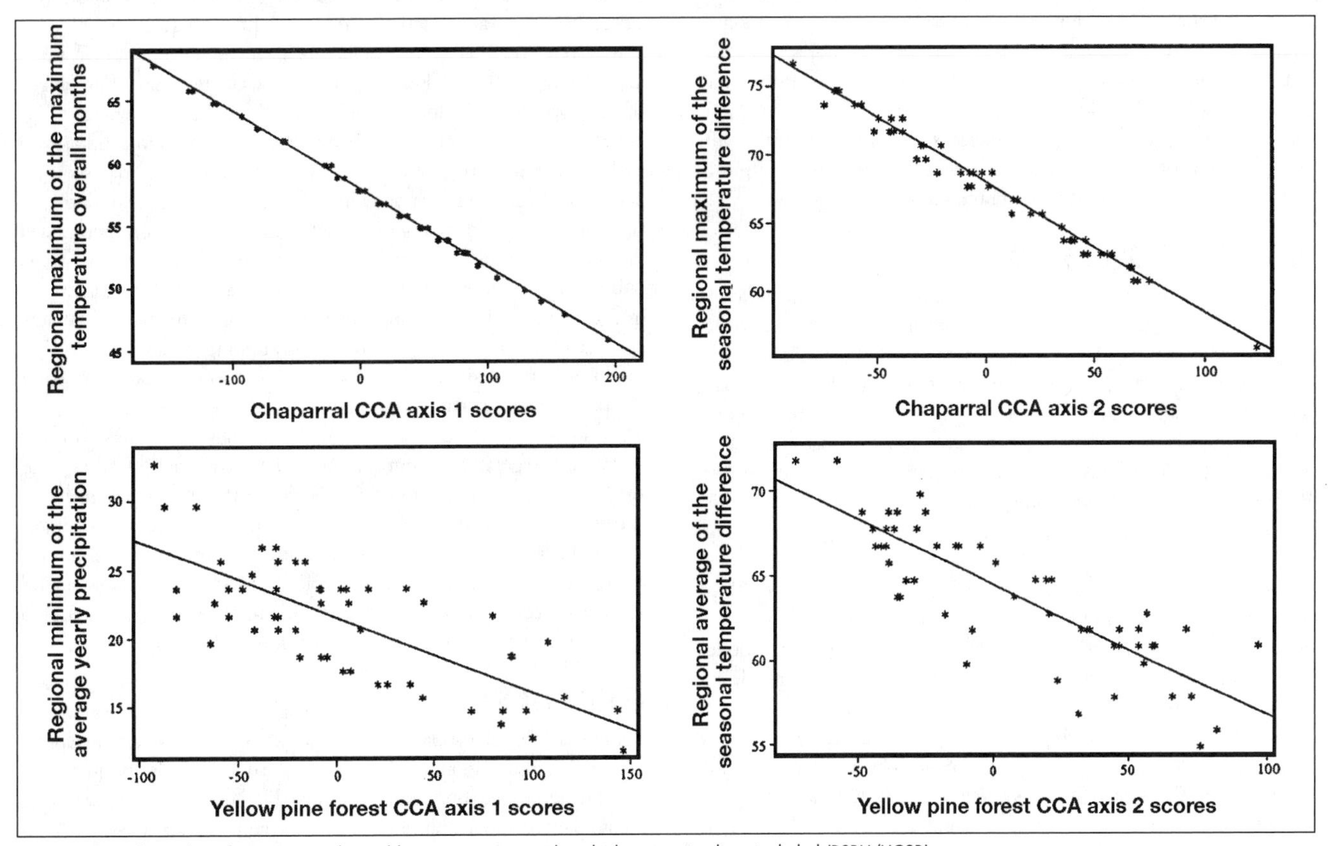

Figure 25-3. Relationship of environmental variables to axis scores with multiple regression lines included (RSRU/UCSB).

environmental variables. The variable correlations indicate the strength of the relationship between the environmental variables and the ordination axes. One may infer the relative importance of each environmental variable for prediction of species composition from the eigenvalue and correlation scores.

The current approach to predicting the possible effects of a doubling of CO_2 is to examine the output of models of global climate. Grid cells for the Goddard Institute for Space Sciences (GISS) model span 10° of longitude and 7.83° of latitude; cells for the Geophysical Fluid Dynamics Laboratory (GFDL) model span 7.5° of longitude and 4.44° of latitude. Both global models differ in some respects (e.g., solar constants, levels of CO_2, inclusion of diurnal cycle), but despite the model limitations, the two models predict rather similar initial temperatures and temperature changes for January in the California region (4°–6.5°C-rise). For July, the models predict a 2°–5°C-rise. Both models also predict an increase in January precipitation of about 0.3–3.9 cm from southern California northward, and an increase of July rainfall by 0.06–0.6 cm. Generally the cloud cover is expected to increase, but at present, there is more confidence in projections for temperature than for precipitation. These data were supplied to Westman and Malanson (1992) by the Data Support Section, National Center for Atmospheric Research.

The Lum region map was recoded to match the values of climate variables selected by CCA analysis. These four climate variables were then modified to reflect potential climate changes in order to graphically assess the sensitivity of the vegetation communities. The projected increases in temperature, seasonal temperature difference, and precipitation are taken from the GISS and GFDL circulation model outputs previously mentioned. The increase of temperature used in this study is 5°C, increase of seasonal temperature difference is 2°C, and the increase in precipitation is 4 cm. The new canonical trend surface maps present the change in the environmental gradients in relation to the original multiple regression coefficients, which are based on the floristic variation and original environmental axes fits. The multiple regression coefficients are like canonical coefficients, which relate to the rate of change in species composition by changing the corresponding environmental variable.

RESULTS AND DISCUSSION

The flora of California are a mixture of northern temperate elements and xeric southern elements and are characterized by a very high degree of endemism (Raven and Axelrod 1978). These floral differences from region to region are generally based on the large mesoscale climatic gradients occurring in California. Climate varies from hot desert and Mediterranean warm summer in the south, to Mediterranean cool summer and highland steppe in the north. Precipitation increases northward in latitude, but topography varies the patterns of temperature and precipitation greatly. The flora are influenced by all three factors. The modified canonical trend surface maps of increasing warmer and wetter climate are compared with present climate conditions. Present and future gradients in species composition in relation to chosen environmental variables as derived from CCA analysis are presented graphically in the green (axis 1) and red (axis 2) color composite maps of Plate 25-1.

An increase in minimum precipitation and average seasonality of temperatures in areas of yellow pine forest produces the most visible changes in the northern Sierra Nevada, Cascade Range, and Modoc Plateau. Tones associated with the modern-day Modoc Plateau have shifted into the Cascade Range. Species associated with mountains of the Modoc highland steppe environment might be driven into the Cascade province by the gradient in seasonality. Tones found in the northern Sierra Nevada also appear to shift southward. Coastal and southern California and regions east of the Sierra Nevada seem to display a smaller shift in environmental characteristics relative to the existing differentiation. This might suggest a smaller change in floristic composition.

An increase in maximum temperatures and the seasonality of temperatures for chaparral produces strong visual changes as well. Tones associated with the Cascade province shift into the Modoc Plateau region and the Siskiyou Mountains. Tones representing environment/species relationships in southern California appear to shift northward through the interior coastal ranges. Once again, the coastal regions seem to display less dramatic shifts in environmental characteristics.

Modification of existing climate data by a single offset value, regardless of location, shifted environmental gradients in marginal areas beyond the limits found in modern-day conditions. Areas of yellow pine forest which are driven beyond existing environmental relationships differ between the two CCA axes. The first CCA axis reaches unusual values in the Warner Mountains of the Modoc Plateau. Extreme predicted values for the second axis occur in the north coastal area. For chaparral, the areas which are extrapolated beyond existing environmental limits in the first axis occur practically along the entire eastern range of the distribution. In axis 2, the southern desert regions and southern Sierra Nevada exceed existing environmental limits. It may be of concern that the southern Sierra Nevada exceeds environmental tolerances for both the first and second CCA axes 1 and 2 of the chaparral community.

Although biotic interactions play significant roles in regulating relative abundances of species, the variability in composition of a vegetation type arises largely from the differing response of each component species to physical factors of the environment (Westman and Malanson 1992). Therefore, as climate changes, the abundances of each species will fluctuate in accord with their individual tolerances, resulting in biotic assemblages that differ both in composition and relative abundance from present conditions. This analysis indicates the species compositions in the two represented communities that might shift with changing climatic gradients. Certain species will fare better than others in the new environmental landscapes; others will more than likely be pushed out of existence. The effects on overall species diversity are extremely difficult to predict as new species may be incorporated from surrounding communities. This might even cause significant changes in the species makeup, despite the persistence of visual dominants by which the community is generally identified. Research at regional and global scales has typically addressed vegetation communities as homogeneous entities. The work presented here demonstrates both that there may be significant species turnover within communities, and that the predicted changes in environment may affect the composition of different communities in different ways. CCA analysis coupled with a GIS provides a useful method of capturing the primary trends in species turnover within vegetation communities relative to environmental variation.

CONCLUSION

Models are simplifications of natural systems, but they can present us with usable information from which to make decisions, derive questions, and further refine our understanding. The use of GIS-based data, manipulation, and graphical analysis can help tremendously in landscape characterization efforts. The methods employed in this paper are an initial effort to study the effects of possible climate change on large-scale biogeographic phenomena using GIS techniques. This effort presents hypotheses regarding climate-change effects on vegetation by identifying the primary climatic gradients that correlate with shifts in species composition of selected vegetation associations. This highlights both potential impacts on species diversity and the need for critical examination of potential errors in treating communities as homogeneous units of study in the analysis of climate change. The specific results of this analysis must be seen at the broadest level of interpretation in order to match the coarse bioclimatic and taxonomic detail which is used. The results are affected by ambiguous Munz categories, generalized species distributions, localized environmental effects (e.g., edaphic), and errors in the environmental data sets. The variable choices made for each community type are strictly of a correlative nature and cannot be tested for accuracy or whether selected variables will be driving factors in climate-change scenarios. Comprehensive analysis of change in communities also demands consideration of the interactive effects of climate, fire, edaphic factors, and anthropogenic effects (e.g., air pollution) on the competitive interactions between individual species. These factors are beyond the scope of this study, but in some cases may be important considerations with respect to climate-change modeling.

As much as anything else, the effort presented here is meant to be a cautionary note to those GIS modelers in the landscape characterization community who treat vegetation communities as homogeneous units. The floristic variation within communities and their relationship to mesoscale environmental factors should be recognized in studies of ecological functioning and spatial patterns. Knowledge that certain compositional groups within a community may move and crowd out preexisting groups within the same community in response to changing environmental gradients may be useful both to those who are concerned with long-term ecosystem functioning and those who are concerned with the character and diversity of natural areas.

Acknowledgments

Funding for the research was provided by NASA grant #NAGW-1743. Peter Richerson and Marc Hoshovsky graciously supplied the floristic data sets used by Richerson and Lum (1980).

References

Averack, R., and M. F. Goodchild. 1984. Methods and algorithms for boundary definition. In *Proceedings of the International Symposium on Spatial Data Handling, Zurich, Switzerland,* vol. 1, 238–50.

Brown, J. H. 1988. Species diversity. *Analytical Biogeography,* edited by A. A. Meyers and P. S. Giller. London: Chapman and Hall.

Davis, M. B., and C. Zabinski. 1992. Changes in geographical range resulting from greenhouse warming: Effects on biodiversity in forests. In *Global Warming and Biological Diversity,* edited by R. L. Peters and T. E. Lovejoy, 297–308. New Haven: Yale University Press.

Earth System Sciences Committee. 1988. *Earth System Science: A Closer View.* Washington, D. C.: National Aeronautics and Space Administration.

Fairbanks, D. H. K. 1993. "Relationship of GIS-based Environmental and Remotely Sensed Data to Floristic Gradients within California Vegetation Communities." Master's thesis, University of California, Santa Barbara.

Flenley, J. R. 1979. *The Equatorial Rain Forest.* London: Butterworths.

Gleason, H. A. 1926. The individualistic concept of plant association. *Bulletin of the Torrey Botanical Club* 53:7–56.

Hanes, T. L. 1988. Chaparral. In *Terrestrial Vegetation of California,* edited by M. G. Barbour and J. Major, 417–69. New York: John Wiley and Sons.

Heusser, C. J. 1974. Vegetation and climate of the southern Chilean lake district during and since the last interglaciation. *Quaternary Research* 4:290.

Jongman, R. H. G., C. J. F. ter Braak, and O. F. R. van Tongeren, eds. 1987. *Data Analysis in Community and Landscape Ecology.* Wageningen: Pudoc.

Keeley, J. E., and S. C. Keeley. 1988. Chaparral. In *North American Terrestrial Vegetation,* edited by M. G. Barbour and W. D. Billings, 165–207. Cambridge: Cambridge University Press.

Kullman, L. 1983. Past and present tree lines of different species in the Handolan Valley, central Sweden. In *Tree Line Ecology,* edited by P. Morisset and S. Payette, 25–42. Quebec: Centre d'etudes nordiques de l'Universite Laval.

Lubchenco, J., A. M. Olson, L. B. Brubaker, S. R. Carpenter, M. M. Holland, S. P. Hubell, S. A. Levin, J. A. MacMahon, P. A. Matson, J. M. Melillo, H. A. Mooney, C. H. Peterson, H. R. Pulliam, L. A. Real, P. J. Regal, and P. G. Risser. 1991. The sustainable biosphere initiative: An ecological research agenda. *Ecology* 72:371–412.

Lum, K. 1975. "Gross Patterns of Vascular Plant Species Diversity in California." Master's thesis, University of California, Davis.

MacArthur, R. H. 1972. *Geographical Ecology: Patterns in the Distribution of Species.* New York: Harper and Row.

Matyas, W. J., and I. Parker. 1980. *CALVEG—Mosaic of Existing Vegetation of California.* San Francisco: USDA Forest Service, Regional Ecology Group.

Miller, W. F., P. M. Dougherty, and G. L. Switzer. 1987. Rising CO_2 and changing climate: Major southern forest management implications. In *The Greenhouse Effect, Climate Change, and U.S. Forests,* edited by W. E. Shands and J. S. Hoffman, 157–187. Washington, D.C.: Conservation Foundation.

Munz, P. A., and D. Keck. 1968. *A California Flora and Supplement.* Berkeley: University of California Press.

Peters, R. L. 1992. Conservation of biological diversity in the face of climate change. In *Global Warming and Biological Diversity,* edited by R. L. Peters and T. E. Lovejoy, 15–30. New Haven: Yale University Press.

Peters, R. L., and T. E. Lovejoy, eds. 1992. *Global Warming and Biological Diversity.* New Haven: Yale University Press.

Raven, P. H., and D. I. Axelrod. 1978. Origins and relationships of the California flora. *University of California Publications in Botany* 72:11–34.

Richerson, P. J., and K. Lum. 1980. Patterns of plant species diversity in California: Relation to weather and topography. *American Naturalist* 116:504–36.

Rundel, P. W., D. J. Parsons, and D. T. Gordon. 1988. Montane and subalpine vegetation of the Sierra Nevada and Cascade ranges. In *Terrestrial Vegetation of California,* edited by M. G. Barbour and J. Major, 559–99. New York: John Wiley and Sons.

Smith, J. B., and D. Tirpak, eds. 1989. The *Potential Effects of a Global Climate Change on the United States: Draft Report to Congress.* Washington, D.C.: Environmental Protection Agency.

Smith, T. M, R. Leemans, and H. H. Shugart. 1991. Sensitivity of terrestrial carbon storage to CO_2-induced climate change: Comparison of four scenarios based on general circulation models. *Climate Change* 21:367–84.

Soulé, M. E., and K. A. Kohm, eds. 1989. *Research Priorities for Conservation Biology*. Washington, D. C.: Island Press.

ter Braak, C. J. F. 1986. Canonical correspondence analysis: A new eigenvector technique for multivariate direct gradient analysis. *Ecology* 67:1,167–79.

Thorne, R. F. 1988. Montane and subalpine forests of the Transverse and Peninsular ranges. In *Terrestrial Vegetation of California*, edited by M. G. Barbour and J. Major, 537–57. New York: John Wiley and Sons.

U.S. Congress, Office of Technology Assessment. 1987. *Technologies to Maintain Biological Diversity*. Washington, D. C.: GPO.

Westman, W. E. 1991. Measuring realized niche spaces: Climatic response of chaparral and coastal sage scrub. *Ecology* 72:1,678–84.

Westman, W. E., and G. P. Malanson. 1992. Effects of climate change on Mediterranean-type ecosystems in California and Baja California. In *Global Warming and Biological Diversity*, edited by R. L. Peters and T. E. Lovejoy, 258–76. New Haven: Yale University Press.

Whittaker, R. H. 1977. Evolution of species diversity in land communities. *Evolutionary Biology* 10:1–67.

Woodward, F. I. 1992. A review of the effects of climate on vegetation: Ranges, competition, and composition. In *Global Warming and Biological Diversity*, edited by R. L. Peters and T. E. Lovejoy, 105–23. New Haven: Yale University Press.

Dean Fairbanks received his B.A. and M.A. degrees in geography from the University of California, Santa Barbara (UCSB). He is currently with the CSIR's Division of Forest Science and Technology in South Africa. His interests lie in environmental monitoring/modeling, natural resource assessment, and applied research addressing the influence of anthropogenic effects on environmental systems using GIS and remote sensing.
Remote Sensing Research Unit
Department of Geography
University of California
Santa Barbara, California 93106-4060
Phone: (805) 893-3845
Fax: (805) 893-3703
Email: *dean@geog.ucsb.edu*
Environmental Information Technology Programme
Council for Scientific and Industrial Research
Division of Forest Science and Technology
PO Box 395
Pretoria 0001
Republic of South Africa

Kenneth C. McGwire is currently acting manager of the Remote Sensing Research Unit (RSRU) at UCSB.

Kelly D. Cayocca is currently working on an M.A. degree in geography at UCSB.

Jeffrey P. LeNay is currently working on an M.A. degree in geography at UCSB.

John E. Estes is the director of RSRU and is currently a visiting senior scientist at the USGS National Mapping Division Headquarters.
Remote Sensing Research Unit
Department of Geography
University of California
Santa Barbara, California 93106-4060

Introduction Part 2

Environmental Modeling Linked to GIS

Part II focuses on efforts to link GIS and environmental modeling software. This approach, which combines the functionality of two independently designed software modules and links them through common files, has been quite successful over the last two decades. It combines the power of environmental modeling software to model environmental processes with the power of GIS to perform input, output, and basic housekeeping functions, such as preliminary resampling or transformation of data. In such combinations, GIS usually performs functions that can be characterized as preprocessing or postprocessing. Preprocessing includes coordinate transformation and projection change, resampling and conversion between data models and structures, windowing and clipping to fit study areas, and analysis and modeling of uncertainty. Postprocessing includes cartographic and visual display, simple spatial analysis of results, verification, and visual analysis of outliers and residuals. One of the major impediments to such linkage lies in the lack of common data models, structures, and common interfaces, such that modelers must frequently resort to writing modules in source code to create workable linkages. Poor performance is frequently encountered if the GIS architecture has not been optimized for real-time linkage with other modules.

When we arranged the chapters in Part II, we deliberately chose not to cluster all chapters on a given environmental process, hoping instead to emphasize the common theme that runs through the chapters, the importance it has on policy development, and the need to focus on linkages between processes rather than their separability. These chapters describe projects that have taken the route of integrating GIS and environmental modeling.

First, two review chapters focus on the two areas of environmental modeling where linkage with GIS has arguably had the greatest success in recent years. Hydrologic Modeling and GIS, written by the late Ian Moore, covers in detail how underlying GIS data models influence the characterization of hydrologic processes, a common thread in many of the chapters in this section. Carol Johnston, Josef Cohen, and John Pastor emphasize the importance of the dynamic component of ecological models and options for representing temporal change in GIS.

George Leavesley and colleagues describe the development of a linked environment in which environmental processes are modeled using a selection of modular components, coupled to a GIS through a simple interface. Chapter 29 describes a coupled modeling environment for the hydrology of the Gunnison River basin in Colorado; Chapter 30 discusses a comparable approach to coupling a hydraulic model and GIS in Hong Kong for flood risk assessment.

The next four chapters focus on landscape dynamics. J. R. Krummel et al. discuss the role of spatial patterns in ecological process and make interesting use of the spatial analytic capabilities of GIS. David Mladenoff and colleagues describe research with a model of forest landscape disturbance and succession linked with a raster-based GIS. The linked model of forest dynamics is presented by Miguel Acevedo and colleagues in Chapter 33, wherein the role of GIS is that of a postprocessing tool for display and analysis of model results. Brendan Mackey et al. illustrate the complex software environment needed for integrated modeling of ecosystems.

Chapter 35, written by Philip Emmi and Carl Horton, examines the data accuracy theme explored in Part I in the context of the assessment of seismic risk and the sensitivity of GIS-based assessments to uncertainty in the underlying data. The theme of environmental risk also underlies the following chapter, in which Deborah French and Mark Reed discuss the coupling of an environmental impact model and GIS for monitoring the impact of oil and chemical spills.

Chapters 37 through 40 return to the theme of hydrologic modeling, specifically of distributed and nonpoint systems linked with GIS. Baxter Vieux and colleagues describe a modeling environment constructed around the GRASS-GIS for modeling stormwater runoff. Bahram Saghafian also uses GRASS to implement a distributed hydrologic model in Chapter 38. Piotr Jankowski and Gregory Haddock describe the integration of the AGNPS nonpoint source pollution model with PC ARC/INFO, while Chapter 40 by R. Srinivasan et al. describes integration of the SWAT nonpoint source model with GRASS.

The use of a multiscale GIS to monitor and model environmental processes, emphasizing the importance of understanding process on a range of scales, is discussed by Claude Duguay and Donald Walker. Stephen Brandt and colleagues explicitly describe spatial models of the aquatic habitat in Chapter 42, followed by David Miller's exploration of the use of knowledge-based approaches coupled with GIS for modeling processes in ecological systems.

The next three chapters focus on models of couplings between atmospheric and other processes. Lauren Hay et al. describe a modeling environment for the coupling of climatic and hydrographic processes. Tsengdar Lee and Roger Pielke focus on the coupling of atmospheric processes with the land surface, using digital elevation models and characterizations of surface cover. In Chapter 46, George Malanson and colleagues use a modified JABOWA model to examine the effects of spatial structure and climate change on forest dynamics.

Groundwater modeling is the topic of the next few chapters. Deane McKinney and Han-Lin Tsai examine the use of successively coarser grids, or multigrids, to solve partial differential equations in heterogeneous aquifers. Andrew Rogowski describes the use of conditional simulation to model flow in the root zone, coupling a simulation package with the IDRISI GIS. Chi Ho Sham and colleagues couple a three-dimensional groundwater model with GIS to estimate groundwater movement in a drainage basin. In Chapter 50, F. A. D'Agnese et al. discuss the integration of several software modules, including GIS, to model three-dimensional groundwater flow in the Death Valley basin.

The penultimate group of chapters in this section share an increasing concern for decision making and policy formulation. In Chapter 51, Joseph DePinto and colleagues describe GEO-WAMS, a system for support of modeling and management being applied to the Buffalo River watershed. LOADSS, a decision support system based on a GIS for regional planning in the Lake Okeechobee basin, is discussed next. Chapter 53, by Paul Craig and Gerald Burnette, reviews the Environmental Protection Agency's QUAL2E model with GIS for water quality planning. Roger Cronshey and colleagues discuss an interface between GIS and a series of water quality models, followed by a discussion by T. G. F. Kittel et al. of software integration and data issues that emerge from the study of the effects of climate change on terrestrial ecosystems. The damage from army training exercises is the focus of a spatiotemporal model described by Susan Cuddy and colleagues in Chapter 56.

Land surface properties are infinitely variable in space, yet must be generalized for representation in a spatial database. C. S. B. Grimmond discusses this issue in Chapter 57 within the context of atmospheric models. To end this section, Mercedes Lakhtakia et al. emphasize the importance of model integration as environmental modeling moves into the era of abundant data from NASA's Earth Observing System and the vital role that GIS must play in this new era.

Despite obvious diversity, it is possible to identify common themes in Part II. First, there is broad agreement on the importance of GIS as a preprocessing and postprocessing tool. Although some authors discuss the use of GIS for substantial parts of the modeling effort, in all cases, it is emphasized that use must be made of other software modules to perform modeling functions. Second, there is broad awareness of resulting problems from coupling software modules that have not been designed to do so and the consequent degradation of performance that often results.

In some cases it has been possible to design modeling software around the GIS; however, the more ambitious and farsighted alternative—to redesign GIS to better meet the needs of environmental modeling—is the topic of Part III.

26

Hydrologic Modeling and GIS

Ian D. Moore

Five different algorithms for estimating specific catchment area, one of the most hydrologically important terrain attributes, are compared. Results indicate that the traditional D8 *algorithm has significant deficiencies that have been largely overcome with the slope-weighted (*FD8 *and* FRho8*) and stream-tube (*DEMON *and* TAPES-C*) approaches. The impacts of the different algorithms on modeling hydrologic and erosion processes are demonstrated. The results show that the spatial characterization of hydrologic processes or phenomena using GIS is very much methodologically dependent and that these methodological differences should not be ignored in environmental modeling and database development. The* D8 *method of estimating specific catchment area, which is the most commonly used method today, is not recommended.*

INTRODUCTION

Topography plays an important role in the hydrologic response of a catchment to rainfall and has a major impact on the hydrological, geomorphological, and biological processes active in that landscape. If meaningful hydrologic predictions are to be achieved at the landscape scale, the ability to characterize the spatial variability of hydrologic processes in a simple, yet physically realistic way is of major importance. The automation of terrain analysis and the use of digital elevation models (DEMs) have made it possible to quantify the topographic attributes of a landscape. A number of terrain-based indices have been derived and relationships have been sought between these indices and a range of hydrologic processes (Moore et al., *Geographic*, 1993). One topographic attribute has proven to be particularly important in characterizing hydrologic process: specific catchment area, which is an approximate measure of runoff per unit width and the convergence and divergence of flow. Specific catchment area together with other terrain attributes such as slope and profile and plan curvature have been used in different functional forms to describe the spatial distribution of zones of surface saturation, soil water content, runoff, evapotranspiration, erosion and deposition, and catenary soil development.

Previous studies have shown that different methods of estimating terrain attributes produce somewhat different results (Moore et al., *Modelling*, 1993, *Geographic*, 1993). The estimation of specific catchment area was found to be particularly sensitive to the method of estimation. This chapter explores the sensitivity of spatially distributed predictions of specific catchment area as a function of method of computation and DEM structure. Algorithms that use both grid- and contour-based DEM structures are examined. This issue is explored for the primary terrain attribute of specific catchment area, A_s, itself as well as the compound topographic and potential erosion indices, for which specific catchment area is a key parameter.

ESTIMATING FLOW DIRECTION AND SPECIFIC CATCHMENT AREA

Grid-Based DEMs

Aspect, Ψ (measured in degrees clockwise from north), is the direction of steepest descent of a point in a catchment and is the direction in which water would flow from that point. It can be estimated from the directional derivatives of the elevation surface f_x and f_y by:

$$\psi = 180 - \arctan\left(\frac{f_y}{f_x}\right) + 90\left(\frac{f_x}{|f_x|}\right) \tag{26-1}$$

When the gradient is less than some minimum value the aspects computed by equation (26-1) are somewhat arbitrary and the terrain should be classed as flat or as a singular point with undefined aspect (Mitasova and Hofierka 1993).

The primary flow direction for water moving over the land surface, *FLOWD*, is an approximate surrogate for aspect. The simplest method of calculating *FLOWD* is to assume that it is in the direction of steepest gradient to one of the eight nearest neighbor nodes (i.e., from central node 9 to one of the eight nearest neighbor nodes, 1 to 8, in Figure 26-1a). FLOWD can be written as:

$$FLOWD = 2^{j-1} \quad \text{where } j = i \text{ for} \left\{ \max_{i=1,8} \varphi(i) \left| \frac{z_9 - z_i}{\lambda} \right| \right\} \tag{26-2}$$

where $\varphi(i) = 1$ for NSEW neighbors (i.e., for i = 2, 4, 6, 8 in Figure 26-1a), $\varphi(i) = 1/\sqrt{2}$ for N-E, S-E, S-W, N-W neighbors (i.e., for i = 1, 3, 5, 7 in Figure 26-1a), λ is the grid spacing and the z is the elevation. *FLOWD* has values of 1, 2, 4, 8, 16, 32, 64, 128, as shown in Figure 26-1b, or a value of 0 if the central node is a sink. *FLOWD* is an approximate surrogate for aspect, Ψ, where $\Psi \approx 45$ j. For example, a value of *FLOWD*=16 corresponds to an aspect of 225°. This flow-direction identifier can be extended to the case where flow occurs to

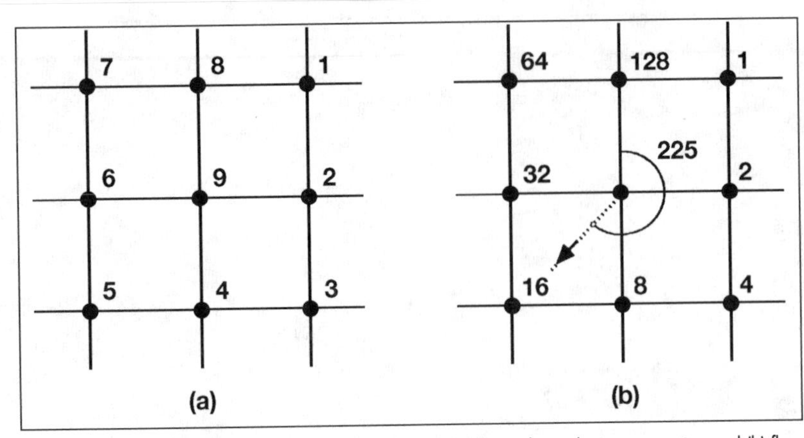

Figure 26-1. 3 × 3 subgrid of a grid-based DEM showing (a) the node numbering convention and (b) flow direction numbering convention.

more than one of the eight nearest neighbors. In such cases a single number is defined by

$$FLOWD = \sum_j 2^{j-1} \qquad (26\text{-}3)$$

that uniquely identifies all possible combinations of flow directions. For example, if flow occurs to nodes 2, 3, 4, and 8 (Figure 26-1a), then $FLOWD=2^{8-1}+2^{4-1}+2^{3-1}+2^{2-1}=128+8+4+2=142$ (Figure 26-1b) (Moore et al., *Modelling*, 1993).

Three different approaches have been used to compute upslope-contributing area based on a drainage-accumulation function determined by the nodal flow or drainage directions described above (Moore et al., *Modelling*, 1993).

D8 *(deterministic-8 node) algorithm:* Developed by O'Callaghan and Mark (1984), *D8* allows flow from a node to only one of eight nearest neighbors based on the direction of steepest descent, as defined by (26-1). This algorithm tends to produce flow in parallel lines along preferred directions, which will only agree with the aspect when the aspect is a multiple of 45°, and cannot model flow dispersion. For example, on a surface with aspects ranging from 0-22.5° the *D8* algorithm will predict a constant flow direction *FLOWD*=128 (Figure 26-1b), which corresponds to a constant aspect of 0° or 360° (i.e., north). Even with these significant limitations, the *D8* remains the most commonly used method for determining drainage areas. It has been incorporated into ARC/INFO version 6.1.

Rho8 *(random-8 node) algorithm:* Developed by Fairfield and Leymarie (1991), *Rho8* is a stochastic version of the *D8* algorithm in which $\varphi(i) = 1/(2-r)$ for N-E, S-E, S-W, N-W neighbors (i.e., i = 1, 3, 5, or 7 in Figure 26-1a) in (26-1), where r is a uniformly distributed random variable between 0 and 1. With this algorithm the expected value of the direction is equal to the aspect. Like the *D8* algorithm, the *Rho8* algorithm cannot model flow dispersion, but it does simulate more realistic flow networks.

FD8 *and* FRho8 *algorithms:* Are modifications of the *D8* and *Rho8* to allow flow dispersion or catchment spreading to be represented (Moore et al., *Modelling*, 1993). The algorithm allows flow to be distributed to multiple nearest-neighbor nodes in upland areas above defined channels and uses the *Rho8* or *D8* algorithms below points of channel initiation. Above channels, the proportion of flow or upslope-contributing area assigned to multiple downslope nearest neighbors is determined on a slope-weighted basis using methods similar to those proposed by Freeman (1991) and Quinn et al. (1991). The fraction of catchment area passed to neighbor i is given by

$$F_i = \frac{Max\left(0, Slope_i^{\upsilon}\right)}{\sum_{j=1}^{8}\left[Max\left(0, Slope_j^{\upsilon}\right)\right]} \qquad (26\text{-}4)$$

where *Slope* is the slope from the central node to the nearest neighbor and υ is a constant. Freeman (1991) found that υ = 1.1 produced the most accurate results for artificial conical surfaces. All of the above algorithms have been incorporated as different options in *TAPES-G* (Terrain Analysis Programs for the Environmental Sciences–Grid version) (Moore 1992).

Recently, an alternative approach that can model flow dispersion and which is conceptually similar to the stream-tube approach used with contour-based DEMs, described below, has been developed by Costa-Cabral and Burges (1993) and Lea (1992). Costa-Cabral and Burges's program is called *DEMON* (digital elevation model network extraction). In *DEMON* flow is generated at each pixel (source pixel) and is followed down a stream tube until the edge of the DEM or a pit is encountered. The stream tubes or flow paths are computed as points of intersection of a line drawn in the gradient direction (aspect) and a grid cell edge. The amount of flow, expressed as a fraction of the area of the source pixel, entering each pixel downstream of the source pixel is added to the flow- or area-accumulation variable (*COUNT*) of that pixel. After flow has been generated on all pixels and its impact on each of the other pixels has been added, the final value of the flow-accumulation variable, *COUNT*, is the total upslope area contributing runoff to each pixel. The *DEMON* algorithm has been incorporated as an option into *TAPES-G*, with some modification. In *TAPES-G*, *DEMON* can be applied to either the original DEM or a derived depressionless DEM; the nodes of the DEM define the centroid of the pixels rather than the vertices; and the flow direction of a stream tube in each pixel is defined by the aspect, calculated using (26-1). A similar algorithm has also been developed by Mitasova and Hofierka (1993) for predicting maximum flow-path lengths. With this algorithm the flow-path curve stops at the cell edge where the gradient is less than some minimum value and the grid cell represents flat terrain or a singular point. This algorithm has recently been incorporated into GRASS version 4.1 (beta version).

Contour-Based DEMs

Vector elevation data, in the form of digitized contour lines, can be used to partition a catchment into a series of interconnected elements using a stream-tube analogy. The method can handle both flow concentration and dispersion. Elements are formed by adjacent contour lines, which are assumed to be equipotential lines, and a pair of adjacent streamlines that are orthogonal to the equipotential lines. As a result it is a "natural" method for hydrological applications because it is based on the way water flows over the land surface. Catchment areas are determined by accumulating element areas down a stream tube. The flow direction is computed as the orthogonal to the contour line, in the downstream direction. An example of the discretization using this approach is presented in Figure 26-2. This method forms the basis of the *TAPES-C* terrain-analysis method (Moore 1992) and an earlier version is described in more detail in Moore and Grayson (1991).

Results

The five methods for determining specific catchment area: (1) *D8*, (2) *Rho8*, (3) *FD8* and *FRho8*, (4) *DEMON*, and (5) *TAPES-C* were applied to the 17-km² Coweeta catchment located in North Carolina (Webster et al. 1992). This catchment is topographically complex with slopes ranging from 0%–156%, with a median of about 38.5%. A

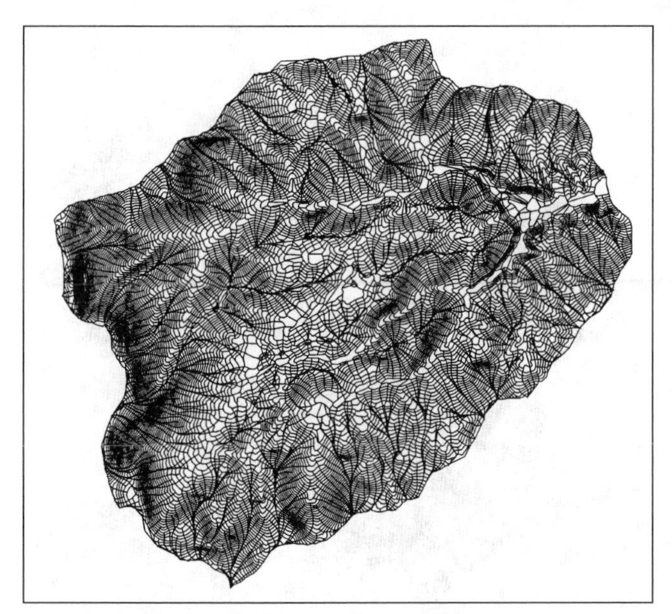

Figure 26-2. Discretization of the Coweeta catchment into a series of interconnected elements using the *TAPES-C* method of analysis (5- and 10-m contour intervals, 70-m average element downslope width).

USGS 30-m × 30-m grid DEM was available for the catchment and provided the primary data for the grid-based analyses. A file of digital contour lines at a contour interval of 10 m was obtained by applying the interpolation program MAPCON (Hutchinson, personal communication 1993) to this 30-m × 30-m grid DEM. This file provided the input data for *TAPES-C* and the discretization of the catchment using this method is presented in Figure 26-2.

Figure 26-3a compares the cumulative frequency distributions of specific catchment area computed by the five different methods. The *D8* and *Rho8* algorithms produce similar results and the distributions have a "stepped" appearance at lower values of specific catchment area. This occurs because in both algorithms the flow- or area-accumulation function, *COUNT*, is always a multiple of the cell area (30 × 30 = 900 m²). Figures 26-4a and 26-4b compare the flow directions, *FLOWD*, and the connectivity of the nodes in the DEM of the *D8* and *Rho8* algorithms, respectively. The *D8* algorithm produces many areas with long linear flow lines and uniform flow directions, whereas the *Rho8* algorithm tends to break up these regions of uniform flow directions. Both algorithms produce virtually the same principal stream network, shown by the thick black lines in Figure 26-4, which was derived making the somewhat simplistic assumption that channel initiation occurs once the upslope contributing area exceeds some critical value. I arbitrarily assumed a critical area of 90,000 m² (9 ha). Although the differences in the computed flow paths for the two methods are not reflected as significant differences in the cumulative specific-area frequency distributions for the Coweeta catchment, other applications (Moore et al., *Modelling*, 1993, *Geographic*, 1993) indicate that generally the breakup of the long linear flow paths produced by the *D8* algorithm and the *Rho8* algorithm occurs at the expense of producing more single-cell drainage areas. Neither algorithm models flow dispersion.

The *FD8*, *FRho8*, and *DEMON* algorithms produce virtually identical cumulative frequency distributions (Figure 26-3a) and spatial distributions (not shown) of specific catchment area. Figures 26-5a and 26-5b contrast the spatial distributions of specific catchment area predicted by the *D8* and *FD8* algorithms, respectively. The strong linear features produced by the *D8* algorithm are more diffuse with the *FD8* algorithm. Figure 26-3a shows that the *D8* and *Rho8* algorithms produce a far greater proportion of small specific drainage areas than the *FD8*, *FRho8*, and *DEMON* algorithms. For example, the median specific catchment areas for the *D8* and *FD8* algorithms were 66 m and 125 m²m⁻¹, respectively. However, for specific catchment areas greater than about 800 m²m⁻¹ all four methods (*D8*, *Rho8*, *FD8* and *FRho8*, and *DEMON*) produce the same cumulative frequency distributions and this part of the regime corresponds to the main drainage lines. The object behind the development of the *FD8*, *FRho8*, and *DEMON* algorithms was to be able to model both flow convergence and divergence. The *FD8* and *FRho8* algorithms are based on a slope weighting approach, whereas the *DEMON* algorithm uses a stream-tube approach. It is interesting (and encouraging) that two quite different approaches to the problem produce virtually identical results. Overall, the slope-weighting approach is easier to implement.

The *DEMON* and *TAPES-C* algorithms use the same conceptual approach, that of the stream tube, but apply it to different DEM data structures. The cumulative frequency distribution functions (Figure 26-3a) are

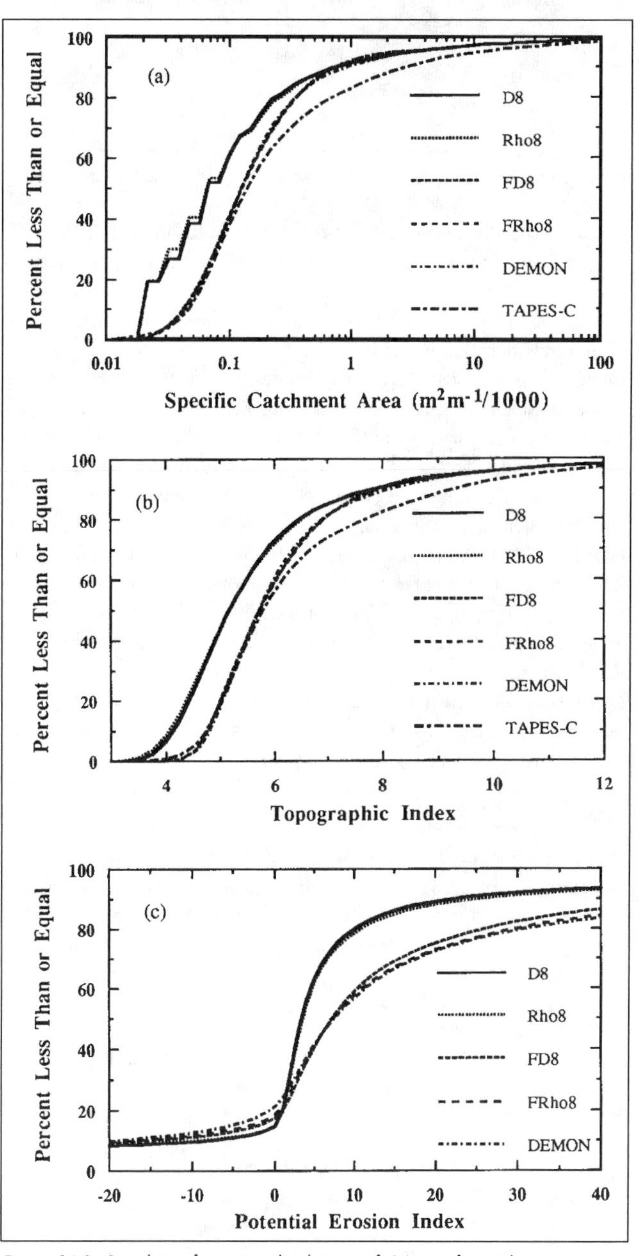

Figure 26-3. Cumulative frequency distributions of: (a) specific catchment area, (b) topographic index, and (c) potential erosion index using *D8*, *Rho8*, *FD8*, *FRho8*, *DEMON*, and *TAPES-C* algorithms.

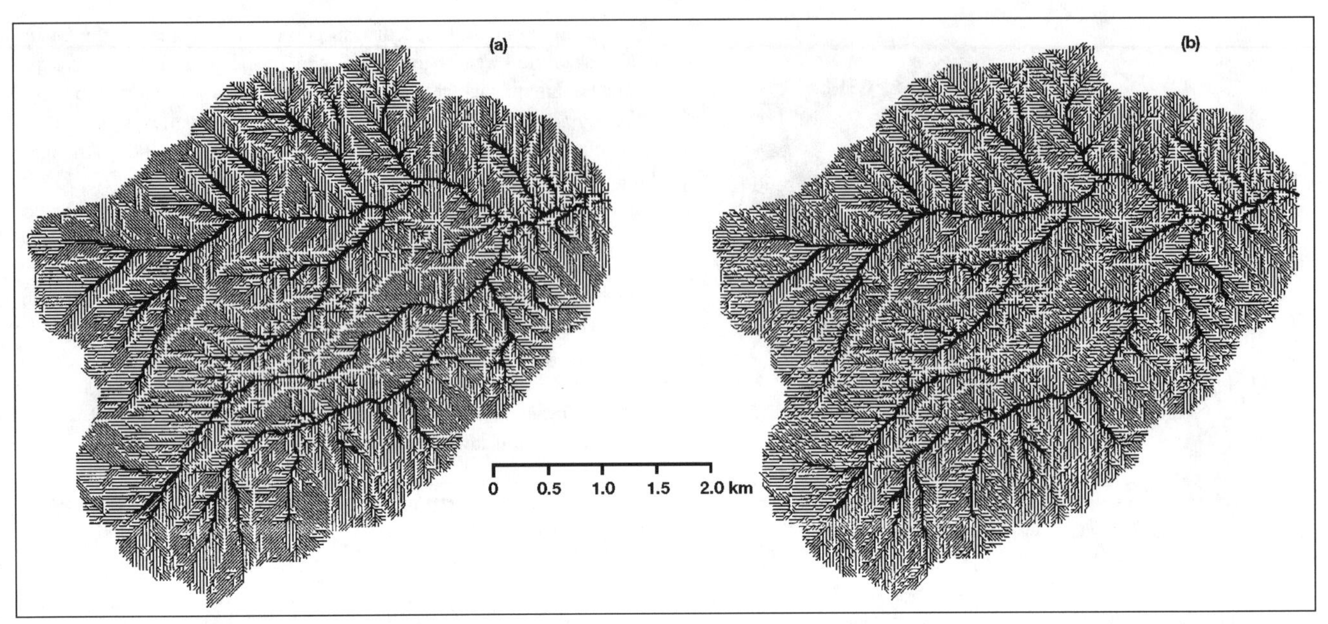

Figure 26-4. Flow directions and flow networks computed using the (a) *D8* and (b) *Rho8* algorithms.

similar for specific catchment areas less than about 120 m^2m^{-1} and the median specific catchment areas are similar, being 124 m and 139 m^2m^{-1}, respectively. However, significant differences occur in the upper half of the cumulative frequency distributions, corresponding to channel or near-channel elements or pixels (i.e., elements or pixels with large specific catchment areas). Channels are long narrow linear features. *TAPES-C* produces considerably larger elements along the main channel lines (see Figure 26-2) compared to the 30-m × 30-m pixels of the grid-based methods. As a result, the *TAPES-C* discretization of the catchment produces a higher proportion (on an area weighted basis) of elements with large specific catchment areas than do any of the grid-based methods. *TAPES-C* was originally developed as a structure for dynamic hydrologic modeling for which the linear channel features are treated separately to the normal land elements. That dissociation does not occur in the simple terrain-analysis application presented here. Furthermore, the data-preparation time for a *TAPES-C* analysis is significantly greater than for the grid-based methods and places major limitations on the suitability of the method for simple terrain analysis applications such as those demonstrated here (as opposed to dynamic hydrologic modeling applications).

APPLICATIONS

Two examples of the use of terrain attributes to model key environmental processes are outlined below. These are used to demonstrate how the differences in the methods of estimating specific catchment area impact on the processes they are attempting to characterize.

Hydrologic Modeling

Beven (1986) and Beven and Kirkby (1979) developed the following equations for predicting the pattern of soil water deficit, S, from a knowledge of topography and soil hydraulic characteristics:

$$\underset{\text{A}}{\frac{1}{m}\left(\bar{S}-S_i\right)}=\underset{\text{B}}{\left(\chi_i-\bar{\chi}\right)}+\underset{\text{C}}{\left(\log\bar{T}-\log T_i\right)} \qquad (26\text{-}5)$$

and

$$\chi_i=\log\left(\frac{A_s}{\tan\beta}\right)_i,\quad \bar{\chi}=\frac{1}{A}\int_A \chi_i,\quad \bar{S}=\frac{1}{A}\int_A S_i,\quad \bar{T}=\exp\left[\frac{1}{A}\int_A \log(T_i)\right] \qquad (26\text{-}6)$$

where A_s is the specific catchment area, β_i is the local slope angle ($\tan\beta$ is an approximation of the local hydraulic gradient), T_i is the local soil transmissivity, $\bar{T}$ is the mean catchment transmissivity, $\bar{S}$ is the mean catchment soil water deficit, χ_i is the local value of the topographic index, $\bar{\chi}$ is the mean catchment value of the topographic index, and m is a parameter that is dependent on the rate of change of hydraulic conductivity with depth. These equations form the basis of the TOPMODEL hydrological model. Equation (26-5) consists of three parts: (1) a soil water-deficit distribution function A, (2) a topographic distribution function B, and (3) a soil-distribution function C. If the spatial variation in soil properties is ignored, then the soil water-deficit distribution can be expressed as a function of the topographic index.

Figure 26-3b presents the cumulative frequency distributions of the topographic index, $\log(A_s/\tan\beta)$, computed using the different methods of estimating A_s. The same finite difference method of computing the element or pixel slope angles, β, was used for all of the grid-based methods (Moore et al., *Modelling*, 1993), but for *TAPES-C* slopes were estimated using the relationship:

$$\beta=\arctan\left[\Delta z\frac{(\omega_u+\omega_d)}{2A_e}\right] \qquad (26\text{-}7)$$

where Δz is the difference in elevation of adjacent contours forming the element, ω_u and ω_d are the upslope and downslope widths of the element, respectively, and A_e is the element area. The only significant differences between the shapes of the cumulative frequency distributions of specific catchment area (Figure 26-3a) and topographic index (Figure 26-3b) are that the stepped nature of the distributions for the *D8* and *Rho8* algorithms has been smoothed out in the computation of the topographic index. The mean topographic index computed for the *D8* and *Rho8* methods was about 5.68, for the *FD8*, *FRho8*, and *DEMON* methods about 6.17, and for *TAPES-C* about 6.49. The corresponding median values were 5.16, 5.71, and 5.79, respectively.

Erosion Modeling

There is a large body of empirical evidence that suggests that the sediment flux, q_{sx}, at any point in a catchment can be approximated by the following relationship:

$$q_{sx} = k_1 q_x^{p+1} (\sin\beta)^n \qquad (26\text{-}8)$$

where q_x is the specific discharge and β is the slope angle at x, and p and n are exponents. The specific discharge $q_x = A_{sx} r_x$, where A_{sx} is the specific catchment area and r_x is the rainfall excess. The sediment loss per unit area is $Y_x = \partial q_{sx}/\partial x$. A useful approximation is $\sin\beta \approx \tan\beta = \partial z/\partial x$, for which the error ranges from 0% at 0% slope to 4.4% at a 30% slope. Also, in many simplified erosion models the rainfall excess per unit area is assumed to be spatially uniform (i.e., $r_x = r$ and $\partial r_x/\partial x = 0$). With these assumptions the soil loss per unit area can be written as

$$Y_x = k_1 A_{sx}^p \left(\frac{\partial z}{\partial x}\right)^{n-1} \left[n A_{sx} \left(\frac{\partial^2 z}{\partial x^2}\right) + (p+1)\left(\frac{\partial z}{\partial x}\right)\left(\frac{\partial A_{sx}}{\partial x}\right) \right] r^{p+1} \qquad (26\text{-}9)$$

(Moore and Burch 1986). This equation represents the effects of topography on soil loss in three-dimensional terrain for the detachment-limiting case and is a more complete form of the equations derived by Foster and Wischmeier (1974). The terms in this equation A_{sx}, $\partial A_{sx}/\partial x$, $\partial z/\partial x$ and $\partial^2 z/\partial x^2$ are topographic attributes that can be readily calculated by terrain analysis. Typical values of n and p are 1.2 and 0.6, respectively (Moore and Wilson 1992).

Figure 26-3c presents the cumulative frequency distributions of the potential erosion index [$Y_x/(k_1 r^{p+1})$ from equation (26-9)] computed by the different grid-based algorithms. A negative value indicates deposition. The values of $\partial z/\partial x$ and $\partial^2 z/\partial x^2$ used in the analyses were the same in all cases; the only differences being in the A_{sx} and $\partial A_{sx}/\partial x$ terms. Once again, the *D8* and *Rho8* algorithms produce similar results, as do the *FD8*, *FRho8*, and *DEMON* algorithms. However, the differences between the methods illustrated in Figure 26-3a appear to be magnified in the computation of the potential erosion index. Because neither the *D8* nor *Rho8* algorithms model flow dispersion, the $\partial A_{sx}/\partial x$ term computed by these algorithms is always positive. For these algorithms deposition (negative value of the potential erosion index) is only computed on concave hillslopes where $\partial^2 z/\partial x^2$ is negative. On the other hand, the *FD8*, *FRho8*, and *DEMON* algorithms all attempt to model flow dispersion, so the $\partial A_{sx}/\partial x$ term can be both positive (flow convergence) or negative (flow dispersion). Therefore, for the *FD8*, *FRho8*, and *DEMON* algorithms, deposition is predicted when is $\partial^2 z/\partial x^2$ is negative and/or $\partial A_{sx}/\partial x$ is negative. Figure 26-6 contrasts the spatial distribution of the potential erosion computed by the *D8* and *FD8* algorithms. Once again, the strong linear features produced by the *D8* algorithm (Figure 26-6a) are evident.

CONCLUSION

The spatial distribution of primary and secondary or compound terrain attributes can be easily calculated using a variety of terrain-analysis methods and GIS using different DEM data structures. One of the most hydrologically important primary attributes is the upslope drainage area or specific catchment area, which is often used as a surrogate for runoff rate per unit width or specific discharge. The most commonly used method of estimating specific catchment area is the *D8* algorithm. This technique has already been incorporated into the grid module of ARC/INFO, for example. However, the *D8* algorithm has significant deficiencies including the production of long, linear flow paths that appear to be unrealistic in many landscapes and an inability to model flow dispersion. In the last two years several alternative algorithms have been proposed to overcome these deficiencies. The predictions from four alternative algorithms are compared.

The results indicate that the *D8* and *Rho8* methods produce similar statistical results, but the *Rho8* algorithm does tend to break up the long, linear flow paths produced by the *D8* algorithm. However, both algorithms allow flow to occur from a node to only one of its eight nearest neighbor nodes and so cannot model flow dispersion. The *FD8*, *FRho8*, *DEMON*, and *TAPES-C* methods all attempt to model flow dispersion, but in quite different ways. The *FD8* and *FRho8* algorithms use a slope-weighted multiple-flow direction algorithm, whereas the *DEMON* and *TAPES-C* algorithms use the concept of "flow tubes." *DEMON* and *TAPES-C* are applicable to grid-based and contour-based DEM data structures, respectively. All of these methods produce virtually identical results, with the exception of *TAPES-C* in the upper half of its range (due to large element areas in channel and near-channel locations). As a result, for applications in environmental modeling where the hydrological behavior exerts significant control over landscape processes, I recommend that any one of these flow-dispersion algorithms be used to estimate specific catchment areas, rather than the traditional *D8* algorithm. The *FD8* and *FRho8* algorithms are simple to implement and easy to use. The *DEMON* algorithm is computationally more demanding, but also easy to use. The *TAPES-C* algorithm is computationally the least demanding of the methods, but requires considerable effort in establishing the elevation data file in the correct format and identifying saddle-points and other topographic features.

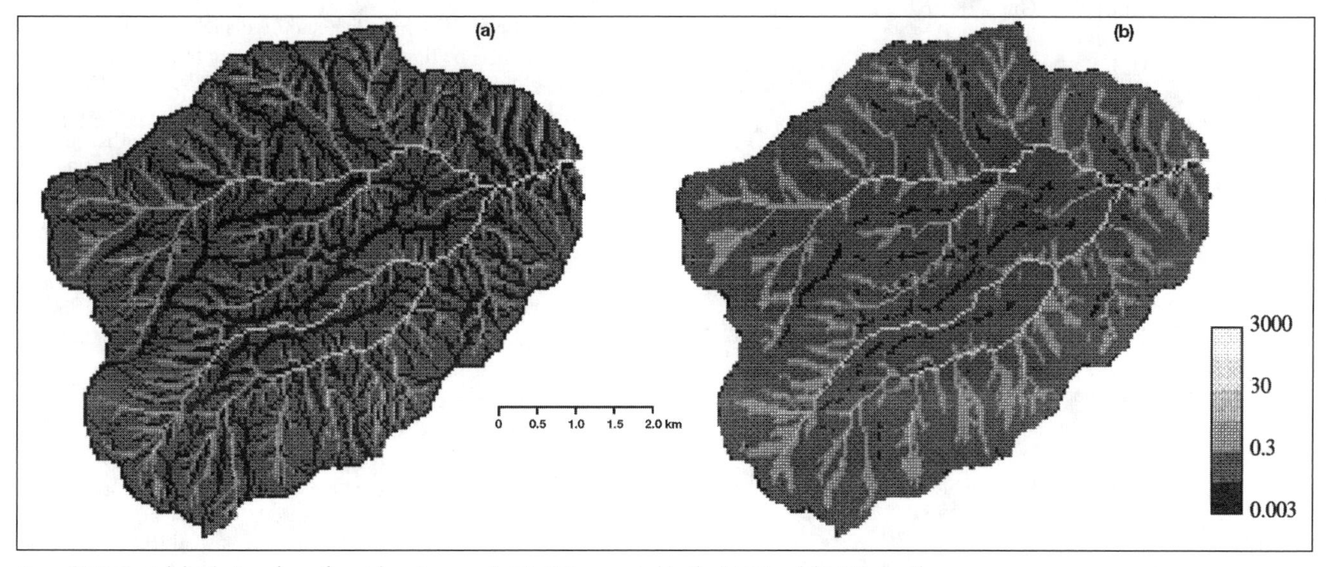

Figure 26-5. Spatial distribution of specific catchment area, $m^2m^{-1}/1{,}000$, computed by the (a) *D8* and (b) *FD8* algorithms.

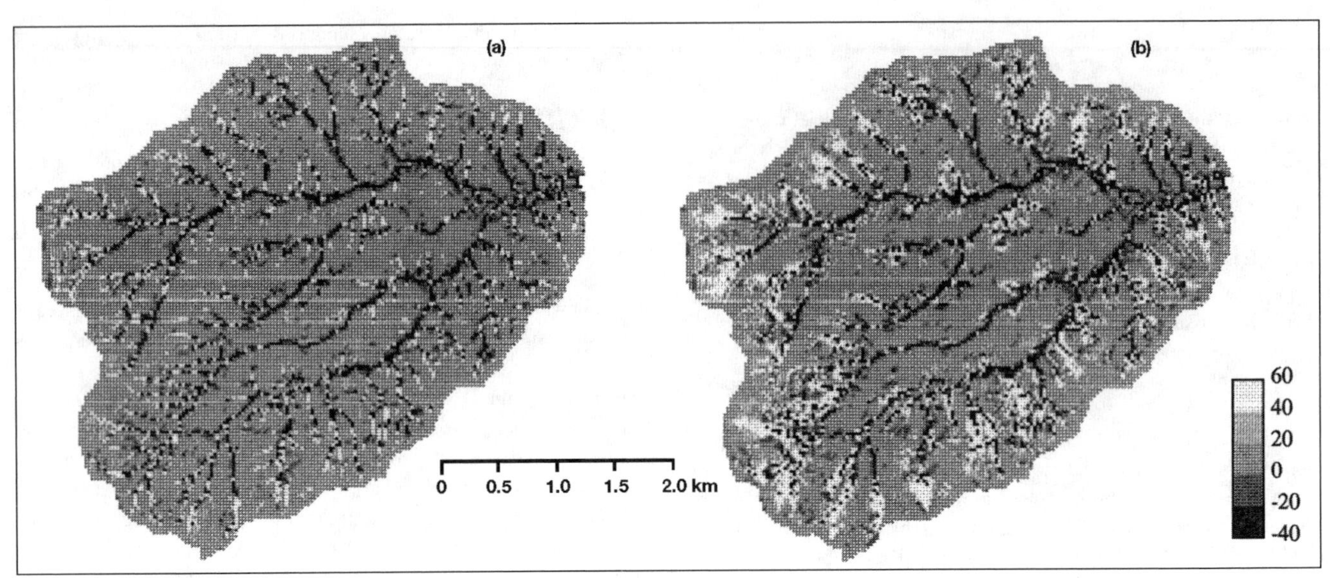

Figure 26-6. Spatial distribution of potential erosion index (+ve erosion, -ve deposition) computed by the (a) *D8* and (b) *FD8* algorithms.

The results show that the spatial characterization of hydrologic processes or phenomena using GIS is very much methodologically dependent. In a related study, Moore et al. (*Modelling*, 1993) have also shown that it is highly scale dependent as well. These methodological and scale-dependent characteristics should not be ignored in environmental modeling and database development.

Acknowledgments

This study was funded in part by grant #1991-92:ANU3 from the Land and Water Resources Research and Development Corporation and by the Water Research Foundation of Australia. The author thanks John C. Gallant for assistance with programming and figure production and Dr. Gunter Bloschl for assistance in developing the grid- and contour-based DEMs for the study area. The DEM was originally obtained from Dr. Murugesu Sivapalan.

References

Beven, K. J. 1986. Runoff production and flood frequency in catchments of order *n*: An alternative approach. In *Scale Problems in Hydrology*, edited by V. K. Gupta, I. Rodriguez-Iturbe, and E. F. Woods, 107–31. Lancaster: D. Reidel.

Beven, K. J., and M. J. Kirkby. 1979. A physically based variable contributing area model of basin hydrology. *Hydrol. Sci. Bull.* 24:43–69.

Costa-Cabral, M., and S. J. Burges. 1993. Digital elevation model networks (DEMON): A model of flow over hillslopes for computation of contributing and dispersal areas. *Water Resource Research* 30(6):1,681–692.

Fairfield, J., and P. Leymarie. 1991. Drainage networks from grid digital elevation models. *Water Resour. Res.* 27:709–17.

Foster, G. R., and W. H. Wischmeier. 1974. Evaluating irregular slopes for soil loss prediction. *Trans. Am. Soc. Agr. Engrs.* 17:305–09.

Freeman, G. T. 1991. Calculating catchment area with divergent flow based on a regular grid. *Computer and Geosciences* 17:413–22.

Lea, N. J. 1992. An aspect-driven kinematic routing algorithm. In *Overland Flow and Erosion Mechanics*, edited by A. J. Parsons and A. D. Abrahams, 393–407. New York: Chapman and Hall.

Mitasova, H., and J. Hofierka. 1993. Interpolation by regularized spline with tension: Application to terrain modeling and surface geometry analysis. *Math. Geology* 25(6):657–69.

Moore, I. D. 1992. Terrain analysis programs for the environmental sciences—TAPES. *Agric. Syst. & Information Tech.* 4:37–39.

Moore, I. D., and G. J. Burch. 1986. Modeling erosion and deposition: Topographic effects. *Trans. Am. Soc. Agr. Engrs.* 29:1,624–30, 1,640.

Moore, I. D., and R. B. Grayson. 1991. Terrain-based catchment partitioning and runoff prediction using vector elevation data. *Water Resour. Res.* 27:1,177–91.

Moore, I. D., A. Lewis, and J. C. Gallant. 1993. Terrain attributes: Estimation methods and scale effects. In *Modelling Change in Environmental Systems*, edited by A. J. Jakeman, B. Beck, and M. McAleer. Chichester, England: John Wiley and Sons.

Moore, I. D., A. K. Turner, J. P. Wilson, S. K. Jenson, and L. E. Band. 1993. GIS and land surface-subsurface process modeling. In *Geographic Information Systems and Environmental Modeling*, edited by M. F. Goodchild, B. O. Parks, and L. T. Steyaert, 196–230. Oxford University Press.

Moore, I. D., and J. P. Wilson. 1992. Length-slope factors for the revised universal soil loss equation: Simplified method of estimation. *J. Soil and Water Cons.* 47:423–28.

O'Callaghan, J. F., and D. M. Mark. 1984. The extraction of drainage networks from digital elevation data. *Computer Vision, Graphics and Image Processing* 28:323–44.

Quinn, P., K. Beven, P. Chevallier, and O. Planchon. 1991. The prediction of hillslope flow paths for distributed hydrological modeling using digital terrain models. *Hydrological Processes* 5:59–79.

Webster, J. R., S. W. Golloday, E. F. Benfield, J. L. Meyer, W. T. Swank, and J. B. Wallace. 1992. Catchment disturbance and stream response: An overview of stream research at Coweeta Hydrologic Laboratory. *River Cons. and Management* 232–53.

Ian Donald Moore was born and educated in Melbourne, earning a B.S. and M.S. in engineering from Monash University. He gained his Ph.D. in agricultural engineering from the University of Minnesota in 1979 and was then appointed an assistant professor in the Department of Agricultural Engineering at the University of Kentucky. In 1983, Ian returned to Australia as a senior research scientist in the then CSIRO Division of Water and Land Resources. He then moved to Minnesota in 1986 as Associate Professor in the Department of Agricultural Engineering and Adjunct Professor of Forest Resources. He returned to Canberra in 1990 to take up the newly established Jack Beale Chair of Water Resources at the Center for Resource and Environmental Studies. He died in September 1993 after a brief illness.

27

Modeling of Spatially Static and Dynamic Ecological Processes

Carol A. Johnston, Yosef Cohen, and John Pastor

Most ecological models predict temporal change at a single location, simulating change based on in situ *feedback mechanisms and/or environmental alteration. Such models have been made two dimensional by using spatially explicit input parameters (e.g., climatic data, soil maps), but they are still spatially static in the sense that the modeled units function independently of each other, with no fluxes of materials or information between them. Spatiodynamic ecological models require incorporation of horizontal (and often vertical) movement, such as dispersion of propagules or organisms. This approach is standard in modeling watershed fluxes but uncommon for modeling organism movement because animals move in ways that are difficult to predict spatially and temporally. The use of two types of ecological models to predict landscape pattern is described: (1) Markov modeling of beaver ponds based on historical trends, and (2) spatial interpolation of a forest simulation model to predict effects of climate change on tree species distribution. The same forest simulation model is used in a spatiodynamic application to animate changes in plant distribution and soil nitrogen over time across an entire landscape. Other spatiodynamic examples described include simulation of moose foraging and a marine fouling community.*

INTRODUCTION

Most ecological models focus on *in situ* processes without regard to the modifying effects of landscape position. The growth of organisms is based on given site parameters, physiological constraints, successional rules, and food web hierarchies. Vertical structure (e.g., subcanopy forest regeneration, lake thermoclines) is commonly considered, as are vertical fluxes of organisms (e.g., diel movement of phytoplankton communities) and materials (e.g., nutrient cycling in forests); however, location in horizontal space is generally ignored (Johnston 1993).

Predicting temporal change at a single location has long been a strength of ecological models (Shugart 1984; Pastor and Post 1988; Cohen and Pastor 1991). A model is run for a simulated time period to predict the results of *in situ* feedback mechanisms and/or changes in environmental parameters. This type of modeling, based on knowledge of current conditions and relationships, can thus be used to predict future changes. Such models have been made two dimensional by using spatially explicit input parameters (e.g., climatic data, soil maps), but they are still spatially static in the sense that the modeled units function independently of each other, with no fluxes of organisms, materials, or information between them.

Spatiodynamic models require incorporation of horizontal (and often vertical) movement, such as dispersion of propagules or organisms. This approach is standard in watershed modeling where movement is unidirectional (downslope) and easily predicted by environmental parameters (topography). Modeling fluxes of organisms are uncommon, however, for one important reason: organisms react to their environment (by modifying their behavior, reproduction, etc.), and therefore move in ways that are difficult to predict spatially and temporally.

In this chapter, we describe the development and use of spatially explicit models to predict environmental change. Some of the model applications do not incorporate horizontal fluxes (spatially static), whereas others do (spatiodynamic). The spatially static models were linked with a commercial GIS, but the spatiodynamic models used custom analysis and display routines due to limitations of commercial GIS packages for this purpose. We suggest ways in which GIS could be improved to meet the needs of innovative developments in ecological modeling.

SPATIALLY STATIC MODELS

Spatial Interpolation of Environmental Simulation Models

Environmental simulation models attempt to duplicate ecological function via coupled differential equations that describe key ecosystem processes (Shugart 1984; Pastor and Post 1986). They are often driven by environmental input parameters that can be derived from satellite image interpretation or existing GIS databases (e.g., leaf-area index, soils, climate), which make them conducive to integration with GIS (Running et al. 1989; Pastor and Johnston 1992).

We used an environmental simulation model (LINKAGES) (Pastor and Post 1985) in combination with a GIS to examine spatial variability of forest response to climate change (Pastor and Johnston 1992). The model simulates annual establishment, growth, and death of individual trees, and the decay of litter from them in a 1⁄12-ha forest plot (Figure 27-1). The model assumes that species migrate to new sites to the extent that temperature and soil water availability are optimal, and that growth is limited by temperature, water, nitrogen, or light, whichever is most restrictive.

Forests were simulated for 200 years at twenty points across eastern North America using average monthly temperatures and precipitation from long-term climatic data sets. These climatic properties were then altered linearly for the next 100 years to reach a simulated climate corre-

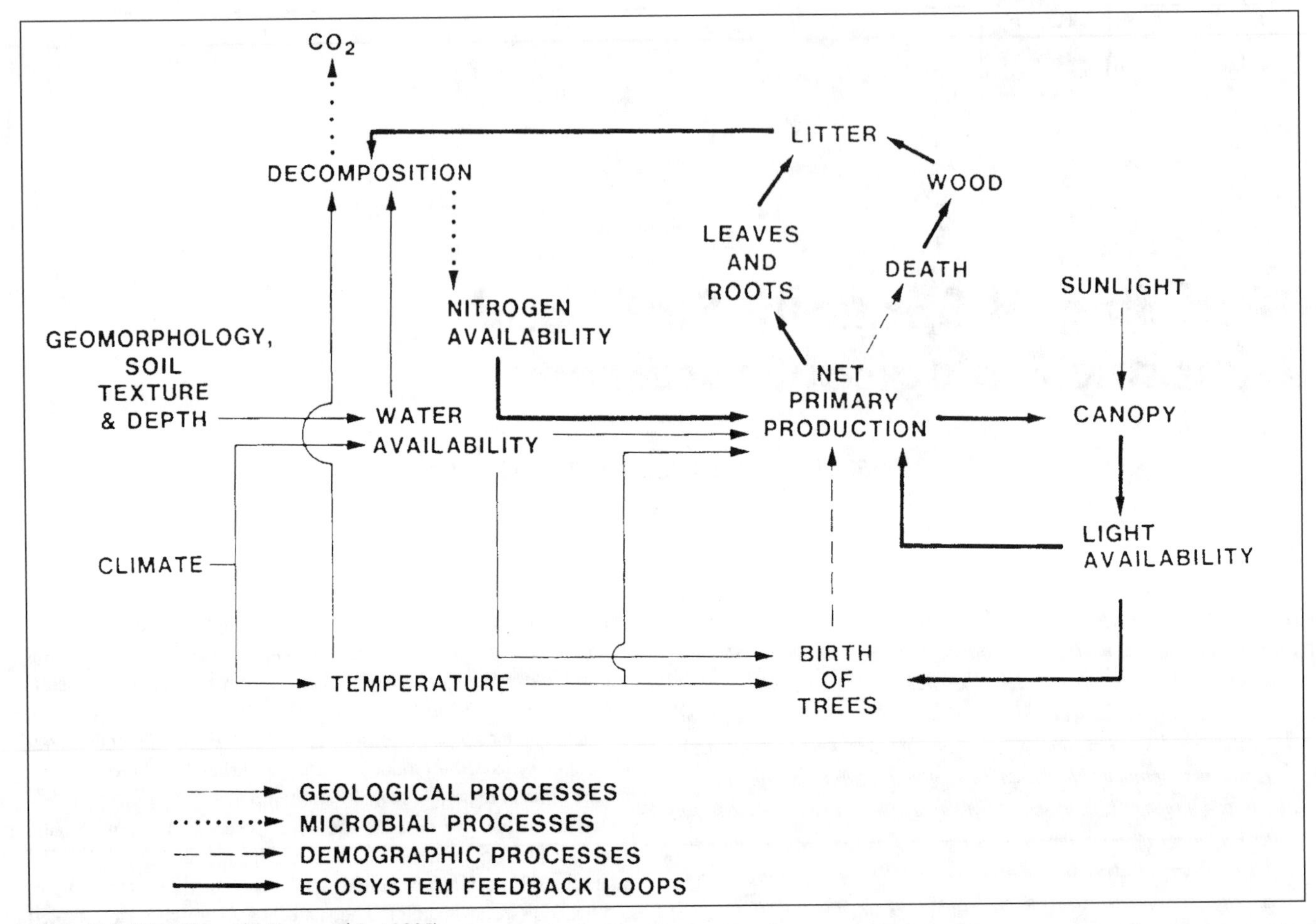

Figure 27-1. LINKAGES model (Pastor and Post 1985).

sponding to a 2x current CO_2 concentration for each site, followed by 200 more years under the new climate. Species biomass predictions at the network of points were interpolated to the subcontinent east of the 100th meridian using the ERDAS-GIS (Figure 27-2).

Under current climatic conditions, modeled species-distribution ranges corresponded well to independent distribution range maps, despite the paucity of data points simulated. The predicted climate changes resulted in migration of tree species over time (Figure 27-2). Note that even though this modeling approach did not explicitly incorporate horizontal processes (e.g., via seed dispersion), changes in species distribution were predicted due to changes in the spatial distribution of the climatic driving variables that influenced species presence or absence at a particular location.

Markov Modeling

One of the major motifs of American plant ecology in the twentieth century, and most recently landscape ecology, is that vegetation patterns are not random but emerge from the interactions of organisms with their environment. Vegetation maps depict the arrangement of those patterns, but quantification of those patterns was only recently made practical with the advent of GIS. To evaluate changes in those patterns over time, two or more databases showing vegetation on different dates are intersected to compute the percentage of total land area in all possible transition categories. When assembled in a matrix and used to generate a temporal series, known as a Markov chain, these transition probabilities form a simulation model of changes in areas of different cover types over time (Pastor et al. 1993; Acevedo et al. this volume).

The eigenvalues and eigenvectors of a Markov matrix are useful mathematical properties from which ecological processes can be inferred. These satisfy the equation:

$$A\mu = \lambda \mu \tag{27-1}$$

where A is the matrix of transition probabilities and λ is a particular eigenvalue associated with the eigenvector μ. The dominant eigenvector of a Markov matrix is the steady-state distribution of areas among the various states, and the rate of approach to steady state can be used to predict the time required for convergence to steady state (Caswell 1989).

We used Markov modeling to evaluate historical landscape changes caused by beaver pond building on the 298 km² Kabetogama Peninsula of Voyageurs National Park in northern Minnesota for the five decades since 1940, a period during which the beaver population was increasing in numbers and expanding in range. Aerial photos were used to map and classify "floodable areas," all land that was impounded by beavers at any time during the fifty-year period, into four classes, of which the first three describe soil moisture conditions within beaver ponds: (1) moist, (2) seasonally flooded, (3) flooded, and (4) not ponded. A GIS was used to compute transition probabilities among these four moisture classes (Johnston et al. 1993).

The modeling revealed that although the steady state distribution among the different hydrologic classes differed by decade (Figure 27-3), the time required for 95% convergence of the landscape to steady state fell within a narrow range, 30–300 years, regardless of spatial scale (entire peninsula or subwatershed thereof), particular watershed, antecedent soil type (upland, mineral soil wetland, or peatland), or cohort of pond creation (Pastor et al. 1993).

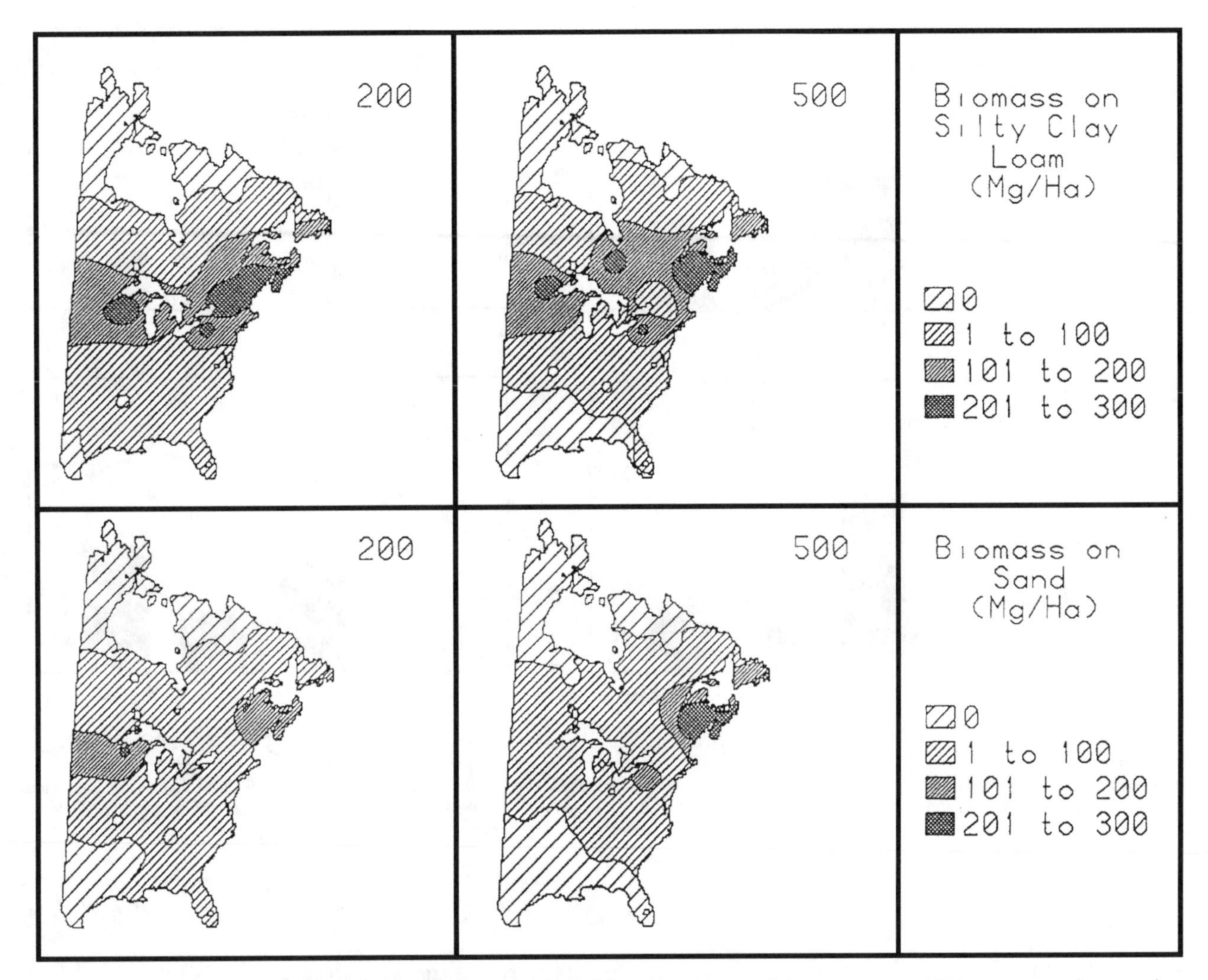

Figure 27-2. Simulated distribution of maple after 200 years under current climate followed by 100 years of climatic warming and 200 more years under a greenhouse climate. (Reprinted with permission from *New Perspectives in Watershed Management.* Copyright 1992 by Springer-Verlag New York, Inc. "Using Simulation Models and Geographic Information Systems to Integrate Ecosystem and Landscape Ecology"; J. Pastor and C. A. Johnston, pp. 324–346, figure 11.4.)

Based on the eigenvector results, the general behavior of beaver ponds was summarized by three rules:

1. The portion of the dominant eigenvector associated with "not ponded" was always 0. That is, all floodable area was permanently converted to some form of beaver pond, and "not ponded" was a transient state.
2. The portion of the dominant eigenvector associated with seasonally flooded pond areas was small (generally < 0.2) and fluctuated randomly from one decade to the next. Seasonally flooded ponds were therefore a minor and nondirectional component of the landscape dynamics.
3. The portions of the eigenvector associated with flooded and moist pond areas dominate the dynamics of the landscape at all scales and were inversely related to one another regardless of watershed, antecedent conditions, or cohort of pond creation. This inverse relationship between the stable areas of flooded and moist soils is a type of attractor along which the landscape moves over time, with tendency toward domination by flooded ponds during beaver occupation and by moist meadows during abandonment (Figure 27-4). The attractor itself is invariant, but the eventual position of the landscape along it depends on decade, geomorphology, antecedent conditions, and cohort of pond creation (Pastor et al. 1993).

This work illustrates the utility of Markov modeling for evaluating patterns of spatial change across the landscape. The use of GIS with digitized maps provides more precision in determining transition probabilities over different portions of the landscape at different times, removing many of the difficulties experienced by earlier workers in parameterizing Markov matrices.

However, the usefulness of this approach in predicting potential changes in the spatial distribution of plants is limited for the following reasons:

1. The transition probabilities are derived statistically: the actual mechanisms such as reproduction, overcrowding, and seed dispersal that result in plant migration are not considered. Extrapolations can be made only if one assumes that the system is measured in equilibrium, and that the extrapolation is applied to a similar system in equilibrium.
2. Markov chains, as all other classical statistical analyses, spatial or not, are linear. By linear we mean the following:

$$f(cS) = cf(S), \text{ and } f(S1+S2) = f(S1) + f(S2) \qquad (27\text{-}2)$$

where f is some transformation on a system S, and c is a constant. This definition does not preclude nonlinear equations (such as polynomials) in

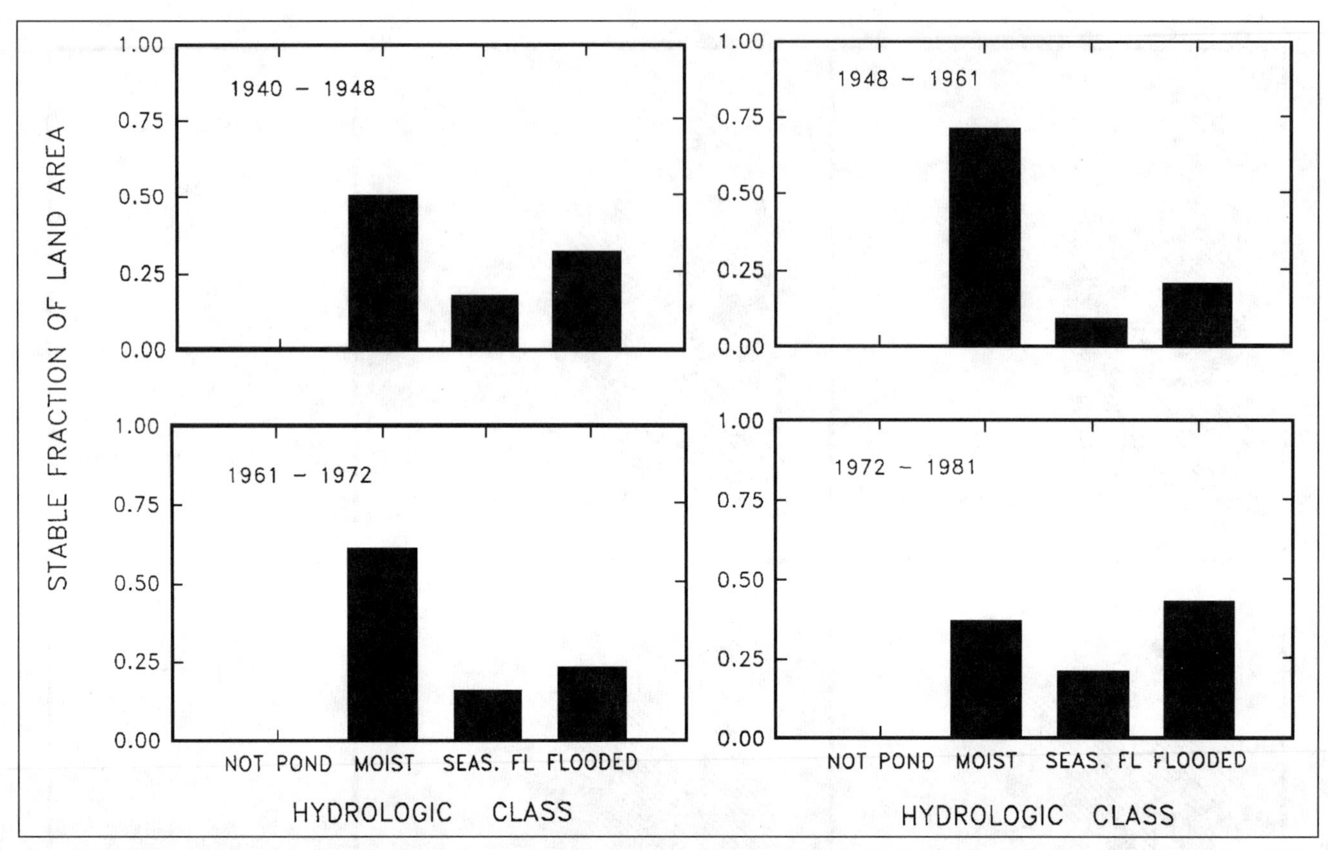

Figure 27-3. Stable areal distributions of four hydrologic classes predicted by dominant eigenvectors of Markov matrices for floodable areas of Voyageurs National Park. (Reprinted with permission from *Lectures on Mathematics in the Life Sciences*, Vol. 23, by permission of the American Mathematical Society. "Markovian Analysis of the Spatially Dependent Dynamics of Beaver Ponds," by J. Pastor, J. Bonde, C. A. Johnston, and R. J. Naiman.)

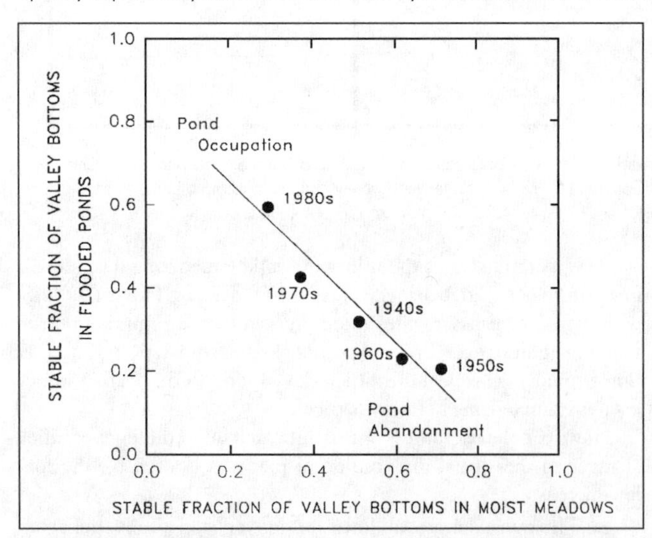

Figure 27-4. The portion of the dominant eigenvector associated with flooded ponds varies with that associated with moist meadows. (Reprinted with permission from *Lectures on Mathematics in the Life Sciences*, Vol. 23, by permission of the American Mathematical Society. "Markovian Analysis of the Spatially Dependent Dynamics of Beaver Ponds," by J. Pastor, J. Bonde, C. A. Johnston, and R. J. Naiman.)

modeling the system. Biological systems include feedbacks. These, along with other biological processes introduce nonlinearities to the system. It is well known that even the simplest nonlinear systems (e.g., the logistic growth equation) can display very complex dynamic behavior (May 1976).

These problems plague many current GIS applications. Next, we discuss our attempts to circumvent these difficulties by incorporating mechanisms into our models as much as possible. It must be realized, however, that mechanistic and descriptive models are two extremes along a continuum or mix of both, and that all models (including ours) include descriptive (statistical) mechanisms (equations) at some level.

SPATIODYNAMIC MODELING

Dynamic Forest Ecosystem Simulation

We have demonstrated the use of the LINKAGES model in a spatially static application. Recently, we developed a spatial version of the model, called S_L (Spatial Linkages). In S_L, a landscape of a prespecified size is simulated. Plant species grow, reproduce and distribute seeds, and die on the landscape according to their tolerance to light, water, nitrogen conditions, seed production, and life-history characteristics. These tolerances and characteristics are derived from known plant-species properties and their geographic distribution. Simulations exhibit successional changes that are expected. These changes are produced mostly by competition for light and nitrogen, the availability of seeds, and the conditions for seed germination. The competition, however, is not simulated explicitly: it is a consequence of the tolerances of plant species to various environmental conditions, shading, independent extraction of resources from the soil, and differing optimal conditions for seed germination.

S_L presents challenging development problems. Because of the model's computational intensity, efforts to implement it in a GIS software framework failed: the runs were unacceptably slow. A typical run on a 250-m × 250-m landscape lasts one to three days on a SUN Sparcstation 2. We thus had to code the spatiotemporal changes and their animation from "scratch." The model is coded in C++, and the code is highly optimized. One of the model outputs is a dynamic animation of changes in plant distribution and soil nitrogen over the landscape (e.g., Figure 27-5). These animations were channeled to a VCR, and are thus available for wide distribution.

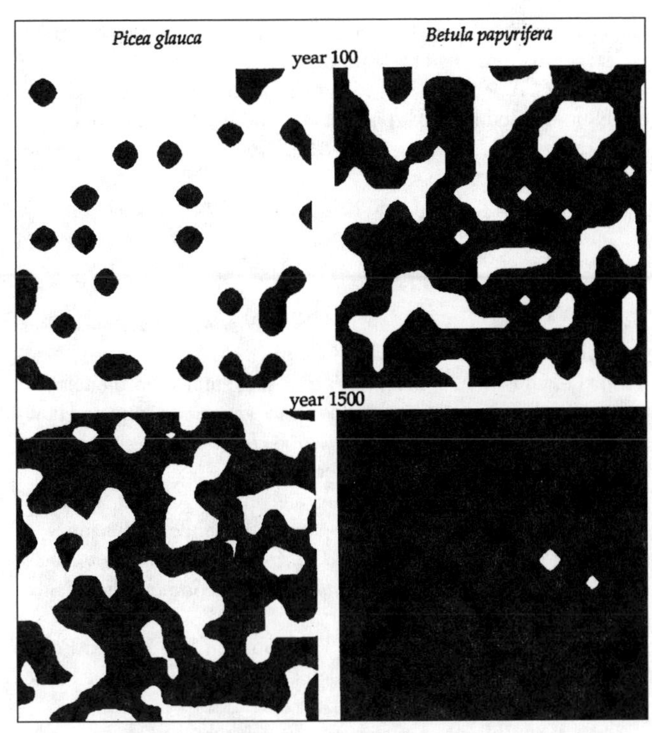

Figure 27-5. Representative frames of the spatial dynamics of two species biomass on a 160 × 160 landscape. White represents values close to 1, maximum biomass; black close to 0.

One of the insights gained from the animation (the results of spatiotemporal runs) was the realization that there are sudden surges in nitrogen concentrations over the landscapes. Although small, these surges permeate the whole landscape, and last for relatively short periods of time (five to fifteen years). Such surges would not have been obvious without direct observations of the animation.

Potential advancements in parallel computing will allow us to run larger landscapes at shorter simulation times. But even with such developments in computer hardware, we do not expect to implement the model through existing GIS software packages: these packages have been developed primarily for static analysis, with its inherent limitations, as discussed above. Yet, GIS software can be potentially useful (see Conclusion).

Moose-Foraging Simulation

In this model, we simulate the landscape that emerges as a result of moose foraging. The simulation begins with a user-adjustable initial landscape, containing prespecified "moose-units," simulated as moving entities on the landscape. The moose movements are implemented as a series of "decisions" the moose makes with regard to foraging. These decisions are based on: (1) the energetic conditions of the moose (whether it has or has not supplemented its requirements at a particular time of the year); (2) the history (in years) of the moose-population energetic conditions; (3) the history of plant distribution, as it depends on plant growth and its reaction to the intensity of browsing; and (4) the optimal foraging rules that dictate the moose-foraging behavior. All of these rules and parameters are adjustable. With time, one can observe the consequences of optimal moose-foraging rules on the emerging landscape (i.e., changes in the density and spatial distribution of plant species over time).

A fundamental working hypothesis in animal optimal foraging theory is that animals forage according to the marginal value theorem: an animal forages in a patch until its marginal return from the patch reaches that of the mean return from the animal's habitat (Charnov 1976). In our model we "allow" the moose to forage according to this rule. Other rules implemented in the model are: (1) the so-called Markovian rule—the animal forages in a patch until no food is available, and then moves on to the next best patch; and (2) the random rule—the animal exhausts the food in an existing patch and moves on to a different patch, randomly. It turns out that the emerging landscapes are different based on the foraging rule.

One conclusion reached from this spatiodynamic simulation is that an animal cannot forage according to the marginal value theorem unless there are external disturbances to the habitat. Without such disturbances (e.g., fire), all patches will be reduced to the mean marginal return value of the habitat, and the animal winds up standing in a patch, not knowing where to move next. These decision rules do not imply that the animal has conscious decision-making abilities, but are a consequence of evolution; that is, those animals that happen to forage according to these rules survive and leave more progeny than others, and thus their genes (which include the "foraging-rule template") become more frequent with time. Although in hindsight, such a conclusion seems trivial, we did not think about this possibility of failure of the marginal value theorem until we observed the simulated foraging moose stopping, for no apparent reason (for a long time we thought that there was a bug in the code).

As with S_L, the model could not be implemented through an existing GIS software package: runs were unacceptably slow. The model is currently coded in C++.

Spatiodynamic Modeling of a Marine Fouling Community

Stationary marine surfaces are subject to so-called biofouling. The process begins with a clean surface. Microalgae and barnacles settle on the surface and stick to it. With time, sessile organisms (e.g., algae) settle on the barnacles and the surface and eventually cover it. As expected, this process annoys the shipping industry: the fouling community sticks to ship hulls and slows their movement considerably. The power-producing industry suffers similar consequences: the community develops on discharge grills and clogs them. To combat this phenomenon, antifouling paints were developed. These paints are highly toxic, containing one of the most polluting marine compounds, tribolytin.

In an effort to solve some of these problems, we developed a spatiotemporal biological-control model. It simulates the development of the fouling community. In experiments, we demonstrated that limpets, who might settle on such surfaces, crawl over it, eat the sessile algae, and bulldoze barnacles off the surface. They thus may keep the surface clean. The model was developed to examine the potential cleaning effect of limpets and to design an optimal biological control program. Thus, the model simulates the growth and succession of the biofouling community and the foraging behavior of limpets. By regulating the number of limpets, we can calculate the minimum number needed to keep the surface clean based on the rate of development of the fouling community.

The model is spatiotemporal: (1) fouling organisms arrive at a particular rate (according to the Poisson process) to particular locations on the surface, then settle and grow; and (2) limpets crawl on the surface, according to prespecified rules, and thus keep it clean (Figure 27-6). Therefore, given a specific number of limpets on the surface, one may observe their movements and conclude that they are territorial. The fact is that one may reach such a conclusion even when the limpets move according to purely random rules. This demonstrates that care must be taken in deriving conclusions about territoriality and potentially other animal behavioral traits.

As with S_L and the moose-foraging model, GIS software could not be used; simulations were much too slow. We thus developed the model in C++, and runs are satisfactorily fast.

CONCLUSION

The previous examples presented us with some common problems. When it comes to simulating "real-time" dynamics (i.e., observing landscape changes during simulation time), GIS software was not helpful. These software packages were not designed and implemented

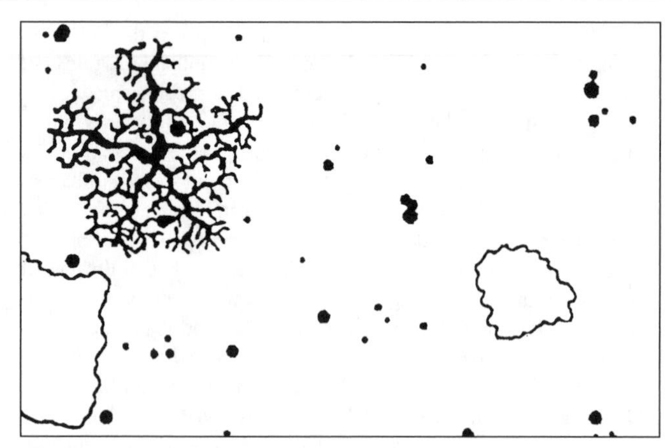

Figure 27-6. A typical surface produced by a simulation on a 35- × 35-cm² panel with 15 limpets of size 18 mm. Hydrozoa and bryozoa growth is simulated with diffusion percolation.

with such considerations in mind. Yet, the state of the art and the ever-increasing demands from simulation models require spatiodynamic approaches that are mechanistically based (as much as possible).

In order to help, GIS software developers may consider the following:

1. Supply a large part of the software in the format of an object library. Such libraries should include a host of functions (sometimes called subroutines or procedures). The tasks that such functions perform should be well defined and as independent of other tasks as possible. Each function should perform a single specific task.
2. The software should provide "hooks" for dynamic modeling. Such hooks should be as modular as possible and yet allow implementation of dynamic processes. For example, a single routine might ask the user to supply a location, a trait, and equations that describe the dynamic changes of this trait. This routine, coupled with others (e.g., contour smoothing, color rendering, and three-dimensional smoothing) should calculate new values and render them on the monitor.
3. The tasks of spatial analysis using GIS software must be simplified considerably. Given the current technology (e.g., Graphical User Interface, the so-called GUI) such simplifications are feasible.
4. Establish open architecture, and agreed-upon standards. By open architecture we mean that the algorithms (not the code) of various functions are well described and access to isolated parts of the software is easy (as discussed in 1 above). This will simplify software development and increase user friendliness (at the "expense" of secrecy and simplifications). A good case in point is the computer hardware industry that only recently has been forced to adopt open architecture and cooperation.
5. Implement sophisticated tools, such as algorithms, to solve partial differential equations, and to assist in decision making and risk-analysis software (e.g., Bayesian statistics).
6. Integrate multimedia capabilities with the GIS software.

Acknowledgments

Research support from the National Science Foundation (#BSR-8906843, #BSR-9009169, and #DEB-9119614) is gratefully acknowledged.

References

Caswell, H. 1989. *Matrix Population Models.* Sunderland: Sinauer Associates.

Charnov, E. L. 1976. Optimal foraging: The Marginal Value Theorem. *Theoretical Population Biology* 9:129–36.

Cohen, Y., and J. Pastor. 1991. The responses of a forest model to serial correlations of global warming. *Ecology* 72:1,161–65.

Johnston, C. A. 1993. Introduction to quantitative methods and modeling in community, population, and landscape ecology. In *Environmental Modeling with GIS,* 276–83. New York: Oxford University Press.

Johnston, C. A., J. Pastor, and R. J. Naiman. 1993. Effects of beaver and moose on boreal forest landscapes. In *Landscape Ecology and Geographical Information Systems,* 236–54. London: Taylor and Francis.

May, R. M. 1976. *Theoretical Ecology: Principles and Applications.* Philadelphia: W. B. Saunders Company.

Pastor, J., and C. A. Johnston. 1992. Using simulation models and geographic information systems to integrate ecosystem and landscape ecology. In *New Perspectives in Watershed Management,* 324–46. New York: Springer-Verlag.

Pastor, J., J. Bonde, C. A. Johnston, and R. J. Naiman. 1993. Markovian analysis of the spatially dependent dynamics of beaver ponds. *Predicting Spatial Effects in Ecological Systems,* 5–27. Vol. 23 of *Lectures on Mathematics in the Life Sciences.* Providence: American Mathematical Society.

Pastor, J., and W. M. Post. 1985. Development of a Linked Forest Productivity-Soil Process Model. ORNL/TM-9519. Oak Ridge: Oak Ridge National Laboratory.

———. 1986. Influence of climate, soil moisture, and succession on forest carbon and nitrogen cycles. *Biogeochemistry* 2:3–27.

———. 1988. Response of northern forests to CO_2-induced climate change. *Nature* 334:55–58.

Running, S. W., R. R. Nemani, D. L. Peterson, L. E. Band, D. R. Potts, L. L. Pierce, and M. A. Spanner. 1989. Mapping regional forest evapotranspiration and photosynthesis by coupling satellite data with ecosystem simulation. *Ecology* 70:1,090–1,101.

Shugart, H. H., Jr. 1984. *A Theory of Forest Dynamics.* New York: Springer-Verlag.

Carol A. Johnston is a senior research associate at the Natural Resources Research Institute of the University of Minnesota where she directs the Natural Resources GIS Laboratory. She holds a Ph.D. in soil science from the University of Wisconsin-Madison. Johnston serves on the Minnesota Governor's Council on Geographic Information and was on the organizing committee for the Second International Conference/Workshop on Integrating Geographic Information Systems and Environmental Modeling.

John Pastor is a senior research associate at the Natural Resources Research Institute of the University of Minnesota where he conducts research on boreal forest ecosystems. He holds a Ph.D. from the University of Wisconsin-Madison.
Natural Resources Research Institute
University of Minnesota
Duluth, Minnesota 55811
E-mail: *cjohnsto@sparkie.nrri.umn.edu*

Yosef Cohen is a professor in the Department of Fisheries and Wildlife at the University of Minnesota and conducts research on mathematical modeling. He received a Ph.D. from the University of California, Berkeley.
Department of Fisheries and Wildlife
University of Minnesota
St. Paul, Minnesota 55108
E-mail: *yc@turtle.fw.umn.edu*

28

MMS:
A Modeling Framework for Multidisciplinary Research and Operational Applications

G. H. Leavesley, P. J. Restrepo, L. G. Stannard, L. A. Frankoski, and A. M. Sautins

The Modular Modeling System (MMS) is an integrated system of computer software that is being developed to provide the research and operational framework needed to support development, testing, and evaluation of physical-process algorithms and to facilitate integration of user-selected sets of algorithms into operational physical-process models. MMS uses a module library that contains compatible modules for simulating a variety of water, energy, and biogeochemical processes. A model is created by selectively coupling the most appropriate modules from the library to create an "optimal" model for the desired application. Where existing modules do not provide appropriate process algorithms, new modules can be developed. A geographic information system (GIS) interface is being developed for MMS to facilitate model development, application, and analysis. This interface permits application of a variety of GIS tools to characterize the topographic, hydrologic, and biologic features of a physical system for use in a variety of lumped- and distributed-parameter modeling approaches. MMS display capabilities permit visualization of the spatial distribution of model parameters and of the spatial and temporal variation of simulated state variables during a model run.

INTRODUCTION

The interdisciplinary nature and increasing complexity of environmental and water-resource problems require the use of modeling approaches that can incorporate knowledge from a broad range of scientific disciplines. Selection of a model to address these problems is difficult given the large number of available models and the potentially wide range of study objectives, data constraints, and spatial and temporal scales of application. Coupled with these problems are the problems of study area characterization and parameterization once the model is selected. Guidelines for parameter estimation are normally few, and the user commonly has to make decisions based on an incomplete understanding of the model developer's intent.

To address the problems of model selection, application, and analysis, a set of modular modeling tools, termed the Modular Modeling System (MMS), is being developed. The approach being applied in developing MMS is to enable a user to selectively couple the most appropriate process algorithms from applicable models to create an "optimal" model for the desired application. Where existing algorithms are not appropriate, new algorithms can be developed and easily added to the system. This modular approach to model development and application provides a flexible method for identifying the most appropriate modeling approaches given a specific set of user needs and constraints.

A major component of MMS is a geographic information system (GIS) interface to facilitate model development, parameterization, application, and analysis. This interface permits application of a variety of GIS tools to lumped- and distributed-parameter modeling approaches. These tools permit development and testing of a variety of objective characterization and parameterization techniques. They also permit visualization of the spatial distribution of model parameters and simulated state variables at a variety of temporal and spatial scales.

The linking of modeling and GIS tools provides a common framework in which to focus multidisciplinary research and operational efforts to provide improved understanding of complex water, energy, and biogeochemical processes. Current model development within MMS is focused on hydrologic processes. The purpose of this chapter is to provide (1) an overview of the system, (2) an introduction to the concepts and capabilities of MMS with a discussion of currently available (1993) system components, and (3) a description of the GIS interface.

MMS OVERVIEW

The conceptual framework for MMS has three major components: preprocess, model, and postprocess (Figure 28-1). The preprocess component includes the tools used to input, analyze, and prepare spatial and time-series data for use in model applications. The model component includes the tools to develop and apply models. The postprocess component provides a number of tools to display and analyze model results and to pass results to management models or other types of software. The model component is currently the most fully developed of the three components. However, a number of the preprocessing and postprocessing tools are being developed, tested, and made available for linkage to the model component.

A system supervisor, in the form of an X-Windows graphical user interface (GUI), is proposed to provide user access to all the components and features of MMS. The present framework has been developed for UNIX-based workstations and uses X-Windows and Motif for the GUI. The GUI provides an interactive environment for users to

Any use of trade, product, or firm names in this chapter is for descriptive purposes only and does not imply endorsement by the U.S. Government.

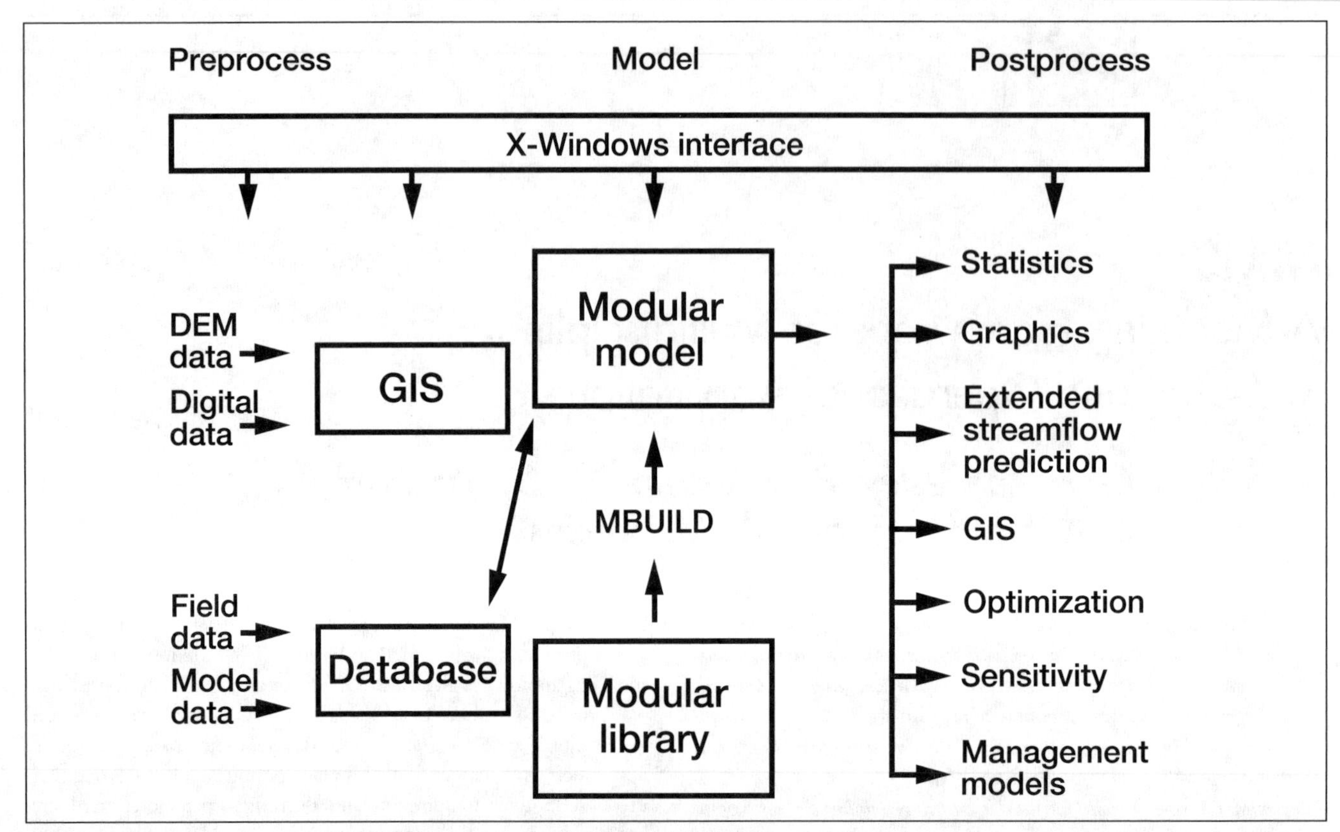

Figure 28-1. A schematic diagram of the components of the Modular Modeling System (MMS).

access model component features, apply selected options, and graphically display simulation and analysis results. The current GUI is being expanded and enhanced into the full system supervisor, incorporating the linkages needed to access features in all the system components.

Preprocess Component

Preprocess component functions include all data preparation and analysis functions needed to meet the data and selected parameterization requirements of a user-selected model. A goal in the development of the preprocess component is to take advantage of the wide variety of existing data preparation and analysis tools and to provide the ability to add new tools as they become available. Spatial data analysis is accomplished using GIS tools that the users have installed on their computer system. Such analyses could include segmentation and characterization of a watershed into subareas for distributed-parameter modeling. Digital elevation model (DEM) data and selected digital databases that include information on soils, vegetation, geology, and other pertinent physical features would provide the data on which to develop such characterizations.

Time-series data from existing databases as well as from field instrumentation are prepared for use in selected model applications by generating and combining these data into a single flat ASCII file. Additional tools are being developed to detect bad or missing values, replace bad or missing data via selected statistical procedures, aggregate data to longer time steps, disaggregate data to shorter time steps, and apply transform functions to produce a new time series. Methods to create simulated time series from model output or from the analysis and extrapolation of measured data to unmeasured points or gridded fields are also being developed.

The database(s) used to store the spatial and time-series data provide(s) the interface between the preprocess and model components. The goal is to enable the use of a variety of databases, dependent on user preference or prescribed needs. Interface to the model component can be provided by the use of Structured Query Language (SQL) commands or filter programs that access the database.

Model Component

The model component is the core of the system and includes the tools to selectively link process modules from the module library to build a model and to interact with this model to perform a variety of simulation and analysis tasks. These interactions are provided using a variety of X-Windows and graphical techniques. The model component is discussed in more detail in a later section titled "Modular Model Concepts and Capabilities."

Postprocess Component

The tools to analyze model results are included in the postprocessing component. These include a variety of statistical and graphical tools as well as the ability to interface with user-developed special purpose tools. Statistical and graphical analysis procedures provide a common basis for comparing module performance and can be used to aid in making decisions regarding the most appropriate modeling approach for a given set of study objectives, data constraints, and temporal and spatial scales. A GIS interface provides the visualization and analysis tools to display spatially distributed model results and to analyze results within and among different simulation runs.

Two of the currently available postprocessing capabilities interact with the model component. The parameter-optimization and sensitivity-analysis tools are provided to optimize selected model parameters and evaluate the extent to which uncertainty in model parameters affects uncertainty in simulation results. A modified version of the National Weather Service Extended Streamflow Prediction Program (ESP) (Day 1985) provides forecasting capabilities using historic or synthesized meteorological data.

MODULAR MODEL CONCEPTS AND CAPABILITIES

Modules

A major feature of the model component of MMS is the module library which contains a variety of compatible modules for simulating water, energy, and biogeochemical processes. The library can contain several modules for a given process, each representing an alternative conceptualization or approach to simulating that process. The user, through an interactive model-builder interface (MBUILD), selects and links modules to create a specific model. Once a model has been built, it may be saved for future use without repeating the MBUILD step. This capability allows "canned" versions of models to be provided to end users.

Initial modules in the library were derived from the U.S. Geological Survey Precipitation Runoff Modeling System (PRMS) (Leavesley et al. 1983). Additional modules have been included using selected process algorithms from the National Weather Service River Forecast System (NWSRFS) model (Anderson 1973), the Streamflow Synthesis and Reservoir Regulation (SSARR) model (U.S. Army 1989), and TOPMODEL (Beven and Kirkby 1979). New modules for channel transport of solutes and sediment also have been developed and included. Additional modules can be added to the library as research and operational applications expand MMS use.

The ability to link modules developed by a variety of users is provided by the use of a standardized module structure. A module is composed of a minimum of four functions: declare, initialize, run, and main. The declare function is used to specify parameters and variables that are being declared in this module. The initialize function is used to initialize parameter and variable values used in the module. The run function contains the algorithm code that simulates the specific process. The main function directs system calls to the declare, initialize, and run functions of a module. A module can be written in either the FORTRAN or C programming language.

Communication among modules and between a module and the MMS system is accomplished using specific MMS function statements that a module developer uses in the module code. Parameter and variable data structures are created by the "declparam" and "declvar" function statements respectively. When MMS is executed, the declare function of all modules is executed once to obtain the needed information from the declparam and declvar statements to build the MMS parameter and variable databases. During the execution of the initialize and run functions of a module, related MMS function statements "getparam" and "getvar" are used to read current values of parameters and variables from these databases. Values are written to the variables database automatically by MMS when the variable is modified in the module where it was declared. To modify a variable from a module where the variable was not declared, a "putvar" system function is used.

Arguments in the declparam and declvar function statements include a definition of the parameter or variable and the units in which it is expressed. The definition is made available to the model user through a help feature during MMS execution. The definition describes the parameter or variable, provides guidance on the estimation of its initial value, and lists any other pertinent information. This feature provides a mechanism for module developers to imbed their knowledge and expertise within the module and have this information available to all users. The units argument is used in the MBUILD process to ensure the compatible linkage of module parameters and variables.

In the MBUILD procedure, a comparison is made among selected modules to ensure that all getparam and getvar statement functions are satisfied by a declparam or declvar statement for the specified parameter or variable. The declparam or declvar statement may be in the same module or another module.

The ability to compare parameter and variable names among modules requires that a consistent terminology be used in their naming. A dictionary of currently used terms and their definition is included in MMS to provide information to system users as well as to maintain consistency among module developers. This is one of the more difficult system concepts to develop. However, the value in improved communication within the modeling community and the establishment of some consistency in process parameter and variable definition are seen as major needs in being able to compare modeling approaches.

While there is a requirement for consistency in the external naming convention used for module linkage, there is no immediate need to change the terminology in the already existing code that is converted to a module. Two arguments included in the declparam and declvar function statements are an external system name and an internal module name. These are made equivalent when incorporated into MMS.

Models

Modules are linked using MBUILD in a user-defined sequence to create a model. The modules are executed sequentially in a time-based loop whose time step may be variable and is defined by the input data stream. When a model is executed within MMS, one pass is made through the initialize functions of all the modules to initialize all parameters and user-specified variables. Each subsequent pass through the modules is directed to the run function to execute the process algorithms. Groups of processes that have feedback must currently be included in the same module in order to accommodate these mechanisms. However, the ability to group a set of modules into a feedback unit is currently being developed to provide more system flexibility.

During module compilation, the MMS X-Windows graphical user interface (GUI) is combined with the modules to provide the links between the selected model and the MMS support functions. A series of pull-down menus in the GUI provide the links to a variety of system features. These include the ability to (1) select and edit parameter files and data files; (2) select a number of model execution options such as a basic run, a run with a graphical and/or a spatial visualization output, a run with ESP, an optimization or a sensitivity analysis; and (3) select a variety of statistical and graphical analyses of simulation output.

Reviewing and editing of model parameters is accomplished through the use of a spreadsheet-type interface. Spreadsheets are selected as a function of the indices used in defining the spatial or temporal distribution of groups of parameters. For example, one spreadsheet might use the index of months to display the parameters that are defined as varying by month. Another spreadsheet might use the index of spatial basin subareas to display the parameters that are defined for each subarea. Spreadsheets are also available to support two-dimensional parameters that, for example, might change monthly for each basin subarea. Pointing and clicking on the parameter name in the spreadsheet opens a help window that displays the definition information that was included in the declparam statement for the selected parameter.

One MMS graphical tool permits the use of up to four graphical display windows in which the user can display, during a model run, any of the variables that have been declared in a declvar statement. As many as ten variables can be displayed in each window and plotted results can be output in HPGL or PostScript formats either to a digital file or to a printer.

GIS INTERFACE

A GIS interface is being developed to provide tools for the analysis and manipulation of spatial data in the preprocess, model, and postprocess components of MMS. The GIS currently being used is the Geographical Resources Analysis Support System (GRASS) developed by the U.S. Army Corps of Engineers (1991). This interface permits the coupling of a variety of GIS tools to characterize the topographic, hydrologic, and other physical features of a watershed for use in a variety of lumped- or distributed-parameter modeling approaches.

In the preprocess component, the GIS interface is accessed through a GRASS shell screen which the user operates through a set of pull-down menus. The menus provide access to frequently used GRASS tools and other GIS tools developed specifically for a variety of modeling tasks. The functions provided include the ability to (1) display raster, vector, and site maps, (2) create slope, aspect, and elevation maps, (3) delineate watershed subbasins, (4) develop model input parameters and a parameter characterization table, and (5) generate an MMS input parameter file. The GRASS shell also provides flexibility to the interface by allowing the user to execute any GRASS command.

The delineation of basin subareas and the estimation of a range of physical and hydrologic parameters for these subareas are major functions of the preprocess GIS tools. These procedures use DEM data for the delineation of watershed boundaries and the generation of slope, aspect, and elevation maps. Digital databases for soils, vegetation, geology, and other physical and hydrologic features provide additional information that can be overlayed on the topographic characterization for use in the estimation of parameter values for delineated basin subareas. Estimated parameter values can be written to MMS model parameter files for direct use in a selected model.

While the GIS tools provide the mechanisms for delineating basin subareas and estimating parameters for these subareas, much work remains to be done on the development, testing, and identification of the specific parameterization algorithms that are most appropriate for each of the water, energy, and biogeochemical processes that may be simulated using MMS. As these algorithms are defined for selected modules, they will be linked to the module. When a model is built or selected for execution, the parameterization procedures linked with each module will be made available to the user for automated execution if the spatial databases needed are available.

Within the model component, the GIS interface enables the visualization of the spatial and temporal variation of simulated state variables during a model run. The basin subarea characterization developed in the preprocess component is displayed to the user in an X-Window. At run time, any of the state variables being simulated for each subarea can be selected for an animated display during the run. The capability to store selected images from this animation for user-defined time periods is currently being developed.

The GIS interface in the postprocessing component will provide the capabilities to display and analyze model results. Again, the basin subarea characterization developed in the preprocess component will provide the base map for the display of the spatial and temporal variation of selected state variables. One capability currently being developed is the ability to display the images of simulated snow-covered areas taken from the run time animation with remotely sensed data provided by the National Weather Service. The ability to compare simulated and measured spatial and temporal variation in snow-covered areas, or other stated variables, provides important additional independent measures of distributed-parameter model performance.

FUTURE ENHANCEMENTS

A number of additional system enhancements and capabilities are being designed to facilitate model development, application, and analysis. The ability to couple a variety of resource-management and risk-analysis models with user-selected water, energy, and biogeochemical process models is being included for use in evaluating alternative resource-management policies and in developing operational short- and long-term resource-management plans. Interfaces are also being developed to import and export data and model results from and to other external data management and analysis systems.

CONCLUSION

MMS is an integrated system of computer software that has been developed to provide the research and operational framework needed to support the development, testing, and evaluation of physical-process algorithms and to facilitate the integration of user-selected sets of algorithms into an operational model. MMS provides a common framework in which to focus multidisciplinary research and operational efforts. Researchers in a variety of disciplines can develop and test model components to investigate questions in their own areas of expertise as well as work cooperatively on multidisciplinary problems without each researcher having to develop the complete system model. Results that demonstrate improved simulation performance can be used to modify or enhance current operational models for application within the same framework.

Continued advances in physical and biological sciences, GIS technology, computer technology and data resources will expand the need for a dynamic set of tools to incorporate these advances in a wide range of interdisciplinary research and operational applications. MMS is being developed as a flexible framework in which to integrate these activities.

References

Anderson, E. A. 1973. National Weather Service river forecast system, snow accumulation, and ablation model. *NOAA Technical Memorandum NWS Hydro-17.* Silver Spring: U.S. Department of Commerce.

Beven, K. J., and M. J. Kirkby. 1979. A physically based variable contributing area model of basin hydrology. *Hydrological Sciences Bulletin* 24:43–69.

Day, G. N. 1985. Extended streamflow forecasting using NWSRFS. *Journal of Water Resources Planning and Management* 111:157–70.

Leavesley, G. H., R. W. Lichty, B. M. Troutman, and L. G. Saindon. 1983. Precipitation-runoff modeling system: User's manual. U.S. Geological Survey *Water Resources Investigations Report 83-4238.*

U.S. Army. 1989. *Users Manual, SSARR Model.* Portland: U.S. Army Corps of Engineers.

U.S. Army Corps of Engineers. 1991. *GRASS Version 4.0 User's Reference Manual.* Champaign: USA-CERL.

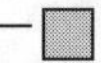

George H. Leavesley received his Ph.D. in watershed sciences from Colorado State University, Fort Collins, Colorado, in 1973. Since that time, he has been employed by the Water Resources Division (WRD) of the U.S. Geological Survey. He is presently project chief of the WRD National Research Program's Precipitation-Runoff Modeling Project. He is conducting research on the development of water, energy, and biogeochemical process models to simulate the effects of climate and land-use change.

G. H. Leavesley and *L. G. Stannard*
U.S. Geological Survey
Box 25046, MS 412
Denver Federal Center
Lakewood, Colorado 80225

P. J. Restrepo
1115 Hancock Drive
Boulder, Colorado 80303

L. A. Frankoski and *A. M. Sautins*
University of Colorado
Center for Advanced Decision Support for Water and Environmental Systems, Department of Civil Engineering
Boulder, Colorado 80309

29

Using GIS to Link Digital Spatial Data and the Precipitation Runoff Modeling System:
Gunnison River Basin, Colorado

William A. Battaglin, Gerhard Kuhn, and Randolph Parker

The U.S. Geological Survey Precipitation Runoff Modeling System, a modular distributed parameter watershed modeling system, is being applied to twenty smaller watersheds within the Gunnison River basin. The model is used to derive a daily water balance for subareas in a watershed, ultimately producing simulated streamflows that can be input into routing and accounting models used to assess downstream water availability under current conditions, and to assess the sensitivity of water resources in the basin to alterations in climate. A geographic information system (GIS) is used to automate a method for extracting physically based hydrologic response unit (HRU) distributed parameter values from digital data sources and for the placement of those estimates into GIS spatial data layers. The HRU parameters extracted are: area, mean elevation, average land surface slope, predominant aspect, predominant land cover type, predominant soil type, average total soil water-holding capacity, and average water-holding capacity of the root zone.

INTRODUCTION

The Gunnison River basin in southwestern Colorado (see Figure 29-1) has a drainage area of 7,930 square miles and elevations that range from 4,520 to 14,300 feet. The basin is a major source of water for the Colorado River system, providing more than 40% of the river's streamflow at the Colorado–Utah state line (Ugland et al. 1990). Water from the Colorado River is used by more than twelve million people and irrigates more than one million hectares of agricultural land (Mueller and Moody 1984). Most streamflow in the Gunnison River basin is derived from snowmelt from the mountainous areas around the perimeter of the basin.

As a part of the U.S. Geological Survey Gunnison River Basin Climate Study, the Precipitation Runoff Modeling System (PRMS) (Leavesley et al. 1983), a modular distributed parameter watershed modeling system, is being applied to twenty smaller watersheds in the Gunnison River basin. Figure 29-1 shows an example watershed. Simulated streamflow from the watersheds will be input into routing and accounting models that are used to assess downstream water availability under current conditions and to assess the sensitivity of water resources in the basin to alterations in climate (Kuhn and Parker 1992).

The PRMS model is used to derive a daily water balance for subareas in a watershed. The outflows from these subareas are combined to produce simulated streamflows (Leavesley et al. 1983). The subareas are called Hydrologic Response Units (HRUs). Each HRU is assumed to be homogeneous in its hydrologic response. Traditionally, parameter values for each HRU were manually determined from various source materials and manually entered into PRMS. Although this system was tractable for situations where the modeled area was small or contained a small number of HRUs, the procedure is cumbersome, slow, and susceptible to human bias and error when large watersheds containing many HRUs are modeled. The number of HRUs in a watershed can range from less than thirty to more than 100, making it undesirable to manually determine HRU parameter values and manually enter those values into PRMS.

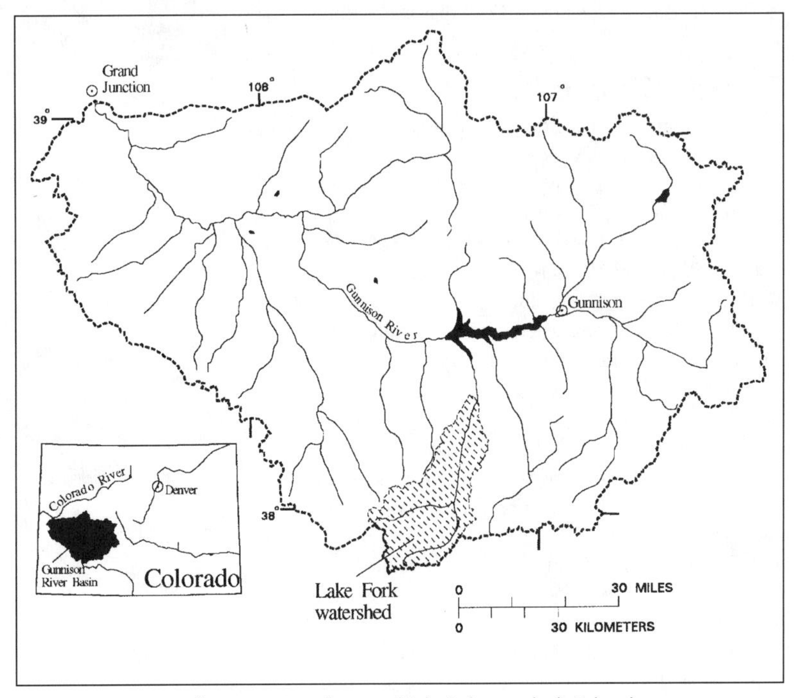

Figure 29-1. Location of Gunnison River basin and Lake Fork watershed, Colorado.

In addition, some watersheds have few data for particular parameters and values must be transferred from other areas. Testing the transfer value of modeled parameters from nearby basins can be accomplished by allowing the easy insertion and alteration of transfer functions. Such a system needs to accommodate relations among various parameters and the physical attributes of the HRUs (Kuhn and Parker 1992).

In a snowmelt-dominated watershed like the Gunnison River basin, values for about forty model parameters need to be estimated for each HRU. These parameter values fall into three categories: (1) regional climatic parameters, (2) physically based HRU distributed parameters, and (3) parameters distributed by subsurface and groundwater reservoirs (Kuhn and Parker 1992). This chapter focuses on the derivation of estimates for the physically based HRU distributed parameters.

GIS APPLICATIONS

A series of GIS programs were developed to automate the transfer of parameter values from GIS spatial data layers (coverages) into digital representations of HRU boundaries. The HRU parameters transferred are: area, mean elevation, average land surface slope, predominant aspect, predominant land cover type, predominant soil type, average total soil water-holding capacity, and average water-holding capacity of the root zone. Once transferred from the digital sources into the HRU coverage, parameter values are written out in a format that can be read directly into PRMS.

In this chapter, HRUs are delineated as 25-km^2 grid cells (see Figures 29-2 and 29-3) (Kuhn and Parker 1992), but the GIS programs that were developed also worked with irregularly shaped HRUs. When a basin boundary crossed a grid cell HRU, the HRU was defined as the portion of the grid cell that fell within the basin. Therefore, HRU area was 25 km^2, unless a drainage basin boundary crossed a grid cell, in which case, HRU area was defined as the area of the grid cell that was within the drainage basin.

Processing Elevation Data

GIS procedures were written to automate the estimation of mean elevation, average land-surface slope, and predominant aspect within HRUs. These parameters were derived from 30-arc-second digital elevation model (DEM) data (point spacing of about 900 m). Techniques described in this chapter also work with other DEM resolutions.

Mean elevation was calculated by first identifying all DEM points within each HRU. If the portion of the 25-km^2 grid cell HRU that was within a particular basin was so small that no DEM points were within that HRU, then the mean elevation value of all DEM points within the 25-km^2 grid cell was assigned to that HRU. For all other HRUs, the mean elevation was calculated as the mean value of the DEM points within the HRU. An example of mean HRU elevations for the Lake Fork watershed is shown on Figure 29-2a.

Average slope was calculated from a coverage of preprocessed slope and aspect information. This coverage was created by making a TIN (triangulated irregular network) of all DEM elevations values in and around the Gunnison River basin, and then converting that TIN into a polygon coverage that contains slope and aspect values for each TIN facet polygon. HRU boundaries were then overlain on this slope and aspect coverage and the area-weighted average slope value was calculated from all TIN facet polygons and portions of TIN facet polygons within each HRU. An example of average HRU slopes for the Lake Fork watershed is shown in Figure 29-2b.

Predominant aspect was calculated from the preprocessed slope and aspect coverage. The aspect data in the preprocessed coverage are given as the compass direction of the maximum rate of descent across each facet from the TIN. The aspect values for each TIN facet polygon were classified into one of six classes: (1) flat class (no aspect), (2) north-facing class (aspect values greater than 337.5° or less than 22.5°, (3) south-facing class (aspect values greater than 157.5° and less than 202.5°), (4) east- and west-facing class (aspect values between 22.5° and 67.5° or between 247.5° and 292.5°), (5) northeast- and northwest-facing class (aspect values between 22.5° and 67.5° or between 292.5° and 337.5°), and (6) southeast- and

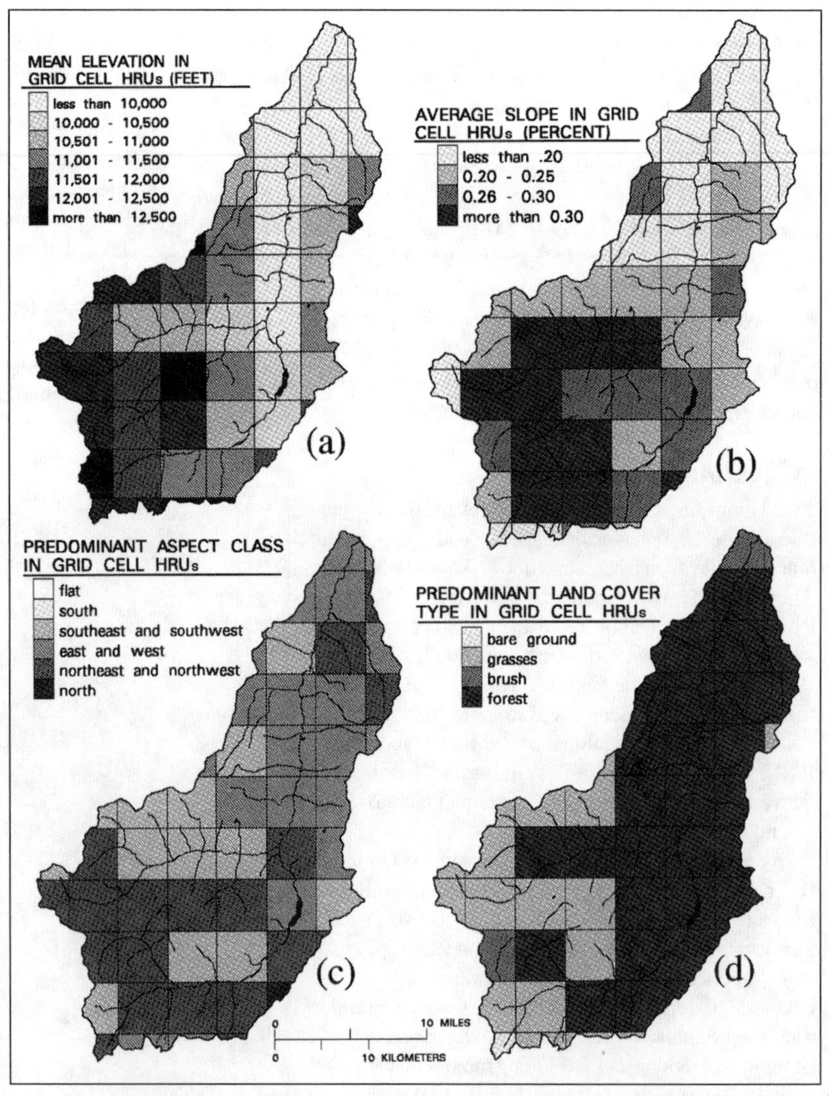

Figure 29-2. Estimates in grid cell HRUs of: (a) mean elevation, (b) average land surface slope, (c) predominant aspect class, and (d) predominant land cover type; for the Lake Fork near Gateview watershed of the Gunnison River basin, Colorado.

southwest-facing class (aspect values between 112.5° and 157.5° or between 202.5° and 247.5°). Aspect values for east- and west-, northeast- and northwest-, and southeast- and southwest-facing facets were grouped together because the aspect values are used in the calculation of potential evapotranspiration, and these aspects are similar in terms of exposure to incident solar radiation. Predominant aspect was calculated for each HRU by summing the areas of all TIN facet polygons and portions of TIN facet polygons that were classified in each of the six aspect classes and then picking the class with the largest area within the HRU. An example of predominant HRU aspects for the Lake Fork watershed is shown on Figure 29-2c.

Processing Land Cover Data

Estimates of predominant land cover were derived from a preclassified land-use coverage. The source material used for this procedure is a digital land-use/land cover database of Anderson's level II land-use categories (Anderson et al. 1976, Fegeas et al. 1983). Four aggregate classes of land cover are required by PRMS: (1) bare ground, (2) grasses, (3) brush, and (4) forest. Table 29-1 shows which level II land-use categories were aggregated to establish the four land cover classes required for this study.

Predominant land cover was calculated by overlaying HRU boundaries on the preclassified land-use coverage. Areas of land cover polygons in each of the four classes in each HRU are calculated and the class with the largest area within each HRU is picked. An example of predominant HRU land cover for the Lake Fork watershed is shown on Figure 29-2d.

Processing Soils Data

Each STATSGO map unit (soil-association polygon) (see Figure 29-3) can have multiple components (soils series), and each component can have multiple layers. Data about the soil association are organized in a series of relatable tables. The STATSGO component table specifies the percentages of each soil series present in a soil association. The STATSGO layer table gives more detailed information about each soil series in a soil association including the number of layers in the soil, layer thicknesses, soil type, and available water-holding capacities (Soil Conservation Service 1991).

Generalized estimates of predominant soil type, average total water-holding capacity, and average water-holding capacity of the root zone are required by PRMS. A series of AWK (Aho et al. 1988), FORTRAN, and GIS programs were written to compute predominant soil type, soil depth, and average available water-holding capacity for each soil series in each soil association. Then area-weighted estimates of predominant soil type, soil depth, and average available water-holding capacity were computed for each soil association using the information in the component table (Battaglin et al. 1992).

PRMS requires three classes of soil type: (1) sand, (2) loam, or (3) clay. The American Association of State Highway and Transportation Officials (AASHTO) group classification values in STATSGO were used to determine soil type (American Association of State Highway and Transportation Officials 1982; Das 1990). The AASHTO system classifies soils into seven groups based upon the soil's particle size distribution and plasticity (Das 1990). For application in the PRMS model, soils with AASHTO classifications A-1, A-2, or A-3 are classified as sand; those with classifications A-4 or A-5 are classified as loam; and those with classifications A-6 or A-7 are classified as clay. Predominant soil type is calculated by summing the areas of each soil type in each HRU and then selecting the class that has the largest area. An example of predominant HRU soil type for the Lake Fork watershed is shown on Figure 29-3a.

Table 29-1. Aggregation of Anderson's[1] level II land-use/land-cover categories into land-cover classes required by PRMS.

PRMS aggregated land cover classes	Anderson's level II land-use categories[2]	
Bare ground	11	Residential
	12	Commercial
	13	Industrial
	14	Transportation, communications, and utilities
	15	Industrial and commercial complexes
	16	Mixed urban or built-up land
	17	Other urban or built-up land
	51	Streams and canals
	52	Lakes
	53	Reservoirs
	73	Sandy areas other than beaches
	74	Bare rock below tree limit
	75	Strip mines, quarries, and gravel pits
	76	Transitional land
	83	Bare ground
	91	Perennial snow fields
Grasses	21	Cropland and pasture
	22	Orchards, nurseries, horticultural areas
	23	Confined feeding operations
	24	Other agricultural land
	31	Herbaceous rangeland
	62	Nonforested wetland
	82	Herbaceous tundra
	84	Wet tundra
	85	Mixed tundra
Brush	32	Shrub and brush rangeland
	33	Mixed rangeland
	61	Forest and shrub wetland
	81	Shrub and brush tundra
Forest	41	Deciduous forest land
	42	Evergreen forest land
	43	Mixed forest land

[1] Anderson et al. 1976.

[2] Numbers correspond to class codes on land-use overlay maps.

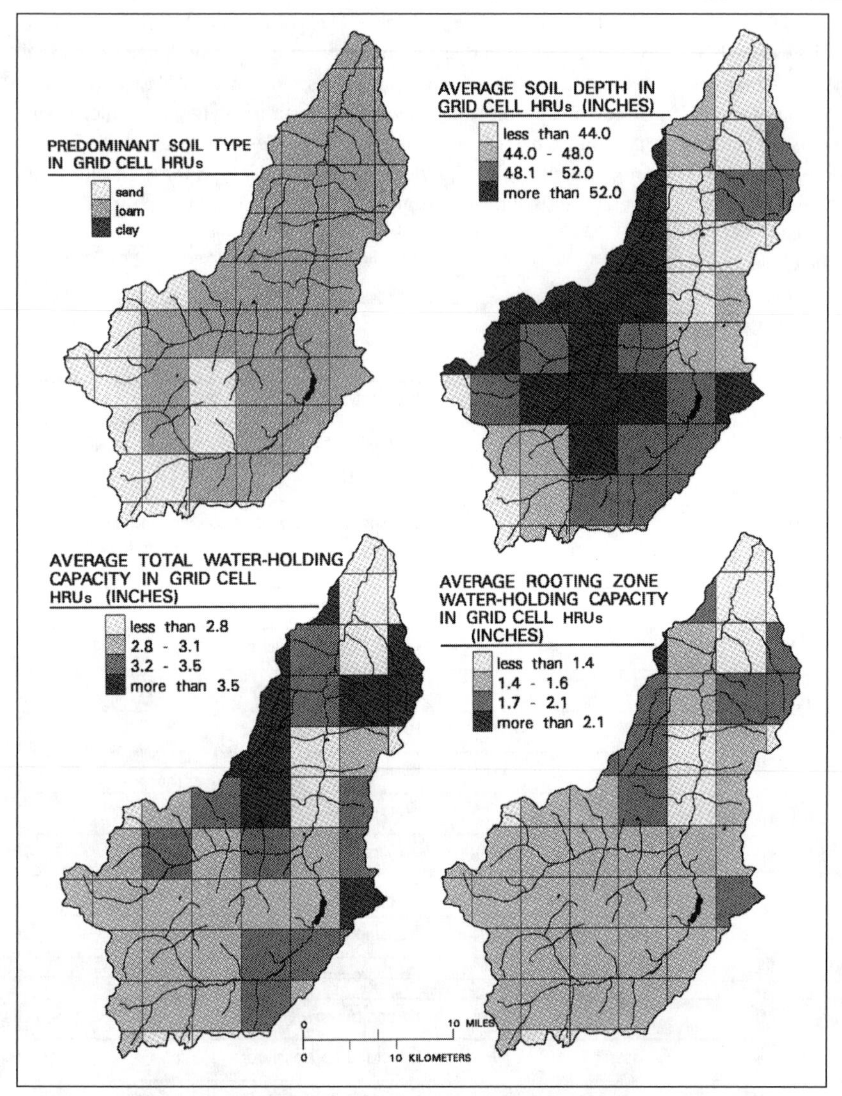

Figure 29-3. Estimates in grid cell HRUs of: (a) predominant soil type, (b) average soil depth, (c) average total water-holding capacity, and (d) average water-holding capacity of the root zone; for the Lake Fork near Gateview watershed of the Gunnison River basin, Colorado.

Estimates of average total water-holding capacity and average water-holding capacity of the root zone are calculated from soil depth and available water-capacity estimates in the STATSGO layer table. Average soil depth is computed by calculating the soil depth for each soil series in an association and computing an area-weighted estimate of soil depth for each soil association using the information in the component table. Available water capacity expressed as inches per inch is also calculated by computing an average value for each soil series and then computing an area-weighted average estimate of available water capacity for each soil association. Estimates of total water-holding capacity for each soil association are calculated as the product of average soil depth and average available water capacity. Estimates of average water-holding capacity of the root zone are calculated as the smaller of 50% of the total water-holding capacity or 2.5 inches of water. Estimates of average total water-holding capacity and average water-holding capacity of the root zone for each HRU are calculated as the area-weighted average value from the soil associations within each HRU. Examples of estimated HRU soil depth, average total water-holding capacity, and average water-holding capacity of the root zone are shown on Figure 29-3b–d.

PROCESSING DATA FOR INPUT INTO PRMS

The result of the data processing just described is a polygon coverage containing estimated values for the eight physically based HRU distributed parameters for all HRUs in a user-specified watershed. The parameter values are output from the HRU coverage into an ASCII file and reformatted using AWK to produce a data file as shown in Table 29-2.

A FORTRAN program was developed to write the GIS-derived HRU parameter values to a PRMS input file. In addition to the parameter values listed in Table 29-2, values for eight other physically based HRU parameters are calculated by the program for inclusion in the PRMS input file. The additional parameters are: (1) summer and (2) winter cover density for the predominant vegetation type; (3) summer and (4) winter interception storage for rain; (5) interception storage for snow; (6) solar radiation transmission coefficient for the vegetation canopy over the snowpack; (7) elevation-based coefficient for calculation of evapotranspiration; and (8) solar-radiation plane.

The FORTRAN program contains simple algorithms that assign values to the additional parameters on the basis of the HRU's mean elevation, average land surface slope, predominant aspect class, and predominant land cover class (Table 29-2). The algorithms were developed primarily from analysis of topographic and orthophoto maps, climatological data, and other watershed modeling studies in Colorado. The algorithms can be modified for calculation of different parameter values as needed.

Use of the FORTRAN program in conjunction with the GIS programs for derived parameter values substantially reduces the time required to create the parameter files needed for input to the PRMS model; if a watershed has a large number of HRUs (>50), the time saved may be one to three days. Thus, additional time is available for doing model simulations and evaluating how parameter values affect model results.

CONCLUSION

An automated method for deriving estimates of physically based parameters required for a distributed parameter watershed model (PRMS) has been used on twenty smaller watersheds in the 7,930 square-mile Gunnison River basin, in southwestern Colorado. The method uses a GIS to quantify parameter values in hydrologic response units that have been defined by the system user. By use of this method, a set of parameters for hydrologic response units in any watershed can be quickly calculated and a file prepared for input to PRMS. This method allows PRMS to be rapidly calibrated to a series of gaged watersheds. In addition, the use of fixed algorithms to manipulate calculated model parameters helps in the transferability of the model to ungaged watersheds. Finally, automation of data entry provides the hydrologist with more time for model calibration, verification, and evaluation of the sensitivity of model results to changes in model parameter values.

Table 29-2. Example of formatted data file used to transfer HRU parameter values from the GIS to the PRMS.

HRU Number	HRU area (acres)	Mean elevation (feet)	Average slope (percent)	Aspect class[1]	Land cover class[2]	Soil type[3]	Soil water-holding capacity (in) total	root zone
1	119	9800	0.12	3	3	1	5.4	2.5
2	1887	8714	0.15	3	2	2	7.8	2.5
3	5	8680	0.05	5	2	2	8.0	2.5
4	3370	9537	0.17	2	3	2	5.9	2.5
5	6134	8565	0.13	3	2	2	6.1	2.5
6	1057	8871	0.12	2	2	2	4.2	2.1
7	9	9545	0.18	3	2	2	8.0	2.5
8	5634	9961	0.18	4	2	2	7.5	2.5

[1] Aspect classes defined as follows: 1=north, 2=northeast and northwest, 3=east and west, 4=southeast and southwest, 5=south, and 6=flat.

[2] Land cover class defined as follows: 0=bare ground, 1=grasses, 2=brush, and 3=forest.

[3] Soil type defined as follows: 1=sand, 2=loam, 3=clay.

References

Aho, A. V., B. Kernighan, and P. J. Weinberger. 1988. *The AWK Programming Language.* New York: Addison-Wesley.

American Association of State Highway and Transportation Officials. 1982. *AASHTO Materials*, part I: *Specifications.* Washington, D.C.

Anderson, J. R., E. E. Hardy, J. T. Roach, and R. E. Witmer. 1976. *A Land-Use and Land Cover Classification System for Use with Remote Sensor Data.* U.S. Geological Survey Professional Paper 964.

Battaglin, W. A., L. E. Hay, R. S. Parker, and G. H. Leavesley. 1992. Application of a GIS for modeling the sensitivity of water resources to alterations in climate in the Gunnison River basin, Colorado. In *Managing Water Resources During Global Change,* edited by R. Herrmann, 741–50. Reno: American Water Resources Association.

Das, B. M. 1990. *Principles of Geotechnical Engineering.* Boston: PWS-KENT.

Fegeas, R. G., R. W. Claire, S. C. Guptill, K. E. Anderson, and C. A. Hallam. 1983. *Land-Use and Land Cover Digital Data.* U.S. Geological Survey Circular 895-E.

Kuhn, G., and R. S. Parker. 1992. Transfer of watershed model parameter values to noncalibrated basins in the Gunnison River basin, Colorado. In *Managing Water Resources During Global Change,* edited by R. Herrmann, 741–50. Reno: American Water Resources Association.

Leavesley, G. H., R. W. Lichty, B. M. Troutman, and L. G. Saindon. 1983. *Precipitation-Runoff Modeling System—User's Manual.* U.S. Geological Survey Water-Resources Investigations Report 83-4238.

Mueller, D. K., and C. D. Moody. 1984. Historical trends in concentration and load of major ions in the Colorado river system. In *Salinity in Watercourses and Reservoirs—Proceedings of the 1983 International Symposium on State-of-the-Art Control of Salinity,* edited by R. H. French, 181–92. Boston: Butterworth Publishers.

Soil Conservation Service. 1991. *State Soil Geographic Database (STATSGO) Data User's Guide.* U.S. Department of Agriculture Miscellaneous Publication Number 1492.

Ugland, R. C., B. J. Cochran, M. M. Hiner, R. G. Kretschman, E. A. Wilson, and J. D. Bennett. 1990. *Water Resources Data for Colorado 1990,* volume 2: *Colorado River Basin.* U.S. Geological Survey, Water-Data Report CO-90-2.

William A. Battaglin is a hydrologist for the U.S. Geological Survey, Water Resources Division. He received a B.S. in geology from the University of Colorado, Boulder, in 1984, and a Masters of Engineering from Colorado School of Mines in 1992. He is currently working on studies that use GIS to investigate the fate and transport of agricultural chemicals in the midwestern United States and the effects of climate change on water resources in the Gunnison River basin, Colorado.

U.S. Geological Survey
Water Resources Division
Box 25046, MS 406
Denver Federal Center
Lakewood, Colorado 80225
Phone: (303) 236-5939
E-mail: *wbattagl@gisrcolka.cr.usgs.gov*

30

Linking GIS with Hydraulic Modeling for Flood Risk Assessment:
The Hong Kong Approach

Allan J. Brimicombe and Jonathan M. Bartlett

Hong Kong's northern lowland basins have undergone substantial urban and suburban development over a twenty-year period. This has been associated with worsening recurrent flood problems. An approach has been adopted whereby hydraulic modeling has been used in conjunction with geographic information systems (GIS) to produce 1:5,000 scale basin management plans (BMPs). GIS has a dual role: (1) in data integration and quantification as an input to hydraulic modeling; and (2) in data interpolation, visualization, and assessment of flood hazard and flood risk using the outputs from the hydraulic modeling. By using current land-use and various development scenarios to be modeled over a range of rainstorm events, "what if" decision support can be used in devising BMPs. Linking with hydraulic modeling requires a different approach to GIS data modeling than the more traditional linkage with hydrological modeling only. The methodology developed in Hong Kong is presented as a case study.

INTRODUCTION

Drainage basins undergoing rapid urbanization are complex, dynamic, human–environment systems. The carefully balanced rural coexistence with the river and its floodplain, developed by trial and error (often over centuries), is invariably disrupted by the geomorphic response to urban development. Flooding can become both more frequent and more severe in terms of depth and duration. The consequences can be costly with periodically disrupted livelihoods, damage to property and essential services, and even loss of life.

FLOODING IN HONG KONG

Hong Kong has a land area of 1,070 km^2 of which approximately 80% is mountainous terrain. With a population of six million and a vibrant economy, the pressures for urban development on the narrow coastal plains and river floodplains are intense. The climate is monsoonal with an average annual rainfall of 2,225 mm. During thunderstorms and typhoons, rainfall can reach intensities of 90 mm/hr. The steep terrain leads to rapid runoff concentration and flash flooding in the lowland basins. High storm surges along the coast (associated with typhoons) occasionally combine with high rainfall to produce critical floods. Planned new towns, ad hoc development, and changes in agriculture in the northern part of the territory have substantially modified both runoff characteristics and drainage configuration and have caused a socially unacceptable increase in flood hazard. The Hong Kong case is well documented (United Nations 1990; United Nations Development Program 1991).

The Drainage Services Department (DSD) of the government of Hong Kong was established in 1989 to take overall responsibility for flood control and storm water drainage. To provide statutory powers for river maintenance, a land drainage bill is being drafted for which BMPs and designation of water courses into maintenance classes are required. A town planning ordinance also is to be amended, requesting drainage impact assessment studies from developers. DSD has commissioned consultants to formulate appropriate strategies and to prepare 1:5,000 scale BMPs for the five largest and most severely flood-prone drainage basins. GIS were an integral part of the studies. Details of how the BMPs are formulated are given in Townsend and Bartlett (1992).

SPATIAL DECISION SUPPORT FOR FLOOD MITIGATION

GIS is an effective tool for storing, analyzing, and visualizing spatial information. Their strength lies in providing syntheses of spatially complex data sets. Supporting decision making in a spatial context is implicit in the use of GIS. However, if GIS is to move beyond simplistic levels of decision support based on "data retrieval and sifting" (Rhind 1988), then analytical capability must be extended to include dynamic modeling of spatial phenomena to include "what if"-type queries (Densham and Goodchild 1989; Hazelton et al. 1992). Modeling of phenomena could be based on deterministic or stochastic simulation to provide multiple, optional outcomes as a sound basis for decision making. While many GIS offer some capability for mathematical and statistical modeling through macrolanguages or internal functionality, the specialist requirements of civil engineering (in areas such as slope instability, liquid/smoke/sediment dispersion, noise, transportation, aquifer management, and drainage design) are frequently more sophisticated than the tools provided. Besides, for most of these requirements, proven and benchmarked proprietary software are already widely used, though graphics functions remain rudimentary by comparison to GIS. One way to move forward is to provide data links between such software and GIS so as to weld their capabilities.

The use of GIS in flood modeling and mitigation has, to date, taken two forms: (1) graphic visualization/communication of flood hazard information and (2) hydrological modeling. An example of the former is the work of the U.S. Federal Emergency Management Agency

(Cotter and Campbell 1987), which uses GIS to integrate map data from a number of sources for the production and dissemination of flood hazard maps. The GIS in this type of application is neither involved in flood modeling nor in establishing hazard potential—that is left to engineers and geomorphologists to establish by traditional non-GIS means. The main role of GIS is as a digital cartographic tool.

Hydrological modeling has been widely associated with GIS for applications in water quality, nonpoint pollution, erosion, and flooding. These applications are founded predominantly on grid cell digital elevation models (DEMs). Other parameters for modeling the watershed, such as land use and soils, are then stored as raster layers registered to the DEM. The catchment area is thus discretized into unit elements with parameters controlling inputs and outputs stored for each element. Gradient and aspect derived from the DEM allow determination of flow direction and sequencing of flow from drainage divide to basin outlet. The hydrological response is computed for each element within the proper time and spatial sequence and accumulated toward the outlet. Thus, in modeling overland flow, unit hydrographs can be developed. GIS-based hydrological modeling is summarized by Johnson (1989) and Smith and Brilly (1992); recent applications are described by Bitters et al. (1991) and Wu and Chen (1992).

Hydraulic modeling is used to simulate runoff along a drainage network in response to specified rainfall events. Models are constructed using true channel details, aggregated runoff coefficients, floodplain storage mechanisms, and significant structures such as bridges and weirs. Thus hydraulic modeling gives a truer simulation of channel capacities and their flows than hydrological modeling. Apart from studying the existing situation, hydraulic modeling is widely used as a design tool for remedial measures. Whereas hydrological modeling is based on a tessellation of the drainage basin, hydraulic modeling is based on nodes linked into a topological network. Nodes may be located on drainage channels at typical cross sections or may be used to represent locations of flood storage. The hydrological characteristics of each node's subcatchment area are aggregated and used as attributes of the node. Thus data modeling within the GIS for hydraulic applications is necessarily different from that used in hydrological applications. No applications of GIS and hydraulic modeling have been found in the literature.

By linking proprietary GIS and hydraulic modeling software into a spatial decision support tool, the study team was able to assess flood hazard for current and projected land-use scenarios for a range of rainfall events (1:2- through 1:200-year return periods) and for a variety of mitigatory measures. On the basis of these multiple outcomes, well-founded decisions could be made regarding appropriate proposals and options for the BMPs.

PREPARATION OF GIS DATA INPUTS TO HYDRAULIC MODELING

Figure 30-1 shows the data integrated in the GIS. The range of data is wide since the GIS has a number of functions: (1) preparation of inputs to hydraulic modeling, (2) extrapolation and analysis of modeling outputs, and (3) cartographic production for reports.

Data Resolution

The DSD consultancy brief specified 1:5,000 scale BMPs. Another influence on data resolution is the sensitivity or resolving power of the hydraulic modeling. Hong Kong drainage basins are typically small, but the intensive complex pattern of land use demands a high level of attention to detail rarely found elsewhere. However, given the time and cost of data collection, resources would be wasted if the data had a higher resolution than that discernible in the mathematical modeling. Hydraulic models may be accurately calibrated for stream levels and flows, but once the flow spreads into the floodplain, considerable approximations are made. Nevertheless, simulation of local backwater effects and features, perhaps a constriction at a bridge, may be essential to characterize the flooding of a basin. Therefore different aspects of the data were collected by different techniques over a range of scales depending upon their significance in the modeling.

Base Maps

For coregistration of all the data, existing 1:5,000 scale topographic maps were reprinted as contour and detail separates (excluding text and symbols). These were scanned, vectorized, imported into the GIS, and edge-matched. The base was neither cleaned, labeled, nor had the topology been built, as it would mainly be used as a backdrop for other data. A few key items such as drainage lines were picked off into separate layers and edited.

Hydrological Characteristics

The engineers located the nodes to be modeled by inspecting the drainage pattern, general topography, and other site-specific characteristics. Coordinates of each node and the length of connecting drainage lines can be determined by the GIS, thus avoiding errors inherent using paper maps. Cross-sectional channel geometry and the dimensions and nature of structures were collected by site survey and stored directly in the modeling software. For each node, key hydrological parameters needed to be mapped and aggregated in the GIS. These were the subcatchment boundaries for each node, topographic break of slope, and land cover. These were interpreted from aerial photographs.

The Hong Kong terrain is characterized by a break of slope between the mountainous upland areas and the gently sloping valley floors. Upland and lowland areas need to be distinguished because stream reaches must be modeled separately if realistic unit hydrographs are to be developed. Eighteen classes of land cover were mapped to reflect runoff conditions; therefore, each could be

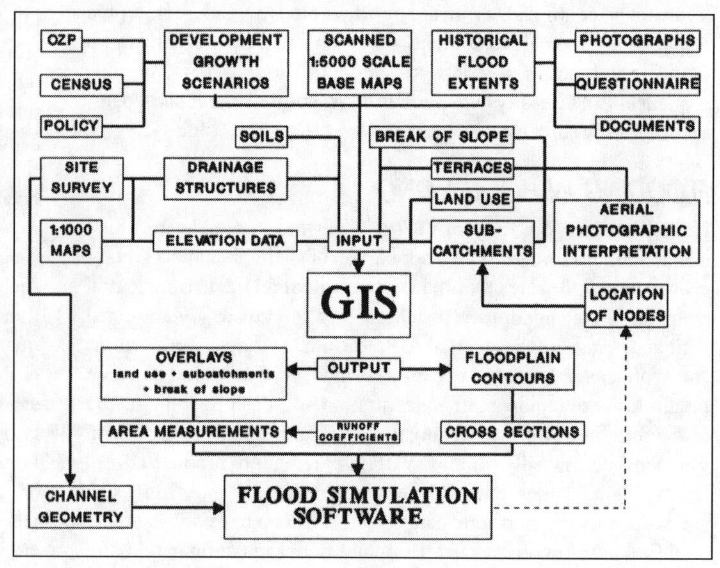

Figure 30-1. Data integration and data preanalysis for input to hydraulic modeling.

matched with a SCS runoff curve number (CN). Soils did not feature strongly as a separate hydrological characteristic. The break of slope approximates the boundary between upland *in situ* decomposed soils (often with a colluvial veneer) and lowland alluvial soils. For the high-intensity rainfall associated with flood events, land cover in Hong Kong is a more important determinant of runoff than soil infiltration capacity and is a proven approach for Hong Kong conditions.

Land use, subcatchments, and break of slope were overlaid and area tables prepared. These summarized for each node the area of each land use within the upland and lowland portions of its subcatchment. By equating land uses with CN values, area-weighted average CN values for upland and lowland portions could be passed to the hydraulic modeling. The same procedure was also carried out for a number of basin-development scenarios stored in the GIS so that both current and projected future land-use patterns could be used in the modeling.

Floodplain Topography

A detailed topography of the floodplain up to the break of slope was required, mainly for further processing of the outputs from the hydraulic modeling. As an input, however, some of the stream channel cross sections used in the modeling required extending to the edge of the floodplain, and this could be done from the floodplain contours generated in the GIS. The floodplain geometry is quite complex with levees, terraces, reclamations, and pond bunds. From observations of flooding, these features are known to have an effect on flood routing and creation of backwaters. A dense pattern of spot heights was digitized from existing 1:1,000 scale survey sheets and supplemented where necessary from site survey records and as-built drawings. Terrace scarps are, in some areas, important geomorphic features likely to restrict the lateral extent of flooding. Where these were identified, they had to be included in the terrain model but were poorly identifiable from the 1:1,000 scale sheets. Relevant terrace scarps were interpreted and mapped from aerial photographs. Elevations along the top and bottom of the scarps were extrapolated using either linear or quadratic trend surfaces to model the terraces from the available spot heights. Spot heights and breaks were triangulated to interpolate 0.5-m contours. These contours were then rasterized to form a 10-m cell DEM.

Historical Flood Events

Two well-documented flood events (1988 and 1989) were mapped from oblique photographs taken by helicopter and from village surveys. These were stored in the GIS with certain and uncertain boundaries, and although they represented the floods as partially receded, they were an important means of calibrating the outputs of the hydraulic modeling. Another flood event occurred in 1992 during the course of the study and was similarly documented.

GIS PROCESSING OF DATA OUTPUTS FROM HYDRAULIC MODELING

The hydraulic modeling was pseudo-two dimensional, solving the full St. Venant equations to simulate variations of flow in space and time. While flow in reaches were one dimensional using mean velocities, overflow into floodplain storage and flow could also be simulated. Structures such as bridges, weirs, and dams were fully described. Modeling outputs of height and velocity values relate only to the nodes, and while the software can generate long profiles, the outputs are essentially alphanumeric. For the hazard to be assessed, the data need to be given added spatial dimensionality.

Flood hazard is usually determined as a function of both depth and velocity of the water; thus a 0.2-m flood at 2-m/sec velocity may be considered a low hazard, whereas a 1-m flood at 1-m/sec velocity would be considered a high hazard. Such classifications appear more sensitive to flood height. Although the outputs from the hydraulic modeling include both height and velocity, it is difficult to make assumptions about the behavior of velocity and therefore extrapolate it over the floodplain. Depth is more easily dealt with, and therefore depth values extrapolated over the floodplain were used as an indication of hazard.

A typical flood has a subtle topography partly reflecting the gross characteristics of the underlying floodplain, the ability of tributaries to drain into the main valley, and the damming effects of structures. Thus, the flood surface cannot be assumed to be flat like a lake but has a series of gradients. In extrapolating the maximum flood depths calculated for each node over the floodplain, the flood "topography" must be modeled. The objective of this extrapolation is to determine the lateral extent of flooding from the nodes and the depth of flooding in all areas from which to assess hazard. This important postprocessing of the hydraulic modeling outputs is carried out using the GIS as flowcharted in Figure 30-2. By assuming that the height of flooding is constant perpendicular to the direction of flow at each node, it is possible to introduce dummy nodes at the edge of the floodplain, triangulate and contour the flood surface. The contours are rasterized and processed with the floodplain topography using "maximum" and "subtract" functions to derive a raster map where each cell has a depth value (values of 0 denote no flooding). In previous projects elsewhere by the same consultants,

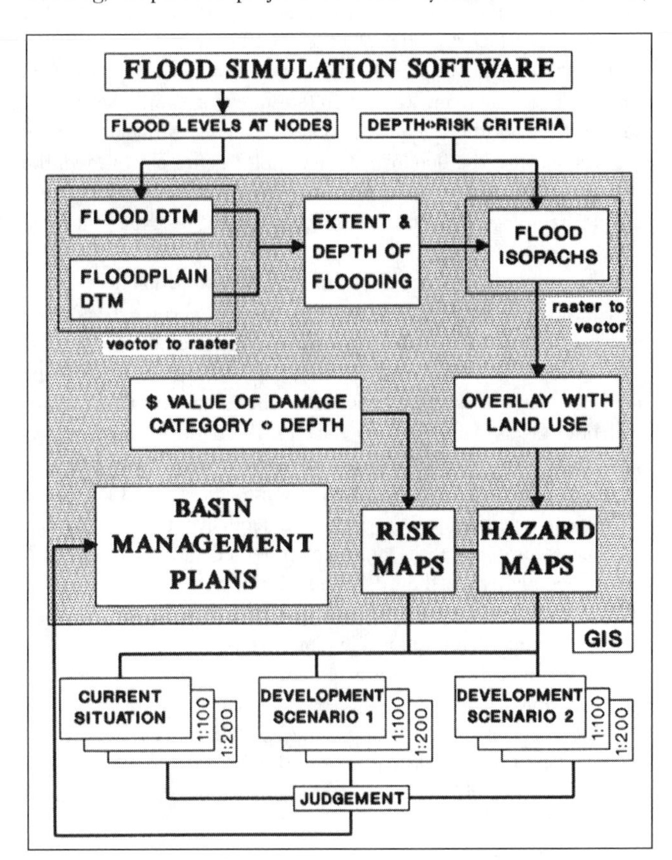

Figure 30-2. GIS postprocessing of data output from hydraulic modeling.

proprietary CAD surface modeling software had been used to achieve the extrapolation by directly intersecting the two TIN surfaces (floodplain and flood) to generate an isopachyte model (Thompson and Bush 1992).

The flood extents simulated for known storm events can be superimposed with the historical records for model calibration. An exact match cannot be expected as records show partially receded floods and may have inaccuracies. Floods for a range of design return periods (1:2–1:200 years) can be superimposed with relevant land uses to objectively assess the hazard. Where dollar values can be estimated for damage resulting from a range of flood depths for given land uses, flood risk maps could be developed. The iterative approach that can be used with the GIS and hydraulic modeling to produce flood hazard maps reflecting changing land uses, the introduction of mitigatory measures, and different rainfall events provided true spatial decision support for the engineers producing the BMPs.

CONCLUSION

A twenty-four-month project precluded the development and testing of a fully integrated GIS/flood simulation system. The particular proprietary hydraulic modeling used was the preference of the DSD as an appropriate modeling and design tool for Hong Kong. The GIS challenge lay in how best to interact with and support the hydraulic modeling so as to set up a decision-support loop, again, using off-the-shelf software. Data transfer was through ASCII files that were reformatted using short FORTRAN programs.

The choice of hydraulic modeling with a hydrological submodel rested with the need to simulate channel flow and, in particular, the effects of structures more realistically. The use of GIS for the preparation of data inputs for the hydrological submodel required that the data be viewed as attributes of nodes in a network rather than as a regular partitioning of space. Nodes were positioned at a typical cross section for a reach, and were therefore rarely at confluences. This tended to make the catchment boundaries more difficult to define especially across the floodplain. Nevertheless, once established, the use of a GIS allowed the benefits of an automated approach in quantifying hydrological parameters to be achieved as experienced in raster approaches elsewhere (Ragan 1991).

The output of node attribute data from the simulation necessitates extrapolation if decisions are to be made in a spatial context. As noted above, CAD surface modeling software has been used for this purpose thus avoiding the tedious and uncertain task of doing it manually. The main advantage of using a GIS lies in the ability to integrate the results with other layers of information required as part of the decision-making process. Although a vector approach was most appropriate for quantifying the hydrological parameters, a raster approach was found to be more effective in determining flood extent and depth over the floodplain.

The color encoded hazard maps provided the engineers with a new perspective of their data and added to their understanding of the modeling outputs. For example, they could easily identify areas of unexpected deep flooding where parameters in the model (channel friction, bridge dimensions) required refinement or correction.

The decision-support loop enabled a wide range of options to be developed in the GIS and passed through the simulation. The results, together with the base data, could be efficiently reproduced in a number of formats for reports using the cartographic capability of the GIS.

Acknowledgments

The authors are grateful to the Drainage Services Department, Hong Kong government, for permission to present this paper.

References

Bitters, B., P. J. Restrepo, and M. R. Jourdan. 1991. Using geographic information systems to predict the effects of flooding on the Han River, South Korea. *Proceedings ACSM-ASPRS Annual Convention*, vol. 4, 11–20. Baltimore.

Cotter, D. M., and R. K. Campbell. 1987. Concepts for a digital flood hazard database. *Proceedings 25th URISA Annual Conference*, vol. 2, 156–70. Fort Lauderdale.

Densham, P. J., and M. F. Goodchild. 1989. Spatial decision support systems: A research agenda. *Proceedings GIS/LIS'89*, vol. 2, 707–16. Orlando.

Hazelton, N. W. J., F. J. Leahy, and I. P. Williamson. 1992. Integrating dynamic modeling and geographic information systems. *URISA Journal* 4(2):47–58.

Johnson, L. E. 1989. MAPHYD—A digital map-based hydrologic modeling system. *Photogrammetric Engineering & Remote Sensing* 55:911–17.

Ragan, R. M. 1991. *A Geographic Information System to Support Statewide Hydrologic and Nonpoint Pollution*. Maryland Department of Transportation.

Rhind, D. 1988. A GIS research agenda. *International Journal of Geographical Information Systems* 2:23–28.

Smith, M. B., and M. Brilly. 1992. Automated grid element ordering for GIS-based overland flow modeling. *Photogrammetric Engineering & Remote Sensing* 58:579–85.

Thompson, G., and I. Bush. 1992. Digital ground models in hydro-environmental engineering. *Proceedings 3rd International Conference on Floods and Flood Management.* Florence.

Townsend, N. R., and J. M. Bartlett. 1992. Formulation of basin management plans for the northern new territories of Hong Kong. *Proceedings 3rd International Conference on Floods and Flood Management*, 39–47. Florence.

United Nations. 1990. *Urban Flood Loss Prevention and Mitigation.* New York: United Nations.

United Nations Development Program. 1991. *Manual and Guidelines for Comprehensive Flood Loss Prevention and Management.* New York: United Nations.

Wu, B., and S. Chen. 1992. A flood disaster information system and its application. *Proceedings GIS/SIG'92*, 728–38. Ottawa.

Allan J. Brimicombe
Department of Land Surveying and Geo-Informatics,
Hong Kong Polytechnic
Hong Kong
E-mail: *lsajbrim@hkpcc.hkp.hk*

Jonathan M. Bartlett
Senior Hydraulics Engineer
Binnie & Partners
Grosvenor House, Redhill
Surrey
United Kingdom

31

A Technology to Analyze Spatiotemporal Landscape Dynamics:
Application to Cadiz Township (Wisconsin)

J. R. Krummel, C. P. Dunn, T. C. Eckert, and A. J. Ayers

As landscape ecology has matured, it has gone beyond description of land-use changes to examining the functional relationships between spatial patterns of landscapes and ecological processes. Attempts to describe these relationships at larger scales or in complex landscapes have been hampered by the lack of spatially explicit distributed parameter models linked dynamically to geographical information systems (GIS). This chapter describes developments we have made to link such models to GIS and to develop visualization methods (a graphical interface) that permit the user to readily manipulate large element files containing model parameters. We then present preliminary results illustrating the effects of pattern (in an agricultural landscape) on water and material flow across a heterogeneous landscape composed of multiple watersheds. These dynamics are driven in large measure by the location, size, and number of forest patches. By use of soil, hydrologic, and vegetation data from a real landscape, the effects of spatial relocation of vegetation on water and sediment dynamics are explored through the model-GIS combination. The spatiotemporal modeling approach described here could be useful in effectively managing ecosystem restoration and rehabilitation at the landscape scale.

INTRODUCTION

Landscape ecology has been described as the discipline that studies linkages between spatial and temporal patterns and ecological processes at broad scales (Turner 1989). Because spatial pattern is explicitly considered in landscape ecology, quantitative methods have been developed to examine spatial heterogeneity, or pattern, of landscape elements (e.g., vegetation, disturbance, populations). Although no single index of landscape pattern has been universally adopted, many have been proposed and most have been shown to have considerable utility in quantifying some aspect of landscape pattern (Krummel et al. 1987; Milne 1988; O'Neill et al. 1988). However, quantifying landscape pattern is only part of the effort to link pattern with process. To date, very few studies have explicitly linked pattern and process at the landscape scale, with the possible exception of the relationship between landscape pattern and such disturbances as fire and insect infestations (Turner 1987; Leitner et al. 1991).

While a great deal of work has been conducted on nutrient dynamics in watersheds and ecosystems (Likens et al. 1977; Peterjohn and Correll 1984; Bartell and Brenkert 1991), very little research has explicitly addressed the relationship between nutrient and sediment dynamics and spatial pattern of landscapes (Turner 1989). Landscape ecology research should address two key areas to provide knowledge about natural phenomena that operate at broad geographic scales: (1) the role of landscape pattern in the redistribution of nutrients and energy (Turner 1989), and (2) moving beyond description to an understanding of dynamics that allows projections of potential outcomes from natural or anthropogenic disturbances (Westoby 1987).

Projection requires an explicit linkage between landscape pattern and ecosystem processes (e.g., nutrient dynamics). Process models must explicitly incorporate spatial heterogeneity so that pattern is not a "black box" component of system analysis. For example, Kesner and Meentemeyer (1989) analyzed the effect of the spatial pattern of nitrogen application for crop production and projected nitrogen storage and loss in a southern Georgia watershed.

The development of spatially explicit catchment models (Beasley and Huggins 1981; Beven and Kirkby 1979; O'Loughlin 1981) provides an opportunity to link landscape patterns to ecosystem processes. These distributed-parameter models have been designed to forecast water and material movement over single watersheds or catchments. The basic modeling approach is to develop a set of spatial elements (or cells) for the watershed in which all process events (e.g., interception and infiltration) in an individual element are uniform. However, between elements, parameter values vary in an unrestricted manner to reflect a heterogeneous landscape. With further development and modification of this basic modeling approach, it is possible to begin to quantify the effect of known pattern changes on known process fluxes across defined landscapes.

While our primary objective is to model the effect of landscape pattern on process dynamics through a spatiotemporal approach, technology development is required to efficiently simulate complex spatial phenomena. The technology development consists of linking a distributed-parameter computer code, geographic information system (GIS) database, and analytical tools to construct the element file, and visualization methods (i.e., a graphical user interface) to efficiently manipulate large arrays that contain model parameters. In this chapter we discuss the technology, especially that related to methods of incorporating spatial information in simulation models, and the application of the technology to forecast changes in ecosystem processes as a function of changing landscape patterns. We then use a database developed for an intensively studied southern Wisconsin agricultural landscape (Shriner and Copeland 1904; Curtis 1956; Sharpe et al. 1987; Dunn et al. 1991; Leitner et al. 1991) to illustrate how technology can be used to address fundamental problems in landscape ecology.

TECHNOLOGY

The Model

A modified version of the Areal Nonpoint Source Watershed Environment Response Simulation (ANSWERS) (Beasley and Huggins 1981) provides the basic algorithms required to simulate and evaluate landscape pattern–process interactions. The use of ANSWERS as a surface water and material transport model has been validated for watersheds in Indiana, Oklahoma, Ohio, South Carolina, and Texas (Beasley et al. 1980).

ANSWERS simulates the flow of water and material over a watershed with hydrologic parameters distributed in an array of elements or cells. These elements are connected via topographic conditions that affect surface, storm water, and tile flow. Water movement, as a function of a rainfall event, erodes, transports, and deposits materials and nutrients within channels (i.e., streams) on the landscape. Hydrological processes that occur within an individual element include interception, infiltration, surface detention, surface retention, subsurface drainage, and sediment and nutrient transport (specific nutrient reactions supplied by additional subroutines) (Figure 31-1). Equations describing these processes use parameters such as surface roughness, rainfall rate, vegetation type and cover, soil type and erodability, and slope. The primary output from an ANSWERS simulation is the amount of sediment lost or deposited from each element and a hydrograph quantifying the dynamics of water and sediment movement out of the watershed.

Correlations and spatial statistics are calculated by comparing input data with soil and nutrient movement. Thus, ANSWERS allows pattern and spatial statistics to be computed and compared for both sediment and nutrient transport and their governing factors. In essence, explicit spatial information is dynamically connected to ecosystem processes.

To address landscape ecology issues, we have modified how the spatial boundaries of the model are defined, allowing the user to incorporate multiple watersheds and surface outflows when bounding a study landscape. By removing the constraint of a single watershed, we can address those sets of problems controlled by human-dominated boundary conditions (e.g., land ownership, public lands), as well as landscapes defined by natural drainage features.

GIS Interface

Because the parameters controlling the dynamics of water flow and material transport are site specific, a critical aspect of model development is constructing the input data set (i.e., the element file). For many landscape system boundaries, thousands of elements are required to approximate the scale of surface flow dynamics. For example, a 100-km^2 landscape with an element resolution of 0.25 ha will have 40,000 elements, each of which contains georeferenced parameter values that are a function of soil type, land cover (e.g., deciduous forest, pasture, concrete), and topographic conditions that vary over the landscape. Thus, while the model can be a powerful tool to analyze process–pattern interactions, an efficient, cost-effective method is required to develop the data input file.

File transfer algorithms have been incorporated to use the spatial database capabilities of a GIS. The GIS provides an efficient method to construct the DEM and to input land cover and soil information from other digital databases (e.g., Landsat images) or cartographic products (e.g., aerial photographs or topographic maps). Because the element file is a gridded array of cells, it is important to be able to convert vector data into raster data. For example, topographic and soil data are not usually available in digital format, and therefore polygon (vector) cartographic information must be digitized and converted to raster (cell) information for the element file.

Construction of the topographic data file provides an example of how GIS technology improves the efficiency and accuracy of the modeling technology (Figure 31-2). Digital terrain data are available for some areas of the United States, but in many areas topographic maps must be digitized to produce slope and aspect information required for the model. To accomplish this task we: (1) digitize topographic maps with AutoCad software, (2) send vector files to ARC/INFO software and use the TIN algorithm to produce a digital elevation model, (3) use the GRID algorithms in ARC/INFO to produce an cell file of slope and aspect, and (4) send the cell file to the modified ANSWERS code to initialize the surface drainage. With this process, accurate and precise cartographic data, which is fundamental to the dynamic process, is efficiently transferred from a map to the model.

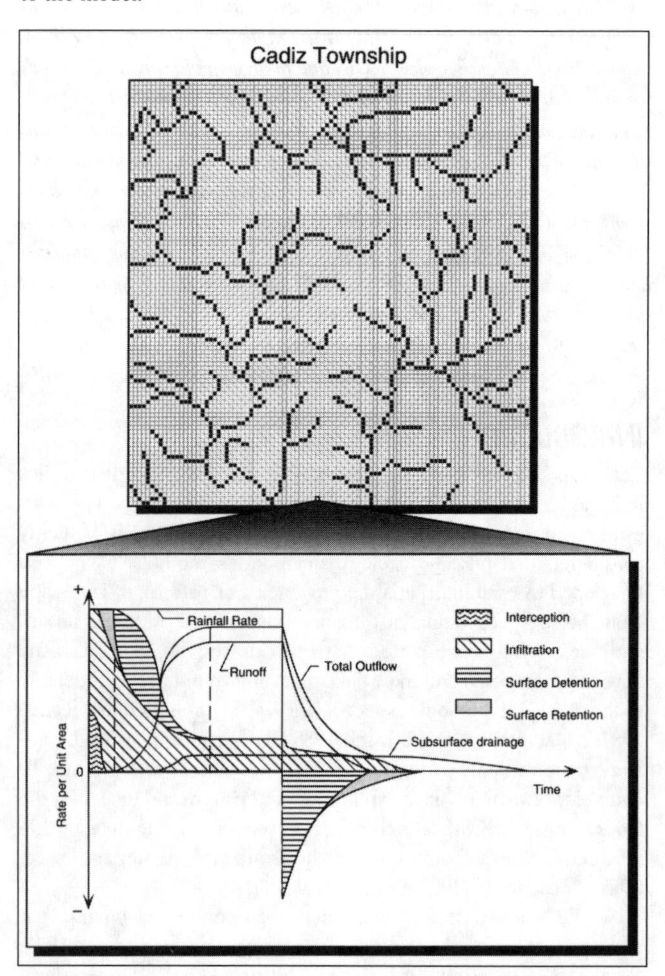

Figure 31-1. The Cadiz Township element file, with the stream channel elements, consists of 10,000 elements. Each individual element contains the water and sediment movement relationships (Beasley and Huggins 1981).

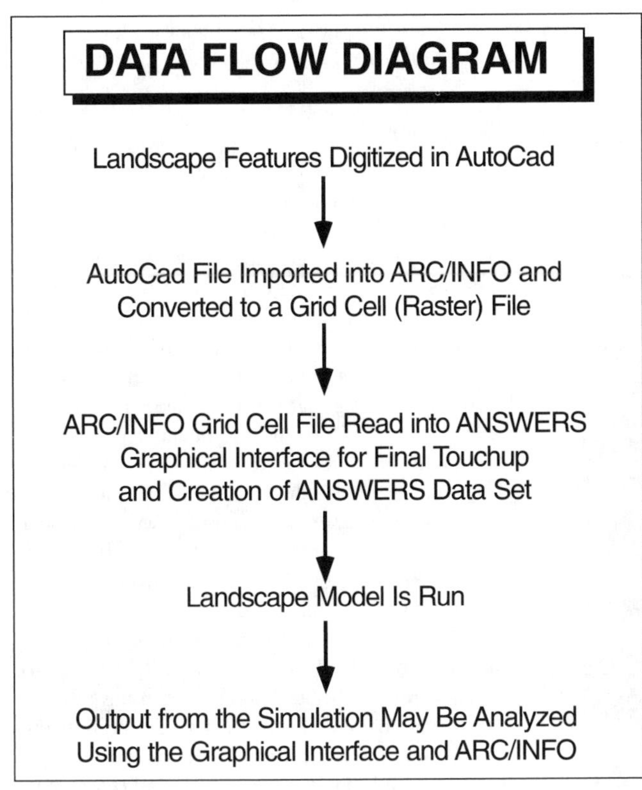

Figure 31-2. The efficiency of spatial data input is increased by connecting a GIS to the model interface.

Visualization Tools

In conjunction with the use of a GIS to construct the data input file, we have developed a set of visualization tools that allow the analyst to edit the data layers on the computer screen (Figure 31-3). The package of tools is written in the C language and runs under X-Windows, version 11, revision 4 or higher. These tools graphically represent the major layers of input data for ANSWERS. All data are stored in the ARC/INFO SVF format. Program functions allow the user to make minor editing changes to the input data. Once the simulation has been run, the sediment transport may be viewed and analyzed with the visualization tools and compared with input parameters such as land cover.

RESEARCH PROBLEM

In this chapter, the modeling technology is used to analyze the effect of changing land cover conditions on soil/sediment movement and surface hydrology in Cadiz Township, Wisconsin. Studies of historic deforestation and fragmentation patterns, disturbance effects on presettlement vegetation, current vegetation–site relationships, and models of future landscape patterns have analyzed the structural components of this landscape. This work has produced a GIS data set of soil, vegetation, elevation, and hydrological parameters for each of 10,000 1-ha grid cells on the 10-km × 10-km Cadiz Township landscape. We can use the data contained in each grid cell to simulate water and soil flux with the distributed parameter model. The modeling technology allows us to address a series of questions:

1. How do rates of water and soil movement for the current landscape pattern compare with the presettlement landscape pattern?
2. Does the current land cover pattern result in long-term, nonsustainable conditions (i.e., nonequilibrium)?
3. What levels and pattern of reforestation are necessary to return disrupted processes to presettlement (e.g., forested) conditions?

Study Area

Data are from Cadiz Township (Green County, Wisconsin) (Figure 31-1), which has been the subject of studies on deforestation and fragmentation pattern (Shriner and Copeland 1904; Curtis 1956; Sharpe et al. 1987), presettlement vegetation–site–disturbance interactions (Leitner et al. 1991), present-day vegetation–site relationships (Sharpe et al. 1987), and models of future landscape pattern assuming revegetation scenarios motivated by the Conservation Reserve Program (CRP) (Dunn et al. 1991, 1993). Early in this century, Shriner and Copeland (1904) quantified the pattern of postsettlement forest fragmentation in this township and related that pattern to changes in regional streamflow and sedimentation.

Cadiz Township contains several drainage networks, the major water course being the Pecatonica River that runs north to south in the western portion of the township. The river served as a fire break during presettlement times. Fires west of the river resulted in xeric oak (*Quercus macrocarpa*) forest, savanna, and prairie. To the east of the river, the lack of major or frequent fire resulted in the development of mesic oak (*Q. alba, Q. rubra*), ironwood (*Ostrya virginiana*), and maple (*Acer saccharum*) forest (Leitner et al. 1991).

Following European settlement in the early 1830s, the original vegetation was rapidly fragmented and fires were eliminated. By 1978, total forest cover had been reduced from 8,724 ha to 473 ha allocated

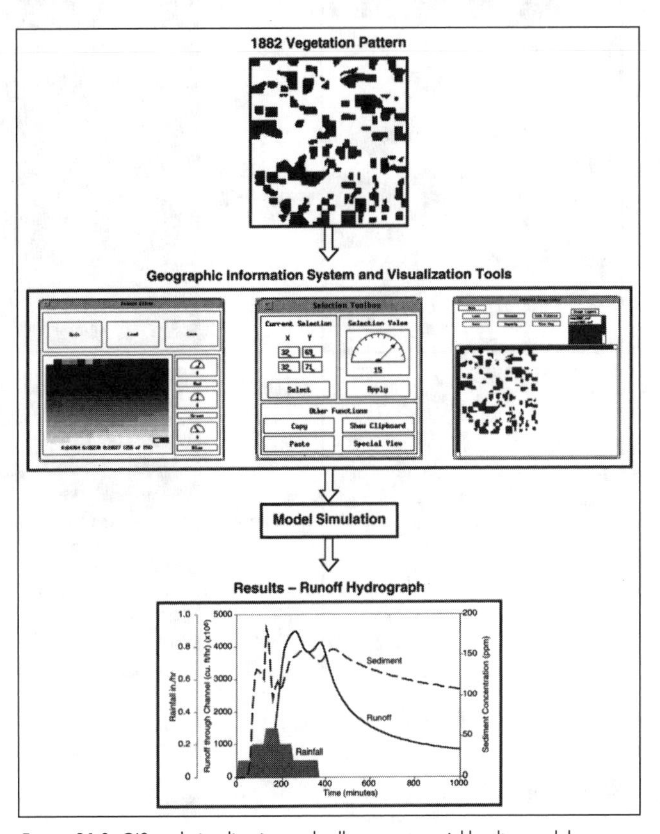

Figure 31-3. GIS and visualization tools allow one to quickly alter model parameters to evaluate different scenarios of landscape pattern and evaluate the results of the simulation.

among eighty-four forest patches, each averaging 6 ha (Dunn et al. 1991). Cleared land has been converted largely to agriculture.

This landscape is approximately 10 km × 10 km and has been divided into nearly 10,000 1-ha grid cells. The spatiotemporal database includes land cover types and environmental data for each grid cell. The presettlement landscape was assumed, for the purposes of this chapter, to be in equilibrium, and results from all simulations for other dates were compared with this baseline. Further details of the database are described by Sharpe et al. (1987) and Dunn et al. (1991).

Results and Discussion

As expected, potentially erosive areas (defined by slope, soil, and elevation) lose substantially more soil when converted to agriculture than when left in forest (Figure 31-4). As forest cover is converted into agricultural land, soil erosion becomes an increasingly important perturbation to the landscape system. As a measure of sustainable conditions, the magnitude of the loss relative to the presettlement baseline is strongly dependent on the type of cover (e.g., agriculture or mesic forest).

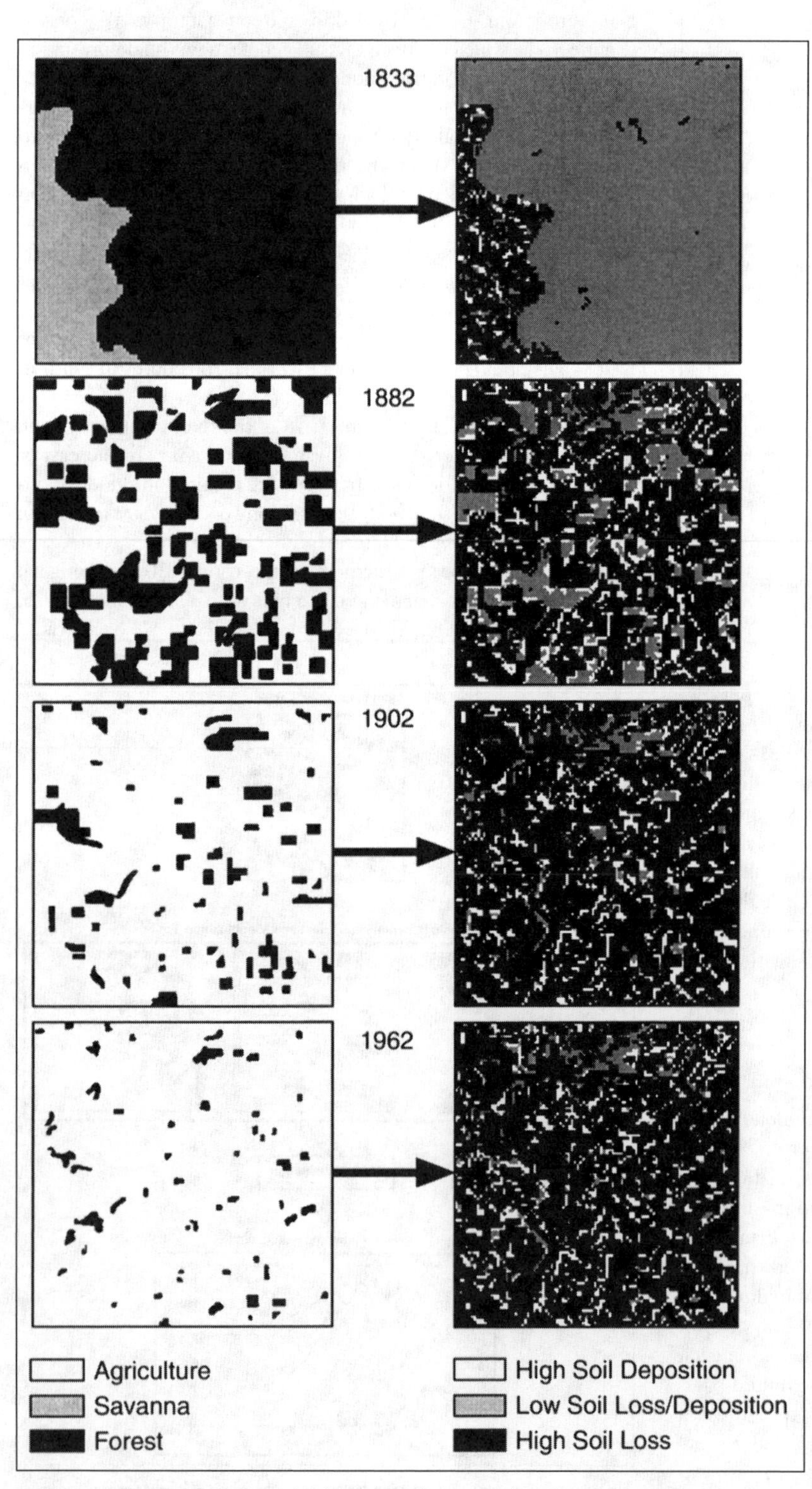

Figure 31-4. The left column shows the changing landscape with loss of natural vegetation in Cadiz Township. The right column shows the increase in sediment transport over the landscape as the vegetation changed.

The rates of loss and the quantity of sediment exported are different for each natural vegetation type and each watershed (Figure 31-4). The conversion from a presettlement landscape without agriculture to one dominated by agriculture has resulted in new rates of erosion. As land was converted to agriculture and the number and size of the forest patches decreased, the pattern of erosion and deposition mirrored the land cover changes (Figure 31-4). Clearly, pattern dictates process, but in this case, the man-made pattern represents a severe perturbation to natural ecosystem functions. These rate changes can be directly applied to measures, indices, or models of sustainability.

In addition to measuring past and present disturbance, landscape restoration and rehabilitation scenarios can also be quantified in order to project future conditions. Figure 31-5 shows how scenarios that increase forest cover in a portion of Cadiz Township can alter water and sediment movement. Both scenarios have equal increases in the area of forest cover, but with different patterns of forest cover distribution. In scenario one, a 1-ha forest perimeter has been added around each existing forest patch. In scenario two, the amount of forest area has been used to connect the existing forest patches. Scenario two mimics the establishment of forest corridors. Results show that both scenarios reduce sediment movement, but the larger, individual forest patches are more effective than the forest corridors in restricting soil loss.

Shriner and Copeland (1904) described the pattern of forest fragmentation in Cadiz Township from presettlement (1831) to 1902, by which time forest cover was reduced to 10% of the original. The authors noted that as forest cover was reduced, there was a reduction in streamflow, with many streams becoming completely dry. They speculated that increasing forest cover to 27% of the original would result in the return to presettlement streamflow. We anticipate that as a sufficient acreage of erodible land is enrolled in the Conservation Reserve Program (CRP), erosion will be substantially reduced. In fact, if all eligible CRP land were eventually converted to forest, the landscape would be 27% forested (Dunn et al. 1991).

The effects of erosion on aquatic habitats and agricultural economics are well known. However, ero-

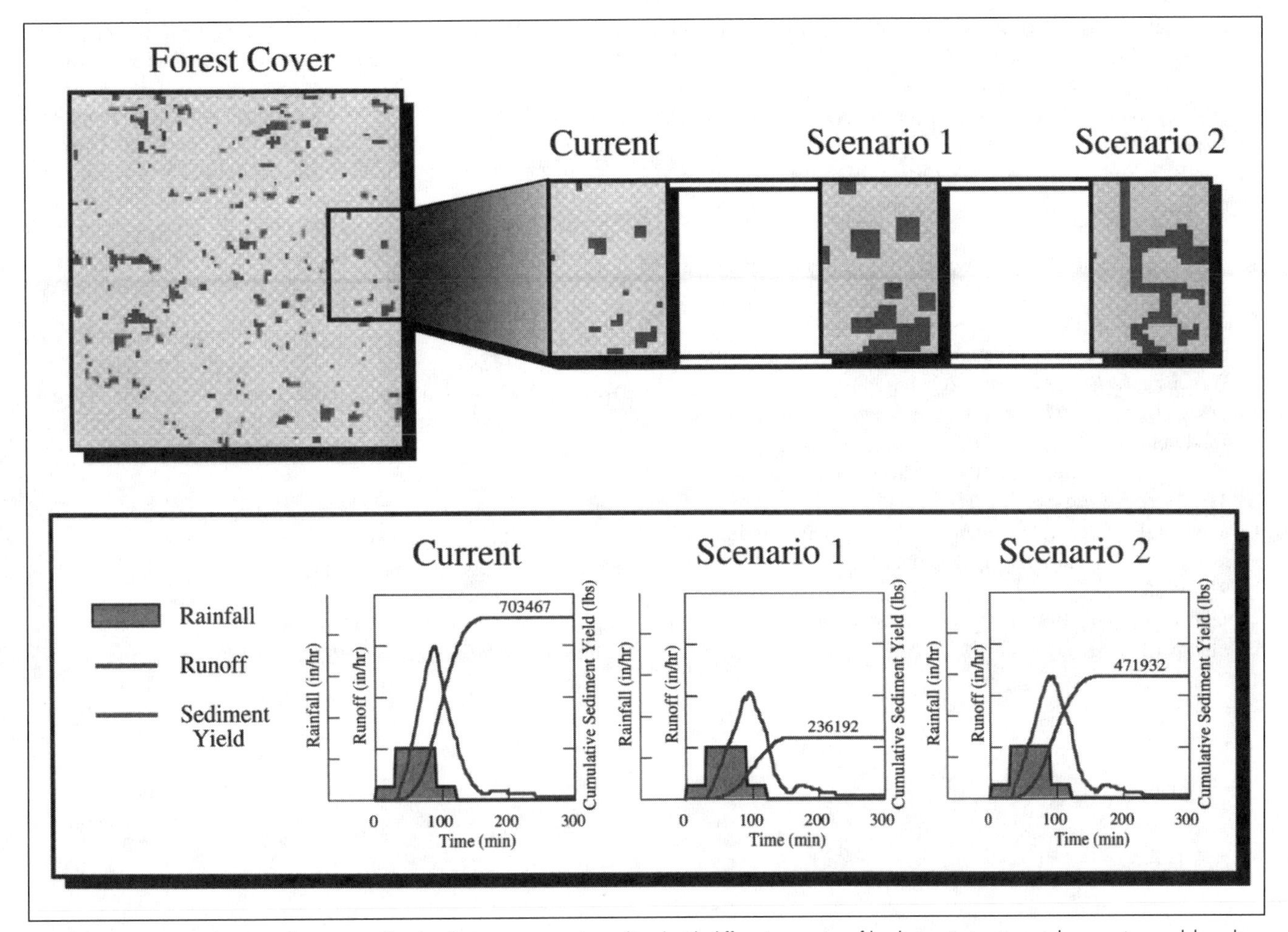

Figure 31-5. Current and projected water runoff and sediment movement is analyzed with different scenarios of landscape restoration and comparing model results.

sion is not the only negative consequence of fragmentation. Other ecological processes are disrupted spatially and temporally. Decreasing size and increasing distances among forest patches lead to changes in forest composition, stand structure, species dispersal patterns, altered disturbance regimes, local and regional species loss, increase in exotic species, and changes in the incidence of plant pathogens (Franklin and Forman 1987; Dunn and Stearns 1993).

CONCLUSION

Our work with the technology of modeling complex spatial phenomena shows that landscape pattern–process interactions can result in the projection of past and future conditions. These projections can be used to analyze sustainable conditions or restoration scenarios. Moreover, linking of simulation models to GIS tools provides a powerful technology to manipulate and visually demonstrate spatially complex phenomena. Indeed, we argue that a spatially explicit modeling approach is the only way to show the environmental degradation that occurs from the cumulative effects of minor, but additive, changes that operate over long time periods.

Acknowledgments

This work was supported by the U.S. Department of Energy, Assistant Secretary for Environmental Restoration and Waste Management, under contract #W-31-109-Eng-38.

References

Bartell, S. M., and A. L. Brenkert. 1991. A spatiotemporal model of nitrogen dynamics in a deciduous forest watershed. In *Quantitative Methods In Landscape Ecology*, edited by M. G. Turner and R. H. Gardner, 379–98. New York: Springer-Verlag.

Beasley, D. B., et al. 1980. ANSWERS: A model for watershed planning. *Transactions of the ASAE* 23:938–44.

Beasley, D. B., and L. H. Huggins. 1981. *ANSWERS—User's Manual.* EPA-905/9-82-001. Chicago: U.S. Environmental Protection Agency.

Beven, K. J., and M. J. Kirkby. 1979. A physically based, variable contributing area model of basin hydrology. *Hydrologic Science Bulletin* 24:43–69.

Curtis, J. T. 1956. The modification of midlatitude grasslands and forests by man. In *Man's Role in Changing the Face of the Earth*, edited by W. L. Thomas, 721–26. Chicago: University of Chicago Press.

Dunn, C. P., et al. 1991. Methods for analyzing temporal changes in landscape pattern. In *Quantitative Methods in Landscape Ecology*, edited by M. G. Turner and R. H. Gardner, 173–98. New York: Springer-Verlag.

———. 1993. Ecological benefits of the Conservation Reserve Program. *Conservation Biology* 7:132–39.

Dunn, C. P., and F. Stearns. 1993. Landscape ecology in Wisconsin: 1830–1990. In *Wisconsin Vegetation Ecology: John T. Curtis and His*

Legacy, edited by J. Fralish, R. P. McIntosh, and O. L. Loucks. Madison: Wisconsin Academy Press.

Franklin, J. F., and R. T. T. Forman. 1987. Creating landscape patterns by forest cutting: Ecological consequences and principles. *Landscape Ecology* 1:5–18.

Kesner, B. T., and V. Meentemeyer. 1989. A regional analysis of total nitrogen in an agricultural landscape. *Journal of Landscape Ecology* 2:151–63.

Krummel, J. R., et al. 1987. Landscape patterns in a disturbed environment. *Oikos* 48:321–24.

Leitner, L. A., et al. 1991. Effects of site, landscape features, and fire regime on vegetation patterns in presettlement southern Wisconsin. *Landscape Ecology* 5:203–17.

Likens, G. E., et al. 1977. *Biogeochemistry of a Forested Ecosystem.* New York: Springer-Verlag.

Milne, B. T. 1988. Measuring the fractal geometry of landscapes. *Applied Mathematics and Computation* 27:67–79.

O'Loughlin, E. M. 1981. Saturation regions in catchments and their relations to soil and topographic properties. *Journal of Hydrology* 53:229–46.

O'Neill, R. V., et al. 1988. Indices of landscape pattern. *Landscape Ecology* 1:153–62.

Peterjohn, W. T., and D. L. Correll. 1984. Nutrient dynamics in an agricultural watershed: Observations on the role of a riparian forest. *Ecology* 65:1,466–75.

Sharpe, D. M., et al. 1987. Vegetation dynamics in a southern Wisconsin agricultural landscape. In *Landscape Heterogeneity and Disturbance,* edited by M. G. Turner, 137–55. New York: Springer-Verlag.

Shriner, F. A., and E. B. Copeland. 1904. Deforestation and creek flow about Monroe, Wisconsin. *Botanical Gazette* 37:139–43.

Turner, M. G., ed. 1987. *Landscape Heterogeneity and Disturbance.* New York: Springer-Verlag.

Turner, M. G. 1989. Landscape ecology: The effect of pattern on process. *Annual Review of Ecology and Systematics* 20:171–97.

Westoby, M. 1987. Soil erosion as a landscape ecology phenomenon. *Trends in Ecology and Evolution* 2:321–22.

J. R. Krummel, C. P Dunn, T. C. Eckert, and *A. J. Ayers*
Argonne National Laboratory
Environmental Assessment Division
9700 S. Cass Avenue
Argonne, Illinois 60439
E-mail: *krummelj@smtplink.eid.anl.gov*

32

LANDIS:

A Spatial Model of Forest Landscape Disturbance, Succession, and Management

David J. Mladenoff, George E. Host, Joel Boeder, and Thomas R. Crow

LANDIS is a stochastic, spatially explicit model of forest landscape disturbance and succession. It is designed to simulate the forests of the northern lakes states. LANDIS is raster-based and programmed in C++ with both imperative code and hierarchical, object-oriented data structures. LANDIS simulates succession semiquantitatively as tree species age classes. This approach allows concentration of model complexity on algorithms that simulate landscape-scale spatial interactions such as seed dispersal on a matrix of land types with differing disturbance regimes. LANDIS contains interacting windthrow and fire disturbance regimes, that we believe have not previously been modeled. The model links dynamically with GIS by operating in an ERDAS raster file format and is implemented for both MS-DOS and UNIX operating systems. The model includes a graphical interface and its own routines for spatial analysis and calculation of various indices of landscape pattern, and graphical and map output. The model has been developed to analyze changes in landscape structure in response to fire and windthrow disturbance regime combinations and forest harvest levels and patterns.

INTRODUCTION

Objective

We are conducting research to understand landscape-scale forest ecosystem dynamics in the northern lakes states region (Mladenoff et al. 1993, 1994). This region is dominated by highly altered and extensive second-growth forests that followed destructive logging in the past. In this context, our goal has been to develop a model of forest disturbance and succession for research into landscape-scale processes and applications in the northern lakes states. Traditional research methods that incorporate experimental methods and replication are seldom possible at landscape scales. Simulation models provide a tool to conduct experiments and examine results over large spatial and temporal domains. Similarly, landscape models provide a tool for managers and policy makers to test and evaluate management applications.

Background

Ecological landscape models, including forest and landscape disturbance models, have recently been reviewed and classified (Baker 1989; Sklar and Costanza 1991; Turner and Dale 1991). Here, we are interested particularly in spatially explicit forest landscape models that include dynamic processes of disturbance (including management) and succession and that also have a functional link with geographic information systems (GIS).

Forest ecosystem models that simulate single-plot species interactions, "gap" models of the JABOWA/FORET type, have been in use for some time in simulating forest change (Botkin et al. 1972; Shugart 1984). Such approaches are valuable for understanding ecosystem dynamics at within-stand scales, but even a single-stand level simulation (10s–100s ha) using such a method would challenge state-of-the-art computer capabilities. The challenge in forest landscape models, as in others, is to balance resolution, detail, and practical functionality. This means that no model is likely to meet all needs. We need to assure that the perceived need in model resolution is real for the anticipated tasks and not an artifact of our inability to adjust our traditional conception of informational needs from a fine-grained, site-specific scale to the landscape as a whole.

A few spatial models of forest fire disturbance exist (Kessell et al. 1984) and can simulate the spread of a single disturbance based on landscape characteristics, but they do not have the ability to simulate multiple, repeated dynamic changes (Baker et al. 1991).These models have a recovery or successional component that is strictly deterministic and lacks spatial interaction. A spatial model of fire disturbance that incorporates climatic driving variables has been developed for the Boundary Waters Canoe Area (BWCA), a wilderness area in northeastern Minnesota (Baker 1992). The model simulates recovery of patch age classes and therefore landscape structure as influenced by climate change and fire suppression effects on fuel and disturbance susceptibility. However, the model does not include an explicit forest-succession routine.

Roberts (1994a, 1994b) combined fire disturbance and forest succession in a spatial model of landscape dynamics in the southwestern United States. His model (VAFS/LANDSIM) is polygon based and operates on species life history characteristics or vital attributes (Noble and Slatyer 1980), fire susceptibility and response, and site characteristics. The model balances the need for computational efficiency while including dynamic processes between disturbance and recovery. Its limitations are the fixed polygons and spatial processes that are driven by the order of polygon neighborhoods rather than actual distances. Although LANDSIM contains some highly generalized processes, the model realistically simulates the forest landscape dynamics of the southwestern United States. Roberts (1994a) also examined the behavior of the model algorithm using sensitivity analysis based on varying the main ecological parameters.

MODEL DESCRIPTION

General Characteristics

Our model of landscape disturbance and succession (LANDIS) is an elaboration of the approach of Roberts (1994a, 1994b). It is similar in concept, temporal resolution (ten-year time step), and basic model algorithm. The basic conceptual structure of LANDIS is also similar to the forest gap models, where succession is based on interactions between species life history characteristics (Table 32-1), site conditions, and disturbance regime or management. With LANDIS, we have attempted to retain more of the ecological dynamics of gap models with the capability to spatially model larger areas. LANDIS simulates forest change semiquantitatively by modeling tree species as ten-year age classes, not as individual stems as in gap models. Succession and disturbance contain stochastic, spatially dynamic elements.

Table 32-1. List of species life history parameters that drive the model.

Long	Species longevity (years)
Mature	Age of sexual maturity (years)
Shade	Shade tolerance class (1–5)
Fire	Fire tolerance class (1–5)
Wind	Windthrow tolerance class (1–5)
Effseed	Effective seed dispersal distance
Maxseed	Maximum seed dispersal distance
Vegprob	Vegetative reproduction probability
Sprout	Maximum sprouting
Estab	Species establishment coefficient (by land type)

However, LANDIS differs in several ways from the approach of Roberts (1994b):

1. Our model operates in raster mode, which allows complex dynamics to be more easily modeled than in vector format, such as dissolution and aggregation of patches.
2. Spatial interactions such as seed dispersal and disturbance spread are based on distances instead of polygon neighborhoods.
3. Model algorithms are programmed to allow operation at different scales of resolution by modifying cell size.
4. The model is programmed in C++ using hierarchical, object-oriented data structures (Figure 32-1) (Boeder et al., *Spatially*, 1993). Although LANDIS is more detailed than LANDSIM, the object-oriented structure and raster format maintain computational efficiency.
5. The model has a user interface and a free-standing spatial analysis package (APACK) that allows rapid calculation of landscape summaries and indices without exporting to a GIS or statistical package (Boeder et al., *Spatially*, 1993; *Spatial Analysis*, 1993).
6. LANDIS reads and writes ERDAS raster files, making GIS links rapid. Real-time operation and linkage with GIS to interact with other data layers can be done with appropriate hardware and software. Map output also can be easily converted to other formats such as ARC/INFO and Postscript.
7. Development of the model for the northern lakes states required designing both windthrow and fire disturbance routines and interaction between them. We are not aware of other models that incorporate such disturbance dynamics.

Successional Dynamics

The model successional algorithm is based on that of Roberts (1994a, 1994b), with several additions. Succession is a competitive process driven by species life history parameters (Table 13-2). Pattern of seed dispersal, species establishment, and shade tolerance, along with disturbance (fire and windthrow) susceptibility and response, determine vegetation change on different kinds of sites (Figures 32-2 and 32-3). The values for life history parameters are derived from published literature (Curtis 1959; Burns and Honkala 1990; Loehle 1988). Our approach differs most from Roberts (1994a) in the disturbance and seeding algorithms. The large number of tree species in the region exhibit a great variety of seed longevity and dispersal modes (Pastor and Mladenoff 1992). Seed dispersal is parameterized as actual dispersal distance curves, rather than by polygon neighborhood as in Roberts (1994a). This allows for more complex dispersal and establishment dynamics to be modeled. Using actual dispersal distances also allows the model to be used at variable scales or cell sizes.

The seeding algorithm is the slowest component of the model. The seeding algorithm can be switched to simpler modes in the model if desired, including no dispersal and uniform dispersal.

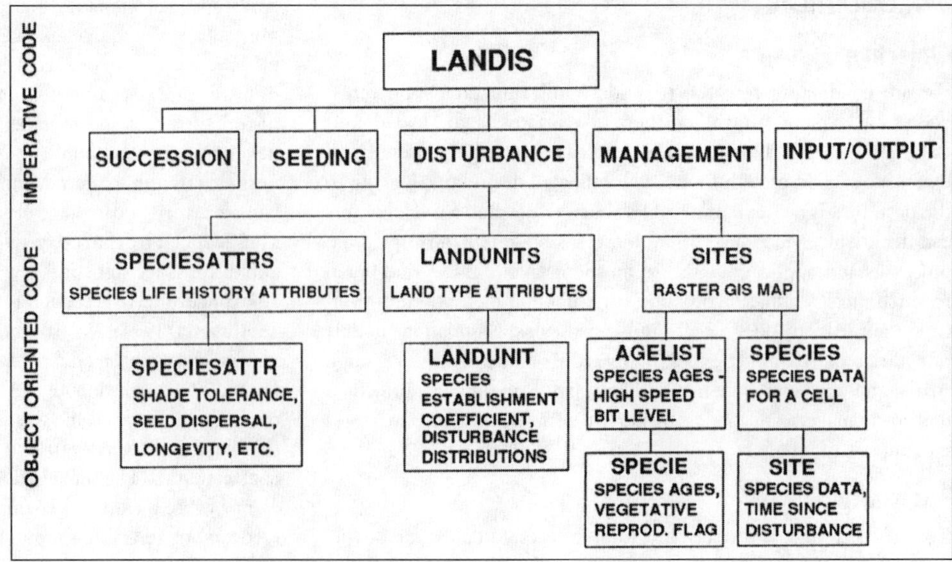

Figure 32-1. General structure of LANDIS model code. The model contains both more traditional imperative code and object-oriented structures.

Table 32-2. Analysis methods and indices currently calculated in APACK module. Algorithms and sources are in Boeder et al. (1993b).

Fractal dimension	Edge ajacency
Perimeter: area ratio	Electivity index
Patch area statistics	Connectivity (distance)
Patch perimeter (edge) statistics	Percolation ratio
Landscape diversity/dominance	Angular second moment

However, we feel that these spatial seeding dynamics are some of the most important characteristics of the model. These stochastic processes allow simulation of local species recolonization following disturbance, more realistic successional change over time on large landscapes, and the potential for modeling species migration in response to major large-scale events such as global climate change.

Site Factors

Site characteristics are input to the model as a map layer consisting of ecologically defined land types. Ecological land units can be identified at multiple scales by dominant controlling factors (Barnes et al. 1982; Host et al. 1987). Land units in this landscape are defined in terms of major landscape geomorphic features. Ecologically, in terms of a soil moisture/nutrient gradient, these correspond roughly to mesic, dry mesic, and xeric (Curtis 1959). Any practical number of such classes are possible at a selected resolution.

For each species an establishment coefficient summarizes the probability of that species establishing on each land type, without competition (Roberts 1994a). These values are estimates based on data in Curtis (1959) and Burns and Honkala (1990). The establishment parameters encapsulate what is known about environmental constraints on each species in the region, primarily moisture and nutrient requirements as expressed by site conditions. These land type differences are important because they affect species establishment and relative success. Site characteristics also influence fuel accumulation and persistence (productivity and decomposition), which is important in the interaction of the fire and windthrow disturbance regimes and variation in disturbance susceptibility with stand age (Clark 1991; Frelich and Lorimer 1991a).

Disturbance

Modeling succession over large landscapes in the northern lakes states requires an integration of windthrow and fire disturbances (Mladenoff and Pastor 1993). Xeric sites in the region usually support vegetation that is both more fire prone and requires fire for reestablishment. Such interactions on real landscapes are complicated by human changes, such as fire suppression, which modify natural disturbance regimes, succession, and fuel loads (Clark 1988; Baker 1992). Fire disturbance is implemented in the model similar to Roberts (1994a), where fires are categorized within five severity classes according to time since last fire. Fuel accumulation also varies according to site characteristics. As a result of these factors, fuel accumulation and decomposition curves are mediated by land type in the model. These curves are estimates based on examples and patterns in the literature for different forest and site types (Bormann and Likens 1979; Gore and Patterson 1985). These factors interact with the fire tolerance parameter of the tree species. The combination of fire severity class and species fire tolerance determines which species age classes are killed (Figure 32-2).

Windthrow is a top-down disturbance, and susceptibility increases primarily with tree size (age class). Windthrow is implemented in the model within five severity classes, based on percent of canopy removal. This differs from fire, which is a bottom-up disturbance, where smaller (younger) age classes are killed first, according to disturbance severity and species susceptibility. Both types of disturbance are important in the northern lakes states, with windthrow important in all types of forest (Canham and Loucks 1984; Mladenoff 1987). Fire is more important on drier sites, typically with pine (Heinselman 1973). Besides the differing modes of these two disturbances, their temporal interactions produce complex combinations and patterns of forest ecosystems on the landscape (Figure 32-2). For example, fire following a windthrow disturbance will have fuel loads different from a fire in the same landscape without a preceding disturbance.

Both fire and windthrow disturbances are generated in the model by selecting randomly from the appropriate disturbance-size distributions. These are estimated negative exponential distributions based on the published literature for the region (Frelich and Lorimer 1991a, 1991b; Canham and Loucks 1984; Baker 1991, 1992). The disturbance frequency and rotation period for the landscape (by land type) are estimates based on Canham and Loucks (1984) and Frelich and Lorimer (1991a). For a given disturbance event, a potential disturbance size is selected from the distribution, but realized size or ultimate spread is based on local susceptibility. The realized disturbance-patch spread must be contiguous for fire, but for windthrow it need not be.

Model Output

The model produces maps based on landscape composition at specified time steps (Plate 32-1). The maps can be produced as cover types or age classes. Several forest classification algorithms may be selected (Boeder et al., *Spatially*, 1993), based on order of dominants, successional stage, or fuzzy community membership (Roberts

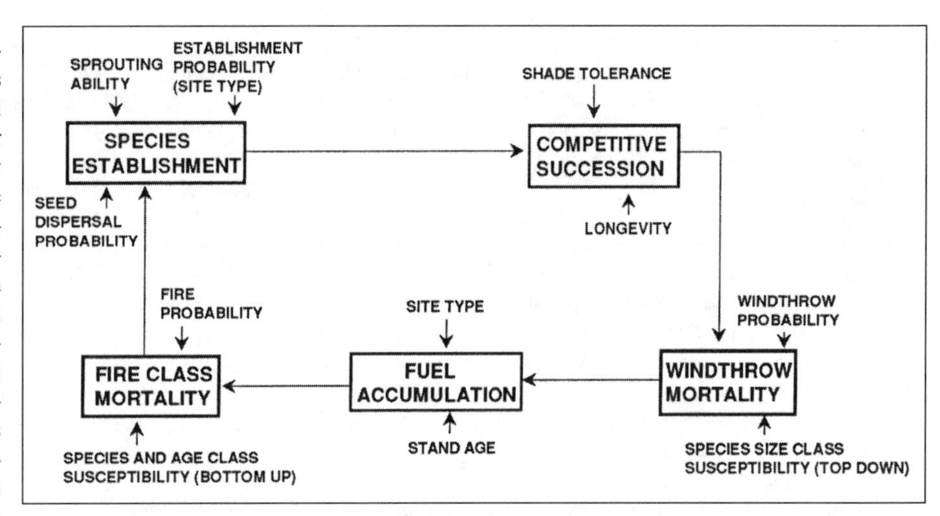

Figure 32-2. Succession (upper half) and disturbance (lower half) dynamics of the LANDIS model. Disturbances (windthrow and fire) can occur in any order, or singly on the landscape, with future disturbances susceptibilities modified accordingly for a given site.

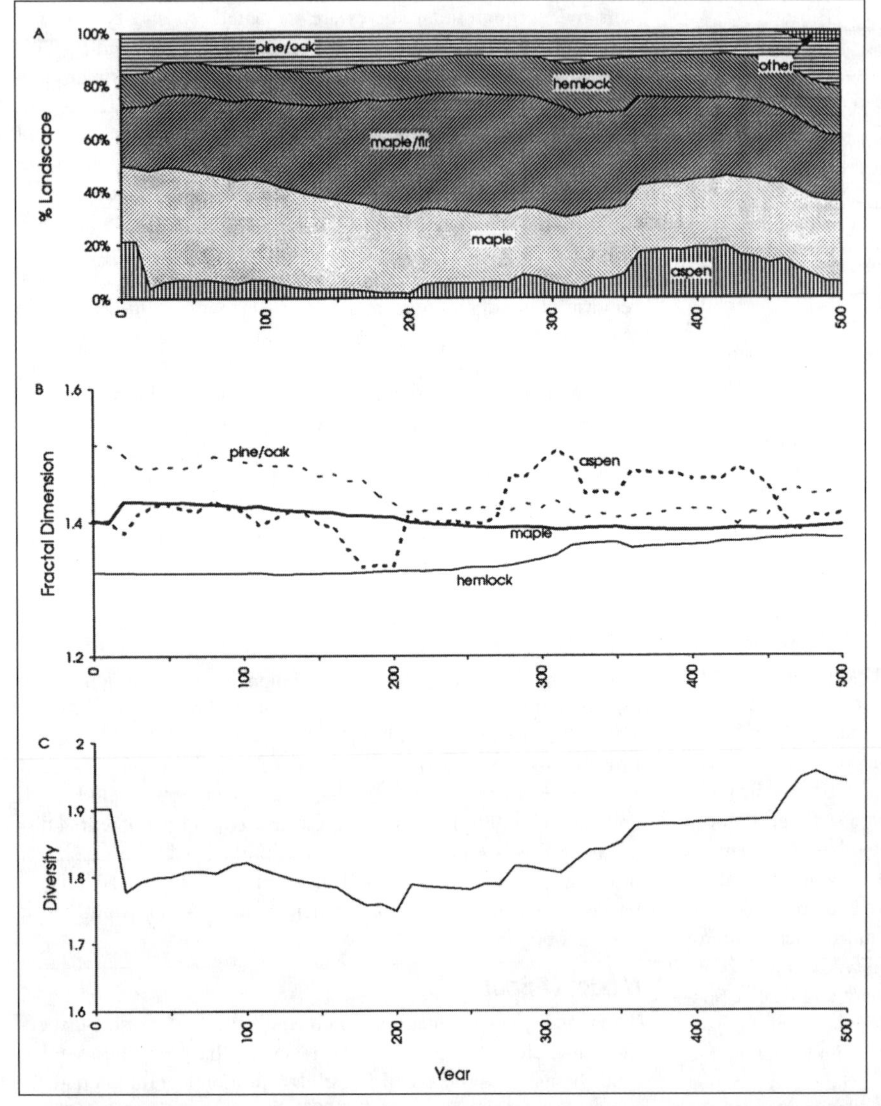

Figure 32-3. Example of selected analytical and landscape index output from APACK plotted through the 500-year simulation: (a) landscape composition by dominant species, (b) fractal dimension of patch types by dominant species, (c) landscape diversity.

1994a). Disturbance and age class maps are also produced at desired time steps, as well as a cumulative disturbance map at the end of a simulation. Spatial analyses and indices are produced as specified in the appropriate output format (table, graph, or chart).

Sample Results

In the example simulation (Plate 32-1), a landscape of approximately 10,000 ha is simulated for 500 years at 60-m resolution (approximately 37,500 cells). Landscape succession responds to disturbances occurring at different rates on the three land types. The starting map (1980s) is dominated by early successional forest following widespread logging and fire before the middle of the century (White and Mladenoff 1994). Gradual inward spread of late successional species like maple (*Acer saccharum*) and balsam fir (*Abies balsamea*) accelerate from recolonization nodes that establish by longer distance seed dispersal.

Landscape pattern indices describe change over the simulation period. Fractal dimension is more variable among the early successional patch types dependent upon higher-intensity disturbances. Landscape diversity decreases over time as the landscape becomes more dominated by late successional types but increases as the disturbance susceptibility and realized disturbances again increase the area of successional types, such as aspen (*Populus tremuloides*). Codominant or age class maps provide additional insights. Running the same simulation as shown without the natural disturbance rates results in maple dominance 150 years sooner. Once maple is dominant, windthrow changes the age class of patches, but without fire, most areas remain dominated by maple (Plate 32-1).

CONCLUSION

The spatially dynamic disturbance and dispersal algorithms are the core of LANDIS model behavior and allow it to reasonably simulate the dynamics and composition of the northern lakes states' forests for various land types. The model shows promise in examining forest landscape dynamics in relation to disturbance and management (cutting) and in analyzing changing landscape pattern. The interaction of disturbance regimes (windthrow and fire) in the model is a novel implementation which we have only begun to explore.

Routines are being implemented that incorporate landscape structure effects on disturbance spread and succession, such as edge effects on windthrow susceptibility and probabilistic selection of windthrow directionality. The object-oriented structure of LANDIS (Figure 32-1) (Boeder et al., *Spatially*, 1993) is intended as the base for development of other rule-based management applications of the model, such as forest management and habitat utilization (Mladenoff and Host 1994).

Acknowledgments

This work has been funded by a U.S. Forest Service Ecosystem Management cooperative agreement with D. J. Mladenoff and G. E. Host, NRRI-UMD; and T. R. Crow, U.S.F.S. This work is contribution #115 of the Natural Resources Research Institute and #22 of the Natural Resources GIS Laboratory, University of Minnesota, Duluth. Development of LANDIS benefited from conversations with William Baker, Lee Frelich, Robert Gardner, Glenn Guntenspergen, Andrew Hansen, Bud Heinselman, John Pastor, and Forest Stearns. We are especially grateful to Dave Roberts for his assistance. Mark White provided technical assistance throughout the project. We appreciate the comments we received on the manuscript from William Baker, John Pastor, Dave Roberts, and two anonymous reviewers. This chapter is dedicated to Miron L. (Bud) Heinselman (1920–1993), in acknowledgment of our debt to his pioneering work in forest disturbance and landscape ecology.

References

Baker, W. L. 1989. A review of models of landscape change. *Landscape Ecology* 2:111–33.

———. 1992. Effects of settlement and fire suppression on landscape structure. *Ecology* 73:1,879–87.

Baker, W. L., S. L. Egbert, and G. F. Frazier. 1991. A spatial model for studying the effects of climatic change on the structure of landscapes subject to large disturbances. *Ecological Modelling* 56:109–25.

Barnes, B. V., K. S. Pregitzer, T. A. Spies, and V. H. Spooner. 1982. Ecological forest site classification. *Journal of Forestry* 80:493–98.

Boeder, J., D. J. Mladenoff, and G. E. Host. 1993. A spatially explicit model of forest landscape disturbance, management, and succession (LANDIS). *User's Guide and Technical Documentation.* NRRI Technical Report. Duluth: Natural Resources Research Institute, University of Minnesota.

Boeder, J., Y. Xin, D. J. Mladenoff, and G. E. Host. 1993. Spatial analysis package (APACK). *User's Guide and Technical Documentation.* NRRI Technical Report. Duluth: Natural Resources Research Institute, University of Minnesota.

Bormann, F. H., and G. E. Likens. 1979. *Pattern and Process in a Forested Ecosystem.* New York: Springer-Verlag.

Botkin, D. B., J. F. Janak, and J. R. Wallis. 1972. Some ecological consequences of a computer model of forest growth. *Journal of Ecology* 60:849–72.

Burns, R. M., and B. H. Honkala, coordinators. 1990. *Silvics of North America.* Washington, D.C.: Forest Service, USDA.

Canham, C. D., and O. L. Loucks. 1984. Catastrophic windthrow in the presettlement forests of Wisconsin. *Ecology* 65:803–09.

Clark, J. S. 1988. Effect of climate change on fire regimes in northwestern Minnesota. *Nature* 334:233–34.

———. 1991. Disturbance and population structure on the shifting mosaic landscape. *Ecology* 72:1,119–137.

Curtis, J. T. 1959. *The Vegetation of Wisconsin.* Madison: University of Wisconsin Press.

Frelich, L. E., and C. G. Lorimer. 1991a. Natural disturbance regimes in hemlock-hardwood forests of the upper Great Lakes region. *Ecological Monographs* 61:145–64.

———. 1991b. A simulation of landscape-level stand dynamics in the northern hardwood region. *Journal of Ecology* 79:223–33.

Gore, J. A., and W. A. Patterson III. 1985. Mass of downed wood in northern hardwood forests in New Hampshire: Potential effects of forest management. *Canadian Journal of Forest Research* 16:335–39.

Heinselman, M. L. 1973. Fire in the virgin forests of the boundary waters canoe area, Minnesota. *Quaternary Research* 3:329–82.

Host, G. E., K. S. Pregitzer, C. W. Ramm, J. B. Hart, and D. T. Cleland. 1987. Landform-mediated differences in successional pathways among upland forest ecosystems in northwestern lower Michigan. *Forest Science* 33:445–57.

Kessell, S. R., R. B. Good, and A. J. M. Hopkins. 1984. Implementation of two new resource management information systems in Australia. *Environmental Management* 8:251–70.

Loehle, C. 1988. Tree life history strategies: The role of defenses. *Canadian Journal of Forest Research* 18:209–22.

Mladenoff, D. J. 1987. Dynamics of nitrogen mineralization and nitrification in hemlock and hardwood treefall gaps. *Ecology* 68:1,171–180.

Mladenoff, D. J., and G. E. Host. 1994. Ecological applications of remote sensing and GIS for ecosystem management in the northern lakes states. In *Forest Ecosystem Management at the Landscape Level: The Role of Remote Sensing and GIS in Resource Management Planning, Analysis and Decision Making,* edited by V. R. Sample. Washington, D.C.: Island Press.

Mladenoff, D. J., and J. Pastor. 1993. Sustainable forest ecosystems in the northern hardwood and conifer region: Concepts and management. In *Defining Sustainable Forestry,* edited by G. H. Aplet, J. T. Olson, N. Johnson, and V. A. Sample, 145–80. Washington, D.C.: Island Press.

Mladenoff, D. J., M. A. White, T. R. Crow, and J. Pastor 1994. Applying principles of landscape design and management to integrate old-growth forest enhancement and commodity use. *Conservation Biology* 8(3):752–62

Mladenoff, D. J., M. A. White, J. Pastor, and T. R. Crow. 1993. Comparing spatial pattern in unaltered old-growth and disturbed forest landscapes. *Ecological Applications* 3:293–305.

Noble, I. R., and R. O. Slatyer. 1980. The use of vital attributes to predict successional changes in plant communities subject to recurrent disturbances. *Vegetatio* 43:5–21.

Pastor, J., and D. J. Mladenoff. 1992. The southern boreal–northern hardwood forest border. In *A Systems Analysis of the Global Boreal Forest,* edited by H. H. Shugart, R. Leemans, and G. B. Bonan, 216–40. Cambridge: Cambridge University Press.

Roberts, D. W. 1994a. Modeling forest dynamics with vital attributes and fuzzy systems theory. *Ecological Modelling.*

———. 1994b. Landscape vegetation modeling with vital attributes and fuzzy systems theory. *Ecological Modelling.*

Shugart, H. H. 1984. *A Theory of Forest Dynamics: The Ecological Implications of Forest Succession Models.* New York: Springer-Verlag.

Sklar, F. R., and R. Costanza. 1991. The development of dynamic spatial models for landscape ecology: A review and prognosis. In *Quantitative Methods in Landscape Ecology,* edited by M. G. Turner and R. H. Gardner, 239–88. New York: Springer-Verlag.

Turner, M. G., and V. H. Dale. 1991. Modeling landscape disturbance. In *Quantitative Methods in Landscape Ecology,* edited by M. G. Turner and R. H. Gardner, 323–51. New York: Springer-Verlag.

White, M. A., and D. J. Mladenoff. 1994. Old-growth forest landscape transitions in the northern lakes states from pre-European settlement to present. *Landscape Ecology* 9(3):191–205.

David J. Mladenoff has been a research associate for the past six years at the Natural Resources Research Institute (NRRI), University of Minnesota, Duluth. He received his Ph.D. in forest ecology from the University of Wisconsin-Madison in 1985. His research program is in the area of forest ecosystems and landscape ecology.

Joel Boeder was an applications programmer at NRRI while working on this work; he currently is with Datamap, Inc.

George Host is a research associate at NRRI in forest and landscape ecology. He earned his Ph.D. at Michigan State University in 1987.
Natural Resources Research Institute (NRRI)
University of Minnesota-Duluth
Duluth, Minnesota 55811
Phone: (218) 720-4279
Fax: (218) 720-4219
E-mail: *dmladeno@ua.d.umn.edu*

Thomas Crow heads the Landscape Ecology Program with the U.S. Forest Service, North Central Forest Experiment Station.
U.S. Forest Service
North Central Forest Experiment Station
Forestry Sciences Laboratory
Rhinelander, Wisconsin 54501

33

Landscape Scale Forest Dynamics: GIS, Gap, and Transition Models

Miguel F. Acevedo, Dean L. Urban, and Magdiel Ablan

This development illustrates the integration of state transition models and GIS spatial capabilities to analyze forest dynamics at the landscape scale. The transition model (MOSAIC) is semimarkovian with probabilities, distributed lags, and discrete time lags estimated from simulation runs of a gap model (ZELIG) at the plot scale. This parameter estimation procedure assures consistency in the change of scale. Further simplification is achieved by using a limited number of typal species representing functional roles as states of the transition model. Environmental factors are stored as GIS files and transferred to MOSAIC to adjust parameters for simulation; values for the states at each landscape cell are generated by MOSAIC and transferred to the GIS for display and analysis. The capabilities of the model to answer management questions at the landscape scale are demonstrated.

INTRODUCTION

The benefits of using a compact model of forest dynamics for landscape analysis, based on patch transitions among several functional roles, are enhanced by the use of a GIS for display and spatial analysis of model output as well as for model input management. As many of the other chapters in this book demonstrate, the linkages of environmental models to GIS greatly enhance their capabilities for decision support systems (Lam and Swayne 1991; Fedra 1990). For the purposes of this work, the landscape is assumed to be composed of a collection of a large number of cells (units, ecotypes, or tesseras [Naveh and Lieberman, 1984]). Each one of these cells is modeled as a mosaic of smaller, gap-scale plots; its cover type, or state, is given as the proportion of total area in each of several cover types. Landscape dynamics are simulated as changing proportions of within-cell cover types.

Cover types are defined according to dominance by a limited number of species roles or functional types. These roles are based on the requirement of canopy gaps for regeneration and the capacity to create canopy gaps by mortality (Shugart 1984, 1987). Transition probabilities among the states and holding times in each transition lead to semimarkovian analytical calculations of the state probabilities and to simple simulations. The cell-level model has been described in detail in Acevedo, Urban, and Shugart (1993).

The model makes feasible the direct exploration of hypotheses and the possibility of fast computation from closed-form solutions and formulae, which can be a premium in applications requiring repetitive numerous simulations, as often occurs in landscape dynamics.

Detailed forest simulators have linked environmental parameters to demographics and growth using the JABOWA-FORET approach (Botkin et al. 1972; Shugart and West 1977) and its many variants (Urban and Shugart 1992). This approach has limited application for landscape and regional scale analysis due to the computer time and memory required for a large number of model plots.

Modeling the dynamics of landscape-scale forest patterns can be simplified by aggregating species according to some adequate criteria; for example, "pioneer" or "gap-requiring" species and "climax" or "shade-tolerant" species (Acevedo 1981a, 1981b; Whitmore 1989; Swaine and Whitmore 1988). This classification based on regeneration requirements leads to a dynamic interpretation of a forested landscape as an ever-changing mosaic of patches cycling through defined successional phases (Watt 1947; Oldeman 1978; Whitmore 1989; Whittaker and Levin 1977; Bormann and Likens 1979). Even though this classification may be simplistic for many reasons (Denslow 1987; Barton 1984; Hubbell and Foster 1986; Smith et al. 1992; Lieberman et al. 1989; Brokaw and Scheiner 1989), it allows for a practical modeling methodology equipped to answer questions related to coarse-scale dynamics of the forest mosaic. Other aggregation schemes are possible; for example, definition of cover states according to successional status (Shugart et al. 1973), definition of types according to shade and drought tolerance (Smith and Huston 1989), and identification of several tree species' roles based on patterns observed in FORET-type simulations (Shugart et al. 1981; Shugart 1984, 1987). This last approach served as the basis of the model described in Acevedo, Urban, and Shugart (1993) and is used here for the cell-level dynamics.

Combining the gap-creating properties of trees derived from the mortality process, with the gap-requiring properties derived from the regeneration process, Shugart (1984, 1987) proposed four main roles (groups) of tree species: (1) gap-creating and gap-requiring (large, shade-intolerant); (2) gap-creating and nongap-requiring (large, shade-tolerant); (3) nongap creating and gap-requiring (small, shade-intolerant); and (4) nongap-creating and nongap-requiring (small, shade-tolerant). This scheme associates size with gap creation and therefore is similar to the one proposed by Swaine and Whitmore (1988) for tropical forests. Shugart and Urban (1989) have simulated forests with single species representing the four roles to infer typical dynamic patterns.

The advantages and disadvantages of using markov processes and differential equations to model plant succession have been discussed by many authors (e.g., Weinstein and Shugart 1983; Leps 1988; Usher

1992; van Hulst 1980; Hobbs and Legg 1983; van Dorp et al. 1985). The use of a semimarkov model, which makes the transitions dependent on the time spent in a given state, can resolve many of the disadvantages (Acevedo 1981a; Acevedo, Urban, and Shugart 1993). Semimarkov models have not been used extensively in ecological modeling research (Marcus et al. 1979; Matis et al. 1992).

CELL-LEVEL DYNAMICS

Forest dynamics at the cell level are simulated with a semimarkov model for the transitions of a forest plot among the four roles. This section summarizes the model, which is described in more detail by Acevedo, Urban, and Shugart (1993).

A gap-size forest plot is assumed to make transitions among several states defined by the dominance of one of the four roles. The total area covered by a collection of gap-size plots making up a cell will be distributed among the four roles according to proportions $X_i(t)$ which are approximately equal to the probabilities $p_i(t)$ that a plot is dominated by role i at time t. The model is specified by the transition probabilities p_{ij}, from role j to role i, and the holding-time densities h_{ij}; that is, the probability densities for the time spent in making the transition from role j to i for every pair of roles i and j. A convenient form to use for the holding-time density is a gamma density (Lewis 1977; Acevedo 1981a), which has two parameters: d_{ij}, and k_{ij}. The first one—d_{ij}—is a first-order rate, and the second—k_{ij}—is an integer representing the order of the function. Both parameters define the mean k_{ij}/d_{ij} of the $h_{ij}(t)$ density.

Statistics of interest (e.g., occupancy probabilities, entrance probabilities, transit time) can be calculated from semimarkov theory (Howard 1971). In particular, the steady-state occupancy probabilities can be readily calculated to represent the fraction of the cell that will be occupied by each role after a sufficient time period has passed. A large value for the stationary state of one role can be due to a long holding time in the transition from that role, or, to a low probability of transition from that role.

A set of first-order differential equations emulating the gamma function (Acevedo 1981a; McDonald 1978; Matis et al. 1992) are integrated numerically. The transition from state j to state i can be considered as a sequence of transitions among intermediate states, at the rate d_{ij}. However, when a long latency is needed in the holding-time density with a large value for the mean and a low value for the variance, it is desirable to add a fixed time delay to the gamma distributed holding-time density. Therefore, for each one of the valid transitions, four parameters are needed: (1) the rate for each one of the distributed time lags, (2) the order for each one of the distributed time lags, (2) the order of the fixed lag, and (3) one transition probability. The resulting structure is schematically depicted in Figure 33-1.

The parameters used in the semimarkov model can be estimated from a series of runs of a gap model simulator using species that correspond to the four roles. Counting the frequency of occurrence and timing of each transition pair allows an estimation of the transition probabilities and the holding-time densities. For example, four hypothetical species were constructed in the ZELIG gap model (Urban and Shugart 1992) to represent the four contrasting functional roles, exhibiting a classic successional pattern over several hundred years: Role 3 asserts an early importance but quickly declines; Role 2 reaches its greatest dominance in the first century; Role 2 increases steadily into the next centuries; and Role 4 is the most common component of the understory in older stands but does not have the stature to dominate the canopy. Discrete forest plot states were classified by assigning each plot to the role with the greatest basal area. The time course of the frequency of occurrence of a transition from state i to state j allows estimation of the holding-time density for this transition. Fixed latencies can be extracted from these time traces and the estimated mean of the gamma densities is obtained by subtracting the latency from the observed mean. Results for 400 plots over 500 years of simulation with ZELIG were produced in this case. Simulation results of the transients with the estimated parameters, using an integration time step of one year, are shown in Figure 33-2.

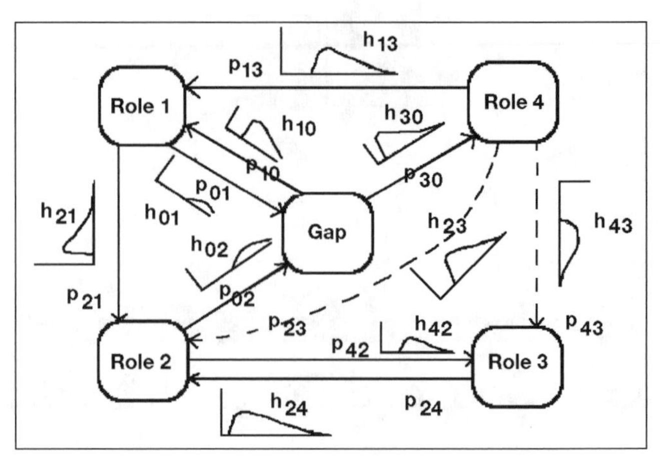

Figure 33-1. Semimarkov model of forest dynamics based on aggregation of species in four main groups according to their dynamic role in regeneration and mortality and gaps. Dashed lines represent low probability. Cycles are contained in this structure. Only the major transitions are shown.

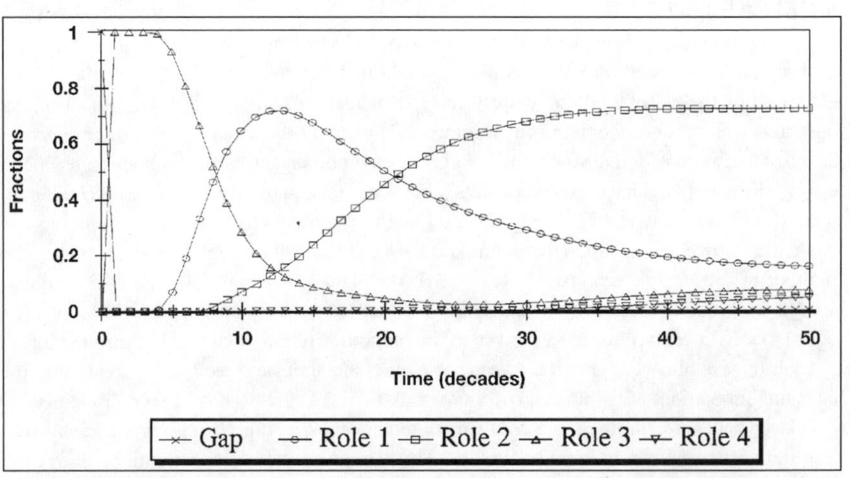

Figure 33-2. Semimarkov simulation results at the cell level with parameters adjusted by ZELIG simulation. Typical successional dynamics are exhibited by this simulation. All empty-gap plots are assumed for the initial condition.

LANDSCAPE DYNAMICS

The transition model can be used to run a simulation of the landscape dynamics, by repetitive calculations for each cell at each time step. We will refer to this version of the model as MOSAIC. A systematic method is needed to extract the transition model parameters directly from cell environmental conditions, such as soil moisture, nutrients, temperature, and so on.

Due to the Monte Carlo nature of gap-type simulators, their use for landscape applications requires multiple runs for every plot constituting a cell, whereas MOSAIC only requires one run per cell. For example: an area of 10 ha per cell would require runs of a gap model for 100 plots of 1/10 ha each, which implies a reduction of approximately a factor of 100 when using the semimarkov simulation (Figure 33-3).

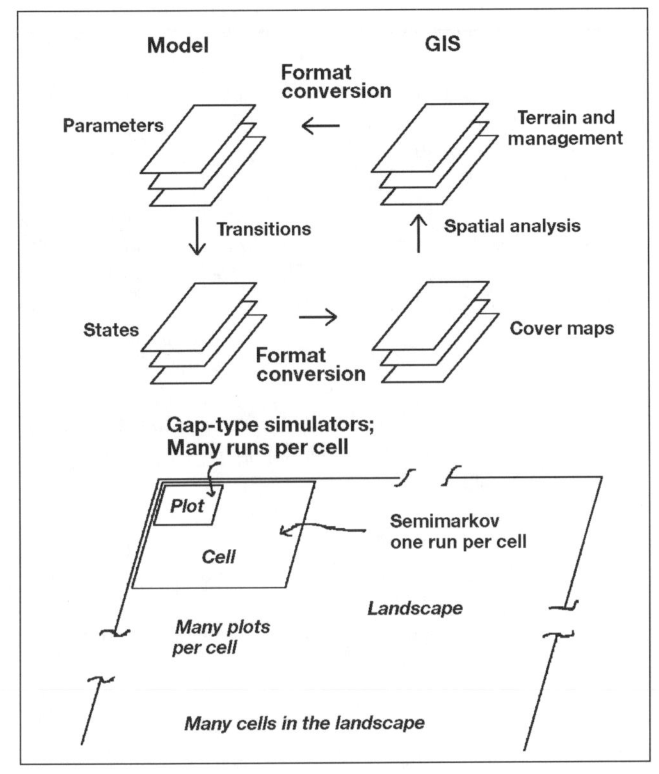

Figure 33-3. Landscape dynamics simulation; schematic diagram. Comparison of the MOSAIC semimarkov approach with gap models.

The potential of this model to simulate landscape dynamics can be enhanced by linking it to a GIS containing files for terrain characteristics (soil moisture, slope, aspect, elevation, fertility, etc.) and to other environmental conditions for every cell. The parametric dependence of successional models on different soil conditions has been shown to be important for landscape modeling by many authors (e.g. Weinstein and Shugart 1983; Shugart 1989; Urban and Smith 1989). The interaction of these variables deals with slope/aspect effects on radiation, elevation effects on temperature and precipitation, and terrain effects on hydrology via topographic convergence.

Neighboring cell interactions could also be included to develop spatial patterns of the dynamic mosaic due to contagious disturbance (Knight 1987; Sprugel and Bormann 1981), seed dispersal, or other neighborhood effects (Turner 1987). A similar approach has been recently reported by Wissel (1992) to study the forest mosaic cycle applying the cellular automata method and considering only the dominant species.

The MOSAIC model prototype is conveniently executed in the UNIX environment and linked to the GRASS-GIS (U.S. Army Corps of Engineers 1991). Model output files in ASCII are converted to raster format using GRASS commands. This operation, plus the opening of monitors, display of raster files, and spatial analysis commands are programmed within a shell script running in GRASS. This shell script also converts raster files for terrain data and invokes the executable simulation program, which calculates values for the parameters and the state variables. An example of model output is shown in Figure 33-4.

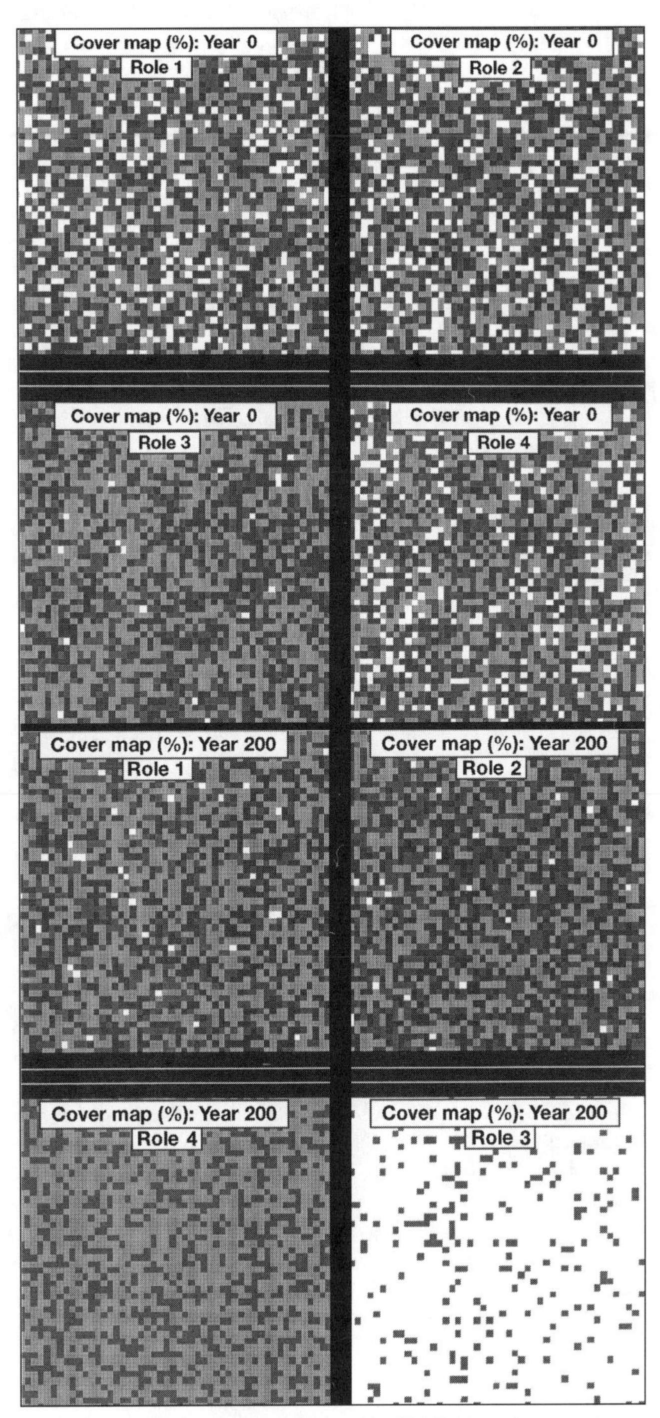

Figure 33-4. Output from MOSAIC displayed by GRASS. Four roles at simulation years 0 and 200. Increasing intensity of gray in a cell corresponds to increasing values of the proportion of that role in the cell.

Computer resource utilization and model performance depend on the number of states, the order of the delays, and the number of nonzero transition probabilities. To illustrate model performance: MOSAIC runs in approximately twenty minutes, occupying about 30

MB of memory when executed in a SPARC 2 workstation with 32 MB memory and 38 MB swap, for a 500-year simulation of a landscape of 10,000 ha, approximately 1,000 cells of 10 ha each, five states (four roles plus empty gaps), 10 nonzero transition probabilities, maximum gamma-delay order of 5, and maximum fixed delay of 90 years (parameter values from Acevedo, Urban, and Shugart 1993). By contrast, a 500-year-simulation run will take about three hours for a small watershed (approximately 50 ha) using the current version of ZELIG. Simulating a large number of cells would degrade MOSAIC's performance due to increased swapping.

Current work includes:

1. Benchmarking computer resource demands by this program in its alternative linkages to GRASS.
2. Improvements in code to reduce memory and time requirements.
3. Improvements of GRASS linkages.
4. Development of linkages of terrain effects and soil characteristics to MOSAIC parameters, incorporating among-cell interactions and neighborhood effects.
5. Potential use of remotely sensed data for model calibration and validation.

CONCLUSION

This approach can play a role in the understanding of ecosystem dynamics at the landscape scale. Questions at this large scale can be addressed while a consistent conceptual and empirical basis is maintained with finer-scale ecological detail through its correspondence with gap models. Consistency across scales is greatly needed in landscape ecology (Risser 1987; Meentemeyer and Box 1987).

This framework provides new capabilities for land-use and management strategies by making it possible to obtain fast simulation results for the interaction of time and space heterogeneity of land management, natural disturbances, and successional dynamics (Turner and Bratton 1987; Remillard et al. 1987; Godron and Forman 1983). For example, the model could be used to design a management strategy for timber harvest—as a temporal schedule and distribution of areas to be cut—taking into account spatial distribution of fires, tree falls, landslides, and other landscape features, to satisfy a given criterion. The objective might be a stable distribution of forest-age classes, or a high but sustainable proportion of Role 1 stands. The spatial and temporal characteristics of the management practice are defined as a set of maps, one for each discrete intervention event. Each map has polygons representing areas of common intensity of the intervention. Disturbances and other landscape features have a similar set of maps. The model responds dynamically to these, giving a change of values for the states in each cell. Spatial analysis on the set of maps is examined for the value of the criterion or objective function. Changes in the management practice are developed according to the result. In addition to simplifying the simulations, transition semimarkovian models in landscape dynamics can be extremely helpful by providing analytical guidance to the simulations and in permitting direct exploration of hypotheses via closed-form solutions and formulae.

References

Acevedo, M. F. 1981a. Electrical network simulation of tropical forest successional dynamics. In *Progress in Ecological Engineering and Management by Mathematical Modeling,* edited by D. M. Dubois, 883–92. Liege: CEBEDOC.

———. 1981b. On Horn's markovian model of forest succession with particular reference to tropical forests. *Theor. Pop. Biol.* 19(2):230–50.

Acevedo, M. F., D. L. Urban, and H. H. Shugart. 1993. Models of forest dynamics based on roles of tree species. *Ecological Applications.*

Barton, A. M. 1984. Neotropical pioneer and shade-tolerant tree species: Do they partition tree fall gaps? *Trop. Ecol.* 25:196–202.

Bormann, F. H., and G. G. Likens. 1979. *Pattern and Process in a Forested Ecosystem.* New York: Springer-Verlag.

Botkin, D. B., J. F. Janak, and J. R. Wallis. 1972. Some ecological consequences of a model of forest growth. *J. Ecol.* 60:849–73.

Brokaw, N. V. L., and S. M. Scheiner. 1989. Species composition in gaps and structure of a tropical forest. *Ecology* 70(3):538–41.

Denslow, J. S. 1987. Tropical rain forest gaps and tree species diversity. *Ann. Rev. Ecol. Syst.* 18:431–51.

Fedra, K. 1990. A computer-based approach to environmental impact assessment. *EUR 13060 EN*:11–39.

Godron, M., and R. T. T. Forman. 1983. Landscape modification and changing ecological characteristics. In *Disturbance and Ecosystems: Components of Response,* edited by H. A. Mooney and M. Godron, 12–28. New York: Springer-Verlag.

Hobbs, R. J., and C. J. Legg. 1983. Markov models and initial floristic composition in heathland vegetation dynamics. *Vegetatio* 56:31–43.

Howard, R. A. 1971. *Dynamic Probabilistic Systems: Semi-Markov and Decision Processes, Vol. 2.* New York: John Wiley and Sons.

Hubbell, S. P., and R. B. Foster. 1986. Canopy gaps and the dynamics of a neotropical forest. In *Plant Ecology,* edited by M. J. Crawley, 77–96. Blackwell Scientific.

Knight, D. H. 1987. Parasites, lightning, and the vegetation mosaic in wilderness landscapes. In *Landscape Heterogeneity and Disturbance,* edited by M. G. Turner, 59–83. New York: Springer-Verlag.

Lam, D. C. L., and D. A. Swayne. 1991. Integrating database, spreadsheet, graphics, GIS, statistics, simulation models, and expert systems: Experiences with the RAISON system on microcomputers. *NATO Series* G26:429–59.

Leps, J. 1988. Mathematical modeling of ecological succession—A review. *Folia Geobotanica et Phytotaxonomica* 23:79–94.

Lewis, E. R. 1977. Linear population models with stochastic time delays. *Ecology* 58:738–49.

Lieberman. M., D. Lieberman, and R. Peralta. 1989. Forests are not just Swiss cheese: Canopy stereogeometry of nongaps in tropical forests. *Ecology* 70(3):550–52.

Marcus, A. H. 1979. Semimarkov models in ecology and environmental health. In *Compartmental Analysis of Ecosystem Models,* edited by J. H. Matis, B. C. Patten, and G. C. White, 261–78. International Cooperative Publishing House.

Matis, J. H., W. E. Grant, and T. H. Miller. 1992. A semimarkov process model for migration of marine shrimp. *Ecol. Model.* 60:167–84.

McDonald, N. 1978. *Time Lags in Biological Models.* Springer-Verlag.

Meentemeyer, V., and E. O. Box. 1987. Scale effects in landscape studies. In *Landscape Heterogeneity and Disturbance,* edited by M. G. Turner, 15–34. New York: Springer-Verlag.

Naveh Z., and A. S. Lieberman. 1984. *Landscape Ecology: Theory and Application.* New York: Springer-Verlag.

Oldeman, R. A. A. 1978. Architecture and energy exchange of dicotyledonous trees in the forest. In *Tropical Trees as Living Systems,* edited by P. B. Tomlinson and M. Zimmermann, 535–59. Cambridge University Press.

Remillard, M. M., G. K. Gruendling, and D. J. Bogucki. 1987. Disturbance by beaver and increased landscape heterogeneity. In

Landscape Heterogeneity and Disturbance, edited by M. G. Turner, 103–22. New York: Springer-Verlag.

Risser, P. G. 1987. Landscape ecology: State of the art. In *Landscape Heterogeneity and Disturbance*, edited by M. G. Turner, 3–14. New York: Springer-Verlag.

Shugart, H. H. 1984. *A Theory of Forest Dynamics: The Ecological Implications of Forest Succession Models*. New York: Springer-Verlag.

———. 1987. Dynamic ecosystem consequences of tree birth and death patterns. *Bioscience* 37(8):596–602.

———. 1989. The role of ecological models in long-term ecological studies. In *Long Term Studies in Ecology: Approaches and Alternatives*, edited by G. E. Likens, 90–109. New York: Springer-Verlag.

Shugart, H. H., T. R. Crow, and J. M. Hett. 1973. Forest succession models: A rationale and methodology for modeling forest succession over large regions. *For. Sci.* 19(3):203–12.

Shugart, H. H., and D. L. Urban. 1989. Factors affecting the relative abundance of forest tree species. In *Toward a More Exact Ecology*, edited by P. J. Grubb and J. B. Whittaker, 249–73. Blackwell Scientific Publications.

Shugart, H. H., and D. C. West. 1977. Development of an Appalachian deciduous forest succession model and its application to assessment of the impact of the chestnut blight. *J. Env. Management* 5:161–79.

Shugart, H. H., D. C. West, and W. R. Emanuel. 1981. Patterns and dynamics of forests: An application of simulation models. In *Forest Succession Concepts and Applications*, edited by D. C. West, H. H. Shugart, and D. B. Botkin, 74–94. New York: Springer-Verlag.

Smith, A. P., K. P. Hogan, and J. R. Idol. 1992. Spatial and temporal patterns of light and canopy structure in a lowland tropical moist forest. *Biotropica* 24(4):503–11.

Smith, T. M., and M. Huston. 1989. A theory of the spatial and temporal dynamics of plant communities. *Vegetatio* 83:49-69.

Sprugel, D. G., and F. H. Bormann. 1981. Natural disturbance and the steady state in high-altitude balsam fir forests. *Science* 211:390–93.

Swaine, M. D., and T. C. Whitmore. 1988. On the definition of ecological species groups in tropical rain forests. *Vegetatio* 75:81–6.

Turner, M. G. 1987. Spatial simulation of landscape changes in Georgia: A comparison of three models. *Landscape Ecol.* 1:29–36.

Turner, M. G., and S. P. Bratton. 1987. Fire, grazing, and the landscape heterogeneity of a Georgia barrier island. In *Landscape Heterogeneity and Disturbance*, edited by M. G. Turner, 85–101. New York: Springer-Verlag.

U.S. Army Corps of Engineers. 1991. *GRASS. User's and Programmer's Manual for the Geographic Resource Analysis Support System*. CERL ADP Report N-87/22.

Urban, D. L., G. B. Bonan, T. M. Smith, and H. H. Shugart. 1991. Spatial applications of gap models. *For. Ecol. Mgmt.* 42:95–110.

Urban, D. L., and H. H. Shugart. 1992. Individual-based models of forest succession. In *Plant Succession: Theory and Prediction*, edited by D. C. Glenn-Lewin, R. K. Peet, and T. T. Veblen, 249–92. Chapman & Hall.

Urban, D. L., and T. M. Smith. 1989. Extending individual-based forest models to simulate large-scale environmental patterns *Bull. Ecol. Soc. Amer.* 70:284.

Usher, M. B. 1992. Statistical models of succession. In *Plant Succession: Theory and Prediction*, edited by D. C. Glenn-Lewin, R. K. Peet, and T. T. Veblen, 215–48. Chapman & Hall.

van Dorp, D., R. Boot, and E. van der Maarel. 1985. Vegetation succession on the dunes near Oostvoorne, the Netherlands, since 1934, interpreted from air photographs and vegetation maps. *Vegetatio* 58:123–36.

van Hulst, R. 1980. Vegetation dynamics or ecosystem dynamics: Dynamic sufficiency in succession theory. *Vegetatio* 43:147–51.

Watt, A. S. 1947. Pattern and process in the plant community. *J. Ecol.* 35:1–22.

Weinstein, D. A., and H. H. Shugart. 1983. Ecological modeling of landscape dynamics. In *Disturbance and Ecosystems: Components of Response*, edited by H. A. Mooney and M. Godron, 29–45. New York: Springer-Verlag.

Whitmore, T. C. 1989. Canopy gaps and the two major groups of forest trees. *Ecology* 70(3):536–38.

Whittaker, R. H., and S. A. Levin. 1977. The role of mosaic phenomena in natural communities. *Theor. Pop. Biol.* 12:117–39.

Wissel, C. 1992. Modeling the mosaic cycle of a middle European beech forest. *Ecol. Model.* 63:29–43.

Miguel Acevedo is an electrical engineer and biophysicist, whose main research interests are in ecosystem modeling and environmental instrumentation. He earned his Ph.D. at the University of California-Berkeley.

Magdiel Ablan is a systems engineer, currently pursuing a Ph.D. in environmental science.

Miguel F. Acevedo and Magdiel Ablan
Department of Geography and Institute of Applied Science
University of North Texas
Denton, Texas 76203
Phone: (817) 565-2694
Fax: (817)-565-4297
E-mail: *acevedo@unt.edu; mablan@sol.acs.unt.edu*

Dean Urban is an ecologist whose focus is forest modeling and landscape ecology. Dean received his Ph.D. from the University of Tennessee.
Department of Forest Science
Colorado State University
Fort Collins, Colorado 80523
Phone: (303) 491-7529
Fax: (303) 491-2339
E-mail: *deanu@populus.cfnr.colostate.edu*

34

Spatial Analysis of Boreal Forest Ecosystems: Results from the Rinker Lake Case Study

B. G. Mackey, R. A. Sims, K. A. Baldwin, and I. D. Moore

An integrated ecosystem modeling project underway in northwestern Ontario at Rinker Lake is described in this chapter. A critical first step is to develop spatially reliable estimates of the landscape processes that control the availability and distribution of energy, moisture, and mineral nutrients. These provide a basis for developing spatially referenced predictive models of plant/vegetation-environment response. The case study is imbedded within a parallel regional/provincewide modeling framework. Preliminary results are presented where soil moisture regime is predicted on a 20-m grid over a 900-km² area as a function of the topographic wetness index and the class of geological substrate. Future work aims at refining this model and developing similar spatial models for the mineral nutrient regime. These form the basis for modeling the vegetation component of boreal forest ecosystems at both operational and strategic scales.

INTRODUCTION

The Rinker Lake case study aims to develop space/time models of the ecological resources of boreal forest ecosystems at scales relevant to both strategic and operational resource decision making. By "operational scale," we mean generating spatial information that can be incorporated into harvest schedule plans where mapped information is presented at a scale of 1:20,000. The Rinker Lake case study encompasses 900 km² of boreal forest and is located in northwestern Ontario, Canada. It is explicitly multi- and interdisciplinary and involves collaboration between a number of federal and provincial research and resource-management agencies. The case study is designed to promote the integrated analysis of the physical environment, vegetation, wildlife, and timber supply. Computer-based studies are being underpinned by three summers of field survey.

Ecologically sustainable development of forest ecosystems requires reliable spatial data about the distribution of ecological as well as traditional timber resources. Spatial analyses present two problems: (1) many ecological and timber resources cannot be directly sensed through either satellite imagery or air photography but must be modeled or derived from other primary data; (2) information about causal processes is at least as important as data about extant landscape patterns. The natural distributions of individual plant species and vegetation associations are the result of complex interactions between the genetically controlled response of plants to both landscape and disturbance processes. Landscape processes determine the availability of energy (radiative, thermal), moisture, and mineral nutrients, that is, the primary environmental regimes (PERs). Disturbance processes (such as fire, disease, insects) serve to facilitate successional changes in the vegetation. As a first step, predictive (though static) vegetation models can be built based on empirical relations between the extant distribution of plants and the physical landscape processes that control the distribution and availability of the PERs. The crucial step then becomes to develop reliable spatial estimates of the PERs. The relevant processes operate at a range of scales. Estimating available soil water, for example, requires integration of climatic inputs (precipitation, evaporation) and topography and soils (in redistributing soil water across the landscape).

Analysis of plant–environment response must include the cross-scale processes noted above. Consequently, the Rinker Lake case study (at a nominal scale of 1:20,000) is imbedded within a regional/provincewide study (about 1:250,000 scale) based on the same environmental modeling framework. A new digital elevation model (DEM) for Ontario has been developed using the ANUDEM procedure (Hutchinson 1989). Mathematical surfaces of long-term mean monthly climate have been fitted to a network of available weather stations for the province using the ANUSPLIN procedure (Hutchinson 1987).

Using these techniques, plus ancillary surface interrogation programs, estimates of various climatic and derived bioclimatic variables (e.g., degree days above a specified base temperature) can be generated at any point in the province. Also, by resolving the coefficients at each grid intersection, a regular grid of the climatic variable can be generated. The context for the provincewide DEM and climate-modeling work is discussed by Mackey and McKenney (1993), and results will be presented in a series of forthcoming papers (Mackey and Sims 1993). These digital models enable the climatic response of taxa (based on either floristic or structural characteristics) to be empirically determined and the potential climatic domain spatially predicted (Nix 1986). These spatial data and predictions are being coupled to data from remotely sensed sources, mapped geological/substrate data, and terrain analysis based on the DEM, to provide improved estimates of the PERs, extant land cover, and hence more precise spatial predictions about potential plant response (Mackey et al. 1989; Mackey 1993).

The role of the PERs in boreal forest ecosystems was noted by Sims et al. (1989) who demonstrated that the distribution of mature vegetation associations in the forest ecosystems of northwestern Ontario are strongly influenced by the moisture and mineral nutrient regimes. Vegetation types were configured in a two-dimensional ordination whose axes were interpreted to represent increasing values of moisture and nutrients. The configuration reflected increasing complexity in the physiognomic structure of the vegetation (which in boreal forest ecosystems is strongly correlated with floristics). The two regimes were considered the major environmental determinants of the vegetation gradient. The capacity to generate reliable spatial estimates of the PERs is therefore a prerequisite to modeling plant/vegetation response at both the strategic and operational scales. Hence an important first stage of analysis is to integrate climate, terrain, and substrate data to obtain better spatial estimates of soil water. In this chapter, results are presented of preliminary analyses from the Rinker Lake case study where we demonstrate a method for generating gridded spatial estimates of an index of soil water called the soil moisture regime (SMR [an annualized estimate of the available soil water at a given forest site and derived in the field using a simple lookup table based on a range of attributes including soil texture, abundance of mottles/gley at depth, and slope] Sims et al. 1989). The method utilizes a distributed model of catchment hydrology based on the so-called topographic wetness index and incorporates the local effects of soil parent material.

CHARACTERIZING THE SPATIAL DISTRIBUTION OF SOIL WATER CONTENT

The location of variable source areas of runoff generation and the distribution of soil water are influenced by (1) soil characteristics, (2) topography, (3) vegetation, and (4) weather (Moore et al. 1993). Beven (1986) and Beven and Kirkby (1979) developed the following equation for predicting the pattern of soil water deficit, S_i, from a knowledge of topography and soil hydraulic characteristics

$$S_i = \overline{S} - m[\lambda_i - \overline{\lambda} + \ln(\overline{T}/T_i)] \qquad (34\text{-}1)$$

and

$$\lambda_i = \ln\left(\frac{A_s}{\tan\beta}\right), \overline{\lambda} = \frac{1}{A}\int_A \lambda_i, \overline{S} = \frac{1}{A}\int_A S_i, \overline{T} = \exp\left[\frac{1}{A}\int_A \ln(T_i)\right] \qquad (34\text{-}2)$$

where A_s is the specific catchment area (i.e., the upslope contributing area per unit width of contour), β_i is the local slope angle ($\tan\beta$ is an approximation of the local hydraulic gradient), T_i is the local soil transmissivity (the depth integrated hydraulic conductivity = $\int K(z)dz$, where $K(z)$ is the saturated hydraulic conductivity at depth z, $\overline{T}$ is the mean catchment transmissivity, $\overline{S}$ is the mean catchment soil water deficit, λ_i is the local value of the topographic index, $\overline{\lambda}$ is the mean catchment value of the topographic index, and m is a parameter that is dependent on the rate of change of hydraulic conductivity with depth. A reasonable assumption is that $T_i = K_s/f$ and $m = \Delta\theta/f$, where K_s is the saturated hydraulic conductivity at the soil surface, $\Delta\theta$ is the soil water content deficit (m^3/m^3) and f is a parameter that typically varies from 1.0 to $13m^{-1}$ (Beven 1986). The parameter m (or f) is usually assumed to be spatially invariant and is often treated as a fitted parameter. Equation (34-1) consists of two parts: a topographic component $[\lambda_i - \overline{\lambda}]$ and a soils component $[\ln(\overline{T}/T_i)]$.

The soil water deficit can be written as a simple linear function of the form:

$$S_i = b_0 - m[\lambda_i + \ln(\overline{T}/T_i)] = b_0 - m\ln\left(\overline{T}\frac{A_s}{T_i\tan\beta}\right) = b_0 - m\ln(\varpi_i) \qquad (34\text{-}3)$$

where b_0 is a constant and ϖ_i is what, in an independent derivation, O'Loughlin (1986) called the wetness index. The soil is saturated at the surface when $S_i \leq 0$. The underlying assumptions on which equations (34-1) and (34-3) are based are that all points with the same topographic soil index $[A_s/(T\tan\beta_i)]$ are hydrologically similar and transient conditions can be approximated by successive steady state conditions (Beven 1986). Recently Barling et al. (1993) modified equation (34-3) by substituting the effective specific catchment area A_e for A_s where A_e is calculated from topographic (element slope, area, and connectivity) and soil properties (saturated hydraulic conductivity, saturated soil thickness, effective porosity), and an effective drainage time using a subsurface time-area routing approach. O'Loughlin et al. (1989) and Moore et al. (*Geographic*, 1993, *Application*, 1993) have recently expanded equation (34-3) to account for the effects of vegetation and climate on the pattern of soil water deficit and soil water content in topographically complex catchments by modeling spatially variable net radiation, and hence evaporation.

In both equations (34-1) and (34-3), a spatially varying transmissivity can be represented. However, such data are rarely, if ever, available except for small experimental catchments. Normally only topographic attributes are used to characterize soil water distribution. One rationale for only using topographic attributes to predict soil water content is that in many landscapes pedogenesis of the soil catena occurs in response to the way water moves through the landscape, so that the spatial distribution of the topographic index, λ_i, that characterizes these flow paths inherently captures the spatial variability of soil properties at the mesoscale as well (Moore and Hutchinson 1991). However, the development of soils in northwestern Ontario is dominated, like all the Canadian shield, by series of continental glaciations, the last of which occurred approximately 10,000–22,000 years ago. Many landform features in northwestern Ontario are the result of glacial and postglacial processes. Different modes of deposition led to variation in the depth and texture of the unconsolidated materials overlaid on the bedrock. Soil profiles are therefore relatively young and have developed largely *in situ*. The implication is that spatial variation in soil water content is likely to be as strongly influenced by the soil parent material, and associated soil attributes, as by the topographic wetness index. Here we make the assumption that soil attributes relevant to defining transmissivity are spatially covariant with the major classes of soil parent material. Thus we examine the hypothesis that soil water can be predicted as a function of (1) the upslope contributing area, (2) local slope, and (3) the class of soil parent material.

METHODS

A DEM was generated at a 20-m grid based on digital topographic data from the 1:20,000-scaled Ontario base map series. The contour and streamline data were interpolated using the ANUSPLIN procedure (Hutchinson 1989). A preliminary map of the quaternary geology of Rinker Lake was produced from interpretation of aerial photographs. The mapped polygon boundaries were digitized and then rasterized to a 20-m grid spacing. Twelve classes of geology were recognized and then ranked according to their relative texture, with peatlands and exposed bedrock assigned to opposite ends of the scale. The geology map was reclassified to reflect this ranking. A field survey was undertaken in the summer of 1992, which built upon a previous field survey in the region done by R. A. Sims and colleagues. Measurements were taken at fifty forest plots of the SMR, upslope contributing area, plot slope, and the class of geological substrate (these attribute data are a subset of a more extensive ecological survey conducted at each 10-m^2 plot). SMR was estimated using the field estimation method of the Ontario Institute of Pedology (Sims et al. 1989) where the soil profile is allocated to one of eleven ranked classes.

The upslope contributing area (USCA) was calculated by delineating the immediate watershed boundary through walking upslope from the plot. Boundary points were flagged based on the visual interpretation of the likely course of surface water flow as a function of slope. A chain was then walked from the plot to key points on the boundary, and distances and angles were recorded such that the USCA could later be calculated. The slope of the forest plot was observed with a hand-held inclinometer. A hand auger or 1-m^2 pits were used to sample the soil profile for determining SMR and soil parent material class. The topographic wetness index, λ_i, was calculated for each plot using

$$\lambda_i = (A_s / S) \tag{34-4}$$

where A_s (the specific catchment area) is given by A/w (A is the USCA, w the plot width) and S is the plot slope.

The Monomax algorithms (Bayes and Mackey 1991) were used to examine the relation between variation in the observed SMR as a function of (1) λ_i and (2) the ranked geological class. A goodness of fit for this model of about 70% was returned. Monomax delivers maximum-likelihood estimates using dynamic programming subject to assumptions of a monotonic relation between the dependent and independent variables. The dependent variable can be a ranked categorical variable as is the case with the SMR. A PC-based program TREEWHERE (Intkhab and Mackey, unpublished) was used to generate the spatial predictions of SMR. Gridded versions of the topographic wetness index and surficial/bedrock geology files were read into TREEWHERE. The matrices created by the Monomax analysis then became look-up tables, such that each grid cell was located within the matrices based on their wetness index and geological values and assigned the corresponding predicted probability. The results of the Monomax analysis used here, together with a complete methodology description and results of subsequent vegetation analysis, are given in a forthcoming paper by the authors.

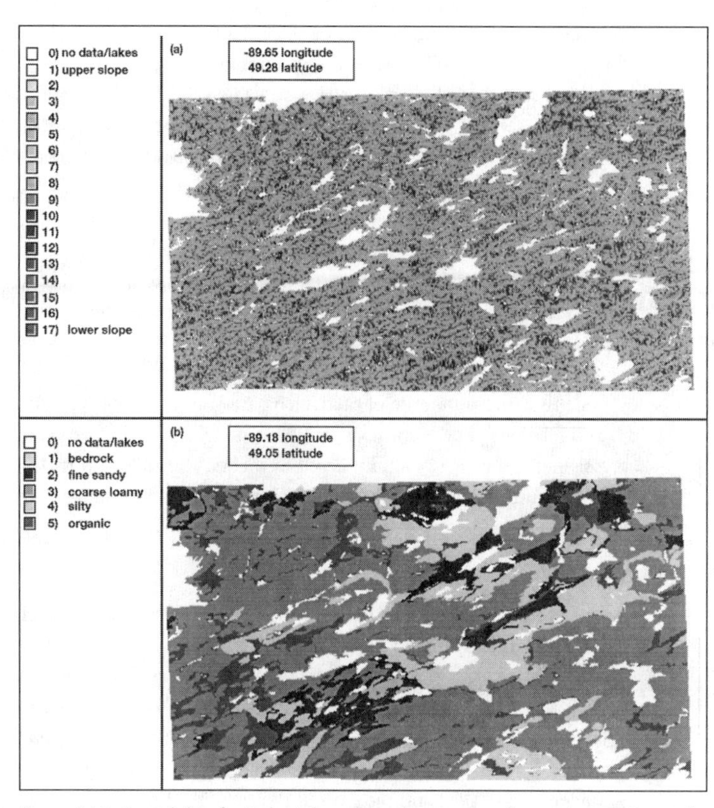

Figure 34-1. Spatial data for Rinker Lake case study area georeferenced on a 20-m grid: (a) topographic wetness index, (b) geology ranked for texture.

CONCLUSION

The gridded estimates of λ_i and the geological texture rankings are shown in Figures 34-1a and 34-1b. The predicted probabilities for the SMR classes are given in Figure 34-2a—interactions between the topography and geology are evident. In some places geology dominates while elsewhere drainage patterns are apparent. The variables can combine to either reinforce or negate the SMR trend. However, the available data proved limited on a number of accounts. First, the field data were insufficiently representative. Hence, only five grouped classes are recognized in Figure 34-2a as it was not possible to account for the variation in all eleven SMR classes (see key for Figure 34-2a; SMR 1 and 2 are "dry"—SMR 9, 10, and 11 are "wet"). Also, no observations were taken from landscapes with lower slope positions and shallow tills on bedrock. Second, the geological map was derived from interpretation of aerial photographs unsupported by field validation. Of particular concern is the inability of air-photo interpretation to distinguish between coarse and fine-grained till/glacial fluvial deposits; consequently, these areas are all inaccurately mapped as "coarse loamy." The result of these data limitations is that extensive areas of Figure 34-2a are predicted as relatively homogeneous areas of map classes 1 and 4. These areas are actually more diverse as illustrated by the range in the predicted probability of occurrence for map class 4 shown in Figure 34-2b. A field survey is being conducted over 1993/1994 to rectify these data limitations. Also, additional data are being collected so that the method described here can be applied to develop a spatial model for soil mineral nutrients.

The results illustrate the complex system of analysis needed to undertake modeling at landscape scales. Field survey, computer-based

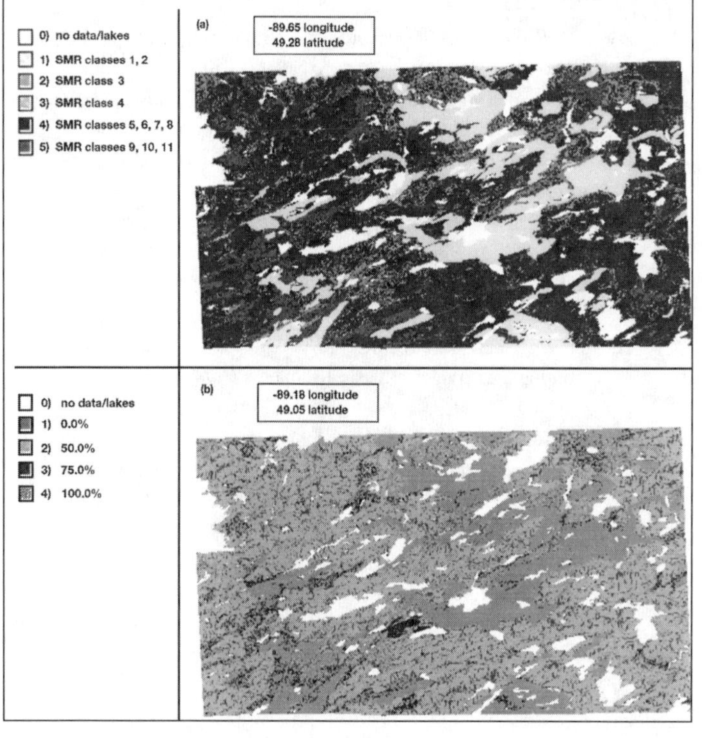

Figure 34-2. Predicted spatial data for Rinker Lake case study area georeferenced on a 20-m grid: (a) soil moisture regime (SMR), (b) probability of occurrence for map class 4 (SMR classes 5, 6, 7, 8).

simulation, and spatial statistics must all be drawn upon and integrated within a common computer-based environment. Our main interest in generating spatial predictions of soil water is its value as a predictor of vegetation response. When coupled with similar data about the mineral nutrient regime, and regional-based analyses of plant/vegetation–climate/environment relations, these techniques provide the basis for developing reliable spatial estimates of both the ecological and wood production potentials of boreal forest ecosystems. Landscape units can then be defined on the basis of shared-potential plant response. These spatial data also provide a basis for implementing more dynamic models of forest succession and ecosystem process. Of utmost importance is the need to develop a generic framework that can be applied across entire ecosystems with a minimum of calibration based on empirically (locally) derived relations. The wetness index is a good example of the type of analysis needed as it is soundly based in theory; it enables (in its full form) climatic, terrain, and substrate variables to be spatially integrated; and it provides a quantitative context for utilizing field data.

Acknowledgments

Funds for this research were provided by a Collaborative Research Agreement with the Ontario Forest Research Institute (OMNR) and the Northern Ontario Development Agreement and Forestry Canada Green Plan. B. M. Mackey's contribution was part of the Bio-environmental Indices Project. Logistic support for the field program was generously provided by Gerry Racey of OMNR. I. D. Moore received additional funding under grant #90/82 and the Special Project 1991–1992, ANU3 from the Land and Resources Research and Development Corporation, and by the Water Research Foundation of Australia. The authors gratefully acknowledge the technical assistance of Norm Sczyrec, Cheryl Widderfield, Kevin Lawrence, Andrew Batchelor, and Intkhab Ali.

References

Barling, R. D., I. D. Moore, and R. B. Grayson. 1993. A quasi-dynamic index for characterizing the spatial distribution of zones of surface saturation and soil water content. *Water Resources Research.*

Bayes, A. J., and B. G. Mackey. 1991. Algorithms for monotonic functions and their application to ecological studies in vegetation science. *Ecological Modelling* 56:135–59.

Beven, K. J. 1986. Runoff production and flood frequency in catchments of order *n*: An alternative approach. In *Scale Problems in Hydrology,* edited by V. K. Gupta, I. Rodriquez-Iturbe, and E. F. Wood, 107, 131. Lancaster: D. Reidel.

Beven, K. J., and M. J. Kirkby. 1979. A physically based variable contributing area model of basin hydrology. *Hydrological Sciences Bulletin* 24:43–69.

Hutchinson, M. F. 1987. Methods for generation of weather sequences. In *Agricultural Environments: Characterization, Classification and Mapping,* edited by A. H. Bunting, 149–57. Wallingford, United Kingdom: CAB Int.

———. 1989. A new procedure for gridding elevation and streamline data with automatic removal of pits. *J. Hydrol.* 106:211–32.

Mackey, B. G. 1993. A spatial analysis of the environmental relations of rain forest structural types. *Journal of Biogeography.*

Mackey, B. G., and D. W. McKenney. 1993. *The Bioenvironmental Indices Project. A Northern Ontario Development Agreement Report.* Forestry Canada.

Mackey, B. G., H. A. Nix, J. A. Stein, S. E. Cork, and F. T. Bullen. 1989. Assessing the representativeness of the wet tropics of Queensland world heritage property. *Biological Conservation* 50:279–303.

Mackey, B. G., and R. A. Sims. 1993. A climatic analysis of selected boreal tree species and potential responses to global climate change. *World Resources Review.*

Moore, I. D., J. C. Gallant, L. Guerra, and J. D. Kalma. 1993. Modeling the spatial variability of hydrological processes using GIS. In *Application of Geographic Information Systems in Hydrology and Water Resources Management, International Association of Hydrological Sciences Conference,* edited by K. Kovar and H. P. Nachtnebel, 161–69. International Assoc. of Hydrological Sciences Conference, Wien, Austria, AHS Pub. no. 211.

Moore, I. D., and M. F. Hutchinson. 1991. *Spatial Extension of Hydrological Process Modeling.* Proc. Int. Hydrology and Water Resources Symposium, Inst. Engrs. Aust. Pub. no. 91/22, 803–08.

Moore, I. D., A. K. Turner, J. P. Wilson, S. K. Jenson, and L. F. Band. 1993. GIS and land-surface–subsurface process modeling. In *Environmental Modeling with GIS,* edited by M. F. Goodchild, B. O. Parks, and L. T. Steyaert. 196–230. New York: Oxford University Press.

Nix, H. A. 1986. A biogeographic analysis of Australian elapid snakes. In *Atlas of Australian Elapid Snakes*, edited by R. Longmore, 4–15. Australian Fauna and Flora Series, No. 7. Canberra: AGPS.

O'Loughlin, E. M. 1986. Prediction of surface saturation zones on natural catchments by topographic analysis. *Water Resources Research* 22:794–804.

O'Loughlin, E. M., D. L. Short, and R. W. Dawes. 1989. *Modelling the Hydrological Resources of Catchments to Land-Use Change.* Hydrology and Water Resources Symposium, 335–40.

Sims, R. A., W. D. Towill, K. A. Baldwin, and G. M. Wickware. 1989. *Field Guide to the Forest Ecosystem Classification for Northwestern Ontario.* Forestry Canada/Ontario Region and Ontario Ministry of Natural Resources.

Brendan G. Mackey has a Ph.D. in Plant Ecology from the Australian National University and specializes in spatial analyses. He has worked for the Australian Commonwealth Scientific and Industrial Research Organization Division of Water and Land Resources and the Center for Resource and Environmental Studies, Canberra. He is presently working as a research scientist with the Canadian Forest Service, Ontario. His current research is focused on the development of spatially distributed models of the ecological resources of Ontario's boreal forests as part of the Bio-environmental Indices Project.

B. G. Mackey, R. A. Sims, K. A. Baldwin
Forestry Canada Ontario Region
PO Box 490 Sault Ste. Marie
Ontario, Canada P6A 5M7
E-mail: *bmackey@%soo.dnet@cedar.pfc.forestry.ca*

I. D. Moore (deceased)

35

Seismic Risk Assessment, Accuracy Requirements, and GIS-Based Sensitivity Analysis

Philip C. Emmi and Carl A. Horton

The purpose of this chapter is to demonstrate the use of Monte Carlo simulation to define the sensitivity of seismic risk-assessment results to random perturbations in the boundaries between earthquake ground-shaking intensity zones. A risk-assessment accuracy standard is established. An algorithm is deployed for inducing within prescribed limits random perturbations in intensity zone boundaries. The effects of boundary perturbations are noted as perturbations are allowed within an increasingly wide corridor surrounding the original boundary alignment. Of particular interest is the width of the corridor at which the risk-assessment accuracy standard is violated. The width of this corridor gauges the degree of accuracy needed in the alignment of intensity zone boundaries to get reliable risk-assessment results.

INTRODUCTION

Natural and technological hazards vary geographically. The media through which hazardous effects are propagated possess physical properties that vary spatially. The populations that might be exposed are also spatially distributed. Risk is inherently a spatial phenomenon. Increasingly, the objective of risk assessment is to define not only the degree of risk but also its spatial variation.

Risk assessment requires data on the components of risk. The spatial assessment of risk requires component data that are mapped or geographically referenced. The quality of risk assessment is increasingly dependent upon the quality of spatially referenced data.

Confidence in risk-assessment results can be compromised by measurement error and by model specification error. These two forms of error are especially amplified by the multiplicative and exponential functions used within the mathematical models that link physical hazards to human exposure and harm (Alonso 1968). Relying on geographic information systems (GIS) to map spatial variation in risk adds a form of error uniquely associated with cartographic representation. Included are digitizing error, map scale error, errors in digital overlay analysis, and vector-to-raster conversion error.

Sensitivity analysis is a generalized methodology for measuring the response of a model's output to variation in its various inputs. Here, sensitivity analysis is used to gauge the ways that assessed risk varies in response to change in selected components of risk. It is a helpful tool when trying to judge the degree of confidence one should have in risk-assessment results. It is particularly helpful when the data quality and the reliability of model specifications are unknown.

Monte Carlo simulation is a statistical technique used to identify the effects of error in input data on risk-assessment results. Two general approaches are possible. The first focuses on the propagation of errors with known statistical distributions. The second approach—the approach used here—identifies the input error that would be permissible given an output accuracy standard. No prior information is required concerning the likely distribution of input error. Instead, a set of successively increasing ranges of possible error is defined. Within each range, random perturbations to input data are induced, and their effects on risk-assessment results are traced. The range of error is extended until the effects on assessment results violate accuracy standards.

This second approach is applied to probabilistic seismic risk assessment (SRA) for Salt Lake County, Utah (population = 725,600, see Figure 35-1 for study-area map). Findings from the Salt Lake

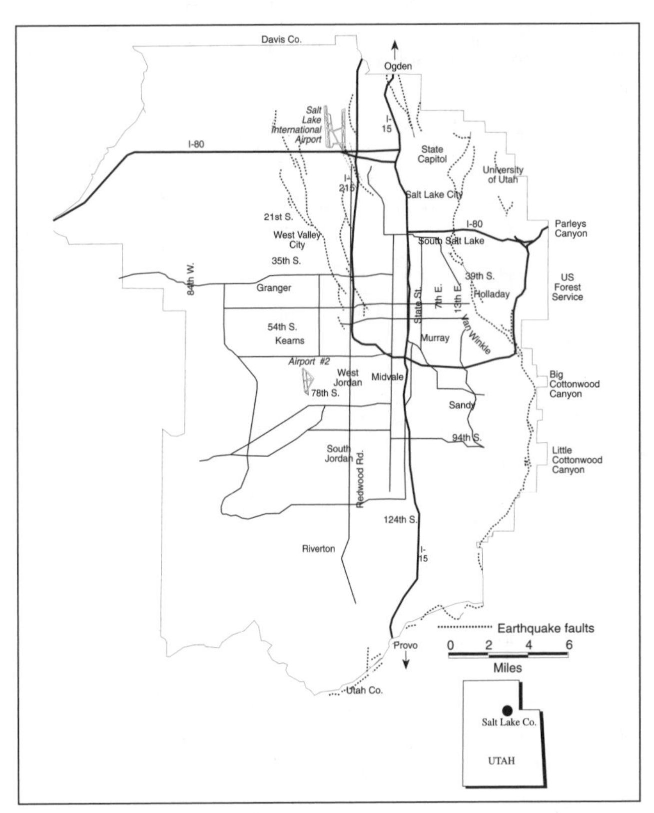

Figure 35-1. Salt Lake Valley, Utah: earthquake faults, major roads, and cities.

SRA are based on (1) a microzonation of earthquake ground-shaking hazard using the Modified Mercalli Intensity Scale (MMI); (2) an inventory of buildings by use, value, and structural frame type; (3) earthquake engineering damage functions relating the seismic performance of buildings to ground-shaking intensities; (4) data on the density of residential and employee populations; and (5) earthquake engineering casualty functions relating casualty risk to the intensity of building damage. The analysis is supported by the algebraic combination of thematic layers within a GIS.

Uncertainty in the location of boundaries between earthquake ground-shaking intensity zones undermines confidence in Salt Lake County seismic risk-assessment results. An accuracy standard for boundary alignments is established to include any random realignment that yields expected residential property loss estimates that are within 5% of original estimates 95% of the time. Random vertex densification and line generalization algorithms are used to induce random perturbations in boundary alignments. Weed tolerances are used to contain boundary perturbations within prescribed ranges surrounding original zonal boundaries. Monte Carlo simulation is used to define the effects that perturbations within each range have on risk-assessment results. It shows that ground-shaking intensity boundaries can vary at random within a 1.5-km-wide corridor before the permissible accuracy standard is violated.

ELEMENTS OF EARTHQUAKE RISK ASSESSMENT

Seismic risk is a blanketing phenomenon. It can be visualized as a three-dimensional surface whose *x*, *y*, and *z* axes are defined by the three conceptual components of a probabilistic risk assessment shown in Figure 35-2. The *x* axis defines the length of the prospective time period over which the study site and its population are exposed to risk. The *y* axis defines the probabilities with which various event intensities are exceeded over various exposure periods. The *z* axis defines event intensities given exceedance probabilities and exposure times. A probabilistic assessment of seismic risk requires the simultaneous treatment of exposure time, exceedance probabilities, and event intensities.

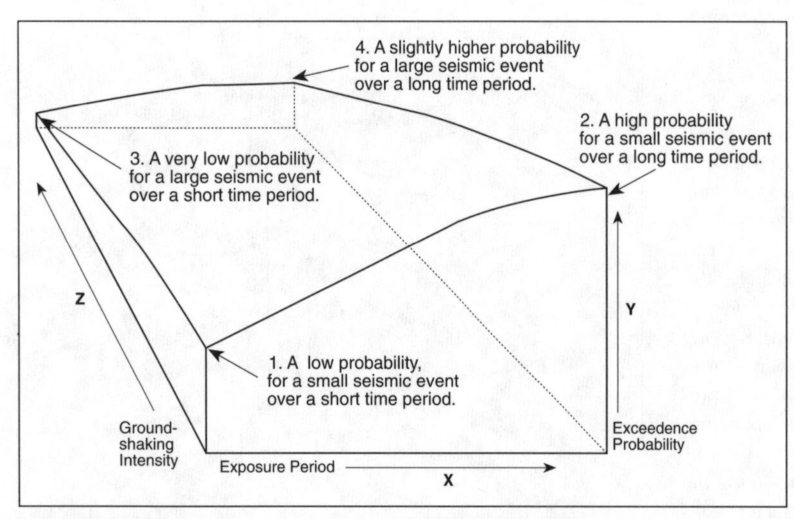

Figure 35-2. Conceptual components of a probabilistic seismic risk assessment.

Because of risk's three-dimensional nature, risk communication is difficult. Practical limits on communication require that risk be reconceptualized as a two-dimensional phenomenon. This can be done by holding constant one of risk's three dimensions and considering a "slice" through the surface. Although any of three slices through this surface could be used, the field of seismic risk assessment has established a practice of referring to seismic risk in terms of the ground-shaking intensities that have a 10% chance of being exceeded over 10-year, 50-year, and 250-year exposure periods.

The intensity of ground shaking is measured on the MMI, a qualitative scale keyed to a variety of human perceptions, structural damage indicators, and geophysical indicators of ground-shaking intensity (Wood and Neumann 1931). Intensities of less than 7 usually produce little damage to most ordinary structures, intensities of 7 to 9 produce considerable damage to poorly built structures, while intensities of 10 or higher result in considerable damage to most structures.

In practice, this approach generates perspective statements such as: "At the Salt Lake International Airport, there is a 10% chance of seismic ground shaking exceeding an intensity level of 10 on the MMI scale over any 50-year exposure period." Nonetheless, when one refers to a given ground-shaking intensity for which there is, for example, a 10% chance of exceedance over a 50-year period, this statement implicitly includes other probabilities and exposure periods on the risk curve for events of that intensity.

DATA, METHODOLOGY, AND FINDINGS

An assessment of regional ground-shaking hazard defines by location the likelihood of various ground-shaking intensities expected over various exposure periods. This hazard represents the surface on which expected damage is to occur. Distributed over this surface are various types of residential and commercial structures and their associated populations. The physical links between the ground-shaking hazard, structures, and local populations are defined by earthquake engineering damage and casualty functions. The damage functions specific to each structural frame type define the expected proportion of a building's value lost as a function of ground-shaking intensity. Casualty functions define the probabilities of injury and death as a function of the intensity of structural damage. Data on residential populations, employee populations, and diurnal patterns of structure use indicate the number of persons at risk from expected damage to residential and commercial structures. Ground-shaking intensities, inventories of structures, damage functions, population densities, and casualty functions are basic to estimates of earthquake property damage and associated casualties.

This report is based on a mapped probabilistic assessment of the ground-motion hazard in Salt Lake County by Emmi (1993) and an assessment of earthquake property damage and casualty risk by Emmi and Horton (1993). The latter study uses the GIS overlay and calculate functions to create the algebraically weighted combinations of thematic data layers needed to assess expected property and casualty losses. The volume and spatial distribution of losses are defined for ground-shaking intensities having a 10% chance of being exceeded over 10-, 50-, and 250-year exposure periods. Figure 35-3 is a visual schematic of the process used to estimate expected damage to structures and associated casualties to persons occupying the damaged structures. Details are given in Horton (1993).

The analysis of the earthquake ground-shaking hazard used in these studies combines a probabilistic assessment of seismic sources and an analysis of seismic attenuation from sources to sites with two models of earthquake ground-motion amplification (Emmi 1993). The chaining of multiple models amplifies uncertainty in the resulting microzonation of the ground-shaking hazard. The range of ground-shaking intensities found does conform to expectations established by earlier studies and by expert opinion. But it is difficult to judge whether the boundaries between ground-shaking intensity zones are accurately located.

It is also difficult to assess the effects that error in the location of boundary lines between intensity zones might have on the accuracy of assessment results. Needed is a technique that randomly varies the spatial alignment of boundaries between ground-shaking zones and examines the effects of this random variation on original risk-assessment results. The objective is not to refine the model's conclusions, but to see if the same conclusions about risk can be drawn when uncertainty in microzonation boundary alignments is included.

Lodwick et al. (1990) suggest the use of sensitivity analysis to assess the effects of error in databases on analytical outcomes. This is done by inserting perturbations in selected data items and then analyzing their effects on analytical outcomes. Openshaw et al. (1991) demonstrate the use of Monte Carlo simulation as a method for implementing sensitivity analysis on spatially referenced data for which the likely distribution of errors is known *a priori*. In this project, sensitivity analysis by Monte Carlo simulation is used to examine the effect on seismic loss estimates of uncertainty in the boundaries of ground-shaking intensity zones when little prior knowledge exists about the distribution of errors in the location of these lines. The continuous nature of the ground-shaking zonation is suited to a method that randomly perturbs MMI zonal boundaries within successively larger constraints. The random disturbance of boundary lines alters the ground-shaking intensities to which specific structures are exposed. If the data sets with perturbed boundaries are then processed through the risk-assessment procedure outlined in Figure 35-3, a relationship between boundary location error and risk-assessment results can be ascertained.

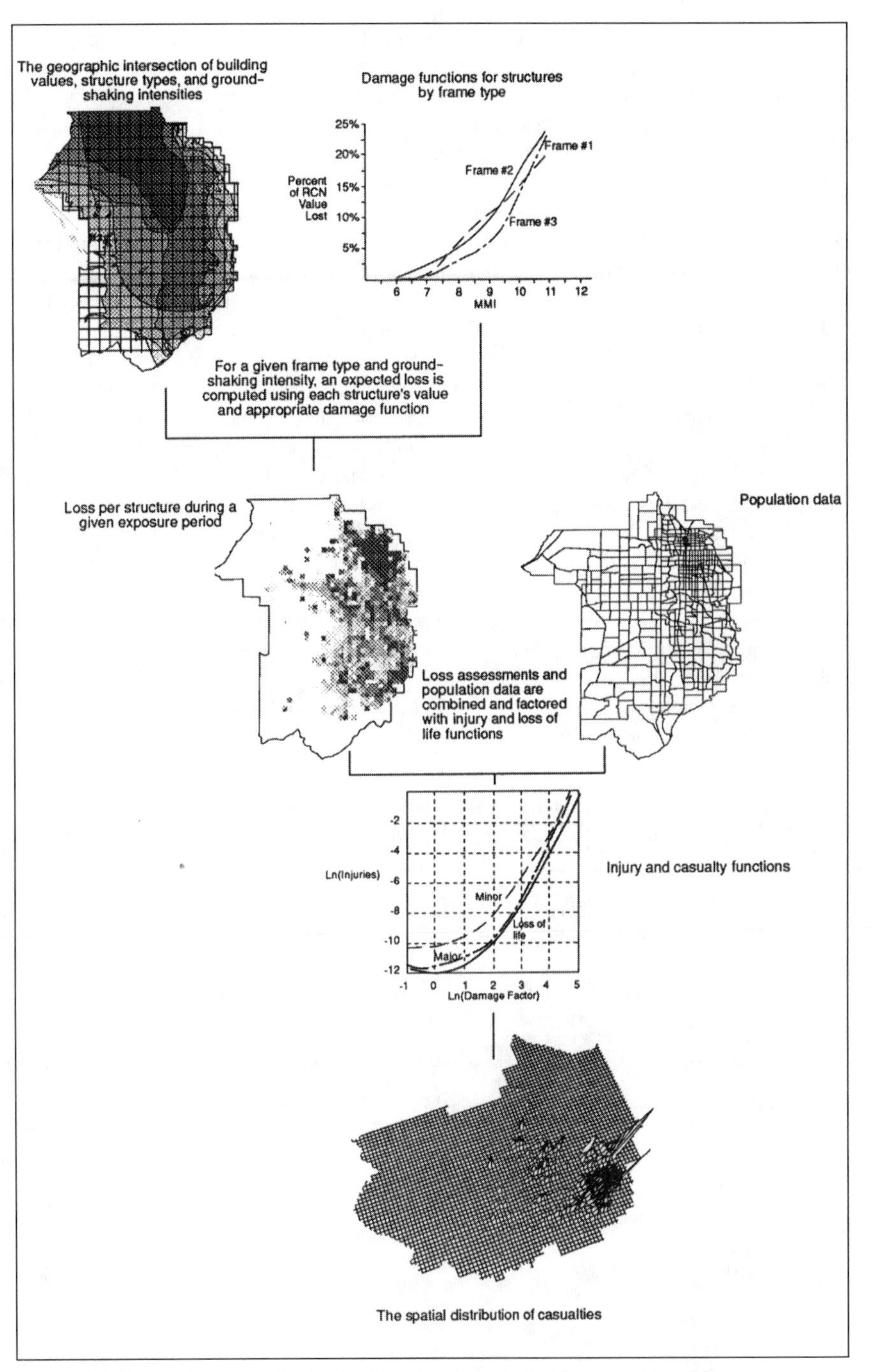

Figure 35-3. The use of data and functional relationships in a seismic risk-assessment model.

A risk-assessment accuracy requirement can be stated operationally as the physical range within which ground-shaking boundaries can fluctuate and still yield risk-assessment results that are within 5% of their original value 95% of the time. The map depicting the ground-shaking intensities to which Salt Lake County is subject over a 50-year exposure period is used as the basis from which subsequent perturbations are made. The model of seismic risk through which the effects of perturbations are assessed is the model of structural damage to residential properties only. Boundaries are perturbed randomly within bounds of 100, 250, 500, and 1,000 m of their currently mapped locations. These distances allow transitional boundaries to fluctuate within corridors that range up to 2,000 m.

Implementation of the Monte Carlo simulation is coupled with a vertex densification/line generalization algorithm. First, vertices are added at random intervals along the boundary line of each MMI zone. Vertex densified boundaries are then generalized by a line-generalization technique that maintains their approximate shape within user-defined constraints. Line generalization depends on the spacing and location of each boundary's vertices. The degree to which each boundary can vary spatially is constrained by the vertex distance value, or "weed tolerance," used in the line generalization process. Repeated application of the vertex densification/line generalization process produces a map series where the boundaries on each map are modestly different from one another and from the original map.

Weed tolerances of 100, 250, 500, and 1,000 m are specified prior to line generalization. This has the effect of allowing each boundary line to be moved randomly within preset tolerance limits. Once a cycle of random boundary perturbation is complete, the property damage model is run. New loss values are computed and compared with the original estimate. The entire process is run repeatedly: each iteration uses a new, randomly selected vertex densification interval. When enough iterations are completed, descriptive statistics can be used to compare the effect of boundary changes on risk-assessment results.

Vertex densification does not change the shape of a boundary line, but line generalization does. Each randomly densified boundary line is generalized according to a weed tolerance set first at 100 m and then at 250, 500, and 1,000 m. The Douglas–Poiker algorithm (Douglas and Poiker 1973) is used as the weed algorithm. This routine maintains the general shape, yet the prior densification allows minor fluctuation in the critical bends of each boundary. Each resultant boundary map is limited in variation away from the original boundary by the weed tolerance. Each map of similar weed tolerance is different because of randomly chosen vertex densities. The advantage of this approach is that the general shape of each line is retained, yet within prescribed limits the process randomly shifts boundary shape and location.

Fifty maps for each of the four weed tolerances are created. These serve as the basis for 200 residential property-loss estimates. This sampling is sufficient to create a randomly distributed set of risk-assessment results.

Figure 35-4 shows the results of our Monte Carlo simulations. Of particular interest is how variation in risk-assessment results increases as weed tolerance increases. The coefficient of variation in risk-assessment results is only 0.02%, when MMI boundary lines are allowed to wander from their original location by no more than 100 m. When boundary fluctuations are allowed up to 1,000 m, the coefficient of variation increases to 4.5%.

Our objective is to identify the range over which the boundary lines between ground-shaking intensity zones might wander while estimates of expected loss to residential structures remain within 5% of their original values 95% of the time. The lower portion of Figure 35-4 explains how this range is defined. There, a second-order polynomial equation is fitted statistically to four observed data points generated by Monte Carlo simulation. Each point identifies the coefficient of variation in loss assessments associated with a predefined weed tolerance. The resulting polynomial equation is then used to define a positional accuracy requirement for the location of boundary lines between MMI zones.

The coefficient of variation that corresponds to our accuracy standard is 5% divided by 1.96 standard deviations, or 0.025. The polynomial regression in Figure 35-4 shows that this value corresponds to a weed tolerance of 780 m. Thus, the positional accuracy requirement for the location of MMI boundary lines is ±780 m. In other words, these boundary lines may wander at random within a corridor 1,560-m wide and still yield residential property-loss estimates that vary from original estimates by no more than 5%, 95% of the time. This suggests that great precision in the location of ground-shaking intensity boundaries is not needed for an accurate assessment of expected risk to residential real estate.

The variation in expected loss to residential structures when the location of boundaries between ground-shaking intensity zones wander randomly within 100, 250, 500, and 1000 m of their original locations.

Location error (weed tolerance)	Mean loss (billions $)	Std. deviation (millions $)	Coefficient of variation
100 meter	$1.158	$0.276	0.02%
250 meter	$1.159	$5.521	0.44%
500 meter	$1.164	$10.011	0.90%
1000 meter	$1.171	$52.659	4.50%

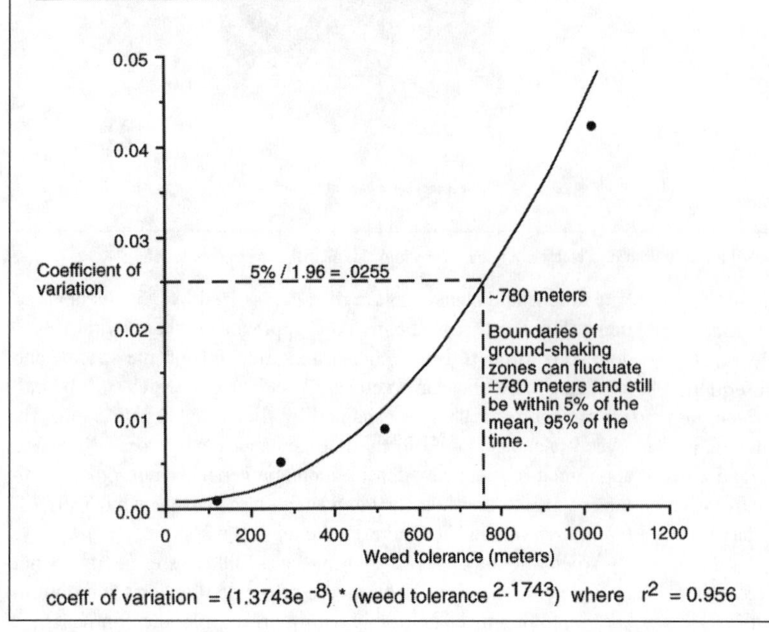

coeff. of variation = $(1.3743e^{-8}) * (\text{weed tolerance}^{2.1743})$ where $r^2 = 0.956$

Figure 35-4. The relationship between error in boundary locations and variation in loss estimates.

CONCLUSION

Risk assessment requires data on the components of risk. The spatial assessment of risk requires spatially referenced component data. The quality of risk analysis increasingly depends upon the quality of spatially referenced data.

The purpose of this study is to explore the sensitivity of a GIS-based assessment of property damage due to seismically induced ground shaking within urban Salt Lake County, Utah. Sensitivity analysis measures the response of a risk assessment to variation in its various data inputs. It helps to establish the degree of confidence one could have in risk-assessment results. It is particularly helpful when the data quality and the reliability of model specifications are uncertain. Monte Carlo simulation is used to identify the effects of error in the alignment of ground-shaking zonation boundaries. The precise location of MMI boundary lines appears not to be as important as first impressions might suggest. Assuming accurately calibrated levels of ground-shaking intensity, accuracy standards for residential property-loss assessments are not exceeded until boundaries are misplaced by more than 780 m on either side of their original location.

The increased use of digital cartographic products in public policy deliberations is cause for celebration. GIS has clearly shown its ability to produce maps of considerable significance to the quality of public discourse. Yet, the increased use of GIS products has created a need for clearer information on map accuracy. With GIS's ability to create analytically complex maps comes a commensurate complexity in the nature of map error.

In some policy issues, map error is an insignificant matter. But in issues like the siting of a radioactive waste dump or the assessment of earthquake risk, spatial variation in map accuracy is important. By identifying the effects of errors in the components of risk on assessment results, sensitivity analysis helps determine where future risk research dollars are most profitably spent. In short, information of the spatial variation in errors around mapped analytical results may be as relevant to policy as the results themselves (Rejeski 1991). The complexity of error in risk assessments and its increased relevance to policy deliberations further confirm the need for additional research into these issues.

Acknowledgments

The authors would like to thank the U.S. Geological Survey for its financial support of this research. The U.S. Geological Survey is not responsible for any errors of fact or method as these are the responsibility of the authors.

References

Alonso, W. 1968. Predicting best with imperfect data. *American Institute of Planners Journal* 7:248–55.

Douglas, D., and T. Poiker. 1973. The number of points required to represent a digitized line or its caricature. *The Canadian Cartographer* 10(2):112–22.

Emmi, P. C. 1993. A mapping of ground-shaking intensities for Salt Lake County, Utah. In *Applications of Research from the U.S. Geological Survey Program, Assessment of Regional Earthquake Hazards and Risk along the Wasatch Front, Utah,* edited by P. L. Gori, 91–120. U.S. Geological Survey Professional Paper 1519. Washington, D.C.: U.S. GPO.

Emmi, P. C., and C. Horton. 1993. A GIS-based assessment of earthquake property damage and casualty risk: Salt Lake County, Utah. *Earthquake Spectra* 9(1):11–35.

Horton, C. 1993. "Earthquake Risk: A Geographic Information System-Based Model and Sensitivity Analysis." Ph.D. diss., University of Utah, Salt Lake City, Utah.

Lodwick, W., W. Monson, and S. Svoboda. 1990. Attribute error and sensitivity analysis of map operations in geographical information systems: Suitability analysis. *International Journal of GIS* 4:412–28.

Openshaw, S., M. Charlton, and S. Carver. 1991. Error propagation: A Monte Carlo simulation. In *Handling Geographical Information: Methodology and Potential Applications,* edited by I. Masser and M. Blakemore, 78–102. Avon: Bath Press.

Rejeski, D. 1993. GIS and risk: A three-culture problem. In *Environmental Modeling with GIS,* edited by M. F. Goodchild, B. O. Parks, and L. T. Steyaert, 318–31. New York: Oxford University Press.

Wood, H. O., and F. Neumann. 1931. Modified Mercalli Intensity Scale of 1931. *Seismological Society of America Bulletin* 24(4):277–83.

Philip C. Emmi holds degrees in economics from Harvard (1967) and in urban and regional planning from the University of North Carolina at Chapel Hill (1979). He is currently the chair of the Urban Planning Program Committee with teaching responsibilities in the urban planning curriculum and the GIS curriculum. His research has focused on models in urban planning analysis and on the integration of GIS and environmental models.
Department of Geography
University of Utah
Salt Lake City, Utah 84112
E-mail: *pcemmi@geog.utah.edu*

Carl A. Horton is a 1993 graduate of the geography department's doctoral program and is currently employed as a GIS analyst with the Earth Sciences and Resources Institute of the University of South Carolina.

36

Integrated Environmental Impact Model and GIS for Oil and Chemical Spills

Deborah P. French and Mark Reed

A computer model of the physical fates, biological effects, and economic damages resulting from oil and chemical spills has been developed by Applied Science Associates, Inc. (ASA) to be used in Type A natural resource damage assessments under the Comprehensive Environmental Response, Compensation, and Liability Act of 1980 (CERCLA). Natural Resource Damage Assessment Models for Great Lakes Environments and Coastal and Marine Environments (NRDAM/GLE and NRDAM/CME) will support NOAA's damage-assessment regulations under the Oil Pollution Act of 1990. The physical and biological models are three dimensional. Direct mortality from toxic concentrations and oiling, impact of habitat loss, and food-web losses are included in the model. Estimation of natural resource damage is based on both the last value of the injured resource and on the cost to restore or replace the resource. A coupled geographical information system (GIS) allows gridded representation of coastal boundaries, bathymetry, shoreline types, and biological habitats. The user may view and edit these data, and so change the physical fates, ecological effects, and damages calculated by the model. Tools are available to import geographical data from commercially available GIS packages, such as ARC/INFO.

INTRODUCTION

There is an increasing need for quantitative and objective assessment of environmental impact (injury) and natural resource damage resulting from the release of toxic substances. Accidental spills, chronic releases, and continued contamination from historical dumping all need to be assessed for effective planning and decision making in order to minimize environmental impact and natural resource damage.

The need to compensate the public for damages resulting from chemical spills prompted the United States government to pass the Comprehensive Environmental Response, Compensation, and Liability Act of 1990 (CERCLA, also known as Superfund). This act directs that compensation be made by the responsible party for costs incurred from spills and the resulting damage to natural resources.

To assess large spills, or significant loading, numerous field studies have been conducted to measure injury. However, event-specific field efforts are expensive, time consuming, and difficult to coordinate because a number of researchers are needed in several fields. Additionally, the results are often difficult to interpret or do not accurately reflect the nature of the injuries because of a lack of baseline information and a high degree of natural variability in both time and space that can obscure pollutant-induced changes.

CERCLA provides two types of damage-assessment methodologies: (1) Type A, standard simplified procedures requiring minimal field observations; and (2) Type B, more complex and detailed studies for conducting assessments in situations that warrant further attention. For Type A assessments, numerical models developed by ASA are used: NRDAM/CME and NRDAM/GLE.

ASA has further developed the model system and developed other model components that may be incorporated into the system for other applications. These additional components address spill-response management, spill hindcasting, risk assessment, error estimation, and detailed prediction of oil–shoreline interactions. This chapter will focus on the two NRDAM systems for Type A assessments.

TYPE A ASSESSMENT MODELS

The model system contains several major components (Figure 36-1):

- physical fates
- biological effects
- compensable value
- restoration
- user interface

Physical Fates Model

The physical fates model estimates distribution (as mass and concentration) of contaminant on the water surface, on the shoreline, in the water column, and in sediment. If the density of a spilled chemical (i.e., oil) is less than or equal to that of water, the model estimates surface spreading, slick transport, entrainment into the water column, and evaporation to determine trajectory and fate at the surface. Surface slicks interact with shorelines, depositing and releasing material according to shoreline type. For modeling purposes, crude oils and petroleum products are represented by four components: two aromatic fractions considered toxic to organisms, a volatile relatively insoluble fraction, and a nonvolatile insoluble (residual) fraction. A contaminant heavier than water is modeled as a convective plume which is transported by currents and randomized dispersion. All contaminants may adsorb to particulates and settle.

Biological Effects Model

The biological effects model uses habitat-specific and seasonally varying estimates of the abundance of fish, shellfish, birds, mammals, and the productivity of plant and animal communities at the base of the food chain in the environment of the spill to determine biological injury. It is assumed that a portion of the wildlife in a surface-slick area will die, based on probability of encounter with the slick and mortality once oiled. Estimates for these probabilities are derived from information on behavior and on field observations of mortality under similar circumstances. Fish and their eggs and larvae are affected by dissolved contaminant concentration (in the water or sediment). For oil and petroleum products, dissolved aromatic concentrations are used. The movements of biota, either active or by current transport, are accounted for in determining time and concentration of exposure and resulting mortality. Organisms killed are integrated over space and time by habitat type to calculate a total kill.

Lost production of plants and animals at the base of the food chain is also computed. Lost production of fish, shellfish, birds, and mammals due to reduction in food supply is estimated using a simple food-web model. In addition to direct-kill and food-web losses (starvation) of eggs and larvae, fish young-of-the-year may be lost via habitat disruption by lethal concentrations or oiling. Losses are assumed proportional to the habitat loss. Fish and wildlife population losses are calculated as the number absent in present and future years as the result of the spill after natural and harvest mortality are subtracted. Loss of harvest and of viewing are also calculated.

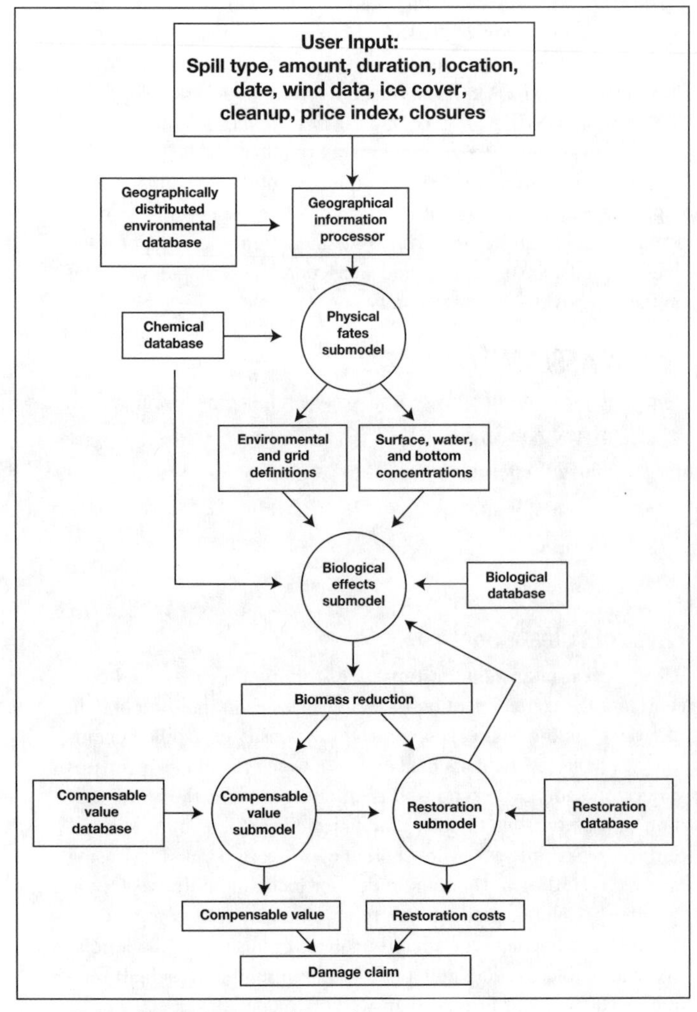

Figure 36-1. Schematic overview of the model system.

Compensable Value Model

The compensable value model utilizes consumptive-use values for fish and wildlife based on the value of the resource set in place (e.g., price at the dock or participation costs of fishing or hunting). Nonconsumptive-use values, such as viewing and photography, are also based on the cost of the activity. Damages resulting from injury to biological resources are the sum of these values, multiplied by the number of unavailable animals due to the spill. Additional damages result from the loss of beach use and recreational boating. These damages are based on the cost of participation and the participation rate (users per day) in the affected area.

Restoration Model

Direct restorative action with habitats and restocking of fish and wildlife are assumed performed if they reduce injuries resulting from the spill. The net injury represents services lost due to the spill. The value of these lost services is the compensable value portion of the total damages. Restoration costs are added to compensable values to estimate total damages.

User Interface

The model system is implemented on an IBM PC with a VGA color monitor. A user-friendly menu system (both keyboard and mouse driven) allows the user to view and enter data, run the model, and view model output. Geographical data and model output are mapped and animated on the color screen and can be printed; tabular information can be scanned. Data are entered into pull-up forms or are mouse driven for graphic information. Locations of site-specific information (such as critical habitats or response equipment) are indicated by colored dots or icons on a map. This information may be interrogated and updated by the user using a windowing system into a textual database.

CONCLUSION

ASA's model system can be used to quantify physical fates, biological impacts, and economic damages resulting from the release of pollutants into aquatic environments for a variety of purposes: (1) to focus on response efforts after a spill, (2) to estimate maximum liability for a spill, (3) to evaluate management strategies and results obtained, and (4) to educate the public about the impact of a chemical spill.

Deborah French has been a senior scientist with Applied Science Associates, Inc., since receiving her Ph.D. in biological oceanography from the University of Rhode Island in 1984. Dr. French was coprincipal investigator in the development of the first CERCLA type A damage assessment model for marine waters and is the designer of the biological effects and restoration components of the second generation GIS-based version of this model, for both the Great Lakes and United States coastal waters (i.e., an updated version of the original marine model). She has managed numerous projects, with primary emphasis on realistic, dynamic simulation of biological behaviors juxtaposed with physical fates and effects models.

Applied Science Associates, Inc.
70 Dean Knauss Drive
Narragansett, Rhode Island 02882
Phone: (401) 789-6224

37

Integrated GIS and Distributed Storm Water Runoff Modeling

Baxter E. Vieux, Nadim S. Farajalla, and Nalneesh Gaur

The research presented herein reviews recent efforts at integrated modeling of stormwater runoff and the errors propagated in hydrograph response due to attribute errors. Spatially variable infiltration parameters in the Green and Ampt infiltration equation must be captured by a given grid cell resolution. A soil map of the Little Washita watershed in southwest Oklahoma is used to investigate the effects of grid cell resolution and error propagation on distributed modeling of infiltration. Spatial distribution of attribute error gives insight into the uncertainty associated with the Green and Ampt infiltration parameters. Hydrographs are simulated using the model r.water.fea, *which is a finite element simulation model integrated with GRASS, a raster GIS. Distributed flow depth maps are simulated using parameter maps of the hydraulic conductivity and other infiltration parameters derived from soil properties. A four-fold increase in peak discharge results from a 50% decrease in the values of hydraulic conductivity contained in a distributed map. Distributed parameter maps ensure that the relative contributions of clayey and sandy regions are present in the simulation. Thus, distributed modeling using GIS is particularly useful when modeling spatially variable rainfall.*

INTRODUCTION

Hydrologic simulation at the river basin scale has historically lumped rainfall, infiltration, and other hydraulic parameters. With the advent of distributed modeling, the basin is subdivided into computational elements at a scale smaller than the basin. Storm runoff is composed of subprocesses affected by many spatially and temporally variable parameters. Without suitable information management techniques, distributed models quickly become cumbersome in terms of the number of input parameters. Distributed simulation models must address management of the voluminous input parameters necessary for topographically based models.

Geographic information systems (GIS) are becoming an important part of many professions, including hydrology, as an enabling technology. Because natural resource data are commonly available in raster format, for example, digital elevation models (DEMs), the focus of this chapter will be on the application of the raster GIS to distributed simulation of storm water runoff. Geographic Resources Analysis Support System (GRASS), a raster-based GIS, developed by the U.S. Army Construction Engineering Research Laboratory (USA-CERL) is used in this study.

Background

A fully integrated simulation model and GIS allow the user to simulate spatially variable parameters without lumping. This allows the utilization of spatially and temporally variable rainfall fields coupled with soils, land use, and digital elevation data for flood prediction. Given the complexities of storm water runoff and other hydrologic processes, highly efficient computer storage and information management techniques are essential to distributed modeling at the river basin scale.

Vieux (1988) and Vieux et al. (1990) presented a finite element simulation of overland flow and applied it to modeling direct surface runoff from a subbasin without channel routing. Vieux (1991) presented this application using a triangulated irregular network (TIN) to supply land surface slope to the finite element model. The advantages of integrating distributed numerical models with a GIS include calculation and display of runoff flow depths distributed throughout the subbasin. The integration of this finite element solution with GRASS resulted in the model, *r.water.fea* (Vieux and Gaur 1994). Internal integration allows "seamless" simulation of hydrograph response using flow networks derived from DEMs.

Assessing the largest cell size that captures the spatial variability of each model parameter is important to distributed modeling of hydrologic processes. The idea "smaller is better" may cause wasted computer storage and computational time when applied to grid cell resolutions in a GIS. A larger cell resolution may suffice if it produces essentially the same results as smaller-sized cells. The following section explores the effects of error related to grid cell size in spatially variable parameter maps and the effects on hydrologic modeling.

Entropy

Shannon and Weaver (1964) first introduced the concept of entropy and information content for digital signal processing. Entropy or information content applied to a map becomes a measure of the spatial variability. The rate of change of entropy with different resolutions is the Hurst coefficient. Entropy is a measure of how many categories are present in a parameter map at a given sampling interval or grid cell resolution. If the resolution is too large, categories will drop out of the ensemble affecting the hydrologic and infiltration model results. The rate at which entropy changes (Hurst coefficient) with cell resolution depends on the spatial variability. Thus, the

Hurst coefficient is an indication of adjacency. That is, as the grid cell resolution becomes coarser, if the same categories are found, then the parameter surface is not spatially variable.

Leopold and Langbein (1962) discussed how the thermodynamic entropy, concerned with order and disorder, could be used to describe the probability and improbability of an observed state. They applied the second principle of thermodynamics to longitudinal profiles of rivers and stream networks. Huang and Turcotte (1989), using spectral image analysis techniques, compared the topography of Arizona to synthetic images and found that the topography over very large areas is nearly Brownian (H = 0.48).

Errors may be propagated due to resampling and other spatial filtering of topography. Vieux (1993) studied the impacts of smoothing and aggregation on a DEM and the resulting error in modeling surface runoff. The loss of entropy was related to hydrograph errors in the finite element solution of the kinematic wave equations. Furthermore, for a catchment with H = 0.68, the logarithm of error and relative entropy loss were found to have a linear relationship for a small ungauged watershed near Rapid City, South Dakota. This error was due primarily to a flattening of the apparent watershed slope caused by spatial filtering (smoothing). Vieux and Farajalla (1994) analyzed the effects of smoothing hydraulic roughness data on overland flow modeling. Because of the uniformly random distribution of field-measured values, resampling at larger resolutions did not change the ensemble of values. Therefore, only small errors were propagated. For this kind of parameter distribution, any sampling interval or grid cell resolution captures the essential spatial variability.

In order to test the ability of entropy loss to predict error propagated in infiltration parameters, Farajalla and Vieux (1994) measured the change in entropy with grid cell resolution. Each Green and Ampt parameter derived from the soils map was determined at increasing grid cell resolution. The graph of entropy versus grid cell resolution is shown in Figure 37-1. Two regions are evident, one at small resolution where information content does not change and another where information content decreases rapidly. The boundary between these two regions, though not sharp, is a critical resolution beyond which significant entropy is lost.

The spatial variability of maps may be assessed by computing a spatial variability measure (SVM). A maximum entropy parameter surface, where I = log(N), decreases in information content with coarser cell resolution because the number of cells decreases. This change does not represent a loss of information due to missing categories of parameters, but simply because the number of cells, N, decreases. The SVM is determined by normalizing the entropy of the various maps by the logarithm of the number of cells, and then subtracting this from one (log(N)/log(N)) to obtain the departure from an equal probability plane. The SVM differentiates between the loss of entropy that is due to a decrease in the number of cells, as is the case for the equal probability plane, and the loss in entropy that is due to an actual loss in the spatial variability. The spatial variability measure, SVM, is calculated by:

$$1 - \frac{\sum_{i=1}^{\beta} P_i \log(P_i)}{\log(N)} \quad (37\text{-}1)$$

where N is the number of cells; β is the number of categories; and P_i is the probability of each ith category. A value of 0 indicates that the surface represents an equal probability plane; and a value of 1 occurs in the limit as $N \rightarrow \infty$.

As can be seen in Figure 37-2, the SVM of an equal probability plane is constant (0) while the DEM and soil maps are the most spatially variable. This is also supported by the Hurst coefficient values calculated between 1,209-m and 1,500-m resolution (which represents the beginning of the steep decline in entropy). The map of infiltrated depth is shown in Figure 37-3 at several grid cell resolutions. The critical grid cell resolution of this parameter is ≈1,200 m. At grid cell sizes larger than ≈1,200 m, the mode and mean values of infiltrated depth decrease as entropy is lost. By different methods, Wood et al. (1988) have found that catchment response variance is minimal at the critical resolution of 1 km^2. They referred to this resolution as the "representative elementary area" (REA).

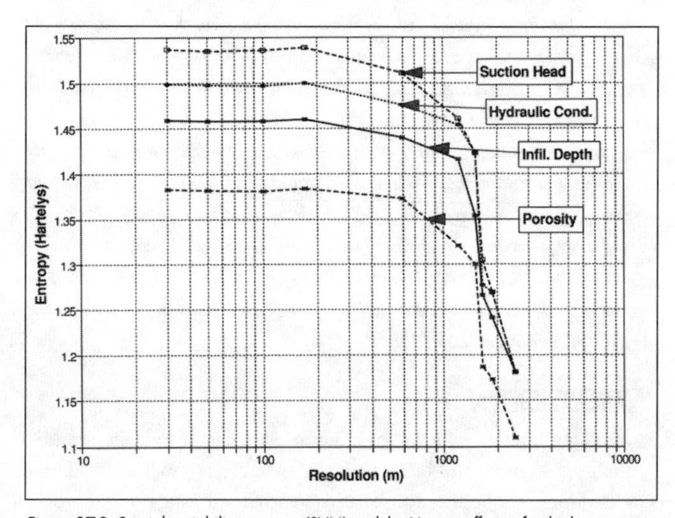

Figure 37-2. Spatial variability measure (SVM) and the Hurst coefficient for the base maps.

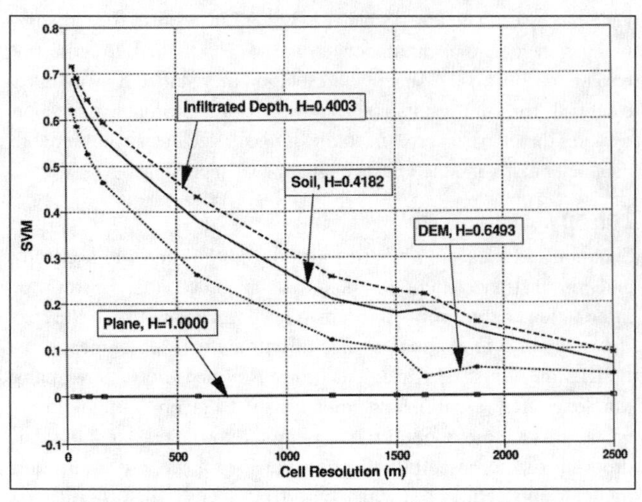

Figure 37-1. Entropy versus grid cell resolution for infiltration parameters.

Infiltration affects storm response in river basins, and its distribution pattern may control storm response at various grid cell resolutions.

Infiltration

Green and Ampt (1911) developed an equation for determining infiltration that is based on physical considerations and assumes the wetting front to be an abrupt interface between wetted and dry soil:

$$f = K\left(1 + \frac{n\, y_f}{F}\right) \quad (37\text{-}2)$$

where f is the infiltration rate, K is the hydraulic conductivity, F is the cumulative infiltrated depth, n is soil-moisture deficit, and y_f is capillary pressure head at the wetting front. Rawls et al. (1983) developed a procedure to determine Green and Ampt infiltration parameters from data available

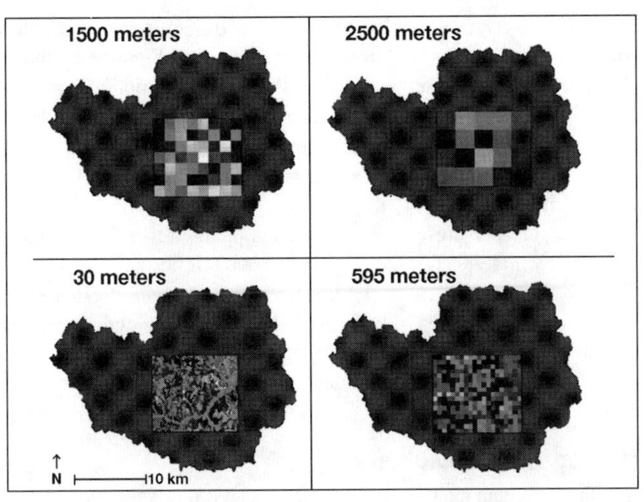

Figure 37-3. Map of infiltrated depth at various grid cell resolutions.

in published soil surveys. They constructed charts from which the Green and Ampt parameters could be determined from soil properties such as percent clay, percent sand, percent organic matter, and bulk density. Alberts et al. (1989) then presented equations to estimate the Green and Ampt parameters from the same soil properties used by Rawls et al. (1983). Using these techniques, Brakensiek et al. (1985) found that parameter estimations predicted infiltration rates and amounts to within one standard deviation of the mean calculated from field measurements. Thus, the Green and Ampt equation may be applied using soil maps and soil properties available from published soil surveys or field investigations. The methods for identifying the influence of attribute accuracy and propagated error on infiltration and hydrograph simulation are discussed next.

METHODOLOGY

The error in each map attribute or parameter propagates in the hydrograph response of a distributed model. Infiltration maps are produced from soils maps using empirical equations that relate soil properties to Green and Ampt parameters. The propagation of error is predicted using the standard error equation. The effect of this propagated error is demonstrated using the following methodology for hypothetical storms producing runoff in a large river basin.

The Little Washita River basin located in southwest Oklahoma covers 610 km^2 (Allen and Naney 1991). Published soil surveys of the watershed made by the USDA Soil Conservation Service indicate 64 different soil series with 162 soil phases reflecting differences, within the series, in the surface soil texture, slopes, stoniness, degree of erosion, and other characteristics affecting land use. Land use is mainly cropland and rangeland with small areas of timber and water bodies interspersed throughout the watershed (Loesch 1988).

Infiltration Parameter Estimation

The Green and Ampt equation is used to determine the cumulative infiltrated depth by integrating the infiltration rate with time. The Green and Ampt equation expressed in terms of cumulative infiltration, F is:

$$K_s = F - N_s \ln(1 + \frac{F}{N_s}) \tag{37-3}$$

where K_s is the saturated hydraulic conductivity, N_s is the effective matric potential, and t is time. Ahuja et al. (1984, 1989) used the generalized Kozeny-Carmen equation to determine the saturated hydraulic conductivity K_s:

$$K_s = B\,\phi_e^n \tag{37-4}$$

where ϕ_e is the effective porosity, and B and n are constants that vary with each soil. Ahuja (personal communication, 1993; Ahuja et al. 1989) recommended the following values for B and n in (37-4) for calculating K_s in cm/hr:

$$K_s = 764.5\,\phi_e^{3.29} \tag{37-5}$$

The effective porosity, in percent, was approximated by Rawls et al. (1983) as:

$$\phi_e = \phi_t - \theta_r \tag{37-6}$$

where ϕ_t is total porosity and θ_r is residual soil moisture as a fraction. The wetting front suction head ψ (in meters) as determined by Rawls and Brakensiek (1983) is:

$$\psi = e^b \tag{37-7}$$

where the coefficient b is:

$$\begin{aligned} b = {} & 6.531 - 7.33\,\phi_e + 15.8\,Cl^2 + 3.81\,\phi_e^2 + 3.40\,Cl\,Sa \\ & - 4.98\,Sa\,\phi_e + 16.1\,Sa^2\,\phi_e^2 + 16.0\,Cl^2\,\phi_e^2 \\ & - 14.0\,Sa\,Cl - 34.8\,Cl^2\,\phi_e - 8.0\,Sa^2\,\phi_e \end{aligned} \tag{37-8}$$

where Sa is the fraction of sand and Cl is the fraction of clay in the soil.

Therefore, the soil characteristic maps of clay, bulk density, sand, and organic matter are the basis from which the Green and Ampt parameters are derived. In determining propagated error, partial derivatives of the Green and Ampt parameters with respect to the soil characteristic maps were symbolically differentiated as described in the following section.

Standard Error Analysis

The standard error equation applied to mathematical combinations of maps provides a means for assessing the propagation of attribute accuracy. Provided there is no correlation between the map layers, the standard error equation is applicable to arithmetic operations in GIS. Burrough (1986) presented expressions for error propagation for arithmetic map combinations. The standard error equation for the resulting map, u is:

$$\sigma_u^2 = \left(\frac{\partial u}{\partial x}\right)^2 \sigma_x^2 + \left(\frac{\partial u}{\partial y}\right)^2 \sigma_y^2 + \left(\frac{\partial u}{\partial z}\right)^2 \sigma_z^2 + \ldots \left(\frac{\partial u}{\partial n}\right)^2 \sigma_n^2 \tag{37-9}$$

where $\sigma_u, \sigma_x, \sigma_y, \sigma_z, \sigma_n$ are the standard errors of each of the map layers; and the derivatives are with respect to each map/variable in the arithmetic combination. Though simple in form, this equation is quite complex in application because the derivatives are not spatially invariant. That is, because derivatives vary across the map, there is no single standard error formula for the map combination. In order to generate the derivatives, Mathematica® is used to symbolically differentiate equations (37-5), (37-6), and (37-7) for hydraulic conductivity, porosity, and wetting front suction head, with respect to sand, clay, organic matter, and bulk density. The derivatives were then placed in a rules file for use by *r.mapcalc* (a GRASS module) in order to produce a standard error map. The results of this analysis applied to the Little Washita follow.

RESULTS AND DISCUSSION

This section describes the propagation of error estimated for the parameters in the Green and Ampt equation and in the simulated hydrograph response. Error in the hydrograph response may be due to attribute uncertainty predicted by the standard error equation; it may also be due to grid cell resolution as measured by entropy loss. The following treats these two components of the distributed error.

Soil Reclassification

Based on soil properties, the soil map is reclassified into four distinct maps, each showing the percent clay, percent sand, percent organic matter, and bulk density distributions. For example, several different soils that have the same percentage of clay are lumped together into one category when the soils map is reclassified into a map showing the distribution of clay in the watershed. This reduction in variability results in a decrease in the maximum value of the entropy of the different soil properties. At a cell resolution of 30 m, the entropy of the original soil map is 1.60, which decreases to 1.41 for sand, 0.96 for clay, 0.76 for bulk density, and 0.62 for organic matter. Thus, when the maps of soil properties (bulk density, clay, sand, and organic matter) are reclassified into model parameters, the spatial variability and entropy decrease at a given resolution. This is due to little or no difference between soil mapping units in terms of clay content or other properties. The cell resolution (30 m) is well below the critical resolution. Thus, entropy loss is not expected to propagate error in the hydrograph response.

Standard Error Maps

Classification error is expected wherein at any particular location, the soil may not have the percent clay, sand, organic, or bulk density as indicated by the soil survey. This error propagates in the infiltration and hydrograph simulations. The standard error equation is used to find the propagation of attribute error in the estimation of the Green and Ampt parameters. The resulting maps, using the techniques described, are shown in Figure 37-4. It is apparent that given a standard error of unity in each parameter, the resulting propagated error

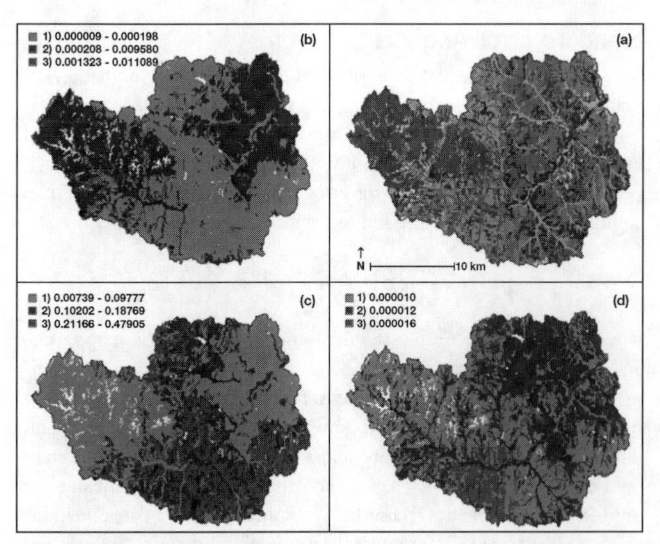

Figure 37-4. Error maps: (a) soils map, (b) wetting front suction head, (c) hydraulic conductivity, (d) porosity.

for hydraulic conductivity, which ranges from $\sigma^2(K_s) = 0.01$ to 0.5, is much larger than that for the wetting front suction head ($\sigma^2(\psi) = 0$ to 0.01). The error in porosity is the least in magnitude and variability with $\sigma^2(\phi_e) = 0.00001$ to 0.000016. The hydraulic conductivity map shows a wide variation of values with higher errors occurring in sandy areas. On the other hand, higher errors in wetting front suction tend to be in clayey areas. In terms of relative significance, $\sigma^2(K_s)$ is up to 50 larger than σ^2_{ψ}. The distributed maps in Figure 37-4 indicate the spatial distribution and the relative contribution of error by each Green and Ampt parameter. Thus, the propagated error reflects the relative significance of each parameter.

The advantage of using a distributed parameter map over lumped methods is that the relative contributions of clayey and sandy regions are present. Errors in each parameter will propagate into the hydrograph simulation. The next phase in the error analysis is to test the effects on hydrographs for a hypothetical storm. The spatial distribution of each parameter can be maintained while adjusting the map values by some percentage through calibration. To demonstrate, the K_s map is decreased 25% and 50% and hydrographs are compared.

Hydrograph Simulation

The finite element model routes the rainfall excess via overland and channel flow to the outlet of the basin. Thus, the hydrograph is a composite of the effects of infiltration and associated errors. To illustrate the effects of error in the hydraulic conductivity map, hydrographs are simulated using the K_s produced by applying (37-5) to the Little Washita soils map. A spatially constant and temporally varying rainstorm with a maximum intensity of 6.5 cm/hr is used to produce the hydrographs. The rainstorm used in this simulation is 6.5 cm/hr for forty-five minutes and 3.0 cm/hr for fifteen minutes. Channel hydraulic characteristics are given as equivalent trapezoidal shapes. Once these parameters and reasonable basin parameters for hydraulic roughness and channel characteristics are entered, two simulations take place. The first simulation is for hydrographs for each of the selected subbasins. The second takes the hydrographs from the subbasins and routes them downstream using the kinematic wave analogy.

The effect on the discharge hydrograph of varying K_s is investigated by varying the values contained in the K_s map. A total of three simulations were performed. Only the values in the map are altered for every simulation while other parameters are kept constant. The hydrograph at the basin outlet for the original K_s simulation is shown in Figure 37-5. In the second simulation, K_s values in all grid cells were reduced by 25%. The hydrograph for this simulation is shown in Figure 37-6. Similarly, in the third simulation, values in all grid cells were reduced by 50% of the original. The hydrograph at the basin outlet for this simulation is shown in Figure 37-7.

The distributed information content produced by *r.water.fea* offers the advantage of knowing where, when, and how much runoff occurs throughout the basin. The distributed map of flow depths is solved as a part of the finite element/finite difference solution and is valuable in depicting areas of high runoff and erosion potential. *R.water.fea* produces maps depicting the spatial distribution of flow depths. Flow depth maps are shown in Figure 37-8 for all three simulations at 115 minutes after the beginning of simulation. Dark areas indicate lower flow depths. The 50% reduction of K_s produces more overland flow and a more well-defined flow network. In general, a 50% decrease in the original K_s values results in over a four-fold increase in the peak of the simulated hydrographs. Besides the error in estimating infiltration parameters from soil properties, K_s has a large impact on peak discharge and runoff volume. K_s is a very sensitive parameter, as evidenced by the hydrograph simulations and has the highest error propagated due to attribute uncertainty.

CONCLUSION

The maps of propagated error indicate regions in which simulations may be in error due to attribute uncertainty. Efforts can then be

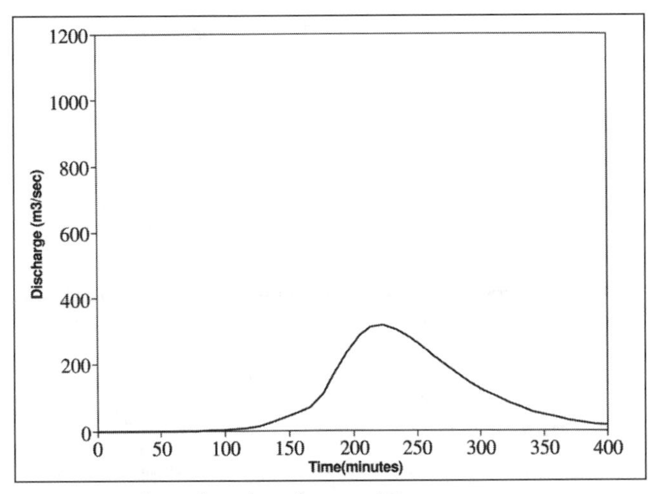

Figure 37-5. Hydrograph simulation for original K_s.

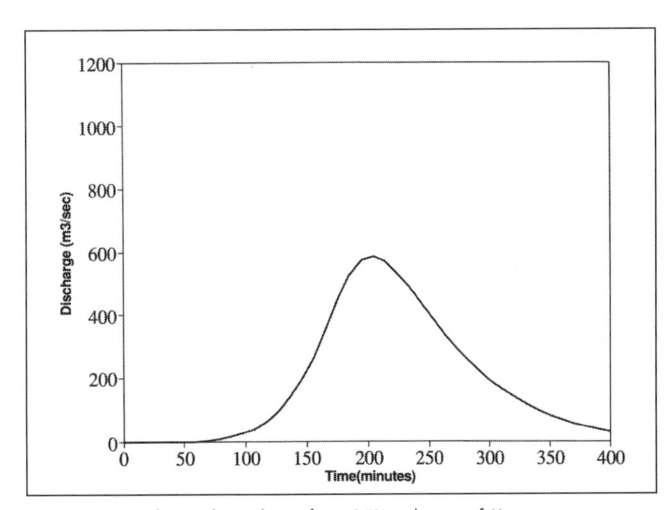

Figure 37-6. Hydrograph simulation for a 25% reduction of K_s.

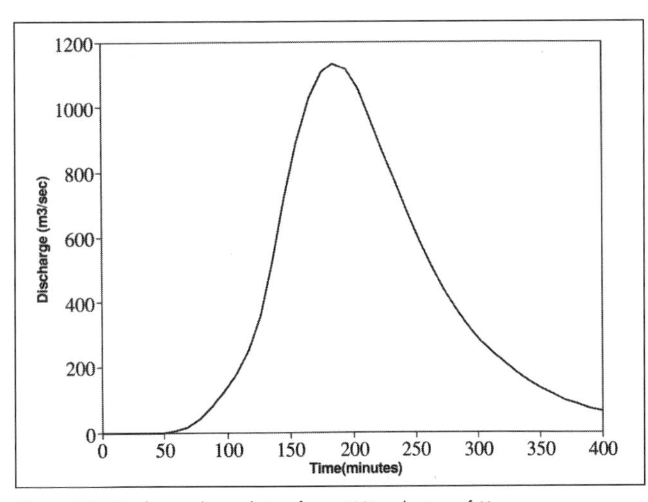

Figure 37-7. Hydrograph simulation for a 50% reduction of K_s.

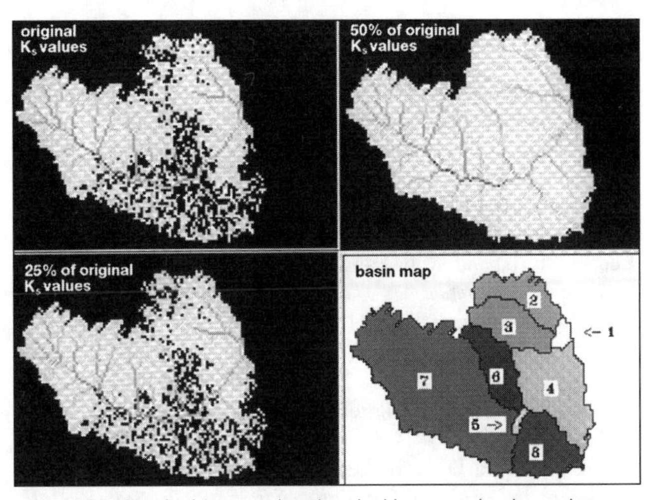

Figure 37-8. Flow depth maps and numbered subbasins used in the simulation.

applied in those regions to improve the accuracy and reduce propagated error. The magnitude and distribution of each derivative indicate whether attribute errors grow or diminish. If the derivative is larger than one over significant areas, then errors will magnify substantially in these areas. Thus, the benefit of using distributed parameter maps is that the relative contributions of clayey and sandy regions are present in the solution.

Two levels of error are present: those inherent in the uncertainty of the infiltration parameters, and those present in the hydrograph simulations. K_s is a sensitive parameter that affects the behavior of the runoff hydrograph as is evident from this investigation. K_s can be used effectively to calibrate a storm runoff model because of its sensitivity. The hydrograph is the composite of all the distributed parameters affecting the process. As such, the important question of, "What is the effect of distributed parameter uncertainty on the simulated hydrograph?" can be answered.

References

Ahuja, L. R., D. K. Cassel, R. R. Bruce, and B. B. Barnes. 1989. Evaluation of spatial distribution of hydraulic conductivity using effective porosity data. *Soil Science* 148(6):404–11.

Ahuja, L. R., J. W. Naney, R. E. Green, and D. R. Nielsen. 1984. Macroporosity to characterize spatial variability of hydraulic conductivity and effects of land management. *Soil Science Society of America Journal* 48(4):699–702.

Alberts, E. E., J. M. Laflen, W. J. Rawls, J. R. Simanton, and M. A. Nearing. 1989. Soil component. In *Water Erosion Prediction Project Manual.* USDA-Agricultural Research Service, 6.1–6.15.

Allen, P. B., and J. W. Naney. 1991. *Hydrology of the Little Washita River Watershed, Oklahoma: Data and Analyses.* Durant: U.S. Department of Agriculture, Agricultural Research Service.

Bogard, V. A., A. G. Fielder, and H. C. Meinders. 1978. *Soil Survey of Grady County, Oklahoma.* USDA–Soil Conservation Service.

Brakensiek, D. L., W. J. Rawls, and C. A. Onstad. 1985. Evaluation of Green and Ampt infiltration parameter estimates. In *Proceedings of the Natural Resources Modeling Symposium,* edited by D. G. DeCoursey, 339–44. Pingree Park.

Burrough, P. A. 1986. *Principles of Geographical Information Systems for Land Resources Assessment.* Monographs on Soil and Resources Survey No. 12, 103–35. Oxford Science Publications.

Farajalla, N. S., and B. E. Vieux. 1994. Capturing the essential spatial variability in distributed hydrological modeling: Infiltration parameters. *Hydrological Processes* 9(1):55–68.

Green, W. H., and G. A. Ampt. 1911. Studies in soil physics: The flow of air and water through soils. *J. of Agricultural Science* 4:1–24.

Huang, J., and D. L. Turcotte. 1989. Fractal mapping of digitized images: Application to the topography of Arizona and comparison with synthetic images. *J. of Geophysical Research* 94:7,491–95.

Leopold, L. B., and W. B. Langbein. 1962. *The Concept of Entropy in Landscape Evolution.* U.S. Geological Survey Professional Paper 500-A. Washington, D.C.: U.S. GPO.

Leosch, T. N. 1988. "The Use of Satellite Remote Sensing and Geographic Information Systems in a Rural Watershed Hydrologic Model." Master's thesis, University of Oklahoma, Norman.

Rawls, W. J., and D. L. Brakensiek. 1983. A procedure to predict Green and Ampt infiltration parameters. *Proceedings of the American Society of Agricultural Engineers Conference on Advances in Infiltration*, 102–12.

Rawls, W. J., D. L. Brakensiek, and N. Miller. 1983. Green and Ampt infiltration parameters from soil data. *Journal of Hydraulic Engineering* 109(1):62–70.

Rawls, W. J., D. L. Brakensiek, and B. Soni. 1983. Agricultural management effects on soil water processes: Soil water retention and Green and Ampt infiltration parameters. *Transactions of the American Society of Agricultural Engineers* 26(6):1,747–52.

Shannon, C. E., and W. Weaver. 1964. *The Mathematical Theory of Communication.* Urbana: University of Illinois Press.

Vieux, B. E. 1988. "Finite Element Analysis of Hydrologic Response Areas Using Geographic Information Systems." Ph.D. diss., Department of Agricultural Engineering, Michigan State University, East Lansing.

———. 1991. Geographic information systems and nonpoint source water quality and quantity modeling. *Hydrological Processes* 5(1):101–13.

———. 1993. DEM aggregation and smoothing effects on surface runoff modeling. *Journal of Computing in Civil Engineering: Special Issue on Geographic Information Analysis* 7(3):310–38.

Vieux, B. E., V. F. Bralts, L. J. Segerlind, and R. B. Wallace. 1990. Finite element watershed modeling: One-dimensional elements. *J. of Water Resources Planning and Management* 116(6):803–19.

Vieux, B. E., and N. S. Farajalla. 1994. Capturing the essential spatial variability in distributed hydrological modeling: Hydraulic roughness. *Hydrological Processes* 8(3):221–36.

Vieux, B. E., and N. Gaur. 1994. Finite element modeling of storm water runoff using GRASS-GIS. *J. of Microcomputers in Civil Engineering.*

Wood, E. F., M. Sivapalan, D. Thongs, K. Bevan, and L. Band. 1988. A DEM-based model for catchment storm response using catchment morphology. *Eos Trans. AGU* 69:1,224.

Baxter E. Vieux
University of Oklahoma
202 West Boyd Street
Norman, Oklahoma 73019
Phone: (405) 325-3600
Fax: (405) 325-7508
E-mail: *vieux@chief.ecn.uoknor.EDU*

38

Implementation of a Distributed Hydrologic Model within GRASS

Bahram Saghafian

A number of strategies in conjunctive use of hydrologic models and GIS are outlined. By using database structure and functions of GRASS-GIS, this particular effort follows the "full-integration" strategy by reprogramming CASC2D distributed parameter hydrologic model within GRASS-GIS. Several advantages gained in fully integrating these two components are highlighted. While major features of the hydrologic model CASC2D are briefly described, the input requirements and output options in relation to GRASS are discussed in more detail. Some general issues in conjunctive use of hydrologic models and GIS are also addressed.

BACKGROUND

Nowadays, hydrologists have turned their attention to geographic information systems (GIS) for assistance in studying some of the natural and man-made processes that describe the movement of water within the hydrologic cycle. Hydrologic models, particularly those distributed in nature, deal with the spatial distribution of numerous watershed and rainfall characteristics. As such, GIS can play a significant role in facilitating the treatment of the spatial dimension.

Several of the advantages gained by using GIS in the pre- and postprocessing of spatial data to and from hydrologic models have been exploited in a number of applications. The influence of spatial aggregation on runoff has been reported by Mancini and Rosso (1989), who investigated the spatial variability of the SCS curve number using a raster GIS and a distributed model. Leavesley and Stannard (1991) took GIS outputs defining the boundaries and associated parameter values of hydrologic response units (HRU) as input to the Precipitation Runoff Modeling System (PRMS). Remotely sensed snow cover data were also used in conjunction with the GIS to verify PRMS results predicting snowmelt processes. Steube and Johnston (1990) compared the performance of GIS-derived results with manually crafted methods in runoff simulation using the SCS curve number approach. GIS was found to be an acceptable alternative, greatly simplifying data preparation and handling.

In a review of water quality and quantity modeling and GIS applications in water resources, Vieux (1991) used a GIS-based Triangulated Irregular Network (TIN) to process the terrain of a small watershed for further application of the finite element solution to the kinematic overland flow equations. It was stressed that GIS can combine the model results with other map coverages to allow comparison of cause-and-effect relationships. Recently, Rewerts and Engel (1993) developed a hydrologic toolbox in which spatial data needs of ANSWERS (Areal Nonpoint Source Watershed Environmental Response Simulation) model were served by GRASS (Geographic Resources Analysis Support System) GIS. A project manager program was designed as a user interface to simulation scenario management and visualization.

Maidment (1993) identified four different levels of hydrologic modeling in association with GIS: (1) hydrologic assessment, (2) hydrologic parameter determination, (3) hydrologic modeling inside GIS, and (4) linking GIS and hydrologic models. In the assessment level, hydrologic factors pertaining to some situations are mapped in GIS. Hydrologic parameter determination involves the analysis of terrain and land-cover features to yield the parameters for hydrologic models. In his description of hydrologic modeling inside GIS, Maidment (1993) limited such operations to steady state processes. It was suggested, however, that with developing space-time data structures in GIS, it would be realistic to begin thinking about performing numerical modeling "within GIS." This study attempts to cover some aspects of hydrologic modeling within the GIS framework.

APPROACH

Conjunctive Use of Hydrologic Models and GIS

A number of strategies can be inferred from available literature on conjunctive use of hydrologic models and GIS. Although the distinction may not readily be apparent, alternatives in conjunctive use can be categorized as follows:

1. Development of an independent spatial data structure to be used by a distributed hydrologic model.
2. Model input/output processing using a GIS.
3. Linkage of a model and a GIS through an interface.
4. Full integration of a model "within" a GIS.

In the first category, the mere development of a distributed parameter hydrologic model requires construction of a spatial data structure for storage, manipulation, and even visualization of input/output, such as model input parameters and output discharge quantities. Thus a poor-man GIS is built as an integrated part of the model. Recent extensive use of standard GIS in determination of hydrologic model parameters, output analysis, and visualization may be classified in the second category. Such effort involves the preprocessing of spatial input data by a GIS in a format suited for the model and occasionally the subsequent analysis of spatial output results. Determination of slope, aspect, watershed boundaries, and stream network are some examples of the preprocessing tasks being carried out using GIS. Category 3, describing the linkage of models and GIS through interfaces, is gaining popularity among hydrologists. In fact, the linkage performed in this category is an advanced user-friendly implementation of the second category. The ANSWERS interface (Rewerts and Engel 1993) is one example of establishing linkage between a hydrologic model (ANSWERS) and a GIS (GRASS). Spatial Decision Support Systems (SDSS) could be considered the ultimate product in category 3. Category 4 represents a full integration and involves programming of a hydrologic model "within GIS." This strategy encourages the model to adopt the spatial data structure of the GIS. While the model and the GIS are separate units in categories 2 and 3, GIS is a subunit of the model in category 1, whereas the opposite is true in category 4.

Except in category 1, the hydrologic model of conjunctive use could be either a lumped parameter or a distributed one. Watersheds and their properties are major spatial components of hydrologic modeling. While lumped parameter models must rely on different levels of aggregation prior to their use, the distributed watershed modeling can take full advantage of the GIS spatial operation capabilities for more accurate analysis that provides meaningful and verifiable spatial output results such as runoff discharge, surface depth, and sediment concentration. The time-varying nature of hydrological processes, however, needs to receive more attention in a GIS environment.

CASC2D-GRASS Integration

The model of this study, CASC2D, is a distributed raster-based hydrologic model and has been previously used in conjunction with GRASS-GIS to study the hydrological impacts of soil disturbances created by training exercises (Doe and Saghafian 1992). A five-step methodology was used in that study: (1) disturbance scenario development, (2) spatial characterization of the watershed in GRASS, (3) linkage of GRASS spatial data to the CASC2D hydrologic model, (4) hydrologic simulation of the scenario in CASC2D, and (5) linkage of model output to GRASS for spatiotemporal analysis.

This current effort revolves around offering a direct dynamic modeling capability to the GIS by adding a time-varying hydrologic model component (CASC2D) to the existing GIS modules, thus providing the GIS toolbox with an environmental simulation capability. While GIS benefits from such full integration by adding to its growing assets, the performance of the hydrologic model is improved by: (1) facilitating spatial data management; (2) taking advantage of data sets already prepared to perform other environmental studies using GIS; (3) gaining direct access to the GIS library for data resampling, manipulation, etc.; and (4) employing the GIS visualization modules for display purposes. The GIS umbrella for this effort is GRASS.

It is of significance to mention that the CASC2D-GRASS full integration is aimed at preparing the foundation for sediment transport simulations.

GRASS is a public domain, image processing GIS, written in the C programming language that runs under the UNIX operating system. GRASS was originally developed by researchers at USACERL to assist land managers at military installations. At present, GRASS is used by a wide variety of public and private agencies, some of which have developed and contributed a number of GRASS programs. Data in raster format is of particular interest because GRASS is well developed to handle such data.

There are several commands in GRASS for direct terrain analysis that may be useful for hydrologic models. The raster command *r.slope.aspect* takes digital elevation models (DEM) as input and generates raster map layers of slope and aspect. *R.watershed* is a program that delineates subwatersheds and stream network in a given geographic region. *R.drain* can trace a flow line through an elevation raster map. *S.surf.tps* is a site program which interpolates site data, computes tangential and profile curvatures, and generates a raster map layer. The interpolation is accomplished using spline with tension. This program may be particularly useful in DEM generation.

HYDROLOGIC MODEL CASC2D

General Model Features

CASC2D is a physically based rainfall runoff watershed model. Square grid elements are used to represent the distributed watershed and rainfall domains. Although spatial variability is allowed from one element to the next, each element is assumed to be a homogeneous unit. The primary features of the current version of the model include an advanced soil moisture accounting, primarily based on the Green–Ampt infiltration model, and a two-dimensional diffusive wave overland flow routing coupled with a one-dimensional diffusive wave channel routing. The numerical technique used to solve the continuity equation and the diffusive form of the momentum equation is an explicit finite difference method.

The existing formulation of the point infiltration component allows for the continuous simulation of multistorm events by computing soil moisture content at all times. The main input parameters used to activate the infiltration module are the saturated hydraulic conductivity, the capillary suction head, the effective porosity, and the initial soil moisture. Other input data may be necessary depending upon whether one wishes to model the soil desaturation during low-intensity and/or no-rainfall periods. At each time step iteration, prior to the first surface ponding, the soil moisture is incremented by the prorated soil moisture deficit at the present time. The rating factor is based on the ratio of the time step duration to the estimated ponding time that corresponds to the current soil moisture and rainfall intensity. Once ponding is reached, the original Green–Ampt equation is applied to compute the infiltration rate while the surface soil moisture stays at saturation. If and when the rainfall intensity falls below the saturated hydraulic conductivity, the water profile is redistributed to account for soil desaturation. This redistribution is performed based on the application of Darcy's law as indicated by Smith et al. (1993). Once a new burst of storm arrives and the rainfall intensity increases again, a new water profile is formed within the soil in addition to the still redistributing first-water profile. This second profile is fed by a rate given by the Green–Ampt equation in which the moisture deficit is calculated relative to the moisture of the first profile. These two profiles join when the depth of the second profile exceeds that of the first profile.

Once the excess surface water depth at any given time step is determined at all grid cells by removing infiltration losses, the overland surface depth is routed in two dimensions depending upon water surface slope. The model allows the simulation of runon, which occurs when surface runoff from upstream cells infiltrates in a pervious downstream cell. Where the overland surface runoff meets a channel segment embedded in a cell, it becomes part of the channel flow. The model can also simulate overbank flow by connecting the flow over the floodplain to the overland flow routing component. A complete description of flow routing in CASC2D can be found in Julien and Saghafian (1991) and Saghafian (1992).

Input Description

The input data necessary for model simulations can be classified into three major categories: (1) spatial, (2) temporal, and (3) parametric. While influenced by the selected grid resolution, most of the spatial data are stored and manipulated in GRASS. Such data include the following items:

1. Watershed boundary (mask) map: This is optional; however, failure to provide the shape map would result in extra computations for the elements outside the desired watershed boundary. The *r.watershed* program in GRASS may be used to delineate the watershed of interest.

2. Elevation map in DEM form: Where raster DEM is not available, digitized contours can be used to generate the DEM using *s.surf.tps* in GRASS. Noise and depression removal are generally not required because the model can handle water accumulation and backwater effects, if any, due to its diffusive wave nature. It is recommended, however, that unreal depression areas be filled to prevent excessive water accumulation that otherwise would contribute to the discharge at the watershed outlet.

3. Surface roughness map: If proper correlation between the vegetation cover map and surface roughness is established, *r.reclass* of GRASS may be used to reclassify the vegetation map and generate the roughness map. If no such map is provided, the watershed is assumed to be uniform with respect to overland surface roughness, requiring the input of a single roughness value.

4. Soil infiltration parameter maps: To compute infiltration losses using the original Green–Ampt equation, four maps are necessary: saturated hydraulic conductivity, capillary suction head at the wetting front, effective porosity, and antecedent soil moisture content. The soil textural maps can assist in construction of the first three maps using available tables (Rawls et al. 1983). As such, the *r.reclass* command in GRASS may be used to produce soil parameter maps from the soil textural maps. The maps of soil pore-size distribution index and residual soil moisture content are also required if soil desaturation is to be simulated.

5. Channel network and cross-sectional properties: Channel network information should describe network spatial layout and connectivity. Cross-sectional geometry data and a surface roughness value must be assigned to each channel segment. *R.watershed* of GRASS can assist in delineating the stream network.

6. Rain gauge network: The position of each recording rain gauge in a network in or nearby the watershed must be provided if simulation of such rainfall data is desired.

7. Outlet location: This is specified in terms of easting and northing of the outlet in the UTM coordinate system.

The temporal data mainly include the rain gauge rainfall data which are provided via an ASCII file. The parametric model inputs are computational time step duration, rainfall duration, and total simulation time. For simulating uniform watersheds and/or uniform rainfall events, the parametric input data may also include, wherever appropriate, the uniform values of overland surface roughness, soil infiltration parameters, and rainfall intensity.

Output Description

The user can select several computed variables to be saved as output results for further analysis and visualization. These outputs may include discharge hydrograph at the outlet and time series of raster maps describing surface depth, rainfall intensity, infiltration depth, infiltration rate, and soil moisture content at the surface. The raster maps in a given time series have consecutive extensions. For further analysis, the time series of raster maps can be visualized and put into animation using certain display capabilities of GRASS. Advanced three-dimensional multisurface visualization and animation are also possible in GRASS using the SG3D module, a specially developed graphics program for Silicon Graphics computers. When utilizing SG3D, various surfaces including rainfall intensity, surface depth, and infiltration depth may be displayed while aligned with one another in vertical position.

Model Improvements

Addition of a major option for channel routing computations is currently being implemented. An implicit channel routing algorithm would enhance the capability of the hydrologic model in simulating internal boundary conditions such as bridges, culverts, etc. The trend in further development of the model is being directed toward simulation of sediment transport.

GENERAL ISSUES

There are a number of issues which need to be addressed in conjunctive use of hydrologic modeling and GIS in general and in attempting to fully integrate distributed hydrologic models with GIS in particular. Most of the issues are related to the current weaknesses of GIS, while others deal with numerical techniques used in the hydrologic models, and still others root from the errors in spatial data. One may express these issues as follows:

1. Elevation data quality: The quality of USGS DEM has recently caused some concern in the field of distributed hydrologic modeling. In particular, the error in DEM may generate unrealistic depression areas which kinematic-based models, for instance, cannot handle. The existence of error in a DEM will also introduce inaccuracy in delineation of subwatersheds and, most importantly, stream network. However, it is important to realize that some depression areas are real and do exist in watersheds; therefore, the choice of hydrologic model to simulate such depressions must be at least a diffusive-based model.

2. Transition from raster to vector operations: Two-dimensional overland flow routing using the finite difference approach is well suited to raster environments, while one-dimensional channel routing may best be performed in a vector domain. The transition from raster to vector operations needs special attention in models that operate as such. In the current version of the CASC2D model, the channel network is known by identifying the raster cells that contain a channel segment. The length of this segment is assumed to be equal to cell size, but work is underway to incorporate the specified actual length of segment in each cell as a cell attribute.

3. Temporal variability: Currently most standard GIS, including GRASS, have no straightforward way of dealing with temporal variations. Distributed hydrologic models can typically generate numerous spatial map layers through time. Storing all such map layers in the same spatial database may be confusing unless some way of identifying the time attribute is established.

4. Conjunctive ground and surface water simulations: Conjunctive distributed surface and ground water simulations within GIS require the existence of a multilayer (or multidimensional) data handling capability. Development of a temporally varied multilayer database in a GIS greatly facilitates the integration of linked surface and subsurface models with GIS.

5. *Float/zero data handling:* This particular problem pertains to GRASS, which currently is not capable of handling floating point input/output data. Therefore, ASCII maps containing floating point values must be converted to integer maps by multiplying the maps by certain factors before they can be imported into GRASS. The multiplication factors may have to be known in advance to GRASS modules operating on converted data. Moreover, in producing output maps containing floating point variables, still other multiplication factors must be applied. This method of handling floating point data is rather tedious and sometimes confusing, considering the number of maps involved in distributed hydrologic modeling. Also zero value data in GRASS are generally interpreted as no data. Fortunately, both floating point and zero data management capabilities are being designed for future GRASS releases.

CONCLUSION

A number of strategies may be implemented to assist hydrologists in conjunctive use of hydrologic models and GIS. One of the strategies revolves around the full integration of distributed hydrologic models with GIS by programming a model within GIS and thus making the model a member of the GIS toolbox. The effort in this chapter capitalizes on the raster-based nature of the hydrologic model CASC2D as a suitable candidate for integration with a raster GIS such as GRASS. The model CASC2D is briefly described and its input data requirements in relation to GRASS are discussed. Some of the issues related to conjunctive use of hydrologic models and GIS are also outlined.

Acknowledgments

This work has been supported in part by an appointment to the Research Participation Program at the U.S. Army Construction Engineering Research Laboratories (USA-CERL) administered by the Oak Ridge Institute for Science and Education through an interagency agreement between the U.S. Department of Energy and USA-CERL.

References

Doe, W. W., and B. Saghafian. 1992. Spatial and temporal effects of army maneuvers on watershed response: The integration of GRASS and a two-dimensional hydrologic model. *Proceedings, 7th Annual GRASS GIS User's Conference.* Denver.

Julien, P. Y., and B. Saghafian. 1991. *CASC2D User's Manual: A Two-Dimensional Watershed Rainfall Runoff Model.* Civil Engineering Report No. CER90-91PYJ-BS-12, Fort Collins: Colorado State University.

Leavesley, G. H., and L. G. Stannard. 1991. *Application of Remotely Sensed Data in a Distributed Parameter Watershed Model.* Denver: USGS.

Maidment, D. R. 1993. GIS and hydrologic modeling. In *Environmental Modeling with GIS,* edited by M. F. Goodchild, B. O. Parks, and L. T. Steyaert, 147–67. New York: Oxford University Press.

Mancini, M., and R. Rosso. 1989. Using GIS to assess spatial variability of SCS curve number at the basin scale. *New Direction for Surface Water Modeling,* 435–44. IAHS Publication No. 181.

Rawls, W. J., D. L. Brakensiek, and N. Miller. 1983. Green–Ampt infiltration parameters from soils data. *J. of Hydraulic Engineering* 109(1):62–70.

Rewerts, C. C., and B. A. Engel. 1993. ANSWERS on GRASS: Integrating of a watershed simulation with a geographic information system. In *Conference Agenda and Listing of Abstracts, 8th Annual GRASS GIS User's Conference and Exhibition.* Reston.

Saghafian, B. 1992. "Hydrologic Analysis of Watershed Response to Spatially Varied Infiltration." Ph.D. diss., Colorado State University, Fort Collins.

Smith, R. E., C. Corradini, and F. Melone. 1993. Modeling infiltration for multistorm runoff events. *Water Resources Research* 29(1):133–44.

Steube, M. M., and D. M. Johnston. 1990. Runoff volume estimation using GIS techniques. *Water Resources Bulletin* 26(4):611–20.

Vieux, B. 1991. Geographic information systems and nonpoint source water quality and quantity modeling. *Hydrological Processes* 5(1):101–13.

Bahram Saghafian
Research Associate
U.S. Army Construction Engineering Research Laboratories (USA-CERL)
2902 Newmark Dr.
Champaign, Illinois 61826
Phone: (217) 352-6511
Fax: (217) 373-7222
E-mail: *bahram@zorro.cecer.army.mil*

39

Integrated Nonpoint Source Pollution Modeling System

Piotr Jankowski and Gregory Haddock

The major problems in using computer models for simulating the consequences of nonpoint source pollution are the intensive data requirements and time involved in compiling the model input file. These problems can be overcome by integrating nonpoint source pollution models with Geographic Information Systems (GIS). This chapter presents a computerized system developed by integrating GIS PC-ARC/INFO and a dynamic, event-based nonpoint source pollution model. Using the PC-ARC/INFO macrolanguage, Pascal, and batch programming, a menu-driven system was developed that integrates the Agricultural Nonpoint Source Pollution Model (AGNPS) with PC-ARC/INFO. Running on a DOS platform with 640 kilobytes of memory, the integration prototype converts a set of PC-ARC/INFO coverages into an AGNPS input data file. The output from the AGNPS model is converted back into PC-ARC/INFO coverage and can be displayed as a map using the user-selected format.

INTRODUCTION

In the last decade, several watershed modeling programs predicting the amount and content of eroded material were developed for personal computers. These applications incorporate databases holding attributes of the watershed with algorithms that generate soil-loss data. Due to the extensive data requirements, however, these programs are generally suitable for small watersheds if fine resolutions are needed.

GIS are designed to store, organize, and manipulate large amounts of spatial data such as soil type, land usage, and topography as separate georegistered coverages. Spatial and attribute data handling capabilities of a GIS are well suited for the geographically complex nature of nonpoint source pollution modeling. To increase the efficiency and extent of a nonpoint source model, vast amounts of geographical data must be synthesized and organized for processing. The synthesis of GIS and nonpoint source models is therefore needed. This chapter describes the design and implementation of an integrated nonpoint source pollution modeling system.

DESIGN GUIDELINES FOR AN INTEGRATED NONPOINT SOURCE POLLUTION MODELING SYSTEM

The main objective of this project was to create an integrated system based on a seamless link between the GIS software PC-ARC/INFO (ESRI 1990) and Agricultural Nonpoint Source Pollution Model (AGNPS) (Young et al. 1989). The integrated system should allow the user to create many input selection files for simulating different storm events and subwatersheds. It should be possible to select any coverage, including the model output coverage, and display it on the CRT monitor using the system's user interface. The architecture of the system, demonstrating linkages between the database, PC-ARC/INFO, and AGNPS model is presented in Figure 39-1 (Haddock 1992). The PC-ARC/INFO environment is the central

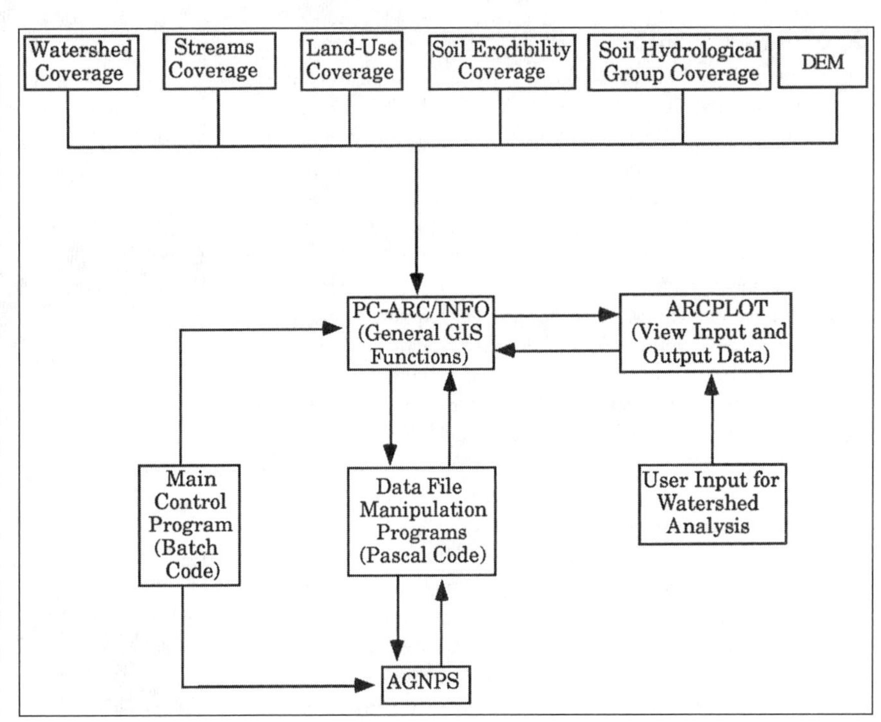

Figure 39-1. Architecture of the integration prototype and data set.

component of the prototype. The approach adopted in this project uses the AGNPS model for generating nonpoint source data output whereas all data storage, manipulation, retrieval, and display operations are carried out in PC-ARC/INFO. The PC-ARC/INFO environment is used to link the data with the AGNPS model, using the ARCPLOT interface for user access. Although PC-ARC/INFO is capable of exporting coverages to a variety of formats, there is no set of commands that will produce a data file suitable for direct AGNPS input. Separate utility programs are necessary for detailed data manipulation of PC-ARC/INFO output data. To achieve a transparent operation, these programs need to be executable within the PC-ARC/INFO environment. The utility programs can be coded in any programming language. In this project, the Pascal programming language was used to write these programs. Pascal was chosen for its structured approach to handling many data items. To create an input data file for the AGNPS model, several coverages are required. These include watershed delineation, streams, soils data, and land-use coverages. These coverages, with the digital elevation model (DEM) file, form the data set for the integration prototype. To match the coordinates within the DEM, each coverage should be in the UTM projection. The integration prototype converts these coverages to a raster-based file for cell-by-cell analysis and manipulation.

SYSTEM IMPLEMENTATION

The implementation of the system prototype consists of several programs linking the AGNPS model with PC-ARC/INFO. These programs are written in SML, Turbo Pascal version 6.0, and DOS 5.0 batch code directives. The system prototype was implemented by writing several linking programs for three reasons. First, more than one programming language was used. Second, a modular programming design makes the programming and debugging process less complicated. Third, large programs cannot be executed within the ARC/INFO environment. The batch programs call PC-ARC/INFO and AGNPS, which cannot run simultaneously.

Data Viewing Interface

The integrated system provides an interface to the ARCPLOT module of PC-ARC/INFO for viewing the coverages. This is also the system's user interface where the user examines the coverages and selects the subwatershed for AGNPS analysis (Figure 39-2). Following the guidelines of the prototype integration, this interface is menu-driven. Several options are available within this interface:

1. System Interface Draw and Clear Options: The Draw option provides basic mapping functions, which include drawing the coverages on the graphic screen. The user has the option of zooming in and out of the drawing by changing the scale interactively. The color selection also can be changed from the default settings. The polygon coverages can be shaded with the Draw option through an interactive menu system. This consists of selecting the coverage, selecting desirable attributes, and choosing the color for filling in each polygon selected. Using this option, other polygon coverages may be shaded, including coverages not listed in the master file.

Once the AGNPS analysis has been completed, the user also may view the output as a map coverage. The same Draw option, including shading, is available for the output coverage as for the other coverages. The Draw option provides the means of graphically overlaying the output coverage on the input coverage, such as the K factor coverage. This would allow the user to see which K factors yield higher soil losses.

The Clear option resets the graphics screen. This permits the user to clear everything drawn to the screen. The map scale, any changed colors, and all drawn features are also reset to the default values.

2. System Interface List and Query Options: The List option searches the attribute data of a user-specified coverage and writes the information to a text file. The file is then displayed within another window on the screen. Using this option, the user can view the entire content of the coverage attribute.

With the Query option, the user interactively selects one line or polygon drawn on the screen. The interface identifies which feature was selected and searches the coverage attributes for the information specific to this feature. Similar to the List option, the information on this feature is displayed within a window.

The List and Query options take advantage of powerful GIS functions. The Query option is more advantageous than the List option. Although it only displays a small portion of the attribute data, the Query option demonstrates the direct relationship between the geographic and attribute data.

3. System Interface AGNPS Settings Option: Several model parameters are defined from the coverage and DEM data set. Other parameters, however, are specific to the analysis and are specified by the user before running the model. This is accomplished in the AGNPS Settings option. Several alternatives are available within this option. The user can load another master file if more than one is present. The user also may create different input situations for AGNPS analysis. If there are missing coverages in the data set, the user must provide the assumed values before creating an input file.

When creating an input file, the user selects a subwatershed and enters the cell resolution. The system interface asks for several other specific details necessary for AGNPS parameter generation. This information is stored in input files linked to its parent master file. This is accomplished by creating a file with a similar name to the master file that lists the input files created by the system interface. This prevents mixing the input files with separate master files. When an input file is created within the AGNPS option, it is given the status of "currently loaded."

Once an input file is loaded, the integration prototype is ready to assemble the AGNPS input data file and run the model. This is performed with the Run AGNPS selection. The entire process of creating the AGNPS input, running the model, and creating an output coverage is done automatically.

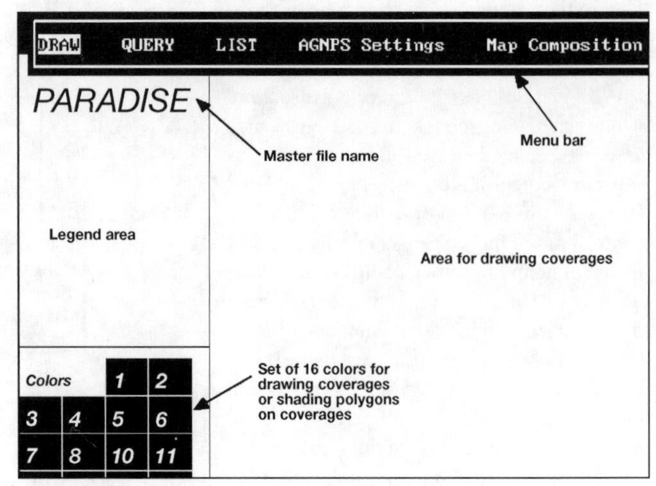

Figure 39-2. Reduced screen capture of the system interface.

4. System Interface Map Composition Option: The ARCPLOT module of PC-ARC/INFO provides several useful cartographic tools. Using the System Interface Map Composition option, the user creates map composition files from the coverages. These files may be used to create hardcopy maps or to import the graphic images into other software packages.

Using a menu, the interface prompts the user to select the coverage for mapping. The items of this coverage are presented in another menu. The user may select only one item for the map composition. The attributes that are associated with this item can be used to create thematic map compositions. Thematic maps are produced by dividing the numeric data into several classes and shading the polygons accordingly. The classes are derived from a user-specified classification technique. This includes two automatic methods to create equal-interval or equal-frequency classes and an interactive manual selection of break points. The shading patterns for each class are selected by the user interactively.

With the Map Composition option, several map elements are available. They include a title, a key, and a map scale. Using a predefined map layout, the user only needs to enter the requested information in the menu-driven interface. Given the width of the desired output page, the system automatically calculates the page height and map scale. With a prescribed page size, the user may decide to diminish the scale, but may not augment it.

Other coverages may be added to the map composition. Although shading is not possible with additional coverages, the coverage features can be drawn over the existing map. This option may provide a reference to the map audience, such as streams or roads.

Running AGNPS from the System Interface

Before running AGNPS, all missing parameters must be substituted with assumed constants. This is done under the AGNPS option. The Master File Functions option may also be used to change the current master file within the system interface. Once these considerations are taken care of, an input file can be created with the Create Input option.

The system will request that a subwatershed be pointed to with the mouse or the cursor-pointer device. The area of the watershed or subwatershed is then calculated and reasonable bounds of cell sizes will be calculated. The user then needs to enter the desired resolution. The larger the number entered (cell width and height, in meters), the coarser the resolution; hence, the less overall cells for the analysis. The minimum number of cells is thirty.

The other data prompted with the menu series is the USLE C-Factor, the USLE P-Factor, soil texture, fertilization level, fertilization availability, precipitation level, and energy-intensity level.

When the input file menu series has ended, the input file is created and ready to run. The Run AGNPS option under the AGNPS Settings option is used to run the currently selected input file. Selecting this option will invoke the model simulation. An AGNPS data file will be created with the information supplied by the user and will be used by the AGNPS model. The integration prototype uses the output data to create a coverage that can be viewed using the interface Draw option.

CONCLUSION

The crucial component of any integrated modeling system is the user interface. A well-designed user interface encourages operation through an ease-of-system use. An important side effect of the proper interface design is integration efficiency. The present system prototype provides an opportunity to use PC-ARC/INFO coverages in preparing AGNPS input data. The seamless and automatic nature of the integration allows for multiple AGNPS analysis within a short period. Finally, the menu-driven user interface grants an ease-of-use approach to watershed modeling.

In program testing, approximately 1,100 cells were processed by the integration prototype in about twenty-two minutes. This was performed on an 80386-25 Mhz machine. In one study, Wang (1991) asserts that manually compiling the data for only one AGNPS cell may take twenty minutes in itself.

The time required for the manual compilation of the data provided a strong incentive for developing the integrated modeling system. It may take several weeks to create an AGNPS data file using manual techniques. Digitizing the coverages necessary for the prototype may only take a few days. Once the coverage database is assembled, the entire process can be performed in a short time.

The effectiveness of the integration prototype also lies in its ability to model best management practices (BMPs). By making small changes in the database or AGNPS input file using the system interface, management comparisons can be made. This can help the user model several practices and predict the response from the watershed. In summary, the usefulness of this integration prototype lies in its efficiency and flexibility.

References

ESRI. 1990. *Understanding GIS—The ARC/INFO Method, PC Version.* Redlands: Environmental Systems Research Institute.

Haddock, G. 1992. "Integrating Nonpoint Source Watershed Modeling with a Geographic Information System." Master's thesis, University of Idaho, Moscow.

Wang, Y. 1991. "Application of a Nonpoint Source Pollution Model to a Small Watershed in Virginia." Master's thesis, Virginia Polytechnic Institute.

Young, R. A., C. A. Onstad, D. Bosch, and P. Anderson. 1989. AGNPS: A nonpoint source pollution model for evaluating agricultural watersheds. *Journal of Soil and Water Conservation* 2:168–73.

Piotr Jankowski is an assistant professor in the Department of Geography, College of Mines and Earth Resources, University of Idaho. His current professional activities include teaching and externally funded research in the areas of watershed modeling and spatial decision support systems for environmental management.

Gregory Haddock is a Ph.D. student in the Department of Geography, College of Mines and Earth Resources, University of Idaho. His current research is in the area of integrated hydrologic modeling at the watershed scale.

Department of Geography
College of Mines and Earth Resources
University of Idaho
Moscow, Idaho 83843
E-mail: *Piotr@IDUI1.Bitnet*

40

Hydrologic Modeling of Texas Gulf Basin Using GIS

R. Srinivasan, J. Arnold, W. Rosenthal, and R. S. Muttiah

Geographic information systems (GIS) have been successfully integrated with distributed parameter, continuous time, nonpoint source pollution model (NPS), and Soil and Water Assessment Tool (SWAT). The integration has proven to be effective and efficient for data collection and to visualize and analyze the input and output of simulation models. The SWAT-GIS system is being used to model the hydrology of eighteen major river systems in the United States as part of a project called the Hydrologic Unit Model for the United States (HUMUS). This chapter focuses on the integration of SWAT (basin scale hydrologic model) with the Geographical Resources Analysis Support System (GRASS-GIS) and a relational database management system. The system is then applied to the Texas Gulf River basin. Input data layers (soils, land use, and elevation) were collected at a scale of 1:250,000 from various sources. The average monthly simulated and observed streamflow records from 1970–1979 are presented for the six-digit basins defined by the United States Geological Survey (USGS) in the Texas Gulf basin.

INTRODUCTION

The Texas Gulf basin covers more than 80% of Texas (68.32 million ha). Ninety-seven percent of the state is nonfederal land; of this, rangeland is the largest at 61%. The terrain and climate features are diverse: desert mountains in the west have precipitation rates of 254 mm per year; the forested sections in the east have rainfall rates of 1,524 mm per year. In an average rainfall year, it is estimated that about 42% of the precipitation falling on Texas evaporates directly back into the atmosphere and about 47% is lost through plant transpiration. Only a little more than 1% of the precipitation that falls actually recharges aquifers, and the remaining 10% runs off to become streamflow in rivers and tributaries (Texas State Soil and Water Conservation Board 1991). Domestic, industrial, recreation, power generation, fish industries, and rural agriculture water demands depend on the freshwater supply from streams, reservoirs, and groundwater. Given the relatively high cost of distributing water from reservoirs and limited water supplies, agriculture most often relies on groundwater for irrigation or depends totally on rainfall.

There are fifteen major river basins and eight coastal basins in Texas, of which eighteen contribute their water yield into the Gulf of Mexico. There are approximately 3,700 streams and tributaries and 128,000 linear km of streambed. The United States Geological Survey (USGS) has divided the eighteen basins that flow into the Gulf into approximately twenty-two subwater resource regions (Figure 40-1) called six-digit hydrologic unit areas (HUA). For this study, only eighteen of the twenty-two subwater resource regions were selected. The four others were located along the coast and had inadequate detail to meet the model input requirements. Because of the importance of freshwater, it is necessary to understand how potential alterations in climate, land use, and other hydrometeorological parameters may affect water resources.

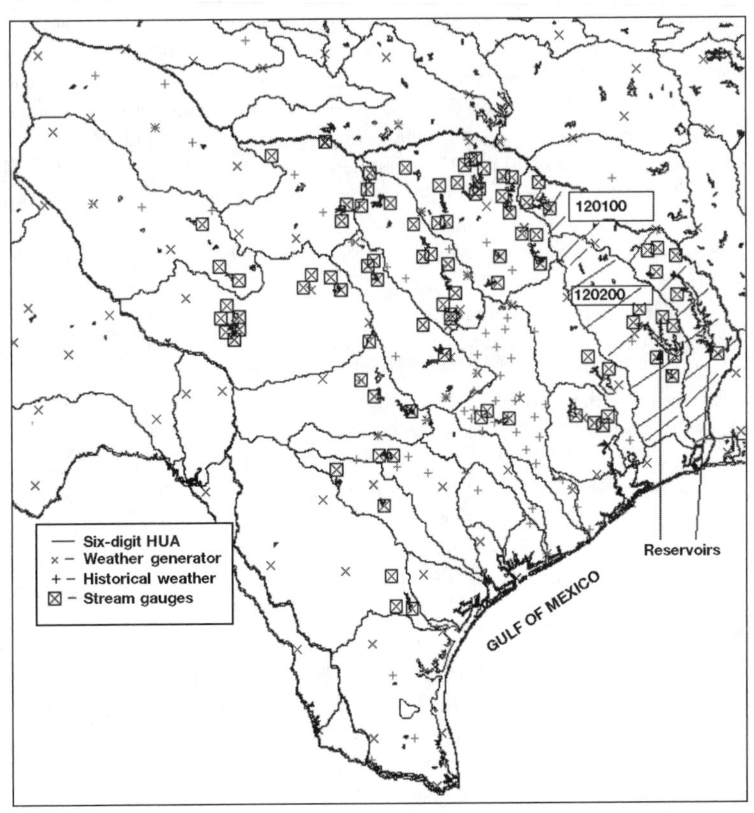

Figure 40-1. Texas Gulf six-digit HUA layer with water bodies, weather, and stream gauge locations.

The Resources Conservation Act (RCA) of 1977 requires the Department of Agriculture to appraise the status, condition, and trends in the uses and conservation of nonfederal soil and water-related natural resources. This current study addresses some of the issues related to the RCA appraisal of 1997 through the Hydrologic Unit Model of the United States (HUMUS) (Srinivasan et al. 1993).

In the past, erosion and runoff estimates were predicted using empirically derived equations, including the Universal Soil Loss Equation (USLE) (USDA 1992) and SCS curve number methods (USDA 1972). More recently, runoff, soil erosion, and chemical movement models have been based on the major processes of soil erosion and water movement such as the detachment and transport of particles by rainfall and runoff (Beasley et al. 1980; Young et al. 1989). Existing soil erosion models such as Erosion Productivity Impact Calculator (EPIC) (Williams et al. 1984); Chemicals, Runoff, and Erosion from Agricultural Management Systems (CREAMS) (Knisel 1980); Water Erosion Prediction Project (WEPP) (Foster and Lane 1987); Areal Nonpoint Source Watershed Environment Response Simulation (ANSWERS) (Beasley et al.); Agricultural Nonpoint Source Pollution Model (AGNPS) (Young et al. 1989); Simulator for Water Resource Rural Basin (SWRRB) (Arnold et al. 1990); TOPMODEL (Beven and Kirkby 1979); and Soil and Water Assessment Tool (SWAT) (Arnold et al., *Proceedings*, 1993) provide users with analytical tools that allow them to predict runoff and erosion characteristics of slopes, fields, watersheds, and channels. These models also allow evaluation of management practices that influence certain factors contributing to runoff and erosion and provide significant insight into the processes of soil erosion. However, they have a number of limitations that restrict their use.

The limiting factors of simulation models as management tools include: large data and input parameter requirements, parameters that are difficult to estimate or obtain, uncertainty in inputs, and lack of technical assistance to analyze the overwhelming amount of model outputs. Researchers have successfully shown that integration of simulation models with spatial databases and expert systems can significantly reduce the time and resources required to develop input and interpret output from simulation models (Arnold and Sammons 1989; Heatwole 1990; Srinivasan 1992; Srinivasan and Arnold 1993). Further, they have used several forms of graphical tools including GIS to visualize spatially and/or temporally varying data such as runoff and sediment yield (Srinivasan 1992).

GISs are designed to collect, manage, store, and display spatially varying data. Several nonpoint source simulation models including ANSWERS, AGNPS, TOPMODEL, and SWAT have been integrated/interfaced with GIS to enhance the use and utility of the models (Srinivasan 1992; Rewerts and Engel 1991; Texas State Soil and Water Conservation Board 1991; Chariat and Delleur 1993). This chapter describes an application of an integrated SWAT-GRASS (Geographic Resources Analysis Support System) (Shapiro et al. 1992) model to the Texas Gulf River basin. The results were reported at six-digit hydrologic units (Figure 40-1).

THE SWAT MODEL

SWAT was developed to predict the effect of alternative management decisions on water, sediment, and chemical yields with reasonable accuracy for ungauged rural basins. The model was developed by modifying the SWRRB model for application to large and complex river basins. Major changes from SWRRB involved: (1) expanding the model to allow simultaneous computations on several hundred subwatersheds, and (2) adding components to simulate lateral flow, groundwater flow, reach-routing transmission losses, and sediment and chemical movement through ponds, reservoirs, streams, and valleys. SWAT operates on a daily time step and is capable of simulating 100 years or more. Major components of the model include hydrology, weather, sedimentation, soil temperature, crop growth, nutrients, pesticides, groundwater and lateral flow, and agricultural management.

The SWAT model offers significant advantages over the combined SWRRB/ROTO (Arnold 1990) model. SWRRB routed from subbasin outlets directly to the basin outlet for simplicity. The new routing structure in SWAT is required to allow large basins to be simulated, provide more realistic routing, allow for more subbasins to be easily added, and simplify GIS linkages and database management. A set of commands is used to control the channel routing, which route and add flows through the watershed via reaches and reservoirs. The model reads each command and performs the given hydrologic command.

Total streamflow from large basins is the sum of surface runoff and groundwater flow. Groundwater flow volumes and recession periods must be simulated to accurately predict streamflow, sediment concentrations, and chemical concentrations in the streamflow. Water percolating past the root zone is assumed to recharge the shallow aquifer. Shallow aquifer components include recharge, groundwater evaporation, flow to the stream, percolation to the deep aquifer, and pumping withdrawals. The shallow aquifer also interacts directly with the streams and reservoirs through transmission losses and seepage. A detailed description of the model and model input descriptions can be found in Arnold (*Proceedings*, 1993).

Since SWAT was developed for large basins, a component to simulate water transfer between subbasins was developed. Given the reach-routing command structure, it is relatively easy to transfer water within a basin. This can account for irrigation flow paths and could provide a management tool for irrigation management districts and other agencies concerned with irrigation water rights. The algorithm developed here will allow water to be transferred from one reach or reservoir to any other in the watershed. It will also allow water to be diverted and applied as irrigation directly in a subwatershed.

THE SWAT-GIS INTEGRATED SYSTEM

The GIS tool chosen was GRASS (Shapiro et al. 1992), a public domain raster GIS designed and developed by the Environmental Division of the U.S. Army Construction Engineering Research Laboratory (USA-CERL). GRASS is a general purpose modeling and analysis package that is highly interactive and graphically oriented (both two dimensional and three dimensional). It provides tools for developing, analyzing, and displaying spatial information. GRASS is used by numerous federal, state, and local agencies as well as by private consultants.

A toolbox (modular approach) rationale was utilized in providing a collection of GIS programs to assist with the data development and analysis requirements of the SWAT model. The SWAT-GRASS input interface programs and other tools are written in C language and are integrated with the GRASS libraries. The SWAT model is written in FORTRAN 77 language, and both the interface and model run under the UNIX environment. The input-interface tools assist with preparation and extraction of data from the GIS database for use in the SWAT model (Figure 40-2). The input interface (Srinivasan and Arnold 1993) consists of three major divisions: (1) the project manager; (2) tools to extract and aggregate inputs for the model; and (3) tools to view, edit, and check the input for the model. The function of the project manager is to interact with the user to collect, prepare, edit, and store basin and subbasin information to be formatted into a SWAT input file.

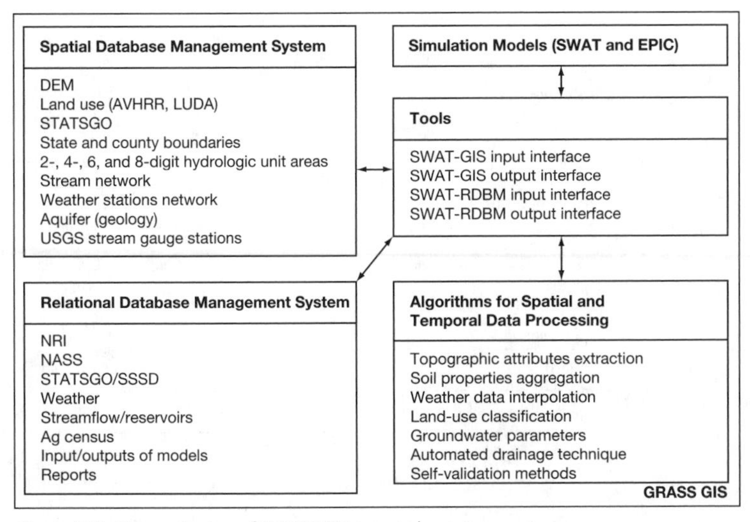

Figure 40-2. Schematic view of SWAT-GIS integrated system.

The extract and aggregate step uses a variety of hydrologic tools (Srinivasan 1992). The GIS layers that are required at this step include: subbasin, soils, elevation, land use, pesticide application, and weather network. In addition, the reservoirs, inflow, pond, and lake data can be collected directly from the user. In the third step, the user can either view, edit, or check the data extracted from the previous phase by using a subbasin number as input. There are about fifteen different data forms that can be modified by the user. The developed interface reduces the data collection and manipulation phase of watershed simulations (Rosenthal et al. 1993). The interface allows rapid modification of the various management practices and prepares the data for subsequent model runs. The interface can also be used to perform sensitivity analysis by modifying the GIS data layers and/or choosing different aggregation methods for various input data.

DATABASES

The most critical component of the SWAT-GIS integrated system is the collection of data required to run the simulation models. To model the six-digit hydrologic unit areas of the Texas Gulf, for example, the required information was historical weather, soil properties, topography, natural vegetation, cropped areas, irrigation, state and county boundaries, reservoir (stage flow) data, and agricultural practices (Figure 40-2). The SWAT model data requirement can be classified as spatial and relational. The spatial databases include: topography, land use, soils, state and county boundaries, hydrologic unit area (watershed boundaries), stream network, weather station locations, geology maps, and stream gauge stations (Figure 40-2). The relational databases are: national resources inventory (NRI), national agricultural statistical survey (NASS), state soil survey database (SSSD), weather parameters, streamflow and reservoir operation data, and agricultural census data (Ag Census).

USGS-developed spatial data were used for this study at 1:250,000 scale. The DEM (digital elevation model), LULC (land use and land cover), and stream gauge data were obtained from USGS and processed in albers equal area (AEA) projection for the study area. Several quads of 1° × 1° DEM and 1° × 2° LULC were processed and patched together into one map using several of the GRASS-GIS procedures. The soil layers called STATSGO (USDA 1992) were obtained from SCS, and the attribute databases were loaded into an INFORMIX relational database manager. The DEM, LULC, and STATSGO soil layers are at a scale of 1:250,000. Other relational databases such as NRI, NASS, and Ag Census were analyzed for periodic intervals of five to ten years. Historic streamflow from USGS stream gauge stations and weather information were used for the simulation. When weather data (daily precipitation and temperature) were unavailable, weather parameters were simulated using a stochastic weather generator (Arnold et al. 1990). The streamflow values predicted from SWAT simulations were validated against the historical streamflows using USGS stream gauges at the outlet of each six-digit HUA.

SWAT-GIS APPLICATION ON TEXAS GULF RIVER BASINS

In this chapter, average monthly results from two six-digit HUA covering the Seguin (120100) and Naches (120200) river basins are presented (Figure 40-1). Each river basin spans multiple climatic zones and widely varying soils and land uses. Also, each basin contains major reservoirs. The GIS layers obtained from USGS for land use (LULC, 200-m square grid) and DEM (1-m vertical interval, 3 arc-second data) at a scale of 1:250,000 were assembled in the AEA projection. The STATSGO soils survey layer (1:250,000 scale) was obtained from the SCS, and soil attributes were loaded into an INFORMIX relational database manager. From the USGS water-body layer at a scale of 1:2,000,000, the reservoirs were identified and inputs for the reservoirs were created for the SWAT model using the SWAT-GIS integrated system. In order to use the SWAT-GIS integrated system, the river basin first was subdivided into multiple subbasins, using the DEM layer as an input into the GRASS *.watershed* program. Thus, Seguin and Naches river basins were subdivided into 115 and 116 subbasins respectively.

Using the SWAT-GIS integrated system, the required inputs for each of the subbasins within each basin were extracted and formatted. The extracted information included soils, land use, topographic, weather generator, rain and temperature gauges, reservoir, and groundwater attributes. Table 40-1 gives additional information about the basins. In addition, the routing structure (Arnold et al., *Proceedings*, 1993) needed to run the model was automatically developed using the flow-path data created during the extraction of topographic attributes. This procedure also detects and automates the routing procedures if any reservoir or inflow data exist in any of the subbasins. The system allows the user to edit errors that occur when extracting the routing structure using either the keyboard or through

Table 40-1. Six-digit HUA basin characteristics.

Six-digit number **Name**	**120100** **Seguin River**	**120200** **Naches River**
Drainage area (km^2)	24469.2	25161.0
Length of main channel (km)	604.0	440.7
Average main channel slope (%)	0.0001	0.0001
Average overland slope (%)	0.002	0.002
Number of subbasins	115	116
Number of weather stations	6	6
Number of weather generator stations	11	5
Number of reservoirs	2	1

a graphical user interface. The SWAT-GIS integrated system helps users to model a river basin and saves tremendous time, compared to several man-weeks and months, depending on the size and variability of a basin. Since detailed reservoir operation rules are difficult to simulate, average monthly measured USGS streamflow data from the reservoir outlet were used as input to the model. The SWAT model was then run for nine years in both river basins from 1970–1979, and average monthly outputs were stored from the model for validation.

It is important for simulation models to produce frequency distributions that are similar to measured frequency distributions. Close agreement between means and standard deviations indicates that the frequency distributions are similar. Generally, simulated values compared well with measured values at the outlet of the river basin, with average monthly predicted flows 5% higher than measured flows (Table 40-2) . The standard deviations between measured and predicted compared well (within 2%) (Table 40-2). Figures 40-3 and 40-4 show the close agreement of seasonal trends of average monthly observed and predicted streamflow for 1970–1979 (120 months) for the Seguin and Naches river basins.

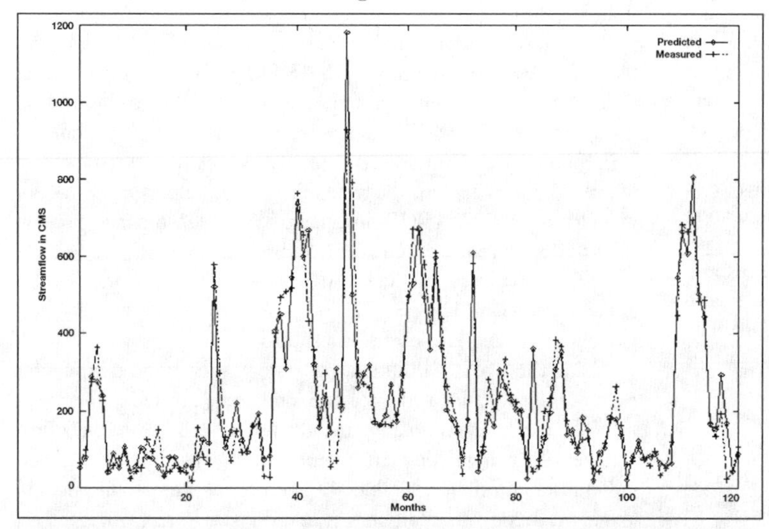

Figure 40-3. Predicted and measured average monthly streamflow for the Seguin River basin for 120 months (1970–1979).

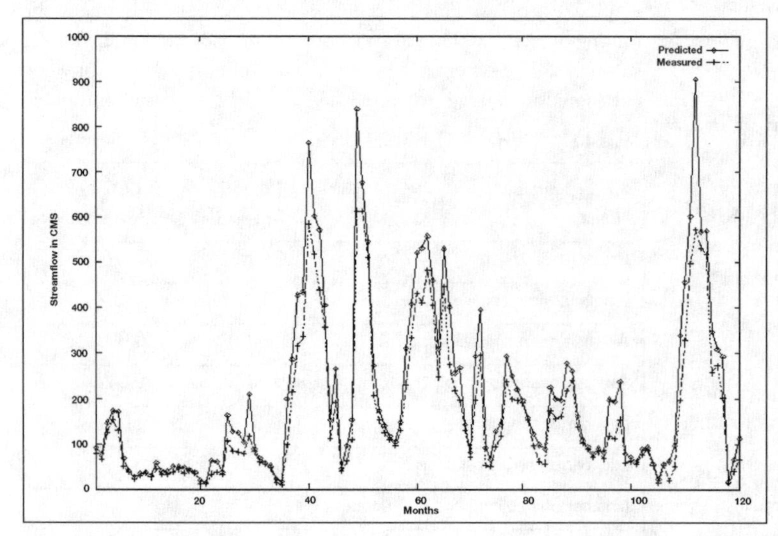

Figure 40-4. Predicted and measured average monthly streamflow for the Naches River basin for 120 months (1970–1979).

Table 40-2. Six-digit HUA basin statistics between observed and predicted average monthly streamflow values at the outlet of the river basins for the period of 1970–1979.

Six-digit number **Name**	**120100** **Seguin River**	**120200** **Naches River**
Measured mean (m^3/sec)	228.89	207.43
Predicted mean (m^3/sec)	230.59	218.45
Measured std. dev.	205.28	192.68
Predicted std. dev.	201.70	194.58
R^2	0.866	0.831
Regression slope	0.947	0.903
Nash Sutcliffe	0.863	0.818
Number of observations	120	120

Since values for the Nash Sutcliffe of 0.863 and 0.818, on an allowable range of -1 to +1, are close to 1, the simulation model performed well. In both basins, the SWAT model does predict close to the observed data (Table 40-2). It is important to note that at the outlet of each reservoir, measured streamflow data were used as input to SWAT, which could partially account for the relatively close agreement of the model results with observed data. However, considering the extreme spatial variability above and below the reservoir, the model was still able to predict streamflow reasonably close to observed values.

CONCLUSION

The SWAT model was integrated with the GRASS-GIS tool to develop a continuous time, distributed parameter modeling tool to assist with management of runoff, erosion, pesticide, and nutrient movement in large basins. The integrated system assists with development of SWAT input from GIS layers. The system is currently being evaluated for several watersheds within the Texas Gulf. Preliminary results suggest that the integrated SWAT-GIS model significantly reduces the time required to obtain input data and simplifies model operation. One of the limitations of the modeling system was its inability to mimic the complex reservoir operation rules, and attempts are being made to improve this in the SWAT model.

The integrated SWAT-GIS was applied to the Texas Gulf USGS defined six-digit HUA. Results from two of the river basins (Seguin and Naches) were reported in this chapter. SWAT model inputs including data on soils, topography, land use, and weather were automatically derived from map layers and associated databases using the integrated GIS system. Simulated average monthly streamflows were in close agreement (within 5%) with observed flows for both the river basins.

Acknowledgments

The authors thank USDA-SCS for supporting part of this work from HUMUS funds. Any use of trade, product, or firm names in this publication is for descriptive purposes only and is not an endorsement by the authors or their institutions.

References

Arnold, J. G. 1990. ROTO: A continuous water and sediment routing model. *ASCE Proceedings of the Watershed Management Symposium*, 480–88.

Arnold, J. G., P. M. Allen, and G. Bernhardt. 1993. A comprehensive surface-groundwater flow model. *Journal of Hydrology* 142:47–69.

Arnold, J. G., B. A. Engel, and R. Srinivasan. 1993. A continuous time, grid cell watershed model. In *Proceedings of Application of Advanced Information Technologies for the Management of Natural Resources.*

Arnold, J. G., and N. B. Sammons. 1989. Decision support system for selecting inputs to a basin scale model. *Water Resources Bulletin* 24(4).

Arnold, J. G, J. R. Williams, A. D. Nicks, and N. B. Sammons. 1990. *SWRRB: A Basin Scale Simulation Model for Soil and Water Resources Management.* College Station: Texas A&M University Press.

Beasley, D. B., L. F. Huggins, and E. J. Monke. 1980. ANSWERS: A model for watershed planning. *Transactions of the ASAE* 23(4):938–44.

Beven, K. J., and M. J. Kirkby. 1979. A physically based, variable contributing area model of basin hydrology. *Hydrological Sciences Journal* 24(1):43–69.

Chairat, S., and J. W. Delleur. 1993. Effects of the topographic index distribution on predicted runoff using GRASS. *Water Resources Bulletin* 29(6):1,029–34.

Foster, G. R., and L. J. Lane. 1987. *User Requirements USDA-Water Erosion Prediction Project (WEPP).* INSERL Report No. 1, National Soil Erosion Research Laboratory.

Heatwole, C. D. 1990. Knowledge-Based Interface for Improved Use of Models as Management Tools. Presented in ASAE 1990 International Winter Meeting, Paper No. 90-2642.

Knisel, W. G., ed. 1980. *CREAMS: A Field Scale Model for Chemicals, Runoff, and Erosion from Agricultural Management Systems.* USDA, Conservation Research Report No. 26.

Rewerts, C. C., and B. A. Engel. 1991. *ANSWERS on GRASS: Integrating a Watershed Simulation with a GIS.* ASAE Paper No. 91-2621.

Rosenthal, W., R. Srinivasan, and J. G. Arnold. 1993. A GIS watershed hydrology model link to evaluate water resources of the lower Colorado River in Texas. *Proceedings, Application of Advanced Information Technologies for the Management of Natural Resources.*

Shapiro, M., J. Westervelt, D. Gerdes, M. Larson, and K. R. Brownfield. 1992. *GRASS 4.0 Reference manual.* Champaign: USA-CERL.

Srinivasan, R. 1992. "Spatial Decision Support System for Assessing Agricultural Nonpoint Source Pollution Using GIS." Ph.D. diss., Agricultural Engineering Department, Purdue University, West Lafayette.

Srinivasan, R., and J. G. Arnold. 1993. Basin scale water-quality modeling using GIS. *Proceedings, Application of Advanced Information Technologies for Management of Natural Resources.*

Srinivasan, R., J. G. Arnold, R. S. Muttiah, C. Walker, and P. T. Dyke. 1993. Hydrologic unit model for United States (HUMUS). *Proceedings, Advances in Hydro-Science and Engineering.*

Texas State Soil and Water Conservation Board. 1991. *A Comprehensive Study of Texas Watersheds and Their Impacts on Water Quality and Water Quantity.* Temple: Texas State Soil and Water Conservation Board.

USDA. 1972. Hydrology. *National Engineering Handbook*, Section 4. Washington, D.C.

———. *STATSGO—State Soils Geographic Database.* Soil Conservation Service, Publication Number 1492. Washington, D.C.

Williams, J. R., C. A. Jones, and P. T. Dyke. 1984. A modeling approach to determine the relationship between erosion and soil productivity. *Transactions of the ASAE* 27(1):129–44.

Wischmeier, W. H., and D. D. Smith. 1978. *Predicting Rainfall Losses—A Guide to Conservation Planning.* USDA Agricultural Handbook No. 537.

Young, R. A., C. A. Onstad, D. D. Bosch, and W. P. Anderson. 1989. AGNPS: A nonpoint source pollution model for evaluating agricultural watersheds. *Journal of Soil and Water Conservation* 44(2):168–73.

R. Srinivasan, J. Arnold, W. Rosenthal, and *R. S. Muttiah*
Blackland Research Center
Temple, TX 76502
E-mail: *srin@iiml.tamu.edu; arnold@iiml.tamu.edu; rosentha@iiml.tamu.edu;* and *muttiah@iiml.tamu.edu*

41

Environmental Modeling and Monitoring with GIS:
Niwot Ridge Long-Term Ecological Research Site

Claude R. Duguay and Donald A. Walker

Research at the Niwot Ridge Long-term Ecological Research (LTER) site is concentrating on the ecological effects of altered snowpack and changes in precipitation and temperature regimes on alpine plant and animal communities. This is accomplished through the use of: (1) a hierarchic geographic information system (HGIS), (2) a snow-fence experiment to examine altered patterns of snowpack, and (3) analysis of remotely sensed data. The HGIS database is helping us study the links between species patterns at the level of plots, landscape patterns of plant communities, and regional patterns seen on satellite images. In this chapter, an overview of ongoing research activities using the HGIS at the plot, landscape, and regional scale is given. Results obtained during recent satellite remote sensing investigations at the Niwot Ridge LTER are also highlighted. These studies consist of: (1) determining the patterns of greenness as a function of regional climate, (2) evaluating the potential of multiresolution remotely sensed data combined with ancillary topoclimatic data to map tundra vegetation, (3) monitoring and modeling the spatial distribution and temporal patterns of snow cover, and (4) modeling the radiation and energy balance in the alpine.

INTRODUCTION

Remote sensing imagery and ancillary data from geographic information systems (GIS) are important sources of information for input to ecological and climatic models of seasonal and long-term environmental change. One of the goals of the United States Long-term Ecological Research (LTER) program is the systematic monitoring and studying of patterns and controls within a variety of natural ecosystems at various spatial and temporal scales, with an underlying objective to monitor change on the Earth's surface resulting from natural and anthropogenic processes. Within that program, the integration of remote sensing and GIS data sets will be critical toward linking established and detailed ecological studies at plot and landscape levels to regional scale interpretations through ecological simulation and modeling (Walker et al. 1993). Accordingly, all eighteen LTER sites have full remote sensing data sets and GIS capabilities in place for these and other purposes.

These technologies are of particular interest at the Niwot Ridge LTER site in the Colorado Rocky Mountains (Figure 41-1). Alpine environments in general are remote, inaccessible, and often possess complex and extreme environmental gradients that are highly sensitive

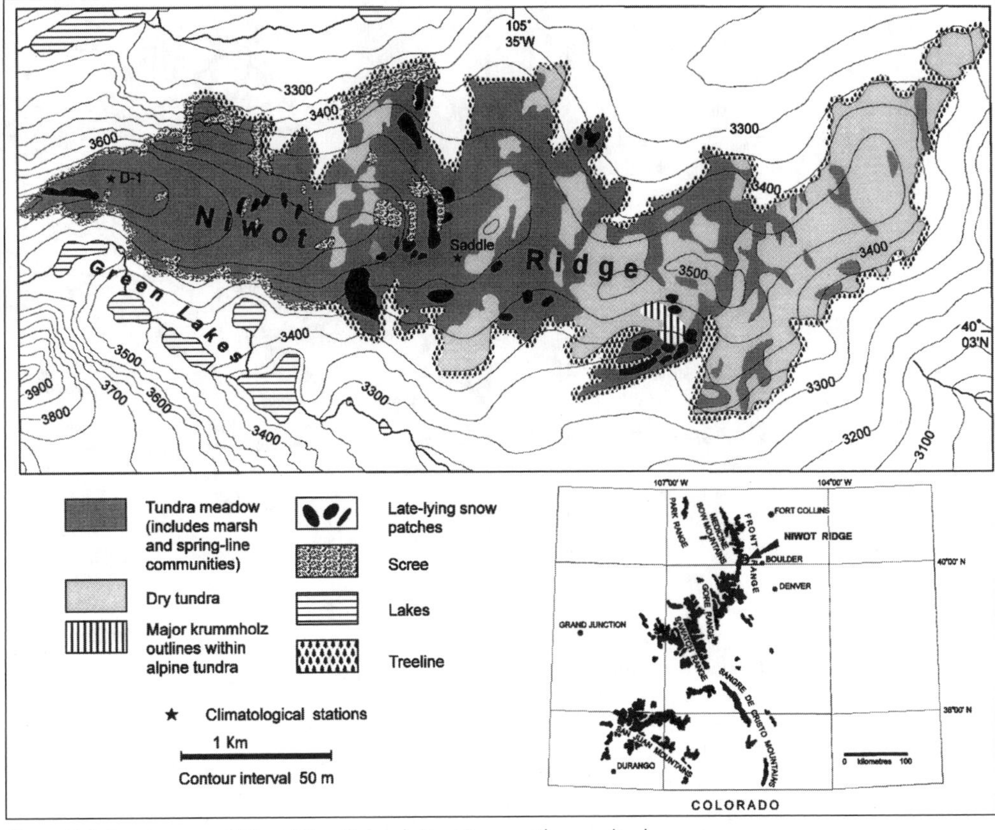

Figure 41-1. Location map of Niwot Ridge, Colorado Front Range, with generalized cover types.

to topographic and climatic controls. For example, in the alpine tundra ecosystem on Niwot Ridge, snow accumulation is an important and complex factor affecting the distribution of alpine vegetation communities. Snow cover insulates plants from severe winter temperatures, wind exposure, and desiccation but serves to shorten the growing season. Snow availability is affected by mountain precipitation regimes modified by strong prevailing westerly winds from the nearby Continental Divide, which force snow removal, redistribution, and accumulation based on topographic orientation. The Niwot Ridge LTER site has been the locale of numerous ecological and climatological studies. However, to monitor evidence of environmental change at different spatial and temporal scales within the Colorado alpine, it is also useful to capture pertinent information about vegetation, topography, and climate from remotely sensed and ancillary information sources to enhance present knowledge and understanding of this complex environment.

A hierarchic GIS database has been developed to analyze the relationships between snow distribution and ecosystem patterns at the plot, landscape, and regional scales in the Colorado alpine. In this chapter, ongoing research activities using GIS at the Niwot Ridge LTER are presented. In particular, we focus our attention on recent remote sensing analysis and monitoring of complex interactions between vegetation and snowpack dynamics in order to provide critical surrogates for assessing the effects of environmental gradients and climatic variability at this midlatitude alpine tundra site.

HIERARCHIC GEOGRAPHIC INFORMATION SYSTEM

The term hierarchic GIS, as used herein, corresponds to a nested set of GIS databases at several spatial scales. Long-term ecological studies often require data collected from a wide range of spatial domain so that, for example, changes observed in species distributions can be linked to changes in regional patterns of spectral reflectance as observed with Earth-orbiting satellites. The three primary goals for establishing the alpine HGIS are to: (1) provide accurate spatial frameworks for studies of ecosystem processes and geobotanical patterns at appropriate scales, (2) develop baselines for long-term observations of natural and anthropogenic change, and (3) provide geographically referenced data for models of ecosystem processes. The hierarchy of disturbances in the alpine ranges from soil heave caused by the formation of needle ice crystals at spatial scales of 10^{-4}–10^{1} m^2 to the disturbance caused by glaciers, which currently cover areas as small as 10^4 m^2 but which covered areas as large as 10^8 m^2 in the Front Range during the Pleistocene (Figure 41-2). The databases are prepared to address questions related to disturbances at plot, landscape, and regional scales.

A conceptual diagram of the HGIS (Figure 41-3) summarizes its tiers, the major topics of research at each scale, the themes that provided the linkage between tiers, and models that provide conceptual integration. A suite of standardized mapping methods makes this approach useful for multiscale and intersite comparisons (see Walker et al. 1993 for a complete description).

MULTISCALE STUDIES OF SNOW–VEGETATION INTERACTIONS

Plot-Scale Studies

At plot-level scales (10^{-2}–10^{2} m^2), the main hypothesis is that plant species react to changes in snowpack in a manner that is predictable from their present-day distribution along snow-depth gradients. We are interested in the plant species dynamics associated with snow distribution. Snowpack indirectly controls the distribution of many plant species by limiting the length of the growing season. Wind-exposed sites have extremely low winter soil temperatures and high moisture stress. The distribution of pocket gophers (*Thomomys talpoides*) is also strongly controlled by snow patterns. Gophers are largely responsible for the fine-scale mosaic of many plant communities; they maintain species diversity by creating gaps in the plant canopy, redistributing nutrients and soil, and suppressing species that would otherwise dominate.

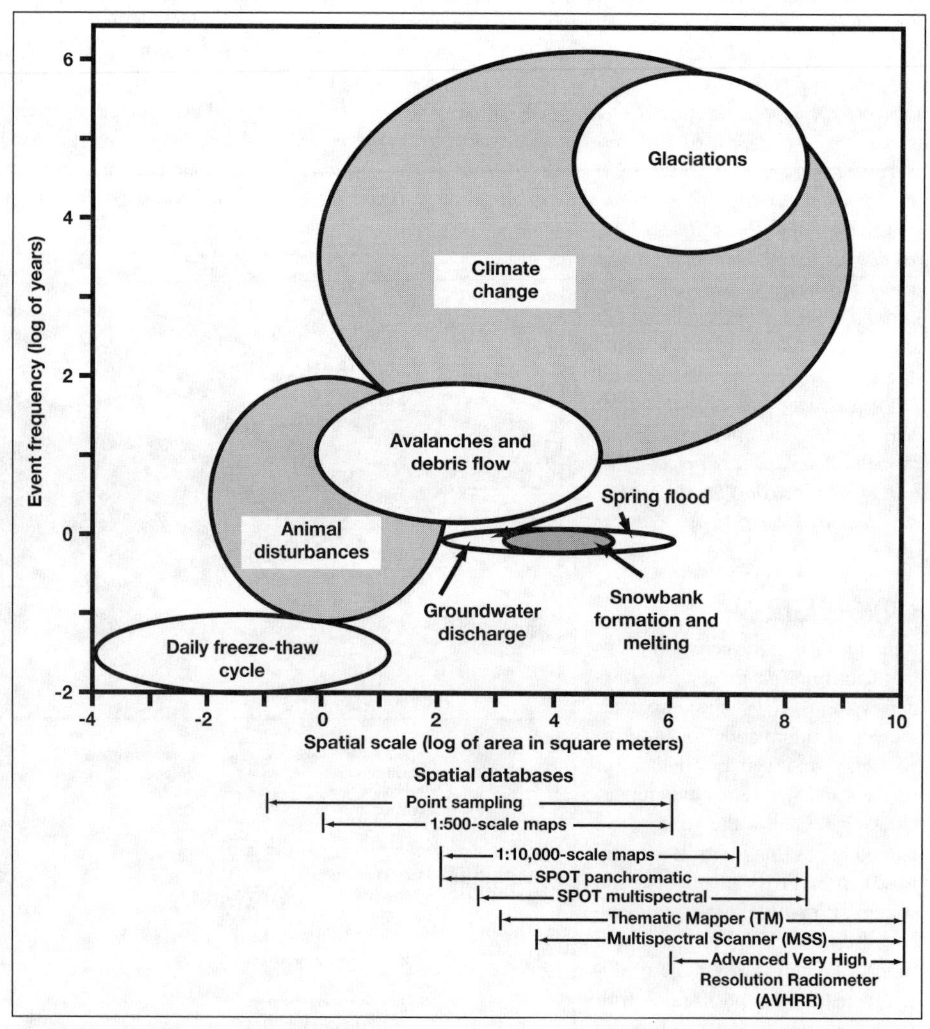

Figure 41-2. Spatial and temporal domains for natural disturbances at the Niwot Ridge LTER site. The shaded ellipses represent disturbance types that are major focuses of study at the site. The available data types for examining various scales of disturbance are shown at the bottom of the figure (Walker et al. 1993).

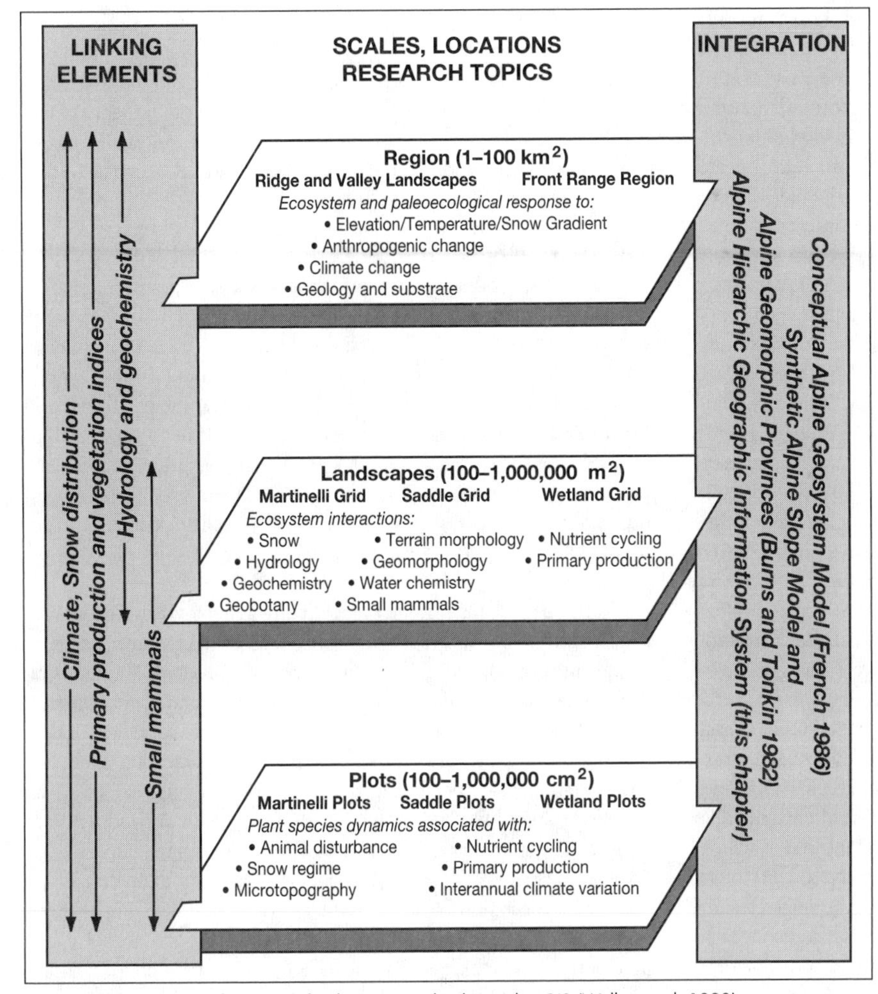

Figure 41-3. Conceptual framework for the Niwot Ridge hierarchic GIS (Walker et al. 1993).

To monitor changes in species composition, we are using a grid of eighty-eight permanent plots that span the snow gradient from windblown sites to areas with over 6 m of maximum winter snowpack. This is the most detailed level of the HGIS. Data at this level allow for detailed examination of the controls of microscale environmental variation because the data are spatially referenced to the UTM coordinate system, both with respect to the 350-m × 500-m Saddle grid and the 1- × 1-m permanent plots. An examination of the distribution of plant species along a snow gradient indicates that most species are found in relatively narrow ranges of snow depth, but a few cosmopolitan species such as *Acomastylis rossii* occur in a broad range of winter snow regimes with its optimal occurrence in 1–2 m of snow.

Landscape-Scale Studies

At the landscape scale (10^2–10^6 m^2), the main long-term research hypothesis is that shifts in snowpack regimes should cause changes to vegetation-community boundaries that are predictable from present-day vegetation–snow relationships. We are examining the patterns of vegetation communities, primary production, and small-mammal distribution associated with hill-slope toposequences, and snow gradients from windblown sites to deep snow patches. Our main study site is the Niwot Ridge, Saddle grid, where we prepared a detailed vegetation map at 1:500 scale and analyzed the vegetation patterns in relation to mean maximum snow depth, slope, and aspect. At this level of the hierarchy, vegetation is mapped using the Braun-Blanquet approach (Komárková and Webber 1978) and six different soil–plant communities (noda) (Webber and May 1977).

The control of snow on the distribution of the dominant alpine species is apparent at the plant-community level. For example, the primary vegetation types on west-facing slopes are either fellfield associations (*Trifolietum dasyphylli* and *Sileno-Paronychieum*) or dry sedge meadow associations (*Selaginello densae-Kobresietum myosuroidis*). In contrast, east-facing slopes have a predominance of deep snow accumulation areas and snow bed associations. Plant communities associated with either windblown sites or snow patches cover a total of 78.6% of the Saddle grid and 77.9% of the alpine area of Niwot Ridge mapped by Komárková and Webber (1978)—an indication of the importance of wind and snow cover in this windy alpine landscape.

Regional-Scale Studies

At the regional scale associated with the entire city of Boulder watershed or the Front Range (10^6–10^8 m^2), remotely sensed data provide an efficient means to examine regional patterns of greenness. The main hypothesis at this scale is that primary production is broadly controlled by gradients associated with changing elevation, but also is influenced by smaller-scale topographic interactions with wind and snow. Currently we are using SPOT HRV and Landsat TM satellite images to examine patterns of greenness along elevation gradients. We are also building a database at a scale of 1:10,000 that covers the city of Boulder watershed and the entire Niwot Ridge LTER site. In addition to remotely sensed data, this database includes vegetation maps (Braun-Blanquet vegetation associations and vegetation noda) and digital elevation models. Surficial geology and surficial geomorphology maps will be added in the future to the analysis of NDVI (Normalized Difference Vegetation Index) in relation to geobotanical features and terrain information.

In addition to linking ground-level observations to remotely sensed imagery through the utilization of NDVI, we are also investigating the use of SPOT MLA as well as Landsat TM and MSS for long-term monitoring and modeling experiments within the alpine tundra zone of Niwot Ridge. In particular, we are conducting studies aimed at: (1) evaluating the potential of multiresolution remotely sensed data, combined with ancillary topoclimatic data to map tundra vegetation; (2) monitoring and modeling the spatial distribution and temporal patterns of snow cover; and (3) modeling the radiation and energy balance within the alpine. These studies provide some of the building blocks necessary for improving simulation experiments of the response of alpine tundra to climatic change. Results of our most recent remote-sensing investigations are briefly described next.

REMOTE-SENSING STUDIES

Spatial Patterns of Green Biomass

Our early studies relating patterns of green biomass to elevation in the Front Range using SPOT HRV (20 m; September 4, 1988) indicate that the fine-scale patterns associated with snow distribution at plot and landscape scales do have an influence on regional patterns of primary production (Komárková and Weber 1978). NDVI generally showed a strong negative correlation with elevation in the Front Range, which is logical because of colder temperatures and longer periods of snow cover at higher elevations. However, this relationship is also affected by strong winds which are responsible for the redistribution of snow. West-facing slopes east of the Continental Divide showed no relationship between elevation and NDVI, an indication of the strong control of wind on production at all elevations.

Based on the preliminary results with SPOT, we are currently investigating the relationship (correlation) between NDVI values derived from Landsat TM (30 m; August 6, 1986) and several topographic and topoclimatic variables for Niwot Ridge alone. One objective of this study is to develop an empirical model of NDVI based on a simple set of topographic/topoclimatic variables. The six topographic variables describe elevation: (1) DEM, (2) slope (SLP), (3) relief (REL), (4) down-slope convexity (DSC), (5) cross-slope convexity (CSC), and (6) angle of incidence (INC). The five topoclimatic variables correspond to: (1) an orogenic precipitation index (OPI), (2) growing degree days (GDD), (3) slope-aspect index (SAI), (4) insolation index (INS), and (5) snow probability (PS). The topoclimatic indices (Plate 41-1) are derived from a sixteen-year Landsat data archive, geomorphometry from a DEM, and meteorological station data. The procedure used to derive these indices is described in a recent paper by Peddle et al. (1993).

Briefly, we define OPI as altitude/distance from the Continental Divide. Higher OPI values indicate greater likelihood for higher relative precipitation amounts and, consequently, greater moisture availability potential for plant growth. GDD refers to the magnitude a given temperature measure exceeds a temperature (5°C) at which plant growth is inhibited. We compute cumulative GDD totals throughout the growing season to gain insight into the total amount of energy available for plant growth. SAI enables us to distinguish between terrain orientations likely to contain ample soil moisture due to snow redistribution versus those that would be exposed, windblown, and dry. High values of SAI are found on steep, leeward (east-facing) slopes where snow accumulation is great and moisture is readily available. INS gives a measure of global insolation computed in terms of global transmission, extraterrestrial solar radiation corrected for the angle of incidence, diffuse sky radiation, and angles of direct beam incidence to sloping and horizontal surfaces. Finally, PS is derived from coregistered Landsat MSS images acquired in different years and at different times in the growing season to capture relevant information pertaining to both the seasonal and annual variability of snow-cover patterns.

Correlations between NDVI and the eleven topographic/topoclimatic variables were computed for Niwot Ridge at various levels of stratification of slope and aspect classes. Only the results that involved stratification into east-facing (67.5°–112.5°) and west-facing slopes (247.5°–292.5°) with gradients 3°–10° are discussed herein. For east-facing slopes, NDVI is highly correlated with OPI (r=0.77), GDD (r=-0.75), and DEM (r=0.75). Correlations are below +0.27 for all other variables. In the case of west-facing slopes, NDVI is also highly correlated with OPI (r=0.78) but only moderately with GDD (r=-0.47) and DEM (r=0.51). In contrast to east-facing slopes, PS (r=0.50) and DSC (r=-0.51) for west-facing slopes are moderately correlated with NDVI. All other correlation coefficients are below +0.18 for these slopes. One possible explanation for the moderate correlations between NDVI and the variables PS and DSC is that some of the west-facing slopes (most probably concave), which are covered by snow during the winter months, provide moister soil conditions for plant growth (i.e., snow-patch communities). These preliminary results should, however, be interpreted with some caution since the relationships between NDVI and the topographic/topoclimatic variables have been derived from a data set at 30-m resolution. For example, fine-scale topographic variations associated with the presence of features such as turf-banked terraces and lobes and their associated drainage ways, which occur mainly on the upper part of Niwot Ridge, may be best captured at spatial resolutions of the order of 2–5 m.

Tundra Vegetation Mapping

Remote-sensing investigations into mapping alpine tundra vegetation in the Saddle area of Niwot Ridge have been carried out with some degree of success using digitized color infrared 1:50,000 scale aerial photography with indicators of topographic context (slope, aspect) (Frank and Thorn 1985) and topoclimatic index values (Frank and Isard 1986). We recently complemented and built upon these studies by exploring the development of a spectral, topographic, and climatic digital data set to test our hypothesis that classification accuracy and precision can be increased by incorporating ancillary climate indices into satellite image analysis of vegetation patterns within the complex alpine tundra ecosystem on Niwot Ridge.

The five topoclimatic indices described in the previous section were tested individually and together with a Landsat TM image and topographic measures from a DEM to assess their significance for increasing the accuracy and precision of maximum likelihood land-cover classification with respect to the hierarchical Braun-Blanquet vegetation classification system. GDD and OPI had the highest classification accuracies among individual topoclimatic indices, with the highest accuracies obtained using all five indices together with several TM bands (74% to 83% at the highest and lowest levels of precision tested, respectively). Experiments showed good correspondence between digital classifications and a vegetation field map at the class and order levels of aggregation of the Braun-Blanquet system. However, greater accuracy and precision might be expected at the alliance and association levels using a higher resolution remote sensing (i.e., airborne imaging spectrometers) and DEM data sets together with more sophisticated classification algorithms (i.e., evidential reasoning).

Snow-Cover Variability

As shown earlier in this chapter, snow is one of the principal factors controlling alpine vegetation at all levels of investigation. For example, an increase in snowfall affects plant processes in several ways, the most important being to shorten the growing season, thus reducing the time available for photosynthesis and reproductive development. Under a climate scenario of increased winter snowfall on Niwot Ridge, dry exposed ridges receive little, if any, additional snow, whereas traditional snowbed sites are more susceptible to impact. Over several decades, weather modification resulting in an increase or decrease in winter snowfall would most probably alter the proportional composition of alpine community classes. Satellite remote sensing provides the capabilities necessary for monitoring changes in the spatial/temporal patterns of snow cover. In addition, when combined with a DEM and climate data, remote-sensing imagery may be used to estimate snow depth, an important input parameter to runoff forecasting.

Snow-covered areas, expressed as a percentage of the total area of Niwot Ridge, have been extracted from a Landsat MSS image archive (1973–present) and correlated to temperature and precipitation records from two climate stations on the Ridge to see if snow cover can be utilized as one of the many indicators of climate variability/change at the Niwot LTER (Plate 41-2). Preliminary results show a good correspondence between snow-covered areas, derived from Landsat MSS during the major period of seasonal runoff, and climate conditions (average monthly temperatures and total snowfall calculated from October through May). Long-term monitoring of seasonal snow cover with Landsat MSS (and higher resolution sensors) will complement another experiment in which we are artificially altering winter snow regimes with a large snow fence (100-year snow fence experiment). Because alpine plant species are long-living perennials, we expect that many ecosystem responses will be slow. Therefore, we plan to continue the experiment until we no longer see changes in species patterns or soil properties. The experiment is likely to provide many insights to the ecosystem consequences of climate-altered snowpack regimes.

Radiation and Energy-Balance Modeling

The accurate determination of the net radiation flux on the Earth's surface is an important objective in climate-related research on all spatial and temporal scales. Net radiation is a fundamental measure of the energy available at the earth–atmosphere interface since it regulates evapotranspiration, sensible heat flux, soil heat flux, and photosynthesis. A knowledge of the spatial distribution of net radiation is therefore of considerable importance to geoecological research on Niwot Ridge. We have recently proposed a model to estimate the radiation balance over alpine snow fields under clear-sky conditions (Duguay 1993) and presented an approach to derive surface albedo from Landsat TM imagery and digital terrain data (Duguay and LeDrew 1992). Results were in good agreement with data acquired during field measurement campaigns. We now plan to use the approach of Duguay and LeDrew (1992) to monitor changes in surface albedo at the Niwot LTER on an ongoing basis. In addition, the net radiation model is currently being revised to consider cloudy conditions and provide further capabilities to estimate both the radiation balance and evapotranspiration over the alpine tundra zone of Niwot Ridge.

CONCLUSION

In this chapter we have presented an overview of ongoing research using a hierarchic geographic information system database at the Niwot LTER site. Thus far, much of our effort has been directed at examining the relationships between patterns and processes at a variety of spatial scales. The hierarchic analysis of alpine vegetation patterns has demonstrated that plant species, community, and green-biomass patterns are largely controlled by snow distribution, which varies greatly with topography and wind patterns. Environmental monitoring and modeling experiments within the alpine tundra zone of Niwot Ridge have mostly been conducted at the regional scale using remotely sensed imagery and ancillary GIS data.

In the future, high-resolution remote sensing and digital elevation data should be acquired in order to address more fully the question of scale. With respect to the development of an empirical model of NDVI or a model of snow accumulation and distribution, there is a need to collect high-resolution data (2–5 m) to capture fine-scale topographic variations that control snow redistribution. Finally, high-resolution sensors could provide some important land-surface parameters for initializing and validating alpine tundra simulation models, such as the one recently presented by Grant and French (1990).

Acknowledgments

This work was supported by the Niwot Ridge Long-term Ecological Research project (#DEB 9211776). Additional funding for this research was provided by a Natural Sciences and Engineering Research Council of Canada research grant to Claude Duguay.

References

Burns, S. J., and P. J. Tonkin. 1982. Soil–geomorphic models and the spatial distribution and development of alpine soils. In *Space and Time in Geomorphology*, edited by C. E. Thorn, 25–43. London: Allen and Unwin.

Duguay, C. R. 1993. Modeling the radiation budget of alpine snowfields using remotely sensed data: Model formulation and validation. *Annals of Glaciology* 17:288–94.

Duguay, C. R., and E. F. LeDrew. 1992. Estimating surface reflectance and albedo from Landsat-5 Thematic Mapper over rugged terrain. *Photogrammetric Engineering & Remote Sensing* 58(5):551–58.

Frank, T. D., and S. A. Isard. 1986. Alpine vegetation classification using high-resolution aerial imagery and topoclimatic index values. *Photogrammetric Engineering & Remote Sensing* 17:381–88.

Frank, T. D., and C. E. Thorn. 1985. Stratifying alpine tundra for geomorphic studies using digitized aerial imagery. *Arctic and Alpine Research* 52:179–88.

French, N. R. 1986. Hierarchical conceptual model of the alpine geosystem. *Artic and Alpine Research* 18:133–46.

Grant, W. E., and N. R. French. 1990. Response of alpine tundra to a changing climate: A hierarchical simulation model. *Ecological Modelling* 49(3-4):205–27.

Komárková, V., and P. J. Webber. 1978. An alpine vegetation map of Niwot Ridge, Colorado. *Arctic and Alpine Research* 10:1–29.

Peddle, D. R., C. R. Duguay, and G. Deschamps. 1993. Use of ancillary climate data in satellite image analysis of an alpine tundra ecosystem, Front Range, Colorado Rocky Mountains. *Proceedings of the 16th Canadian Symposium on Remote Sensing and 8e Congrès de l'Association Québécoise de Télédétection*, 215–20.

Walker, D. A., J. C. Halfpenny, M. D. Walker, and C. A. Wessman. 1993. Long-term studies of snow-vegetation interactions. *BioScience* 43(5):287–301.

Webber, P. J., and D. E. May. 1977. The distribution and magnitude of below-ground plant structures in the alpine tundra of Niwot Ridge, Colorado. *Arctic and Alpine Research* 19:155–66.

Claude Duguay is an associate professor in the Department of Geography at the University of Ottawa and director of the Laboratory for Earth Observation and Information Systems. He is also coinvestigator of the NASA-EOS CRYSYS (Cryospheric System to Monitor Global Change in Canada) project. His general research interests include the derivation of geophysical parameters required in climate analysis from remotely sensed data and climate modeling and monitoring in arctic and alpine environments.
Laboratory for Earth Observation and Information Systems
Department of Geography
University of Ottawa
and
Ottawa-Carleton Geoscience Center
Ottawa, Ontario K1N 6N5
Canada
E-mail: *duguay@acadvm1.uottawa.ca*

D. A. (Skip) Walker is codirector of the Joint Facility for Regional Ecosystem Analysis at the Institute of Arctic and Alpine Research, University of Colorado, Boulder, Colorado. He is also an adjunct professor in the Department of Environmental, Population, and Organismic Biology at the University of Colorado.
Joint Facility for Regional Ecosystem Analysis
Institute of Arctic and Alpine Research
and
Department of Environmental, Population, and Organismic Biology
University of Colorado
Boulder, Colorado 80309
E-mail: *swalker@aimyr.colorado.edu*

42

Spatial Modeling of Fish Growth:
Underwater Views of the Aquatic Habitat

Stephen B. Brandt, Doran M. Mason, Andrew Goyke, Kyle J. Hartman, James M. Kirsch, and Jaingang Luo

Large aquatic systems often provide a spatially complex physical and biological habitat for fishes. Such spatial patterning in the environment can have profound and nonlinear effects on ecological processes and predator–prey interactions. To examine the functional relationships between spatial patterning in underwater habitats and ecological processes, we have developed a new approach that incorporates the spatial environment into a modeling framework. The approach combines the strengths of: (1) underwater acoustics to measure fish density and size with fine-scale spatial resolution, (2) bioenergetics models to accurately simulate fish growth, and (3) spatial modeling to link the physical and biological structure of the habitat and maintain their spatial integrity. Data visualization techniques are used to display spatial maps of fish growth rates. The spatial modeling approach has direct applications to ecological studies and fisheries management.

INTRODUCTION

As in terrestrial landscapes, the underwater world is typified by a spatially complex biological and physical habitat (Haury et al. 1978; Steele 1978; Legendre and Demers 1984; Bennett and Denman 1985; Frost et al. 1988; Wiens 1989; Nero et al. 1990). This spatial patterning in the environment is often ignored in ecological models that try to evaluate predator–prey interactions and fish production because of the difficulty of measuring biological patchiness underwater. Biological processes that occur at relatively small spatial scales can significantly affect production at the system level (Possingham and Roughgarden 1990; Kotliar and Wiens 1990), and predator–prey interactions are clearly scale-dependent (Cox et al. 1982; Kareiva and Anderson 1988; Koehl 1989; Rose and Leggett 1990; Kotliar and Wiens 1990; Chesson and Rosenweig 1991). For example, Lasker (1978) showed that survival and growth of larval anchovies depended on the existence of food patches, and that average values of prey density were meaningless to the predator. Moreover, the thermal structure of aquatic systems sets up a mosaic of environments that control the rates of processes that regulate fish growth. Most of the physical and biological factors that regulate biological rates are nonlinear. Therefore, the use of values that are spatially averaged over a given body of water may lead to serious errors.

Spatially explicit modeling is one approach that incorporates the spatial environment into a modeling framework (Brandt et al. 1992) and can be used to assess the functional relationships between spatial patterning of the environment and ecological processes (Legendre and Demers 1984; Carpenter 1988; Hoffman 1988; Powell 1989; Turner and Gardner 1991; Magnuson et al. 1991). The basic approach in spatial modeling is to subdivide a heterogeneous habitat into small homogeneous subunits or cells. Data input and output are distributed in cells across a spatial grid. Process-oriented simulation models of the same model structure are run in each cell but are parameterized differently according to the specific habitat conditions in each cell. One key advantage of this approach is that if cell sizes are sufficiently small, environmental conditions within each cell can be assumed to be homogeneous, thus simplifying the structure of the ecological models.

One of the difficulties of implementing spatially explicit modeling is that high-resolution spatial data are required to parameterize models in each cell. The study of spatial processes in terrestrial ecology (Turner and Gardner 1991) has made significant progress in recent years, largely because remote-sensing techniques have provided sufficient spatial information to allow hypotheses to be formulated and tested (Roughgarden et al. 1991; Ustin et al. 1991). Progress has been slower in aquatic systems because it is difficult to measure biological heterogeneity under the surface of the water (Steele 1978). For our approach, we take advantage of the ability of underwater acoustics to provide high-resolution, continuous data on the spatial distributions of fishes across large bodies of water (Nero et al. 1990). This technology permits us to view fish distributions in midwater at new spatial scales, thus providing a template for spatially explicit ecological modeling. Acoustic techniques have been applied routinely for fish stock assessment (Brandt et al. 1991; MacLennan and Simmonds 1992), but the spatial information inherent in acoustic data has largely been ignored in ecological modeling. Recent developments in computer technology and data visualization capabilities provide the means to more fully exploit acoustic information.

SPATIALLY EXPLICIT MODELS OF FISH GROWTH

Fish Growth-Rate Potential

Fish growth rate is an important consideration in fisheries management because it directly affects fish production, survival, and reproductive success (Zastrow et al. 1991; Monteleone and Houde 1990). The growth rate of an individual fish depends on the innate growth potential of the species and on local environmental conditions, such as prey availability and water temperature, which can vary across space and time. Thus fish growth rates should be affected by spatial patterning in the environment.

As a first application of spatial modeling in aquatic systems, we developed the concept of fish growth-rate potential. Fish growth-rate potential defines a potential functional response; that is, the expected growth rate of a predator if placed in a particular volume (i.e., cell) of water having known physical and biological characteristics. Fish growth-rate potential is derived from the integration of ecological models with spatially explicit data on food availability and habitat structure and is thus a spatially varying attribute of the environment. Fish growth-rate potential is a measure of the habitat suitability for the predator and is independent of actual predator distributions (Brandt et al. 1992).

Conceptual Model

Fish growth-rate potential is calculated using spatial modeling techniques. High-resolution acoustic data on prey sizes and densities form the template to generate a two-dimensional array of prey availability. The prevailing physical and biological data from each spatial cell are incorporated into a species-specific, physiological-based bioenergetic model that estimates the growth rate of an individual predator, if that predator is located in that particular cell. The result is a spatial map of growth-rate potentials.

Space is modeled as an explicit feature of the environment by subdividing the water column into small volumes (V_i) of water (i.e., cells) that define a grid with axes of horizontal distance and water depth (Plate 42-1). The size of the cells is determined by the scale of observation, and the number of cells is determined by water depth and horizontal coverage of the sampling. Each cell is treated individually and is characterized by a measured prey density (D_i), prey size (S_i), and water temperature (T_i). A predator placed in a cell for a set interval will have a growth rate specified by the biological and environmental conditions in the cell and the innate growth potential of the predator. A foraging model (f) estimates the consumption rate of the predator and is a function of prey densities and sizes. The growth model (g) estimates predator growth rate based on consumption rate, predator physiology, and water temperature. We assume that environmental conditions in each cell remain constant over the modeling time interval.

By running foraging and growth models in every cell, cross-sectional maps of fish growth rates that would be achieved in each cell will result if the predator occupied that particular cell for a specified unit of time. The model is run on a UNIX workstation using Interactive Data Language software (Research Systems 1990). Recent developments in interactive data visualization are used to display and evaluate spatial patterns of fish growth rates (Platt and Sathyendranath 1988; Manley and Tallet 1990). Brandt et al. (1992) provide a general introduction and discussion of this approach.

Acoustic Measures of Prey Density and Size

Acoustic systems sample the water column by sending short (e.g., 0.2–1.0 m/sec), repetitive (e.g., 5 pulses sec^{-1}) pulses of high-frequency sound (e.g., 12–420 kHz) in a directed beam, downward throughout the water column as the survey vessel moves across the surface. When the sound wave encounters an acoustic scatterer, an echo propagates radially outward from the target and is received at the surface. Echoes contain information on prey location, size, and numerical abundance. The strength of an echo can be related to the biomass of the target using various techniques (Burczynski and Johnson 1986; Foote et al. 1986; Holliday et al. 1989; Pieper et al. 1990; Peterson et al. 1976; Clay 1983; Stanton and Clay 1986). General reviews of acoustic techniques can be found in Clay and Medwin (1977), Forbes and Nakken (1972), Greenlaw and Johnson (1983), Urick (1983), Thorne (1983), and MacLennan and Simmonds (1992). Since sound travels in seawater at about 1,500 m/sec, the entire water column can be quickly sampled, and a continuous, detailed map of prey densities and sizes can be obtained. Acoustic data are converted into cells by defining the depth intervals and horizontal distances over which the continuous data are pooled.

Foraging Model

Various models of foraging have been developed to define the relationship of prey size and density to predator consumption (Holling 1966; Pyke 1984; Wright and O'Brien 1984; Stephens and Krebs 1986). Generally, foraging is considered a random process whereby the predator encounters prey randomly and prey are assumed to be randomly distributed. Normally, the latter only occurs in a small local volume (cell). To simplify the problem of prey spatial patchiness, we reduce the spatial scales of observation to a volume of water (or cell) sufficiently small to assume random distribution of prey. Thus, predator consumption (C_i) becomes a function of the encounter rate (E = the number of prey encountered per unit time) and the probability that the predator ingests the prey once encountered (Fuiman and Gamble 1989). Encounter rates between predator and prey depend on predator reaction distance (R = the distance at which an individual prey is recognized), swimming speeds of predator (v) and prey (u) and prey densities. For a specific volume of water with a given prey density, prey size, and temperature, a number of common foraging models can be applied (e.g., Gerritsen and Strickler 1977; Gibson and Ezzi 1990; Persson and Greenberg 1990). In this chapter, we use the encounter rate model of Gerritsen and Strickler (1977) which defines encounter rate in cell i with prey density D_i as:

$$E_i = \frac{\pi R^2}{3} \cdot \frac{3v^2 + u^2}{v} \cdot D_i \qquad (42\text{-}1)$$

Bioenergetics Model

Bioenergetics models of fish growth rate are mass-balance equations that equate energy gained through consumption with growth (somatic and reproductive) and energy loss (respiration, egestion, excretion). Fish growth rates are generally very sensitive to food abundance and water temperature. We adopted a bioenergetics model (Kitchell et al. 1977; Hewett and Johnson 1987, 1992) that has been used widely to model growth responses of fishes to changing environmental conditions and to evaluate predator–prey relationships (Ney 1990; Hill and Magnuson 1990; Hewett and Stewart 1989; Bevelheimer et al. 1985; Stewart and Ibarra 1991; Brandt et al. 1991; Kraft 1993; Hewett 1989). Sensitivity analyses (Stewart et al. 1983; Bartell et al. 1986) and model validation studies (Rice and Cochran 1984; Beauchamp et al. 1989) have shown that these bioenergetic models generally provide accurate and robust estimates of fish consumption and growth (Brandt and Hartman 1993; Hansen et al. 1993).

A bioenergetics model requires species-dependent physiological parameters that are most often derived from laboratory experiments. To apply the model to a particular aquatic system requires environmental inputs such as water temperatures, prey type, caloric density, prey size, and prey abundance. The bioenergetics model is run in each cell to determine the potential growth rate of an individual predator that might occupy that cell. Growth rate (G_i) in cell volume i depends on consumption rate (C_i), metabolic costs (R_i), excretion (U_i), and egestion (F_i):

$$G_i = C_i - (R_i + F_i + U_i) \quad (42\text{-}2)$$

Consumption and metabolism are influenced by water temperature and predator weight. The bioenergetics model is specific to a particular size and type of predator.

Predator consumption rate (C_i) is calculated from the foraging model as an input to the growth model. There are physiological constraints on consumption; an individual fish cannot consume more than its stomach can contain, assimilate, and evacuate. Normally, for a given-sized predator, maximum consumption (C_{max}) increases gradually with temperature to a maximum and then declines sharply (see equations in Kitchell et al. 1977; Hewett and Johnson 1992). Consumption remains constant at prey densities above this threshold level.

CHESAPEAKE BAY EXAMPLE

We illustrate the approach by modeling the growth-rate potential of a four-year-old (1.9 kg) striped bass (*Morone saxatilis*) in the Chesapeake Bay. The striped bass is an important recreational fish in the Chesapeake Bay and feeds primarily on the most abundant pelagic prey fish, bay anchovy (*Anchoa mitchilli*), and Atlantic menhaden (*Brevoortia tyrannus*) (Manooch 1973; Gardiner and Hoff 1982; Hartman 1993). The bioenergetics model (Hewett and Johnson 1992; Moore 1988; Hartman 1993) shows that striped bass maximum consumption increases from about 1% body weight per day at low temperatures (4°–6°C) to a maximum of 3.4% body weight per day at 2°C (Brandt and Kirsch 1993). Maximum consumption is much lower at higher temperatures; striped bass stop feeding at temperatures exceeding 25°C.

The overall relationship of striped bass growth rate to prey density and water temperature is nonlinear (Brandt et al. 1992; Brandt and Kirsch 1993). Any particular growth rate can be derived from a variety of different combinations of temperatures and prey densities. The highest growth rate of striped bass with an unlimited food supply occurs at 19°–20°C. At any particular temperature, growth rate generally increases linearly with prey density but is constant at prey densities exceeding maximum consumption.

To apply the model to the Chesapeake Bay, the spatial distributions of prey densities and sizes were measured along five cross-bay transect lines during April and August 1991. Spatially explicit maps of prey-biomass density and modeled striped bass growth rates are shown in Plates 42-2 and 42-3, respectively. During April, water temperatures ranged from 9°–13°C. Striped bass growth-rate potential was generally highest in regions of high-prey density and in the upper portions of the water column. In contrast, growth rates were poor during August in all regions, despite high densities of prey (Plate 42-3). Poor growth rates occurred in August because water temperatures were too high (24°–29°C) to support feeding.

Brandt and Kirsch (1993) compared the growth rates of striped bass calculated with spatially explicit modeling to growth rates calculated with spatially averaged data. Growth rates estimated from mean conditions exceeded the spatially explicit estimates by 176%–400% in different seasons. The reason for this was that prey densities that exceeded the maximum consumption level of striped bass in any particular cell did not contribute to predator growth potential in the spatially explicit approach. When mean prey densities were used, prey were implicitly assumed to be distributed uniformly across the bay and no cells would contain surplus prey.

MODEL DEVELOPMENT

Spatially explicit models represent a measure of the growth-rate potential of a fish over a specific unit of time and, as such, can be considered to represent only a snapshot of habitat functionality. However, fish can move among cells. How can the dynamics of predator–prey interactions be captured in spatial modeling? One approach to examining the dynamics of fish growth-rate potential is to evaluate a time series of spatially explicit "snapshots" taken seasonally, dielly, hourly, and even in shorter time frames. A second approach is to use dynamic spatial modeling (Sklar and Costanza 1991), whereby conditions within each cell can vary through time and fish (both predators and prey) are allowed to move among cells. Luo and Brandt (1994) have developed prototype software and visual animation capabilities that allow prey to move across cell boundaries in directions and at rates determined by cell attributes and behavioral algorithms of the fish. Model simulations will be used to generate specific hypotheses on predator–prey interactions in a spatially complex and time-varying environment.

CONCLUSION

Most effort in evaluating physical/biological coupling in the sea has dealt with planktonic organisms that respond to the advective processes operating at different scales (Gower et al. 1980; Legendre and Demers 1984; Powell 1989). In contrast, fish are often very capable of behaviorally selecting their habitat. Our approach evaluates the functional quality of the habitat from the perspective of the fish as well as the different consequences to growth of behavioral responses (Brandt 1993). By coupling spatially explicit acoustic data within an ecological-modeling framework, we go beyond the simple (but essential) correlation of biological and physical structure to model the functional response field of fishes to their physical and biological habitats. Our approach provides a unique framework for assessing and predicting fish growth rates and system production at various time and space scales and their dependency on predator behavior and physiology and on the spatial patterning within the environment. We hypothesize that spatial complexity is critical in evaluating predator–prey interactions and fish growth potential and analyses that consider this complexity could provide counterintuitive predictions to those obtained from analyses of system averages and mean trends (Wiens 1989). The application and, indeed, the motivation for our research is to find and reliably measure such functional features that ultimately will help characterize ecosystem production.

Acknowledgments

This research was funded by the National Science Foundation, Biological Oceanography Program (#OCE-911607 and #OCE-9415740). Additional support was provided by the Environmental Protection Agency, Chesapeake Bay Program (#CB-993197-01) and the Maryland Sea Grant Program (#NA 90AA-D-S6063 and #NA 46RG0091). We thank F. Younger for the preparation of final figures.

References

Bartell, S. M., J. E. Breck, R. H. Gardner, and A. L. Brenkert. 1986. Individual parameter perturbation and error analysis of fish bioenergetic models. *Can J. Fish. Aquat. Sci.* 43:160–68.

Beauchamp, D. A., D. J. Stewart, and G. L. Thomas. 1989. Corroboration of a bioenergetics model for sockeye salmon. *Trans. Am. Fish. Soc.* 118:597–607.

Bennett, A. F., and K. L. Denman. 1985. Phytoplankton patchiness: Inferences from particle statistics. *J. Mar. Res.* 43:307–35.

Bevelheimer, M. S., R. A. Stein, and R. F. Carline. 1985. Assessing significance of physiological differences among three esocids with a bioenergetics model. *Can. J. Fish. Aquat. Sci.* 42:57–69.

Brandt, S. B. 1993. The effect of thermal fronts on fish growth: A bioenergetics evaluation of food and temperature. *Estuaries* 16:142–59.

Brandt, S. B., and K. J. Hartman. 1993. Innovative approaches with bioenergetics models: Future applications to fish ecology and management. *Trans. Am. Fish. Soc.* 122:731–35.

Brandt, S. B., and J. Kirsch. 1993. Spatially explicit models of striped bass growth potential in Chesapeake Bay. *Trans. Am. Fish. Soc.* 122:845–69.

Brandt, S. B., D. M. Mason, and E. V. Patrick. 1992. Spatially explicit models of fish growth rate. *Fisheries* 17(2):23–35.

Brandt, S. B., D. M. Mason, E. V. Patrick, R. L. Argyle, L. Wells, P. Unger, and D. J. Stewart. 1991. Acoustic measures of the abundance and size of pelagic planktivores in Lake Michigan. *Can. J. Fish. Aquat. Sci.* 48:894–908.

Burczynski, J. J., and R. L. Johnson. 1986. Application of dual-beam acoustic survey techniques to limnetic populations of juvenile sockeye salmon (*Oncorhynchus nerka*). *Can. J. Fish. Aquat. Sci.* 43:1,776–88.

Carpenter, S. R. 1988. *Complex Interactions in Lake Communities.* New York: Springer-Verlag.

Chesson, P., and M. Rosenweig. 1991. Behavior, heterogeneity, and the dynamics of interacting species. *Ecology* 72(4):1,187–95.

Clay, C. S. 1983. Deconvolution of the fish scattering PDF from the echo PDF for a single transducer sonar. *J. Acoust. Soc. Amer.* 73:1,989–94.

Clay, C. S., and H. Medwin. 1977. *Acoustical Oceanography: Principles and Applications.* New York: John Wiley and Sons.

Cox, J. L., L. R. Haury, and J. J. Simpson. 1982. Spatial patterns of grazing-related parameters in California coastal surface waters, July 1919. *J. Mar. Res.* 40:1,127–53.

Foote, K. G., A. Aglen, and O. Nakken. 1986. Measure of fish target strength with a split-beam echo sounder. *J. Acoust. Soc. Am.* 80:612–21.

Forbes, S. T., and O. Nakken. 1972. Manual of methods for fisheries resource survey and appraisal: The use of acoustic instruments for fish detection and abundance estimation. *FAO Manual Fish. Sci. S.*

Frost, T. M., D. L. De Angelis, T. F. H. Allen, S. M. Bartell, and D. J. Hall. 1988. Scale in the design and interpretation of aquatic community research. In *Complex Interactions in Lake Communities,* edited by S. R. Carpenter, 229–58. New York: Springer-Verlag.

Fuiman, L. A., and J. C. Gamble. 1989. Influence of experimental manipulations on predation of herring larvae by juvenile herring in large enclosures. *Cons. Int. Explor. Mer.* 191:359–65.

Gardiner, M. N., and J. B. Hoff. 1982. Diets of striped bass in the Hudson River estuary. *NY Fish. Game J.* 29:152–65.

Gerritsen, J., and J. R. Strickler. 1977. Encounter probabilities and community structure in zooplankton: A mathematical model. *J. Fish Res. Bd. Can.* 34:73–82.

Gibson, R. N., and I. A. Ezzi. 1990. Relative importance of prey size and concentration in determining the feeding behavior of the herring (*Clupea harengus*). *Marine Biology* 107:357–62.

Gower, J. F. R., K. L. Denman, and R. J. Holyer. 1980. Phytoplankton patchiness indicates the fluctuation spectrum of mesoscale oceanic structure. *Nature* 288:157–59.

Greenlaw, C. F., and R. R. Johnson. 1983. Multiple-frequency acoustical estimation. *Bio. Ocean.* 2:227–52.

Hansen, M. J., D. Boisclair, S. B. Brandt, S. W. Hewett, J. F. Kitchell, M. C. Lucas, and J. J. Ney. 1993. Applications of bioenergetics models to fish ecology and management—Where do we go from here? *Trans. Am. Fish. Soc.* 122:1,019–30.

Hartman, K. J. 1993. "Striped Bass, Bluefish, and Weakfish in the Chesapeake Bay: Energetics, Trophic Linkages, and Bioenergetics Model Applications." Ph.D. diss., University of Maryland, College Park.

Haury, L. R., J. A. McGowan, and P. H. Wiebe. 1978. Patterns and processes in the time-space scales of plankton distributions. In *Spatial Patterns in Plankton Communities,* edited by J. H. Steele, 277–327. New York: Plenum Press.

Hewett, S. W. 1989. Ecological applications of bioenergetics models. *Am. Fish Soc. Symp.* 6:113–20.

Hewett, S. W., and B. L. Johnson. 1987. A Generalized Bioenergetics Model of Fish Growth for Microcomputers. University of Wisconsin Sea Grant Technical Report. No. WIS-SG-92-250.

———. 1992. Fish Bioenergetics Model. University of Wisconsin, Sea Grant Institute, Technical Report. No. WIS-SG-92-250.

Hewett, S. W., and D. J. Stewart. 1989. Zooplanktivory by alewife in Lake Michigan: Ontogenetic, seasonal, and historical patterns. *Trans. Am. Fish. Soc.* 118:581–96.

Hill, D. R., and J. J. Magnuson. 1990. Potential effects of climate warming on the growth and prey consumption of Great Lakes fishes. *Trans. Am. Fish. Soc.* 119:265–75.

Hoffman, E. E. 1988. Plankton dynamics on the outer southeastern U.S. continental shelf: A coupled physical-biological model. *J. Mar. Res.* 46:919–46.

Holliday, D. V., R. E. Pieper, and G. S. Kleppel. 1989. Determination of zooplankton size and distribution with multifrequency acoustic technology. *J. Cons. Int. Explor. Mer.* 46:52–61.

Holling, C. S. 1966. The strategy of building models of complex ecological systems. In *Systems Analysis in Ecology,* edited by K. E. F. Watt, 195–214. New York: Academic Press.

Kareiva, P., and M. Andersen. 1988. Spatial aspects of species interactions: The wedding of models and experiments. In *Community Ecology,* edited by A. Hastings, 38–54. New York: Springer-Verlag.

Kitchell, J. F., D. J. Stewart, and D. Weininger. 1977. Applications of bioenergetics model to yellow perch (*Perca flavescens*) and walleye (*Stizonstedion vitreum*). *J. Fish. Res. Bd. Can.* 34:1,922–35.

Koehl, M. A. R. 1989. Discussion: From individuals to populations. In *Perspectives in Ecological Theory,* edited by J. Roughgarden, R. M. May, and S. A. Levin, 39–53. Princeton: Princeton University Press.

Kotliar, N. B., and J. A. Wiens. 1990. Multiple scales of patchiness and patch structure: A hierarchical framework for the study of heterogeneity. *OIKOS* 59:253–60.

Kraft, C. E. 1993. Phosphorus regeneration by Lake Michigan alewives in the mid-1970s. *Trans. Am. Fish. Soc.* 122:749–55.

Lasker, R. 1978. The relation between oceanographic conditions and larval anchovy food in the California current: Identification of factors contributing to recruitment failure. *Cons. Int. Explor. Mer.* 173:212–30.

Legendre, L., and S. Demers. 1984. Toward dynamic biological oceanography and limnology. *Can. J. Fish. Aquat. Sci.* 41:2–19.

Luo, J., and S. B. Brandt. 1993. Bay anchovy (*Anchoa mitchilli*) production and consumption in mid-Chesapeake Bay based on a bioenergetics model and acoustic measures of fish abundance. *Mar. Ecol. Prog. Ser.* 98:223–36.

———. 1994. Virtual reality of planktivores: Are fish really size selective? International Council for the Exploration of the Sea.

MacLennan, D. N., and E. J. Simmonds. 1992. *Fisheries Acoustics.* London: Chapman and Hall.

Magnuson, J. J., T. K. Kratz, T. M. Frost, C. I. Bowser, B. J. Benson, and R. Nero. 1991. Expanding the temporal and spatial scales of ecological research and comparison of divergent ecosystems: Roles for

LTER in the United States. In *Long-term Ecological Research*, edited by P. G. Risser, 45–70. New York: John Wiley and Sons.

Manley, T. O., and J. A. Tallet. 1990. Volumetric visualization: An effective use of GIS technology in the field of oceanography. *Oceanography* 3(1):23–29.

Manooch, C. S. 1973. Food habits of yearling and adult striped bass (*Morone saxatilis*, Walbaum) from Albemarle Sound, North Carolina. *Ches. Sci.* 14:73–86.

Monteleone, D. M., and E. D. Houde. 1990. Influence of maternal size on survival and growth of striped bass (*Morone saxatilis*) Walbaum eggs and larvae. *J. Ecp. Mar. Biol. Ecol.* 140:1–11.

Moore, C. M. 1988. "Food Habits, Population Dynamics, and Bioenergetics of Four Predator Fish Species in Smith Mountain Lake, Virginia." Ph.D. diss., Virginia Polytechnical Institute and State University, Blacksburg.

Nero, R. W., J. J. Magnuson, S. B. Brandt, T. R. Stanton, and J. M. Jech. 1990. Fine-scale biological patchiness of 70-kHz acoustic scattering at the edge of the Gulf Stream—Echofront 85. *Deep Sea Res.* 37(6):999–1,016.

Ney, J. J. 1990. Trophic economies in fisheries: Assessment of demand-supply relationships between predators and prey. *Rev. Aquat. Sci.* 2:55–81.

Persson, L., and L. A. Greenberg. 1990. Optimal foraging and habitat shift in perch (*Perca fluviatilis*) in a resource gradient. *Ecology* 71:1,699–713.

Peterson, M. L., C. S. Clay, and S. B. Brandt. 1976. Acoustics estimates of fish density and scattering function. *J. Acoust. Soc. Amer.* 60:618–22.

Pieper, R. E., D. V. Holliday, and G. S. Kleppel. 1990. Quantitative zooplankton distributions from multifrequency acoustics. *J. Plank. Res.* 12(2):433–41.

Platt, T., and S. Sathyendranath. 1988. Oceanic primary production: Estimation by remote sensing at local and regional scales. *Science* 241:1,613–20.

Possingham, H. P., and J. Roughgarden. 1990. Spatial population dynamics of a marine organism with a complex life cycle. *Ecology* 71:973–85.

Powell, T. M. 1989. Physical and biological scales of variability in lakes, estuaries, and the coastal ocean. In *Perspectives in Ecological Theory*, edited by J. Roughgarden, R. M. May, and S. A. Levin, 157–76. Princeton: Princeton University Press.

Pyke, G. H. 1984. Optimal foraging theory: A critical view. *Ann. Rev. Ecol. Syst.* 15:523–75.

Research System, Inc. 1990. *IDL User's Guide: Interactive Data Language Version 2.0.* Boulder.

Rice, J. A., and P. A. Cochran. 1984. Independent evaluation of bioenergetics model for largemouth bass. *Ecology* 65:732–39.

Rose, G. A., and W. C. Leggett. 1990. The importance of scale to predator-prey spatial correlations: An example of Atlantic fishes. *Ecology* 71:33–44.

Roughgarden, J., S. W. Running, and P. A. Matson. 1991. What does remote sensing do for ecology? *Ecology* 72:1,918–22.

Sklar, F. H., and R. Costanza. 1991. The development of dynamic spatial models for landscape ecology: A review and prognosis. *Ecological Studies Analysis and Synthesis* 82:238–88.

Stanton, T. K., and C. S. Clay. 1986. Sonar echo statistics as a remote-sensing tool: Volume and seaflow. *IEEE J. Oceanic Eng.* 11:79–96.

Steele, J. H. 1978. *Spatial Pattern in Plankton Communities.* New York: Plenum Press.

Stephens, D. W., and J. R. Krebs. 1986. *Foraging Theory.* Princeton: Princeton University Press.

Stewart, D. J., and M. Ibarra. 1991. Predation and production by salmonen fishes in Lake Michigan. *Can. J. Fish. Aquat. Sci.* 48:909–22.

Stewart, D. J., D. Weininger, D. V. Rottiers, and T. A. Edsall. 1983. An energetics model for lake trout, *Salvelinus namaycush*: Application to the Lake Michigan population. *Can. J. Fish. Aquat. Sci.* 40:681–98.

Thorne, R. E. 1983. Assessment of population abundances by hydroacoustics. *Biol. Oceanogr.* 2:253–62.

Turner, M. G., and R. H. Gardner. 1991. Quantitative methods in landscape ecology: The analysis and interpretation of landscape heterogeneity. *Ecol. Studies,* vol. 82. New York: Springer-Verlag.

Urick, R. J. 1983. *Principles of Underwater Sound,* 3rd ed. New York: McGraw-Hill.

Ustin, S. L., C. A. Wessman, B. Curtiss, E. Kasischke, J. Way, and V. C. Vanderbilt. 1991. Opportunities for using the EOS imaging spectrometers and synthetic aperture radar in ecological models. *Ecology* 72(6):1,934–45.

Wiens, J. A. 1989. Spatial scaling in ecology. *Funct. Ecol.* 3:385–97.

Wright, D. I., and W. J. O'Brien. 1984. The development and field test of a tactical model of the planktivores feeding white crappie (*Pomoxis annularis*). *Ecol. Mono.* 54:65–98.

Zastrow, C. E., E. D. Houde, and L. G. Morin. 1991. Spawning, fecundity, hatch-date frequency and young-of-the-year growth of bay anchovy, *Anchoa mitchilli*, in mid-Chesapeake Bay. *Mar. Ecol. Prog. Ser.* 73:161–71.

Stephen B. Brandt is currently Director of the Great Lakes Center and Professor of Biology at SUNY College at Buffalo, New York. Brandt received his Ph.D. in Oceanography and Limnology from the University of Wisconsin, Madison (1978); has held faculty positions at SUNY-ESF (Syracuse) and the University of Maryland; and was a senior research scientist at CSIRO Marine Laboratories in Australia. His research interests include fish ecology in marine and freshwater systems, trophic relationships, spatial processes, and the application of underwater acoustics to aquatic systems.

Stephen B. Brandt, Kyle J. Hartman, and *Jaingang Luo*
Great Lakes Center
Buffalo State College
1300 Elmwood Avenue
Buffalo, New York 14222
E-mail: *brandtsb@snybufaa.cs.snybuf.edu*; *hartmakj@snybufaa.cs.snybuf.edu*; and *luoj@snybufaa.cs.snybuf.edu*

Doran M. Mason
Center for Limnology
University of Wisconsin
7680 North Park St.
Madison, Wisconsin 53706-1492
E-mail: *mason@limnosun.limnology.wisc.edu*

Andrew Goyke
Great Lakes Indian Fish & Wildlife Commission
PO Box 9
Odanah, Wisconsin 54861

James M. Kirsch
Center for Environmental and Estuarine Studies
Chesapeake Biological Laboratory
University of Maryland
PO Box 38
Solomans, Maryland 20688
E-mail: *kirsch@cbl.umd.edu*

43

Knowledge-Based Systems for Coupling GIS and Process-Based Ecological Models

David R. Miller

This chapter describes the bases for the coupling of a process-based vegetation succession model with GIS, using a knowledge-based system and the design of "knowledge models" representing the formalization of available information on ecological factors, spatial analyzes, and data handling. Knowledge of the components of the vegetation model and of spatial data handling are encoded to facilitate the graphical and statistical expression of possible changes in distribution of a vegetation type (Bracken, Pteridium aquilinum*). A frame structure is used for the representation of knowledge and attributes of spatial data. An assessment of the rapidity of change and the possible complexity of changes in land cover and a prediction of the future land cover of an area can be made. The importance of explanation, in addition to descriptions of a system, its operation, and the results are discussed.*

INTRODUCTION

The rapid development of GIS has enabled the incorporation of modeling of different forms: geographic, inductive (Aspinall 1993), rule-based, and knowledge-based (Fedra 1993). Coupling of models and GIS is a major area of required study (Nyerges 1993). The GIS itself becomes an integral part of a model by "providing both the mapped variables and the processing environment" (Berry 1993). The complexity of geographically referenced data, the (potentially) large databases to be manipulated, and the diversity of application areas make GIS a candidate for the application of artificial intelligence techniques (Leung and Leung 1993a, 1993b). Tasks and methods may be classified based on various qualities and characteristics. Recognition of "task characteristics" and the "problem-solving method" of the expert produces a methodological and (less idiosyncratic) development of a knowledge base.

This chapter uses the mapping of vegetation and prediction of vegetation change as the bases for discussing some issues associated with using knowledge-based systems (KBS) for coupling GIS and process-based models. Similar work is being done in other areas of modeling, for example, in hydrology (MacKay et al. 1993). In the mapping of vegetation, GIS may be employed in this twin role as presenting both spatial data and aspatially represented knowledge (Lowell and Astroth 1989).

Models of vegetation change may be based upon knowledge of past states at a location for the prediction of future changes. A synthesis of techniques for spatial data handling applying a process-based understanding of vegetation dynamics may provide the means for monitoring changes in land cover (Lowell and Astroth 1989). In particular, a methodology for what and where to observe may be developed for monitoring those seminatural classes which have been least successfully mapped. However, there needs to be a partitioning of the variability into component parts of successional trend, climatic fluctuations, cyclic changes, and the associated spatial patterns that are emphasized by several authors (Watt 1947; Austin 1981).

Changes in vegetation cover can be predicted from mapped vegetation and knowledge of environmental factors influencing vegetation distribution. Satellite imagery, digital terrain data, and small-scale soil and landform data provide spatially expressed inputs, whereas measured vegetation relationships with ecological variables, vegetation dynamics, and vegetation succession are specified aspatially (Miles 1988; Watt 1947). Techniques are available for the analysis of spatial data and integration of different types of models with spatial data using GIS (Aspinall 1993). The latter data need not be precise or quantifiable; their form may be as knowledge of processes or relevant hypotheses (Buchanan and Shortliffe 1984).

Using a knowledge of associations formed by species and their requirements, the group, or groups, of species likely to occur within an area can be established. In this chapter, an expert system approach is presented, which synthesizes environmental data expressed spatially with knowledge on vegetation dynamics and vegetation succession in particular (temporal models of vegetation types, which have different degrees of reliability, with analysis of "static" spatial data), coupling GIS to understanding of process-based models.

KNOWLEDGE-BASED SYSTEMS AND GIS

Coupling KBS with a GIS requires two knowledge bases to be linked: spatial data in GIS and application-specific knowledge. These may be separated between the "knowledge base" and "inference engine" (Skidmore 1989): (1) knowledge base being a series of rules (quantitative or qualitative) or relationships, and (2) inference engine referring to software and hardware that link the user's questions to the knowledge base and instruct the user on why the system is taking a particular approach.

Ripple and Ulshoefer (1987) discuss the potential of expert systems for improving user-friendliness and efficiency of GIS. However, their argument is split between the need for appropriate data models for efficient representation, retrieval, and analysis of geographical data and the value of expert systems for carrying out these tasks. The reliability of knowledge and data, whether originating from a graphic map, such as probabilities of class membership from satellite image analysis, can be used to determine the ordering of rules within a rule-based system (Mason et al. 1988). In the process of image segmentation described by Mason et al. (1988), each segmented region is assigned a confidence level that can be used in subsequent analysis.

A further limitation of existing systems is that customization of software for a particular task reduces the generic application of the system. It often appears that the GIS provides the graphical interface, the value residing largely in displaying a model and the spatial changes in its state through time for dynamic models. Implementations of expert systems for applications in environmental modeling have kept domain knowledge and problem-solving rules in separate models: the first model is the GIS, the second is the expert system (Skidmore 1989).

In order to effectively manage, share, and use spatial data, it is important to have many types of information about the data, various structures, and often results from many complex processing steps that depend on specific processing parameters, judgments, and objectives of the developer (Clancey 1992). Information such as data statistics can be computed directly from the basic spatial data and may or may not be stored. Metadata include the history of the processing on the data and the history of requests the system has had to tackle.

Bill (1992) lists ten types of knowledge and their implicit or explicit description in GIS. Procedural and prototypical knowledge are implicitly described, but declarative, inexact, incomplete, and uncertain knowledge are explicitly described. A value of KBS is the potential to apply areas of expertise in the absence of complete data where explanation is in terms of analogy and precedent.

What level of explanation of presence or absence and changes in vegetation cover is appropriate? In compiling a facility for the provision of information relevant to studies of changes in vegetation cover, the key components are: (1) applicability to task, (2) reliability indicators on content and results, and (3) pointers to sources of additional information. Other important factors include minimization of redundant information (unless desired) and effectiveness of communication of results. The data used above have subsequently been used in conjunction with the knowledge base to describe the presence of vegetation locally and to then explain it in terms of land use, soils, and topographic factors.

The role of expert systems or knowledge-based systems in vegetation science was described by Noble (1987). He notes that a potentially valuable subject area is applied ecology, where ecological expertise is needed across a wide range of disciplines, such as land management. The impact of ecological theories will be dependent on the degree to which deep knowledge is used in formulating knowledge bases. The value of expert or knowledge-based approaches is of significant benefit, if the interrelationships between ecological processes are stated clearly and linked in a way that can provide advice (that is, predict) (Noble 1987).

KNOWLEDGE-BASED SYSTEMS

Coupling an expert system or knowledge-based system enables a GIS facility (data, software, hardware) to be knowledgeable about its own constraints and potential as well as sensitive to data source, context, and use. Data representation in the KBS is in a frame structure within slots that hold the elemental information relating to data structure: cell, vector (arc), point, or a user-defined object. Treating all data structures as (potential) components of an object provides a facility for flexible analysis of data by KBS and GIS.

The coupling takes the form of exchanging files between GIS and an environmental model where the input, display, and output stages are the only users of the GIS capability. Any close coupling requires a sufficiently open GIS architecture to provide all the linkages and interfacing necessary. The hooks and links between the environmental, the remote sensing and the GIS models, and the GIS functions must be designed into the data model at the outset of the project.

The knowledge of bracken and mapping techniques is encoded in an example schematic inquiry of the knowledge-based system, to follow through the decision-making process involved in mapping bracken and predicting where it will spread. Frequently it is impossible to decide which of an array of ecological differences between two species is responsible for the expression of a particular behavioral or morphological difference. Different environmental conditions lead to different plant characteristics (the plasticity of the plant). Therefore, the selection of the expert needs to be assessed as valid. An alternative solution may require additional or different functions.

Information is relevant data in the context of a set of questions. Therefore, contexts are used for selecting rules and context masks—from experts—used on a posterior basis. The content and priorities of the expert routines and their rules are inherently a representation of knowledge. Their procedures and rule content are the encoding of what to do to satisfy the goal. The priorities are a representation of when to initiate a goal-satisfying routine. Finally, the methodological frames and the encoded goals within experts provide the metaknowledge of how to satisfy the goals.

This greatly enhances the system and its opportunities. The process of backward chaining (which unwinds steps back to a particular decision node) includes cross referencing to other domains of knowledge (i.e., production technology or particular user specifications).

To analyze metadata and record processing operations more generally, a data history model for both quantitative and qualitative spatial metadata is designed and implemented. The components of metadata used here build on that of the United States Spatial Data Transfer Standard (Federal Geographic Data Committee 1992). Access is required to metadata for query and update and for use in data processing. Any operations on a data set preserve the data set's metadata.

The inclusion of information on data sources provides one linkage between data input and processing for cartographic presentation. If the data are raster scanned, the scale of capture is taken from the empirical relationship between scale and maximum resolution. A rule-of-thumb to follow is that the scale of maximum valid use of data follows a guideline of 2.5 times the source scale. Other metadata include the nature of the data source (classified from satellite imagery as opposed to raw data or vector digitized and rasterized). Data currency, which relates to the estimated length of time the data are relevant for and the routine history data linked to the absolute and relative error slots, allows error tracking and lineage reporting (Lanter 1990).

MODELS OF VEGETATION GROWTH

The extensive research undertaken into all aspects of bracken has produced factual information that is either quantifiable or descriptive of the physiological processes and mechanisms involved in bracken's growth cycle. Some knowledge of bracken (where to look and what to look for) provides a basis for its mapping. The knowledge base is of ecological data on bracken presence: when, where, where not, and why not. "When," relates to the growth cycle of bracken; "where," to the environmental niches that provide a likely habitat for bracken; "where not," to those environmental niches bracken will definitely not inhabit; and "why not," to land use and bracken ecology that determine bracken presence (Table 43-1).

Fluctuations and cyclic regeneration are present in observations of long-term trends in the distribution of a species. Appreciation of which types of temporal change are being observed is the basis of applying the observations in a predictive mode. Similarly, spatial patterns may denote the extent and location of changes in a vegetation stand. The role of nucleation (a core patch) was studied by Watt (1947), who identified its role in determining spatial pattern and temporal variability of the community. This concept combined the expanding spatial front of a building phase and the existence of a persistent clonal patch. The patch of the dominant species shows four zones: pioneer, building, mature, and degenerate. This pattern, the associated plant density around the edge of the stand, interrelates with the earlier observations on vegetation continua (Foody 1992).

Table 43-1. A hierarchy of successional causes (modified from Pickett et al. 1987).

General causes of succession	Contributing processes or conditions	Defining factors
Site availability	Coarse-scale disturbance	Size, severity, time, dispersion
Differential species availability	Dispersal	Landscape configuration
	Propagule pool	Dispersal agents

The model of change is based upon vegetative reproduction of bracken, and, as such, requires: (1) a bracken presence, and (2) changes along the boundaries of that existing bracken. Contextual information of the vegetation cover may be lost with interpretation of polygonal information without boundary information. The mature hinterland to bracken locally shows signs of degeneration, which may be as low growth or discoloration similar to deficiency disease. Regression in bracken may be due to patch aging, climate, or pollution; and locally, to soil conditions and microclimate. These changes become obvious when bracken cover reveals gaps.

PROTOTYPE KNOWLEDGE-BASED SYSTEMS

An expert system shell (SBS) (Baldock et al. 1987) is being used as a framework within which to develop the approach to vegetation analysis. Separation of types of knowledge into different groups or sources helps modularity but complicates control and communication within the system. Using a blackboard system (Engel et al. 1990), each source may have its own type of knowledge and processing. The output of that source is recorded on the blackboard, which acts as a common database. A scheduler is used to select the knowledge source appropriate to the current task. This shell uses the language POP-11 for the scheduling of expert routines, which retain systemwide information (i.e., current estimates of accumulated errors). Information specific to individual experts (i.e., soil type details) is held in a frame structure, which manipulates the object-referenced data but not with an object-oriented code. Priorities assigned to each expert may be changed during a system run. The experts operate opportunistically on the current goal according to their goal, priorities, and prerequisites. Access to the raster format spatial data sets is by means of FORTRAN routines initiated by expert routines. The expert routines and the POP-11 database provide the vehicles for transferring information between the goal list and the GIS functions in the form of subgoals, function parameters, and variable values.

Frame-based representations of objects allow class–subclass, is–a, and has–a relations (Usery 1988). Attributes are assigned to these objects which, if changed, may have a trigger system for recalculating other associated attribute values within other objects. Thus, changes in the area of one object will necessitate a change in the area of neighboring objects. The frame is used to represent a group of entities with attendant facts (1988). For example, a frame contains slots for each type of soil map unit represented and used in the spatial data set: slots in turn are frames containing details of individual soil map units.

In most blackboard systems, each expert routine functions without knowledge of any other. In this prototype we include representation of the other expert routines, their goals and prerequisites, within a frame structure (Usery 1988), and the KBS has the potential to call on other aspects of knowledge and expertise. The system executes its goals by means of matching information between database entries and current problem details. This is a mechanism that contains a list of data items which may in itself be a list. Accompanying this information is a degree of confidence that the relationship and error assessments for one class will have consequential impact on error assessments of classes which it is adjacent to or confused with (Veregin 1989).

Use of a daemon procedure within a frame provides a summary of the local topographic, soil, and land-use environment. Daemon procedures are executed on receipt of a request for the value of the slot in a frame. The daemon triggers POP-11 code, which in turn calls FORTRAN code. The same sequence of procedures is triggered for decoding the classified land cover classes and their probabilities.

The data model for locational information provides for the probability levels associated with each pixel in a raster format. This element of the data model is embedded in the frame structure and inspectable in the data set from a FORTRAN routine that decodes the class value and probability level into either raster format data sets or information on single pixels. Three forms of information are utilized within the system, which may be quantitative or qualitative. The former output option would be used when all the *a priori* settings and the thresholds for the probability levels are decided upon. The latter option provides the representation of the change in the vegetation types along the continuum of vegetation types.

Feature extraction or manipulation is of three types: (1) spectral (intensity or leaf-area index), (2) spatial (with horizontal or horizontal and vertical components), or (3) temporal (dynamics, human interference). The rate of change of difference in likelihood between classes provides information on the nature of the boundary between the two most likely classes. Where the rate of change of the difference is slow, the extent of intergradation between the two classes is high, and the boundary becomes increasingly diffuse until only isolated patches or fronds of bracken are visible. In locations where the rate of change of the difference is high, the nature of the boundary is observed to be more distinct and narrower in dimension.

The stages of assessment of likelihood of spread involve checking the following:

1. Bracken heterogeneity as measured by the number of bracken polygons in an area minus the number of holes in those polygons (the Euler number). This is calculated at different probability levels to give a measure of the susceptibility to bracken spread, according to the rules in the POP-11 database.
2. Adjacent environment to that occupied by bracken. The altitude, slopes, aspect, soil types, exposure levels, and land cover types adjacent to existing bracken are summarized, and each bracken pixel is assigned a score of suitability of its neighboring environment to bracken spread.
3. Appreciation of which types of temporal change are being observed to identify active and static edges to the bracken stands.
4. Vegetation succession model (Miles 1988), which allocates the likelihood of spread according to the neighboring vegetation type.
5. A classification matrix is applied at each location around the boundary.

The KBS scheduler triggers tests on the possible ecological meaning of interpretations of spatial data at any particular state. For example, is there any value in looking at the topology of the second most likely class? Matching knowledge of successional causes with the spatial pattern of classification output may indicate the likely existence of rhizomes (roots) or a local source of spores.

The first description is a listing of the system routines used and the decisions made by the scheduler. The second description uses data stored in the frames of what can be determined from frames or expert routines being accessed and what that data implies for changes in bracken distribution. This is the value of the KBS approach (Buchanan and Shortliffe 1984)—the coupling of process-based understanding of the subject to the spatial organization of its components. The extent of the explanation depends on the depth of the knowledge base, the credibility depends on the sources of knowledge, and the reliability depends on inferences.

CONCLUSION

The extent to which expert system tools can be readily built into or upon GIS applications depends largely upon the nature of future users.

If the users' backgrounds in environmental applications become more diverse, then the degree to which technology and user expertise can be formally coupled becomes a more valuable area of research. If the remit of users develops more narrowly along the lines of specialization of, for example, cartography, image processing, or forestry and agriculture, then the expert system coupling may be less useful. In either case, the need for theoretical and developmental work into the nature and form of data models will continue to increase in importance.

The effectiveness of the communication to the user of a coupled knowledge base and GIS for describing a resource (specifically, selected vegetation types) would be enhanced by the use of, for example, multimedia facilities. The inherent use of displays of multiple types of data would be well suited to description, illustration, and explanation of the basis and nature of different types of knowledge being brought to bear on a subject. Finally, the representation of ecological knowledge using knowledge-based approaches provides a means of building the linkages between the skill of the ecologist and spatial analyst and the needs of the user (and therefore the specification of the map product). This is not for technological fulfillment; rather, the value lies in the answers it provides to questions not easily answered by one discipline on its own.

Acknowledgments

The author would like to thank the Scottish Office Agriculture and Fisheries Department for its funding of the work from which this chapter is derived. Acknowledgment is also due to Richard Aspinall, for his detailed discussion on context and objectives of this approach; and to Jane Morrice, Alistair Law, and Matt Wells for their advice on running POPLOG and SBS. Thanks, also, to the Medical Research Council and Population Cytogenetics Unit at the University of Edinburgh for use of the SBS expert system shell.

References

Aspinall, R. J. 1993. Exploratory spatial analysis in GIS: Generating geographical hypothesis from spatial data. In *Proceedings of GIS Research UK*. Keele: University of Keele.

Austin, M. P. 1981. Permanent quadrats: An interface for theory and practice. *Vegetatio* 46(11):1–10.

Baldock, R. A., J. Ireland, and S. J. Towers. 1987. *SBS User Guide*. Edinburgh: Medical Research Council and Population Cytogenetics Unit.

Berry, J. K. 1993. Cartographic modeling: The analytical capabilities of GIS. In *Environmental Modeling with GIS*, edited by M. F. Goodchild, B. O. Parks, and L. T. Steyaert, 58–74. New York: Oxford University Press.

Bill, R. 1992. On the acquisition, representation, and application of knowledge in geo-information systems. Washington: ASPRS.

Buchanan, B. G., and E. H. Shortliffe. 1984. *Rule-Based Expert Systems, The MYCIN Experiments of the Standard Heuristic Programming Project*. New York: Addison-Wesley.

Clancey, W. J. 1992. Model construction operators. *Artificial Intelligence* 53(1):1–115.

Engel, B. A., D. B. Beasley, and J. R. Barrett. 1990. Integrating expert systems with conventional problem-solving techniques using blackboards. *Computers and Electronics in Agriculture* 4:287–301.

Federal Geographic Data Committee. 1992. *Information Exchange Forum on Spatial Meta-Data*. Reston: USGS.

Fedra, K. 1993. GIS and environmental modeling. In *Environmental Modeling with GIS*, edited by M. F. Goodchild, B. O. Parks, and L. T. Steyaert, 35–50. New York: Oxford University Press.

Foody, G. M. 1992. A fuzzy-sets approach to the representation of vegetation continua from remotely sensed data: An example from lowland heath. *Photogrammetric Engineering and Remote Sensing* 58(2):221–25.

Lanter, D. P. 1990. Lineage in GIS: The problem and a solution. National Center for Geographical Information and Analysis Technical Paper, No. 90.

Leung, Y., and K. C. Leung. 1993a. An intelligent expert system shell for knowledge-based geographical information systems. I. The tools. *International Journal of Geographical Information Systems* 7(3):189–99.

———. 1993b. An intelligent expert system shell for knowledge-based geographical information systems. II. Some applications. *International Journal of Geographical Information Systems* 7(3):201–13.

Lowell, K. E., and J. H. Astroth. 1989. Vegetative succession and controlled fire in a glades ecosystem—A geographical information system approach. *International Journal of Geographical Information Systems* 3(1):69–81.

MacKay, D. S., V. B. Robinson, and L. E. Band. 1993. An integrated knowledge-based system for mapping spatiotemporal ecological simulations. *AI Applications* 7(1):29–36.

Mason, D. C., D. G. Corr, A. Cross, D. C. Hogg, D. H. Lawrence, M. Petrou, and A. M. Tailor. 1988. The use of digital map data in the segmentation and classification of remotely sensed images. *International Journal of Geographical Information Systems* 2(3):195–215.

Miles, J. 1988. Vegetation and soil change in the uplands. In *Ecological Change in the Uplands*, edited by M. B. Usher and D. B. A. Thompson. Special publications series of the British Ecological Society, No. 7.

Noble, I. R. 1987. Expert systems in vegetation science. *Vegetatio* 69(2):115–22.

Nyerges, T. L. 1993. Understanding the scope of GIS. In *Environmental Modeling with GIS*, edited by M. F. Goodchild, B. O. Parks, and L. T. Steyaert, 75–93. New York: Oxford University Press.

Openshaw, S. 1993. A concepts-rich approach to spatial analysis for GIS. In *Proceedings of GIS Research UK*. Keele: University of Keele.

Pickett, S. T. A., S. L. Collins, and J. J. Armesto. 1987. A hierarchical consideration of causes and mechanisms of succession. *Vegetatio* 69(2):109–14.

Ripple, W. J., and V. S. Ulshoefer. 1987. Expert systems and spatial data models for efficient geographic data handling. *Photogrammetric Engineering and Remote Sensing* 53(10):1,431–33.

Skidmore, A. K. 1989. An expert system classifies eucalypt forest types using thematic mapper data and a digital terrain model. *Photogrammetric Engineering and Remote Sensing* 55:1,449–64.

Usery, L. E., P. Altheide, R. R. P. Deister, and D. J. Barr. 1988. Knowledge-based GIS techniques applied to geological engineering. *Photogrammetric Engineering and Remote Sensing* 54(11):1,623–28.

Veregin, H. 1989. A taxonomy of error in spatial databases. National Center for Geographical Information and Analysis Technical Paper, No. 89.

Watt, A. S. 1947. Pattern and process in the plant community. *Journal of Ecology* 35:1–22.

Yee L., and K. S. Leung. 1993. An intelligent expert system shell for knowledge-based geographical information systems: The tools. *International Journal of Geographical Information Systems* 7(3):189–200.

David Miller is a graduate of the Departments of Geography and Topographic Science at the University of Glasgow and the Department of Mathematics at the University of Aberdeen. He has been working at the Macaulay Land Use Research Institute for the last ten years on environmental modeling using spatial data. His current research includes knowledge-based systems for use with ecological models, dynamic classification of landscape for visual impact assessments, and the impact of acidification and eutrophication on freshwater quality.

GIS and Remote Sensing Group
Macaulay Land Use Research Institute
Aberdeen, Scotland
Phone: 44 224 318611 x 2240
Fax: 44 224 311556
E-mail: *d.miller@mluri.sari.ac.uk*

44

Integrating GIS, Scientific Visualization Systems, Statistics, and an Orographic Precipitation Model for a Hydroclimatic Study of the Gunnison River Basin

Lauren Hay, Loey Knapp, and Janet Bromberg

As part of the U.S. Geological Survey's Gunnison River Basin Climate Study, hydroclimatic models are used to assess the potential effects of climate change on water resources. Complex hydroclimatic-modeling problems commonly involve overlapping data requirements, as well as massive amounts of one- to four-dimensional data in multiple scales and formats. Geographic Information Systems (GIS) and Scientific Visualization Systems (SVS), combined with advanced statistical capabilities (STAT), are powerful tools for developing and analyzing complex hydroclimatic models. In this chapter, a four-component system is presented in which a hydroclimatic model, GIS, SVS, and STAT are all accessed from a graphical user interface, providing a tool for spatial data management and manipulation, model parameterization, visual data interpretation, and model verification.

INTRODUCTION

Background

Much of the available software for scientific analysis of data have been developed along functional lines: GIS perform analytical functions on spatial data; SVS render data in a variety of ways; and STAT perform classical and modern statistics. In general, the software development in each of these areas has followed a set of assumptions derived from different primary users, creating a lack of coherence relative to scientific data. GIS tend to use only two-dimensional data because they were initially developed for land management and utility problems. SVS render *n*-dimensional scientific data, but lack data management and analytical functions because they were developed to visualize modeled data. STAT tend to use nonspatial data because they were developed for business applications.

Solutions to these scientific problems are difficult and tedious because they require a cross section of these functions that are working interactively in a modeling application. As a result, scientists have been forced to take on the role of data administrators, commonly reformatting their data to fit the various input/output requirements before completing their analyses. Using currently available software, scientists spend 30% or more of their time on data manipulation, adversely affecting the timeliness and cost of projects and the pace of scientific progress.

An integrated scientific platform consisting of GIS, SVS, and STAT that could be easily linked with models would provide scientists with statistical and visual representations of phenomena across space and time. The ease of data access and the ability to develop flexible methods for the quantification of spatial variables over discrete areas make GIS an integral tool to modelers (Hay et al. 1992; *Proceedings*, 1993 and *Environmental*, 1993) using these methods. GIS have become popular for environmental analysis, partly because of their display capabilities; but these display capabilities generally are limited to the creation of static, fixed-color maps (Knapp 1993). SVS tend to be used exclusively for the display of complex images through time. In general, GIS and SVS do not have the exploratory data analysis, graphics, and statistical tools found in many STAT. However, STAT does not have the advanced rendering capabilities of SVS nor the spatial analytical capabilities of GIS. A system that incorporates the capabilities of GIS, SVS, and STAT would facilitate model development, calibration, and verification by eliminating the time and effort for transfer of data across systems. Improved communication of model results would lead to a broader understanding of the model assumptions and results. This chapter describes a structure for such an integrated analysis and visualization system and its application to a hydroclimatic modeling effort in the Gunnison River basin.

Gunnison River Basin Climate Study

The objectives of the U.S. Geological Survey's Gunnison River Basin Climate Study are to identify the sensitivity of water resources in the basin to reasonable scenarios of climate change and to develop techniques useful in assessing the sensitivity of water resources to changes in climate. One of the techniques being developed is the use of GIS, SVS, and STAT as tools for spatial data management and manipulation, visual data interpretation, and model calibration and verification.

The Gunnison River basin in southwestern Colorado has a drainage area of 20,530 km^2 and elevations ranging from 1,410–4,400 m. The basin is geologically and hydrologically diverse and provides a challenge in defining the spatial distribution of various components of the hydrologic system (e.g., precipitation, temperature, and evaporation). The spatial distribution of precipitation in the Gunnison River basin is variable and complex because of orographic effects. Snow is the principal source of available water in the basin, with seasonal accumulation and storage located above 2,800 m. However, no long-term precipitation station exists above this elevation. Evaluation of hydrologic response to climate variability and change depends on accurate estimates of winter snowpack at these higher elevations. For the Gunnison River basin study, an orographic precipitation model was chosen to estimate the spatial and temporal distribution of winter precipitation within the basin.

THE SYSTEM

Hay and Knapp (1993) demonstrated the use of a three-component system that uses a precipitation model, GIS, and SVS to aid in spatial data management and manipulation, model parametrization, visual interpretation, and model verification. They concluded that there is a need for a system that more thoroughly integrates the display capabilities of SVS and analytic functions of GIS.

Continued work on this integration has resulted in a system that is currently being developed at the U.S. Geological Survey in collaboration with IBM. The system, illustrated in Figure 44-1, is comprised of the following four components: (1) a GIS–ESRI ARC/INFO; (2) a MODEL–RHEA-Colorado State University (RHEA-CSU) orographic precipitation model; (3) a STAT–StatSci S-PLUS; and (4) an SVS–IBM Data Explorer.

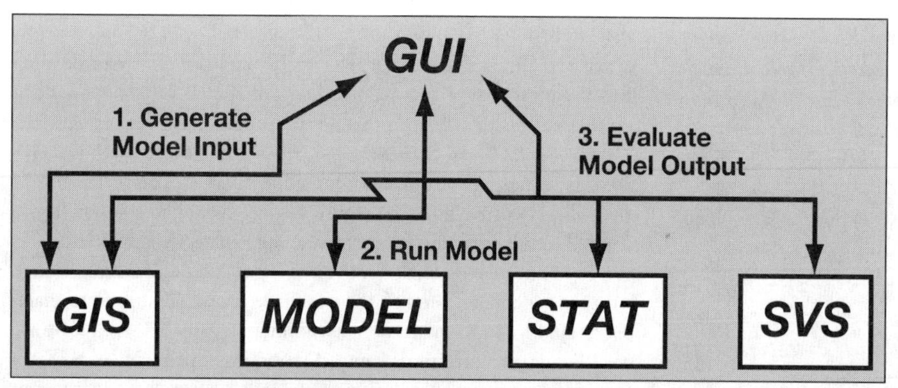

Figure 44-1. Conceptual view of the input-output relations between a GUI, GIS, STAT, and SVS.

Communication between the various systems is handled by the graphical user interface (GUI), which is modular in design and provides users with a building-block type of framework that allows a sequence of instructions to be applied as a macro that can be stored, reused, or modified. In this system, model input is generated using GIS. Model output is then accessed by the GIS, SVS, and STAT for visualization and analytical interpretation.

Orographic Precipitation Model

As part of the Gunnison River Basin Climate Study, the RHEA-CSU model is being used to estimate precipitation on a daily basis at a variety of scales (Hay et al., *Environmental*, 1993). To assess the effects of climate change on the water resources within the Gunnison River basin, the RHEA-CSU model is applied within a larger modeling framework in which general circulation and mesoscale general circulation models are linked to the RHEA-CSU model and a watershed model to produce possible scenarios of climate change (Leavesley et al. 1992; Hay et al. 1992; Kuhn and Parker 1992). Before using this approach, the accuracy and "optimal" scale of the RHEA-CSU model need to be assessed.

The RHEA-CSU model was developed in the late 1970s (Rhea 1977). The model has been updated and converted from FORTRAN to C programming language to allow for user-designated resolution with no code modifications. The model is steady state, multilayer, and two dimensional; one dimension is along the prevailing 700-mb wind direction and the other is vertical. The model requires as input: (1) twice daily soundings from nearby or surrounding upper-air stations, and (2) gridded elevation data. The soundings provide measurements of atmospheric pressure, temperature, relative humidity, wind direction, and wind speed at various altitudes. Nine elevation grids are generated using DEM point values, one for every ten degrees of rotation from 0°–80°; grids at complementary and supplementary angles are derived from these nine. An elevation grid is selected that corresponds to the 700-mb wind direction at the center of the study area, rounded to the nearest 10°. Upper-air data are interpolated to the upwind edge of the selected elevation grid. Precipitation estimates are calculated at each point of the selected rotated grid and interpolated to the nonrotated inner grid using the inverse distance squared from the four nearest grid points. The interpolated inner grid is used for model output and covers an area common to all of the rotated grids. Figure 44-2 shows examples of two rotated grids and the common interpolated inner grid.

The RHEA-CSU model simulates the interaction of air layers with the underlying topography by allowing vertical displacement of the air column while keeping track of the resulting condensate or evaporation. The lifting process is assumed to be moist adiabatic. The lift due to large-scale vertical motion is linearly additive to the topographic lift, but buoyant convection is not treated. The orographic component is proportional to the terrain height and depends on the vertical stability in the interpolated soundings. As the layers flow across the region, part of the condensate precipitates. The precipitation amount for a given layer is estimated as a temperature-dependent fraction of the total condensate (precipitation efficiency), while the remainder of the condensate is advected downwind to the next grid cell. In the model, individual layers can moisten by: (1) dry or vertical displacement as a result of topography and large scale vertical motion field, and (2) precipitation that falls from higher layers. Vertical layer computation of moisture conditions for each grid cell starts at the highest layer and works down. Evaporation of falling precipitation into unsaturated lower layers moistens these strata and decreases the precipitation reaching the ground.

Geographic Information Systems—GIS

GIS are used to establish a common database for individuals working on different aspects of the Gunnison River Basin Climate Study, and to develop methods for acquiring, generating, managing, and displaying spatial data required for modeling efforts. Additionally, GIS provide several means for verifying model results and enhance the flow of information and ideas between project personnel with different specialties (Hay et al., *Proceedings*, 1993). GIS provide input to the RHEA-CSU model, SVS, and STAT, and analyze output from the RHEA-CSU model.

While the primary function of GIS within the system described in this chapter is to provide the input files for the model, it also provides spatial data for the SVS bounded to the correct domain. Development of the elevation grids required by the RHEA-CSU model were automated using GIS: this eliminated the most labor-intensive step involved in applying the model to a new area (M. D. Branson, Colorado State University, Department of Atmospheric Science, oral communication 1991). Elevation grids required by the RHEA-CSU model are generated from DEM values, making it pos-

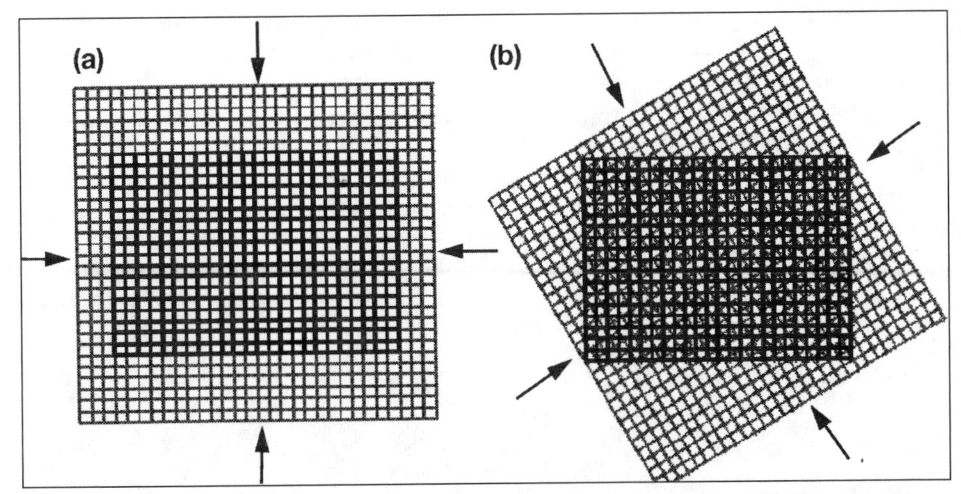

Figure 44-2. RHEA-CSU rotated grid and interpolated inner grid examples: (a) 0° rotation, and (b) 60° rotation. Arrows represent four possible wind directions for each rotated grid.

sible to simulate precipitation over a range of spatial scales and enabling the user to choose the method of topography characterization (e.g., mean, maximum, or minimum elevation) (Hay et al. 1992, *Environmental*, 1993 and *Proceedings*, 1993; Battaglin et al. 1993). Routines written in ARC Macro Language (AML) are used to generate the rotated elevations grids at a user-specified resolution.

The analytic capabilities of GIS are used to calculate the input parameters used in the RHEA-CSU model's interpolation scheme to the inner grid (Figure 44-2). The model uses the inverse distance squared from the four closest rotated grid cells when interpolating to the inner grid. GIS can be used to generate alternative input files for interpolation. For example, instead of using inverse distance squared, the percent area of each rotated grid cell that falls within the domain of the nonrotated inner grid can be easily calculated within GIS and used for model interpolation.

Scientific Visualization Systems—SVS

SVS are used to animate sequences of images displaying model results over space and time. SVS allow the user to interact with the data in a flexible manner. The user can: (1) change the control display characteristics in real-time (i.e., opacity, value, and hue); (2) determine the number of variables to be displayed; and (3) provide multiple views of the data (statistical as well as spatial). As indicated in Figure 44-1, SVS are used to evaluate output from the model, GIS, and STAT by effectively displaying data through space and time, a function not found in either GIS or STAT.

Several techniques were used in this study to determine the optimal scale to execute the RHEA-CSU model. Concurrent animation of model output at different scales over space and time allowed direct examination of the effect of scale. Additionally, grid subtraction indicated the effects of a change in scale. For example, precipitation values were simulated at 10-, 5-, and 2.5-km resolutions. In the SVS, 10-km elevation grids refined to 5- or 2.5-km can be subtracted from the 5- and 2.5-km grids and precipitation differences can be animated through both space and time.

Statistics—STAT

STAT are used in this study for exploratory data analysis and graphics. The primary function of STAT within this system is the evaluation of model output through statistical and graphical analysis (Figure 44-1). Statistical analysis, including regression and time-series analyses, is accessible through STAT. Graphical output, such as x–y scatterplots, boxplots, and time-series plots, can be employed to better understand the relations between model results and data. In the Gunnison River Basin Climate Study, STAT is used to identify outliers, perform residual analysis, and determine the relation between grid size and model output.

CONCLUSION

A four-component system was described that is being used in hydroclimatic modeling application from the Gunnison River Basin Climate Study. The system will be used to investigate the effects of spatial scale on precipitation processes, as well as to aid in model calibration and verification. Results will provide a basis for developing intra- and interregional variation in these processes, transferring these results to other mountainous regions, and improving the understanding of orographic precipitation processes in mountainous regions.

The system consists of a model, a scientific visualization system (SVS), a geographic information system (GIS), and a statistical package (STAT), which are all accessible from a graphical user interface (GUI). The modular design of the GUI provides a building-block framework for the system. The SVS visualization capabilities allow for the rendering of complex images through time and space and allow more flexibility and breadth of display. The GIS is an integral part of the system, facilitating the automation of input data generation for the orographic precipitation model and the analysis of model output. The STAT provides the statistical and advanced graphical capabilities. Together the system forms an integrated analysis tool to aid in modeling applications.

This integrated scientific platform provides scientists with statistical and visual tools needed to model phenomena across space and time, and facilitates model development, calibration, and verification by eliminating the time and effort for transfer of data across systems. Improved communication of model results leads to a broader understanding of model assumptions and results. The application presented in this chapter is a demonstration of how such a system might operate. Analysis of this system provides a basis for the transferability of these results to other models for development of similar systems. The utility of such a system could be expanded/modified to incorporate other models.

References

Battaglin, W. A., L. E. Hay, R. S. Parker, and G. H. Leavesley. 1993. Application of a GIS for modeling the sensitivity of water resources to alterations in climate in the Gunnison River basin, Colorado. *Water Resources Bulletin* 29(6):1,021–28.

Branson, M. D. 1991. "An Historical Evaluation of a Winter Orographic Precipitation Model." Master's thesis, Colorado State University, Department of Atmospheric Sciences, Fort Collins.

Hay, L. E., W. A. Battaglin, R. S. Parker, and G. H. Leavesley. 1993. Modeling the effects of climate change on water resources in the Gunnison River basin, Colorado, using GIS technology. In *Environmental Modeling with GIS*, edited by M. F. Goodchild, B. O. Parks, and L. T. Steyaert, 173–81. New York: Oxford University Press.

Hay, L. E., W. B. Battaglin, M. D. Branson, and G. H. Leavesley. 1993. Application of GIS in modeling winter orographic precipitation, Gunnison River basin, Colorado. *Proceedings from International Conference on Application of Geographic Information Systems in Hydrology and Water Resources Management,* 491–500.

Hay, L. E., M. D. Branson, and G. H. Leavesley. 1992. Simulation of precipitation in the Gunnison River basin using an orographic precipitation model. *Proceedings from AWRA 28th Annual Conference and Symposia,* 651–60.

Hay, L. E., and L. Knapp. 1993. Visualization techniques for hydrologic modeling. *Proceedings of Symposium on Hydrologic Modeling Demands in the 90's.*

Knapp, L. 1993. Task analysis and geographic visualization. *Proceedings from the American Society of Photogrammetry and Remote Sensing.*

Kuhn, G., and R. R. Parker. 1992. Transfer of hydrologic model parameters from calibrated to noncalibrated basins in the Gunnison River basin, Colorado. *Proceedings from AWRA 28th Annual Conference and Symposia,* 741–50.

Leavesley, G. H, M. D. Branson, and L. E. Hay. 1992. Investigation of the effects of climate change in mountainous regions using coupled atmospheric and hydrologic models. *Proceedings from AWRA 28th Annual Conference and Symposia,* 691–700.

Rhea, J. O. 1977. "Orographic Precipitation Model for Hydrometeorological Use." Ph.D. diss., Colorado State University, Department of Atmospheric Science, Fort Collins.

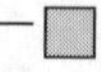

Lauren Hay is a hydrologist for the U.S. Geological Survey, Water Resources Division. She received her M.S. in hydrology from the University of Arizona, Tucson, in 1986 and is working on her Ph.D. at the University of Colorado, Boulder. She is presently researching the effects of climate change on water resources in mountainous regions.
U.S. Geological Survey
Water Resources Division
MS 412, Box 25046, DFC
Denver, Colorado 80225
E-mail: *lhay@lhay.cr.usgs.gov*

Loey Knapp and *Janet Bromberg*
International Business Machines Corporation
6330 Spine Rd.
Boulder, Colorado 80301
E-mail: *knappl@bldfvm9.vnet.ibm.com*; *jbromberg@bldfvm9.vnet.ibm.com*

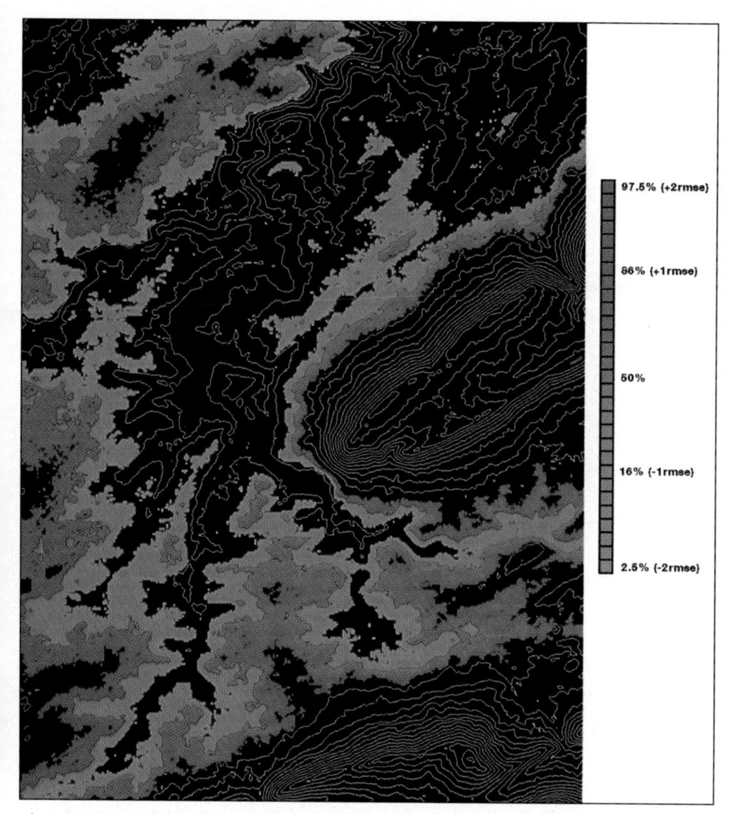

Plate 3-1. Visualization of uncertainty in the position of the 350-m contour for the State College, Pennsylvania, 7.5-minute quadrangle. Shown are the 95% confidence limits on the contour position, based on the standard 7-m root mean square error in ground surface elevation and an assumed Gaussian distribution. (Source: U.S. Geological Survey 30-m DEM.)

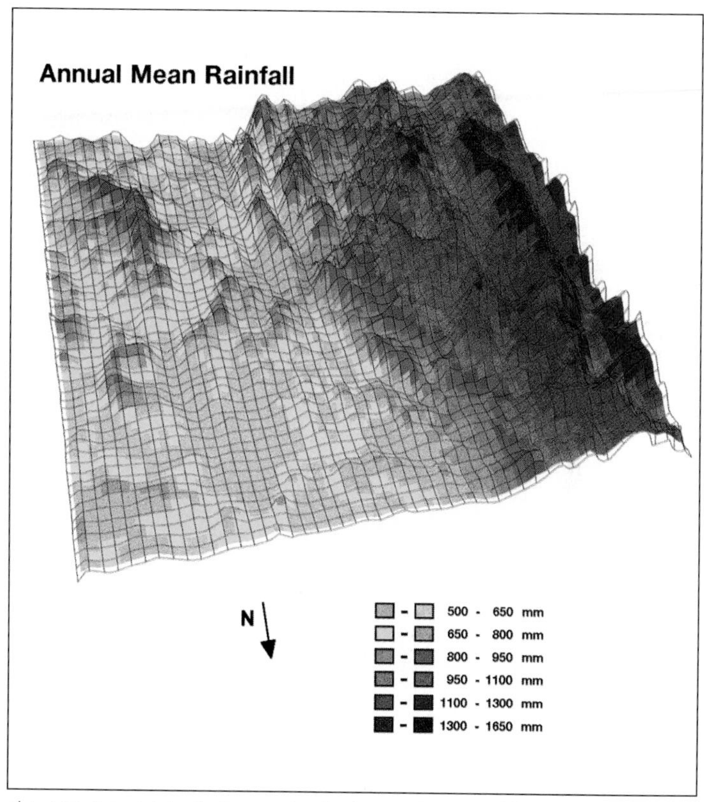

Plate 16-1. Estimated standard error surface for the annual mean rainfall plotted in Figure 16-4.

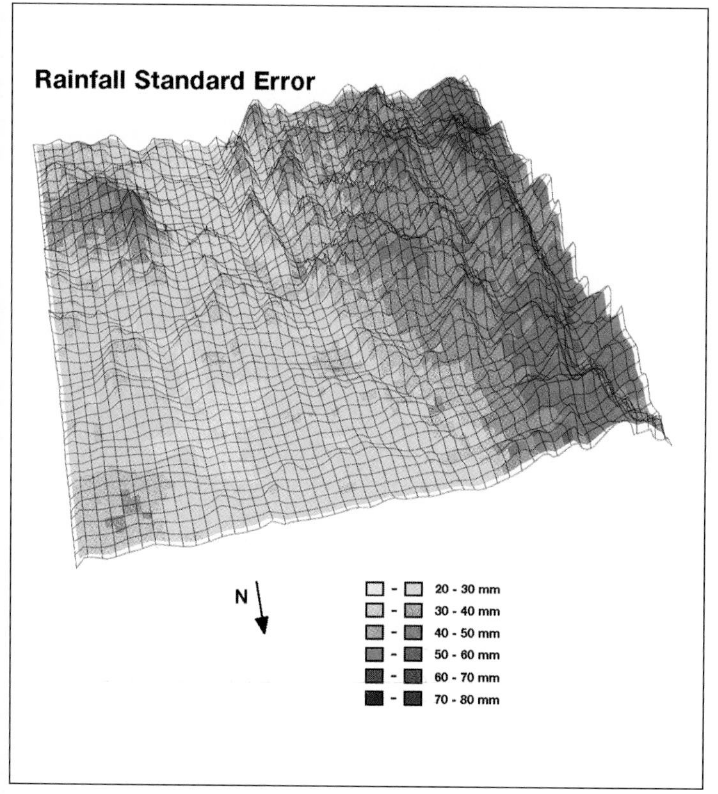

Plate 16-2. Logarithm of rainfall variance versus logarithm of annual mean rainfall.

Plate 18-1. Montana digital elevation grid, 1-km horizontal resolution, 6-m vertical resolution, with markers indicating the location of weather stations used in this analysis.

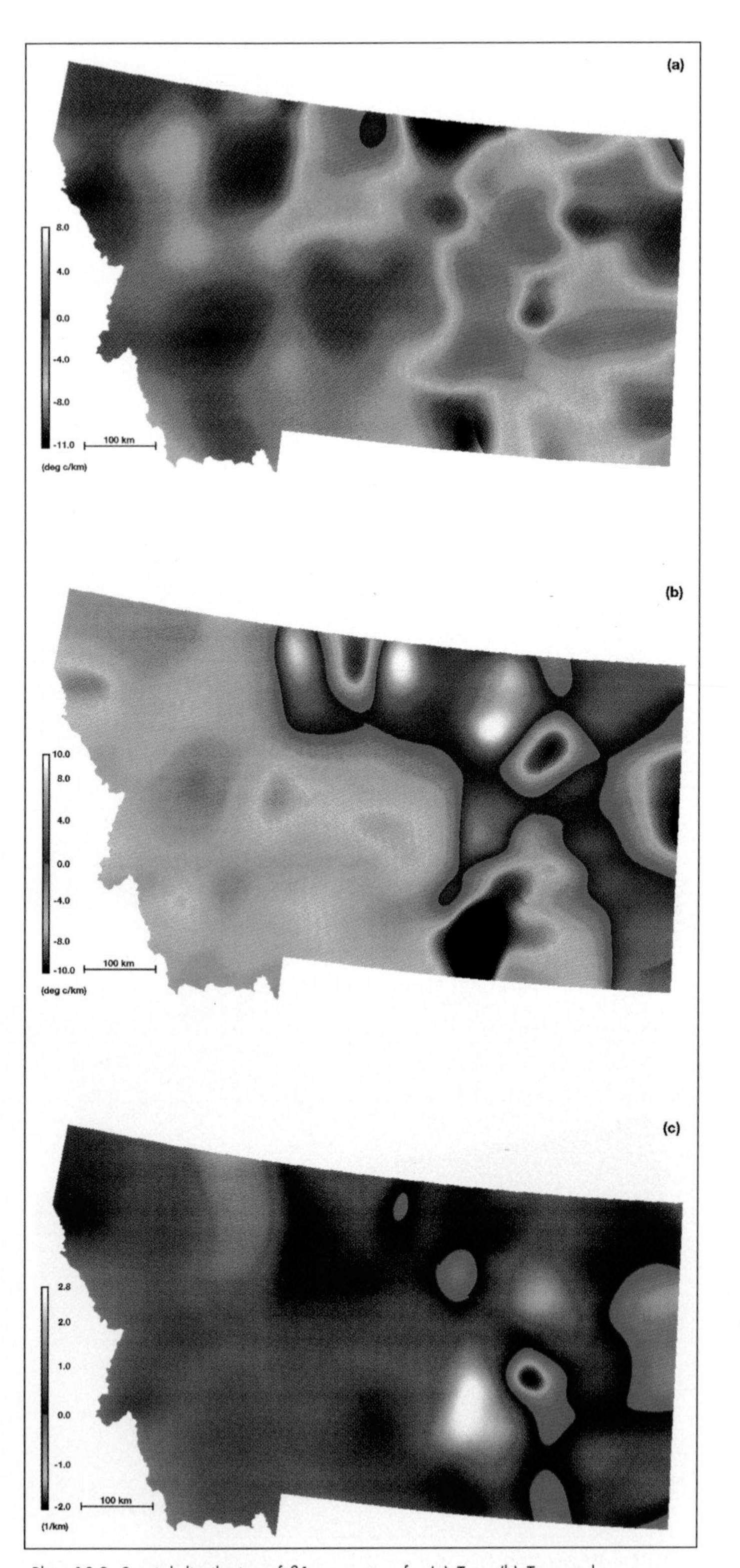

Plate 18-2. Spatial distribution of *ß1* parameters for (a) T_{max}, (b) T_{min}, and (c) *Prcp* regressions, averaged over the entire year at each grid cell.

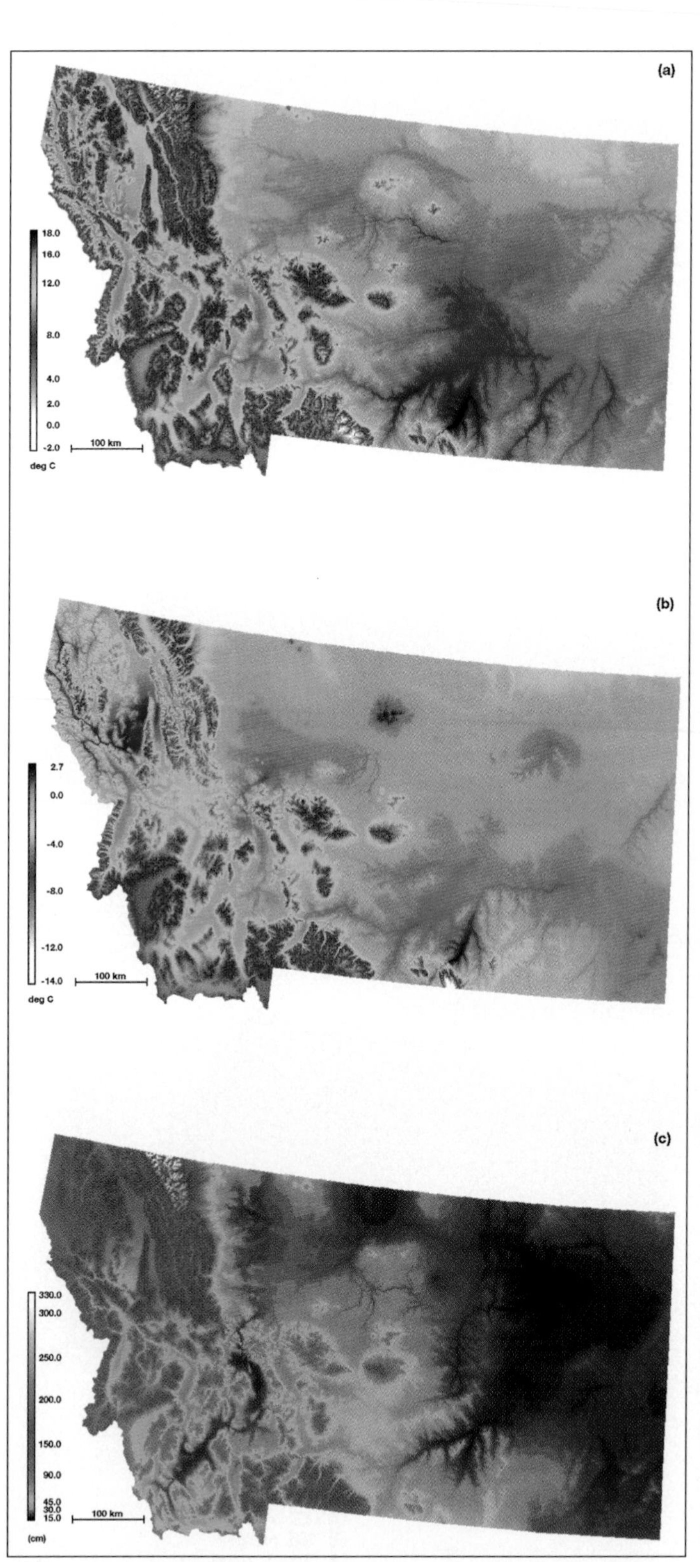

Plate 18-3. Spatial distribution of 1990 annual average for (a) T_{max}, (b) T_{min}, and (c) annual total *Prcp*.

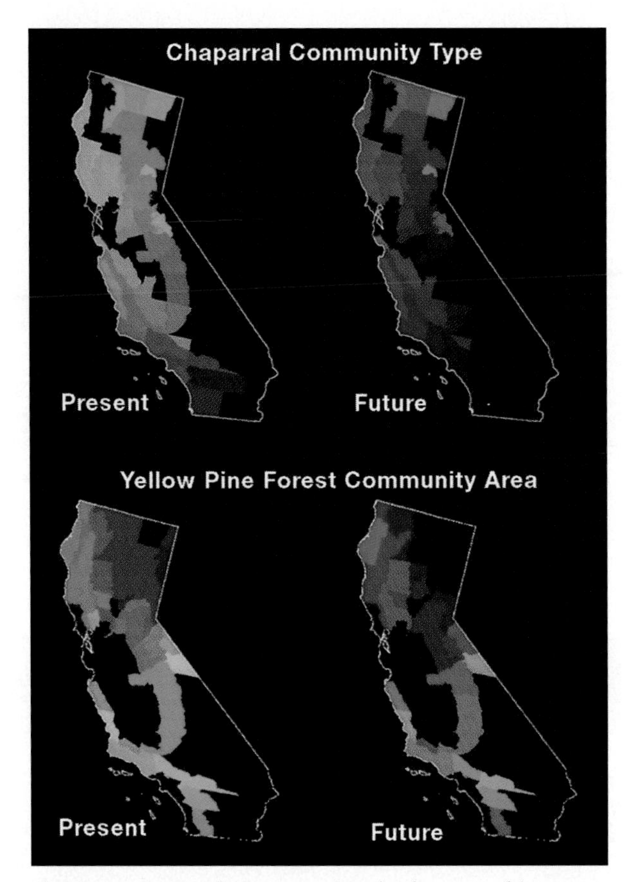

Plate 25-1. Present and future canonical trend surface maps of the main environmental gradients for the chaparral and yellow pine forest Munz community types (RSRU/UCSB).

Plate 32-1. Map output at selected time steps from a 500-year LANDIS simulation of a 10,000-ha landscape at 60-m cell size (approximately 37,500 cells): (a) initial state, (b) 50 years, (c) 300 years, and (d) cumulative disturbance over the 500-year model run.

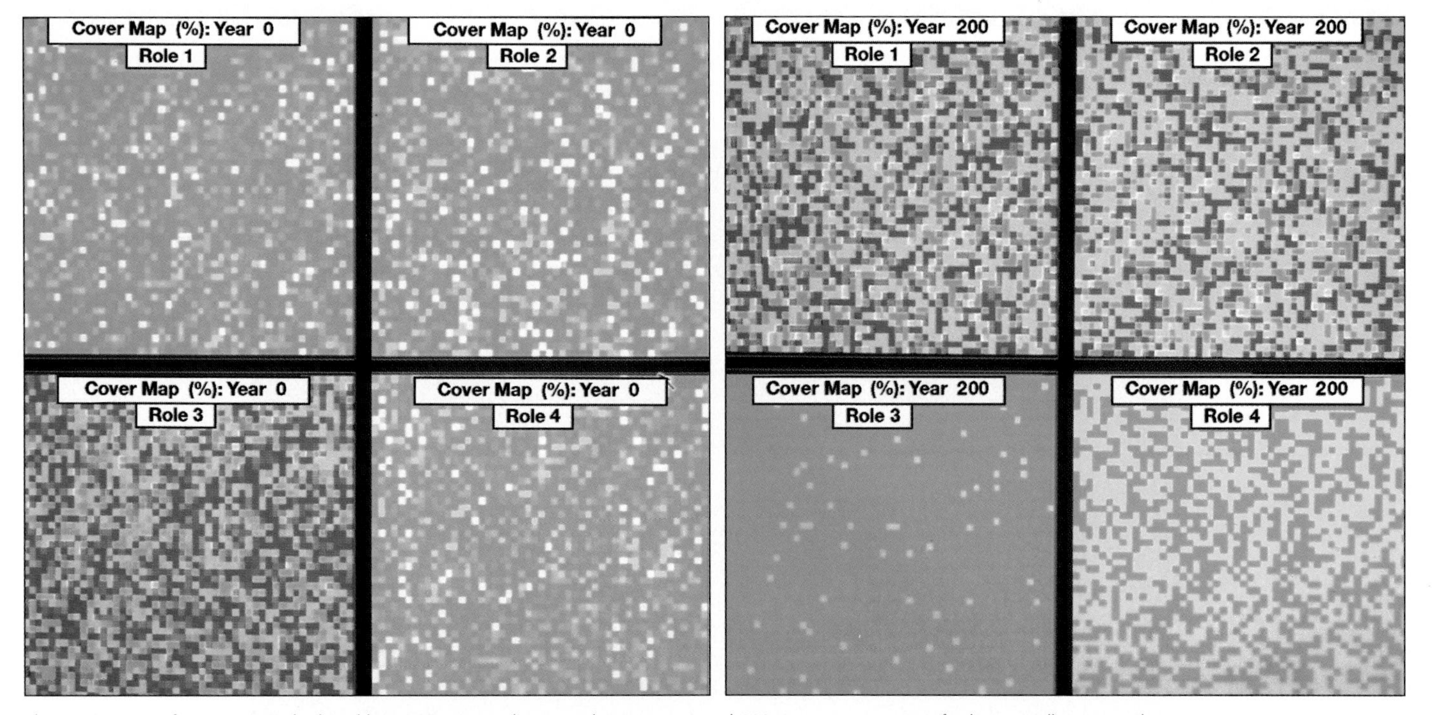

Plate 33-1. Output from MOSAIC displayed by GRASS. Four roles at simulation years 0 and 200. Increasing intensity of color in a cell corresponds to increasing values of the proportion of that role in the cell.

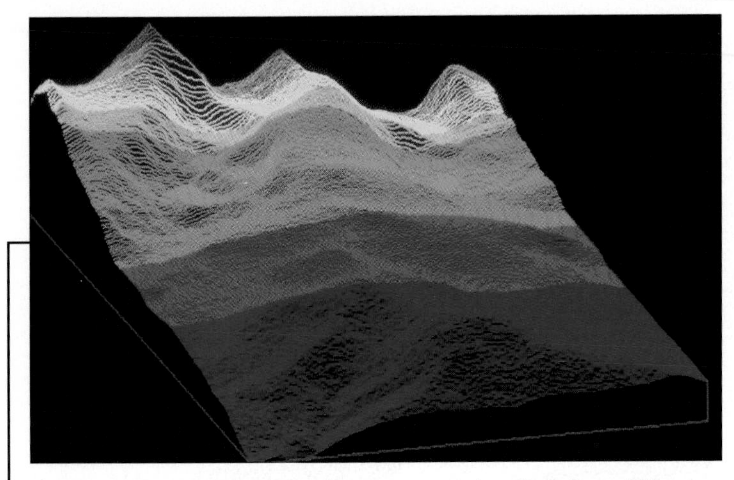

Plate 41-1a. Orogenic precipitation index image computed as altitude from a DEM divided by distance from the Continental Divide (west of the site). Minimum value is 0.32 (dark blue); maximum value is 1.97 (red).

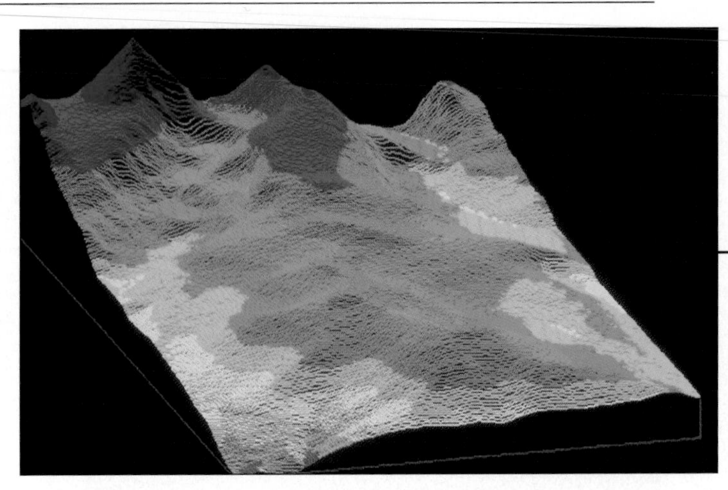

Plate 41-1b. Growing degree days image computed from daily temperature grids derived using environmental lapse rates and meteorological station data. Minimum value is 6.26 (dark blue); maximum value is 317.77 (red).

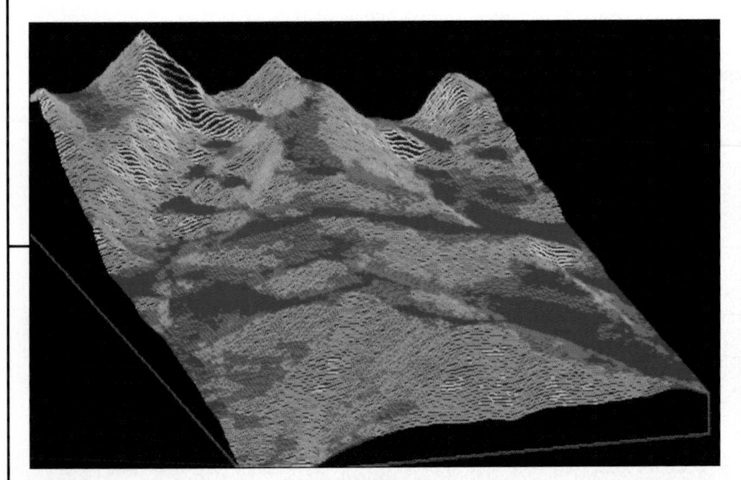

Plate 41-1c. Topoclimatic slope-aspect index image. Minimum value is 0.00 (dark blue); maximum value is 148.11 (red).

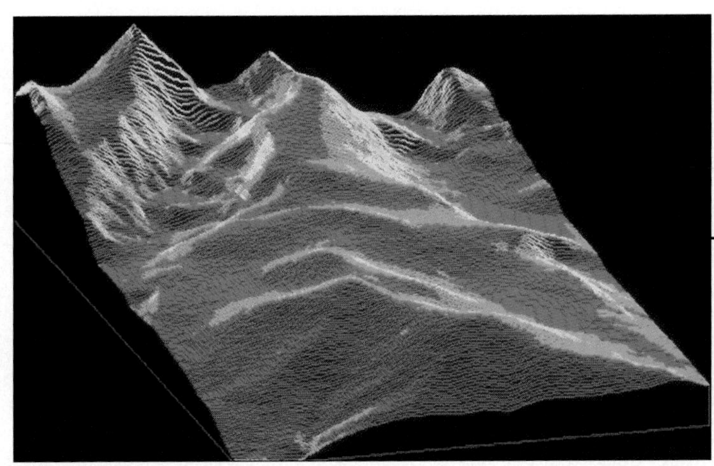

Plate 41-1d. Insolation index image. Minimum value is 10.39 (dark blue); maximum value is 25.79 (red).

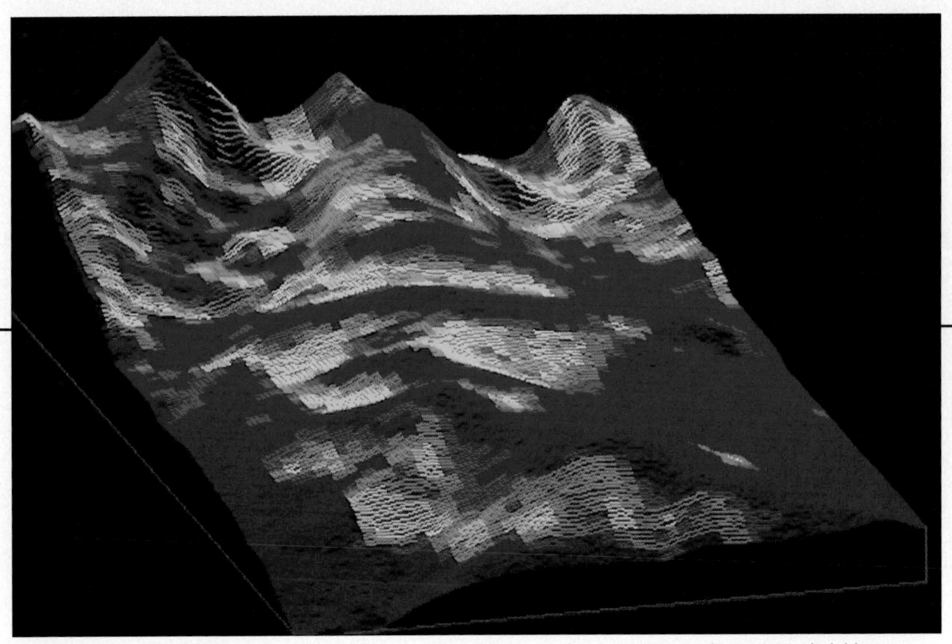

Plate 41-1e. Snow probability image derived from Landsat MSS data archive. Minimum value is 0.00 (dark blue); maximum value is 1.00 (red).

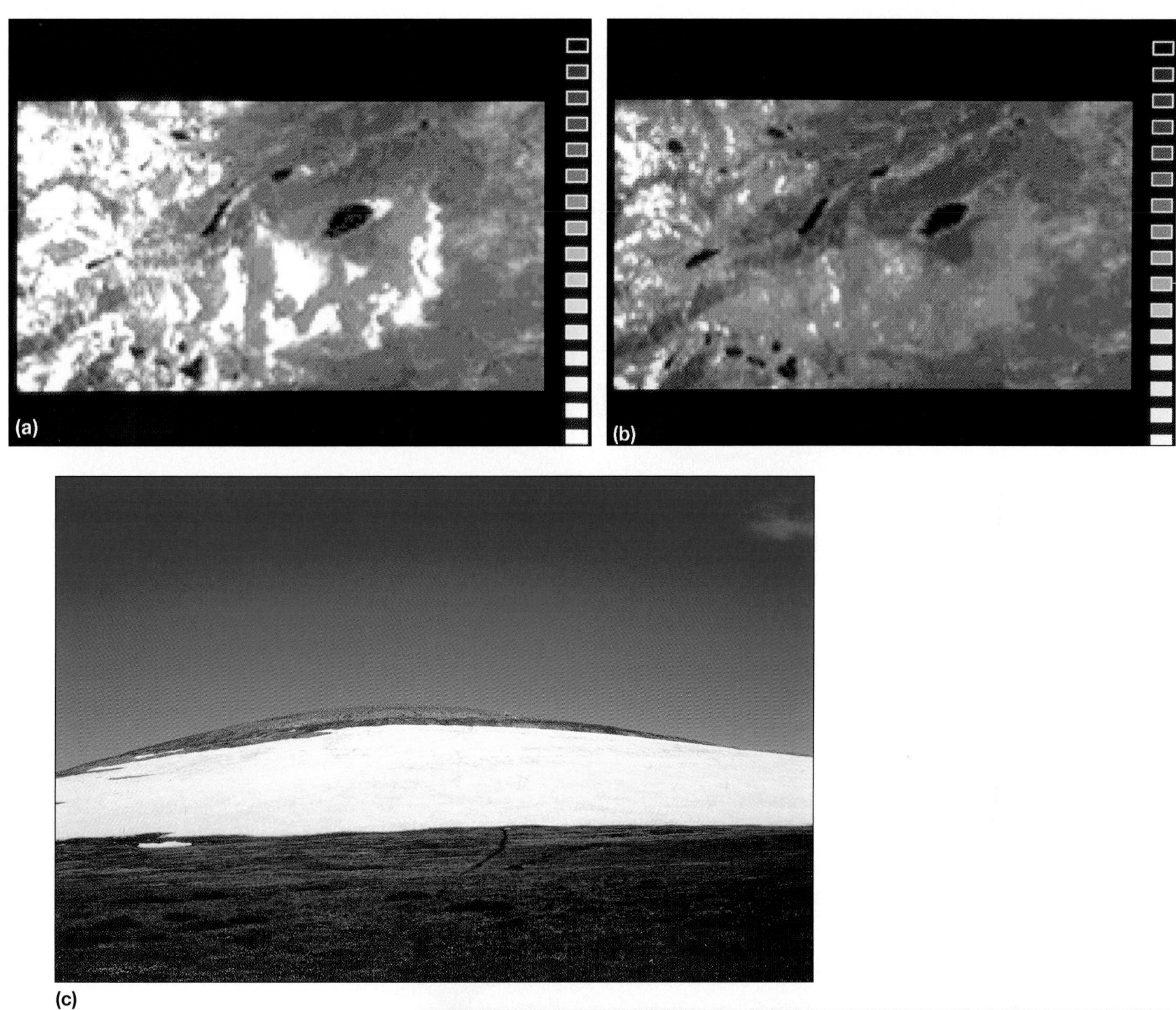

Plate 41-2. Landsat MSS band 7 images of the Niwot Ridge LTER acquired on (a) June 19, 1986; (b) June 22, 1987. Landsat MSS images are used to monitor year-to-year and seasonal variations in the spatial extent of snow cover. Photographs of the Saddle snowfield taken during (c) June 1986 and (d) June 1987 field campaigns show the effect of a decrease in snowfall on plant growth.

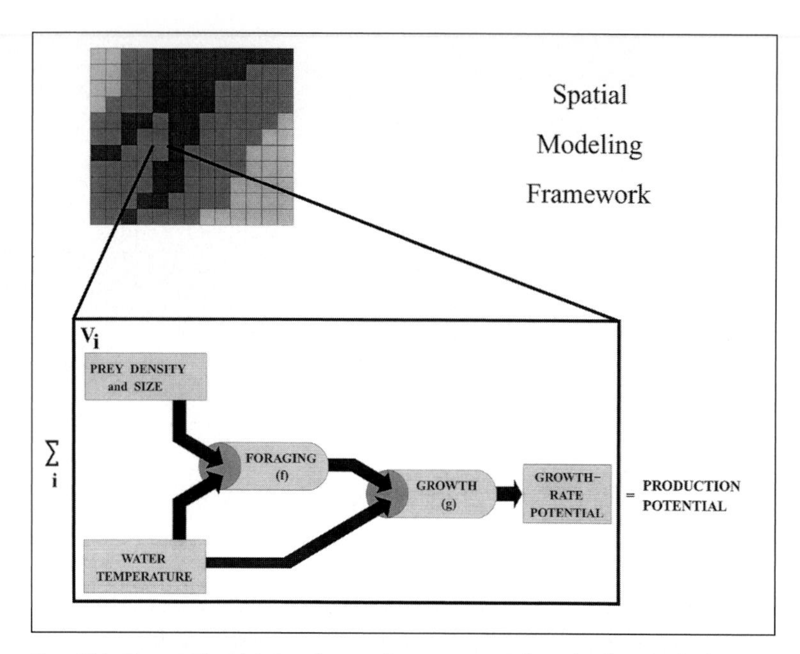

Plate 42-1. Conceptual model of predator production in a spatially explicit framework. The water mass is divided by columns and rows into a grid of cells, depicted here in the vertical and horizontal dimension. Spatial heterogeneity is indicated by shading of the cells. Cell *i* has volume V_i. Prey density, prey size, and water temperature are used as inputs to a foraging model (*f*) and a growth model (*g*) to produce the growth-rate potential for the predator in that volume of water. The integration of growth-rate potential over all water volumes produces an estimate of system production potential.

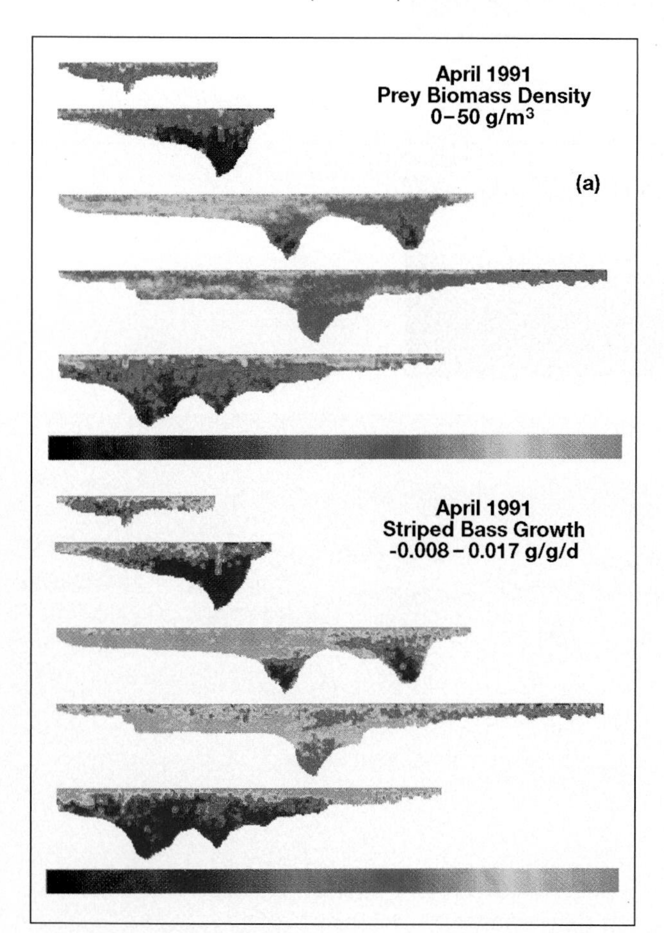

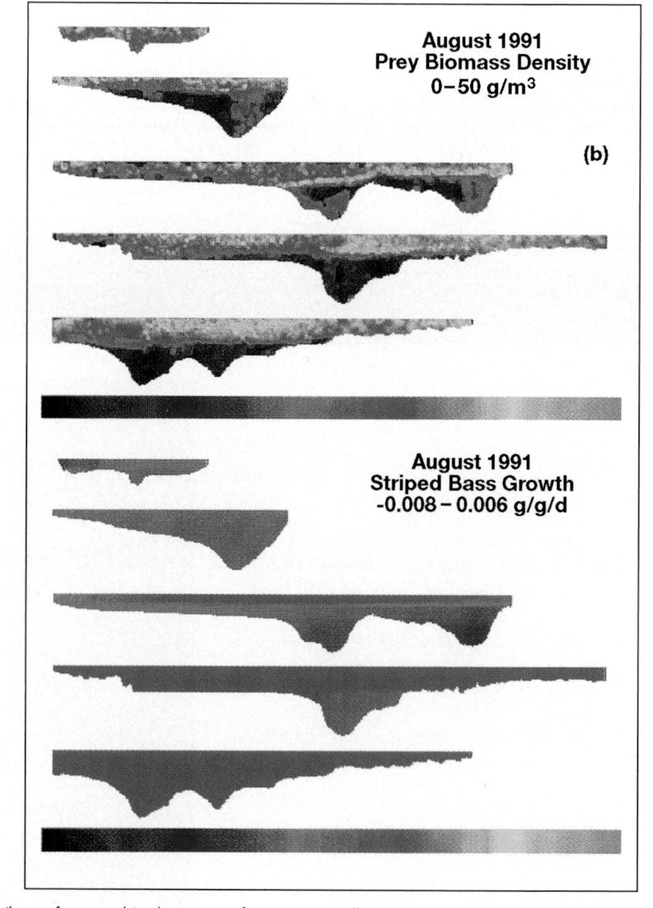

Plate 42-2. Prey biomass density (upper five panels) and striped bass growth-rate potential (lower five panels) taken across five geometrically separated transects (approximately 25–45 km between transects) stretching from Still Pond in the north (top panel) to Smith Point in the south (bottom panel). Data were collected at night during (a) April 1991, and (b) August 1991. Maximum bottom depth is about 34 m. Cell size is 0.5 m in the vertical dimension and 30 m in the horizontal dimension.

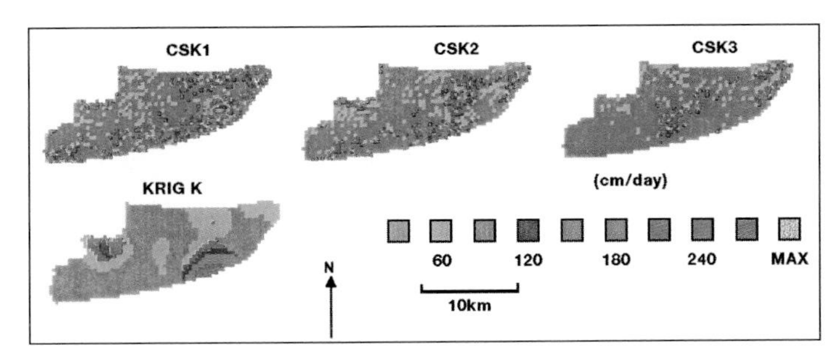

Plate 48-1. Example of three conditional simulations (CS) and a kriged (KRIG) distribution of hydraulic conductivity on the Mahantango Creek watershed using conditioning data given in Table 48-1.

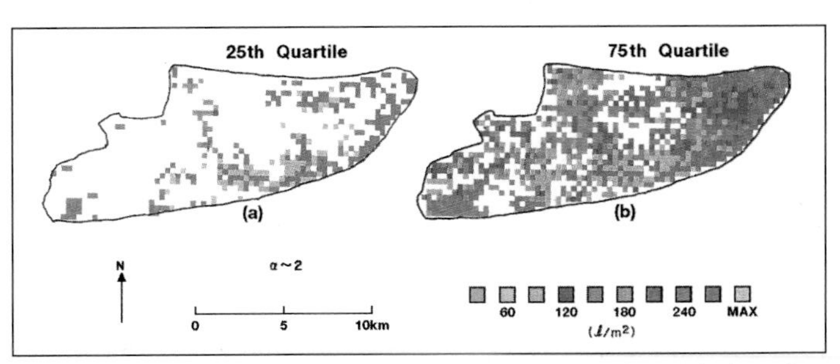

Plate 48-2. Distribution of recharge flux associated with (a) the 25th, and (b) the 75th quartiles of travel time to the water table.

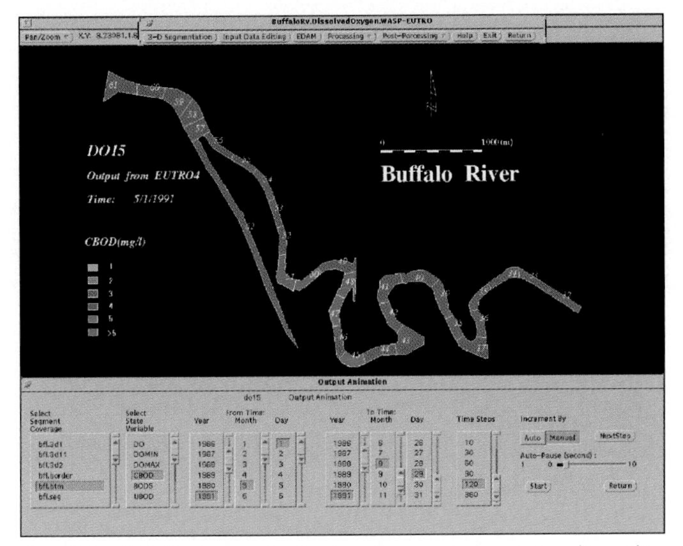

Plate 51-1. GEO-WAMS screen showing three-dimensional segmentation editing utility with a segmentation scheme for the Buffalo River depicted in the edit windows.

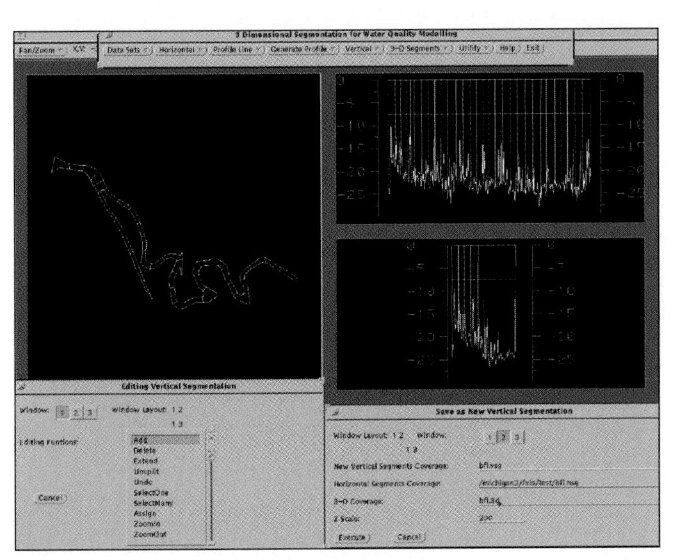

Plate 51-2. Snapshot of GEO-WAMS model output animation module display of dissolved oxygen in the lower waters of the Buffalo River.

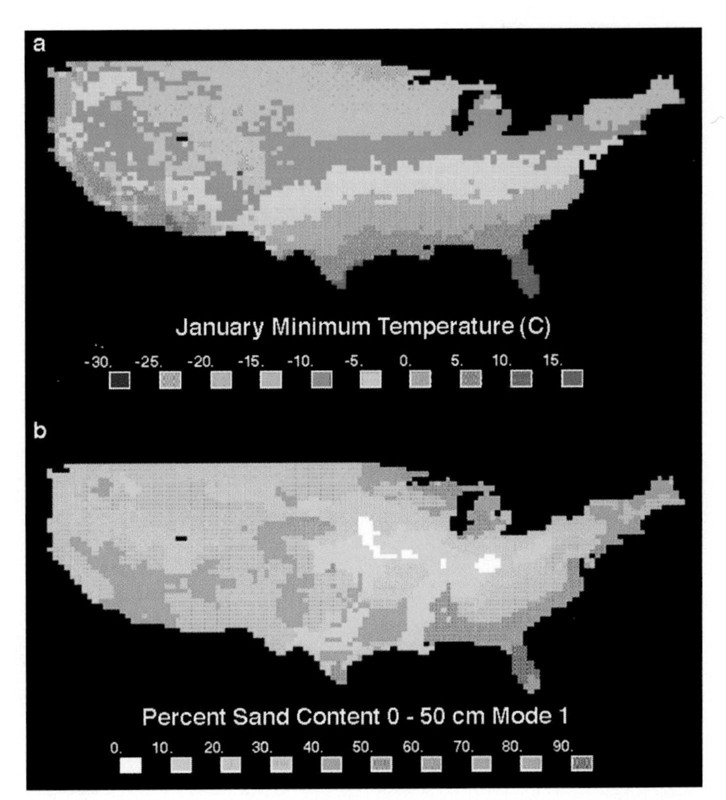

Plate 55-1. (a) Mean January minimum temperature, spatially interpolated with adiabatic adjustment from surface station means. (b) Sand content (%) of the top 0–50-cm soil layer for the dominant soil type (first modal soil). Grid interval is 0.5° latitude by 0.5° longitude.

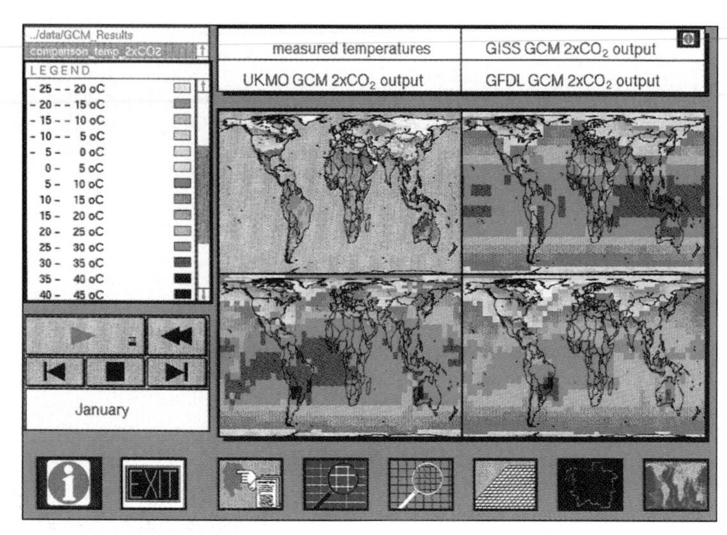

Plate 74-1. CLIMEX: Comparison of thematic maps.

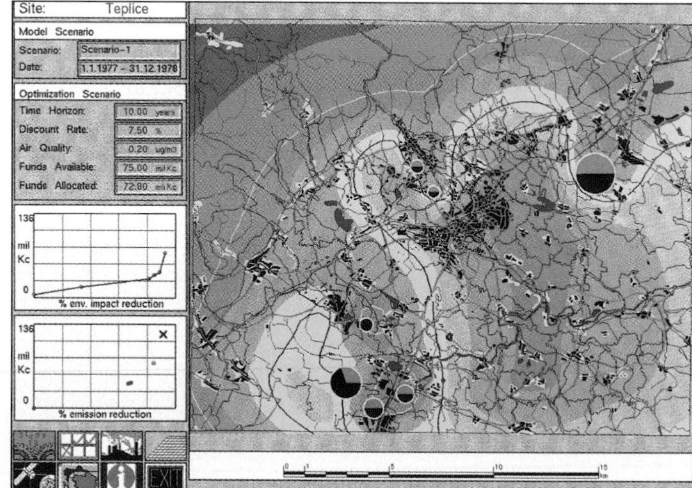

Plate 74-2. ISC: Abatement strategy optimization module.

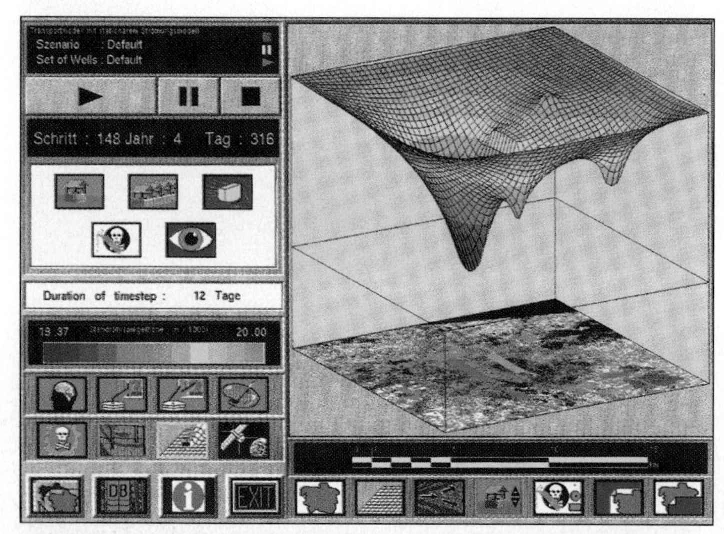

Plate 74-3. XGW: Three-dimensional display of groundwater contamination.

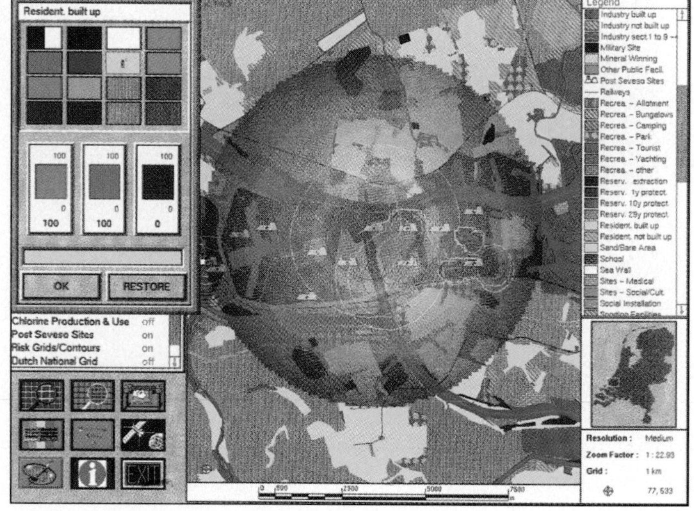

Plate 74-4. XENVIS: Visualization of risk contours.

45

GIS and Atmospheric Modeling:
A Case Study

Tsengdar J. Lee and Roger A. Pielke

Through a series of examples, we demonstrate how GIS can be used in atmospheric modeling in terms of initializing the model. We also have demonstrated the potential usage of GIS in three-dimensional analyses. Challenged by the needs of atmospheric-, hydrologic-, and ecologic-coupled climate modeling, the application of GIS techniques has become more and more critical.

INTRODUCTION

The recent development of geographic information systems (GIS) and computer technologies has created a new era of environmental modeling. Larger and faster computers have made running atmospheric models at fine (~10 km) spatial scale possible. In addition, the awareness of possible human-caused climate changes has made environmental modeling challenging. In order to understand the function of the climate system, a state-of-the-art general climate model (GCM) should consist of several submodels, including an atmospheric model, an oceanic circulation model, a hydrological model, and an ecological model. Obviously, the use of GIS has become essential in providing surface and/or boundary conditions to these various models.

As described in Lee et al. (1993), current applications of GIS in atmospheric modeling provide land surface boundary and initial conditions that are necessary to perform model simulations. The initial conditions usually include soil temperature and moisture, while the boundary conditions generally consist of land cover, land use, and elevation. However, the use of GIS in atmospheric modeling should not be limited to processing the surface data. Due to the four-dimensional nature of atmospheric modeling, we should extend the concept of GIS to include temporal variations of three-dimensional spatial data.

In the following sections, through the use of the Colorado State University (CSU) Regional Atmospheric Modeling System (RAMS), we: (1) show how station data are interpolated onto a model grid and how spatial resolution is defined in an atmospheric model; (2) introduce numerical filters due to the difficulties in handling the fine-scale features in an atmospheric model; (3) show how a digital elevation model (DEM) is applied in atmospheric models with the application of a "silhouette-averaging" technique; (4) demonstrate the application of the USGS land surface cover data in the CSU RAMS; and (5) discuss the difficulties in the application of "landscape classifications" within the model.

INTERPOLATION OF STATION DATA ONTO A MODEL GRID

Most of the analyses of the atmospheric state use a grid system. The irregularly distributed observational meteorological station data must, therefore, first be interpolated onto the grid system. Recall that since the atmosphere is a three-dimensional system, the interpolation must take both horizontal and vertical variation into account, which is very difficult. Fortunately, for dry atmospheric processes (processes that do not change the phase of water substance), the atmospheric motion follows a constant potential temperature surface; this physical constant is utilized in a technique developed by Tremback (1990). Potential temperature is the temperature a parcel of dry air would have if brought adiabatically from its initial state to the (arbitrarily selected) standard pressure of 1,000 mb. Using potential temperature as the vertical coordinate, atmospheric dry motion becomes two dimensional on this surface, and the interpolation can be completed on this surface. This coordinate transformation technique effectively converts three-dimensional interpolation into two-dimensional interpolation, which is a lot easier to perform. Another advantage to performing analyses using an isentropic (constant potential temperature) coordinate is that it permits higher resolution near a frontal zone where potential temperature gradient is large. Figure 45-1 shows an example of a potential temperature north-south vertical cross section along the 100° W meridian and centered at 39° N on June 6, 1990. A

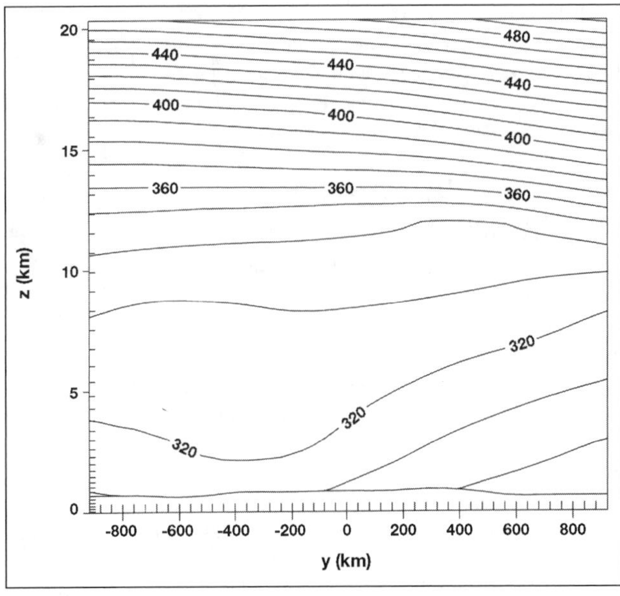

Figure 45-1. Potential temperature (K) along 100° W meridian and centered at 39° N on June 6, 1990.

front, which is indicated by the leading edge of the slanted isentropic surfaces, is visible in this coordinate system. The two-dimensional interpolation is performed on the slanted surfaces in this example. Notice, however, this technique is not applicable everywhere in the atmosphere. Very close to the surface, in the convective boundary layer, for example, the isentropic coordinate system provides no resolution, and other coordinate systems must be used to supplement the vertical resolution near the surface (Benjamin 1989; Benjamin et al. 1991). (This is beyond the scope of this chapter.)

SPATIAL RESOLUTION AND NUMERICAL CONSTRAINT

Having the technique to interpolate station data to an arbitrary model grid, the natural next step is to determine the size of the grid increment. A grid system can only resolve features that are at least 2 grid increments. However, numerical methods used to integrate the equations that govern atmospheric motion forward in time cannot handle features as small as 2 grid increments (Pielke 1984). Thus, the effective spatial resolution of a numerical atmospheric is no smaller than 4 grid increments. For this reason, a numerical filter is introduced when interpolating station data to the model grid. A commonly used interpolation scheme, the Barnes' analysis (Barnes 1964, 1973), is very wavelength selective. The user can input parameters that will determine the characteristics of the response function. Features that are larger than the specified wavelength are retained at higher amplitude, while features that are smaller than the specified wavelength are retained at a lower amplitude (not completely removed). The steepness of the response function at lower wavelength determines how selective the filter is. Figure 45-2 shows the response function as a function of wavelength. The response function is chosen to retain 90% amplitude at a 250-km wavelength. Figure 45-3 shows an antecedent precipitation index (API) map created for a 60 × 47 model grid with a 40-km grid increment and using the same response function shown in Figure 45-2. API is needed in a model in order to define the rainfall over a simulation domain prior to beginning a model simulation and to estimate soil moisture.

THE USE OF A DEM IN AN ATMOSPHERIC MODEL

Perhaps the most important surface boundary condition is the topography. Since the air near the surface must flow over or around a surface obstacle, the treatment of underlying topography has a major impact on the near-surface flow. Past studies have shown that the atmosphere should respond to the "envelope topography" in which the highest topography in a DEM should be used in the so-called "silhouette-averaging" process (Bossert 1990). Consider a case in which terrain features equal to or longer than 16 km should be retained while using a 1-km DEM. Since the smallest wavelength is 16 km, the model grid system should have a grid increment of 8 km. The silhouette-average scheme first makes a 16-km increment DEM, which effectively removes terrain features smaller than 16 km. The 16-km silhouette-average DEM is created by finding the mean height of the silhouette, as viewed from the east or west of the set of 16 × 16 pixels in 1 coarse DEM cell, and the silhouette of the same points as viewed from the north or south, and then averaging the 2 silhouette heights together. The final step is to interpolate the 16-km DEM onto the 8-km grid.

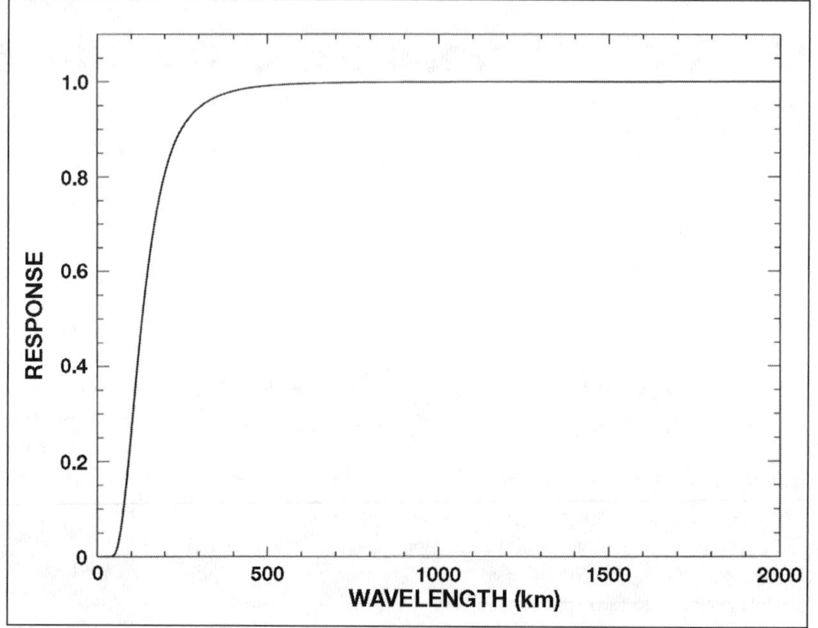

Figure 45-2. Barnes' amplitude response as a function of wavelength. The response function is chosen to retain 90% magnitude for a 250-km wavelength feature.

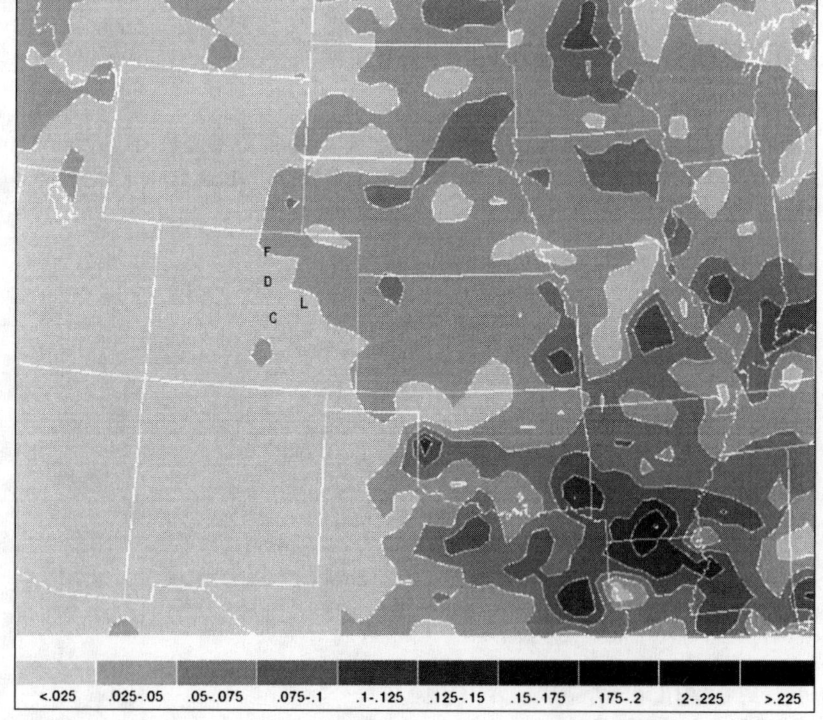

Figure 45-3. An antecedent precipitation index (API) map for a 60 × 47 model grid with a 40-km grid increment. The Barnes' response function shown in Figure 45-2 is used to produce the map. Contour interval is 1 in and starts at 0.

Having developed techniques to interpolate station data to a model grid system, Figure 45-4 illustrates the topography near Denver, Colorado, as "seen" by the atmospheric model. A thirty-second-interval DEM of the United States is used to produce the figure. As demonstrated in Figure 45-4a, the original topography has many fine-scale features. Removing 4 grid increment waves by silhouette averaging (the finest-resolution atmospheric grid, in this case a grid increment of 1 km) results in the topography in Figure 45-4b. Figures 45-4c and 45-4d compare the difference between the regular running-averaged topography and the silhouette-averaged topography. Features shorter than 16 grid increments have been removed for the ease of demonstration. As shown in Figures 45-4c and 45-4d, the silhouette-averaged topography retains the maximum height of the topography.

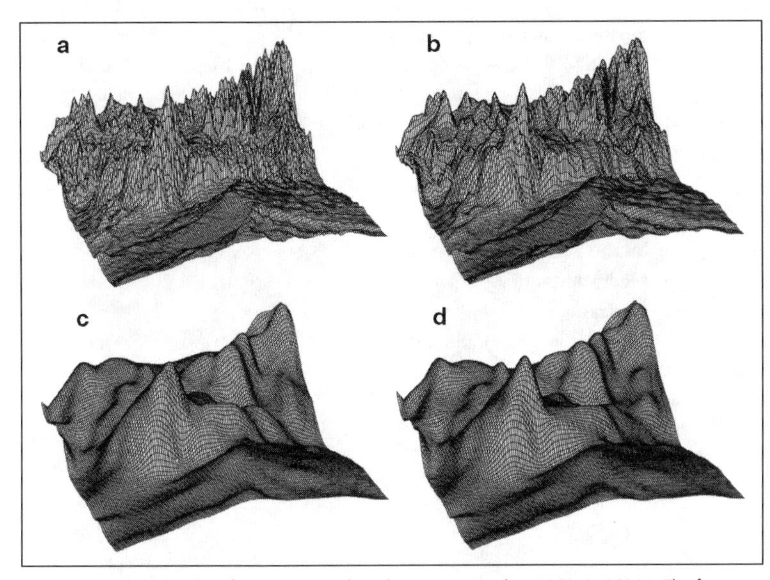

Figure 45-4. Topography of a 140 × 140 km² domain centered at 105° W 39° N. The four panels are: (a) 1-km original data; (b) envelope terrain with a wavelength of 4 km; (c) regular running-average terrain with a wavelength of 16 km; and (d) envelope terrain with a wavelength of 16 km. The perspective of view is from the same point in each figure.

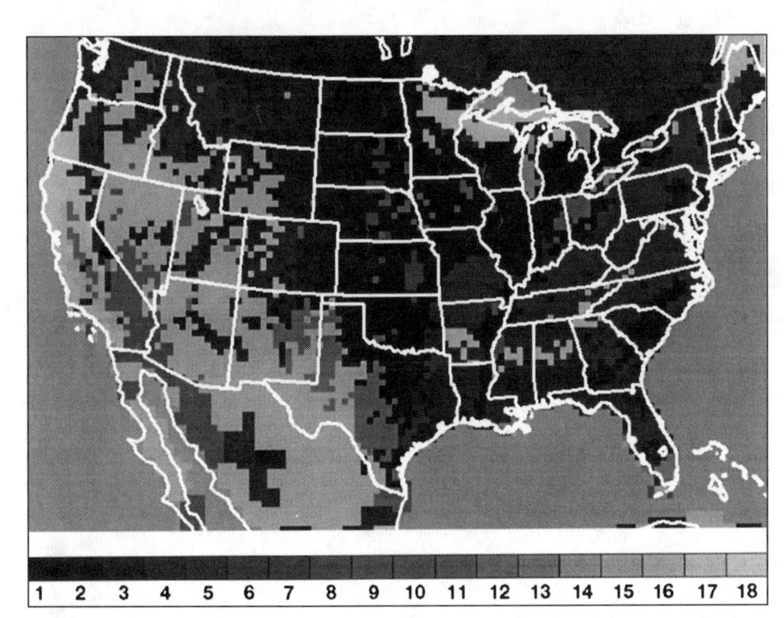

Figure 45-5. USGS land-cover data over the United States using a 120 × 80 RAMS grid with 40-km grid increments. The dominant USGS land-cover class has been converted to a BATS classification. (See text for more detail.)

THE USE OF USGS LAND COVER DATA IN RAMS

Many of the recent atmospheric models include a surface module that calculates land surface hydrology. Having the topography data ingested into the model, the next task is to obtain land-use and land cover data for insertion into the model. Unlike the topography data, in which a numerical filter can be applied to remove undesired, small-scale features, the classification of land-use and land cover categories cannot be straightforwardly averaged or filtered in the form in which this information is normally available. Although we can average the fundamental parameters that were used to develop the classifications, the current application of land-use data in RAMS has been to utilize the most frequent classification that is observed in a grid box. Figure 45-5 shows the land cover data over the United States using a 120 × 80 RAMS grid with 40-km grid increments.

The USGS land cover data (Loveland et al. 1991; Brown et al. 1993) was first converted to a BATS classification (Dickinson et al. 1986) and then the dominant class for 1,600 pixels in a grid box was chosen to represent the grid box. The percentage of water coverage (not shown) in a grid box, which is also a land cover parameter used by RAMS, can also be obtained simply by counting the pixels that are classified as water. Finally, the Leaf Area Index (LAI) used by RAMS can be obtained from the USGS NDVI data (Eidenshink 1992) using a similar technique.

CONCLUSION

Through a series of examples implementing CSU RAMS, we have shown that by using a coordinate transform, two-dimensional interpolation can be applied to interpolate three-dimensional atmospheric data onto a model grid system. Due to the constraint of numerical techniques and limitations in sampling, spatial resolution is defined in an atmospheric model as no smaller than 4 grid increments. Since it is difficult to handle the fine-scale features in an atmospheric model, numerical filters are introduced to remove subgrid-scale features in the analysis. It has been demonstrated how a DEM is applied in atmospheric models with the application of a "silhouette-averaging" technique, with which the highest topography in a defined window (usually associated with the resolution of the grid system) is used to represent the topography in that window. Finally, the use of the USGS conterminous United States land cover surface data in CSU RAMS has been demonstrated. The difficulties in the application of "landscape classifications" in the model are also discussed. We also have demonstrated the potential application of GIS in three-dimensional analyses. Challenged by the needs of atmospheric-, hydrologic-, and ecologic-coupled climate-system modeling, the application of GIS techniques has become more and more critical.

Acknowledgments

This research was supported by the United States Geological Survey, Department of the Interior, under assistance #ATM-14-08-0001-A0929 and the National Aeronautics and Space Administration under grant #NAG5-2078. We thank Lou Steyaert and Tom Loveland for their continuing professional input to this work.

References

Barnes, S. L. 1964. A technique for maximizing details in numerical weather map analysis. *J. Appl. Meteor.* 3:296–409.

———. 1973. *Mesoscale Objective Map Analysis Using Weighted Time-Series Observations*. NOAA Technical Memorandum ERL NSSL-62. Norman: National Severe Storms Laboratory.

Benjamin, S. G. 1989. An isentropic mesoscale analysis system and its sensitivity to aircraft and surface observations. *Mon. Wea. Rev.* 117:1,586–603.

Benjamin, S. G., T. L. Smith, P. A. Miller, D. Kim, T. W. Schlatter, and R. Bleck. 1991. Recent improvements in the MAPS isentropic-sigma data assimilation system. *Proceedings Ninth Conference on Numerical Weather Prediction, Denver, Colorado.* American Meteorological Society.

Bossert, J. E. 1990. "Regional-Scale Flows in Complex Terrain: An Observational and Numerical Investigation." Ph.D. diss., Colorado State University, Fort Collins.

Brown, J. F., T. R. Loveland, J. W. Merchant, B. C. Reed, and D. O. Ohlen. 1993. Using multisource data in global land cover characterization. Concepts, requirements, and methods. *Photo. Eng. Rem. Sens.* 59:977–87.

Dickinson, R. E., A. Henderson-Sellers, P. J. Kennedy, and M. F. Wilson. 1986. *Biosphere–Atmosphere Transfer Scheme for the NCAR Community Climate Model.* Technical Report NCAR/TN-275+STR, Boulder, Colorado, NCAR.

Eidenshink, J. C. 1992. The 1990 conterminous U.S. AVHRR data set. *Photo. Eng. Rem. Sens.* 58:809–13.

Lee, T. J., R. A. Pielke, T. G. F. Kittel, and J. F. Weaver. 1993. Atmospheric modeling and its spatial representation of land surface characteristics. In *Environmental Modeling with GIS*, edited by M. F. Goodchild, B. O. Parks, and L. T. Steyaert, 108–22. New York: Oxford University Press.

Loveland, T. R., J. W. Merchant, D. O. Ohlen, and J. F. Brown. 1991. Development of a land cover characteristics database for the conterminous United States. *Photo. Eng. Rem. Sens.* 57:1,453–63.

Pielke, R. A. 1984. *Mesoscale Meteorological Modeling.* New York: Academic Press.

Tremback, C. J. 1990. "Numerical Simulation of a Mesoscale Convective Complex: Model Development and Numerical Results." Ph.D. diss., Department of Atmospheric Science, Colorado State University, Fort Collins.

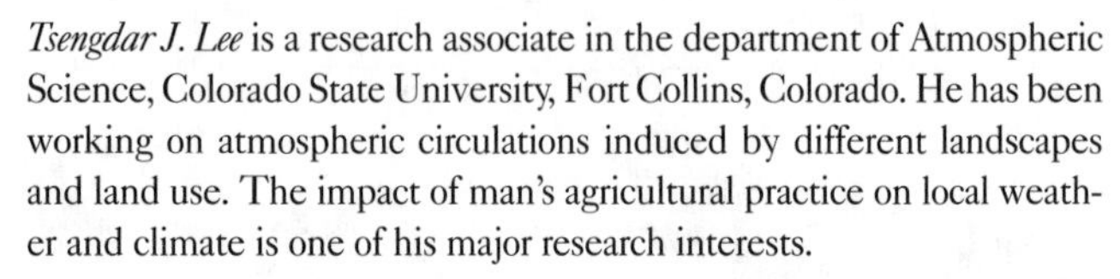

Tsengdar J. Lee is a research associate in the department of Atmospheric Science, Colorado State University, Fort Collins, Colorado. He has been working on atmospheric circulations induced by different landscapes and land use. The impact of man's agricultural practice on local weather and climate is one of his major research interests.

Roger A. Pielke is a professor in the department of Atmospheric Science, Colorado State University, Fort Collins, Colorado. He was formerly employed by NOAA's Experimental Meteorology Lab and by the University of Virginia.

Department of Atmospheric Science
Colorado State University
Fort Collins, Colorado 80523
E-mail: *lee@tachu.atmos.colostate.edu*

46

Fragmented Forest Response to Climatic Warming and Disturbance

George P. Malanson, Marc P. Armstrong, and David A. Bennett

The objective of this research is to show how forest stands are affected by fragmentation, barriers, and corridors in the landscape during periods of climatic change. We use a computer simulation model of the dynamic processes of establishment, growth, and death of forest trees in a spatially explicit framework. A modified version of the JABOWA-FORET model is used to elucidate differences in three general factors that affect forest dynamics and structure: climatic range, dispersal abilities, and spatial structure. At the scales used, dispersal is a small factor in limiting response to climatic change, and fragmentation has no additional effect. The results of our simulation experiments indicate that model scale must be addressed in more detail, possibly using large parallel systems.

INTRODUCTION

Several research teams have concluded that the global climate will warm during the next century, in large part because of human activities. This anthropogenic climatic change will have far-reaching outcomes, and one particularly important impact is its effect on the ecology of forests. Though computer simulation models have projected significant changes in the productivity, biomass, and species composition of forests in eastern North America as a consequence of the greenhouse effect (Malanson 1993), most studies have failed to consider the spatial isolation of forest stands. Levin (1992) argued that spatial pattern and scale were the core problems in ecology, and it has been argued that they are at the heart of geography (Meentemeyer 1989).

In this research we analyze the spatial dynamics of plant communities during periods of climatic change. The specific objective of the research is to investigate how forest stands are affected by fragmentation, barriers, and corridors in the landscape. We use a computer simulation model of the dynamic processes of establishment, growth, and death of forest trees in a spatially explicit framework to show that the spatial configuration of landscapes controls their reproduction. We also examine interaction among species competing for space. These simulation experiments are designed to address some of the issues raised by Malanson (1986) about the possible effects of spatial autocorrelation on the community structure of vegetation.

The relatively detailed fossil record of responses to climatic warming among plant species in the Holocene reveals two key points (Malanson 1993). First, species have responded individualistically to climatic change; though species have lived in a variety of assemblages that are no longer extant, they are now living in other assemblages. Second, while species distributions came into equilibrium with climate over a period of a millennium (Webb 1992), exceptions occurred where the landscape was naturally fragmented (Cole 1985; Davis et al. 1986). The use of Holocene records to project vegetation change in response to the greenhouse effect is a tenuous enterprise, however. Future rates of climatic change in response to increased concentrations of carbon dioxide will likely differ from rates of past climatic change, and thus, transfer functions based on fossil records for the Holocene are unlikely to be true indicators of future responses.

Two versions of a computer simulation model (JABOWA-FORET) have been used by several researchers to examine the dynamics of forest stands in response to climatic change (Shugart et al. 1992). The major shortcoming of the JABOWA-FORET model is that it has oversimplified a key factor that did not affect its original intended use, but which does affect its use for modeling response to aspects of climatic change. Since the model incorporates an assumption of ubiquitous dispersal, the propagules of any species in the array of species modeled can arrive on any site at any time, and the species can establish as a seedling if other environmental conditions are met. To state an extreme case, if the simulated climate of northern Canada warms drastically, tupelo can establish there immediately. This assumption is clearly untenable. Consequently, we have developed a spatially explicit version of this model that allows us to test for the effects of different dispersal abilities and different degrees of landscape fragmentation (Hanson et al. 1989, 1990). In a limited range of tests, we found that when limits to dispersal are included, forest stands become less diverse and some species become extinct.

METHODS

Computer Simulations

Our modification of the JABOWA-II simulation model, the Module of Spatially Explicit Landscapes (MOSEL), is used as the principal vehicle for conducting simulation runs. The JABOWA-FORET class of simulators has been extensively examined and tested, studied in a variety of situations relevant to the work described here, and details of the model are given in widely available references (Shugart et al. 1992). It is based on functional relationships between tree growth and climate and has been modified for many forest types. Growth, based on growing-degree days, is now modified by shading in all versions and by soil moisture and nitrogen in ours. If growth falls to a low level, mortality is likely. Mortality also occurs randomly, simulating windfall, for tall trees. Establishment is limited differentially by the available light on a site, and light varies as trees grow and die.

Several factors are held constant throughout the series of simulation runs. The initial conditions and the amount of climatic change are constant, except for the case of no climatic change. Other parameters, which are outside the scope of our research, are the effects of CO_2 fertilization and feedbacks from vegetation change to climatic change.

The standard inputs for tree species adaptations are used in a modified form. The original model represented a single 1/12-ha stand of trees. We use these 1/12-ha stands to represent a surrounding area of forest. For this research 200 stands are defined in a 10-column × 20-row grid (Figure 46-1). Each simulation is run for 1,000 years and includes seventy-five of the dominant tree species of eastern North America.

Over this grid we superimposed an initial north-south gradient of growing-degree days (GDD; summed above 4.4°C base) and an east-west gradient of moisture. The GDD are calculated based on data for Virginia, Minnesota, representing row 6. Based on the algorithms used in JABOWA-II, for a latitudinal change of approximately 0.8° latitude per row, we calculated an average step of 245 growing-degree days for each row. The coldest row begins with 413 GDD, while the warmest is at 5,306. The west-east gradient is given by the simulation parameter WILT (WILT = [AET-PET]/AET), which we change over the 10 columns from 0.10 to 0.01 in 0.01 decrements (e.g., growth and establishment are multiplied by 0.0 to 0.99 in 0.20 increments across this gradient for a moisture-sensitive species such as *Nyssa aquatica*). While the moisture remains constant, the GDD are increased by one each year over the first 500 years of the simulation. By raising the GDD by 500, we effectively increase the annual average temperature by 1.67°C in the south where the growing season is in the range of 300 days, and by 5°C in the north where the growing season is in the range of 100 days.

Dispersal Functions

In a spatial model, dispersal can be treated as absent, unlimited, or as a spatially varying function. Perhaps because plant-dispersal ability is difficult to document (van der Pijl 1982), in many past models it has either been assumed to be unlimited (Solomon 1986) or has been ignored. In spite of the difficulty involved, however, dispersal can be modeled in a general way. Johnson et al. (1981) provided a useful formalization of this process and a variation of their dispersal model has been operationalized (Hanson et al. 1990). Our method uses a series of dispersal rules that apply to the major dispersal types identified by van der Pijl (1982).

Ubiquitous dispersal: This is the assumption implemented in most previous modeling efforts. In such simulations the probability of seed from every species falling on every plot is 1.0 every year. Establishment is then limited by light conditions and random selection. We refer to these runs as UCC and UNCC, with and without climatic change.

In our modified version, however, species are dispersed by several methods: gravity, water, wind, mammal, scrub-bird, and bluejay.

Gravity: To simulate dispersal by gravity we assume that any species producing seed on a site will have potential replacement on a site (p=1.0) and a 0.10 probability of dispersal to adjacent cells if they are of equal or lower elevation. This assumption means that if a species is present on a site and is producing seeds, it has the same chance for establishment as under the ubiquitous-dispersal assumption.

Water: In this version of the model we have developed a cursory implementation of hydrochory—we simply calculate a 0.01 probability of dis-

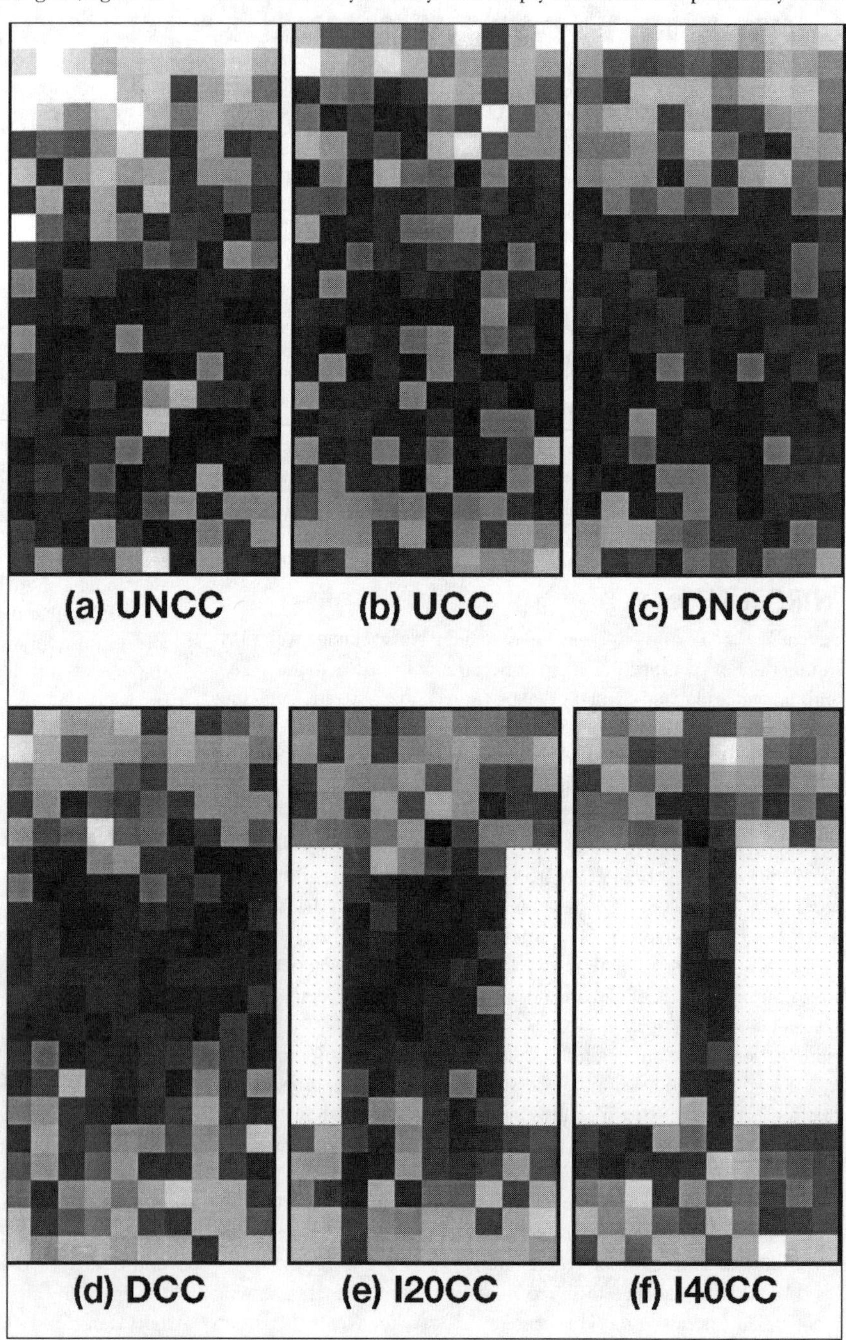

Figure 46-1. The 10 × 20 grid used in the simulation shows the basal area (increasing with the gray scale); areas of initial deforestation (stippled) represent fragmentation to a corridor. All patterns shown are for the end of the simulation period after 1,000 years.

persal by a hydrochore to any cell. We have developed, but have not yet debugged, code for simulating hydrochory in which species in topologically linked river cells can disperse downstream to cells along the river. *Wind:* Anemochory is modeled as a negative exponential decay away from each source cell using a Monte Carlo type simulation popularized in geography in the 1960s (Hagerstrand 1965; Gould 1969). To implement the model we use a grid of dispersal probabilities that allows dispersal for up to 3 rows and/or columns in any direction in the grid. The dispersal probability field is biased toward the east in this implementation to account for prevailing wind patterns. As above, we have developed, but have not yet debugged, code for wind-dispersed species that (1) calculates wind fields for each dispersal season, and (2) creates wind-dispersal grids based on wind direction and speed and four classes of dispersal type.
Mammal, scrub-bird, and bluejay: Zoochory is also modeled using a probability given by a negative exponential decay away from source cells. For each category we have a grid of dispersal probabilities that allows dispersal for up to 3 rows and/or columns in the grid in any direction, but in this case the dispersal-probability field is symmetrical. Dispersal by bluejay is calculated as double the probability of that for mammals and scrub-birds and is limited to sites of low total biomass.

Dispersal modes for seventy-five common tree species of eastern North America have been identified. The propagule morphology of lesser-known species was determined by comparing propagule morphology with that of species for which better information is available. For each plot, the simulation model calculates whether a seed source is extant based on the presence of individuals of reproductive age by species and the probability of a seed crop in a given year. The model then calculates the seed rain to surrounding plots.

The seed rain on a plot is assessed independently for each of the means available to a species and for each of the source stands and then summed. We create a dispersal grid for each species that accumulates the dispersal events to each cell in the grid. Thus a species that disperses by both mammal and bluejay may have two successful dispersal events from each of several cells to a target cell in a single year. The number of successful dispersal events to a given cell by any species could be high, but in practice, it seldom exceeds four. More than one successful dispersal linearly increases the probable number of saplings established on the site, which is a randomized function of observed sapling densities and is limited by thresholds of available light.

Spatial Structure

The basic spatial configuration used in the simulation experiments has all cells available for forest (Figure 46-1, a–d). In subsequent experiments, conditions with variable amounts of forest fragmentation are simulated. Those cases in which all cells are forested are coded as DCC and DNCC: with and without climatic change. A second set of simulation experiments examines the effects of corridors, in which blocks of contiguous cells are defined as nonforested. These are examined for cases where 20% and 40% of the cells are forested. We examined one type of corridor, which we call the "I-formation." In the I-formation, the pattern of forested cells forms an "I" with the post and bars varying in width (Figure 46-1, e and f); we coded these cases as I20CC and I40CC.

Analyses of Projections

Calculated output for each species includes absolute and relative density, basal areas, frequency, a composite importance value, biomass, and net primary productivity, all of which can be subdivided among classes of basal diameter. In order to test the effects of dispersal abilities and spatial configuration on forest response to climatic change, we analyzed the stand-level variable of total basal area. However, the explanations for variations in stand-level variables often lie in the dynamics of individual species (Liu and Malanson 1992), and this will be examined in future work.

An especially important aspect of the application of GIS capabilities to this research is the integration of spatial modeling with display technology. The fusion of these components has allowed us to explore various aspects of spatial problems by enabling us to evaluate visually the results of modeling efforts and to then revise the model to incorporate information gained in previous iterations. This type of problem solving involves the use of heuristic information and, with geostatistics, forms the basis for a rapidly growing branch of GIS research involving the design of decision support systems (Armstrong and Densham 1990). In addition, visualization capabilities enable us to examine the patterns of change that occur among the simulation experiments. Such scientific visualization techniques have made it possible for the scientific community to gain new insight into the nature of problems (Upson et al. 1989). For example, we have constructed time series of the simulation model results which have been captured on videotape. As these images are displayed, the animated sequence of results provides additional insight into the spatial dynamics of simulated environmental processes.

RESULTS

We compared the projections for the grid after 1,000 annual iterations. At this point we report only the gross results of total basal area and the number of empty cells.

Tabular Results

Some differences between the assumptions of dispersal, fragmentation, and climatic change are shown in Table 46-1. The mean basal area for the top and bottom rows of the simulation grid reveal that for the cases of ubiquitous dispersal, the values of basal area are slightly higher than when dispersal is simulated, even when no fragmentation has occurred, but these differences may not be significant. Increases in the amount of fragmentation had no additional effect in these simulations. The effects of climatic change are seen in that for the northernmost row, if the climate does not change, the basal area is smaller; for the southernmost row, if the climate does not change, the basal area is larger. These changes may indicate an increase in forest growth in northern regions and a decline in southern areas where the temperature is too high.

Table 46-1. The average basal area for the northernmost and southernmost rows at year 1,000 of the simulation (cm^2).

Case	Climate Change	Fragment	Row 0	Row 19
UCC	yes	no	7,032	10,457
UNCC	no	no	661	12,241
DCC	yes	no	8,887	8,632
DNCC	no	no	4,841	11,875
I20CC	yes	yes	9,518	8,159
I40CC	yes	yes	8,781	8,300

Visual Results

A common trend in all simulations is that peak basal areas increase. The initial conditions have few cells with exceptionally high basal area. After 1,000 years, many of the cells have high total basal area. It is not possible to distinguish particular patterns among the different scenarios at this point.

DISCUSSION AND CAVEATS

An initial interpretation of our results would indicate that limitations of dispersal and forest fragmentation do not increase forest decline. This interpretation may or may not be correct, however. It may be correct if, over the course of 1,000 years, fragmentation leads to protection of some species from competition and so balances the effects of increased isolation, but it may be incorrect because of scaling artifacts in the simulation.

Inertia: While fragmentation tends to decrease the probability that a species can successfully move its range northward before the climate changes beyond its ability to exist, it may also favor inertia. Cole (1985) noted that species at isolated sites were able to persist in a climate beyond their equilibrium climatic range because their competitors had not yet arrived. The climatic range of a species is a function of its fundamental niche and interaction with competitors. The dispersal abilities of species and the spatial structure of the landscape contribute to the isolation of species from potential competitors. If the direction of climatic change is to a less stressful condition (e.g., warmer but not drier), however, then spatial isolation can protect a species from competitors. Inertia assumes that a competitive hierarchy (Keddy 1989) exists on the climatic gradient and that the fundamental niche is larger than the realized niche (Hutchinson 1957). In this case, a species may be able to persist longer, giving itself a greater probability of eventual successful dispersal. With a rapidly changing climate on a fragmented geographical field, nonequilibrium responses may be critical. Inertia is likely to occur when isolation and the difference between the fundamental niche and the realized niche are both significant (Malanson et al. 1992).

Scale: When dispersal is simulated, a species must continue to exist on the grid in at least one location; a seed source must be extant in order for establishment to occur. One problem with our present approach is that each row represents a climate defined by a number of growing-degree days, but there is a large jump from one row to the next. This discontinuity means that at some times what would be the optimum climate for a species is found between the rows of the grid, and thus, that species has no place in which its growth and reproduction are favored. For example, when row 6 has 1,675 GDD and row 7 has 1,920 GDD, a species that has its optimum conditions at 1,800 GDD will have no place where it is favored. Because of the consequences of extinction, the discontinuity in growing-degree days is critical when dispersal is simulated, but it is only a temporary effect when ubiquitous dispersal operates. The overall number of cells is also important. A species must persist on one of the cells in which climate is favorable. If, by chance, it becomes extinct, it can never recover. A grid of more cells would decrease the probability of chance extinction. More importantly, the dispersal distances need to be scaled to the distances between grid cells. In our simulation, the distances are set as if the cells were in fact contiguous and do not truly represent the probability of dispersal across many kilometers from one cell to the next or farther. A difficulty with accurate scaling is that empirical studies of plant dispersal have focused on short-range events, and little is known about the mechanisms of long-range dispersal (Portnoy and Willson 1993). It may be necessary to simulate a grid several orders of magnitude larger using large parallel processing systems in order to establish a basis for comparison.

CONCLUSION

The development of MOSEL contributes to basic knowledge and the advancement of research in three areas: (1) environmental modeling—by including spatial processes in a spatially differentiated multicell simulation model; (2) GIS—by linking spatial databases to environmental simulation models for dynamic visualization and spatial analysis; and (3) ecological impact of climatic change—by assessing the importance of spatial structure of the landscape (e.g., fragmentation) on the structure and composition of forests.

The most important result of our study is the further delineation of the importance of scale effects in incorporating spatial pattern and process in environmental models. While the effects of scale within single cells have been recognized, some aspects of multicell models require additional investigation (Moloney et al. 1992). A simple expansion to more cells, each representing a smaller area, will solve some of the problems with discontinuity of climate gradients that we encountered in our simulation experiments. An additional area of investigation will be in altering the microclimate of individual cells depending on the vegetation, or lack thereof, on neighboring cells (Kupfer and Malanson 1993). Allen and Hoekstra (1992) have made a convincing argument that community patterns are hierarchically constrained by landscape patterns. They noted that one could examine the communities at a coarse spatial scale (e.g., our grid) or at a fine spatial scale (e.g., one of our cells), and that one can examine landscape by examining either scale, but that it should be most informative to examine the realm of constraint, which we interpret to be the effects of grid (landscape) pattern on the community composition of cells.

MOSEL is designed to use a raster-based GIS input that is scaled to match our cell size. The inclusion of additional environmental factors will improve the level of realism in the simulation model. At actual simulation cell size, 1/12 ha, the cells can be matched to Landsat TM 30-m resolution data. These data plus digital elevation models can provide the ancillary information needed for each cell: elevation, slope, aspect, and percent rock surface; while other GIS coverages can provide information on soil depth, nitrogen (which can also be internalized), and depth to water table. This information, and the models developed, will improve our ability to assess changes in forests given projections for global climate warming. The linkage of pattern at this scale to regional pattern is a central part of this research and is fundamental to a hierarchical modeling framework (Brown et al. 1993). These contributions are founded on a tenet of physical geography: the assumed importance of space in mediating the interaction of physical and biological process-response systems. The future development of this model will address spatial questions of theoretical interest and allow more realistic projections to evolve as information about species responses to change accumulates through field studies.

Acknowledgments

This research was funded in part by a grant from the DOE Midwest Center for Global Environmental Change Research. We thank Amy Ruggles, Rajesh Krishnamurthy, and John Knaack for assistance with the visualization process.

References

Allen, T. F. H., and T. W. Hoekstra. 1992. *Toward a Unified Ecology.* New York: Columbia University Press.

Armstrong, M. P., and P. J. Densham. 1990. Database organization strategies for spatial decision support systems. *International Journal of Geographical Information Systems* 4:3–20.

Brown, D. G., D. M. Cairns, G. P. Malanson, S. J. Walsh, and D. R. Butler. 1993. Remote sensing and GIS techniques for spatial and biophysical analyses of alpine treeline through process and empirical models. *NSF Symposium Proceedings.*

Cole, K. 1985. Past rates of change, species richness, and a model of vegetational inertia in the Grand Canyon, Arizona. *American Naturalist* 125:289–303.

Davis, M. B., K. D. Woods, S. L. Webb, and R. P. Futyama. 1986. Dispersal versus climate: Expansion of *Fagus* and *Tsuga* into the upper Great Lakes region. *Vegetation* 67:93–104.

Gould, P. R. 1969. *Spatial Diffusion.* Washington, D.C.: Association of American Geographers.

Hagerstrand, T. 1965. A Monte Carlo approach to diffusion. *European Journal of Sociology* 6:43–67.

Hanson, J. S., G. P. Malanson, and M. P. Armstrong. 1989. Spatial constraints on the response of forest communities to climate change. In *Natural Areas Facing Climate Change,* 1–23. The Hague: SPB Academic.

———. 1990. Landscape fragmentation and dispersal in a model of riparian forest dynamics. *Ecological Modelling* 49:277–96.

Hutchinson, G. E. 1957. Concluding remarks. *Cold Spring Harbor Symposium in Quantitative Biology* 22:414–27.

Johnson, W. C., D. M. Sharpe, D. L. DeAngelis, D. E. Fields, and R. J. Olson. 1981. Modeling seed dispersal and forest island dynamics. *Forest Island Dynamics in Man-Dominated Landscapes,* 215–39. New York: Springer-Verlag.

Keddy, P. A. 1989. *Competition.* New York: Routledge, Chapman, and Hall.

Kupfer, J. A., and G. P. Malanson. 1993. Observed and modeled directional change in riparian forest composition at a cutbank edge. *Landscape Ecology* 8(3):185–99.

Levin, S. A. 1992. The problem of pattern and scale in ecology. *Ecology* 73:1,943–67.

Liu, Z-J. and G. P. Malanson. 1992. Long-term cyclic dynamics of simulated riparian forest stands. *Forest Ecology and Management* 48:217–31.

Malanson, G. P. 1986. Spatial autocorrelation and the distribution of plant species on environmental gradients. *Oikos* 45:278–80.

———. 1993. Comment on modeling ecological response to climatic change. *Climatic Change* 23:95–109.

Malanson, G. P., W. E. Westman, and Y-L. Yan. 1992. Realized versus fundamental niche functions in a model of chaparral response to climatic change. *Ecological Modelling* 64:261–77.

Meentemeyer, V. 1989. Geographical perspectives of space, time, and scale. *Landscape Ecology* 3:163–73.

Moloney, K. A., S. A. Levin, N. R. Chiariello, and L. Buttel. 1992. Pattern and scale in a serpentine grassland. *Theoretical Population Biology* 41:257–76.

Portnoy, S., and M. F. Willson. 1993. Seed dispersal curves: Behavior of the tail of the distribution. *Evolutionary Ecology* 7:25–44.

Shugart, H. H., T. M. Smith, and W. M. Post. 1992. The potential for application of individual-based models for assessing the effects of global change. *Annual Review of Ecology and Systematics* 23:15–38.

Solomon, A. M. 1986. Transient response of forests to CO_2-induced climate change: Simulation modeling experiments in eastern North America. *Oecologia* 68:567–79.

Upson, C., T. Farlhaber, D. Kumins, D. Laidlaw, D. Schlegel, S. Vroom, R. Gurmitz, and A. Van Dam. 1989. The application visualization system: A computational environment for scientific visualization. *IEEE Computer Graphics and Applications* 9:30–42.

van der Pijl, L. 1982. *Principles of Dispersal in Higher Plants.* Berlin: Springer-Verlag.

Webb, T. 1992. Past changes in vegetation and climate: Lessons for the future. *Global Warming and Biological Diversity,* 59–75. New Haven: Yale University Press.

George P. Malanson is an associate professor at the University of Iowa. He earned his Ph.D. at UCLA in 1983.

Marc P. Armstrong is an associate professor at the University of Iowa. He earned his Ph.D. at the University of Illinois in 1986.

David A. Bennett is an instructor at Southern Illinois University. He is a doctoral candidate at the University of Iowa.

Department of Geography and
Center for Global and Regional Environmental Research
316 Jessup Hall
The University of Iowa
Iowa City, Iowa 52242
Phone: (319) 335-0151
Fax: (319) 335-2725
E-mail: *george-malanson@uiowa.edu*

47

Solving Groundwater Problems Using Multigrid Methods in a Grid-Cell-Based GIS

Daene C. McKinney and Han-Lin Tsai

Multigrid methods use a series of successively coarser grids to accelerate the solution of equations arising from the numerical approximation of partial differential equations. A grid-cell-based geographic information system (GIS) with a map algebra language is capable of displaying and manipulating spatial data and attributes and handling the data arrays and results arising in groundwater simulation problems. Using a grid-cell-based system, we have performed groundwater modeling directly within a GIS without going outside the system for model solution. We have applied the multigrid method to solve steady state groundwater flow problems in heterogeneous aquifers. Execution times for the GIS-multigrid method exceed those of traditional groundwater simulation models due to the creation of temporary grids for intermediate calculations during the simulation. Efficient handling of boundary conditions is another difficult problem in GIS grid-cell-based modeling. The GIS multigrid modeling approach and its application to solving groundwater flow problems are presented.

INTRODUCTION

Multigrid methods are iterative methods for the solution of algebraic equations, especially those arising from the numerical solution of multidimensional boundary-value problems on discrete spatial domains. A unique characteristic of multigrid methods is the use of a series of coarser grids to accelerate the convergence of iterative methods of solving linear systems of equations. The combination of fine- and coarse-grid solutions provides optimal resolution of all spatial frequency components comprising the solution. In recent years, multigrid methods have been used extensively to solve engineering problems; however, comparatively little use of multigrid methods has been made in solving groundwater flow and mass transport problems.

One of the major difficulties encountered in implementing multigrid methods in a conventional programming environment is the creation and manipulation of data structures that support the computations at different levels of discretization. Raster- or grid-cell-based GISs provide a simple and versatile method for managing information on different grids and performing the intergrid data transfer operations necessary to implement the multigrid method. GISs are well suited to displaying and manipulating the spatial data and attributes common to many groundwater modeling problems. With the advent of GIS grid-cell-based macroprogramming languages, finite difference and multigrid solutions of groundwater problems can be computed directly within the GIS without the need to access external simulation models.

GISs were originally conceived, designed, and programmed by cartographers and geographers to produce maps and perform spatial analysis on static two-dimensional data sets. As a result, the current generation of GISs tend to have low computational efficiency and speed. However, the solution of groundwater and environmental problems normally entails the solution of dynamic, three-dimensional partial differential equations using spatially distributed data. Thus, the need arises to enhance the computational capability and speed of GISs to perform simulations of these systems.

We have applied the multigrid approach to the finite difference solution of steady groundwater flow problems using a grid-cell-based GIS environment. The development of the GIS multigrid method and its application to groundwater flow examples are presented below.

MULTIGRID METHODS

The simulation of flow in aquifers often requires the solution of large sets of linear equations. The linear solver portion of a typical groundwater model can consume the majority of computer processing time. Multigrid methods have been successful in reducing the percent of CPU time devoted to the linear solver. In this section, we discuss linear solvers and their implementation in multigrid methods.

Iterative Methods

Consider a system of n linear equations in n unknowns

$$Ax = b \qquad (47\text{-}1)$$

where A $(n \times n)$, b $(n \times 1)$, and x $(n \times 1)$ are a coefficient matrix, a vector of right-hand sides, and the exact solution of the system, respectively. Given an initial guess x_o, an iterative method is defined as

$$Mx_k = Nx_{k-1} + b \qquad k = 1, 2, \ldots \qquad (47\text{-}2)$$

where $A = M - N$, M is diagonal or lower triangular for either the Jacobi or Gauss Seidel methods, respectively, and x_k is the current estimate of the solution to (47-1). After k iterations, the residual is

$$r_k = b - Ax_k \qquad (47\text{-}3)$$

and the error in the approximate solution is

$$e_k = x - x_k \tag{47-4}$$

The error satisfies the residual equation

$$Ae_k = r \tag{47-5}$$

An iterative method can be applied to (47-5) to approximate the error and improve the approximate solution, or

$$x_{k,new} = x_{k,old} + e_k \tag{47-6}$$

A major difficulty with iterative methods for solving equation (47-1) is that convergence is often extremely slow due to long wavelength (low frequency) Fourier components of the error e. Using standard iterative methods, short wavelength error components damp in a few iterations, but long wavelength components can take hundreds of iterations to be eliminated (Hirsch 1988). This has led to the development of acceleration techniques for iterative methods such as multigrid. In the multigrid method the long wavelength error components contained in the residual r on a fine grid become short wavelength error components on a coarser grid and are quickly damped by iteration on the coarser grid.

Coarse Grid Correction

Let x^h be the solution of the system of equations resulting from the discretization of a partial differential equation on a uniform grid of representative mesh size $h = \Delta x$, then x^h satisfies

$$A^h x^h = b^h \tag{47-7}$$

where A^h is the matrix resulting from the discretization, finite difference say, and b^h is the right-hand side containing boundary condition and source/sink information. The solution to the system of equations on this mesh can be represented as

$$x^h = x_k^h + e_k^h \tag{47-8}$$

where x_k^h is the approximate solution after k iterations of an iterative method applied to (47-7) and e_k^h is the error after k iterations. The short wavelength components of the error are eliminated by the first few iterations. The long wavelength error components can be eliminated by transferring the residual,

$$r^h = b^h - A^h x_k^h \tag{47-9}$$

to a coarser grid (mesh size $2h = 2\Delta x$) to obtain

$$r^{2h} = I_h^{2h} r^h \tag{47-10}$$

where I_h^{2h} is a fine-to-coarse grid transfer operator defined below. This coarse-grid residual becomes the right-hand side of the residual equation (47-5) on the coarse grid. Several iterations are then applied to the residual equation on the coarse grid to eliminate the fine-grid long wavelength error components, which are now the coarse-grid short wavelength components, or

$$A^{2h} e^{2h} = r^{2h} \tag{47-11}$$

After k iterations, the error term e_k^{2h} is then transferred back to the fine grid, that is

$$e_{new}^h = I_{2h}^h e^{2h} \tag{47-12}$$

where I_{2h}^h is a coarse-to-fine grid-transfer operator defined below. The fine-grid solution from (47-8) is then updated

$$x_{new}^h = e_{old}^h + e_{new}^h \tag{47-13}$$

Multigrid Method

In the multigrid method, the coarse-grid correction technique is used recursively on a series of M increasingly coarse meshes, where the mesh spacing is $(2^{m-1})h$, $m = 1, ..., M$ and h is the fine-grid mesh spacing. The basic V-cycle multigrid method proceeds as follows (Briggs 1987; Wesseling 1992).

1. Starting with $m = 1$ and using x_{old}^m as an initial guess on the finest grid, apply n iterations of an iterative method on the fine grid to smooth the short wavelength error components. Then project the residual onto the next coarser grid (m+1) by a restriction of the residual $r_n^m = b^m - A^m x_n^m$ from the finer grid, that is,

Iteration: $A^m x^m = b^m$ n times (47-14)

Projection: $b^{m+1} = I_m^{m+1}(b^m - A^m x_n^m)$ (47-15)

where I_m^{m+1} is a fine-to-coarse grid-projection operator, and x_n^m is the approximate solution on the mth grid after n iterations.

Update: $x_{old}^m = x_n^m$.

2. Continue this iteration-projection process until the coarsest grid ($m = M$) is reached.
3. Solve the system of equations on the coarsest grid (M) to convergence.

$$A^M x^M = b^M \tag{47-16}$$

4. Perform interpolation from the coarser (m+1) grid to the next finer grid (m) and correct x_{old}^m on the finer grid. Then perform n iterations of a relaxation method using x_{new}^m as an initial guess, that is,

Interpolation: $x_{new}^m = x_{old}^m + I_{m+1}^m x_n^{m+1}$ (47-17)

Interation: $A^m x^m = b^m$, n times

where I_{m+1}^m is coarse-to-fine grid-interpolation operator to be discussed below.

5. Continue the interpolation-relaxation process until the finest grid (m=1) is reached.

This basic multigrid V-cycle is repeated as needed until convergence of the approximate solution on the finest grid, x^1, is obtained. This cycle is illustrated in Figure 47-1.

Intergrid Transfer Operators

To implement the multigrid method we need two intergrid-transfer operators: one to go from a fine grid to a coarser grid, the projection oper-

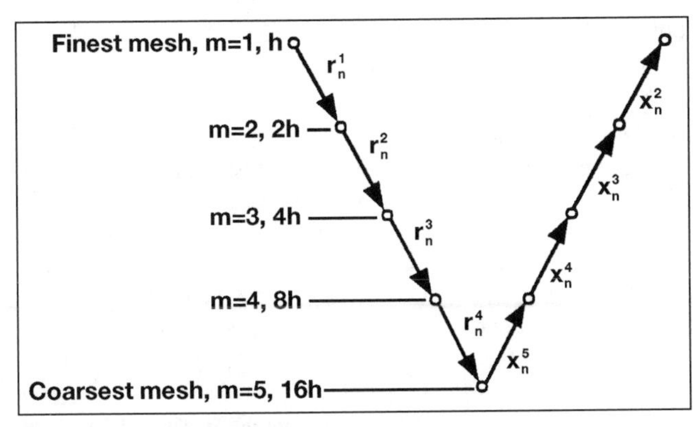

Figure 47-1. V-cycle multigrid cycle.

ator I_m^{m+1}, and one to go from a coarse grid to a finer grid, the interpolation operator I_{m+1}^m. For the purposes of intergrid transfer of information, simple linear interpolation is very efficient and effective (Briggs 1987). Various intergrid transfer operators have been derived for multigrid methods (Briggs 1987; Hackbusch 1985; Wesseling 1992). The majority of these operators are useful for mesh-centered finite difference methods and only a few are appropriate for block-centered finite difference approximations. The operators used here are for the block-centered finite-difference method of approximating two-dimensional porous media flow equations discussed below. Only block-centered methods are considered since this is the mesh topology used in most GIS grid-cell-based systems and they have attractive mass-conserving properties. The bilinear projection and interpolation operators for operators on two-dimensional block-centered grids are illustrated in Figure 47-2.

FLOW IN A CONFINED AQUIFER

We are concerned with determining the effectiveness and efficiency of modeling steady state, two-dimensional flow in saturated porous media using a grid-cell-based GIS. In this section we present the governing equations for flow and the discretization method that is used in the GIS model.

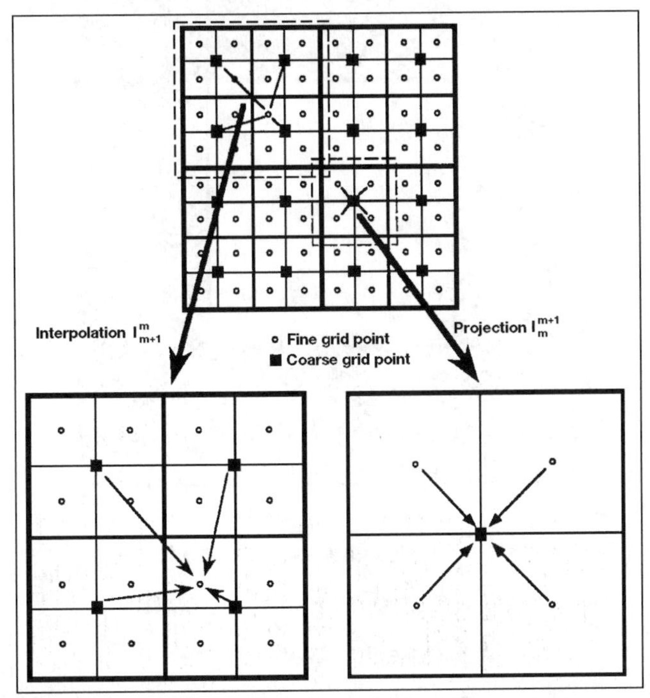

Figure 47-2. Intergrid projection and interpolation transfer operators.

Governing Equation

The governing partial differential equation for steady state, two-dimensional, confined flow in saturated porous media is

$$\frac{\partial}{\partial x}\left(T_x \frac{\partial h}{\partial x}\right) + \frac{\partial}{\partial y}\left(T_y \frac{\partial h}{\partial y}\right) + Q = 0 \qquad (47\text{-}19)$$

where T is the heterogeneous and anisotropic transmissivity [L^2/T], h is the hydraulic head [L], and Q is a source/sink term [$L^3/T/L^2$]

Equation (47-19) is approximated with a second-order accurate finite difference method using the block-centered finite difference grid shown in Figure 47-3.

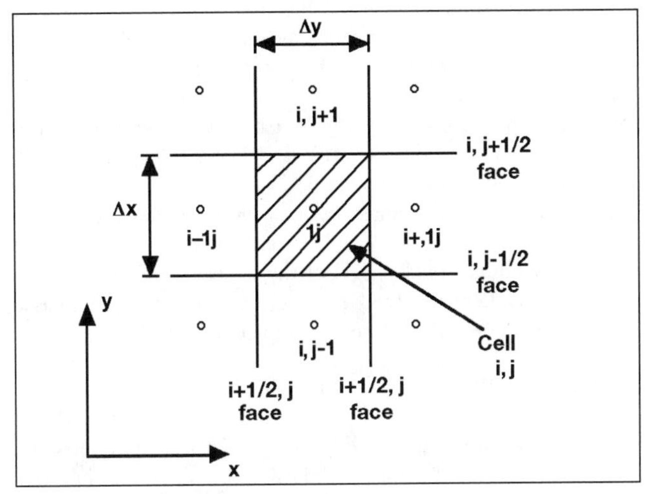

Figure 47-3. Block-centered finite difference grid in two dimensions.

Assuming a uniform cell size, $\Delta^2 = \Delta x \Delta y$, typical of GIS grid-cell-based systems, we obtain (Bear and Verruijt 1987)

$$-E_{ij}h_{ij} + A_i h_{i-1,j} + B_j h_{i,j-1} + C_i h_{i+1,j} + D_j h_{i,j+1} + \Delta^2 Q_{ij} = 0 \qquad (47\text{-}20)$$

where

$$A_{ij} = T_x^{i-1/2,j},\ B_{ij} = T_y^{i,j-1/2},\ C_{ij} = T_x^{i+1/2,j},\ D_{ij} = T_y^{i,j+1/2},$$
$$\text{and } E_{ij} = (A_i + B_j + C_i + D_j) \qquad (47\text{-}21)$$

The transmissivities in (47-21) are evaluated at the cell faces using a harmonic average; for example,

$$T_x^{i+1/2,j} = \frac{2T_x^{i,j}T_x^{i+1,j}}{T_x^{i,j}+T_x^{i+1,j}} \qquad (47\text{-}22)$$

When (47-20) is written for each cell of the grid, a system of linear equations similar to (47-7) results

$$Ah = b \qquad (47\text{-}23)$$

where A (n x n) is a pentadiagonal matrix of coefficients involving transmissivities, h (n x 1) is a vector of unknown cell head values, b (n x 1) is a vector of right-hand sides containing boundary and source information, and n is the number of cells in the domain. The solution of equation (47-23) in the grid-cell-based GIS system using the multigrid method is discussed in the next section.

APPLYING MULTIGRID METHODS IN A GIS ENVIRONMENT

Grid-Based Modeling

We are interested in modeling flow in porous media using a grid-cell-based GIS. This type of modeling environment has recently become available and its use in environmental engineering and groundwater modeling has not been investigated. While graphic environments for groundwater modeling have been available for some time now (McKinney and Loucks 1986), the majority of them have been special-purpose graphic pre- and postprocessing systems linked to a specific model or application and provide only the most rudimentary data management and analysis capabilities. The approach used here allows the definition of the model directly within the grid-cell-based GIS and does not rely on external programs for simulation.

The GIS used here consists of a combined vector- and grid-cell-based topological data model in which spatial features are represented as grid cells, points, lines, or polygons, and the attributes of those features are stored in a tabular or relational database (Menon et al. 1991; Morehouse 1985). Region boundaries, sources or sinks, and aquifer property zones can be defined easily in the vector-based portion of GIS. The vector-based data are then transferred into grid-cell-based data for groundwater modeling. Upon the completion of the modeling, the resulting information can be transferred back to the vector-based portion of the GIS and displayed or postprocessed as needed.

Multigrid in the Grid Cell System

We have implemented the multigrid method of solving (47-23) in the GIS grid-cell-based modeling environment. The geometry and properties of an aquifer are defined using the vector-based portion of the GIS. Vector coverages are used to represent wells, aquifer geometry, and property zones. Once coverages are stored in the GIS, they are converted to raster grids with the appropriate grid sizes for multigrid implementation. Individual grids are created that represent the geometry, the well locations, aquifer properties, and so on, for each grid resolution needed in the multigrid computation. This makes the task of data management for the multigrid method much simpler and more convenient than with conventional programming techniques. After the grids have been constructed, the multigrid computation discussed above is carried out in the GIS to solve the flow problem.

To apply the multigrid method to solve the groundwater flow equations, an iterative method is used to determine the value of head in a particular grid cell as a function of the values of its neighboring cells. A Jacobi iterative method has been used here to solve the linear system of equation (47-23). The grid-cell-based system and its associated macrolanguage are used to construct iteration k for cell (i,j) on grid level m as

$$h^m_{ijk}=\frac{1}{E_{ij}}(A_{ij}h^m_{i-1,j,k-1}+B_{ij}h^m_{i,j-1,k-1}+C_{ij}h^m_{i+1,j,k-1}+D_{ij}h^m_{i,j+1,k-1}+\Delta^2Q_{ij}) \quad (47\text{-}24)$$

where the matrices A, B, C, D, and E are defined above, and h^m_{ijk} and $h^m_{i-1,j,k-1}$ are the head values at the current (k) and previous (k-1) iterations, respectively.

All of the quantities needed in equation (47-24) are stored as grids; this is one of the advantages using a grid-cell-based method. As discussed above, the multigrid method requires the transfer of information between grid levels. Several intergrid transfer operations are available in the grid-cell-based system: (1) nearest neighbor, (2) bilinear interpolation, and (3) bicubic interpolation. Bilinear interpolation is used here as the intergrid projection and interpolation transfer operations since it is the obvious choice for block-centered finite difference methods.

RESULTS

The grid-cell-based multigrid method has been implemented using the GIS macroprogramming language and its grid-cell-based system. The major function of the macrolanguage is to control program input/output, execution of the grid cell program, and error detection. The functions and commands of the grid-cell-based system handle the execution of the multigrid algorithm, that is, the iterative method and the intergrid transfers.

In the example of the GIS multigrid method shown here, Jacobi iteration is used to solve steady state flow problems with constant head boundaries in a confined aquifer system. The input grids needed for defining a flow problem consist of: (1) an initial head guess, (2) sink/source locations and strengths, and (3) the transmissivity values. The output grid is the hydraulic head distribution after the simulation.

The example consists of steady flow of groundwater in a 16×16 grid in a homogeneous and isotropic-confined aquifer with constant head boundaries. In this example, the constant head boundaries are handled by adding a buffer layer of additional columns and rows to the flow-domain grid to maintain the constant head boundary values as information is transferred from one grid to another. In the examples shown here, we use four grid levels (m=4) in a V-cycle multigrid method, so eight additional columns or rows of grid cells are needed for the constant head boundaries. As a result, the actual input and output grid size is 32×32 with a total of 1,024 grid cells. Figure 47-4 shows the relationship between the buffer layer of grid cells outside the boundary and the interior solution domain. The domain we are interested in is a 16×16 grid area encompassed

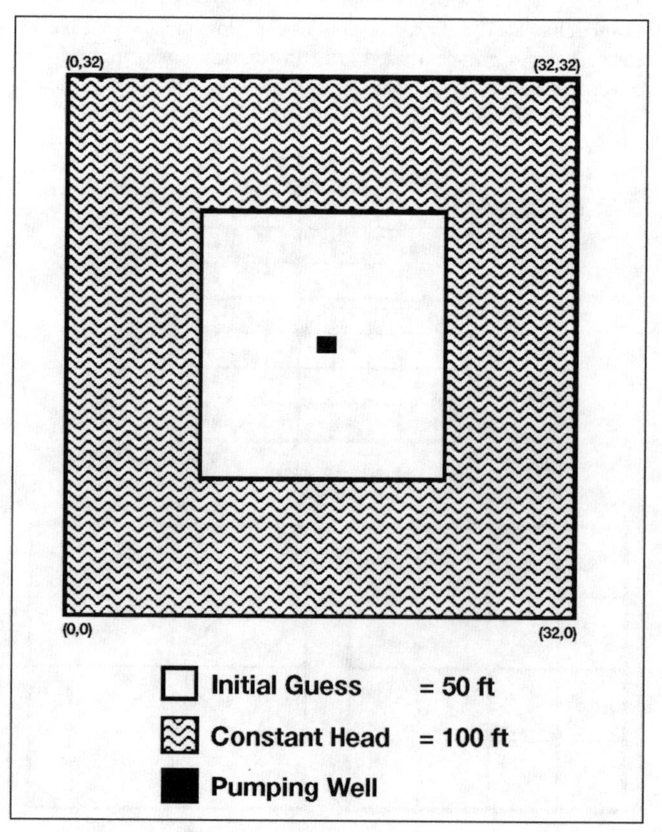

Figure 47-4. Head input grid for multigrid solution.

Figure 47-5. Head contours (ft.) for the GIS-multigrid solution.

by constant head boundaries with a value equal to 100 feet. A pumping well is located inside the grid area at cell (16,16) with pumping rate equal to 30 ft/min. Two different input grids are needed: one for the initial guess of the head distribution, and another for the sink or source.

The solution of this example problem took ten multigrid V-cycles to meet the convergence criteria. A contour plot of the resultant head distribution is shown in Figure 47-5. For each V-cycle, five iterations were required for the solution of the system of equations on the coarsest grid to meet the convergence criteria. The Sun SPARC 2 CPU time was one hour and fifty-three minutes.

CONCLUSION

The GIS used here consists of a combined vector- and grid-cell-based topological data model in which spatial features are represented as grid cells, points, lines, or polygons, and the attributes of those features are stored in a tabular or relational database. The vector-based GIS component is used as a database management system for the groundwater modeling performed in the grid-cell-based system. In this way, an aquifer's geometry, properties, wells, and other features are entered, stored, analyzed, and manipulated prior to being discretized and transferred to the grid-cell-based modeling environment. Upon the completion of modeling, the resulting information can be transferred back to the GIS and displayed or post-processed as needed or desired.

Some conclusions can be drawn from this work:

1. Steady state groundwater flow problems can be successfully solved using the grid-cell-based GIS.
2. The multigrid method is an efficient technique for accelerating the iterative solution of systems of equations. The advantage of using the multigrid method in the grid-cell-based system is that the intergrid error and correction transfers are accomplished by a set of built-in functions of the GIS.
3. The time needed for this approach exceeds that of typical FORTRAN simulation models. The major factor causing the excess CPU time required for the solution is that the GIS records all data from temporary grids onto hard disk during program execution.
4. A limitation of the grid cell system used here is that updated cell values cannot be immediately used in grid calculations making efficient iterative methods such as Gauss Seidel or successive-over-relaxation difficult to implement.
5. A more efficient method of handling boundary conditions in intergrid transfer operations other than the expansion of the grid domain is needed.

References

Bear, J. 1972. *Dynamics of Fluids in Porous Media.* New York: Elsevier.

Bear, J., and A. Verruijt. 1987. *Modeling Groundwater Flow and Pollution.* Dordrecht: D. Reidel.

Briggs, W. L. 1987. *A Multigrid Tutorial.* Philadelphia: SIAM.

Golub, G. H., and C. F. van Loan. 1989. *Matrix Computations.* Baltimore: Johns Hopkins University Press.

Hackbusch, W. 1985. *Multi-Grid Methods and Applications.* Berlin: Springer-Verlag.

Hirsch, C. 1988. *Numerical Computation of Internal and External Flows,* vol. 1. *Fundamentals of Numerical Discretization.* Chichester: John Wiley and Sons.

McKinney, D. C., and D. P. Loucks. 1986. Interactive modeling for groundwater management. In *Finite Elements in Water Resources,* edited by A. Sa da Costa et al. Berlin: Springer-Verlag.

Menon, S., P. Gao, and C. Zang. 1991. *GRID: A Data Model and Functional Map Algebra for Raster Geo-Processing.* Draft Report. Redlands, Calif.: Environmental Systems Research Institute, Inc.

Morehouse, S. 1985. ARC/INFO: A georelational model for spatial information. *Proceedings Auto Carto* 7, 388–97.

Wesseling, P. 1992. *An Introduction to Multigrid Methods.* New York: John Wiley and Sons.

Daene C. McKinney and *Han-Lin Tsai*
Department of Civil Engineering
University of Texas
Austin, Texas 78712
E-mail: *daene_mckinney@cemailgate.ce.utexas.edu*

48

Conditional Simulation of Percolate Flux Below a Root Zone

Andrew S. Rogowski

*A multiple indicator conditional simulation program (ISIM-3) and a raster-based geographic information system (GIS-IDRISI) were used to construct overlays of percolate flux below a root zone. The approach called for a substitution of conditionally simulated values of hydraulic conductivity (*K*) into a simplified gravity flow equation to give spatial estimates of flux below a root zone. These estimates were subsequently used to construct quantity-arrival-time distributions of recharge flux at the water table. The procedure is illustrated with an example from a Pennsylvania watershed. Sampling positions, where soil *K *was measured with a Guelph permeameter, were identified in the field with a global positioning system (GPS), while GIS modeling was utilized to manipulate, combine, and display the spatial flow properties. Conditional simulation of flow parameters, when coupled with currently available GIS and GPS technology, appeared to successfully account for spatial variation of field-measured soil attributes and helped to identify potential groundwater recharge zones within a watershed.*

INTRODUCTION

Groundwater is the source of drinking water in many rural areas, but the specific recharge zones that influence groundwater availability and quality are usually difficult to identify. Because highly fractured, very pervious bedrock underlies shallow soils in parts of central Pennsylvania (Urban and Gburek 1988), rate and quality of water discharge from below a root zone controls both the potential recharge rate at the water table and the quality of the groundwater at a scale of a watershed. Consequently, identifying spatial distributions of discharge from below a root zone may help delineate possible zones of groundwater recharge and contamination.

Distribution of soil water discharge from below a root zone and subsequent recharge to the groundwater varies in space and in time. Contributing areas depend on the amount, timing, and distribution of precipitation; their position relative to the water table; land use and soil properties; and underlying geology. A key soil property that controls the flow within an area is the distribution of soil hydraulic conductivity (K). Unfortunately K is a highly variable soil property that may differ greatly from point to point and vary appreciably in time.

The objectives of this study were to: (1) measure spatial distribution of K over an area; (2) simulate alternate equiprobable distributions of K values conditioned on measured data; and (3) demonstrate how a variable K would affect discharge from below a root zone, recharge at the water table, and distribution of contributing areas on a 123-km^2 watershed in east-central Pennsylvania.

FLUX BELOW ROOT ZONE

Spatial distributions of flux below a 0.6-m depth and approximating the depth of the root zone were computed from field-measured distributions of soil hydraulic conductivity (K) utilizing gravity flow approximation (Jury et al. 1991). If K can be written in the exponential form as a function of water content (θ),

$$K(\theta) = K_s \exp [\alpha(\theta - \theta_s)] \tag{48-1}$$

and the change of soil water content, or pressure with depth, approaches 0 (unit hydraulic gradient), water will drain uniformly from all soil depths and a simplified drainage equation (Davidson et al. 1969; Warrick et al. 1977) will describe the flux below the root zone:

$$q = K_s / (1 + \alpha K_s t/L) \tag{48-2}$$

where q (cm/day) is the flux at some depth L (cm) below the root zone, α is a constant relating K to the soil volumetric water content (θ), subscript "s" denotes field saturation, and t is time in days. The profile depth L_r is the depth of the root zone. Taking $L_r = 0.62$ m, $\alpha = 2.1$, at $t = 1$ day, and integrating, equation (48-2) becomes:

$$CUM.FLUX \sim 10 \times \Sigma q\, \Delta t \sim 10 \times 1 \times [K_s / (1 + 0.0337 K_s)] \tag{48-3}$$

where the $CUM.FLUX$ is in 1/m^2 and the total duration of precipitation Δt during spring, when expressed in a compressed form (Rogowski 1990), does not exceed one day.

Assuming that soil controls the rate of flow in the vadose zone, the recharge flux to the groundwater can be made numerically equal to the discharge flux from the bottom of the root zone. Under these circumstances, the residence time in the root zone (T_r) and the travel time to the groundwater (T_g), can be computed as a function of depth to the groundwater L_g, and effective porosity P through which the flow takes place.

$$T_r = L_r P_r / q \tag{48-4}$$

$$T_g = L_g P_g / q \tag{48-5}$$

QUANTITY-ARRIVAL-TIME DISTRIBUTIONS

To evaluate adverse impacts of any surface activity on groundwater, we should know an approximate extent of outflow boundary and be able to estimate contaminant arrival time and concentration at the water table. Nelson (1978) has proposed a method to compute contaminant arrival-time distributions. In Figure 48-1 sample calculations for one location and a set of input parameters given in the caption are illustrated.

In the first row of Figure 48-1, time, as reckoned from the start of an event, is listed. In the second row q, the flux rate below the root zone, is calculated from (48-2). Values of q are then used in (48-4) and (48-5) to compute the residence time in the root zone (T_r) and the travel time to the groundwater (T_g), shown in the next two rows. The following row gives an estimate of a cumulative recharge flux (1/m^2) to the groundwater, shown as the dashed line in Figure 48-1b. Assuming an effective twenty-four-hour event, the first (t=initial) and the last (t=1.0 days) cumulative recharge fronts are shown as dashed paired curves in Figure 48-1b, arriving at the water table after 32.5 days for the initial front and after 35 days for the final one. The two are of the same magnitude but offset by 2.5 days corresponding to the water residence time in the root zone. The difference between the two curves, expressed as a q-pulse (1/m^2/d) and shown by a solid curve in Figure 48-1b, is an effective (Rogowski 1990) recharge flux pulse arriving at the water table at 32.5 through 155 days. It is computed by subtracting the lower dashed curve from the upper and dividing by 32.5. Hence, the scale on the right of Figure 48-1b is equal to the cumulative flux term on the left divided by the earliest (32.5 days) estimated arrival time of the recharge front at the water table.

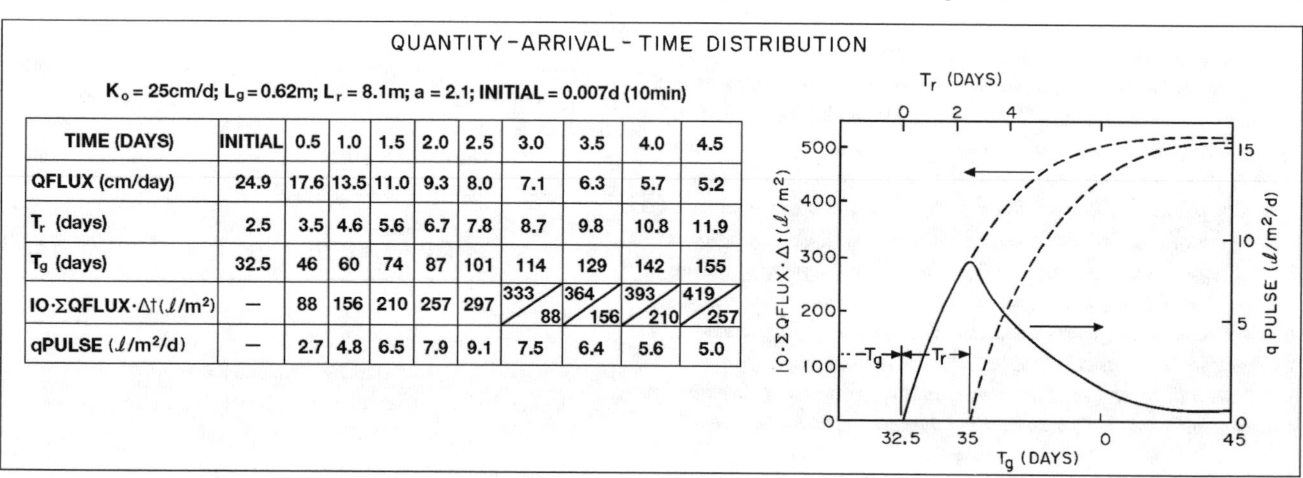

QUANTITY-ARRIVAL-TIME DISTRIBUTION

K_o = 25cm/d; L_g = 0.62m; L_r = 8.1m; a = 2.1; INITIAL = 0.007d (10min)

TIME (DAYS)	INITIAL	0.5	1.0	1.5	2.0	2.5	3.0	3.5	4.0	4.5
QFLUX (cm/day)	24.9	17.6	13.5	11.0	9.3	8.0	7.1	6.3	5.7	5.2
T_r (days)	2.5	3.5	4.6	5.6	6.7	7.8	8.7	9.8	10.8	11.9
T_g (days)	32.5	46	60	74	87	101	114	129	142	155
IO·ΣQFLUX·Δt (ℓ/m²)	—	88	156	210	257	297	333 / 88	364 / 156	393 / 210	419 / 257
qPULSE (ℓ/m²/d)	—	2.7	4.8	6.5	7.9	9.1	7.5	6.4	5.6	5.0

Figure 48-1. (a) Tabular and (b) schematic, computation of the quantity-arrival-time distribution.

During each time period considered here, there were from three to five major storm events and a number of lesser ones. A single "compressed" front was assumed to reach the water table, carrying with it all of the percolating water for that time period.

CONDITIONAL SIMULATION

Conditional simulation of water flux from below a root zone constitutes a spatial response function model requiring spatially distributed soil hydraulic conductivity (K) as input. Equation (48-2) acts as such a transfer function: that is, it translates a spatial distribution of point measurements of K into a spatially distributed estimate of water flux below the root zone. Transfer functions operating over a large area require attribute data to be defined for a whole watershed. Because such exhaustive data sets are generally not available, I resorted to simulation and constructed GIS overlays of flux below the root zone from conditionally simulated, equiprobable realizations of K processed through (48-2).

A three-dimensional multiple indicator conditional simulation program (ISIM-3) (Gomez-Hernandez and Srivastava 1990) was implemented in two dimensions to create a number of alternate equiprobable realizations of soil K on a watershed, based on field-derived data. Details of the underlying theory can be found in Journel (1987).

The model implementation was as follows: An IDRISI (Eastman 1989) raster grid, with a pixel size of 0.3 × 0.3 km, and consistent with desk computer limitations, was imposed on the watershed. Field-measured values of K falling anywhere within a pixel grid cell were then assumed to apply to a whole pixel. All field-measured soil hydraulic conductivity values were converted into a set of 0s and 1s by the indicator transformation and spatially discretized into several thresholds. The thresholds corresponded to the Q1 (25th percentile), Q2 (median), and Q3 (75th percentile) moments of the statistical distribution. Numerical values were assigned a magnitude of 0, if greater or equal to the threshold, and 1 if less. Based on their distribution over an area, variogram and covariance models of spatial dependence were created.

In the next step, an algorithm for adding a new value at an unsampled location was defined. The algorithm was consistent with prior information and a pertinent threshold covariance model. The first step involved estimation of a conditional probability that a new value would be less than a given threshold. This was done by kriging each unsampled location using surrounding indicator values. The resulting estimate of conditional probability distribution (CPDF) was between 0 and 1.

The actual simulation of a corresponding indicator value at an unsampled grid location was accomplished by drawing a random number between 0 and 1 from a uniform distribution (Monte Carlo). If this number was less than or equal to a kriged value, a simulated indicator value of 1 was assigned to that location. If it was more, a value of 0 was assigned (Alabert 1987). The new value became a part of the conditioning data set, and the procedure was repeated at another location until all unsampled locations in a watershed grid were filled.

ILLUSTRATIVE EXAMPLE

Field Measurements

The example discussed here is based on a field study conducted on the 123-km² Mahantango Creek watershed in Pennsylvania (Figure 48-2a). The watershed is situated in the valley and ridge physiographic province. In the valleys, land use is predominantly cropland, with areas of permanent pasture, truck crops, small woodlots, and orchards. The ridges are generally forested.

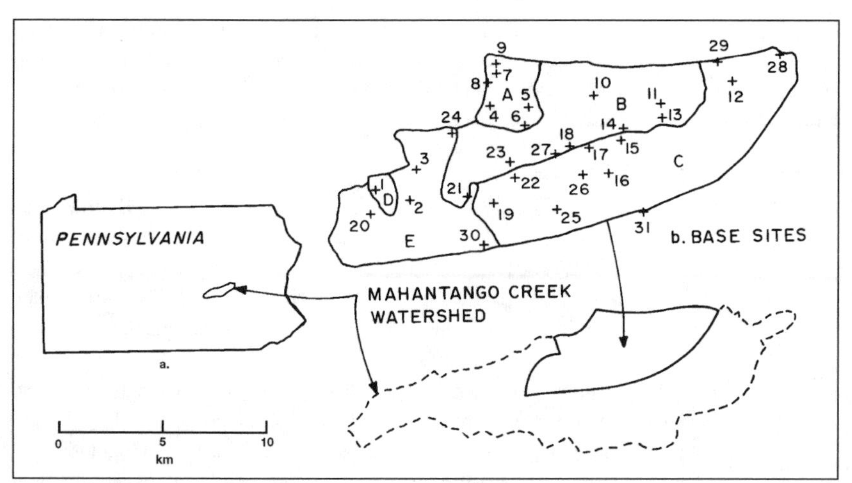

Figure 48-2. (a) Location of the Mahantango Creek watershed, and (b) location of the thirty-one base sites.

Field-measured values of soil hydraulic conductivity (K) were used to compute water flux below the root zone. The root zone depth was on the average 0.62 m. The K values were measured at 0.3-m depth, on thirty-one base sites (Figure 48-2b) with the Guelph permeameter (Elrick et al. 1989) at a prevailing ambient water content and pressure. These values were assumed to represent effective hydraulic conductivity of the root zone layer. Concurrent readings of soil water content and pressure also were made and used for the computation of the coefficient alpha in (48-1) and (48-2). Base-site locations were identified in UTM (Universal Transverse Mercator projection) coordinates with a commercial GPS.

Model Applications

Figure 48-3 shows a flowchart of the GIS model and illustrates some of the intermediate results obtained. Individual inserts represent component overlays for spring of 1990. Lines and arrows indicate internal relationships and the computational chain within the model. "KRIG" means that individual grid values have been kriged; "CONDITIONAL SIMULATION" means individual grid values have been simulated using the ISIM-3 program (Gomez-Hernandez and Srivastava 1990); otherwise, an appropriate algorithm used to process an overlay, pixel by pixel, is indicated. Spatial variability was retained in the model by utilizing kriged or conditionally simulated distributions of input variables.

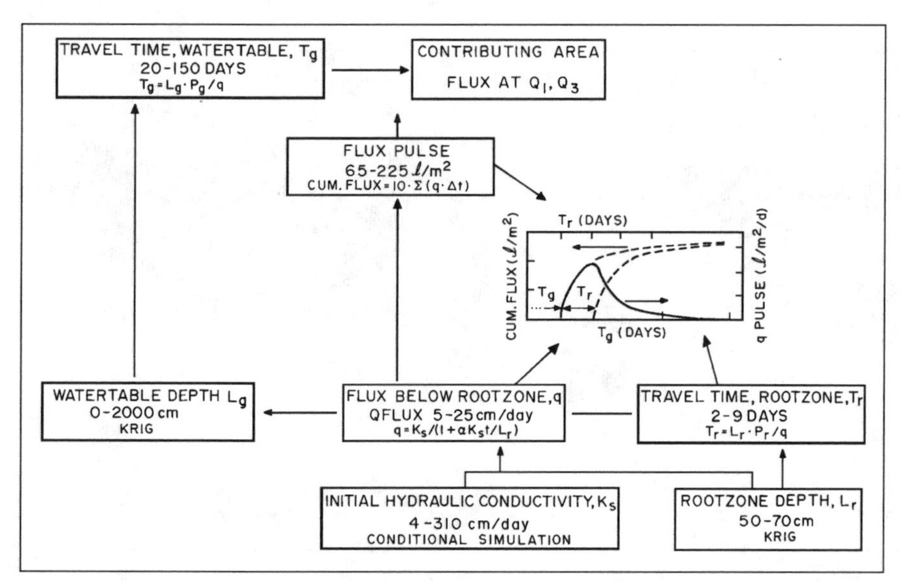

Figure 48-3. A generalized flowchart of a GIS model.

An example of typical univariate statistics for three sets of conditional simulations and conditioning data are given in Table 48-1. Probability (*pdf*) and cumulative density functions (*cdf*) for one of the simulations and the conditioning data are illustrated in Figure 48-4. Three conditional simulations of hydraulic conductivity (CSK1, CSK2, and CSK3) are shown in Figure 48-5.

Inspection of Table 48-1 and Figure 48-4 suggests that conditional simulation of flow parameters does account for spatial variability of field-measured soil attributes. The conditionally simulated values appear to preserve spatial data structure necessary to construct alternate, equiprobable spatial distributions of soil K over the watershed.

When conditional simulations of soil hydraulic conductivity in Figure 48-5 are compared with their interpolated (KRIG) equivalents, a smoothing effect of interpolation and a rather dispersed distribution of conditionally simulated values are readily apparent. Part of the reason is that usually, as in Figure 48-5 and Table 48-1, only a few simulations can be shown at any one time. However, if twenty-five or more simulations are averaged, a spatial continuity pattern becomes evident. Interpolated estimates of a highly variable attribute such as K may give a false impression of homogeneity and continuity. But the apparent smoothness reflects primarily the lack of data. In contrast, conditionally simulated distributions may be more representative because they are able to reflect the prevailing patterns of local variability.

Table 48-1. Univariate statistics for selected conditionally simulated (CS) and conditioning data (COND) distributions of hydraulic conductivity on the Mahantango Creek watershed.

PARAM	$\bar{X}$	SD	MODE	MAX	Q3	MED	Q1	MIN
			(cm/day)					
CSK1	75	91	5	310	128	25	10	4
CSK2	51	73	4	307	41	20	10	3
CSK3	34	57	4	308	25	16	9	4
COND	39	57	13	311	49	23	13	4

Contributing Areas

Figure 48-6 gives the primary contributing areas associated with Q1 and Q3 quartiles of the travel time to underlying water table. The distribution of travel time was obtained by dividing the simulated flux below the root zone overlay into the kriged depth to water table overlay (48-4). When the result was intersected with the CUM.FLUX overlay (48-3) (Figure 48-3), distribution of recharge flux at the water table was obtained. Subject to the assumptions of this study, the figure shows potential magnitude and distribution of recharge flux between 35 and 105 days following the event. It also suggests that 50% (Q3-Q1) of recharge flux could be associated with the amounts and areas delineated in Figure 48-6.

CONCLUSIONS

The approach discussed in this chapter may be useful for evaluating the potential impact of agriculture on underlying groundwater in the context of soil properties that affect water yield and quality. The inputs and outputs can be readily handled within the framework of a GIS. This allows application of results to a variety of practical field problems. Since the input data, model operations, and output are all georeferenced in the same GPS coordinates, comparisons of properties associated with a given area can readily be made. Potential applications involve a delineation of primary recharge contributing areas within a watershed.

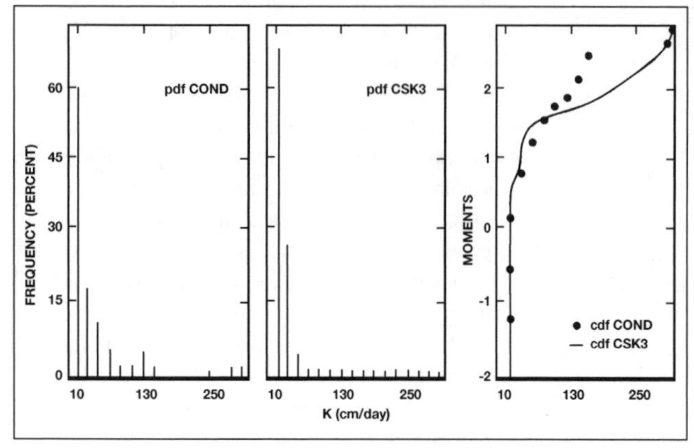

Figure 48-4. Probability density functions (pdf) and cumulative density functions (cdf) for a selected conditional simulation (CSK3) and conditioning (COND) data.

References

Alabert, F. G. 1987. "Stochastic Imaging of Spatial Distributions Using Hard and Soft Information." Master's thesis, Stanford University, Stanford.

Davidson, J. M., L. B. Stone, D. R. Nielsen, and M. E. Larue. 1969. Field measurement and use of soil-water properties. *Water Resources Res.* 5(6):1,312–21.

Eastman, J. R. 1989. *IDRISI: A Grid-Based Geographic Analysis System.* Worcester: Clark University Cartographic Service.

Elrick, D. E., W. D. Reynolds, and K. A. Tan. 1989. Hydraulic conductivity measurements in the unsaturated zone using improved well analyses. *Groundwater Monitoring Review* Summer:184–93.

Gomez-Hernandez, J. J., and R. M. Srivastava. 1990. ISIM3D: An ANSI-C three-dimensional, multiple indicator conditional simulation program. *Computers & Geosciences* 16(4):395–440.

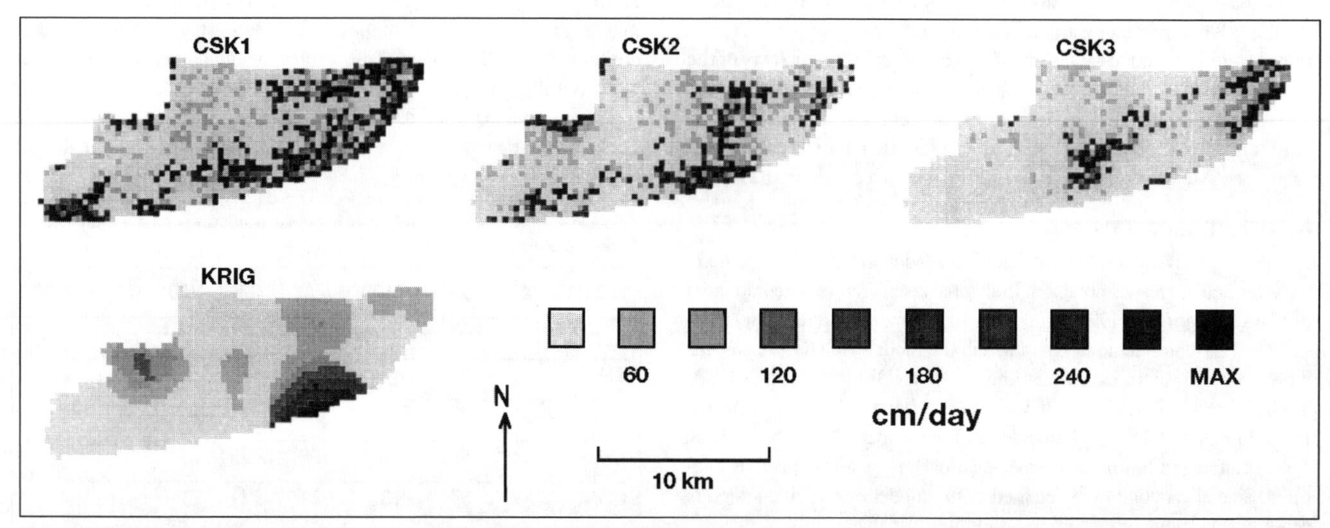

Figure 48-5. Example of three conditional simulations (CS) and a kriged (KRIG) distribution of hydraulic conductivity on the Mahantango Creek watershed using conditioning data given in Table 48-1.

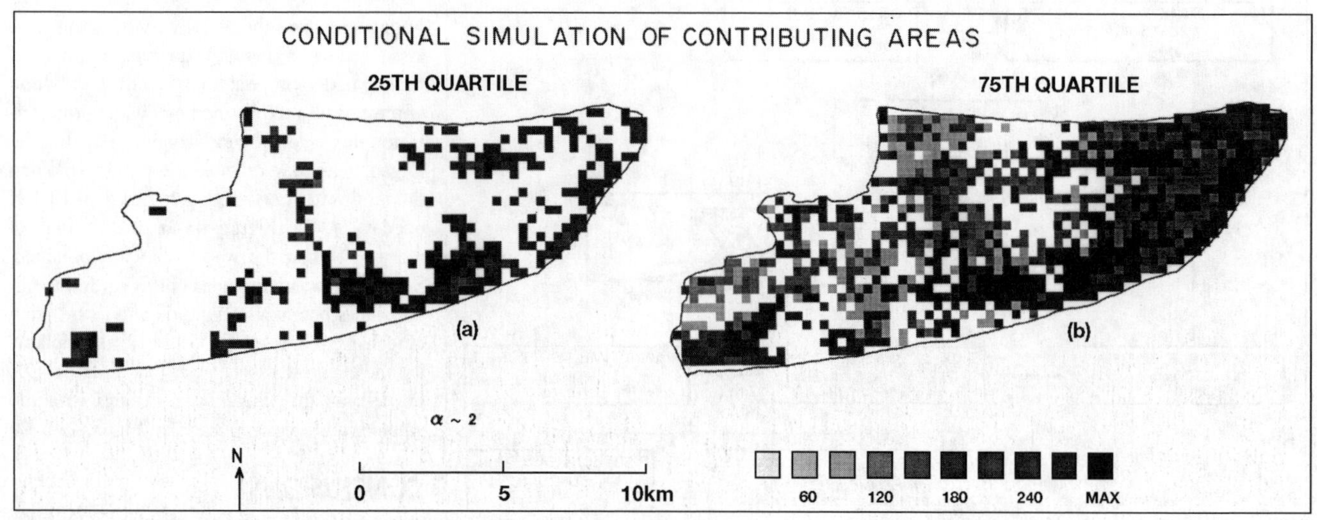

Figure 48-6. Distribution of recharge flux associated with (a) the 25th, and (b) the 75th quartiles of travel time to the water table.

Journel, A. G. 1987. Imaging of spatial uncertainty: A non-Gaussian approach. In *Geostatistical Sensitivity and Uncertainty Methods for Groundwater Flow and Radionuclide Transport Modeling,* edited by B. E. Buxton, 585–99. Columbus: Battelle Press.

Jury, W. A., W. R. Gardner, and W. H. Gardner. 1991. *Soil Physics.* New York: John Wiley and Sons.

Nelson, R. W. 1978. Evaluating the environmental consequences of groundwater contamination: Obtaining location/arrival time and location/outflow quantity distributions for steady flow systems. *Water Res. Research* 14(3):416–28.

Rogowski, A. S. 1990. Estimation of the groundwater pollution potential on an agricultural watershed. *Agricultural Water Management* 18:209–30.

Urban, J. B., and W. J. Gburek. 1988. A geologic and flow-based rationale for groundwater sampling. In *Groundwater Contamination: Field Methods,* ASTM STP 963, edited by A. G. Collins and A. J. Johnson, 468–81. Philadelphia: Am. Soc. for Testing and Materials.

Warrick, A. W., G. J. Mullen, and D. R. Nielsen. 1977. Predictions of the soil water flux based upon field-measured soil-water properties. *Soil Sci. Soc. Am. J.* 41:14–19.

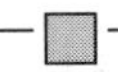

Andrew S. Rogowski is a soil scientist with the U.S. Department of Agriculture, Agricultural Research Service at University Park, Pennsylvania, and Adjunct Professor of Soil Physics at Pennsylvania State University. After joining the Agriculture Research Service, he investigated spatial variability of soil properties, nature of water flow in sloping soils, response of reclaimed mine soils to precipitation and acid mine drainage, and hydrologic behavior of compacted clay liners. His current research interests include distribution of recharge on agricultural watersheds, flow of water in the root zone, and applications of geostatistical methods in the GIS and GPS environment.
USDA-ARS-PSWM
University Park, Pennsylvania 16802-4709

49

Analyzing Septic Nitrogen Loading to Receiving Waters:
Waquoit Bay, Massachusetts

Chi Ho Sham, John W. Brawley, and Max A. Moritz

Waquoit Bay, a shallow bay on Cape Cod, Massachusetts, is exhibiting symptoms of eutrophication, largely attributed to septic nitrogen inputs in the drainage basin. This study assessed septic nitrogen inputs by linking the following components: a three-dimensional groundwater model, a geographic information system (GIS), and a customized spatiotemporal nitrogen loading calculation program. The groundwater model provided estimates of groundwater movement, delineated as annual "time bands" across the drainage basin. Using GIS statistical functions and land parcel data, we derived the spatial distribution of residential housing within the various groundwater travel time bands and generated input for the nitrogen loading calculation program. Temporal characteristics, such as when building occurred on a parcel, were incorporated into the loading calculation. Since 1940, residential development in the Waquoit Bay drainage area has increased about fifteenfold. Due to the slow speed of groundwater movement, the bulk of septic nitrogen entering the bay lags behind development by nearly a decade. If residential building is held at the 1990 level, nitrogen level input from septic systems will increase by 36% over current levels. At full residential build-out, septic nitrogen loading will increase to more than twice the current levels.

INTRODUCTION

There is growing concern about nitrogen loading to coastal waters, predominately because primary production rates in coastal waters are nitrogen limited (Howarth 1988). Coastal waters are increasingly at risk of eutrophication due to anthropogenic nutrient loading to ground and surface water (Valiela and Costa 1988; Giblin and Gaines 1990). Waquoit Bay is a shallow coastal bay located within the towns of Falmouth and Mashpee, Massachusetts, on the southwestern edge of Cape Cod (Figure 49-1). In 1989, Waquoit Bay was chosen as one of five Land Margin Ecosystems Research (LMER) sites by the National Science Foundation. The Waquoit Bay LMER project team conducted interdisciplinary investigations into the relationship between changes in land use and changes in the nutrient dynamics and primary productivity of shallow estuarine ecosystems.

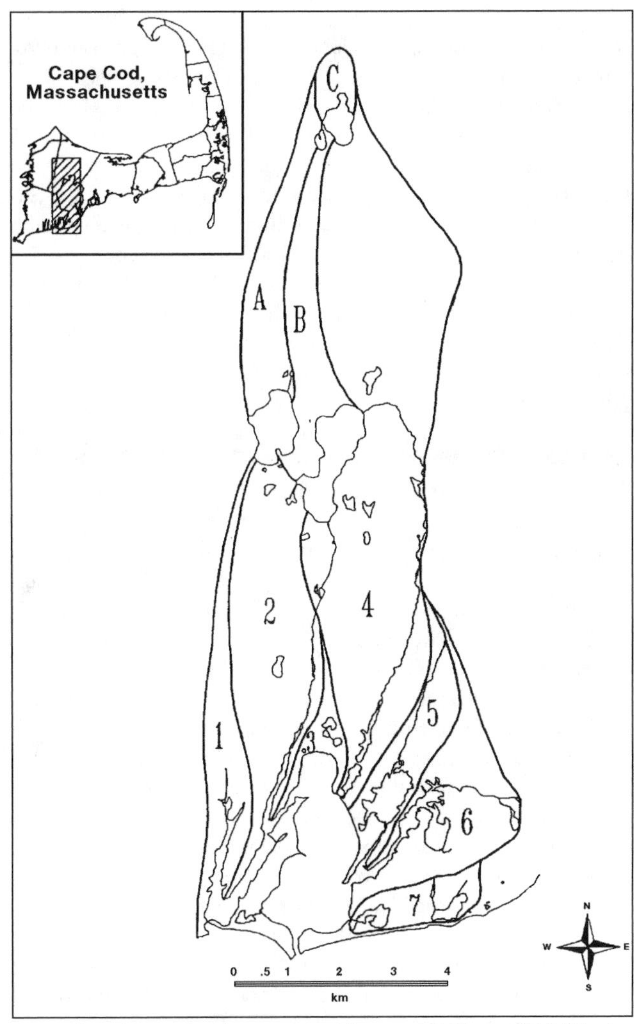

Figure 49-1. Waquoit Bay drainage area delineations. Subbasins: (1) Eel Pond, (2) Childs River, (3) Head of the Bay, (4) Quashnet River, (5) Hamblin Pond, (6) Jehu Pond, and (7) Sage Lot Pond. Kettle Pond recharge areas: (A) Ashumet Pond, (B) Johns Pond, and (C) Snake Pond.

Residential septic systems are one of the principal sources of nitrogen to Waquoit Bay, as the drainage area is heavily populated and largely unsewered. Other nitrogen sources include atmospheric deposition and lawn and agricultural fertilizers; however, these sources are not being addressed in this chapter. Because the geology of the drainage area consists of highly permeable, unconsolidated sand and gravel of glacial and marine origin (LeBlanc and Guswa 1977), rapid percolation of precipitation is favored, making surface runoff a negligible contribution to receiving waters. As a result, nutrients from terrestrial sources, such as septic systems, are transported primarily by groundwater, through groundwater-fed streams or direct seepage along the coast.

The Waquoit Bay drainage area covers approximately 50 km² and is subdivided into seven subbasins and three kettle pond contributing areas (Figure 49-1). Residential housing density varies widely in the area, with a higher concentration along the coast. Increased nitrogen concentrations delivered to the bay are associated with at least three major ecological alterations: increased seaweed growth, reduced eelgrass growth, and adverse effects to the food web (Costa 1988; Valiela et al. 1990). Because of these ecological changes and the danger of further eutrophication, it is important to assess and quantify the rate of anthropogenic nitrogen loading to Waquoit Bay over time. Quantification is essential to gain a better understanding of the ecological response to increasing nitrogen loading and to implement responsible development strategies for the future in other similar coastal environments.

This chapter describes the methods, assumptions, and results of a unique approach to quantifying nutrient loading. This study integrates both spatial and temporal characteristics of nitrogen loading to yield a more accurate quantification when compared with results from simpler, steady-state models. Due to the spatial variability of housing density and the relatively slow movement of groundwater, we estimate nitrogen input rates to the bay using both groundwater travel time and the distribution of residential development. A groundwater model has been used to delineate annual "time bands" of groundwater flow, indicating the approximate number of years it takes for groundwater to move from the point of recharge to the receiving waters of Waquoit Bay. Land parcel information stored in a GIS is coupled with groundwater time bands to estimate the temporal and spatial distribution of nitrogen loading across the drainage area. Using a customized loading calculation program written in C, this distribution is then processed to quantify the total amount and trend of nitrogen loading to Waquoit Bay.

METHODS

Groundwater Modeling and Delineation of Time Bands

The finite difference, three-dimensional flow model MODFLOW was used to analyze groundwater flow characteristics within the Waquoit Bay drainage area. MODFLOW and its particle-tracking component, MODPATH, were developed by the U.S. Geological Survey (USGS) and are widely used. Processing-MODFLOW software by Chaing and Kinzelbach (1991) was used for pre- and post-processing model inputs and results.

Using the groundwater models, specific cells occurring in streams and along the shorelines of estuaries, the bay, and ponds were "marked" for particle tracking. Backward flow paths with one-year increments delineated along each flow line were generated, output in digital exchange format (DXF), transferred to a UNIX workstation, and imported into ARC/INFO (developed by Environmental Systems Research Institute). Once geographically referenced, the flow lines and their annual flow markers were incorporated into a base map of the drainage area.

From this new flow line base map, annual groundwater travel time bands could be derived and manually digitized. The time bands were delineated from this base map in conjunction with the December 1991 water table map developed by the Cape Cod Commission (Cambareri et al. 1993). Figure 49-2a represents the Childs River subbasin with its annual groundwater travel time bands. The time band number indicates the length of time in years that it will take groundwater recharge to reach the bay from that particular point.

Quantifying Residential Loading Rates

To estimate the flux of nitrogen through the aquifer and into receiving waters, the rate of nitrogen input from each home must be specified. The value derived for this study was 29 mg/l for each household, which coincides with the value determined from a literature search of nitrogen concentration of septic effluent reaching the water table (Valiela et al. 1993).

Site-specific data on nitrogen removal during transport through the aquifer are not available for the Waquoit Bay area. Nitrogen transport through this type of aquifer is believed to be very conservative (Frimpter et al. 1990). Although it is possible that some attenuation of nitrogen is occurring, 100% conservative transport of nitrogen through the saturated zone of the aquifer is assumed until more conclusive data are available.

GIS Analysis of Time Bands and Land Parcel Data

Digital 1989 land parcel data for the entire drainage area were acquired from the towns of Falmouth, Mashpee, and Sandwich and imported into the GIS. Figure 49-2b shows land parcel polygons for the Childs River subbasin. Attributes associated with the land parcel layer include: (1) land use by property type (e.g., agricultural, developed and developable residential, and public utility); (2) land parcel area; and (3) residential year-built data. Year-built data are necessary for calculating nitrogen loading over time in this drainage area because large-scale, detailed land use data are limited and aerial photos are only available on an infrequent basis.

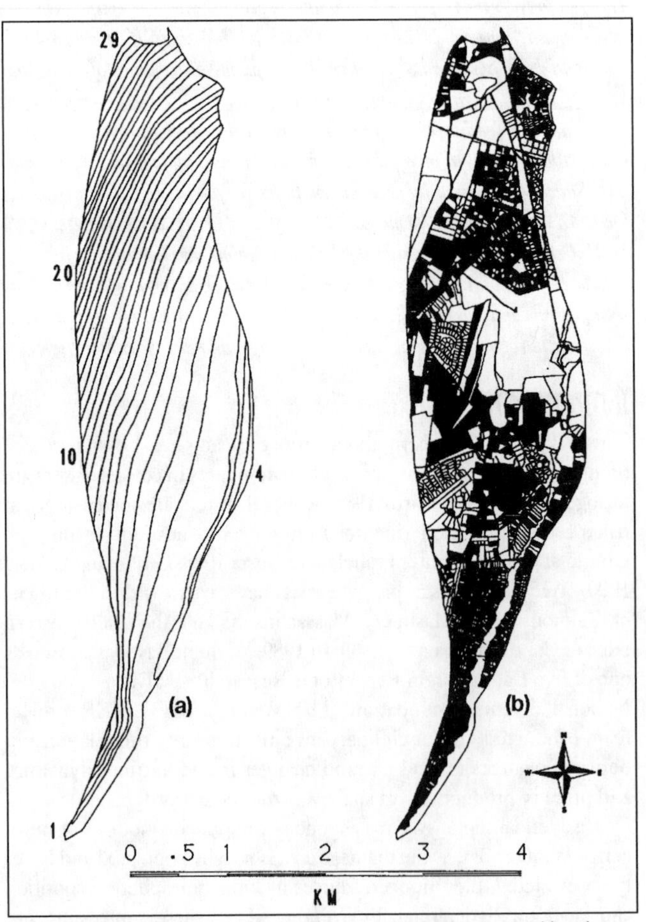

Figure 49-2. (a) Childs River subbasin with annual groundwater travel bands, and (b) land parcel coverage. The shaded polygons represent developed residential housing units as of 1990. Unshaded parcels are undeveloped and classified as vacant land, agricultural land, or institutional land.

Using the land parcel database and statistical functions in the GIS, an historical account of housing development was produced. A limitation of the year-built data is that they may be skewed toward the present, due to the way parcel information is maintained at the local level. At least one town in the study area, Mashpee, updates year-built data for a parcel when any structural modifications are made on that parcel. For this reason, the land parcel database will tend to underestimate the number of houses built in earlier decades and overestimate the number of houses built in later decades. In terms of septic nitrogen loading to receiving waters, this could result in an underestimation of current levels, but this skewing will lessen as future nitrogen loading is modeled. This practice is believed to be limited; therefore, the error is probably small. If such a bias is causing some error, the error is temporal and not one of overall magnitude.

To determine the temporal effect of residential development on nitrogen loading to groundwater, the land parcel data layer was overlaid with the annual groundwater travel time bands. Through this overlay procedure, it was possible to determine when a specific parcel started to contribute nitrogen to the groundwater and how long it will take for this specific contribution to enter the bay.

To automate the repetitive process of "backcasting" (i.e., selecting only those residential parcels that existed in a particular year and generating statistics for them), a set of macroprograms was used to determine the number of developed residential parcels in each time band over the last 300 years. The number of developable residential parcels in each time band (as of 1989) was also extracted for projections of future growth scenarios. These statistics were extracted from the GIS as ASCII files and transferred from the UNIX workstation to a personal computer. With the time- and site-specific statistics, it was possible to estimate nitrogen loading from residential septic systems to the water table and subsequently to the receiving bay.

Total Loading Calculation Model

The total loading calculation model was written in the C programming language and ran on an IBM-compatible 486 personal computer. To compute loading for the whole drainage area, the model started at the northern end of the drainage area and worked south toward the bay in order to quantify the nitrogen loading to the ponds first. The calculations for the ponds had to be completed before the assessment of the down-gradient areas receiving groundwater discharging from the ponds. Due to the large amount of data being manipulated for each subbasin and recharge area (e.g., over 100,000 array elements), loading for each contributing area was computed separately and then summed. Loading for a subbasin in a particular year was calculated as total input into the subbasin's first annual time band during the prior year, plus the total input into the second annual time band two years prior, and so on. This procedure is analogous to that of repeatedly summing down the diagonals of a spreadsheet with time bands as columns and year-built data as rows.

For pre-1990 years, the statistics files indicated the number of residential parcels associated with each time band in the drainage basin. For post-1990 years, the model calculated residential growth in each time band according to various hypothetical growth rates. Because the statistics files also indicated the number of developable parcels in each time band in 1989, growth was allowed to continue up to the maximum number of developable parcels to simulate full residential build-out. The output of the total loading calculation model was a series of ASCII text files that listed the total nitrogen loading reaching Waquoit Bay over time. The files were imported into a spreadsheet for graphing and further analysis.

RESULTS AND DISCUSSION

Between 1940 and 1990, residential parcels across the entire drainage area increased approximately fifteenfold to approximately 4,230 parcels. While the bulk of nitrogen entering Waquoit Bay lags behind residential nitrogen input by about a decade, nitrogen from some portions of the drainage area may take several decades to reach the bay—up to a maximum of about 115 years. Because such a discrepancy exists between the level of nitrogen input to the aquifer and nitrogen output to the bay for a given year, it is clear that this methodology provides a more accurate assessment of nitrogen loading over time when compared to a steady state analysis.

In addition to holding the number of residential parcels constant at the 1990 level (0% growth), the post-1990 scenarios included annual uniform residential growth of 1%, 2%, and 5%. For the 0% growth scenario, the number of residential parcels and the nitrogen loading reaching the bay over time are shown in Figure 49-3. Nitrogen continues to increase sharply over the next decade, finally leveling off more than 100 years from now at approximately 18,000 kg/yr, which is about 36% higher than the current level of approximately 13,200 kg/yr. It is interesting to note the sharp increase in the rate of nitrogen loading to receiving waters beginning around 1950, which corresponds to the decrease in eelgrass beds over the same period, as noted by Costa (1988) and Valiela et al. (1990).

Figure 49-4 shows the growth scenarios for nitrogen loading reaching the bay. Regardless of the annual growth rate, it is apparent that nitrogen loading from septic sources continues to increase for many decades. The difference between the 1990 residential development level and full residential build-out in the Waquoit Bay drainage area is approximately 2,280 parcels. Total nitrogen loading

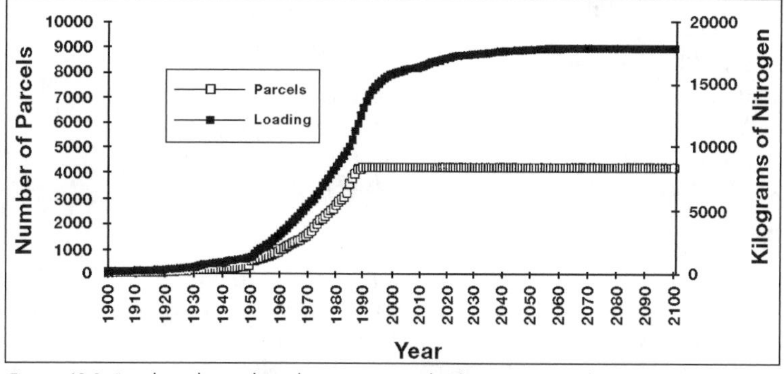

Figure 49-3. Residential parcels and septic nitrogen loading over time. This scenario assumes no growth in residential development after 1990. Note that loading lags development and continues to increase long after development is held at a constant level.

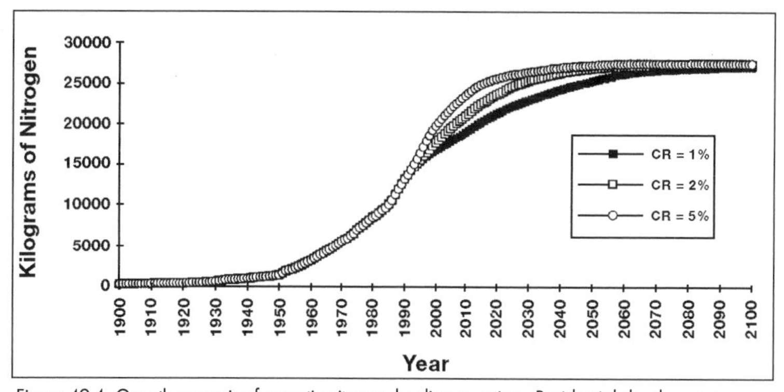

Figure 49-4. Growth scenarios for septic nitrogen loading over time. Residential development grows at the rates indicated; the graph shows the loading associated with these rates. Regardless of the growth rate chosen, loading will eventually reach the same level at full residential build-out.

from septic sources increases to approximately 28,100 kg/yr at full residential build-out—more than twice the current level.

The significance of these findings is highlighted when one considers that the ecological changes are largely attributed to the increased nitrogen inputs affecting Waquoit Bay. There is concern that the frequency of anoxic and hypoxic events is increasing, and higher nitrogen inputs can only raise the likelihood of these events. Based on our analysis, it is apparent that septic nitrogen loading is increasing and will affect Waquoit Bay for many decades to come.

CONCLUSION

Quantifying septic nitrogen loading to Waquoit Bay, where septic systems are significant sources of nitrogen in the drainage area, is crucial to understanding ecological changes over time. Given our assumptions, this analysis indicates that septic nitrogen inputs to the bay have increased sharply since 1950 as the ecological health of the bay declined, and they will continue to increase.

In this analysis, it has been demonstrated that it is necessary to incorporate the spatial and temporal characteristics of groundwater flow and land use data to accurately quantify nutrient loading in a drainage area such as the Waquoit Bay. A static analysis of nutrient inputs and outputs may lead to errors in past, present, and future loading estimates, unless groundwater is insignificant or the drainage area is largely sewered.

Incorporation of these results with other research being conducted in the Waquoit Bay LMER will result in a detailed model of the inputs, outputs, and biogeochemical processes that are occurring in the Waquoit Bay estuarine system. Currently, there are projects concerning the quantification of atmospheric deposition and vegetative uptake of nutrients, the nutrient transformation occurring in different types of ecosystems, and several other related topics.

Acknowledgments

We thank Ivan Valiela, Jim Kremer, Brad Seely, and Glynnis Collins for helpful reviews of the manuscript. We also thank Eric Ettlinger for GIS and graphical assistance. This work was supported by NSF/EPA grant #OCE-8914729 and NOAA grant #NA 90AA-H-CZ131.

References

Cambareri, T. C., E. M. Eichner, and C. A. Griffeth. 1993. *Characterization of the Watershed and Submarine Groundwater Discharge: Hydrologic Evaluation for the Waquoit Bay Land Margin Ecosystem Research Project.* Final Report submitted under DEP 319, Contract #91-05 and EPA Grant #C9001214-90-0.

Chaing, W. H., and W. Kinzelbach. 1991. *Processing MODFLOW.* Washington, D.C.: Scientific Software Group.

Costa, J. M. 1988. "Distribution, Production, and Historical Changes in Abundance of Eelgrass (*Zostera marina*) in Southeastern Massachusetts." Ph.D. diss., Boston University.

Frimpter, M. H., J. J. Donohue, and M. V. Rapazc. 1990. *A Mass-Balance Nitrate Model for Predicting the Effects of Land Use on Groundwater Quality.* U.S. Geological Survey Open File Report 88-493.

Giblin, A. E., and A. G. Gaines. 1990. Nitrogen inputs to a marine embayment: The importance of groundwater. *Biogeochemistry* 10:309–28.

Howarth, R. W. 1988. Nutrient limitation of net primary production in marine ecosystems. *Annual Review of Ecology and Systematics* 19:89–110.

LeBlanc, D. R., and J. H. Guswa. 1977. *Water Table Map of Cape Cod, Massachusetts.* U.S. Geological Survey Open File Report 77-419.

Valiela, I., G. Collins, J. Kremer, J. Brawley, K. Lajtha, C. H. Sham, and B. Seely. Nitrogen loading from coastal watersheds to receiving waters: Evaluation of methods, n.d.

Valiela, I., and J. E. Costa. 1988. Eutrophication of Buttermilk Bay, a Cape Cod coastal embayment: Concentrations of nutrients and watershed nutrient budgets. *Environmental Management* 12(4):539–53.

Valiela, I., J. Costa, K. Foreman, J. M. Teal, B. Howes, and D. Aubrey. 1990. Transport of groundwater-borne nutrients from watersheds and their effects on coastal waters. *Biogeochemistry* 10:177–97.

Chi Ho Sham is Director of GIS Operations with The Cadmus Group, Inc. He is a graduate of the University of Regina, Saskatchewan, Canada (B.A.) and the State University of New York at Buffalo (M.A. and Ph.D.).

The Cadmus Group, Inc.
135 Beaver St.
Waltham, Massachusetts 02154
Phone: (617) 894-9830
Fax: (617) 894-7238

John W. Brawley and *Max A. Moritz* (at the time of this research) were analysts with The Cadmus Group, Inc. Currently, John W. Brawley is a doctoral student at the Center for Environmental and Estuarine Studies, University of Maryland at Solomons; Max A. Moritz is a doctoral student in the Department of Geography of the University of California at Santa Barbara.

50

Using Geoscientific Information Systems for Three-Dimensional Regional Groundwater Flow Modeling:

Death Valley Region, Nevada and California

F. A. D'Agnese, A. K. Turner, and C. C. Faunt

Three-dimensional groundwater flow modeling of the complex Death Valley hydrologic basin requires the application of a number of geoscientific information system (GSIS) techniques. This study, funded by the U.S. Department of Energy as a part of the Yucca Mountain Project, focuses on a 100,000 km² area in Nevada and California. The geologic conditions are typical of the basin and range province; a variety of sedimentary, igneous, and metamorphic intrusive and extrusive rocks have been subjected to both compression and extension deformation.

GSIS techniques allow the synthesis of geologic, hydrologic, and climatic information gathered from numerous sources. Integration of these data facilitates the development of detailed spatial models that can be used for distributed parameter modeling. Construction of a three-dimensional hydrogeologic framework model and definition of surface and subsurface hydrologic conditions are possible with the combined use of software products, including traditional GIS products and sophisticated contouring, interpolation, visualization, and numerical modeling packages.

INTRODUCTION

Yucca Mountain on the Nevada test site in southwestern Nevada is being studied as a potential site for a mined geologic repository for the long-term storage of high-level radioactive waste. In cooperation with the Department of Energy, the U.S. Geological Survey (USGS) is evaluating the site as part of the Yucca Mountain Project. Because of the potential for radionuclides to be transported by groundwater from the repository to the accessible environment, studies are being conducted to characterize the Death Valley regional groundwater flow system, of which Yucca Mountain is a part (Bedinger et al. 1989).

This chapter describes the methods used by the USGS to develop both conceptual and numerical groundwater flow models of the Death Valley groundwater basin using three-dimensional data management and visualization. Hydrogeologic and hydrologic modeling applications of this kind require representation of subsurface and surface conditions; characterization of the geologic, hydrologic, climatic, and environmental controls to water behavior; and linkages to various data manipulation procedures. Therefore, more geologically oriented geoscientific information systems, or GSIS, were used (Turner 1991; Turner et al. 1991). The term "GSIS" is used to differentiate these systems from the more common two-dimensional GIS products (Raper 1989).

The study area, which is defined by the Death Valley regional groundwater flow system boundaries, lies within the area bounded by latitude 35° and 38° N and longitude 115° and 118° W. (Figure 50-1). The model includes about 100,000 km² and extends to depths of more than 10 km. The study area has a semiarid to arid climate and is within

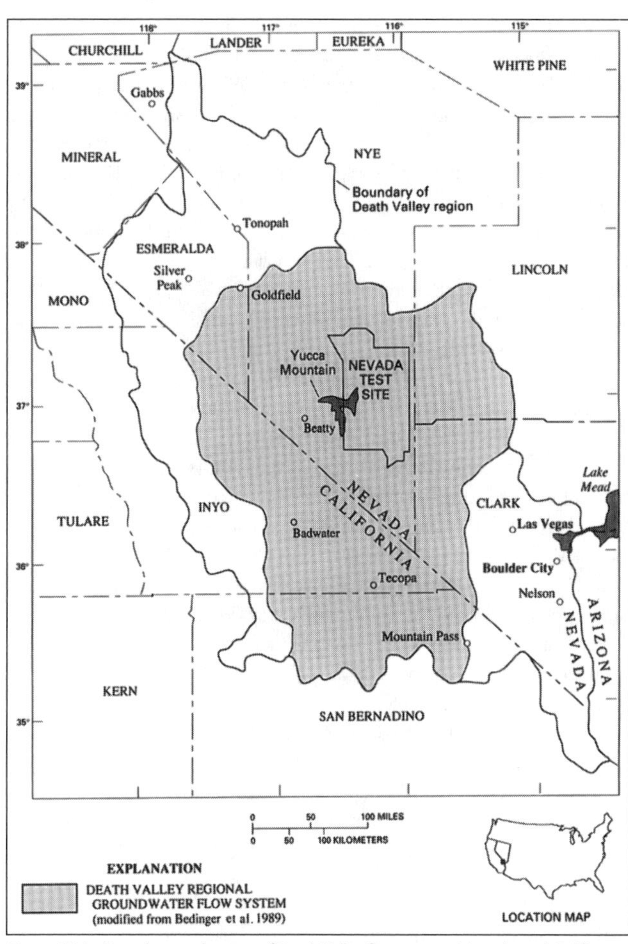

Figure 50-1. Map showing location of Death Valley flow system, Nevada and California.

Any use of trade, product, or firm names in this chapter is for descriptive purposes only and does not imply endorsement by the U.S. Government.

the southern Great Basin, a subprovince of the basin and range physiographic province. Geologic conditions are typical of the basin and range geologic province, with a variety of intrusive and extrusive igneous, sedimentary, and metamorphic rocks that have been subjected to several episodes of compressional and extensional deformation throughout geologic time. Elevations range from 90 m below sea level to 3,600 m above sea level; thus the region includes a variety of climatic regimes and associated recharge/discharge conditions. These complex geologic and hydrologic conditions require different computer-based techniques to develop conceptual and numerical models.

REGIONAL GROUNDWATER FLOW SYSTEM MODELING

Regional groundwater flow systems have been theoretically described by Toth (1963, 1972) and Freeze and Witherspoon (1966, 1967). These investigators developed descriptions and models of idealized flow characteristics of regional groundwater systems. However, conditions for such idealized flow are rarely found in natural hydrologic environments. This is especially true in geologically, climatically, and ecologically complex terrain such as the Death Valley region. Therefore, to correctly interpret this complex natural flow system, many deviations from the ideal need to be considered (Fiero et al. 1974). As with many other mountainous areas, the Death Valley region has historically been difficult to conceptualize and model because it is data sparse, deeply circulating, and structurally complex (Forster and Smith 1988).

Previous numerical modeling efforts of the groundwater systems in the Death Valley region have relied on two-dimensional, lumped parameter methods (Waddell 1982; Czarnecki and Waddell 1984; Rice 1984; Sinton 1987). These investigators utilized simplifying assumptions to simulate the complex, three-dimensional system in two dimensions. These simplifications involved considerable abstractions of the natural environment and became highly dependent on modeling system parameters that were prone to estimation error. These investigators concluded that their lumped parameter representations prevented accurate simulation of the three-dimensional nature of the system including the occurrence of vertical flow components, subbasinal groundwater flux, large hydraulic gradients, and physical subbasin boundaries.

In contrast, three-dimensional distributed parameter representations allow examination of the internal, spatial, and process complexities of the hydrologic system. These models require, in addition to numerical values of the various hydrologic properties, an understanding of the processes affecting these properties and their spatial distribution (Domenico 1972). The use of these modeling methods introduces several concerns resulting from: (1) the large quantity of data required to describe the system, (2) the complexity of the spatial and process relations involved in the system, and (3) the complex mathematics used to balance flow-system equations. The problems of data management, spatial and process complexity, and computational power usually encountered when using this approach are alleviated by the use of GSIS techniques.

Three-dimensional regional groundwater flow modeling efforts at Yucca Mountain utilize GSIS at each of the following seven stages:

1. Conversion and integration of two- and three-dimensional data sets into a GSIS.
2. Development of a hydrogeologic framework model of the Death Valley region that characterizes the three-dimensional subsurface geologic structures and materials.
3. Analysis of selected framework model components to define the physical boundaries and flow parameters of the system.
4. Evaluation of the mechanisms of regional groundwater recharge, discharge, and flow to characterize hydraulic boundaries and flux conditions.
5. Development of a series of numerical model input arrays using an interface between the GSIS database and the numerical model.
6. Numerical simulation of groundwater flow and evaluation of the model predictions through GSIS visualization.
7. Repetition of the above stages to achieve numerical model calibration.

DATA CONVERSION AND INTEGRATION

Extensive conversion and integration of regional-scale data are required to characterize the hydrologic system, including traditional hydrologic data used in groundwater modeling and other spatial data such as geologic maps and sections, soil surveys, vegetation maps, surface water maps, spring localities, meteorologic data, and remote-sensing imagery. Data are converted into a consistent digital format using the two-dimensional ARC/INFO GIS (Figure 50-2). Some data sets are available in digital formats and only require extraction and transformation. Other generalized digital data types, such as remote-sensing images, require processing and classification into suitable thematic layers that describe vegetation or geomorphologic properties as being useful in the modeling process. Manuscript map data are scanned using a raster-to-vector Tektronix scanner, and the resulting vector files are further processed to remove artifacts of the scanning process, transformed to a convenient

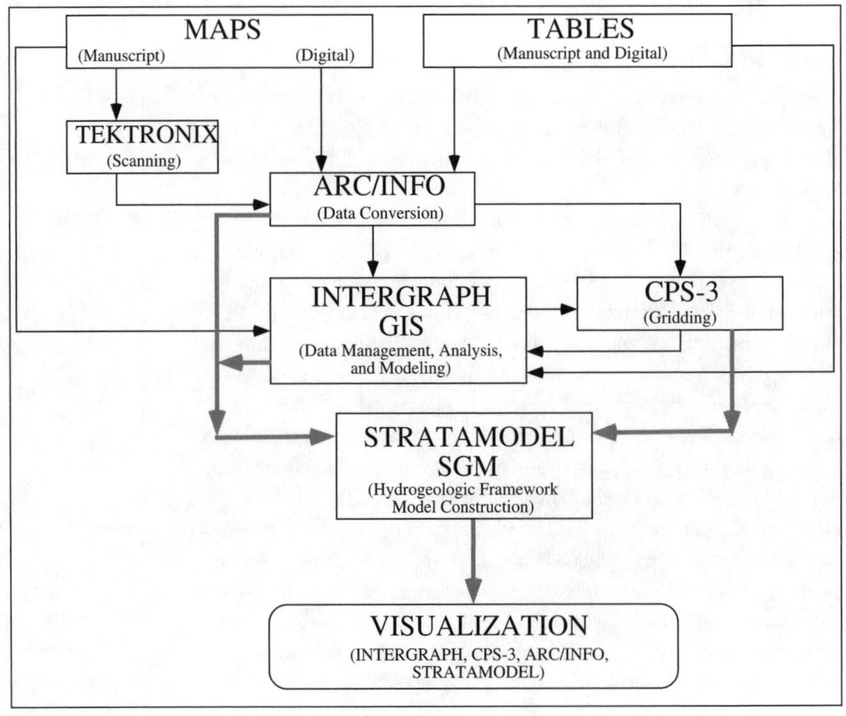

Figure 50-2. Flowchart showing systems used for modeling.

geographic coordinate system, and edited to achieve accurate topology. The procedures utilize traditional vector-based ARC/INFO capabilities and raster-based capabilities of ARC/GRID. The digital files are then moved to the Intergraph Corporation Modular GIS Environment (MGE), which is used to store, integrate, analyze, and visualize raster and vector three-dimensional data sets (Figure 50-2).

CHARACTERIZATION OF THE HYDROLOGIC SYSTEM COMPONENTS

The hydrologic system is characterized by defining the three-dimensional hydrogeologic framework and the surface and subsurface hydrologic conditions. These components are used to conceptualize the system.

Defining the Three-Dimensional Hydrogeologic Framework

The hydrogeologic framework defines the physical geometry and composition of the subsurface materials through which the groundwater flows. It defines the fundamental nature of the groundwater flow system and includes a definition of both the hydrostratigraphy and the hydrogeologic structures of the area. A hydrostratigraphic unit is "a geologic unit that has considerable lateral extent and has reasonably distinct hydrologic properties because of its physical (geological and structural) characteristics" (Maxey 1964). Hydrogeologic structures include faults, fractures, or joints that alter the physical framework of a region and, thus, influence groundwater flow.

Construction of a true three-dimensional hydrogeologic framework model is possible only with the combined use of many available software products, including traditional GIS and sophisticated contouring, interpolation, and visualization packages (Figure 50-3). Digital geologic maps and sections are accurately placed in their correct three-dimensional spatial relations, and their features are attributed using Intergraph Corporation's MGE. Radian Corporation's CPS-3 gridding system and Full Fault Modeling System (FFMS) interpolate fault planes and geologic horizon surfaces between existing cross sections and boreholes. These gridded surfaces are used to construct true three-dimensional solid models in several software products that then are displayed, or visualized, graphically. The geometrical model components are supported by a complex sequence of attribute information that describes their hydrogeologic properties.

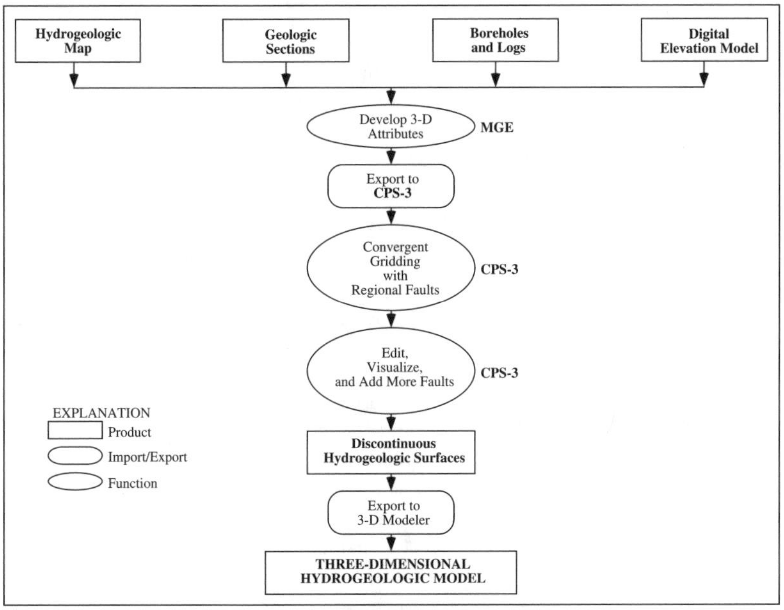

The GSIS procedures allow the hydrogeologic framework model to be repeatedly analyzed and interpreted, permitting investigators to query alternative hypotheses concerning geologic structures and potential characteristics of groundwater moving through this complex terrain. Various configurations of structure and hydrostratigraphy can be developed that are alternative interpretations of relatively sparse data. These multiple hypotheses help the investigators during conceptualization to: (1) determine the most feasible interpretation of the system given the available database, (2) determine the location and type of additional data that will be needed to reduce uncertainty, (3) select potential physical barriers to groundwater flow, and (4) develop hypotheses about preferred regional flow paths.

Hypothesis testing with this framework model is critical in the numerical flow modeling process. Alternative strategies for evaluating the hydrologic systems within the faulted terrain can be explored: for example, they could be analyzed and interpreted as either heterogeneous porous media or equivalent porous media. In each case, the distribution and values of numerical model parameters are determined directly from the hydrogeologic data contained in the database. The heterogeneous distribution of the flow parameters could be estimated using stochastic procedures and probability distributions (Haldorsen et al. 1988; Haldorsen and Damsleth 1993).

Defining the Surface and Subsurface Hydrologic Conditions

The regional groundwater flow system reflects interactions among all the natural and man-induced mechanisms controlling how water enters, flows through, and exits the system. The inflows to the system can include subbasinal flux or groundwater recharge. Outflows can include spring flow, base flow to streams, evapotranspiration, pumping, and subbasinal flux (Anderson and Woessner 1992). Inflow volumes are estimated using data that describe system components controlling precipitation, evapotranspiration (ET), surface water runoff, and subbasinal flux. Outflow volumes are estimated using data that describe system components controlling springs discharge, ET, and groundwater use (Anderson and Woessner 1992). Groundwater flow paths are delineated using a combination of integrated data sets that describe regional gradient, hydrochemistry, hydrogeologic structure, and source/sink locations.

Maps describing average annual precipitation, topography, land use, vegetation, surface hydrography, diffuse groundwater discharge areas, and soils are developed and integrated using GSIS techniques (Figure 50-4). These maps are overlaid to develop a comparative model that describes how surface conditions control groundwater recharge and discharge mechanisms. Each surface class represented in this model is compared with previous point or areal estimates. The resulting recharge and discharge model estimates are compared to empirical estimates developed by Eakin et al. (1951), water-balance estimates developed by Rice (1984), and distributed parameter estimates suggested by Leavesley et al. (1983). These comparisons ultimately result in the selection of "most appropriate" recharge and discharge models that are incorporated into the integrated GSIS database (Figure 50-4).

Maps that show water levels, surface hydrography (lakes, springs, and perennial streams), groundwater

Hydrogeologic Map
Geologic Sections
Boreholes and Logs
Digital Elevation Model
Develop 3-D Attributes — MGE
Export to CPS-3
Convergent Gridding with Regional Faults — CPS-3
Edit, Visualize, and Add More Faults — CPS-3
Discontinuous Hydrogeologic Surfaces
Export to 3-D Modeler
THREE-DIMENSIONAL HYDROGEOLOGIC MODEL
EXPLANATION
Product
Import/Export
Function

Figure 50-3. A true three-dimensional hydrogeologic framework model.

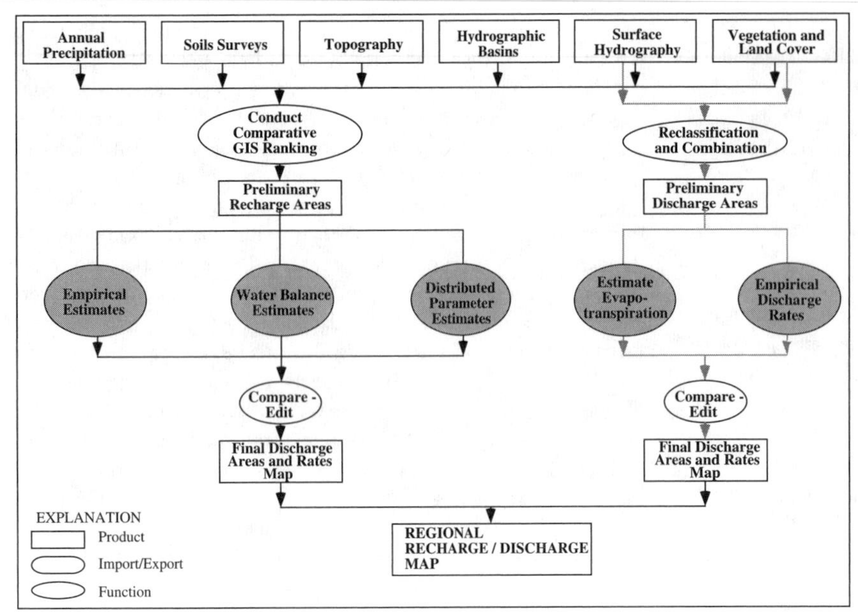

Figure 50-4. "Most appropriate" recharge and discharge models incorporated into the integrated GSIS database.

discharge areas (wetlands and wet playa-lakes), vegetation, and surface water basins are developed and integrated using GSIS techniques (Figure 50-5). Altitudes of lakes, perennial streams, regional springs, and wetland areas, interpreted as areas where the potentiometric surface intercepts the land surface, are combined with water levels from a regional well database. Since drainage basin boundaries commonly coincide with groundwater basin boundaries, they are used to control a sophisticated gridding algorithm and appropriately place local maxima or minima in the regional potentiometric surface map. After initial gridding, a series of interactive editing steps ensure that the gridded result conforms to hydrologic constraints. The resulting grid and contour map then is compared to existing maps. Existing maps might not correctly describe the true potentiometric surface; nevertheless, discrepancies are noted and evaluated. A final optimal potentiometric surface map and grid model for the region are stored in the integrated GSIS database (Figure 50-5).

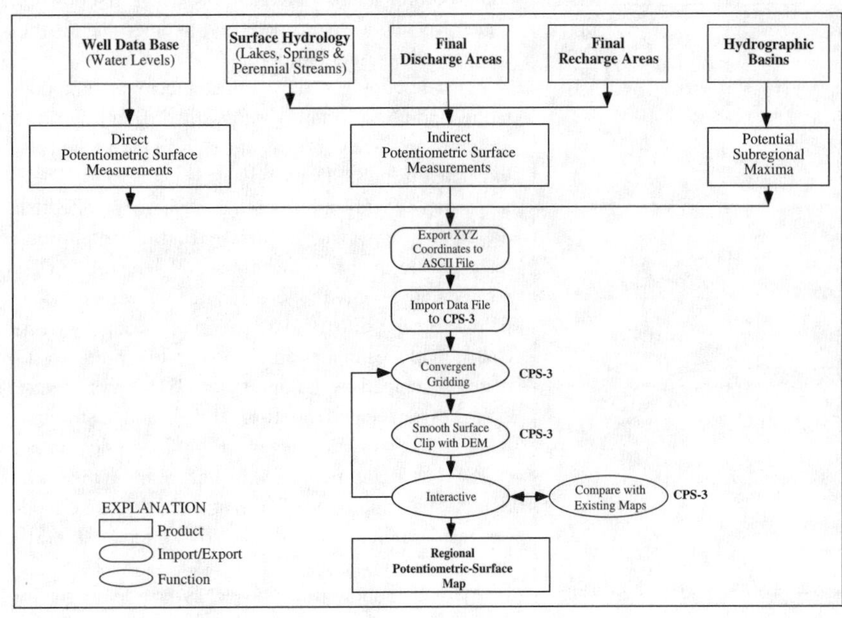

Figure 50-5. Optimal potentiometric surface map and grid model.

CONCEPTUALIZATION OF THE REGIONAL FLOW SYSTEM

Prior to numerical simulation, conceptualization of the regional flow system is undertaken by the investigators to select a single coherent representation from the many possible representations of the natural flow system. This conceptual model integrates the data and includes: (1) a detailed water budget, (2) descriptions of the physical framework of the hydrogeologic system, (3) descriptions of physical and hydraulic boundary conditions, (4) estimates of the hydraulic properties of the hydrogeologic units, (5) estimates of groundwater source and sinks, and (6) hypotheses about regional and subregional flow paths (Anderson and Woessner 1992).

The conceptual model allows the investigators to distinguish between what is known about the system on the basis of available data from the results of numerical simulations that result from simplifying assumptions and system abstractions. GSIS techniques aid modelers in making these determinations by displaying data control on interpreted map products, by rendering multiple likely scenarios, and by conducting statistical evaluations of model estimates.

NUMERICAL MODEL ARRAY CONSTRUCTION

By use of the Intergraph Corporation's ERMA interface to the MODFLOW numerical model, selected maps or grids that describe model inputs are developed, or extracted from, the integrated GSIS database, resampled to the desired model resolution, and converted to ASCII model arrays (Figure 50-6). GSIS capabilities enhance model design and construction by allowing modelers to develop numerous grid configurations, different boundary-type designations, and multiple source/sink representations.

Once the arrays are developed, various combinations of model runs are executed. The model results are returned to the GSIS database and visualized. The numerical model is then calibrated to water levels and/or regional discharge fluxes by making modifications to the input arrays on a node-by-node basis or by revising interpreted characterization components. Although it is understood that the calibrated model is one of many appropriate configurations, numerical methods that are based on inverse techniques and/or sensitivity analysis help to quantify the significance of the conceptual model.

CONCLUSION

Available GSIS technology assists in modeling the complex natural hydrologic system in the Death Valley region. Three-dimensional geological modeling and visualization allow characterization of the "data sparse" subsurface, while integrated image processing and hydrologic process modeling using traditional GIS techniques support surface-based process modeling. However, these models are hindered by the use of existing lumped parameter models in describ-

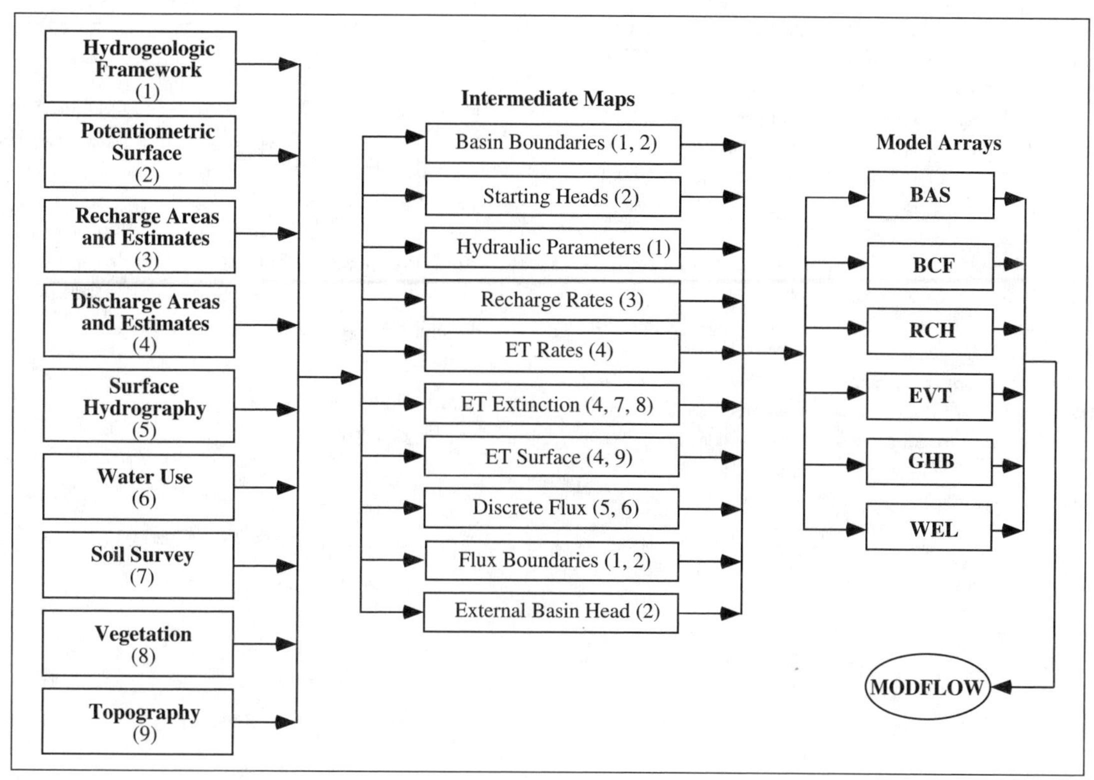

Figure 50-6. MODFLOW numerical model converted to ASCII model arrays.

ing selected hydrologic and hydrogeologic phenomena. The resulting models are hybrid representations of lumped parameter models and distributed parameter models. Further developments in GSIS technology offer a means for developing fully three-dimensional, spatially accurate, distributed parameter process models.

References

Anderson, M. P., and W. M. Woessner. 1992. *Applied Groundwater Modeling: Simulation of Flow and Advective Transport.* New York: Academic Press.

Bedinger, M. S., K. A. Sargent, and W. H. Langer. 1989. *Studies of Geology and Hydrology in the Basin and Range Province, Southwestern United States, for Isolation of High-Level Radioactive Waste: Characterization of the Death Valley Region, Nevada and California.* U.S. Geological Survey Professional Paper 1370-F.

Czarnecki, J. B., and R. K. Waddell. 1984. *Finite-Element Simulation of Groundwater Flow in the Vicinity of Yucca Mountain, Nevada–California.* U.S. Geological Survey, Water Resources Investigations Report 84-4349.

Domenico, P. A. 1972. *Concepts and Models in Groundwater Hydrology.* New York: McGraw-Hill.

Eakin, T. E., G. B. Maxey, T. W. Robinson, J. C. Fredericks, and O. J. Loeltz. 1951. *Contributions to the Hydrology of Eastern Nevada.* Water Resources Bulletin No. 12. Nevada: Office of the State Engineer.

Fiero, G. W., Jr., J. R. Illian, G. A. Dinwiddie, and L. J. Schroder. 1974. *Use of Hydrochemistry for Interpreting Groundwater Flow Systems in Central Nevada.* U.S. Geological Survey Report 474-178.

Forster, C. B., and L. Smith. 1988. Groundwater flow systems in mountainous terrain: Numerical modeling techniques. *Water Resources Research* 24(7):999–1,010.

Freeze, R. A., and P. A. Witherspoon. 1966. Theoretical analysis of regional groundwater flow: Analytical and numerical solutions to the mathematical model. *Water Resources Res.* 2:641–56.

———. 1967. Theoretical analysis of regional groundwater flow: Effect of water table configuration and subsurface permeability variations. *Water Resources Res.* 3(2):641–56.

Haldorsen, H. H., P. J. Brand, and C. J. MacDonald. 1988. Review of the stochastic nature of reservoirs. In *Mathematics of Oil Production,* edited by S. Edwards and P. R. King, 109–210. New York: Oxford University Press.

Haldorsen, H. H., and E. Damsleth. 1993. Challenges in reservoir characterization. *Amer. Assoc. Petroleum Geologists Bulletin* 77(4):541–51.

Leavesley, G. H., R. W. Lichty, B. M. Troutman, and L. G. Saindon. 1983. *Precipitation Runoff Modeling System: User's Manual.* U.S. Geological Survey Water Resources Investigations Report 83-4238.

Maxey, G. B. 1964. Hydrogeology of desert basins. *Ground Water* 6(5):10–22.

Raper, J. F. 1989. The three-dimensional geoscientific mapping and modeling system: A conceptual design. In *Three-Dimensional Applications in Geographic Information Systems,* edited by J. F. Raper, 11–19. London: Taylor and Francis.

Rice, W. A. 1984. Preliminary Two-Dimensional Regional Hydrological Model of the Nevada Test Site and Vicinity. Pacific Northwest Laboratory, SAND83-7466.

Sinton, P. O. 1987. "Three-Dimensional, Steady State, Finite Difference Model of the Groundwater Flow System in the Death Valley Groundwater Basin, Nevada–California." Master's thesis, Department of Geology and Geological Engineering, Colorado School of Mines, Golden.

Toth, J. 1963. A theoretical analysis of groundwater flow in small drainage basins. *Journal of Geophysical Research* 68(16):4,795–812.

———. 1972. Properties and manifestations of regional groundwater movement. *Twenty-Fourth International Geological Congress*, Section 11, 153–63.

Turner, A. K. 1991. Applications of three-dimensional geoscientific mapping and modeling systems to hydrogeological studies. In *Three-Dimensional Modeling with Geoscientific Information Systems*, edited by A. K. Turner, 327–64. Vol. 354 of NATO ASI Series C: Mathematical and Physical Sciences, Dordrecht, the Netherlands: Kluwer Academic Publishers.

Turner, A. K., J. S. Downey, and K. E. Kolm. 1991. Potential applications of three-dimensional geoscientific mapping and modeling systems to hydrogeological assessments at Yucca Mountain, Nevada. *ASPRS-ACSM Annual Convention Proceedings*.

Waddell, R. K. 1982. *Two-Dimensional, Steady State Model of Groundwater Flow, Nevada Test Site and Vicinity, Nevada–California*. U.S. Geological Survey Water Resources Investigations Report 82-4085.

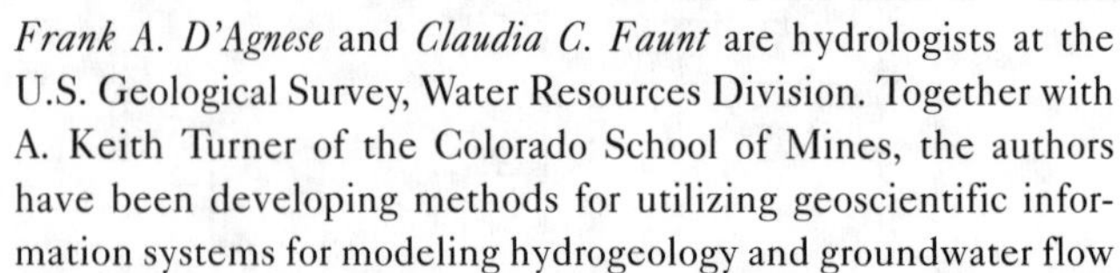

Frank A. D'Agnese and *Claudia C. Faunt* are hydrologists at the U.S. Geological Survey, Water Resources Division. Together with A. Keith Turner of the Colorado School of Mines, the authors have been developing methods for utilizing geoscientific information systems for modeling hydrogeology and groundwater flow in complex geologic and hydrologic environments. They are currently applying these methods to the Death Valley region as part of the ongoing Yucca Mountain site-characterization studies.

U.S. Geological Survey, Yucca Mountain Project

MS 421, Box 25046

Lakewood, Colorado 80225

Phone: (303) 236-5195

51

Development of GEO-WAMS:

A Modeling Support System to Integrate GIS with Watershed Analysis Models

Joseph V. DePinto, Joseph F. Atkinson, Hugh W. Calkins, Paul J. Densham, Weihe Guan, Hui Lin, Frank Xia, Paul W. Rodgers, and Tad Slawecki

In order to facilitate the site-specific–problem-specific development and application of water-quality models in the Great Lakes watershed, a modeling support system that links water quality models with a GIS (ARC/INFO) has been developed. This system, which is called GEO-WAMS (geographically based watershed analysis and modeling system), automates such modeling tasks as: (1) spatial and temporal exploratory analysis of system data; (2) model scenario management; (3) model input configuration; (4) model input data editing and conversion to appropriate model input structure; (5) model processing; (6) model output interpretation, reporting and display; (7) transfer of model output data between models; and (8) model calibration, confirmation, and application. The design of GEO-WAMS and its feasibility and utility are demonstrated by a prototype application to the Buffalo River (Buffalo, New York) watershed. The prototype provides a modeling framework for addressing management questions related to the spatial and temporal distribution of dissolved oxygen in the Buffalo River. It includes a watershed loading model, a groundwater contaminant transport model, as well as a modified version of the EPA's WASP4 model for simulation of dissolved oxygen in the river.

INTRODUCTION

For almost thirty years, the management of water quality in the Great Lakes basin has been aided by the development and application of process-oriented water-quality models. From chloride to nutrient/phytoplankton and, more recently, toxic substances, deterministic, mass balance models have been used to simulate the concentration of important materials in various compartments of the aquatic ecosystem as a function of the loadings of these materials to the system.

Regardless of the level of sophistication, application of these water quality models on a site-specific basis requires efficient acquisition, storage, organization, reduction, and analysis of model input data accompanied by manipulation, interpretation, reporting, and display of model output data. Among these site-specific input data are: system geometry, external loading of modeled constituents, land-use activities within the system of interest, hydrometeorological forcing data (e.g., solar radiation, precipitation, wind velocity, etc.), basic physical and chemical properties of the system (e.g., air, land, water, and bottom sediments) that govern transport and transformation processes, and initial conditions of modeled constituents. Also required are data from field observations in space and time of model state variables and process rates that are to be compared with model output for calibration, verification, and system response projections. Since most of these data have a spatial context, the premise of this project is that performing these functions for water quality models applied to the Great Lakes watersheds can be greatly facilitated by linking models with a geographic information system (GIS) database management system.

To accomplish the coupling of GIS with water-quality models, we have undertaken the development of a modeling support system that will link a geographically based data management system (GDMS) with an integrated watershed modeling system (IWMS) to create a geographically based watershed analysis and modeling system (GEO-WAMS).

The modeling support system facilitates the job of the water quality modeler in accomplishing the various data-interactive tasks necessary to develop and apply one or a series of site-specific–problem-specific, process-oriented mathematical models. The modeling support system provides the following benefits:

Problem identification/specification: Use GIS database for exploratory analysis to identify where and what potential impacts are.

Model enhancement: Greatly expand the modeler's ability to consider more information and a wider variety of interrelationships in addressing environmental problems so that models can be applied with a higher spatial, temporal, and kinetic interaction resolution.

Model input development: Ultimately reduce the labor involved in acquisition, management, analysis, aggregation, and input of data each time a model is applied to a new ecosystem or in a different problem context within the same ecosystem.

Model calibration, confirmation, diagnostic use: Facilitate model calibration, confirmation, and diagnostic applications through enhanced visualization of model output overlaid on system observations; passes model output back through the GIS for display in a spatial context.

Model sensitivity and alternative management analysis: Facilitate running whole-system model sensitivity and component analyses and testing alternative management strategies (i.e., conducts "what if?" scenarios such as changing land-use pattern, agricultural practices, implementation of point or nonpoint source controls, etc.).

Uncertainty analysis: Facilitate evaluation of model uncertainty (i.e., Monte Carlo analyses) that results from data error, parameter estimation error, aggregation error, and so on.
Visualization of future conditions: Allow efficient access and display of all existing and future observations on a specific site along with model predictions, which will expedite the documentation of the outcome of system management actions as well as postauditing of model predictions of these actions.
Facilitate data storage, retrieval, archiving: Create a database that can be accessed by any interested parties (i.e., modelers, managers, planners, regulatory agencies, etc.), which can serve as a repository for new data and refined knowledge on the system.

In order to achieve these benefits, GEO-WAMS contains a library of utilities (some generic and some application-specific) that perform a wide variety of preprocessing, processing, and postprocessing model development and application functions.

While GEO-WAMS is intended to be a generic modeling support system for use by relatively expert aquatic system modelers, we felt it was important to develop a prototype system that could demonstrate the feasibility and utility of a coupled GIS/database process modeling system such as GEO-WAMS. For a number of reasons, we selected dissolved oxygen in Buffalo River as our prototype problem. To demonstrate the spatial-temporal data and model linkages that are possible with GEO-WAMS, we included two models in this prototype application: (1) a conceptually simple but spatially complex watershed loading model (WLM), and (2) a version of WASP4-EUTRO4 (a generic surface water quality modeling code supported by EPA), which we call BREUTRO. The watershed loading model demonstrates the significant utility of GIS data for modeling the environmental behavior of a watershed. BREUTRO demonstrates the capability of GEO-WAMS to manage the relatively complex and diverse data processing functions necessary to configure a generic water quality model for a site-specific–problem-specific application. Taken together, the two models demonstrate the ability of our system to transfer model output that requires the data as part of its input data set from one model (WLM) to another model (BREUTRO).

This chapter presents the design of GEO-WAMS and describes the modeling support capabilities built into the Buffalo River watershed prototype.

GEO-WAMS DESIGN

Modeling Support Needs

GEO-WAMS was designed with careful consideration of the process a modeler goes through in developing and applying a surface water quality model on a site-specific basis. Generally, this process involves a series of steps that include: problem specification, theoretical/conceptual construct, numerical construct, model code development and implementation, model calibration, model confirmation, and diagnostic or management application.

One immediately recognizes that throughout this process there is significant interaction with process experimental and field observation data. The types of model input and output data are somewhat problem- and site-specific; however, almost all model applications deal with the following types of input data: (1) system geometry, (2) external loading of modeled constituents, (3) land-use activities within the system of interest, (4) hydrometeorological forcing data (e.g., solar radiation, precipitation, wind velocity, etc.), (5) basic physical and chemical properties of the system (air, land, water, and bottom sediments) that govern transport and transformation processes, (6) rate coefficients and other process parameters, and (7) initial conditions of modeled constituents. Common output data are: (1) spatial and temporal distributions of model state variable concentrations, (2) process rates, and (3) mass fluxes.

Analysis of pollutant impact in an aquatic system almost always requires utilization of a series of disparate models. These models are usually linked in an *ad hoc* way and are often run by different groups without regard for linkages. This observation led to our desire to create a system with a common database that would link a series of models in a way that was transparent to the user and would permit a single user ("user" in this case refers to an experienced water-quality modeler) to accomplish a complete analysis of pollutants in aquatic systems from original source to ultimate sink.

Based on this analysis, it was determined that the functional capabilities and linkages desired in GEO-WAMS could be grouped into three major categories, based on their chronology in a typical model application: preprocessing, processing, and postprocessing. Following is a list of envisioned tasks to include in GEO-WAMS (listed by category):

Preprocessing

- site specification and problem definition
- input database generation—existing data and user-provided data
- computation of segment-dependent data—volumes, interfacial areas
- assignment of parameters—global, spatially variable, time-dependent forcing
- prescription of boundary conditions
- prescription of initial conditions
- input data reduction and integration

Processing

- model selection, linkage, and execution
- spatial and temporal animation of results during model run
- model interrupt/restart capability
- parameter reassignment capability during a run—global, spatially variable, forcing functions, boundary conditions

Postprocessing

- basic display of primary state variables—spatial variation, temporal variation
- calibration to field data
- model scenario comparison for such tasks as sensitivity analysis or remediation alternative evaluation
- computation and display of secondary variables (e.g., fluxes)—spatial and temporal variation
- segment/system mass balance diagrams
- uncertainty analysis
- data export for use in other models—space averaged, time averaged, other necessary manipulations.

Another crucial aspect of a modeler's job is to keep track of the input data associated with a given model's run for a potentially large list of model runs throughout the entire process. With this in mind, we built into GEO-WAMS what we call the "scenario manager." The scenario manager keeps track of all model input and output data for a given site-specific–problem-specific application. It allows retrieval, editing, and saving of any model scenario, thereby greatly facilitating such modeling tasks as calibration, confirmation, management-alternative evaluation, sensitivity analysis, postaudit, and so on.

System Components and Integration

To meet the modeling support needs defined above, GEO-WAMS was designed and implemented to include five major components:

1. Spatial/temporal database: A database management system that allows modelers to input, store, analyze, retrieve, and display all spatially and temporally referenced data.
2. Data model management interface (DMMI): A software interface for data conversion between spatial database and process models and for user access to the database, process models, and tool kit utilities.
3. User interface (UI): A window and screen menu program written in macrolanguage so the user can visually examine the spatial and temporal data sets and manage and analyze them interactively through the DMMI. A series of help/explanation windows are also part of this interface.
4. Process models: A group of existing and/or newly developed mathematical models for aquatic system analysis and management. These models could range from relatively simple conservative substance transport models to complex, high-resolution ecosystem food web models.
5. Analyst toolkit utilities: A library of utilities used for data manipulation, data analysis, model development, and model application.

(These components and their interactions are depicted in Figure 51-1.)

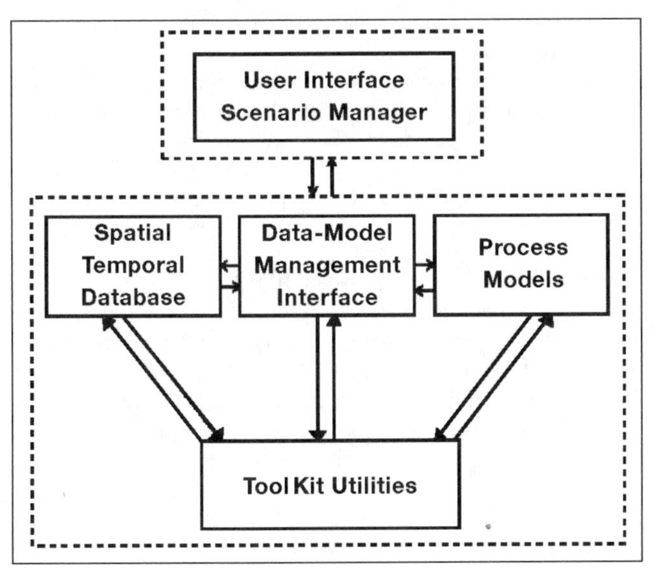

Figure 51-1. Conceptual design and component integration of GEO-WAMS.

System integration is the key to the development of GEO-WAMS. As an integrated system, GEO-WAMS can be seen as a collection of software programs that communicate by passing data, model parameters, and solutions among themselves. A variety of issues must be addressed in designing linkages among diverse pieces of software because they may use different representations of space and spatial objects, have multiple, inconsistent user interfaces, and several mechanisms for importing and exporting data. In GIS literature, three perspectives on GIS-model linkages can be identified: technical, functional, and conceptual (Burrough et al. 1988; Densham 1991; Fedra 1993; Nyerges 1993).

Technical perspective: This perspective is concerned with data. ARC/INFO, the GIS used for GEO-WAMS, interacts with the modeling software by passing files through the data model management interface (DMMI), which reformats files to ensure that object differences are resolved. Because we can directly modify the data input and export filters of the modeling software, we are able to pass data directly from one model to another. Model outputs are also passed back to ARC/INFO via the data–model management interface. These results are stored in the database and a series of records are built to track who used which data and/or models, when, and to what effect (function of scenario manager).

Functional perspective: This perspective on integration is provided by examining the forms of human–system interaction supported by the linked system. Pertinent issues range from data availability to the forms of user interface supported. Burrough et al. (1988) identify numerous issues from their experience in linking GIS and land management models. GEO-WAMS has essentially two types of user interface. The main GEO-WAMS interface is built within ARC/INFO using AML. A series of screens, each with pull-down menus and push buttons, provide access to all system functions, including all external modeling programs. The flexibility of AML enables users to hide much of the system's underlying complexity behind this interface, simplifying its use and enabling them to concentrate on the problem at hand. A second interface is associated with each of the models. Although we tried to maximize the amount of user interaction via the main interface, there are some stages during modeling that require users to interact directly with individual models. In concert, these two types of interface support all user interaction with GEO-WAMS, which can be said to exhibit a high degree of integration from the functional perspective.

Conceptual perspective: This perspective is concerned with the purpose of the linkage and how it affects system architecture. Extending the work of Fedra (1993), GIS has four levels of integration that can be used: (1) as a spatial database system to support modeling, (2) as a mapping tool to display output from models, (3) as both data manager and display generator for one or more models, and (4) to incorporate modeling functions. GEO-WAMS falls somewhere between the third and fourth levels of integration. Some of the model support tools are embedded, being built entirely from GIS functions. The external models correspond more to the third level of integration, although interaction with these models is minimized so that they appear to be embedded.

Overall, GEO-WAMS exhibits a fairly high degree of integration. The diverse nature of the numerous system components and the complexity of the linkages among them are largely transparent to the user, hidden behind a consistent graphical user interface.

MODELING SUPPORT TOOLS

Presented in this section is a brief description of some of the more important modeling support tools that are included in GEO-WAMS. These capabilities can be placed in context of the entire system by referring to the functional flow diagram of GEO-WAMS for a typical application (Figure 51-2).

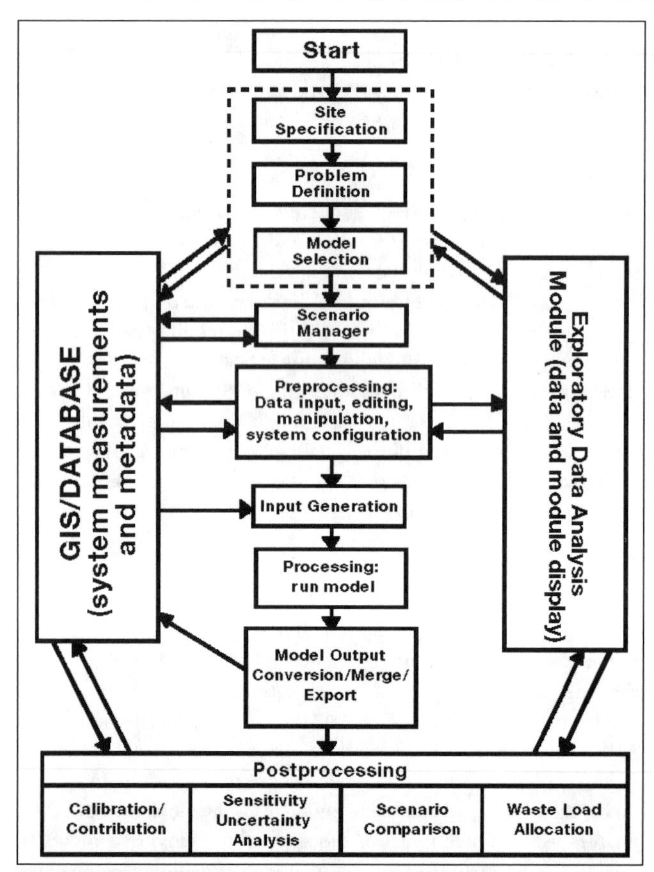

Figure 51-2. Functional flow diagram for a typical model application of GEO-WAMS.

Scenario Manager

Possibly one of the more valuable utilities of a coupled GIS modeling system is the scenario management utility built into GEO-WAMS. The purpose of the scenario manager is to ensure the integrity of scenarios generated by system users by tracking what a user does to the system's data sets and with what effects. The proposed system contains a large number of data sets dealing with hydrology, land use, soil properties, water quality, and so on; these data sets and their derivatives are analyzed by the models in GEO-WAMS. Because a given application may employ a series of models and may entail running the models a number of times, it is important to record which data sets are used and how they may have been modified to form a scenario (e.g., changing land use to simulate a development project); to keep track of what models are used and any intermediate results that are important to save; and, of course, to relate model output to all the input information that defines a scenario. The scenario manager provides this tracking system. By recording which data sets and intermediate results are associated with each user and each scenario, users are able to adopt and modify scenarios (either their own or those of others) without regenerating the entire scenario. From a user's perspective, the scenario manager simplifies the whole process of generating and modifying scenarios, and it prevents users from deleting or modifying "base" data sets and intermediate results associated with existing scenarios. In addition, it reduces storage requirements by ensuring that, where possible, data sets and intermediate results are reused rather than regenerated and stored every time they are required. In a multiuser, multiscenario system, the scenario manager is a crucial element in database management, user interaction, and integrated system operation.

Three-Dimensional System Segmentation

Based on the finite segment modeling approach of Thomann (Thomann and Mueller 1987), spatial discretization (segmentation) is a fundamental step for using WASP4 (Ambrose et al. 1987). It defines a natural water body as a series of linked segments. Each segment is considered as a completely mixed reactor with kinetic reactions occurring within each segment and transport process occurring between segments. The design of the segmentation scheme usually depends on physical characteristics of the water body, major water quality gradients in the system, hydrodynamic features of the system, and the spatial and temporal resolution requirements of the problem being investigated. Segmentation of a given aquatic system requires an experienced modeler who can apply both experience and theory to develop a workable segmentation scheme.

A special support utility has been developed in GEO-WAMS that greatly facilitates the process of segmentation and specification of segment-specific model parameters for WASP4. Using AML, this program allows the modeler to generate a new segmentation scheme or to edit an existing scheme for any water body for which surface coverage and bathymetric data are available. In this approach, the user defines all segment boundaries in the ARCEDIT environment. Background display of water-quality information from sample stations, water flow, waste loading, and bottom geometry data are options available as references in the segmentation process. Users are allowed to select, move, delete, or add segment boundaries as arcs on the surface plan view of the water body. The third dimension is provided by defining vertical segment boundaries on a vertical profile as a separate coverage in ARC/INFO. Once a vertical boundary is added, the interface program generates a new coverage recording the subsurface layer of the segments, and overlays the surface segmentation boundaries on the subsurface layer to generate horizontal and longitudinal segment boundaries of the subsurface segments. A view of the editing screen for this utility, showing a two-dimensional (longitudinal and vertical) segmentation of the Buffalo River, is presented in Plate 51-1.

After segment boundaries are determined, it is necessary to calculate the volumes of each segment and the exchange surface areas between pairs of segments and between segments and boundaries, as well as other system geometry parameters. This tool significantly reduces the effort required by the user as it automatically recalculates these variables whenever the segmentation scheme is changed.

Model Input Data Entry and Editing

Many of the input parameters for WASP4 are potentially spatially and temporally variable. Spatial variability is managed by permitting segment-specific parameterization. Temporal variability is managed by permitting specification of a time series (with various options for temporal interpolation) for each parameter that varies in time (e.g., seasonally varying intersegment dispersion rates). GEO-WAMS contains another utility that allows the user to input or edit any WASP4 model input value through what is called a modeling data editing routine. This utility, which is written in AML and operates within the ARCEDIT environment, has a user interface that looks much like the one depicted in Plate 51-2 for the segmentation editing. For segment-specific parameters, the user can display the system segmentation with segment numbers specified on the coverage. The user can then select the data block to edit from a list that includes intersegment dispersion, input advective flow, flow routing, boundary conditions, initial conditions, loading, time functions, parameters, constants, simulation variables, and scaling factors. Depending on which data block is selected for editing, an appropriate editing window will be displayed at the bottom of the screen. The exception is the "loading" selection, which calls the geographically based loading model (or other loading model available within GEO-WAMS). (The watershed loading model is discussed in the Intermodel Data Transfer section.)

Each data block that is specified or edited is saved as a separate coverage or INFO file and appropriately logged within the scenario manager so that it becomes part of the scenario currently being edited. If a given data block file has not been edited in modifying a given scenario, then the scenario manager will not rewrite that file but merely "point" to the old file as part of the new scenario. One of the features of the scenario manager is its ability to determine which individual files may be deleted when a full scenario is deleted and which must be saved because they are part of another resident scenario.

Data Model Management Interface (DMMI)

WASP4 requires a specific format for each input variable. If done manually, data entry and formatting are tedious and time consuming; a WASP4 input file in excess of 1,000 lines is not unusual. The DMMI of GEO-WAMS interacts with the user interface, the scenario manager, the GIS database, the utility tool kit, and the process models to automatically convert ARC/INFO files to an ASCII data file that meets the input format requirements of WASP4. (This interface is model specific; there is a separate C code for each model included in the system.) The conversion involves data selection, extraction, merging, and reformatting.

Following the processing of a WASP4 model run, the model output (ASCII text) is reconverted and transferred back to the ARC/INFO database. The DMMI rearranges the data matrices to extract data for each state variable into a separate file. The processed matrix file has segment polygons as rows and simulation variables as columns with incremented time-step series. The processed files are converted to INFO files in the GIS database. The AML interface programs select data of certain time and certain variables defined by the user and relate them to the specified coverage for display, calibration, or animation.

Intermodel Data Transfer

As mentioned earlier, one of the motivations for developing GEO-WAMS was the need to link several models in a series in order to develop a complete modeling framework for a given watershed. To automate the appropriate transfer of data from one model to the next, GEO-WAMS works through its DMMI and its GIS database to convert each individual model's output to a common structure and format within the GIS database. Then, when the next model in the series requires output from the model before its input file, the DMMI can simply access the appropriate input file from the GDMS and convert it to input for that model. This data transfer approach allows for the possibility that a given model may be able to utilize output data from several different "upstream" models.

In our GEO-WAMS prototype, the example of intermodel data transfer is the transfer of output from the watershed loading model (WLM) to provide loading input to BREUTRO. As mentioned previously, when the "loading" data block is selected in editing BREUTRO input data, the system initiates WLM. The watershed loading model in our GEO-WAMS prototype was developed to take advantage of ARC/INFO's capabilities. Its input data include a series of coverages, such as land-use polygons, soil association polygons, hydrologic unit polygons, stream course arcs, point source location points, and rain gauge location points. Other time functions and model parameters to be used in the model are also managed in the INFO database.

The output of WLM consists of time series for each delivery location, with date, flow, and loading of pollutants. The delivery location may be a point on the river shore, or an arc as a section of the river shoreline. The output files are also managed directly in the INFO database and can be read directly by the DMMI to generate segment-specific, time-variable loading of state variables necessary for the WASP4 input deck.

Postprocessing Support Capabilities

In the GEO-WAMS prototype, visualization of data (either field observation data or model output data) is performed in one of two ways. First, we developed a spatiotemporal exploratory data analysis module which, as depicted in Figure 50-2, can be accessed at several points in the functional flow of GEO-WAMS. It can either be used to display system data when making decisions about model selection or model configuration. It can also be accessed after a model run to perform various postprocessing tasks that require simultaneous display of both observation data and model output data (calibration, confirmation, postaudit, uncertainty analysis) or simultaneous display of more than one model scenario output (scenario comparison, sensitivity analysis, remediation, or regulatory alternative evaluation). (Because of space constraints, it is impossible to do justice to this system within this chapter, so this useful module of GEO-WAMS is described elsewhere [Lin et al. 1993].)

Another postprocessing capability of GEO-WAMS that takes advantage of GIS capabilities of the system is the model output spatial animation module. This AML program allows time-dependent simulation model output to be transferred to the GIS for display in two-dimensional map form on the screen. At each time step, the spatial variation of any given parameter may be displayed. Dynamic changes with time can then be illustrated by sequentially displaying maps at different time steps. This provides a valuable means of viewing model output.

In the case of BREUTRO, the animation program in the GEO-WAMS program takes animation time period, time steps, and segmentation coverage as user input, reads in the specified EUTRO4 output data, and develops automated sequential shading of the segment polygons for any given water-quality variable in a time series. Preset default shading patterns or user-specified shading patterns may be used. This process generates a false run time animation of the water quality in the studied river. An example of the display screen at a given time step for dissolved oxygen in the Buffalo River is presented in Plate 51-2.

CONCLUSION

The development of the GEO-WAMS prototype demonstrates the feasibility and utility of GIS as a tool for developing and applying surface water-quality models on a site-specific basis. In the case of the Buffalo River watershed application, we are able to effectively use high-resolution spatial data to develop input data necessary to drive a watershed loading model and a river water quality model and to display the model output. In addition to the database model integration, the two models in the prototype have been coupled to the extent that output from one model automatically is converted to input for the next model in the chain.

Future directions for the continued evolution of GEO-WAMS include the following enhancements:

- addition of new utilities to the analyst toolkit for better visualization of data and model output and for additional exploratory and statistical data and model analysis;
- incorporation of more Great Lakes process models addressing other issues (such as hydrodynamics and sediment transport, groundwater transport of pesticides and hazardous substances, nutrient dynamics and eutrophication, toxic contaminant exposure and effects, ecosystem food web interactions, etc.) at various levels of sophistication; and
- application of GEO-WAMS to other Great Lakes watersheds, including expansion to regional and whole-lake watershed applications.

The concept of developing a GIS for the entire Great Lakes basin, with the potential to apply GEO-WAMS to any part of the system, is an attractive long-term goal. In this way, data model integrated systems such as GEO-WAMS will open the door to the next generation of environmental models, a generation in which new data and knowledge can rapidly and functionally be incorporated into analysis and modeling frameworks, thus keeping pace with the need for greater sophistication in water resource management.

Acknowledgments

This work was funded through Environmental Protection Agency cooperative agreement (#CR818560) issued through the EPA, Environmental Research Laboratory–Duluth, Large Lakes and Rivers Research Station, Grosse Ile, Michigan. William L. Richardson, LLRS Branch Chief, is the project officer for this project. We would also like to recognize several agencies and individuals who provided Buffalo River watershed data for use in building our database: U.S. EPA (Region II, LLRS, and the Great Lakes National Program Office); New York Department of Environmental Conservation; U.S. Geological Survey, Soil Conservation Service; Buffalo Sewer Authority; National Climatic Data Center HydroNet database; Jill Singer and Kim Irvine at Buffalo State College; Cornell Laboratory for Environmental Applications and Remote Sensing (CLEARS); TIGER files; and U.S. Army Corps of Engineers, Buffalo office.

References

Ambrose, R. B., Jr., T. A. Wool, J. L. Martin, J. P. Connolly, and R. W. Schanz. 1987. *WASP4: A Hydrodynamic and Water-Quality Model—Model Theory, User's Manual, and Programmer's Guide* (Revision for WASP4.3x). Athens: Environmental Research Laboratory, Office of Research and Development, U.S. Environmental Protection Agency.

Burrough, P. A., W. van Deursen, and G. Heuvelink. 1988. Linking spatial process models and GIS: A marriage of convenience or a blossoming partnership? *Proceedings, GIS/LIS '88.*

Densham, P. J. 1991. Spatial decision support systems. In *Geographical Information Systems: Principles and Applications*, edited by D. J. Maguire, M. F. Goodchild, and D. W. Rhind, 403–12. London: Longman.

Fedra, K. 1993. GIS and environmental modeling. In *Environmental Modeling with GIS*, edited by M. F. Goodchild, B. O. Parks, and L. T. Steyaert, 35–50. New York: Oxford University Press.

Lin, H., F. Xia, and J. V. DePinto. 1993. GIS exploratory spatiotemporal data analysis: A study of water-quality data of the Buffalo River, New York, USA. *Proceedings, Geo-Information Beijing '93 International GIS Workshop*. Beijing, China.

Nyerges, T. 1993. GIS for environmental modelers: An overview. In *Environmental Modeling with GIS*, edited by M. F. Goodchild, B. O. Parks, and L. T. Steyaert, 75–93. New York: Oxford University Press.

Thomann, R. V., and J. A. Mueller. 1987. *Principles of Surface Water-Quality Modeling and Control*. New York: Harper and Row.

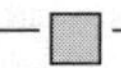

Joseph V. DePinto is currently Professor of Civil Engineering and Director of the Great Lakes Program at the State University of New York at Buffalo. Dr. DePinto received his B.S. degree in physics from Miami University in 1967. He has an M.S. in physics (1970), an M.S. in environmental engineering (1972), and a Ph.D. in environmental engineering (1975) from the University of Notre Dame. After sixteen years on the faculty in the Department of Civil and Environmental Engineering at Clarkson University, he joined the faculty at the University at Buffalo in 1991. Dr. DePinto's current research interests are focused on process experimentation and mathematical modeling of fate and transport of contaminants in aquatic systems.
Great Lakes Program
State University of New York at Buffalo
207 Jarvis Hall
Buffalo, New York 14260
Phone: (716) 645-2088
Fax: (716) 645-3667
E-mail: *depinto@superior.eng.buffalo.edu*

52

LOADSS:
A GIS-Based Decision Support System for Regional Environmental Planning

B. Negahban, C. Fonyo, K. L. Campbell, J. W. Jones, W. G. Boggess, G. Kiker, E. Hamouda, E. Flaig, and H. Lal

LOADSS (Lake Okeechobee Agricultural Decision Support System) was developed to help address problems created by phosphorus runoff into Lake Okeechobee. It was designed to allow regional planners to alter land uses and management practices in the Lake Okeechobee basin, then view the environmental and economic effects resulting from the changes. The Lake Okeechobee coverage incorporates information about land uses, soil associations, weather regions, management practices, hydrologic features, and political boundaries for approximately 1.5 million acres of land and consists of close to 7,000 polygons. LOADSS runs on SUN SPARC stations using ARC/INFO GIS software and requires 80 mb of hard disk storage. It is a completely mouse- and menu-driven user interface that can perform the following functions:

1. Create maps and reports detailing existing features (land uses, soil associations, weather regions, roads, hydrography, basin and county boundaries, etc.).

2. Change land uses and management practices on polygons selected using a mouse or logical criteria (for example, a particular land use or soil association).

3. Calculate phosphorus runoff, phosphorus assimilation along streams and canals, and the final phosphorus loading to Lake Okeechobee for a particular regional plan.

4. Create maps and reports detailing material imports and exports, economic indices, and environmental effects of selected land use and management practices. Compare the net effects of different regional plans.

5. Pass information to and call external simulation models, incorporate simulation results back into LOADSS databases, and display results in reports and maps. It displays time-series output of simulation runs.

6. Select water quality, quantity, and weather stations in the Lake Okeechobee basin and display time-series charts and graphs of monitoring data.

INTRODUCTION

Phosphorus concentrations in Lake Okeechobee water increased 2.5 times over a fifteen-year period from an annual average of 0.049 mg/L in 1973–1974 to a peak of 0.122 mg/L in 1988. Repetitive occurrences of extensive blooms of blue-green algae were viewed as an additional sign that the lake was receiving excessive amounts of nutrients, primarily phosphorus, which threatened the overall health of the lake's resources (SFWMD 1989). The lake serves as a major source for aquifer recharge for east coast municipal well fields; irrigation supply for annual $1.5 billion crops of sugarcane, rice, and winter vegetables; inflow for the ecologically unique wetland ecosystem comprised of the three Everglades water conservation areas and Everglades National Park; and potable water supply for five lakeside municipalities. Most agricultural activities in the Lake Okeechobee basin are net importers of phosphorus. The largest sources of net phosphorus imports to the basin are improved pasture (45.9% of the total), followed by sugar mills (14.9%), dairies (14.3%), sugarcane fields (13.5%), and truck crops (6.9%) (Fonyo et al. 1991).

The South Florida Water Management District (SFWMD), which is responsible for managing Lake Okeechobee, has initiated numerous projects to develop effective control practices to reduce the level of phosphorus in agricultural runoff as part of the Lake Okeechobee SWIM (Surface Water Improvement and Management) plan. These projects, numbering over thirty, have been designed to develop information on the control and management of phosphorus within the lake basin and to determine the costs and effectiveness of selected management options. There are three types of control options being studied: (1) nonpoint source controls, such as pasture management; (2) point source controls, such as sewage treatment; and (3) basin scale controls, such as aquifer storage and retrieval. With the completion of the majority of these research efforts, the need arose for a comprehensive planning tool that could integrate the results for all three classes of phosphorus control practices (PCPs). In response to these needs, design and implementation of a decision support system was initiated with the following objectives:

1. Organize spatial and nonspatial knowledge about soils, weather, land use, hydrography of the lake basin, and PCPs under a GIS environment.
2. Develop and implement algorithms for modeling nonpoint-source, point source, and basin scale PCPs.
3. Develop and implement algorithms for evaluating the technical feasibility and cost effectiveness of alternative combinations of PCPs applied to the basin.
4. Design and develop a user interface that would facilitate the use of the system by noncomputer experts.

The Lake Okeechobee Agricultural Decision Support System (LOADSS) is a GIS-based decision support system aimed at evaluating the effectiveness of different PCPs in the Lake Okeechobee basin for reducing phosphorus loads to the lake. LOADSS was designed to assist regional planners in decision making by generating reports and maps on regional land attributes, called external hydrologic simulation models, and by displaying historical water quality and quantity monitoring station data.

LOADSS DESIGN

LOADSS consists of three primary components:

1. Regional scale GIS-based model used to develop and manipulate regional plans aimed at reducing phosphorus loading to Lake Okeechobee in a cost-effective manner.
2. Interactive dairy model (IDM) (used to develop field-level management plans for dairies and run the field hydrologic and nutrient transport model (FHANTM) to estimate nutrient transport on individual dairy fields.
3. Lake Okeechobee graphic interface (LOGIN) used to access and display water quality and quantity monitoring station data.

Although these components can run independently, they are fully integrated in the LOADSS package and can exchange information where necessary. A design schematic of LOADSS is given in Figure 52-1.

Regional Scale GIS-Based Model

LOADSS serves both as a decision support system for regional planning and as a graphical user interface for controlling the different components. An example of the user interface is shown in Figure 52-2. One consideration in the design of LOADSS was the size of the database that was being manipulated. Since the land use database consisted of nearly 7,000 polygons, it was decided that running the simulation models interactively would not be feasible. Thus, the CREAMS-WT (Heatwole et al. 1987) runoff model was prerun for different levels of inputs and management for each land use, soil association, and weather region (Kiker et al. 1992). Depending on the land use and its relative importance as a contributor of phosphorus to the lake, anywhere from one (background levels of inputs to land uses like barren land) to twenty-five (dairies, beef pastures) levels of inputs were selected. Each set of inputs to a particular land use was given a separate PCP identification code. A CREAMS-WT simulation was performed for each PCP, on each soil association, and in each weather region. This resulted in some 2,600 simulation runs. Annual average results were computed for use in LOADSS.

The imports, exports, and costs of each PCP are based on a per-production-unit basis. Depending on the type of polygon, the production unit can be acres (pastures, forests, etc.), number of cows (dairies), or millions gallons of effluent (waste-treatment plants and sugar mills). To develop the regional plan for the Lake Okeechobee basin, a PCP identification code was assigned to each one of the polygons. Accessing the results of a regional plan is a process of multiplying the production unit of each polygon by its appropriate database import, export, or cost attribute and summing the resulting values over all polygons in the Lake Okeechobee basin. LOADSS runs in the ARC/INFO version 6.01 GIS software on SUN SPARC stations.

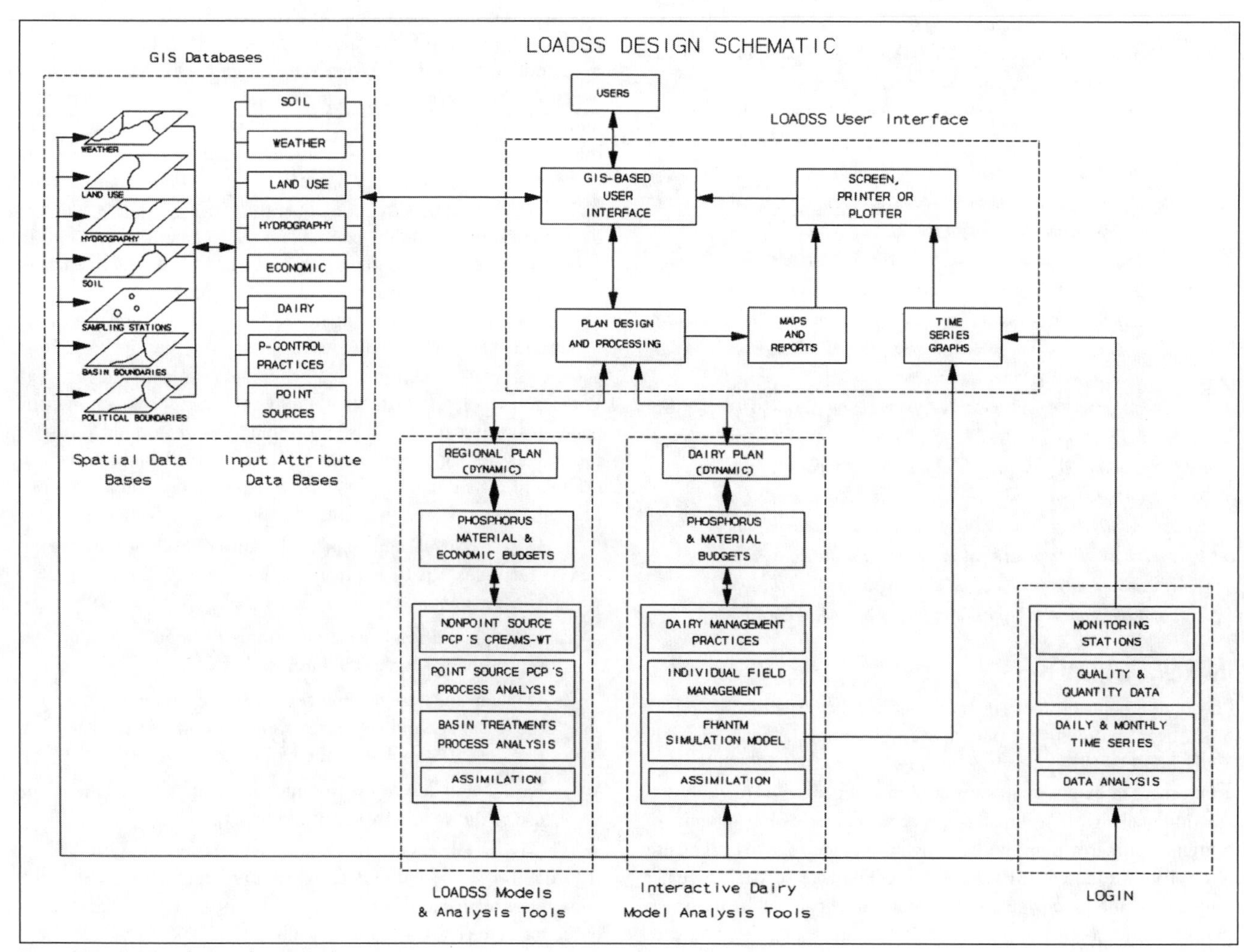

Figure 52-1. LOADSS design schematic.

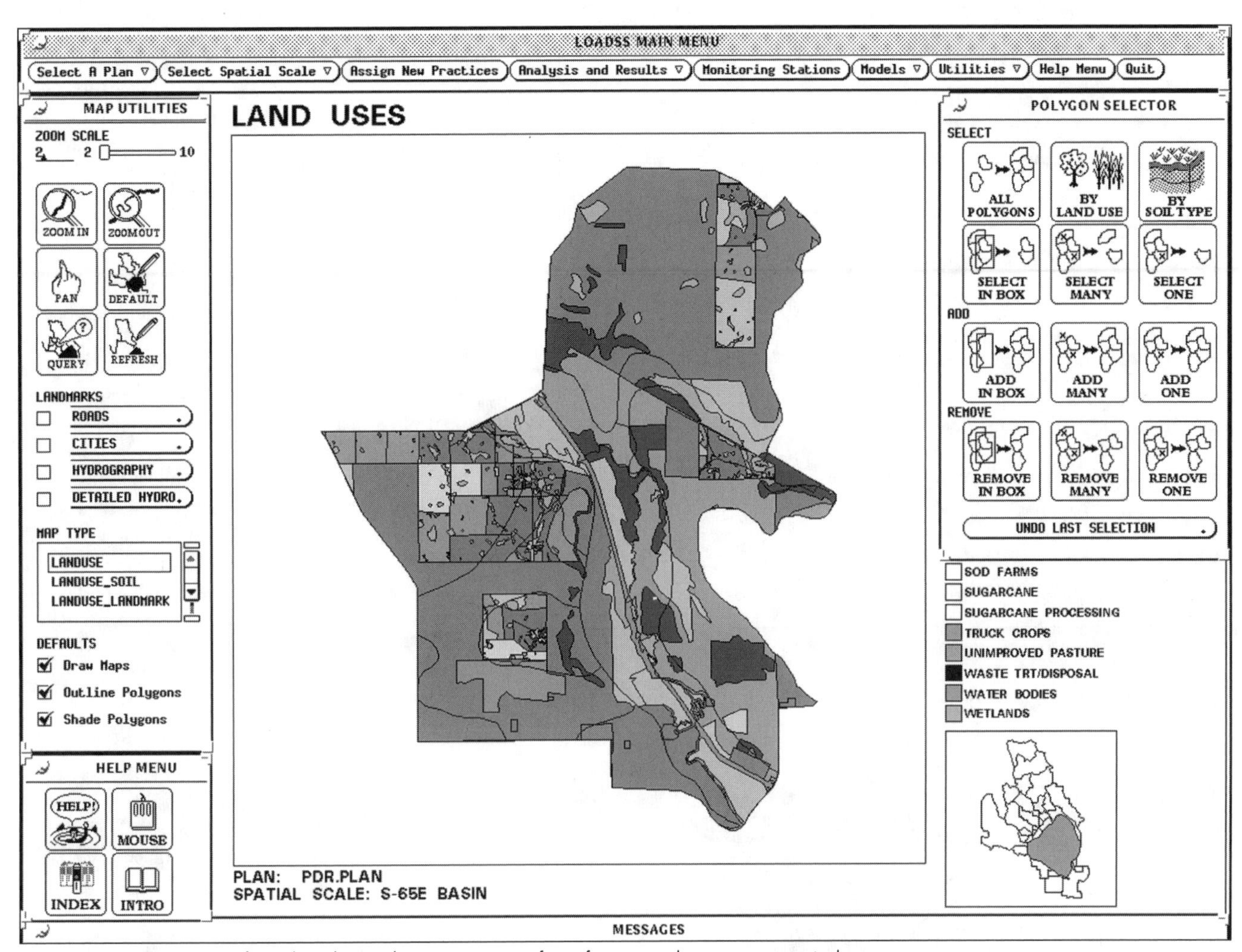

Figure 52-2. LOADSS user interface. The polygon selector menu is one of over forty menus that appear as required.

CREAMS-WT provides an average annual estimate (based on simulated runoff from twenty years of historical weather data) of phosphorus runoff from each polygon. Phosphorus assimilation along flow paths to Lake Okeechobee is estimated as an exponential decay function of distance traveled through canals and wetlands (SFWMD 1989).

A typical session in LOADSS consists of:

1. Selecting a regional plan to provide initial base conditions.
2. Selecting a spatial scale (basin, subbasin, county, region, or entire lake basin) for modification and analysis.
3. Changing land uses and PCPs on selected polygons within the specified spatial scale to develop a new regional plan.
4. Displaying and analyzing the effects of the changes made to create the new regional plan.

Over 100 different maps and reports relating to the following topics can be created:

Plan details—Descriptions and acreages of different PCP types assigned to polygons in the lake basin (reports only).

Materials budgets—Material imports and exports (fertilizer, sugarcane, milk, etc.) into the lake basin (maps and reports).

Phosphorus budgets—Phosphorus imports and exports, runoff, assimilation, and lake loading for polygons in the lake basin (maps and reports).

Economic summaries—Private and public investment, operation and maintenance, regulatory, and secondary cost comparisons (reports only).

Reports and maps are generated for only those polygons that are selected at the time the display option is invoked. This allows the user to create reports and maps for any set of polygons that are of interest. An example report generated in LOADSS is presented in Figure 52-3. Examples of maps generated in LOADSS are presented in Figure 52-4.

Interactive Dairy Model (IDM)

The LOADSS model was designed primarily to assist in regional environmental planning rather than for detailed field-level analysis of management alternatives. However, since dairies are one of the larger and most concentrated sources of phosphorus runoff into the lake and have been required to implement rather sophisticated phosphorus management systems, the district requested that LOADSS incorporate the capability for detailed, field-level analysis of management practices on dairies. IDM was developed as a GIS-based interactive hydrologic/water-quality simulation model specifically for Lake Okeechobee basin's dairy operations. The model operates on an individual dairy basis and incorporates a field-scale model (FHANTM) to provide detailed water and phosphorus transport information for specific fields of a dairy in addition to the large-scale spatial planning available through the main menu of LOADSS.

The FHANTM model was derived from DRAINMOD 4.03 (Skaggs 1989) with modifications to allow simple surface water routing and phosphorus release rates in surface runoff and groundwater

11/16/92 PAGE 1

REPORT TYPE: PHOSPHORUS BUDGET (ASSIMILATION)
GEOGRAPHIC AREA: S-65E BASIN
EXAMPLE.PLAN: CURRENT (MODIFIED)
PCPS INCLUDED: ALL PCPS
AGGREGATION(S): POLYGONS, SOILS

LAND USE	PCP ID.	P RUNOFF (lbs)	P ASSIM. (lbs)	P AT BASIN OUTLET(lbs)
BARREN_LAND	NBL01000	30	26	4
CITRUS	NCT02000	89	5	84
MILK_HERD_PASTURE	PDR20MHP	571	554	17
MILK_HERD_PASTURE	PDR50MHP	1,688	1,557	131
MILK_HERD_PASTURE		2,259	2,111	148
.	.	.	.	.
.	.	.	.	.
.	.	.	.	.
TRUCK_CROPS	NTC01000	3,336	1,803	1,533
UNIMPROVED_PAS	NUP01000	2,564	1,298	1,266
WETLANDS	NWL01000	0		
S-65E		58,367	44,348	14,019

Figure 52-3. Sample analysis report created in LOADSS (portions deleted).

(Campbell and Tremwel 1992). FHANTM was developed specifically for the sandy spodosols of south Florida. The model is area-independent within the limits of homogeneous soil, cultural, and weather conditions. It simulates surface and subsurface flow, phosphorus transport, and phosphorus uptake by plants. FHANTM is run interactively as required by the IDM.

Users are provided the opportunity to alter numerous management parameters through the GIS menu system for six different dairy land uses: spray field, milking herd pasture, other pasture, solids spreading area, hay field, and high-intensity area. Digitized coverages of these individual land uses were provided for each of the dairies in the basin. The IDM allows for the development and evaluation of detailed dairy-management plans that otherwise would be impossible at a regional scale.

While LOADSS only provides average annual results, IDM is able to display daily time-series simulation results to the user. An example time-series graph can be seen in Figure 52-5. IDM utilizes the same assimilation algorithm and can produce the same phosphorus budget maps and reports as in LOADSS. IDM runs under the ARC/INFO version 6.01 GIS software on SUN SPARC stations.

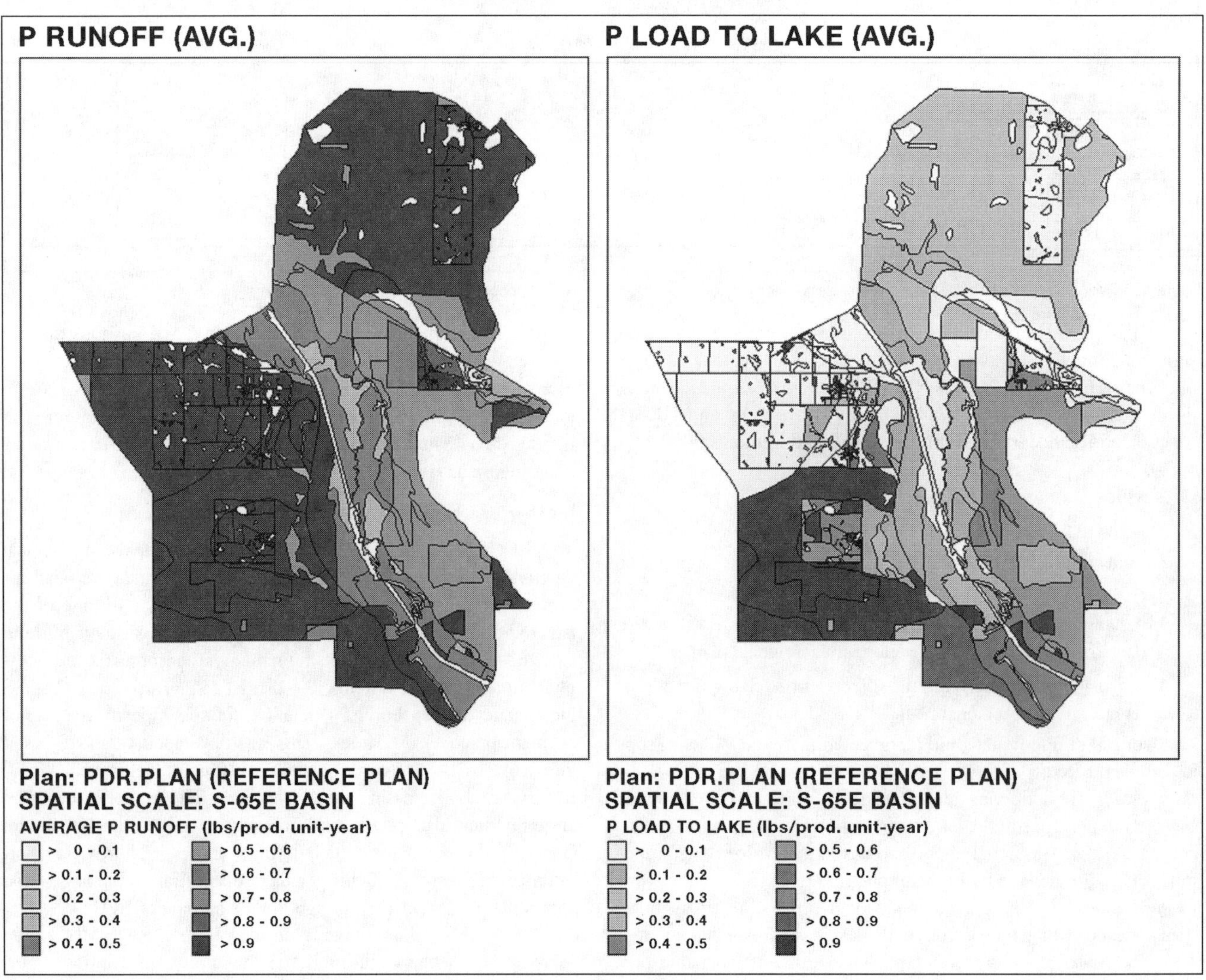

Figure 52-4. Sample analysis maps created in LOADSS.

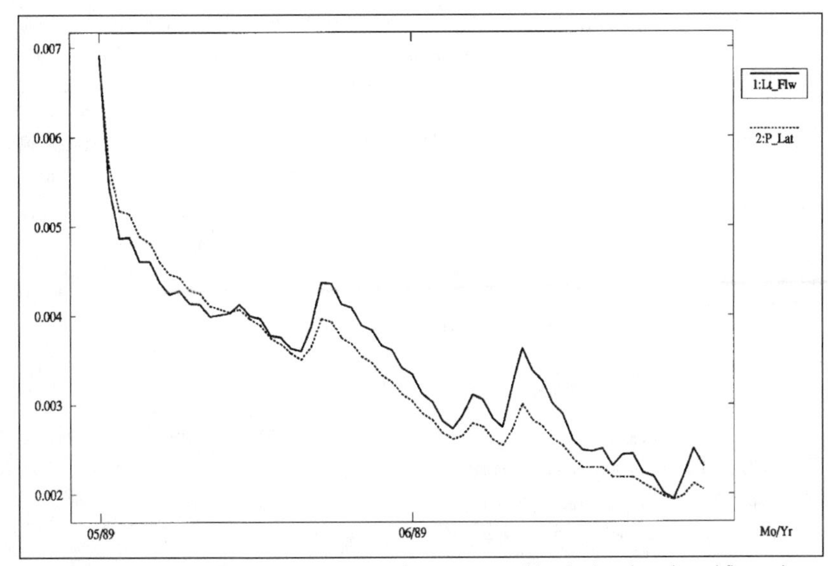

Figure 52-5. Sample simulation time-series output from IDM. Variables displayed are lateral flow volume and phosphorus load.

A typical session in IDM consists of:

1. Selecting a dairy plan for modification and analysis. A dairy plan consists of management plans for all dairies in the Lake Okeechobee basin.
2. Selecting a particular dairy for analysis.
3. Changing land uses and management practices on selected fields to develop a new plan for the dairy.
4. Running the FHANTM simulation model for selected fields in the dairy.
5. Analyzing and displaying the effects of the changes made to the dairy plan.

Monitoring Station Data Graphic Interface

LOGIN was developed to access and display measured data from water quality and quantity monitoring stations in the Lake Okeechobee basin. LOGIN allows the user to select monitoring stations from a basin map on the screen and then view time-series data and summarized sampling data for those stations (Figure 52-6). This tool allows users to examine historical data for trends or unusual values and to compare actual and simulated data in the LOADSS interface. LOGIN runs under the ARC/INFO version 6.01 GIS software and PV-WAVE version 4.0 data-analysis software on SUN SPARC stations.

INITIAL RESULTS

The model was used to estimate loads to Lake Okeechobee under current regulations and management practices. Total loads are estimated at 235 tons per year, nearly a 40% reduction from the adjusted SWIM load estimate. The estimate of current loads is 63 tons greater than the adjusted SWIM target. However, fourteen of the twenty-two basins meet or exceed the target reductions. Three basins (S-191, S-154, and S-65D) account for 78% of the target shortfall, despite the fact that these basins are estimated to have achieved a 50% average reduction in load. These three basins are all important dairy producing areas.

Alternative means for obtaining additional reductions in lake loads can be evaluated using the three components of the LOADSS system. First, the IDM can be used in consultation with district specialists and dairy operators to examine alternative phosphorus management practices for particular fields on individual dairies. Second, LOGIN can be used to access the district's current water quality data and compare it to the LOADSS estimates to identify problem areas. Finally, the regional LOADSS planning model can be used to evaluate the cost effectiveness of additional nonpoint, point, and basin-scale control practices for individual basins or for the entire drainage area.

CONCLUSION

Initial results from LOADSS indicate that current regulations and PCPs have achieved approximately 75% of the target load reduction. The remaining 25% reduction is likely to be more difficult and expensive. The LOADSS system provides a means for "what if?" evaluations of the cost effectiveness of additional PCPs applied at various scales within the drainage area. As is generally the case, initial results obtained from the system provide insights into additional research needs. In this particular case, it appears that additional research to refine the assimilation algorithm is warranted.

In a more general sense, the LOADSS system provides a versatile set of components useful in regional environmental planning. The system was designed to be as flexible as possible in order to allow for additional modifications to the capabilities of the system as well as to facilitate application to other regions and problems.

Acknowledgments

This decision support system was developed as a joint project by the Institute of Food and Agricultural Sciences, University of Florida and the South Florida Water Management District, as part of Southern Region Project S-249 of the USDA-CSRS. Approved as Florida Agricultural Experiment Station Journal Series #N-00798.

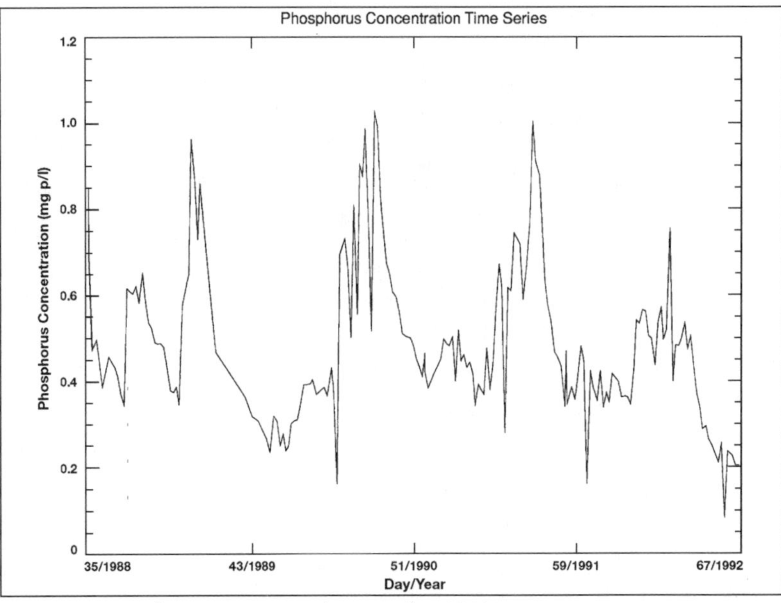

Figure 52-6. Example monitoring station data output from LOGIN.

References

Campbell, K. L., and T. K. Tremwel. 1992. *Biogeochemical Behavior and Transport of Phosphorus in the Lake Okeechobee Basin: FHANTM Users Manual.* West Palm Beach: South Florida Water Management District..

Fonyo, C., R. C. Fluck, W. Boggess, C. Kiker, H. Dinkler, and L. Stanislawski. 1991. *Biogeochemical Behavior and Transport of Phosphorus in the Lake Okeechobee Basin, Area 3: Final Report.* Gainesville: Department of Agricultural Engineering and Food and Resource Economics, IFAS, University of Florida.

Heatwole, C. D., K. L. Campbell, and A. B. Bottcher. 1987. Modified CREAMS hydrology model for coastal plain flatwoods. *Transactions of the ASAE* 30(4):1,014–22.

Kiker, G. A., K. L. Campbell, and J. Zhang. 1992. CREAMS-WT linked with GIS to simulate phosphorus loading. *ASAE Paper No. 92-9016.* St. Joseph: American Society of Agricultural Engineers.

SFWMD (South Florida Water Management District). 1989. *Interim Surface Water Improvement and Management (SWIM) Plan for Lake Okeechobee.* West Palm Beach: South Florida Water Management District.

Skaggs, R. W. 1989. DRAINMOD User's Manual. Interim Technical Release. Fort Worth: USDA-SCS, South National Technical Center.

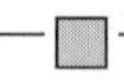

B. Negahban, C. Fonyo, K. L. Campbell, J. W. Jones, W. G. Boggess, G. Kiker, E. Hamouda, E. Flaig, and *H. Lal*
Agricultural Engineering Department
Food and Resource Economics Department
University of Florida
PO Box 110570
Gainesville, Florida 32611-0570
Phone: (904) 392-8534
Fax: (904) 392-4092
E-mail: *klc@agen.ufl.edu*

53

Basinwide Water-Quality Planning Using the QUAL2E Model in a GIS Environment

Paul M. Craig and Gerald A. Burnette

QUAL2E is an EPA-developed water-quality model that has gained wide acceptance as a planning tool. It is a comprehensive one-dimensional, steady state, streamwater-quality model that can simulate up to fifteen constituents including dissolved oxygen, biochemical oxygen demand, temperature, ammonia, nitrogen, and coliforms. It allows for multiple waste discharges, withdrawals, tributary flows, and incremental inflow and outflow. QUAL2E can be an excellent tool for studying the impact of waste loads on in-stream water quality. Because of these characteristics, QUAL2E is an ideal candidate for use in a GIS environment. The ability to examine basin geometry and query a computerized database from within a single framework is a tremendous aid in building a complete, reliable model data set. Furthermore, posting the results of a particular QUAL2E model run back to the database facilitates the type of "what if?" analysis that is so vital in effective planning. This chapter examines the issues relevant in utilizing the QUAL2E model for water-quality planning in a GIS environment. Throughout the discussion, examples are cited from a case study involving the investigation of a stream's capacity to assimilate discharge from a sewage treatment plant.

INTRODUCTION

Our nation's waters have historically received wastes and waste byproducts from sanitary and industrial activities. In the early 1900s it became clear that the wastes entering our natural water systems often degraded the water quality in these streams beyond acceptable limits. As a result, regulatory and planning agencies became involved in trying to minimize the impact of industrial activities on water quality by applying various technologies to wastewater discharges. As waste treatment technologies and mathematical modeling techniques developed, the focus began to shift from simply setting waste load targets and measuring attainment to predicting the impact of changes in waste load on water quality. A key element that facilitated this change in focus was the water-quality planning study (including waste assimilative capacity modeling), during which measurements of water quality were made and used as a baseline against which the impact of proposed waste loads could be compared. Of course, since the creation of the Environmental Protection Agency and subsequent state water-quality organizations, these water-quality planning studies have become even more prevalent.

In analyzing applications for wastewater discharge permits, many of these water-quality planning agencies have compiled large databases of water-quality information on individual streams and associated watersheds. Sometimes these data are used as input to an assimilative capacity study (ACS). An ACS is a comprehensive study of the water quality in a stretch of river. These studies combine the historical data on a stream with new data collected during intensive water-quality surveys. These data are then used for calibration and validation of a predictive mathematical water-quality model of the study reach. Many such models have been developed over the years, but for routine water-quality modeling, the EPA has developed and maintained the QUAL2E model (USEPA 1987). This model is particularly applicable to water bodies during steady low-flow conditions, during which streams are typically the most stressed from point-source loadings. This makes QUAL2E a very common planning tool for evaluating point-source loadings to receiving streams. (More description of the model will be provided later.)

QUAL2E principally uses spatially referenced data and therefore provides an ideal match to the capabilities of a geographic information system (GIS). The model also outputs spatially referenced results. The opportunity for the development of a graphically based water-quality planning tool in a geographically referenced environment becomes clear.

SCOPE

This chapter describes some of the work involved in developing a GIS application using Intergraph hardware and software—specifically, coupling the EPA QUAL2E model to the GIS platform. To test the capabilities of this new application, a previously conducted ACS was revisited using the application. The results of the study are published in Eiffe et al. (1991).

GIS/QUAL2E INTEGRATION

This section examines various issues that arise during the integration of the QUAL2E model into the GIS environment. Not all the issues have been completely resolved. (The North Mouse Creek study in Eiffe et al. [1991] offers more insight into the complications.)

Hardware and Software

ECE's GIS platform consists of an Intergraph 2400 series graphics workstation with 32 Mb of RAM, a 486 Mb hard disk drive, and a 19" high-resolution monitor. A 36" × 48" digitizing table is attached. The workstation runs a variation of the UNIX operating system. The baseline graphics software is Intergraph's Microstation. Practically all of Intergraph's high-end application software use Microstation as a basis on which to build customized capabilities. In the area of basic GIS software, Intergraph has a project-management-oriented shell called Modular GIS Environment (MGE). For more advanced analysis, Intergraph has the MGE Graphics Analyst (MGA) and other tools. Microstation and subsequent products are capable of interfacing seamlessly with several database management products. The sample study used the Informix database management system (DBMS).

QUAL2E Model Description

QUAL2E is a comprehensive one-dimensional streamwater-quality model that can simulate up to fifteen constituents including dissolved oxygen, biochemical oxygen demand, temperature, ammonia, nitrogen, and coliforms. It is applicable to dendritic streams that are well mixed. It assumes that the major transport mechanisms, advection and dispersion, are significant only along the main direction of flow (longitudinal axis of the stream or canal). It allows for multiple waste discharges, withdrawals, tributary flows, and incremental inflow and outflow. It also has the capability of computing required dilution flows for flow augmentation to meet any prespecified dissolved oxygen level.

Hydraulically, QUAL2E is limited to the simulation of time periods during which both the streamflow in river basins and input waste loads are essentially constant. QUAL2E can operate either as a steady state or dynamic model. When operated as a steady state model, it can be used to study the impact of waste loads (magnitude, quality, and location) on in streamwater-quality and in conjunction with a field sampling program to identify the magnitude and quality characteristics of nonpoint source waste loads. By operating the model dynamically, the user can study the effects of diurnal variations in meteorological data on water quality (primarily dissolved oxygen and temperature) and diurnal dissolved oxygen variations due to algal growth and respiration. However, the effects of dynamic forcing functions, such as headwater flows or point loads, cannot be modeled in QUAL2E.

Porting and Verifying the Model

The first step in linking the QUAL2E model to the GIS environment was to get the model to run on the workstation. Since the model is maintained by the EPA, it was desirable to keep the changes in the source code to a minimum. QUAL2E, as is typical of environmental models that have been developed over a number of years, is written in FORTRAN. The EPA offers versions for personal computers, VAXs, and IBM mainframe systems. All of these versions are maintained using one set of source files with separate sections that are specific to the operating system. By commenting out the unneeded portions of these sections and recompiling, a version of the model for the new platform can be prepared quickly. Porting the model to the new UNIX-based platform was eased somewhat by this fact. An examination of the code revealed which supported environment contained system calls that most closely resembled the UNIX environment's system calls. Preparing the code as if it were to be compiled on the closest supported system produced a model that was almost acceptable. A few additional modifications were required in order to complete the conversion and produce a model that ran correctly on the GIS platform.

The EPA supplies several example data sets with the model that serve to familiarize the user with the model's capabilities. Since the examples contain both input and output data, they can be used to ensure that any new version of the model is completely compatible with the EPA authorized versions. Several of these example data sets were run on the newly ported model. The resulting output files were compared to the ones supplied by the EPA to confirm a correct porting. In all cases, the only discrepancies noted were on the order of 0.0001%, which is easily attributable to differences in round off within the machines.

System Overview

The desired fundamental capability to run a water quality model in a GIS environment is to associate model parameters, both input and output, to graphic elements. The QUAL2E model appears to be amenable to such capabilities: hydrologic and hydraulic parameters are input to describe reaches, reaches are composed of one or more computational elements, and results are output for each computational element. There is, however, a subtle complication that makes direct linkage of the model to the graphics more difficult. In the model input data set, reaches are defined by specifying the upstream and downstream river miles for that reach. The computational element length is constant throughout the model, so that defining a reach produces a fixed number of computational elements. Elements are numbered in a downstream fashion, beginning with the most upstream point of the main stream. When a junction is encountered, numbering continues with the most upstream point of the tributary. The element of a receiving stream that is immediately downstream of the tributary is assigned the next number after the last element in the tributary. The modeler then specifies a numeric code, identifying the element type (headwater, junction, standard, etc.), for each element in sequence.

The numbering scheme causes problems when trying to post modeling results to the database. The model produces an ASCII file containing values for all the modeled parameters for each computational element. In order to associate this information with specific graphic elements, knowledge of which graphic elements represent the modeled computational elements is required. The model outputs no information that will identify the element number with a particular reach. The solution to this problem is to add another layer to the flow of information: a reach description file that contains the information regarding which graphic lines represent reaches that were modeled and the correspondence between computational elements and reaches. This file acts as an interpreter between the model and the graphics, pointing model results to the appropriate database rows by matching the computational element to the graphic element that represents the appropriate reach.

The model interface that ECE is developing follows a logical breakdown in the work flow for performing a QUAL2E run. The three distinct phases are preprocessing, running the model, and postprocessing and interpretation.

Preprocessing and data entry: The first step in performing the QUAL2E modeling study is to prepare a baseline graphic (map) of the stream system to be modeled. The most common map depiction of streams, particularly small ones, is simple line work. The base map will, therefore, contain linear elements to represent the stream system. The Intergraph software, like most GIS packages, only allows database entries to be associated with polygons or other closed forms. This means that no database rows may be associated with the line work that represents the stream system. The result is that the user is forced to use polygons rather than lines to depict the stream system. A routine has been developed to generate the polygons based on the single lines. This has the advantage of providing polygons that are sufficiently wide to allow attribution, yet sufficiently narrow so they do not artificially dominate other features on the base map or any overlaid scanned images.

After the base map is prepared and polygons depicting the computational elements are created, other graphic elements are placed that denote sampling points or other reference locations. These reference points will have database entries attached to them that define the values of reach, river mile, flow, dissolved oxygen concentration, sediment oxygen demand, and other model inputs. The values in the database can come from an ACS, historical data, or other sources. Ideally, there should be only one such reference point within each reach. An exception to this is that the most upstream reach of each water body may have two reference points: one should be at the upstream end of the reach to set the upstream boundary conditions. If there is another point within the first reach, it may be used to supply initial conditions for the reach. Likewise, if model downstream boundary conditions are to be specified, the last reach in the model might have two reference points.

A graphical form-based preprocessor has been developed to assist the user in entering the database information from which the necessary QUAL2E input files are generated. Figure 53-1 shows the initial screen of this preprocessor, which contains control information. When prompted to generate the input files, the preprocessor checks to ensure that all necessary parameters have been supplied; missing information causes the preprocessor to prompt the user for input. The graphical nature of the interface allows the user to quickly repeat model runs during the calibration process and allows "what if?" analysis for facility siting.

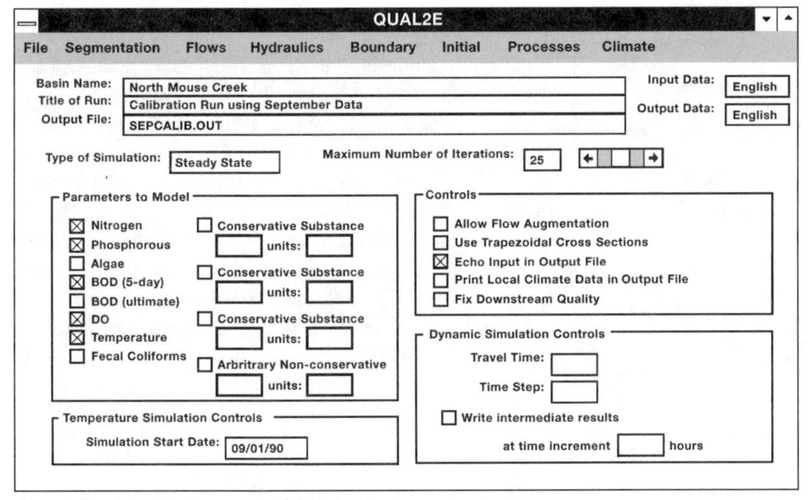

Figure 53-1. The opening screen of the QUAL2E preprocessor.

Running the model: The first step in a model run is to define the system to be modeled. Definitions of reaches should be chosen to allow simulation of the study area. Parameters for the reaches are determined from database entries tied to the reference location(s) within the reach. After a title for the run is supplied, some fundamental model options are specified (what to model, maximum number of iterations, etc.), and the model run configuration file is written. This file is used to help post results to the appropriate graphic elements and can also be used to reproduce model runs with different parameters. This file contains a physical description of the modeled system, which is a combination of the descriptions used by the graphics and the model. The model only needs to know the beginning and ending river miles for each reach and where tributaries join other reaches. The graphics file (including the underlying database) will contain information on the modeled reaches, but might also contain information on stream sections that are outside the modeled area.

Once the configuration file has been created and the database entries are complete, the modeler must calibrate the model before it can be used for predictive analyses and planning purposes. The pre- and postprocessors help the modeler conduct this process quickly and efficiently.

Postprocessing and interpretation: Postprocessing consists of posting model results to the database. One table receives the values stored in the ASCII file the model outputs. That table contains columns for the element number and all the parameters the model is capable of simulating. Any parameters that are not modeled for a particular run are flagged as missing with null values. The postprocessor also reads the reach description file in order to determine which graphic elements correspond to which computational elements, and therefore where to post the data.

Interpretation of modeling results is made easier by the GIS environment. Interpretive actions that are possible are much like those of most other GIS systems. The modeler can, for instance, color code the computational elements according to ranges of values of one or more constituents.

AN EXAMPLE APPLICATION

In order to test the QUAL2E GIS application, the study conducted on the North Mouse Creek (located in southeastern Tennessee) was selected. The following subsections provide a brief overview of the study site, the water-quality parameters necessary to the ACS study, and a brief description of how some of those parameters were obtained.

Watershed Description

North Mouse Creek is a small stream with a total drainage area, at its confluence with the Hiwassee River, of 93.7 square miles. The study area was between creek mile (CM) 18.4 and CM 26.6. Figure 53-2 provides a general site map of the area noting the features of interest.

North Mouse Creek watershed is a long, narrow basin in the valley and ridge province of east Tennessee. It is underlain by shales of the Conasauga group and limestone and dolomite of the Knox group. The soils in the drainage basin are silt loams and silt clay loams with chert and shale present in many locations. These soils result in a relatively low infiltration rate, which in turn results in a relatively low base flow in the stream. Land use is principally agricultural, with one existing wastewater treatment plant, the Niota plant, and one wastewater discharge planned for the city of Athens.

ACS

An ACS was conducted on upper North Mouse Creek to collect water-quality data for the modeling and planning of a proposed wastewater treatment plant. An ACS consists of two or more intensive water-quality surveys to obtain the data necessary to calibrate and validate the model. These studies generally are conducted in one of two ways. The first method, which was employed in this study, follows a particular slug of water as it moves downstream and measures its water-quality parameters. All significant inputs and withdrawals are noted and measured during the study. This results in sufficient information to develop a mathematical model of that portion of the stream. The second method consists of collecting an "instantaneous snapshot" of the water-quality parameters at many locations up and down the stream and, at selected points, source loadings and withdrawals. This method works best in steady state systems.

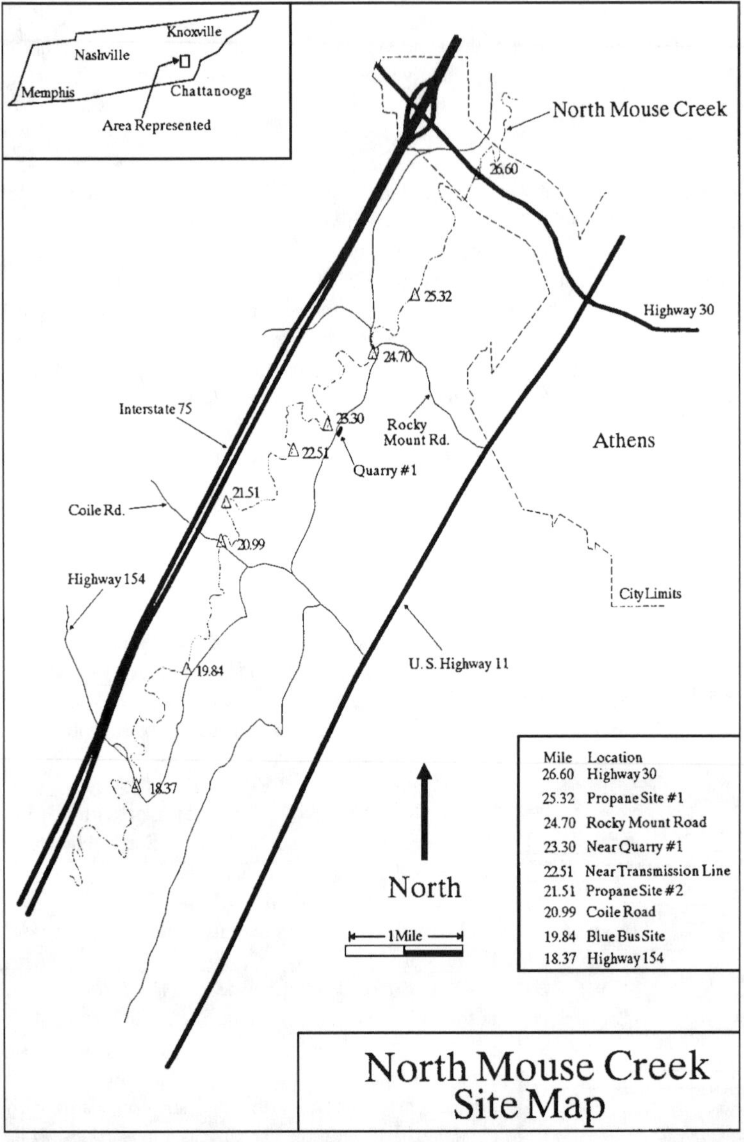

Figure 53-2. The study area of North Mouse Creek, December 1989.

Data Sources

Little water quality information existed for North Mouse Creek at the beginning of this study. The United States Geological Survey (USGS) had developed regional regression analysis of low flow in local streams, primarily as a function of drainage area (Bingham 1986). The State of Tennessee had mandated that the regulatory low-flow target would be set by the statistical flow determined from a flow frequency analysis. Three-day, twenty-year low flow (3Q20) can be determined from this analysis. The 3Q20 at North Mouse Creek, mile 24.7 (the proposed plant location) was 7.2 cfs. From this regional regression equation, the base flow along the studied stream reach was computed.

The remaining information that was collected as part of the ACS included hydraulic information necessary to determine the time taken for a slug of water to travel downstream. Time of travel is a function of flow in the stream; therefore, two to three time-of-travel studies at different flow rates were necessary. In addition, water quality parameters such as dissolved oxygen, pH, nitrogen series, phosphorus series, temperature, reaeration rate, deoxygenation rate, sediment oxygen demand, and carbonaceous biochemical oxygen demand were required. These water-quality parameters were collected from a slug of water as it moved downstream by tracing a fluorescent dye wave as it passed. Significant tributaries and discharge points were also sampled for the various parameters. These data were imported into the QUAL2E GIS application.

Model Setup and Execution

The North Mouse Creek study area was modeled for this application as a single stream system. The stream was divided into seven reaches based on physical and hydrologic characteristics. The computational element length was 0.2 miles. The ACS gathered data on a subregion of this creek; therefore, data were estimated for this test application for the remaining portions of the stream based upon the actual field measurements. Dissolved oxygen was the primary constituent of interest. After calibration, the oxygen balance in North Mouse Creek was modeled in a steady state condition at the 3Q20 in the stream with wastewater discharges from the proposed Athens plant. The model was run using waste load discharges of 1, 2, and 3 million gallons per day. Modeling results were compared to the state mandated lower limit of 5 mg of dissolved oxygen per liter of water. The modeling results were reviewed in conjunction with the design engineers, allowing determination of optimum wastewater discharge permit limits.

CONCLUSION

A general water-quality modeling and planning tool has been developed for use in an Intergraph GIS environment. A graphical user interface to the standard QUAL2E model assists the user in entering and validating data. After the data have been entered and the model calibrated, the user has a rapid assessment tool for quickly and efficiently evaluating different water quality stresses to a stream segment. The modeler can easily explore the effects on water quality of additional wastewater discharge from existing and/or proposed plants. Proposed plants can be placed interactively anywhere along the river system to determine their potential impact. In a similar fashion, withdrawals can be evaluated quickly and easily to determine what impact they may have on downstream water quality.

References

Bingham, R. H. 1986. Regionalization of low-flow characteristics of Tennessee streams. In *United States Geological Survey Water Resources Investigations Report.* 85-4191.

Eiffe, M. A., M. L. Gentry, and P. M. Craig. 1991. *A Study of the Assimilative Capacity of North Mouse Creek Near Athens, Tennessee.* Knoxville: Environmental Consulting Engineers, Inc.

United States Environmental Protection Agency (USEPA). 1987. *The Enhanced Stream Water-Quality Models QUAL2E and QUAL2E-UNCAS: Documentation and User Manual.* Athens: USEPA, Environmental Research Laboratory.

United States Geological Survey. 1980. Discharge measurements at gaging stations. *Techniques of Water-Resources Investigations of the United States Geological Survey,* vol. 3.

Paul M. Craig and *Gerald A. Burnette*
Environmental Consulting Engineers, Inc.
PO Box 22668
Knoxville, Tennessee 37933

54

GIS Water-Quality Model Interface:
A Prototype

Roger G. Cronshey, Fred D. Theurer, and R. L. Glenn

The Soil Conservation Service (SCS) has water-quality responsibilities that require the use of comprehensive computer models. In order to carry out these responsibilities, SCS is adopting several Agricultural Research Service (ARS) water-quality (WQ) models. In addition, they are building an interface to connect geographically referenced watershed data (both spatial and tabular) to the models via control of a model user. While this project eventually will allow SCS personnel at many levels to access water-quality analysis, the prototype is limited to use by eight volunteer data collection test sites (Iowa, Michigan, Missouri, New York, Tennessee, Texas, Washington, and West Virginia) and developing offices. The test sites have collected data on a hydrologic unit area to meet the input requirements for the four ARS WQ computer models (AGNPS, SWRRB, EPIC, and GLEAMS) that will be included in the interface. The first two models are watershed scale and the latter two are field scale. Relational database attribute data for the models will be associated with eight GIS data layers (soils, field, elevation, watershed boundary, subwatershed, geomorphic, stream network, and point data). Many model input parameters need to be derived from the source attributes collected using GIS data, attribute data, and mathematical relationships. Additionally, some model parameters will be required at run time. The prototype interface will allow easy use of the models as well as generation and modification of input parameters for evaluation of the water-quality models themselves.

INTRODUCTION

The Soil Conservation Service (SCS) has the responsibility to provide landowners with technical assistance to prevent agricultural nonpoint source pollution. To accomplish this, SCS is adopting several Agricultural Research Service (ARS) water quality (WQ) models and is building an interface to connect geographically referenced watershed data (both spatial and tabular) to the models via control of a model user. The scope of this effort is to examine water quality at the hydrologic unit (HU) level, but portions of HUs also may be examined. No new WQ models will be developed as part of this project, although the evaluation of the models that the interface addresses will be included.

A prototype interface (using the Geographical Resource Application Support System [GRASS] GIS where practical) is currently being developed as the initial step in the long-term development of a fully attributed GIS-WQ model interface (Theurer and Geter 1993). The prototype's purpose is to quickly provide: (1) a functional version of an interface to evaluate the data collection and data aggregation requirements, (2) automated input to the models on a limited scale, and (3) a method of model input for model component evaluation. An operational version of the interface subsequently will be developed with the user more in mind than the quick and less friendly approach of the prototype. The data requirements of the prototype input generation to the four models is the subject of this chapter, which is an update on progress reported in a paper presented at the Eighth Annual GRASS User Conference in March 1993 (Cronshey et al. 1993).

ARS Models Addressed by the Interface

The four ARS WQ computer models selected are:

- Agricultural Nonpoint-Source Pollution Model (AGNPS) (Young et al. 1987, 1993)
- Simulator for Water Resources in Rural Basins (SWRRB) (Arnold et al. 1990, 1991)
- Erosion Productivity Impact Calculator (EPIC) (Williams et al. 1990; Dumesnil 1993)
- Groundwater Loading Effects of Agricultural Management Systems (GLEAMS) (Knisel 1978; Knisel et al. 1992a, 1992b)

The first two models are watershed scale models and the latter two are field scale models. AGNPS is a single event model while the others provide continuous simulation over a user-defined period of time. AGNPS divides the watershed by superimposing a grid of square cells over the entire basin and evaluating each grid cell, routing results from one cell to another. SWRRB allows the model user to divide the watershed into a maximum of ten subbasins, based on smaller drainage pattern boundaries. Each subbasin is evaluated and independently routed to the watershed outlet. EPIC and GLEAMS, which consider only a single field, have limited routing capability and do not subdivide the field being analyzed. An assumption of homogeneity is made for each model view (cell, subbasin, or field) as appropriate. This assumption can be difficult on large drainage area watersheds, especially for SWRRB, because of the limitation on the number of subbasins.

DATA CATEGORIES

While most categories of data and some specific parameters required by each of these models are the same, many are different. The difference may be in the way data are entered (maximum and minimum temperature versus mean temperature), units (English versus metric), or the need for special parameters due to a specific subcomponent used in the model; for example, feedlot data are only considered in AGNPS.

Data can be divided into five categories: GIS layer information, source data, reference data, run time data, and derived data.

GIS Layer Information Data

GIS layer information data are the geographically referenced data that generally include an identifier that will point to other attributes. Depending on the GIS layer, an identifier can represent an area, a line, or a point. Seven required layers and one optional layer are used as basic data by the interface. The layer name, data type (area, point, or line), and identifier name(s) are:

- soils (area)—SCS soil mapping unit ID
- field (area)—field ID
- elevation (point)
- watershed boundary (area)
- subbasin or subwatershed (area)—subbasin ID
- geomorphic (area)—geomorphic region ID
- stream network (line)
- point data (point)—reservoir, pond, feedlot, point source, gully, terrace system, or sinkhole ID (optional)

The identifiers on the GIS layers are pointers to various parameters in the source and reference data categories.

Source Data

Source data include most watershed-specific data that are associated with one of the GIS layers. Source data are stored in a relational database (Informix 1990a, 1990b) keyed to the appropriate GIS layer identifier. Table 54-1 includes a list of the source data tables referenced to the GIS layers.

Table 54-1. Source data tables.

GIS Layer	Table
Soils	Soil surface data
	Soil layer data
Field	Land use, field characteristics, and supplemental table identifiers
Elevation	(No tables associated with layer)
Watershed boundary	Watershed characteristics
Subbasin or subwatershed	Subwatershed characteristics
Geomorphic	Geomorphic channel parameters
Stream network	Transportation system (future table) Bridges and culverts (future table) Diversions and waterways (future table) Local geomorphic channel parameter override (future table)
Point data	Feedlot data Point sink data Point source data Gully erosion Terrace system Pond data Reservoir data

The field attributes contain most of the data collected. Five of the attributes contained in the field data are identifiers (pointers) for supplemental tables that can be used on multiple fields in a watershed. These pointers are: noncropland, cropland operations, fertilization, pesticide, and irrigation. As the tables associated with these identifiers are not directly geographically referenced, they are treated as part of the next data category.

Reference Data

Reference data include information connected to the watershed only through reference to some parameter in the source or run time data. These data are not site specific as they can be referenced in several locations within a watershed or in several watersheds. Table 54-2 lists the reference tables and the key elements (in brackets) used to extract data from them.

Table 54-2. Reference data tables.

Channel type codes (AGNPS) [by channel type]
Climate reference data [by simulation period]
Cropland operations (rotation) [by cropland operations identifier][1]
Acceptable crop names
Crop name equivalence (EPIC and GLEAMS) Crop reference data [by crop name]
Fertilization application table [by fertilization identifier][1]
Fertilizer application methods Fertilizer (nutrient) reference [by fertilizer name]
Animal nutrient and COD data (AGNPS) [by animal name]
Irrigation application table [by irrigation identifier][1]
Land use parameters [by land use name]
Length slope (LS factor) [by length and slope]
Noncropland operations [by noncropland identifier][1]
Pesticide application table [by pesticide identifier][1]
Pesticide application methods Pesticide reference data [by pesticide name]
SCS runoff curve numbers [by land use, treatment, hydrologic condition, and soil group]
Soil reference data [by soil mapping unit ID]
Terrace P factors (AGNPS) [by terrace slope]
Acceptable tillage operations
Tillage operation equivalence (EPIC and GLEAMS) Tillage operation reference data [by tillage equipment or operation]
Planting method names
[1]Pointer from field source data, not directly geographically represented.

Several tables that are associated via reference to the field ID in the field characteristics table provide data that can be used repeatedly with a watershed. These tables are developed locally and could be considered as part of source data, but they are not directly geographically referenced. For this reason they are included as part of the reference data. Types of data (with examples) include:

- pesticide application (pesticide name, date applied, application method, and amount)

- fertilizer application (fertilizer name, date applied, application method, and amount)
- cropland operations (crop name, operation date, operation name, seeding rate)
- irrigation application (manual—date applied and amount; automatic—define conditions to apply, maximum annual amount)

Run Time Data

Run time data are parameters that vary so often that inclusion as part of the source data is not practical. Types of parameters include: which model to generate input for, starting date and period of simulation, initial conditions that may be dependent on chosen starting date, and so on. These types of data should be kept to a minimum, thus avoiding repeated entry of redundant information for consecutive interface executions. Depending on the water-quality model, one to five screens will be used to collect these data.

Derived Data

Derived data include all parameters that are calculated, aggregated over area, length, or time, or have units converted. Most parameters that are input to the models are derived. Calculated parameters use one or more of the source or reference attributes and a mathematical relationship. Data aggregation occurs over the evaluation area of each model (grid cell, subbasin, or field). The aggregation is generally a weighted area average. For some parameters, a weighted average does not make sense, so a dominant value is used; for example, the development of a synthetic crop to represent all the crops that may be growing in a subbasin at one time. The synthetic crop must be classed as either an annual or perennial. A discussion follows of some specialized aggregations that mostly involve SWRRB. Units conversions from English and metric or vice versa are made as needed.

Universal Soil Loss Equation (USLE) Parameters—USLE (Wischmeier and Smith 1978) is used by the models that the interface accommodates to predict soil loss. USLE parameters are derived from different GIS base layers: (1) soil erodibility factor (K) is based on the soils layer; (2) cover and management factor (C) is based on the field layer; (3) support practice factor (P) is also based on the field layer; (4) topographic factor (LS) is based on the elevation layer; and (5) rainfall and runoff factor (R) is a constant for the watershed based on the climate data used. All of the factors except R are aggregated based on potential erosion production. Using the GIS-based values for each of the parameters, an erosion layer is created and used to individually weight each of the parameters.

Cropping—Only three crops are allowed by the SWRRB model for the simulation period. A synthetic crop is generated based on aggregating the crop parameters (harvest index, potential heat units, water stress yield factor, etc.) associated with each field in the subbasin for a given year. Each year will probably have a different synthetic crop; therefore, only the first three years of a crop rotation can be handled by SWRRB.

Fertilizer—Only five applications are allowed by the SWRRB model during a year. As the model operates on a watershed subbasin and fertilizer applications are defined on a field basis, aggregation must take place over area (subbasin) as well as time. Each year is divided into five time periods: preplanting, postharvest, and three equal-length intervals during the growing season. The application date is weighted on the amount of nitrogen (N) and phosphorus (P) applied on a given date. Field-applied N and P are individually summed over the time period and adjusted for the entire subbasin area for each of the five application periods.

Pesticides—Only five pesticides are allowed by the SWRRB model during a year. Only the five pesticides with the largest applied amounts in the subbasin for a year are considered. Similar to fertilizer, each of these will be aggregated (along with date applied) over the five time periods.

Irrigation—Three types of irrigation can occur: manual by field (apply a certain amount of water on a set date); automatic (apply when plants are water stressed); or none. Only one type can be used on a subbasin at a time. As the field in the subbasin can be a mixture of all three types, some criteria are needed for which type to apply to the subbasin. The dominant type (manual or automatic) is determined for the irrigated fields in the subbasin. If manual is dominant, use it. If automatic is dominant, and 75% of the subbasin is irrigated, use it. Otherwise, there is no irrigation applied. Once the type is determined, the parameters associated with the type must be weighted by area.

Climate data—Daily values (temperature, precipitation, solar radiation, relative humidity, and wind speed) are retrieved based on the starting and ending dates for the simulation period. Monthly values (temperature, precipitation, solar radiation, relative humidity, dew point, and wind velocity) are based on the entire period of record for a nearby climate station. Note that many parameters are needed for both daily and monthly values, depending on the model selected. Some parameters are required for both time periods within a model.

Flow direction—The AGNPS model requires that flow direction be determined for each of the grid cells in the watershed. The automated flow direction process can be time consuming to execute and not 100% accurate. Inaccuracies that can be encountered are: collisions (two cells that flow toward each other); circularity (cells that form a flow loop); and multiple outlets from the watershed. Each of these situations must be addressed (and corrected) to establish a logically comprehensive grid cell flow network. Also the flow direction will generally not change unless either the elevation layer or the size of the AGNPS grid cell changes. As flow direction does not need to be regenerated each time AGNPS is run, it is handled as a preprocessor to the interface.

A Gaussian elimination approach is used to provide a plane of best fit for each cell. The slope plane generated (represented as northing and easting) is converted from an uphill slope direction to a downhill flow aspect and then ultimately to the AGNPS flow direction code. AGNPS uses one of eight directions representing the eight cells that surround the current cell for the code. Initially the conversion from aspect to flow direction code will allow 2° to represent the

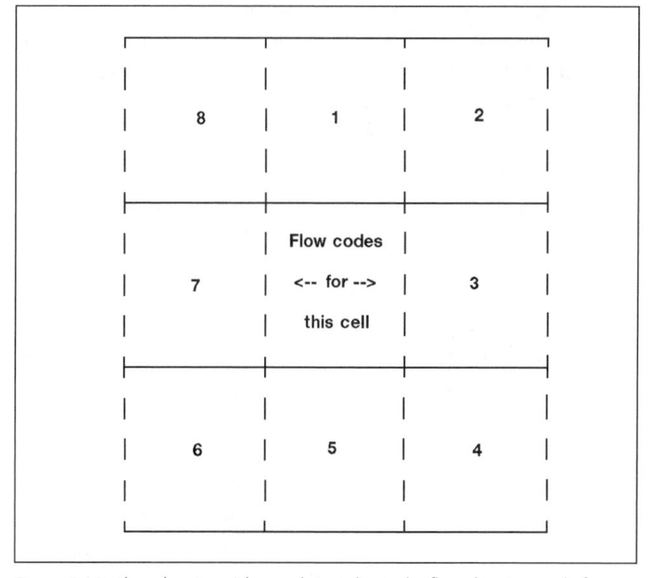

Figure 54-1. Flow direction. (The numbers indicate the flow direction code from the center cell.)

flow into diagonally oriented cells, and 88° for the horizontal and vertical located cells (Figure 54-1).

The approach for establishing flow direction used in the prototype is based on the land slope direction. Upland flow direction is generally in the direction of the land slope but once flow concentrates in a stream, it is more likely at right angles to the land slope. Thus, this procedure can lead to indicated directions that are 90° off from the true direction of flow. The flow direction determination will be reexamined, incorporating a stable, automated stream network generation that will eliminate this situation.

OPERATION OF THE INTERFACE

Initially it was expected that the basic GIS tool (GRASS) would be the "workhorse" in generating the derived parameters for the interface. Most of the GRASS commands used were in conjunction with the flow direction preprocessor. Commands and their uses:

- *r.volume*—extract watershed and subwatershed centroid
- *g.remove*—remove temporary maps
- *v.to.sites*—convert contour (vector) map to point elevation values
- *s.surf.tps*—elevation surface (raster) interpolation from point elevation values
- *r.mapcalc*—determine change (%) from original to edited version of flow map
- *d.vect*—display a vector map
- *d.rast*—display a raster map

While several GRASS commands are used, the bulk of interface processing is done by internal functions developed for extracting data directly from the Informix database. The approach taken was C language access to the database, generally using a GIS label or value to extract a single value, multiple values, or complete table records. This was only practical because GRASS is in the public domain; the source code is available to any user. Developed functions:

- *a.rast.edit*—Modification of *d.rast.edit* to include flow type error tracking codes (collisions, circularity, and multiple outlets) as part of the flow direction raster value. Also allows redisplay of revised flow direction while still in editor and can overlay stream and/or contour layers.
- *a.rast.arrow*—Modification of *d.rast.arrow* to strip out the error tracking codes in the raster values created by *a.rast.edit* to understand flow direction.
- *a.rast.zoom*—Modification of *d.rast.zoom* to fix apparent bug in the zoom-out function.
- *r.slope.aspect*—Portions of the GRASS code (read elevations and then generate percent slope parts of MAIN) used to create a percent slope layer. This layer in turn was used along with Informix table to generate a USLE LS factor layer.
- *r.reclass*—Some of the GRASS modular code (*parse*, *add_rule*, and *reclass*) with this command name was used internally for a reclass operation. Reports written in Informix Standard Query Language (ISQL) were used to develop reclassification rule tables. The rule logic includes simple mathematical functions and assignment of integer values. A large portion of the interface involved developing rule sets for various parameters and using this function.
- *get_stat*—Callable function to evaluate data attributes assigned to a "reclassed" layer. The layer is temporarily created for the evaluation of values within watershed boundary and the current model view. Several options are available in the function for returning a single value. One of the following can be returned: weighted average value, average, mode (dominant), or count (number of rasters) for a data class. This new function incorporates the functionality of *r.mask* and *r.stats* GRASS commands. The *get_stat* function was developed after experimenting with repeating a series of calls: *r.reclass*, *r.mask*, and *r.stats* for each AGNPS grid cell over many input parameters. The time required to accomplish this was too long considering the number of cells (1,000+) and parameters (20+) that normally would be used.
- *get_list*—Callable function to identify and accumulate a list of point items within the watershed boundary and the model view. A specific point data category (e.g., feedlot, gully pond, etc.) is specified for this function. A list of the values for the identified point category is returned.

USER VIEW OF INTERFACE

The interface operates in an active or background mode when generating the model input. The input generation is time-consuming (especially for the watershed models), and the background mode will free the computer terminal for other tasks. The interface computer will notify the user when the input has been generated. Active mode provides continuous on-screen display as to what step in the input generation the interface is currently working on.

The purpose of the prototype is to be functional and accurate, not glamorous or fancy. Thus, a text interface was used with simple fill-in-the-blank fields for data. The operational version will vastly improve in this area and will provide output display capability. The following are the steps the user must take to operate the prototype system.

1. Enter *Rice* to start interface.
2. From the WQ model list, select the model for input generation.
3. From the utility function list (*create*, *run*, *modify*, and *rename*), select create.
4. Provide the names for each of the seven basic GIS layers.
5. Enter model run time information (includes one or more screens, depending on model).
6. Select active or background mode of operation; input generation will begin.

CONCLUSION

By the time this book is published, Beta testing at the data collection sites should be well under way for the interface and the four models. Reports on the testing experience were due in September 1993. An operational version of the interface was anticipated in early 1994.

References

Arnold, J. G., J. W. Williams, R. H. Griggs, and N. B. Sammons. 1991. SWRRB—*A Basin Scale Model for Assessing Management Impacts on Water Quality*. Temple: Agricultural Research Service, Grassland, Soil and Water Research Laboratory.

Arnold, J. G., J. W. Williams, A. D. Nicks, and N. B. Sammons. 1990. SWRRB—*A Basin Scale Simulation Model for Soil and Water Resources Management*. College Station: Texas A&M University Press.

Cronshey, R. G., F. D. Theurer, and R. L. Glenn. 1993. *Water Quality Computer Model—GRASS Interface Prototype*. Washington, D.C.: RIGIS Division, Soil Conservation Service.

Dumesnil, D., ed. 1993. *EPIC User's Guide—Draft Version 2275*. Temple: U.S. Department of Agriculture, Agricultural Research Service.

Informix. 1990a. *INFORMIX SQL Reference Manual,* Version 4.0. Menlo Park: Informix Software, Inc.

———. 1990b. *INFORMIX SQL User's Guide,* Version 4.0. Menlo Park: Informix Software, Inc.

Knisel, W. G., ed. 1978. CREAMS: *A Field Scale Model for Chemicals, Runoff, and Erosion from Agricultural Management Systems.* U.S. Department of Agriculture, Conservation Research Report No. 26.

Knisel, W. G., R. A. Leonard, and F. M. Davis. 1992a. *The GLEAMS Model Plant Nutrient Component, Part I: Model Documentation.* Tifton: U.S. Department of Agriculture, Agricultural Research Service, Southeast Watershed Research Lab.

———. 1992b. *GLEAMS Version 2.0, Part III: User Manual.* Tifton: U.S. Department of Agriculture, Agricultural Research Service, Southeast Watershed Research Lab.

Theurer, F. D., and F. Geter, eds. 1993. *WQ Computer Model–GRASS Interface: Spatial and Attribute Data Needs.* Washington, D.C.: U.S. Department of Agriculture, Soil Conservation Service.

U.S. Army Corps of Engineers. 1991. *GRASS Version 4.0 User's Reference Manual.* Champaign: USA-CERL.

Williams, J. R., P. T. Dyke, W. W. Fuchs, V. W. Benson, O. W. Rice, and E. D. Taylor. 1990. EPIC—Erosion productivity impact calculator: User's manual. In *U.S. Department of Agriculture Technical Bulletin 1768,* edited by A. N. Sharpley and J. R. Williams. Tifton: U.S. Department of Agriculture, Agricultural Research Service.

Wischmeier, W. H., and D. D. Smith. 1978. Predicting rainfall erosion losses—A guide to conservation planning. In *U.S. Department of Agriculture, Agricultural Handbook No. 537.*

Young, R. A., C. A. Onstead, D. D. Bosch, and W. P. Anderson. 1987. *AGNPS, Agricultural Nonpoint-Source Pollution Model: A Watershed Analysis Tool.* U.S. Department of Agriculture, Conservation Research Report 35.

———. 1993. *Agricultural Nonpoint Source Pollution Model Version 4.0.* Computer program.

Roger G. Cronshey and *Fred D. Theurer*
Soil Conservation Service
U.S. Department of Agriculture
PO Box 2890
Washington, D.C. 20013

R. L. Glenn
U.S. Department of Agriculture
2625 Redwing Rd., Suite 110
Fort Collins, Colorado 80526

55

Model GIS Integration and Data Set Development to Assess Terrestrial Ecosystem Vulnerability to Climate Change

T. G. F. Kittel, D. S. Ojima, D. S. Schimel, R. McKeown, J. G. Bromberg, T. H. Painter, N. A. Rosenbloom, W. J. Parton, and F. Giorgi

Our ability to implement and analyze geographic and time-dependent model simulations is crucial for assessing the potential response of ecosystems to altered climate and CO_2 forcing over larger domains. The integration of ecosystem models, geographic information systems (GIS), and statistical analytical software can provide essential tools for managing and evaluating ecosystem model experiments. Needed capabilities are: (1) spatial and statistical tools to develop physically consistent data sets of boundary conditions and driving variables that cover large domains, yet realistically capture gradients and represent subgrid information; (2) methods to facilitate transfer and storage of model inputs and outputs that include model GIS interfacing, time-referenced storage in GIS, and time-series decomposition storage techniques; and (3) analytic techniques to statistically evaluate both the geographical and temporal nature of modeled ecosystem responses.

INTRODUCTION

Terrestrial ecosystem processes, including biogeochemical dynamics and vegetation structure, are potentially sensitive to rapid change in climate and elevated atmospheric CO_2 (Bazzaz 1990; Melillo et al. 1990; Neilson 1993). Understanding the vulnerability of ecosystems to altered forcing is important because of potential impacts on human systems and feedback to the climate system (Dickinson 1985; Jäger and Ferguson 1991; Pielke and Avissar 1990).

Crucial to such assessments is the ability to simulate transient ecological responses over regional to global domains. This process is limited in part by the inherent difficulty in handling large data sets in the implementation and evaluation of model runs. Integration of ecosystem models, GIS, and analytical software provides an essential tool to facilitate: (1) development of input data sets, (2) transfer and storage of model inputs and outputs, and (3) analysis of results.

In this chapter, we discuss these three processes in the context of linking models, GIS, and statistical applications. We start with a discussion of scientific issues regarding global change and ecosystems to illustrate the need for geo- and time-referenced modeling, data handling, and analytical capabilities. In the discussion of model input development, we present examples from the creation of data sets for simulation of United States ecosystems. In the following sections, we discuss innovations for data transfer and storage and address the requirements for statistical processing of model results. In chapter 77 of this volume, Bromberg et al. present the integration of a regional ecosystem model (CENTURY) (Parton et al. 1987, 1994) with a GIS and a statistical software package.

SCIENTIFIC ISSUES: NATURE OF ECOLOGICAL RESPONSES

Ecosystems are potentially sensitive to rapid changes in climate and elevated CO_2 through changes in ecosystem function and structure. Functional responses include altered rates of plant biomass production, evapotranspiration, soil decomposition, and nitrogen mineralization (Melillo et al. 1993; Ojima et al. 1993; Running and Nemani 1991). Structural changes include changes in vegetation height and density, altered species composition, and more extensive changes in dominant vegetation life-form (e.g., grassland to shrubwood or forest) (Neilson 1993; Urban et al. 1993; Webb 1987). Spatially explicit simulations are key to understanding the nature of these responses over continental or global domains. We expect the pattern of ecosystem response to be a function of the spatial distribution of altered forcing, initial state (such as ecosystem type and carbon pool sizes), and boundary conditions, including soil type (Pastor and Post 1988).

The response of some ecosystem components will likely lag others. Lags arise from: (1) interactions between soil biogeochemical and plant physiological processes with different response times, ranging from nearly instant to multidecadal time scales (Schimel et al. 1990); and (2) slow changes in vegetation composition that in turn influence biogeochemical dynamics (Pastor and Post 1988). These nonlinear interactions give rise to complex transient dynamics that have little resemblance to a linear path to a new equilibrium state. Time-explicit simulations and the ability to handle and analyze time-referenced output are crucial for understanding potential ecosystem trajectories. Transient ecological responses will depend both on internal system dynamics and the course of altered climate and CO_2 forcing.

The need to understand geographic dependence and time dependence in ecosystem responses gives rise to specific requirements for regional and global change modeling studies. These required elements are: (1) physically consistent model input data, (2) the ability to store and rapidly exchange large amounts of data between GIS and the model, and (3) spatial and temporal analytical capabilities.

Model Input Database Development

Accurate representation of the spatial distribution of driving variables and boundary conditions is needed for simulation of regional to global ecological dynamics. This can be difficult to achieve because of the paucity of extensive regional data and the coarse resolution needed to cover large domains. In addition, simulations require physical and spatial consistency across input data layers, such as among climate, vegetation, and elevation fields. This can be accomplished in part by using standard GIS functions (e.g., reprojection, resampling) to spatially coregister data sets. However, because the data are often from different sources—reflecting nonuniform sampling techniques, spatial resolution, and accuracy—physical consistency is not always assured.

For simulations covering the conterminous United States, we developed an integrated database of monthly climate, soils, and land cover (Kittel et al., n.d.). The grid used for this data set was a 0.5° latitude × 0.5° longitude interval grid covering the lower forty-eight states. We used two approaches to create spatially consistent inputs that capture important regional features of model boundary conditions and forcing variables. These are: (1) topographically based spatial interpolation, and (2) statistical representation of subgrid scale heterogeneity.

Topographically based interpolation of climate data—Because of the limited number of weather stations in mountainous regions, spatial interpolation techniques that account only for the horizontal placement of stations do not adequately create gridded fields of climate variables. This is because climate variables such as temperature and precipitation are strongly affected by topography. We applied alternative approaches that account for the effects of elevation on temperature fields and orography on precipitation.

To create a 0.5° grid of long-term monthly mean minimum and maximum temperatures for the United States, we used station means for a thirty-year period (1961–1990) from NCDC (1992). We adiabatically adjusted temperatures from station elevations to sea level using algorithms of Marks and Dozier (1992) to remove elevational effects in the data, creating fields that were smoother and more readily gridded. We gridded the sea level adjusted temperatures using inverse distance-square interpolation to the 0.5° grid and adiabatically readjusted these temperatures back up to grid point elevations (Plate 55-1a).

We based 0.5° grids of mean monthly precipitation on a 10-km gridded data set developed by Daly et al. (1994) using the PRISM model. PRISM spatially interpolates station data by: (1) stratifying topography (as represented in a digital elevation model) into facets with similar aspect, and (2) developing precipitation–elevation regressions for each facet. The same topographic stratification is used to extrapolate precipitation data throughout the domain on like facets. We degraded the 10-km data to the 0.5° grid by averaging. The 0.5° averages of the 10-km PRISM output are more representative of grid precipitation than if PRISM had been implemented with 0.5° terrain elevations because of the high degree of topographic heterogeneity that is missed on a 0.5° grid. This approach allows coarse gridded fields of precipitation to reflect subgrid scale topographic forcing on precipitation generation.

For climate inputs, scenarios of potential climate change are needed in addition to data sets of current climate. Scenarios for regional ecological change studies are generally derived from general circulation model (GCM) climate sensitivity experiments. The usual approach for generating scenarios is to modify current climatologies based on spatially interpolated changes in surface variables between control and altered (e.g., 2× CO_2) climate runs. While this approach can form the basis for evaluating the sensitivity of ecosystems to climate change, the GCM grid interval is too coarse for the scenarios to adequately reflect regional climatic sensitivity. Regional effects on climate sensitivity are mostly a concern where the land surface is highly variable, for example, in mountain regions, along coastlines, and in the vicinity of large lakes. In these areas, we expect that local climate change will be strongly influenced by regional processes.

An approach to account for regional effects is the nesting of regional atmospheric model runs within the coarse mesh of a GCM (Giorgi et al. 1994; Copeland et al. 1994). Giorgi et al. (1994) implemented a limited area regional climate model (RegCM, based on the PennState/NCAR mesoscale model MM4) on a 60-km interval grid covering the United States. Nested within the GCM grid, RegCM was initialized and driven by updated boundary conditions from a set of 1× CO_2 and 2× CO_2 GCM experiments (Thompson and Pollard 1995a, 1995b). Because it incorporates the mesoscale effects of orography and large lakes on regional climate, the nested model "dynamically" interpolates the large-scale forcing from the GCM runs to a spatial resolution more appropriate for assessing regional change.

Statistical representation of subgrid heterogeneity in soil and land cover type—Boundary conditions, such as soil and land cover type, commonly exert nonlinear control over ecological processes. When these variables vary in space at scales finer than the simulation grid, grid averages may not adequately represent existing conditions. For example, spatial discontinuities in soil texture occurring across short distances can have significant effects on system state and response tendencies (Burke et al. 1990; Pastor and Post 1988). When two distinct soil types occur in the same grid cell, averaging creates a set of soil properties that is not characteristic of any soil type present. An alternative is to represent the soil in a cell by dominant ("modal") soil types. Model experiments can be run either with the dominant soil or with a suite of dominant types. In the former case, model runs are driven by soil properties representative of an actual soil type. In the latter, weighting model outputs by the relative areal coverage of each soil type will produce simulation results representing the range of soils present.

We created a 0.5°-grid modal soils data set for the United States from Kern's (1994a, 1994b) 10-km gridded data based on Soil Conservation Service national level (NATSGO) polygons. We sampled all 10-km elements within each 0.5° cell and extracted the two to four dominant soil types and the percent area coverage of each type with a cell. While this process was accomplished external to a GIS, using C code linked to S-Plus with S+Interface/CASIM (StatSci, Inc.), it illustrates where tight coupling between GIS and statistical software could facilitate input data set development. Properties of the dominant soil type (Plate 55-1b) show important differences compared to average fields in the vicinity of edaphically distinct regions such as the Sand Hills of Nebraska, the Gulf coastal plains in the southeastern United States, and mountain zones. In source data sets where soils types are better resolved than the coarse NATSGO set, differences between modal and average representations should be more critical. The same process can be applied to land cover data based on such high-resolution cover data as the 1-km set of Loveland et al. (1991).

A consequence of independently choosing the dominant vegetation and soil types for a grid cell is that the selected vegetation–soil combination may not exist. To avoid this situation, identification of actual vegetation–soil type pairs can be made using bivariate histograms of the occurrence of vegetation and soil types within each cell (Figure 55-1). Simulations can be made for dominant vegetation–soil combinations and the results combined on an area-weighted basis. Such statistical treatment of spatial heterogeneity and spatial coherence among variables that nonlinearly control ecosystem processes allows implicit inclusion of subgrid information without explicitly increasing the resolution of model runs, which would increase the already high computational requirements of large domain simulations.

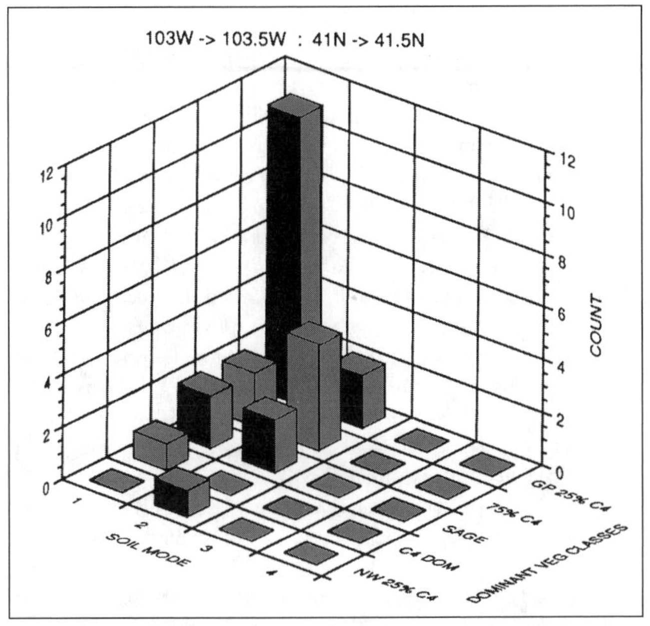

Figure 55-1. Bivariate histogram of soil and land cover types for a 0.5° cell centered on 41.25°N and 103.25°W in the vicinity of the Nebraska Sand Hills. The vertical axis is the number of 10-km cells with a given combination of soil and vegetation in the 0.5° cell. Texture of mode 1 soil is 32% sand and 47% silt; mode 2 is 41% sand and 37% silt. There were no soil modes 3 and 4 for this cell. Land cover types are: C4 DOM = grassland dominated by C4 (photosynthetic pathway) grasses; 75% C4 = grassland consisting of roughly 75% C4 and 25% C3 grasses; GP 25% C4 and NW 25% C4 = grasslands consisting of roughly 25% C4 and 75% C3 grasses; and SAGE = sagebrush shrub steppe. Land cover types are aggregated classes from Loveland et al. (1991).

Data Transfer and Storage Innovations

System level interface between GIS and models—Seamless integration of software responsible for storage of model data layers with model code is needed to manage the cumbersome amount of input and output data for regional and global simulations. Tight coupling of these two research tools, GIS and models, allows automated and rapid exchange of inputs and outputs between application and database management software. This is accomplished through a systems level linkage between applications, a graphical user interface (GUI), and underlying data structures. Bromberg et al. present the architecture of such linkages in Chapter 77.

Time-referencing GIS capability—The need to evaluate transient results also impacts data transfer, storage, and analysis requirements. For example, the manner in which time-dependent information is stored affects how easily results can be analyzed. Temporal statistical operations on model results stored as separate map coverages for each point in time, as is common for GIS, are awkward at best. Temporal analyses can be expedited if data layers are time referenced by a GIS in a manner comparable to geographic referencing. Temporal referencing treats maps of the same coverage that vary in time as a continuum, so that time functions, such as time averaging and interpolation, can be performed on them as a set (Beller et al. 1991; Bromberg et al. Chapter 77).

Output time-series decomposition—If an output series consists of repeated and overlying temporal patterns, time-series decomposition techniques can reduce output storage requirements (Schimel et al. 1994). Output time series from, for example, multidecadal runs (50–500 years or longer) can be decomposed into mid- and long-term trends (e.g., 5-year moving averages), seasonal cycle, other spectral components, and a residual series. This approach reduces output to be stored, if residuals are considered unimportant. Time series can be reconstructed from stored components for later analysis, and individual components such as the interdecadal trend can be analyzed separately.

Analysis of Model Results

Geographical analytical capabilities—Methods to evaluate spatial relationships in ecosystem model results include display of mapped responses, map differencing, map comparison statistics (e.g., Kappa Index of Agreement), and animation. Both visualization and statistical techniques are valuable for revealing and evaluating responses in terms of the spatial distribution of altered climate forcing, the system's initial state, and boundary conditions (e.g., vegetation and soil types). For example, stratification by vegetation type permits separate analysis of responses within an ecoregion (e.g., eastern temperate deciduous forest, northern mixed grass prairie). Such stratification can be used to determine average responses by vegetation type and to address questions such as whether responses to uniform forcing (e.g., elevated CO_2) are even across a region or dependent on initial water or nutrient limitation.

Some of these spatial functions are included in GIS. Other more complex statistical tasks can be accomplished externally through export of map files that are then brought into analytical software. However, this process becomes burdensome for large files and repeated cases and when GIS file structures are proprietary. This task is made more efficient through the enhancement of spatial analytical tools within a GIS or through two-way linkage of GIS and statistical software, such as S+GISLINK for ARC/INFO (StatSci, Inc.) and as described in Chapter 77.

Temporal analytical capabilities—Effective temporal methods for analysis of model output range from determination of time mean averages to more complex techniques for time signal decomposition (as discussed in the previous section) and visualization, including animation of a map series. Power spectra, autoregression, and trend analyses are useful methods to evaluate short- and long-period changes in model output time series. Multivariate techniques such as analysis of spectral coherence and phase can evaluate lag/lead relationships between variables, including those between forcings and system processes or between system components.

CONCLUSION

The integration of ecological models, GIS, and statistical software provides capabilities for implementation and analysis of large domain, transient ecosystem simulations. Crucial tasks facilitated by closer ties between these tools are: (1) development of physically consistent model input data sets, (2) smooth transfer of data between GIS and applications, and (3) analysis of model results in both geographical and temporal contexts. Use of climate interpolation schemes and statistical tools increases the spatial accuracy and consistency of data sets by accounting for the effects of topography and spatial heterogeneity in their representations of model inputs. Key GIS enhancements needed for coping with and exchanging large data sets are: (1) seamless interfaces to user-supplied models and statistical analytical applications, and (2) time-referencing capabilities comparable to georeferencing. Visualization and statistical techniques, ranging from determination and mapping of means and trends to more complex analyses of spatial and temporal relationships within and among model outputs, are valuable tools in evaluation of complex responses of terrestrial ecosystems under multivariate (CO_2 and climate) forcing. Improvements in, and integration of, these software systems will strengthen their application to environmental modeling studies in general and to global change research in particular.

Acknowledgments

This work was supported by an IBM Independent Research and Development project; Vegetation/Ecosystem Modeling and Analysis Project (VEMAP) sponsors (NASA Mission to Planet Earth, Electric Power Research Institute, USDA Forest Service Southern Region Global Change Research Program); UCAR's Climate System Modeling Program (with funding from NSF and DOE); the Model Evaluation Consortium for Climate Assessment (MECCA); the USGS-USFS TERRA Laboratory; and Genasys II. We are also grateful for assistance from StatSci, Inc. The National Center for Atmospheric Research (NCAR) is funded by the National Science Foundation. We thank Gary Bates, Aaron Beller, Donna Beller, Rob Braswell, Susan Chavez, Monica Engle, Hank Fisher, Melannie Hartman, Loey Knapp, Brian Newkirk, and Christine Shields for their valuable assistance. We gratefully acknowledge Roy Barnes, Chris Daly, Roy Jenne, Dennis Joseph, Jeff Kern, John Kineman, Thomas Loveland, Danny Marks, Dennis Shea, and Louis Steyaert for access to data sets and model output; Gaylynn Potemkin for help with manuscript preparation; and NCAR Climate and Global Dynamics Division for computer systems support.

References

Bazzaz, F. A. 1990. The response of natural ecosystems to the rising global CO_2 levels. *Annual Review of Ecology and Systematics* 21:167–96.

Beller, A., T. Giblin, K. V. Le, S. Litz, T. Kittel, and D. S. Schimel. 1991. A temporal GIS prototype for global change research. In *GIS/LIS '91 Proceedings*. Vol. 2.

Burke, I. C., D. S. Schimel, W. J. Parton, C. M. Yonker, L. A. Joyce, and W. K. Laurenroth. 1990. Regional modeling of grassland and biogeochemistry using GIS. *Landscape Ecol.* 4:45–54.

Copeland, J. H., T. Chase, J. Baron, T. G. F. Kittel, and R. A. Pielke. 1994. Impacts of vegetation change on regional climate and downscaling of GCM output to the regional scale. In *Regional Impacts of Global Climate Change: Assessing Change and Response at the Scales that Matter.* Proceedings of the 32nd Hanford Symposium on Health and the Environment.

Daly, C., R. P. Neilson, and D. L. Phillips. 1994. A statistical topographic model for mapping climatological precipitation over mountainous terrain. *Journal of Applied Meteorology* 33(2):140–58.

Dickinson, R. E. 1985. Climate sensitivity. *Advances in Geophysics* 28A:99–129.

Giorgi, F., C. S. Brodeur, and G. T. Bates. 1994. Regional climate change scenarios over the United States produced with a nested regional climate model. *Journal of Climate* 7(3):375–99.

Jäger, J., and H. L. Ferguson, eds. 1991. *Climate Change: Science, Impacts and Policy*. New York: Cambridge University Press.

Kern, J. S. 1994a. Spatial patterns of soil organic carbon in the contiguous United States. *Soil Science Society of America Journal* 58:439–55.

———. 1994b. Geographic patterns of soil organic carbon in the contiguous United States. *Soil Science Society of America Journal* 59(4):1,126–33.

Kittel, T. G. F., N. A. Rosenbloom, T. H. Painter, D. S. Schimel, and VEMAP Modeling Participants. The VEMAP integrated database for modeling United States ecosystem/vegetation sensitivity to climate change. *Global Ecology and Biogeography Letters*, n.d.

Loveland, T. R., J. W. Merchant, D. O. Ohlen, and J. F. Brown. 1991. Development of a land cover characteristics database for the conterminous United States. *Photo. Eng. Rem. Sensing* 57:1,454–63.

Marks, D., and J. Dozier. 1992. Climate and energy exchange at the snow surface in the alpine region of the Sierra Nevada: 2. Snow cover energy balance. *Water Resources Research* 28:3,043–54.

Melillo, J. M., T. V. Callaghan, F. I. Woodward, E. Salati, and S. K. Sinha. 1990. Effects on ecosystems. In *Climate Change: The IPCC Scientific Assessment*, edited by J. T. Houghton, G. J. Jenkins, and J. J. Ephraums, 283–310. New York: Cambridge University Press.

Melillo, J. M., A. D. McGuire, D. W. Kicklighter, B. Moore III, C. J. Vorosmarty, and A. L. Schloss. 1993. Global climate change and terrestrial net primary production. *Nature* 363:234–40.

National Climatic Data Center (NCDC). 1992. 1961–1990 Monthly Station Normals Tape. U.S. Department of Commerce, Data Tape TD 9641.

Neilson, R. P. 1993. Vegetation redistribution: A possible biosphere source of CO_2 during climatic change. *Water, Air, and Soil Pollution* 70:659–73.

Ojima, D. S., W. J. Parton, D. S. Schimel, J. M. O. Scurlock, and T. G. F. Kittel. 1993. Modeling the effects of climatic and CO_2 changes on grassland storage of soil C. *Water, Air, and Soil Pollution* 70:643–57.

Parton, W. J., D. S. Schimel, C. V. Cole, and D. S. Ojima. 1987. Analysis of factors controlling soil organic levels of grasslands in the Great Plains. *Soil Sci. Soc. Am. J.* 51:1,173–79.

———. 1994. A general model for soil organic matter dynamics: Sensitivity to litter chemistry, texture, and management. In *Quantitative Modeling of Soil Forming Processes*, edited by R. B. Bryant and R. W. Arnold. SSSA Special Publication. Madison: ASA, CSSA, and SSSA.

Pastor, J., and W. M. Post. 1988. Response of northern forests to CO_2-induced climate change. *Nature* 334:55–58.

Pielke, R. A., and R. Avissar. 1990. Influence of landscape structure on local and regional climate. *Landscape Ecology* 4:133–55.

Pielke, R. A, D. S. Schimel, T. J. Lee, T. G. F. Kittel, and X. Zeng. 1993. Atmosphere–terrestrial ecosystem interactions: Implications for coupled modeling. *Ecological Modelling* 67:5–18.

Running, S. W., and R. R. Nemani. 1991. Regional hydrologic and carbon balance responses of forests resulting from potential climate change. *Climatic Change* 19:349–68.

Schimel, D. S., T. G. F. Kittel, D. S. Ojima, F. Giorgi, A. Metherall, R. A. Pielke, C. V. Cole, and J. G. Bromberg. 1994. Models, methods, and tools for regional models of the response of ecosystems to global change. In *Proceedings, Sustainable Land Management for the 21st Century*, edited by R. C. Wood and J. Dumanski, 227–38. Lethbridge, Canada.

Schimel, D. S., T. G. F. Kittel, and W. J. Parton. 1991. Terrestrial biogeochemical cycles: Global interactions with the atmosphere and hydrology. *Tellus* 43AB:188–203.

Schimel, D. S., W. J. Parton, T. G. F. Kittel, D. S. Ojima, and C. V. Cole. 1990. Grassland biogeochemistry: Links to atmospheric processes. *Climatic Change* 17:13–25.

Thompson, S. L., and D. Pollard. 1995a. A global climate model (GENESIS) with a land surface transfer scheme (LSX). Part 1: Present climate simulation. *Journal of Climate.* 8(4):732–61.

———. 1995b. A global climate model (GENESIS) with a land surface transfer scheme (LSX). Part 2: CO_2 sensitivity. *Journal of Climate.* 8(5):1,104–121.

Urban, D. L., M. E. Harmon, and C. B. Halpern. 1993. Potential response of Pacific northwestern forests to climatic change: Effects of stand age and initial composition. *Climatic Change* 23:247–66.

Webb, T., III. 1987. The appearance and disappearance of major vegetational assemblages: Long-term vegetational dynamics in eastern North America. *Vegetatio 69*:177–87.

Timothy Kittel is Deputy Project Scientist at UCAR and Research Associate, Natural Resource Ecology Laboratory (NREL), Colorado State University.

Dennis Ojima is Research Scientist at NREL.

David Schimel is a scientist at NCAR and an NREL senior scientist.

Rebecca McKeown is Associate Scientist at NCAR and Research Associate at NREL.

William Parton is Senior Scientist at NREL.

Filippo Giorgi is a scientist at NCAR.
University Corporation for Atmospheric Research (UCAR)
National Center for Atmospheric Research (NCAR)
Box 3000
Boulder, Colorado 80307-3000
Fax: (303) 497-1695
E-mail: *kittel@ncar.ucar.edu*
and
Natural Resource Ecology Laboratory (NREL)
Colorado State University
Fort Collins, Colorado 80523

Nan Rosenbloom is a programmer at NCAR and graduate student, Department of Geology and Institute for Arctic and Alpine Research, University of Colorado, Boulder.
Department of Geology and Institute for Arctic and Alpine Research
University of Colorado
Boulder, Colorado 80309
Fax: (303) 497-1695

Jan Bromberg was a member of the GIS Center, IBM Federal Systems Company in Boulder and is now with Evolving Systems, Englewood, Colorado.

Thomas Painter was a programmer at NCAR; he is currently a graduate student in the Department of Geography, University of California, Santa Barbara.

56

Integrating Time and Space in an Environmental Model to Predict Damage from Army Training Exercises

Susan M. Cuddy, J. Richard Davis, and Peter A. Whigham

An integrated land management advice system (LMAS), using ARC/INFO GIS, a soil moisture model, and an expert system was developed and installed at Puckapunyal Army Range, Australia, to predict the environmental damage from training exercises. Both temporal and spatial modeling were required. It was not possible to develop LMAS entirely within the GIS for both efficiency and functionality reasons. This chapter describes the system design and identifies some shortcomings of current GIS that prevent truly embedded temporal and spatial modeling suitable for such environmental management tasks.

INTRODUCTION

The development of integrated software packages that combine different technologies to provide diverse features is not new. Management information systems and decision support systems, to name two common classes of software, fit this description. GIS—a relatively new environment for developing integrated software packages—provides powerful macrolanguages and databases for linking system components. However, techniques to achieve an integrated look within a GIS are often cumbersome and commonly rely on inelegant solutions such as using external files to pass data and/or commands between components.

This chapter traces the development of an "integrated" land management advice system (LMAS) that helps the Australian Army manage their premier armored training range. The rationale for the system design is given, together with descriptions of the components and how they are linked. We identify the spatial modeling facilities that were required but that were not available within the ARC/INFO GIS used for the project. (As this software is one of the leaders in the GIS world, we have assumed that such features are not generally available in commercial vector-based GIS).

Background

By the late 1960s, much of the Australian Army's armored vehicle training range at Puckapunyal, Victoria, was untrafficable. The heavy vehicle use combined with the generally poor condition of the land (from earlier clearing, rabbit plagues, and overgrazing) resulted in waterlogging, compaction, and extensive gully and tunnel erosion.

Committed to retaining the range for armored vehicle training, the army embarked on a major restoration program in 1970. This work, which included tree planting, pasture establishment, battering and filling gullies, building dams, and associated runoff controls, took almost fifteen years.

To sustain this resource (and protect their investment), the army contracted the CSIRO Division of Water Resources to research and develop software to assist with management of the range. Discussions with range management staff focused attention on the need for a model that could predict, at the time of booking, the likely environmental damage from future exercises and the cost of repairing that damage (Cuddy et al. 1990).

Scope of the Problem

Training exercises are typically booked up to eighteen months in advance. The number and type of vehicles participating are often not known at the time of booking, though these can be estimated from extent of area booked and level of exercise (troop, brigade, etc.). A brigade exercise may have in excess of 300 vehicles (tanks, armored personnel carriers [APCs], supply trucks, etc.) and may last up to ten days. Because of the logistics required to organize such exercises, they normally proceed, regardless of the weather or the condition of the range.

Typical training exercises have five phases: advance, withdrawal, defense, attack, and harboring/resupply. An advance, the most common phase, consists of many tracked vehicles (tanks and APCs) moving forward through a landscape corridor. This corridor bounds the movement of vehicles. It also defines the land features that will be traversed (or avoided). Vehicles proceed along an axis of advance across the width of the corridor, undertaking tactical maneuvers that provide protection against possible enemy action. Standard descriptions of these maneuvers are described in a tank driver's handbook. Figure 56-1 shows an exercise with vehicles using hills, vegetation, and other landscape features for tactical maneuvers that provide maximum safety through the corridor. Vehicle movements that are known to lead to environmental damage are of four types: (1) traversals (moving through an area as quickly as possible), (2) jockeying for observation positions, (3) hill ascent/descent to reconnoiter, and (4) creek crossings. Decisions on the most appropriate maneuver rely on identifying critical landscape features and noting their location relative to their surroundings. For example, if a vehicle is approaching a hill and there is no other hill within tactical range, then traversing to the hill and jockeying on the side of the hill may be the most appropriate maneuver.

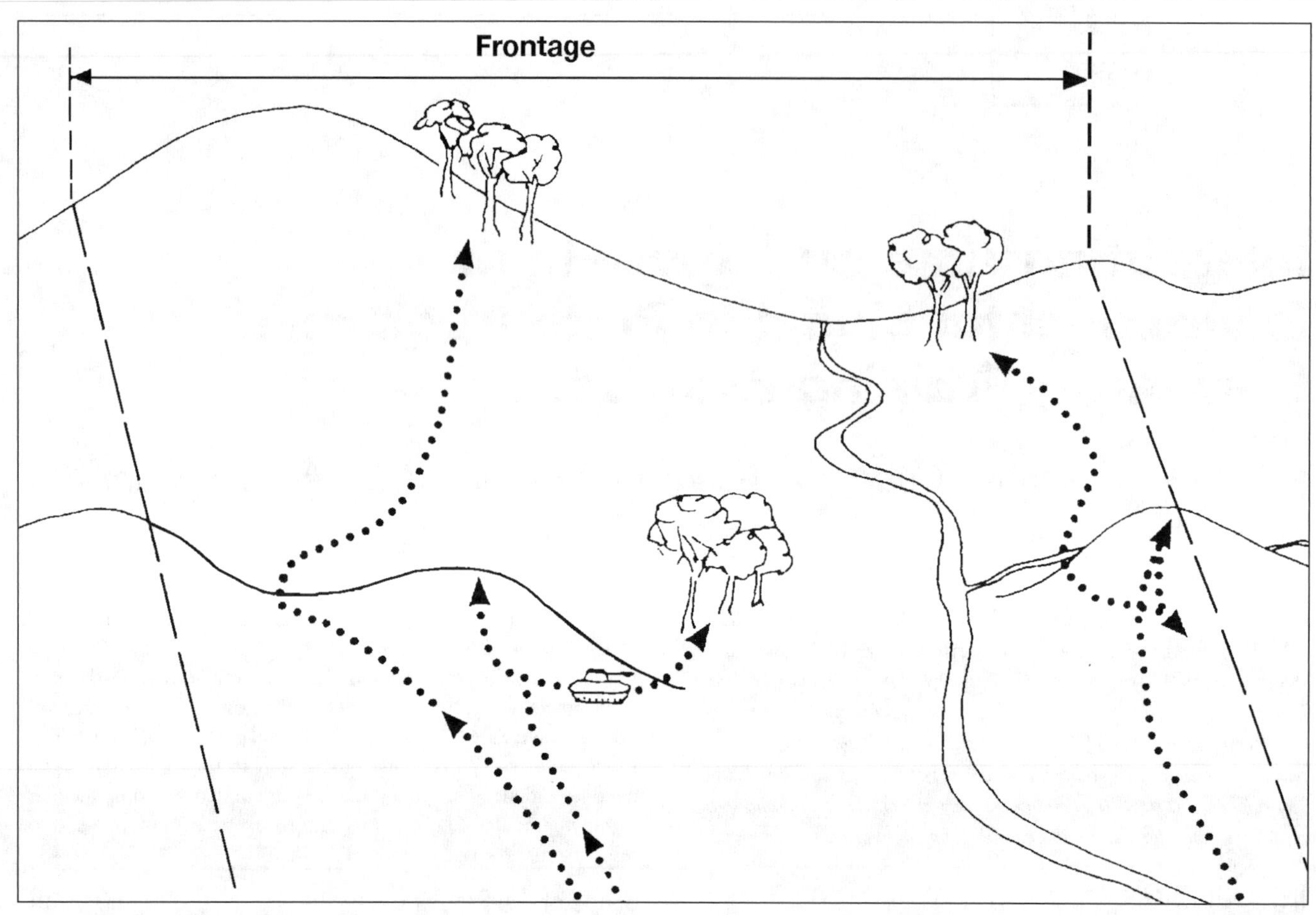

Figure 56-1. Example of vehicle tactical maneuvers within an advance corridor.

Vehicle maneuver decisions also rely on identifying features that change over time, such as the extent of saturated swampy areas. For example, a valley floor may be ideal for traversal in dry conditions; however, if waterlogged, it should be avoided.

The impact of an exercise, particularly during or after wet weather, can be severe. Loss of vegetation, streambank collapse, and severe rutting of the surface from wheel and tank tracks are common. Management in the aftermath of an exercise involves survey of damage and planning appropriate (and cost-effective) repair, with most attention being given to areas damaged by vehicle tracks. Severe damage must be repaired as soon as possible to minimize gully formation. Moderately damaged areas can wait until the following spring to be plowed and reseeded. Managers intuitively categorize damage after an exercise by the work required to repair the range.

Analysis of the problem showed that a computer system that could predict the extent and severity of ruts from tracked and wheeled vehicles and the consequent repair and cost would assist range managers in balancing the short-term needs of exercise commanders and the longer-term need of maintaining the range for training.

SYSTEM DESIGN

Design Constraints

The system design was influenced by the needs and experience of the end users. These were the range managers—career army officers with no scientific training. In 1988 when the project commenced, there was not a single PC installed at range control. Thus, the first design criterion was to develop a package that was easy to use and interesting enough to encourage exploration of both data and predictions by noncomputing professionals. The second criterion was that existing models should be used for prediction if possible. The Australian Army provided a model to predict soil moisture and soil strength for different soil texture classes (soil moisture/soil strength prediction [SMSP] model) (Rush and Kennedy 1982), and an engineering algorithm (Duell, personal communication) to predict rut depth as a function of soil strength from single or multiple passes of tracked and wheeled army vehicles. Both programs were written in FORTRAN 77. Third, the army purchased ARC/INFO version 5.1 for running on a SUN Sparc 1 workstation. This was the GIS that the prediction model had to access for spatial data, storage, and display.

Design Components

The LMAS comprises a number of components (Figure 56-2). The rainfall component generates rainfall scenarios by combining actual local rainfall with synthetic rainfall patterns. The SMSP model takes a selected rainfall scenario and calculates soil moisture and soil strength for the twenty-four soil types mapped on the range on a daily time step up to the date of the exercise. After the user has described an exercise (location of phases and number and type of vehicles involved), likely vehicle movements are modeled. A rut depth model takes these predictions of vehicle movement, together with the underlying soil strength calculated by SMSP, to predict depth and extent of ruts. The system assesses the damage and recommends appropriate repair action. Final outputs are a map of the damage and reports recommending repair action and likely costs.

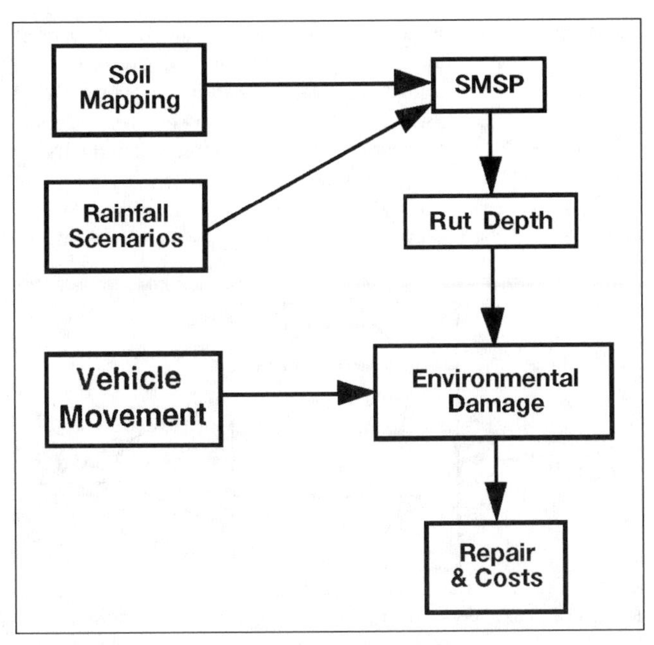

Figure 56-2. Components of the LMAS structure.

Topography and topology are important for modeling soil moisture and vehicle movement. The 42,400 ha of the range were subdivided into about 4,000 polygons (units), which were delineated at 1:50,000 scale. Soil type, land form, slope, wetness potential, and drainage were then recorded as (homogeneous) attributes of these units in the GIS.

Some parts of the problem were clearly qualitative—particularly vehicle movements—and these were best modeled using an expert system. ARX, a spatial expert system (Whigham and Davis 1989), has been developed specifically to represent spatial knowledge and infer conclusions from such knowledge. Environmental models may then be expressed as a set of spatial rules. For example, a spatial rule to model the tactical maneuver of a vehicle approaching a hill could be:

IF near a hill

AND vegetation cover is sparse

AND there are no other hills within 500 m

THEN move to hill.

The movement of vehicles along the axis of advance can be simulated by "slicing" the corridor into sections, each representing the field of opportunity of a vehicle commander. Likelihood of a maneuver could then be modeled as a function of the landscape features within the next slice (usually 100 m) and distance since last occurrence of the maneuver. Thus, a typical rule governing maneuvers could be:

IF near a hill

AND vegetation cover is sparse

AND there are no other hills within 500 m

AND (time since last jockey is 10 or more minutes

OR distance since last jockey is between 200 m and 500 m)

THEN move to hill and perform jockey.

The GIS can provide the spatial operations needed to instantiate the spatial operations in such rules.

Range managers need to know the location, size, and duration of saturated sites to assist with rescheduling exercises if conditions are unsuitable on booked sites. For example, if the system were to predict that damage would be excessive at a particular date, then it would be necessary to analyze the underlying soil moisture conditions. The SMSP model predicts soil moisture for specific soil types on a daily time step. To satisfy the rescheduling requirements, it was necessary that the computer system predict the temporal and spatial extent (in addition to the daily status) of soil moisture.

GIS Limitations

In this section we describe the three LMAS requirements that ARC/INFO software did not satisfy. These were spatial movement of vehicles, spatial extent of soil moisture, and temporal extent of soil moisture.

The spatial movement of vehicles can be implemented in ARC/INFO, though not in real time. Each field of opportunity (e.g., a "slice") can be represented as a rectangular polygon forward along the axis of advance from the current position (Figure 56-3). Then the land form and vegetation features that lie within the slice can be examined for suitable tactical position and maneuver. This requires the dynamic construction of rectangles and their overlay with two coverages (land form and vegetation). These overlays (and there are many—one for each vehicle for each slice) are transient during the modeling phases. Such frequent dynamic spatial overlay is slow in a GIS and impractical in a real-time system.

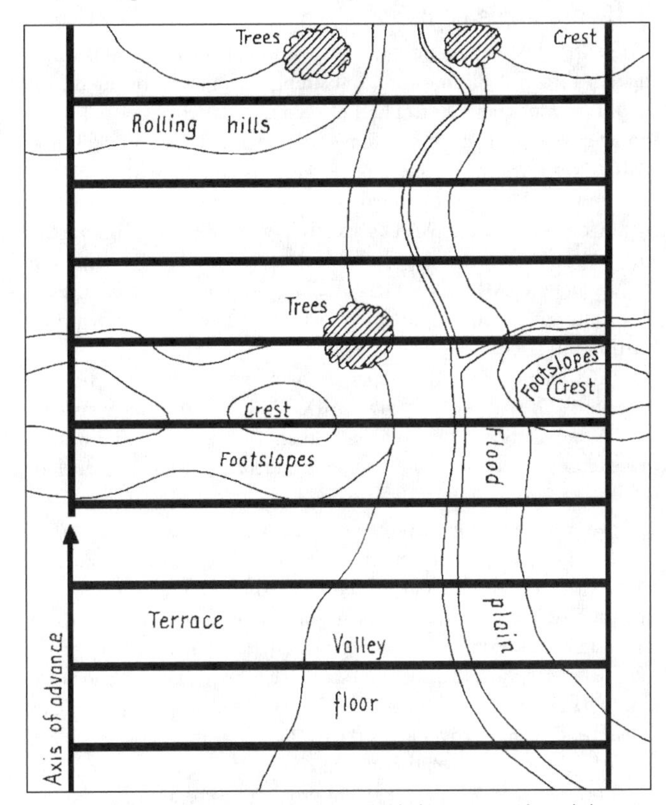

Figure 56-3. "Slicing" technique used to constrain vehicle movement. The underlying terrain is a two-dimensional GIS representation of the advance corridor of Figure 56-1.

As described earlier, soil moisture status is an important factor in scheduling training exercises. Predictions from the SMSP model need to be available for 4,000 polygons for each of potentially 540 days (if an exercise is booked eighteen months in advance). These moisture status values could be stored in 540 tables. However, analysis of the spatial extent of areas with similar moisture status would require these values to be mapped to the 4,000 units and an ARC/INFO DISSOLVE operation carried out to create a new "soil moisture unit" coverage for each day (with adjoining units with the same status forming new, larger units). It would not be feasible to generate this many coverages or perform so many DISSOLVE operations in real time.

In addition, knowledge of the occurrence and persistence of status for the original 4,000 units is required. The spatial and temporal extents of the soil moisture are tightly coupled, as areas of similar status would change shape over time because different soil types "wet up," or drain, at different rates.

In practice, only certain phenomena may be of interest (e.g., severely waterlogged areas); for example, the length of time that an area stays waterlogged may be of more interest than the length of time that it took to wet up. To answer such queries would require knowledge of persistence across coverages. ARC/INFO does not easily provide this facility.

Since range managers are really interested in the location, areal extent, and persistence of these "interesting" phenomena, they may be better modeled as spatial/temporal objects. Some authors (Gahegan and Roberts 1988) discuss object-oriented technology (OOT) for such temporal/spatial data. There are two issues in doing this: (1) storing existing spatial/temporal data in a way that allows the analyst to identify separate objects, and (2) dynamically recognizing objects from the output of models (fuzzy data sets).

Software Implementation

The entire LMAS was designed on a process basis from rainfall prediction through soil moisture, soil strength, vehicle movement, rut depth, to repair and cost. The final system combined several technologies—a spatial expert system (ARX), mathematical models, and ARC/INFO. Mathematical models were used as provided, on the assumption that they provided superior speed and precision compared to expert systems. The expert system was used where the information was essentially qualitative and unsuited to mathematical representation. ARC/INFO was used to record the spatial data, carry out some spatial operations, and provide the graphical and mapping display interface.

ARX readily allowed the spatial movement of vehicles to be represented. The rule syntax within ARX allows the dynamic construction of polygons and the rapid assessment of features from different coverages. Rules consist of a condition and a conclusion part, each of which use a quadruplet structure:

<parameter> <relation> <expression> <spatial expression>

The first three elements are as usually defined for a conventional rule-based expert system. The *<spatial expression>* identifies where the premise clause is to be evaluated or the conclusion clause is to be asserted, and its syntax supports target (spatial domain where resultant parameters are selected) and source (where spatial operations are applied) domains (coverages). In the previous example of an advance, the target coverage is the dynamic rectangle coverage, and the land form and vegetation are the source coverages.

The spatial and temporal extent problems were not overcome in the final system. Soil moisture and strength were predicted for each of the twenty-four soil types occurring on the range for the start date of a training exercise and stored in a table. This table was then linked as an external INFO table and related to the 4,000 polygons to give the spatial distribution of moisture for one day only. Thus, for an exercise scheduled up to eighteen months in advance, the SMSP soil moisture predictions were calculated daily but transferred to the GIS "soil moisture" coverage only at the time of the exercise. Thus, this communication occurs only once per run. This is a simple requirement and is handled adequately by the GIS, but it does not provide the spatial/temporal extent of wet areas.

By way of contrast, the communications between ARX and ARC/INFO are more frequent and complex. Three types of communication are involved (Figure 56-4): (1) some of the spatial searches required by the ARX rule syntax are performed by calling ARC routines; (2) parameter values in rules are instantiated by inspecting the INFO database (e.g., testing whether "vegetation is sparse" in a particular polygon is carried out by querying the appropriate record in the INFO database; and (3) when the goal of the expert system (e.g., assessment of damage) has been determined for all the polygons of interest, ARX calls on ARCPLOT routines to color and display the output map of damage.

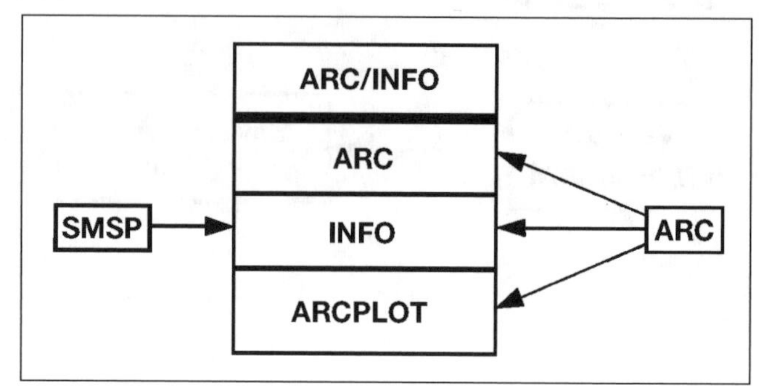

Figure 56-4. LMAS software structure.

ARX and ARC/INFO were integrated at source code level. SMSP and ARC/INFO were integrated via exchange of data and command files. ARC/INFO provided the graphical menu interface, database storage and manipulation, and map display capabilities. Because the ARCPLOT module of ARC/INFO version 5.1 did not have multiple active windows, this feature was provided by ARX. Most of the LMAS front end was written in the ARC/INFO macrolanguage (AML) with the FORTRAN and ARX programs executed as external programs.

CONCLUSION

Though GIS, as represented by ARC/INFO, has many features that make it a useful environment for implementing highly graphical packages that combine spatial and nonspatial data, it falls short in providing the range of tools needed for environmental modeling.

We have looked at the system design of a complex LMAS program developed within a GIS environment for managers of an army training range. The components have been described and their spatial and temporal modeling and storage needs identified. The use of other technologies, such as a spatial expert system, to address these needs has been explored. Of particular concern are the limitations of GIS to handle three fundamental requirements of the LMAS: (1) dynamic production and overlay of polygons, (2) spatial extent of "interesting" features, and (3) the temporal extent of these features.

However, GIS was a powerful environment for software development and provided all the tools necessary to develop an attractive program with links to external models handled transparently to the end user. The installation of LMAS at the range has provided management with a new tool to assist them in overseeing the range and its activities.

Acknowledgments

We wish to thank the Australian Army for funding this project and Terry Duell of the Mobility Terrain and Analysis Group, Engineering Development Establishment of the Australian Army, for providing the SMSP and rut depth models. We also wish to thank ESRI for their assistance in providing advice and code for the integration of ARX with ARC/INFO.

References

Cuddy, S. M., P. Laut, J. R. Davis, P. A. Whigham, J. Goodspeed, and T. Duell. 1990. Modeling the environmental effects of training on a major Australian army base. *Mathematics and Computers in Simulation* 32:83–88.

Gahegan, M. N., and S. A. Roberts. 1988. An intelligent, object-oriented geographical information system. *International Journal of Geographical Information Systems* 2(2):101–10.

Rush, E. S., and J. G. Kennedy. 1982. *Updated Soil Moisture–Soil Strength Prediction Methodology*. Vicksburg: U.S. Army Corps of Engineers Waterways Experimental Station.

Whigham, P. A., and J. R. Davis. 1989. Modeling with an integrated GIS/expert system. *Proc. ESRI User Conference.*

Susan Cuddy has a B.S. in pure mathematics from Queensland University and postgraduate qualifications in computer science from Canberra University. She worked as a mainframe database analyst before joining CSIRO in 1984 as a programmer. The Puckapunyal project commenced in 1988, and she provided project management and GIS and soil moisture model development and programming. Her current work is directed toward developing decision support systems for water resource managers.

Richard Davis earned his Ph.D. in physics from the Australian National University in 1973 and a B.S. from the ANU in 1975. He joined CSIRO in 1976 and worked on the early development of GIS and land use planning methodologies. Since then he has developed software packages for rural land use planning, wildfire management, and catchment management using techniques from artificial intelligence. His current work is focused on integrating models, GIS, and expert systems for a range of environmental management problems.

Peter Whigham obtained a B.S. in 1983 in computer science at the Australian National University. He joined CSIRO in 1987 and undertook the substantial design and development of the ARX spatial expert system. He has been involved in applications of ARX to remote area trafficability and land management of army bases. He is currently studying for a Ph.D. in computer science at the University of New South Wales.

CSIRO Division of Water Resources
GPO Box 1666, Canberra ACT
Australia 2601
Fax: 61 6 246 5800
E-mail: *sue@cbr.dwr.csiro.au*

57

Dynamically Determined Parameters for Urban Energy and Water Exchange Modeling

C. S. B. Grimmond

A methodology is presented to link meteorological source area models to geographic information systems (GIS). The objective is to define surface parameters for site descriptions for measurement and modeling studies in an objective way to ensure spatial consistency between measured and modeled data domains. The approach is illustrated with two applications in Sacramento, California. The surface properties of the areas influencing radiosonde and tower-based eddy correlation energy balance measurements are determined, along with variable surface properties of the latter, and are documented with measurements taken during a seven-day period in the summer of 1991. The results and their implications are considered for site description and model input.

INTRODUCTION

The active surface of any system is one of its most important climatic determinants because it is the primary site of energy, mass, and momentum transfer and transformation. Climatological and meteorological measurement and modeling studies both require the surface datum to be defined and described for the following reasons: (1) to characterize the site where measurements have been conducted, (2) to provide input for numerical models, or (3) to ensure spatial consistency between measured and modeled data. In model evaluations, it is essential that surface parameters (the model domain) represent the same surface area as that for which the measurements were conducted (the measurements' source area).

The conventional approach in meteorological studies of urban areas to surface description is static. One description is used to describe a measurement site or to provide input to a numerical model for all periods of measurement and modeling. Typically, this description is based on the mean parameter for a circle around the measurement site. This approach assumes that the surface is spatially homogeneous and/or that over time, variations in wind direction will create spatial averaging and that variability at time scales less than this period are not of interest. In reality, there are preferred wind directions and meteorological conditions and hour-to-hour variations in the fluxes, and thus the areas they represent are of interest. This becomes even more important as the surface becomes more complex. Consequently, the surface area contributing to meteorological measurements at any point at a given time is of interest.

The source area for meteorological measurements is dependent on the physical process involved, the instrumentation used, and the meteorological conditions under which the measurements occurred. For radiant fluxes, the source area is fixed in time by the field of view of the instruments, that is, by geometry. This source area can be determined using procedures outlined by Reifsnyder (1967) and Schmid et al. (1991). For the turbulent fluxes, the source area is not fixed but varies through time: as a sensitive function of sensor height, atmospheric stability, and surface roughness, in that order of importance. A number of numerical models, based on boundary layer diffusion theory, have been developed to determine the dimensions, weighting, and areal extent of the source area of turbulent measurements (Gash 1986; Schuepp et al. 1990; Leclerc and Thurtell 1990; Schmid and Oke 1990; Horst and Weil 1992).

This chapter presents a methodology to link source area models for turbulent fluxes to a surface database within a GIS. It is illustrated with respect to Sacramento, California, the site of an intensive urban climate measurement program (Grimmond et al. 1993). In the Sacramento study there are two main objectives underlying the linkage of the source area model to the GIS: (1) to provide a basis for interpreting the flux measurements in terms of the surface features influencing them, to assess their spatial representativeness, and to provide a basis for combining different measurement techniques (in this case, eddy correlation energy balance and radiosonde-based estimates of surface energy and water exchanges); and (2) to provide model input for determining surface parameters that are spatially consistent with the measured data used to evaluate numerical boundary layer models.

Two examples of the application of the source area GIS methodology are presented here:

1. The surface properties of the source areas of eddy correlation energy balance measurements and the implications of their variability through time considered in the context of site descriptions and model inputs.
2. A comparison of the surface properties of the source areas of eddy correlation energy balance flux measurements and radiosonde-based estimates for the same times.

Full descriptions of the meteorological measurements, instrumentation, and site are presented in Grimmond et al. (1993) and Cleugh and Grimmond (1993).

METHODOLOGY

Source Area Model for the Fixed-Tower Measurements

A full description of the source area model (SAM) used for the turbulent measurements is provided by Schmid and Oke (1990). This model was selected because it is a three-dimensional model, originally developed for the scale of urban areas. SAM is based on a probability density function plume dispersion model. Simply, it can be visualized using Figure 57-1a. Instead of calculating the dispersion from a point (e.g., a smoke stack) to an area, the "reciprocal" is calculated, that is, the area influencing that point (in this case the instruments on the measurement tower). The SAM output consists of ten weighted ellipses aligned upwind of the measurement tower along the mean wind direction for the period of interest (Figure 57-1b). The surface area between each ellipse has equal influence on the measurements (Figure 57-1c). All source area dimensions increase as the stability changes from unstable to neutral to stable conditions (Pasquill 1972).

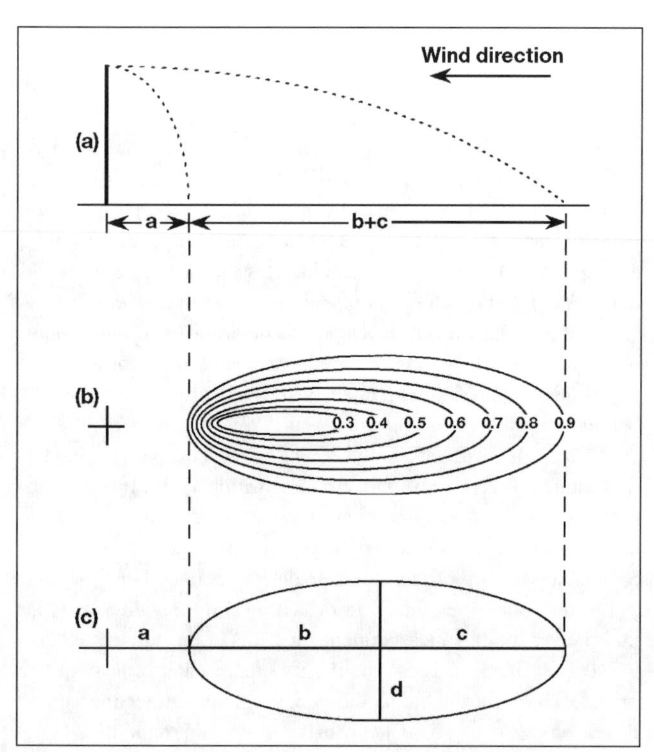

Figure 57-1. Schmid and Oke (1990) source area model output of weighted ellipses: (a) side view, (b) and (c) plan view. (Note: not all ellipse weighting levels are illustrated.)

The inputs necessary to determine the weighted source areas from the model are: the measurement height of the instruments; the site conditions (roughness length, z^0 and zero plane displacement, *d*); the meteorological scaling parameters (Obukhov stability length, *L*, and friction velocity, u^*); and the wind direction, its standard deviation, and mean wind speed.

Source Area for the Radiosonde-Based Measurements

In the Sacramento study, sensible and latent heat fluxes were also estimated from temperature and humidity profiles measured in the convective boundary layer from radiosondes (Cleugh and Grimmond 1993). Of interest is the similarity between these and the tower-based flux measurements and the spatial representativeness of such estimates. The source area of the radiosondes is aligned with the wind direction. The upwind length dimension is determined from Raupach's (1991) length scale of convective boundary layer (CBL) adjustment, based on horizontal and convective velocities and CBL depth. The width of the source area is based on the standard deviation of the wind direction. Here the radiosonde source areas are only considered while they remain in the CBL.

The Surface Database: Sacramento, California

The georeferenced database was developed at three spatial scales (regional, local, and micro) illustrated in Figure 57-2. The full methodology is outlined by Grimmond and Souch (1994). The spatial scale of interest in the Sacramento study is the local scale (Oke 1984): length dimension 10^2–10^4 m; time dimension 10^3–10^4 s.

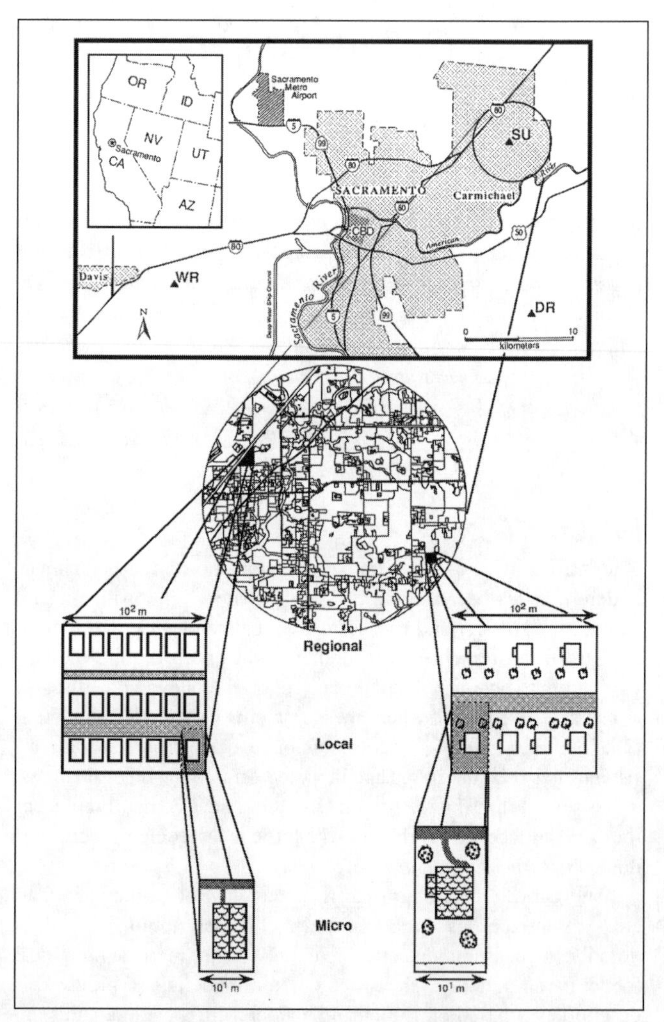

Figure 57-2. Schematic representation of the nested database methodology and the location of the Sacramento study area.

The study area is centered on latitude 38°29' 33.28" N, longitude 121°19' 2.01"W; the site of a pneumatic tower used to collect micrometeorological flux measurements (Grimmond et al. 1993). Land use for a 5-km radius circle around the pneumatic tower was mapped using aerial photography (1:12,000, Geonex Sacramento, flown April 7, 1991). The primary criteria for differentiating between land use units were building dimensions/density and tree cover dimensions/density and surface properties significant in surface water and energy flux partitioning. A total of fifty land use classes were mapped, which produced 1,345 polygons. The map was digitized using ARC/INFO.

More detailed "local scale" (Figure 57-2) information was collected on surface cover and building densities using randomly located 200-m × 200-m grid squares on the aerial photographs. Based on replicates within each land use category, mean densities and percent plan area cover were calculated for each land use category. Field surveys provided additional information on surface cover (e.g., building heights, roof materials, tree species, etc.) at the scale of the individual lot. All three levels of information are linked to provide information on the spatial distribution of surface attributes across the study area; for example, building heights, surface materials, tree cover, etc. Figure 57-3 presents an example of the spatial distribution of percent plan area impervious (concrete, asphalt, roofs, etc.).

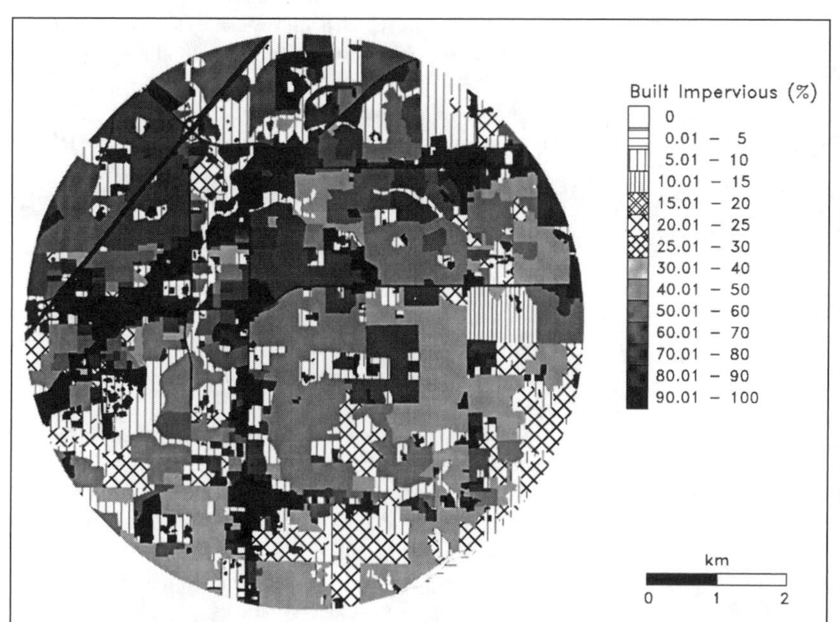

Figure 57-3. Percent plan area impervious across the study region, Sacramento, California, based on land use polygons and surface attribute information from detailed aerial photograph analyses.

Linking the Source Area Model to a GIS

Additional subroutines were added to the Schmid and Oke (1990) model to output the actual ellipse boundaries, aligned along the mean wind direction, in universal transverse mercator (UTM) grid coordinates. This format can be imported into Atlas GIS (with the aid of Atlas Import/Export) or GRASS, the software used in subsequent analyses. The cord length (the distance between two consecutive points on the ellipse) was set between 10 and 20 m. The shorter distance was preferable because it allowed each source area ellipse to be more accurately defined and prevented ellipse polygon boundaries crossing each other, especially in the vicinity of the measurement tower. However, the shorter cord length generated polygons that frequently had too many coordinate pairs for a single Atlas GIS polygon. This was particularly true for stable/neutral conditions when larger source areas occur; thus, a 15-m–20-m cord length was most commonly used. This is not a constraint in GRASS.

Using a series of Atlas GIS or GRASS commands, the land use layer polygons were split with the ellipse layer polygons to create a new layer so that the parameters of interest could be determined. Figure 57-4 illustrates three source areas for the fixed tower superimposed on the land use polygons. The influence of changing wind direction and atmospheric stability can be seen clearly.

RESULTS

Variability in Average Surface Attributes of Source Areas for the Fixed-Tower Measurements

During August 1993, intensive micrometeorological flux measurements were conducted at the suburban site, Carmichael, in Sacramento (Grimmond et al. 1993) (Figure 57-2). Figure 57-5 contains the results from the linkage of the Schmid and Oke source area model to the Sacramento GIS for the periods when radiosondes were released. The results presented are average surface properties (proportion of plan area cover): buildings, vegetation (trees, grass, and shrubs), pavement (roads, sidewalks, and parking lots), open water (lakes, rivers, and pools), and other (sand, scrub, and dirt) in the source area of the tower-based energy balance measurements. The measurements, taken over a seven-day period, represent a range of meteorological (wind and stability) conditions. The variability evident in the results suggests the potential for significant changes in surface controls on flux partitioning through this period, which must be considered when these data are analyzed. Furthermore, when these measured data, which are representative of varying surface conditions, are used to evaluate numerical boundary layer models, changes in the model surface parameters need to be incorporated.

Comparison of Source Area Properties for Fixed-Tower and Radiosonde-Based Measurements

Figure 57-6 illustrates the source area for the tower-based eddy correlation and radiosonde-based measurements for August 24 at 1130. At this time the predominant wind was from the south. The atmospheric conditions were unstable, hence

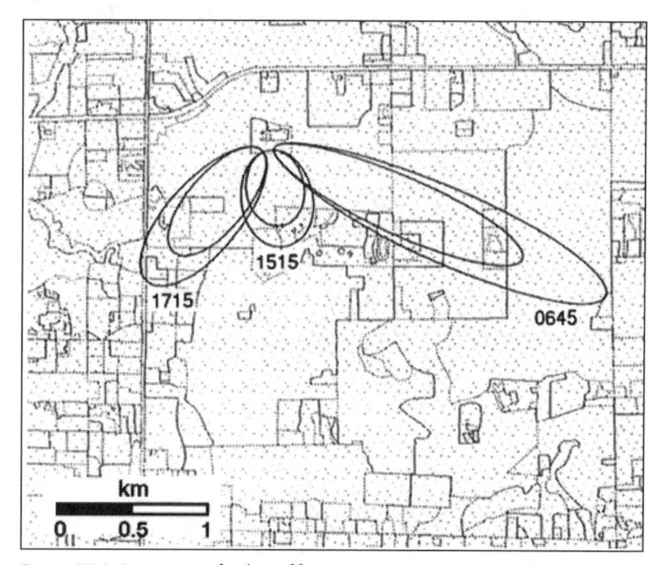

Figure 57-4. Source areas for three, fifteen-minute measurement periods, August 28, 1991. The times of the measurements each source area represents are indicated.

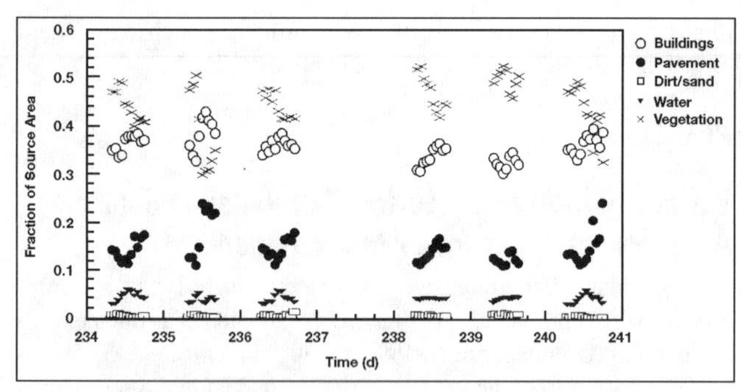

Figure 57-5. Average surface attributes (proportion of plan area cover) in eddy correlation source areas within 5 km of the measurement tower for radiosonde release periods (August 22: 234–August 28: 240).

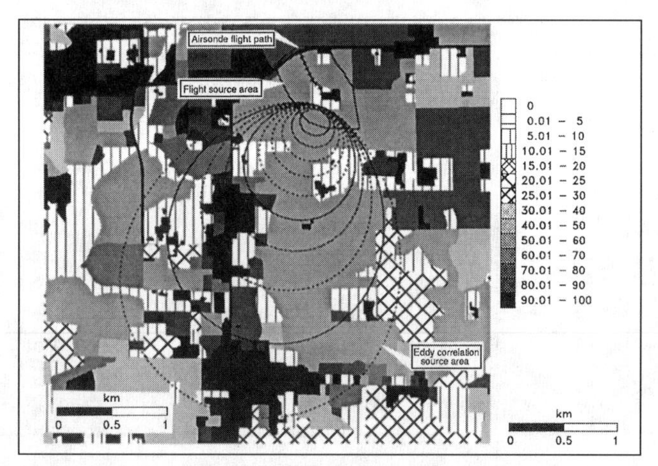

Figure 57-6. Source areas for fixed tower-based and radiosonde measurements at 1130, August 24, 1991. The radiosonde track is shown with the railway tie symbol, the radiosonde source area is delimited with a solid line, and the ten source area ellipses with dotted lines. These are superimposed on the percent plan area impervious map.

the source areas calculated for both the fixed-tower-based measurements and the radiosonde before it left the boundary layer are small. Surface attributes calculated for the two source areas indicate 52% and 42% vegetated, and 48% and 58% built for the tower-based and radiosonde measurements respectively. In this case, the surface properties of the source areas influencing the measurements are quite different, and the flux measurements should be interpreted accordingly.

CONCLUSION

The linkage of meteorological source area models to GIS provides an objective method for site description for both measurement and modeling studies. Here, this methodology has been used to determine both the properties of source areas for different measurement approaches and the variability in the surface properties influencing one set of measurements through time. The results of such analyses can be used to explain differences and similarities in measured fluxes through time and to permit the assignment of appropriate surface parameters for modeling studies.

Acknowledgments

Special thanks to Dr. T. R. Oke (UBC, Canada) and Dr. H. A. Cleugh (Macquarie University, Australia) for their assistance with the fieldwork and their support of the research; Dr. C. Souch (IU, Indianapolis) for assistance with the database and useful discussion; Dr. H. P. Schmid (ETH, Switzerland) for providing version 2 of SAM; Ms. A. Johnson for assistance with aerial photography mapping; and Mr. J. Pitts for assistance with program development. This research was supported by an IU summer faculty fellowship.

References

Cleugh, H. A., and C. S. B. Grimmond. 1993. A comparison of measured local scale suburban and areally averaged urban heat and water vapor fluxes. *Exchange Processes at the Land Surface for a Range of Space and Time Scales.* IAHS Publ. 212:155–63.

Gash, J. H. C. 1986. A note on estimating the effect of a limited fetch on micrometeorological evaporation measurements. *Boundary Layer Meteorol.* 36:409–14.

Grimmond, C. S. B., T. R. Oke, and H. A. Cleugh. 1993. The role of "rural" in comparisons of observed suburban–rural flux differences. *Exchange Processes at the Land Surface for a Range of Space and Time Scales.* IAHS Publ. 212:165–74.

Grimmond, C. S. B., and C. Souch. 1994. Surface description for urban climate studies: A GIS-based methodology. *Geocarto International* 9:47–59.

Horst, T. W., and J. C. Weil. 1992. Footprint estimation of scalar flux measurements in the atmospheric surface layer. *Boundary Layer Meteorol.* 59:279–96.

Leclerc, M., and G. W. Thurtell. 1990. Footprint predictions of scalar fluxes using a Markovian analysis. *Boundary Layer Meteorol.* 52:247–58.

Oke, T. R. 1984. Methods in urban climatology. In *Applied Climatology*, edited by W. Kirchofer, A. Ohmura, and W. Wanner, 19–29. Zurich: Zurcher Schriften.

Pasquill, F. 1972. Some aspects of boundary layer description. *Quart. J. Roy. Meteor. Soc.* 98:469–94.

Raupach, M. R. 1991. Vegetation–atmosphere interaction in homogeneous and heterogeneous terrain: Some implications of mixed layer dynamics. *Vegetatio* 91:105–20.

Reifsnyder, W. E. 1967. Radiation geometry in the measurement and interpretation of radiation balance. *Agricultural Meteorol.* 4:255–65.

Schmid, H. P., H. A. Cleugh, C. S. B. Grimmond, and T. R. Oke. 1991. Spatial variability of energy fluxes in suburban terrain. *Boundary Layer Meteorol.* 54:249–76.

Schmid, H. P., and T. R. Oke. 1990. A model to estimate the source area contributing to turbulent exchange in the surface layer over patchy terrain. *Quat. J. Roy. Meteorol. Soc.* 116:965–88.

Schuepp, P. H., M. Y. Leclerc, J. I. McPherson, and R. L. Desjardin. 1990. Footprint prediction of scalar fluxes from analytical solutions of the diffusion equation. *Boundary Layer Meteorol.* 50:355–74.

Sue Grimmond is an assistant professor in the Climate and Meteorology Program at Indiana University. She earned her B.S. at the University of Otago, New Zealand, and her M.S. and Ph.D. at the University of British Columbia, Canada. Her research interests are in measurement and modeling of energy and water exchanges in heterogeneous terrain, in particular, urban areas.

Climate and Meteorology Program
Department of Geography
Student Building 120
Indiana University
Bloomington, Indiana 47405
Phone: (812) 855-7971
Fax: (812) 855-1661
E-mail: *grimmon@ucs.indiana.edu*

58

GIS as an Integrative Tool in Climate and Hydrology Modeling

Mercedes N. Lakhtakia, Douglas A. Miller, Richard A. White, and Christopher B. Smith

The current generation of climate and hydrology models requires new approaches to the management, analysis, and visualization of model input and output. GIS provides an integrative framework that meets most of these requirements.

CLIMATE AND HYDROLOGY MODELS AND GIS IN THE EARTH-OBSERVING SYSTEM ERA

Humanity's concern about global climate change and its impact on society are well documented. Perhaps in no other component of the earth system are these concerns more critical than in the area of water resources. How will a changing climate impact a growing population that requires water? What information will be available upon which to base decisions regarding the allocation of water resources on local, regional, and global scales? In this context, the scientific community seeks a unified understanding of coupled climate and hydrology processes at all scales.

Climate and hydrology models, coupled with observations, are important tools for developing a clearer understanding of the controlling elements in the earth system. These models are becoming more complex and require increased amounts of information about the nature and properties of the earth system. The launch of the NASA-sponsored Earth Observing System (EOS) in the late 1990s will result in nearly 15×10^{15} bytes of new data. Much of this information will be used directly to initialize climate and hydrology models. Greater strides in linking GIS must be made if the scientific community is to be successful in creating realistic models of climatic and hydrologic processes.

As a part of EOS, the Earth System Science Center (ESSC) at Penn State is conducting an interdisciplinary investigation of the global water cycle. This research program focuses on the processes controlling the movement of water in the earth system from a global, to a regional, to a local scale. A series of study sites, in areas ranging from humid temperate regions like the northeastern United States to New Zealand in the South Pacific, have been selected for the fifteen-year research program. The Susquehanna River basin (SRB), a 62,419-km^2 watershed covering portions of New York, Pennsylvania, and Maryland, has been chosen as the study site for the initial experiment (also known as SRB experiment, or SRBEX).

The goal of SRBEX is to link and test a hierarchy of environmental models (a subset of these models is shown in Figure 58-1), using data from field studies for calibration and verification, and then, in conjunction with EOS observations, to produce information on physical and biological variables and process rates. The key objective is to document the water cycle (sources, sinks, and flux rates) to understand its variation in the past and to predict future conditions.

THE ROLE OF GEOGRAPHIC INFORMATION SYSTEMS

Context

The movement of water and energy in the earth system is controlled to a large extent by the distribution of land forms, surface cover, and soils. These features change in space and time. Climatologists and hydrologists have long recognized that in order to realistically simulate the controlling processes, information on the characteristics of these elements is required. Prior to the development of GIS, the modeling community developed much of its own software to process the sparse spatial information available. These "preprocessing" programs were generally used to reformat existing data to meet model

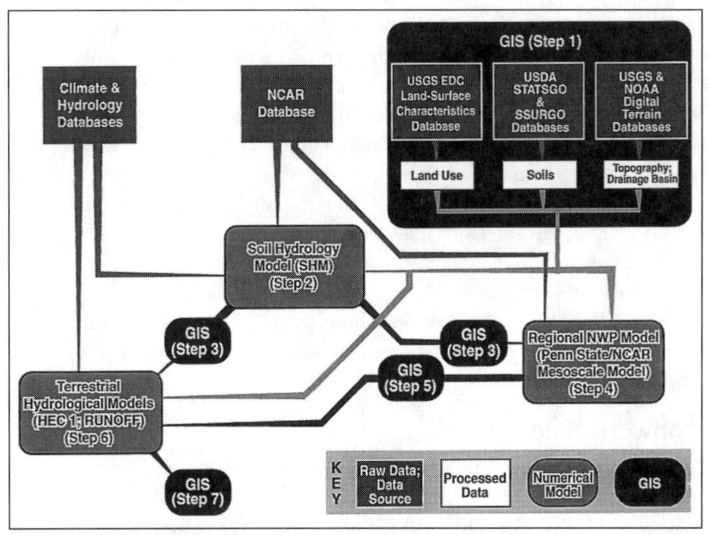

Figure 58-1. Subset of SRBEX databases and environmental models.

requirements. GIS research and development in the early 1980s made available, for the first time, software (both public domain and private sector products) for management and analysis of spatial data. The capabilities of these software packages far outstrip the "bare bones" approach taken by model developers.

Some members of the climate and hydrology modeling community have embraced GIS more fully than others. A number of factors appear to be responsible for this, including computing environments, model complexity, and idiosyncrasy.

Interdisciplinary science teams, who want to study a system like the water cycle in an interactive way, find themselves at various stages of expertise with regard to GIS. The key in this regard is to discover the best way to integrate models using a GIS environment without wasting time and resources on items such as duplicated software development and incompatible data exchange formats. In the following sections, the outline of the current SRBEX research program is presented, and some of the ways in which these problems are being approached are discussed.

Approach

Since the very beginning of SRBEX, GIS has been acknowledged to be central to the implementation and linkage of the models. In fact, GIS is the "medium," or environment within which the models can be linked in one-way and ultimately two-way interaction scenarios.

GIS provides spatial data management capabilities (data ingest, map projection definitions, etc.) that can provide a standard framework for tracking and coordinating numerous, large spatial data sets, which may include primary spatial data (soils, topography, land cover, etc.) and model outputs (soil moisture fields, wind fields, temperature fields, etc.). The spatial analysis components found in the typical GIS perform functions (aggregation, statistical summarization, locational operations) not available in most environmental models. These functions add important capabilities for evaluating the sensitivity of the models to changes in the structure of the spatial input data.

The GIS environments needed by the SRBEX investigators are varied. The models are currently grid cell-based and interface quite well with the standard raster structure found in GIS. Some of the model components in the program rely heavily on input from remotely sensed data. Again, the raster GIS data structure is a common denominator with the added requirement of image processing capability. However, most of the models also rely on multiple attribute information about various elements of the land surface. Soils information is a good example in this case. Typically, soils data are captured in a vector format with multiple attributes associated with each digitized soil polygon. In this case, a vector-based data structure coupled with a relational database management package is required.

Finally, some of the SRBEX research requires the development of GIS-based software for "nonstandard" analysis. Thus, the GIS packages must have the flexibility to allow for both the development of modules and the link with existing routines. This will most likely be accomplished within public domain software. The overall requirement is, therefore, for a heterogeneous software environment that will meet multiple needs.

Software Suite

A GIS software environment has been developed that includes four primary components: LAS (NASA Land Analysis System), ARC/INFO, ERDAS, and GRASS. The combination of these commercial and public domain packages provides the required capabilities.

LAS and ERDAS are primarily used for image processing, although some limited GIS capabilities are found in LAS, and more extensive GIS functionality can be found in the ERDAS package. LAS, developed at NASA in the 1980s, is a well-documented and well-engineered software package. It provides extensive library source codes and a rigid set of coding standards for developing user-implemented software. It has the ability to handle floating point raster operations—a critical element in many of the SRBEX applications. ERDAS provides a user-friendly environment and a gentle learning curve for new users.

The GRASS and ARC/INFO packages have also been incorporated into the working environment. These popular packages are complementary. ARC/INFO provides a vector-based data structure with links to relational database management implementations. GRASS provides a growing, "open" development environment for UNIX-based GIS.

THE SRBEX DATABASE

Management

Given the requirements for GIS and spatial data already discussed, data management becomes an issue of prime importance. With nearly two dozen investigators and graduate students in the EOS research program at Penn State, the potential for chaos is high. The approach to managing data resources has been to concentrate on providing safe storage of all primary data sources, easy access to these data and all "derived data products" (e.g., topographic characteristics like slope, aspect, etc.), and linkages between GIS packages to facilitate data interchange in the research process. In addition, it is possible to develop and maintain metadata concerning the database elements.

To meet these needs, a data storage and archival mechanism is being developed, using the CRAY YMP/2E housed at ESSC. The extensive on-line data storage of the CRAY, coupled with an IBM 3490E D42 tape cartridge system, provides a useful tool for managing the extremely large volumes of data required for SRBEX. The CRAY is closely integrated with the Penn State and ESSC networks used by the research group, ensuring easy access to the database.

In order to facilitate this access, an integrated data management and browse facility is under development, which will allow a user to access metadata about the SRBEX database. This management system will be structured as a server that can be accessed from a variety of PCs and workstations located anywhere on the Internet. The goal is to develop appropriate tools for the SRBEX environment that will minimize support overhead, yet provide full on-line documentation of the data resources.

The SRBEX database will hold a wide range of information related to the climate and hydrology of the SRB, including spatial data on topography, land cover, and soils. Other elements will include climate and hydrology data sets from a wide range of government programs and agencies, information about infrastructure (e.g., USGS digital line graph [DLG] data), and remotely sensed data from aircraft and satellite platforms. Data related to climate and hydrology from ongoing field programs in the SRB also will be included. The database will continue to evolve as the research program moves toward the 1998 EOS launch date.

Following is an overview of three major data sets that are currently being used in the set of climate and hydrology models shown in Figure 58-1. An example of the way in which GIS is used to link these models is also discussed.

Topographic data: The topographic components of the SRBEX database (Table 58-1) are complete down to the 3-arc-sec. U.S. Geological Survey (USGS) digital elevation model (DEM) scale. The USGS 30-m DEMs for the SRB are in the process of being obtained and archived. Approximately 500 of these cover the whole basin, out of which about 250 are already archived in-house. In addition to the raw data layers, all "derived" topographic characteristics products developed in the research (i.e., drainage basin boundaries, flow paths, geomorphometrics) are being archived.

Table 58-1. Digital topographic data layers for SRBEX.

SOURCE	RESOLUTION
NCAR/GCM	4 × 7
NGDC—Navy Charts	10 min.
NGDC—Navy Charts	5 min.
NGDC—DMA Charts	30 arc-sec.
USGS—1:250,000	3 arc-sec.
USGS—1:24,000	30 m

Land cover data: Land cover characteristics are an important component of the modeling effort within SRBEX. The land surface characteristics database developed at the USGS EROS Data Center (EDC) (Loveland et al. 1991) has been obtained and utilized in preliminary studies. This database provides land surface cover information at a 1-km resolution for the entire conterminous United States. For the current applications, the 167 land cover classes in this database are reduced to the eighteen vegetation/surface cover types used in the biosphere–atmosphere transfer scheme (BATS) (Dickinson et al. 1993).

Soils data: Soils information is also a key component of the SRBEX modeling effort. Through cooperation with the National Cooperative Soil Survey Center in Lincoln, Nebraska, the latest U.S. Department of Agriculture Soil Conservation Service (USDA SCS) STATSGO soils database for the SRB area is being acquired. STATSGO is a soils database that is generalized from the detailed county level soil surveys and created specifically for regional scale analyses. Also, through a cooperative relationship with the USDA SCS in Pennsylvania, the digitized soils information is being required from the detailed county-level soil surveys for the Mahantango Creek watershed where some of the SRBEX investigators have ongoing field experiments. It is hoped that by the 1998 EOS launch date, a complete digital database of the detailed county-level soil surveys for the SRB will be part of the SRBEX database.

NUMERICAL MODELS

Penn State/NCAR Mesoscale Model

The Penn State/NCAR mesoscale model (MM) is a versatile three-dimensional, limited-area meteorological numerical model (Anthes and Warner 1978) that has been applied to the study of a variety of atmospheric phenomena (Anthes 1990). The nonhydrostatic version of the model (Dudhia 1993) is being used in SRBEX. It has a double-nested horizontal grid as shown in Figure 58-2. The outer, intermediate, and inner grids have a grid increment of 36, 12, and 4 km, respectively. All three grids have 61 × 61 horizontal grid points and twenty-five irregularly spaced vertical levels. A high-resolution planetary boundary layer (PBL) model developed by Blackadar and described by Zhang and Anthes (1982), is utilized in this version of the MM. The PBL model utilizes a modified version of BATS (Lakhtakia and Warner 1994) as a sophisticated surface physics/soil hydrology parameterization module.

Soil Hydrology Model

The soil hydrology model (SHM) utilized in SRBEX is a one-dimensional, diffusion gravitation model (Capehart 1992), which is driven by conventional meteorological data (i.e., precipitation, cloud cover, surface air temperature, moisture, and wind speed and direction), as well as by information on topography, soils, and land cover. It produces a vertical profile of the soil water content for each of the grid elements in the MM grid (Figure 58-2) (Smith 1993). These soil water content fields are then used as part of the initialization of the mesoscale and the terrestrial-hydrology models.

The Terrestrial-Hydrology Model

The terrestrial-hydrology model (THM) being developed as a part of SRBEX will be a modular format terrestrial-hydrologic model that will incorporate a Penn State-developed digital topographic data processing package for a geomorphic feature extraction (GEOMORPH) package (a lumped parameter flood hydrograph package [HEC-1] developed by the U.S. Army Corps of Engineers Hydrologic Engineering Center in Davis, California), and a detailed distributed-parameter hydrologic model currently under development at Penn State (RUNOFF). Pending full development of RUNOFF, HEC-1 has been restructured to accept direct input from GEOMORPH, the MM, and the GIS.

GIS/MODEL INTEGRATION EXAMPLE

One example of GIS/environmental model integration within SRBEX is presented in the schematic in Figure 58-3. There, the SHM provides the initial soil water content to the MM and the THM. In turn, the MM provides the simulated precipitation data to the THM. The GIS is used at every step of the process to provide model parameters on a specific model grid, to convert output data from one model into input data for another model, and for analysis and visualization.

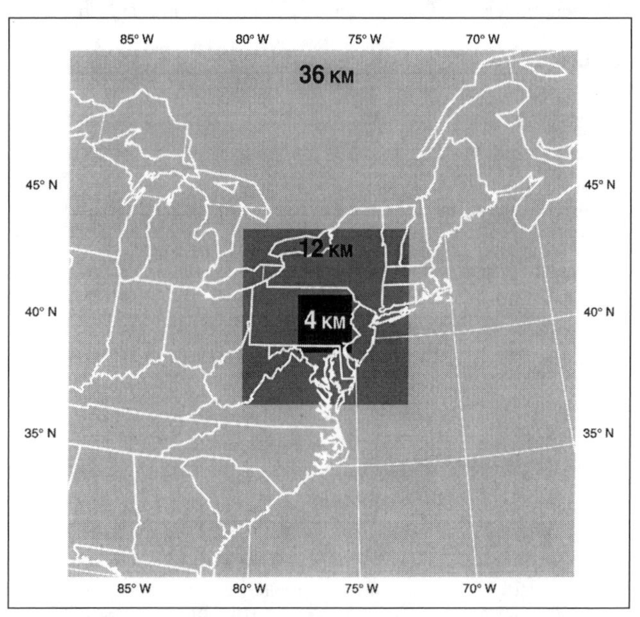

Figure 58-2. The Penn State/NCAR mesoscale model grids utilized in SRBEX.

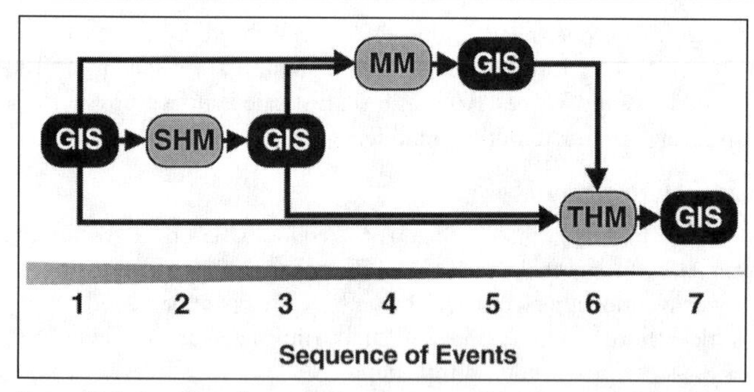

Figure 58-3: Sequence of events in the integration of GIS/environmental models.

In Step 1 (Figures 58-1 and 58-3), the GIS is used to register the data obtained from a particular database (i.e., the USGS EDC land surface characteristics, the USDA SCS soils, and the USGS 3-arc-sec. DEM databases) to a specific model grid. In this particular case there are two different types of horizontal grids involved (i.e., the SHM/MM grids and the THM grid). The GIS provides the means for model grid registration. Step 2 represents the execution of the SHM for the particular experiment. The output from the SHM migrates into the GIS environment (Step 3) for regridding, in order to provide part of the initial conditions for the THM. This SHM output is also utilized as part of the initial conditions for the MM, which is then executed in Step 4. The spatial and temporal distributions of the precipitation field simulated by the MM is piped into the GIS environment, where it is regridded (Step 5) for use by the THM as a forcing term in the surface hydrology balance. The THM is then executed in Step 6 and its output is migrated into the GIS (Step 7).

CONCLUSION

The climate and hydrology modeling community is beginning to recognize the importance of GIS and its role in providing an environment for linking models of the various components of the hydrologic cycle. The requirements of programs like EOS and interdisciplinary science investigations like SRBEX will continue to foster the growth of linkages between GIS and environmental models. Here, an example of the way in which these relationships are beginning to evolve is provided. As this and other research programs develop, the role of GIS and the links between models and GIS will grow in importance and sophistication.

References

Anthes, R. A., and T. T. Warner. 1978. Development of hydrodynamic models suitable for air pollution and other mesometeorological studies. *Mon. Wea. Rev.* 106:1,045–78.

Anthes, R. A. 1990. Recent applications of the Penn State/NCAR mesoscale model to synoptic, mesoscale, and climate studies. *Bull. Amer. Meteorol. Soc.* 71:1,610–29.

Capehart, W. J. 1992. "Constructing of a Meteorologically Driven Substratum Hydrology Model." Master's thesis, Department of Meteorology, Pennsylvania State University.

Dickinson, R. E., A. Henderson Sellers, P. J. Kennedy, and M. F. Wilson. 1986. Biosphere–atmosphere transfer scheme (BATS) for the NCAR Community Climate Model. *NCAR/TN-275 + STR.* Boulder: National Center for Atmospheric Research.

Dudhia, J. 1993. A nonhydrostatic version of the Penn State/NCAR mesoscale model: Validation tests and simulation of an Atlantic cyclone and cold front. *Mon. Wea. Rev.* 121:1,493–513.

Lakhtakia, M. N., and T. T. Warner. 1994. A comparison of simple and complex treatments of surface hydrology and thermodynamics suitable for mesoscale atmospheric models. *Mon. Wea. Rev.* 122(5):880–96.

Loveland, T. R., J. W. Merchant, D. O. Ohlen, and J. F. Brown. 1991. Development of a land cover characteristics database for the conterminous United States. *Photogrammetric Engineering & Remote Sensing* 57:1,453–63.

Smith, C. B. 1993. "Initialization of Soil Water Content for Regional-Scale Atmospheric Prediction Models." Master's thesis, Department of Meteorology, Pennsylvania State University.

Zhang, D., and R. A. Anthes. 1982. A high-resolution model of the planetary boundary layer—sensitivity tests and comparisons with SESAME-79 data. *J. Appl. Meteor.* 21:1,594–609.

Mercedes N. Lakhtakia received a *Licenciatura en Ciencias Meteorológicas* (1981) from the University of Buenos Aires, Argentina, and an M.S. (1985) and Ph.D. (1991) in meteorology from Pennsylvania State University. Her area of expertise is mesoscale numerical weather prediction, in particular, soil–vegetation atmosphere transfer schemes and surface forcing of the lower atmosphere. She is a research associate for the Earth System Science Center (Penn State) working on a NASA EOS project. Before joining Penn State, she was a research assistant for the Argentinean Navy Weather Service for over two years and taught for a year at the University of Buenos Aires.

Douglas A. Miller received his B.S. in Earth science (1981) and M.S. in agronomy (1987) from Pennsylvania State University. He is currently a Ph.D. candidate in soil science at Penn State. He has over ten years of experience in the application of remote sensing and geographic information systems to a wide range of natural resource management problems. He is currently the project manager for a NASA EOS-sponsored investigation of the global water cycle being conducted by an interdisciplinary team of scientists at Penn State's Earth System Science Center.

Richard A. White received a B.A. in physics (1957) from Oberlin College and a Ph.D. in physics (1962) from the University of Wisconsin. In his current position as a senior research associate in the Earth System Science Center at Pennsylvania State University, he is helping to integrate multiscale remotely sensed imagery, geographic data, and atmospheric and hydrologic models. Before joining Penn State in 1989, he was with Computer Sciences Corporation for eighteen years, primarily supporting development of remote-sensing software and systems under a series of NASA contracts. From 1963 to 1969, he taught physics at Makerere University, Kampala, Uganda.

Earth System Science Center
248 Deike Building
Pennsylvania State University
University Park, Pennsylvania 16802
Phone: (814) 863-9540
Fax: (814) 865-3191
E-mail : *lakht@psumeteo.psu.edu*

Christopher B. Smith received a B.A. in sociology (1987) from Gordon College in Wenham, Massachusetts; a B.S. in meteorology (1991) from the University of Lowell (now the University of Massachusetts at Lowell); and an M.S. in meteorology (1993) from Pennsylvania State University. He currently works at the Massachusetts Institute of Technology Lincoln Laboratory as a research meteorologist in the Weather Sensing Group.

Weather Sensing Group
MIT Lincoln Laboratory
Lexington, Massachusetts 01273

Introduction Part 3

Building Environmental Models with GIS

In this final section, we turn our attention to recent research efforts to enhance the capabilities of GIS so that it can perform the role of a modeling environment or language, obviating the need to couple GIS and modeling software. This is an ambitious goal, in part because much of the impetus for the development of GIS over the past decades has come not from environmental modeling but from less sophisticated applications in the areas of facilities management and inventory. The idea that GIS might be a tool to support sophisticated spatiotemporal modeling is still far from being broadly accepted.

If GIS is to perform this role, it is clear that its capabilities must be significantly enhanced. Thus, the chapters in Part III focus on impediments in current GIS, such as the difficulty of handling the temporal dimension, of linking data at different scales, and the need for improved programming languages for spatial modeling.

The introductory review chapter by David Maidment emphasizes the need to model *within* GIS, rather than couple to it. It examines the representation of space and time within environmental models and concepts such as vector fields, which are commonly used in modeling but are not largely supported in GIS.

Chapters 60 and 61 focus on cell-based models, a widely recognized framework for spatiotemporal modeling. Peng Gao and colleagues review cell-based modeling in GIS, and Keith Clarke presents an implementation of a cellular automata model for wildfire propagation.

In Chapter 62, Karen Kemp proposes a new level of user interaction with spatial databases that would largely hide the need to be concerned with details of representation, allowing the user to focus instead on the manipulation of continuous fields. This theme is echoed in Chapter 63, where Helena Mitasova et al. discuss the objectives of Open GIS and describe a series of general tools for surface manipulation.

Chapters 64 through 67 focus on decision support and the construction of integrated systems for modeling that can be used to help solve poorly structured problems. Dean Djokic devotes Chapter 64 to the design of expert system shells that can surround GIS and other special-purpose modules. Steven Frysinger et al. examine the requirements of environmental decision support, particularly with respect to the user interface and its implications for system architecture. Peter Keller and James Strapp explore the use of application programming interfaces for rapid development of decision support systems for specialized applications. Finally, Chapter 67, by Stephen Kessell, looks at the design of a decision support system for bush fire management in Australia.

GIS were designed to manipulate certain types of spatial objects, and whether or not these are the same types of objects needed by environmental modelers remains open to question. In Chapter 68, Terence Smith and colleagues explore the modeling concepts needed to describe the process of modeling, arguing that GIS designs have been more concerned with describing the phenomena being modeled, particularly their locations and attributes. Peter Crosbie takes a different approach to the same issue by looking at the more advanced data modeling concepts that have recently emerged within GIS. In Chapter 70, Jonathan Raper and David Livingstone take a specific example of the modeling of coastal geomorphic processes to argue the advantages of object-based approaches to data modeling. Dale White and Matthias Hofschen examine in detail the modeling of the components of a hydrologic system. Finally, T-S. Yeh and B. de Cambray discuss the representation of time and its integration into spatial databases as a conceptually continuous dimension.

Chapter 73, by Perry LaPotin and H. L. McKim, introduces recent research in the field of image processing and the use of neural nets and distributed processors to test the results of pattern recognition and feature identification.

The next four chapters focus on the all-important topic of system integration. Kurt Fedra reviews alternative integration strategies in Chapter 74, with examples of tight coupling—a strategy that avoids many of the problems associated with loose coupling strategies discussed in Part II, by effectively sharing common data structures. David Lam and Christian Pupp discuss the integration of GIS with expert systems. D. D. Cowan and colleagues describe the conceptual design of an integrated environmental information system in Chapter 76. Finally, J. G. Bromberg et al. discuss the integration of GIS, the CENTURY model, and other components into a scientific GIS.

Techniques that allow the environment to be represented and modeled in hierarchical fashion are the subject of Chapters 78 through 80. Hierarchical structures allow explicit links to be made between data at different scales and resolutions; permit models of process links between scales to be implemented; and support the aggregation and disaggregation of data and models. Ferenc Csillag looks at recent research in the hierarchical representation of spatial data. In Chapter 79, David Bennett and colleagues describe a system for managing object-oriented models and representations of environmental systems. David Stoms et al. discuss the use of hierarchical concepts in modeling species distributions in Chapter 81, followed by Alan Johnson, who examines the role of hierarchical concepts of space and time in ecological theory and modeling.

Effective data modeling of environmental systems can be a time-consuming and elaborate exercise. In Chapter 82, René Reitsma describes an innovative project to automate the construction of data models to support studies of river basin hydrology.

The final four chapters focus on visualization. Many of the traditional cartographic techniques for display of spatial data have been implemented in GIS, but the digital environment provides a more versatile environment for new techniques and also creates a need for more efficient methods, given the vast amounts of data and modeling results that potentially require display. First, Barbara Buttenfield examines the current state of visualization capabilities in GIS and presents a useful formalism as the basis for more advanced methods. Loey Knapp and colleagues describe the scientific visualization techniques being used in the Gunnison River Basin. Sidey Timmins and Carolyn Hunsaker devote Chapter 85 to a review of tools for visualizing patterns of land cover. Lastly, in Chapter 86, Len Wanger and Peter Kochevar review the visualization tools of the Sequoia 2000 project.

The chapters in Part III collectively point the way to future development of GIS for better support of environmental modeling. In some fields it will be possible to model spatiotemporal processes directly in the language of advanced GIS, particularly those evolving from the cellular automaton framework. In other fields, tight coupling using shared data models may be more practical, at least in the interim. It is clear, however, that efforts to integrate GIS and environmental modeling will continue to stimulate novel approaches, on both sides of the house. Widespread use of GIS is causing modelers to rethink the importance of software integration and shared data and the role of data models and representations in defining models of process. On the other hand, the interest of environmental modelers in GIS is stimulating the development of more advanced modeling tools that move GIS progressively away from the static, two-dimensional, limited vision that it inherited from its cartographic roots and that still characterizes many of its applications.

59

Environmental Modeling within GIS

David R. Maidment

Principles describing the motion of water, air, and the substances they transport are considered in the context of GIS data structures and functions. Some finite difference and response function models can be solved with existing GIS capabilities, but extensions are necessary to achieve more complete simulations. These include modeling of the flow and transport of water and air using vector functions and the analytical solution of the governing equations of important environmental processes. The time–area diagram for watershed runoff and the advection–dispersion equation of groundwater transport are used as examples showing how intrinsic environmental functions are created within GIS using velocity fields. Atmospheric moisture motion is used as an example to illustrate some of the common operators of vector calculus for dealing with such fields and to show how the Reynolds Transport Theorem can be used as a general approach to link quantities flowing through continuous and discrete space domains.

INTRODUCTION

Geographic information systems (GIS) describe the spatial environment. Environmental modeling simulates the functioning of environmental processes. Such simulations require data about the environment within which the processes occur, and simulation results provide additional data to enrich environmental description. Thus GIS and environmental modeling are synergistic, and GIS can serve as a common data and analysis framework for environmental models. But GIS and environmental modeling grew up separately, so their computer programs have very different data structures, functions, and methods for inputting and outputting spatial information. This makes it difficult to link existing GIS systems and environmental models, or even to link one environmental model with another. Some of these difficulties can be overcome by rewriting environmental models into a form in which they can be imbedded within a GIS and thus make use of data directly from GIS data structures (Ragan 1991).

But beyond the linkage of software components, useful as it is, there lies a deeper issue—how best to use GIS to change the way environmental modeling is being done. How can the fundamental basis of environmental modeling be reconsidered in the light of GIS data structures and functions? Is it possible to use or adapt existing GIS functions to carry out environmental modeling within a GIS? What new data structures and functions are needed to accomplish this goal? What might some of the components of the environmental GIS of the future look like? This chapter explores these questions.

In developing an environmental GIS, there is no need to reinvent science. The basic laws governing the motion of water and air as fluids have been known for more than a century. How quantities are transported in water and air, how the atmosphere circulates, how rainfall is transformed to runoff, and how biological processes function have been elucidated in the intervening years. The key is to be able to isolate from all this knowledge those fundamental principles upon which all else rests, and to adapt or build GIS data structures and functions to represent those principles. Since the scope of environmental modeling is very broad, the goal of isolating its fundamental principles is beyond the knowledge of any single individual and has to be approached gradually by a synthesis of knowledge of specialists in many fields. In particular, terrestrial and aquatic ecosystems are so complex and contain so many random and uncertain elements that they do not lend themselves to generalization by a set of basic laws, as is the case for fluid flow and transport of constituents within water and air.

For these reasons, the scope of this chapter is limited to some aspects of hydrologic modeling within GIS. Ideas are presented in the latter part of the chapter by which some of the methods described earlier for water flow and transport can also be applied to atmospheric systems and, perhaps, to biological systems. At present, environmental modeling within GIS is necessarily simplified, but as environmental GIS develops, its power and scope will increase.

Abstract Models

The construction of any representation of a real system requires the development of an abstract model, which includes a set of basic constructs or variables and a language for manipulating them. In GIS, this abstract model rests first on the representation of geographic features within a spatial domain by data connected to layers of points, lines, polygons, or grid cells, and then on software functions that process existing data layers to produce new ones. As a GIS user becomes familiar with a particular system, its data structures and functions become an analysis system with which complicated data manipulations can be carried out. A principal measure of value of a GIS database is how accurate and complete its data are. Do they faithfully represent what is in the environment and where it is located?

Environmental modeling focuses on different questions such as:

- What are the important environmental processes implicit in a given situation?
- What are their governing equations?
- How will these equations be solved?
- How will the parameters of the equations be determined?
- How can the model results be checked against observations?

Ultimately, all applications of environmental models focus on a particular environment, but the nature and solution of their governing equations are the most critical features, rather than the descriptions of the environment within which the environmental processes occur. It follows that the abstract models used for environmental simulation are quite different from those used in GIS. They use the language of mathematics, particularly algebra and calculus, instead of the language of GIS functions and data structures. Performing environmental modeling within GIS requires the expression in GIS language of the mathematical solutions of the governing equations of environmental processes.

ANALYSIS DOMAIN

An environmental model is applied over an analysis domain, which is a region of space and time within which the variables will be simulated. Important distinctions exist as to how this region is subdivided.

Space

Within GIS, one can create a continuous space or a discrete space model (Goodchild 1991). Vector GIS is a discrete space model in which the data layers contain an arrangement of points, lines, or areas in which each point, line, or area is an individual or discrete entity described by a data record. By overlaying existing data layers, more complex data layers with more extensive descriptive data tables are created (ESRI 1992). Points and lines can lie anywhere in space within the domain and areas can be of any shape. Continuous space models are represented by a surface over the analysis domain whose horizontal plane is subdivided into regularly shaped areas, usually rectangular cells or triangles. Raster or grid GIS are continuous space models, as are the triangulated irregular networks (TINs) often used in digital terrain modeling. This type of model usually has a single associated attribute, such as the cell value in a grid system or the vertex z-coordinate in a TIN system. It is possible to create a discrete space representation in a continuous space model by assigning the same attribute value to all spatial elements in a particular zone (Tomlin 1990), as in the identification of stream links as lines of cells and their contributing areas as zones of cells in grid-based watershed delineation.

Lumped Models

The distinction between discrete and continuous space modeling in GIS is important because a similar distinction exists in environmental modeling between lumped models and distributed models. A lumped model is one in which processes function within a system of discrete spatial objects, and the model solution describes the input and output of each object without attempting to determine the precise spatial distribution of the processes within the object. Lumped models are widely used in rainfall–runoff modeling in which the watershed is treated as a spatially averaged entity, whose input is rainfall over the watershed and whose output is runoff at the stream outlet. This output becomes input to flow in the channel downstream of the outlet, whose output is joined to the outputs of downstream subwatersheds by means of a flow network connecting the subwatersheds and channels in the landscape. The result of the simulation is a hydrograph of discharge versus time at each outlet node in the network. The properties of each subwatershed and stream channel in the network are lumped or spatially averaged so that each spatial object in the model is represented by a single set of descriptive properties, like the attribute values in the tables attached to vector GIS data layers.

A critical fact about lumped models is that they can be represented by ordinary differential equations; that is, by differential equations whose dependent variables are a function of a single variable, which in this case is time. For example, the continuity equation for water flow in a lumped model is written as:

$$dS/dt = I - Q \qquad (59\text{-}1)$$

where S is the amount of water stored in a particular object, I is the inflow rate, Q is the outflow rate, and dS/dt is the rate of change of storage with respect to time in response to the inflow and outflow.

Distributed Models

By contrast, a distributed environmental model operates over a continuous space on which the solution is determined for each spatial element. The solution is one, two, or three dimensional, depending on how many spatial dimensions are used to describe the model variables. For example, the passage of a flood wave down a river can be described by a one-dimensional distributed time varying model in which the discharge and water surface elevation are determined at regular intervals along the course of the river as a function of time (Fread 1993). The response of the water table to well pumping is modeled by two-dimensional time varying groundwater models that determine the water table elevation and the magnitude and direction of the groundwater flow field on a mesh laid over the horizontal domain of the aquifer (Bear 1979). A three-dimensional general circulation model of the atmosphere determines atmospheric properties and the horizontal wind field at regular vertical intervals (Peixoto and Oort 1992). For a two-dimensional model, a map or shaded surface of the solution can be drawn at each time point for each variable determined by the model.

Distributed models are based on partial differential equations, which are differential equations whose dependent variables are a function of two or more variables. For example, the continuity equation for steady two-dimensional groundwater flow without sources and sinks is written as:

$$\frac{\partial Q}{\partial x} + \frac{\partial Q}{\partial y} = 0 \qquad (59\text{-}2)$$

where the partial differentials $\partial Q/\partial x$ and $\partial Q/\partial y$ describe the variation or gradient of the groundwater discharge Q in the horizontal x and y directions, respectively.

Time

Ideally, an environmental simulation would proceed continuously over space and time, but practical constraints of the computer time and memory required to solve systems of differential equations and limitations of available environmental data mean that some reduction in the dimensions of the problem are necessary. In GIS this is particularly true because it is a static system in which there are few procedures for explicitly labeling data with their time of occurrence.

Several approximations to a continuous time model are available. The first and simplest approximation is to consider a steady state or steady flow system in which there are no time variations. This eliminates all time derivatives from the governing equations, such as dS/dt in (59-1), and considerably simplifies their solution. Steady flow assumptions are useful for systems whose inputs change very slowly with time, such as deep groundwater systems being slowly recharged by surface infiltration. In this case, a constant input that represents the mean annual recharge rate is used. A variation on a single steady flow model is to have a sequence of steady flows, such as simulating a system in monthly time steps using monthly average inputs. The result then represents January average conditions, February average conditions, and so on.

A second approximation is an *event* model, where the response of the system to a single input is simulated through time. Hydrologic design for severe storms uses this approach, where the flood hydrograph resulting from a single storm is determined separately from the effects of preceding rainfalls whose effect is represented by a baseflow existing in the river prior to the storm and by the condition of soil moisture in the watershed at the time rainfall began. The event model limits the number of time steps to be considered to perhaps 10–100.

The most complete representation is to have a temporal data description over the whole time domain. It is fairly common, for example, to simulate watershed hydrology with daily or even hourly time steps over a period of 10–20 years. It does not seem productive to try to accomplish such simulations by intrinsic GIS functions; rather the simulation model should be kept separate from the GIS and the GIS used just as a spatial data source. The point is that if the number of time steps can be restricted, simulation of time variation within GIS can be accomplished. Time averaging or single event modeling can be used for this purpose.

MODELING FLOW AND TRANSPORT PROCESSES

Fluid mechanics is the engineering science that describes the motion of fluids such as water and air (Aris 1962; Roberson and Crowe 1990). Although water and air seem like very different fluids, their governing equations of motion are very similar, differing only in the respect that air is a compressible fluid while water in most situations is not. In fluid mechanics, a distinction is drawn between flow processes, which are processes related to the motion of the fluid itself, and transport processes, which are processes related to the substances transported by a fluid, such as sediment or dissolved contaminants in water and particulate matter in air. This distinction is important because describing the flow of a fluid often requires a simultaneous solution for the flow properties over the whole domain, while transport of constituents can be treated in a more localized fashion. For example, in determining a groundwater flow field, pressure effects are transmitted very quickly through an aquifer so that a change in flow at one point can affect flow at some distance away quite quickly, as shown by the propagation of tidal effects upstream through groundwater systems that discharge into the ocean. But groundwater transport occurs at a very slow rate related to the seepage velocity, so an injection of contaminant at a particular location may take years to flow and disperse over even a few hundred meters from the injection point.

For this reason, and when dealing with continuous space models, as is appropriate for groundwater and air systems, it is reasonable to begin to incorporate transport modeling as an intrinsic GIS function before flow modeling because transport can be handled in some circumstances by local rather than by global functions. For surface water hydrology, discrete space or lumped modeling of water flow is a standard approach, and it is more reasonable to develop intrinsic GIS functions to describe both flow and transport.

Solution of Equations of Motion

The equations of motion of a fluid are those equations describing the action of the physical laws governing the way the motion occurs. These equations are based on three physical laws: (1) conservation of mass (continuity equation), (2) Newton's second law of motion (momentum equation), and (3) the first law of thermodynamics (energy equation). These laws can be applied to both discrete space and continuous space models, which lead to ordinary and partial differential equations, respectively. Such equations can be solved analytically or numerically. An analytical solution is one in which the differential equations are integrated over time and space and the result is a mathematical expression, usually in the form of an infinite series of terms whose evaluation yields the values of the dependent variables everywhere over the solution domain. A numerical solution is one in which the dependent variables are calculated on a finite mesh of points over the domain using an approximate form of the governing equations, the approximation being of their differential form when the finite difference method is used and of their integral form when the finite element method is used.

Finite difference and finite element models are continuous space models whose solutions are specific to the particular set of equations being solved and are difficult to generalize. It is possible to set up such solutions within the map algebra language in GIS grid systems, but it appears that the numerical burden of these solutions is so great as to require modification of the software engineering techniques supporting the map algebra languages to make that a practical option. In particular, it is common for GIS computations to be very disk dependent because the result of each computational step is written to disk before the next step is initiated. Numerical analysis is best accomplished by continuously holding all the data within the core memory and writing to disk at infrequent intervals. It may be that at the present stage of environmental GIS development, numerical solutions over flow domains should be accomplished in programs linked to GIS rather than as intrinsic GIS functions, although the distinction between what is linked and what is intrinsic is blurred in a system like GRASS, whose open architecture makes it easier to bind in numerical analysis codes than is the case with other GIS.

It appears that for intrinsic GIS function development, analytical solutions offer a promising development route. This is so for several reasons. First, the analytical solution of a particular equation is an encapsulated, completely defined quantity that can be described precisely in mathematics. Second, an analysis procedure can be developed in which such functions can be sequenced so that a more complex result can be built up from a progressive series of explicit steps, like the progressive development of data layers in a GIS spatial analysis. Third, analytical solution of differential equations has a long scientific history and was the main way these equations were solved before the advent of computers, so extensive lists of solutions have been compiled (Carslaw and Jaeger 1959). The analytical solution within GIS of the advection–dispersion equation of groundwater transport is described later in this chapter.

Response Functions

Response functions are a class of analytical solutions. They are solutions of systems described by differential equations when the inputs have particular properties in time and space. For lumped models in discrete space, the impulse response function describes the output from the system as a function of time in response to an instantaneous input of unit amount. For example, the instantaneous unit hydrograph is the theoretical discharge hydrograph at the watershed outlet which would occur if a unit depth (e.g., 1 in or 1 mm) of rain fell uniformly and instantaneously over the watershed and all of the rainfall became runoff. Two extensions of this idea, the step response function and the pulse response function, describe the response of a system to a continuous input at unit rate per unit time and to a discrete input of unit amount over a specified time interval, respectively.

For distributed models in continuous space, the corresponding response functions are called Greens functions (Zwillinger 1992), which are more complex because they need to specify the values of the variables in both time and space rather than in time alone as for lumped models. For example, a Greens function can be used to calculate the concentration of contamination as a function of space and time throughout an aquifer which results from the instantaneous injection of a unit mass of contaminant at a particular point within the aquifer. If the governing differential equations of the system are known, the mathematical form of the response function can often be obtained analytically, as is the case for many problems in heat flow, for example (Crank 1956). Such solutions are easiest to obtain when the governing equations are linear differential equations (terms involving the dependent variables are added and not multiplied together) with constant coefficients (the parameters of the equations are constant in time and uniform in space).

Linear Systems

Linear systems are systems described by linear differential equations with constant coefficients. These can be either ordinary or partial differential equations. Linear systems have two properties that make them attractive for modeling: additivity and proportionality. Additivity (or superposition) means that the response to two separate inputs can be obtained simply by summing the responses to each input considered individually. In groundwater flow, the drawdown of the water table of pumping from two wells is found by summing the drawdowns produced by each well; in surface water hydrology, the runoff response to two pulses of rainfall occurring one after the other is found by summing the responses to each pulse. Proportionality means that the magnitude of the output is directly proportional to the magnitude of the input. Thus, the drawdown is doubled if the pumping rate is doubled in a well and the runoff hydrograph produced by two inches of excess rainfall has at each time point twice the discharge as the hydrograph resulting from one inch of excess rainfall. While there are few processes in nature that are truly linear, this assumption is sufficiently realistic to serve as a basis for many components of environmental models. Because the mathematical basis of linear systems is well understood and the ideas are widely used, it seems appropriate that intrinsic GIS functionality be created for linear systems.

It is sometimes the case that the governing differential equations are not known precisely; rather, what is available is a series of measured data showing inputs and outputs from the system from which the form of the response function is inferred directly. This is the case in watershed hydrology, in which the unit hydrograph is the pulse response function of a watershed to a unit amount of rainfall falling uniformly over the watershed in a specified interval of time.

Two examples of intrinsic GIS functions that can be used for environmental modeling are now described. Both examples are implemented in version 7.0 of ARC/INFO, presently in beta test release.

TIME–AREA DIAGRAM

Watersheds and stream networks can be delineated from a digital elevation model (DEM) of land surface terrain by allowing water to flow from each grid cell to one of its eight neighboring cells, thus creating a grid of flow direction. If a corresponding grid of flow velocity is added, the spatial velocity field for flow over a watershed is specified, as shown in Figure 59-1a. The flow direction grid can be used to trace out from any cell in the watershed the path of flow to the outlet. Since the velocity of flow is known in each cell of the path, the time of travel to the outlet can be computed. If a time interval Δt is selected, the watershed time of travel grid can be divided into zones of area A_i, $i = 1, 2, \ldots, n$, in increments of Δt thus producing the time–area map, as shown in Figure 59-1b. The boundaries of these zones are lines of equal time of travel to the outlet called isochrones. By plotting the cumulative histogram of the incremental areas A_i the time–area diagram is created.

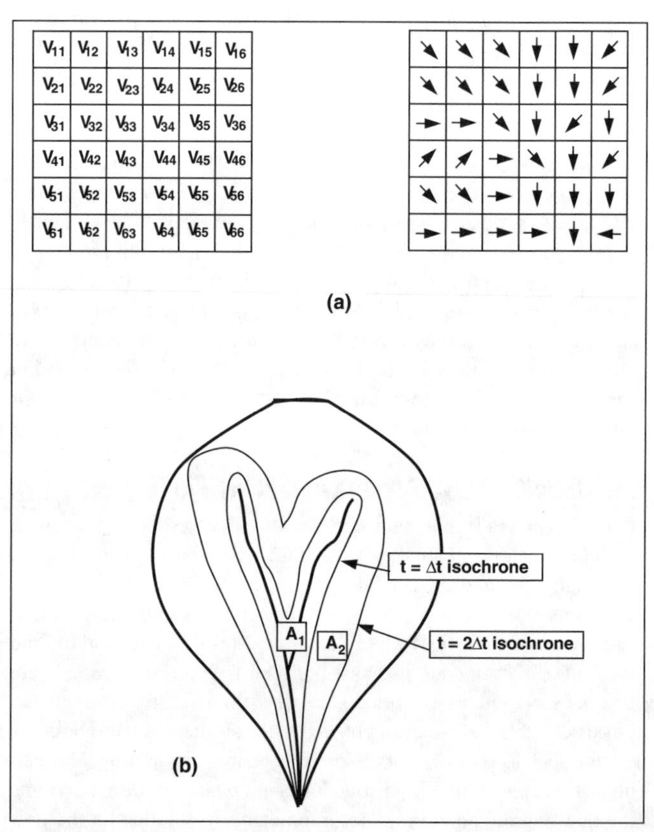

Figure 59-1. Watershed time–area relationships: (a) a velocity field specified by the magnitude and direction of flow velocity; (b) a watershed isochrone map drawn by classifying a grid of time of flow to the outlet.

The motivation for forming the time–area diagram is that if a rainfall begins and continues indefinitely at a constant rate and if production of runoff is uniform over the watershed, the successive ordinate values of the diagram represent the increase in drainage area contributing to flow at the outlet in each time period until all the watershed is contributing (the time of concentration). The concept of the time–area diagram is a very old one in hydrology, originating in the 1920s, but it has not been practical to use this concept because its computation was so laborious. With the widespread availability of digital elevation data and with corresponding functionality in GIS, this constraint can be removed.

The time–area diagram can be implemented in ARC/INFO Grid 7.0 by using a new function called Flowlength. This function takes as an argument the grid of flow direction and optionally a grid of cell weights. If the weights are set to unity, the value computed in each cell is simply the cumulative distance of flow to the outlet. If the weights of the stream cells are set to zero, the distance recorded in the cells on the remaining land surface of the watershed is the distance to the stream. By this means, distance–area diagrams for the watershed can be created, which are an interesting summary of its topographic properties that merit further study. If the weights are set equal to the inverse of the velocity $1/V$, the product of distance d and inverse velocity gives the time of travel t in each cell, $t = d/V$ and the time–area map is created, from which, following a grid classification into finite time intervals, the time–area diagram is found.

The unit hydrograph characterizes the response of watershed runoff to rainfall. It is specified as a series of discharge ordinates occurring at the outlet in response to a unit amount of excess rainfall (1 in or 1 mm) occurring over the watershed. Maidment (1993) showed that this sequence of ordinates is given by $U_i = A_i/\Delta t$, that is, the ordinates of the unit hydrograph in units of discharge per unit of rainfall depth are equal to the incremental areas of the time–area diagram divided by the time interval into which the grid is classified, which must be equal to the specified duration of the rainfall of the unit hydrograph. In this way, a spatial analysis of the runoff velocity field can yield the response function at the outlet.

With GIS grid capabilities for rainfall mapping, the uniform spatial rainfall distribution is no longer necessary so that two subscripts are needed to characterize rainfall, P_{ij}, where P_{ij} is the average excess rainfall over all cells in isochrone zone i during time interval j. Direct runoff Q_n at time $t = n\Delta t$ is given by summing the runoff contributions from each of the applicable isochrone zones suitably lagged in time:

$$Q_n = \sum_{i=1}^{n} \frac{P_{ij} A_i}{\Delta t} \quad \text{where } j = n-i+1 \qquad (59\text{-}3)$$

It appears that the unit hydrograph assumptions of constant time base and linearity of rainfall runoff response are equivalent to assuming that regardless of the amount of rainfall, runoff always follows the same paths with the same average travel time, that is, that the runoff velocity field implied by the unit hydrograph is spatially variable but time and discharge invariant.

To determine the correct time–area diagram for a watershed requires specification of its velocity field. Sircar et al. (1991) have shown how this can be done using a velocity function of the form $V = aS^b$, where S is land surface slope, and a and b are coefficients related to land use taken from McCuen (1982), which are based on procedures of the USDA Soil Conservation Service. The author is presently engaged in a more direct method of determining the required velocity field by using a known unit hydrograph at the outlet and finding that velocity field which will yield the required result. The watershed is divided into two zones, a stream zone and a land surface zone. Velocity is determined as a function of distance from the outlet along the stream channel and as a function of distance from the channel for the land surface. By using regression or nonlinear optimization, the appropriate parameters of these velocity field functions can be determined.

GROUNDWATER CONTAMINANT ADVECTION AND DISPERSION

Dispersion of a substance over a domain can be described graphically in GIS by using spatial averaging functions. But dispersion in the physical sense is a term describing a class of mixing processes in which a constituent in a fluid is spread over a region of space surrounding its centroid, which is being advected or transported at the mean fluid velocity. The governing equation for this process, called the advection–dispersion equation, applies in air (Nuclear Regulatory Commission 1983; Seinfeld 1986), in surface water (Thomann and Mueller 1987; Huber 1993), and in groundwater (Freeze and Cherry 1979; Mercer and Waddell 1993) although the magnitude of the dispersion coefficient is determined differently in each environment.

If the fluid is moving with velocity U in the longitudinal direction x_L and dispersion occurs both longitudinally and transversely, the advection–dispersion equation is written:

$$\partial C/\partial t = D_L\, \partial^2 C/\partial x_L^2 + D_T\, \partial^2 C/\partial x_T^2 - U\, \partial C/\partial x_L \qquad (59\text{-}4)$$

where C is the concentration of the contaminant, D_L is the longitudinal dispersion coefficient, and D_T = the transverse dispersion coefficient in the direction x_T. Equation (59-4) assumes that the constituent is conservative, so additional terms are added to the equation to account for adsorption, decay, and other processes when necessary.

Solution of (59-4) requires initial conditions, which are a distribution of C when t is 0, and boundary conditions which are values C around the boundary of the flow domain (commonly assumed to be at zero concentration). The boundary conditions can be constant or a function of time. All solutions of the advection–dispersion equation have in common the fact that dispersion is a random phenomenon made up of a very large number of small movements. For stationary, homogeneous processes, the statistical accumulation of these movements produces a normal or Gaussian bell curve for concentration distribution in both the x_L and x_T directions where the variance of the normal distribution is related to the dispersion coefficient by: $\sigma^2 = 2Dt$. One common solution of (59-4) is for an instantaneous spill of mass M at the origin at time zero. The solution in this case is:

$$C(x,y,t) = \frac{M}{2\pi\sigma_L\sigma_h} \exp\left(- \frac{(x_L - Ut)^2}{2\sigma_L^2} - \frac{x_T^2}{2\sigma_T^2}\right) \qquad (59\text{-}5)$$

Values of $C(x,y,t)$ at particular points in time and for particular values of C can be mapped as ellipses as shown in Figure 59-2. The centroid of the distribution is advected with the flow field and the dispersion occurs around the centroid. This is called a Lagrangian solution because it follows the mass of the contaminant as it moves, as contrasted to an Eulerian solution in which the focus is on the individual grid cells and the mass that moves between each cell and its neighbors.

This solution of the advection–dispersion equation for groundwater contaminant transport is implemented in the beta version of ARC/INFO Grid 7.0 by three functions: Darcyflow, Particletrack, and Porouspuff (Tauxe et al. 1992). Darcyflow takes as inputs grids of piezometric head, porosity, wetted thickness, and transmissivity and produces grids of Darcy flux magnitude and direction (the seepage velocity U is equal to the Darcy flux divided by the porosity); Particletrack describes the motion of a discrete mass released into the flow field as shown in Figure 59-2a; Porouspuff describes the mapping of the dispersion of the contaminant mass around the centroid as in Figure 59-2b.

VECTOR FIELDS

The examples just described rely on the velocity field of fluid motion. Velocity is a vector quantity because at each point in space it possesses both magnitude and direction. In GIS a vector data struc-

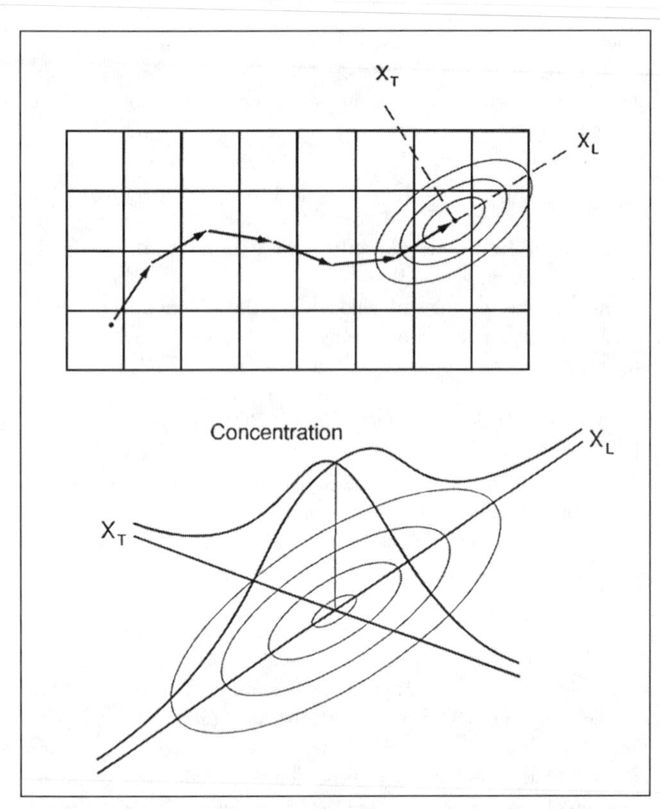

Figure 59-2. Solution of the advection–dispersion equation on a grid (from Tauxe et al. 1992).

ture is one based on point and line descriptions. A vector can be drawn from the origin of a coordinate system to each point in the space. If a line joins two points, its vector is found as the difference between the vectors to its end points. A vector space is a region containing vectors of this type (Marcus and Minc 1965). A vector field is a special type of vector space in which at each point in the space a vector quantity is defined. For example, flow velocity V is a vector defined by three components (u, v, w) in the coordinate directions. Since discharge, mass flux, and other flow-related quantities are all dependent on velocity, vector fields can be developed for them as well. Vector calculus is the mathematical language for manipulation of vector fields (Marsden and Tromba 1988).

Three functions are central to vector calculus: gradient, divergence, and curl (Schey 1992). The gradient is a familiar quantity in GIS, such as when land surface gradient is defined by the magnitude and direction of the line of steepest descent down a slope. In mathematical terms, the gradient of a scalar h is given by the vector $\nabla h = (\partial h/\partial x, \partial h/\partial y, \partial h/\partial z)$. The divergence of a vector field Q is given by $\nabla \bullet Q = \partial Q/\partial x + \partial Q/\partial y + \partial Q/\partial z$ and • indicates a vector dot product. The divergence is a measure of the amount added or subtracted from the vector at a point in the field. Thus, (59-2) can be written as $\nabla \bullet Q = 0$ because the groundwater flow does not have sources or sinks. The curl of a vector field is a measure of its rotation, and is given by $\nabla \times Q$ where × indicates a vector cross product. Atmospheric moisture motion is an example of a vector field where these functions are useful.

Atmospheric Moisture Motion

Figure 59-3 shows a vector field for atmospheric moisture flow over North America. The moisture content is specified by the specific humidity q, and when multiplied by the west–east (u) and south–north (v) components of wind velocity, the vector field of moisture motion $qV = (qu, qv)$ is created. These data are taken from the condition of the atmosphere computed by a general circulation model for weather forecasting operated at the U.S. National Meteorological Center and include information on air mass soundings and other forms of atmospheric measurement. The values of qV are integrated vertically through the atmosphere and averaged over June to August to produce the vector field shown in Figure 59-3. The contours shown in the figure are the divergence of this field $\nabla \bullet qV$ which is equal to E-P, the net rate of addition of vapor to the flow field by evaporation E less precipitation P. Negative divergence is denoted by dashed contours and shows regions where precipitation is extracting moisture from the atmospheric moisture field while solid contours show positive divergence where evaporation is adding moisture.

The concepts of vector fields and their gradient, divergence, and curl operators are a natural extension of the grid data structure in raster GIS, and in developing an environmental GIS, these functions would be very useful. Kemp (1993) has suggested that a special data structure called a field data type should be set up for describing quantities distributed over continuous space, which may be scalars such as elevation or vectors such as velocity.

REYNOLDS TRANSPORT THEOREM

Returning again to the distinction made at the beginning of this chapter between discrete and continuous space models, it is evident that vector calculus is concerned with continuous spaces but that many problems are well described in discrete space. For example, it has always been difficult to relate data derived from general circulation models of the atmosphere to watershed hydrology because the GCM data are determined for large grid cells while watersheds are irregular polygon shapes whose boundaries don't coincide with the GCM grid cells.

This general problem of relating the properties of a fluid to the space through which the fluid flows is central to defining the operating equations of fluid mechanics. A mathematical mechanism has been created to deal with this, called the Reynolds Transport Theorem (Aris 1962). This theorem uses the Eulerian view of motion of looking at a fixed reference frame called the control volume whose boundary is the control surface through which the fluid passes. Physical laws, such as the conservation of mass, Newton's second law of motion, and the first law of thermodynamics, describe the time rate of change of the properties of a discrete mass of fluid moving through space. Using the Reynolds Transport Theorem, these laws can instead be applied to the fluid flowing through a control volume at a fixed location.

If B is the amount of a fluid property, β is the amount of this property per unit mass of the fluid, ρ is the fluid density, and dv is an element of volume within the control volume, the Reynolds Transport Theorem can be written for a control volume of fixed size, such as that shown in Figure 59-4e, as:

$$\frac{dB}{dt} = \frac{d}{dt} \iiint_{c.v.} \beta \rho \, dv + \iint_{c.s.} \beta \rho V \bullet da \qquad (59\text{-}6)$$

where the term dB/dt is the material derivative of the fluid property or its total rate of change with respect to time, the first term on the right-hand side is the time rate of change in the amount of this property stored within the control volume, and the second term is the net outflow of the property across the control surface. For water vapor transport in air, the term B = mass of water vapor W, β = mass of water vapor per unit mass of moist

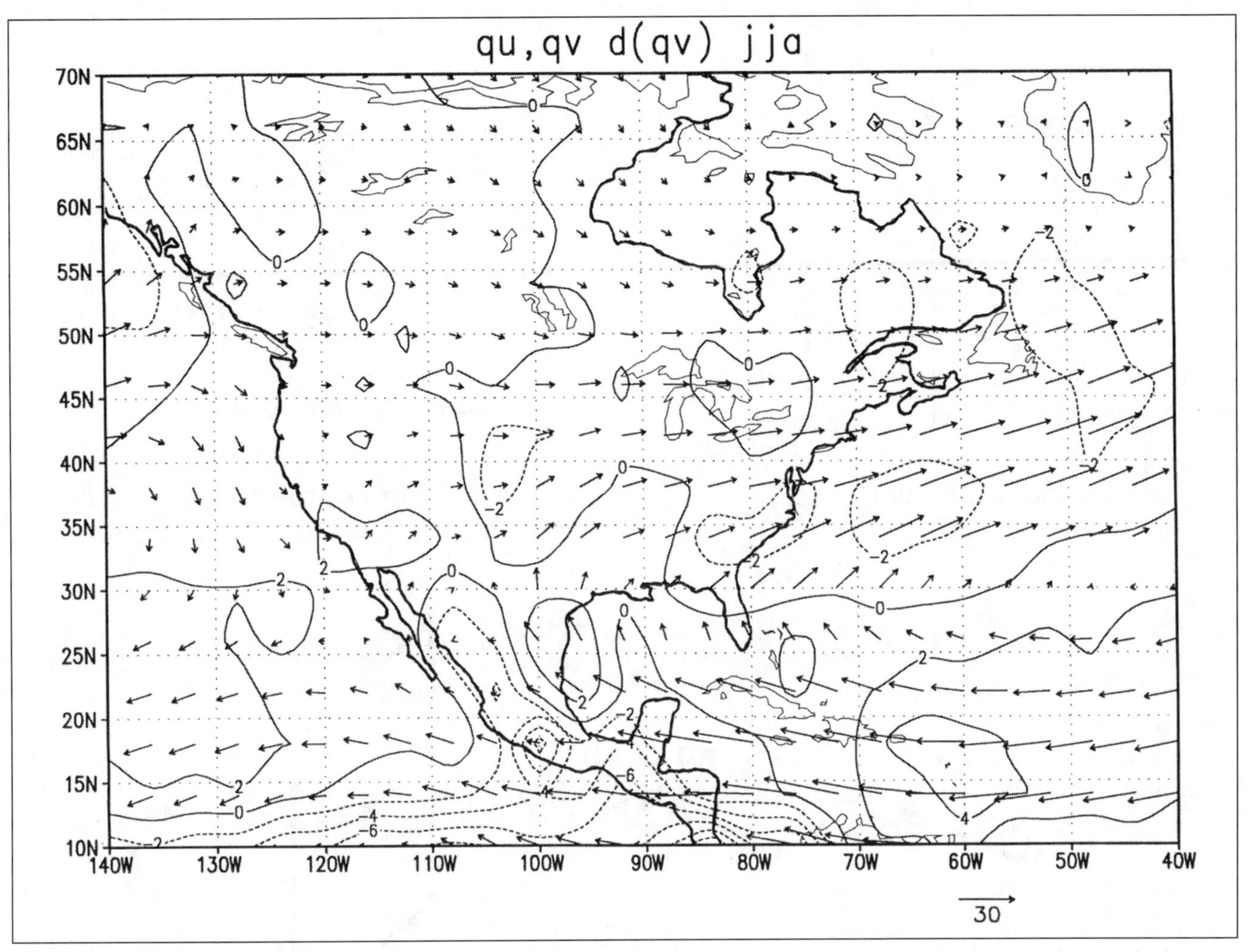

Figure 59-3. Atmospheric moisture motion over North America, averaged over June–August. Arrows indicate moisture flux in gm–cm/s x 10^2, and contours the divergence of moisture flux in mm/day. (Figure courtesy of Eugene M. Rasmusson, University of Maryland.)

air (specific humidity, q), ρ is the density of moist air and dB/dt is the rate of extraction or addition of vapor to the air stream by precipitation or evaporation, respectively. Figure 59-4 shows how the boundaries of a watershed can conceptually be projected vertically and closed above and below with planes to form a control volume, how the velocity can be interpolated from the vector field onto each vertical plane of the control volume boundary, how the moisture flow Q_a through each plane can be found as the vector product $qV \bullet d\mathbf{A}$ where $d\mathbf{A}$ is the area vector of the plane. By substituting all these quantities into (59-6), an equation for computing the atmospheric water balance over a watershed is written

$$\frac{dB}{dt} = \frac{d}{dt} \iiint_{c.v.} q\rho \, dv + \iint_{c.s.} q\rho V \bullet d\mathbf{A} \qquad (59\text{-}7)$$

in which dW/dt is the rate at which water vapor is being added to the air stream by net evaporation (E-P), the first term on the right-hand side is the time rate of change of moisture in the air over the watershed, the second term on the right-hand side is the net outflow of water vapor across the watershed boundaries, and the nomenclature *c.v.* and *c.s.* refer to the control volume and control surface, respectively.

The merit of using the Reynolds Transport Theorem lies in the fact that it is a completely generally defined mathematical operator that applies to all fluid flow situations, and it provides a means of moving from continuous to discrete space representation of fluid properties. The procedure just described is not limited to accounting for atmospheric moisture motion over a watershed, it also can be applied to account for energy balances on the surface and in the atmosphere, to water flow on the watershed, and to other types of water and air flows (Chow et al. 1988). For example, the water balance (59-1) is a special case of the Reynolds Transport Theorem.

CONCLUSION

The synthesis of GIS and environmental modeling has the potential to create a new base for environmental simulation that is different and more powerful than those presently existing. To accomplish this by intrinsic environmental GIS functions requires that the mathematical language of the equations of environmental processes be translated into the programming language of the GIS. This can be accomplished by means of discrete space or lumped models, or by continuous space or distributed models. At the present stage of environmental GIS development, it seems best to focus on developing GIS functions for analytical solutions of particular environmental processes, as illustrated by the solution to the advection–dispersion equation for groundwater transport, and for characterization of spatial response functions such as by using a time–area diagram.

As environmental GIS develops, it would be very helpful to have functions to analyze vector fields of motion; in particular, to be able to calculate the divergence, gradient, and curl of such fields, as illus-

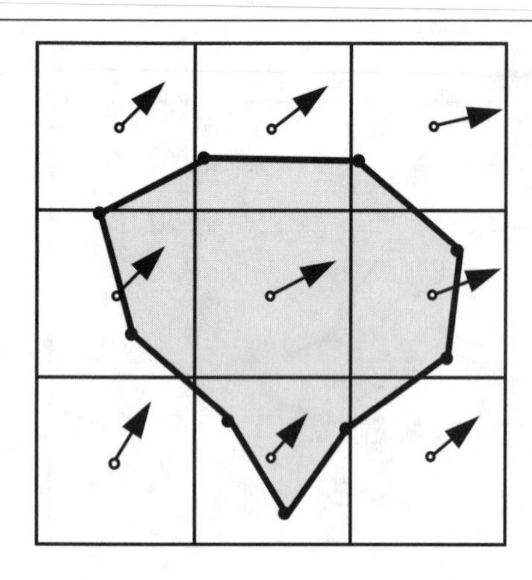

(a) A watershed overlayed on a GCM grid

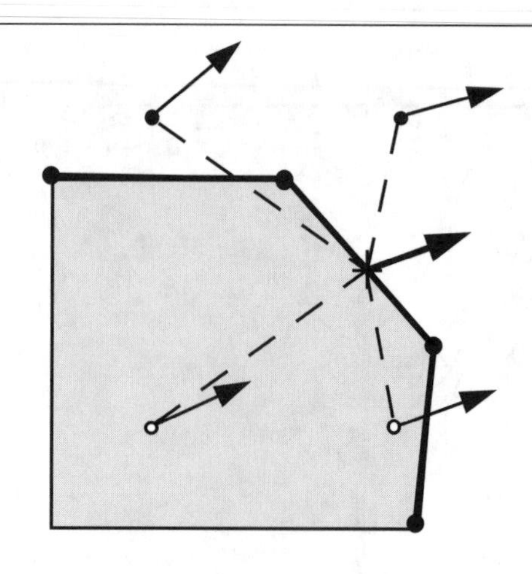

(b) Interpolation of the moisture flux vector onto the watershed boundary

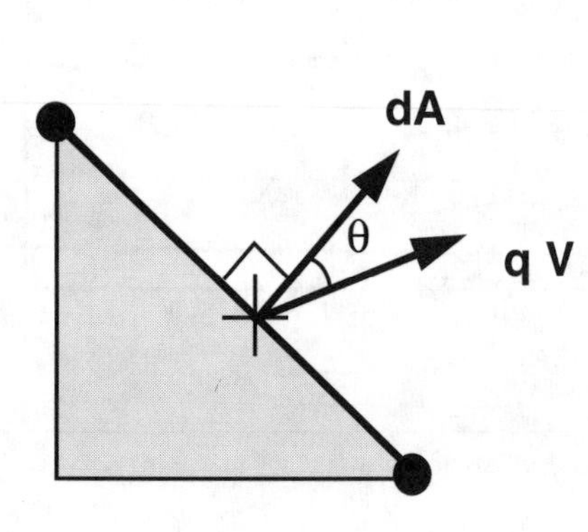

$$dQ_a = q\,V \cdot dA$$
$$= q\,V\,dA\,\cos\theta$$

(c) Computation of the moisture flow through the boundary

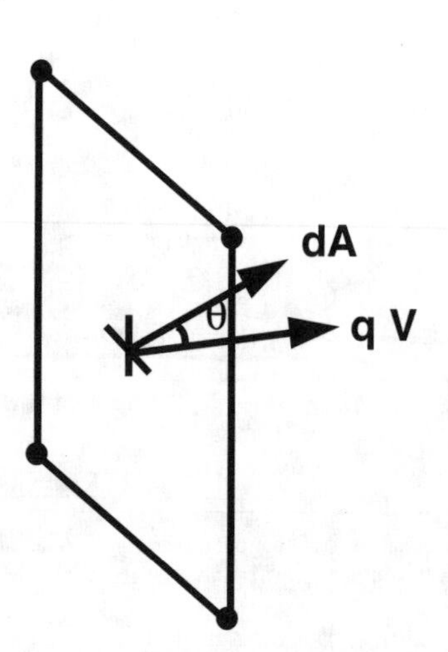

(d) The boundary as a vertical plane in the atmosphere

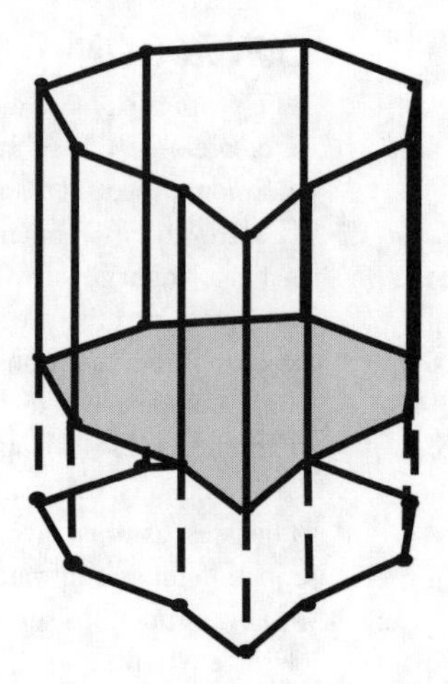

(e) The watershed as a 3-D control volume

Figure 59-4. The computation of atmospheric water balance over a watershed using a three-dimensional control volume defined by vertical projection of the watershed boundary.

trated by considering the vector field of atmospheric moisture motion whose divergence is the rate of evaporation minus the rate of precipitation. An extension of this idea to link vector fields on continuous space with mass and energy balances on a discrete space domain is to use the Reynolds Transport Theorem applied to a fluid control volume defined by projecting vertically the boundaries of a two-dimensional space such as a watershed.

Acknowledgments

The research reported here has been supported in part by grants from the Environmental Protection Agency, the Environmental Systems Research Institute, the Hydrologic Engineering Center of the U.S. Army Corps of Engineers, and by the U.S. Department of Agriculture. The cooperation of Dean Djokic, Tom Evans, Witold Fraczek, Mason Hewitt, Sud Menon, Jon Pickus, Gene Rasmusson, and John Tauxe is gratefully appreciated.

References

Aris, R. 1962. *Vectors, Tensors and the Basic Equations of Fluid Mechanics*. New York: Dover.

Bear, J. 1979. *Hydraulics of Groundwater*. New York: McGraw-Hill.

Carslaw, H. S., and J. C. Jaeger. 1959. *Conduction of Heat in Solids*, 2nd ed. New York: Oxford University Press.

Chow, V. T., D. R. Maidment, and L. W. Mays. 1988. *Applied Hydrology*. New York: McGraw-Hill.

Crank, J. 1956. *The Mathematics of Diffusion*. Oxford: Clarendon Press.

ESRI. 1992. *Understanding GIS, the ARC/INFO method*. Redlands, California: Environmental Systems Research Institute.

Fread, D. L. 1993. Flow routing. In *Handbook of Hydrology*, edited by D. R. Maidment. New York: McGraw-Hill.

Freeze, R. A., and J. A. Cherry. 1979. *Groundwater*. Englewood Cliffs: Prentice Hall.

Goodchild, M. F. 1991. Spatial Analysis and GIS. Course notes for ESRI Summer Institute, Redlands, California.

Huber, W. C. 1993. Contaminant transport in surface water. In *Handbook of Hydrology*, edited by D. R. Maidment. New York: McGraw-Hill.

Kemp, K. 1993. Environmental modeling and GIS: Dealing with spatial continuity. *Int. Ass. Sci. Hydrol. Publ.* 211:107–15.

Maidment, D. R. 1993. Developing a spatially distributed unit hydrograph by using GIS. *Int. Ass. Sci. Hydrol. Publ.* 211:181–92.

Marcus, M., and H. Minc. 1965. *Introduction to Linear Algebra*. New York: Dover.

Marsden, J. E., and A. J. Tromba. 1988. *Vector Calculus*. New York: Freeman.

McCuen, R. H. 1982. *A Guide to Hydrologic Analysis Using SCS Methods*. Englewood Cliffs: Prentice Hall.

Mercer, J. W., and R. K. Waddell. 1993. Contaminant transport in groundwater. In *Handbook of Hydrology*, edited by D. R. Maidment. New York: McGraw-Hill.

Nuclear Regulatory Commission. 1983. *Radiological Assessment: A Textbook on Environmental Dose Analysis*, Report No. NUREG/CR-3332, Washington, D.C.

Peixoto, J. P., and A. H. Oort. 1992. *Physics of Climate*. New York: American Institute of Physics.

Ragan, R. M. 1991. *A Geographic Information System to Support Statewide Hydrologic and Nonpoint Pollution Modeling*. Department of Civil Engineering, University of Maryland.

Roberson, J. A., and C. T. Crowe. 1990. *Engineering Fluid Mechanics*. Boston: Houghton-Mifflin.

Schey, H. M. 1992. *Divergence, Gradient, Curl and All That*. New York: Norton.

Seinfeld, J. H. 1986. *Atmospheric Chemistry and Physics of Air Pollution*. New York: John Wiley and Sons.

Sircar, J. K., R. M. Ragan, E. T. Engman, and R. A. Fink. 1991. A GIS-based geomorphic approach for the digital computation of time–area curves, *Proc. Am. Soc. Civ. Engr. Symposium on Remote Sensing Applications in Water Resources Engineering*.

Tauxe, J. T., D. R. Maidment, and R. J. Charbeneau. 1992. Contaminant transport modeling using new grid operators. *Proc. 12th Annual Users Conference, Redlands, CA, Env. Sys. Res. Inst.* 1:57–62.

Thomann, R. V., and J. A. Mueller. 1987. *Principles of Surface Water Quality Modeling and Control*. New York: Harper and Rowe.

Tomlin, C. D. 1990. *Geographic Information Systems and Cartographic Modeling*. Englewood Cliffs: Prentice Hall.

Zwillinger, D. 1992. *Handbook of Differential Equations*. San Diego: Academic Press.

David R. Maidment is Professor of Civil Engineering at the University of Texas at Austin where he has been on the faculty since 1981. He received his B.S. in Agricultural Engineering from the University of Canterbury, Christchurch, New Zealand; his M.S. and Ph.D. degrees in Civil Engineering from the University of Illinois at Urbana-Champaign. He is a hydrologist by specialization. He is editor in chief of the *Handbook of Hydrology* (McGraw-Hill 1993), editor of the *Journal of Hydrology*, and coauthor of *Applied Hydrology*. In 1989 he began a teaching and research program in GIS applied to hydrology and water resources at the University of Texas, where GIS is being used to support spatial hydrologic modeling.

Center for Research in Water Resources
Department of Civil Engineering
University of Texas
Austin, Texas 78712
E-mail: *maidment@batza.crwr.utexas.edu*

60

An Overview of Cell-Based Modeling with GIS

Peng Gao, Cixiang Zhan, and Sudhakar Menon

This chapter presents an overview of raster GIS functionality for cell-based modeling with emphasis on the following areas: (1) map algebra, (2) distance mapping, (3) topographic feature extraction and surface description, and (4) surface interpolation. For each area, available algorithms and different implementations are discussed along with sample applications. It will be shown that together, these GIS tools provide modeling capabilities ranging from the evaluation of analytic expressions over a spatial domain, Euclidean distance mapping, watershed and stream network delineation, and simple weighted average interpolation, to fire spread modeling, directional path distance mapping, geostatistical interpolation, and flow length and unit hydrograph calculation.

INTRODUCTION

The cell- or raster-based data model is now widely used for spatial analysis and modeling. Its strengths include a uniform location-based representation of space and the ability to represent both continuous and categorical spatial variables. Cell-based models are used in a number of areas including topographic and hydrologic analysis, solar radiation modeling, landscape ecology, wildlife habitat analysis, predictive mapping, and dispersion modeling. In this chapter we provide an overview of raster GIS functionality for cell-based modeling, emphasizing the key areas of map algebra-based analytic processing, distance mapping, topographic feature extraction and modeling, and surface interpolation. We consider two-dimensional GIS that partition space into square or rectangular cells. A spatial database is implemented in these systems as a set of map layers, one for each variable of interest. A cell in a layer stores the value of the variable at the center of the cell. This value may apply to the whole cell for categorical data or may be interpolated at other locations within the cell for continuous data. For environmental modeling, both integer and floating point data types are desirable, together with support for cells with missing data values. Our focus here is not on the integration of existing process models with raster GIS but on the generic GIS tools that allow models to be written within the GIS.

MAP ALGEBRA

The process of spatial analysis and modeling using a cell-based GIS usually involves the application of a sequence of processing operations to a set of input map layers or grids. Each operation takes as input one or more grids and generates as output a new grid. In general, operations do not modify their inputs. The structural framework within which the processing operations in a system are organized is often referred to as a map algebra (Tomlin 1990). A common classification of processing operations on raster map layers is based on the set of cells in the input grid(s) that participate in the computation of a value for a cell in the output grid: operations are classified as per-cell or local, per-neighborhood or focal, per-zone or zonal, and per-layer or global.

Although the map algebra was first introduced in a natural languagelike form (Tomlin 1990), most implementations of the map algebra use function-based modeling languages designed to support the flow of data from one processing operation to the next (ERDAS 1992; Menon et al. 1991; Shapiro 1993). All processing operations are made available in the language as either functions or operators. A parser interprets and evaluates expressions consisting of functions, operators, and grids as well as yields outputs that are saved as new grids. A sequence of processing operations can be implemented as either a single expression or a sequence of expressions.

Most systems support a rich set of per-cell analytic functions including arithmetic, trigonometric, exponential, logarithmic, statistical, tabular reclassification, and conditional evaluation functions. The map algebra allows a user to conveniently represent and evaluate an analytic formula that references a number of input spatial variables, over a specified spatial domain. Many models can be represented as a sequence of analytic expressions or as an iterative application of a sequence of expressions. An example of the use of a GIS map algebra to implement a step in such a model is shown below.

Length-Slope Factor

The length-slope (LS) factor, as a component of the Universal Soil Loss Equation, accounts for the effects of topography on erosion (Moore et al. 1991). The LS factor is defined as:

$$LS = (n + 1) \left(\frac{A_s}{22.13}\right)^n \left(\frac{\sin\beta}{0.0896}\right)^m \qquad (60\text{-}1)$$

where $n = 0.4$, $m = 1.3$, A_s is the specific catchment area, and β is the slope. Using the map algebra, this equation can be easily expressed as:

*LS = 1.4 * pow ((AS / 22.13), 0.4) * pow ((sin (SL) / 0.0896), 1.3)* (60-2)

where *LS*, *AS*, and *SL* are grids corresponding to the output length-slope factor, the specific catchment area and slope respectively, and *pow* and *sin* are per-cell functions available in the language. The map algebra processor evaluates the above expression at each cell in the spatial domain of interest. The flexibility of the map algebra in expressing mathematical equations eliminates the need for writing a new program for each new kind of equation. A cell-based map algebra thus provides an effective way to express and evaluate mathematical equations and to perform overlay modeling.

Cell-based GIS languages can also permit access to neighboring cells using a neighborhood notation. With conditional processing, neighborhood notation and support for iteration, many spreading and diffusive processes can be modeled. Neighborhood notation allows one to make a reference to cells that are neighbors of the current processing cell, using the syntax *layer* (*col,row*), where *layer* is the name of a grid and (*col,row*) is a neighborhood subscript. The two integer constants *col* and *row* are offsets in the column (map *x*) and row (map *y*) directions. Column offsets are positive in the direction of increasing *X*, row offsets are positive in the direction of decreasing *Y* (Figure 60-1). A simplified version of the fire growth model is shown below to illustrate the use of neighborhood notation in a map algebra.

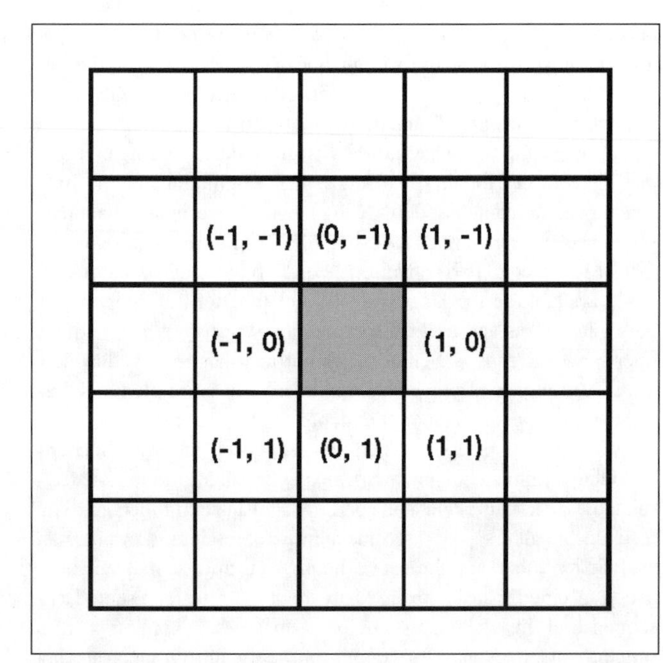

Figure 60-1. Neighborhood notation cell position.

Fire Growth Model

Dispersion over a cost or friction surface is basic to the modeling of many environmental phenomena. The spread of fire over a surface of burnability ratings is one of the examples. The fire spread can be modeled using iterative neighborhood analysis where iterations correspond to steps in time (Johnston 1991). At each iteration, the next state of each nonburning cell is computed by applying a set of rules to the neighborhood of the cell. The rules take into account the state and spatial configuration of the neighboring cells with respect to the prevailing wind direction. Additional factors considered are the effects of the burning fire on the kindling and vegetation in neighboring cells and the length of time a fire will burn before consuming itself:

```
if (old_fire > 15)
        new_fire = 0          /* Burnt Out
else if (old_fire > 10)
        new_fire = old_fire + 1   /* Continue Burning
else if (old_fire (1,0) > 10 or old_fire (1,1) > 10 or old_fire (0,1) > 10)
        new_fire = 10         /* Begin Burning
else
        new_fire = 0          /* No change
endif
```

where a grid of fire intensity at time *t2* (*new_fire*) is derived from a grid of fire intensity at time *t1* (*old_fire*) given a prevailing wind from the southeast. At the end of execution, *new_fire* could be renamed to *old_fire* and the model rerun.

A map algebra-based language can also provide facilities for building models that require decision making and iteration at the individual cell level, based on the value of the cell, before proceeding to the next cell. These facilities are provided by the flow of control and iteration statements made available to the user in the cell-based modeling language. The following example shows how these facilities are used to obtain a new grid *Y* from a given grid *X* where the derivation of *Y* from *X* requires the solution of an implicit equation for each cell, that is, $X + Y + exp(X * Y) = 0$:

```
DOCELL
        yl := -50.0
        yh := 0.0
        y := ( yl + yh ) / 2.0
        t := x + y + exp(x * y)
        while ( abs(t) > 0.01 ) {
                if (t < 0.0)
                        yl := y
                else
                        yh := y
                y := (yl + yh) / 2.0
                t := x + y + exp(x * y)
        }
        outgrid = y
END
```

Note that the *if* statement is used to control the flow of execution and the while statement provides the iteration, at each cell location.

A Rich Suite of Built-In Functions

A raster GIS platform for modeling will usually provide the user with a rich set of built-in per-cell, per-neighborhood, per-zone, and per-layer functions that can be used to generate output layers that form components of complex models. One example of such a built-in function is a per-layer function that generates the catchment area for each cell in the domain of interest given an input elevation surface. Such a function can be used to generate the grid named "AS" used as an input in the expression for the length-slope factor described above.

The set of functions available in raster GIS today includes functions for multisource Euclidean distance mapping, topographic shading analysis, watershed analysis, weighted distance mapping, surface interpolation, multivariate regression, clustering and classification, visibility analysis, and more. Using these basic tools and the map algebra, a range of models can be implemented directly within the GIS. Examples of raster GIS-based models implemented using cell-based languages include SOLARFLUX, an isolation model that computes total direct radiation over a specified time interval (Hetrick et al. 1993), and RHINEFLOW, an integrated GIS water balance model for the river Rhine (Van Deursen and Kwadijk 1993). The remainder of this chapter overviews the functionality available in the specific areas of distance mapping, topographic feature extraction and analysis, and surface interpolation.

DISTANCE MAPPING

Distance is a fundamental variable in spatial operations involving movement or regionalization based on spatial proximity. Distance mapping includes calculation of Euclidean distance, isotropic cost distance, and directional path distance, respectively, from or to a set of source locations.

Euclidean distance mapping is used to calculate the Euclidean distance from each cell in a spatial domain to the nearest cell among a given set of source cells. The set of source cells usually represents different spatial objects where a spatial object consists of a group of usually contiguous cells that share the same cell value (object identifier). It is often used to buffer zones around objects such as wells, roads, pipelines, and lakes. It can also be used to allocate regions of space to a set of objects based on proximity (generating raster Thiessen or Voronoi diagrams). Various algorithms for Euclidean distance mapping are available to efficiently compute true Euclidean distance in discrete raster space (Danielsson 1980; Tomlin 1987; Eastman 1988; Borgefors 1986). Sequential approaches are generally preferred, in which the input data are scanned several times in different directions. During each scan each cell propagates its current distance variables to its neighbors in a section along the scan direction, and each cell receives the distance variables propagated to it if they are smaller than its current one. The algorithm by Danielsson (1980) requires only two scans and a total of ten comparisons at each cell. It also produces a true Euclidean distance map with negligible and quantifiable errors.

Isotropic cost-distance mapping assumes travel over a cost surface in which each cell stores a per-unit-length cost value for travel within the cell. The cost values are isotropic with respect to travel direction. Cost-distance mapping on raster data structures assumes that travel occurs along links that connect any cell to its immediate neighbors. A cost distance can be assigned to any path between two cells based on summing the costs of travel along the individual links along the path. The shortest path from a cell to any other cell is the path with minimum cost distance over all possible paths between the two cells. The shortest cost distance between any two cells is the cost distance corresponding to the shortest path. Assuming that source objects are represented as collections of cells, cost-distance mapping assigns to each cell in the domain the distance that is smallest among the shortest cost distances from the cell to all cells in the closest object. The building block in the mapping is the cost distance between two neighboring cells. This is modeled as the cost distance of travel between the centers of the two cells and is calculated from the values stored in the input per-unit-length cost grid for the two cells (Menon et al. 1991; Zhan et al. 1993). Although it requires more computation, this method is conceptually clear and handles abrupt spatial cost changes much better.

The growth model (used in MAP) (Tomlin 1986), sequential scanning (used in IDRISI) (Eastman 1988), and graph-based approach (used in GRASS) (Dijkstra 1959) are all methods used for isotropic cost-distance mapping, which differ mainly in the method of propagation from source cells to others and the complexity of the cost surfaces that they can handle. Dijkstra's approach is capable of handling any complex cost surface and demands the least computation when all the data can be held in memory for random access. The algorithm treats each cell as a node, and links exist between neighboring cells only. At each step, the cell with the shortest cost distance so far is allowed to propagate to its neighbors, whose cost distances are updated if the newly propagated cost distances are shorter than their current ones. The algorithm ends when every cell is allowed to propagate to its neighbors. For large data sets that cannot be held in memory, the random access required in Dijkstra's approach uses much more computer input/output time than the real distance computation, in which case the Bellman–Ford algorithm (1958) can be used. This algorithm iteratively propagates the path distance from each node to all nodes along its links and updates the distances of those nodes until no node changes. An implementation of the above algorithm using a divide and iterate paradigm (Gao 1991) minimizes disk access and leads to more rapid convergence.

To make the model of distance mapping more general, Zhan et al. (1993) proposed using three additional factors in the distance measure in addition to the isotropic cost or friction surface. The first is the terrain surface factor, which converts the planimetric distance between two cells to surface distance. The surface factor is nondirectional since the surface distance between two cells does not depend on the travel direction. An elevation grid can be used to calculate the surface factor. The second is the value gradient factor, which considers the effect of the value gradient from a cell to its neighbor on the cost of travel between two neighboring cells. Fields of temperature, particle concentration, and elevation can create gradients between cells, which affect the heat flow, particle dispersion, and energy consumption, respectively. The gradient factor is directional since the gradient from cell A to cell B is generally different from that from cell B to cell A. The third is the horizontal direction factor, which accounts for the possible effects of a preferred local horizontal direction at a cell upon the cost of travel within the cell. The cost of travel from a cell to its neighbors is dependent on the angle between the preferred local horizontal direction for the cell and the moving direction from the cell to its neighbor. Wind and water flow fields can be used to create horizontal direction factor grids that can be used in the distance computation.

A general path distance model is established by combining these factors as multiplicative factors in distance measure. Assume the distance traveling from cell F to neighboring cell T is to be computed. Denoting planimetric distance between cells F and T as D_0, the friction of cell F as F_f, the friction of cell T as F_t, the surface factor between cells F and T as S_{ft}, the value gradient factor from cell F to cell T as V_{ft}, and the horizontal direction factors of cells F and T in the moving direction m as H_{fm} and H_{tm}, respectively, the distance from cell F to cell T is given by:

$$D_{ft} = 0.5\,(H_{fm} F_f + H_{tm} F_t)\, V_{ft} S_{ft} D_0 \qquad (60\text{-}3)$$

The friction and surface factors are extracted from the input friction grid and elevation grid. A value gradient function and a horizontal direction function, which can be selected from existing functions in the program or provided by users in forms of lookup tables, establish the mappings from the value gradient computed from a values surface grid to the value gradient factor and from the local horizontal direction and moving direction to the horizontal direction factor, respectively. Any input variable in the formula can be selected or dropped to make submodels for particular applications. A path-distance function corresponding to the above distance model has been implemented in GRID (ESRI 1992). The distance model can be used to select optimal routes and simulate dispersion if proper calibration is provided.

During isotropic cost-distance mapping and directional path-distance mapping, the final propagation link between cells can be saved to create a back link grid, which can be used for path search. The code of the source object, from which a cell obtains the shortest distance, can also be saved at the cell to create an allocation grid which allocates regions to objects based on nearest-neighbor proximity under the distance model used.

TOPOGRAPHIC FEATURE EXTRACTION AND SURFACE DESCRIPTION

Topographic features such as watersheds and ordered stream networks, and topographic surface descriptors such as flow accumulation and flow length can be computed from an input DEM using functions available in current raster GIS. These features and surface descriptors are the primary input to most surface hydrologic models. These models are used for such things as determining the height, timing, and inundation of a flood, as well as locating areas contributing pollutants to a stream or predicting the effects of altering the landscape. Automatic extraction of these topographic features and descriptors is very useful in GIS.

While there are many separate approaches for delineating stream networks (Band 1986; Lay 1991; Peuker and Douglas 1975; Seemuller 1989; Smith et al. 1990; Yoeli 1984) and watershed (Marks 1984; Martz and Jong 1988; Vincent and Soille 1991), the general approach proposed by Marks et al. (1984) and Jenson (1988) holds greater potential as it is more likely to yield a set of connected stream networks, and more importantly, provides a framework for extracting many topographic features. Figure 60-2 is the flowchart showing the functional flow of extracting topographic features and surface description layers from a DEM (ESRI 1992).

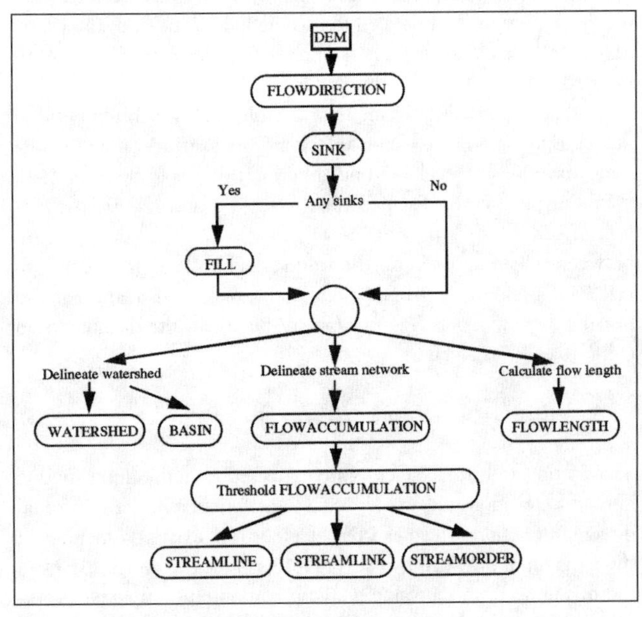

Figure 60-2. Functional flow of topographic feature extraction.

Starting with a given DEM, the flow direction is first calculated. The flow direction is defined, for each cell, as the direction of steepest descent to its neighboring cell. According to Jenson (1988), only one flow direction is allowed for each cell. When there is more than one direction with equal nonzero maximum elevation drop, the flow direction is determined as the representative direction for all via a lookup table. On flat areas where there exist multiple directions with zero maximum drop, the flow direction of a cell is defined by one of its neighboring cells which has a valid flow direction. The flow direction grid is a key component in this framework, and once calculated, it is used as the basis for all other feature extraction and surface descriptions.

The flow accumulation value of a cell is the number of upstream cells that flow to it. An optional weight can be assigned to each cell. In this case, the flow accumulation value of a cell equals the accumulated weight of all cells flowing into it. Cells with a high flow accumulation value are areas of concentrated flow and may be used to identify stream networks. By using a threshold value, stream networks can be delineated as those cells with flow accumulation values exceeding the threshold. It is suggested that a proper threshold value should be selected so that the stream networks are extracted at the appropriate drainage density (Tarboton et al. 1991).

Flow accumulation calculation is still considered a computationally expensive procedure even on powerful workstations. There are basically three approaches for calculating flow accumulation. One approach visits each cell and follows it downslope to the edge of the DEM or to a sink (Mark 1983). All cells along the way have one unit of flow accumulation added to the flow accumulation already present. Since flow accumulation from any cell could potentially flow across the entire data area, the flow direction (or elevation) and the flow accumulation counters for the entire area must be available at all times. Also, since the procedure follows paths in the downslope direction, random access is required. Random access on disk is very expensive; therefore, this approach is only suitable for small data sets that can be completely held in main memory.

A second approach is iterative (Marks et al. 1984). The flow accumulation matrix is first initialized to value one and a temporary matrix is initialized to the number of drainage inputs to a cell. On each iteration, only those cells with zero values in the temporary array are examined. The accumulation values for these cells are added to those of the neighbor to which it drains, and the value for that neighbor in the temporary array is decremented by one. The procedure terminates when no cells with zero values remain in the temporary array. Even though it was first implemented using matrices in main memory, this approach can be easily modified to use secondary storage for large data sets. However, in this case, the number of iterations over the elevation (or flow direction) and the flow accumulation grids will be proportional to the number of cells in the longest drainage path. Unfortunately, this number can be very large.

A third approach is essentially iterative, too. However, a technique called "divide and iterate" is used to reduce the number of disk accesses on the flow direction (input) and the flow accumulation (output) grids (Gao 1991). In this approach, instead of iterating over the entire area, the flow direction and the flow accumulation grids are divided into small blocks. The size of a block is selected such that it is small enough to be held in main memory. The procedure begins with a block that is flagged for update (initially only those blocks containing or adjacent to cells with zero number of drainage inputs are flagged for update). For each cell, with the initial value of 0, within a block, a candidate flow accumulation value is first calculated as:

$$\sum_k (WT(k) + FA(k)) \quad (60\text{-}4)$$

where $WT(k)$ and $FA(k)$ are the weight and the flow accumulation value of the kth neighbor, and the summation is over all neighboring cells that flow into the center cell. The candidate flow accumulation value is then compared with the old *FA* value at the center cell. If the old value is smaller, then the old *FA* value is replaced by the candidate value. If the center cell is on the block boundary, the neighboring blocks are flagged for further update. This process iterates on a block until no change can be made. The procedure continues with the next flagged block and terminates when there is no block that is flagged for update. Since this approach continuously updates the flow accumulation values within a block until the values converge to a stable state while the block is in the main memory, it effectively reduces the number of disk accesses over the flow direction and the flow accumulation grids. This approach has been implemented as the FlowAccumulation function within GRID (ESRI 1992).

Flow length is another primary topographic attribute. The upstream flow length, also called flow path length for a cell, is defined as the maximum distance of water flow to the cell from its farthest upstream source cell (Moore et al. 1991). The downstream flow length is the distance from the cell downhill to the outlet. Both the upstream and downstream flow length can be calculated similar to the flow accumulation except that the maximum linear distance is calculated instead of accumulated flow. While there are algorithms based on constructing flow lines (Mitasova 1993), an iterative approach is described here (Gao 1991). On each iteration, for each cell, with the initial value of 0, within a block, a candidate flow length value is calculated as:

$$MAX\,(C(k)\,(W_0+WT(k))+FL(k)) \qquad (60\text{-}5)$$

for upstream flow length, where $WT(k)$ and $FL(k)$ are the weight and the flow length value of the kth neighbor, W_0 is the weight of the center cell, $C(k)$ is the factor used to adjust the distance for the diagonal cells, and the maximization is on all neighboring cells that flow into the center cell. For the downstream flow length, a candidate flow length for a cell is computed as:

$$C(k_0)\,(W_0+WT(k_0))+FL(k_0) \qquad (60\text{-}6)$$

where k_0 denotes the neighboring cell it drains into. The candidate flow length value is then compared with the old *FL* value at the center cell. If it is smaller, the old *FL* value is replaced by the candidate value, and the change flag is set properly for further update.

A watershed is the upslope area contributing flow to a given location and can be delineated from a DEM. Technically, there is a difference between the watersheds for specified source locations and the watersheds, or basins, for sinks. The basin can be simply computed using the connected labeling technique where the connectivity is based on flow direction (Marks et al. 1984). However, it is relatively more difficult to delineate watershed for any specified locations. One approach is similar to the iterative ones for the flow accumulation. Initially, only locations corresponding to given source locations are assigned with valid labels and the rest are marked as undefined. On each iteration only those cells with the undefined label are examined. For each cell, if the downslope cell has a valid label, the current cell is assigned with that label. The procedure stops when all cell locations have been assigned with valid labels. It is easy to see that this approach also can be improved using the divide-and-iterate technique (Gao 1991).

Table 60-1 lists the results of a performance test on the flow accumulation implemented using the first and third approaches described above. All time measures are in CPU seconds. The test is run on a SUN SPARC 1+ in a networked environment. Although nothing can be concluded from this simple test, one can observe that the computing time of the first approach increases dramatically as the size of the data set increases, while that of the third approach increases slowly in nearly linear relationship with the size of the data set.

Table 60-1. A performance test on the flow accumulation.

Size (Cells)	Approach 3 (cpu seconds)	Approach 1 (cpu seconds)
67,280	49	132
269,120	192	1,544
1,076,480	800	19,772
2,422,080	2,241	N/A
4,317,992	4,396	N/A
9,715,482	6,856	N/A

Unit Hydrograph Estimation

Given a DEM for a watershed, a grid of flow direction is defined from each cell to one of its eight neighboring cells in the direction of the maximum downhill slope. From this a grid of flow direction distance can be computed for the watershed by tracing from each cell downstream to the lowest cell and storing in each cell its flow distance to the outlet. By assigning a velocity of flow, which takes the form: $V=aS^b$, where S is the land surface slope and a and b are coefficients related to the land use, to each cell, a grid of flow times, *ft*, to the outlet can similarly be computed. The time–area diagram for the watershed, $A(t)$, is then defined as:

$$A(t) = \sum_{i,j} A_t(i,j);\; A_t(i,j) = \begin{cases} a_0 & ft(i,j) \le t \\ 0 & \textbf{Otherwise} \end{cases} \qquad (60\text{-}7)$$

where i, j are row and column numbers and a_0 is the cell area. It is pointed out that the slope of the time–area diagram is the unit hydrograph with the dimensions of L^2/T (Maidment 1993). Using the tools described in this section, the flow time grid and time–area diagram can be determined. The following script calculates the flow times, given a DEM, for a watershed:

```
fd = flowdirection (dem)
fv = a * pow (slope (dem), b)
ft = flowlength (fd, fv, DOWNSTREAM)
```

where a and b are coefficients related to the land use and *dem* is the DEM for the watershed. For a given time step, t, the time–area value, $A(t)$, can be calculated as the following:

```
DOCELL
    at += ft <= t
END
```

where *ft* is the flow time grid calculated from the last step. The time–area diagram can be compiled by calculating $A(t)$ at different time steps. Once the flow time–area diagram is constructed, the estimation of the unit hydrograph is done by estimating the slope of the time–area diagram.

SURFACE INTERPOLATION

There is a rich set of surface interpolation methods available. Starting with trend surface analysis, the simplest way to describe gradual long-range variations is to model them by polynomial regression. The idea is to fit a polynomial surface by least squares through the data points (Burrough 1986). It is assumed that the spatial coordinates X, Y are the independent variables and that Z, the surface property of interest, is the dependent variable. Finding the coefficients of a polynomial is a standard problem in multiple regression, and therefore easy to accomplish. Code is also available in FORTRAN or C (Press et al. 1988). However, it is warned that polynomial regression is an ill-conditioned least squares problem that needs careful numerical analysis, due to the extremely large value differences among the terms ($X, Y, X^2, Y^2, XY, X^3, \ldots$) in a polynomial. It is recommended that all terms be rescaled in the range $-1<0<1$ to avoid some numerical problems.

Inverse distance weighting (IDW) interpolation is another widely used method. IDW is based on the intuitively appealing notion that values at nearby data points are more significant than distant observations when an estimation is to be made at an unknown location. The influence of an observation is taken as being inversely related to its distance from the unknown location (Burrough 1986). IDW is a simple procedure, thus easy to implement. However, if the k nearest data points are desired for the interpolation, the task of efficiently finding these k nearest data points from a given location is not a trivial one. A spatial search structure is needed to facilitate the task. To correct the problems associated with clustered data, a modification has been made to the original IDW (Heine 1986). Rather than select the k nearest points, the search space is divided into quadrants and an equal number of points are selected from each quadrant. This ensures that the interpolated value will depend on data points from all directions and will be more independent of the cluster effect.

Triangle-based methods are a class of interpolation techniques that explicitly use the spatial ordering established by the triangulation. Linear interpolation, where planar facets are fitted to each triangle, generates a continuous surface coinciding with all the data points. The surface property z value can be easily calculated for any point within a triangle by a weighted average of the three observations. It is a fast method and most useful for initial investigation where smooth isolines are not important. To obtain a smoother surface, higher-order interpolating functions should be used. Akima (1978) suggested using a fifth-degree polynomial in x and y defined in each triangle cell which has projections of three data points in the x–y plane as its vertices. The z value at any location is interpolated as:

$$z(x, y) = \sum_{i=0}^{5} \sum_{j=0}^{5-i} q_{ij} x^i y^j \tag{60-8}$$

where q_{ij} are coefficients. The values of the function and first-order and second-order partial derivatives given at each vertex are used to estimate the twenty-one coefficients.

Kriging has been highly recommended as a method of interpolation for GIS (Griffith 1992; Oliver and Webster 1990). Kriging interpolation is considered to be optimal in the sense that the interpolation weights are selected so as to optimize the interpolation function to provide a best unbiased estimate of the value of a variable at a given point, provided that the conditions, stationarity of difference and variance of differences are satisfied. Kriging is a two-step process including semivariance estimation and interpolation. The semivariance can be estimated from sample data using the following formula:

$$\hat{\gamma}(\vec{h}) = \frac{1}{2n} \sum_{i=1}^{n} \{Z(\vec{x}_i) - Z(\vec{x}_i + \vec{h})\}^2 \tag{60-9}$$

where $\vec{x}$ and $\vec{h}$ are two-dimensional location and distance and n is the number of pairs of data points separated by distance h, or by distance range $(h\text{-}d/2,\ h\text{+}d/2)$, if a distance interval, d, is used. In an actual implementation, the distance interval should be controllable by the end user so that a satisfactory semivariogram can be found. Since the estimated semivariance only exists at discrete distance intervals, and is often scattered due to imperfection in sample data, a mathematical model is fitted in order to describe the way in which the semivariance changes continuously with the lag. There are many models available including linear, circular, spherical, exponential, and Gaussian (McBratney and Webster 1986) and procedures to fit the estimated semivariogram to the models (Press et al. 1988).

Once the semivariance estimation step is finished, the fitted semivariance can be used to determine the weights needed for interpolation. The interpolated value at an unvisited location is given by:

$$\hat{Z}(\vec{x}_0) = \sum_{i=1}^{n} \lambda_i Z(\vec{x}_i) \tag{60-10}$$

The weights λ_i are chosen so that the estimate $\hat{Z}(\vec{x}_0)$ is unbiased, and so that the estimation variance is less than for any other linear combination of the observed values and can be found by solving the following system of equations:

$$\left\{\begin{array}{l} \sum_{i=1}^{n} \lambda_i = 1 \\ \sum_{j=1}^{n} \lambda_i \gamma(\vec{x}_i, \vec{x}_j) + \mu = \gamma(\vec{x}_i, \vec{x}_0)\ ; i = 1, 2, 3, \ldots, n \end{array}\right\} \tag{60-11}$$

There are $n+1$ equation and $n+1$ unknowns (λ_i, μ). This system of equations needs to be constructed and solved for every cell. This form of kriging is often called ordinary point kriging (Burgess and Webster 1980a). There are also some other forms of kriging in use, such as ordinary block kriging and universal kriging, which are designed to perform interpolation of an average value on a block (Burgess and Webster 1980b) and interpolation in the presence of local trends (Webster and Burgess 1980; Royle et al. 1981), respectively.

CONCLUSION

GIS have matured from their early stage where only a limited set of primitive functions were available, and where one had to resort to programming to solve any real modeling problem. Today, most raster GIS provide facilities to express almost any mathematical equations in the form of the map algebra and other tools such as distance mapping, topographic feature extraction, and surface interpolation, to either generate input variables for modeling or perform the modeling itself.

References

Akima, H. 1978. A method of bivariate interpolation and smooth surface fitting for irregularly distributed data points. *ACM Trans. Mathematical Software* 4:148–59.

Band, L. E. 1986. Topographic partition of watersheds with digital elevation models. *Water Resources Research* 22:15–24.

Bellman, R. 1958. On a routing problem. *Quarterly of Applied Mathematics* 16(1):87–90.

Borgefors, G. 1986. Distance transformations in digital images. *Computer Vision, Graphics and Image Processing* 34:344–71.

Burgess, T. M., and R. Webster. 1980a. Optimal interpolation and isarithmic mapping of soil properties I: The semivariogram and punctual kriging. *Journal of Soil Science* 31:315–31.

———. 1980b. Optimal interpolation and isarithmic mapping of soil properties II: Block kriging. *Journal of Soil Science* 31:333–41.

Burrough, P. A. 1986. *Principles of Geographical Information Systems for Land Resources Assessment.* New York: Oxford University Press.

Danielsson, P. 1980. Euclidean distance mapping. *Computer Graphics and Image Processing* 14:227–48.

Dijkstra, E. W. 1959. A note on two problems in connection with graphs. *Numerische Mathematik* 1:269–71.

Eastman, J. R. 1988. Pushbroom algorithm for calculating distances in raster grids. *Auto-Carto 9* 288–97.

ERDAS. 1992. *Writing Models in ERDAS IMAGINE Using the Spatial Modeler Language*. ERDAS, Inc.

ESRI. 1992. *Cell-Based Modeling with GRID 6.1*. ESRI, Inc.

Gao, P. 1991. *GRID Hydrologic Tools*. ESRI Internal Technical Report. ESRI, Inc.

Griffith, D. A. 1992. Which spatial statistics techniques should be converted to GIS functions? *Annals of Regional Science*.

Heine, G. W. 1986. A controlled study of some two-dimensional interpolation methods. *COGS Computer Contributions* 2:60–72.

Hetrick, W. A., P. M. Rich, and S. B. Weiss. 1993. Modeling insolation on complex surfaces. *Proceedings of the Thirteenth Annual ESRI User Conference* 2:447–58.

Jenson, S. K., and J. O. Domingue. 1988. Extracting topographic structure from digital elevation data for geographic information system analysis. *Photogrammetric Engineering and Remote Sensing* 54(11).

Johnson, K. 1991. Using a GIS to model dispersion: A fire-spread example. In *Proceedings of the First International Conference/Workshop on Integrating Geographic Information Systems and Environmental Modeling*, Boulder, Colorado.

Lay, J. 1991. Terrain feature extraction from digital elevation models: A multiperspective exploration. *GIS/LIS '91 Proceedings*. Atlanta, Georgia.

Maidment, D. R. 1993. Developing a spatially distributed unit hydrograph by using GIS. *HydroGIS '93: Application of Geographic Information Systems in Hydrology and Water Resources*. Vienna, Austria.

Mark, D. M. 1983. Automated detection of drainage networks from digital elevation models. *Auto-Carto 6 Proceedings*. Ottawa, Canada.

Marks, D., J. Dozier, and J. Frew. 1984. Automated basin delineation from digital elevation data. *Geo-Processing* 2:299–311.

Martz, L. W., and E. D. Jong. 1988. CATCH: A FORTRAN program for measuring catchment area from digital elevation models. *Computers & Geosciences* 4:627–40.

McBratney, A. B., and R. Webster. 1986. Choosing functions for semivariograms of soil properties and fitting them to sampling estimates. *Journal of Soil Science* 37:617–39.

Menon, S., P. Gao, and C. Zhan. 1991. GRID: A data model and functional map algebra for raster geoprocessing. *GIS/LIS '91 Proceedings*. Atlanta, Georgia.

Mitasova, H. 1993. Surfaces and modeling. *Grassclippings* 7:18–19.

Moore, I. D., R. B. Grayson, and A. R. Ladson. 1991. Digital terrain modeling: A review of hydrological, geomorphological, and biological applications. *Hydrological Processes* 5:3–30.

O'Callaghan, J. F., and D. M. Mark. 1984. The extraction of drainage networks from digital elevation data. *CGVIP* 28:323–44.

Oliver, M. A., and R. Webster. 1990. Kriging: A method of interpolation for geographical information systems. *Int. J. GIS* 4:313–32.

Peuker, T. K., and D. H. Douglas. 1975. Detection of surface-specific points by parallel processing of discrete terrain elevation data. *Computer Graphics and Image Processing* 4:375–87.

Press, W. H., B. P. Flannery, S. A. Teukolsky, and W. T. Vetterling. 1988. *Numerical Recipes in C: The Art of Scientific Computing*. Cambridge University Press.

Royle, A. G., F. L. Clausen, and P. Frederiksen. 1981. Practical universal kriging and automatic contouring. *Geo-Processing* 1:377–94.

Seemuller, W. W. 1989. The extraction of ordered vector drainage networks from elevation data. *CVGIP* 47:45–58.

Shapiro, M. 1993. R.mapcalc: Raster map layer data calculator. *GRASS Reference Manual*.

Smith, T., C. Zhan, and P. Gao. 1990. A knowledge-based two-step procedure for extracting channel networks from noisy DEM data. *Computers and Geosciences* 16(6):777–86.

Tarboton, D. G., J. C. J. Bras, and I. Rodriguez-Iturbe. 1991. On the extraction of channel networks from digital elevation data. *Hydrological Processes* 5:81–100.

Tomlin, C. D. 1986. *The IBM Personal Version of the Map Analysis Package*. The Laboratory for Computer Graphics and Spatial Analysis, Harvard University, Cambridge.

———. 1987. Three cartographic distance-weight interpolation techniques. *Proceedings, the First Latin American Conference on Geographical Information System*. Costa Rica.

———. 1990. *Geographic Information Systems and Cartographic Modeling*. Englewood Cliffs: Prentice Hall.

Van Deursen, W. P. A., and J. C. J. Kwadijk. 1993. RHINEFLOW: An integrated GIS water balance model for the river Rhine. *HydroGIS '93: Application of Geographic Information Systems in Hydrology and Water Resources* 1:507–18.

Vincent, L., and P. Soille. 1991. Watersheds in digital spaces: An efficient algorithm based on immersion simulations. *Trans. on Pattern Analysis and Machine Intelligence* 13:583–98.

Webster, R., and T. M. Burgess. 1980. Optimal interpolation and isarithmic mapping of soil properties III: Changing drift and universal kriging. *Journal of Soil Science* 31:505–24.

Yeoli, P. 1984. Computer-assisted determination of the valley and ridge lines of digital terrain models. *International Yearbook of Cartography* 24:197–206.

Zhan, C., S. Menon, and P. Gao. 1993. A directional path distance model for raster distance mapping. *COSIT '93*.

Peng Gao, *Cixiang Zhan*, and *Sudhakar Menon*
Environmental Systems Research Institute, Inc.
380 New York St.
Redlands, California 92373
E-mail: *pgao@esri.com*
E-mail: *czhan@esri.com*
E-mail: *smenon@esri.com*

61

Refining a Cellular Automaton Model of Wildfire Propagation and Extinction

Keith C. Clarke and Greg Olsen

A cellular automaton-based model previously was designed to simulate the propagation and extinction behavior of wildfire. Research on the model has moved from prototyping and calibration to refinement, using more sophisticated data from remote sensing, field data collection with GPS, and extension to forest as well as range fires. Data collected in 1993 were used to classify and model the fuel properties of an extensive area in the Santa Cruz Mountains in California. The area includes several state forests and habitat protection districts, contains extensive stands of uncut Douglas fir and coastal redwood, and is particularly susceptible to fire. Accurate fuel characterization, coupled with model-based estimates of fuel moisture conditions, were used as input to the Monte Carlo version of the fire model. Data have been classified, geometrically and radiometrically corrected, and ground truth checked and integrated with the fire model using a suite of GIS-based and other software tools. The result has been a set of fire risk maps and assessments that can be used to assist in planning for fire risk management in the Santa Cruz Mountains. Issues related to model limitations, data accuracy, data integration, and fire risk management are discussed. The integration of data, both real and modeled, with the cellular automaton prediction to make risk maps is seen as an effective and inexpensive means by which intelligent wildland management can be coupled with remote sensing to save life and property and to protect biological diversity and land resources.

INTRODUCTION

Fires are important for understanding global change because: (1) they decompose significant amounts of organic biomass into minerals such as nitrates and silicates, and into carbon-based gases such as carbon dioxide and carbon monoxide; (2) they substantially alter the species and biogeochemical geography of affected areas, especially vegetation and soils, both in the short and long terms; and (3) they have an impact on fauna, both in terms of habitat and population. Fires kill and injure people and animals, destroy property, and cost a great deal of money to combat. Fires are a major problem in areas that have dry climates, in areas that experience periodic or seasonal drought, and in areas where ignition sources are regularly present. While many fires are "natural," the dominant ignition sources in the United States are lightning, accidents, and arson, and a significant number of fires in the United States have their behavior modified by countermeasures.

Wildland fires, mostly in coniferous forest, chaparral, and grassland, are most likely to involve significant areas, longer periods of burning, and less fire fighting. Fire management policy varies by land use and ownership, varying from "fight every fire" to "let-it-burn" approaches. In general, national fire management policy changed significantly as a result of the catastrophic Yellowstone fires of 1988, with the result that human intervention in the natural fire cycle of fire-prone landscapes is now usually "work with" rather than "work against."

Working with fire implies data collection and fire modeling, and significant research literature addresses these issues. Fire data collection has been significantly improved by the use of data from remote sensing. Continued sensor improvements have led to some major breakthroughs in measuring fire and smoke parameters in real time. Similarly, fire modeling has continuously built upon the classical fire models to improve the prediction of fire parameters.

In an earlier work (Barbour et al. 1987), a fire model was proposed that used a cellular automaton to simulate fire behavior. The model worked on a regular grid and used data from remote sensing, digital elevation models (DEMs), and local environmental conditions to simulate fire spread and extinction spatially and thermally over time. One outcome of the model is probabilisitic estimates of fire risk over space, based on a Monte Carlo implementation of the cellular automaton. The model was calibrated using data for the Lodi Canyon fire in the San Dimas experimental forest in 1986 and was found to provide useful and reasonably accurate estimates of aggregate fire risk.

More recently, the model has been extended. First, two of the static spatial parameters that were scalar estimates in the earlier calibration application have been replaced with spatially distributed layers. The first parameter is the critical fuels layer, which has been replaced by values from a land cover classification of a Thematic Mapper Simulator layer. Published species data from wind tunnel experiments and estimates of aboveground dry weight biomass allow the conversion of vegetation cover to a fuel layer. The second parameter is the fuel moisture, which has been estimated by distributed parameter hydrologic modeling using the GRASS-GIS.

This chapter discusses both classical fire models and the cellular automaton model and its refinements and investigates some of the issues related to using a GIS to integrate data and models for fire risk assessment. The Conclusion points to further necessary model refinements, to additional useful fire data sets that are suitable for applications of the model, and to a set of recommendations for the next necessary stages if the results of this research are to be used for effective fire management in the field. Increased integration between GIS and programming tools necessary for modeling is a clear need for future work, a trend that is already evident in research literature.

CLASSICAL FIRE MODELING

The pioneering physical model of fire spread was by Rothermel (1972). Rothermel's model is the basis of all United States-based fire behavior simulations by computer programs such as FIRECAST (Cohen 1986) and BURN (Albini 1976). The model consists of a set of flux equations for the physical and chemical reactions within fires. Spread is determined by measuring differences in fluxes. Most of the physical and chemical constants for the equations, such as the thermal content of various fuels and the relationship between spread rates and fuel densities, have been determined experimentally by more than fifty years of cumulative work.

Anderson (1983) developed the currently accepted model of combustion geometry, and his ideas have been incorporated into fire models such as BURN. Anderson built upon some classic results by Fons (1946), which included the results of 198 wind tunnel combustion chamber experiments, and a number of well-studied wildfires. By least squares fits of log regressions of wind speed and amount of forward spread, Anderson produced a set of equations that describe the fire geometry as a bicentered ellipse, allowing the computation of fire area and total perimeter as well as maximum fire width, since the length-to-width ratio of the fire is a function of wind speed. Thus, as long as fire remains (1) as discrete, single-centered, uncorrelated spread sequences; (2) unaffected by major differences in fuel; (3) in unchanged wind; and (4) uninfluenced by topography other than forward spread gradient, Anderson's geometric model (and by implication, Rothermel's general model), will provide a good estimate of fire behavior.

Spatially, however, few real fires conform to these behavioral constraints. Pyne (1984), for example, stated, "For mathematical analysis, the shape can be considered as a double ellipse. The stronger the wind or slope, the closer the total fire shape will approximate the ellipse. When the fire burns slower, influenced by several factors, more rugged shapes result, for which common terms are necessary." Fire managers and firefighters use the "common" terminology of experience and observation to describe features of a fire such as islands, spots, and pockets. Photographs of fire scars show extreme variation in shape, and even topology, and that fire scars are irregular, fingerlike, have islands and detatched outliers, and show crenulated edges.

THE CELLULAR AUTOMATON MODEL

Basis of the Model

The basis of the fire model developed is that fire spread is a process closely resembling that of diffusion limited aggregation (DLA). DLA is an aggregation process that has been used to explain the growth of snowflakes, mineral crystals, and cities (Fotheringham et al. 1990; Meakin 1983; Mullins and Sekerka 1963). This body of theory has followed from that of fractals, and DLA has been shown to produce objects that are self-affine and scaling—both properties of fractals (Mandelbrot 1985).

DLA is a process-oriented model and assumes that an object will grow and change shape due to a large number of essentially random events, each of which is influenced by the events that preceded it. In the context of fires, DLA can be thought of as a process by which "firelets" are sent out one at a time from a fire source. The process has been implemented as a cellular automaton, in which the application is to elements of a regular grid, which coincides with the resolution of the DEM. Each stage of change consists of the fire status at a time, plus a set of rules for change and survival. Ignition consists of a single moving firelet. If the firelet finds fuel, it ignites it and moves in a direction determined by the fire environment. If there is no fuel at its new location, or if the firelet has moved too far from its source, the firelet stops. The next firelet then moves out from the fire center. If this firelet finds a cell that has already been burned, the firelet continues on its journey. Upon finding virgin fuel, the firelet ignites it and stops.

Clearly, this process would result in a fire that always goes out after burning all the fuel within a circle of the source. Three other factors come into play to alter this balance: (1) fires with many successful firelets are allowed to breed new centers, each of which resembles the original; (2) when firelets go out, the fire center itself moves in the direction of the last successful burn; and (3) to counteract the propagation, after each "wave" of firelets from all centers, the fire is "aged" by allowing all burning cells to consume an amount of fuel determined by the local "heat" level. This final factor allows old fires to die and normally eventually extinguishes the fire.

In the cellular automaton, a computer program reads the fire environment layers (the DEM and a fuels map) and sets several scalar environment parameters such as air temperature and relative humidity. The program then ignites a fire at a location given by the user so that any number of fires can be started simultaneously or at set times and places. The eight-cell neighbors of any given pixel are then numbered off in octal. These neighbors start with weights assigned on the basis of the wind direction and magnitude. These weights are modified by changing the weights to reflect the topography; upslope aspects are weighted by the magnitude of the slope. A second modification again changes the weight to reflect the fuel load, taking into account the current status of the fire. A random number is then drawn to determine direction of movement. The new fire center is then "burned," and the fire moves on. A "run" will stop when the image edge is reached, when no fuel remains, or at random. Each center continues to generate random fire runs of a length that reflects the fuel moisture and preheating conditions until it has no unburned fuel. When successive runs find no new fuel, the fire center goes out.

In addition, fire spreading centers (unlike the classical DLA model) are permitted to "migrate" toward the active fire front, with the direction and magnitude of the movement determined partly at random and partly to reflect the location of the most active fire front. To further stimulate spread, when the fire intensity reaches a set level, new fire centers are permitted to form at the ends of firelets, each of which functions as a new independent fire, therefore making the process recursive and self-affine. The maximum number of simultaneous secondary fires is user-determined and is a function of total intensity over all fire centers and the preheating of the fuel.

To simulate the combustion process, fires can age in two ways. First, as a run burns anew across a pixel, the fire consumes some of the fuel at this location. At the end of a spreading phase, all burning pixels are aged by allowing additional fuel consumption. This simple mechanism makes the interior of a burning area eventually go out, leaving the active new fire at the front. The mechanism also forces the fire to go out if fire conditions are unfavorable, such as too little wind, not enough fuel, and so on. The burning pixels in the model are assigned a temperature level of between zero and eleven, with zero used for unburned and eleven for extinguished.

Data Requirements

The computer version of the model uses a pseudorandom number generator seeded by the system process identification number and therefore produces similar but never identical fires in successive runs. Using a given number, however, then allows the same fire to be repeated with different conditions. Input variables for the fire consist of a fuel map, a terrain map in the form of a DEM, a wind table, the wind magnitude and direction, the air temperature and relative humidity, and the fuel moisture content. The wind table is a matrix of probabilities of fire movement at different compass directions and wind magnitudes. The default table is determined by taking measurements from a figure in Anderson (1983).

Many other factors are controlled in the model, their values having been set by calibration with the Lodi Canyon data. These include rates of spread, maximum permitted number of new fires and the rate of self-replication, weighting factors for terrain slope, and extinction conditions. Each of these factors is partially linked, however, to one or more of the input variables and is therefore partly under user control. Also, the model has no smoke component, which is a severe limitation since smoke affects preheating, the flaming front, shadows, etc. Finally, the model assumes that ignition is determined by the user or at random and has no process model to simulate lightning or any other fire cause.

Fire factors not yet included in the model are specific calorimetric values for fuels, details of fuel thermal behavior, fuel packing ratios, and wind variability. Currently, no fires are allowed to be subjected to dynamic variations in conditions, that is, once a run is started, conditions must remain fixed until the end.

Two versions of the model have been programmed. The first uses the X-Windows graphical user interface to present the user with three windows: a fire control window, a fire status window, and a map display. The user moves sliders and sets values in the control window and then activates the simulation by pressing a button. As the simulated fire burns, its physical properties are displayed on a moving graph in the status window, while the thermal and spatial patterns are draped over a map of the topography in the map window. This version was designed as a prototype "trainer" for fire managers, to allow experiments with fire control parameters for the Lodi Canyon test data, and was used extensively in the calibration phase of the model.

The second version of the model was designed without graphics and allows the user to repeat one or more fires under similar or differing conditions many times, in each case accumulating the likelihood that a pixel is burned. This Monte Carlo version of the model was used to compute probability estimates and can do so either for constant or variable environment conditions. Thus, for example, the program can be set to repeat random fires under the same conditions or allow wind, temperature, fuel moisture, and so on, to vary, either at random or within constraints.

Testing and Calibration

A test version of the model was parameterized using the X-Windows implementation until satisfactory linkages for the environmental constraints were achieved. The model was then calibrated against the Lodi Canyon test data both spatially and temporally. In spite of some limitations of the test data, particularly due to sensor saturation in the infrared at higher temperatures, the calibration allowed the duplication of the real data using only the input data, the ignition points, and the behavior rules. Correspondence between predictions and observations was good. For example, with 100 iterations under constant conditions, of the pixels estimated by the model to have a greater than even chance of burning, 77.9% were actually burned in the Lodi test burn. With randomly determined fire conditions, 82.5% of the actually burned pixels were correctly identified by the simulations (Clarke et al. 1994).

EXTENDING THE MODEL

Two major shortcomings in this research exist due to the circumstances under which the prototype model was developed. First, as far as data were concerned, the model's calibration would be greatly enhanced by using actual fuel estimates as a data layer rather than the nominal layer assumed for the Lodi Canyon data. Second, the model assumes wind to be uniform in magnitude and direction, whereas, in fact, wind interacts with both topography and the fire and has vertical and horizontal structure. The Lodi test was exceptional in that: (1) it was ignited by a helicopter-carried napalm torch with many ignitions spread widely over the test burn in close to calm wind conditions; and (2) the fuel had been prepared by cutting to maximize drying since the goal was to maximize combustion temperature. The present work addresses the former point only. In short, we have sought to find a study area for which extensive information is available, both from fieldwork and remote sensing, to provide meaningful fuel load estimates as quantitative assessments of biomass. The region chosen was part of the Santa Cruz Mountains in California.

Extended Test Data Set

A new test data set has been compiled with the assistance of NASA's Ames Research Center, USGS National Mapping Division's Western Mapping Center, and with fieldwork by the authors using GPS. The input data layers added to the model consist of four mosaicked digital elevation models (DEMs) produced by interpolation from the digital line graph hypsography (Big Basin, Mindego Hill, Franklin Point, and LaHonda 7½ minute quadrangles) and an image from NASA's Ames Daedalus Thematic Mapper Simulator taken on March 19, 1988, by Ames's ER-2 high-altitude aircraft. At an altitude of 19,800 m, the Daedalus TMS gives a nominal ground resolution of 25 m at nadir. The field data consist of twenty-five ground control points surveyed on foot and located using a Magellan NAV 5000 handheld GPS receiver. At each point, the dominant species canopy contribution and canopy closure were noted.

Land Cover, Fuel Load, and Biomass

Species composition affects the combustibility of fuel, requiring that the vegetated landscape be stratified into dominant species classes. Fire fuel parameters for most species are available from published combustion chamber experiments. Species composition was extracted from the remotely sensed data radiometrically, based on the spectral signature of representative sites. Each class must be analyzed separately to estimate the amount of aboveground vegetation, because each overstory vegetation type will have a different set of allometric relationships and a different understory species composition. Any individual location will have ecological community type with a particular species composition. Canopies can be multitiered and can be composed of varying percentages of subdominant species, but the relationships between species in a community are relatively constant (Barbour et al. 1987). There can be significant overlap between communities, however, which requires a certain subjectivity in map generalization (Kuchler and Zonneveld 1988).

The classification of the imagery was done by a "clusterbusting" algorithm (Narumalani et al. 1993) applied to a set of band ratios. The band ratios were chosen based on their sensitivity to floristic variation in the canopy cover (Peterson et al., *IEEE*, 1986).

The band ratios chosen were the thematic mapper equivalent bands 4/3, 5/4, and 2/3. Band ratios were used rather than the raw spectral bands in order to reduce the effects of topography. An isodata algorithm was applied to the three band ratios, which provided a set of spectral signatures that were then input into a maximum likelihood classifier. The signatures that could readily be identified as a given class were merged and the classification was run again until all clusters could be accounted for. The resulting classes were grassland composed of wild oat (*Avena* spp.) and wild rye (*Elymus* spp.); hard chaparral composed primarily of coyote brush (*Baccharis pilularis*), chamise (*Adenostoma fasciculatum*), deer weed (*Lotus scoparius*), and California sage (*Artemisia californica*); oak woodland composed of coast live oak (*Quercus agrifolia*), canyon oak (*Q. chrysolepis*), interior live oak (*Q. wislizenii*), bay laurel (*Umbellularia californica*), and madrone (*Arbutus menziesii*); mixed evergreen composed of coast live oak, bay laurel, Douglas fir (*Pseudotsuga menziesii*), and madrone; and mixed coniferous composed of Douglas fir, coastal redwood (*Sequoia sempervirens*), and tan oak (*Lithocarpus densiflorus*). The oak woodland, mixed evergreen, and mixed coniferous classes all had canopy closure of 80%–100%, and the Douglas fir component of mixed evergreen contributed less than 50% to the canopy. Percent cover types for the entire image were grassland 16.7%, chaparral 5.7%, oak woodland 15.1%, mixed evergreen 25.3%, and mixed coniferous 37.2% (Figure 61-1).

Figure 61-1. Land cover for the Santa Cruz Mountains. Daedalus TMS image, classified by supervised classifier.

Not only is the vegetative cover complex in its species composition but also in its forest structural characteristics. Forest structure has an impact on the relative combustibility of vegetation as fuel. Estimates of forest structure are based on relationships derived from linear regression analysis. A vegetation index such as the TM 4/3 simple ratio may be regressed against sites of known forest structure (Peterson et al., *Remote*, 1986; Spanner et al. 1984). To extract estimates of aboveground dry weight biomass, TM 4/3 ratios within each class were regressed against estimates of aboveground dry weight biomass from a forest survey done in 1986.

Topography

Topography in the Santa Cruz Mountains is as extreme but more varied than in the San Dimas experimental forest. Elevations in the four quadrangle area vary from sea level to over 650 m, with steep slopes and only a few valley bottoms. Most topography is canyon and ridge type, with more rolling hills to the north and more rugged topography toward the Big Basin area (Figure 61-2). Given the impact of topography upon fire, there is adequate variation in the study area for the demonstration of significant topographic effects.

Figure 61-2. Topography for four quadrangles in the Santa Cruz Mountains.

Topography also has a strong impact upon species cover and microclimate. Significant variations in soil moisture and species were observed due to differences in slope aspect, since almost all moisture in this area comes from the ocean to the west and falls as fog, especially in summer. Field observation of wind magnitude and direction also showed some significant local topographic effects.

Soil Moisture

Another major limitation of the cellular automaton model is the assumption of uniform fuel moisture. Depending on species, fuel moisture can vary from about 4% in dead fine fuel to over 200% in living wood and leaves. In many models, at a fuel moisture content of about 30% the conditions favoring ignition are effectively reduced to zero, placing an upper limit on fire spread. For the enhanced data set, fuel moisture was also estimated as a continuous value, but by modeling rather than measurement. Hillslope hydrology was simulated by using a grid-based model that incorporates data layers of position in a watershed (ridge top, stream, or somewhere in between on a relative index); slope angle; insolation based on slope angle, slope aspect, and local horizon conditions at the appropriate latitude; and soil hydrologic properties. An automated basin delineation algorithm (Band 1986) was used to generate the position in the watershed from the DEM. Slope angle and slope aspect were calculated. Insolation was calculated from the duration of direct beam solar radiation, slope angle, and slope aspect.

Duration was calculated from local horizons that are determined from the DEM (Dozier 1989) (Figure 61-3).

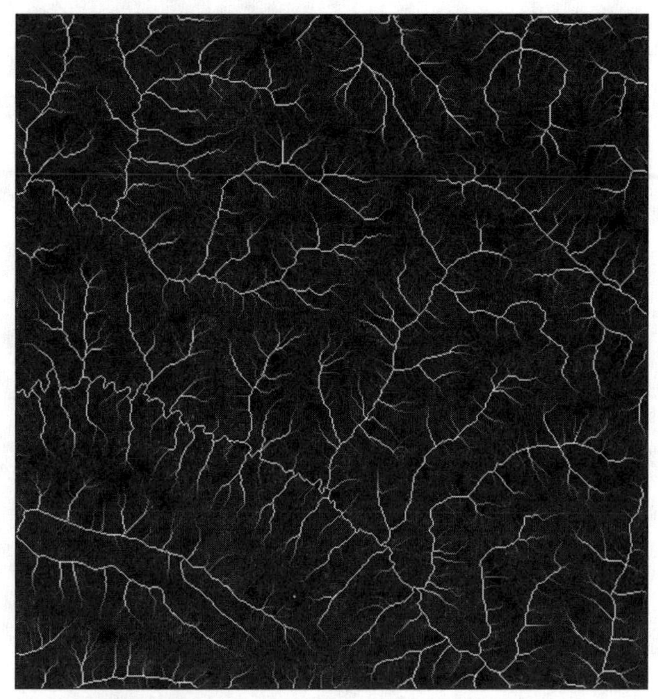

Figure 61-3. Soil moisture derived from GRASS hydrographic model.

CRITICAL ISSUES

Model Limitations

Extension of the cellular automaton model has gone far toward addressing some of the identified limitations. The primary remaining limitation is the simple way in which wind is addressed. An improved model would include: (1) three-dimensional wind structure, (2) air cells in three dimensions, each of which influences its neighbors, (3) a model to have wind interact with topography, (4) a model to have wind interact with the fire's thermal plume, and (5) a model for the distribution of smoke, which by shadowing, reduces the preheating conditions downwind. These refinements will constitute the next phase of the research. In addition, new suitable data sets for application of the model will be identified and used for further testing. Eventually, once parameters are calibrated fully, any area of choice for which input data are available could be used.

Data Accuracy

Each of the data sets has inherent accuracy. The DEMs are believed to be the most accurate, as they are derived from the DLG hypsometry, and meet national map accuracy standards. For this reason, this layer was chosen to give the final model resolution. The remotely sensed data had need of significant geometric and radiometric enhancement and was eventually reinterpolated to 30 m. The likely combined error here is about one to two pixels. This compared favorably with the ground resolution of the GPS receiver, believed to be about 40–50 m horizontal during the fieldwork. The overall classification accuracy of about 82% is comparable to many remote-sensing-based vegetation classifications and compared favorably to a species and timber survey of the area conducted by NASA Ames in 1986. There was a fair amount of confusion between mixed evergreen and mixed conifer classes due to classic mixed pixel problems. The tan oak understory of the mixed conifer class has a significant effect on radiances. The effect is particularly pronounced in the Big Basin redwoods where there has not been a fire in several decades and the understory is quite dense. Furthermore, as the spatial resolution of the sensor becomes finer, more high-frequency variation is introduced into the scene as forest structure (crown size, tree density) becomes more apparent and species composition less apparent (Spanner et al. 1984). Finally, there is no way of testing the accuracy of the fire risk predictions, other than by waiting for fires to take place. Fire risk maps will be mailed to fire stations in the region, with invitations to annotate actual fires and their extents for return to the authors.

Data Integration

Many of the practical problems of this research were related to the fact that data came in as: (1) point samples in latitude and longitude, (2) grids at 30 m in UTM in DEM format, and (3) raw Daedalus scanner data. Data integration involved file format conversion, mosaicking (for which custom software had to be written), classification, and registration. The image was ground-registered to UTM using the 7½ minute quadrangles with some difficulty, especially in the coniferous areas where few identifiable features were visible. As in many GIS applications, the layer-to-layer registration was essential and difficult.

Furthermore, GIS software involved was ERDAS (both 7.5 and Imagine), GRASS for the hydrologic modeling, and the in-house software packages read_dem (Clarke 1990), terrapin, and mosaic. Model runs were conducted using the cellular automaton, which is a C language program using the above layers as inputs. Output images were displayed using ERDAS and converted between formats using Image Alchemy.

Clearly, accounting for the myriad data processing decisions that resulted in the model predictions is complex. On the other hand, every step of the process is duplicable with the same or additional data, and the same results are to be expected given the same data conditions.

Fire Risk Management

A map of fire risk is only one possible outcome of this model (Figure 61-4). Placing this layer back into the GIS allows further modeling. For example, the roads layer, in conjunction with the locations of high ridges and peaks, could be used to provide an acceptable spatial model for ignition. Further model runs from these, rather than random, ignition points could provide better real-risk estimates. Second, the risk maps are only one component of a risk management system. People make complex decisions based on risk. The Big Basin ranger station, for example, is accessible via two roads, one an emergency road suitable only for an alternative escape route in particular fire conditions. Similar ideas could be applied to roads, communications lines, and to the locations of buildings. Fire risk maps could be used to plan land use, development, and in the construction of passive fire breaks and location of fire stations.

Recent work at NASA Ames has produced a means by which real-time data can be telemetered down to Earth for real-time processing. The model outlined here could provide the heart of a system for the real-time management of fire, allowing cheaper and more effective fire fighting and the ability to update the risk map as events unfold.

CONCLUSION

The authors have extended the cellular automaton fire model by the provision of superior data and the more integrated use of remotely sensed and GIS modeled data. As a result, fire risk maps of the study area in the Santa Cruz Mountains have been produced, and a mechanism built by which the estimates can be made to respond to changes in the control parameters, such as weather and fuel conditions. While the model is already pro-

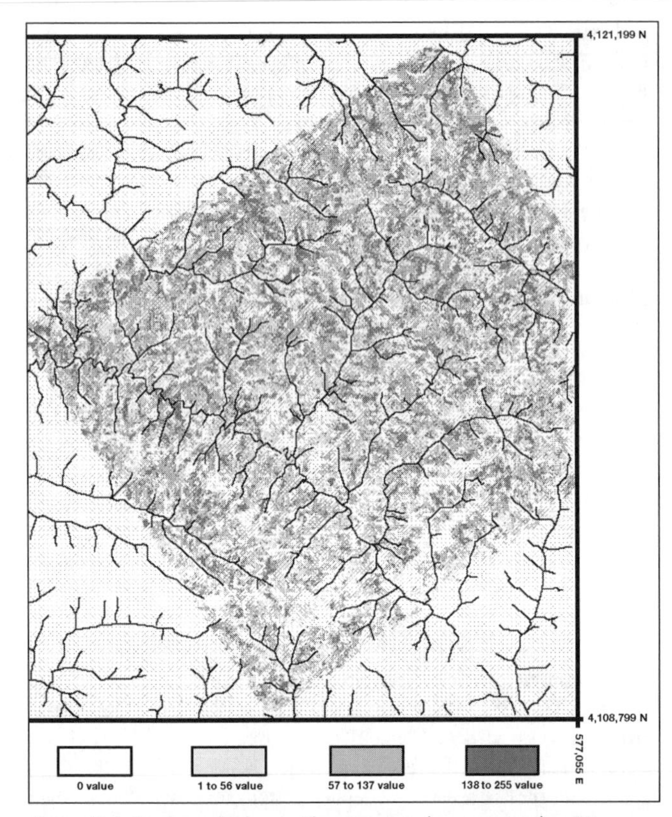

Figure 61-4. Fire hazard risk map. The minimum value represents the minimum fire risk, and the highest value represents the highest fire risk on this map.

ducing some interesting results, the authors seek to extend the model to use a more subtle accounting for variations in wind direction and magnitude as a function of local effects. In addition, the model is extendable to other regions for which data are available, to other scales, and to real-time applications. Any of these new applications could require different calibration, which will necessarily reflect the time and space scaling properties of the physical environmental variables in question.

Our understanding of fire as an environmental factor has improved as a result of this and many other modeling studies. Similarly, GPS, GIS, and remote-sensing technology have much to offer fire management and fire science. On the other hand, people will continue to face unnecessary environmental risk from fire, and the expansion of urban settlements continues to cross the border between fire safety and disaster.

References

Albini, F. A. 1976. *Computer-Based Models of Wildland Fire Behavior: A User's Manual.* USDA Forest Service unnumbered publication. Intermontain Forest and Range Experiment Station, Ogden, Utah.

Anderson, H. E. 1983. Predicting Wind-Driven Wildland Fire Size and Shape. USDA Forest Service, Research Paper, INT-305, Intermountain Forest and Range Experiment Station, Ogden, Utah.

Band, L. E. 1986. Topographic partition of watersheds with digital elevation models. *Water Resources Research* 22:15–24.

Barbour, M. G., J. H. Burk, and W. D. Pitts. 1987. *Terrestrial Plant Ecology,* 2nd ed. Menlo Park: Benjamin/Cummings Publishing Co.

Clarke, K. C. 1990. *Analytical and Computer Cartography.* Englewood Cliffs: Prentice Hall.

Clarke, K. C., J. A. Brass, and P. J. Riggan. 1994. A cellular automaton model of wildfire propagation and extinction. *Photogrammetric Engineering and Remote Sensing* 60(11):1,355–67.

Cohen, J. D. 1986. *Estimating Fire Behavior With FIRECAST: User's Manual.* U.S. Department of Agriculture, Forest Service, General Technical Report PSW-90, Pacific Southwest Forest and Range Experiment Station.

Dozier, J. 1989. A faster solution to the horizon problem. *Computers and Geosciences* 7:145–51.

Fons, W. T. 1946. Analysis of fire spread in light forest fuels. *Journal of Agricultural Research* 73(3):93–121.

Fotheringham, A. S., M. Batty, and P. A. Longley. 1990. Diffusion-Limited Aggregation and the Fractal Nature of Urban Growth. Annual Meeting of the Association of American Geographers, Toronto, Ontario.

Kuchler, A. W., and I. S. Zonneveld, eds. 1988. *Vegetation Mapping.* Dordrecht, the Netherlands: Kluwer Academic Publishers.

Mandelbrot, B. B. 1985. Self-affine fractals and fractal dimension. *Physical Scripta* 32:257–60.

Meakin, P. 1983. Diffusion-controlled cluster formation in two, three, and four dimensions. *Physical Review* A-27:604–07.

Mullins, W., and R. Sekerka. 1963. Morphological stability of a particle growing by diffusion or heat flow. *Journal of Applied Physics* 34:323–29.

Narumalani, R., J. R. Jensen, J. Michel, M. Hayes, and T. Montello. 1993. Remote Sensing for Habitat Assessment of Dawhat Al Musallamiyah and Ad Daffi in Saudi Arabia Prior to the Gulf War Oil Spill. Annual Meeting of the Association of American Geographers, Atlanta, Georgia.

Peterson, D. L., M. A. Spanner, S. W. Running, and K. B. Teuber. 1986. The relationship of thematic mapper simulator data to leaf area index of temperate coniferous forests. *Remote Sensing of Environment* 22:323–41.

Peterson, D. L., W. E. Westman, N. J. Stephenson, V. G. Ambrosia, J. A. Brass, and M. A. Spanner. 1986. Analysis of forest structure using thematic mapper simulator data. *IEEE Transactions on Geoscience and Remote Sensing* GE-24(1):113–21.

Pyne, S. J. 1984. *Introduction to Wildland Fire: Fire Management in the United States.* New York: John Wiley and Sons.

Rothermel, R. C. 1971. A Mathematical Model for Predicting Fire Spread in Wild Land Fuels. USDA Forest Service, Research Paper INT-115, Intermountain Forest and Range Experiment Station, Ogden, Utah.

Spanner, M. A., J. A. Brass, and D. L. Peterson. 1984. Feature selection and the information content of the thematic mapper simulator data for forest structural assessment. *IEEE Transactions on Geoscience and Remote Sensing* GE-22(6):482–89.

Keith C. Clarke holds a B.A. with honors from Middlesex University, London, and an M.A. and Ph.D. from the University of Michigan, specializing in Analytical Cartography. Since 1982, Dr. Clarke has been located in New York, and is currently professor and chairman of the department of Geology and Geography at Hunter College and in the Earth and Environmental Sciences program at the Graduate School and University Center of the City University of New York. Dr. Clarke is the current North American editor of the *International Journal of Geographical Information Systems*, serves on the editorial board of *Cartography and GIS*, and is series editor for the *Prentice Hall Series in Geographic Information Science.* He is the author of the textbook *Analytical and Computer Cartography* (Prentice Hall, 1990).

Department of Geology and Geography
Hunter College and CUNY
Graduate School and University Center
695 Park Ave.
New York, New York 10021
E-mail: *kclarke@everest.hunter.cuny.edu*

62

Managing Spatial Continuity for Integrating Environmental Models with GIS

Karen K. Kemp

An important problem that is preventing full integration of GIS with environmental modeling is the incompatibility of the spatial models used in each of these two enterprises. While environmental models generally depend upon continuous mathematics to represent continuous processes acting on spatially continuous phenomena, GIS, as digital technology, can only work with discrete mathematics and data. Recognizing this incompatibility is important if environmental modelers are to be successful in linking GIS and their models. This chapter discusses this incompatibility and offers a strategy to alleviate the problem.

INTRODUCTION

For most environmental modeling projects, GIS are seen as convenient and well-structured databases for handling the large quantities of spatial data demanded of them. Traditional GIS tools such as overlay and buffering are important for developing derivative data sets that serve as proxies for missing variables or for characterization (a process in which multiple aspects of the environment are combined into single variables). As better spatial analysis methods become incorporated into GIS and programming languages, GIS will also become important in model building, validation, and operation.

However, one significant incompatibility still hampers the true integration of GIS and environmental models: While environmental models generally deal with dynamic and continuous phenomena, current GIS manage only static and discrete data. Digital spatial data are discrete representations of continuous reality, and the form and basic assumptions inherent in these representations affect the outcome of the models. It is essential that environmental modelers recognize this fundamental dichotomy, understand the assumptions and limitations of the discrete representations of continuous reality that are used in both the models and the databases, and develop strategies to deal with continuity in a discrete computing environment. This chapter briefly reviews these first two issues and offers suggestions for dealing with the third.

SPATIAL CONTINUITY IN ENVIRONMENTAL MODELING

Since environmental models generally depend upon physical principles, the mathematics of these models are often in the form of differential equations. These equations implicitly recognize the continuity of space and constantly changing values of the independent variables. Likewise, the phenomena being represented by these models are often, though not always, spatially continuous. In order to implement these models in digital computers, it is necessary to develop techniques for discretizing the spatial continuity of both the models and the phenomena.

Modeling and Computing Spatially Continous Processes

The majority of physically based environmental models are based largely upon variations of a few fundamental laws of physics, many of which are expressed as partial differential equations. For example, many soil and groundwater models incorporate Laplace's equation for steady state flow. In two dimensions, Laplace's equation is:

$$\frac{\partial^2 u}{\partial x^2} + \frac{\partial^2 u}{\partial y^2} = 0 \qquad (62\text{-}1)$$

where u is the quantity being measured (i.e., heat, water). To solve such equations, for example, to find the value of u at any place, it is necessary to isolate this variable on one side of the equation. Ideally, a solution can be found by analytical means. However, such solutions are generally only attainable for models that are simple or that have some special form. Since most models of natural systems are complex, they normally must be solved numerically (Jeffers 1982).

There are several different types of numerical solutions. Finite difference numerical solutions discretize time and space into small units. These stepped algebraic solutions for the governing differential equations are calculated for each time and space unit, and a final solution is achieved by "integrating" (generally through simple addition) the results across the entire study area and time period (Gerald and Wheatley 1989). Finite element solutions divide the study area into units that are homogeneous in ways that allow some terms of the governing equations to be simplified. Analytical solutions can then be determined for each element, and the total solution is determined through simultaneous solution of the set of simplified equations (Desai 1979). Some global climate models are solved through the use of spectral analysis. Here, instead of discretizing space, the response spectrum itself is dissected into a set of ordinary differential equations for which solutions can be found (Bourke 1988). It is important to note that all of these numerical solutions require the discretization of continuous processes and/or space. Continuous mathematics generally require discrete solutions.

Modeling and Representing Spatially Continuous Phenomena

Spatially continuous phenomena are best described by the concept of the *field*. A physical field traditionally is defined as an entity distributed over space, whose properties are functions of space coordinates and, in the case of dynamic fields, of time:

$$z = f(x,y) \text{ or } z = f(x,y,t) \quad (62\text{-}2)$$

Scalar fields are characterized by a function of position and possibly time, with value at each point that is scalar, while the value at any location in a vector field is a vector (i.e., wind fields where the value at a location has both magnitude and direction).

Since they are continuous, physical fields are particularly distinguished by their extremely high degree of spatial autocorrelation. Thus, while we cannot measure the values of continuous phenomena everywhere, we know that locations near those we can measure will have very similar values. Knowledge of spatial autocorrelation, however, gives us little information about how rapidly and erratically the values change between locations at which we know the value. In order to represent and manipulate fields for mathematical modeling, we must have some way of linking the continuous variation of the field as it is observed in nature, to the individual numbers or letters stored in the computer, as representations of the values of the field at specific locations. In a few special cases, values and variation in space can be represented by an equation such as:

$$z = x^2 + xy + y^2 \quad (62\text{-}3)$$

where x, y are horizontal Cartesian coordinates and z is the value of the phenomenon at any (x,y) location. However, since surfaces in reality are rarely this smooth, the linkage between continuous reality and its representation in the computer is achieved by first dividing continuous space into discrete locations for which discrete values can be measured and recorded, and then establishing a rule for interpolating unknown values between these locations. The first of these is discretization, which was previously introduced with respect to the discretization of space for the solution of continuous mathematics; the second is accomplished through the use of spatial data models.

Spatial Data Models

In a literal sense, just as a hydrological model represents hydrology and a plant growth model represents plant growth, the term "data model" suggests a formal representation of data, not of reality. It is important to recognize that the process of developing spatial data models of a specific reality (data modeling), involves the discretization of the spatial variation of that reality. Unfortunately, data modeling is often confused with issues of data structure and becomes mired in questions of how points, lines, and areas should be represented. In fact, this confusion of terms may be why a lack of understanding exists regarding the fundamentally different ways these data models represent reality. Each one embodies one or more important assumptions about the form of the reality represented. These assumptions critically affect how the data model can be manipulated mathematically.

For field representation, six different spatial data models currently are available: (1) cell grids, (2) polygons, (3) TIN, (4) contour models, (5) point grids, and (6) irregular points (Goodchild 1992). Goodchild suggested that these six models represent two distinct ways of exploiting the spatial autocorrelation of fields: piecewise models and sample models.

Piecewise models—(cell grids, polygons, and TIN) dissect the surface into contiguous regions. A value is defined at every location on the surface. The continuous variation of the value of the phenomenon within each region is described by a simple mathematical function of the coordinates. In two models—cell grids and polygons—this mathematical function is a constant, while in the TIN model, the function is linear. Thus, if the values of the phenomenon represented are drawn as a third dimension, cell grid and polygon models produce a stepped surface of horizontal regions, while the regions of the TIN model are sloping planes, with the edges of each region coincident to those of their neighbors. The crucial assumption in all piecewise models is that the value or function assigned to each region is representative of the average value or general trend of the surface in the region. While each individual point may not be represented precisely, it is assumed that the integral of the values over this surface will produce the value or linear function assigned.

Sampled models—(contours, point grids, and irregular points) use a very different approach. In these models, the phenomenon is sampled precisely at a number of different points. Sampling is done either at points, as in point grids and irregular point models, or along lines, as in contour models. Sampling schemes may be unbiased (point grids) or biased (contours and some irregular point models). Also, as is the case with irregular points, samples may be taken at critical points (i.e., at the peaks and changes of slope in an elevation surface) or at representative points (i.e., at rain gages for which sites have been chosen as representative of conditions in the region around them). Note that the differences between models based on critical versus representative points are not apparent in the data model itself and should be regarded as necessary metadata for the data set as they affect how that data set should be manipulated.

In sampled models, no values are assigned to locations that have not been sampled, and, except in the limited case of contour models, no information is provided about the variation in the value of the phenomenon between sample sites. In order to represent the continuous surface between these sample locations, an assumption must be made that variation between these points can be described by a mathematical function. However, unlike piecewise models, this function is not always clearly defined. Frequently, linear functions are used, although other forms are also common (e.g., higher-order functions that fit a surface exactly to points in the 3×3 window of a point grid). The interpolation function chosen may vary for a single data set used in different applications. Also, the accuracy with which the value of a given point on the surface can be predicted varies, depending upon its distance from a sampled site since, in general, the value of a point close to a sampled site can be predicted with greater accuracy than the value of a location a great distance away.

Thus, we have two groups of spatial data models with varying basic assumptions: While piecewise models provide a generalized representation of the continuous phenomenon, sampled models provide precise data at a limited number of locations. Moreover, while there has been considerable discussion as to how humans perceive the world (Couclelis 1992), how we develop conceptual spatial data models based upon those perceptions (Burrough 1992), and how we construct abstractions from them in the form of spatial databases (Nyerges 1991), for the most part, construction of database models of reality is out of the hands of environmental modelers. Databases often exist before the modeling task is conceived. Thus, the question is not how to represent reality, but rather *how* to *understand* and *work with* a given database's representation of reality.

MODELING WITH FIELDS

Since there are many ways to represent fields, and these representations affect how fields can be manipulated, it is useful for the mathematical operations that are to be performed on data about fields to be conceptually separated from issues related to the form of spatial discretization used to represent those phenomena in the computer. This separation can be achieved through specifying field variables. Field variables are the logical, or functional, representations of the concept of fields. These variables are spatially continuous and represent values of the field at a single slice in time. Like other types of variables, field variables can be represented with symbols; for example, the field, "temperature," may be indicated by symbols like "T" or "TEMP." For any field variable, it is possible to determine the specific value at any location (i.e., T[*x*,*y*]) and these specific values will differ from location to location within the same field variable. Once field variables have been identified, the entire field and its related data set may be referenced by a simple variable and incorporated directly into the mathematical statements in such a way that the simple concept of the field can be used during the mathematical modeling effort.

At the implementation stage, the computer is incapable of mathematically manipulating two continuous fields to produce a third continuous field. All fields must be reduced to simple finite numbers before mathematical manipulation can proceed. This is achieved through the use of spatial data models. However, in order to manipulate two fields simultaneously (as in addition or multiplication), the locations for which there are simple finite numbers representing the values of the fields must correspond. For example, to add field *A* to field *B*, the value of *A* must be added to the value of *B* at the same location. Different spatial data models express location in ways that are generally incompatible. Thus, in order to perform mathematical operations on data in various spatial data models, estimates of values for locations in one field variable for which we have data in another field variable first must be extracted.

Fortunately, algorithms that can be used to convert one spatial data model to others are widely available and implemented in many different forms (e.g., stand-alone software modules, contour modeling packages, full-scale GIS). However, in some cases there are several different ways one representation of reality can be converted to another. The choice of which procedure to use should be determined by the reality being represented rather than by the form of the representation itself. This implies that the appropriate conversion procedure should be determined by the characteristics of data sets themselves rather than by the modeling environment. This, in turn, supports the contention that the mathematical operations of a mathematical model should be treated separately from issues related to the form of spatial discretization.

In most cases, the correct procedure to convert spatial data from one specific representation to another can be determined simply by considering the characteristics of each spatial data set (Kemp 1993). These characteristics include a description of the spatial data model used and other critical properties, such as the spatial density of the data and the measurement system used (i.e., categorical or numerical). However, since similar spatial data models can be used to represent very different phenomena, in a few cases it is desirable that additional information be supplied that cannot be deduced directly from the characteristics of the digital representation. In particular, since point models do not explicitly include rules or assumptions about how values vary between the points where values are known, knowledge about the reality being represented should be used to determine appropriate interpolation procedures. For example, in the case of irregular point data models discussed above, if the values at the points included in the data set are representative of the neighborhood of each point, a kriging approach may be appropriate; if the points are the critical points in the field, a triangulation procedure might be used. Since such information cannot be deduced from the representation itself, it should be attached to data sets at the time of their creation, since this is the only time when reality and its representation can be directly compared.

These data model characteristics and their related properties, such as appropriate interpolation procedures, can be declared within the GIS/environmental modeling system in a number of different ways depending upon how the field concept is implemented. In most cases, sufficient information can be provided so that no further input is required when manipulation of these variables is called for. In this manner, the objective of separating the manipulation of spatial data from aspects of its representation can be achieved.

Field Functions and Special Field Variables

Once the concept of field variables has been established, it is possible to formulate new functions and variables designed specifically for fields. Some of these include:

- *Integration*—a function specifically for continuous variables that has wide application in models handling fields. While algorithms to approximate integration are relatively easy to design for field spatial data models, each data model requires a different solution.
- *Slope and aspect*—variables that are important in many environmental models that can be implemented as field functions. While these functions are currently implemented in traditional forms in many GIS, their expression as field functions available in the mathematical modeling environment would be useful.
- *Statistical operations*—certain statistical operations, like *mean* and *standard deviation*, have special meanings when performed on fields, but their implementation is simply a matter of defining appropriate algorithms for the various data models.
- *Latitude and longitude*—two reserved field variables, *lat* and *long*, can be identified. Like other fields, they can be determined for any location. How the specific value of these variables is calculated for any location is determined by the spatial data model of other field variables in the mathematical statements.

Vector Fields

If we accept the notion that fields can be handled like ordinary scalar variables, then the possibility of working with vector fields within GIS can be considered. Vector fields can be represented, using the separable components of vectors, by specifying the length of unit vectors along each of the relevant axes, in two dimensions, as (x,y), and in three dimensions, as (x,y,z), or by specifying direction and magnitude (α,r). Vector fields can be expressed using current data models by permitting multivalued attributes at each location included in the data set. This approach is most suitable for the point-based spatial data models since the vectors then describe the direction and magnitude of the flow of the phenomenon at each point location. Cell grids and even polygons can be used to identify the location of vectors; however, like scalar fields, the vector must be assumed to represent the average flow over the entire area included in the spatial element.

How might vector fields be used in environmental models? If vectors are recorded in the (x,y) or (x,y,z) form, vector algebra easily can be performed on the vector elements. Analytical, finite difference, and finite element solutions of complex equations may proceed in the normal scalar fashion where solutions are found for each spatial element and integrated for the whole. Vector fields and related operations such as divergence and gradient provide useful new ways to conceptualize interactions in continuous phenomena. However, since vector fields have not been available in spatial databases, their use has been restricted to a small group of highly specialized and technical mathematical models that have little in common with current GIS. Before vector fields and related operations can become widely used by environmental modelers using spatial data stored in GIS, vector fields must become accessible and manageable within available software. Vector field variables provide a first step in this direction.

Implementation in Scientific Programming Languages

Existing programming facilities that allow us to define abstract or user-defined data types and to establish standard procedures, either as subroutines or encapsulated operations, provide immediate means by which field variables can be incorporated into mathematical models of environmental phenomena.

In standard procedure-oriented languages such as C or FORTRAN, field variables can be specified as user-defined data types. The declaration of field variables establishes not only the type of data contained within the variable, but how it can be manipulated. Declaration statements can establish certain functionality constraints and options particular to a specific data type. Properties can be stored as flags or values and accessed when needed to signal the appropriate subroutine or to insert the appropriate value in a routine. Specialized functions and subroutines using these user-defined data types can be created. With procedural languages, details of the implementation of field data model conversions and related procedures do not need to be handled within the program itself, but may be developed as standard subroutine libraries accessed through operations on field variables. These standard libraries may be written to interface directly between various programming languages and specific GIS. It is possible to envision a full range of such subroutines contained in a standard GIS library developed for each programming language.

With object-oriented programming languages, the six field data models may be defined as abstract data types with associated methods to determine appropriate conversion routines. Specific field variables can then be defined as instances of these object classes. Appropriate conversion procedures will be inherited. Instance-specific overloading of operators can be enabled by creating subclasses of the six object classes.

USING THE FIELD CONCEPT

There are a number of ways the field concept can be used and several different reasons for doing so. At its most fundamental level, this concept can be used as a thinking tool—to conceptualize the relationships of the model in a procedural fashion. By isolating issues related to the manipulation of the spatial data models, the difficulty of dealing with data model conversions is eliminated, and the modeler can work directly with the variables of the model.

Pedagogically speaking, this approach emphasizes the fundamental differences between the representation and the reality that is being modeled. While reality can be modeled as abstract scalar and field variables, the variables themselves must be declared and defined as specific spatial data models, as must conversion procedures and other special operations as well. This separation of spatial operations from modeling operations helps clarify limitations and restrictions inherent in working with discrete representations of a continuous world.

On a more practical level, this field concept can be used to map out how a linkage between a model and GIS can be implemented. There are several different levels at which this linkage can be achieved (Fedra 1993):

1. The lowest level of integration, simple manual file exchange between the environmental model and GIS, is not addressed by this strategy since it can be achieved by solutions such as map algebra, where all data must be in the same cell grid spatial data model before modeling is begun.
2. The next level of integration makes use of special interface programs between the model and GIS that manage the file format conversions in such a way that file sharing is transparent to the user. At this level, the field concept could be implemented through a compiler that converts the calls for manipulation of spatial data sets into calls to various data management systems and data model conversion subroutines, and integrates these with the standard mathematical operations of the program. This solution suggests that implementation using a traditional procedural language is suitable.
3. The highest level of integration—when the model becomes one of the analytical functions inside a GIS or the GIS becomes an option in the file management and output components of the model—suggests another possible approach to implementation of the field concept. Here, the various field functions and data model conversion procedures can be "unbundled" and made available as individual tools in GIS or a model toolbox. The model itself calls these functions directly. This particular integration approach requires that the specification of field variables be an integral part of the model. This solution may be most appropriate for implementation in an object-oriented environment.

CONCLUSION

Linking a mathematical model to GIS through the field concept requires the conscious development of a rigorous, structured declaration of how reality is digitally represented and how its various representations should be manipulated. By recognizing a distinction between the fundamental continuous mathematics of the model and the implementation of that mathematics on discrete spatial data, the necessary operations that need to be performed become explicit. The field concept provides a conceptual foundation that can be exploited as a design framework for the interface between the model and GIS.

The strategy presented here allows a flexible approach to handling variables representing continuous phenomena. Modelers can express as much or as little as they prefer about how these variables are to be represented and manipulated. If a modeler is unfamiliar with the types of operations that may or should be performed on spatial data, information about spatial data model conversion procedures and even field variable declarations can be omitted from the modeling code. These elements can be specified when the model is implemented with the GIS by other experts who are more familiar with the characteristics of spatial databases and spatial functions.

An important aspect of this approach is that it does not assume a homogeneous database. Data can be stored in any form, preferably one closest to that in which they have been collected. Conversion to the spatial data model required by a particular mathematical model is performed only when necessary. This helps to keep to a minimum the number of conversions performed on a particular spatial data set, thus reducing the inevitable information loss that occurs with each conversion.

Directions for the Future

Implementation of the field concept merits exploration. Implementation in a number of different environmental modeling domains would provide considerable opportunity for refinement of the generic concepts discussed here. The development of interfaces between various programming languages and GIS that interpret statements about field variables and translate them into computable elements is highly desirable. The creation of libraries of field variable functions would be useful extensions to existing programming languages and GIS.

There are also several opportunities for extensions to the field concept. Certainly three-dimensional fields should be considered as well as other types of continuous variables, such as tensors and statistical fields. In addition, fields exist in areas outside environmental modeling domains and should be explored. For example, models of human activities frequently make use of density variables modeled as fields and thus may benefit from treatment similar to that given to models of physical processes. The continuity of time also suggests that a similar approach to recognizing the distinction between its continuous real form and its discrete representations may assist in the development of a general theoretical framework for handling time data.

Acknowledgments

This research was completed at the National Center for Geographic Information and Analysis (NCGIA), University of California, Santa Barbara. The support of the NCGIA by the National Science Foundation under Grant #SES-10917 is gratefully acknowledged. As well, direction and critique from Dr. Michael Goodchild, Director of NCGIA, was instrumental in the completion of this work.

References

Bourke, W. 1988. Spectral methods in global climate and weather prediction. In *Physically Based Modeling and Simulation of Climate and Climatic Change*, edited by M. E. Schlesinger, 169–20. Kluwer Academic Publishers.

Burrough, P. A. 1992. Are GIS data structures too simple minded? *Computers & Geosciences* 18:395–400.

Couclelis, H. 1992. People manipulate objects, but cultivate fields: Beyond the raster–vector debate in GIS. In *Theories and Methods of Spatiotemporal Reasoning in Geographic Space*, edited by A. U. Frank, I. Campari, and U. Formentini, 65–77. New York: Springer-Verlag.

Desai, C. S. 1979. *Elementary Finite Element Method.* Englewood Cliffs: Prentice Hall.

Fedra, K. 1993. GIS and environmental modeling. In *Environmental Modeling and GIS*, edited by M. F. Goodchild, B. O. Parks, and L. T. Steyaert, 35–50. New York: Oxford University Press.

Gerald, C. F., and P. O. Wheatley. 1989. *Applied Numerical Analysis*, 4th ed. New York: Addison-Wesley.

Goodchild, M. F. 1992. Geographical data modeling. *Computers and Geosciences* 18:401–08.

Jeffers, J. N. R. 1982. *Modeling.* New York: Chapman and Hall.

Kemp, K. K. 1993. Environmental Modeling with GIS: A Strategy for Dealing with Spatial Continuity. Technical Report 93-3. Santa Barbara, CA, National Center for Geographic Information and Analysis, Department of Geography, University of California, Santa Barbara.

Nyerges, T. L. 1991. Geographic information abstractions: Conceptual clarity for geographic modeling. *Environment and Planning A* 23:1,483–99.

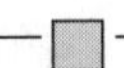

Karen Kemp received her B.S. in Calgary, her M.S. in Victoria, and her Ph.D. in California. She taught at a regional college in British Columbia before joining the NCGIA in 1988. There, she was coordinator of education programs and coeditor, with Dr. Michael Goodchild, of the NCGIA Core Curriculum in GIS. In 1992 she left NCGIA to work on education programs at the Technical University Vienna, and then joined Longman GeoInformation, Cambridge, United Kingdom, as their distance learning development manager. In addition to building an international reputation in GIS education, Dr. Kemp's research focuses on the integration of environmental models with GIS from pedagogic and theoretical perspectives. She returned to NCGIA in 1994.

National Center for Geographic Information and Analysis
University of California
Santa Barbara, California 93106-4060
Phone: (805) 893-7094
Fax: (805) 893-8617
E-mail: *kemp@ncgia.ucsb.edu*

63

Modeling Spatial and Temporal Distributed Phenomena:

New Methods and Tools for Open GIS

H. Mitasova, L. Mitas, W. M. Brown, D. P. Gerdes, I. Kosinovsky, and T. Baker

The concept of Open GIS has created a favorable environment for integration of process-based modeling and GIS. To support this integration, a new generation of tools is being developed in the following areas: (1) interpolation from multidimensional scattered point data, (2) analysis of surfaces and hypersurfaces, (3) modeling of spatial processes, and (4) three-dimensional dynamic visualization. A general interpolation and approximation method, using smoothing spline with tension, has been derived for d *independent variables, and its special cases for* d = 2,3,4 *are implemented in GIS. High-accuracy surface and hypersurface geometry analysis using the direct computation of partial derivatives from the interpolation function are also incorporated. Interactive visualization is an integral part of modeling. It currently supports the visualization of multiple phenomena through combination of multiple surfaces, color, transparency, point and vector data, the display of data in a spherical coordinate system, and the animation necessary for modeling of spatial dynamic processes. New capabilities are being added to support volumetric visualization. Examples of two applications are given: spatial and temporal modeling of erosion and deposition, and multivariate interpolation and visualization of nitrogen concentrations in Chesapeake Bay.*

INTRODUCTION

In environmental sciences, various phenomena can be modeled at a certain level of approximation by smooth functions: terrain, climatic phenomena, surface water depth, concentrations of chemicals, and so on. Multivariate smooth functions are also important for modeling processes described by scalar fields and their associated vector fields representing fluxes. Current GIS support modeling with two-dimensional fields and some systems also support three-dimensional fields and temporal data sets (IBM GIS 1991). We anticipate that with environmental models incorporating the study of interacting systems, and with the increasing need for analysis of multiparametric models, higher-dimensional tools become necessary. Open GIS is a suitable approach for efficient development and testing of the new generation of multidimensional GIS tools.

MULTIVARIATE SPLINE INTERPOLATION

One of the basic tools needed for multidimensional modeling in environmental sciences is interpolation and approximation from d-dimensional scattered point data to d-dimensional grids. Although numerous interpolation methods, especially for bivariate case, are available, significant progress recently has been achieved using methods obtained as solutions to variational problems, commonly known as splines (Franke 1987; Hutchinson 1989, 1991; Hutchinson and Gessler 1993; Wahba 1990; Mitas and Mitasova 1988; Mitasova and Mitas 1993). We recall some of the basic principles of this approach and present a general expression for d-variate smoothing splines with tension and special cases implemented in GIS.

Spline methods are based on the assumption that the approximation function should pass as closely as possible to the data points and, at the same time, should be as smooth as possible. These two requirements can be combined into a single condition of minimizing the deviation from the measured points and the smoothness seminorm of the function, which is formulated as follows. Given N values of the studied phenomenon $z^{[j]}$, $j=1,\ldots,N$ measured at discrete points $\mathbf{x}^{[j]} = (x_1^{[j]}, x_2^{[j]},\ldots,x_d^{[j]})$, $j = 1,\ldots,N$ within a certain region of a d-dimensional space, find a function $S(\mathbf{x})$ which fulfills the condition:

$$\sum_{j=1}^{N} |z^{[j]} - S(\mathbf{x}^{[j]})|^2 w_j + w_0 I(S) = minimum \qquad (63\text{-}1)$$

where w_j, w_0 are positive weighting factors and $I(S)$ is the smooth seminorm. The solution of this problem can be expressed as a sum of two components (Talmi and Gilat 1977):

$$S(\mathbf{x}) = T(\mathbf{x}) + \sum_{j=1}^{N} \lambda_j R(\mathbf{x}, \mathbf{x}^{[j]}) \qquad (63\text{-}2)$$

where $T(\mathbf{x})$ is a "trend" function and $R(\mathbf{x},\mathbf{x}^{[j]})$ is a radial basis function with an explicit form that depends on the choice of $I(S)$. For our choice of the smooth seminorm (Mitasova and Mitas 1993) the "trend" function $T(\mathbf{x})$ is a constant:

$$T(\mathbf{x}) = a_1 \qquad \text{for an arbitrary } d \qquad (63\text{-}3)$$

and the general form of radial basis function for d-variate case is given by:

$$R_d(\mathbf{x}, \mathbf{x}^{[j]}) = R_d(|\mathbf{x} - \mathbf{x}^{[j]}|) = R_d(r) = \psi^{-\delta}\,\gamma(\delta, \psi) - \frac{1}{\delta} \qquad (63\text{-}4)$$

where $\delta = (d\text{-}2)/2$, and $\psi = (\varphi r/2)^2$. Further, φ is a generalized tension parameter, $r^2 = \sum_{i=1}^{d} (x_i - x_i^{[j]})^2$, and finally, $\gamma(\delta,\psi)$ is the incomplete gamma function (Abramowitz and Stegun 1964). For the special cases d=2,3,4, equation (63-4) can be rewritten as:

$$R_2(r) = -\left[\mathrm{E}_1(\psi) + \ln\psi + C_E\right] \tag{63-5}$$

$$R_3(r) = \sqrt{\frac{\pi}{\psi}}\,\mathrm{erf}\,(\sqrt{\psi}) - 2 \tag{63-6}$$

$$R_4(r) = \frac{1 - e^{-\psi}}{\psi} - 1 \tag{63-7}$$

where C_E = 0.577215. . . is the Euler constant, $\mathrm{E}_1(.)$ is the exponential integral function and erf(.) is the error function (Abramowitz and Stegun 1964). The coefficients $a_1,\{\lambda_j\}$ are obtained by solving the following system of linear equations:

$$a_1 + \sum_{j=1}^{N} \lambda_j \left[R(\mathbf{x}^{[i]}, \mathbf{x}^{[j]}) + \delta_{ji} w_0/w_j\right] = z^{[i]}, \quad i = 1, \ldots, N \tag{63-8}$$

$$\sum_{j=1}^{N} \lambda_j = 0. \tag{63-9}$$

Besides high accuracy (Mitasova and Mitas 1993), the derived functions have several useful properties. They have regular derivatives of all orders everywhere that make them suitable for direct analysis of surface and hypersurface geometry. The generalized tension φ controls the distance over which the given point influences the resulting surface or hypersurface. For the bivariate case, tuning the tension can be interpreted as tuning the character of the resulting surfaces between membrane and thin plate. The proper choice of smoothing and tension parameters is important for successful interpolation or approximation. The optimal smoothing parameter for smoothing splines can be found via ordinary or general cross-validation scheme (Wahba 1990; Hutchinson and Gessler 1993). We have used Wahba's formulation of ordinary cross-validation to find both optimal tension and smoothing. If we assume that $\{w_j\}$ are the same for all data points, then the smoothing parameter is $w = w_j/w_0$. Let $S^{[k]}_{N,\varphi,w}$ be the smoothing spline with tension using all the data points except the kth. We can take the ability of $S^{[k]}_{N,\varphi,w}$ to predict the missing data point z_k as a measure of predictive capability of the spline with the given tension φ and smoothing w, which can be represented by a mean error V_0:

$$V_0(\varphi, w) = \frac{1}{N}\sum_{k=1}^{N}\left[S^{[k]}_{N,\varphi,w}(\mathbf{x}^{[k]}) - z^{[k]}\right]^2 \tag{63-10}$$

then the optimal values of φ_0, w_0 are found by minimizing $V_0(\varphi, w)$. The example for Chesapeake Bay nitrogen concentration data (Figure 63-1) shows that for smoothing w = 0 the predictive error sharply increases for tension $\varphi < \varphi_0$. A small value of smoothing can significantly reduce the predictive error for the low values of tension. The cross-validation error computed for each point also can be used for the analysis of spatial distribution of predictive error and location of areas where this error is high and denser sampling is needed.

Due to the solution of the system of linear equations (63-8) and (63-9), computational demands for the presented method are proportional to N^3 and their application to large data sets becomes problematic. To overcome this limitation, an algorithm for segmented processing has been developed. Segmentation is based on the division of the given region to (hyper)rectangular segments. The size of each segment is adjusted to the density of points in the given subregion using 2^d-trees (quad-trees for two dimensional, oct-trees for

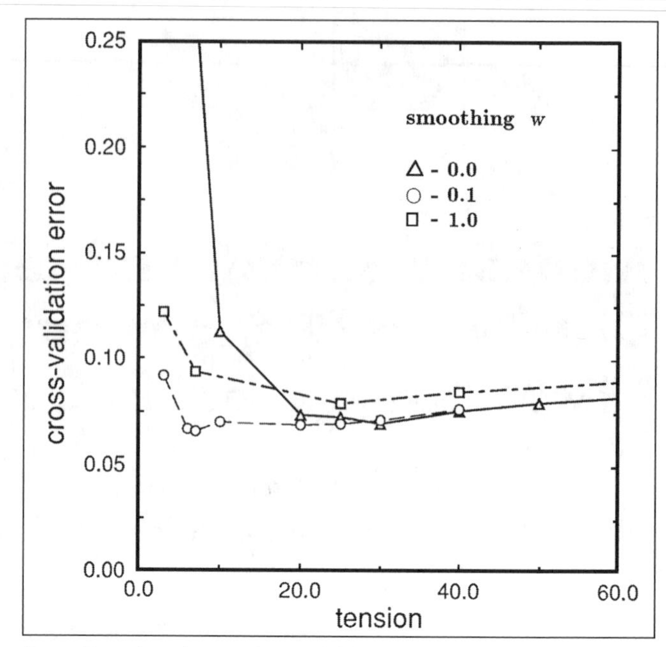

Figure 63-1. Relation between the cross-validation error, tension, and smoothing parameters for bivariate interpolation of DIN concentrations in the Chesapeake Bay.

three dimensional, etc.). Then, the interpolation function is computed for each segment using the points from this segment and from its neighborhood, located in the window which adjusts its size to the density of points in the neighborhood of each segment (Figure 63-2). This segmentation is dimension-independent, applicable to data with heterogeneous spatial distribution and makes the computation proportional to N.

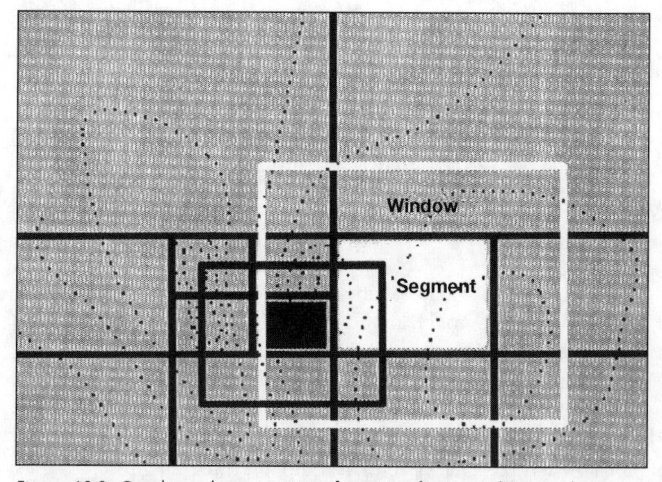

Figure 63-2. Quad-tree decomposition of a region for interpolation with segmented processing.

Multidimensional interpolation is also a valuable tool for incorporating the influence of an additional variable into interpolation; for example, for interpolation of precipitation with the influence of topography or concentration of chemicals with the influence of the environment where it is distributed. We have implemented the approach proposed by Hutchinson (1991) to support this type of modeling. To model the spatial distribution of d-dimensional phenomenon $u = F(x_1, ..., x_d)$ from the given point data $(x^{[i]}_1, ..., x^{[i]}_d, u^{[i]}) = (\mathbf{x}^{[i]}, u^{[i]})$, i = 1, ..., N influenced by another

d-dimensional phenomenon $z = G(x_1, ..., x_d)$, we can take the values of the "influencing" phenomenon as a d+1 coordinate $x^{[i]}_{d+1} = c.G(\mathbf{x}^{[i]})$ where c is a rescaling factor, so that we get the points $(x^{[i]}_1, ..., x^{[i]}_{d+1}, u^{[i]}) = (\mathbf{x}^{*[i]}, u^{[i]})$. The values of $x^{[i]}_{d+1}$ can be either given with data or can be interpolated from the function $z = G(\mathbf{x})$. Now we can interpolate from these points using the (d+1)-variate function $S^*(\mathbf{x}^*)$ and compute the resulting $S(\mathbf{x})$ as an intersection $S^*(\mathbf{x}) = S^*(\mathbf{x}^*) \cap G(\mathbf{x})$. The ($d$+1)th variable can be either a real surface like terrain or geological layer, its modification, or an abstract surface.

SURFACE GEOMETRY ANALYSIS

For process-based modeling it is often necessary to compute the geometrical parameters of surfaces and hypersurfaces directly related to the fluxes in landscape. We have used d-dimensional differential geometry (Thorpe 1979) to formulate the algorithms for d-dimensional topographic analysis. Assume that the studied phenomenon is modeled by a d-dimensional surface (d-surface), in a (d+1)-dimensional space, represented by a smooth d-variate function $z = g(x_1, \ldots, x_d)$. If we set $z = x_{d+1}$ we can represent this function in an implicit form as $f(x_1, \ldots, x_{d+1}) = x_{d+1} - g(x_1, \ldots, x_d)$ and then use the general expressions for d-surface analysis, applicable also to d-surfaces that are not functions (e.g., isosurfaces). Associated with each function $f(\mathbf{x})$ is a smooth vector field called the gradient ∇f, defined by:

$$\nabla f = \left(\frac{\partial f}{\partial x_1}, \ldots, \frac{\partial f}{\partial x_{d+1}}\right) = \left(-\frac{\partial g}{\partial x_1}, \ldots, -\frac{\partial g}{\partial x_d}, 1\right) \quad (63\text{-}11)$$

Such vector fields often arise in geosciences as velocity fields or fluid flows. Scalar fields associated with gradient are gradient magnitude $||\nabla f||$ and gradient direction $(\alpha_1, \ldots, \alpha_d)$

$$\|\nabla f\|^2 = \sum_{i=1}^{d+1} \left(\frac{\partial f}{\partial x_i}\right)^2, \quad \cos \alpha_i = \frac{1}{\|\nabla f\|} \frac{\partial f}{\partial x_i} \quad i = 1, \ldots, d \quad (63\text{-}12a, b)$$

For the bivariate case, in most applications a horizontal projection of ∇f: $\nabla f' = (\partial f/\partial x_1, \partial f/\partial x_2, 0) = (-\partial g/\partial x_1, -\partial g/\partial x_2, 0)$ is being used, and gradient magnitude is measured by the slope angle β, $\tan \beta = ||\nabla f'||$ or $\cos \beta = (\partial f/\partial x_1)/||\nabla f||$. The direction of $\nabla f'$ is used as aspect. Besides these scalar fields, associated with a gradient field is a family of curves called flow lines. To generate the flow lines and their related parameters we have developed an improved flow tracing algorithm based on a vector-grid approach (Mitasova and Hofierka 1993; Mitasova et al. 1995). The points defining the flow line are computed as the points of intersection of a line constructed in the $\nabla f'$ direction and a grid cell edge. Flow lines can be generated in the direction of $\nabla f'$ (downhill) for the computation of upslope contributing areas (Moore and Wilson 1992) and extraction of streams because downhill flow lines merge in valleys (Figure 63-3a). Flow lines generated in the direction of $-\nabla f'$ (uphill) are useful for computation of flow path length and extraction of ridge lines, as these flow lines merge on ridges (Figure 63-3b). A modified version of this algorithm has been developed for tracing the flow lines in three dimensions by Zlocha and Hofierka (1992).

To study the rate of change in the velocity of flow due to the shape of d-surface, the Weingarten map or *shape operator* $L_p(\mathbf{v})$, which measures the change of the normal direction $\mathbf{N} = \nabla f / ||\nabla f||$ when moving in the direction $\mathbf{v}$ on a d-surface, is defined as:

$$L_p(\mathbf{v}) = -\nabla_{\mathbf{v}} \mathbf{N} \quad (63\text{-}13)$$

where $\nabla_{\mathbf{v}}$ denotes the directional derivative in the direction given by a vector $\mathbf{v} = (v_1, ..., v_{d+1})$ which belongs to the tangent space of the modeled d-surface. Normal component of acceleration at a given point in the direction $\mathbf{v}$ is measured by a normal curvature $k(\mathbf{v})$ computed directly from the Weingarten map as:

Figure 63-3. (a) Flow lines generated downslope, merging in valleys; (b) flow lines generated upslope, merging on ridges.

$$k(\mathbf{v}) = L_p(\mathbf{v}).\mathbf{v} = -\frac{1}{\|\nabla f\|} \sum_{i,j=1}^{d+1} \frac{\partial^2 f}{\partial x_i \partial x_j} v_i v_j \quad (63\text{-}14)$$

For the bivariate case, when we consider flow over a 2-surface in a three-dimensional space, we get the profile curvature related to the flow acceleration for the vector $\mathbf{v}$ representing the tangent to the surface in the direction of flow. If $\mathbf{v}$ represents the tangent to the surface in the direction of a tangent to the contour, we get the tangential curvature related to the flow convergence/divergence (Mitasova and Hofierka 1993). Figure 63-4 shows the importance of using an appropriate method for the estimation of first- and second-order derivatives. While local polynomial interpolation applied to the noisy data with insufficient resolution produced unacceptable results, tangential curvature computed from smoothing spline reflects the basic structure of ridges and valleys.

Various other measures related to acceleration and shape of the d-surface can be derived from Weingarten map (e.g. mean, Gauss–Kronecker, and principal curvatures) and used not only in process-based models, but for enhancing the visualization as well.

MULTIDIMENSIONAL DYNAMIC VISUALIZATION

For the implementation and applications of the presented approach within GIS, a highly interactive and efficient tool for visualization of data and the resulting surfaces was important. This was needed both for the visual evaluation of the functionality of methods, algorithms, and programs as well as for the communication of application results. To support the use of visualization as both an analytical and communication tool, a program called *SG3d* was developed and fully integrated with the data structure of the GIS. The program enables users to view all types of currently supported geographic data (raster, point, vector) in the same three-dimensional space and

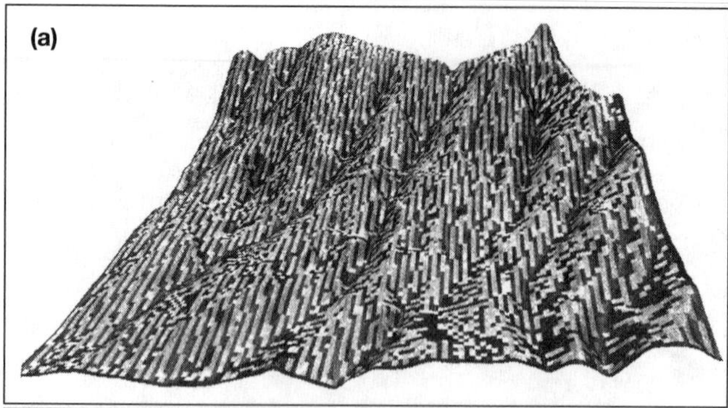

Figure 63-4. (a) Tangential curvature computed from the standard 30-m DEM using local polynomial interpolation; (b) tangential curvature computed from the same data as in Figure 63-4a, using spline for resampling to 10-m resolution, smoothing, and computation of partial derivatives.

allows scripting for producing dynamic visualizations via animation. The support for the volumetric visualization of three-dimensional data sets (three-dimensional raster, three-dimensional point, and three-dimensional vector) not currently supported by the GIS, is being integrated with *SG3d*. *SG3d* has the following capabilities (Brown and Gerdes 1992):

1. Visualization of two-dimensional raster data as 2-surfaces in three-dimensional space, using one data set for surface topography and a second one for surface color.
2. Interactive, easy positioning, zooming, and z-scaling, important for the efficient use in analysis of the results of models.
3. Interactive lighting with adjustable light position, color, intensity, and surface reflectivity, useful for detecting noise or small errors in data or models and enhancing the three-dimensional perception of the surface topography.
4. Draping vector and point data with the "glyphs" option for point data to facilitate the visualization of data with uncertainty (Figure 63-6).
5. Animation capabilities with two options: scripting for automatically generating animations from temporal data, and key frame animation to establish a path for moving the viewpoint through the data and automatically rendering and saving a user-defined number of frames.
6. Display and animation of multiple related surfaces (e.g. for hydrologic modeling, rainfall intensity surface, terrain with water surface depth draped as a color, and infiltration depth surface), displayed and animated in one, three-dimensional space.
7. Fast data query.
8. Display in spherical coordinate system for rendering the surfaces draped over sphere from latitude/longitude data.

A new version of the visualization program that integrates the capabilities of *SG3d* with programs for volumetric visualization of three-dimensional data sets is being developed. This program combines the standard 2-surfaces, vectors, and points with isosurfaces, solids, slices, and points representing three-dimensional phenomena, in the same three-dimensional space, while preserving the flexibility and efficiency of *SG3d*.

APPLICATION EXAMPLES

Spatial and Temporal Distribution of Soil Erosion/Deposition

By analyzing several currently used erosion models and sediment flux equations, Moore and Wilson (1992) derived the dimensionless index of sediment transport capacity T representing the influence of terrain on soil erosion:

$$T = \left(\frac{A}{22.13}\right)^m \left(\frac{\sin\beta}{0.0896}\right)^n \qquad (63\text{-}15)$$

which is the unit stream power-based LS-factor proposed by Moore and Burch (1986). In this equation, A is the upslope contributing area per unit contour width ($m^2\ m^{-1}$) and β is slope. The index becomes unity for the case when the upslope area A=22.13 ($m^2\ m^{-1}$), and the slope is 9%.

By considering the sediment transport limiting case, the topographic potential of landscape for erosion or deposition E can be computed as a change in the sediment transport capacity in the direction of flow (Figure 63-5). As an alternative to the finite difference approach used by Moore et al. (1992), E can be computed as a directional derivative of the surface $T = g(x_1,x_2)$ representing the sediment transport capacity (Mitasova et al. 1995):

$$E = \frac{\partial g}{\partial s} = \frac{\partial g}{\partial x_1}\cos\alpha + \frac{\partial g}{\partial x_2}\sin\alpha \qquad (63\text{-}16)$$

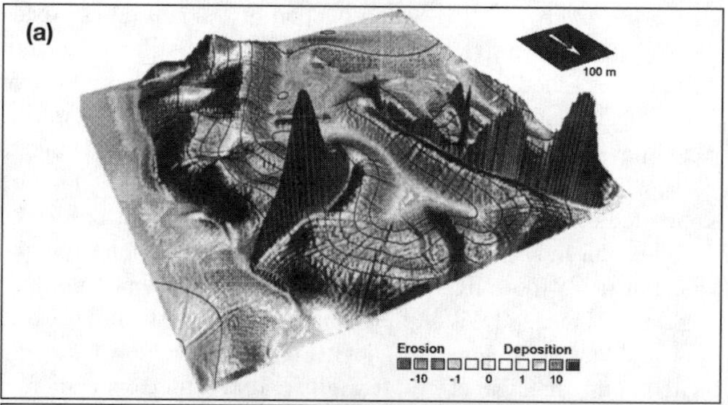

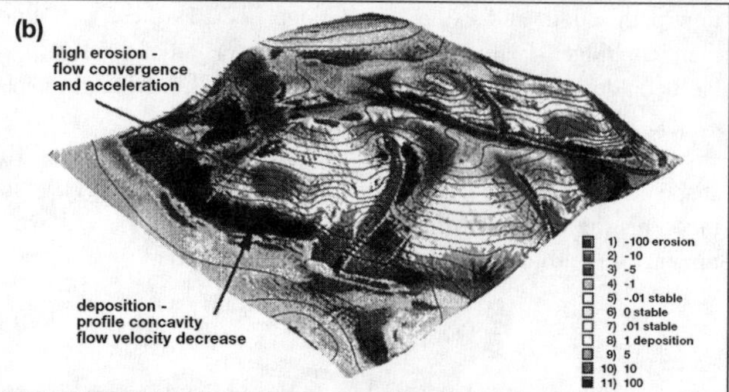

Figure 63-5. (a) Sediment transport capacity T visualized as a surface, with high values in valleys due to water-flow accumulation. (b) Topographic potential for erosion and deposition E computed as a directional derivative of surface in (a) draped over terrain.

where (x_1,x_2) are the georeferenced coordinates, and α is aspect (direction of flow) computed from the elevation surface $z = f(x_1,x_2)$. The smoothing spline with tension is used for the estimation of partial derivatives in (63-16). Equation (63-16) incorporates the influence of slope angle, shape of terrain in flow direction (through derivative in the direction of flow) as well as water flow convergence/divergence (through upslope area). Because the unit stream power-based LS-factor T has known relation to the revised universal soil loss equation (RUSLE) (Moore and Wilson 1992), the influence of rainfall intensity, soil properties, and land cover can be incorporated using the RUSLE R, K, and C factors respectively.

We have used this approach to model the spatial and temporal distribution of erosion and deposition at a military installation undergoing significant changes in land use during the year. The high-resolution (5 m) DEM with slope and aspect was created from digitized contours using the smoothing spline with tension. Upslope contributing area for each grid cell was computed by the vector–grid-based flow tracing algorithm. Unit stream power-based LS-factor T and erosion/deposition potential E were computed using the raster map calculator *r.mapcalc* (Shapiro and Westervelt 1992). The series of raster maps representing the spatial and temporal distribution of potential soil loss L during the year was then computed as:

$$L(x_1,x_2,t_i) = R(t_i).C(x_1,x_2,t_i).K(x_1,x_2).T(x_1,x_2),\; i = 1,\ldots,12 \quad (63\text{-}17)$$

where R is the rainfall factor uniform over the area, but significantly changing during the year; C is the cover factor, which varies in space and time; and K is the soil factor. The study has shown that both spatial and temporal distribution of erosion are important for land-use management to minimize the soil loss. Spatial analysis allowed us to propose to locate the intensively used areas in regions least susceptible to erosion, and to create protective buffers with natural vegetation along the streams in areas of high erosion and deposition. Temporal analysis showed that rescheduling the timing of intensive use so that it does not coincide with the highest rainfall intensity can further reduce soil erosion risk.

Multivariate Interpolation of Nitrogen Concentrations

The presented methods have been applied to interpolation, analysis, and visualization of the concentrations of dissolved inorganic nitrogen (DIN) in the Chesapeake Bay. DIN is measured at forty-three stations, at two or four depths, at each station approximately twice per month. It takes about three days to collect the data from all stations (Chesapeake Bay Program 1992). If we consider time to be an independent variable, then the data represent scattered points in four-dimensional space, and the distribution of DIN in space and time can be modeled by a function $u = S(x_1,x_2,x_3,x_4)$ where the fourth variable x_4 is proportional to time. To compare the differences between interpolation at various dimensions, we have applied bivariate interpolation to model the spatial distribution of DIN at the surface layer. Trivariate interpolation was used to model the concentrations of DIN in the volume of water for each set of measurements with time varying for no more than three days. Finally, the four-variate function was used to model both spatial and temporal distribution of DIN concentrations.

SG3d was used to visualize the dynamics of spatial distribution of DIN concentrations at the surface layer of the bay during the year, with surface and color representing concentrations of DIN, and glyphs representing sampling sites, colored and sized according to the cross-validation error (Figure 63-6). The integrated *SG3d* and volume visualization tool were used to create the animations from the results of trivariate and four variate interpolation. Although trivariate interpolation and volume visualization provided good representation of spatial distribution of DIN in the volume of water (Figure 63-7), the animations created from a series of three-dimensional grids interpolated by trivariate interpolation failed to capture the dynamic character of DIN movement. However, application of four variate interpolation and computation of three-dimensional grids with seven-day steps resulted in an excellent animation and proved the importance of using time as a fourth independent variable when processing the time series of data measured in three-dimensional space.

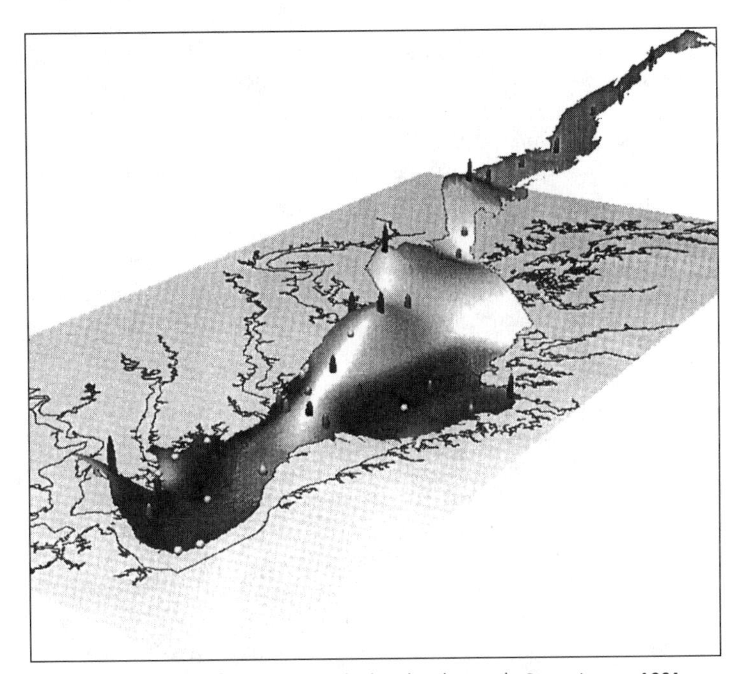

Figure 63-6. DIN concentrations at 0.5-m depth in the Chesapeake Bay in January 1991, visualized as a surface, with cross-validation error represented by glyphs.

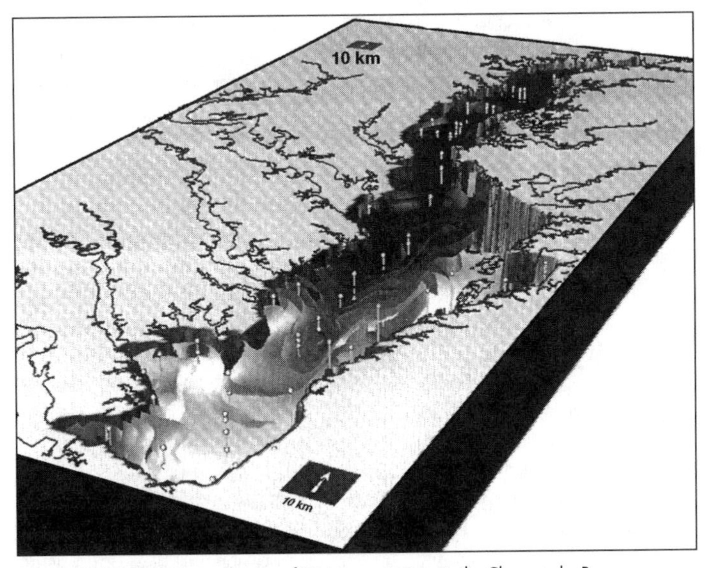

Figure 63-7. Volumetric visualization of DIN concentrations in the Chesapeake Bay.

CONCLUSION

We have used GRASS-GIS as an environment for implementing the presented system for multidimensional modeling, analysis, and visualization for environmental applications. We were able to fully integrate the tools for bivariate case with GRASS as commands *s.surf.tps* for interpolation and topographic analysis from scattered point data; *r.resample.tps* for smoothing, resampling, and topographic analysis of noisy raster data; *r.flow* (contributed by J. Hofierka and M. Zlocha) for the construction of flow lines, computation of flow path lengths and upslope contributing areas; and the powerful interactive visualization tool *SG3d* (U.S. Army Corps of Engineers 1993). We were able to handle time series quite efficiently using shell scripts with GRASS commands; however, more appropriate data structure is desirable. The programs supporting the higher-dimensional interpolation and visualization (*s.surf.3d, s.surf.4d, SG4d*) are consistent with GRASS and linked to it through the geographic region. Because the standard GRASS-GIS does not support three-dimensional and four-dimensional data, we have used our own data structure, which potentially may be incorporated into GRASS.

The presented approach to the integration of multidimensional modeling and GIS using the open GIS concept creates the basis for the new generation of GIS technology that supports the study of complex interacting systems needed for environmental modeling.

References

Abramowitz, M., and I. A. Stegun. 1964. *Handbook of Mathematical Functions.* New York: Dover.

Brown, W., and D. P. Gerdes. 1992. *SG3d—Supporting Information.* Champaign, Illinois: U.S. Army Corps of Engineers, Construction Engineering Research Laboratories.

Chesapeake Bay Program. 1992. *Trends in Nitrogen in the Chesapeake Bay (1984–1990).* Report CBP/TRS 68/92, Chesapeake Bay Program.

Franke, R. 1987. Recent advances in the approximation of surfaces from scattered data. In *Topics in Multivariate Approximation*, edited by Chui, Schumaker, and Utreras, 79–98. Academic Press.

Hutchinson, M. F. 1989. A new procedure for gridding elevation and streamline data with automatic removal of spurious pits. *Journal of Hydrology* 106:211–32.

———. 1991. The application of thin plate smoothing splines to continent-wide data assimilation. In *Data Assimilation Systems*, edited by J. D. Jasper, 104–13. BMRC Research Report no. 27. Melbourne: Bureau of Meteorology.

Hutchinson, M. F., and P. E. Gessler. 1994. Splines—More than just a smooth interpolator. *Geoderma* 62(N1-3):45–67.

IBM GIS. 1991. *What Is Scientific GIS?* Conference Proceedings, IBM GIS, Federal Sector Division, Boulder, Colorado.

Mitas, L., and H. Mitasova. 1988. General variational approach to the interpolation problem. *Computers and Mathematics with Applications* 16:983–92.

———. n.d. *D*-variate interpolation for scattered data. *Journal of Computational Physics.*

Mitasova, H., and J. Hofierka. 1993. Interpolation by regularized spline with tension: II. Application to terrain modeling and surface geometry analysis. *Mathematical Geology* 25:657–69.

Mitasova, H., J. Hofierka, M. Zlocha, and R. L. Iverson. n.d. Modeling topographic potential for erosion and deposition using GIS. *Int. Journal of Geographical Information Systems.*

Mitasova, H., and L. Mitas. 1993. Interpolation by regularized spline with tension: I. Theory and implementation. *Mathematical Geology* 25:641–55.

Mitasova, H. et al. 1995. Modeling spatially and temporally distributed phenomena—New methods and tools for GRASS-GIS. *International Journal of Geographical Information Systems* 9(4):433–46.

Moore, I. D., and G. J. Burch. 1986. Physical basis of the length-slope factor in the universal soil loss equation. *Soil Science Society of America Journal* 50:1,294–98.

Moore, I. D., and J. P. Wilson. 1992. Length-slope factors for the revised universal soil loss equation: Simplified method of estimation. *Journal of Soil and Water Conservation* 47:423–28.

Moore, I. D., J. P. Wilson, and C. A. Ciesolka. 1992. Soil erosion prediction and GIS: Linking theory and practice. *Proceedings of Geographic Information Systems for Soil Erosion Management Conference*, Taiyuan, China.

Shapiro, M., and J. Westervelt. 1992. *R.mapcalc, an Algebra for GIS and Image Processing.* Champaign: U.S. Army Corps of Engineers, Construction Engineering Research Laboratories.

Talmi, A., and G. Gilat. 1977. Method for smooth approximation of data. *Journal of Computational Physics* 23:93–123.

Thorpe, J. A. 1979. *Elementary Topics in Differential Geometry.* New York: Springer-Verlag.

U.S. Army Corps of Engineers. 1993. *GRASS 4.1 Reference Manual.* Champaign: U.S. Army Corps of Engineers, Construction Engineering Research Laboratories.

Wahba, G. 1990. *Spline Models for Observational Data.* CNMS-NSF Regional Conference Series in Applied Mathematics 59. Philadelphia: SIAM.

Zlocha, M., and J. Hofierka. 1992. Computation of three-dimensional flow lines and their application in geosciences. *Acta Geologica Universitatis Comenianae* 48/I, Comenius University, Bratislava, Slovakia.

Helena Mitasova graduated from the Slovak Technical University, Bratislava, Czechoslovakia, in 1981 and received her Ph.D. in geodetic cartography in 1987. She worked at the Department of Physical Geography and Cartography at Comenius University in Bratislava until 1990, when she came to the Illinois Natural History Survey as a visiting research scientist. Since 1991, she has been involved in the development of methods for surface interpolation, analysis, and erosion modeling for GRASS-GIS at U.S. Army CERL. At present, she is an affiliate at the Department of Geography, University of Illinois at Urbana–Champaign.

Spatial Modeling and Systems Team
U.S. Army Construction Engineering Research Laboratories
PO Box 9005
Champaign, Illinois 61826-9005
and
Department of Geography
University of Illinois at Urbana–Champaign
Urbana, Illinois 61801
E-mail: *helena@ginko.cecer.army.mil*

Lubos Mitas graduated from Slovak Technical University in 1983 and received his Ph.D. in physics in 1989. He worked at the Institute of Physics in Bratislava, Czechoslovakia, until 1990 when he came to the Department of Physics at the University of Illinois at Urbana-Champaign as a postdoctoral research associate. Since 1991, he has been working at the National Center for Supercomputing Applications on Quantum Monte Carlo computations of electronic structure. In addition to his work in physics he has been developing the interpolation methods for geoscientific applications since 1985.

National Center for Supercomputing Applications
University of Illinois at Urbana-Champaign
Urbana, Illinois 61801

William M. Brown graduated from the University of Illinois, Urbana-Champaign in 1981 with a B.S. in biology and received an A.A.S. in Visualization Computer Graphics from Parkland College, Champaign in 1992. He completed an internship at the National Center for Supercomputing Applications in 1991 and has since been a research programmer for the Modeling and Visualization Group, Spatial Analysis and Systems Team, at U.S. Army CERL. His work at CERL has concentrated on developing tools for multiple dimension visualization and analysis of geographic data.

David Gerdes graduated with a B.S. in math and computer science from the University of Illinois in 1984. He has since worked at the U.S. Army CERL as a developer of the GRASS-GIS system and the leader of GRASS visualization research. He currently is designing the next GRASS vector system.

Irina Kosinovsky received her B.S. in computer science and mathematics from University of Illinois in 1991 and her M.S. in applied mathematics from University of Illinois in 1993. Her graduate work was in the area of optimization and computation. She currently works at U.S. Army CERL doing research and development in spatial modeling and analysis.

Terry Baker graduated from the University of Illinois in 1989 with a B.S in psychology and received an A.A.S in visualization computer graphics from Parkland College, Champaign in 1991. She completed a graphics internship at the National Center for Supercomputing Applications in 1991 and has since worked with the Modeling and Visualization Group, Spatial Analysis and Systems Team, at U.S. Army CERL, most recently developing tools for three-dimensional volume visualization. She is currently a graduate student in the Department of Computer Science, University of Illinois.

Spatial Modeling and Systems Team
U.S. Army Construction Engineering Research Laboratories
PO Box 9005
Champaign, Illinois 61826-9005

64

Toward a General-Purpose Decision Support System Using Existing Technologies

Dean Djokic

In this chapter, a case is made for the use of existing, commercially available software as tools for creating a general-purpose spatial decision support systems (SDSS) "shell." By creating interfaces between geographic information systems (GIS), expert systems (ES), numerical models (NM), and utility software, all the capabilities of respective technologies become available in a unified computer environment. Both GIS and ES software commercially available are mature enough to provide, in conjunction with engineering numerical modeling software, most of the capabilities required of a generic SDSS. Developing interfaces, a one-time effort, requires much less energy than customizing or writing a new code for existing GIS, ES, or NM features of an SDSS for each specific task at hand. Using an SDSS shell, effort can be concentrated on developing the application instead of developing the computer environment. The integration of ARC/INFO (GIS), Nexpert Object (ES), and HEC-1 (NM) is provided as an example of an SDSS shell used to develop a computerized tutor that helps select HEC-1 methods based on available data and prepares the input for HEC-1.

INTRODUCTION

Spatial decision support systems (SDSS) are computerized technology used to support complex decisions based on some kind of spatially distributed information. They are used for semi- or ill-structured problems that do not lend themselves to the rigid, structured problem-solving techniques usually implemented in a computerized environment. SDSS can be viewed as a collection of tools and techniques that are used for data interpretation, manipulation, and analysis.

The major requirements of an SDSS are that they are designed to be interactive and easy to use; to have access to all data and analysis procedures; and to have a solution procedure (developed interactively by the user) by creating a series of alternative solutions and the ability to select the most viable one (Wright and Buehler 1990). Walsh (1993) provides a discussion on the SDSS framework within the water resources area.

Works by Loucks and Fedra (1987), Loucks, Kindler et al. (1985), Loucks, Taylor et al. (1985), and Fedra and Loucks (1985) provide in-depth discussions about emerging trends in water resources analyses and the need for more interactive methods. Although they do not explicitly call it SDSS, what they describe as needed in water resources modeling is exactly what SDSS are about.

> The essential difference between what is happening now and the more traditional off-line, noninteractive approaches is that instead of generating solutions to specific problems, model developers are now beginning to deliver, in a much more useful and user-friendly form, computer-based turnkey systems for exploring, analyzing, and synthesizing plans or policies. Such tools permit the user to evaluate alternative solutions based on his or her own objectives and subjective judgment in an interactive learning and decision-making process (Loucks and Fedra 1987).

Published research papers dedicated to SDSS applications show that there has indeed been a switch toward more powerful, complex, and interactive methods for decision support. Their scope and the ways in which the applications have been developed, vary from simple demonstration projects to complex regional systems, but the overall trend indicates that the complexity of problems to be solved is increasing and covers much more than pure technical problems that water resources engineers concentrated on in the past.

SDSS SHELL—THE TOOLBOX APPROACH

By definition, SDSS are tools that can be used in user-friendly environments to analyze data, with the ultimate goal of developing solutions for particular problems. In order to create general-purpose SDSS, integration of factual modeling, reasoning, and decision making has to be accomplished. Moreover, the user needs to have the capability to direct the flow of information and modeling effort toward a desired goal. The SDSS computer environment provides functionality to develop specific SDSS using general tools, that is, without the need to write special code that performs the necessary operations. It can be viewed as an SDSS "spreadsheet," where the user creates an application using spreadsheet functionality, but never has to write the functions themselves.

A solution approach to a complex environmental problem can be subdivided in several major tasks: data collection and preprocessing, numerical analyses, alternative evaluation, and solution presentation. Traditionally, each of the above-mentioned tasks has been implemented in a computer environment as a separate technology. Database management systems (DBMS) for tabular information and GIS for spatially distributed information have been used for data storage and manipulation. Expert system shells (ES) have been developed to emulate the reasoning process, while a number of general-purpose and specialized programs exist that can perform numerical modeling tasks and present the results.

An SDSS shell can be put together by combining individual technologies in a single computer environment, where each technology is aware of the others; that is, they can share the data and control the execution of the overall solution process. Viewed this way, SDSS are a collection of larger toolboxes that are used for a group of specific tasks. Each toolbox has a number of tools, used within the narrow scope of a single question or task (Figure 64-1).

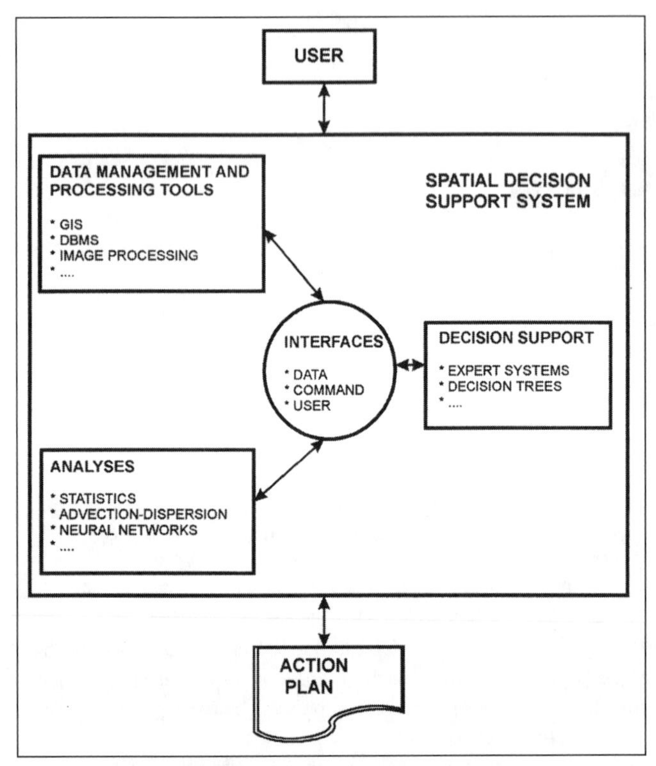

Figure 64-1. Schematics of a spatial decision support system shell.

There are two basic approaches for creating an SDSS shell:

1. Develop it from the ground up, by writing the whole code for desired functionality from scratch. This approach allows tight integration and the most efficient operation since the programmer has total control of all aspects of the program. The problem with this approach is that such a shell is complex and basically reinvents the code for most required functionality. This is a daunting task; as a consequence, implementation of such a system has not even been attempted.
2. Create a shell by putting existing applications together that provide the necessary tools (Djokic 1991). In this approach, most of the elements of an SDSS shell are tools that have already been implemented in a computer environment by someone else. The implementation, though, can take several forms. It can be in the form of a full-featured, commercially available software package, a set of subroutines that the user has to put together, or a public (shareware) domain software. The most difficult part of creating the shell using this approach is data and command control, that is, how to make different applications within the shell talk to each other.

Software Component Requirements of an SDSS Shell

There are three main requirements that a software package has to have to be viable for inclusion in an integrated computer application such as an SDSS shell:

1. The software has to be powerful enough to cover at least the current needs for a specific type of analysis. In principle, if there is a choice, a more powerful package should be considered for inclusion in the shell.
2. Each application has the "open" structure. Open in this context means two things: first, that the database for each application is accessible and easily modifiable by other elements in the shell (either directly or by some utility program), and that any application can be controlled by an input stream coming from a variety of sources. For example, consider statistical analyses. A statistical package that can be controlled only through its graphical user interface (GUI) will not be as useful within the shell as a package that also can be driven by a script (macro) language. The reason is the flexibility in the way the particular tool can be used. Let us say that in the course of execution of a large process we need to perform some statistical computations. The GUI package requires user intervention, even if all the parameters of the statistical operation are known. The other package can be run without user intervention by preparing the script from the application that determined what kind of statistical computation is needed (e.g., expert system), to run the script and then use the results to continue with the larger process. An example of this type of operation (expert system writing GIS macrocode) is presented by Djokic and Maidment (1993).
3. The applications that provide the desired functionality coexist in the same computer environment (platform). Today, this is more of a "desire" stemming from the reality of computer technology than a "hard requirement." With the trend toward open systems, networking, and downsizing, it is possible to build complex computer systems that have different components handled not only by different software packages, but that also have these packages residing on different computer platforms. Today there are many organizations that keep the data on a mainframe, analyze them on a workstation, and produce presentations of the analysis on a Macintosh or a PC.

Although conceptually attractive for its streamlined and efficient operation, such a system is highly susceptible to malfunction due to malfunction of any of its components. For SDSS, where interactive and user-friendly use is one of the prime operational considerations, multiplatform systems are not yet feasible (although they could be soon).

Building Interfaces

The burden of developing an SDSS shell using existing software packages is on creating interfaces that allow data transfer (data bridge) and command control (command bridge) between various applications. A lot of work can be saved by careful selection of shell elements. For example, if all the software components have intrinsic interfaces to a common database format (dBase, Oracle, Ingress, etc.), then there is no need to develop the data bridge because there is a common data format all the packages can understand and exchange the data with.

If that capability is lacking, simple data bridges can be programmed by using flat, text file structure (ASCII) as an intermediate step. Although this complicates the interface somewhat, it provides flexibility in terms of support in case a particular software changes its native data structure.

User interfaces used to be a major component of any custom application development. With current emphasis on the graphical user interfaces and standard windowing environments (Windows, Motif, Open Look, etc.), most of the more complex packages provide the tools for some kind of GUI development. They can be developed in GIS, ES, or DBMS, or written using specialized GUI tools. This feature provides the flexibility to interact with the user where the interaction is needed, and it reduces the effort required to develop a single application that handles all interaction with the user.

Professionally developed command bridges between different applications are very rare. This task is usually the most difficult to implement. Again, with careful selection of the software components ("openness") and operating system (multitasking), many difficulties can be avoided. Although initially command interfaces seem complex, most required functionality can be accomplished using simple methods (Djokic and Maidment 1993). If a data bridge exists, it is possible to develop a limited command bridge by using the common database as the "blackboard" on which applications look for tasks to execute (Sumic 1990), thus avoiding a "real" command bridge altogether.

Benefits and Pitfalls of the Approach

Using existing technologies as building blocks for an SDSS shell has several advantages and disadvantages. Often, individual components of the shell are already in use in an organization that is considering developing an SDSS. This means that the learning curve for implementation of the shell is not going to be as steep as with a new product, and that most of the effort can be placed on the overall interaction of the components as opposed to their individual capabilities.

Existing technologies (especially if they are widely used products) usually have proven track records in terms of user support, quality control, data availability, and flexibility. The burden of maintenance, development, and support of individual components is on the software producers, so the user can concentrate on specific shell implementation. Large user base means that there are already many applications developed for a specific product, so some solutions will not need to be developed, simply integrated. It also means that the dissemination of the product and its results are easier since more people have access to the technology.

There are several pitfalls as well. Even if individual elements of the shell are used in the same company, they might be used by different groups that do not have much interaction. This can sometimes hinder the implementation of the shell because different groups could have noncompatible agendas (an unfortunate reality).

Sometimes, the needs for a particular SDSS are much less than what the shell offers; consequently, there is substantial overhead and unnecessary complexity due to shell implementation. If the implementation of the data and command bridges does not allow full functionality of all the components, or the components are lacking necessary tools, the shell will not be as useful, yet still will create high overhead. Using commercially available tools can be very expensive, so start-up costs can be high, especially if training is required.

APPLICATION

A shell consisting of ARC/INFO (ESRI 1992), GIS, and Nexpert Object (Neuron Data 1991) expert system (Evans et al. 1993) has been applied to the analysis and preparation of input data for HEC-1 rainfall runoff model. Many of the methods for HEC-1 parameter extraction are empirical. These empirical expressions have been developed for specific regions and geographic conditions, which limit their use to watersheds of similar characteristics. This is often overlooked, so the idea is to create an application that will know the conditions for which these expressions were developed and help the user decide which are the correct methods applicable in his or her particular case.

The basic mode of operation has five main steps:

1. Basic data collection and preprocessing (ARC/INFO).
2. Combination of data from different GIS databases into a subwatershed-based object representation (Nexpert Object).
3. Application of rules to determine applicability of empirical parameter extraction methods and selection of appropriate methods (Nexpert Object).
4. Computation of parameters and their storage in the GIS database (Nexpert Object).
5. Formatting of basin data in HEC-1 format (custom C program).

The basic GIS data and ES object relationship is presented in Figure 64-2.

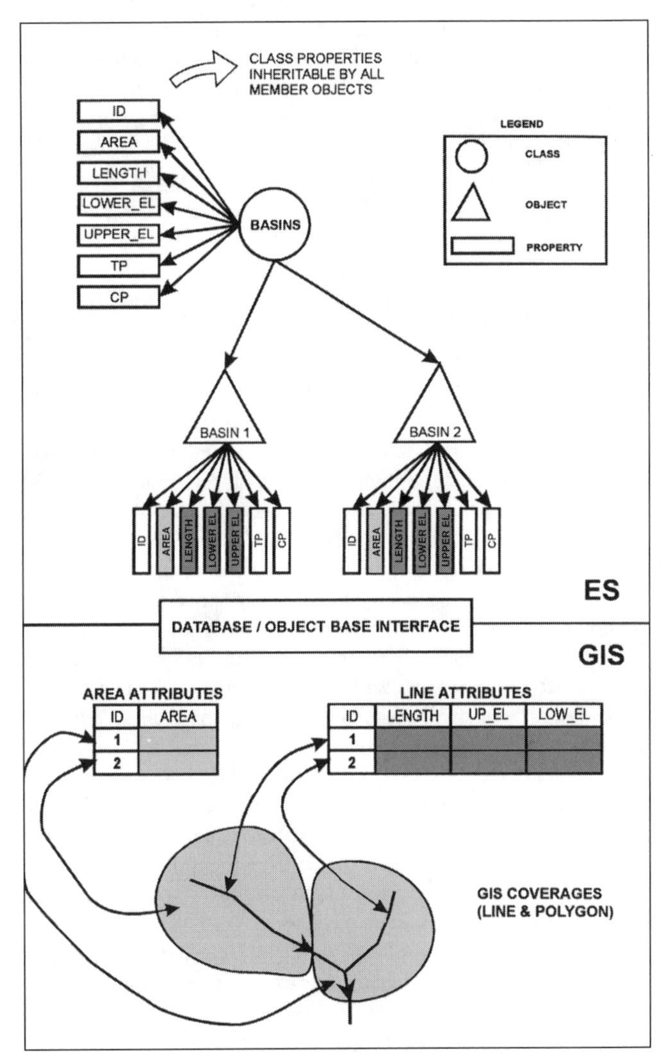

Figure 64-2. GIS databases and expert system object base interaction.

The basic geographic information for a watershed, its subwatersheds, and streams is developed and stored in ARC/INFO. There are two basic coverages: (1) the subwatershed coverage, which is a polygon coverage; and (2) a stream coverage, which is a line coverage. A number of GIS tools are used to develop these databases, requiring most of the features that ARC/INFO provides. The coverages usually will contain more information than required for HEC-1 analysis, but that is one of the advantages of a GIS system—different applications can share the common database and use only parts of it.

After spatial analysis is finished and coverages are completed, Nexpert Object extracts information it requires from the subwatershed coverage (e.g., AREA) and the stream coverage (e.g., LENGTH, UP_EL, and LOW_EL) to create objects representing individual subwatersheds. The rules are then applied to these objects to determine which parameter extraction methods are applicable. Once the method is selected for a subwatershed, additional parameters for HEC-1 rainfall runoff are computed (e.g., TP and CP). All of these operations are performed in computer memory (as opposed to disk for most database operations), increasing the efficiency and speed of execution and minimizing redundancy.

The newly computed HEC-1 parameters are stored back into GIS, mostly for quality control and interface consistency. The necessary data from the GIS database are then extracted and formatted according to HEC-1 input specifications. At this stage of application development, each subwatershed has a single file that the user has to put together in proper HEC-1 sequence. Eventually this step will be automated as well.

This organization of data and processing allows each technology to be used for what it does best: ARC/INFO for spatial data processing, Nexpert Object for rule-based reasoning on object-oriented data, and HEC-1 for rainfall runoff modeling.

CONCLUSION

Creation of a general-purpose spatial decision support systems shell is a viable undertaking. The major tasks in solving complex engineering or scientific problems—spatial data handling, numerical analyses, and reasoning—can be implemented in the computer environment using existing, commercially available software. The software products available today are fully featured and stable enough to provide most capabilities required for complex analyses.

By using commercially available software as components of the SDSS shell, development, support, and distribution are the responsibility of the software companies making the components and not the application developer. The individual software components, with their full capabilities, can also be used for other tasks that do not involve application of SDSS, so the investment in purchase, training, and maintenance can be shared by many users.

Interfaces that allow flow of data and control between different software packages are not difficult to implement, especially if careful consideration is given to the "openness" of the components and the computer environment in which to develop the SDSS shell. These interfaces are much easier to develop than rewriting major portions of individual applications.

An SDSS shell is not a solution for all problems, and there could be high costs associated with building or using one. The full capabilities of individual packages could be too advanced for the particular problem at hand, or the needs could be too specialized for a general-purpose package to handle. The computing overhead of many packages running at the same time could make the application too slow for interactive use. In general, SDSS shell technology has to be evaluated on a case-by-case basis, as any other complex technology and its computer implementation.

The presented application shows how a combination of off-the-shelf GIS and expert system packages can be utilized to efficiently construct a system that provides help in the selection of numerical methods for specific numerical modeling purposes.

Acknowledgments

This research has been supported in part by the Texas Low-level Radioactive Waste Disposal Authority.

References

Djokic, D. 1991. "Urban Stormwater Drainage Network Assessment Using an Expert Geographical Information System." Ph.D. diss., The University of Texas at Austin, Austin.

Djokic, D., and D. R. Maidment. 1993. Creating an expert geographical information system: The ARC/INFO–Nexpert Object interface. In *ASCE Monograph on Integration Issues in Expert Systems Technology*. New York: ASCE.

Environmental Systems Research Institute, Inc. 1992. *ARC/INFO Data Model: Concepts & Key Terms*. Redlands: ESRI.

Evans, T., D. Djokic, and D. R. Maidment. 1993. Development and application of an expert geographic information system. *J. Comput. Civ. Eng.* 7(3):339–53.

Fedra, K., and D. P. Loucks. 1985. Interactive computer technology for planning and policy modeling. *Water Resour. Res.* 21(2):114–22.

Loucks, D. P., and K. Fedra. 1987. Impact of changing computer technology on hydrologic and water resources modeling. *Rev. Geophysics* 25(2):107–12.

Loucks, D. P., J. Kindler, and K. Fedra. 1985. Interactive water resources modeling and model use: An overview. *Water Resour. Res.* 21(2):95–102.

Loucks, D. P., M. R. Taylor, and P. N. French. 1985. Interactive data management for resources planning and analysis. *Water Resour. Res.* 21(2):131–42.

Neuron Data. 1991. *Nexpert Object Functional Description*. Palo Alto: Neuron Data.

Sumic, Z. 1990. "The Concept and Feasibility of Automated Electrical Plat Design Via an Intelligent Decision Support System Approach." Ph.D. diss., University of Washington, Seattle.

Walsh, M. R. 1993. Toward spatial decision support systems in water resources. *J. Water Resource Plng. Mgmt.* 119(2):158–69.

Wright, J. R., and K. A. Buehler. 1990. *A GIS Proving Ground for Water Resources Research*. Tech. Rept. 189. West Lafayette: Water Resources Research Center, Purdue University.

Dean Djokic is a research associate and a lecturer at the Center for Research in Water Resources, the University of Texas at Austin. He got his B.S. and M.S. degrees in civil engineering from the University of Zagreb, Croatia; his graduate degree in hydrology from IHE in Delft, the Netherlands; and his Ph.D. from the University of Texas at Austin. Before moving to Texas in 1987, he worked as a consultant on hydrologic and hydraulic analyses pertinent to development of hydropower. He specializes in application of GIS and other advanced computer methods in water resources and plans to continue his career in the academic environment.

Center for Research in Water Resources
University of Texas
Austin, Texas 78712
Phone: (512) 471-1807
E-mail: *dean@batza.crwr.utexas.edu*

65

Environmental Decision Support Systems (EDSS):

An Open Architecture Integrating Modeling and GIS

Steven P. Frysinger, David A. Copperman, and Joseph P. Levantino

The information-intensive nature of environmental management has led to increased interest in a class of computer applications called environmental decision support systems (EDSS). EDSS are designed to assist users make complex environmental management decisions by employing multiple technologies. As defined here, EDSS focus on specific rather than generic problems to provide a user interface that is sufficiently rich, yet simple enough to encourage casual users. This is in sharp contrast with the general-purpose character of software systems like geographic information systems (GIS), although the spatial nature of environmental management problems virtually dictates the use of GIS technology in any EDSS. Therefore, extensible software architectures supporting customization toward various decision problems and integration of various support tools (including GIS) are required if EDSS are to meet our environmental stewardship imperative. We briefly describe such an architecture, with special emphasis on mechanisms by which a variety of mathematical modeling programs are integrated with GRASS-GIS and a graphical user interface (GUI) to yield a single coherent system, easily adapted to a wide variety of decision problems.

WHAT IS AN ENVIRONMENTAL DECISION SUPPORT SYSTEM?

Environmental decision support systems (EDSS) are computer systems that help humans make environmental management decisions. They facilitate "natural intelligence" by making information available to the human in a form that maximizes his or her cognitive decision processes. EDSS bring to bear multiple technologies to accomplish this goal, and a number of such systems recently have been reported (Fedra 1993; Frysinger et al. 1993; Fürst et al. 1993).

While the term "EDSS" can broadly refer to virtually any sort of computer application that assists an analyst in arriving at an environmental management decision (Guariso and Werthner 1989), we choose to limit the definition to those systems that integrate multiple supporting technologies and that generally provide explicit support for the spatial character of environmental management problems. A further restriction in this definition is that EDSS focus on a particular decision problem and decision maker. Thus, they are not general purpose tools with which anything can be done (if only you knew how to do it). Rather, they are particularly tailored to a problem facing an analyst, and they offer a user interface that is optimized for that problem.

The focused nature of EDSS helps simplify the user's interface to the computer system, allowing the user to concentrate on the problem at hand and the information and tools needed to solve it. It also dictates a software architecture that facilitates the development of sibling systems embracing different decision problems with an essentially common user and data interface.

In order to effectively use spatial and aspatial data in decision making, analysts often use modeling tools that permit predictive analyses. These modeling tools generally are cumbersome, offer poor human interfaces, and little or no means for interaction with GIS data. There is a definite need for EDSS to provide a human interface to such computational models, ideally incorporating robust support for the analyst while he or she defines a conceptual model of the problem at hand (Heger et al. 1992). Furthermore, uncertainty in this conceptual model often results in a need to evaluate the model stochastically. This generally necessitates executing the modeling codes multiple times within a Monte Carlo simulation loop, with each iteration sampling distributions describing the uncertainty of the input parameters.

If the user adds to these features the following: (1) access to empirical data, (2) exploratory data analysis tools to analyze and interpret these data, and (3) a carefully designed user interface, he or she has created a tool that can significantly enhance the quality of the environmental decision maker's working environment. Integration of these facilities into a single "workbench" that minimizes clutter and maximizes ease of use becomes the single most important task of the EDSS designer. Such an interface can best be defined by employing the task analysis approach of human factors engineering, wherein the specific decision process is analyzed and characterized in terms of the needs of the decision maker.

Once having defined the requirements for an EDSS, one is faced with the technical challenge of integrating the software tools needed to support the required interface. The following section will address this topic in some depth, with special emphasis on the software interface between environmental models and a GIS-based user interface.

THE SOFTWARE PATCH PANEL

An open architecture for EDSS has been proposed and prototyped (Frysinger 1994). In order to simultaneously maximize software portability and extensibility, this architecture is hosted by UNIX workstations under X-Windows (Figure 65-1). UNIX, while not in the public domain, is available from various vendors and is widely used in high-performance workstations as well as on personal computer hardware. Its multitasking capability allows us to adopt a very flexible software architecture. X-Windows is a public domain client server architecture that supports windows-based graphical user interfaces on a variety of hardware and software platforms. The approach takes full advantage of the multitasking support of the UNIX operating system, to produce a truly open architecture.

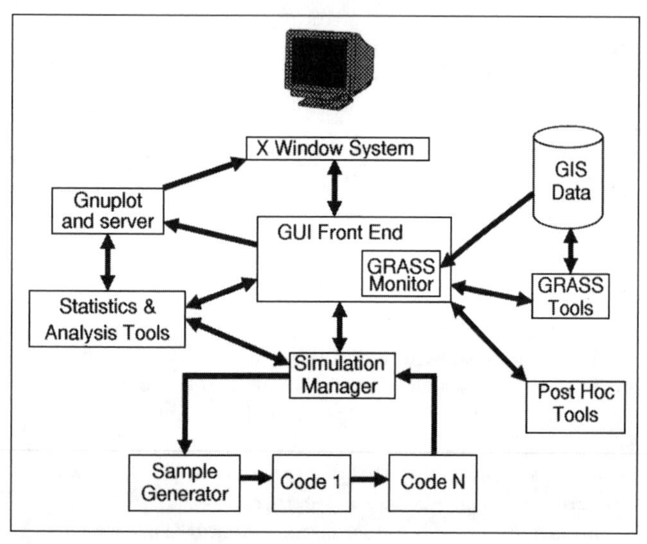

Figure 65-1. An open architecture for EDSS.

Individual modeling codes are maintained as stand-alone programs managed by shell scripts. Data import and export for these codes are accomplished by providing data to the UNIX standard input file and trapping results from the standard output file. Alternatively, codes that use named files for input and output are provided with names of temporary files for this purpose. These data can then be reformatted using standard UNIX tools (e.g., Awk and Sed) to conform to the requirements of subsequent codes. This "arms-length" relationship with computational codes has two significant advantages:

- Existing codes, such as those from federal agencies, can generally be used "as is," with no modifications. Apart from cost reduction, this preserves the sanctity of certified codes.
- New codes can be developed and tested independently and only integrated into EDSS when their correctness is assured. This minimizes the errors caused by the integration process and reduces the burden of regression testing.

These properties result in an architecture that is open, with respect to computational models for a given decision problem, and facilitate its extension to other decision problems with different models.

The interface between this multitasking simulation architecture and the software that manages the user interface is, to the maximum extent possible, data driven. For example, the set of data plots available for display is defined by a text file and ultimately displayed by a slightly modified version of the public domain Gnuplot program (which runs as a stand-alone UNIX process). If an additional plot is to be added to the offering, it can be accomplished through modification of shell scripts. The new views of the data can be provided through invocation of UNIX-based statistics and data processing tools, and the new data set (with its menu entry string) passed to the "front end" user interface process as data.

Likewise, the architecture incorporates the GRASS-GIS (a public domain UNIX-based GIS developed system, maintained by the Army Corps of Engineers' Construction Engineering Research Laboratory [CERL]). GRASS analytical tools retain their normal status as stand-alone processes, with an interprocess communication interface to the front end control software responsible for the GUI. GRASS's monitor has been modified to remove window decorations and has been incorporated into the GUI front end to support rapid image refreshing. While it provides the primary display window of the EDSS, all user interactions (such as mouse management, pop-ups, etc.) are managed by the GUI front end, thus preserving a coherent "look and feel" in the EDSS.

GRASS analytical tools are invoked through UNIX's system call or its equivalents, with data from files moved between the GIS domain and the modeling codes via shell scripts under a data passing protocol. Spatial animation (such as contaminant plumes or hypothetical wells) is drawn onto the GRASS window without the intervention of GRASS native code, yielding a much more interactive spatial display than provided by conventional GIS.

While the architecture just described is in fact a collection of diverse, independent software tools, the user interface is crafted in such a way that the analyst always has the impression that he or she is interacting with a single, coherent system.

CASE STUDIES

A case study of such an EDSS architecture may be found in the Sandia Environmental Decision Support System (SEDSS), a family of EDSS that tackle monitor well network design, waste facility evaluation, and risk analysis, with more applications planned. A similar approach is underway to address fisheries management decision making by New Jersey's Division of Fish, Game, and Wildlife.

Sandia Environmental Decision Support System (SEDSS)

The Sandia environmental decision support system (SEDSS) is a family of workstations being developed jointly by Sandia National Laboratories and AT&T Bell Laboratories and is representative of this technology. SEDSS is a family of software applications with strong family traits, but with significant individuality. Sharing a common user interface philosophy and software architecture, the members of the SEDSS family are designed to facilitate human decision making with respect to the hydrological aspects of hazardous waste management. SEDSS is based upon the open architecture advocated here and elsewhere (Frysinger 1994). Its prototype was, in fact, the proving ground for many concepts implicit in that architecture.

Sandia National Laboratories, as the operator of chemical waste facilities regulated under the Resource Conservation and Recovery Act (RCRA), needed to determine the number and location of monitoring wells to be installed around these sites. RCRA specified a minimum of one up-gradient background well and three down-gradient monitor wells, but it did not provide a quantitative basis for determining the actual number and placement of wells to be drilled and monitored. As a result, Sandia (like other waste site operators) was compelled by regulators to install more and more wells, without the benefit of a rational placement strategy or beforehand knowledge of the total number of wells required.

In response to this circumstance, Sandia used the mathematical models of site geohydrology to determine likely flow paths and placed wells accordingly (Parsons and Davis 1991). Using computational models of flow in the unsaturated and saturated zones and applying Monte Carlo simulation techniques to sample the uncertain parameters of a conceptual model of the site's hydrology and geology, it was possible to evaluate the performance of an actual or proposed monitoring well network and determine where to best locate additional wells to achieve the desired probability of plume detection (should a leak occur).

This analytical methodology employs a scoring approach, taking the plume realizations generated by Monte Carlo simulation as representative of the population of all possible plumes. For each plume realization, the plume is considered detected if it intersects at least one well in the monitoring network (which can be composed of either existing wells, hypothetical wells, or some combination of the two). The fraction of plume realizations detected is taken as the probability of detection of any plume by the network. Conventional optimization methods can then be employed to determine combinations of well locations that achieve a specified threshold detection probability with a minimum number of wells.

Unfortunately, any such modeling effort requires the development of a rather extensive set of assumptions about the hydrology and geology of a site, and uncertainties about these assumptions are not easily conveyed to the regulators when presenting modeling results. Furthermore, a credibility issue arises when the regulators themselves cannot manipulate the parameters of the conceptual model to evaluate the sensitivity of the modeling solution.

The remedy for these difficulties was to develop an interactive geohydrological modeling platform to share with the regulators so they can control the assumptions behind modeling operations and make their own decisions based upon actual data. Thus, the monitor well network designer (MWND) was born—the first "sibling" of the SEDSS family (Frysinger and Parsons 1992; Frysinger et al. 1992).

Since the cost of monitoring wells is significant, MWND serves an important function in helping to ensure that contaminant plumes are likely to be detected while minimizing the number of wells installed to achieve the desired detection goal. MWND allows the analyst to: (1) interactively manipulate the parameters of the hydrological conceptual model (including consideration of uncertainty), (2) use various computational methods to simulate contaminant plumes, (3) invoke linear programming optimization algorithms to suggest well locations that meet the desired detection probability, and (4) use interactive visualization techniques to conduct exploratory data analyses. Since it offers a "what if?" capability, MWND allows analysts, who traditionally distrust modeling results, to develop their own results based upon their own conceptual model, and to interactively test the sensitivity of these results to changes in the conceptual model.

The waste facility risk analyzer (WFRA) is the second member of the SEDSS family, having been prototyped after the completion of the MWND prototype. It is designed to help the Nuclear Regulatory Commission (NRC) staff execute low-level waste facility performance assessment methodology (Kozak et al. 1990) for the evaluation of license applications for low-level radioactive waste disposal facilities. The impact of the proposed facility location and design on the neighboring population is estimated through modeling of various facility failure scenarios and exposure pathways. A variety of models are employed to simulate the movement of radionuclides from the facility to the hypothetically exposed individual, with the goal of calculating the probable exposure of that individual to radiation leaving the facility. The development effort for this prototype was approximately one-tenth that for MWND, due to the advantage of having the MWND architecture as a starting point.

Marine Environmental Decision Support System (MEDSS)

In a different vein, EDSS architecture is being applied to the New Jersey Division of Fish, Game, and Wildlife's Ocean Stock Assessment Program. Their needs in coastal fisheries management include plotting and spatial analysis of species sampling surveys and water quality studies. While many of their requirements could be considered classic GIS applications, the opportunity to integrate EDSS with predictive models of harvest, weather, or pollution impacts suggests some advantages to applying the architecture we describe. An additional advantage will accrue in that the GIS and modeling functions will integrate into a user-friendly system that encourages biologists who are not experienced GIS users to benefit from its capabilities.

The effort to design such a system has been undertaken as a community service to the division, complementing other related volunteer activities. However, it also will be a useful test of the applicability of the EDSS architecture to a completely different application and could result in a more generalized tool for coastal management.

TECHNOLOGICAL CHALLENGES

Despite these and other very encouraging successes in the integration of environmental models with GIS to form environmental decision support systems, there remain some significant technical challenges.

Three-Dimensional Modeling

The most fundamental limitation of GIS in EDSS applications is the two-dimensional nature of the conventional GIS data model. While current GIS technology provides "two-and-one-half-dimensional" capabilities, such as perspective surface plots, there is no true three-dimensional raster structure available in common GIS. It is possible to simulate, for data storage purposes, a three-dimensional raster by using multiple layers of two-dimensional rasters, but access to data so stored is clumsy at best, and analyses can only be performed by tools custom written for the purpose. In contrast, most hydrological models are inherently three dimensional, with two- or one-dimensional simplifications undertaken only where necessary for reasons of computational complexity or data sparsity. Recent work in geoscientific information systems (GSIS) shows promise in resolving this shortcoming (Battaglin 1993; Faunt et al. 1993; Fisher 1993; Schenk et al. 1993).

Advanced Spatial Data Representation

Another challenge receiving less attention is associated with the way in which spatial data are displayed. GIS workers have long noted that their multicolor map displays often misrepresent the data by omitting the element of spatial uncertainty. For example, one could get the impression from viewing a soils map that the boundaries between soil types are well defined and well known.

When one integrates environmental models with GIS, this problem becomes even more acute. These models involve parameters that are, to varying degrees, spatial and often uncertain. While it is straightforward to represent, say, moisture content as a raster layer in a GIS, it is more difficult to represent the uncertainty associated with these values, especially when the uncertainty is itself a spatial quantity.

One visual technique that may help resolve this is a dithering algorithm that defines multipixel cells of homogeneous value. For each such cell, the color assigned is a function of its value, as is usual. However, the fraction of member pixels displayed in that color depends on the degree of certainty in the value. If the value is completely certain, all pixels in the cell are given the color; if there is less certainty, fewer pixels receive that color (the others receive the background color). This results in a dithering, or washing out, which can indicate the degree of certainty of a spatial parameter.

Alternatively, nonvisual displays may be used to communicate such extraspatial data to the analyst. Auditory displays (Frysinger 1990; Kramer 1993) offer some possibilities here, some of which are just being explored in a spatial context (Weber personal communication, 1993; Krygier personal communication, 1993). An obvious example of the use of sound would be to design an auditory display for certainty (e.g., let the harmonic content of a tone increase with increasing uncertainty), and then use this mapping to display the uncertainty associated with the point in space to which the mouse is pointing. The uncertainty of two or three variables could be simultaneously represented this way by assigning each a voice of a different pitch. While such a notion may seem bizarre to those used to working with strictly visual tools, the nature of our display requirements and the history of auditory data representation research suggest such studies are worth undertaking.

Empirical Measures of Effectiveness of EDSS

Finally, there exists a pressing need to measure the effectiveness of the decision support tools we offer environmental managers if we are to distinguish useful enhancements from attractive nuisances. While this necessity exists throughout the universe of software tool design, it is especially crucial for EDSS technology because the stakes are high, the users are drawn in many directions, and the general level of computer expertise is low. Psychophysics experiments evaluating the merits of such elements as data representation are fairly easy to design and conduct (Mezrich et al. 1984). However, it will prove much more challenging to design and validate experiments that test the efficacy of an EDSS with respect to total task performance. In part, this is due to the difficulty of knowing when an environmental decision is "correct"—how does one balance economic cost against risk, for example. Nonetheless, we must begin the difficult process of defining objective metrics for our decision support tools and replace anecdotal successes with measures of worth. This effort will be made easier by incorporating systematic methods during EDSS requirements specification and design. Such methods will help us understand the task to which we address our systems and against which we must measure them.

CONCLUSION

The architecture described herein, demonstrated through the cited case studies, offers a clean, open mechanism for integrating environmental models and GIS. On this foundation, it will be possible to expand the Sandia Environmental Decision Support System family at a greatly reduced cost and with consistent, friendly human interfaces, and to develop new EDSS to satisfy the disparate needs of environmental managers. Ultimately, we hope that this leads to better environmental management decisions and corresponding improvements in environmental stewardship.

Acknowledgments

The SEDSS project involved a team of Sandia and Bell Lab scientists and engineers, including Steve Conrad, Roger Cox, Paul Davis, Tom Feeney, Matt Kozak, Bob Knowlton, Jim McCord, Alva Parsons, Dick Thomas, and Tony Zimmerman. In addition, the U.S. Army CERL contributed substantially to this effort through their development and support of GRASS. Finally, the budding MEDSS project is being coordinated with New Jersey marine biologist Don Byrne under the auspices of the Frank Jewett Chapter of Telephone Pioneers.

References

Battaglin, W. 1993. Use of volume modeling techniques to estimate agricultural chemical mass in groundwater. In *Applications of Geographic Information Systems in Hydrology and Water Resources Management—Proceedings of HydroGIS '93*, edited by K. Kovar and H. P. Nachtnebel, 591–99. IAHS Press.

Faunt, C., F. D'Agnese, and A. K. Turner. 1993. Development of a three-dimensional hydrogeological framework model for the Death Valley region, southern Nevada and California, USA. In *Applications of Geographic Information Systems in Hydrology and Water Resources Management—Proceedings of HydroGIS '93*, edited by K. Kovar and H. P. Nachtnebel, 227–34. IAHS Press.

Fedra, K. 1993. Models, GIS, and expert systems: Integrated water resources models. In *Applications of Geographic Information Systems in Hydrology and Water Resources Management—Proceedings of HydroGIS '93*, edited by K. Kovar and H. P. Nachtnebel, 297–308. IAHS Press.

Fisher, T. R. 1993. Integrated three-dimensional geoscientific information systems (GSIS) technologies for groundwater and contaminant modeling. In *Applications of Geographic Information Systems in Hydrology and Water Resources Management—Proceedings of HydroGIS '93*, edited by K. Kovar and H. P. Nachtnebel, 235–41. IAHS Press.

Frysinger, S. P. 1990. Applied research in auditory data representation. In *Extracting Meaning from Complex Data—Proceedings of the SPIE/SPSE Symposium on Electronic Imaging.*

———. 1994. An open architecture for environmental decision support. *International Journal of Microcomputers in Civil Engineering* 10(2):119–26.

Frysinger, S. P., R. Cox, P. Davis, and A. Parsons. 1992. Environmental decision support systems and GRASS: A Sandia monitor well case study. In *GRASS User's Conference.*

Frysinger, S. P., and A. M. Parsons. 1992. A decision support system for evaluating the performance of a monitor well network. In *Environmental Geotech Symposium—American Society of Civil Engineers.*

Frysinger, S. P., R. P. Thomas, and A. M. Parsons. 1993. Hydrological modeling and GIS: The Sandia Environmental Decision Support System. In *Applications of Geographic Information Systems in Hydrology and Water Resources Management—Proceedings of HydroGIS '93*, edited by K. Kovar and H. P. Nachtnebel, 45–50. IAHS Press.

Fürst, J., G. Girstmair, and H. P. Nachtnabel. 1993. Application of GIS in decision support systems for groundwater management. In *Applications of Geographic Information Systems in Hydrology and Water Resources Management—Proceedings of HydroGIS '93*, edited by K. Kovar and H. P. Nachtnebel, 13–21. IAHS Press.

Guariso, G., and H. Werthner. 1989. *Environmental Decision Support Systems.* Chichester, England: Ellis Horwood Books.

Heger, A. S., F. A. Duran, S. P. Frysinger, and R. G. Cox. 1992. Treatment of human–computer interface in a decision support system. In *IEEE International Conference on Systems, Man, and Cybernetics*, 837–41.

Kozak, M. W., M. S. Y. Chu, and P. A. Mattingly. 1990. *A Performance Assessment Methodology for Low-Level Waste facilities.* NUREG/CR-5532, SAND90-0375.

Kramer, G., ed. 1993. *Auditory Display: Proceedings of the First International Conference on Auditory Display*. Sante Fe Institute Proceedings Series. New York: Addison-Wesley.

Mezrich, J. J., S. P. Frysinger, and R. Slivjanovski. 1984. Dynamic representation of multivariate time-series data. *Journal of the American Statistical Association* 79:34–40.

Parsons, A. M., and P. A. Davis. 1991. *A Proposed Strategy for Assessing Compliance with the RCRA Groundwater Monitoring Regulations, Current Practices in Groundwater and Vadose Zone Investigations*, edited by D. M. Nielsen and M. N. Sara, ASTM STP 1118, Philadelphia: American Society for Testing and Materials.

Schenk, J., K. Kirk, and E. Poeter. 1993. Integration of three-dimensional groundwater modeling techniques with multidimensional GIS. In *Applications of Geographic Information Systems in Hydrology and Water Resources Management—Proceedings of HydroGIS '93*, edited by K. Kovar and H. P. Nachtnebel, 243–49. IAHS Press.

Steve Frysinger holds a B.S. in environmental studies/physics, M.S. degrees in computer science and applied psychology, and a Ph.D. in environmental sciences. He is Associate Professor at James Madison University in the College of Science and Technology. He also is on part-time status at Bell Laboratories as a Member of Technical Staff (MTS).

James Madison University
College of Science and Technology
Harrisonburg, Virginia 22807
E-mail: *frysinsp@jmu.edu*
and
AT&T Bell Laboratories
8209 Valley Pike
Middletown, Virginia 22645
Phone: (540) 568-2710
Fax: (540) 568-2761
E-mail: *frysinger@att.com*

Dave Copperman has been an MTS at AT&T Bell Laboratories for ten years, holding a B.S. in geography and an M.S. in computer science.

Joe Levantino has been an MTS/technical manager at AT&T Bell Laboratories for over fifteen years, holding a B.S. in electrical engineering and an M.S. in computer science.

AT&T Bell Laboratories
101 Crawfords Corner Rd.
Holmdel, New Jersey 07733
Phone: (908) 949-7596
Fax: (908) 949-6029
E-mail: *spf@hoqaa.att.com*

66

Multicriteria Decision Support for Land Reform Using GIS and API

C. Peter Keller and James D. Strapp

Numerous applied multicriteria land related problems require decision making. These problems are not solvable by conventional mathematics; they require logical search procedures leading to feasible solutions of acceptable compromise. Commercially available GIS offer an appropriate technology for data inventory, routine manipulation, and visualization, but they lack the necessary advanced analytical capabilities and search procedures. Necessary solution procedures can be designed as spatial decision support systems (SDSS). The challenge becomes how to customize GIS to incorporate individual SDSS designs efficiently and at a low cost. One promising solution is to customize a GIS using an application programming interface (API). This chapter reports on the formalization of a land related decision-making problem into a decision support framework and the subsequent customization of GIS into an SDSS. The problem concerns agricultural land consolidation.

INTRODUCTION

Farm fragmentation occurs where landholdings of individual farmers consist of small and widely scattered parcels of land. Fragmented land ownership exists in many parts of the world. Where fragmented land ownership dominates the pattern of holdings, it has been identified as a form of ownership that controls agricultural practices and a way of life.

Opinions differ concerning potential merits and drawbacks of a fragmented system of landholdings. Some point out long-presumed costs and identify significant benefits derived from fragmentation, such as ecological diversity for an individual farmer (Bentley 1990). Others contend farm fragmentation is the single greatest deterrent to modern agricultural development, creating inefficiencies in movement of labor and machinery, hindering large-scale mechanization of production processes, and increasing administration expenses, given the complexity of the cadastre and rights-of-way (Farmer 1960; Agarwal 1972; Bonner 1987). The authors advocate elimination of farm fragmentation through a process called "land consolidation."

Land consolidation defines the process of changing land ownership by redistribution so that individual farmers own fewer, more compact, larger land parcels. The economic rationale underlying land consolidation has been recognized by many governments. The potential for improved efficiency and competitiveness has led politicians and planners to favor land consolidation schemes in many parts of the world.

Conducting a land consolidation scheme has proven to be an extremely complex and political process given the many criteria that must be considered when reallocating land and the importance most societies and individuals place on land ownership. Experience has shown that voluntary consolidation schemes rarely have been successful since a small minority can effectively create havoc in the overall planning process. Most schemes, therefore, have required varying degrees of enforced land consolidation following a complex planning process.

The steps required to conduct a planned land consolidation scheme are well understood. Implementing these steps in an efficient and effective manner, however, has proven difficult for several reasons:

1. A fair land consolidation process requires access to considerable volumes of data about the land and its occupants. Efficient technology to support storage, manipulation, and visualization of these data has been lacking until the advent of digital geographic information systems (GIS). GIS are beginning to be applied in land consolidation schemes and promise to be an enabling technology to facilitate the inventory, query, and communication components of land consolidation.
2. The multicriteria nature of the consolidation process implies that a straightforward logical or mathematical solution procedure to consolidate land ownership does not exist. Indeed, in the past, the actual process of shuffling land ownership to yield more compact and larger parcels has been an ad hoc and often a subjective process. The latter has resulted in a call for research to examine and formalize the land consolidation process (King and Burton 1983) and to provide tools to implement it. While limited progress has been made on formalizing the land consolidation process (Kik 1990), a response to this call for research has been slow, given the lack of mathematical formality and available technology to allow realistic implementation of a formalized process. Commercially available GIS do not offer a solution since they lack the necessary mathematical solution procedures. However, recent advances in tools to develop enhancements to commercial GIS appear to offer technology to support implementation of a formalized definition of the land consolidation process.

Given the above, we set out to explore the feasibility of combining the inventory and data manipulation capabilities of GIS with a formalized definition and mathematical solution procedure of a multicriteria decision support system for land consolidation. This would require creating and adding a number of unique analytical tools to the usual suite of functional capabilities of GIS and to modify the user interface so that these tools easily could be used and understood.

Several commercial GIS vendors offer or are about to offer application programming interfaces (API) for their systems. In their simplest form, these API libraries of code allow users to access the GIS's spatial database directly from external programs. More advanced routines may be available to control GIS user interfaces and to access external attribute databases. The result of creating a program with an API is an executable file that, depending upon the vendor's implementation, may be distributed without the parent GIS. API frees programmers to concentrate on adding analytical capabilities to a spatial database management system instead of recreating GIS functions. Consequently, API promises to be a powerful solution to GIS customization, and the combination of GIS and API appears to be well suited to building a spatial decision support system (SDSS) (Strapp and Keller 1992). We argue that

GIS + API = SDSS Toolbox.

A secondary purpose of our research efforts, therefore, was to test the advantages of using API to customize GIS into an SDSS. In this chapter we summarize our efforts to build an interactive land consolidation decision support system by programming a formalized mathematical solution procedure for land consolidation into a commercially available GIS using API.

In the next section we discuss objectives and steps involved in a land consolidation process, which leads to the introduction of a formalized definition of the land consolidation procedure in the third section. After that, we summarize our efforts to program the formalized definition into GIS using API. Finally, we comment on our experience and offer suggestions for future work.

SOLVING AGRICULTURAL LAND CONSOLIDATION

Once agreement is reached in principle that land consolidation should proceed, the following general consolidation process can begin:

1. Determine ownership and produce cadastre.
2. Determine value of land.
3. Circulate value for comment.
4. Repeat 2 and 3 if deemed necessary.
5. Give opportunity to state preference (optional).
6. Conduct consolidation process.

In order to conduct the above steps, a number of issues must be addressed. They include identification of what criteria will determine land value and wealth, what constraints are to be imposed on the consolidation process, and what consolidation solution procedure is to be applied in step 6 above.

These issues can be addressed, in part, by identification and examination of the objectives underlying the consolidation initiative. The following appear to be typical motives and arguments for consolidation.

Improve Efficiency	**Maintain Equity**
• Parcel efficiency (shape/size)	• Individual equity ("How did I do?")
• Spatial efficiency (contiguity/proximity)	• Community equity ("How did the others do?")
• System efficiency (infrastructure)	

From an efficiency perspective, planners will seek to create large parcels shaped to suit mechanized farming—notable rectangles or squares. Contiguity and proximity are surrogates for accessibility: an individual farmer's land holdings should be close to each other and close to the homestead.

Consolidation schemes often include plans to improve road infrastructure and to build irrigation systems; therefore, system efficiency will dictate boundaries along proposed new roads and irrigation canals. Wherever possible, maintenance of existing fences, walls, and hedges, etc., to avoid extra economic and environmental costs, is a second system efficiency consideration.

Given this, our opinion is that the formal consolidation scheme should include the opportunity for planners to modify the existing field pattern interactively using GIS, allowing them to tessellate space into newly defined parcels large enough to give individual plots of land a simpler geometry to support farm mechanization: The parcels would abide by boundaries imposed by new roads and irrigation canals; existing structures would be maintained as much as possible. This tessellation process should still yield a set of parcels large enough to support equitable redistribution.

Consolidation is concerned with the redistribution of land, not the redistribution of wealth. No one involved in the consolidation scheme should lose or benefit more than another; ideally, the scheme would result in an equal net benefit for every participant. However, defining costs and benefits experienced by farmers as a result of consolidation is problematic given the many noncommensurable criteria that define land value, the lack of a universally accepted yardstick by which to place value on land, and the fact that the relative value for different classes of land may change considerably through time. The following is a list of some of the criteria that ought to be considered when setting out to derive a fair assessment of land value:

Naturally endowed value	**Value added**	**Nonagricultural value**
• slope	• fencing	• real estate
• soil	• orchards	• sentiment
• topography	• vineyards	• burial grounds
• sun exposure	• access to roads	• political agendas
• precipitation	• irrigation	• heritage

It should become obvious that these criteria cannot all be measured in the same way, and that much of the information will be subjective and subject to uncertainty and bias. The land consolidation process, therefore, must accommodate many noncommensurable criteria and objectives. Planners must accept that a farmer's perception of value may be quite different from a mathematically or logically derived value assessment, however defined.

With this in mind, we feel that the formalized version of the land consolidation process should avoid trying to combine noncommensurable assessments of land. We favor the multicriteria budget approach. Individual farmers begin by evaluating parcels of land independently on a number of agreed-upon criteria. At the outset of consolidation, the decision support system calculates how much of each criterion each individual farmer owns under the fragmented system. This becomes each farmer's starting budget consisting of a multicriteria account balance. Farmers should be able to assess how individual components of their budget change as ownership of land is reallocated in the consolidation process.

Ideally, each individual's multicriteria account statement at the end of consolidation should match exactly the starting account balance. This obviously is an unrealistic expectation. Farmers and planners must therefore negotiate what constitutes acceptable and fair trade-offs between individual noncommensurable criteria. The consolidation process ought also to be able to enforce threshold limits on individual components of the multicriteria budget. In other words, a farmer's gain or loss for a specific item in the budget may not exceed a predetermined percentage or actual number. Given the above, the land consolidation decision support system must monitor each farmer's multicriteria budget carefully, flag any accounting discrepancies, and produce an output of individual budgets on request.

Some consolidation schemes in the past have allowed farmers to participate in the consolidation process by giving them the opportunity to state their preferences for what land they wish to retain, what land they are willing to lose, and what land they would like to acquire. We felt that a land consolidation decision support system should support this option; the decision support system should allow the set of all individual farmers m to view and classify a map of land parcels to be consolidated, breaking all parcels W_{ij} into one of four options:

$W_{kj}=1$	parcel k to be retained by farmer i
$\sum_{i=1}^{m} W_{ij}=0, \quad W_{kj}=0$	parcel j not wanted by any farmer i
$\sum_{i=1}^{m} W_{ij}=1, \quad W_{kj}=0$	parcel j wanted by one farmer i
$\sum_{i=1}^{m} W_{ij}>1, \quad W_{kj}=0$	parcel j wanted by more than one farmer

Where k is the farmer owning the land before consolidation and $W_{ij}=1$ if farmer i wants parcel j, 0 otherwise.

FORMALIZATION OF PROPOSED CONSOLIDATION MODEL

Land consolidation is a messy multicriteria problem (e.g., location, allocation, assignment), to which an elegant, efficient solution procedure, yielding a single optimal answer does not exist. We strongly suspect that any solution procedure is intractable (NP hard), although we have not set out to prove this mathematically.

We conclude that an appropriate solution procedure for land consolidation should:

- be sufficiently automated to produce alternative solutions in a reasonable time.
- be able to consider multiple structured objectives explicitly.
- account for multiple unstructured objectives by supporting interaction with solution processes.
- be sufficiently generic to be of use in many different consolidation programs.
- be able to produce more equitable solutions by better representing land value through retention of innate attributes of a parcel.
- provide a mechanism for the direct input of participants' preferences.

A preferable solution is to conduct the consolidation process by following a logical sequence of steps using a land consolidation SDSS built on GIS. The steps are as follows:

1. Reach consensus on what criteria should be included in the consolidation process.
2. Collect data set criteria and build a digital inventory using GIS.
3. Examine the old field pattern and interactively modify it where necessary to meet geometric concerns.
4. Calculate individual farmer's multicriteria budgets.
5. Allow farmers to state ownership preferences.
6. Enforce threshold limits on individual components of the multiple criteria budget.
7. Implement a heuristic to reallocate the land subject to threshold limits defined in step 6.
8. Allow farmers to respond to results and interactively repeat from step 5.

Two issues arise: (1) a search heuristic needs to be designed and coded into the GIS to reallocate land, and (2) the GIS needs to be customized to support a user interface and routine output generation that easily can be understood and operated by farmers and planners.

The Heuristic

A search heuristic was designed consisting of five modules that can be implemented in any sequence. The modules are:

Reassign: reallocates parcels back to original owners because owners want to retain them.

Uncontested: identifies and allocates parcels wanted only by one farmer.

Adjacency: assigns parcels to the farmer owning the contiguous parcel with the longest common boundary.

Deficit: allocates a parcel to the farmer with the largest deficit in a chosen attribute.

Unwanted: reassigns parcels that no farmers want to original owners.

Each module is designed to consider each farmer's multicriteria budget and threshold constraint. Additional detail concerning design of the heuristic can be found in Strapp (1992).

The heuristic requires access to seven related data tables. A spatial adjacency table is built at the outset by extracting and reformulating topological relationships maintained in the spatial database of the GIS. This table is built at the outset to reduce time-consuming searches of the GIS database when running the heuristic procedures. The other data tables contain detail about individual farmers and parcels, multicriteria budgets, threshold constraints, and graphics housekeeping.

System Customization Using Application Programming Interfaces (API)

The next logical challenge is to combine the land consolidation data inventory stored in GIS with the heuristic, packaging and customizing the lot into a user-friendly and consistent decision support system using a screen language familiar to farmers and planners—not only to GIS experts.

Application programming interfaces (API) offer a promising method of integrating the heuristic into GIS. An API for GIS consists of a library of routines that allows the user to access and integrate most functional capabilities of the GIS in a standard programming language. The idea is to allow a user to write an analytical program in a standard programming language and place calls to the GIS via API functions to handle spatial database management, graphic display, and user interaction.

In early 1992 when we conducted this work, API for GIS was still in the development stage. The beta releases we examined at that point still lacked cohesion and complete documentation. We selected an API supplied by Digital Resources Systems (DRS) of Nanaimo, British Columbia, for their TerraSoft GIS. The TerraSoft API offers C language routines to access spatial data directly, to build a user interface containing both graphical and textual screens, and generic routines to access attribute data contained in different external database management systems.

While the TerraSoft API was the foundation of our SDSS development, it was clear it would not meet our needs for fast, complex processing of the attribute database. Because the API was based on a standard programming language, we were able to take advantage of another specialized API—Code Base (Sequiter Software, Edmonton, Alberta, Canada). Code Base is a C function library that replicates dBASE (Borland International, Inc.) commands and enables data to be stored in a dBASE-compatible format.

Given the preliminary release and lack of documentation for the TerraSoft API, we had to proceed by trial and error. However, by using two API, it proved possible for an individual without a computer science degree to program the heuristic and integrate it into the TerraSoft GIS using less than 2,000 lines of code. Users can initiate assignment heuristics through a graphical user interface and see the solution portrayed cartographically as it emerges. The complete system consists of an executable file, three TerraSoft system files, data files containing the map and theme information, and several database files.

CONCLUSION

Customization of the TerraSoft GIS into a land consolidation SDSS using an API proved feasible. The SDSS was tested for several scenarios using data obtained from a consolidation scheme in Spain. The SDSS was found to perform to expected analytical specifications (Strapp 1992).

Preliminary research did not allow us to give actual decision makers a chance to use the software to conduct a land redistribution scheme, although one of the authors is now involved in such work in the Czech Republic. The performance and success of an SDSS, of course, cannot be judged properly without applying it in a real world decision-making process; therefore, we cannot yet report on the success of our formalization of the consolidation process.

However, we can make the following observation: Numerous multicriteria land-related problems require decision making. These problems do not tend to be solved by conventional mathematics; rather, they require logical search procedures, leading to feasible solutions of acceptable compromise. Commercially available GIS offer appropriate technology to handle data inventory, routine data manipulation, and data visualization, but they lack the necessary advanced analytical capabilities and search procedures. After all, search procedures and analytical steps are unique for each applied land-related decision-making problem.

What we need is the capability to customize commercially available GIS into specialized SDSS at a relatively low cost. Our experience has demonstrated that API offers this capability. We conclude that considerable work is required to advance the linkage of GIS and API to yield a user-friendly package.

Our vision is that of an API that will allow a programmer to treat the GIS like a graphics library and spatial database server. While generic graphics libraries can be used to create GIS interfaces, the commonality of display and interaction with cartographic data is better satisfied with a specific GIS function library. By approaching GIS as a database server, vendors are not forced to meet the individual needs of an ever-expanding group of users, but instead, offer an efficient method of storing and retrieving spatial data that can satisfy all applications. A few vendors are now offering spatial database servers, and as enhancements are made to database technology and standard query language (SQL) access, more should follow.

Acknowledgments

Support for this research was received from the British Columbia Science Council, Digital Resources Systems, and the Natural Science and Engineering Research Council of Canada, grant #OGP0006533.

References

Agarwal, S. K. 1972. *Economics of Land Consolidation in India.* New Delhi: S. Chand.

Bentley, J. L. 1990. Economic and ecological approaches to land fragmentation: In defense of a much-maligned phenomenon. *Annual Review of Anthropology* 16:31–67.

Bonner, J. P. 1987. *Land Consolidation and Economic Development in India: A Study of Two Haryana Villages.* Riverdale: Riverdale.

Farmer, B. H. 1960. On not controlling the subdivision of paddy lands. *Transaction of the Institute of British Geographers* 28:225–35.

Kik, R. 1990. A method for reallotment research in land development projects in the Netherlands. *Agricultural Systems* 33:127–38.

King, R. L., and S. P. Burton. 1983. Structural change in agriculture: The geography of land consolidation. *Progress in Human Geography* 6(4):475–94.

Strapp J. D. 1992. "A Spatial Decision Support System for Agricultural Land Consolidation." Master's thesis, Department of Geography, University of Victoria, British Columbia.

Strapp J. D., and C. P. Keller. 1992. Integrating spatial analysis and GIS through application program interface. *Proceedings, GIS'92.* Vancouver.

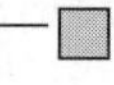

Peter Keller is an associate professor at the University of Victoria. His research interests are in spatial decision making using GIS and multicriteria analysis. He is codirector of the University of Victoria's Spatial Science Laboratories.
Department of Geography and
School for Earth and Ocean Sciences
University of Victoria
Victoria, British Columbia V8N 3P5
Canada

James Strapp completed his M.S. at Spatial Science Laboratories and is now a GIS consultant with Coopers & Lybrand, where he continues to work on land reform optimization using GIS.
The Coopers & Lybrand Consulting Group
Victoria, British Columbia
Canada V8W 1E3

67

The Integration of Empirical Modeling, Dynamic Process Modeling, Visualization, and GIS for Bushfire Decision Support in Australia

Stephen R. Kessell

The development of bushfire decision support systems that link both dynamic modeling and GIS has been a major research area in Australia over the past decade. These systems are now being used routinely in the management of approximately ten million hectares of fire-prone rural land. While installations have been customized for different organizations, most share several important linked components:

1. *Physical processes, including fire behavior and its environmental effects, are simulated by dynamic process models rather than by simple empirical or statistical models.*
2. *Complex environmental patterns and mosaics are represented with the aid of* n*-dimensional artificial space models rather than by simple classifications.*
3. *Advanced visualization software has been developed and implemented to assist with the construction of these models.*
4. *Simple user interfaces allow the ready access of very complex models by a range of technical, managerial, and administrative staff.*
5. *All of these modules build upon and/or are linked with major commercial GIS packages.*

I argue that the successful development and use of these systems have resulted from their effective integration of research results from several traditionally quite distinct disciplines. The chapter briefly reviews the development of such systems, with emphasis on the interdisciplinary linkages. It also explores how such generic techniques are being applied to three quite different spatial modeling problems, including optimization of urban fire fighting resources, metropolitan and rural health-care facilities, and gold exploration. It concludes with a conceptual model of how process and environmental gradient models can be integrated more fully.

CONVERGING TRADITIONS

Once upon a time, GIS was the province of Harvard's School of Design, ecological modelers were driven by IBP funding and were part of biology departments at places like Cornell, PC workstations of today were the stuff of science fiction, and nobody seemed to talk to anyone outside their own discipline. As a result, spatial analysis and modeling tools developed twenty or more years ago are only now being "discovered" by a new generation of GIS practitioners. I am more perplexed than cynical about this curious state of affairs. While I appreciate that the hardware required to integrate dynamic models with large GIS data sets was prohibitively expensive in the past, I find it odd that very useful generic tools are only now being "dusted off" and applied to current environmental modeling problems.

I would like to support this rather audacious introductory statement by outlining four such tools: (1) the integration of dynamic process models with GIS; (2) the linkage of real space with abstract (*n*-dimensional) artificial resource space models; (3) the application of simple statistical/ordination/visualization software in developing such models; and (4) the provision of user interfaces that are not hostile to mere mortals.

I also will discuss briefly how these tools—whose application and integration have been restricted largely to terrestrial environmental GIS applications in North America and Australia—are now being used in the development of three quite different applications. Finally, recent and promising work that attempts to link process and environmental gradient models to describe the distribution and abundance of organisms as a consequence of energy balances is discussed.

SOME USEFUL GENERIC TOOLS

Using Dynamic Process Models

Many applications of GIS to environmental analysis are limited by the restricted features provided by commercial packages (the "If you can't do it by reclassification, overlay, and buffering, it can't be done" mentality). A further restriction results from the common use of "empirical models"—collect some data and run lots of regressions—to represent complex environmental processes. (Until recently, empirical models were commonly used to estimate bushfire behavior in Australia: divide the wind speed in km/hr by the relative humidity in percents; then multiply by the number you thought of in the first place. Such models tend to crash when utilized outside the domain of the data from which they were constructed and usually perform inadequately even within that domain [Kessell 1990].) There are better ways.

The North American Fire Behavior Model developed by Rothermel (1972) two decades ago was first implemented in GIS for national park management in 1975 (Kessell 1976). A refined version of the Rothermel model was used in a more advanced forest management GIS (FORPLAN), developed for the USDA Forest Service

(Potter et al. 1979). Nearly another decade passed until the same linkages were available in Australia (Kessell 1990). I use the incorporation of the Rothermel fire model within GIS as an example of both the opportunities and problems involved.

Unlike previous empirical regression fire models, the Rothermel model is based upon principles of thermodynamics and combustion physics. Initially even its author deemed it too complex to link with GIS (Rothermel 1974, personal communication). But such a linkage is possible if all inputs required by the model are:

1. available as coverages (at an appropriate level of resolution) in the GIS,
2. derivable from other coverages in the GIS,
3. entered by the user at the time of model execution, or
4. sufficiently unimportant that they can be ignored.

Integrating the Rothermel fire model with a raster GIS, first at Glacier National Park in Montana (Kessell 1979) and later across eastern Australia (Kessell 1990), involved all four items: (1) Many inputs (slope, aspect, vegetation structural type) were available as existing coverages; (2) other inputs (fuel loadings, packing ratios, heat content) could be derived using secondary models from existing coverages (vegetation, terrain data, fire history); (3) weather inputs were either input at run time (for real-time simulations) or extracted from historical data sets (for planning and hazard assessment simulations); and (4) sensitivity analysis demonstrated that some model inputs simply could be ignored by setting an appropriate constant (Kessell 1990).

An important secondary model in this application was of fine fuel accumulation, previously modeled empirically (or, quite frankly, "guestimated"). But a very elegant litterfall decomposition model (Walker 1979), using simple inputs available from fire history and vegetation coverages, was found to be vastly superior to static lookup tables in estimating this parameter (Kessell et al. 1982).

A final example was the incorporation of the "vital attributes/multiple pathways" succession (vegetation dynamics) model (Noble and Slatyer 1977; Cattelino et al. 1979; Kessell and Potter 1980). Rather than forcing the vegetation response to follow a strictly deterministic (Clemensian) pattern (in a lookup table), this approach allows a range of responses as a function of each component species' life history parameters (age to reach maturity, reproduction seed dispersal method, tolerance, etc.) and interdisturbance periodicity. The species' characteristics are held in lookup tables, and a simple IF-THEN-ELSE structure programs the varied responses (Kessell and Potter 1980).

In each example, a research model developed outside of the GIS context was used to improve the GIS decision support system.

Mapping Real Space onto Abstract Space

Gradient analysis attempts to describe and understand the distribution of vegetation in response to one or more environmental, resource, and/or temporal gradients (Whittaker 1973). It contrasts sharply with traditional classification methods in that it seeks to understand continuous variation within an *n*-dimensional abstract space. An example is shown in Figure 67-1 where the distribution of two tree species in Glacier National Park is shown in response to two major gradients.

Gradient modeling links such abstract space models with real space via GIS (Kessell 1979). That is, a point-by-point mapping of real space onto the abstract space is provided by various GIS coverages (in the example shown in Figure 67-1, required coverages are

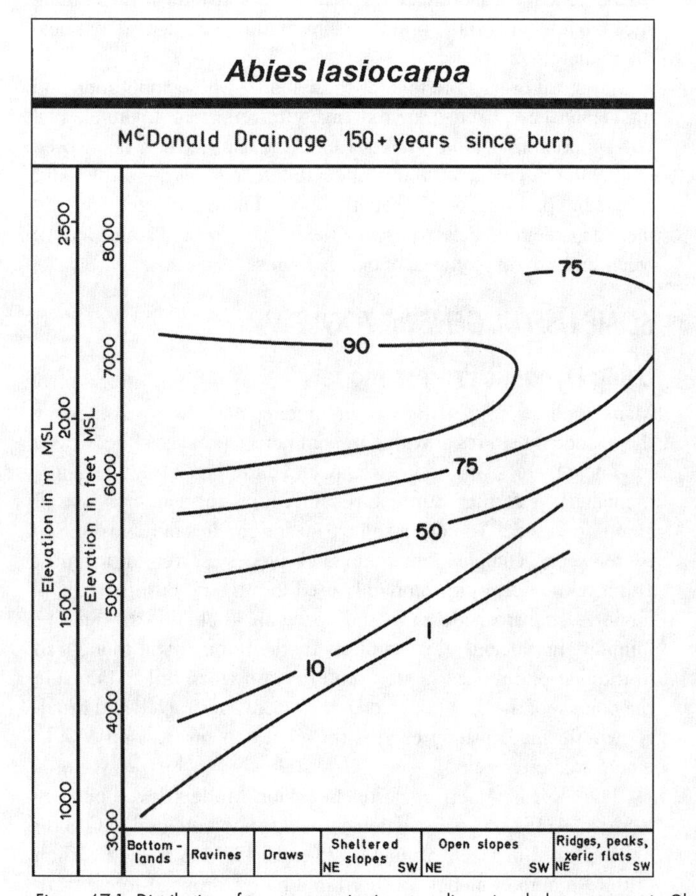

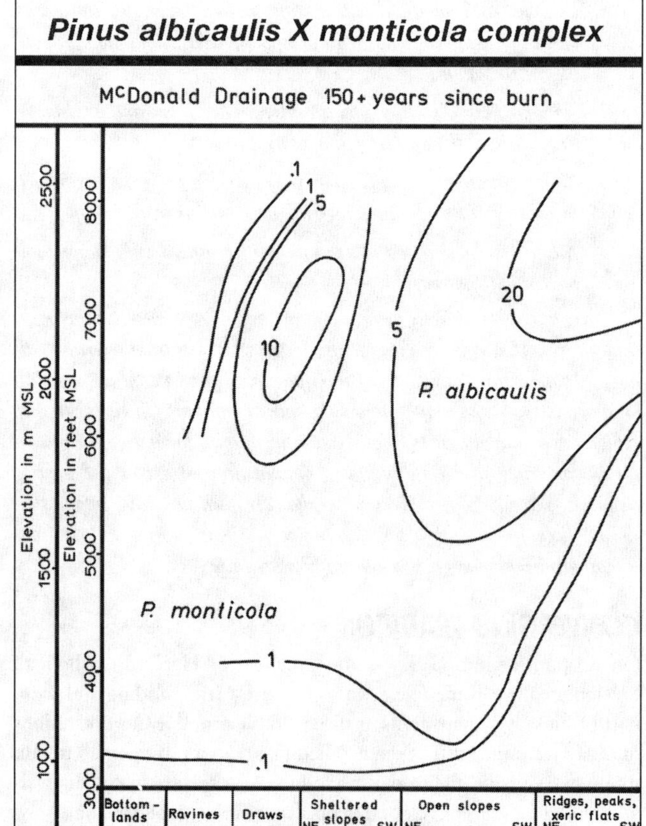

Figure 67-1. Distribution of two tree species in a two-dimensional abstract space in Glacier National Park, Montana (from Kessell 1979).

elevation, aspect, and topographic position). The abstract space models are empirical and basically predictive (rather than "explanatory"); however, they also offer some insight into "what grows where" (this species prefers higher elevation, wetter sites, etc.).

The approach offers two important improvements over traditional classification (and the subsequent digitization of the resulting thematic maps). First, the specification of any real-space coordinate (northing, easting) links, via the coverages just noted, to a single point in the abstract space. From this, one may estimate, for example, the density of every tree species (providing more information than that available from a traditional classification). Second, the method provides automated mapping at a species' level. That is, every pixel may be linked to a point in the abstract space, allowing the production of a wide range of derived maps (i.e., individual species' densities, dominance/codominance, and diversity maps).

This serves as a very good check on the abstract space models themselves. When the initial derived maps are checked against reality, discrepancies are observed, leading to successive refinements of the models as more data are collected.

While gradient modeling largely has been limited to predicting the distribution of vegetation and its derivatives (flammable fuel, wildlife habitat, etc.), the method can in fact be applied to the distribution of any "resource."

Visualization Software as a Model Development Tool

Burrough (1992) noted the need for more "intelligent" GIS, and, in particular, the need for a better linkage between models and GIS. Goodchild (1992) similarly deplores "the lack of integration of GIS and spatial analysis." A promising approach to more intelligent GIS is the linkage of spatial models with improved visualization techniques (McCormick et al. 1987).

Haslett et al. (1990) were inspired by a "rare example of software for dynamic graphics" developed for the Apple Macintosh by Velleman and Velleman (1988) called DataDesk; it provides a "linked view" of multivariate data sets. Haslett et al. noted that:

> Each view provides a different representation of the data. The views are *active*; the user can select and highlight individuals or subsets of individuals in each view. . . . The views are *linked;* highlighting in one view causes the corresponding cases to be highlighted immediately *in all other views*. Each view of the data is an equally valid point from which to query the underlying database. The potential of such a system lies in the development of informative views, especially for high-dimensional data [emphasis added] (1990).

Haslett et al. (1990) then went on to develop SPIDER (SPatial Interactive Data ExploRer) as a query-mode GIS, applying and expanding the features of DataDesk. SPIDER provides "point and click" access to a range of linked coverages, allowing simple "plot this against that," "show me histograms of these," spatial displays. MacDougall (1991) also developed a prototype program for exploratory data analysis and dynamic statistical visualization that attempts to extend the features of programs such as SPIDER.

Three research students at Curtin University took the work a step further. Following a project by Buckleton (1991) to improve automated map generation from the abstract space gradient models, Tuffin (1992) developed a new visualization package in Turbo Pascal that operates on IBM PCs and clones (the other two systems cited above run only on the Macintosh). A significant new feature of Tuffin's package was the ability to provide a range of transformations (log, square root, reciprocal, etc.) to any spatial variable; this often results in a clearer understanding of the relationships and, therefore, better statistical models.

More recently, Hunter (1993) further developed this approach with a more refined linked display package called GISVIZ (GIS VIsualiZation). Written in Visual Basic, it operates under Windows on IBM PCs and clones using data dumped from OSU-Map. It, too, provides transformations (log, square root, reciprocal) for all variables; it also provides a nice user interface and is more flexible than its predecessors. Figure 67-2 plots elevation versus density of *Abies lasiocarpa* in western Glacier National Park (Figure 67-1), after a square root transformation has been applied to the latter. The transformation significantly improves the goodness of fit of the regression. GISVIZ also offers the automated linkage to and mapping from the abstract *n*-dimensional space models; Figure 67-3 plots the density of *Pinus albicaulis* for a portion of the McDonald drainage in Glacier National Park.

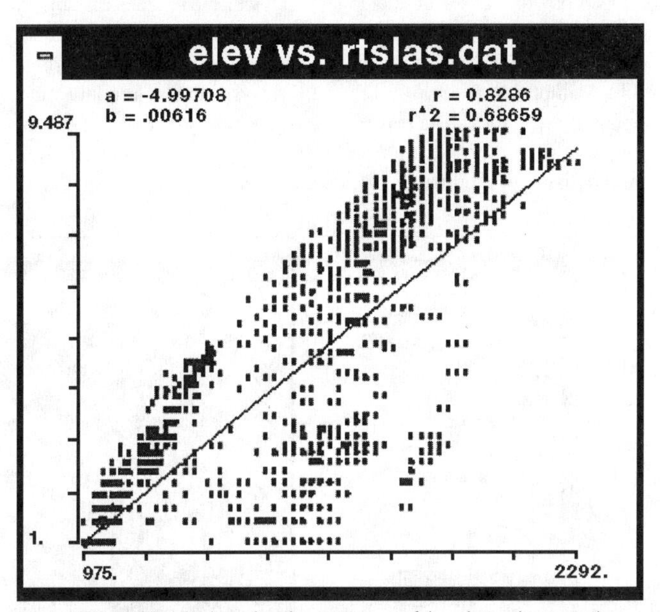

Figure 67-2. GISVIZ display: plots the square root of the relative density and percent of *Abies lasiocarpa* against elevation (m).

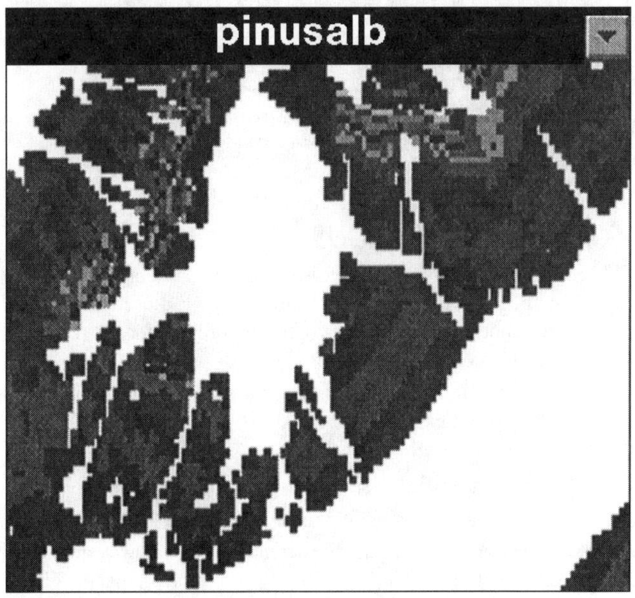

Figure 67-3. By linking the DEM of a study area to the abstract space species distribution models, GISVIZ plots a map showing the relative density of *Pinus albicaulis*.

I expect such visualization tools will play an increasingly important role in the development of both predictive and explanatory spatial models in the future.

User Interfaces

An extremely versatile natural language parsing, command-line-style user interface was introduced in the USDA Forest Service FORPLAN system over a decade ago (Potter et al. 1979); it is better than the command line interfaces seen on some expensive commercial GIS today. More recently, systems developers have seen the advantage of utilizing graphics user interfaces (GUI) as a means of providing easy access to complex packages.

Beck (1989) and Beck and Kessell (1992) have gone a step further with WIMS (Wildfire Incident Management System), a decision support GIS and modeling system developed by Curtin University and the Western Australia Department of Conservation and Land Management. WIMS provides a "visual situation report," which combines base maps, overlays, model results, and resource icons superimposed as required. For example, Figure 67-4 shows the deployment of resources on a forest fire. A simple "click" on a 4 WD's icon, displays a range of attribute information; any of these data may temporarily or permanently be updated.

As systems become more and more complex, better presentation and delivery of the final product is essential.

The second project, in cooperation with a minerals exploration company, is attempting to develop better spatial models for precious metal exploration in the Australian outback.

The third project, with a government public health authority, is attempting to optimize the location and provision of varying health services as a function of demographics in both metropolitan and rural areas.

Despite their very different subjects, all three projects include the need to:

1. Visualize and reduce very large data sets which possess both a significant temporal component and significant "noise."
2. Be more than empirical, overlay-type predictive modeling; all three projects will require the development of explanatory process-type spatial models.
3. Incorporate existing GIS features (such as networks to determine the time response of fire engines) with new analysis tools.
4. Identify and optimize resources (e.g., fire control forces, gold deposits, or aged-care nursing) in response to social, demographic, and/or environmental gradients.
5. Provide a sophisticated final product in a form that may be used easily by noncomputing specialists.

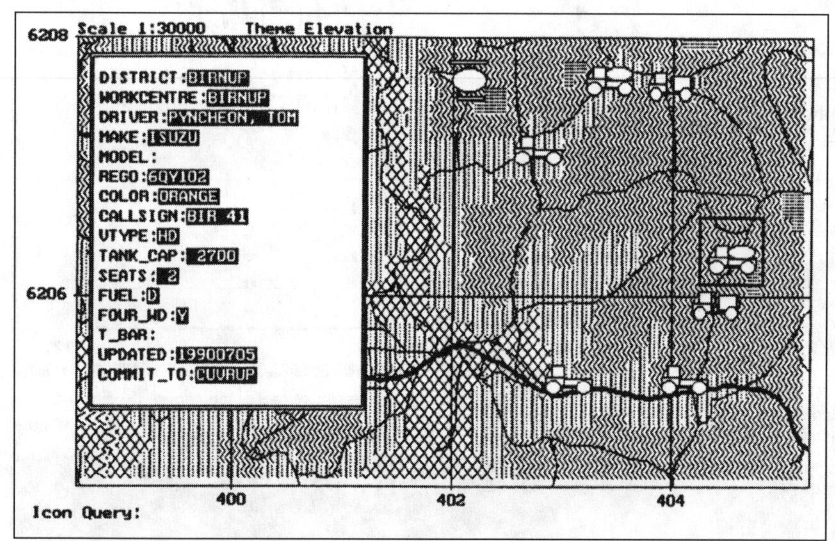

Figure 67-4. WIMS display showing the resources allocated to a particular fire. The query option has been used to display the data affiliated with a particular icon (from Beck and Kessell 1992).

SOME NEW APPLICATIONS

Three new and superficially disparate spatial analysis projects recently undertaken at Curtin attempt to exploit these basic visualization and modeling techniques. The first project, for metropolitan fire service seeks to determine:

1. The optimum location of new fire stations and the possible relocation of existing fire stations, so that any fire incident can be reached within seven minutes of notification
2. The identification of sites where a "standard response" would be inadequate (gas stations, chemical warehouses, etc.)
3. The stages at which the fire services of small towns should be upgraded from volunteer to various professional levels in response to population growth.

CONCLUSION

There remains a large and frustrating gap between stand-alone explanatory or process models and our ability to incorporate them effectively into GIS. For example, Hall and Day (1990 [originally published in 1977]) present a range of environmental modeling case studies—many involving energy and nutrient transfer models—that have never been linked successfully to GIS. Similarly, there is a large gap between the abstract space (essentially empirical, predictive) models described previously, and the selected process models that were incorporated in the fire management GIS. While it is recognized that the abstract space models reflect real responses to environment and resources, the precise nature of these relationships invariably defies prediction.

Hall et al. (1992) provided what well may be a conceptual solution to the problem. Briefly, they argue that the observed species distributions (in *n*-space) relate directly to species-specific energy costs and gains (energy balance) in response to the *n*-dimensional environmental and resource gradients. Despite alarm bells sounding in my head whenever someone suggests a "common-currency" approach to environmental modeling, I agree with these authors on a couple of major points: (1) competition, predation, and other biotic interactions principally operate by increasing energy costs to a species population (and thus can be included as additional gradients of energy costs); and (2) the environment itself (i.e., any particular point in *n*-space) poses an energy cost—the further away from its optimum habitat an individual occurs, the greater the metabolic energy costs for maintenance. The ultimate success of a population depends on both factors.

Population dynamics thus are seen to "reflect the cumulative effect of environmentally induced shifts in organizational energy balance" (Hall et al. 1992). Such an approach has the potential to synthesize theoretical ecology at several levels. The important questions are whether our understanding of the system is sufficiently complete, and whether it is possible to obtain such *n*-dimensional "coverages" with sufficient detail and resolution, so that such a synthesis can be implemented in GIS. No such synthesis and implementation have yet been achieved. However, if we can find a way of accomplishing this for a "real" (in contrast to "toy") environmental system, environmental modeling could well take a quantum leap forward.

References

Beck, J. A. 1989. A geographic information and modeling system for the management of wildfire incidents. In *Australian Urban and Regional Information Systems Association Conference,* 474–84. Perth, Western Australia.

Beck, J. A., and S. R. Kessell. 1992. Tracking suppression resources via an Australian fire management system: In principle and in practice. *Proceedings, GIS '92,* Vancouver, Canada.

Buckleton, L. H. 1991. "Predicting the Limits of Prediction." Honor's thesis, School of Computing, Curtin University, Perth, Western Australia.

Burrough, P. A. 1992. Development of intelligent geographical information systems. *Int. J. GIS* 6:1–11.

Cattelino, P. J., I. R. Noble, R. O. Slatyer, and S. R. Kessell. 1979. Predicting the multiple pathways of plant succession. *Environmental Management* 3:41–50.

Goodchild, M. F. 1992. Geographical information science. *Int. J. GIS* 6:31–45.

Hall, C. A. S., and J. R. Day, eds. 1990. *Ecosystem Modeling in Theory and Practice: An Introduction with Case Histories.* Boulder: University Press of Colorado.

Hall, C. A. S., J. A. Stanford, and F. R. Hauer. 1992. The distribution and abundance of organisms as a consequence of energy balances along multiple environmental gradients. *Oikos* 65:377–90.

Haslett, J., G. Wills, and A. Unwin. 1990. SPIDER—An interactive statistical tool for the analysis of spatially distributed data. *Int. J. GIS* 4:285–96.

Hunter, D. 1993. *Statistical Visualization and GIS.* Computing Project Report. Curtin University, Perth, Western Australia.

Kessell, S. R. 1976. Gradient modeling: A new approach to fire modeling and wilderness resource management. *Environmental Management* 1:39–48.

———. 1979. *Gradient Modeling: Resource and Fire Management.* New York: Springer-Verlag.

———. 1990. An Australian geographical information and modeling system for natural area management. *Int. J. GIS* 4:333–62.

Kessell, S. R., R. B. Good, and M. W. Potter. 1982. *Computer Modeling in Natural Area Management.* Australian National Parks and Wildlife Service Special Publication No. 9. Canberra, Australia.

Kessell, S. R., and M. W. Potter. 1980. A quantitative succession model for nine Montana forest communities. *Environmental Management* 4:227–40.

MacDougall, E. B. 1991. Dynamic statistical visualization of geographic information systems. *Proceedings, GIS/LIS '91,* 158–65. Atlanta, Georgia.

McCormick, B. H., T. A. Defanti, and M. D. Brown. 1987. *Visualization in Scientific Computing.* Washington: National Science Foundation.

Noble, I. R., and R. O. Slatyer. 1977. Postfire succession of plants in Mediterranean ecosystems. *Proceedings, Symposium on the Environmental Consequences of Fire and Fuel Management in Mediterranean Climate Ecosystems.* USDA Forest Service General Technical Report WO-3, 27–36. Washington, D.C.: USDA.

Potter, M. W., S. R. Kessell, and P. J. Cattelino. 1979. FORPLAN—A forest planning language and simulator. *Environmental Management* 3:59–72.

Rothermel, R. C. 1972. *A Mathematical Model for Predicting Fire Spread Rate and Intensity in Wildland Fuels.* USDA Forest Service Research Paper INT-115. Washington, D.C.: USDA.

Tuffin, M. 1992. *A Facility for Exploration of Both Raw and Derived Coverages of a GIS.* Computing Project Report. Curtin University, Perth, Western Australia.

Velleman, P. F., and A. Y. Velleman. 1988. *DataDesk Professional.* Northbrook: Odesta Corp.

Walker, J. W. 1979. Fuel dynamics in Australian vegetation. In *Fire and the Australian Biota,* edited by A. M. Gill, R. H. Groves, and I. R. Noble, 107–27. Australian Academy of Science, Canberra, Australia.

Whittaker, R. H., ed. 1973. *Handbook of Vegetation Science, 5: Ordination and Classification of Communities.* The Hague: Junk.

Steve Kessell is Associate Professor and Department Head of the Department of Geographic Information Systems, Curtin University, Perth, Western Australia. Educated at Amherst College and Cornell University, he has lived in Australia since 1979. Prior to joining the Curtin faculty in 1987, he worked for a variety of private and public sector organizations, developing GIS modeling and simulation systems for environmental management decision support.

Department of Geographic Information Systems
School of Computing
Curtin University of Technology
GPO Box U1987
Perth, Western Australia 6001
Phone: 61 9 351 7297
Fax: 61 9 351 2819
E-mail: *kessell@cs.curtin.edu.au*

68

Computational Modeling Systems to Support the Development of Scientific Models

Terence R. Smith, Jianwen Su, Amitabh Saran, and Anuradha M. Sastri

Computational modeling systems (CMS) are designed to resolve many of the shortcomings associated with systems currently employed in providing support for a wide range of scientific modeling applications. We have designed and developed a CMS—Amazonia—that supports the construction of computationally useful representations of an unlimited array of modeling concepts, including those that describe the process of modeling itself as well as the phenomena being modeled. It is based on a very general but powerful conceptual model of the scientific modeling activity. We introduce a simple and largely declarative computational modeling language, used to express modeling operations at the conceptual level of the scientific investigator. Amazonia has been implemented in a layered architecture, incorporating three key components: (1) a modeling support system that supports an appropriate environment for the construction and testing of models, (2) a tool management system that provides support for the interoperability of heterogeneous tools, and (3) a distributed access system that supports access to, and manipulation of, "data" and "services" in a distributed environment.

INTRODUCTION

In many areas of scientific and engineering research, there is a fundamental need for comprehensive and integrated computational support for the development, evaluation, and application of symbolic models (Requicha 1980) of a broad range of phenomena. Activities that require computational support during the iterative development of symbolic models (Robertson et al. 1991; Hardisty et al. 1993) of phenomena range from the acquisition and manipulation of raw data to the construction and evaluation of complex sets of mathematical equations (Chu 1993; Dozier 1990; Hachem et al., *AAAS*, 1993 and *Proceedings*, 1993; TCDE 1993; Long et al. 1992; Ordille and Miller 1993; Silberschatz et al. 1990; Wolf 1989; Gunther and Schek 1991; Abel 1993). Current computational support for these activities typically involves a heterogeneous collection of tools supporting limited aspects of scientific modeling activities, such as numerical computations, data management, and image processing. Rarely is such support integrated into a single comprehensive system, and there is typically little or no support for the modeling enterprise as a whole. It is our observation that in some Earth science research projects, for example, a significant proportion of effort is focused on computational issues that are largely irrelevant to the scientific research. Significant increases in efficiency are possible for applications in which computation is a major activity.

Integrated computational support for scientific modeling activities is a difficult goal to achieve for a variety of reasons. The construction of scientific models of phenomena is a complex, highly interactive, and highly iterative process. The modeling process is typically distributed over a number of conceptually and physically distinct environments and subenvironments. Scientific modeling activities typically involve a large and evolving body of concepts related both to the phenomenon of interest and to the phenomenon of the modeling process itself, and there are frequently multiple representations of these concepts. Finally, scientists typically are relatively conservative in adopting "new approaches" until they understand the net advantage of the scientific research.

The lack of comprehensive and unified computational environments that hide scientifically irrelevant computational issues leads to many difficulties and inefficiencies (Smith et al. 1993a, 1993b). One instance is the management of scientific data. In essence, current database support has not been designed on the basis of any deep or general characterization of scientific modeling activities. In particular, it does not take into account the fact that it is almost impossible, from a scientific point of view, to separate "database support" activities from other modeling activities, except to the degree that scientists find it of value to employ the concept of "data" in relation to a certain class of scientific entities and a certain class of operations. Hence, database environments that support the handling and management of "data" are typically separated from and unrelated to the programming language environments and other environments that are better able to support the scientific modeling process. They provide little or no explicit support for the fact that scientific modeling typically occurs in a variety of environments, and that there are frequently multiple representations for many entities and relationships.

Computational modeling systems (CMS) are intended to provide scientific investigators with computational support, allowing them to achieve their goals more efficiently. In particular, CMS should provide a unified computational environment in which scientific investigators are provided with sets of tools that support a significant range of activities relating to the processes of constructing, evaluating, and applying symbolic representations of phenomena. In general, a CMS may be viewed as the fusion of three relatively distinct components:

(1) a computational environment that is designed on the basis of a comprehensive and consistent model of scientific modeling activity; (2) a knowledge base that may be tailored for any specific domain of scientific investigation; and (3) persistent programming support, that combines compiler, database, and other technology and is transparent with respect to scientifically irrelevant computation details.

We believe that a CMS should be based on a comprehensive and appropriate model of scientific modeling activities. Such a model should include a characterization of the development, representation, and evaluation of the concepts employed by scientists in their modeling of both the phenomena of interest and the process of modeling itself. In developing and evaluating the concept of a CMS, we have interacted closely with multidisciplinary groups of scientists, since we believe that support for scientific modeling activities must be based on a deep understanding of scientific activity.

The main contributions of the research reported in this chapter include:

1. The development of a conceptual model of the process of scientific modeling. This model has been developed in collaboration with multidisciplinary groups of scientists and includes the development and "definition" of a large number of concepts relating to those aspects of the process of scientific modeling that may be provided with significant computational support.
2. The development of a framework that represents a translation of our conceptual model of scientific activity as a simple, unified, computational framework that we term a CMS.
3. The definition of a simple computational modeling language (CML) with which the modeling concepts that are definable in our computational framework may be constructed and manipulated in a simple and uniform manner.
4. The design and implementation of a specific CMS, Amazonia, that is intended to support scientific research in the area of Earth science investigations.

We believe that a particularly significant contribution of our work is the development of a strong conceptual and theoretical basis for CMS, which clarifies a large number of issues related to scientific modeling activity and its computational support. We note that other approaches to the development of such support are typically based on evolutionary extensions of existing systems (Adams and Solomon 1993; Jones 1990, 1991; de Hoop and van Oosterom 1992; Medeiros and Pires 1994; Paul et al. 1987; van Oosterom and Vijlbrief 1991; Wolf and Schek 1989; Wolf 1989, 1990).

We structure the chapter as follows: In the next section, we provide a simple example of scientific modeling drawn from the earth sciences. We use this example throughout the chapter to illustrate many of the issues that are raised and discussed. We then describe in some detail the conceptual model of scientific modeling activities we have developed during the course of our research. In the third section, we present the language CML, and we end the chapter with a discussion of systems support for a CMS, the system architecture, and the current implementation of Amazonia.

COMPUTATIONAL MODELING SYSTEMS

In this section, we briefly discuss the concept of a computational modeling system which provides scientists with an integrated environment for managing and accessing data sets and developing computational models. We first provide a much simplified example from the EOS/Amazon project and then characterize the computational activities of scientific investigators. Based on this characterization, we then present a conceptual data model.

A Motivating Example

We briefly describe a particular example of a set of computational modeling activities from the geological and hydrological sciences. For illustrative purposes, we focus on a highly simplified version of a problem under investigation by a group of Earth scientists who are building models of various aspects of the hydrology and geomorphology of the Amazon river basin.

The exemplary problem upon which we focus our attention involves the construction, testing, and application of water flow models within complex river systems. We assume that an acceptable model in a natural drainage basin is one in which we are able to:

1. Construct, evaluate, and store acceptable representations of observed land surfaces (including networks of river channels), over which we wish to model the flow of water, observed rainstorm events, and observed flows of water at a given set of locations on the surface.
2. Construct representations and properties of flows over the observed surface in response to observed rainstorms given a representation of some flow generating process.
3. Compare appropriately (representations of) properties of the modeled flows with real-world observations of flows.

A first set of activities includes a variety of operations involving representations of "observations" on the phenomena of interest (i.e., involving "data sets"). For current purposes, it suffices to consider digital elevation models (DEM), rainfall measurements, and river stage observations. It is often the case that sequences of procedures are applied to such instances, with some procedures being relatively complex. In much of this type of processing, it is typical for scientists to work in an iterative mode that involves computation, visualization, and procedure modification. In Figure 68-1, we show the data sets and operations that provide a simplified view of the problem. The computation sequence involves first choosing DEM of interest, by intersecting them with an appropriate area (the region around Manaus), and then combining them into a single DEM. Rainfall data from points inside this area and for specific time periods are retrieved and interpolated over the DEM. The DEM is used to generate a slope map that models the flow of water from each point in the DEM to the mouth of the river. Hydrography is a domain of representation of the discharge of water at the basin mouth as a function of time.

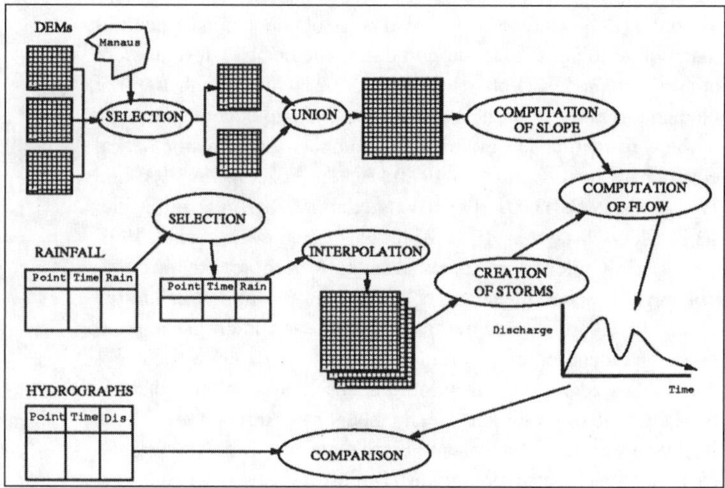

Figure 68-1. Illustration of operations in a subproblem of Amazonia.

A critical task is to select a set of concepts to be employed in the modeling process and to select representations for these concepts. For example, in modeling the flows over land surfaces, one might use the concept of a surface point flow and a representation in terms of a triple of a location, a time, and a flow vector. It may be the case that appropriate concepts or appropriate representations of concepts may not be available. In this case, a major task involves the discovery of both concepts and representations. Given a satisfactory representation of the flow generating process, it is then necessary to select and apply solution procedures to representations of the approximating equations. The application of these procedures involves representations of the surfaces and rainstorms, as well as representations of initial flow conditions over the surfaces. Various transformations may then be applied to the outputs of this processing. For example, it may be desirable to compute the representations of hydrographs at various locations on selected channel segments.

As in the previous processing, it is typically the case that the indicated modeling procedures are highly iterative and involve computation, visualization, and modification. Furthermore, the computational tasks in the process are typically accomplished with the use of a variety of computational tools, such as visualization tools and equation solvers. Consequently, communications between such tools are unavoidable. Without a comprehensive support, investigators have to operate at a relatively low operating system level. However, for computational ease, it is critical that the communications occur at a much higher level. We integrate these techniques with CMS concepts under Amazonia.

Scientific Modeling: Concepts and Representations

This discussion is a characterization of the iterative process by which scientific models of phenomena are constructed. It provides a set of concepts to describe scientific modeling activities and a set of representations for these concepts. It was our goal to provide a basis for those aspects of the scientific modeling process that can and must be supported by CMS.

It is generally accepted that the most fundamental goal of science is the discovery and investigation of appropriate symbolic models (or representations) of phenomena. The concepts that form the basis for most scientific modeling activities typically possess some interpretation in terms of: (1) identifiable entities in a domain being modeled, (2) transformations between such entities, or (3) relationships between entities. We define a modeling concept to be: A domain of entities and relationships, a set of symbolic representations of such entities and relationships in a given language, and an interpretation of mapping representations to entities and relationships. We emphasize that a concept cannot be defined apart from a symbolic representation. The scientific value of any set of modeling concepts is usually closely related to the nature of the representation of the concepts.

There appears to be an important distinction between representations that contain extractable information about the entities they represent and those that contain no extractable information. We term the latter "nominal" representations. The power and success of symbolic modeling activities in science frequently involve the discovery, evaluation, and application of "clever" representations that fall within the former category, such as the place value representation for numbers. We note, however, that nominal representations permit scientists to make efficient reference to arbitrarily complex entities.

As discussed previously, modeling concepts involve representations of classes of identifiable entities, of processes that operate on such entities, and of relationships among the various sets of entities and processes. In this spirit, it is of great value, both theoretically and practically, to consider in detail the representation of a concept in terms of entities and relationships. A representational structure (R-structure) for a concept includes a set of representations of all entities (the representational domain [R-domain]); a set of transformations (operations) on the representations; a set of relationships over the representations; and a finite set of particular instances of representations given in explicit form. We may associate with an R-structure a set (possibly empty) of constraints about the representations. For example, we may specify the representation for the concept polygon in terms of R-structure polygons with an R-domain where any polygon is represented as a sequence $p_1, \ldots, p_n$ of points from the R-domain of the R-structure points. This is illustrated in Figure 68-2.

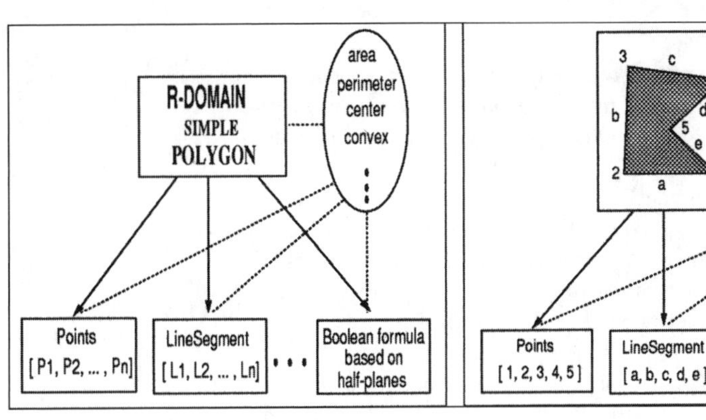

Figure 68-2. Illustrating the concept of R-structures.

The transformations on these representations would include, for instance, area that maps a representation of a polygon into a representation of its area. Relationships between these representations might include intersects, which is true for two polygons if they intersect. In order to make the representation reasonable, it is necessary to associate the R-structure with some constraints, such as $p_1 = p_n$, and that no two edges of the boundary intersect, except at their end points.

We may view the R-structures employed in representing a given class of entities as being composed of two components. These components are: (1) a concrete R-structure that possesses a concrete R-domain, in which the representations of the entities encode some information about the entities they represent; and (2) an abstract R-structure with an abstract R-domain, whose representations are purely nominal. It is therefore not possible to infer any information about the entity on the basis of the representations in the R-domain of an abstract R-structure. In particular, an abstract R-structure may be viewed as an abstraction over a set of concrete R-structures that are equivalent in the sense that they each represent exactly the same set of entities. Therefore, we may view concrete R-structures as containing representations of entities and transformations on entities that may be manipulated to extract further information about entities. Abstract R-structures, however, largely contain nominal information about the entities and transformations that is useful during the process of manipulating the representations.

In particular, abstract R-domains are of value in the support of scientific modeling for at least five reasons:

1. The have the ability to give a name to a particular entity, which often is of value in scientific research (e.g., numerous special classes of polygons, such as triangles, quadrilaterals, parallelograms, and squares).
2. They provide a simple, "representation-free" method for formulating definitions of the representations in new concrete R-domains.
3. They encapsulate corresponding equivalent concrete representations.
4. They provide a simple mechanism for specifying the inheritance of concrete representational structures, transformations, relationships, and constraints from superdomains to subdomains.
5. They serve as useful computational functions in the sense of object-oriented systems and facilitate the management of the efficient computation, if one views the nominal representations in an abstract R-domain in terms of globally unique object identifiers.

In general, we construct representations in a new R-domain by: (1) applying constructors to a finite set of representations that may be selected from any subset of previously defined R-structures, including any subset of R-domain elements, any subset of transformations and relationships, and any subset of the instances; or (2) by constraints on the representations of the new R-domain elements when all new representational components are taken from previously defined R-domains or primitive R-domains, typically including Booleans, integers, real, and strings. Examples of such constructors are set, tuple, sequence, and finite function.

There may be more than one representation in an R-domain that corresponds to an entity in the domain of application. In the case of polygons, for example, it is clear that a cyclic permutation of the point sequence that represents a polygon also will represent the same polygon. There may be equivalent R-domains for a given set of entity representations since it is frequently possible to construct different representations of the same set of entities. A typical reason for constructing different but equivalent representations is that it may be more convenient to compute certain properties of the phenomena using one representation rather than another. In the case of polygons, for example, we may construct additional R-structures in which we employ sequences of line segments or sets of half planes to represent a polygon.

Transformations map representations from a subset of R-domains into R-domain elements. To a significant degree it is the set of transformations and relationships involving R-domain elements that provide semantics to R-structures, since different phenomenological entities may possess structurally equivalent representations in the associated R-domain. For example, the transformation area associated with polygons differentiates the representation $p_1, \ldots, p_n$ of a polygon from a structurally equivalent representation for a chain of line segments.

Using the notion of R-structure, one may view the process of constructing symbolic models of phenomena as one in which an enormous space of potential R-structures is incrementally explored and evaluated in terms of the interpretations provided with the associated set of concepts. Consequently, scientific modeling activities are processes in which scientists: (1) construct, evaluate, and apply collections of R-structures of value in relation to modeling phenomena in specific domains of application and in relation to modeling the process of modeling itself; (2) construct specific instances of R-domain elements and apply sequences of specific transformations to sets of instances of R-domain elements; and (3) construct, evaluate, and apply various statements about representations of domain elements, transformations, and relations in an R-structure.

The Hydrological Example Revisited

We briefly illustrate the manner in which R-structures may be employed in the support of scientific modeling in a CMS by listing a subset of the R-structures that may be used in constructing a computational model of the hydrological problem described previously.

Typical R-structures for this application include those representing relatively low-level concepts, such as sets of geometrical entities (*Points*, *Line_segments* and *Rectangular_Grids*) and basic scientific concepts (such as *Elevations*, *Rainfalls*, *Flows* and *Surface_flow_vectors*). Higher-level R-structures that have value in the hydrological modeling context include five clusters of R-structures: (1) classes of models of the flows of water over a surface (e.g., *Surface_flow_models*, *Surface_flows*); (2) various aspects of the actual surfaces over which the flows occur (e.g., *DEM*, *Basels*, *Drainage_Basins*); (3) various aspects of the rainfall inputs that drive the flows of runoff (e.g., *Rainfall_observations*, *Rainfall_maps*, *Storms*); (4) various observations on observed flows over the surfaces (e.g., *Hydrograph_observations*); and (5) the modeling activities of the scientist (e.g., *Flow_modeling*). The transformations associated with this domain include *solve_method_i*, which takes as input one element from each of *Surface_flow_models*, *Drainage_basins* (representation of the land surface over which the water flows), *Rainstorms* and *Surface_flows* (representation of the initial flow pattern over the surface).

In terms of the representations that we may employ in concrete versions of these various R-structures, the R-structure *DEM*, for example, may have representations in the form of sets of *Points* and *Elevations* pairs *{[p, e]}*. One may transform these basic representations of ground surfaces into various other representations of complex drainage surfaces (e.g., *Channel_links*, *Link_drainage_areas*, *Basels* and *Drainage_basins*). The R-structure *Rainfalls* may be represented as tuples of values of rainfall as measured at a certain location and a certain time: *{[p, t, r]}*. We associate a variety of transformations with this structure, including interpolation (producing from a rainfall data set a set of estimated rainfalls at each point of some [rectangular] grid).

During the modeling process, one may envisage the scientist to be involved in a variety of operations, including the construction and evaluation of particular R-structures, and, in particular, the construction of: representations of transformation and relationships; specific instances and applying sequences of transformations; and the construction, evaluation, and application of various statements about the representations of domain elements, transformations, and relations in an R-structure and associating them with the R-structure.

A HIGH-LEVEL COMPUTATIONAL MODELING LANGUAGE (CML)

In order to carry out the operations involved in constructing R-structures and their components and to manipulate the associated representations during the course of modeling, it is important to have a simple language to represent these various operations. A simple and largely declarative computational modeling language (CML) that is closely related to our conceptual model of scientific activity, may be used to express a wide range of scientific modeling operations. CML is intended to express such operations at the conceptual level of the scientific investigator. It permits easy and natural expression of most operations employed in iterative model development, while hiding irrelevant computational issues.

CML is based on the conceptual model of R-structures, as defined previously. The primary functions of CML include: (1) the definition, creation, manipulation, and storage of new R-structures and their constituent parts; (2) the application of transformations to R-domain elements in general and to R-domain instances in particular; and (3) the search for transformations and specific R-domain elements that satisfy appropriate constraints. CML includes a small set of simple commands: create, delete, modify, access, store (R-structures, R-domains, transformations, relationships, and instances), and apply (transformations to R-domain elements). CML has been extended appropriately to incorporate virtual structures that form the link between the CML and the tool management system. For convenience, we apply the term "data set" to any of the explicit representations of R-domain elements. While this convention is not equivalent to semantics scientists give to the term data set (including the fact that the representation is partly based on direct observation or measurement), any data set in the scientific sense is represented as an element of some R-domain here. The total collection of data sets constitutes a database in a CMS.

The create command in CML permits the construction of abstract and concrete R-structures and the components of such structures. It is necessary to indicate the corresponding abstract R-structure and the constraints on the R-domain representations. For example, DEM is created in a CMS database (where *peg* represents *point_elevation_grid*) by the following:

```
CREATE R-STRUCTURE DEM.peg
    WHERE SUPER R-STRUCTURES { Rectangular_Grid_Maps.peg }
        R-DOMAIN =          r-domain(DEM.peg)
        TRANSFORMATIONS     transformations(DEM.peg)
        RELATIONS           relations(DEM.peg)
        INSTANCES           instances(DEM.peg)
```

DEM.peg indicates that the newly created (concrete) R-structure "implements" the abstract R-structure DEM. Abstract R-structures provide "external" specifications of concrete R-structures in a manner similar to abstract data types (Birtwistle et al. 1973).

The user may "name" the elements by values of string type. Although these names play the same role as object identifiers, this provision in CML provides flexibility and ease in scientific modeling activities. For example, if *Y* holds an identifier of a DEM element, the command *CREATE Manaus = Y IN r-domain(DEM)* creates a new name, *Manaus*, for the element. While each element in an R-structure can have zero, one, or more user-defined names, they must be unique in an R-structure and all its substructures, that is, consistent with the inheritance hierarchy.

The R-domain of a concrete R-structure specifies representations of all the entities, which are a structural component and a value component, specified in terms of a set of constraints on the subcomponents in the structural representation. The following example illustrates the creation of the concrete domain of the R-structure *DEM.peg* in which elevation data are represented by a set of *Points, Floats* pairs:

```
CREATE DEFAULT r-domain(DEM.peg)
    WHERE STRUCTURE = [name:string, resolution:integer,
                      location:[L1:point, L2:point, L3:point,
                      L4:point], P_E:set of [Location:point,
                      Elevation:real]]
        CONSTRAINTS = ...
```

In CML, each abstract domain has exactly one default concrete domain. The default domain is used whenever the concrete domain is neither explicitly specified, stated, nor inferred. This representation is used to avoid nondeterminism and to simplify implementation.

Transformations may be created in CML by the create command, aggregating previously defined transformations into new transformations. This is important in the construction of new R-structures, since the new R-domain elements frequently involve aggregate constructors. Transformations not defined in CML are viewed as binary executables or scripts. Each transformation includes a name, a list of parameters and their types, the output and type, and a sequence of CML commands. The following is a simple example of a transformation *rain_extract*. It retrieves a certain subset of relevant rainfall measurements from any *Rainfall_Map* representation.

```
CREATE TRANSFORMATION rain_extract(yy:int,mm:int,dd:int,-
    hh:int,area:polygon)
    RETURN rr:SET OF [year:int, month:int, day:int, hour:int,
    loc:point]
    BEGIN
    rr = ACCESS { t:[year:int, month:int, day:int, hour:int, loc:point]
    FOR o IN RAINFALL-MAP
    WHERE t IN TPR(o) AND t.year = yy AND T.month = mm AND
    T.date = dd
    AND t.hour = hh AND T.P IN area } ;
END
```

A key operation in CML is the application of transformations to elements from R-domains. Such applications may be expressed in CML in terms of the apply statement. Suppose the variable *Y* holds a set of DEM element identifiers. The command *APPLY DEM.union TO Y* results in a (new) element of type DEM; it also returns the identifier of the new element, which can be stored in another variable to be used later. The apply command has a large number of important applications, which include the creation of data sets in an R-structure. The following example shows how we may create a new explicit instance of an R-domain element of the R-structure *DEM_SLOPES* using the transformation *DEM.compute_slope:*

```
CREATE element IN instances(DEM_SLOPES)
    WHERE VALUE = APPLY DEM.compute_slope TO Y
```

CML provides an important but simple command access for querying about R-structures, their four main components and elements of their components. For example, suppose *Rainfall_Map* is an abstract R-structure representing rainfall data with a (default) concrete structure *Rainfall_Map.tpr*, whose elements possess the representation *{[year:Int, month:Int, date:Int, hour:Int, P:Point, rain:Rainfall]}*. Then, the next query extracts all rainfall data sets for the Manaus region within a particular period. (The function map applies *rain_extract* to each element in R and returns the union.)

```
R = ACCESS { S IN Rainfall_Map
        FOR T IN Rainfall_Map.tpr(S)
        WHERE T.year = 1989 AND T.month = 1 AND T.date = 21
          AND 1 <= T.hour <= 12 AND T.P IN spatial_projec-
          tion(Manaus) }
W = APPLY map TO Rainfall_Map.rain_extract, R
```

Scientific modeling is a complex, highly interactive, and highly iterative process, which is typically distributed over a number of conceptually and physically distinct environments and subenvironments. CMS provides scientific investigators a uniform computational environment consisting of a set of tools that supports a significant range of activities. These tools may be software packages, such as

Mathematica, MatLab, and Khorus. Thus, for a CMS, the uniform, consistent, and transparent integration of all such software tools is clearly necessary.

We have exploited the concept of R-structures for building a conceptual framework for an integrated computational environment. Virtual R-structures, as they are referred to, are used specifically to tie external software code segments within the realm of the CMS modeling activity. In this section, we discuss these conceptual-level constructs and how they have been incorporated in CML.

At the conceptual level it is possible to create virtual (concrete) R-structures whose R-domain elements have a hidden representation. They are treated as black boxes and are managed as plain (binary or ASCII) files. Such virtual R-structures are extremely useful in allowing operations from other software modules to be incorporated and used with the CMS environment. As an example, the following creates a virtual concrete R-domain for the R-structure DEM in a format that is interpretable by MatLab.

```
CREATE r-domain(DEM.MatlabFormat)
                    WHERE STRUCTURE = FILE
                    SUPER DOMAIN = NONE
```

As we shall see, it is also possible to introduce transformations from MatLab, such as *display_dem,* into the set of transformations associated with the R-structure DEM. Hence one may obtain a uniform view from the integration of software packages.

In CMS, a special R-structure *Tools* is used to "register" external software tools. The description about each tool includes the general information about the tool and specific instructions (in UNIX shell script) of how it can be started. The following CML operation adds a new tool MatLab into the system.

```
CREATE ELEMENT IN Tools
   WHERE NAME = MatLab
         DESCIPTION = "Tool for mathematical computation"
         INIT = "/cms/tools/matlab.init"
         EXEC = "/cms/tools/matlab"
         PARAMETERS = INTERACTIVE, . . .
```

We now discuss how transformations may be implemented with the use of other languages and tools. We illustrate this functionality in terms of a MatLab function that displays a DEM.

```
CREATE TRANSFORMATION display_dem (e:DEM)
   SOFTWARE: MatLab
   FILE = MatlabFormat(e)
   BEGIN
   load FILE.dat
   plot(FILE)
   END
```

Here *MatlabFormat* is an isomorphic translation into the representation used by MatLab, which is a virtual domain (returns a file name). The body includes actual MatLab commands.

Note that the transformations are implemented both inside and outside of CML commands. Suppose *Spatial* and *Temporal* are two R-structures representing spatial and temporal objects and that the two transformations *Spatial.projection* and *Temporal.projection* are defined on each respectively. The following defines a transformation *Surface_Flow.compute* that computes, taking the form of a hydrograph, the flow of surface water over some area of interest.

```
CREATE TRANSFORMATION Surface_Flow.compute
   (o:Spatial, t:Temporal)
   RETURN H:Hydrograph
   BEGIN
   X = ACCESS { D IN DEM
      WHERE Spatial.projection(D) INTERSECT
      Spatial.projection(o) }
   Y = APPLY DEM.union TO X
   APPLY Display_dem TO Y
   Z = APPLY DEM.slope TO Y
   A = APPLY Rainfall_Map.extract_by_range TO
      TEMPORAL.projection(t), SPATIAL.projection(o)
   B = APPLY Rainfall_Map.interpolation TO A
   C = APPLY Rainfall_Map.find_storm TO B
   D = ACCESS S { IN Surface_Flow
      WHERE Spatial.projection(S) INTERSECT
      Spatial.projection(o)
        AND Temporal.projection(S) = Temporal.projection(t) }
   H = APPLY SURFACE_FLOW.compute.solve_method to Z, C, D
   APPLY Display_dem TO H
   END
```

Thus, CML provides a basic set of high-level operations intended to serve a significant proportion of the scientific database users' needs.

THE AMAZONIA ARCHITECTURE

Amazonia is based on the preceding model of scientific activity in general, and on CML in particular. Amazonia supports modeling in large-scale Earth science research and, in particular, data intensive and numerically intensive modeling activities in large-scale hydrologic research.

Amazonia is an extensible system, based on a comprehensive and consistent model of R-structures. It provides a simple, uniform framework for the architecture and use of the system (Figure 68-3). The system incorporates three key components that are required to provide adequate computational support for modeling activities: (1) a modeling support system that supports an appropriate environment for the construction and testing of models; (2) a tool management system that provides support for the interoperability of heterogeneous modeling tools; and (3) a distributed access system that supports access to, and manipulation of, data and services in a distributed environment.

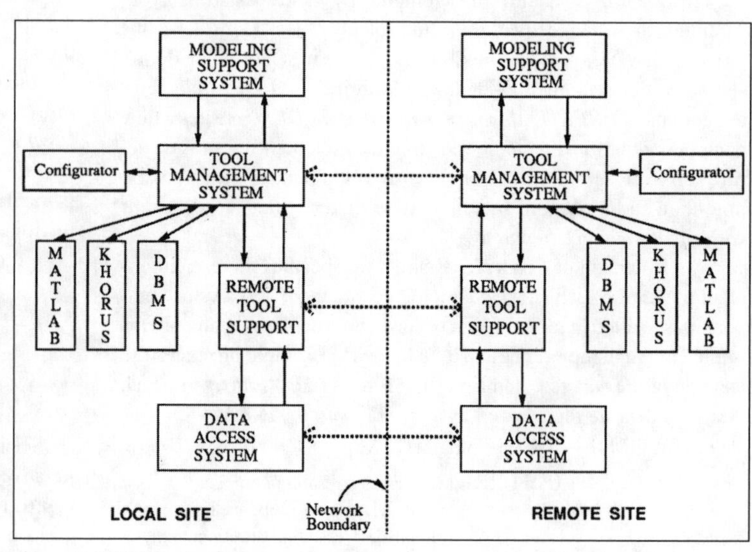

Figure 68-3. High-level architecture of Amazonia.

Modeling Support System

Two types of entities—objects and processes—correspond directly to two key components in our model of scientific activity. Objects are used to represent data sets and abstractions of data sets; processes represent transformations between data sets and their abstractions. A process takes one or more objects as input and produces one or more objects as output. The exact transformation from input to output is included as a procedure call that specifies the operation and its execution details. These abstractions are then translated into proper language constructs to be interpreted by the tool manager, which addresses the corresponding commands to the various tools involved.

Objects are used for two purposes: (1) to provide a unique identifier for the totality of information related to a data set, be it the actual data, associated references, or information related to the correct interpretation of the data; and (2) to abstract the details of format and location, presenting a homogeneous access point to data sets, regardless of their type. Objects contain attributes that provide the basic functionality needed for object abstraction. These attributes provide users with the information necessary for lineage management, change propagation, and similar features. Another important feature provided by these attributes is that several objects can reference the same data set, thus permitting the construction of different views of the same data set.

Processes are used to represent data transformations, and they play a role similar to that of the objects (e.g., provide multiple views of models and a homogeneous interface for data processing, regardless of the execution environment). In Figure 68-1, the interpolation model, the slope model, and the storm model are examples of processes.

A process is implemented as a set of input objects, a transformation applied to those objects, and a list of the resulting output objects. This information is stored as a set of attributes: a list of object names for the input attribute, a transformation name for the transformation attribute, and a list of object names for the output attribute. The transformation can be of two types—primitive or composite.

In its current form, the modeling level encompasses several databases used to store the objects, processes, and models. The metainformation about the models (i.e., the objects and processes implementing models) will be stored in the storage subsystem of Amazonia, which is currently the POSTGRES database system. Such a database provides high-level support for modeling activities, such as query capabilities on the nature and contents of the objects, their attributes, the processes that use them, and the models (Alonso and El Abbadi 1994). In the Amazonia architecture, the tool management system is in charge of executing operations, using either internal or external tools. Thus the modeling subsystem will only specify the transformations to execute, the appropriate inputs (as objects), and the order of execution. The tool manager and the data access system will resolve all the references, find the data sets corresponding to the objects, find the transformations corresponding to the processes, locate the tools required to execute each particular transformation, and generate the commands that will trigger the execution. The results of the execution are stored in the database as data sets, if the user so specifies. This information is then passed on to the modeling subsystem that creates the corresponding objects and references at the modeling level.

The Tool Management System

The tool management system can be divided into two conceptually and functionally distinct levels: an upper-level toolbox and a lower-level tool handler. The toolbox component stores information about each software package, including the name, description, and necessary information and files for execution. This information is provided by the user when each software package is "registered" into the system. It also serves as an interface between other tools and Amazonia. Simple constructs are used for registering the tools that a scientist wishes to use in modeling. Once a tool has been registered in Amazonia, it handles the parsing of internal commands for the tool, processes tool-specific data and language format conversions, and then issues requests to the tool process for executing the commands; or, if the process has not been running, it starts the tool process. This last function is performed with the help of the tool handler. The two levels basically operate using the client–server paradigm, with the toolbox level requesting services on a communication link established by the tool handler. The client and server communicate by means of two simple primitives :

Send_command: Sends various commands for execution by the tool handler. The commands are distinguishable and initiate different primitives in the tool handler. For example, in the case of MatLab, two types of commands could actually start MatLab or load a file.

Receive_output: Gets the output of the command sent to the handler. Again, these can be different kinds, for example, system error/warning messages, flags for successful execution of the command, or some form of resulting data.

The tool handler provides the actual interprocess communication between Amazonia and the software tool. It is here that the "registering" information of a tool is used to initialize and set up the environment for its subsequent execution. Extensive processing is involved in communicating with a software tool. Constant user interaction, as a process running in the background, is required.

A small set of primitives exist that are used to communicate with the external tool:

Start: Sets up the execution environment for the tool and starts it. It returns the communication link identifier to the toolbox.

Send: Sends commands on a link to a tool. In the case of MatLab, it might send the load command.

Receive: Retrieves the result of the command sent previously. Thus, in the case of MatLab, the status (successful/error) of the load command execution in MatLab is received.

Diagnostic: Provides information at the systems level, of the tool process status. System crashes and part of this construct.

Format: Takes care of any changes in the format of the data returned by the tool before it is passed to the subsystem.

The current implementation of the tool management system provides the ability to run tools in the background with the appropriate run-time environment set up. Access to the tools is on a demand-driven basis. Parameters are provided to tune the tool performance. We have integrated, with relative ease, front end tools of DBMS, such as POSTGRES and Coral (Ramakrishman et al. 1993), and mathematical tools, such as MatLab.

Our current implementation uses the relational DBMS POSTGRES for database support. Our approach is to keep Amazonia as loosely bound to POSTGRES as possible, to provide easy portability of Amazonia to other database platforms. The configurator was designed to dynamically configure Amazonia to the underlying DBMS, according to the specifications provided by the user. The configurator is used to translate CML constructs to the appropriate underlying database constructs. To integrate various database systems, the configurator requires specifications for mapping primitive constructs to the constructs of back end databases.

The current implementation of the configurator has run-time directives for translating CML commands to constructs of relational languages like POSTQUEL (UC–Berkeley 1993) and SQL. However, the design allows for the change of directives to translate CML into nonrelational languages such as O2 (Altair 1989).

Distributed Access System

The distributed access system is the lowest layer in Amazonia's hierarchy that provides a homogeneous modeling support environment on top of a collection of heterogeneous data and services distributed over a network.

Traditionally, distributed databases provide remote access to data by implementing a layer on top of existing database systems. This approach is not prevalent since it assumes a homogeneous database environment. The alternative, more popular approach, is to hide the distribution at the lower levels (e.g., at the level of the file system). With this approach, a database product that uses the services of a UNIX file system easily can be ported to any environment consisting of a UNIX-compatible file system, such as the SUN Network File System (NFS) or others. Due to the popularity of UNIX platforms in academic and scientific environments, we have chosen the latter for distribution of data and services. In particular, the fact that the tools, services, and data are distributed in the network remains transparent to the users. Users continue to access data (or files at the lowest level) and services as if they existed on site.

Although a network file system (NFS) can deal with distributed data and services, we chose not to implement it for a variety of reasons:

1. One of the main drawbacks of an NFS is that it assumes that data (i.e., files) are under a single administrative domain in a local area network.
2. Access to distributed services has a different interface, typically remote procedure calls (RPC). Instead, much of scientific data are stored at repositories across the Internet and are accessible via anonymous FTP. Using NFS to make this data accessible locally would entail copying the data locally, which would give rise to a host of complications, ranging from data consistency to data currency.
3. Accessing services is not consistent with that of data in terms of the interface available to the user.

Recently, a file system called Prospero (Neuman 1992) has been developed that can integrate UNIX files and directories scattered across the Internet. We extended the Prospero file system to provide data handling and remote service capabilities in Amazonia. The limitation of Prospero is that most of its features are restricted to file directories. For example, Prospero provides a notion of "union" that can be applied only to file directories. This feature can be used to view physically distinct directories as a logically centralized entity. Prospero also provides a "selection" on a file directory to extract only those files that satisfy a certain selection criterion. In the context of Amazonia, these notions have been extended so that they can be used at the file level. For example, a data set "Rainfall" may consist of several files that are distributed spatially. However, a user may want to view the union of these files as the logically centralized data set "Rainfall." Similarly, a user may be interested in looking at the rainfall for a particular year, which can be extracted from the data set by applying the notion of selection. These modifications have been made to Prospero by changing the underlying network protocols between Prospero clients and servers. (Further details about these protocols appear in Sastri 1994.)

In order to provide access to remote tools and data, we require support for remote execution (i.e., users requesting the remote service must have adequate access permission). Once this assumption is made, access to remote service can be integrated very easily with Prospero. Instead of fetching data by copying the remote file, the data (in reply to a service request) are obtained by executing a program on a demand-driven basis. We designed the functional and protocol extensions to Prospero that are required to support the functionality in the context of Amazonia.

CONCLUSION

The computational modeling system (CMS) Amazonia is designed to resolve many shortcomings associated with systems currently employed in providing support for a wide range of scientific modeling applications. Amazonia is based on a general and powerful model of scientific modeling activity. A key component is the concept of an R-structure, which is intended to capture the notion of a scientific concept, in general, and the representations and transformations of such a concept, in particular. A computational modeling language (CML) that is based on the notion of R-structures may be employed to construct computationally useful representations of an unlimited array of scientific modeling concepts, including those that describe the process of modeling itself and the phenomena being modeled. Amazonia has been implemented in terms of a layered architecture incorporating three key components to provide a consistent and simple framework for modeling activities. These components include a modeling support system, a tool management system, and a distributed access system. The strength of Amazonia lies in its ability to provide an integrated environment in which investigators may construct, run, and evaluate their models; integrate heterogeneous tools into the system "on the fly"; and access services and data in a distributed environment. The system currently is being applied in solving large-scale hydrologic problems.

References

Abel, D. 1993. *Advances in Spatial Databases: Third International Symposium.* Lecture Notes in Computer Science Series. New York: Springer-Verlag.

Adams, P., and M. Solomon. 1993. *An Overview of the CAPITL Software Development Environment.* Technical Report TR-1143. Computer Science Department, University of Wisconsin–Madison.

Alonso, G., and A. El Abbadi. 1994. Cooperative modeling in applied geographic research. In *Proceedings of the International Conference on Cooperative Information Systems, CoopIS'94.* Toronto, Canada.

Altair Corp. 1993. *O2 Query Language Prototype Version 1.0—User's Reference Manual.* France: Altair Corporation.

Birtwistle, G. M., O. J. Dahl, B. Myhrhaug, and K. Nygaad. 1973. *SIMULA Begin.* Philadelphia: Auerbach Press.

Chu, W., ed. 1993. *Proceedings of the NSF Scientific Database Projects.* AAAS Workshop on Advances in Data Management for the Scientist and Engineer. Boston.

de Hoop, S., and P. van Oosterom. 1992. Storage and manipulation of topology in POSTGRES. In *Proceedings of EGIS'92,* 1,324–36. Munich, Germany.

Dozier, J. 1990. Looking ahead to EOS: The Earth-observing system. *Computer in Physics.*

Gunther, O., and H. J. Schek, eds. 1991. *Advances in Spatial Databases: Second Symposium, SSD '91, Zurich, Switzerland.* Lecture Notes in Computer Science Series. New York: Springer-Verlag.

Hachem, N., M. Gennert, and M. Ward. 1993. The Gaea system: A spatiotemporal database system for global change studies. In *AAAS Workshop on Advances in Data Management for the Scientist and Engineer*, 84–89. Boston.

Hachem, N., K. Qiu, M. Gennert, and M. Ward. 1993. Managing derived data in the Gaea scientific DBMS. In *Proceedings of the 19th International Conference on Very Large Databases*. Dublin, Ireland.

Hardisty, J., D. M. Taylor, and S. E. Metcalfe. 1993. *Computerized Environmental Modeling: A Practical Introduction Using Excel*. London: John Wiley and Sons.

Jones, C. V. 1990. An introduction to graph-based modeling systems, I: Overview. *ORSA Journal on Computing* 2(2):136–51.

———. 1991. An introduction to graph-based modeling systems, II: Graph-grammars and their implementation. *ORSA Journal on Computing* 3(3):180–206.

Long, D., et al. 1992. REINAS: *Real Time Environmental Information Network and Analysis System. Concept Statement*. Technical Report UCSC-CRL-93-05. Baskin Center for Computer and Information Sciences, University of California–Santa Cruz.

Medeiros, C. B., and F. Pires. 1994. Databases for GIS. *SIGMOD Record* 23(1):107–15.

NASA. 1990. *EOS: A Mission to Planet Earth*. Washington, D.C.: NASA.

Neuman, B. C. 1992. Prospero: A tool for organizing Internet resources. *Electronic Networking: Research, Applications and Policy* 30(7):28–32.

Ordille, J. J., and B. P. Miller. 1993. Database challenges in global information systems. In *Proc. ACM SIGMOD Int. Conf. on Management of Data, 1993*.

Paul, H., H. Schek, M. Scholl, G. Weikum, and U. Deppisch. 1987. Architecture and implementation of the Darmstadt database kernel system. *Proceedings of the SIGMOD International Conference on Management of Data*, 196–206. San Francisco.

Ramakrishnan, R., D. Srivatsava, S. Sudarshan, P. Seshadri. 1993. Implementation of CORAL deductive database system. *ACM SIGMOD Record* 22(2):167–76.

Requicha, A. A. G. 1980. Representations for rigid solids: Theory, methods, and systems. *Computing Surveys* 12(4):437–64.

Robertson, D., et al. 1991. *Eco-Logic: Logic-Based Approaches to Ecological Modeling*. Cambridge: MIT Press.

Sastri, A. M. 1994. Experiences in heterogeneous service and data access using distributed approaches. Computer Science Department, University of California–Santa Barbara.

Segev, A. 1993. Processing heterogeneous data in scientific databases. In *Proceedings of the NSF Scientific Database Projects, AAAS Workshop on Advances in Data Management for the Scientist and Engineer*, edited by W. Chu. Boston.

Silberschatz, A., M. Stonebraker, and J. D. Ullman. 1990. Database systems: Achievements and opportunities. *ACM SIGMOD Record* 19(4):6–22.

Smith, T. R., J. Su, D. Agrawal, and A. El Abbadi. 1993a. Database and modeling systems for the earth sciences. *IEEE Bulletin on Data Engineering* 16(1).

———. 1993b. MDBS: A modeling and database system to support research in the earth sciences. In *Proceedings of the NSF Scientific Database Projects, AAAS Workshop on Advances in Data Management for the Scientist and Engineer*, edited by W. Chu. Boston.

Smith, T. R., J. Su, A. El Abbadi, G. Alonso, and A. Saran. 1994. *Computational Modeling Systems: Support for the Development of Scientific Models*. Technical Report TRCS94-11. Department of Computer Science, University of California–Santa Barbara.

Technical Committee on Data Engineering (TCDE). 1993. *Bulletin of the Technical Committee on Data Engineering* 16(1).

UC–Berkeley. 1993. *POSTGRES Database Management System Version 4.1—Reference Manual*. Computer Sciences Division, University of California–Berkeley.

van Oosterom, P., and T. Vijlbrief. 1991. Building a GIS on top of the open DBMS POSTGRES. *Proceedings of EGIS '91*, 775–87. Brussels, Belgium.

Wolf, A. 1989. *Design and Implementation of Large Spatial Databases: First Symposium SSD '89*, Santa Barbara, California, edited by A. Buchmann et al. Lecture Notes in Computer Science Series. Berlin; New York: Springer-Verlag.

———. 1990. How to fit geo-objects into databases—An extensibility approach. *Proceedings of the First European Conference on GIS*. Amsterdam.

Wolf, A., and H. J. Schek. 1989. The DASDBS GEO-Kernel—An extensible database system for GIS. In *Three Dimensional Modeling with Geoscientific Information Systems*, edited by A. K. Turner, 67–88. NATO ASI Series. Dordrecht; Boston: Kluwer Academic Publishers.

Terence R. Smith is Professor of Geography and of Computer Science at the University of California, Santa Barbara. He served as Chair of the Department of Computer Science from 1986–1990; Associate Director of the National Center for Geographic Information and Analysis from 1988–1990; and currently is Director of the Center for Computational Modeling and Systems. His current research interests focus on the design, development, and testing of spatial and scientific databases, and the design and development of computational systems that support the modeling of complex phenomena.

Jianwen Su is Assistant Professor of Computer Science at the University of California, Santa Barbara. His research is focused on principles and theory of databases; data modeling and database design; query, transaction, and programming languages for databases; distributed databases; and database methodology for software engineering.

Amitabh Saran is a graduate student researcher in the Department of Computer Science, University of California, Santa Barbara. His research interests include system design and engineering, language and system support for modeling scientific applications and distributed systems. He currently is working on providing computational support for modeling large-scale scientific applications that will provide unified support for interoperability with heterogenenous modeling tools.

Anuradha M. Sastri is a master's candidate in the Department of Computer Sciences, University of California, Santa Barbara. Her research interests include database kernel, distributed systems, networks, and system design. Currently she is working on providing efficient support for large data sets in modeling large-scale scientific applications.

Department of Computer Science
University of California
Santa Barbara, California 93106

69

Object-Oriented Design of GIS:

A New Approach to Environmental Modeling

Peter Crosbie

Geographic information systems' (GIS) technology boasts tremendous potential for the analysis and modeling of spatial data. However, modeling within a GIS environment is presently at an unsophisticated level. One reason is the difficulty, and therefore cost, of integrating modeling procedures. Current literature suggests two levels of dealing with the problem: loose coupling of existing models and GIS through software interfaces, and tight integration of existing models within a GIS framework. Loose coupling, the current approach, is unsatisfactory because models are often system-specific and not reusable. Also, this approach requires a high level of technical expertise beyond that of the average user. Tight integration requires a high initial cost outlay for software development by GIS vendors. Market demand for individual applications of this nature has not been high enough to drive much development in this area (Dangermond 1993). What is needed is a more efficient approach to integrating models with GIS software. Object-oriented data systems provide this opportunity. The unification of attribute geometry and actions on a single object or feature allows for the development of more sophisticated functionality for advanced operations. A third option, therefore is the design of a GIS that has the advanced capability to allow simplified, low-cost user generation of integrated models within the GIS environment.

INTRODUCTION

This chapter proposes a new approach to modeling by building on the notion that analysis capability within GIS is presently far below its potential. It suggests new tools to bridge the gap between loose coupling and tight coupling possibilities. These tools provide the analyst with increased capability for implementing existing models within a GIS environment. More significantly, they facilitate the creation of a new generation of models that take advantage of the latent potential of spatial analysis and GIS for more accurate representation of real-world events. It will describe how the object data structure allows the development of such a system and will explain some of the tools themselves, including triggers, user-defined functions, interactive topology, and a C programming interface. Collectively, these tools allow the user to manipulate the workspace in a highly interactive manner, and, more significantly, control the reactivity of the data. They encourage the generation of relationships between themes and features within the data set to any level of complexity. They allow the user to define how a feature or a class of features will react to user-defined stimuli, and they allow these reactions to be propagated in a cascading fashion. In addition, the C programming interface allows the dynamic coupling of existing models to the GIS database, so that the GIS itself acts as a responsive database to the controlling program. These tools provide the user with complete control over his or her environment in a way not possible in the traditional georelational system, and as such, provide the ability to set up a more realistic and effective modeling environment than was previously possible.

THE MODELING DILEMMA

What is a model? Specifically, what is an environmental model? Look up the definition of "model" in a dictionary and you might find something like: "A small object built to scale that represents another, often larger object." In the case of an environmental model, the "often larger object" refers to the enormously complicated interactions of natural or physical systems. The "small object" is the hydrologic, atmospheric, ecological, land surface, or subsurface model to be created. The problem is how to simplify the large object, in such a way as to make the small object manageable, yet useful. It is a problem compounded by the uncertainty of the interactions involved. A GIS needs to be versatile enough for model generation under these terms.

General consensus is that there is not enough support for this kind of analysis in GIS. Though the potential of GIS for spatial analysis has been widely recognized (Goodchild 1988) so, unfortunately, has its lack of fulfillment. Goodchild described a GIS as: "A database containing a discrete representation of geographical reality in the form of static two-dimensional geometric objects and associated attributes, with a functionality largely limited to primitive geometric operations to create new objects or to compute relationships between objects and to simple query and summary descriptions" (1991a).

Potential of GIS

Today, GIS are frequently used simply as display devices for already calculated results. Though this is useful and valid, it ignores the role GIS can play in generating those results.

GIS provide a workspace for data integration, that is, the ability to handle multiple models and convert between them, for representation of topological relationships important for most spatial analysis

(Chou and Ding 1992) and modeling. Goodchild (1991b) argues this is the greatest advantage of GIS.

More than this, however, GIS provide spatial input into models and allow for consideration of environment, where appropriate, in models otherwise lacking in this dimension. This exposes hidden opportunities for developing new, more effective environmental models utilizing connections to be made based on geographic proximity, which is vital to problem understanding, but often missing without GIS.

Current Problems

Presently, consensus is arranged around the following general assertions:

- GIS currently are not well adapted to environmental modeling;
- user interfaces need to be more carefully designed (Goodchild 1991b);
- commercial software packages lack depth (Nyerges 1993); and
- current systems have limited capacity for spatial analysis and modeling (Chou and Ding 1992).

More specifically, current GIS are characterized by restrictive data models, the lack of error measurement, proprietary file structures and algorithms, and hooks that allow the use of advanced programming tools within GIS (Kemp 1992). Others point to a lack of temporal analysis capacity and poor handling of three-dimensional scenarios.

Required Features of GIS

What features are needed to support environmental modeling in GIS? Goodchild (1993) argues for the following:

1. Support for efficient methods of data input, including import from other digital systems.
2. Support for alternative data models, particularly layer models and conversions between them using creative methods of spatial interpolation.
3. Ability to carry out a range of standard operations, e.g., calculate area, perimeter, and length.
4. Ability to generate new objects on request, including objects created from simple geometric rules from existing objects.
5. Ability to assign new attributes to objects based on existing attributes and complex arithmetic and logical rules.
6. Support for transfer of data to and from analytic modeling packages, e.g., statistical packages and simulation packages. These should take the form of hooks into existing modeling packages rather than the modeling capabilities themselves.

Of these, the first two, and to some degree the third, are already provided for. Missing is support of the fourth, fifth, and sixth; also a seventh: the addition of dynamics and continuity to our understanding of spatial data and spatial interaction, and functionality to the environmental models (Kemp 1992).

One potential solution has been achieved with the implementation of object-oriented concepts that allow for complex feature definition, temporal analysis, and raster–vector integration. It also empowers the concept of dynamic GIS and, finally, the ability to more easily couple process models to GIS.

OBJECT-ORIENTED DATA MODEL

Most current GIS implementations are based on the relational or hybrid model (Egenhofer and Frank 1992). There are a number of reasons why conventional databases are inadequate for GIS applications. The most significant limitation cited in the literature refers to a lack of expressive modeling power (Joseph et al. 1991; Maguire et al. 1990). This lack of power is a result of the simplicity of the relational data model and its inability to handle the complexity of natural data in GIS (Frank 1988, 1984; Egenhofer and Frank 1989, 1992; Maguire et al. 1990). Egenhofer and Frank (1992) contend that the relational data model does not match the natural concept humans have about spatial data. They claim users are required to artificially transform their mental models into a restrictive set of nonspatial concepts. Object orientation offers significant possibilities for GIS applications by addressing the problems of the relational systems listed previously (Egenhofer and Frank 1992).

Complex Feature Modeling

Object orientation describes the basic idea that the world is often perceived as consisting of objects that interact in specific ways. It focuses on modeling objects as humans perceive them in reality. Object orientation places the emphasis on modeling the data rather than on the actions to be performed on the data (Meyer 1988).

The ability to handle complex features is one of the capabilities object-oriented concepts bring to GIS. Complex feature support allows real-world features, such as drainage basins, which can be recursively split into component parts, to be more accurately represented to include all necessary subfeatures. Operations can be performed within and between these composite features taking into account all the inherent natural complexities. This richer data modeling capability is one of the major benefits object orientation and complex feature support brings to the GIS user (Egenhofer and Frank 1992).

Raster–Vector Integration

Many processes are more efficiently modeled in a grid system but often require vector interaction. GIS need to be able to access data in the most appropriate format and act upon that data, regardless of format and without requiring complex conversion routines (Compton et al. 1992).

Temporal Analysis

The problem of handling temporal data in GIS has received a lot of attention. It is vital for many modeling work flows. It is a problem that has been poorly addressed in the commercial software environment. Many researchers point to the architecture of the object-oriented data model as the most promising solution to this dilemma. Present solutions involve taking "snapshots" in time. Advances on this limited implementation involve the ability to store history information on each individual feature, thus allowing the user to query how a feature or group of features has changed over time. The next step would be to expand this solution to take advantage of the potential of object orientation.

DYNAMIC GIS

The current generation of GIS software concentrates on a static view of space occupied by passive objects, and it offers little in support of dynamic interactions analyses (Coucelis 1989).

Support for dynamic GIS is vital to environmental modeling work flows. Dynamic GIS refers to the ability of the system to automatically react in a user-defined fashion to modification of the GIS database. This includes automatic updates of attributes and geometry. This can be achieved through the use of interactive topology, dynamic attributes, user-defined triggers, and software interface.

Interactive Topology

Interactive topology frees the user from concerns of complicated and time-consuming system requirements of topology maintenance. It allows the user to investigate and experiment with this data in a non-process-bound environment. For example, changes to the geometry of a drainage basin feature (whether it be a boundary, modification, addition, or deletion of features, etc.) should be immediately incorporated into the database without requiring user-initiated topological updates. In this way, different analyses options can be investigated under drastically reduced time requirements (Owen 1993).

Dynamic Attributes

Dynamic attributes provide user control of his or her environment at the relationship level. They are extremely powerful modeling tools. They allow the user to define attributes of a feature as a function rather than as a static *char, float, int,* etc. This makes it possible to define a drainage basin attribute so that it calculates the return periods for design storms using the familiar Q= R·A·I equation, where

R=Runoff coefficient

A=Area

I=Rainfall Intensity.

These attributes can be updated automatically whenever any of the input variables change. So, for example, a change in the area of any of the subbasins within the drainage basin under study would trigger an automatic recalculation of flood potential. The significance is that the runoff coefficient is determined by land use, and the drainage basin includes several subbasins for which the *R* and *I* must be calculated separately. The final result involves taking the values calculated for the subbasins and weighting them based on the ratio of their area to the drainage basin area, and combining these weightings to arrive at the solution.

There are a number of dynamic attributes delivered with the system, including a live link to an external relational database. The user must have the ability to define his or her own attribute functions.

User-Defined Triggers

Triggers give complete control of the GIS environment at a reactive level by allowing the user to dictate how features will respond to environmental changes. A simple example would be to automatically place a bridge feature whenever digitizing a road across a river. This easily could be extended to any user-required reaction. In effect, triggers allow the construction of rules to determine the interaction of model input features.

Software Interfaces

The user must be able to take complete control of data input and manipulation through sophisticated software interfaces to the GIS, essentially supplanting or extending the delivered user interface. This gives complete command-level control of data input and manipulation.

With these tools, the user is provided an extended set of functionality for implementing environmental models within the GIS environment. Modeling is supported at the command, relationship, and reactive levels. The final level of support is the database level, which adds the enhanced ability to couple existing models to the GIS.

COUPLING PROCESS MODELS

The provision of a C programming interface allows the user to treat the object space as a slaved database, loading data to and retrieving data from the space to stand-alone C programs. This especially enhances the value of the GIS to the owners of GIS modeling programs needing GIS input and output. Not only is the object space a data repository, but all user-defined functions and triggers can be fully operational while the C program is in control, thus allowing the C program to drive the model and the GIS to react appropriately to the input changes

CONCLUSION

The object-oriented design of GIS allows for the inclusion of several advanced concepts that enhance the functionality of existing systems. The integration of attributes and geometry, interactive topology, dynamic attributes, and dynamic functions, implemented in an object-oriented environment, free the analyst to explore data in a nonprocess-bound environment (Glenn 1990; Owen 1993). In doing so, the value of existing data is intensified through the power of object orientation in implementing a "dynamic GIS." MGE Dynamic Analyst is a commercial implementation of these concepts. It is optimized to handle ad hoc queries and "what-if?" analyses on spatial data by minimizing manual process steps and maximizing dynamics. This is achieved in several ways:

1. "Triggered" processes are used to detect changes in the data model and to automatically maintain defined relationships.
2. Spatial, attribute, and thematic relationships can be defined and left for the features themselves to maintain as the input of new data or the editing of old data continues.
3. Spatial relationships can be studied through a delivered set of spatial operators or by simply defining a new operator.
4. Attribute relationships can be added, edited, disabled, and reactivated at any time. An attribute can even be dynamically related to any table in the relational database; if needed, entire tables of data can be added and modeled as "joined" or related features.
5. Thematic relationships can be defined and edited for a feature class, for an attribute for any feature class, on a query basis, or for any subset or grouping of data desired.

Again, these are relationships that once defined or edited are dynamically maintained (Glenn 1990). This set of tools enables the analyst to take advantage of the power of GIS, not only to more effectively couple existing models in GIS, as is currently the case, but to implement models entirely within the GIS environment and, ultimately, to suggest extensions to existing models to include spatial relationships with dependent features.

References

Abel, D. J., S. K. Yap, G. Walker, M. A. Cameron, and R. G. Ackland. 1992. Support in spatial information systems for unstructured problem solving. *Proceedings of the 5th International Symposium on Spatial Data Handling, Charleston, South Carolina,* 434–43.

Chou, H-C., and Y. Ding. 1992. Methodology for integrating spatial analysis/modeling and GIS. *Proceedings of the 5th International Symposium on Spatial Data Handling, Charleston, South Carolina,* 514–23.

Compton, W., D. Glenn, and P. Crosbie. 1992. Integration of spatially referenced data independent of source format. In *Proceedings of GIS '92, Calgary, Alberta.*

Coucelis, H. 1989. *Geographically Informed Planning: Requirements for a Planning-Relevant GIS.* Paper presented at the North American

Meeting of the Regional Science Association, Santa Barbara, California.

Dangermond, J. 1993. The role of software vendors in integrating GIS and environmental modeling. In *Environmental Modeling with GIS*, edited by M. F. Goodchild, B. O. Parks, and L. T. Steyaert, 51–56. New York: Oxford University Press.

Egenhofer, M. J., and A. U. Frank. 1989. Object-oriented modeling in GIS: Inheritance and propagation. *Proceedings of Auto-Carto 9, Baltimore, Maryland, April 1989*, edited by E. Anderson, 588–98. Falls Church: ACSM, ASPRS.

———. 1992. Object-oriented modeling for GIS. *URISA Journal* 4(2):3–19.

Frank, A. U. 1984. Requirements for database systems suitable to manage large spatial database systems. In *International Symposium on Spatial Data Handling, Zurich, Switzerland, August 1984.*

———. 1988. Requirements for a database management system for a GIS. *Photogrammetric Engineering and Remote Sensing* 54:1,577.

Glenn, D. 1990. An Open-Ended Approach to GIS Analysis. Unpublished paper.

Goodchild, M. F. 1988. A spatial analytic perspective on geographical information systems. *International Journal of Geographical Information Systems* 1:327–34.

———. 1991a. Spatial analysis with GIS: Problems and prospects. *Proceedings, GIS/LIS '91*, vol. 1, 40–48. Bethesda: American Congress on Surveying and Mapping.

———. 1991b. Integrating GIS and environmental modeling at global scales. *Proceedings, GIS/LIS '91*, vol. 1, 117–27. Bethesda: American Congress on Surveying and Mapping.

———. 1993. The state of GIS for environmental problem-solving. In *Environmental Modeling with GIS*, edited by M. F. Goodchild, B. O. Parks, and L. T. Steyaert, 8–15. New York: Oxford University Press.

Herring, J. H. 1992. Tigris: A data model for an object-oriented geographic information system. *Computers and Geosciences* 18(4):443–52.

Kemp, K. K. 1992. Spatial models for environmental modeling with GIS. *Proceedings of the 5th International Symposium on Spatial Data Handling, Charleston, South Carolina*, 524–33.

Maguire, D. J., M. F. Worboys, and H. M. Hearnshaw. 1990. An introduction to object-oriented geographical information systems. *Mapping Awareness* 4(2).

Meyer, B. 1988. *Object-Oriented Software Construction*. Englewood Cliffs: Prentice Hall.

Nyerges, T. L. 1993. Understanding the scope of GIS: Its relationship to environmental modeling. In *Environmental Modeling with GIS*, edited by M. F. Goodchild, B. O. Parks, and L. T. Steyaert, 75–93. New York: Oxford University Press.

———. 1992. Coupling GIS and spatial analytic models. *Proceedings of the 5th International Symposium on Spatial Data Handling, Charleston, South Carolina*, 534–42.

Owen, P. K. 1993. Dynamic function triggers in an on-line topology environment. In *Proceedings of Third European Conference and Exhibition on Geographical Information Systems, Genova, Italy, March 1993.*

Polzer, P. L., B. J. Hartzell, R. H. Wynne, P. M. Harris, and M. D. Mackenzie. 1991. Linking GIS with a predictive model: A case study in a southern Wisconsin oak forest. *Proceedings, GIS/LIS '91*, vol. 1, 49–59. American Congress on Surveying.

Peter Crosbie received his B.S. in cartography from California State University in Chico in 1988. He was a GIS analyst at the Center for Geographic Planning and Analysis from 1988–1992. He received his M.S. in December 1992, with an emphasis in computer applications in geography. Currently he is a senior marketing analyst at Intergraph Corporation, in charge of object-oriented modules of the MGE product line.

One Madison Industrial Park
Huntsville, Alabama 35894-0001
Phone: (205) 730-7534
Fax: (205) 730-6750
E-mail: *ptcrosbi@ingr.com*

70

High-Level Coupling of GIS and Environmental Process Modeling

Jonathan Raper and David Livingstone

This chapter suggests that high-level coupling of GIS with environmental modeling can best be achieved using fully integrated approaches. A coastal geomorphological example is described as background to the study. After reviewing the "compromises" in the spatial representation employed by most position-based GIS, it is suggested that "object" approaches are likely to offer better solutions. In particular, object approaches allow data modeling at a much higher level than the spatial representations employed, and they offer tools for the integration of spatial representation and spatial models. Object data modeling and handling of time/behavior are discussed and shown not to be obstacles to an integrated object solution.

INTRODUCTION

For those researchers concerned with monitoring and analysis of environmental phenomena and systems, the synthesis of spatial data representations and environmental models has become the new Holy Grail. On one hand, this project is driven by the mechanics of time-consuming and wasteful data transfer between the GIS and environmental model; on the other, by concern about the representational scope of separate and combined components.

In order to progress, research in this area clearly requires a strong focus on the coupling between GIS and model components (Raper 1993). Some believe coupling should be achieved at the "problem," or task, level (Fedra 1993) in order to transcend the contrasting paradigms of GIS (spatial location and interrelationship) and environmental modeling (system state and dynamics). Others maintain that the entire representational tool set should be replaced by new languages (Smith 1993) designed to offer support for spatial representations within a modeling framework. Although considerably different, these two approaches share the view that integration is required at a new (higher) level. Although also taking that view, this chapter asks the question, "What structure in the environmental problem domain should drive the implementation mechanisms selected?"

Perhaps since the problem domain each environmental modeler faces is subtly different, no consensus can ever emerge on this subject, and each application will generate its own implementation mechanisms. However, the authors believe that in many cases the implementation options for coupling are chosen by default and without a fresh assessment of the characteristics of the environmental problem domain. In particular, the demands of sampling frames, the nature of matrix data storage, the structure of external data sources (e.g., remote-sensing imagery), and the use of planar-enforced geometry all dictate data models, data structures, and analysis methods employed.

If integration of model and spatial representation is the highest objective, then it is implied that new holistic methods need to be employed. If this contention is correct (NB many continue to argue the merits of low-level coupling), then the debate surely has shifted to the nature of such holistic methods. This chapter describes the particular problem domain driving this analysis and proposes an integrated modeling and representation approach.

THE PROBLEM DOMAIN

Environmental systems are among the most complex and poorly understood in science. This is due to the difficulty of identifying the relevant functional objects in space or time, their temporally and spatially nested nature (Schumm and Lichty 1965), the existence of system thresholds and feedbacks creating complex response to external forcing (Chorley and Kennedy 1971), the unknown characteristics of the external energy inputs, and the precise impact of human intervention and management (Cooke and Doornkamp 1990). These characteristics of environmental systems render the problem domain extremely complex; attempting to hold any one of these factors steady while varying the others is only possible in the laboratory (where scaling problems are then involved), or, with luck, in the field, when a system state change occurs during monitoring.

The problems associated with modeling a coastal geomorphological system are used to illustrate the problem domain for modeling environmental systems using GIS and, therefore, some of the constraints on implementation. Work on Scolt Head barrier island in North Norfolk, England (Bristow et al. 1993; Raper et al. 1993), describes the environment of an evolving coastal barrier backed by sand dunes and salt marshes and terminated in spits and ebb tidal deltas. In this coastal system, the real-world entities are dunes, spits, beaches, marshes, etc. However, the spatial expressions of these entities are tightly coupled with their behaviors: after any storm, their positions will have changed, making it difficult to represent them with fixed geometric data structures. In this environment, GIS are needed for mapping change (which is rapid), for morphology (which is complex), and for modeling future development based on wave energy inputs and sediment budgets. It is this set of self-evolving fuzzy-edged landforms that provides the environmental complex for which an integrated system is required for the ongoing geomorphological research of the authors.

OPTIONS FOR REPRESENTATION

In our research on coastal environments, we rapidly encountered the following major difficulties in translating observations on the barrier island environment into formal GIS representations: (1) the need to discriminate between active and relict features; (2) there is no implicit hierarchy to drive a functional decomposition of features (e.g., salt marsh creeks experience bidirectional flows and, often, anastamose); (3) landforms have fuzzy boundaries and alternate between subaerial and subaqueous states; (4) composition of landforms is internally heterogeneous; and (5) landforms (remaining functionally identical) change their positions and topological relationships on a time scale of days. In this situation (perhaps not uncommon in environmental analysis), finding representations that are useful in the correlation of process observations with the mode of change measured are exceptionally difficult.

The options for representation in this situation include nonspatial approaches based upon one-dimensional models (French 1993), and spatial approaches based upon GIS (Raper et al. 1993). However, when using most GIS, a number of representational compromises need to be made.

Representational Compromises

The software architecture of the most common commercially available systems approximates the shape/position of features in geographic space by vector or raster "geometry," and descriptive "attributes" are recorded using standard alphanumeric data types (Raper and Maguire 1992). Vector systems of this kind have been called "position-based systems" (Herring 1992), since they are organized by the number and type of positions stored in the geometry store. Position-based systems usually are structured by the creation of nodes at line intersections through the "planar enforcement" process (Goodchild 1992). A consequence of applying planar enforcement is that the polygons formed by this process are nonoverlapping, and every locatable point is assigned to one, and only one, polygon. Raster systems also are explicitly "position based," since the cells are mutually exclusive and their value determines whether they are grouped with other cells. Land ownership is well represented by planar enforced geometry (however, note the problem of high buildings); fuzzy environmental phenomena are not.

Storage of overlapping areas or nonintersecting lines forces the use of separate layers. Commercially available position-based vector GIS contain tools to build structure-describing topological relationships such as connectivity and adjacency within layers, to permit the overlay of different layers graphically and logically, and to update the links between the geometry and the attributes. Raster GIS are also "layered" since each cell may only contain one value, making it necessary to store different geometries in different raster maps. In both raster and vector cases, there are normally one-to-one relationships between pieces of geometry, such as polygons, and records containing alphanumeric attributes.

Position-based GIS offer powerful functionality for applications where the layers can be readily defined and where they contain nonoverlapping features that are static. These conditions are met in many application areas such as fixed asset management and modeling statistical data over administrative areas. However, many organizations handling environmental data find that position-based GIS are inadequate for the storage and modeling of features such as river systems or coastlines that are fuzzy and undergo rapid temporal change. Such systems cannot store complex features, which are aggregations of geometry (without using grouping operations on the attributes), and must store features in their different states, which are changing positions, by placing each state in a separate layer. It is also difficult or impossible to derive change factors for the differences between feature states because, by definition, each different geometry is treated as a different feature. Such restrictions apply equally to vector and raster geometric data alike as it is the one-to-one nature of the linkage between the geometric and the nongeometric attributes that is the limiting factor. These limitations have significantly restricted the use of GIS in hydrological, geomorphological, and marine applications (Raper 1993).

These representational compromises, required when using position-based GIS, have both defined and widened the divide between GIS (as a methodology for handling spatial location and interrelationships) and environmental modeling (as a methodology for handling system state and dynamics). Perhaps it can be argued that modeling has (methodologically) pulled away from GIS specifically as a result of these compromises—compromises that have not harmed the applicability of GIS in the facilities management sector (from which the greatest amount of software development funding has come). It can therefore be argued that since these representational compromises force modelers to abandon GIS or separate it from modeling, it is necessary to articulate the requirements of a holistic approach involving fewer (or no) such compromises and an integrated approach.

Hence, the key questions can be defined as, "What implementation mechanism is appropriate?" (if not position-based GIS), and "How should the integrated mechanism be chosen?" Object approaches are clearly potential new implementation mechanisms offering powerful concepts and procedures that relate well to the target problem domain. Their scope and potential are outlined next, along with an assessment of how well they handle some of the representational compromises.

Representations Based on Objects

The arrival of object approaches to GIS offers some powerful new solutions to the limitations of position-based GIS (Maguire et al. 1990; Worboys 1992). Object approaches in GIS are focused on the types of features or objects being represented. This means that the database is constructed from sets of related object classes, where an object class is defined as a phenomenon having defining characteristics that may be grouped to higher levels and broken down into lower ones. These characteristics define each object class in the system data model, and instances (objects) of the object class are stored in the system database. In an object GIS the descriptive characteristics are not differentiated between geometric and nongeometric types at the interface level. Different object GIS employ different mechanisms to unify storage of the geometric and nongeometric attributes; some systems employ object databases with unified storage, others use an object-oriented language to manipulate the data storage in a relational database.

Object GIS have flexible representational tools, hence objects in any object class can have associated geometries that overlap each other to any extent necessary. Interaction of objects in and between object classes is defined by interaction rules that may be governed by topological criteria. Operations that act only on the object class can also be stored with the class, making it possible to manipulate particular object classes in a variety of ways. Object GIS, therefore, model space as a property of the objects defined, making it possible to organize spatial phenomena by other organizing principles while still maintaining their spatial indexation and topological relationships.

The object-oriented paradigm incorporates concepts that can be applied to both the description of real-world systems and to the definition of computer systems through a programming language. This appears to offer good opportunities to represent concepts in both of these types of systems. The object-oriented systems incorporate concepts that translate well into the language of environmental systems; for example, subclassing, inheritance, encapsulation of data and behavior, and aggregations. These concepts are analogous to the terms used to describe the nature and classification of physical systems. Hence, in environmental systems we perceive such things as hierarchical relationships when using typologic and taxonomic schema (see Oertel 1985 for the coastal system) and when identifying aggregation relationships (features made up of agglomerations of smaller features). Therefore, a data model that also incorporates these concepts is a promising vehicle for the storage and processing of environmental data.

The object-oriented data model offers potentially more powerful semantic concepts than the relational model, which is fundamentally mathematical in nature. However, unlike the relational model, there is no set of "commandments" akin to Codd's principles (Codd 1970) that defines the object-oriented model. This problem is being approached by standards bodies such as the Object Management Group (Soley 1990); but, in the meantime, the effect has been to limit the uptake and development of object-oriented databases. The lack of a standard terminology is a limitation on the efficacy of object-oriented data modeling because one of the major advantages of an object-oriented approach is the tendency to dissolve the boundaries between programs and databases, allowing structures and parameters for the numerical models to be persistent objects in a managed database. This is exactly the kind of situation we require for a fully integrated GIS-based environmental modeling system.

In order to design an object-oriented system, in the absence of a definitive object-oriented model, a consistent set of semantics needs to be adopted; this is often based on models used by particular programming environments such as Smalltalk, C++, and Smallworld Magik. However, from the standpoint of system analysis and design, it is not desirable to design a system on the basis of the mode of implementation, and so a more generic set of semantics is required. Booch (1991) describes a set of semantics and methodologies for the design of object-oriented systems which draws upon the histories of object-oriented programming languages, some of which have been available for over twenty years. The usages and descriptions of object-oriented concepts in this chapter are based upon Booch terminology.

OBJECT ORIENTATION AS AN INTEGRATED APPROACH

From this assessment of the representational compromises in position-based GIS and a review of the more powerful semantic concepts available within object approaches, it can be argued that (at least from the perspective of the coastal system) there is a close relationship between real-world concepts discovered in practical field research and the representational approach offered by object GIS. However, it is now necessary to examine the object approach more closely to determine its suitability for modeling purposes.

Data Modeling and Spatial Integrity

In order to be able to define an environmental data model in an integrated system, the definition of spatial concepts must be made at the highest level of abstraction. This permits the modeling of geographical space as a property of entities, and not the other way around (Nunes 1992). However, the data model must now provide or eschew spatial integrity; for example, deal with the situation whereby different instances of the same entity exist as different representational types in the same place, or whereby different instances of different entities occupy the same space. Clearly, since spatial analytic techniques seek to examine spatial patterns and interactions, the system data model must provide information on class structure and spatial relations. Although this arrangement may make it impossible for many existing spatial analysis procedures to work unaltered, that is not a reason to reject object approaches. It is simply a time to recognize that the real world is, in fact, complex in exactly this fashion, and approaches need to be found to analyze these new spatial models.

One of the problems with a project like the coastal project described previously is the continual temptation to try to define precisely a metaphysical model of the universe because of the complexity and scale of environmental processes. These issues must be faced and opinions formed about what might be termed "environmental metaclasses." Ultimately, decisions about metaclasses will be the most important in regard to the flexibility and durability of the system, since these classes will determine what objects it will be possible to represent. In object-oriented parlance, metaclasses are those classes whose instances are classes in themselves, and they can be used to define attributes and methods that are common to all the instances of the classes they create. The instances in metaclasses are models of environmental entities, using whichever definition and representation is available or required—in raster or vector form. Data model fields (e.g., stored as run length codes) can provide the spatial representation just as well as vector data because these concepts no longer govern the overall data model.

Handling Time and Behavior

Object approaches to spatial representation also offer a means to handle temporal change. Time can be handled such that it is a dimension where "events" are defined as "instants in time when objects are changed" (objects are made in time, and time is absolute); or time can be a property of the objects (time is made of successive objects, and time is relative). While the former approach has been implemented using position-based GIS (Langran 1992), the latter approach is well suited to object approaches. This "relative" approach can be implemented such that objects have versions, or that the time-invariant characteristics are stored at higher levels in the class hierarchy.

Object approaches can also integrate behavior through the methods encapsulated with objects. These methods can be written in languages like C and invoked by messages that are acted upon only by certain objects. Little or no work has been done in spatial modeling to develop spatial analytic methods that utilize this message-passing facility.

CONCLUSION

Impatient in the face of the rising tide (literally and metaphorically), the authors suggest that the way forward for GIS and environmental modeling depends on the high-level coupling of representations and models. If the difficult challenges of environmental problems are to be faced, then nothing less than full integration of spatial representations and system state will do. Object systems seem to offer an approach to integration that has the required scope. They have few disadvantages, other than the potential for (over)complexity and a lack of methodologies for development. Let geographic information scientists be among the first who define how such methodologies take shape.

References

Booch, G. 1991. *Object-Oriented Design with Applications.* Redwood City: Benjamin-Cummings.

Bristow C. S., J. F. Raper, and H. M. Allison. 1993. Sedimentary architecture of recurved spits on a macrotidal barrier island, Scolt Head Island, Norfolk, England. In *Tidal Clastics '92,* 14. Courier Forschungsinstitut Senckenberg.

Chorley, R., and B. Kennedy. 1971. *Physical Geography: A Systems Approach.* London: Prentice Hall.

Codd, E. F. 1970. A relational model of data for large shared data banks. *Communications of the ACM* 13(6):377–78.

Cooke, R. U., and J. C. Doornkamp. 1990. *Geomorphology in Environmental Management: A New Introduction,* 2nd ed. Oxford: Clarendon.

Fedra, K. 1993. GIS and environmental modeling. In *Environmental Modeling with GIS,* edited by M. F. Goodchild, B. O. Parks, and L. T. Steyaert, 35–50. New York: Oxford University Press.

French, J. R. 1993. Numerical simulation of vertical marsh growth and adjustment to accelerated sea level rise, North Norfolk, England. *Earth Surface Processes and Landforms* 18:63–81.

Goodchild, M. F. 1992. Geographical data modeling. *Computers and Geosciences* 18:401–08.

Herring, J. 1992. The mathematical modeling of spatial and nonspatial information in geographic information systems. In *Cognitive and Linguistic Aspects of Geographic Space,* edited by D. M. Mark and A. U. Frank, 313–50. Dordrecht: Kluwer.

Langran, G. 1992. *Time in GIS.* London: Taylor & Francis.

Maguire, D. J., M. F. Worboys, and H. Hearnshaw. 1990. Object-oriented data modeling for spatial databases. *International Journal of Geographical Information Systems* 4:369–84.

Nunes, J. 1992. Geographic space as a set of concrete geographic entities. In *Cognitive and Linguistic Aspects of Geographic Space,* edited by D. M. Mark, and A. U. Frank, 9–34. Dordrecht: Kluwer.

Oertel, G. F. 1985. The barrier island system. *Marine Geology* 63:1–18.

Raper, J. F. 1993. Geographical information systems. *Progress in Physical Geography* 17(4):507–16.

Raper J. F., C. S. Bristow, D. Livingstone, and A. Köppen. 1993. Reconstructing three-dimensional sand body geometry with IVM: A case study from a barrier island spit bar. *Dynamic Graphics Applications Journal* 1:43–51.

Raper, J. F, and D. J. Maguire. 1992. Design models and functionality in GIS. *Computers and Geosciences* 18(4):387–94.

Schumm, S. A., and R. W. Lichty. 1965. Time, space, and causality in geomorphology. *American Journal of Science* 263:110–19.

Smith, T. R. 1993. Integrating high-level modeling of spatial phenomena into GIS. *Proc. GISRUK '93,* 1–13.

Soley, R. M., ed. 1990. *Object Management Architecture Guide.* Object Management Group Document 90.9.1.

Worboys, M. F. 1992. A generic model for planar geographic objects. *International Journal of GIS* 6:353–72.

Jonathan Raper is a lecturer in GIS in the Department of Geography at Birkbeck College, University of London. He gained his Ph.D. from Queen Mary College, London, in GIS techniques applied to geomorphology. His research has concentrated on two areas: the optimum use and understanding of GIS through the development of multimedia spatial data exploration tools and the design of GIS user interfaces, and the development of models for the study of geomorphological environments in three dimensions and through time, particularly at the coast.

David Livingstone is a research fellow in GIS in the Department of Geography at Birkbeck College, University of London. His Ph.D. research concerns the implementation of data models for spatial data based upon object approaches. He also has carried out research in multimedia spatial data exploration techniques and three-dimensional visualization.

Department of Geography
Birkbeck College
University of London
7-15 Gresse Street
London W1P 1PA
United Kingdom
E-mail: *raper@earth.ge.bbk.ac.uk* and *davel@earth.ge.bbk.ac.uk*

71

Spatial Analysis of Nutrient Loads in Rivers and Streams

Dale A. White and Matthias Hofschen

Methods that are more objective and consistent are needed to assess water quality over large areas. A spatial model that capitalizes on the topologic relationships among spatial entities, to aggregate pollution sources from upstream drainage areas was developed for land surfaces having point-source and nonpoint-source water pollution effects. An infrastructure of stream networks and drainage basins, derived from 3-arc-sec. digital elevation models (DEM), defines the hydrologic system in this spatial model. The spatial relationships between point and nonpoint pollution sources and measurement locations are referenced to the hydrologic infrastructure using network analysis and relational database management techniques within the geographical information system (GIS) ARC/INFO. A maximum branching algorithm for traversing directed graphs was developed to simulate the effects of cumulative time of travel from a pollutant source to an arbitrary downstream location. The spatial model was then applied to approximately 11,000 stream segments and 3,000 drainage partitions in the state of New Jersey for estimating concentrations of total phosphorus and total nitrogen. Comparisons of the predicted to actual concentrations (using monitoring station information observed from 1983–1987) showed a relatively good fit for both total phosphorus ($RMSE_p$ = 0.28 mg·l^{-1}) and for total nitrogen ($RMSE_n$ = 1.02 mg·l^{-1}). Model fit improved slightly when the spatial variability of nutrient export coefficients and chemical decay rates (i.e., nonconservative transport) was considered. Stratification of monitoring stations by point-source and nonpoint-source-dominated drainage basins improved the model fitting process because either loads from overland flow or pollutant discharges dominated in each stratification subset ($RMSE_{p,nps}$ = 0.09 mg·l^{-1} and $RMSE_{n,nps}$ = 0.9 mg·l^{-1}).

INTRODUCTION

A major problem in assessing the effects of anthropogenic impacts on regional surface water quality has been the difficulty in obtaining reliable information on cause-and-effect relationships for a representative sample of stream segments in a region (Cohen et al. 1988). Cokriging has been one method used to assess regional streamwater quality (Jager et al. 1990). However, for kriging techniques to be applicable on a regional scale, measurement data should exhibit spatial autocorrelation in multiple dimensions and for distances that are as large as the smallest two or three lag intervals. Johnston et al. (1988) performed a linkage between wetland position in the drainage basin to water quality. However, hydrologic process representation in their spatial model was limited to the use of stream order at the wetland mouth, to order at the basin mouth as a method for weighting the positional impact of each wetland on water quality. Other investigators have employed regression techniques to relate upstream drainage basin characteristics to downstream point measurements and/or indices of water quality (Smith et al. 1992; Steedman 1988; Osborne and Wiley 1988). Each application has offered various methods for characterizing the spatial distribution of predictor variables and the inclusion of physical process mechanisms.

Regional water-quality assessment further would benefit the preparation of statewide water-quality summaries like those required under Section 305(b) of the U.S. Clean Water Act. Presentation of these statewide summaries still is beset by problems of inconsistent reporting of total stream length and number of assessed waters in each state (U.S. Environmental Protection Agency 1991). Assessed waters are those for which a state is able to make support decisions about water use and include both "evaluated waters" and "monitored waters." Evaluated waters are bodies of water for which the use support decision is based on information other than current site-specific ambient data, such as data on land use, location of pollution sources, and predictive modeling using estimated input variables (U.S. Environmental Protection Agency 1991). The model development and application explored in this investigation provide necessary tools and evaluative information to improve the consistency and objectivity of regional water quality inventories and assessment.

Objectives

A spatial model for downstream aggregation of distance- or time-weighted water-quality predictor variables was developed and validated by White et al. (1992). This modeling framework was designed to: (1) encompass large and diverse areas, (2) combine point and diffuse (nonpoint) sources of pollution, (3) explicitly link pollution impacts and sinks via stream topology and geometry, and (4) employ existing inventories of water-quality data. The primary objective of this investigation is to apply and determine the success of the spatial model for predicting nutrient concentrations (total phosphorus and total nitrogen) in nontidal rivers and streams within New Jersey. A secondary objective is to show that model application to an actual domain can serve as a test for the spatial variability of input parameters (e.g., export coefficients from nonpoint-pollution sources).

Nontidal rivers and streams in New Jersey have a total length of 9,157 km, and, for application of the spatial model, they were segmented into 10,916 reach segments. The model covers a total area of 15,385 km². River segments and drainage basin boundaries were stored as vector topology within ARC/INFO. Attributes of these entities are stored in a relational database.

METHODS

Model Description

Traversal of a stream network yields a cumulative time-of-travel term that causes a nonconservative pollutant to decay from entry into the stream system to its point of measurement. The assumption is that cumulative time of travel is the basis for calculating predictors of water quality. (The model is briefly reviewed here; for further elaboration see White et al. [1992].)

The point- and nonpoint-source pollution impacts situated upstream of an initial node (e.g., point-of-measurement or monitoring station) comprise the initial loads of various pollutants. By combining distance effects and initial pollutant load, the formulation of a water quality predictor variable for an individual pollutant, s_{j0} at location $j0$, takes the form:

$$s_{j0} = Q_{j0}^{-1} \cdot \Sigma_i \Sigma_j (l_{ij} \cdot W_{ij}) \qquad (71\text{-}1)$$

where

i: index of pollutant source type, including individual point sources (e.g., wastewater treatment plants) and nonpoint sources (e.g., agricultural and diffuse urban activity);

j: reach or subbasin index; $j = (1,2,3...,n)$ of n reaches or subbasins that lie upstream of the initial node located at $j0$;

s_{j0}: pollutant concentration [ML^{-3}] at initial node located at $j0$;

Q_{j0}: cumulative flow or discharge [L^3T^{-1}] of reach at location $j0$;

l_{ij}: load of a pollutant [MT^{-1}] originating from source i in subbasin j; and,

W_{ij}: weighing factor representing the nonconservative transport of a pollutant.

The l_{ij} for point-source pollutants are assumed to be known. l_{ij} for nonpoint-sources are estimated as:

$$l_{ij} = A_{ij} \cdot Y_i \qquad (71\text{-}2)$$

where

A_{ij}: area [L^2] of nonpoint-source or diffuse activity (e.g., the area of agricultural land within a drainage subbasin); and

Y_i: yield [$ML^{-2}T^{-1}$] of the nonpoint-source pollutant type i.

Assuming nonconservative downstream transport of the pollutant according to a first-order reaction (Thomann and Mueller 1987),

$$W_{ij} = exp\,(-K_{ij} \cdot t_j) \qquad (71\text{-}3)$$

where

K_{ij}: decay constant [T^{-1}] for pollutant i at location j; and

t_j: time of travel [T] from the jth reach or subbasin to location of the initial node at $j0$.

Note that parameters having subscript "j" imply that the parameter can vary spatially.

Assuming stream velocity to be constant, a first-order reaction is one in which the rate of loss of the substance is proportional to its concentration at any time. The decrease in nutrient concentrations with increasing time of travel is assumed to be caused by sedimentation within the water column and by biological uptake (mainly attached aquatic macrophytes).

Establishment of Initial Values for Input Parameters

Calculation of the cumulative time of travel (t) was made by implementation of the spatial model (White et al. 1992) for incremental time of travel for each reach segment. The incremental value for the 10,916 reach segments was formulated from the known length and an empirical estimate of velocity for each reach. Velocity was estimated from the hydraulic geometry relationship of Leopold and Maddock (1953):

$$v = k \cdot Q^m \qquad (71\text{-}4)$$

where Q is mean annual stream discharge and v is mean velocity. Values of $k = 0.38$ and $m = 0.24$ were determined for New Jersey rivers and streams based on 21,031 samples of v [$ft \cdot s^{-1}$] and Q [$ft^3 \cdot s^{-1}$] that were measured over short time intervals (i.e., instantaneous) for the period 1975–1981 ($r^2 = 0.68$). Using Equation (71-4) and values for k and m, v was derived for each reach from an estimated Q for that reach. Q was estimated from an isoplethed map of discharge per area based on fifty gauging stations. Once basin area was determined, incremental Q for each subbasin was aggregated using the maximum branching algorithm. The resultant estimates of time of travel were compared to eleven locations distributed over New Jersey where dye tracer tests were conducted. The comparison showed that time-of-travel estimates employed here were underestimated (factor of 0.57). Assignment of higher values of pollutant decay (K) than those reported in the literature would accommodate the anticipated statewide underestimation of time of travel.

Initial values for pollutant decay (K) and yields from diffuse sources (Y) were required to initiate the model calibration process. Initial values for K were selected from previous field and model investigations of phosphorus and nitrogen transport (Bowie et al. 1985). Depending on the particular nutrient transformation, decay rates for nitrogen ranged from 0.02–0.5 ($NH_3 \geq NO_2$), 0.03–0.2 ($NH_3 \geq NO_3$), and 0.2–10.0 ($NO_2 \geq NO_3$); and, for phosphorus, 0.1–1.7 (sediment organic $P \geq PO_4$) and 0.1–0.7 (particulate organic $P \geq PO_4$). All decay constants reported are in units of [d^{-1}]. Initial values for Y were selected from generalized nutrient export coefficients reported by Rast and Lee (1983).

Digital elevation data, using DMA DTED-1 3 arc-sec. (U.S. Geological Survey 1983), resampled to a 90-m resolution, were utilized to generate a hydrologic infrastructure. Using the Jenson and Domingue (1988) hydromorphic extraction techniques, the study area was partitioned into a set of 2,893 contiguous drainage basins (herein referred to as subbasins) with a network of 10,916 stream segments. (Each subbasin can be considered a representative elemental unit for which uniform overland flow processes are assumed to occur.) The mean subbasin size was 5.26 km², or 250 grid cells (std dev = 7.12 km²) using flow accumulation (upstream area) and flow confluence thresholds of 250 grid cells each.

Pollutant Source Types

In general, nutrients from point-source entities originate from wastewater treatment plants (WWTP) (mainly human waste and food residues), industry (mainly potato processing and commercial cleaning operations), cattle feedlots, and concentrations of domestic or wild duck populations. Nonpoint-pollution sources of nutrients originate from fertilizer applications on cultivated land (both agricultural and residential), storm water outfalls (especially combined sewer overflows), and, in smaller quantities, from forested (leaf detritus) and idle land.

From a model viewpoint, the significant contributions of nutrient loads from point- and nonpoint-pollution sources within subbasins were determined from four land area predictor variables associated with: (1) agricultural activity, (2) urban activity, (3) natural vegetation, and (4) loads $[MT^{-1}]$ and discharges $[L^{-3} \cdot T^{-1}]$ for WWTP. The first three predictor variables represent rural and urban nonpoint-source inputs to the channel system. These inputs were determined by an unsupervised classification of Landsat Thematic Mapper reflectance data for August 1985 using the method of Jenson et al. (1982). Interpretation of the final classification was done using 1:24,000-scale orthophotographs. The Landsat classification had an overall accuracy of 75.9% using the Kappa coefficient of agreement.

WWTP flow within subbasins was compiled by Robinson et al. (1992) for 148 known locations of discharges occurring in 1986. There were an estimated 1,050 total (WWTP and non-WWTP) effluent discharges in New Jersey. Effluent load (l) for each point-source location was estimated from the compiled flow and the expected effluent concentration of 18.0 $mg \cdot l^{-1}$ for total nitrogen (except when known concentrations [range of 1–72 $mg \cdot l^{-1}$] of total kjeldahl nitrogen were reported) and 7.0 $mg \cdot l^{-1}$ for total phosphorus (Thomann and Mueller 1987). These expected concentrations assume conventional secondary treatment of detergent-containing effluent.

Observations of mean total phosphorus concentration (n = 103) and mean total nitrogen concentration (n = 99) for sampling locations in New Jersey for the period 1983–1987 were used to calibrate model parameters and assess overall model fit. Total nitrogen concentrations were calculated from the sum of separate sampling events for nitrate–nitrite and total kjeldahl nitrogen (organic and ammonia nitrogen) concentrations. Rather than employing an average concentration for a given year, the 1983–1987 mean was used to reduce the variance in observed data. Further, a five-year mean was employed to produce a more representative picture of water quality impacts suggested by predictor variables with observation dates varying between 1985 and 1986.

RESULTS AND ANALYSIS

The strategy for assessing performance and understanding the overall and regional importance of channel and overland flow processes was to: (1) apply the spatial model to the entire domain using spatially uniform parameters, (2) include spatially variable decay constants and nutrient export coefficients, and (3) stratify model calibration to two subsets as a function of the level of point-source inputs in each drainage basin. Model performance and comparisons were made with an unweighted root mean square error (RMSE) to gain further understanding of the significance of input parameters:

$$RMSE = [n^{-1} \cdot \Sigma_i (P_i - O_i)^2]^{1/2} \quad (71\text{-}5)$$

where P and O are the predicted and observed nutrient concentrations at location i, respectively, and n is the total number of locations or initial nodes. RMSE is a difference index described by Willmott et al. (1985).

Overall Model

The initial model runs were based on literature-derived estimates of nutrient yield (generalized values for the United States) and decay. Using both a spatially uniform Y (Rast and Lee 1983) for each land-use class and the midpoint of the range of K (0.44 d^{-1}) identified in Bowie et al. (1985), both phosphorus and nitrogen models performed with a $RMSE_p$ = 0.32 $mg \cdot l^{-1}$ and a $RMSE_n$ = 1.29 $mg \cdot l^{-1}$, respectively. The magnitude of the RMSE was caused by overprediction of observed total phosphorus concentrations and underprediction of observed total nitrogen concentrations. A review of the signs of residuals suggested that decay should be increased (toward nonconservative behavior) for the phosphorus model, but decreased (toward conservative behavior) for the nitrogen model. Following an experimentation with various magnitudes of Y and K, the best model performance produced a $RMSE_p$ = 0.28 $mg \cdot l^{-1}$ and a $RMSE_n$ = 1.02 $mg \cdot l^{-1}$ and suggested optimal values for Y and K (Table 71-1). The optimal model values for total phosphorus yields in Table 71-1 are in agreement with values reported by Rast and Lee (1983), but optimal values derived from the nitrogen model are somewhat higher than those reported.

Table 71-1. Postcalibrated, optimal values of spatially uniform nutrient export and decay for prediction of total phosphorus and total nitrogen.

Nutrient Model	Decay $[d^{-1}]$	Yield $[g \cdot m^2 \cdot y^1]$		
		Agriculture	Urban	Natural Vegetation
Total Phosphorus	0.77	0.03	0.07	0.01
Total Nitrogen	0.2	0.9	0.9	0.5

Spatial Variability of Decay Constant and Nonpoint-Source Yields

Average channel and subbasin slope: An additional degree of physical process representation was incorporated into the model by inclusion of an average gradient for each subbasin and channel segment. For each grid cell in the DEM, the gradient was estimated as:

$$g = [(\partial z/\partial x)^2 + (\partial z/\partial y)^2]^{1/2} \quad (71\text{-}6)$$

The derivatives in the x and y direction were computed from the central difference expressions:

$$\partial z/\partial x = (z_{i,\,j-1} - z_{i,\,j+1})/(2 \cdot \Delta x) \quad (71\text{-}7a)$$

$$\partial z/\partial y = (z_{i-1,\,j} - z_{i+1,\,j})/(2 \cdot \Delta y) \quad (71\text{-}7b)$$

Values of g in the grid domain were averaged within each subbasin (polygon) and reach (line) entity to account for regional and local variability in Y and K, respectively (values not reported due to space limitations). The assumption was that, in the case of channel gradient, lower values of g would increase sedimentation of particulate from the water column and allow more time for biological uptake of nutrients. For subbasin gradient, higher values of g would increase flow velocity and erosion of surface particulate from upland areas, which would subsequently increase nutrient yield.

When decay constants based on channel gradient were combined with the optimized spatially uniform values of nutrient yield, phosphorus model performance improved modestly ($RMSE_p$ = 0.24 $mg \cdot l^{-1}$) and nitrogen model performance showed no improvement ($RMSE_n$ = 1.02 $mg \cdot l^{-1}$). When variable yields, as functions of subbasin gradients, were included in the simulation, the phosphorus model performed slightly better than the overall model ($RMSE_p$ = 0.23 $mg \cdot l^{-1}$), and the nitrogen model performed considerably lower ($RMSE_n$ = 1.18 $mg \cdot l^{-1}$). Subsequent change in the magnitudes of *K* for each channel gradient class did not improve the RMSE.

The relationship of observed and predicted concentrations for each nutrient type shows that model predictions (based on an optimized set of parameters) are of the same order of magnitude and follow the same trend as the observed concentrations (Figures 71-1 and 71-2). The scatter of observed versus predicted total nitrogen concentrations (Figure 71-2) was more compact and linear than that for total phosphorus, indicating better model fit relative to the magnitude of observed concentrations. WWTP loads dominate the magnitude of the predicted estimate, especially in the prediction of total phosphorus. The dominance of WWTP loads in the predicted total concentration of each nutrient suggested that the calibration of yields (*Y*) from nonpoint sources should be done on those sampling locations without point-source-dominated drainage basins (results not reported due to space limitations). The fact that the total nitrogen model underpredicted observed concentrations (Figure 71-2), even with a conservative decay constant (K = 0.2 d^{-1}), suggested that other sources of nitrogen existed that were not accounted for by WWTP loads alone (e.g., sources such as non-WWTP effluent discharges [902 possible locations] and atmospheric fallout).

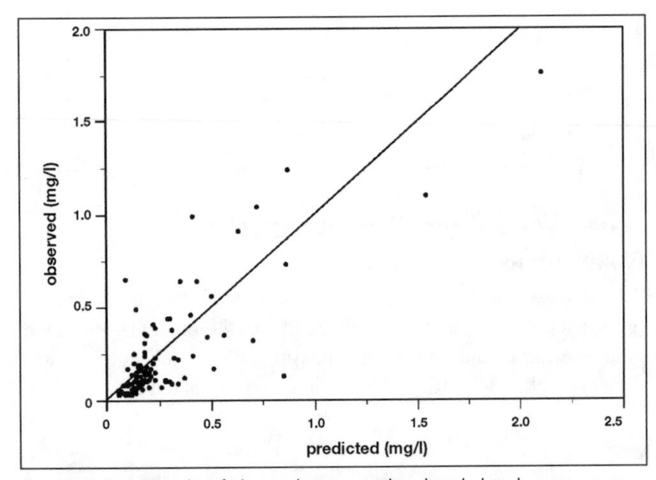

Figure 71-1. Scatterplot of observed versus predicted total phosphorus concentrations (mg/1) for 103 locations using model parameters of uniform Y of 0.13 (agr); 0.14 (urb); and 0.05 (nveg) $g/m^2/y$; spatially variable K based on channel gradient; WWTP LRF=0.6; RMSE=0.174 mg/1.

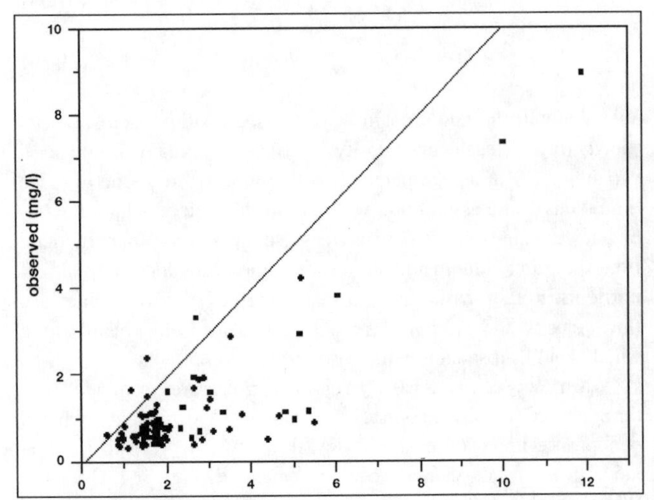

Figure 71-2. Scatterplot of observed versus predicted total nitrogen concentrations (mg/1) for 99 locations using model parameters of uniform Y of 0.9 (agr); 0.9 (urb); and 0.5 (nveg $g/m^2/y$; K=0.11/d; RMSE=1.02/mg/1.

Parameter sensitivity: The importance of nonpoint-source yields (*Y*) and decay rates (*K*) was further examined in a sensitivity analysis of each parameter (values not reported due to space limitations). Both parameters were shown to be influential because the percent change on mean model predictions was at least 15%. This suggested that spatially variable subbasin yields and decay constants would be viable parameters for improving model performance.

In summary, provision of spatially variable decay and subbasin yields did not significantly improve model performance. A plausible explanation for model insensitivity to these parameters is that sufficient spatial variability is accounted for by inclusion of drainage basin attributes upstream of each sampling location alone. Improvement in model performance by accounting for regional factors affecting nutrient inputs and transport was limited by the lack of knowledge in the magnitude of variation over space and their relationship, whether linear or not, to topographic gradient. Further, physiography and county agricultural information were perhaps too summative to be used as predictors of water quality at individual sampling locations.

Model Application

The model fitting process enabled the best selection of decay, subbasin yield, and WWTP load reduction (not described due to space limitations) parameters relative to 103 (99 for total nitrogen) sampling locations. The model was then applied to a sample of 7,213 initial nodes and their upstream drainage basin characteristics (taken from the population of 11,063 reach segments) distributed across New Jersey (Figure 71-3). The objective of this exercise was to compare various nutrient concentration classes of model results with observed concentrations and determine whether monitoring station information was biased upward or downward relative to a true statewide evaluation of water quality. The best phosphorus model produced no consistent difference between per-

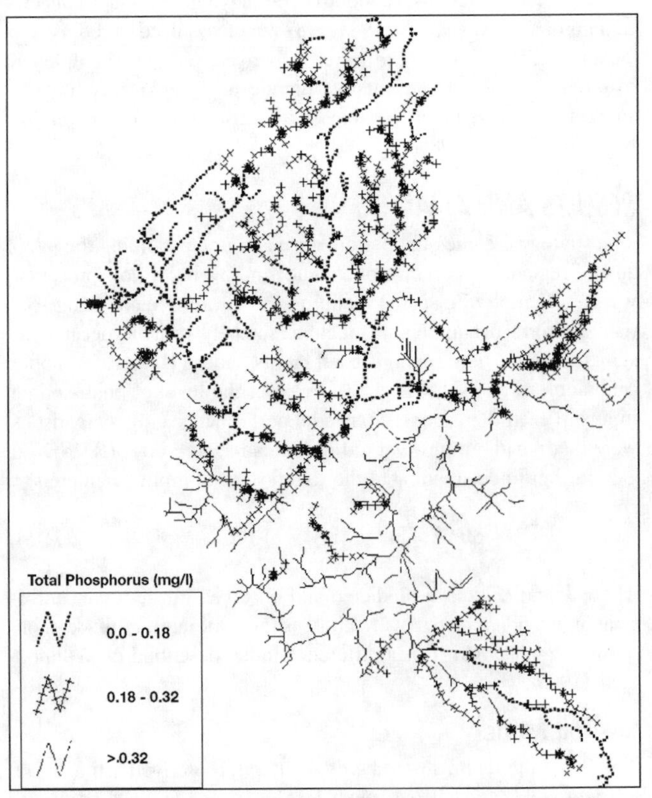

Figure 71-3. Map showing predictions for total phosphorus concentrations for 1,724 reach segments in a drainage basin in northeastern New Jersey. (Refer to Figure 71-1 for identification of model parameters employed.)

centage of samples in each concentration class (Figure 71-4). This model produced a mean phosphorus prediction (n = 7,213) of 0.22 mg·l^{-1} compared to a mean observed concentration (n = 103) of 0.24 mg·l^{-1}. However, there was a slight upward bias in the highest concentration interval (> 0.5 mg·l^{-1}) for the monitoring station samples. The best nitrogen model showed that observed concentrations were severely biased upward in the highest concentration intervals (> 2.0 mg·l^{-1}) relative to model predictions (Figure 71-5). This model produced a mean nitrogen prediction of 1.92 mg·l^{-1} compared to a mean observed concentration of 2.25 mg·l^{-1}. It should be noted that the predicted concentrations of both total phosphorus and total nitrogen were not biased downward.

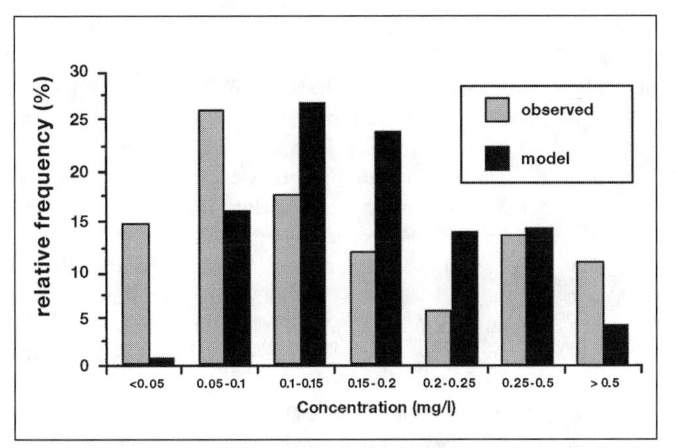

Figure 71-4. Comparison of observed and model-predicted concentrations; total phosphorus model: RMSE=0.17 mg/1; K based on channel gradient; nutrient yields=0.13 (agr); 0.14 (urb); and 0.05 (nveg) g/m^2/y; WWTP LRF=0.6; observed n=103.

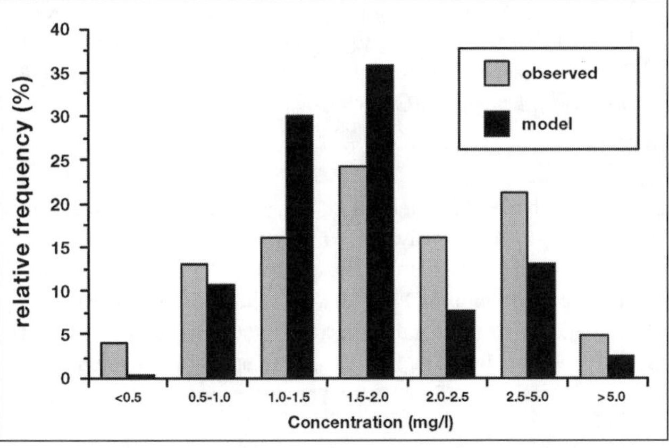

Figure 71-5. Same as Figure 71-4 except total nitrogen model: RMSE=1.05 mg/1; K=0.1/d; nutrient yields=0.9 (agr); 0.9 (urb); and 0.5 (nveg) g/m^2/y; observed n=99.

CONCLUSION

Summary of Results

Based on the results of this modeling investigation, total phosphorus concentrations were generally overpredicted, whereas total nitrogen concentrations were underpredicted. In examining residuals from individual monitoring stations, there were close matches for approximately 90% of these locations. For the remaining locations where phosphorus concentrations were overpredicted, overestimates of WWTP effluent concentrations or the lack of accounting for nutrient sinks (e.g., wetlands) were possible explanations for this difference. For prediction of total nitrogen concentrations, model results consistently showed underestimation. Underestimation was most likely caused by not accounting for point-source (e.g., nitrogen loads from non-WWTP discharges) or nonpoint-source (e.g., nitrogen inputs from atmospheric precipitation and fallout [national average of 1.82 g·m^{-2}y^{1}]) contributions.

The relationships shown in Figures 71-1 and 71-2 indicate that the spatial model implemented in this investigation does account for significant inputs of nutrient sources to a channel system. Although the magnitude of model predictions could be a function of total upstream area from an initial node rather than a reflection of any real physical processes that control nutrient export from the landscape to the stream channel, there was no strong relationship between total upstream area and observed or predicted nutrient concentrations located at the outlet of each drainage basin. Further, model error tends to be associated with smaller drainage basins (not shown due to space limitations). Aside from the large cluster of points near the origin, the relationship indicates that total phosphorus model error increases as basin size decreases. This relationship is even more pronounced for the total nitrogen model. Smaller drainage basins experience more nonlinearities in factors that affect downstream water quality than larger basins. The spatial model explored here is additive and is perhaps better at accounting for the nutrient contributions from upstream land surface characteristics of larger drainage basins.

Differences in the model predicted and the observed concentrations arose from four general sources of error: (1) the highly conceptual structure of the spatial model and not accounting for all significant process mechanisms affecting nutrient transport on the landscape; (2) not accounting for all sources and sinks for each nutrient; (3) the lack of knowledge on site-specific and regional differences in model parameters; and (4) field and laboratory measurements of observed concentrations. For example, in not accounting for all sources and sinks of nutrients, incorporation of point-source nutrient loads was 14% of the total possible inputs. Incorporation of nonpoint-source loads, in terms of land-use activity, was approximately 90% complete.

Overall, the modeling investigation demonstrated that point-source contributions of nutrient pollution tended to dominate the magnitudes of predicted concentrations, especially in drainage basins with high observed levels of total phosphorus and total nitrogen. Nonpoint sources of nutrients played a more important role in drainage basins with low observed levels of total phosphorus. Further, channel decay of nutrients, both uniform and spatially variable representations, were important to accounting for water quality variability over New Jersey.

Importance of Spatial Analysis for Water-Quality Assessment

Information for water-quality assessment is often geographically referenced to point coordinates or river-mile indices, if locational information exists at all. Difficulty exists in accessing information about point, line, and polygon entities that are considered to impact a particular point or line feature (e.g., a sampling location or river segment). Further, there is often a high degree of variability in measuring the water-quality impact of linear or areal features such as the influence of a transportation network or a residential block. Capturing this spatial variability is often dependent on the scale of the modeling application. For example, should one measure the nutrient export rates from individual farm fields or a collection of fields that comprise a drainage basin? The fundamental spatial query in water-quality assessment, when answered, will provide information about upstream land surface characteristics from any arbitrary point entity. Extending this query to isolate the information about intervening drainage areas is also important. In this modeling effort, use of spatial analysis tools (branching algorithm, topological overlay, and a relational data model) were critical for answering the fundamental query of water-quality assessment.

Required Functionality of Spatial Analysis Tools for Water-Quality Modeling

Water-quality modeling further would be served by the inclusion of tools that handle time-dependent input parameters and state variables. This functionality would allow for continuous simulation of water chemistry. Some needed mechanisms include the management of continuous streams of data input and output and the ability for changes in the decision-making process to be incorporated in the model structure. Often these changes are best implemented by their locational attributes (e.g., inclusion of new types of agricultural best-management practices). Incorporation of changing knowledge about parameter estimation and uncertainty would also benefit water-quality modeling. These changes could be efficiently implemented through the use of GIS because these parameters are often geographically referenced. A review of the common analysis tools used in existing spatial models, such as those in hydrology and ecology, would encourage their implementation in commercially available GIS.

Acknowledgments

Support for this project was provided in part by an Intergovernmental Personnel Act Agreement with the U.S. Geological Survey, Water Resources Division (WRD); previous full-time employment by the first author with the School of Natural Resources, Ohio State University (OSU); OSU Office of Research project #8362; and OSU Cooperative Fish and Wildlife Research Unit (for use of computer software and hardware). The authors are thankful for the technical advice and database support given by Curtis V. Price, Richard A. Smith, and Keith W. Robinson, all from the U.S. Geological Survey, WRD.

References

Bowie, G. L., W. B. Mills, D. B. Porcella, C. L. Campbell, J. R. Pagenkopf, G. L. Rupp, K. M. Johnson, P. W. H. Chan, and S. A. Gherini. 1985. *Rates, Constants, and Kinetics Formulations in Surface Water-Quality Modeling.* EPA/600/3-85/040.

Cohen, P., W. M. Alley, and W. G. Wilber. 1988. National water quality assessment: Future directions of the U.S. Geological Survey. *Water Resources Bulletin* 24(6):1,147–51.

Jager, H. I., M. J. Sale, and R. L. Schmoyer. 1990. Cokriging to assess regional stream quality in the southern Blue Ridge province. *Water Resources Research* 26(7):1,401–12.

Jenson, S. K., and J. O. Domingue. 1988. Extracting topographic structure from digital elevation data for geographic information system analysis. *Photogrammetric Engineering and Remote Sensing* 54(11):1,593–600.

Jenson, S. K., T. R. Loveland, and J. Bryant. 1982. Evaluation of AMOEBA: A spectral–spatial classification method. *Journal of Applied Photographic Engineering* 8:159–62.

Johnston, C. A., N. E. Detenbeck, J. P. Bonde, and G. J. Niemi. 1988. Geographic information systems for cumulative impact assessment. *Photogrammetric Engineering and Remote Sensing* 54(11):1,609–15.

Leopold, L. B., and T. Maddock. 1953. *The Hydraulic Geometry of Stream Channels and Some Physiographic Implications.* Geological Survey Professional Paper #252.

New Jersey Department of Agriculture. 1989. *New Jersey Agriculture 1989 Annual Report—Agricultural Statistics.* Circular #524. Trenton: New Jersey Department of Agriculture.

Osborne, L. L., and M. J. Wiley. 1988. Empirical relationships between land use/cover and streamwater quality in an agricultural watershed. *Journal of Environmental Management* 26:9–27.

Rast, W., and G. W. Lee. 1983. Nutrient loading estimates for lakes. *Journal of Environmental Engineering* 109(2):502–17.

Robinson, K. W., C. V. Price, C. Pak, R. A. Smith. 1992. Development of a computerized database of permitted wastewater discharges in New Jersey. *Journal of the Water Pollution Control Federation.*

Smith, R. A., R. B. Alexander, G. D. Tasker, C. W. Price, K. W. Robinson, and D. A. White. 1992. Statistical modeling of water quality in regional watersheds. *Proceedings, Watershed '93.* Alexandria.

Steedman, R. J. 1988. Modification and assessment of an index of biotic integrity to quantify stream quality in southern Ontario. *Canadian Journal of Fisheries and Aquatic Sciences* 45:492–501.

Thomann, R. V., and J. A. Mueller. 1987. *Principles of Surface Water-Quality Modeling and Control.* New York: Harper and Row.

U.S. Environmental Protection Agency. 1991. *Guidelines for the Preparation of the 1992 State Water-Quality Assessments 305(b).* Washington, D.C.: EPA.

U.S. Geological Survey. 1983. *USGS Digital Cartographic Standards, Digital Elevation Models.* U.S. Geological Survey Circular 895-B.

White, D. A., R. A. Smith, C. V. Price, R. B. Alexander, and K. W. Robinson. 1992. A spatial model to aggregate point-source and nonpoint-source water-quality data for large areas. *Computers and Geosciences* 18(8):1,055–73.

Willmott, C. J., S. G. Ackleson, R. E. Davis, J. J. Feddema, K. M. Klink, D. R. Legates, J. O'Donnell, and C. M. Rowe. 1985. Statistics for the evaluation and comparison of models. *Journal of Geophysical Research* 90(C5):8,995–9,005.

Dale A. White has an M.S. and Ph.D. in geography from Pennsylvania State University. Previously, Dr. White was employed by Ohio State University's School of Natural Resources and the U.S. Geological Survey National Mapping Division. Dr. White currently is employed by the Ohio Environmental Protection Agency as an environmental specialist/GIS analyst. His interest lies in spatial modeling of landscape characteristics and their effect on water quality and aquatic biological integrity.
Ohio Environmental Protection Agency
Division of Surface Water
1800 WaterMark Drive
PO Box 163669
Columbus, Ohio 43216-3669
E-mail: *dawhite@cfm.ohio-state.edu*

Matthias Hofschen is presently a graduate student in landscape architecture at Ohio State University. He is completing his Master's thesis on the habitat characteristics and reproductive/migratory behavior of the Oregon spotted owl.
Ohio State University
Department of Landscape Architecture
Columbus, Ohio 43210
E-mail: *hofschen@spowgis.biosci.ohio-state.edu*

72

Time as a Geometric Dimension for Modeling the Evolution of Entities:

A Three-Dimensional Approach

T-S. Yeh and B. de Cambray

Current spatiotemporal databases represent time as a discrete entity; consequently, they lose the intrinsic nature of time. These systems only capture the mutation between two successive states: they represent the history of mutations of entities. To perceive the interactions that occur between living beings and their environment, an environmental database system must represent and process the evolution of its entities. The concept of such a system is not straightforward. Evolution of entities follows continuous laws—information that must not be lost by the model. This requires knowledge of various evolution-law signatures, which are information data associated to entities. We propose the behavioral time sequence (BTS) concept to model these entities. Evaluating spatiotemporal evolution queries is a serious problem due to the high variability of data needed to represent continuous evolution. The proposed model leads us toward unification of spatial and temporal problems by reducing them to geometrical problems and treating time as a geometrical dimension. By transforming a BTS model to a three-dimensional geometrical model, the problem of evaluating spatiotemporal queries is solved through the use of geometric computations.

INTRODUCTION

Our aim is to define a model in a geographical information system (GIS) capable of representing and manipulating the variability of geographical entities. Variability is the domain of the states of an entity; it represents the entity's set of values through time.

GIS users often need answers to questions such as, "Has the cow herd (whose location is known each hour) passed through parcel number three?" Using a common spatiotemporal data model (Langran 1992), which does not capture the notion of *continuous* evolution, such a query will not receive a determinist answer.

Indeed, let us consider the evolution of a cow herd moving through a two-dimensional map (e.g., V1 and V2 plotted in Figure 72-1 are different versions of herd locations at different times). If we only take into account such versions and not all the states between them, we lose the continuity of motion. The only answer that can be given is a positive one, and this is possible only if a version exists that corresponds to the date when the herd passed through the parcel. Therefore, we may not learn when the herd's path intersected the parcel. Of course, it is possible to store the two-dimensional sweeping of motion, but snapshot information is lost.

Hence, a model without the notion of continuous evolution cannot represent entities that are highly variable in form, such as a spreading fire or clouds in motion. These models capture only the changes, called *mutations*, between two successive states (Langran 1992; Snodgrass 1987), representing only the history of entities while losing the essence of continuity. Nevertheless, this problem is less important if the application requires only the modeling of slow phenomena. If this is the case, current spatiotemporal models that represent only discontinuous changes are sufficient.

The representation of evolution in a GIS may lead to highly variable data and, thus, to a large quantity of data. A coarse approach based on the representation of all states is not realistic in computer technology. As a result, existing database technology is inadequate (Egenhofer 1993).

Additional problems concern spatiotemporal interactions between entities. Not only are these spatially localized (interaction between the cow herd and the parcel), they also are localized in time (Dutta 1991). For instance: "When does the herd pass through parcel number three?" (illustrated in Figure 72-1). The system must determine if the shapes of the two entities will intersect during the entities' life spans. Hence, many computations are needed to determine spatiotemporal interactions with a coarse approach because of the large quantity of states due to the continuity of evolution. The BTS model proposed in this chapter overcomes these shortcomings.

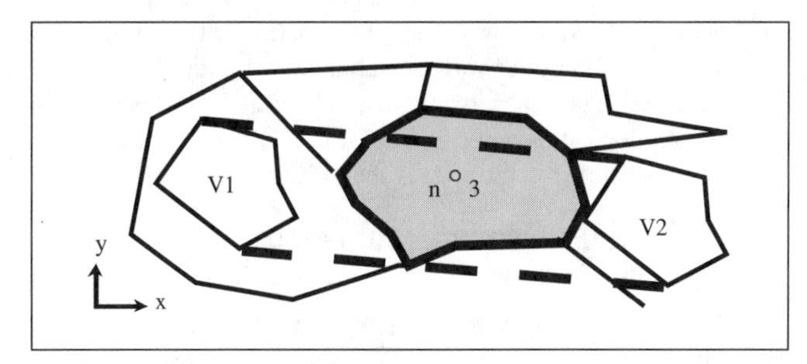

Figure 72-1.

Although little research has been done to solve these problems, and the BTS model is restricted to numerical data (Segev and Shoshani 1987), we present the idea of representing data variability.

Indeed, this model merits particular attention for two reasons: (1) it tackles the problem of variability of numerical data connected to measurements, and (2) it proposes the time sequence (TS) concept. Our purpose is to represent highly variable data as they exist in order to solve numerical quantity measurement problems (i.e., river levels). Although TSs are defined to represent physical phenomena, their use is not restricted to that.

The variability of geographical entities is in the spirit of TS research; however, the data used to describe geographical entities are: (1) geometry, (2) continuity, (3) variability, and (4) the need to consider the entity's variable behavior, which is complex. Entity variable behavior is the way entity modification between two successive states is perceived. For example, temperature behavior may be represented by a spline function. In this way, the BTS representation is not just a simple extension of TS, but it involves an original contribution for resolving these new problems.

In this chapter we present the BTS model for representing the variability of entities. In the second section, we present the variability of geographical entities. We then introduce our BTS model to represent this variability. Consideration of the data types involved led us to a theoretical model of the BTS concept, where time is seen as a geometrical dimension. Thus, these spatiotemporal problems are reduced to strictly geometrical ones.

Typically, environmental systems require more than basic models, which lack the essential notion of variability. Indeed, to perceive interactions that occur between living beings and their environments, such systems must be able to represent and process the evolution of their entities (Cheylan 1993). These interactions may be localized in time and space. The BTS model is particularly adapted to describing these systems.

Entity Variability

In order to produce variability, our approach consists of storing a *description of the data evolution* as it appears to a perception system, as opposed to high-frequency acquisition. This description is therefore generic knowledge (i.e., information), which is depicted with mathematical or algorithmic support, and named *behavioral function*. Then, to produce variability, some data provided by measurements need to be associated to the description to make the behavioral function liable to the entity's evolutionary description. From a machine perspective, the production of variability requires specific mechanisms based on descriptions of behavior and on some measurements of reality. Thus, the pair (*description of behavior* and *measurements*) allow us to produce variability.

All entities do not evolve in an identical way. Basically, in spatial data we have: (1) entities whose shapes vary, (2) entities whose locations vary, and (3) entities with a combination of both. The entities can be real-world objects or beings (e.g., a flood area, the birth rate in a given area [objects may be geometrically represented on a map, such as areas where the birthrate is above 3%]).

Variability in the Geographical World: Specific Behaviors

We think we have a thorough knowledge of variability, but if we consider closely the aspect of geographical variability, it becomes more complicated. The *characteristic behaviors* of an entity are complex because its evolution is subject to factors such as wind, aging, human decisions, etc., which are complex in themselves. To model evolution in a BTS, it is advisable to classify some characteristic features of evolutionary processes.

Behavior expresses the evolution process of some entities. An important example in the real world concerns the displacement of mobile entities. Their characteristic behavior may be classified into several types. The *cloud* (Figure 72-2) is an example of behavior without accidents. A cloud trajectory changes continuously according to the forces applied to it, demonstrating continuous behavior. On the other hand, a vehicle represents noncontinuous behavior—it corresponds to the road's direction or to road conditions (Figure 72-3). The road is mathematically described by a function that, out of necessity, is nonderivable at certain points.

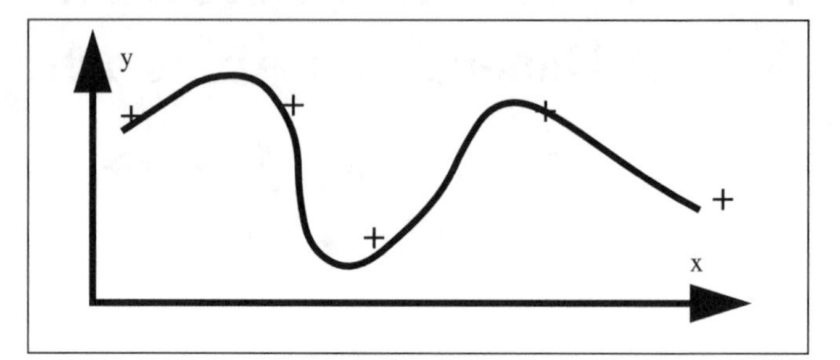

Figure 72-2.

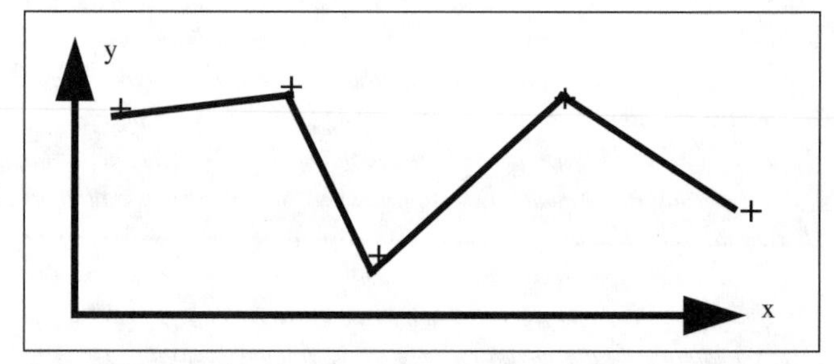

Figure 72-3.

The *regularity* degree of the behavior is a characteristic of the entity's nature. The evolution of sea level is depicted by a curve, whose behavior is a repetition of a pattern with a continuous change during the course of time. The behavior is unique and represents the backward and forward motion of the tide (i.e., a cycle).

On the contrary, the suitability of a road for motor vehicles may be described by several behaviors (Figure 72-4): after a rain it dries and returns to a *normal* state; after a snowfall the ground freezes, and the road then follows another behavior. This type of phenomena is not regular, and so an entity may have a set of different behaviors that are expressed due to irregular variability.

An entity's evolution *speed* is variable, and differences in the evolution speeds of geographical world entities are important. One of the slowest is erosion of a cliff, while one of the quickest is lightning.

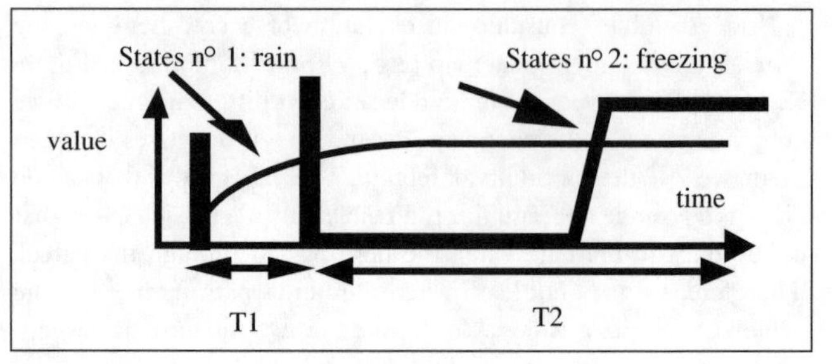

Figure 72-4.

The choice of an observational time scale is important because the same phenomenon can look static at one given time scale and dynamic at a finer time scale. A choice of time scale leads to the selection of a referencing system to observe time. However, within one time scale, observation leads to classifying signatures into two groups: *continuous progressions* (that may be stationary) and *accidents* (that correspond to abrupt changes) (Figure 72-5).

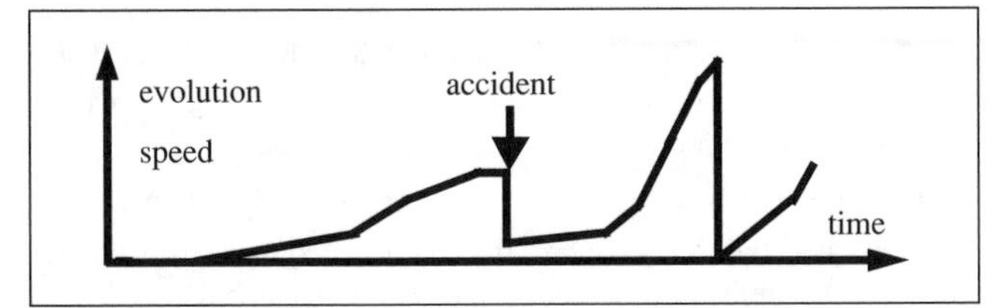

Figure 72-5.

The *continuous* character represents what is uninterrupted during the course of time, since it is a physical characteristic of geographical reality. Variations in the shape of a glacier or the progression of a desert are examples of such evolution. In contrast, an accident is a sudden change and is temporally delimited; that is, an accident has a punctual temporal character and describes an event. This is the case of a forest fire or the division of a plot of land into two parts. Nevertheless, at another time scale, some accidents may appear to be continuous progressions and vice versa. In some cases, the frontier between a continuous evolution and an accidental evolution depends on the time scale. Evolution may be represented by mathematical functions, which may be distinguished by their derivability or by their nonderivability (in a mathematical sense).

Continuity of Identity: The Entity Notion

The evolution manifestation represents the continuity of identity, which is an important notion resulting from both the continuous character of time and the conservation theorem of matter. For example, the cloud displacement represents the cloud sliding, and it is the same entity in different places at different times. In contrast, the *event* evolution indicates a rupture that expresses a deep change in the entity. For example, the event (decision) of dividing a parcel of land into two parts indicates a rupture in the entity evolution (Yeh and Viémont 1992). Each entity is defined on a time interval that represents its life span (Segev and Shoshani 1987). The identity continuity of an entity is assumed during its entire life span.

In information systems, the idea of object identity is essential in order to tie together the successive states of an entity. Object identity is defined during the course of time for the semantic of update. Object identity represents invariance during the course of time (Khoshafian and Copeland 1986) and allows an entity to be uniquely referenced (unicity in time). Thus, relations between different versions constituting an entity are captured.

Such a notion is inadequate for representing some of the evolution processes of real entities, those that cannot be uniquely and unambiguously referred to during the course of time. A flock of sheep splitting into two subherds (diffusion) during the course of time is a typical example. The inverse is true as well. For example, two clouds that collide to then form only one (fusion). Another way to look at it is that when object identity references a unique entity during the course of time, we assume that the creation and the disappearance of the entity are discrete events. However, in the real world, it is not always like that; for example, a dissolving cloud does not permit a date/time to be attached to its disappearance.

THE BTS MODEL DEFINED

1. *DATE* = t and $t \in R$. In the BTS representation, time is considered dense and continuous. Time enables the localization of a state; therefore, a date is a time element whose origin is arbitrarily chosen but must be common for all databases.
2. *SURROGATE* is a value of the objects identifiers domain. A surrogate is an entity identifier in the sense that it is an object identifier for an entity but not for its components. It is a temporal invariant that is unique for a database.
3. *BEHAVIORAL FUNCTION* is an algorithm describing the evolution of a phenomenon. The behaviors of entities are represented by behavioral functions that are mathematical or algorithmic functions. These functions are particularizations of methods used in database applications (Stonebraker 1986). However, a behavioral function follows these binding rules:
   - the signature of the parameter types given to the function is always the same,
   - they are context-free (i.e., all necessary data for processing are communicated by parameter), and
   - the function returns only one value for the date t.
4. *TYPE* is either nonspatial or spatial. A type defines the domain of possible values. Apart from the values domains traditionally provided by databases are nonspatial values (integer, real, string). There are also spatial domains, which enable geometrical objects, such as points, lines, surfaces (and even volumes for three-dimensional space) to be described. In our representation model:
   - a point is represented by its coordinates (x, y) in a two-dimensional map,
   - a line is represented by a set of vertices $\{ ((x, y)\ (x', y')) \}$, and
   - a surface is represented by its boundaries, i.e., by the vertices $((x, y)\ (x', y'))$ bounding it.
5. *UTS* = (s,a,t,c) where $s \in$ *SURROGATE*, a is an alphanumerical or a geometrical value, $t \in$ *DATE*, $c \in$ *BEHAVIORAL FUNCTION*. A unit of time sequence (UTS) represents a measurement or an event reflecting a state of an entity on a given date. A UTS also contains the behavior of the entity between two successive states. It is a 4-uplet defined by a surrogate, a value, a date, and a behavioral function (s,a,t,c). Therefore, a UTS allows us to represent the evolution of an attribute whose value on the given date t is between two successive states (the beginning state corresponding to the one at the date t, and the end state to the date of the next UTS); the evolution is described by the behavioral function c.
6. *BTS* = { *UTS* } = $\{ (s,a,t,c) \}$. A BTS is a set of UTSs. This set represents the variability of an entity during a time interval since it is possible to rebuild the spatiotemporal evolution of the entity from BTS.

Behavioral Function Scope

Each UTS defines its own semantic interpretation of its data through the associated behavioral function. The effect of a behavioral function starts at the date of the UTS that contains it and finishes at the next UTS. This is illustrated in Figure 72-6, which represents a BTS with three UTSs, the behavior $c1$ is defined for the UTS at the date $t1$. Note that $c1$ may differ from the behavior $c2$.

Behavior is a method attached at a UTS level that is a component of the entity. The difference between the domains of object-oriented programming languages and object-oriented databases is to be outlined. In the first case, a method is bound to the object; in the second, the method is bound to the domain/class (Stonebraker 1986).

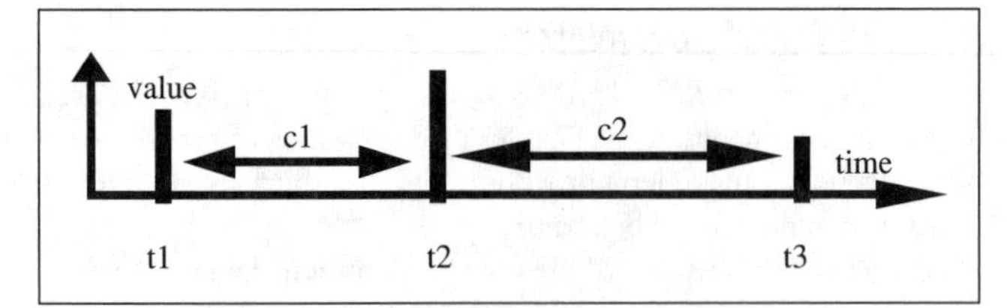

Figure 72-6.

Classes of Behavioral Functions

Behavioral functions can be classified in two groups: law functions and interpolation functions. A law function describes the effects following an event or a state of reality. The variability of values depends only on the UTS representing the event (see Figure 72-7). In this sense, this function reclaims only one UTS as a parameter in order to evaluate the variations which follow. It is, therefore, a prediction function. (Figure 72-7 illustrates a physical phenomenon.)

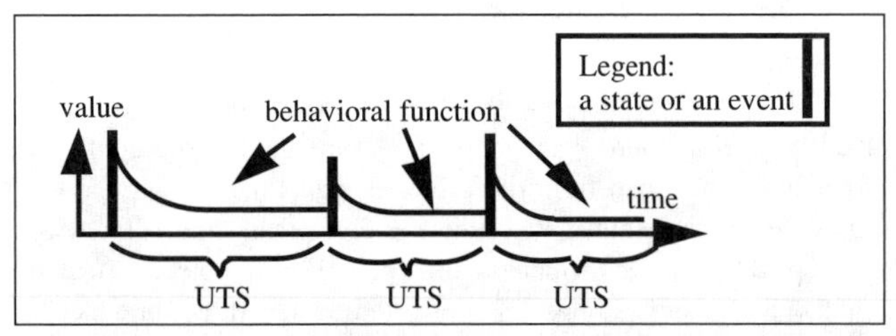

Figure 72-7.

On the other hand, an interpolation function is a mathematical method used to describe the continuous behavior of a changing phenomenon. It is a numerical analysis tool. The goal of the interpolation method is to reproduce, as accurately as possible, the shape that distinguishes particular constraints (the shape representing the original function *F* describing the variability of an entity as a function of time) from a set of measurement points. The interpolation is distinguishable from a law function by the contributions of several stored values (a1,..., a6 in Figure 72-8) of a given entity. (Figure 72-8 shows a linear interpolation.)

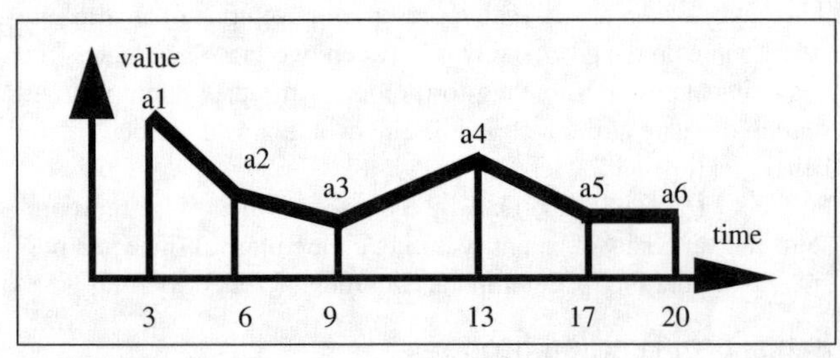

Figure 72-8.

Predefined Behavioral Functions

The model predefines a set of behavioral functions that correspond to the most current interpretations of data. Thus:

- Punctual behavior returns a value if a UTS exists at a given date. It is an event interpretation (i.e., a sale has a punctual semantic).
- Step behavior returns a value *v* of the most recent UTS in relation to a given date. Current temporal database models consider implicitly a *step* semantic, which often corresponds to the validity intervals of data as shown in Figure 72-9.

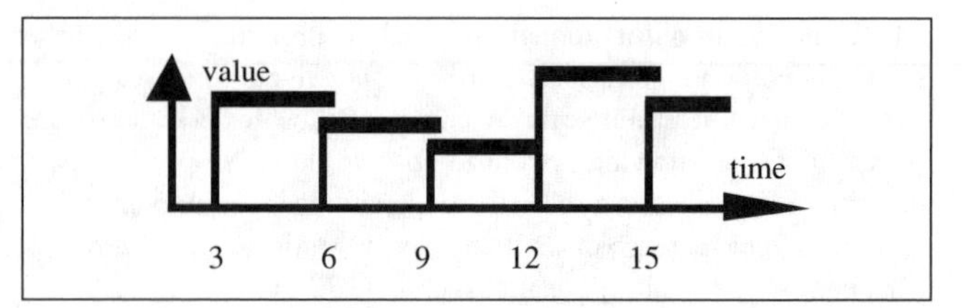

Figure 72-9.

- Linear behavior returns a value obtained by linear interpolation between two successive UTSs (Figure 72-8). For example, vehicle motion may have a linear semantic.
- Spline behavior returns a value describing a spline curve.

Behavior extensibility requires the ability to define behavioral functions in the system. This is done by programming particular methods.

TIME AS A GEOMETRIC DIMENSION

The nature of this highly variable information raises the issue of it begin manipulated by operators integrated in the GIS. The BTS representation associates each value contained in a UTS with the method that defines its semantic. Although a behavioral function is seen as semantic information, it has a dynamic nature, whereas a value is static. It follows that the static and dynamic data are stored in the same way (in a UTS).

Generally, databases define operators that are applied to the data stored in the database (these data are static values); in other words: *operator(data)*. Conversely, operators on BTS are applied on static and dynamic data: *operator(data » behavioral functions)*. The design of such operators is, therefore, more complex.

To solve this problem, time is considered a geometric dimension that possesses properties of localization on the time axis. This assertion is based on the fact that spatial entities are described by their localization. Indeed, dating locates a version in time. This point of view leads us toward a unification for spatial and temporal problems, reducing them to geometric problems. This is done by treating time as a geometrical entity that is more complex than a simple line. Therefore, temporal operators are reduced to geometrical computations.

State Computation

States are values derived from both UTS and behavioral functions. The succession of different states (different values at different dates) expresses data variability. The model indiscriminately represents both events and measurements (low-frequency sampling). Semantic data are contained by the associated behavioral functions; therefore, three information classes coexist: states, events, and measurements (Figure 72-10). UTSs point out occurrences that physically exist in a GIS. Each UTS represents a measurement or a knowledge of the reality at a given date—a description of the state of an entity on a particular date. Considering mathematical aspects, continuity is necessary to produce all intervening states. The problem is then equivalent to applying an interpolation whose support values are complex geometries rather than numerical values.

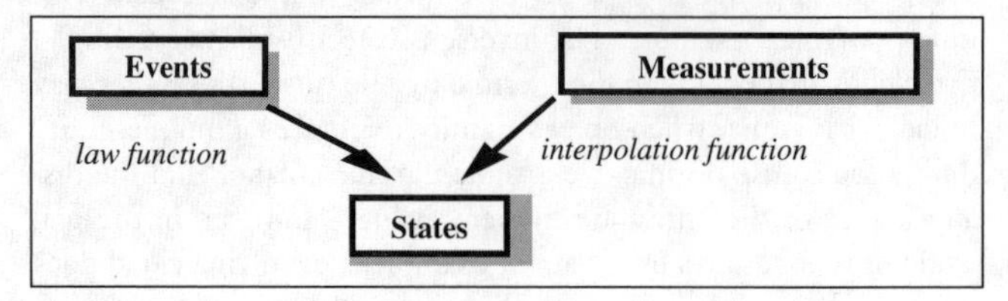

Figure 72-10.

Temporal Sweeping Operator

To represent the evolution of an entity in a continuous way is equivalent to retaining all of its shapes and localizations through time. It is then tantamount to considering a continuous displacement through time: it is a sweeping of the entity representation during the course of time (Figure 72-13) (the perception of time is continuous, and sweeping is a continuous operator), and not a simple union of successive states (Figure 72-12). This temporal sweeping is equivalent to choosing time as the third axis.

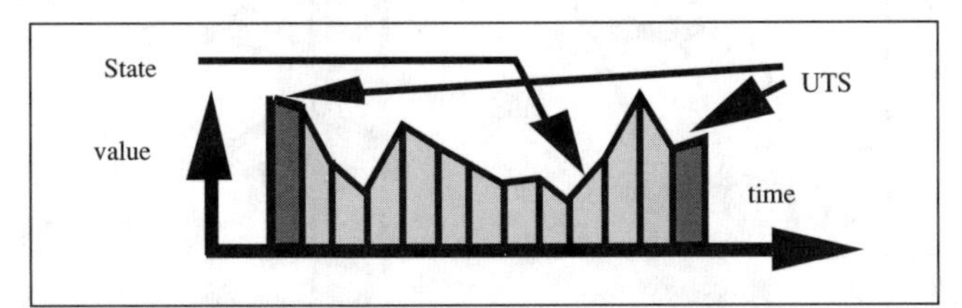

Figure 72-11.

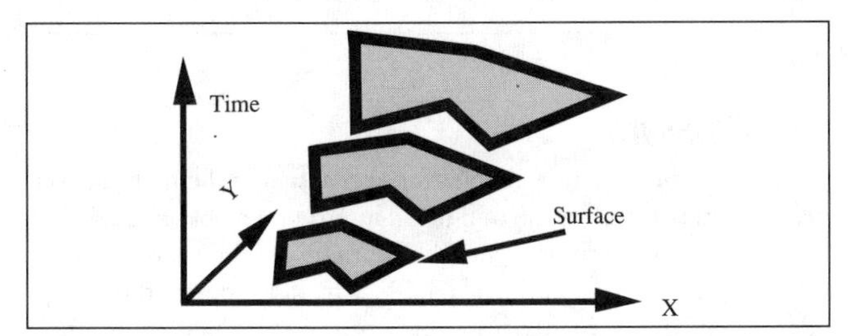

Figure 72-12.

A sweeping representation consists of shifting a surface (or a section) along a spatiotemporal trajectory—the represented solid is thus swept (Figure 72-15) (Foley 1990). The trajectory (*T*) of temporal sweeping corresponds to the entity's center of gravity during the course of time, and the section (*S*) that gets out of shape is the entity; a given section (i.e., for z = t1) is, therefore, the entity's state at t = t1.

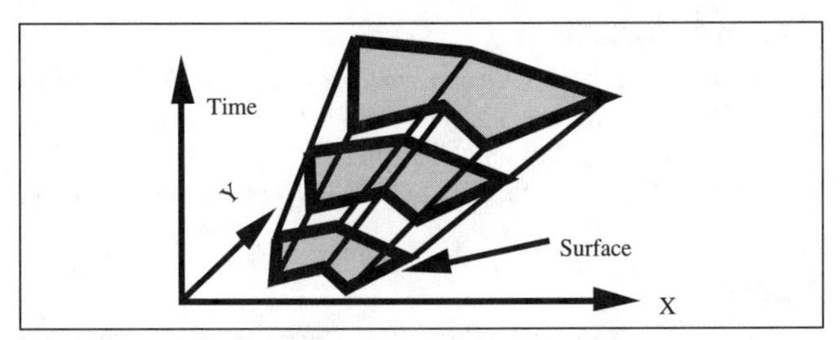

Figure 72-13.

The sweeping shape is depicted by the *behavioral function* associated with the entity and by different intermediate geometrical values that describe the entity's states. The main idea is to get a continuous transformation allowing the first geometrical shape to change into the next shape according to a given *behavioral function.*

Temporally sweeping an entity results in the entity representation increasing by one dimension. Hence, the sweeping of a line gives a three-dimensional surface (Figure 72-14), while sweeping a planar surface gives a three-dimensional volume (Figure 72-13). The temporal sweeping operator transforms the *BTS representation* into a *geometric representation.*

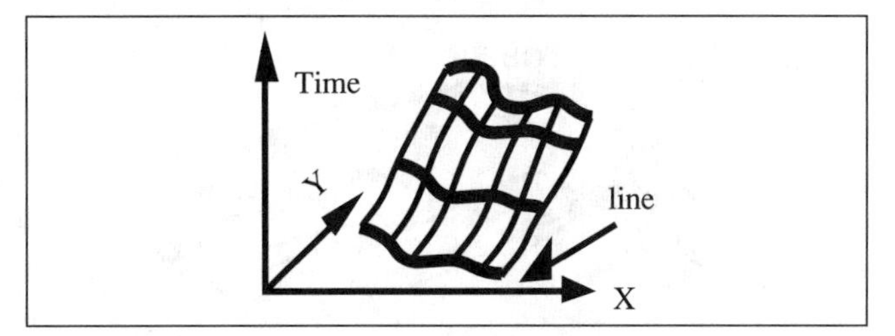

Figure 72-14.

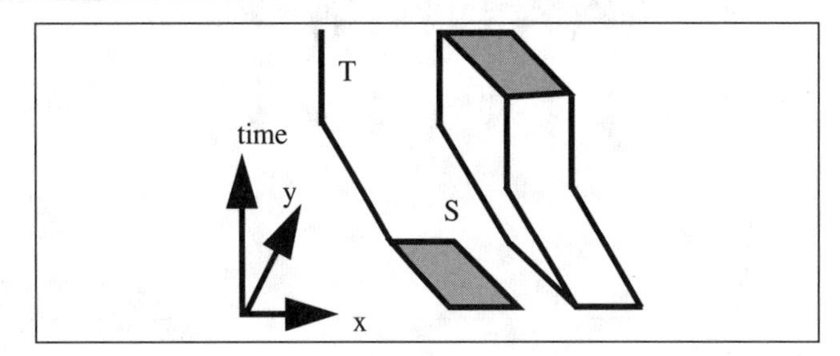

Figure 72-15.

Three-Dimensional Geometric Representation

An entity evolving in time is represented with a four-dimensional shape that is described through a three-dimensional representation. A nonmanifold boundary representation (BR) (Foley 1990) represents a solid in terms of bounding surfaces. They are represented as a union of faces, each of which is defined by its boundaries; for example, the vertices and edges bounding it. (We chose a nonmanifold BR so we could uniformly consider three-dimensional, two-dimensional, or one-dimensional objects, or three-dimensional, two-dimensional, or one-dimensional hybrid objects. Such objects may have dangling faces or dangling edges.) Lists of vertices, segments, and faces are stored. A face is a collection of segments; a segment is described by two vertices (i.e., a segment contains two pointers toward a vertex). Figure 72-17 illustrates the BR of the object of Figure 72-16.

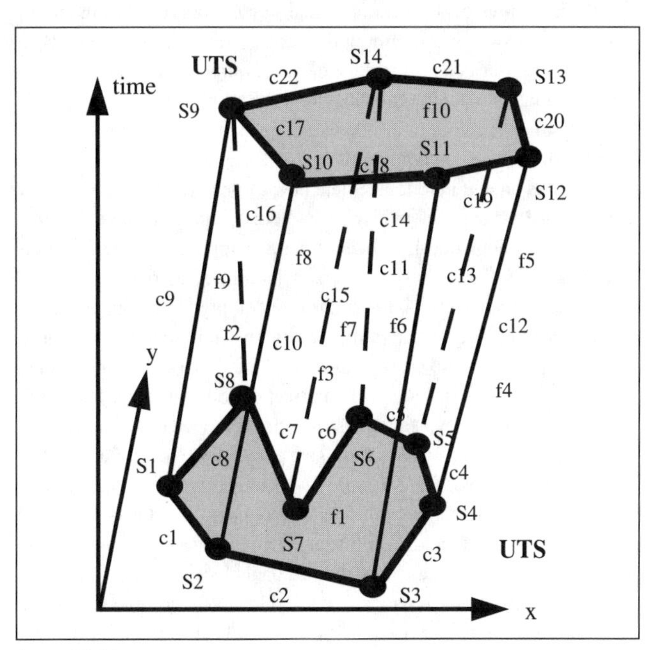

Figure 72-16.

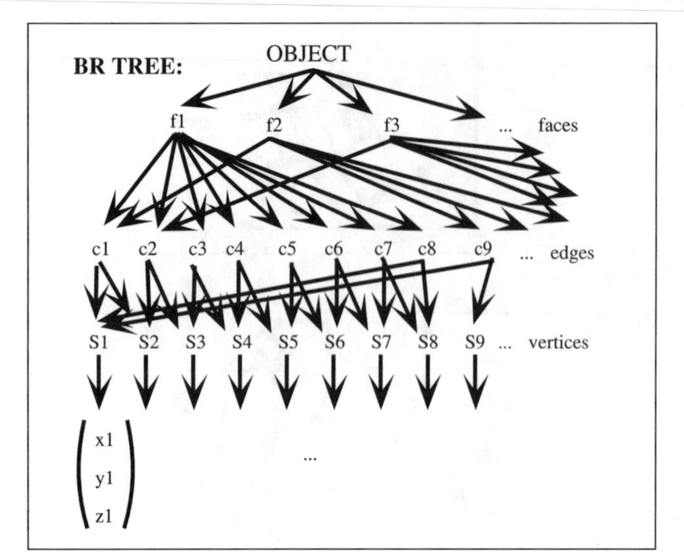

Figure 72-17.

Because these three-dimensional objects correspond to the representation of the variability of entities and, therefore, to the temporal sweeping of entities, they have certain characteristics. First of all, they are built by relating two successive UTSs, which are planar faces (perpendicular to the time axis which is the z-axis, since a UTS is a representation of a two-dimensional entity at a given date). These (shaded patches in Figure 72-16) are joined by segments or by lines that can be described with functions, by tying the vertices of the two sections two-by-two. This is possible because two successive sections have the same number of vertices. If this is not the case, an operator creates pseudovertices to obtain equality between the number of vertices. They then can be described with points, lines, and faces. Lines are approximated by a collection of segments.

DATA MANIPULATION

The manipulation of high-variability data uses geometrical computational techniques on four-dimensional objects previously presented in order to answer spatiotemporal queries. Hence, some answers to requests are traduced in terms of three-dimensional geometrical computations (Preparata and Shamos 1985). Our model defines different spatiotemporal operators, like *cross section*, which provides the states of all entities on a given date.

We focus on one here to illustrate how these operators are used. For example, we want to know if the cow herd has passed through parcel number three. Semantically, it is equivalent to applying the *overlap* operator. In most GIS, the spatial operator is defined as two geometrical shapes that overlap if their intersection is not empty. Hence, an extension of this definition to the spatiotemporal dimension is: two surface shapes that overlap in time and space if their conversion to a three-dimensional geometrical shape has a non-null intersection. So, our query about the cow herd is resolved by using the spatiotemporal overlap operator.

Indeed, a cow herd passes through a parcel if, and only if, it is at the same place on a given date. Figure 72-18 shows two, four-dimensional objects that represent the evolution of a cow herd (light gray area) and of a parcel (hatched area). As illustrated, there is no solution since the intersection between these four-dimensional objects is empty.

Geometrical process is expensive. However, computations are accelerated through the use of different levels of minimal bounding rectangular parallelepipeds of the three-dimensional object (Cambray 1993), allowing computation with only part of the BR.

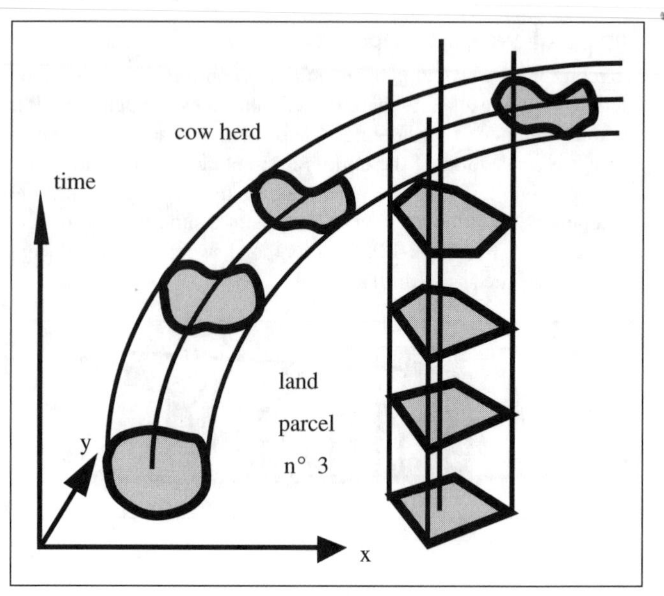

Figure 72-18.

CONCLUSION

The BTS representation enables representation of the high variability of entities as well as the traditional interpretation of temporal data (*step*). This model captures both the continuity of time and the inherent complexity of the entities. The BTS model supports the manipulation of these highly variable entities through geometrical algorithmic representations. The spatial operators for data access are semantically extended to space–time.

GéoSabrina, an existing database system based on an extended relational model and supporting built-in spatial data types, operations, and indices, currently is extended to support the variability. GéoSabrina, developed at the MASI Laboratory, uniformly manages both semantical data and cartographical data (location and geometrical shape) (Larue et al. 1993). New operations are defined to manipulate the entities as they evolve through time.

Acknowledgments

We are deeply grateful to Georges Gardarin and Yann Viémont. We gratefully acknowledge the members of the Cassini research group, especially Jean-Paul Cheylan. We wish to thank Gilles Marechal and John Evens for their useful remarks after reading this work.

References

Cheylan, J. P., and S. Lardon. 1993. Spatiotemporal processes: Toward a conceptual data model and analysis using GIS. *COSIT '93* (September).

de Cambray, B. 1993. Three-dimensional modeling in a geographical database. *Proceedings of Auto-Carto 11.*

Dutta, S. 1991. Topological constraints: A representational framework for approximate spatial and temporal reasoning. *SSD'91*, 161–80.

Egenhofer, M. J. 1993. What's special about spatial? Database requirements for vehicle navigation in geographic space. *Proceedings of the 1993 ACM SIGMOD International Conference on Management of Data*, 398–402.

Foley, J. D., A. van Dam, S. Feiner, and J. Hughes. 1990. *Computer Graphics, Principles and Practice.* Systems Programming Series. New-York: Addison-Wesley.

Gupta, R., and G. Hall. 1992. An abstraction mechanism for modeling generation. *Proceedings of the Eighth International Conference on Data Engineering*, 650–58.

Khoshafian, S. N., and G. P. Copeland. 1986. Object identity. *OOPSLA'86 Proceedings*, 406–16.

Langran, G. 1992. *Time in Geographic Information Systems*. Taylor & Francis.

Larue, T., D. Pastre, and Y. H. Viémont. 1993. Strong integration of spatial domains and operators in a relational database system. *SSD'93*, 53–72.

Preparata, F. P., and M. I. Shamos. 1985. *Computational Geometry, An Introduction*. New York: Springer-Verlag.

Segev, A., and A. Shoshani. 1987. Logical modeling of temporal data. *Proceedings of ACM SIGMOD Conference 1987*, 454–66.

Snodgrass, R. 1987. The temporal query language TQuel. *ACM Transactions on Database Systems* 12(2):247–98.

Stonebraker, M. 1986. Inclusion of new types in relational database system. *Proceedings of International Conference on Data Engineering, Los Angeles, California, February 5–7, 1986*, 262–69.

Yeh, T. S., and Y. H. Viémont. 1992. Temporal aspects of geographical databases. *Proceedings of EGIS'92*, 320–28.

T-S. Yeh and *B. de Cambray* work at the PRiSM (ex-MASI) Laboratory, respectively, on spatiotemporal geographical databases and three-dimensional geographical databases. Their research interests include spatiotemporal models, three-dimensional models, databases, GIS, and CAD. They belong to a research group (Cassini) of the French national research council (CNRS) on GIS and spatial analysis, whose aim is to design the next generation of GIS. The prototype made allows the study of spatial and temporal interactions between flocks of sheep (mobile entities) and their environment. This is an environmental problem and is one of Cassini's research projects.

Laboratoire PRiSM
(CNRS, Université de Versailles-St Quentin)
and
GDR Cassini
45 avenue des Etats-Unis, F-78000
Versailles, France
Phone: 33 1 39 25 40 48
Fax: 33 1 39 25 40 57
E-mail : *(Tsin-Shu.Yeh, Beatrix.De-cambray)@prism.uvsq.fr*

73

Optimal Classification Techniques for Feedforward Parallel Distributed Processors

Perry J. LaPotin and H. L. McKim

In this chapter optimal decision (recognition) tests are presented for use in digital image processing, pattern recognition, and feature identification. The tests are derived for machine architectures that belong to the important class of feedforward neural networks (NN) and feedforward parallel distributed processors (PDP). These architectures are the focus of our analysis since their organization clearly separates sequential processes from parallel processes, and optimal tests for feedforward architectures can be applied to the larger class of recurrent designs. Within the formal decision (recognition) tests, pattern recognition occurs when the routed signal is both consistent and of sufficient strength to exceed critical levels. The tests are derived using a statistical model that yields a precise threshold value for the k=2 *class pattern recognition problem commonly employed for pattern identification. For both single-layer feedforward architectures and multiple-layer feedforward architectures, the derived tests are shown to be optimal in that they approximate the uniformly most powerful (UMP) size* **a** *test for discriminating two pattern classes. Since these tests can be applied within a recursive pattern recognition algorithm, the results can be extended to the* k>2 *class problem using binary trees. Unlike the traditional Minsky and Papert threshold tests, the UMP tests presented in this study yield a systematic optimal threshold value that can be applied to architectures of variable width and depth.*

INTRODUCTION

Neural Networks (NN) and Parallel Distributed Processors (PDP) are the focus of extensive study in the field of pattern recognition and image processing (IP). In the pattern recognition community, the general class of PDP designs has been used to perform difficult mapping and transfer operations with high efficiency and robust behavior (White 1989). In traditional image processing, PDP architectures are employed to perform classification and discriminant analysis on remotely sensed data and have been applied to perform complex filter operations on multispectral data sets (Widrow and Winter 1988). Full-connection PDP architectures are capable of efficient character recognition and pattern classification using recursive feedback techniques, known as back propagation, to update and weight specific elements within the design (Sejnowski and Rosenberg 1986). In addition, the general class of PDP architectures outlined in McClelland and Rumelhart (1989) and the specific subclass of architectures applied to neurobiological systems outlined in Grossberg (1980) have been applied to the problem of video-based pattern recognition using multiple-layer feedforward architectures and hybrid recurrent designs (Sivilotti et al. 1987). The applicability of PDP architectures to multispectral image processing is discussed in McKeown (1987), and specific algorithms for applying PDP techniques to both raster and vector data layers are presented in Kanal (1971) and Anderson and Rosenfeld (1989).

The applicability of using NN and PDP for pattern classification was demonstrated by the early Perceptron architectures of Rosenblatt (1958a, 1958b) and Minsky and Papert (1972, 1988). The first formal logic for using parallel processing in time-varying (state–space) models was outlined by McCulloch and Pitts (1943) and later applied to the early NN and piecewise linear machines by Nilsson (1965). There are three main criticisms that surround the use of NN and PDP for solving mainstream pattern recognition problems (Minsky and Papert 1988): Both NN and PDP methods implicitly assume that (1) a training sequence is available for sensitizing the weights used to modulate each processor in the architecture, (2) a black box transfer function is acceptable for performing the pattern recognition process, and (3) an appropriate architecture is available and known *a priori* for performing the necessary transfer operation.

In both NN and PDP architectures, the training sequence consists of a series of test patterns that are presented to the design prior to the formal pattern recognition experiment. Within the training sequence, the weights assigned to each processor are modified to reward operations that promote the correct classification of the test pattern, and to penalize operations that lead to a false characterization of the test pattern by the architecture and its associated decision (recognition) test. The judicious selection of an appropriate training sequence is vital for the adequate classification of the test pattern (Gibbons and Rytter 1990). When no training sequence is available, as in real-time data acquisition, adaptive methods may be used—provided an appropriate optimization technique is employed in the nonsupervised classification (Kanal and Lemmer 1986). Examples of this method are the simulated annealing approach employed by Hinton (1985), and the partially adaptive architectures described by Fukushima (1988) and Banquet and Grossberg (1986).

In both traditional and adaptive architectures, a weight matrix is generated that is used to perform a specific black box mapping operation. The weight matrix generally contains as many cells (rows × columns) as there are processors in the architecture. Following the pattern recognition experiment (with or without a formal training sequence), the architecture performs a back-propagation sequence that adjusts the magnitude of the cells in the weight matrix in an attempt to optimize the pattern separation and classification. As in any formal simulation model, initial and boundary conditions must be prescribed prior to the pattern recognition experiment.

A final critical element for applying parallel processing techniques to pattern recognition problems centers on the appropriate organization of the processors within the architecture. Under general regularity conditions (Baum and Haussler 1989), recurrent designs systematically outperform the multiple-layer feedforward full-connection architectures (MFF) described by Rumelhart and McClelland (1987). Unfortunately, few algorithms are available for optimizing recurrent designs using single-band or multispectral data sets (Anderson and Rosenfield 1989; Baase 1988). Conversely, specific algorithms may be derived for MFF designs and then applied to the less restricted set of feedforward and recurrent architectures (Hornik et al. 1989). In addition, the general class of MFF architectures can be generalized to nonfeedforward designs, such as recurrent networks: "We have thus far restricted ourselves to feedforward nets. This may seem like a substantial restriction, but as Minsky and Papert point out, there is, for every recurrent network, a feedforward network with identical behavior (over a finite period of time)" (Rumelhart and McClelland 1987).

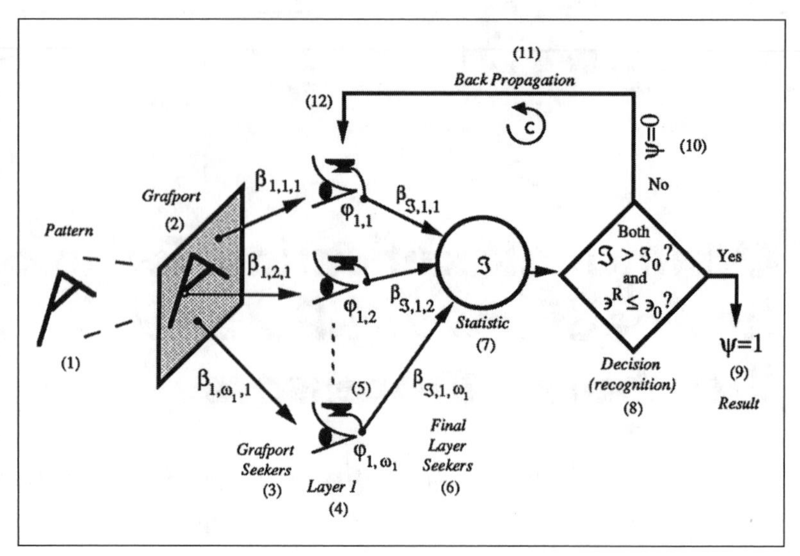

Figure 73-1. Schematic single-layer feedforward PDP. The architecture feeds information forward from the test pattern in step 1 to the final result in step 9. When a test pattern is recognized in step 9, a Boolean result $\psi=1$ is assigned. Otherwise, a Boolean result $\psi=0$ is assigned in step 10, and weights are adjusted within a formal back-propagation sequence shown in step 11. In step 12, each weight in the single layer is modified using the formal weight adjustment algorithm applied in step 11.

Using MFF architectures as the basis for this discussion, we will derive a statistical model and a formal decision (recognition) test for the back propagation sequence that is common to feedforward pattern recognition architectures. While this derivation uses designs with full connection (each processor in layer i+1 is routed a signal from each processor in layer i), the derivation is general and may be applied to nonfull connection architectures and to architectures that apply nontraditional routing techniques. Since MFF architectures provide the foundation for the majority of NN and PDP designs that have been applied to multispectral image classification (Batcher 1980; Barto et al. 1983; Forsyth and Rada 1986), the applicability of these tests to performing optimal pattern recognition can be demonstrated using both raster and vector imagery.

PARALLEL DISTRIBUTED PROCESSOR (PDP)

Pattern recognition is an event controlled by the sampling accuracy of the architecture and the ability of the design to precisely process, weight, and route the information contained within the sampled test pattern. The precision of the sample is generally measured using the raster recognition error (Baillard and Brown 1982), and no architecture with a large sampling error may be considered for pattern recognition. In evaluating the sampling accuracy of the architecture, the machine design uses unprocessed information that is available within the first-layer seekers. The seekers are connective ingredients within an architecture that are used to route the signal between operations. In feedforward architectures, the signal is routed from the image display (grafport) to the first-layer processors. Within the first-layer processors, the signal is evaluated and then weighted before it is then routed to the next subsequent operation (or layer) in the architecture. As an example, consider the single-layer feedforward PDP architecture shown in Figure 73-1. The architecture uses raw pattern data (symbolized by the letter *A* in step 1) to portray the information within the boundaries of a finite two-dimensional image plane referred to as the grafport (step 2). The resolution of the grafport varies with the requirements of the pattern recognition experiment and is sensitive to the granularity of the pattern data for images that are nonperpendicular (nonparallel) to the coordinate axes of the grafport (Hall 1988). The pattern data are sampled by the architecture using the grafport seekers (arrows shown in step 3) that route information from the grafport to the first-layer processors or *atoms* (step 4). The single-layer design contains ω_1 atoms, labeled $\{\varphi_{1,1}, \varphi_{1,2},.., \varphi_{1,\omega_1}\}$, where the term *atom* is used to indicate that the respective processor contains no subprocesses and is itself the simplest processing element within the architecture. In this notation, each atom is denoted using an *eye* symbol to reinforce the fact that (visual) image data are being processed for use within a parallel-pattern recognition architecture. Within Figure 73-1, a single layer of parallel processors is presented between the grafport (step 2) and the statistic (step 7). To simplify the illustration, each atom in layer one is routed a single seeker from the grafport $\{\beta_{1,1,1}, \beta_{1,2,1},.., \beta_{1,\omega_1,1}\}$ where:

$\beta_{i,j,k}$ = the information content within the *k*th seeker pointing to atom j within layer i. (73-1)

For the special case of the final atom layer, exactly one seeker from each final layer atom points to the test statistic $\Im$. These seekers are labeled as:

$\beta_{\Im,1,k}$ = the information content within the *k*th seeker of the final layer of atoms pointing to the test statistic $\Im$ (73-2)

where

$\beta_{\Im,1,k} = \beta_{L+1,1,k}$ is the *k*th seeker pointing to the test statistic (in layer L+1).

The simplification of one seeker per atom is made to reduce the number of connections shown in Figure 73-1. In the actual operation of a single-layer and multiple-layer PDP, many seekers may route information from the grafport test pattern to each first-layer processor. For this reason, the statistical model (described later) does not restrict the architecture to only one seeker per atom in any layer of the architecture.

In step 4 of Figure 73-1, each of the ω_1 atoms yields a Boolean result $\{\varphi_{1,1}, \varphi_{1,2},.., \varphi_{1,\omega_1}\}$ that is based upon the magnitude of the seekers that are routing information from the grafport test pattern. The result of each atom $\varphi_{i,j}$ is combined with a weight $\alpha_{i,j}$ in step 5, to yield a weighted Boolean response. The *anvil* symbol is used to designate the weight, and a single unique weight $\{\alpha_{1,1}, \alpha_{1,2},.., \alpha_{1,\omega_1}\}$ is assigned to each atom in the architecture. The weighted Boolean response is bounded [-1,+1] and is used to measure the response of the architecture to the sampled pattern data. The weighted response $\{\beta_{\Im,1,1}, \beta_{\Im,1,2},.., \beta_{\Im,1,k}, .., \beta_{\Im,1,\omega_1}\}$ is routed to the test statistic in step 6, and the information is formally reduced to yield descriptive statistical measures in step 7. The statistical measures in step 7 are then routed to the formal decision (recognition) test in step 8, and the decision (recognition) test is used to perform the pattern classification.

In Figure 73-1, two results are possible within the decision (recognition) test (i.e. the k=2 class pattern recognition problem). If the magnitude of the statistic $\Im$ exceeds a critical threshold $\Im_0$, and if the spatial accuracy of the sample $\ni^R$ is below critical levels $\ni_0$ as measured by the raster recognition error, then the test pattern is recognized by the architecture. When recognition occurs, a result ψ=1 is assigned in step 9 and the architecture stops the feedforward processing. Conversely, if the decision (recognition) test fails, a result ψ=0 is assigned in step 10, and the architecture performs weight adjustment within a formal back propagation sequence. The back propagation sequence shown in step 11 is used to either reward (penalize) those atoms that contributed to the correct (incorrect) classification of the test pattern. As shown below step 11, an index c is assigned to measure the number of back propagation cycles required for pattern recognition. In step 12, the architecture begins a new cycle by adjusting the respective weights of each architecture in the single-layer design. The weight adjustment is governed by the magnitude of signal received by each processor in the previous sequence c-1, and is also governed by the current magnitude of the individual processors in the current cycle c.

To gauge the significance of the processed information, the output seekers $\beta_{i+1,j,k}$ ($\beta_{i+1,j,k} = \beta_{\Im,j,k}\ \forall$ j,k ; i=L) are used as the basis for a formal decision (recognition) test. The decision (recognition) test examines the magnitude of the seekers that route information from each layer in the architecture to determine the signal strength of the processed information. The test also examines the distribution of the signal routed between each layer as a method to gauge the overall consistency of the processed information. In the formal decision (recognition) test, a pattern recognition event occurs when there is strong evidence to suggest that the routed signal is both consistent and of sufficient strength to exceed critical levels.

STATISTICAL MODEL

In this section, we discuss a decision (recognition) test, derived for use within the general class of MFF architectures, that approximates the Uniformly Most Powerful Size **a** Test—denoted as the UMP test. The derivation of the UMP test is first presented for use in single-layer architectures (as shown in Figure 73-1) and then is extended to multiple-layer designs.

For single-layer architecture, the test statistic is based upon the ω_L seekers that route information from the final layer L to the test statistic $\Im$. In the multiple-layer architecture, the seeker responses produced by each layer are included in the decision (recognition) test to determine the overall signal strength for the design.

For both single- and multiple-layer designs, the symbols **a** and **b** are used to denote the respective Type I and Type II error probabilities, and the power function π notation for the decision (recognition) test is introduced as follows:

$$\pi_0 = \text{Prob}\{\text{reject } H_0 \mid H_0 \text{ is true}\} = \mathbf{a} \text{ and} \quad (73\text{-}3)$$
$$1-\pi_1 = \text{Prob}\{\text{accept } H_0 \mid H_0 \text{ is false}\} = \mathbf{b}$$

The UMP size **a** test is uniformly most powerful since its power function π_1 = Prob{reject H_0 | H_0 is false} is the maximum for all competing tests with size **a** or less. Here "uniformly" is used to denote *all* alternative tests of power **a** or less (Krishnaiah 1969). As a result, this test is desirable within both NN and PDP designs, since among all size **a** tests, it has the greatest likelihood for rejecting the null hypothesis H_0 when it should. These tests are sensitive to the width and depth of an architecture during adaptation and are based upon familiar likelihood ratio principles discussed in Lehmann (1959). The motivation for applying UMP tests to the general class of MFF designs is presented following the definition of the decision (recognition) test for single-layer and multiple-layer architectures.

Single-Layer Architecture

A single-layer architecture contains ω_1 atoms as shown in Figure 73-1. In the single-layer architecture, the magnitude and the distribution of the processed signal can be measured using the individual seekers $\{\beta_{\Im,1,1}, \beta_{\Im,1,2},.., \beta_{\Im,1,k}, .., \beta_{\Im,1,\omega_1}\}$ that route information from the first-layer atoms to the test statistic $\Im$. The magnitude of each seeker $\beta_{\Im,1,k}$ is sensitive to changes in both the full connection term $\alpha_{1,k}\varphi_{1,k}$ and the propagation term $\beta_{1,k,j}$, it also is sensitive to changes in the weight assignment of each atom in the first layer. The decision (recognition) test assumes that the ω_1 seekers define a random sample from an unknown distribution function $f(\mu_\beta,\sigma^2_\beta)$, whose mean population response is μ_b, and whose population variance is σ^2_β. Hence,

$$\beta_{\Im,1,1}, \beta_{\Im,1,2},.., \beta_{\Im,1,k}, .., \beta_{\Im,1,\omega_1} \sim f(\mu_\beta,\sigma^2_\beta) \quad (73\text{-}4)$$

where

$-1 \le \mu_\beta \le +1$, and $\sigma^2_\beta > 0$.

Using the individual seekers $\{\beta_{\Im,1,1}, \beta_{\Im,1,2},.., \beta_{\Im,1,k},.., \beta_{\Im,1,\omega_1}\}$ to compute a mean response for the architecture, a single-sided hypothesis test may be examined of the form:

$$H_0: \mu_\beta \le \mu_0 \text{ versus } H_1: \mu_\beta > \mu_0 \quad ; L=1 \quad (73\text{-}5)$$

where

μ_0 is a critical value for the mean seeker responses in the population, and

μ_β is the mean seeker response for the population as shown in (73-4).

In the hypothesis of (73-5), the test pattern is considered for recognition if the mean response μ_β exceeds the critical level μ_0 and is (otherwise) not recognized when $\mu_\beta \le \mu_0$. For the conditions of the single-sided hypothesis, consider the following test statistic $\Im$ as the basis for the decision (recognition) test:

$$\Im = \Im(\beta_{\Im,1,1}, \beta_{\Im,1,2},.., \beta_{\Im,1,\omega_1}) = \sum_{k=1}^{\omega_1} \beta_{\Im,1,k} \quad ; L=1. \quad (73\text{-}6)$$

The test statistic in (73-6) is equal to the summation of the output seekers produced by the single-layer architecture. From Neyman–Pearson Theory (Bickel and Doksum 1977), a constant $\Im_0$ may be selected to meet Type I error conditions of Prob{$\Im > \Im_0$ | H_0}=**a**. Hence, the test:

$$\text{Reject } H_0 \text{ if } \sum_{k=1}^{\omega_1} \beta_{\Im,1,k} > \Im_0 \quad ; L=1 \quad (73\text{-}7)$$

is an approximate UMP size **a** test for the hypothesis H_0: $\mu_\beta \le \mu_0$ versus H_1: $\mu_\beta > \mu_0$ provided that σ_β^2 is known.

The test $\sum_{k=1}^{\omega_1} \beta_{\Im,1,k} > \Im_0$ for the hypothesis H_0: $\mu_\beta \le \mu_0$ versus H_1: $\mu_\beta > \mu_0$ is of a similar form to the threshold summation test used by Minsky and Papert for the Mark I Perceptron. However, using Neymand–Pearson techniques, a value for $\Im_0$ may be selected that is not arbitrary (as in the Mark I design), but is specifically prescribed to match the size **a** test. For notational purposes, let

$$\phi(\beta) = \frac{1}{\sqrt{2\pi}} \exp\{-\tfrac{1}{2}\beta^2\} \quad \text{and} \quad \Phi(\beta) = \int_{-\infty}^{\beta} \phi(\upsilon)\, d\upsilon \tag{73-8}$$

where

$\phi(\beta)$ is the standard normal distribution, N(0,1) for the random variable β, and

$\Phi(\beta)$ is the cumulative normal distribution for the random variable β.

Using this notation with (73-7), the Type I error is such that:

$$\mathbf{a} = \text{Prob}\{ \sum_{k=1}^{\omega_1} \beta_{\Im,1,k} > \Im_0 \mid H_0\} \approx 1 - \Phi\left(\frac{\Im_0 - \omega_1\mu_0}{\sqrt{\omega_1}\,\sigma_\beta}\right) \tag{73-9}$$

for ω_1 atoms in the single-layer architecture.

In (73-9), the probability is exact for the specific case when the output seekers $\beta_{\Im,1,k}, \beta_{\Im,2,k}, \ldots, \beta_{\Im,1,\omega_1}$ are randomly drawn from an underlying normal population with mean μ_0 and unknown variance σ_β^2 and the error in the approximation declines as the width of the single-layer PDP increases. Writing (73-9) in terms of the corresponding percentile for the standard normal distribution:

$$\frac{\Im_0 - \omega_1\mu_0}{\sqrt{\omega_1}\,\sigma_\beta} \approx Z_{1-\mathbf{a}} \tag{73-10}$$

where

$Z_{1-\mathbf{a}}$ is the (1-**a**)th percentile of the standard normal distribution.

Hence, the test for the hypothesis: H_0: $\mu_\beta \le \mu_0$ versus H_1: $\mu_\beta > \mu_0$ becomes the following:

$$\text{Reject } H_0 \text{ if } \sum_{k=1}^{\omega_1} \beta_{\Im,1,k} > \omega_1\mu_0 + \sqrt{\omega_1}\,\sigma_\beta\, Z_{1-\mathbf{a}}$$

or alternatively:

$$\text{Reject } H_0 \text{ if } \beta > \mu_0 + \frac{\sigma_\beta}{\sqrt{\omega_1}} Z_{1-\mathbf{a}} \tag{73-11}$$

where

$$\beta = \frac{1}{\omega_1} \sum_{k=1}^{\omega_1} \beta_{\Im,1,k} \quad ; L=1 \tag{73-12}$$

is the mean seeker response for the single-layer architecture with ω_1 atoms in the first layer.

As presented in (73-11), the size **a** test has intuitive appeal. The null hypothesis is rejected only if the mean seeker response in the single layer of atoms exceeds the test population mean μ_0 by an amount that is scaled relative to the population variance σ_β^2, and the (1-**a**)th percentile of the standard normal distribution. However, the test also assumes that the population variance for the single layer of seekers σ_β^2 is known. To avoid this problem, a suitable statistic for estimating σ_β^2 must be applied within the size **a** test. The first choice is to select the sample variance for the seekers, s_β^2, as an unbiased estimate for the population variance σ_β^2. For the single-layer PDP, the variance is calculated as:

$$s_\beta^2 = \frac{\sum_{k=1}^{\omega_1} (\beta_{\Im,1,k} - \beta)^2}{\omega_1 - 1} \quad ; L=1 \tag{73-13}$$

where

β = the mean seeker response for the single-layer PDP as shown in Equation (73-12), and

s_β^2 = the variance of the seekers that route data from the single-layer L=1 to the test statistic $\Im$.

If the test statistic $\Im$ is set equal to the pivotal quantity:

$$\Im = \frac{(\beta - \mu_0)}{s_\beta/\sqrt{\omega_1}} \quad ; L=1 \tag{73-14}$$

then $\Im$ tends to be larger when $\mu_\beta > \mu_0$ than for $\mu_\beta \le \mu_0$. As a result, a test may be based on $\Im$ such that H_0 is rejected when $\Im$ is sufficiently large, namely:

Reject H_0 if $\Im > \Im_0$

where

$\Im$ is as shown in (73-14), and $\Im_0$ is the threshold value for the test.

If $\Im = \Im_0$, then $\Im$ has an approximate central t-distribution with ω_1-1 degrees of freedom. Therefore, $\Im_0$ can be determined by setting:

$$\mathbf{a} = \text{Prob}\{\Im > \Im_0 \mid \mu = \mu_0\} \text{ which implies that } \Im_0 \approx t_{1-\mathbf{a},\, \omega_1 - 1} \tag{73-15}$$

where

$t_{1-\mathbf{a},\, \omega_1-1}$ is the (1-**a**)th percentile of the central t-distribution with ω_1-1 degrees of freedom.

Using (73-15) to set the size **a**, the test for the hypothesis: H_0: $\mu_\beta \le \mu_0$ versus H_1: $\mu_\beta > \mu_0$ becomes:

$$\text{Reject } H_0 \text{ if } \Im > \Im_0 = \frac{(\beta - \mu_0)}{s_\beta/\sqrt{\omega_1}}\,. \tag{73-16}$$

where

$\Im_0 = t_{1-\mathbf{a},\, \omega_1-1}$ and $\Im$ =

Rearranging terms, (73-16) may be written as a mean threshold test:

$$\text{Reject } H_0 \text{ if: } \beta > \mu_0 + \frac{s_\beta}{\sqrt{\omega_1}} t_{1-\mathbf{a},\, \omega_1-1} \quad ; L=1 \tag{73-17}$$

or alternatively as a summation threshold test of the Minsky and Papert format:

$$\text{Reject } H_0 \text{ if: } \sum_{k=1}^{\omega_1} \beta_{\Im,1,k} > c_1 \quad ; L=1 \tag{73-18}$$

where

$$c_1 = \mu_0\omega_1 + s_\beta\sqrt{\omega_1}\, t_{1-\mathbf{a},\, \omega_1-1} \text{ and } \sum_{k=1}^{\omega_1} \beta_{\Im,1,k} = \omega_1\beta$$

The test in (73-17) and (73-18) is the general likelihood ratio test for the single sided hypothesis H_0: $\mu_\beta \le \mu_0$ versus H_1: $\mu_\beta > \mu_0$. The test is also an approximate size **a** UMP test for the single-layer NN and PDP pattern recognition architecture that applies feedforward full-connection routing.

Multiple-Layer Architecture

The decision (recognition) test for multiple-layer architectures is an extension of the single-layer test shown in (73-16). In the extension, the signal strength produced by each layer in the architecture is examined with respect to: (1) the mean seeker response β_{i+1}; (2) the seeker standard deviation $s_{\beta_{i+1}}$; and (3) the width and depth of the architecture. The mean seeker response β_{i+1} is calculated using the $\omega_i\omega_{i+1}$ seekers that route information from layer i to layer i+1. To simplify notation, let $W_{i+1} = \omega_i\omega_{i+1}$, and let $\Sigma_{jk} = \sum_{j=1}^{\omega_{i+1}} \sum_{k=1}^{\omega_1}$. The mean seeker response is then calculated as:

$$\beta_{i+1} = \frac{1}{W_{i+1}} \Sigma_{jk}\, \beta_{i+1,j,k} \quad ; i = 1,.., L ; L\geq 2 \qquad (73\text{-}19)$$

where

β_{i+1} is the mean seeker response produced by layer i that is routed to layer i+1, and when i=L,

$\beta_{L+1} = \beta_{\Im}$ is the mean seeker response produced by the final layer L that is routed to the test statistic $\Im$,

and the variance in the seekers (routing information from layer i to layer i+1) is calculated as:

$$s^2_{\beta_{i+1}} = \frac{\Sigma_{jk}(\beta_{i+1,j,k} - \beta_{i+1})^2}{W_{i+1} - 1} \quad ; i = 1,.., L ; L\geq 2 \qquad (73\text{-}20)$$

where

$$\Sigma_{jk}(\beta_{i+1,j,k} - \beta_{i+1})^2 = \sum_{j=1}^{\omega_{i+1}} \sum_{k=1}^{\omega_i} (\beta_{i+1,j,k} - \beta_{i+1})^2.$$

In (73-19) and (73-20), the subscript "i" ranges from 1 to L. When i=1, the mean and variance are calculated using the $W_2=\omega_1\omega_2$ seekers that route information from the first layer to the second layer of the architecture. Hence, the term β_2 refers to the mean seeker response from the first layer that is being routed to the second layer in the architecture. In a similar manner, the term "s_{β_2}" refers to the seeker standard deviation that is calculated using the $W_2=\omega_1\omega_2$ seekers that route information from the first layer to the second layer of the architecture. When i=L, the mean $\beta_{L+1} = \beta_{\Im}$, and the variance $s^2_{\beta_{L+1}} = s^2_{\beta_{\Im+1}}$, are each calculated using the ω_L seekers that route information from the final layer L to the test statistic.

As in the single-layer hypothesis of (73-5), the test pattern is considered for recognition if the mean response $\mu_{\beta_{i+1}}$ exceeds the critical level μ_0 and is (otherwise) not recognized when $\mu_{\beta_{i+1}} \leq \mu_0$. For the multiple-layer PDP, a total of L≥2 single-sided hypothesis tests are examined of the form:

$$H_0: \mu_{\beta_{i+1}} \leq \mu_0 \text{ versus } H_1: \mu_{\beta_{i+1}} > \mu_0 \; ; i = 1,.., L ; L\geq 2 \qquad (73\text{-}21)$$

where

$\mu_{\beta_{i+1}}$ is the mean seeker response for the population produced by layer i and routed to layer i+1, and μ_0 is a critical value for the mean seeker responses in the population.

In (73-21), the hypothesis examines mean seeker responses produced by each layer i = 1,.., L. When i=L, the hypothesis examines the mean seeker responses produced by the final layer of the architecture. The hypothesis does *not* include a mean seeker response from the grafport to the first layer, since β_1 does not contain weighted Boolean responses for the design, but contains only unweighted pixel data.

To examine the mean seeker response from each layer, a total of L≥2 simultaneous single-sided tests are performed for decision (recognition) of the test pattern. In these tests, the null hypothesis is rejected if the test statistic exceeds the critical level $\Im_0$ in all layers of the architecture:

Reject H_0 if:

$$\Im_{i+1} > \Im_0 \text{ in each layer} \quad ; i = 1,.., L ; L\geq 2 \qquad (73\text{-}22)$$

where

$$\Im_{i+1} = \frac{(\beta_{i+1} - \mu_0)}{s_{\beta_{i+1}} / \sqrt{W_{i+1}}} \qquad (73\text{-}23)$$

is the test statistic based upon the W_{i+1} seekers that route information from layer i to layer i+1, and $\Im_0$ is the threshold value for the test.

The test statistic $\Im_{i+1}$ is of a similar form to that shown in (73-14) for the single-layer PDP. The subscript "i+1" is added to the multiple-layer expression to reinforce the fact that a separate pivotal quantity is computed for each layer i = 1,.., L; based upon the W_{i+1} seekers that route information from layer i to layer i+1. The threshold value for the test is defined as follows:

$$\Im_0 = \sqrt{\frac{L(\varpi-1)}{(\varpi-L)} F_{\mathbf{a};L;\varpi-L}} \quad ; \varpi > L \qquad (73\text{-}24)$$

where

L = the number of layers in the PDP architecture,

$\varpi = \langle \frac{1}{L} \sum_{j=1}^{L} \omega_j \rangle$ the integer valued mean width of the architecture,

$F_{\mathbf{a};L;\varpi-L}$ = central F distribution with L and ϖ-L degrees of freedom, and the brackets $\langle x \rangle$ symbolize the greatest integer less than or equal to x $\forall$ real x.

The threshold value $\Im_0$ defines an approximate size **a** test for examining the L≥2 simultaneous single-sided tests shown in (73-22). The integer "ϖ" is used to calculate the total degrees of freedom ϖL, since this produces a conservative acceptance region that is based upon the greatest integer less than or equal to the mean width of the architecture. The threshold value is based upon the central F distribution with L and ϖ-L degrees of freedom, since this produces an acceptance region that is more conservative than the corresponding single-sided t-tests of the form shown in (73-16). If the integer "ϖ" is less than or equal to L, the ordinate $F_{\mathbf{a};L;\varpi-L}$ cannot be calculated in (73-24), and the test pattern is not recognized by the architecture.

Combining (73-22), (73-23), and (73-24) and rearranging terms, the approximate size **a** test for the hypothesis: $H_0: \mu_{\beta_{i+1}} \leq \mu_0$ versus $H_1: \mu_{\beta_{i+1}} > \mu_0$ becomes:

Reject H_0 if:

$$\beta_{i+1} > \mu_0 + \frac{s_{\beta_{i+1}}}{\sqrt{W_{i+1}}} \sqrt{\frac{L(\varpi-1)}{(\varpi-L)} F_{\mathbf{a};L;\varpi-L}} \quad ; i = 1,.., L ; L\geq 2 \qquad (73\text{-}25)$$

or alternatively as a summation threshold test of the Minsky and Papert format:

Reject H_0 if:

$$\Sigma_{jk}\, \beta_{i+1,j,k} > c_{i+1} \quad ; i = 1,.., L ; L\geq 2 \qquad (73\text{-}26)$$

where

$$c_{i+1} = \mu_0 W_{i+1} + s_{\beta_{i+1}} \sqrt{\frac{W_{i+1} L(\varpi-1)}{(\varpi-L)} F_{\mathbf{a};L;\varpi-L}} \quad \text{and } \Sigma_{jk}\, \beta_{i+1,j,k} = W_{i+1}\beta_{i+1} \qquad (73\text{-}27)$$

Equation (73-25) states that the mean weighted Boolean response for each layer (i = 1,.., L) in the architecture must exceed the critical level $\mu_{\beta_{i+1}}$ by an amount that is scaled relative to: (1) the seeker standard deviation $s_{\beta_{i+1}}$; (2) the square root of the number of seekers routing information to the next successive layer $W_{i+1}=\omega_i\omega_{i+1}$; and (3) the critical level for the test $\Im_0$ as shown in (73-24). The inequality in (73-26) is of the same form as the test shown in (73-18) for the single-layer architecture. In each case, the threshold summation test varies with the size of the architecture and the distribution of the seekers within the design.

In the single-layer hypothesis of (73-5) and the multiple-layer hypothesis of (73-21), a conservative value for μ_0 is selected for decision (recognition) of the test pattern. The value is set below the maximum level of $\mu_0 = \mu_{0_{i+1}} = 1$; i = 1,.., L, since this test requires a mean seeker response from each layer β_{i+1} that is equal to unity. Since this condition is too severe for the purpose of pattern recognition using vector data (Hall 1988), a 95% threshold test is applied. In this test, $\mu_0 = \mu_{0_{i+1}} = 0.95$; i = 1,.., L, and the hypothesis test for pattern recognition in the PDP may be stated as:

$$H_0: \mu_\beta \leq 0.95 \text{ versus } H_1: \mu_\beta > 0.95 \quad ; i = 1; L = 1 \quad (73\text{-}28)$$

in the single-layer PDP, and for the multiple-layer architecture:

$$H_0: \mu_{\beta_{i+1}} \leq 0.95 \text{ versus } H_1: \mu_{\beta_{i+1}} > 0.95 \quad ; i = 1,.., L ; L\geq 2 \quad (73\text{-}29)$$

Summarizing the results for the single- and multiple-layer architectures, the tests for the signal strength within the back propagation sequence are:

Accept the test pattern as being recognized if $\mathfrak{z}^R \leq \mathfrak{z}_0$ and if in the single-layer architecture:

$$\beta > 0.95 + \frac{s_\beta}{\sqrt{\omega_1}} t_{0.95,\, \omega_1 - 1} \quad ; i = 1 ; L = 1 \quad (73\text{-}30)$$

and in the multiple-layer architecture:

$$\beta_{i+1} > 0.95 + \frac{s_{\beta_{i+1}}}{\sqrt{W_{i+1}}} \sqrt{\frac{L(\varpi-1)}{(\varpi-L)} F_{0.05;L;\varpi-L}} \; ; \; i = 1,.., L ; L\geq 2 \quad (73\text{-}31)$$

where

$$W_{i+1} = \omega_i \omega_{i+1}, \text{ and } \varpi = \langle \frac{1}{L} \sum_{i=1}^{L} \omega_i \rangle$$

otherwise *reject* the PDP test pattern as being recognized and continue the adaptation process.

CONCLUSION

The decision (recognition) tests shown in (73-30) and (73-31) have distinct advantages over the traditional fixed threshold summation tests of the Minsky and Papert form. First, the threshold value $\mathfrak{I}_0$ varies with the significance level of the test and is sensitive to the width and the depth of the architecture. This yields a more precise critical region for decision (recognition) of the test pattern. Second, the test statistic $\mathfrak{I}_{i+1}$ varies with the magnitude and the distribution of the seekers in the architecture and is separately calculated using the W_{i+1} seekers that route information from layer i to layer i+1; i = 1,.., L ; L≥2. As a result, the precision of the PDP tests increases with the architectural width since the number of seekers used to estimate the mean β_{i+1} and standard deviation $s_{\beta_{i+1}}$ also increases. Third, the test statistic $\mathfrak{I}_{i+1}$ and the threshold value $\mathfrak{I}_0$ define an approximate size **a** test that is based upon the uniformly most powerful likelihood ratio tests. Among all size **a** tests, the UMP tests have the greatest likelihood for rejecting the null hypothesis H_0, whenever they should.

In the analysis of operations within the back propagation sequence, the seekers are the subject of the decision (recognition) test, since their magnitude is affected by the response of both atoms and weights. The decision (recognition) test is based upon the seekers, since this is a more conservative approach when compared to equivalent tests based upon either the atoms or the weights. As shown in (73-30) and (73-31), the condition $\mathfrak{z}^R \leq \mathfrak{z}_0$ is necessary for pattern recognition but is not sufficient since the signal strength may be below critical levels as measured by β_{i+1} and $s_{\beta_{i+1}}$. In addition, the event $\varphi_{i,j} = 1 \; \forall$ j in layer i is necessary for pattern recognition but is not sufficient since the corresponding weights $\alpha_{i,j}$ may be below critical levels to warrant recognition of the test pattern. Examining (73-31), the condition $\varphi_{i,j} = 0 \; \forall$ j in layer i is necessary and sufficient for not recognizing the test pattern since this corresponds to an architecture that has sampled few highlighted pixels from the grafport. When $\varphi_{i,j} = 0$, the full connection term $F_{i,j,k} = \alpha_{i,k}\, \varphi_{i,k}$ is equal to 0, and the maximum routed signal is equal to 0.5 using a standard routing equation: $\beta_{i+1,j,k} = \frac{1}{2}(F_{i,j,k} + P_{i,j,k}) \; \forall$ j,k ; i = 1,.., L ; $-1 \leq \beta_{i+1,j,k} \leq +1$. As a result $\beta_{i+1} \leq 0.95$ in at least one layer i+1 of the PDP test shown in (73-31). Note that in (73-30) and (73-31), the condition $\alpha_{i,j} > 0 \; \forall$ j in layer i is necessary for pattern recognition but is not sufficient since the corresponding atoms $\varphi_{i,j}$ may be equal to 0.

Examining the distribution of the seekers in (73-30) and (73-31), it is evident that the event $\varphi_{i,j} = 1$ and $\alpha_{i,j} > 0.95$; i = 1,.., L; j = 1,.., ω_i is necessary and sufficient for pattern recognition provided $s_{\beta_{i+1}}$ is sufficiently small i = 1,.., L; L≥2. Writing (73-31) in terms of s_β and $s_{\beta_{i+1}}$, the seeker standard deviation in the single-layer PDP must be such that:

$$0 \leq s_\beta \leq \frac{\sqrt{\omega_1}(\beta - 0.95)}{t_{0.95;\omega_1 - 1}} \quad ; \beta \geq 0.95 \quad (73\text{-}32)$$

and in the multiple-layer PDP, the seeker standard deviation $s_{\beta_{i+1}}$ must be such that:

$$0 \leq s_{\beta_{i+1}} \leq \frac{\sqrt{W_{i+1}}(\beta_{i+1} - 0.95)}{\sqrt{\frac{L(\varpi-1)}{(\varpi-L)} F_{0.05;L;\varpi-L}}} \quad ; \beta_{i+1} \geq 0.95; \; i = 1,.., L ; L\geq 2 \quad (73\text{-}33)$$

When $\varphi_{i,j} = 1$ and $\alpha_{i,j} > 0.95$; i = 1,.., L; j = 1,.., ω_i, then $\beta_{i+1} > 0.95$ for i = 1,.., L; L≥2 as required in (73-31). When (73-33) is satisfied, the seeker signal is consistently close to the mean level β_{i+1} for each layer in the design. When this occurs, all seekers are either above the critical level of 0.95, or sufficiently close to the critical level to warrant recognition of the test pattern.

The decision (recognition) tests shown in (73-30) and (73-31) can be used to discriminate both raster and vector patterns into two distinct classes (feature versus background), and may be applied within a recursive design to evaluate test patterns whose discriminant function contains greater than two distinct classes. This is accomplished by nesting the UMP test within a back propagation sequence that repeatedly checks for pattern (recognition) and then classifies those patterns that are recognized within a binary tree. The binary tree is a hierarchy that maps the decision (recognition) event for each class k≥2, shows the most important discriminators (those with maximum separation) at the lowest (root) levels within the tree, and shows the least important discriminators at the highest (branch) levels within the tree. Since the binary tree expands with the number of data layers, the decision (recognition) tests can be applied to both single-band and multispectral image sets without loss of generality. In this application, (73-30) would be applied to simple decision (recognition) problems whose solution is linearly separable, and (73-31) would be applied to complex decision (recognition) problems whose discriminate function requires nonlinear separation.

References

Anderson, J. A., and E. Rosenfeld, eds. 1989. *Neurocomputing: Foundations of Research.* Cambridge: MIT Press.

Baase, S. 1988. *Computer Algorithms—Introduction to Design and Analysis,* 2nd ed. Reading: Addison-Wesley.

Baillard D., and C. Brown. 1982. *Computer Vision.* Englewood Cliffs: Prentice Hall.

Banquet, J. P., and S. Grossberg. 1986. Probing cognitive processes through the structure of event-related potentials during learning: An experimental and theoretical analysis. *Computer Vision, Graphics, and Image Processing.*

Barto, A. G., R. S. Sutton, and C. W. Anderson. 1983. Neuronlike adaptive elements that can solve difficult learning control problems. *IEEE Trans. Syst. Man. Cybern.* SMC-13:834–46.

Batcher, K. E. 1980. Sorting networks and their applications. *Proceedings of the AFIPS, Spring Joint Computer Conference.*

Baum, E. B., and D. Haussler. 1989. What size net gives valid generalization. *Neural Computation* 1:151.

Bickel, P., and K. Doksum. 1977. *Mathematical Statistics: Basic Ideas and Selected Topics.* San Francisco: Holden-Day.

Forsyth, R., and R. Rada. 1986. *Machine Learning: Applications in Expert Systems and Information Retrieval.* Chichester: Ellis Horwood Limited.

Fukushima, K. 1988. A neural network for visual pattern recognition. *Computer* 65–75.

Gibbons, A., and W. Rytter. 1990. *Efficient Parallel Algorithms.* Cambridge: Cambridge University Press.

Grossberg, S. 1980. How does a brain build a cognitive code? *Psychological Review* 87(1):1–45.

Hall, P. 1988. *Introduction to the Theory of Coverage Processes.* New York: John Wiley and Sons.

Hinton, G. E. 1985. Learning in parallel networks. *Byte Magazine* 10(4).

Hogg, R. V., and A. T. Craig. 1978. *Introduction to Mathematical Statistics.* 4th ed. New York: Macmillan.

Hornik, K., M. Stinchcombe, and H. White. 1989. Multilayer feedforward networks are universal approximators. *Neural Networks* 2:359–66.

Kanal, L. 1971. *Generative, Descriptive, Formal and Heuristic Modeling in Pattern Analysis and Classification.* Re. Aerosp. Med. Div. AF System Command, Wright-Patterson AFB, Ohio, Contract F33615-69-C-1571; also Univ. Md., College Park, CSC, Tech. Rep. TR-151.

Kanal, L., and J. Lemmer, eds. 1986. *Uncertainty in Artificial Intelligence.* North Holland: Elsevier.

Krishnaiah, P. R., ed. 1969. *Multivariate Analysis,* vol. II. New York: Academic Press.

Lehmann, E. L. 1959. *Testing Statistical Hypotheses.* New York: John Wiley and Sons.

McClelland, J. L., and D. E. Rumelhart. 1989. *Explorations in Parallel Distributed Processing: A Handbook of Models, Programs, and Exercises.* Cambridge: MIT Press.

McCulloch, W. S., and W. Pitts. 1943. A logical calculus of the ideas immanent in nervous activity. *Bull. of Mathematical BioPhysics* 5:115–33.

McKeown, D. M., Jr. 1987. The role of artificial intelligence in the integration of remotely sensed data with geographic information systems. *IEEE Trans. Geoscience and Remote Sensing* GE-25(3):330–48.

Minsky, M., and S. Papert. 1972. *Perceptrons: An Introduction to Computational Geometry.* 2d ed. Cambridge and London: MIT Press.

———. 1988. *Perceptrons: An Introduction to Computational Geometry.* Cambridge and London: MIT Press.

Nilsson, N. J. 1965. *Learning Machines.* New York: McGraw-Hill.

Rosenblatt, F. 1958a. *The Perceptron: A Theory of Statistical Separability in Cognitive Systems.* Cornell Aeronautical Laboratory Report No. VG-1196-G-1.

Rosenblatt, F. 1958b. The perceptron: A probabilistic model for information storage and organization in the brain. *Psych. Rev.* 65.

Rumelhart, D. E., and J. L. McClelland. 1987. *Parallel Distributed Processing Vol. 1—Foundations.* Cambridge: MIT Press.

Sivilotti, M. A., M. A. Mahowald, and C. A. Mead. 1987. Real-time visual computations using analog CMOS processing arrays, *Advanced Research in VLSI: Proceedings of the 1987 Stanford Conference,* edited by P. Losleben, 295–312. Cambridge: MIT Press.

Sejnowski, T. J., and C. R. Rosenberg. 1986. *NETalk: A Parallel Network that Learns to Read Aloud.* Johns Hopkins University Electrical Engineering and Computer Science Technical Report, JHU/EECS-86/01.

White, H. 1989. Some asymptotic results for learning in single hidden-layer feedforward network models. *Journal of the American Statistical Association* 84(408):1,003–13.

Widrow, B., and R. Winter. 1988. Neural nets for adaptive filtering and adaptive pattern recognition. *Computer* 21(3):25–40.

Perry J. LaPotin
Remote Sensing/GIS Center
USACRREL
72 Lyme Road
Hanover, New Hampshire 03755

H. L. McKim
School of Natural Resources
University of Vermont
George D. Aiken Center
Burlington, Vermont 05405

74

Distributed Models and Embedded GIS:
Integration Strategies and Case Studies

Kurt Fedra

There are several strategies and approaches to integrate spatially distributed environmental models, including expert systems and GIS. They range from simple pre- and postprocessor linkage through shared data files to building models as complex analytical functions into fully functional GIS, or embedding the required GIS functionality into spatially distributed models. Following an overview of integration strategies and their relative merits, this chapter describes a number of applications that use a tight coupling of GIS functionality with environmental models, integrated into a common graphical user interface. Based on a low-level tool kit, these systems exemplify the high degree of flexibility that can be achieved with generic low-level building blocks in an object-oriented design. Databases, expert systems, simulation and optimization models, and GIS functionality, together with hypertext and multimedia elements, can effectively be configured in user-friendly environmental decision support systems (EDSS) that are problem-specific and customized for individual institutions or user groups.

Examples include a global change information and assessment system, air-quality management tools, water resources, and river basin management systems, including surface and groundwater quality and coastal marine models, and systems for environmental impact assessment and technological risk analysis. The examples illustrate principles and strategies of integration and indicate desirable features of next-generation integrated GIS and models.

MODELS AND GIS

Environmental problems usually are also spatial problems. And, as a logical consequence, environmental models are increasingly spatially distributed. This increasing development and use of spatially distributed models replacing simple spatially aggregated or lumped parameter models are also made possible by the availability of more powerful and affordable computers (Loucks and Fedra 1987; Fedra and Loucks 1985).

Geographic information systems (GIS) are tools to capture, manipulate, process, and display spatial or georeferenced data. They contain both geometry data (coordinates and topological information) and attribute data, i.e., information describing the properties of geometrical spatial objects such as points, lines, and areas. In GIS, the basic concept is one of location, of spatial distribution and relationships; the basic elements are spatial objects. By contrast, the basic concept of environmental modeling is one of state, expressed in terms of numbers, mass or energy, of interaction and dynamics; the basic elements are "species," which may be biological, chemical or environmental media such as air, water, or sediment.

The overlap and relationship are apparent, and thus the integration of these two fields of research, of technology, or sets of methods—their paradigms—is an obvious, promising, and widely accepted idea (e.g., Fedra 1993; Fedra and Kubat 1993).

METHODS OF INTEGRATION

The integration of GIS and environmental models can come in many forms. In the simplest case, two separate systems, the GIS and the model, just exchange files. The model obtains some of its input data from the GIS and produces some of its output in a format that allows import and further processing and display with the GIS (Figure 74-1). This seems to be a common approach since it requires little if any software modification. Only the file formats and the corresponding input and output routines, usually of the model, have to be adapted. However, depending on the implementation, a solution based on files shared between two separate applications, usually with a different user interface, is cumbersome and possibly error prone if it involves a significant amount of manual tasks.

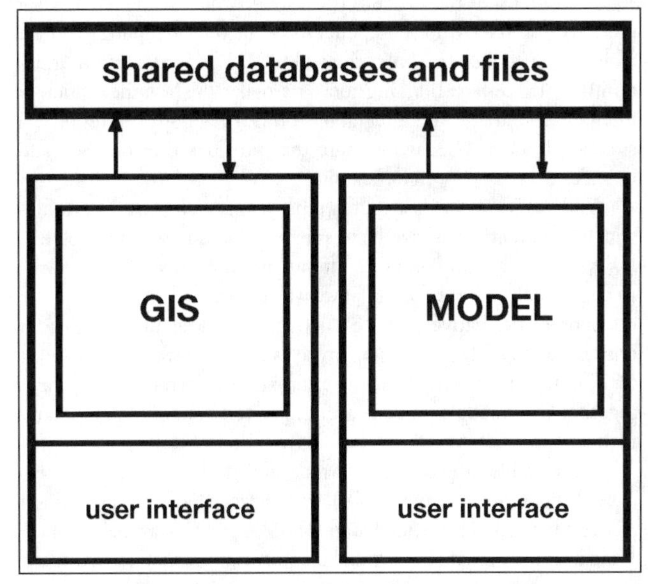

Figure 74-1. Shallow coupling through common files.

Deeper integration provides a common interface and transparent file or information sharing and transfer between the respective components (Figure 74-2). One way to achieve this integration is to use a higher-level application language or the application generators built into the GIS; ARC/INFO's AML or the Subroutine Development Libraries (ARC/SDL) are such approaches (ESRI 1992) and form the basis of numerous integrated applications. Application generators and modeling capabilities with commercial GIS also offer the possibility of tight integration within the limits of the respective package options.

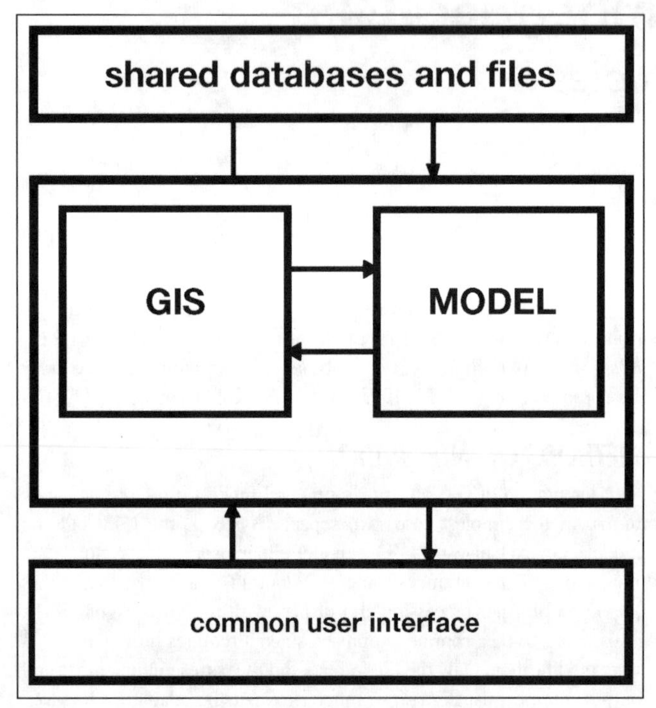

Figure 74-2. Deep coupling in a common framework.

An alternative is to use an open GIS tool kit with a standardized interface, such as GRASS (USA-CERL 1988; Fedra and Kubat 1993). Modules of the overall GIS system (which really are a set of tools with a standardized pipeline-type flexible coupling) can be included in modeling applications. The X-Windows system and a number of interface-building tool kits make this a rather efficient integration strategy. Any integration at this level, however, requires a sufficiently open GIS architecture that provides the interface and linkages necessary for tight coupling.

While efficient for fast prototyping, using such predefined tools and components also can be restrictive: investments in tool kits always carry the temptation to reformulate problems in terms of the available tools rather than the other way around.

Another alternative is to use a do-it-yourself tool kit that provides both customized GIS functionality as well as interface components for simulation models. A recent example of integration that draws together GIS, models, spreadsheet and expert systems in a programmable system is RAISON (Lam and Swayne 1991).

For a problem-specific information and decision support system (DSS), rather than a generic tool for spatial data handling, only a subset of the functions that a full-featured GIS supports may be required. Functions such as data capture and preprocessing and interactive analysis can conveniently be separated: they support different users with different tasks. This layered functionality concept (Figure 74-3) leads to embedded GIS functions, such as map (and model output) display, including animation of dynamic models, and model-related analysis. While providing only a small subset of all the functions of a complete GIS, the selective, limited functionality required in a given application can be integrated efficiently with a minimum of overhead but with great flexibility and performance. A deeper level of integration would merge the two approaches, such that the model becomes one of the analytical functions of a GIS, or the GIS becomes yet another option to generate and manipulate parameters, to input and state variables, to model output, and to provide additional display options.

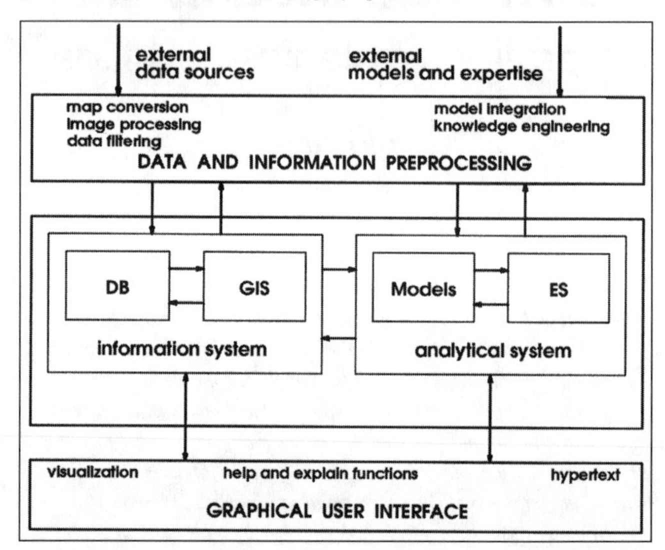

Figure 74-3. An integrated framework for environmental information systems.

However, this requires tools sufficiently modular to allow coupling of software components within one single application with shared memory rather than files and a common interface. Other components such as expert systems, hypertext, optimization, and DSS also are included. Obviously, this most elegant form of integration is also the most costly in terms of development effort. However, if the ultimate goal is to develop a better research tool, more powerful models, and analysis software, as well as aid in the environmental planning and policy-making processes, more than integration of environmental models and GIS technology will be required to successfully integrate these methods.

A DSS Framework

Decision making can be understood (or rather, formalized) as choosing between alternatives. Alternatives can be described in terms of criteria, which are attributes or characteristics of the problem and alternative solutions that can be represented by measurable entities we call descriptors. Since most environmental problems are spatial in nature or involve an important spatial component, these descriptors usually refer to spatial objects.

The criteria relevant to a problem can come in two forms: objectives and constraints. However, objectives and constraints, while well defined in the methods of mathematical programming, are rather fluid and interchangeable in human thinking. For example, we want as much as possible of *Xo* (an objective), but are willing or able to pay at most *Yc* (a constraint) for it. However, we also want at least *Xc*, paying as little as possible (of *Yo*) for it, which interchanges objective and constraint. Also, whether or not certain criteria enter the selection process often depends on the actors involved and their values and preferences rather than on some intrinsic property of the problem.

In other words, any efficient DSS should allow a user to select criteria and arbitrarily decide whether to maximize or minimize them or to use them as a threshold for eliminating alternatives instead. However, the most important part is the generation or design of the alternatives in the first place. If the set to choose from is too small and does not contain satisfactory (or "optimal") solutions, even optimal ranking and selection methods will not lead to satisfactory results.

An Object-Oriented Structure

A common structure for integration can be based on an object-oriented representation. We consider two base classes of objects: spatial objects and thematic objects. Using the CLIMEX system as an example, spatial objects include: the globe, regions, countries, provinces, cities, and observation stations. These form an administrative or political hierarchy. Alternatively, we have major landforms and river basins, which often are independent of national boundaries.

Thematic objects also form a hierarchy (or better, a heterarchy) and include issues, indicators, descriptors, and raw data (e.g., time series of observations). Where spatial (topo)logical and substantive elements meet, we have to consider multiple inheritance. For example, a river reach is a reach in a graph and thus inherits properties from the general arc class of graphs. At the same time, it is a water body and inherits properties from this parent class as well.

The central object in this approach is the descriptor. Derived from raw data, it is used to build indicators, which in turn are grouped into topics or issues. Descriptors are defined by a name, a unit of measurement, a legal range and/or a set of legal values, and a set of methods that can be used to instantiate a specific object in a given context. Methods are defined for each of the spatial object classes for which a descriptor can meaningfully be applied. They are rule sets or functions (including, in many cases as a default, a question to the user). These functions can, for example, derive the average annual rainfall over a river basin or the fraction of the population older than fifty within a country.

The methods that provide the linkage of spatial objects and thematic objects are defined as part of the data, that is, the system's knowledge base. This provides a very high level of flexibility to the approach. And since rules can also be methods, an expert system can coordinate methods as inference trees. This also provides all the related tools, such as recursive explain functions and the knowledge base browser, as well as the hypertext explain files associated with each descriptor.

In this simple view of spatial and thematic objects, models and GIS functions are understood as methods for tasks that become immaterial for a problem structured in the above terms. Integration simply becomes the pragmatic question of which tools can perform the required task most efficiently in a given context.

Embedded GIS

From a problem-oriented viewpoint, a decision support system for spatial problems requires the capability to manipulate, display, and analyze spatial data and models. Alternatives need to be efficiently generated, analyzed, and compared.

By excluding certain basic GIS functions such as data capture (i.e., digitizing) and concentrating on the display and analysis of sets of thematic maps and model output, GIS functionality can be embedded into a DSS framework. The embedded GIS must (1) support the selection and generation of background and thematic maps in various display styles; (2) provide access to spatially distributed data including model input and saved model scenarios; (3) display model output (or time-series data) as animation; and (4) support the comparative analysis of alternative scenarios.

Most analytical functionality is performed at the level of the model, e.g., a two-dimensional finite element code that allows the consideration of very complex and dynamic relationships between the data "layers" and regions (i.e., sets of neighboring elements). However, essential steps, such as preparing model input from "maps" based on rules, may be performed by methods that are proper neither to model nor to GIS, but simply methods available to thematic objects. The distinction between model and GIS becomes blurred and meaningless if the model operates in terms of map overlays, and the analysis of map overlays is done by the model.

APPLICATION EXAMPLES

The following examples illustrate the idea of embedded GIS and their integration with simulation and optimization models and expert systems. The examples are based on recent work at IIASA's Advanced Computer Applications (ACA) project (Fedra 1993; Fedra and Kubat 1993).

Global Change Assessment

CLIMEX is a global change modeling (GCM) and assessment framework designed to support comparative analysis of global change scenarios and to provide a common database and information system for global change impact assessment and modeling. The software system can integrate numerous global data sets relevant to environmental and climate change, including IIASA's global climate database (Leemans and Cramer 1991), as well as base data, such as global elevation, soils, population and land cover, some of which are directly or indirectly derived from satellite imagery and remote sensors. CLIMEX also includes model-generated data sets, such as GCM results or output from IIASA's global agricultural models.

In addition, it integrates selected global models, such as a combined energy and biosphere model (Ahamer 1993), a vegetation and carbon budget model, and RIVM's IMAGE model, in particular, its land-use component (Bouwman et al. 1992). For more regional and local environmental assessment of generic problems, a number of simulation models for air quality, and surface, ground, and coastal marine water quality have been implemented that use embedded GIS integrating vector and raster data, including satellite imagery.

CLIMEX integrates a rule-based expert system that evaluates a given scenario in terms of a set of indicators covering environmental as well as socioeconomic aspects. The expert system in combination with the global GIS and databases offers the possibility to use complex rules for classification and assessment. Both qualitative rules as well as algorithmic components in the form of various models and spatial statistics can be combined to derive spatially distributed indicators.

The data sets in CLIMEX are accessible through a graphical user interface that offers a hierarchical selection of topics and descriptors (discussed previously) and displays the corresponding data set in the form of a topical map, i.e., over a large set of spatial objects. Alternatively, a single spatial object such as a country or an observation station can be selected and all its data holdings can be displayed. The basic mapping and display system uses spatial information prepared or preprocessed by an external GIS such as ARC/INFO or GRASS. Some of the basic features of this dedicated and embedded GIS functionality are described by Fedra and Kubat (1993).

The system handles both vector and raster data simultaneously. Arbitrary zooming is supported, and a special color editor allows the user to modify the colors associated with legend entries. A map editor supports the selection and combination of subsets of features from any map (e.g., a single soil type or group of soils from the global soil map or an elevation band, vegetation class, climate zone, etc.). Alternatively, in the case of GCM results and the corresponding data (temperature, precipitation, cloudiness) from the climate database, a direct comparison of multiple thematic maps is made possible by the simultaneous display of four map windows (Plate 74-1).

In the case of a time series of maps (e.g., population development or monthly climate features), a special tool with a tape deck interface gives the user control over an animation sequence (Plate 74-1).

The expert system in CLIMEX serves a number of purposes. Embedded in the interface of the integrated models, the expert system assists the user in defining consistent model scenarios. Its major role, however, is to derive new, user-defined indicators from available data. A typical example would be an indication of potential water resource problems, i.e., we could formulate a set of rules that include population growth, the current ratio of precipitation and potential evapotranspiration, expected changes according to climate-change scenarios, dependency of the economy on irrigated agriculture, etc. (Kulshreshtha 1993). The resulting set of rules would then be evaluated for a number of spatial objects, depending not only on the degree of spatial resolution required, but also on the nature and resolution of the information used. This might be countries, river basins, or a global grid. The result again would be a global map of this new indicator, possibly for alternative scenarios of population growth, climate change, economic development, etc.

In more theoretical terms, the expert system can reason with indicators that may be spatially distributed, that is, describe spatial objects. These could be points (like measurement stations or even cities on a global scale), polygons (most typically countries, but also supersets and subsets, i.e., regions and provinces, or river basins, etc.), lines or corridors (e.g., areas along rivers or coastlines), and grid cells. The reasoning may involve any or all indicators that apply to this spatial object, which may require sampling or interpolation and aggregation. The inference also may refer to other spatial objects, for example, bordering, or in the "neighborhood" of, the primary object. And since some of the descriptors are dynamic, a historical dimension can be included as well.

The assessment itself is based on a combination of simple models and rules. The models are triggered to provide individual estimates for key variables, for example, the demand for irrigation water based on a set of FAO estimation methods. Using a distributed parallel computing scheme, they are run as an integrated part of the expert system inference mechanism. CLIMEX uses a straightforward syntax, combining an object-oriented design for the descriptors, the basic elements in the inference procedure, with near-natural language rules. For a detailed description of the expert system component, its architecture, and an application for environmental impact assessment, see Fedra et al. (1991).

Air-Quality Modeling

A typical example of a highly integrated environmental decision support system with an explicit spatial dimension and embedded GIS is a set of models for air-quality management developed at IIASA. The system brings together spatial databases (emission inventories, meteorological data, and model scenarios) and hypertext background information, mapping functions, simulation models with different resolution in time and space, an optimization model and multicriteria DSS component, and an expert system for the estimation of point and area source emissions.

Spatially distributed data are used directly for the models (e.g., a DEM or land cover model used to estimate surface roughness and surface temperature differentials). They also are used to (1) derive spatially distributed model input such as wind fields generated from several meteorological stations, plus the above terrain characteristics; (2) derive emission estimates (e.g., urban areas and major traffic arteries used to estimate area and line source emissions); (3) for impact assessment (e.g., vulnerability of differing land uses to various pollutants); and (4) for geographical background data for spatial orientation of the user, the location of emission sources, and as a spatial reference for model results. While used for different purposes in the system, each of these data sets also can be viewed as a thematic map and in conjunction with other overlays in the GIS mapping system. The emission inventories are available either through the map display by picking sources for display and editing their characteristics or from a parallel listing of named sources. Basic source characteristics such as location, emission for various pollutants, stack parameters, and cost functions for alternative pollution abatement technologies are stored in the inventories. An embedded expert system can be used to derive emission estimates from basic technological data, such as fuel consumption and characteristics or production technologies and volumes.

The simulation models of the system include an implementation of EPA's Industrial Source Complex (ISC) model, a Gaussian model that can be run both for short episodes and long-term frequency data (US-EPA 1979). Alternatively, a three-layer finite element model, used conjunctively with a spatially distributed wind field generator, can be used for dynamic short-term runs over a few days. These models describe pollutants such as SO_2, NO_x or dust.

The output of the long-term Gaussian model can be used as an emission and impact scenario for the optimization module (Plate 74-2). Using a source-receptor matrix and a spatially distributed, nonlinear environmental impact function, which can assign different weights to different land use or population zones (transparently retrieved from the GIS), this component finds cost-effective strategies for pollution abatement. Each controlled source has a number of alternative control technologies available, including the option, in some cases, to close a plant. Each option is associated with costs, and for a given overall budget, the model finds the most effective investment strategy (in terms of the environmental impact calculated for the region affected). Varying the budget or the time horizon and discount rate for the cost estimates results in a large number of scenarios that can be further analyzed by a discrete multicriteria optimization tool (Zhao et al. 1985).

An "optimal" emission control scenario can then be used again at the level of the simulation models and tested with a broad range of individual weather scenarios (rather than the frequency data used for the long-term model) and the spatially distributed concentration field, to test the abatement strategy under specific conditions including worse-case assumptions.

Model results are displayed as color-coded overlays over the background maps, as pseudo-three-dimensional displays of pollutant concentration over the DEM, or as a set of symbols representing emission reductions at the source locations in the optimization model.

More Examples

The same basic structure summarized in Figure 74-3 is used in numerous similar and related environmental information systems

(Fedra 1991, 1993; Fedra et al. 1991). They all use maps to facilitate access to spatially referenced data, to provide input to spatially distributed models, to visualize and animate model results in a geographical context, and to add further layers of analysis of model results (Plates 74-3 and 74-4).

CONCLUSION

Integrated environmental information and DSS are tools to support environmental planning and policy making. They are designed to bring the best available knowledge to bear on decision-making processes, reach a broad audience, be easy to use and understand, and help generate and explore a large number of options. They provide direct access to large volumes of data, including as a central element, spatial data, and a set of tools for their analysis and interpretation, including, scenario analysis and forecasts.

Spatially distributed data play an important role in environmental problems, and GIS functionality is an important, common, and expected component in any environmental information system. Levels of integration vary widely, and with them not only the complexity and sophistication, but also the price of the tools. But, to paraphrase a pragmatist's advice on ideological purity versus results, it does not matter whether your integration is deep or shallow, as long as it catches mice.

References

Ahamer, G. 1993. "Influence of an Enhanced Use of Biomass for Energy on the CO_2 Concentration in the Atmosphere." Ph.D. diss., Graz University of Technology.

Bouwman, A. F., L. Van Staalduinen, and R. J. Swart. 1992. *The IMAGE Land-Use Model to Analyze Trends in Land-Use Related Emissions.* National Institute of Public Health and Environmental Protection. Report 222901009. Bilthoven, the Netherlands.

ESRI. 1992. *Understanding GIS: The ARC/INFO Method.* Redlands, California, Environmental Systems Research Institute, Inc.

Fedra, K. 1991. Environmental information systems: State of the art and perspectives. In *ECOLOGIA*, edited by A. Moroni., E. Aloj Tot'aro, and A. Anelli, 12. Atti Del Quarto Congresso Nazionale Della Societ'a Italiana di Ecologia. Arcavacata di Rende, Cosenza.

———. 1993. Models, GIS and expert systems: Integrated water resources models. In *Application of Geographic Information Systems in Hydrology and Water Resources Management*, edited by K. Kovar and H. P. Nachtnebel, 297–308. IAHS Publication no. 211.

———. 1994. GIS and environmental modeling. In *Environmental Modeling with GIS*, edited by M. F. Goodchild, B. O. Parks, and L. T. Steyaert, 35–50. New York: Oxford University Press.

Fedra, K., and M. Kubat. 1993. Hybrid GIS and remote sensing in environmental applications. In *Proceedings of the 25th International Symposium on Remote Sensing and Global Environmental Change*, 657–68.

Fedra, K., and D. P. Loucks. 1985. Interactive computer technology for planning and policy modeling. *Water Resources Research* 21:114–22.

Fedra, K., L. Winkelbauer, and V. R. Pantulu. 1991. *Expert Systems for Environmental Screening: An Application in the Lower Mekong Basin.* Laxenburg: International Institute for Applied Systems Analysis.

Kulshreshtha, S. N. 1993. *World Water Resources and Regional Vulnerability: Impact of Future Changes.* Laxenburg: International Institute for Applied Systems Analysis.

Lam, D. C. L., and D. A. Swayne. 1991. *Integrating Database, Spreadsheet, Graphics, GIS, Statistics, Simulation Models and Expert Systems: Experiences with the Raison System on Microcomputers.* NATO ASI Series, Heidelberg: Springer.

Leemans, R., and W. P. Cramer. 1991. *The IIASA Database for Mean Monthly Values of Temperature, Precipitation and Cloudiness on a Global Terrestrial Grid.* Laxenburg: International Institute for Applied Systems Analysis.

Loucks, D. P., and K. Fedra. 1987. Impact of changing computer technology on hydrologic and water resource modeling. *Review of Geophysics* 25(2).

USA-CERL. 1988. *GRASS 3.0 User's Manual.* Champaign: U.S. Army Construction Engineering Research Laboratory.

US-EPA. 1979. *Industrial Source Complex, ISC: Model User's Guide*, vols. I, II. EPA Report No. EPA-450/4-79-030/311. Research Triangle Park: U.S. Environmental Protection Agency.

Zhao, C., L. Winkelbauer, and K. Fedra. 1985. *Advanced Decision-Oriented Software for the Management of Hazardous Substances. Part VI. The Interactive Decision-Support Module.* Laxenburg: International Institute for Applied Systems Analysis.

Kurt Fedra is head of the Advanced Computer Applications (ACA) project at the International Institute for Applied Systems Analysis (IIASA) and has managed IIASA's computer services since 1989. Dr. Fedra studied at the University of Vienna, where he received his Ph.D. in biology in 1978. He joined IIASA in 1978 and worked until 1982 as a research scholar in the resources and environment area. He returned to IIASA again in 1984 after spending a postdoctoral year at the Massachusetts Institute of Technology. He is an expert in environmental computing applications, and, as project leader of the ACA team, is responsible for the design and development of information and decision support systems in the area of environmental management, development planning, and risk analysis for international governmental and industrial clients. He is the author of more than ninety articles and reports and has contributed to several books on environmental systems analysis and related computer applications.
International Institute for Applied Systems Analysis (IIASA)
A-2361 Laxenburg, Austria

75

Integration of GIS, Expert Systems, and Modeling for State-of-Environment Reporting

David Lam and Christian Pupp

For state-of-environment reporting, information needs to be integrated from environmental, social, economical, agricultural, and many other sources. We propose a framework that, under one system, will integrate various types of information, including GIS, databases, documents, and spreadsheets. The challenge was to develop appropriate information technologies that combine all these components, identify their linkages, and present them as a coherent whole. In addition to traditional modeling techniques, we developed a neural network and the expert system capability to cover both numerical and non-numerical information. To demonstrate the approach, we give examples that fill spatial data gaps and integrate environmental and agricultural data and models.

INTRODUCTION

State-of-environment reporting is an initiative of the federal government of Canada to provide timely, accurate, and accessible information about environmental issues for both public and technical audiences. It requires input from many sources, intelligent interpretation, and communication skills to summarize information to various levels of detail. Many environmental problems are complex, multidisciplinary, and require solution techniques that can integrate data, models, and knowledge across a wide spectrum of research areas. In the case of state-of-environment reporting, we have to summarize, explain, and report these results through the integration of information, not only from environmental domains but also from social, economical, industrial, agricultural, and other domains. It is necessary to find ways to combine all this information. We developed an environmental information system, RAISON, that offers database, map, spreadsheet, and statistical analyses components (Lam and Swayne 1991; Lam et al. 1992) that can accept various types of information such as numerical data, descriptive files or metafiles, GIS maps, documents, models, knowledge rule bases, and so on.

In this chapter, we explore the system's potential for broader application in state-of-environment reporting. The purpose of this study is twofold. First, we use the RAISON system to access different databases and analysis tools to prepare information summaries. Second, we use it to integrate knowledge bases and broaden interpretation capabilities.

APPROACH AND METHODOLOGY

When data or information from different disciplines are combined, there are generally no established rules on how to find their interrelationships. It is necessary to have a system that can access and retrieve any part of the information and try different methods to detect the relationships. In the case of simple numerical data, conventional statistical methods such as regression analysis should be made available and the results made presentable by computer graphics. Not only should all relevant parts of the database be accessible for analyses, but more advanced procedures (e.g., modeling) should be available or programmable through macrolanguage capabilities. The system also should be able to repeat these procedures in batch mode for the entire database when necessary. Where exact relationships are not available, a knowledge-based system component (e.g., knowledge rule bases, GIS attributes) should be available to develop and process the rules and attributes. To acquire the knowledge in a multidisciplinary environment, knowledge representation in a simple format (e.g., a spreadsheet) should be used so that experts from different disciplines easily can enter their knowledge rules and can understand quickly what other experts have entered. In those cases where knowledge data or rules have to be learned or constructed, a neural network capability may be desired and made available. It should have an object-oriented design in the interface so that the network layers and nodes can be entered as objects.

Figure 75-1 shows the schematic of the main components in the RAISON system. RAISON has two levels: basic and advanced. At the basic level, the database, spreadsheet, map, graphics, and statistical analysis are all accessible to each other (Lam and Swayne 1991). For state-of-environment reporting, simple data summary, graphic results, aggregate maps, and annotated documents can be prepared for browsing by a general audience. In the advanced level, the macroprogramming language, modeling, expert system, and neural network are available to the technical user. For example, an expert system can be embedded within another expert system so that when a rule in the first system is executed, it may call for the second expert system to access the database, retrieve information from the GIS, or do some subtasks. This is especially convenient when parts of the expert system need to be called many times and can be treated as a subroutine. Similarly, an expert system can be embedded within a model so that certain model coefficients or boundary conditions can be derived dynamically, for example, from the map information according to a given rule base. Conversely, a model can be nested within an expert system so that the model can be activated during the execution of the expert system, which will then use the model results to determine the outcome of the rules. The RAISON programming language (RPL), a macrolanguage capability in RAISON, can be used to control the embedding and deployment of all basic and advanced components in RAISON and serves as the "glue" that holds them together. Following are some examples.

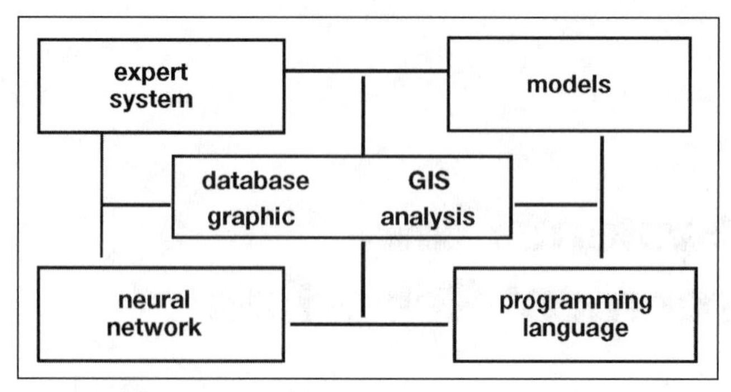

Figure 75-1. The components of the RAISON system.

Environmental Information System

Using the RAISON system, we conducted a pilot study for the Great Lakes lowlands of southern Ontario, which forms part of the mixed wood–plain ecozone in Canada. Over 100 megabytes of data were obtained from federal, provincial, and municipal governments and agencies for air and water quality, meteorology, river flows, ecology, health, population (rural and urban), employment (various economic sectors), social survey, landfill sites, agriculture and land use, livestock and farming practices, etc. Three different levels of accessibility were attainable—raw data, data summaries, and advanced manipulation—through a user interface programmed in RPL. For example, data summaries were prepared by processing the raw data with statistical procedures and storing the results as maps, tables, figures, and documents. Figure 75-2 shows the sequence of menu options used during a presentation session. Figure 75-3 shows that stations easily can be identified with and without pH data on a map from available information in the water-quality database.

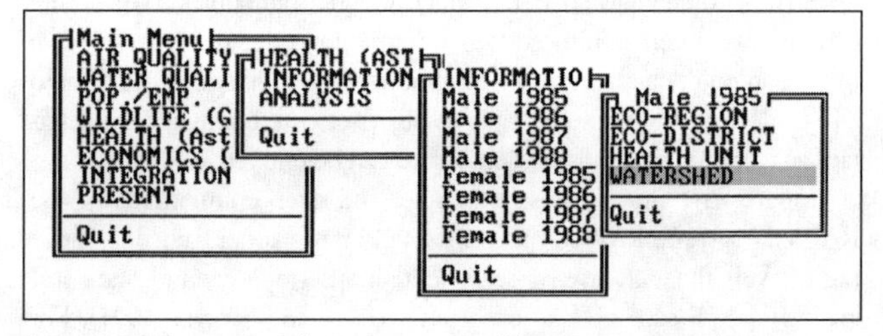

Figure 75-2. A sequence of menu options used during a presentation session.

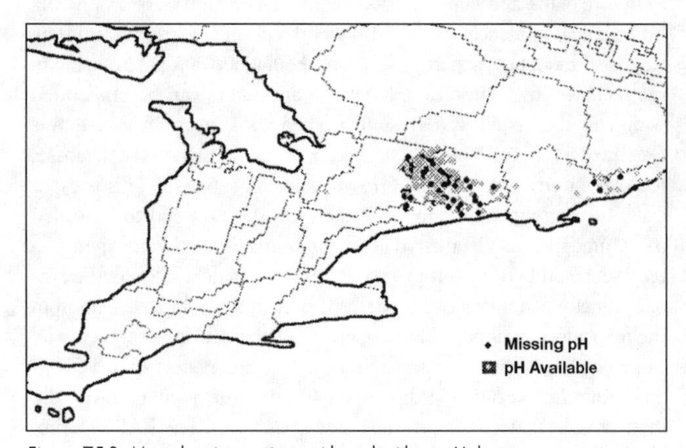

Figure 75-3. Map showing stations with and without pH data.

Neural Network

A common problem in environmental databases is missing data. For example, several parameters were measured in a number of stations, but occasionally one or more parameters were not measured. One possible solution is to use the neural network method, in which all or part of the parameters are used to train the network, which is then used to estimate the missing values. While a number of network paradigms can be used, we chose the multilayer back propagation model because of the relative simplicity and flexibility required to interface it with other modules in RAISON (Eberhart and Dobbins 1990). In this network model, input data are fed into one or more hidden layers and a set of connection weights are continually adjusted under the supervised training mode. The feedforward output state calculation is combined with backward error propagation involving both current and previous corrections (Eberhart and Dobbins 1990).

To illustrate the method, we used a water-quality data set in the Great Lakes mixed wood–plain ecoregion. The parameters measured were sodium (Na), potassium (K), calcium (Ca), magnesium (Mg), sulfate (SO_4), alkalinity (Alk), dissolved organic carbon (DOC), aluminum (Al), and pH. While there were full records of all these parameters for most stations, the values for pH were missing for some (Figure 75-3). A straightforward application of the neural network was to make use of those stations with the full set of data and train the network to estimate pH values from the values of all the other parameters. However, instead of using all parameters, which may or may not have been related to pH, we found that only those that were related should be used. To establish whether one or more parameters were related, we used the cluster analysis method and found that one cluster contained DOC and Al, and the other cluster contained pH and the remaining parameters. Therefore, DOC and Al were excluded from the network for this test case.

Figure 75-4 shows the network topology based on the parameters identified by the clustering analysis. The input layer consists of values of Na, K, Ca, Mg, SO_4, and Alk. The output is the estimated pH value. There are two hidden layers used for back propagation. For comparison purposes, we also trained the network using all parameters (i.e., including DOC and Al). For the same number (one million epochs) of training cycles, the network with clustering produced fewer errors (e.g., a 75th percentile error of 3.6% and a median error of 2.5%) than the network without clustering (a 75th percentile error of 5.5% and a median error of 2.4%). Observed and predicted results with clustering analysis are shown in Table 75-1.

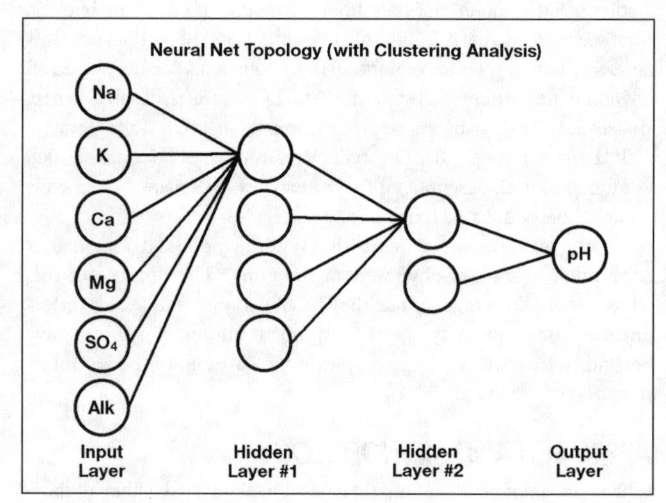

Figure 75-4. Topology of the neural network for estimating pH data.

Table 75-1. Results of pH prediction using neural network.

Record #	Obs. pH	Pred. pH	Rel. Error %
251	6.24	6.23	0.2
252	6.22	6.22	0.0
253	6.04	6.03	0.2
254	5.27	5.46	3.6
255	6.37	6.34	0.5
256	6.05	6.21	2.6
257	5.86	6.11	4.3
258	5.54	5.55	0.2
259	5.35	5.49	2.6
260	6.01	6.03	0.3

Expert System

When different types of data or models are combined, they generally are incompatible in many ways: spatial and temporal scales, degrees of accuracy, and uncertainty, etc. When the data and models are from very different disciplines, it is almost like comparing apples and oranges. One way to overcome this problem is to superimpose or overlay maps by means of GIS techniques, but some control over the incompatibility (e.g., with fuzzy GIS attributes) is required in such an approach. We suggest using the expert system technique (Lam et al. 1992); that is, a rule-based system in which experts from each concerned discipline contribute to the construction of the rules. For example, to evaluate the potential of phosphorus pollution from agricultural inputs (livestock and fertilizer) and from sediment loss (Figure 75-5), we asked the advice of several agricultural experts and environmental hydrologists. In our pilot study, information on livestock, fertilizer, precipitation, soil cover, soil erosion, control measures, and soil slope provided the input data. These input data were obtained from numerical tabulations (e.g., Statistics Canada census information), GIS maps, and descriptive reports.

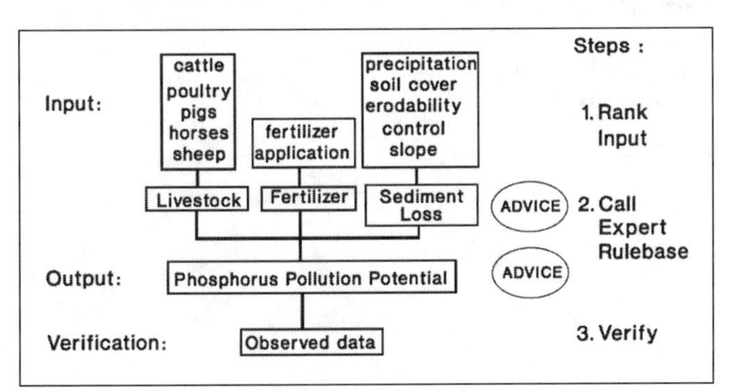

Figure 75-5. An expert system approach to integration of environmental and agricultural data and models for estimating phosphorus pollution.

In the first step (Figure 75-5), numerical values for subareas were transformed into semiquantitative rankings (high, medium, low). In the second step, a rule base was constructed to obtain a ranking for sediment loss (also as high, medium, low). The universal soil-loss equation (USLE) (Wischmeier and Smith 1978) was the prime instrument in generating this rule base. Similarly, the input of phosphorus due to livestock and fertilizer was estimated. Another rule base was generated to produce an overall ranking of the phosphorus pollution potential by combining rankings from livestock, fertilizer, and sediment loss. In the pilot study, this overall ranking scheme was based on limited expert input and some rather crude estimates. It was therefore necessary to provide comments and cautions from the experts during the ranking process. For example, Figure 75-6 shows typical advice issued during a session on the ranking based on the USLE model approach. Nevertheless, the rule bases were applied to the input data associated with the ecodistricts of the study area, and the results of the final ranking for different ecodistricts were presented (Figure 75-7). In the third step, these results were compared to the averages (similarly ranked) of observed phosphorus data measured at major stream outlets. Despite the crude approximation, the agreement between predicted and observed phosphorus pollution was quite satisfactory. For better results, it will be necessary to use a more sophisticated rule base and considerably more expert input.

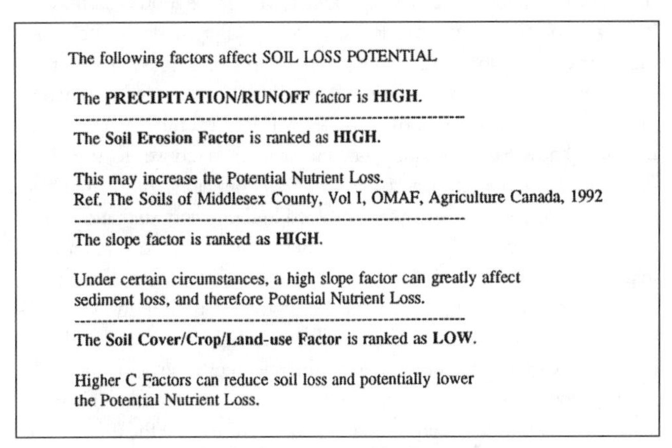

Figure 75-6. An example of advice generated by the expert system.

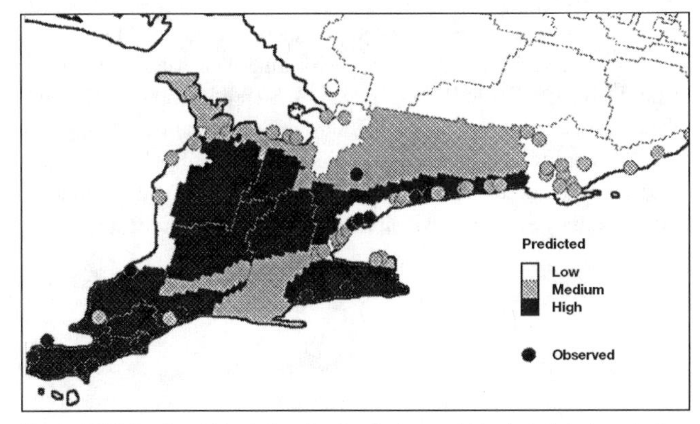

Figure 75-7. Predicted (shaded regions) and observed (shaded circles) results for phosphorus pollution.

One of the main strengths of the expert system approach is the capability to substitute nonrigorous, common-sense rules for rigorous functional relationships when the latter are not available or are difficult to apply. Another advantage of this approach is the added empirical knowledge and so-called metainformation collected from experts and provided as advice and comments during the execution of the expert system. These comments and cautions are valuable for determining where the strengths and weaknesses are for specific outputs and stages during the execution of the rule base.

CONCLUSION

We investigated several techniques to integrate data, GIS, and models for a pilot study on an environmental information system for state-of-environment reporting. We found it most advantageous to link data, maps, documents, graphics, and statistical procedures so that exploratory steps to integrate or summarize the information could be carried out interactively. When necessary, a macrolanguage was used to repeat procedures, control data flow, or run models. The idea of embedding models, expert systems, and neural networks within each other, controlled by macrolanguage programs, was found to be feasible. This idea opens up many new methods of integrating environmental and other information sources.

The strength of this approach is in applying expert knowledge to model coefficients or results during or after the model execution, and also in providing machine-learning capabilities such as filling data gaps in GIS and databases. The drawbacks are that experts are required throughout this approach and in the development of the integration system, and, because experts' backgrounds are often widely different, several iterations may be required to reach a consensus. However, the requirement for expert participation is inherent in the preparation of the state-of-environment report, particularly in the processes of data assemblage, consensus building, knowledge consolidation, and scenario analysis. To speed up these processes, the setup, accessibility, and manipulation of the rule base are made transparent to everyone involved in the project so that they can be debated, tested, and changed with full knowledge given to other partners. Object-oriented interfaces are found to be helpful in simplifying these applications and in making the system more user-friendly. It is anticipated that this expert system approach, when fully developed, will contribute to state-of-environment reporting by providing a platform for storing information (including data, GIS maps, and expert opinion) and as a means for communication among the experts to analyze the information for more timely and accurate interpretation.

Acknowledgments

Thanks are due to Statistics Canada, the Ontario Ministry of Environment, and other agencies for providing their data. The authors thank I. Wong, J. Kerby, P. Fong, D. Kay, A. Storey, J. Storey, D. Swayne, N. Spooner, and other team members for their contributions to this project. This study was supported in part by the Artificial Intelligence Research and Development Fund; by Industry, Science, and Technology Canada; and by State-of-Environment Reporting, Environment Canada.

References

Eberhart, R. C., and R. W. Dobbins. 1990. *Neural Network PC Tools—A Practical Guide.* New York: Academic Press.

Lam, D. C. L., and D. A. Swayne. 1991. Integrating database, spreadsheet, graphics, GIS, statistics, simulation models, and expert systems: Experiences with the RAISON system on microcomputers. In *NATO ASI Series,* Vol. G. 26, edited by D. P. Loucks and J. R. de Costa, 429–59. Heidelberg: Springer-Verlag.

Lam, D. C. L., I. Wong, D. A. Swayne, and J. Storey. 1992. A knowledge-based approach to regional acidification modeling. *Environmental Monitoring and Assessment* 23:83–97.

Wischmeier, W. H., and D. D. Smith. 1978. *Predicting Rainfall Erosion Losses—A Guide to Conservation Planning.* Agricultural Handbook No. 537. Washington, D.C.: U.S. Department of Agriculture.

David Lam is a senior scientist and the project chief of environmental synthesis and prediction of the International Program Group at the National Water Research Institute, Canada Center for Inland Waters, Environment Canada, Burlington, Ontario, Canada. He is also an adjunct professor at the Computer Science Department, University of Waterloo, Waterloo, Ontario, Canada. His research interests include expert systems, neural networks, and GIS applications.
National Water Research Institute, Environment Canada
PO Box 5050
Burlington, Ontario L7R 4A6
Canada
Phone: (905) 336-3916
Fax: (905) 336-4972
E-mail: *david.lam@cciw.ca*

Christian Pupp is a senior science advisor with State-of-Environment Reporting, Environment Canada, Ottawa, Canada. He obtained his Ph.D. in chemistry from the University of Innsbruck, Austria, in 1963 and began working for Environment Canada in 1973. He is currently senior science advisor with the State-of-Environment Reporting, Environment Canada, Ottawa, Canada.
State-of-Environment Reporting, Environment Canada
Ottawa, Ontario K1A 0H3
Canada
Phone: (613) 941-9618
Fax: (613) 941-9650

76

An Integrative Information Framework for Environmental Management and Research

D. D. Cowan, T. R. Grove, C. I. Mayfield,
R. T. Newkirk, and D. A. Swayne

Effective environmental management and research require environmental and socioeconomic data, information, and knowledge to anticipate and predict the impact of development. However, current information, while often extensive, is fragmented, inconsistent, underutilized, and often inaccessible. These factors have led to "information gridlock"—data and information are available but inordinate amounts of time and expertise are required in order for the process of acquisition and assimilation to begin. There is a desperate need for an integrated environmental information system containing both a knowledge base and decision-support tools to help manage the acquisition process. The University of Waterloo, in partnership with several organizations from both the public and private sector, embarked on a large-scale project to develop integrated information systems to manage human use of the environment. This chapter describes some aspects of the system architecture and the implementation approach.

INTRODUCTION

Managing the use of the environment is one of the most important social and political issues that face mankind. Current environmental stresses on the planet are "unsustainable," and fundamental changes in the way we manage our entire habitat are essential.

Environmental concerns include global warming, ozone depletion, contamination of local groundwater, and preservation of fish habitats. The American scientific community has chosen to focus on global climate change (Gershon and Miller 1993), but this leaves many other extremely important environmental issues, such as the management of urban regions, to be addressed. Because half the world's population will be urbanized by the year 2000, sustainable management of urban regions will be a key environmental issue in the next century.

A major obstacle to improving the way we handle environmental problems is the lack of easily accessible and useful environmental and socioeconomic data and information. There is no lack of data and information, but they are inconsistent, stored in diverse locations (on both paper and in various incompatible machine-readable formats), and on many different computer platforms. Even different agencies within the same levels of government often use incompatible data storage technologies and formats. Accessibility is a particularly serious problem in urban regions because environmental and socioeconomic data and information often are held by different and often overlapping levels of government, as well as by various consultants and public interest groups.

This lack of accessibility has led to information gridlock, where large quantities of environmental and socioeconomic data and information are available, but cannot be effectively accessed or processed to yield insight into mankind's impact on the environment. Overcoming information gridlock should be a significant step toward reaching the goal of sustainable development. The result of information gridlock is that much time and effort redeveloping and recreating data are employed at significant expense, rather than efficiently applying existing data and information to environmental problems. The need to resolve this problem has been recognized by the United Nations, the Brundtland Commission, and many national governments.

A BRIEF OVERVIEW OF CURRENT INFORMATION SERVICES

Computer-based tools already exist to assist with organizing and accessing data. Users of the Internet (Krol 1993) are familiar with applications such as FTP and Archie (Emtage 1993), Gopher (Alberti 1992), Veronica (Foster and Barrie 1993), WAIS (wide-area information servers) (Kahle 1989), and the WWW (World Wide Web) (Berners-Lee et al. 1992). Most of these tools require (at least) a moderate degree of computing expertise to be used effectively. For Internet neophytes the learning curve can be overwhelming. Limitations have been identified, and several projects are underway to produce computer-based information systems that focus on the problem of environmental information systems (EIS), including CIESIN (Consortium for International Earth Science Information Network) (CIESIN 1993); ERIN (Environmental Resource Information Network) (Slater 1992); GENIE (Global Environmental Network for Information Exchange) (Newman et al. 1992); and Lamont (Menke 1991). The goals and interim results of these efforts seem to concentrate on accessibility of information (location, client software) and less on information management. For an EIS to be successful, it must also provide information management tools; in many cases, this entails acting as an information filter as well as an information repository.

OBJECTIVES OF THE EIS FRAMEWORK

A fundamental goal of an EIS is to enable its users to locate, acquire, and process information relevant to a problem and then present results in a meaningful fashion. These users range from "domain experts" (planners, environmental engineers) to the general public, and the latter often lack the computing expertise needed to perform the necessary tasks. Important components of EIS are to provide geographic-based mechanisms for finding and obtaining information and to facilitate using that information to solve problems.

To some degree, an EIS functions like a consultant: it gathers, processes, and presents information. In many cases, information gathered by a consultant already exists, it simply needs to be accessed. Informally, EIS strive to provide users the ability to be their own consultants.

Figure 76-1 shows the general framework of an EIS architecture: "Objectives" represent problems to be solved or decisions to be made; "Resources" are inputs to the solution or decision process; "Use" represents actions or transformations to be performed on the resources; and "Location and Access Strategies" represent the components of the EIS that enhance the accessibility of resources and facilitates improved integration of information and processes. The "Feedback" loop is derived information and knowledge that are returned to the resource base for subsequent use. It indicates a mechanism whereby the information framework provided by the EIS can be used to magnify and amplify expertise. As information and knowledge are created, they can be made available as resources to other users.

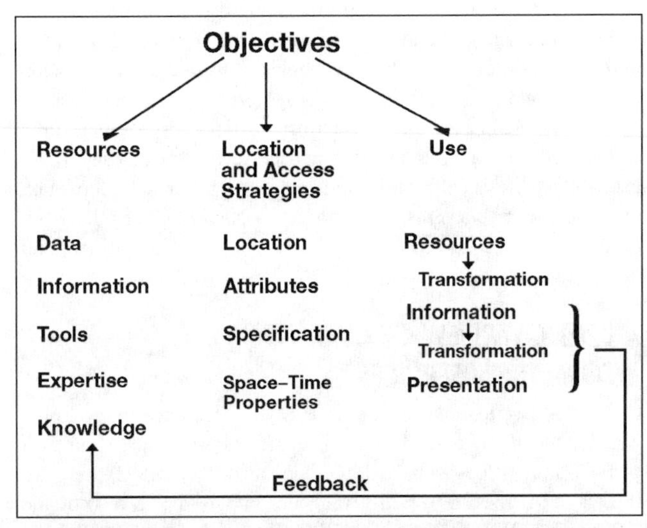

Figure 76-1. An information architecture.

The designer of an EIS must consider two diverse classes of users: (1) experts, who are scientifically sophisticated and have a long-term focus on narrowly defined problems; and (2) integrators (i.e., municipal planners), who have momentary focus on a problem but face many such problems.

Locating and Accessing Resources

It is estimated that 35%–40% of consulting costs are allocated toward acquiring resources such as data, information, expertise, and knowledge (where "knowledge" is viewed as "encapsulated expertise" comprised of encoded data and rules). "Expertise" arises in several categories, including technical (focused knowledge in limited domains), integrative (broad-based general knowledge and experience in its application), and adaptive (cross-disciplinary knowledge and experience). These resources are accessed independently and repeatedly by consultants and practitioners.

Additional costs in information gathering are incurred because of computing expertise that is required to work within the existing *ad hoc* framework. A primary technical goal of the new EIS framework is to reduce this cost by providing a flexible infrastructure to expedite acquisition and utilization of information. The qualitative improvement over raw data repositories will be achieved through the use of partially automated location and access strategies.

Since the fundamental purpose of an EIS is to improve the user's ability to acquire and use information, the access strategies that are embedded into the system should eliminate the need for substantial computing expertise and hide the details of data access.

The User's View

The user's view presented here is solution oriented: the use of the system is motivated by the need to solve a problem. In this context, the goal of an EIS is to improve the "connection" between "Resources" and "Use."

In essence, there are only a few types of questions commonly asked in an environmental context. For discussion purposes, we will focus on the question, "What is the effect of doing *X* on a volume *Y,t?*" (which is representative of commonly asked questions). *X* denotes a physical activity or effect, and *Y* denotes a geographic region with a possible temporal component. Repeated (iterative) application of this class of question commonly is used to determine if regulatory or performance criteria can be met.

An EIS helps answer this kind of question by identifying tools to define and resolve *X*, and by specifying or selecting other resources related to *Y*. For example, *X* might be increased rainfall, and a specification of *Y* might be textual (words or phrases that describe a region) or a polygonal description of an area. If a specific type of study is to be done, the EIS can help locate appropriate information. Conversely, there may exist a data set that must be analyzed; in this case, the EIS can assist in finding an appropriate tool. Oftentimes there is a feedback loop between these two situations: availability of data may motivate the selection of a particular tool that requires additional data, and so on. It is a practical reality, because of expense, that many environmental assessments are done with available data rather than with completely suitable data.

DESIGN OF ACCESS STRATEGIES

Given that there already exist numerous data repositories and collections of tools and modeling systems, the major contribution of the proposed EIS framework is to facilitate access to these items. It is apparent that the existing telecommunications infrastructure initially is sufficient to build some prototypes of an EIS; hence, the access strategies concentrate on the organization of the data and tools and on standardized access methods. Paramount in the organization is the ability to geographically encode queries and data.

Figure 76-2 is a representation of the proposed EIS framework. There are a few components in the EIS framework that require special attention: (1) a dictionary service that contains information about geographic references; (2) a directory service of available databases, tools, and knowledge bases; (3) a trusted-agent mechanism to control access to databases; and (4) a directory index used to resolve database scale and resolution problems. These services are available for use via servers located on a high-speed wide-area network, such as NREN (Aiken et al. 1992), CANARIE (CANARIE 1993), or the Internet. Access mechanisms for a particular item listed in the directory are defined in its directory entry, and although it generally will be accessible via the same network as the framework services, this need not be the case.

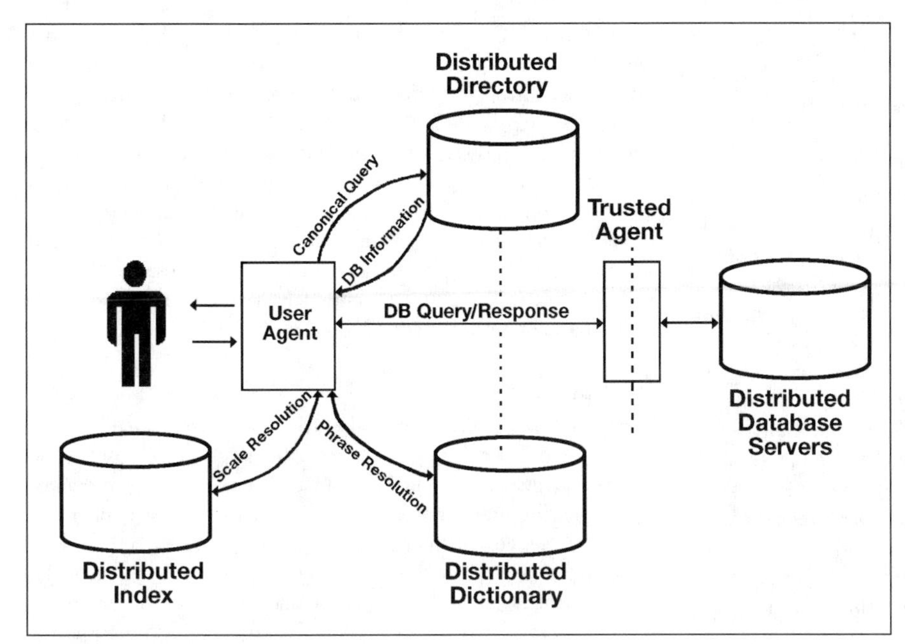

Figure 76-2. Framework for an EIS.

We have organized descriptions of the services by function. In reality, it is likely that the implementation of the framework will be done on an organizational basis. Hence, the design of the framework must encompass fully distributed implementations. The X.500 standard (Bumbulis et al. 1993) for directory services may provide an important standardized base. An important aspect of this standard is its facilities for user authentication, providing flexible access control and security for the databases.

Clearly, the potential scope of the framework described here is global. It is critical that the first stages in the development of the framework concentrate on laying the foundation and creating a suitable "recipe for change." The example of the evolution of the Internet may be instructive in the task. The original sponsoring agency of the Internet (as it came to be known) specified only key protocols and implemented only small prototypes. Many applications were developed independently, and much of the growth was achieved incrementally by improving the availability of developing standards and applications.

Dictionary Service

A dictionary service contains information about geographic references. For example, in the query, "What is the effect of toxic spills in Bloom County?" the reference "Bloom County" would yield a suitable polygon description of the area. "Suitable" is a complex decision in this case. The obvious exact polygon for a particular reference might not be appropriate in some circumstances (for example, computational complexity caused by fractal dimension effects on boundaries that follow geographic features). Therefore, the user should be provided the opportunity to review the "dictionary definition" of an item before its use.

The ideal dictionary service is a knowledge-based system that provides advanced searching and pattern-matching capabilities. However, simpler mechanisms will fit easily into the framework and can be replaced in a modular fashion as new and improved components are developed.

An important aspect of the framework is its ability to capture new knowledge as it is created. This is especially important for the dictionary service. For example, if a new textual geographic reference is defined (by a user), it will be of value to all users if that new reference can be made available. This update facility will require the use of trusted agents or some similar mechanism.

Directory Service

The directory service provides information about databases, knowledge bases, and tools. A query might contain a request for a list of databases that contain information about a geographic region, or a request for a list of tools that pertain to a given keyword. Queries are pattern-matched against entries in the directory, resulting in a list of matching entries. For example, in the previous sample query, the keyword phrase "toxic spill" might be used as the key to search for tools, and the polygon description of "Bloom County" (as provided by the dictionary search) would be the key for the databases search.

It is imperative that a standard representation (a canonical form) for each field in the entry is available. The directory entry indicates whether the canonical form of a field is available; if it is not, the owner of the database is expected to provide transformation functions that can be used to convert the private representation to the canonical form. These functions could include textual descriptions, formulae, or computer programs. Whatever the case, the transformation functions would be made available via the framework. It then becomes the responsibility of the user agent to ensure that a query contains the appropriate data representation.

Directory entries must be sufficiently flexible to reflect the variations in the items that are to be represented. Each entry would have the following types of fields:

Temporal: Some items are applicable to restricted time ranges, and such information should be recorded in the entry.

Spatial: The canonical form (in the directory) for spatial information should be polygons. For a database or knowledge base, this is the area covered by the database; for a tool, it represents any restrictions on the regions for which the tool can be used.

Data specifications and attributes: This information describes the characteristics of the data contained in the database, for example, soil data or groundwater information. The keywords used in the database descriptions are from the same domain as the keywords that describe the tools, thus allowing searches to be restricted to databases that contain information suitable for use by specific modeling tools.

Clearly, items listed in the directory may contain other classes of information of interest to users. The directory services and design must accommodate searching of such indeterminate information. In general, a query should be able to specify matching criteria for any of the fields in the directory entry.

As noted, the directory service probably can be based on the X.500 directory service model. Directory users will forward search requests to a directory user agent who will perform the search and produce a list of items that are (or may be, depending upon search criteria) of interest. Subsequent access to databases can be direct or via a "trusted agent" who accepts and verifies requests. Such a technique provides full-access control to database owners, if they so choose. Thus, the responsibility and control of databases are outside the domain of the EIS framework.

Trusted Agent

Database ownership, security, and access control are of prime importance to data providers. Recognition of the proprietary nature of the intellectual property contained in databases must occur if the commercialization of the framework is to be considered. Therefore, it is important at the outset to incorporate standards for access control, access charges ("pay for service"), and similar issues. The trusted agent mechanism is designed to accommodate such requirements. The trusted agent can act as a "network gateway" for the database, isolating the server from public data networks, tracking usage, and calculating access fees.

The security mechanisms embodied in trusted agent protocols can range from simple (a null process, if the provider wishes to allow unrestricted access) to complex (a *Kerberos*-style authentication system) (Steiner et al. 1988), or to some measure in between these two. In general, the trusted agent is simply a process that intervenes in all accesses to a database and provides the "hook" for all security and access-control mechanisms.

It is envisioned that the trusted agent component of the framework will consist of the protocol for the messages that users transmit to the trusted agent to request access to data. The specification of the protocol is the responsibility of the framework, but the actual implementation of the agent is the responsibility of the data provider. This division of function gives the data provider the assurance that it is his or her code (and only his or her code) that actually is "touching" the database. To simplify the process of implementing the framework, sample implementations of a trusted agent should be made available to data providers.

Processing User Queries

The question, "What is the effect of doing *X* on a volume *Y,t?*" is the user's view of a question. This question can be decomposed into simpler query primitives that are easier to implement with typical information-retrieval and database techniques:

1. List all information about *X* that is contained in *Y,t.*

   Region *Y* may be a jurisdictional name or a geographic polygon (for example, specified with an interactive map-based query tool). If it is a name, it must be transformed into a polygon (i.e., the canonical form for spatial information). The dictionary service will provide such transformations. Once a region reference has been resolved, the directory is searched, using both the area polygon and *X* as keys for the search. *X* is a keyword that describes a geopolitical attribute of a region.

2. List all of the tools or knowledge bases that pertain to *X*.

   The directory contains entries describing known modeling tools and knowledge bases. Associated with each entry is a list of keywords (of the same domain as *X*) that are used to define the applicability of the tool. The result of a query is an enumeration of tools and knowledge bases that may relate to *X* in some fashion.

All user queries can be represented with a logical combination of these or similar primitives, and hence, the EIS framework must support processing of such of primitives.

Directory Index

The search process implied by the first class of question in the previous section yields both intersecting and enclosed polygons. The enclosure case can cause matches with too large an area, leading to false (or meaningless) matches and information "overload." Consider, for example, a directory containing entries for databases about "Bloom County" and "the world." The polygon for "the world" will match a search for "Bloom County," but this result is of little value. Thus, to be effective, the search must be refined in some fashion. Since it is likely that the database described by the worldwide polygon consists of smaller units, a scaling operation can be applied to the search. The directory index provides information about the composition of polygons in the database directory. Using the index, the constituents of an enclosing polygon can be examined to determine if they form a better match for the enclosed (search target) polygon. Constructing an index for an existing database represents an important value-added service.

CONCLUSION

It is a simple task to describe the ideal EIS; however, implementation of such a system is not so easy. Practical realities require that consideration be given to the real-world problem of incorporating existing geographic information into any computer-based EIS (the data legacy problem).

The proposed EIS framework does not specify an end product; rather, it specifies a base that can be achieved with existing technology and provides a standardized path for incremental enhancements to the system. It is important to recognize that there is no ideal system—any successful EIS will, by its very nature, be a dynamic entity.

The most difficult step in the implementation of an EIS is the first step. In order to begin to provide a useful system, existing data owners must be motivated to provide access to their data, and applications must be created to make use of data that are unlikely to be in any standardized form. This is clearly a "chicken and egg" problem: framework-conformant applications cannot be developed until framework-conformant data are available, and such data will not be made available until data owners get something in return for their participation in an EIS. Additionally, potential users of any system are not interested until there is some "critical mass" of available information that justifies the new system's learning curve.

"Jurisdictional overlap" of data, where several independently maintained databases exist for a given geographic region, is already a problem for practitioners. For a computer-based system, this overlap will be compounded as more and more information is made available through the framework. It may be necessary to incorporate some notion of authority into the EIS, in the sense of identifying the primary source versus secondary sources of some particular set of data.

Implementation of the EIS framework is divided into two categories: technical (the computing technology and environmental knowledge), and organizational (the problem of motivating database owners and application providers to adhere to the framework).

Solutions to many technical problems can be obtained with existing technology. However, EIS can and will be improved in the future to provide better technology (faster networks, sophisticated knowledge-based front-end applications to assist searches, better environmental applications, etc.), but the fundamentals of the framework can be designed and implemented now. It also should be noted that it will be feasible to geocode only the directories and not the databases themselves (although database owners should be encouraged to do so).

Solving the organizational problem is not as straightforward. Attracting users to any new system requires that the system provide qualitative and quantitative improvements over the status quo. There are two immediate benefits to EIS users: (1) information about databases and access requirements will be centralized, and (2) the productivity of existing experts will be enhanced. As EIS expand and become more sophisticated, additional benefits will evolve, such as accessibility to "less-than-expert" users.

Motivating information providers is more complex. A small, select group of agencies should be chosen to work as "partners" in the construction of an EIS prototype, using a small test case to prove the concepts. Over time, the framework can be upscaled to involve additional application developers and data proprietors. This user-directed, ground-up approach contrasts the top-down approach that has been used elsewhere. To encourage database providers (and indeed, users), the cost of joining the framework must be minimized; the value of joining must be maximized. Retaining control of data is another key component in motivating data owners.

The approach proposed in this chapter will probably languish in a research laboratory unless key users are motivated to adopt this structure, both internally and as an access method for external users. Finding appropriate partners to adopt the technology is as important as the creation of the original concept.

References

Aiken, R. et al. 1992. *NSF Implementation Plan for Interagency Interim NREN.*

Alberti, B. et al. 1992. *The Internet Gopher Protocol.* Technical report, University of Minnesota, Microcomputer and Workstation Networks Center.

Berners-Lee, T. J. et al. 1992. World-Wide Web: The information universe. *Electronic Networking: Research, Applications and Policy* 2:52–58.

Bumbulis, P. J., D. D. Cowan, C. M. Durance, and T. M. Stepien. 1993. An introduction to the OSI directory services. *Computer Networks and ISDN Systems.*

CANARIE. 1993. *CANARIE: The Canadian Network for the Advancement of Research, Industry and Education.* Ottawa: CANARIE, Inc.

CIESIN. 1993. *CIESIN Mission and Activities Overview.* Briefing prepared for the United Nations Sustainable Development Network.

Emtage, A. et al. 1993. *Archie MAN Page.*

Foster, S., and F. Barrie. 1993. *Frequently Asked Questions about Veronica.*

Gershon, N., and C. G. Miller. 1993. Dealing with the data deluge. *IEEE Spectrum* 30(7):28–32.

Kahle, B. 1989. *Wide Area Information Server Concepts.* Technical Report TMC-202, Thinking Machines Corporation.

Krol, E. 1993. *The Whole Internet: User's Guide and Catalog.* O'Reilly.

Menke, W. et al. 1991. Sharing data over the Internet with the Lamont View-Server System. *EOS* 72:409–16.

Newman, I. et al. 1992. GENIE—The UK global environmental network for information exchange. In *Computing in the Social Sciences '92.*

Slater, W. 1992. *ERIN Concepts.*

Steiner, J.G. et al. 1988. Kerberos: An authentication service for open network systems. In *Winter USENIX.* Dallas, Texas.

D. D. Cowan, T. R. Grove, C. I. Mayfield, and *R. T. Newkirk*
University of Waterloo
Waterloo, Ontario N2L 3G1
Canada
E-mail: *dcowan@csg.uwaterloo.ca; trg@csg.uwaterloo.ca; mayfield@sciborg.uwaterloo.ca* and *newkirk@waterserv1.waterloo.ca*

D. A. Swayne
University of Guelph
Guelph, Ontario N1G 2W1
Canada
E-mail: *dswayne@sciborg.uwaterloo.ca*

77

Integrating GIS and the CENTURY Model to Manage and Analyze Data

J. G. Bromberg, R. McKeown, L. Knapp,
T. G. F. Kittel, D. S. Ojima, and D. S. Schimel

The scientific geographic information system (SGIS) is a systems integration project that provides the ability to analyze temporal data by utilizing commercially available and public-domain software tools. Genamap, S-Plus, and Geo-EAS form the primary core components that provide GIS, statistical, and geostatistical analysis tools. In addition, CENTURY, an ecosystem model, has been linked to the SGIS via a graphical user interface (GUI) to allow model parameters such as land use, soil texture, and climatic data to be retrieved, stored, and analyzed for a variety of simulation experiments. Output from the CENTURY model is organized and stored as temporal continuums and static coverages within the SGIS, where the entire suite of tools provided by the core components is available for data analysis. Following geographical, time series, and geostatistical analysis, data are exported back into the CENTURY model where reiterations of the process fine-tune model parameters to analyze the impact of directional climate change on regional ecosystems.

INTRODUCTION

The 1990s have seen a resurgence of concern over various environmental issues: the threat of global warming, the depletion of resources on a worldwide basis, and an increasingly polluted environment—from atmosphere to subsurface. Research ranging from Earth system modeling to impact analysis is underway on an international basis in an attempt to predict and mitigate potential environmental hazards.

This research has been facilitated by: (1) a widespread interest in man's ability to cope with these problems, (2) funding from the highest levels of government, and (3) advances in technology for data collection and analysis. While, in general, these factors have been positive in nature, they also have been the catalyst behind several characteristics of environmental applications that pose a problem from an information processing perspective. First, data collection techniques, such as remote sensing, are producing massive volumes of data that must be filtered, stored, and analyzed. These data are spatial (two and three dimensional) and temporal. Second, raw data from remote platforms or ground observations are often combined with simulation model output for analysis with geographic information systems (GIS), statistical software, or other models and computational tools. The scientist is ultimately confronted with data of varying spatial and temporal scales and resolutions, uncertain qualities, and a multiplicity of formats. Third, solutions to these problems require interdisciplinary and interagency collaboration for analysis and decision making. This, in turn, mandates systems that account for individual perspectives within an integrated framework.

The large quantities of complex data generated in an interdisciplinary environment can overwhelm research groups, who may spend disproportionate amounts of time on data manipulation and systems integration in order to use the data. To remedy this problem, an effort must be made to provide scientists with integrated tools that support flexible interactive analysis and visualization. The ultimate goal would be to build a system that is so intuitive and easy to use, that it allows the scientist to focus solely on research objectives. Kittel et al. (this volume) discuss the scientific need for integrated systems that tightly link simulation models, GIS, and statistical software.

Our GIS/ecosystem modeling research and development work addresses these issues by examining specific problems and prototyping the requisite building blocks of an integrated analysis and visualization system. This work is based on two assumptions: (1) that such a system must be able to accommodate three- and four-dimensional data; and (2) that a variety of functions are required including science-specific modeling, temporal and spatial statistical analysis, rule-based reasoning, and visualization.

Research has focused on the development of a scientific geographic information system (SGIS), which contains functions for the temporal manipulation of spatial data, a statistical spatial analysis link, and an octree data structure for volumetric data. A prototype of this system has been built using existing software for the GIS and statistical processes. The GIS software is Genasys and the statistics software is S-Plus. In addition, system-level linkage with the CENTURY ecosystem model (Parton et al. 1987, 1994) is implemented as a prototype interface with a user-supplied scientific application. The SGIS is an object-oriented application written mainly in C++ and developed on an IBM RISC System/6000. Currently, it is being ported to Hewlett-Packard workstations. The chapter contains technical descriptions of the SGIS architecture and processes, temporal functions, and linkage to the CENTURY process.

SGIS ARCHITECTURE

SGIS architecture is based on a star topology model. The main driver, called "the process manager," is at the "hub" of the system, serving as a connecting point between various applications, including

SGIS internal modules as well as vendor-supplied or user-supplied applications. The design allows the addition of models and commercial software packages, with minimal disruption to the existing system. Since a new application can be added without recompiling the code base, flexible systems can be designed that change as the scope and focus of research changes. In the current implementation, eight processes are active: the process manager, user interface, GIS (Genamap), temporal analysis, temporal database, animation, statistics (S-Plus), and the CENTURY model.

Process Manager Process

The process manager controls the entire SGIS system. It handles all communication between the processes associated with each application. Messages, called "work units," are distributed among the applications' processes, requesting that various tasks be performed. The work units are fixed-length data structures that contain, among other things, the identification of the sending and receiving processes, the direction of travel (request or response), and a number of request-specific integer and character parameters. Communication between the processes is single-threaded, so only one work unit can be transmitted at a time. For each task, the process manager establishes a two-way socket connection between itself and the appropriate application process until the task is complete or an error is returned.

Communication of data between the processes is achieved via the parameters passed in the work unit. For small amounts of data, such as a file name or map name, this mechanism is sufficient. However, for larger amounts of data, such as a map or the results of calculations on map values, the data are written to a file and the file name is passed to the receiving process. While the process manager controls all communication between the processes, the graphical user interface (GUI) actually determines what actions are to be performed by the SGIS.

User Interface Process

The user interface receives all user requests for action and then passes the necessary information to the process manager, which in turn sends requests to the various subprocesses. These subprocesses include the GIS process, the temporal analysis process, the temporal database process, the animation process, the statistics process, and the CENTURY model process. The initial version of the SGIS was driven by command line instructions. However, since the purpose of the SGIS was to create an intuitive, easy-to-use tool, the need for a more complex interface was recognized early in the development cycle. This resulted in the implementation of a GUI.

The GUI is intended to assist navigation through the SGIS databases; to ease selection of files, maps, data sets, and actions; and to free the user from memorizing command syntax. It is a simple way to access all the SGIS functions, including GIS, statistical, and model control commands. The GUI consists of a window with a menu bar and a series of cascading menus. The menus may have as selections: (1) frequently used SGIS or GIS commands, (2) actions that bring up dialog boxes for browsing through directories, or (3) less frequently used commands. Some Genamap functions are represented by menu selections in the GUI, such as *Shade, Plot,* and *Pixp,* and allow the application of the command to all instances in a time range simply by pressing the "Next" button after the previous map has been processed. Certain dialog boxes brought up by menu items related to Genamap functions allow the entry of a series of commands. Both the Genamap and S-Plus applications have command dialog boxes that contain a history area, which displays previously entered commands and allows replay of these commands with or without modification.

GIS Process

The GIS process is an interface between the SGIS and Genamap, consisting of interprocess communication routines as well as routines that invoke Genamap functions. All the commands available in Genamap for manipulating spatial and attribute data may be invoked to act on temporal data.

Temporal Analysis Process

The temporal analysis process provides internal SGIS functions for time-oriented processing. These include several complex commands ranging from extraction of a time series to aggregation of data from a series of maps. For example, point data obtained from fieldwork can be compared to gridded model output using the "Evaluate" command. The gridded map is retrieved from the GIS database, the point values are located with respect to the gridded map, and the residual values (point data and gridded data) are calculated and sent to the statistical process for display. Additionally, any continuous subset of data can be isolated using the "Extract" command. The resulting time series may be decomposed or displayed by the statistical process. Mean, minimum, or maximum values can be created via the "Aggregate" command. To subset a GIS coverage for further analysis, the "Brush" command can be used to examine areas of interest within a GIS map. The users can select an area of interest by moving the cursor with the mouse. Data associated with the chosen area are sent to S-Plus for display and analysis. The "Brush" command also allows data in an S-Plus scattergram to be selected and sent to Genamap to be stored and displayed as a new map.

Temporal Database Process

The temporal database process manages the SGIS internal database and is responsible for the storage and retrieval of spatial data and attributes stored in Genamap maps. Although the maps themselves are stored by the GIS, the information needed to access maps by date and to organize selected map layers into continuums, sets, and databases is handled by the database process.

Animation Process

The animation process displays specially formatted maps in a sequence, allowing the user to step through up to two series of maps. When two series are examined, they are displayed side by side in order to observe patterns in each series and correlations between the two series. The animation process, once started, displays its own user interface, which is used to select the maps to be displayed and to control the rate and direction of playback.

Statistics Process

The statistics process is the interface between the SGIS and the S-Plus statistical package, providing access to all the S-Plus functions via interprocess communication. A subset of the temporal component of the SGIS is an integration of the Genamap GIS and S-Plus that allows the user to perform statistical analysis using GIS data as input. Conversely, this integration also allows statistical parameters to be input into the GIS for display.

CENTURY Model Process

This process is named after the CENTURY model, which simulates carbon and nutrient dynamics in different types of ecosystems. The design of the process provides control and initialization of the model through the GUI and process manager. The model will be able to request information from the SGIS internal database. For example, CENTURY requires monthly precipitation; monthly average maximum and minimum temperature; percentages of sand, silt, and clay content in soils; and land-use information to simulate the ecosystem dynamics for a particular geographically referenced cell. Maps of each of these parameters may be stored in a temporal map set or as an instant in the SGIS database. The CENTURY process can access this information through its link to the process manager. At the end of a simulation, the output from the CENTURY model is returned to the SGIS for statistical analysis and visualization. This process may be used as a prototype for linking other scientific models to the SGIS.

TEMPORAL SGIS

Temporal Data Terminology

The basic object of the temporal SGIS is to provide a time series of classical GIS maps with the same theme, geographical extent, and grid structure, each representing a specific time. The object is called a temporal map set (TMS) and is associated with a time interval t_0–t_N (Beller et al. 1991). A component map of a TMS is called a time slice or an instant. A series of instants are defined as a continuum. A group of continuums is referred to as a set. Finally, all sets needed in a given project form a database. Any number of databases can be defined in the SGIS.

Temporal Interpolation

The SGIS has temporal interpolation capabilities that provide a TMS with a continuous time interval even though only a discrete set of time slices is recorded. Existing GIS commands are able to process map names specified by a TMS name and date. If the TMS does not contain an existing map time slice for the specified time, interpolation is automatically performed to generate an appropriate time slice.

Transformation

A TMS transformation is any operation that takes one or more TMSs as input and outputs another TMS. The domain of a transformation may be restricted to a subinterval of a TMS by "FROM and/or TO" syntax. In a classical GIS, such a transformation can be classified into point or spatial neighborhood operations that operate on specific time slices. In a temporal GIS, the situation is similar except that neighborhoods can be classified into spatial, temporal, or spatial/temporal (Beller et al. 1991). Note that if the operation is point to point or acts on a purely spatial neighborhood, the transformation can be accomplished by simply distributing the operation across the time slices. The system is able to perform several different types of transformations.

CONCLUSION

As a result of this research, several goals were achieved. Most notable was the development of an integrated system with a hub architecture that allows user-supplied models to be linked with a GIS and statistical software. Additionally, the introduction of a GUI eases system use and results in an integrated system that provides a tool for the efficient management of temporally and spatially referenced data, while allowing flexible interactive model runs, analysis, and visualization.

Acknowledgments

This work was supported by an IBM Independent Research and Development (IRAD) project, the USGS-USFS TERRA Laboratory, and NASA EOS grants. We also appreciate the support of Genasys II as well as StatSci Corporation. We thank Aaron Beller, Steve Metivier, and Bill Parton for their valuable assistance. The National Center for Atmospheric Research is sponsored by the National Science Foundation.

References

Beller, A., T. Giblin, K. V. Le, S. Litz, T. Kittel, and D. S. Schimel. 1991. A temporal GIS prototype for global change research. In *GIS/LIS '91 Proceedings*, vol 2, 752–65.

Parton, W. J., D. S. Schimel, C. V. Cole, and D. S. Ojima. 1987. Analysis of factors controlling soil organic levels of grasslands in the Great Plains. *Soil Sci. Soc. Amer. J.* 51:1,173–79.

Parton, W. J., D. S. Schimel, D. S. Ojima, and C. V. Cole. 1994. A general model for soil organic matter dynamics: Sensitivity to litter chemistry, texture, and management. In *Quantitative Modeling of Soil Forming Processes*, edited by R. B. Bryant and R. W. Arnold. Publ. ASA, CSSA, and SSSA. Madison: Soil Sci. Soc. Amer. Sec.

Jan Bromberg was a member of the GIS Center at IBM Federal Systems Company and is now with Evolving Systems, Englewood, Colorado.
Evolving Systems, Inc.
Englewood, Colorado 80111

Loey Knapp is a member of the GIS Center at IBM and is an adjunct professor in the Department of Geography at the University of Colorado.
IBM Federal Systems Company
Boulder, Colorado 80301

Rebecca McKeown is an associate scientist at the National Center for Atmospheric Research (NCAR) and a research associate at the Natural Resource Ecology Laboratory (NREL), Colorado State University.

Timothy Kittel is Deputy Project Scientist of the Climate System Modeling Program at UCAR and a research associate at NREL.

Dennis Ojima is a research scientist at NREL.

David Schimel is a scientist at NCAR and a senior scientist at NREL.
Climate and Global Dynamics Division
National Center for Atmospheric Research
Box 3000
Boulder, Colorado 80307-3000
and
Natural Resources Ecology Laboratory
Colorado State University
Fort Collins, Colorado 80523
and
Climate System Modeling Program
University Corporation for Atmospheric Research
Box 3000
Boulder, Colorado 80307-3000

78

Variations on Hierarchies:
Toward Linking and Integrating Structures

Ferenc Csillag

Conceptual advantages and computational elegance of hierarchical spatial data structures (HSDS) have stimulated many theories and attracted several applications in geographic information systems (GIS) and environmental modeling. Hierarchical data structures can transparently link fields and objects, and statistically integrate uncertainty in space, time, attributes, and representation. The overview in this chapter suggests that hierarchical data structures provide the ability to focus on interesting and/or meaningful subsets of data.

INTRODUCTION

The first decades of research and development of GIS have been dominated by cartographic tradition (Sinton 1978). Viewing space, time, and attributes as the perpendicular axes of a cube, the primary approach to database development and interaction has been to independently design the characteristics of data modeling by, as Burrough (1986) put it, asking "what," "when," and "where" separately. Data modeling, in this context, refers to the abstraction that formally defines the geometric representation of a set of attributes to be mapped at a given point in time (Langran 1992). These data models, originally called raster and vector, can be synthesized in the conceptual framework of fields and objects (Goodchild 1993). Practically speaking, when one knows what to represent (e.g., mapping counties represented by their boundaries), the object view takes preference, otherwise (e.g., mapping temperature represented by contour lines) the field view is preferred. This apparent dichotomy also is well represented in software tools, which seem to deal with grid cells in the former instance, or with points, lines, and areas in the latter. The difference in preferences primarily determines the relationships between spatial data models and hierarchical data structures.

The primary goal of this chapter is to assess the current status and potential of HSDS, which can be applied for both data models (object view and field view), with regard to interfacing and integrating spatial analysis and environmental models in a GIS. First, HSDS are briefly reviewed, followed by a discussion of the primary issues involved with linking environmental models and these data models based on the nature of environmental (mapping or simulation) units. A major section of the chapter focuses on incorporating accuracy measures into the data structure itself utilizing spatial (or geo-) statistics. The final section summarizes my expectations of developments toward linking and integrating hierarchical structures in the near future.

HIERARCHICAL SPATIAL DATA STRUCTURES

Hierarchy in spatial data structures refers to linkages between (any kind of) spatial entities (in other words, ordering). The primary conceptual and computational advantage of hierarchical data structures is the principle of recursive decomposition of space (Samet 1990a).

Quadtrees

The quadtree, a two-dimensional indexing scheme, was introduced as a cartographic data structure by Klinger and Dyer (1976). It has proven to be efficient in representation and execution times for various databases, and particularly convenient for set operations and spatial search (Samet 1990a). Quadtrees primarily are based on the regular decomposition of a square region into quadrants. The graph showing links of edges between similar quadrants is broken down into a "leaf distribution" on a "tree." Specifically, the graph representation (Figure 78-1) can be characterized as a "directed tree," starting at its root and representing the entire enclosing square, with each link directed from the "parent" node toward its "children" (subquadrants). Each node in the tree has an "out degree" (number of incident edges directed away from the node) of 4, except for "leaf nodes," which are not subdivided further (and hence have an out degree of 0). The depth of the tree is denoted by n, the number of links in the longest path from the root to any leaf node. Thus, the depth determines spatial detail as well, because the level 0 nodes (e.g., cells or pixels) are considered to have side lengths of 1. The full region has a side length of 2^n, whereas a node of level k has a side length of 2^k and represents a quadrant containing 4^k cells, or pixels. (See Samet 1990b for a detailed review of quadtrees and related data structures, and Mark 1986 for a GIS-related summary.)

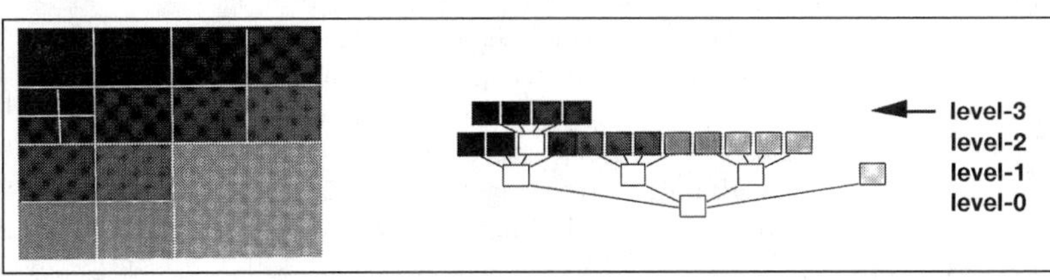

Figure 78-1. An arbitrary quadtree representation of an 8 × 8 region (left) and its leaf distribution (right). Leaf nodes are colored by their corresponding gray tones; nonleaf nodes are white. Decomposition follows a NW, NE, SW, SE order (from left to right on the tree).

Properties of Quadtree Decomposition

While there are many planar decomposition methods, squares are uniquely advantageous because the resulting decomposition yields a partition (1) that has an infinitely repetitive pattern so that it can be used for any size of space; (2) that is infinitely decomposable into increasingly finer patterns; and (3) whose tiles have similar orientation. Nonsquare decompositions of space do not meet all these criteria (e.g., hexagons do not meet the second property; equilateral triangles do not meet the third property).

From the perspective of this discussion, the most relevant geographic property of a quadtree is the principle guiding the decomposition process. The number of levels, the number of nodes, the level of aggregation, or any spatially meaningful criteria can control the decomposition. Some of these criteria are data dependent while others can be defined *a priori*. As we will see, the efficiency and/or usefulness of choosing a particular decomposition rule depends on the nature of features to be represented.

SPATIAL HIERARCHIES AND ENVIRONMENTAL MODELS

There is a well-observed similarity between the conceptual frameworks behind spatial hierarchies and environmental models: aggregation of similar or homogeneous entities. However, HSDS, or more generally GIS, and environmental models (primarily simulation) organize their information in an essentially different way (Hall and Day 1990; Mackay et al. 1992a; Csillag et al. 1992). Clearly, the objects an environmental model has to deal with (e.g., forest stand, hill slope, stream, ecosystem) follow a conceptually different hierarchy than those of a spatial database (e.g., points, lines, polygons, grid cells). Quite extensive examples have been reported as evidence that environmental objects can be derived from spatial objects by various spatial analytical methods and tools (Lawrence et al. 1986; Moore et al. 1993; Hall et al. 1992; Band et al. 1993). Therefore, the fundamental issue of linking and integrating environmental modeling and spatial databases is the way environmentally relevant objects are defined from spatially meaningful objects.

Modeling and Accuracy

HSDS are particularly attractive for studying relationships of, and developing tools for, linking and integrating spatial and environmental objects because of their ability to control the level of aggregation according to "interesting" subsets of the data. Depending on how this hierarchical, or spatially varying aggregation procedure is guided, HSDS can play different roles in environmental modeling. In each case, however, HSDS not only provide potentially high performance in data manipulation, but they also facilitate the integration of accuracy measures into the data structure itself (Tobler 1988; Csillag 1991; Kertész et al. 1993).

When environmental objects (e.g., mapping, or simulation units) are defined *a priori* (i.e., before data processing), HSDS can be important tools in developing global addressing schemes for data conversion. When objects for environmental mapping or simulation are defined *a posteriori* (i.e., after data processing), two distinct cases result: (1) the objects can be formally defined so that a knowledge base communicates between the database of spatial objects and the database of simulation objects, and (2) statistical tools can be applied on spatial objects to determine a feasible partition for environmental units.

Global Addressing Schemes

When spatial units of interest are defined *a priori,* quadtrees (and similar regular decompositions) can serve as spatial addressing schemes. The location of points, lines, or regions then can be encoded in such a way so that the length of their address is a direct measure of their locational uncertainty (Dutton 1988). In the planar case this is simple to follow and implement for either kind of data, but it is interesting to point out that HSDS can support very efficient conversion between them by representing regions, and/or their edges, and/or their centers (Samet 1980). HSDS and data-representation problems also are becoming more popular in supercomputing (especially massively parallel) applications. (Bestul 1991; Mills et al. 1993).

A significant challenge for both GIS and environmental modeling has been the handling of global data sets (Mounsey and Tomlinson 1988). Several regular decomposition schemes recently have been proposed to support global modeling by global hierarchical tesselation. The impossibility of decomposing the earth's surface, similar to the Cartesian case, led Dutton (1984, 1988) and Goodchild and Yang (1992) to project the globe as a sphere onto the surface of an octahedron, and to develop global addressing by regularly decomposing the eight triangular facets (Figure 78-2). This scheme results in slightly varying size facets. Tobler and Chen (1986) used a cylindrical projection instead, where they compromised contiguity (neighborhood) at the poles for equal-area facets.

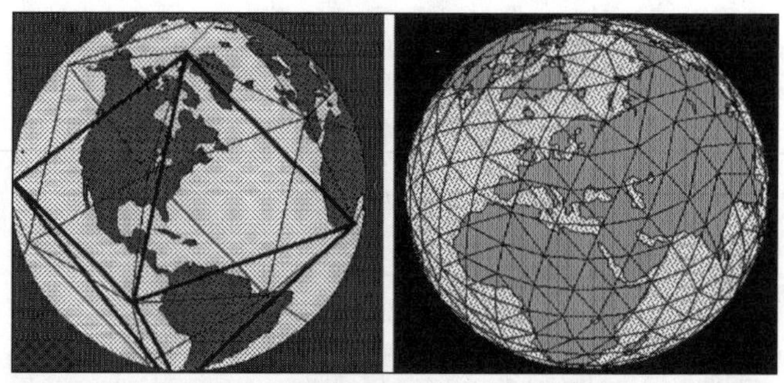

Figure 78-2. Orthographic views of the level 1 and level 3 QTM on a basis octahedron. The 4,398,046,511,104 level 22 facets have a resolution of 1 m, and their address requires 42 code bits. (Courtesy of G. Dutton.)

HIERARCHICAL DECOMPOSITION OF HETEROGENEOUS REGIONS

In most instances, environmental models require the construction of mapping or simulation units during data processing. Data may be available in a variety of forms (e.g., remotely sensed data, digital elevation models, soil maps, etc.), all representing some characteristics of spatial heterogeneity. Measures of heterogeneity (e.g., variance/covariance) can control the level of aggregation.

Knowledge-Based Aggregation

When topography provides a primary guide (e.g., in most forested ecosystems where regional effects dominate the landscape), simulation objects can be derived from simpler spatial units. There are many approaches to a more or less formal definition of simulation objects. Mechanistic aggregation (averaging) along the least-varying variable has been quite frequent, just like extraction of lumped key characteristics (in both time and space) based on personal expertise (Everham et al. 1991; Ollinger et al. 1993). In order to more intelligently handle spatiotemporal variability for watershed level modeling, the hierarchy of streams and hill slopes can be built into a knowledge base that can communicate between spatial objects and simulation objects (Figure 78-3) (Band 1989, 1990; Mackay et al. 1992a, 1992b). Inferences based on this system can thus be evaluated against within-unit variances.

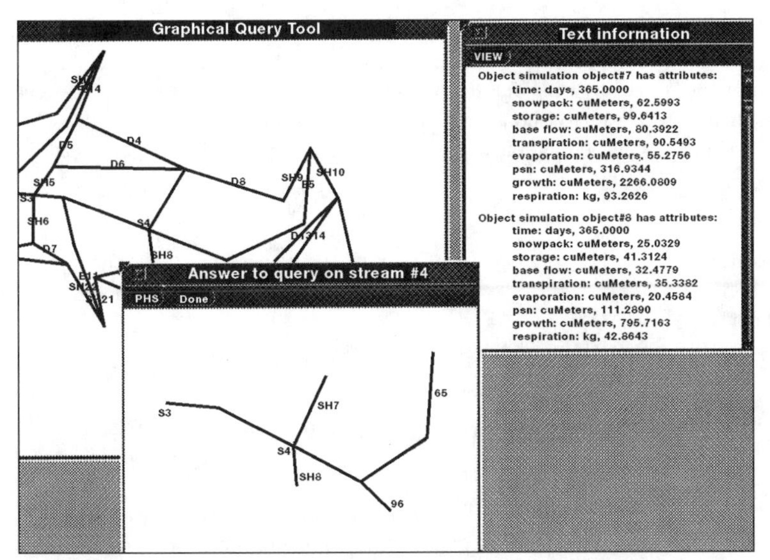

Figure 78-3. An illustration of a knowledge-based geographic information management system. With the graphical query tool, one can query information about streams along hill slopes. The partition of a watershed into hill slopes can be guided by local heterogeneity. (Courtesy of V. Robinson.)

Statistics-Based Optimal Tiling of Surfaces for Sampling and Mapping

In cases where the spatial structure of the phenomenon to be mapped is not known, statistical tools can be used to determine meaningful mapping and simulation units from raw and ancillary (in many cases, remotely sensed) data (Jeon and Landgrebe 1992). Spatial statistical (or geostatistical) measures require fairly strong assumptions about the distribution of variance and the shape of the covariance function (Kashyap and Chellapa 1983), but they are very useful in the design of sampling and efficient in mapping (Webster and Oliver 1990). Recent research and development in HSDS reveals how leaf distributions of quadtrees can be used for statistical inference in cases when no spatial hierarchy can be defined *a priori* (Kummert et al. 1992).

It is rather difficult to determine *a priori* (sampling and mapping) units in any meaningful way (Kertész et al. 1993) when modeling regions where local factors play a dominant role in shaping the surface; for example, in midlatitude grasslands (Hole and Campbell 1985). Testing all possible partitions is obviously impossible. Quadtrees, however, can be thought of as spatial classifications with the advantage that each location not only belongs to one, and only one, leaf, but the location, size, and arrangement of all possible leaves are known. Although this is a strong constraint on the shape, size, and potential number of patches represented by a quadtree as a function of the number of levels, it makes it possible to compare all possible maps during decomposition. Choosing Kullback divergence—an appropriate, nonparametric measure of local heterogeneity (Csiszár 1975; see Kullback Diversion section)—one can partition a heterogeneous field into either a given number of patches (sampling) or into a given threshold in within-unit heterogeneity (mapping). The important property of this partition (or tiling) is that it can adjust to local inhomogeneities so that the size of the units varies while their accuracy stays constant (Kertész et al. 1993; Csillag et al. 1993). It is also worth noting that this method links cost versus quality in a framework where residual heterogeneity is spread over the map as evenly as possible (Figure 78-4). Consequently, at highly variable locations, mapping units are smaller than they are over large homogeneous surfaces. This strategy especially is useful when local variability is significant over a wide range of scales and when contrasts among mapping units are not obvious (Mark and Csillag 1989). It has been tested on a semiarid salt-affected grassland and found to be superior to other point-based interpolation techniques and robust against various levels and spatial structures of noise (Csillag et al. 1993).

A simple transformation of the tile divergence (the total Kullback divergence versus the number-of-leaves function) elucidates the inherent spatial pattern of the phenomenon to be mapped and highlights the relationship between geostatistics and quadtree decomposition. When the tile size equals the grid cell size (i.e., the maximum number of leaves), and when the tile size is 1 (i.e., the root of the quadtree), this function is 0. However, it has a rather characteristic maximum after the first cuts when the number of tiles starts to increase. There will be a cut

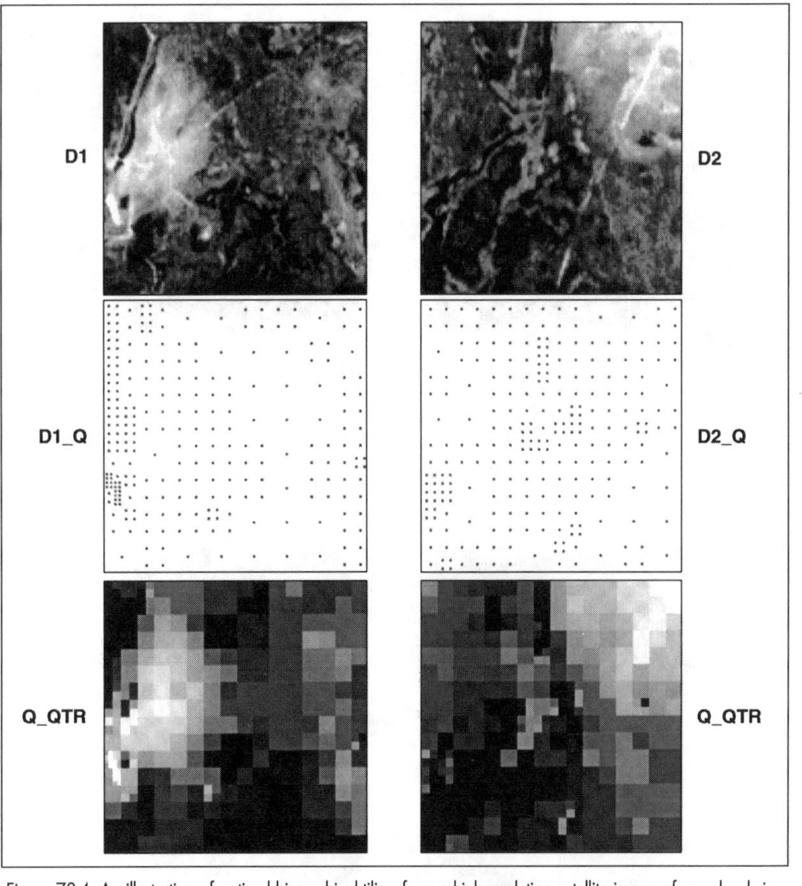

Figure 78-4. An illustration of optimal hierarchical tiling from a high-resolution satellite image of grasslands in east Hungary. The two 128 × 128-pixel test areas (top) sharply differ in spatial characteristics. Their quadtree decompositions (bottom) are the best approximations with 256 homogeneous leaves. The distribution of the centers of the leaves (middle) are to guide the eye.

when the increase in the number of tiles (by 3 in each step) will be compensated for by reduced heterogeneity (Figure 78-5). This maximum, in the case of stationary data sets, corresponds to the spatial correlation length, or range of the semivariogram (Csillag et al. 1993). Furthermore, this maximum is hardly sensitive to deviations from stationarity (e.g., additive noise).

Kullback Divergence

Kullback divergence is closely related to Shanonn's information formula as well as to χ^2. It is a measure of distance between two-dimensional distributions. If a data set is available on a square grid that we would like to approximate with a quadtree (map), then:

$$D(grid \backslash map) = \sum_{i=1} \sum_{j=1} [grid_{ij}/SUM] \, log_2(grid_{ij}/map_{ij}) \quad (78\text{-}1)$$

and

$$SUM = \sum_{i=1}^{I} \sum_{j=1}^{J} grid_{ij} = \sum_{i=1}^{I} \sum_{j=1}^{J} map_{ij}$$

where $grid_{ij}$ is the value for i,jth cell of the grid, map_{ij} the value for i,jth cell of the map, and I and J are the side lengths of the grid to be mapped, and the map in corresponding cells, respectively.

It is obvious from (78-1) that Kullback divergence can measure the similarity between the lattice and the map in a nonspatial manner by assigning one single number to the compared pair. Its real advantage, however, is that for any delineated patch on a map, the contribution of that particular patch to the total divergence can be computed:

$$D(patch_{grid} \backslash patch_{map}) = \sum_{i}^{patch} \sum_{j} [grid_{ij}/SUM] \, log_2(grid_{ij}/map_{ij}) \quad (78\text{-}2)$$

From (78-1) and (78-2) we derive:

$$D(grid \backslash map^{(Q)}) = \sum_{i=1}^{I} \sum_{j=1}^{J} [grid_{ij}/SUM] \, log_2(grid_{ij}/map^{(Q)}{}_{ij}) \quad (78\text{-}3)$$

where $map^{(Q)}$ is the map defined by the Q quadtree $\in$ Q, Q is the set of all quadtrees covering the grid to be mapped; and

$$D(l_{grid} \backslash l_{map}) = \sum_{i=l_{i0}}^{l_{imax}} \sum_{j=l_{j0}}^{l_{jmax}} [grid_{ij}/SUM] \, log_2(grid_{ij}/map^{(Q)}{}_{ij}) \quad (78\text{-}4)$$

where l_{map} is a leaf of the map represented by a quadtree, and l_{grid} is the corresponding lattice area and the sums run for the leaf.

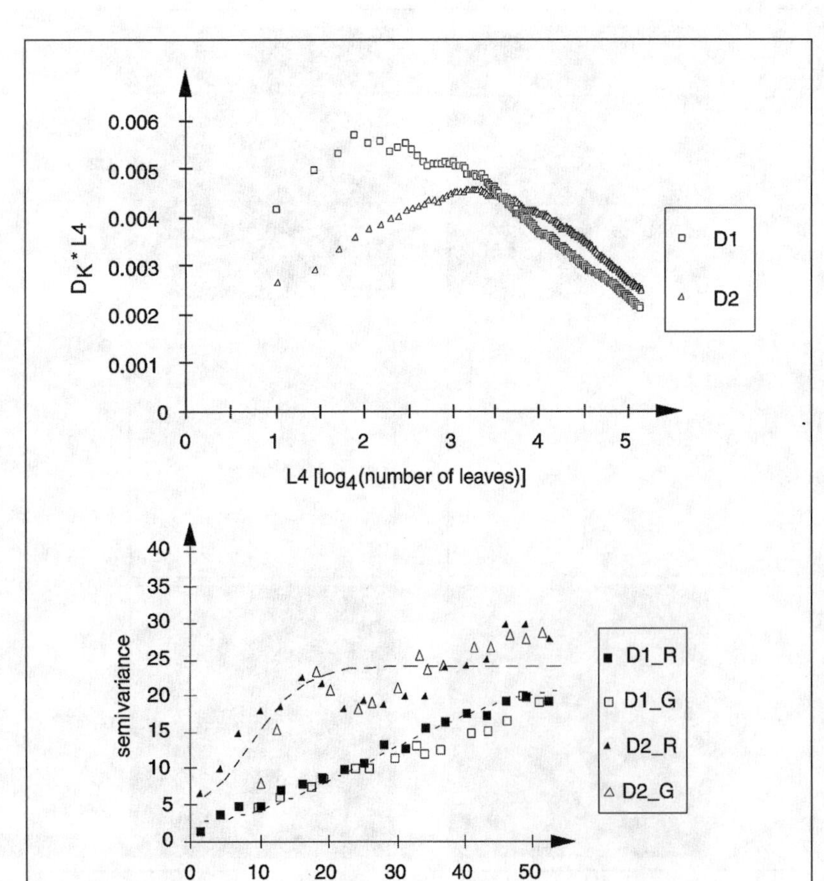

Figure 78-5. Transformed tile divergences for the two test scenes (top), and the corresponding semivariograms estimated from regular and random sampling (bottom). For D1, a range of 35.0 was obtained (roughly 13 equal patches with a size of 35^2); for D2, a range of 13.85 was obtained (roughly 85 patches with a size of 13.85^2).

CONCLUSION

Major challenges and opportunities in linking and integrating environmental models and HSDS are: addressing/ordering in a global GIS, and environmentally meaningful object definitions for regional (and larger-scale) applications either by artificial intelligence, by statistical means, or by both. These issues (i.e., calibration of solutions) require enormous amounts of data, possibly an order of magnitude more than are available today. The most important step toward a "general" solution (if one exists) is to recognize the inherent nature of the phenomena to be mapped and modeled and to adjust our concepts (taxonomies, hierarchies) to them.

Acknowledgments

This research has been funded partially by a grant to Syracuse University from the IBM Environmental Research Program, and by the Program of Science and Technology of the U.S. Agency for International Development. Any opinions, findings, conclusions, or recommendations expressed in this material are those of the author and do not necessarily reflect the views of the sponsors.

References

Band, L. 1986. Topographic partition of watersheds with digital elevation models. *Water Resources Research* 22:15–24.

———. 1989. Spatial aggregation of complex terrain. *Geographical Analysis* 21(4):279–93.

Band, L., P. Patterson, R. Nemani, and S. W. Running. 1993. Forest ecosystem processes at the watershed scale: Incorporating hill slope hydrology. *Agricultural and Forest Meteorology* 63:93–126.

Bestul, T. 1991. A general technique for creating SIMD algorithms on parallel pointer-based quadtrees. In *Proceedings, AutoCarto-10,* 428–44. Bethesda: American Congress on Surveying and Mapping.
Burrough, P. A. 1986. *Principles of Geographical Information Systems for Land Resources Assessment.* Oxford: Clarendon Press.
Csillag, F. 1991. Resolution revisited. In *Proceedings, AutoCarto-10,* 15–29. Bethesda: American Congress for Surveying and Mapping.
Csillag, F., M. Kertész, and A. Kummert. 1992. Resolution, accuracy and attributes: Approaches for environmental geographic information systems. *Computers, Environment and Urban Systems* 16:289–97.
———. 1993. Sampling and Mapping of Two-Dimensional Lattices by Stepwise Hierarchical Tiling Based on a Local Measure of Heterogeneity. (Manuscript under review.)
Csiszár, I. 1975. I-divergence geometry of probability distributions. *Annals of Probability* 3:146–58.
Dutton, G., 1984. Geodesic modeling of planetary relief. *Cartographica* 21:188–207.
———.1988. Modeling locational uncertainty via hierarchical tesselation. In *Accuracy of Spatial Databases,* edited by M. F. Goodchild and S. Gopal, 125–40. London: Taylor & Francis.
Everham, E. M., K. B. Wooster, and C. A. S. Hall. 1991. Forest landscape climate modeling. In *Proceedings, Symposium on Systems Analysis in Forest Resources,* edited by M. Buford. Technical Paper SE-74. Ashville: USFS Southeastern Forest Experimental Station.
Goodchild, M. F. 1993. The state of GIS for environmental problem solving. In *Environmental Modeling with GIS,* edited by M. F. Goodchild, B. O. Parks, and L. T. Steyaert, 8–15. New York: Oxford University Press.
Goodchild, M. F., and Yang Shiren. 1992. A hierarchical spatial data structure for global geographic information systems. *Computer Vision, Graphics and Image Processing: Graphical Models and Image Processing* 54(1):31–44.
Hall, C. A. S., and J. W. Day. 1990. *Ecosystem Modeling in Theory and Practice.* Boulder: University Press of Colorado.
Hall, C. A. S., M. R. Taylor, and E. Everham. 1992. A geographically based ecosystem model and its application to the carbon balance of the Luquillo Forest, Puerto Rico. *Water, Air and Soil Pollution* 64:385–404.
Hole, F. D., and J. B. Campbell. 1985. *Soil Landscape Analysis.* London: Routledge and Kegan Paul.
Jeon, B., and D. A. Landgrebe. 1992. Classification with spatiotemporal interpixel class dependency contexts. *IEEE T. on Geosciences and Remote Sensing* 30:663–72.
Kashyap, R. L., and R. Chellapa. 1983. Estimation of choice of neighbors in spatial-interaction models of images. *IEEE T. on Information Theory* 29:60–72.
Kertész, M., F. Csillag, and A. Kummert. 1993. Mapping Heterogeneous Surfaces by Optimal Tiling. (Manuscript under review.)
Klinger, A., and C. R. Dyer. 1976. Experiments in picture representation using rectangular decomposition. *Computer Graphics and Image Processing* 5:68–105.
Kummert, A., M. Kertész, F. Csillag, and S. Kabos. 1992. *Dirty Quadtrees: Pruning and Related Regular Decompositions for Maps with Predefined Accuracy.* Technical Report 92-3. Budapest: Research Institute for Soil Science.
Langran, G., 1992. *Time in Geographic Information Systems.* London: Taylor & Francis.
Lawrence, G. B., R. D. Fuller, and C. T. Driscoll. 1986. Spatial relationships of aluminum chemistry in the streams of the Hubbard Brook experimental forest, New Hampshire. *Biogeochemistry* 2:115–35.
Mackay, D. S., V. B. Robinson, and L. E. Band. 1992a. Classification of higher order topographic objects on digital terrain data. *Computers, Environment and Urban Systems* 16:473–96.
———. 1992b. Development of an integrated knowledge-based system for managing spatiotemporal ecological simulations. In *Proceedings, GIS/LIS'92,* 494–503. Bethesda: American Society for Photogrammetry and Remote Sensing.
Mark, D. M. 1986. The use of quadtrees in geographic information systems and spatial data handling. In *Proceedings, AutoCarto London,* 517–26. London: Royal Institute of Chartered Surveyors.
Mark, D. M., and F. Csillag. 1989. The nature of boundaries on "area-class" maps. *Cartographica* 26:65–79.
Mills, K., F. Csillag, and M. Kaddoura. 1993. GORDIUS—A data parallel algorithm for spatial data conversion. *Computers & Geosciences* 19(7):1,051–63.
Moore, I. D., A. K. Turner, J. P. Wilson, S. K. Jenson, and L. E. Band. 1993. GIS and land-surface-subsurface process modeling. In *Environmental Modeling with GIS,* edited by M. F. Goodchild, B. O. Parks, and L. T. Steyaert, 196–230. New York: Oxford University Press.
Mounsey, H., and R. Tomlinson, eds. 1988. *Building Databases for Global Science.* London: Taylor & Francis.
Ollinger, S. V., J. D. Aber, G. M. Lovett, S. E. Millham, et al. 1993. A spatial model of atmospheric deposition for the northeastern United States *Ecological Applications* 3(3):459–72.
Samet, H. 1980. Region representation: Quadtrees from boundary codes. *Comm. ACM* 23:163–70.
———. 1990a. *The Design and Analysis of Spatial Data Structures.* Reading: Addison Wesley.
———. 1990b. *Applications of Spatial Data Structures.* Reading: Addison Wesley.
Sinton, D. 1978. The inherent structure of information as a constraint in analysis: Mapped thematic data as a case study. In *Harvard Papers on Geographic Information Systems,* edited by G. Dutton, vol. 7. Reading: Addison Wesley.
Tobler, W. 1988. Resolution, accuracy and all that. In *Building Databases for Global Science,* edited by H. Mounsey and R. Tomlinson, 129–37. London: Taylor & Francis.
Tobler, W., and Z. Chen. 1986. A quadtree for global information storage. *Geographical Analysis* 18:360–71.
Webster, R., and M. Oliver. 1990. *Statistical Methods in Soil and Land Resource Surveys.* Oxford: Oxford University Press.

Ferenc Csillag received his M.S. in geophysics and his Ph.D. in cartography in 1980 and 1986, respectively, from the Eötvös Lóránd University in Budapest, Hungary. After working on digital image processing at the Remote Sensing Center in Budapest, he spent five years with the Soil Research Institute of the Hungarian Academy of Sciences where he focused on geostatistics and GIS. He then taught geographic information analysis at Syracuse University in New York. In 1993 he joined the University of Toronto as an associate professor. His current research interests include the impact of scale and regionalization on spatial data quality.
Institute for Land Information Management
University of Toronto
Erindale College
Phone: (905) 828-3862
Fax: (905) 828-5273
E-mail: *fcs@eratos.erin.utoronto.ca*

79

An Object-Oriented Model Base Management System for Environmental Simulation

David A. Bennett, Marc P. Armstrong, and Frank Weirich

This chapter describes the design and implementation of a geographical modeling system (GMS) that facilitates the construction and use of dynamic environmental simulation models. The GMS provides an operational framework in which spatial knowledge can be stored and managed, theory can be modeled and tested, and alternative resource management strategies can be evaluated. Our goal is to provide users with the materials and tools needed to construct sophisticated geographic models that accurately represent both the structure and behavior of natural systems. To accomplish this goal, we employed object-oriented analysis and design methods to integrate model base management and GIS technologies into a single system.

INTRODUCTION

Commercial geographic information systems (GIS) software do not fully support the study and management of complex environmental systems because they fail to provide users with the capabilities to represent and model dynamic spatial processes. This research provides a new approach to software development in the form of a geographical modeling system (GMS) that is designed to overcome these limitations. The GMS provides a virtual environment in which spatial knowledge can be stored and managed, theory can be modeled and tested, and alternative resource management strategies can be evaluated. Our objective in creating a GMS was to provide users with the materials and tools needed to construct sophisticated geographic models that accurately represent both the structure and behavior of spatial systems. To meet this objective, we integrated model base management techniques with GIS technology using object-oriented analysis, design, and programming techniques. The GMS approach is illustrated by the development of a hydrological model for a mountainous watershed in the San Dimas experimental forest in southern California.

Available Technologies to Represent Environmental Systems

To fully represent an environmental system in the digital domain, both its structure and behavior must be represented. The structure of a system refers to the absolute and relative locations of its constituent components, as well as thematic information relevant to a particular problem or study. The behavior of a system, on the other hand, refers to dynamic physical processes associated with that system (e.g. the flow of water over a landscape). Commercially available GIS software enable researchers to store, retrieve, and manipulate the physical attributes of geographic features. Thus, while GIS software packages provide a suitable environment for the representation and analysis of spatial structure, they provide only limited support for the simulation of spatial behavior. Simulation models are designed to capture such behavior, but they often provide an overly simplified representation of spatial structure. Furthermore, simulation software often lack the database and model base management capabilities needed to modify, analyze, and query data and models. Previous attempts to link the database management capabilities of GIS with simulation models often used a loose coupling (Nyerges 1993) between separate software products. In that approach, GIS software was used to construct input files, which were read by a simulation program. The results of the simulation then were read back into the GIS software for display and analysis.

Although loosely coupled systems can be developed using existing technologies they (1) preclude user interaction and query during a simulation, (2) lack model base management capabilities, and (3) do not effectively capture temporal relationships.

Spatial decision support systems (SDSS) are designed to overcome the limitations of loosely coupled systems by integrating domain-specific knowledge, analytical models, and spatial data-handling techniques (Armstrong and Densham 1990). An SDSS, therefore, represents a more tightly coupled approach to the integration of spatial data handling and computer simulation. Guariso and Werthner (1989) used decision support technology to develop EDSS-1 (an environmental decision support system). Jankowski (1992) used a similar approach to construct a water-quality model (Zeigler 1990). However, both approaches lack support for geographic data structures and, thus, the ability of this software to represent spatial structure is reduced.

While attempts have been made to provide a cohesive seamless environment within which dynamic simulation models can be developed, run, and analyzed, key SDSS modules are not fully developed. Perhaps most importantly, existing SDSS do not possess the model base management capabilities needed to construct, manipulate, and implement dynamic environmental models. Adding model base management capabilities to geoprocessing technologies is one of the primary objectives of this research. As such, this research provides a first step toward the integration of SDSS, GIS, and environmental simulation.

MODELING ENVIRONMENTAL SYSTEMS

A model is a tool used by scientists and decision makers to abstract from reality those components that are most germane to their problem domain. Though models can be conceptual, physical, or numerical, the process of abstraction is intended to reduce the complexity of the real world to a manageable level. Conceptual models diagram the relationships between those components of reality that impact the spatial system under study. For example, ecologists use compartment models, a form of conceptual model, to graphically illustrate the flow of energy and matter through an ecosystem. Physical models are scaled representations of the physical world (e.g., architectural models, maps). These models provide a mechanism through which scientists can observe and manipulate the environment in ways that would not be possible in the field. Numerical models capture the behavior of spatial processes in mathematical form. These models can, for example, represent systems that continually change through time (continuous), change in response to a particular event (discrete), or optimize decision criteria (optimization).

Model base management systems (MBMS) have been developed to support numerical modeling. However, when developing a numerical model, it is often helpful to first organize the relevant components and relationships of the environmental system being simulated in a conceptual model. Therefore, full-featured MBMS should assist users in the construction of conceptual models and, when appropriate, use these representations to facilitate the development of numerical models. The unique characteristics of a physical model can be approximated in MBMS using scientific visualization principles and a graphical user interface (GUI) that allows users to modify the digital environment and emulate the geographic system under study.

Simulation Models as Repositories of Scientific Knowledge

When an environmental simulation model is created, knowledge about landscape form and physical process is captured in digital form. A model base management system that supports environmental simulation must be able to store and manage this knowledge. Armstrong (1991) described three categories of spatial knowledge: geometrical, structural, and procedural. Although this classification scheme was developed to support automated map generalization, these three types of knowledge provide a useful framework for the management of environmental simulation models:

Geometrical knowledge: refers to the dimension, location, and topology of geographic features. When a user selects a cartographic representation (spatial data structure) for a geographic feature, geometrical knowledge is encoded as part of the simulation model.

Structural knowledge: draws from the generating processes that form the landscape. This knowledge is captured by adding thematic information to cartographic objects that represent the form and character of the landscape.

Procedural knowledge: represents an understanding of how environmental processes operate. It is encoded into simulation models as a set of mathematical equations and rules that direct the simulation process.

In most existing simulation models, knowledge is encoded implicitly, and no clear distinction is made between geometrical, structural, and procedural knowledge. This implicit representation of knowledge creates a knowledge base that is difficult to organize and modify. As such, experimentation in a digital environment—one of the primary benefits of the simulation process—is inhibited.

A different approach is promoted in the GMS. Each component of spatial knowledge (geometrical, structural, and procedural), is explicitly represented as a separate construct. Geometrical knowledge is modeled by objects that store spatial data structures. Structural knowledge is represented as an aggregate object composed of a cartographic object and a set of objects that represent problem-specific characteristics. Finally, three forms of procedural knowledge that can be incorporated into the GMS are:

- state dependent—knowledge that describes the behavior of a given environmental process at a particular point in time (e.g., a mathematical representation of sediment-laden streamflow);
- process dependent—knowledge that describes how the behavior of an environmental system can change through time (e.g., streamflow represented as a set of models that capture the full-range flow of behavior [sediment laden, hyperconcentrated, and debris flow]); and
- situation dependent—knowledge that determines how to restructure the simulation model to maintain validity when the use of a particular mathematical representation is no longer consistent with the state of the system being modeled.

Although knowledge in the GMS framework is aggregated into more complex objects that represent geographic features (e.g., a stream segment) and environmental systems (e.g., a watershed), it remains modular. The encapsulation of geometrical, structural, and procedural knowledge into modular objects facilitates experimentation.

GMS DESIGN AND IMPLEMENTATION

GMS are object-oriented software systems (Booch 1991), designed to be flexible and extensible. Few constraints are placed on model design and implementation; as such, users can construct simulation models that best represent the structure and behavior of the particular natural system they wish to simulate. To store cartographic, structural, and procedural knowledge and to support the creation and use of environmental simulation models, the GMS provide the following class hierarchy:

Cartographic objects (class CCartographic): spatial data structures that capture the form and distribution of spatial features (store geometrical knowledge).

Structural objects (class CStructure): encapsulate the attribute information needed to simulate a particular process with a cartographic object that stores the location of a given feature (store structural knowledge).

Behavioral models (class CBehavioralModel): mathematical models that describe spatial processes (store state-dependent and situation-dependent procedural knowledge).

Behavior models (class CBehavior): capture the full range of behavior exhibited by geographic features (store process-dependent procedural knowledge).

Geographic features (class CFeature): composite objects that capture the structural, behavioral, and topological characteristics of geographic features (e.g., a stream segment).

Geographic models (class CGeoModel): geographic objects that capture the structure and behavior of a geographic system (e.g., a stream network).

Flow models (classes CFlow and CMWayFlow): objects that represent the flow of material and energy across the landscape.

Constructor managers: model base managers that assist in the construction of new models.

Implementor managers: model base managers that execute existing models.

New Model Construction

Users interact with the GMS through an icon-based interface. Through this interface, users define the structure and behavior of the natural system that they wish to simulate. Constructor managers accept the design entered by the modeler and create the code needed to implement the model (Figure 79-1). Through this interaction between the user, constructor managers, and interface objects, much of the model creation process is reduced to a drag-and-drop operation.

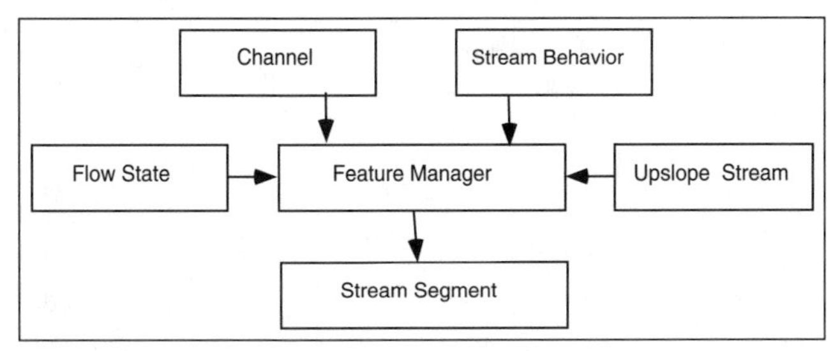

Figure 79-1. Constructor managers interact with the user to construct project class definitions.

Variable Structure Model Management

As suggested, the validity of a mathematical model is often conditional on the physical characteristics of the feature being simulated. Since these characteristics change through time, it may be necessary to represent the behavior of a geographic feature as a set of behavioral models and provide a capability to dynamically change the model used to simulate a feature. This capability, known as variable structure model management, is implemented in the GMS as illustrated in Figure 79-2. Each object of class *CFeature* points to: (1) a behavior (class *CBehavior*) that contains a list of models that describes its full range of behavior, and (2) a behavioral model that best models its current state. Behavioral models (class *CBehaviorModel*) contain a mathematical expression that describes a given process, and knowledge that defines the physical context within which that expression is valid. At each time step during the course of a simulated event, the validity of the model being used to simulate a feature is checked (the *IsConsistent* method). If this check fails, a *FindBestModel* message is sent to the behavior object. In response to this message, an appropriate model is selected from the list of models that comprise the behavior.

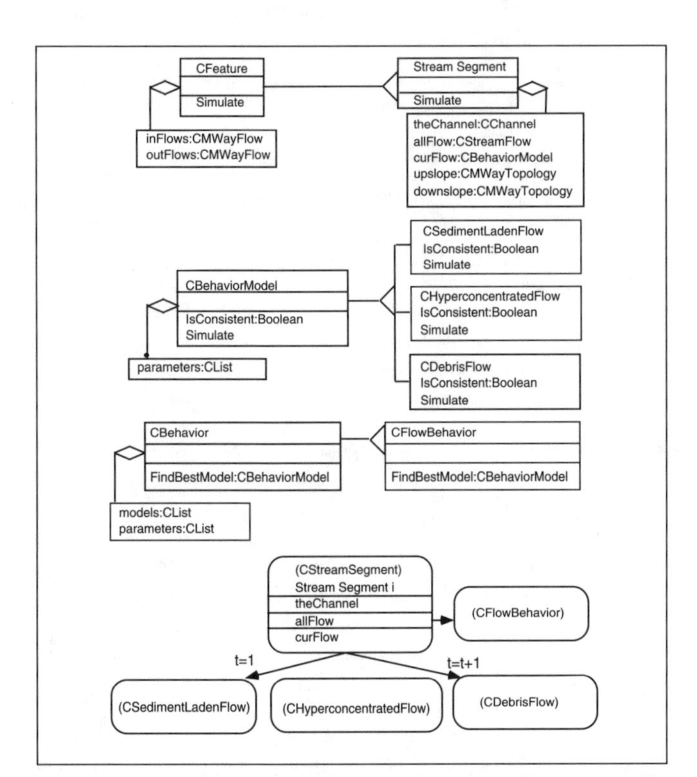

Figure 79-2. Object diagram for geographic features illustrating variable structure model management. At each time interval i, *CStreamSegment* sends a *IsConsistent* message to *curFlow*. If the response to this message is "false," a *FindBestModel* message is sent to *allFlow*, and this method returns a valid (or null) model to *CStreamSegment*.

DESIGN AND IMPLEMENTATION OF A WATERSHED MODEL

To illustrate the use of GMS, a distributed parameter model was developed for a hypothetical watershed that synthesized the characteristics of several basins found in the San Dimas experimental forest in southern California. The following assumptions were used to guide the development of this model:

- stream, overland, and subsurface flow would be simulated;
- flow would be modeled using a distributed parameter approach;
- three streamflow states would be modeled: sediment-laden, hyperconcentrated, and debris flow; and
- information would be gathered relative to stream location (i.e., the system had to support a query, such as "list data for all stream segments upslope of any given stream segment."

These requirements were included to ensure that the system could represent one-, two-, and three-dimensional objects, model polymorphic behavior (behavior simulated by a set of models), and incorporate user-defined topological relationships. To support user needs that are diverse and not wholly known, these capabilities are considered essential.

System Design

The first step in developing an object-oriented software system is to decompose the problem domain into key abstractions. An abstraction "denotes the essential characteristics of an object that distinguish it from all other kinds of objects and thus provides crisply defined conceptual boundaries, relative to the perspective of the viewer" (Booch 1991). This process, referred to as object-oriented analysis, provides input to object-oriented design, which is concerned with the representation of identified abstractions as a cooperative set of objects that capture the structure and behavior of the system under study.

The character of the watershed can be captured by four high-level abstractions: precipitation, throughfall, surface flow, and subsurface flow. These abstractions must be further decomposed into more simple objects until the structure and behavior of each object can be fully defined. In generic form, this hierarchical organization of abstracted features is illustrated in Figure 79-3. A project is composed of models, models are composed of submodels and features, and features are composed of structures and behaviors.

CSurfaceModel: The surface of a watershed can be represented as a stream reach, two hill slopes, and a set of subbasins (Figure 79-4). To

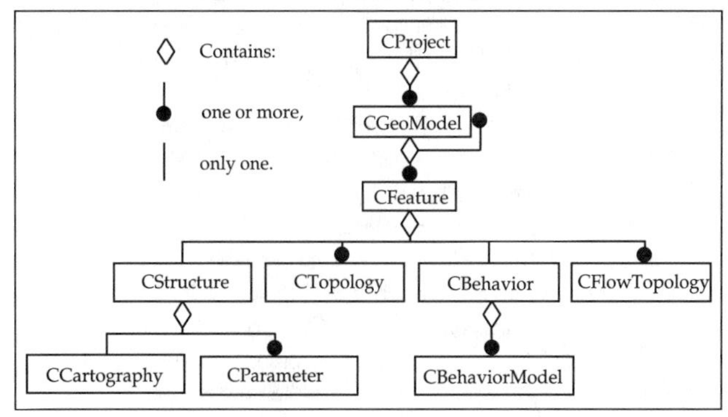

Figure 79-3. Environmental simulation models can be defined as class composition hierarchies.

model surface flow, a *CSurfaceModel* object propagates a *Simulate message* to instances of the features *streamReach, hillslope1, hillslope2,* and *subbasins.* A stream reach, as defined here, is a nonbranching section of stream. However, the flow of water along a stream reach depends on such factors as inflow, outflow, channel slope, channel shape, channel roughness, and water depth. Because these factors may vary along the course of a stream, it is necessary to decompose each stream reach object into a set of stream segments (*CStreamSegment*) that simulate flow behavior. The class *CStreamSegment* is a subclass of *CFeature* and contains an instance of *CChannel* (structure) and *CStreamFlow* (behavior) (see Figure 79-2). In addition, *upstream* and *downstream* topological relationships are included to allow the user to gather information relative to stream location.

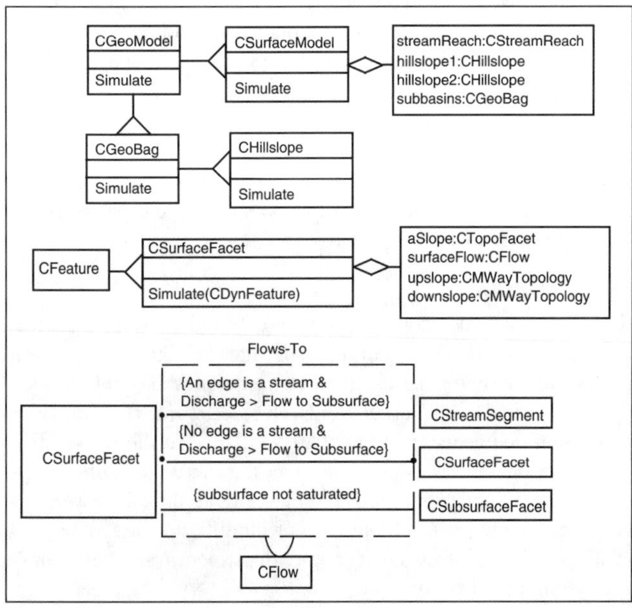

Figure 79-4. Object diagram for class *CSurfaceModel. CFlow* objects in the Flows-To association are stored in the *inFlow* and *outFlow* variables inherited from *CFeature* (see Figure 79-1).

To construct a distributed parameter model it is necessary to decompose hill slopes into tracts of land that are homogeneous with respect to those factors that influence the flow of water and sediment. To store these objects, *CHillslope* is implemented as a subclass of *CGeoBag.* The class *CGeoBag* is a subclass of *CGeoModel* that contains a linked list. For *CHillslope,* each object in this list is an instance of *CSurfaceFacet.* Objects of class *CSurfaceFacet* simulate the flow of water and sediment across the landscape. The procedural knowledge needed to accomplish this simulation is stored in the behavioral class *COverlandFlow.* Flow emitted from *CSurfaceFacet* objects discharge to objects of class *CSurfaceFacet, CStreamSegment,* and *CSubsurfaceFacet.* Since a triangulated irregular network (TIN) is used to tessellate the landscape, the structure of *CSurfaceFacet* is represented by the class *CTopoFacet,* which, in turn, contains an instance of the cartographic class *CTINFacet.*

CSubsurfaceModel: For each instance of *CSurfaceFacet* there is a corresponding instance of *CSubsurfaceFacet* (Figure 79-5). The set of all *CSubsurfaceFacet* objects constitutes the *CSubsurfaceFlow* model. To store these objects, *CSubsurfaceModel* is implemented as a subclass *CGeoBag.* Objects of class *CSubsurfaceFacet* simulate the flow of water through the soil profile. The discharge from each facet can flow to (1) downslope subsurface facets, (2) the surface facet directly above it, or (3) a stream. The

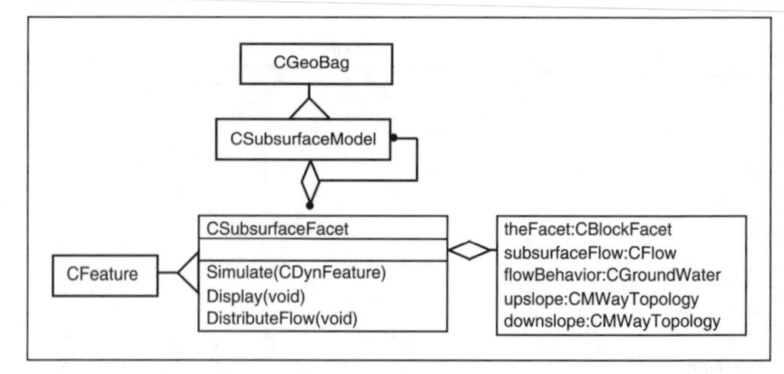

Figure 79-5. Object diagram for *CSuburfaceFacet.*

structure of subsurface facets is captured with class *CBlockFacet,* and subsurface flow behavior is modeled by *CGroundWater.* The *CBlockFacet* contains an instance of the cartographic class *CBlockTINFacet* (a subclass of *CTINFacet* that includes a variable for facet depth).

CVegModel: Throughfall is that part of precipitation that passes through the vegetative cover. As such, each instance of *CVegModel* contains a collection of objects that represents patches of vegetation (*CVegetation*) (Figure 79-6). As implemented here, each patch captures the behavior of one of the common plant community types found in the San Dimas experimental forest (chaparral desert scrub, sagebrush, grassland, and woodland). Unlike most hydrologic models, where vegetation is modeled by attributes stored as part of a data structure that represents land use and land form, vegetation in a GMS is modeled as a unique geographic object possessing structure and behavior (class *CVegetation*). The structure of this class is represented by *CVegPatch* and its behavior is simulated by *CThroughfall.* To track the flow of water through the vegetation submodel, an associative link is constructed between *CSurfaceFacet* and *CVegetation* through a *CFlow* object. Plant communities form a mosaic of irregularly shaped patches across the landscape. As such, the *CVegPatch* class includes an instance of the cartographic class *CPolygon.*

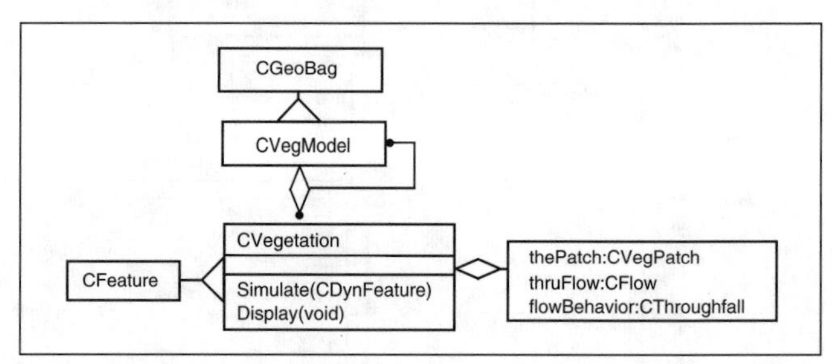

Figure 79-6. Object diagram for *CVegModel.*

CPrecipModel: Precipitation is temporally and spatially variable. A system of rain gauges can be deployed to record this variability; from this sample data, a two-dimensional precipitation pattern can be generated. Thiessen polygons centered on rain gauges often are used to represent the spatial pattern of rainfall. This technique assumes that the nearest empirical data point is the best estimate of rainfall at any given point. For this model, the class *CPrecipModel* is defined as a collection of features (*CPrecipitation*), each of which represents a best estimate of rainfall intensity in a particular region (Figure 79-7). Each *CPrecipitation* object contains an instance of *CPrecipTPoly* (structure) and *CRain* (behavior). Outflow from this feature is modeled by associative links (objects of class *CFlow* stored in a *CMWayTopolgy* object) to *CVegetation* and, in areas where there is no vegetation, to *CSurfaceFacet* objects. Each *CPrecipTPoly* object contains an instance of the cartographic class *CTPoly* (a class representing a Thiessen polygon).

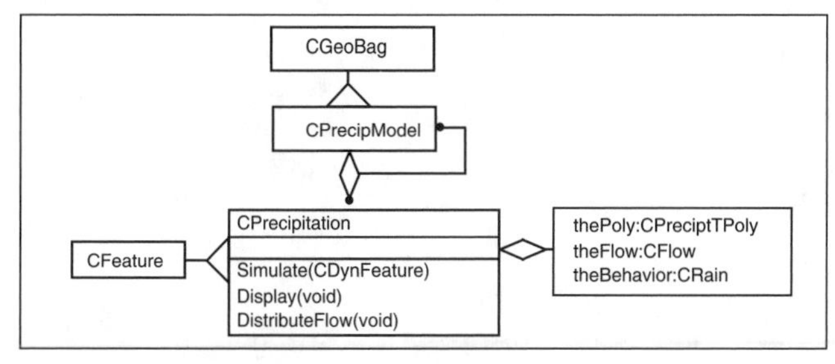

Figure 79-7. Object diagram for *CPrecipModel.*

Results and Future Research

To evaluate the system's model base management capabilities, the watershed model just outlined was constructed and compiled. This model was then used to simulate a hypothetical watershed and rain event. By constructing a hypothetical environment, we had control over its structure and behavior and, thus, were better able to test and track the performance of the model during the simulated event. The system performed as expected, given the rainfall rate and initial conditions selected. This research will now proceed in two directions: (1) the model base and database management capabilities of the GMS will be extended and enhanced, and (2) the watershed model that was developed using the GMS will be applied to an actual watershed.

CONCLUSION

A GMS integrates model base management and GIS technologies into a single integrated software system using object-oriented design and programming techniques. This research develops a new approach to software development that supports the creation and use of environmental models; therefore, it provides a work space within which spatial knowledge can be stored and managed, theory can be modeled and tested, and alternative resource management strategies can be evaluated.

References

Armstrong, M. P. 1991. Knowledge classification and organization. In *Map Generalization—Making Rules for Knowledge Representation,* edited by B. P. Buttenfield and R. B. McMaster, 86–102. New York: John Wiley and Sons.

Armstrong, M. P. and P. J. Densham. 1990. Database organization strategies for spatial decision support systems. *International Journal of Geographical Information Systems* 4:3–20.

Booch, G. 1991. *Object-Oriented Design with Applications.* Redwood City: Benjamin/Cummings.

Guariso, G., and H. Werthner. 1989. *Environmental Decision Support Systems.* New York: Halsted Press.

Jankowski, P. 1992. An architecture for a modeling support system for simulation of environmental processes. *Computers & Geosciences* 18:1,075–93.

Nyerges, T. L. 1993. GIS for enviromental modelers: An overview. In *Environmental Modeling with GIS,* edited by M. F. Goodchild, B. O. Parks, and L. T. Steyaert, 75–93. New York: Oxford University Press.

Zeigler, B. P. 1990. *Object-Oriented Simulation with Hierarchical, Modular Models: Intelligent Agents and Endomorphic Systems.* Boston: Academic Press.

David A. Bennett
Southern Illinois University
Carbondale, Illinois 62901-4514
Marc P. Armstrong and *Frank Weirich*
The University of Iowa
Iowa City, Iowa 52242

80

Hierarchical Representation of Species Distribution for Biological Survey and Monitoring

David M. Stoms, Frank W. Davis, and Allan D. Hollander

Spatial and temporal axes of domain, grain, and sampling intensity can serve as a framework to discuss opportunities for integrating spatial biodiversity data into richer, more complex representations of species distributions. This conceptual framework also highlights many of the problems in integrating data of different spatial, temporal, and thematic properties. A recent analysis of the distribution of the orange-throated whiptail lizard in Southern California is reviewed as an example data set integration. Comparisons of representations resulting from different data sources make biases evident, highlight areas of inadequate sampling, and can lead to new inferences about habitat relationships through convergence of evidence. Improvements in the technology needed to facilitate better integration of distribution models with GIS in the areas of data entry, linkages to tools outside traditional GIS functionality, and new GIS tools to integrate existing data sets are discussed in this chapter.

INTRODUCTION

Increasing support for proactive conservation and sustainable utilization of natural resources has brought into sharp focus the need to consolidate and expand our knowledge of historical, current, and potential distribution of species of plants and animals. Such knowledge generally comes from many sources, such as site observations, range limits, association with mappable environmental variables, review of literature, expert opinion, and models. Yet seldom are all these data integrated into a single GIS database to give a comprehensive representation of species distribution.

Recent advances in automated systems that handle spatial data revolutionized the way we represent distributions of plant and animal species. The range of environmental data to support these modeling efforts also has expanded substantially due to technological advances. These new opportunities have created a host of scientific and technical challenges, defined by the limitations of current ecological theory, data sources, and data-handling techniques. Biological surveys have shown that the distribution patterns of species and the processes responsible for them vary as a function of scale (Wiens et al. 1987; Bergin 1992). Current theory does not provide a unified basis for modeling distribution across scales, and so modeling tends to be exploratory and case-specific. The great expense of inventory work requires that we make the fullest utilization of the limited data we already have and design efficient sampling schemes for collecting new information. Existing database systems are not optimized for the integration of spatial and nonspatial biological data from diverse sources and map scales. Further, their complexity has hindered efforts to import specimen collections and expert opinion into a consistent spatial framework. Therefore, information systems are poorly coupled to analytical tools for modeling, visualization, and uncertainty analysis.

In this chapter, we briefly review the history of digital mapping and modeling of species distribution. Against this background, we outline our vision of hierarchical modeling of species distribution and illustrate with an example using the orange-throated whiptail lizard (*Cnemidophorus hyperythrus*). We conclude with a series of recommendations to better integrate distribution models and GIS in the areas of data entry, linkages to tools outside of traditional GIS, and new GIS tools. We limit the subject here to modeling the presence of a species over its entire extent of occurrence and area of occupancy (Gaston 1991). Other important aspects of distribution, such as home range or territory, abundance, and population dynamics, generally are beyond the scope of this chapter. The term "hierarchical" as used here refers to the nested nature of ecological processes (Allen and Starr 1982) and to the data of different spatial extent and resolution that represent them, and does not necessarily imply use of a hierarchical data structure (e.g., a quadtree). The focus is on spatial biodiversity data, but the importance of models, bibliographic and textual data, graphics, and other nonspatial data readily are acknowledged.

BACKGROUND

The potential advantages of using computer technology for storing and displaying biological survey data were recognized over thirty years ago when *Atlas of British Flora* was published by running punch cards through a line printer (Perring and Walters 1962). Correlations with broad-scale environmental factors such as climate could be inferred by overlaying transparent maps of environmental data on the published atlas sheets. The first dot maps of species observations produced by computer were of plants in Ontario (Soper 1964). Adams (1974) provided a summary of other early efforts in digital distribution mapping.

Many concurrent technological advances have led to a profusion of GIS studies of species distribution. One obvious advance is the vast improvement in computing equipment from punch cards and line printers to modern workstations with their high-speed processing, large data-storage capacities, and color graphics display devices. This has unleashed a parallel advance in GIS technology, such that data on many themes can be more readily captured and compared. Remote sensing has provided the means to produce maps of vegetation cover and other habitat elements over a species' entire range. Global positioning systems (GPS) and radio telemetry have increased the precision at which species locations can be recorded.

Species distribution mapping and modeling basically address two complementary types of research questions: (1) where a species is known or predicted to occur, and (2) what the environmental attributes of those places are. The data used to analyze questions regarding species locations run the gamut, from precisely georeferenced points to crude natural history range maps. If observation data are recorded by date as well, one can examine whether a range is increasing or decreasing in extent (Stoms et al. 1993).

Researchers have taken advantage of the new technology to develop both inductive and deductive models of distributions. The inductive approach derives associations from environmental data at known locations of a species, using a wide variety of statistical, deterministic, or biophysical models. From this approach, one could ask, "What environmental factors or biogeographic barriers are coincident with (and possibly determine) the range limits of a species?" (Root 1988); or, "What environmental factors are associated with this species distribution?" (Price and Endo 1989; Walker and Cocks 1991; Westman 1991; Stoms et al. 1993).

In contrast, the deductive method extrapolates from sampled and unsampled locations on the basis of known habitat associations (Mead et al. 1988; Price and Endo 1989). This habitat-relationships approach usually assumes that spatial factors such as patch size, shape, connectivity, interspersion, and juxtaposition are sufficiently adequate in quality to allow a species to persist in an area (Airola 1988). Recently, spatial components of habitat have been addressed where these assumptions about spatial constraints may not be appropriate or adequate (Pereira and Itami 1991; Aspinall 1992). Besides simple mapping of suitable habitat, one could explore for suitable but unoccupied habitat for the reintroduction of an endangered species (Cogan 1993) or the introduction of a harvest species (Booth et al. 1989). Changes in the area of suitable habitat within the range over time could be identified with multitemporal environmental data. Simulation modeling has been used to predict future distributional changes resulting from climate warming, natural succession, or land-use activities (Benson and Laudenslayer 1986; Busby 1988).

THE INFORMATION HIERARCHY

To accurately analyze species occurrence data, it is important to distinguish the differences between extent of occurrence, data grain, and sampling intensity (fraction of extent sampled at a given grain) (Turner et al. 1989). Figure 80-1 portrays the three spatial axes. Temporal characteristics of spatial data can be described on three similar axes. Our approach presents suites of distribution maps at three different spatial extents: biogeographic, regional, and local (Wiens 1989; Hollander et al. 1994). Together, these three extents create an informative picture of the geography of a species and the ecological factors (i.e., themes) associated with its known distribution. Data assembled from different sources also reveal the limitations and potential biases of existing information. For example, field data are the most concrete data for reconstructing the range of a species. However, these data are often patchy, are of indeterminate grain, and only cover a small fraction of the total extent. In our view, a simple plot of locality data on maps or grids is the most objective but often the least informative representation of distribution. Today, the basic challenge in distribution mapping is to be able to generalize sparse species data originally collected at fine spatial and temporal grains.

Most analyses will be sensitive to the positions data occupy along these axes (Pereira and Itami 1991; Bergin 1992). The solution requires data integration to make the data adequately consistent. Shepherd (1991) explains that inconsistencies tend to be most problematic when data are collected by several organizations, at multiple scales, in several formats, or at different times. Shepherd's list of inconsistencies exactly characterizes species observations and relevant environmental spatial data and is readily described using the framework of axes to explain many of the elements of variation in biodiversity data. Fortunately, there are techniques to reduce some data inconsistencies in GIS and even use them to advantage. For instance, sparse sample data can be used to extrapolate to unsampled areas. Existing information systems are not optimized, however, to deal with all the problems of integrating biodiversity data.

A recent analysis of the distribution of the orange-throated whiptail (Hollander et al. 1994) illustrates the issues involving multiscale distribution data sets, the many steps required for data integration, and the ways in which their interaction can enrich the representation of the distribution of a species. The data sets used in the analysis are characterized in Table 80-1. The exploratory research approach was to work hierarchically. For instance, the local, detailed vegetation map and 432 sightings of orange-throated whiptails were used to test the strength of association of whiptail occurrence with habitat types predicted to be suitable by an existing, coarser-scale habitat relationships model (Airola 1988).

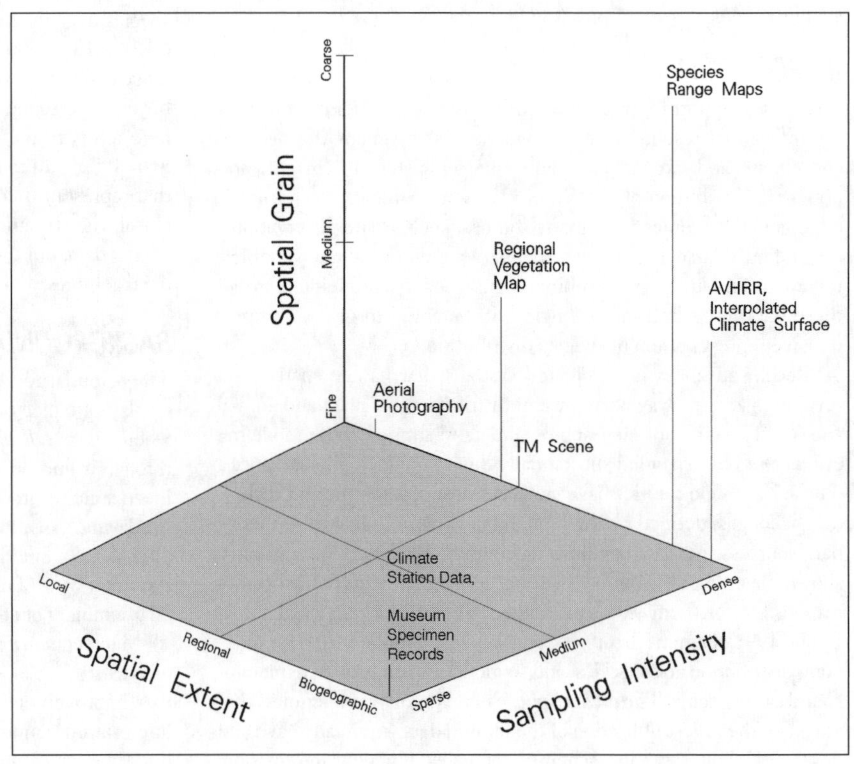

Figure 80-1. Biodiversity data in spatial dimensions of extent, grain, and sampling intensity.

Table 80-1. Data sets in the whiptail analysis, their properties, and integration procedures.

Data Set	Domain	Grain	Intensity	Procedures
Monthly climate	biogeog.	coarse	dense	interpolation, seasonal averaging
Range boundary	biogeog.	coarse	sparse	bound set of points
Museum specimens	biogeog.	fine	sparse	georeference to grid
Digital elevation	biogeog.	moderate	dense	density slice
WHR model and text	biogeog.	moderate	sparse	link tables to spatial objects
Landsat TM scenes	biogeog.	fine	dense	resample, mosaic, cluster
GAP habitat map	regional	moderate	dense	recode to habitats
NDDB sightings	regional	fine	sparse	georeference field data
MSCP habitat map	local	fine	dense	recode to habitats
MSCP sightings	local	fine	moderate	georeference field data, filter by date

The whiptail study highlighted many typical problems associated with species distribution information. Museum specimens potentially provide a valuable record, but geocoding museum locality information presented difficulties similar to those reported by McGranaghan and Wester (1988), such as vague, imprecise place names. The mapping system collection prototype (MSCP) observation data sets were more precisely georeferenced, but were strongly biased toward the urban fringe, where most biological surveys occur during the environmental impact-assessment process. Special habitat elements that generally are not mappable, except at the finest grain over very localized extents or by low-intensity sampling, are often critical factors in determining habitat suitability. For instance, termites are a preferred food of the whiptail, but insufficient data on termite distribution prevents the incorporation of that knowledge as a detailed representation of whiptail distribution. Microclimate is another fine-grain factor that can be critical for some species (Rich and Weiss 1991) but usually is not available over biogeographical extents.

RECOMMENDATIONS

The whiptail study illustrates how different distribution and environmental data at various scales can generate predictive distribution maps and hypotheses about the factors controlling them. None of these representations can be considered definitive, but each encompasses an effective application. The range outline is a coarse-scale representation of a distribution but is useful for analyses at the biogeographic scale. A habitat-based model circumvents the problem of inadequate and biased data collection but calls for habitat requirements to be well known. Point-observation data can be superimposed onto the habitat-based model to show areas of inadequate sampling. Finally, satellite imagery shows a synoptic view of the landscape and provides a useful alternative characterization of the potential habitat. The advantages of each map approach become more apparent when these different representations are considered together.

Unfortunately, existing commercial GIS do not facilitate this sort of interactive work. The heterogeneity of these data sets—composed as they are of vector maps, images, tabular and statistical models, etc.—means much effort must be put into converting data from one form to another. GIS are weak in enabling spatial statistical analyses, but improving the ability to visualize and integrate complex data sets is an active area of development and research in GIS (Shepherd 1991).

This leads us to identifying the functions needed to improve the integration of multiresolution, hierarchical species distribution modeling into GIS. The following list is organized into three categories: facilitating data entry, improving linkages to existing tools outside traditional GIS functionality, and developing new integration tools within GIS.

1. Data entry: Perhaps the greatest need in species modeling is to incorporate the collective knowledge of field biologists. Biologists rightly resist accepting GIS models that identify potential habitat only. Developing a GIS to represent species distribution thus becomes more than just a task of producing a decent interface for the biologist; rather, the expert's opinion should be fed back into the representation of species distribution in the GIS. The complexity of GIS technology, however, obstructs its use by experts most familiar with the species. Ideally the species distribution database should be interactive so that a biologist can view the evidence ascribing the presence of a species to a particular landscape unit as well as modify the ascription based on the expert's own judgment. Thus part of the task of GIS-model integration should be to design an interface to the GIS that frees the biological expert from the need to master the complexity of the software (Mark and Gould 1991). Further, the interface should provide spatial information from the GIS in the form of maps and images as guidelines to tagging locations with species occurrences rather than forcing the biologist to input data as geographic coordinates. A potential interface is shown in Figure 80-2.

This generates a database that is an evolving entity. As each expert adds his or her judgment about the presence (or absence) of a species in a certain location, over time, the distribution database becomes

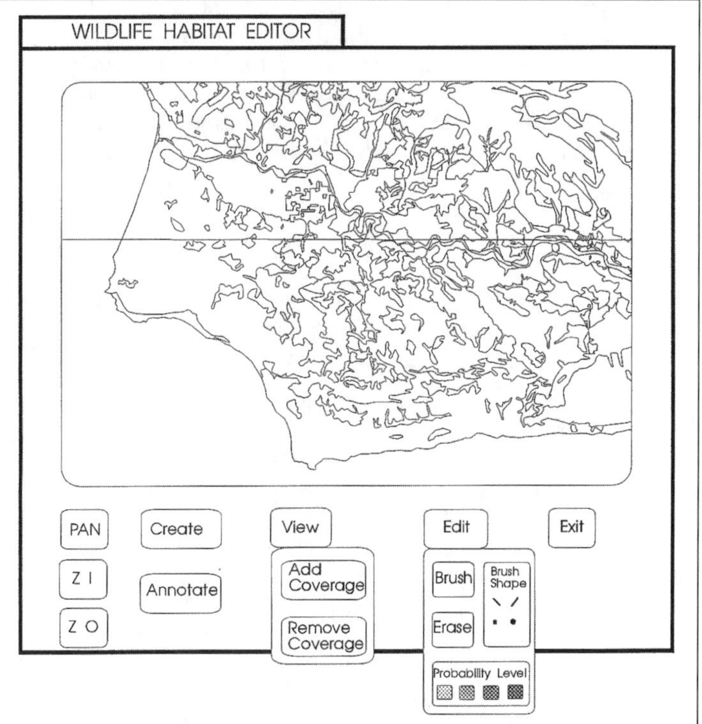

Figure 80-2. Example of an interface editor used by a wildlife expert to designate likelihood of species presence. This editor assumes a raster representation of species distributions, with vector backdrops for context.

richer. Analogous to the authority diagram on earlier paper maps, each landscape region becomes tagged with the identity of the expert on that locality.

Museum collections could be made more accessible and useful for GIS analyses and modeling by redesigning museum voucher forms to provide more precise consistent geographic coordinates (e.g., see the Association of Systematic Collections—Systematics 2000 Agenda) and encouraging the use of GPS to determine locations in the field. It has also been suggested that field data be recorded directly into a data logger using bar-code scanning technology for attributes that recur frequently enough to develop standardized codes (Cogan, personal communication).

2. Linkage to external modeling tools: There are many existing models and tools for analyzing spatial data that do not need to be explicitly incorporated into GIS software. It will be much more efficient, rather, to improve the links between GIS and models in the form of data format conversion, etc. The following tools need to be more efficiently linked to GIS:
   - sensitivity analysis and error simulation
   - statistical confidence limits
   - ordination and classification tools
   - efficient stratified sample design for surveying and monitoring that could reduce map uncertainty (Ferrier and Smith 1990) and optimize routing to sample locations (Cocks and Baird 1991)
   - spatial decision support systems for conservation planning and simulation of alternative future scenarios
   - visualization and exploratory data analysis, and map-guided image classification
3. Development of new GIS tools: The most critical need within GIS functionality is for better tools for integrating biodiversity data with different spatial and temporal properties. Intelligent map-generalization methods need to be developed (Melo and Rebordao 1990) to understand relationships between habitat maps of different spatial grains. Similarly, it is difficult to mosaic local maps from different sources to compile a consistent map of a regional or biogeographic extent. GIS could be used to generate range boundaries based on a combination of observation and habitat data. All three of these examples require simultaneous manipulation of both the geometric and attribute properties of spatial data. In addition, better tools are needed to visualize reliability of distribution maps and how that reliability varies spatially (Aspinall 1992).

CONCLUSION

We envision a mapping environment where the researcher no longer struggles to produce a single map but produces suites of them at will. Data integration is one of the components involved in reaching this goal, but so are flexibility and clarity of the underlying models and multiple images, thereby creating a better representation of the complex reality underlying diverse data sources. The ultimate goal of generating hierarchical representations is to make better habitat models, because for sound management, "why" is more important than "where" in understanding species distribution.

Acknowledgments

We gratefully acknowledge the IBM Environmental Research Program, the National Fish and Wildlife Foundation, the Southern California Edison Company, and the California Department of Fish and Game for their support in the development of the ideas presented in this chapter.

References

Adams, R. P. 1974. Computer graphic plotting and mapping of data in systematics. *Taxon* 23:53–70.

Airola, D. A. 1988. *Guide to the California Wildlife Habitat Relationships System*. Sacramento: State of California, Department of Fish and Game.

Allen, T. F. H., and T. B. Starr. 1982. *Hierarchy: Perspectives for Ecological Diversity*. Chicago: The University of Chicago Press.

Aspinall, R. J. 1992. Bioclimatic mapping—extracting ecological hypotheses from wildlife distribution data and climatic maps through spatial analysis in GIS. *Proceedings of GIS/LIS'92*, 30–39.

Averack, R., and M. F. Goodchild. 1984. Methods and algorithms for boundary definition. *Proceedings of the International Symposium on Spatial Data Handling*, 238–50.

Benson, G. L., and W. L. Laudenslayer. 1986. DYNAST: Simulating wildlife responses to forest management strategies. In *Wildlife 2000*, edited by J. Verner, M. L. Morrison, and C. J. Ralph, 351–55. Madison: The University of Wisconsin Press.

Bergin, T. M. 1992. Habitat selection by the western kingbird in western Nebraska: A hierarchical analysis. *The Condor* 94:903–11.

Booth, T. H., J. A. Stein, H. A. Nix, and M. F. Hutchinson. 1989. Mapping regions climatically suitable for particular species: An example using Africa. *Forest Ecology and Management* 28:19–31.

Busby, J. R. 1988. Potential impacts of climate change on Australia's flora and fauna. In *Greenhouse: Planning for Climate Change*, edited by G. I. Pearman, 387–98. Melbourne: CSIRO.

Cocks, K. D., and I. A. Baird. 1991. The role of geographic information systems in the collection, extrapolation and use of survey data. In *Nature Conservation: Cost Effective Biological Surveys and Data Analysis*, edited by C. R. Margules and M. P. Austin, 74–80. Melbourne: CSIRO.

Cogan, C. B. 1993. "Quantitative Analysis of Habitat Use by the California Condor." Master's thesis, Department of Geography, University of California, Santa Barbara, California.

Ferrier, S., and A. P. Smith. 1990. Using geographical information systems for biological survey design, analysis, and extrapolation. *Australian Biologist* 3:105–16.

Gaston, Kevin J. 1991. How large is a species' geographic range? *Oikos* 61:434–38.

Hollander, A.D., F. W. Davis, and D. M. Stoms. 1994. Hierarchical representation of species distributions using maps, images, and sighting data. In *Mapping the Diversity of Nature*, edited by R. I. Miller. London: Chapman & Hall.

Mark, D. M., and M. D. Gould. 1991. Interacting with geographic information: A commentary. *Photogrammetric Engineering and Remote Sensing* 57:1,427–30.

Mayer, K. E., and W. F. Laudenslayer, Jr. 1988. *A Guide to Wildlife Habitats of California*. Sacramento: California Department of Forestry and Fire Protection.

McGranaghan, M., and L. Wester. 1988. *Prototyping an Herbarium Collection Mapping System*. Technical Papers: 1988 ACSM-ASPRS Annual Convention, vol. 5, 232–38.

Mead, R. A., L. S. Cockersham and C. M. Robinson. 1988. Mapping gopher tortoise habitat on the Ocala National Forest using a GIS. *Proceedings of GIS/LIS'88*, 395–400.

Melo, F. G., and J. M. Rebordao. 1990. Automatic cartographical generalization for Landsat-derived thematical maps. *Proceedings of the Tenth Annual ESRI User Conference.* Redlands: Environmental Systems Research Institute, Inc.

Pereira, J. M. C., and R. M. Itami. 1991. GIS-based habitat modeling using logistic multiple regression: A case study of the Mt. Graham red squirrel. *Photogrammetric Engineering & Remote Sensing* 57:1,475–86.

Perring, F. H., and S. M. Walters. 1962. *Atlas of British Flora.* London: Nelson and Son.

Price, M. V., and P. R. Endo. 1989. Estimating the distribution and abundance of a cryptic species, *Dipodomys stephensi* (Rodentia: Heteromyidae), and implications for management. *Conservation Biology* 3:293–301.

Rapoport, E. H. 1982. *Areography: Geographical Strategies of Species.* Oxford: Permagon Press.

Rich, P. M., and S. B. Weiss. 1991. Spatial models of microclimate and habitat suitability: Lessons from threatened species. *Proceedings of the Eleventh Annual ESRI User Conference.* Redlands: Environmental Systems Research Institute, Inc.

Root, T. L. 1988. *Atlas of Wintering North American Birds: An Analysis of Christmas Bird Count Data.* Chicago: The University of Chicago Press.

Shepherd, I. D. H. 1991. Information integration and GIS. In *Geographical Information Systems, Volume 1: Principles,* edited by D. J. Maguire, M. F. Goodchild, and D. W. Rhind, 337–60. London: Longman Scientific & Technical.

Soper, J. H. 1964. Mapping the distribution of plants by machine. *Canadian Journal of Botany* 42:1,087–1,100.

Stoms, D. M., F. W. Davis, C. B. Cogan, M.O. Painho, B. W. Duncan, J. Scepan, and J. M. Scott. 1993. Geographic analysis of California condor sighting data. *Conservation Biology* 7:148–59.

Turner, M. G., V. H. Dale, and R. H. Gardner. 1989. Predicting across scales: Theory development and testing. *Landscape Ecology* 3:245–52.

Walker, P. A., and K. D. Cocks. 1991. HABITAT: A procedure for modeling a disjoint environmental envelope for a plant or animal species. *Global Ecology and Biogeography Letters* 1:108–18.

Westman, W. E. 1991. Measuring realized niche spaces: Climatic response of chaparral and coastal sage scrub. *Ecology* 72:1,678–84.

Wiens, J. A. 1989. Spatial scaling in ecology. *Functional Ecology* 3:385–97.

Wiens, J. A., J. T. Rotenberry, and B. van Horne. 1987. Habitat occupancy patterns of North American shrubsteppe birds: The effects of spatial scale. *Oikos* 48:132–47.

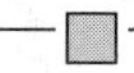

David Stoms is a postdoctoral researcher in the Center for Remote Sensing and Environmental Optics and the Department of Geography at the University of California, Santa Barbara. His current research interests include the use of remote sensing and GIS in the assessment of biodiversity, the effects of scale and resolution in biogeographic analysis, sensitivity and error analysis of GIS data, and database management.

Frank Davis is an associate professor of geography at the University of California, Santa Barbara. His research interests include plant ecology, vegetation remote sensing, and ecological applications of GIS. Several of Davis's recent publications have dealt with the use of GIS for modeling the distribution of species and ecosystems and for conducting biodiversity analyses over large areas.

Allan Hollander is a doctoral student in the Department of Geography at the University of California, Santa Barbara. He has a B.S. from the University of California, Berkeley and an M.S. from Duke University, both in zoology. His research focuses on developing an integrated system for mapping and modeling species distributions.

Department of Geography and Computer Systems Laboratory
University of California
Santa Barbara, California 93106-4060
Phone: (805) 893-7044
Fax: (805) 893-3146
E-mail: *stoms@geog.ucsb.edu*

81

Spatiotemporal Hierarchies in Ecological Theory and Modeling

Alan R. Johnson

Hierarchy theory provides a conceptual framework for the analysis of spatiotemporal aspects of ecological phenomena. It predicts that ecological phenomena will occur in relatively discrete regions of the space–time domain, and that, typically, phenomena that occur at a broader spatial scale will also be characterized by slower or lower-frequency dynamics. As a heuristic guideline, the ecological modeler need typically consider only three hierarchical levels, or spatiotemporal scales, in the construction of an ecological model. The three relevant levels are: (1) a focal level, at which the phenomenon of interest is expressed; (2) a higher level, which imposes constraints; and (3) a lower level, which provides the mechanism that generates the phenomenon of interest. Consideration of more than three levels within a single model is usually unwieldy. Investigation of ecological patterns and processes across a wide range of spatiotemporal scales is often better handled by extrapolation procedures or by the explicit linking of models constructed at differing focal scales.

INTRODUCTION

In recent years ecologists have expressed an increasing awareness of, and interest in, "scale." Scale has become a catchword, used to express both spatial and temporal characteristics of ecological phenomena, of observational or experimental procedures, of data representation and analysis, and of modeling approaches.

The recent interest in scale is driven largely by the recognition that traditional ecological studies have been biased strongly toward particular spatial and temporal scales (a few square meters and less than five years) whereas important environmental problems and management issues are increasingly perceived as being large scale (regional or global) and long term (decades or longer). This leads to two questions: (1) how can ecological studies be designed to address large-scale, long-term phenomena, and (2) can the wealth of small-scale, short-term data be used (i.e., upscaled) to yield information, make inferences or provide testable hypotheses about large-scale, long-term phenomena. Hierarchy theory, especially as it has been applied in the domains of ecosystem ecology and landscape ecology, provides a useful conceptual framework for addressing such scale-related questions.

This chapter first provides an overview of "space–time diagrams" for expressing empirical generalizations about the relevant spatiotemporal scales for various ecological phenomena. Second, the conceptual framework provided by hierarchy theory is explained. Third, the implications of an explicit consideration of spatiotemporal scale(s) for the design and construction of ecological models are considered. Finally, some approaches for extrapolation across scales are described.

SPACE–TIME DIAGRAMS

Many scientists have sought to provide a conceptual ordering of various ecological phenomena by characterizing the dominant spatial and temporal scales associated with such phenomena. A frequently used heuristic device is a graphical representation, or space–time diagram. Many investigators have used space–time diagrams, often borrowing and modifying earlier diagrams. My purpose here is simply to cite a few examples that serve to illustrate the variety of ecological phenomena represented by space–time diagrams.

An early example, which seems to have provided impetus for many subsequent considerations of the space–time domain, is the discussion by Stommel (1963) of oceanographic hydrodynamics. Stommel's diagram schematically decomposes the variability of sea level into its spatiotemporal components. Consider a hypothetical data set in which sea level has been measured at a fixed location every second for the past 100,000 years. If such a time series were available, a spectral analysis would reveal that certain periodicities (i.e., characteristic time scales) are more important than others in contributing to the overall variation with the record. On the time scale of less than a minute, there would be high-frequency oscillations due to ordinary waves (gravity waves in oceanographic parlance). At periodicities of twelve and twenty-four hours, there would be fluctuations due to the tides. Both geostrophic turbulence and synoptic weather patterns will cause fluctuations on the scale of days. Annual fluctuations in sea level also will produce a peak in the spectrum. Finally, the largest amplitude fluctuations will occur at the scale of thousands of years, associated with melting and freezing of ice caps and continental ice sheets in the glacial–interglacial transitions.

Each of these fluctuations also occurs at a characteristic spatial scale. Gravity waves are relatively localized on the scale of tens to hundreds of meters. Sea level changes caused by meteorological events, such as the movement of frontal systems, span hundreds or thousands of kilometers, but are less than global. Tidal fluctuations and glacial–interglacial transitions affect sea level on a global scale.

Spatial and temporal features can be combined in a single diagram by plotting a surface that represents the amount of variance attributable to each combination of space and time scale. In the diagram sketched by Stommel (1963), peaks or ridges in the surface corresponded to the major phenomena outlined above (Figure 81-1). Space and time are represented on logarithmic scales because of the many orders of magnitude spanned by the phenomena under consideration.

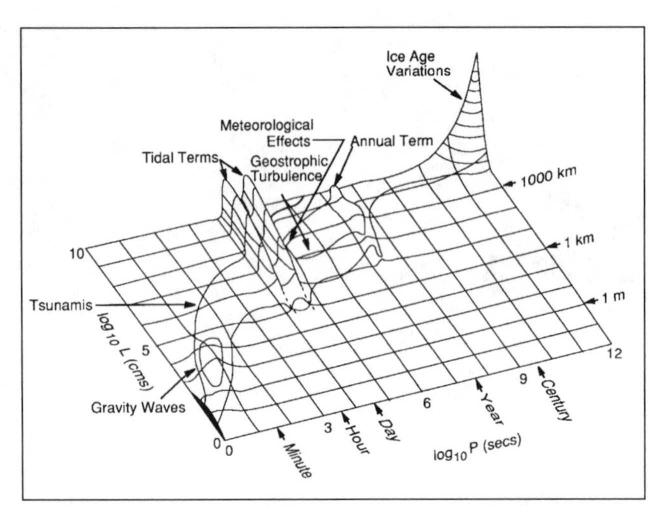

Figure 81-1. Space–time diagram for sea level dynamics. (Redrawn from Stommel 1963.)

Haury et al. (1978) adopted the procedure of Stommel (1963) to represent the spatiotemporal aspects of zooplankton biomass abundance in the oceans (Figure 81-2). Zooplankton exhibits aggregated, patchy distributions at a wide range of spatial and temporal scales. Particularly at broad spatial scales, zooplankton abundance is strongly influenced by hydrodynamic processes (currents, oceanic fronts, etc.) so that there are regions of correspondence in the diagram derived by Haury et al. (1978) for zooplankton that originally presented by Stommel (1963) for sea level. However, zooplankton behavior is also important, leading to peaks corresponding to micropatches (approximately 1 m in size and persisting for a few hours) and swarms (on the order of 1 km and lasting for several days). A ridge in the surface is apparent at a periodicity of twenty-four hours, extending over the range of spatial scales. This corresponds to the diel vertical migrations of zooplankton that occur synchronously over wide distances.

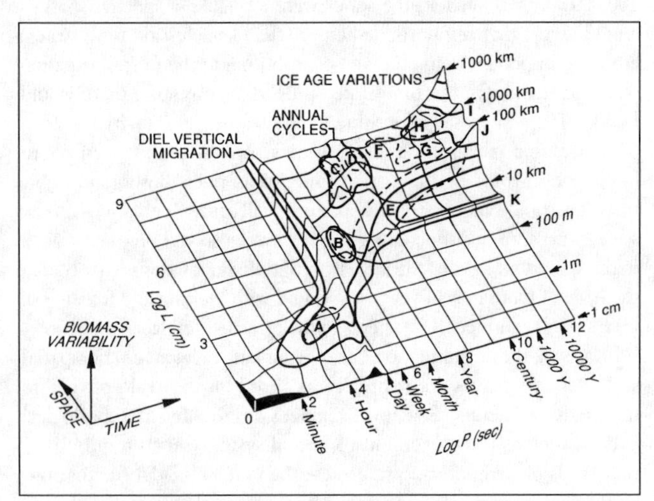

Figure 81-2. Space–time diagram for oceanic zooplankton dynamics. A = micropatches, B = swarms, C = upwelling, D = eddies and rings, E = island effects, F = El Niño events, G = small ocean basins, H = biogeographic provinces, I = length of currents and oceanic fronts, J = width of currents, K = width of oceanic fronts. (Redrawn from Haury et al. 1978.)

Still considering the oceanic environment, Steele (1989) arrayed the major groups of pelagic organisms (i.e., phytoplankton, zooplankton, and fish) along the diagonal of a space–time diagram in which the axes are (1) the body size of the organism, and (2) the doubling time of the population. He also indicates that, at least in upwelling areas, the doubling times of these groups correspond to peaks in the hydrodynamic frequency spectrum. Thus, we see that each of these groups operates at its own characteristic spatiotemporal scale. Since these groups also represent trophic levels, we see that predator–prey interactions are necessarily a multiscale phenomenon.

Terrestrial ecologists also have viewed ecological dynamics within the context of the space–time domain. One prominent scheme is that devised by Delcourt and Delcourt (1991). They partition the spatiotemporal scales studied by ecologists into four domains:

1. The microscale domain (from 1–500 years, 1–10^6m^2) is most familiar to ecologists working with contemporary systems. It is the scale at which processes of population dynamics, productivity, competition, and response to single disturbance events (such as fire or flood) are usually studied.
2. In the mesoscale domain (up to 10^4 years and 10^{10}m^2), research is focused on the events of the last interglacial period (the Holocene) at scales of watersheds or landscape mosaics. Important biotic processes include species invasions, shifts of ecotones or community reassembly in response to climatic shifts. Instead of studying single disturbance events (e.g., fires or floods), the focus shifts to the adaptation to a particular disturbance (e.g., fire or hydrologic) regime.
3. The macroscale domain (up to 10^6 years and 10^{12}m^2) is the realm of Quaternary studies. Biotic processes include displacements of species on a subcontinental scale, with changes in the composition of biomes resulting from differential rates of spread and from extinctions.
4. The megascale domain (greater than 10^6 years and greater than 10^{12}m^2) includes such planetary phenomena as the development of the biosphere, lithosphere, hydrosphere, and atmosphere, along with the entire macroevolutionary history of life on Earth. Reconstruction of this domain is the task of geologists and paleontologists.

Space–time diagrams have frequently been employed in landscape ecology. A prime example is given by Urban et al. (1987), who construct space–time diagrams representing various disturbance regimes, forest processes, environmental constraints on vegetation, and resulting vegetation patterns (Figure 81-3). The latter three space–time diagrams all exhibit a series of identifiable phenomena arrayed along a diagonal from small, short term to large, and long term. These are represented by ellipses, analogous to peaks in Stommel's representation. Disturbance regimes are exceptional in that, although they span a wide range of spatial scales, there is no evident trend of increasing temporal scale (i.e., lower frequency of occurrence) with increasing spatial scale. Thus, although floods can affect much larger regions than tree falls or windthrow, their recurrence intervals are not notably different.

King et al. (1990) employ a space–time diagram to provide the context for their discussion of translating models across temporal scales. These authors consider a range of processes from leaf and whole-tree physiology, up through tree growth, gap dynamics, and forest dynamics. They then illustrate a method for explicitly translating a whole-tree physiology model across temporal scales to arrive at a gap-phase replacement model.

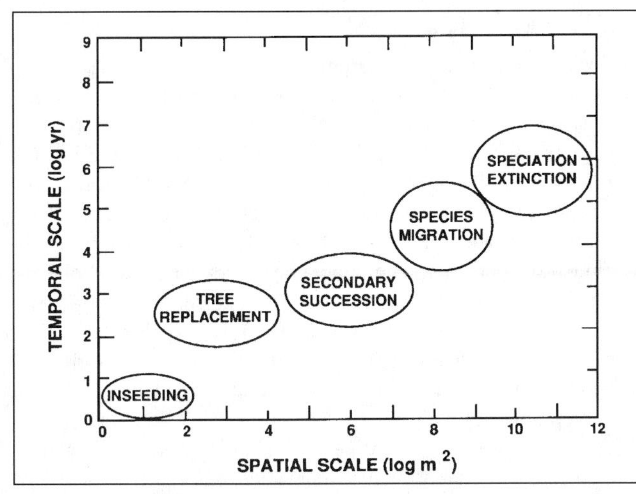

Figure 81-3. Space–time diagram for forest processes. (Redrawn from Urban et al. 1987.)

Our brief review of space–time diagrams demonstrates that although ecological phenomena span a wide range of spatiotemporal scales, identifiable processes generally operate in specific portions of the space–time domain. The characteristic spatiotemporal scale of a process is represented as a peak in diagrams such as that of Stommel (1963) or as elliptical regions in diagrams, such as those of Urban et al. (1987) or King et al. (1990). Moreover, these peaks or ellipses are generally arrayed, at least approximately, along a diagonal line within the space–time domain, such that processes that occur at broader spatial scales also tend to have a longer-term (lower-frequency) temporal scale. Hierarchy theory provides a conceptual framework for explaining these basic trends.

HIERARCHY THEORY

Hierarchy theory arose out of a consideration of the organizational properties of complex systems. Specifically, it has been argued that for systems composed of many interacting components, the evolutionary process will favor a nested hierarchical organization in which the system as a whole can be decomposed into subsystems, each of which can be further decomposed into finer-scale subsystems, and so on (Simon 1962). At each level of decomposition, the components within a subsystem will interact more strongly with each other than with components of other subsystems. In actuality, this nested structure of components is seldom perfect, but hierarchy theory is taken to apply to the set of "nearly decomposable systems" (Simon 1962).

Subsystems within a hierarchical structure can be conceived of as being separated by boundaries or surfaces (Platt 1969; Allen and Starr 1982; Allen et al. 1984). Thus, hierarchical levels are separated by conceptual (and sometimes physical) surfaces. Depending upon the scale of observation, an observer may view a particular hierarchical level from within the surface or from without. Inside the surface, the observed level is seen as being composed of subsystems that interact, subject to environmental inputs that have been filtered by the surface. From outside the surface, the observed level appears as a component of a larger supersystem (Allen et al. 1984).

Surfaces separating hierarchical levels occur where there is a steep gradient in the time constant associated with some characteristic behavior of the system as a function of spatial scale (Allen and Starr 1982). In other words, as long as one focuses on a particular subsystem (i.e., stays within the surface), the time scale of the observed dynamics increases relatively slowly with increasing spatial scale. But, as one includes components from other subsystems (i.e., crosses the surface) the temporal scale increases rapidly with small increases in spatial scale.

Boundaries between hierarchical levels are often associated with steep gradients in other system properties as well as with gradients in characteristic time scales. Thus, a general procedure for identifying hierarchical levels is to find breaks in the scaling of some system property as a function of spatial scale. O'Neill, R. V. et al. (1991) have used breaks in the scaling of the variance of land-use cover types to elucidate the spatially hierarchical structure of landscapes.

In nested hierarchical systems, such as we have been considering here, higher levels contain lower levels, and therefore higher levels are necessarily larger than lower levels. In addition to being larger, higher levels typically operate on a longer, or lower-frequency, time scale. The separation of the temporal scales of hierarchical levels was emphasized by O'Neill et al. (1986) in their discussion of rate-structured hierarchies.

Ecological systems can be viewed as spatially nested, rate-structured hierarchical systems (O'Neill et al. 1986). The requirement that hierarchical systems be nearly decomposable dictates that various levels will be associated with relatively discrete spatiotemporal scales, giving rise to the separate peaks or ellipses commonly exhibited by space–time diagrams of ecological phenomena. Moreover, the fact that higher levels are expected to be both spatially larger and temporally slower (lower frequency) explains why ecological processes tend to follow a positively sloping diagonal in the space–time domain. Thus, the major qualitative features of the empirical space–time diagrams reviewed previously can be seen as natural consequences of the hierarchical organization of ecological systems.

This discussion intentionally has been quite general, without focusing on particular ecological systems or processes. In practical applications, of course, one must focus on the particular. When one does so, a number of different hierarchies may be defined for the same ecological system, depending upon the observational criterion selected (Allen et al. 1984). The objective of hierarchy theory should not be to define the ecological hierarchy, but to define a hierarchy appropriate to a particular problem (Allen et al. 1984; O'Neill et al. 1986).

Various hierarchies have been proposed for the analysis of landscape structure. Noss and Harris (1986) present a scheme designed to account for ecological diversity at multiple scales within a landscape. They propose an approach to conservation that links nodes of high diversity into a functional network by the establishment of multiple-use modules (MUM). Alternatively, Cousins (1993) presents a scheme in which landscapes are partitioned into ecosystem trophic modules (ETM), defined by the areas over which energy is directed toward a particular individual or social group of the top predator(s). Stated more traditionally, ETM correspond to the territorial units of the top predators. When humans enter the picture, a new level of organization is realized, due to the exchange of energy and materials between ETM and to trade. Cousins (1993) considers the representation of landscapes in geographical information systems (GIS) using a combination of ETM and trading network data.

Noss and Harris (1986) and Cousins (1993) present particular hierarchical schemes that may be useful for addressing particular ecological issues. We turn now from the particular back to the general, in order to consider those features of nested, rate-structured hierarchical systems that have implications for the construction of ecological models.

IMPLICATIONS OF HIERARCHY THEORY FOR ECOLOGICAL MODELING

The near decomposability of hierarchical systems is an advantageous feature from the viewpoint of a modeler. The real world displays a variety of structures and dynamics across a wide spectrum of spatiotemporal scales. However, when a model is constructed to address a particular question, the modeler need focus only on those structures and processes that are expressed at the spatiotemporal scale relevant to the question.

Thus, any particular ecological model typically will be keyed to a particular "focal" level within a hierarchy. The dynamics at the focal level of the hierarchy will be generated by the interactions of components (subsystems) at the next lower level. These components are usually represented by state variables within the model. The dynamics generated by these components are restricted by constraints imposed by the level above the focal level. Such constraints are commonly represented by boundary conditions or forcing functions within the model.

Typically, a modeler only needs to consider three hierarchical levels in representing an ecological system: (1) the focal level (n), whose dynamics and structure are to be modeled; (2) the next lower level (n-1) that generates the dynamics and structure of the focal level, and (3) the next higher level (n+1) that imposes constraints on the focal level. Hierarchical levels further removed from the focal level typically are irrelevant in the modeling process. Levels higher than n+1 change so slowly with respect to the focal level that they can be implicitly subsumed into the model structure as constants. Levels lower than n-1 typically vary at frequencies higher than the temporal resolution of the model, so their dynamics need not be explicitly represented, as the integrated results of those dynamics are expressed in the structure of the n-1 level.

The caveat that a modeler typically need only be concerned with three hierarchical levels deserves some mention. This qualification is necessary because exceptions to the rule do occur. Most signals generated by lower levels become rapidly attenuated as they propagate to higher levels (O'Neill et al. 1986). However, certain changes in a lower level, such as the complete loss of a component that performs a vital function, can have effects that propagate across many hierarchical levels. An example of this possibility relating to the ecological role of mycorrhizal fungi is considered by O'Neill, E. G. et al. (1991). Also, it must be noted that at this point in time, many anthropogenic activities occur on anomalous, short time scales, given their large spatial extent. This means that such activities do not fall neatly on the diagonal line in the space–time diagram, and so it may be necessary to examine their effects at multiple hierarchical levels.

Since real ecosystems exhibit many hierarchical levels across a wide spectrum of spatiotemporal scales, it is tempting to represent more than three levels whenever one constructs a model. Such a goal may be achievable in principle, but practical experience suggests that such models are usually too unwieldy to be very useful. The storage of spatial information, which is simultaneously broad in extent and fine-grained in resolution, imposes a demand for computer memory that increases geometrically with the range of scales represented. If temporal dynamics are represented by differential equations, the inclusion of components operating over a broad range of time scales leads to equations which numerical analysts describe as "stiff" because they are notoriously difficult to integrate. Even if a discrete time representation is employed, the coordination of events with multiple time scales can pose a serious technical challenge.

In actual practice, it usually is expedient to construct multiple models, each focusing on a particular spatiotemporal scale to establish clear and explicit links between the various models. We turn now to the problem of linking models constructed at different scales.

EXTRAPOLATION ACROSS SCALES

Ecologists increasingly are turning their attention toward environmental problems that must be addressed at regional and global scales (e.g., acidic precipitation, deforestation, desertification, global climate change). Only recently have such broad-scale patterns and processes been effectively monitored, largely due to the availability of satellite imagery. In contrast to the large-scale nature of the questions, most ecological information gathered by traditional methods is at a much smaller scale. Therefore, most existing ecological models also have been constructed at focal levels that are lower in the hierarchy than appropriate for addressing these large-scale issues. This leads to the question of whether and how information or models originally formulated at one scale can be upscaled to a larger spatial or longer temporal scale.

Any procedure for translating ecological information or models across scales is inherently an extrapolation process. And, like any extrapolation, it is risky. The more fully the ecosystem and the relevant ecological processes are understood, the more likely a successful extrapolation procedure can be devised. But, there is never a guarantee that some unknown process or constraint will not enter at the larger scale, causing the actual ecosystem structure and dynamics at that level to differ from that predicted by the extrapolation from the smaller scale. The wider the range of scales over which an extrapolation must be performed, the greater the chance of unforeseen complications arising.

Still, in the face of pressing large-scale questions, an answer derived from extrapolation, even with the attending uncertainties, may be better than no answer at all. King (1990) identified four general methods of translating ecological models to larger scales: (1) lumping, (2) direct extrapolation, (3) extrapolation by expected value, and (4) explicit integration. (These methods will only be briefly summarized in this chapter.)

1. In extrapolation by lumping, the model structure is unaltered, but the values of state variables and parameters are recalibrated by substitution of average values computed at the larger scale. This is the simplest extrapolation procedure, but it is valid only for linear systems. In systems that display nonlinearity, the bias introduced by lumping can lead to substantial extrapolation errors. Such nonlinearity is common in ecological systems.
2. Direct extrapolation involves explicitly running the small-scale model for a set of discrete elements, scaling the output of each element by the area represented, and then combining the outputs to represent the large-scale system. For independent elements, the combination is often a simple summation. This method is conceptually straightforward. It suffers from being computationally expensive, often prohibitively so if the extrapolation must be made over a wide range of scales.
3. Extrapolation by expected value is essentially a statistical approach. Whereas direct extrapolation requires running the small-scale model for all discrete elements composing the larger system, extrapolation by expected value can be viewed as working from a statistical sample of discrete elements. In practice, one estimates the joint probability distribution for the model arguments (state variables, parameters, forcing variables) and runs Monte Carlo simulations with arguments chosen randomly from the estimated distribution. The accuracy of the method crucially depends on how well the joint probability distribution can be estimated.

4. Extrapolation by explicit integration can be accomplished when the small-scale model can be defined as a function of spatial coordinates. Translating this model to a larger scale then is equivalent to evaluating the integral of this function (analytically or numerically) over the region of interest. Notice that in contrast to the previous extrapolation procedures, explicit integration changes the structure (i.e., functional form) of the model. When feasible, explicit integration is probably the best extrapolation method. The primary limitation of the method is that small-scale ecological models often are procedurally defined rather than explicitly represented in a functional form. Even when a functional representation can be derived, it may be sufficiently complicated as to make integration cumbersome.

An example of extrapolation of ecological models to a larger spatial scale is provided by King et al. (1989). In this study, tundra and boreal forest ecosystem models that had been developed to represent local ecosystems were used to estimate seasonal exchange of CO_2 between the terrestrial biosphere and atmosphere for the circumpolar region from 64°N to 90°N latitude. This was accomplished via an extrapolation by expected value using Monte Carlo simulations.

King et al. (1990) provide an example of extrapolation of an ecological model to a longer temporal scale. They describe a procedure for moving from a physiological tree growth model to a gap-phase replacement model. First the physiological model, with an hourly or shorter time step, is repeatedly run for a year to produce estimates of annual wood production for a factorial of set of input values (initial values of state variables, climatic forcings, etc.). Then a response surface model is derived to represent the statistical relationship between annual wood production and the driving variables. Finally, this statistical relationship is used as a functional component of a gap model that operates on an annual time scale. The integrated dynamics at the focal level (n) of the physiological model become incorporated in the behavior of the components (n-1 level) of the gap model. This general approach also can be applied in the context of extrapolation across spatial scales, as discussed in King et al. (1991).

Thus far our discussion has focused on the extrapolation of ecological models across scales. However, geographic information, which constitutes necessary input data for spatially explicit models, can be represented at various scales of resolution. We turn now to a brief consideration of the problem of scaling up such geographic data when they are represented in a raster format.

Consider a grid of an $N \times N$ array of cells, each of which can exist in one of two states (i.e., black versus white or forest versus grassland). Assume that we wish to rescale this to an $(N/2) \times (N/2)$ array by replacing each 2×2 block of cells in the original array by a single cell in the new array. The question then is how to classify the cells in the new array. Several rules are possible:

1. A presence/absence rule can be used if the 2×2 block contains at least one black (forest) cell—color the new cell black.
2. A majority rule can be used if the new cell adopts the color of the majority of cells in the 2×2 block—ties are assigned a color randomly.
3. A percolation rule can be used if connectivity of the landscape is important and if the black cells span the 2×2 block in an orthogonal direction—color the new cell black.

All such rescalings aggregate the spatial information to a coarser level of resolution. As with any aggregation, a certain amount of aggregation error is to be expected. Renormalization theory provides a method for estimating the amount of error introduced by different aggregation rules (Milne and Johnson, n.d.). As the grid is rescaled, the proportion of cells colored black generally changes. For random maps, the expected change in proportion for any particular aggregation rule can be expressed as a polynomial function of the original proportions of black and white cells (see Milne and Johnson 1991 for a derivation of several such renormalization relations). These renormalization relations also provide at least qualitative information about the aggregation errors introduced when the rules are applied to real (nonrandom) landscapes. Additionally, Milne and Johnson describe an aggregation rule based on preserving the fractal dimension of the map, which typically appears to introduce less aggregation error than other commonly used rules.

CONCLUSION

We have seen that hierarchy theory predicts that ecological systems will be nested, nearly decomposable, rate-structured systems, with higher levels being associated with larger spatial and longer (lower-frequency) temporal scales. The empirical evidence suggests that ecosystems frequently conform to these expectations. This hierarchical organization can be taken advantage of in the construction of ecological models. Models should be designed to represent a single focal level, with dynamics generated by interaction at the next lower level, and constraints imposed by the next higher level.

Practical considerations often lead us to extrapolate models and information from one scale to another. Such extrapolations are not trivial, but several approaches for scaling up ecological models have been proposed and applied, although rigorous testing of the results is, in general, lacking.

Considerations of spatiotemporal scale are likely to be crucial in successfully linking ecological models with GIS. Care must be taken to choose spatial scales in the GIS and temporal scales in the ecological model that are mutually compatible (i.e., correspond to coherent ecological phenomena). Also, successful linkage of a dynamic ecosystem model with a GIS will require representation of temporal change within the GIS data structure (Langran 1992). Continued work on this front promises to be both intellectually challenging and practically rewarding.

Acknowledgments

Thanks to S. J. Turner for his critical review of this manuscript. Many of the ideas expressed here arose in conversations with A. W. King, B. T. Milne, R. V. O'Neill, S. J. Turner, and D. L. Urban. This work was supported by grant #BSR-9107339 from the National Science Foundation.

References

Allen, T. F. H., and T. B. Starr. 1982. *Hierarchy: Perspectives for Ecological Complexity.* Chicago: The University of Chicago Press.

Allen, T. F. H., R. V. O'Neill, and T. W. Hoekstra. 1984. *Interlevel Relations in Ecological Research and Management: Some Working Principles from Hierarchy Theory.* General Technical Report RM-110, U.S. Forest Service.

Cousins, S. H. 1993. Hierarchy in ecology: Its relevance to landscape ecology and geographic information systems. In *Landscape Ecology and Geographic Information Systems,* edited by R. Haines-Young, D. R. Green, and S. Cousins, 75–86. London: Taylor & Francis.

Delcourt, H. R., and P. A. Delcourt. 1991. *Quaternary Ecology: A Paleoecological Perspective.* London: Chapman & Hall.

Haury, L. R., J. A. McGowan, and P. H. Wiebe. 1978. Patterns and processes in the time–space scales of plankton distributions. In *Spatial Pattern in Plankton Communities*, edited by J. H. Steele, 277–327. New York: Plenum Press.

King, A. W. 1990. Translating models across scales in the landscape. In *Quantitative Methods in Landscape Ecology*, edited by M. G. Turner and R. H. Gardner, 479–517. Berlin: Springer-Verlag.

King, A. W., W. R. Emanuel, and R. V. O'Neill. 1990. Linking mechanistic models of tree physiology with models of forest dynamics: Problems of temporal scale. In *Process Modeling of Forest Growth Responses to Environmental Stress*, edited by R. K. Dixon, R. S. Meldahl, G. A. Ruark, and W. G. Warren, 241–48. Portland: Timber Press.

King, A. W., A. R. Johnson, and R. V. O'Neill. 1991. Transmutation and functional representation of heterogeneous landscapes. *Landscape Ecology* 5:239–53.

King, A. W., R. V. O'Neill, and D. L. DeAngelis. 1989. Using ecosystem models to predict regional CO_2 exchange between the atmosphere and the terrestrial biosphere. *Global Biogeochemical Cycles* 3:337–61.

Langran, G. 1992. *Time in Geographic Information Systems*. London: Taylor & Francis.

Milne, B. T., and A. R. Johnson. 1991. Renormalization relations for scale transformation in ecology. In *Some Mathematical Questions in Biology: Predicting Spatial Effects in Ecological Systems*, edited by R. H. Gardner. Providence: American Mathematical Society.

Noss, R. F., and L. D. Harris. 1986. Nodes, networks, and MUM: Preserving diversity at all scales. *Environmental Management* 10:299–309.

O'Neill, E. G., R. V. O'Neill, and R. J. Norby. 1991. Hierarchy theory as a guide to mycorrhizal research on large-scale problems. *Environmental Pollution* 73:271–84.

O'Neill, R. V., D. L. DeAngelis, J. B. Waide, and T. F. H. Allen. 1986. *A Hierarchical Concept of Ecosystems*. Princeton: Princeton University Press.

O'Neill, R. V., R. H. Gardner, B. T. Milne, M. G. Turner, and B. Jackson. 1991. Heterogeneity and spatial hierarchies. In *Ecological Heterogeneity*, edited by J. Kolasa and S. T. A. Pickett, 85–96. Berlin: Springer-Verlag.

Platt, J. 1969. Theorems on boundaries in hierarchical systems. In *Hierarchical Structures*, edited by L. L. Whyte, A. G. Wilson, and D. Wilson, 201–14. New York: American Elsevier.

Simon, H. A. 1962. The architecture of complexity. *Proceedings of the American Philosophical Society* 106:467–82.

Steele, J. H. 1989. The ocean "landscape." *Landscape Ecology* 3:185–92.

Stommel, H. 1963. Varieties of oceanographic experience. *Science* 139:572–76.

Urban, D. L., R. V. O'Neill, and H. H. Shugart, Jr. 1987. Landscape ecology. *BioScience* 37:119–27.

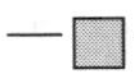

Alan R. Johnson received his B.S. in chemistry from Colorado State University in 1980 and his Ph.D. in environmental toxicology from the University of Tennessee, Knoxville in 1988. He is currently a research assistant professor at the University of New Mexico, where he conducts research in landscape ecology and the modeling of ecological dynamics. He is Secretary of the Theoretical Ecology Section of the Ecological Society of America.

Department of Biology
University of New Mexico
Albuquerque, New Mexico 87131
Phone: (505) 277-2686
Fax: (505) 277-0304
E-mail: *ajohnson@algodones.unm.edu*

82

Bootstrapping River Basin Models with Object Orientation and GIS Topology

René Reitsma

This chapter presents a technique for the automatic generation of river basin simulation and optimization models by using GIS river basin topological information. The technique was derived from an attempt to develop more "data-centered" decision support systems (DSS) for environmental and water resources management. First, traditional model-based and data-centered approaches to DSS for environmental and water resources applications are discussed. Next, an object-oriented technique for modeling river basins is presented. Finally, a method generating these object-oriented models of river basins with the help of GIS topology data is presented and discussed in the context of simulation and optimization.

TOWARD DATA-CENTERED DECISION SUPPORT SYSTEMS (DSS)

Development of DSS for water resources and environmental management traditionally has been centered around the use of existing models. Fundamental to the construction of these systems were mathematical and/or empirical models representing physical phenomena and processes. These models, together with a set of initial conditions, could be used to predict plausible effects of planned impacts on such systems (Loucks et al. 1985; Simonovic and Savic 1989; Reitsma 1990; Camara et al. 1990; Fedra 1990).

Several approaches to the development of large-scale modeling systems have been developed. Each presents a different way to manage the relationships between model and data. The first approach can be denoted "generic" modeling. Generic models reduce physical phenomena to first principles. In order to be applied to a specific case, however, they provide "handles" to which the data of a particular case can be attached. Data describing the system are prepared in files adhering to rigid formatting requirements. The water resources models developed by the Hydrologic Engineering Center (U.S. Army Corps of Engineers 1982) are examples of this model type.

A second approach to solving the model–data problem is the development of completely dedicated models designed for specific cases. Examples of this approach are the Colorado River Simulation Model (CRSM) developed by the U.S. Bureau of Reclamation (Schuster 1987), the Delaware Estuary Model (DOI 1966), and the Potomac River Simulation Model (Hetling 1966). In the implementation of such models, the presentation of physical processes and the representation of the attributes of the system being modeled are not separated. In other words, the attributes and characteristics of, for instance, a river basin, are "hard wired" into source codes for the simulation. Although dedicated models still are being developed, some of the more established ones have undergone a process of generalization to make them easier to apply to other cases.

A third approach has a more hybrid character; it instruments generic models with dedicated data. This typically occurs in the development of what could be called "dedicated DSS." In a dedicated DSS, a generic model is carefully prepared for operation on a very specific situation. This preparation may involve hard wiring of data and connections between data and model, but this hard wiring is not nearly as extensive as in fully dedicated approaches. On the other hand, the resulting DSS is dedicated; it can only be used for that specific situation. If a similar system is required for a different case, significant amounts of reprogramming work are required to refit the model and new data (Fedra 1990, 1991; Reitsma 1990).

Recently, more data-centered approaches to the development of DSS have been developed. The main reasons for this shift toward data-centered systems are both academic and practical. It is inefficient to be restricted to either completely generic or to dedicated models. Similarly, although the application of generic models in a dedicated DSS constitutes an improvement, there is still a significant amount of reprogramming work involved to reapply such a model to new cases. A different reason to look at more data-centered approaches emerges from the organizations that have a definite interest in decision support, but that also have invested substantially in the maintenance of large-scale databases containing their organizational data. In the last ten to fifteen years, there has been a large increase in the amounts of data that organizations collect on their domains (Sutter et al. 1983). These data constitute a powerful resource for modeling. Moreover, since collection and maintenance of these data are fairly costly, it stands to reason that organizations want to make the best use of their data whenever possible. As a result, DSS development, including model development, could center around that data.

Data-centered DSS employ an architecture based on a comprehensive georelational database. Models, data analysis facilities, and other utilities are all "applications" that interact with the database and with each other through data management interfaces (DMI). These are programs that govern the exchange of data between parts of the DSS. The DMI are constructed from a number of basic DMI building blocks. This allows for new applications to be easily added to the DSS.

DATA AS AN IMPLICIT MODEL

A model of a system, such as a river basin, consists of three components:

1. A representation of the system's configuration (e.g., in the form of a network topology [nodes and links]) (Labadie et al. 1986).
2. The attributes of the configuration's components (e.g., head/area volume relationships, historical flows, elevation data, etc.).
3. The first principles governing the system's behavior (e.g., evaporation, power generation, mass balance).

The first two components represent the nongeneric, or "state," components of a model, whereas the third component, the process information, contains the state transition rules (general principles under which the system operates). Generic models concentrate on the process component, whereas dedicated models integrate all three components into a single aggregate application. The dedicated DSS instruments the process component with a set of fixed handles on a set of fixed system data.

For example, when studying an organization's database of a river basin from the perspective of these three model components, it becomes clear that such a database can contain two of the three model components—the system configuration and the attributes of the system's components. This implies that the part of the model representing a system's state is implicitly represented by that database. Of course, the configuration and attribute data are not organized in the framework that a dynamic model might need, but essentially two of the three model components may be implicitly represented in that database.

An example of such a database is a GIS that contains the spatial representation of a river basin, including its topology (Figure 82-1). A tabular (relational) database can contain data on the nonspatial attributes of river basin elements. When integrated, they become complete, albeit implicit, representations of the state variables of a river basin model. One way to view the dynamic component of the model (the state transition rules) is to consider it as just another attribute of the river basin's components. Obviously, when applied to a data structure representing a river reach at time t, it will lead to a new data structure representing the reach at time $t+1$. But since those data structures are stored in the database as system states, application of state transition rules to such a data structure only leads to the addition of another piece of state information to the same database.

From this perspective, it becomes attractive to store the state transition component of a model as another attribute in the database. Object-oriented databases and object-oriented programming offer attractive opportunities for accomplishing this.

OBJECT ORIENTATION

Object-oriented approaches to data representation and modeling offer a number of advantages over non-object-oriented approaches. From a programming point of view, the advantages of object orientation have been well documented (Aho et al. 1983; Ellis and Stroustrup 1990; Lippman 1991; Coplien 1992) so I will not consider them any further here. However, two characteristics of object orientation render the concept quite attractive for data-centered modeling: explicit separation of state information from process information, and inheritance hierarchies.

In object-oriented modeling, as in all modeling, it is common to make a distinction between state information and process information. Object orientation allows both types of information to be related to each other through a common denominator—the object they apply to. For instance, as part of an object-oriented model of a river basin, there is a set of "reservoir" objects. Each of the reservoir objects is equipped with data structures that hold both state and process information. State information is in the form of variables or constant data types; process information is in the form of functions. In object-oriented programming both are tied to the object. As such, an object can monitor its own state and has access to its own set of state transition rules. This, of course, fits nicely on a database structure since database design is often and most naturally concentrated on objects: "things" functioning as information placeholders. For example, it is very likely that a (relational) database for storing information on storage facilities in the Colorado River contains a table for reservoirs. In that table there will be entries for reservoirs such as Lake Mead, Lake Havasu, and Flaming Gorge. Ordering information around objects does not in itself make a database object oriented. On the contrary, most relational database management systems (RDMS) are not object oriented from at least two points of view: (1) they do not allow process information to be stored with the objects, and (2) they do not support inheritance.

This introduces the second advantage object orientation offers for data-centered systems, namely, inheritance. Earlier it was mentioned that state transition rules (process information) constitute the generic component of a model. Object orientation offers preservation of this generality through the mechanism of inheritance. Storing both

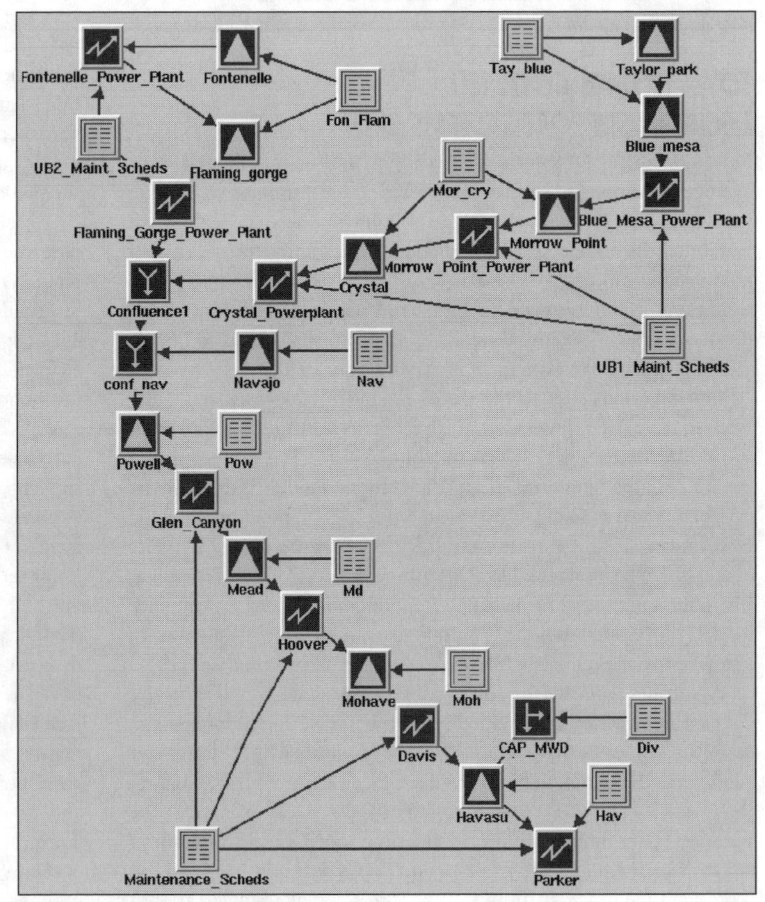

Figure 82-1. The river simulation system (RSS): rivers as networks of objects.

state information and process information on the object does not imply that each object needs to maintain its own copy of the process rules. For that matter, it does not need to store state information that is shared by various objects, nor the way in which this information is stored (internal data management). Inheritance enables the construction of object hierarchies such that objects from anywhere in the hierarchy "inherit" characteristics from objects on higher levels in the hierarchy. These characteristics pertain to both variables and functions. Therefore, the generality of process information can be preserved by storing that process information on a class of object instead of on the object itself. The object then inherits that process information from its class. In this way, class functions can be applied by, and to, the object itself, without it having to maintain its own copy of those functions.

Object orientation offers a powerful tool for modeling in water resources. The elements of a river basin (reservoirs, power plants, diversions, reaches, etc.) show close correspondence with autonomous objects that behave as functions of their states and state transition rules. Their general transition rules can be stored on archetypes or class objects, whereas their specific characteristics can be stored on the objects themselves. Objects then interact with each other by sending each other messages (releases, inflows, demands, etc.), that then are processed by the receiving objects. And, since at least the state part of this information often is stored in an organization's database, it must be possible to almost automatically create the state information of such an object-oriented model directly from the organization's database.

RIVER SIMULATION SYSTEM (RSS): SIMULATION

The River Simulation System (RSS) (Figure 82-1) constitutes software for the rapid construction of (object-oriented) river simulation models. The system was developed by the Center for Advanced Decision Support for Water and Environmental Systems (CADSWES) in cooperation with the U.S. Bureau of Reclamation (USBR). River basin simulation models constructed with RSS are truly object oriented in that the river basin configuration is not explicitly represented in the model. Instead, the elements that make up a configuration (reservoirs, streams, power plants, etc.) maintain links with each other across which they can send information, such as quantities of water or demands for water. River basin configurations are thus implicitly represented through sets of autonomous objects that communicate with each other through sets of information links. In RSS, the instrumentation of a model occurs by the addition of nonphysical objects to the object network. Examples of such objects are time-series handlers, plotter objects, reporting objects, and objects that can conduct multicriteria evaluation analysis.

In RSS, all objects are treated as information processors. A reservoir, for instance, processes inflow data into evaporation, bank storage, storage, and outflow data. Likewise, a plotter object can process the same inflow or outflow data, but converts it into a plot. A multicriteria evaluation object, in its turn, conducts its operations on the same data or other data provided by other objects. Since all objects in RSS share this method of operation, common processing modes can be defined on a class hierarchy of objects. In RSS, two main classes, each subdivided into many more subclasses, exist: the water class and the table class. All physical objects (reservoirs, streams, power plants, etc.) are members of the water class. These objects, just like real river basins and their parts, do not have any memory. They know about their current state and they know how to respond to an external stimulus such as an inflow. Once they have undergone a state transition, however, they forget everything about their previous state. Therefore, if RSS users desire to monitor attributes of physical objects over time, they need to instrument the network with objects that can do so; these are the objects of the table class. Table class objects can be hooked into the network just like water class objects, but although they have no "knowledge" that allows them to perform hydrologic or hydraulic behavior, they do know how to provide data from a time series to other objects, and how to collect data from other objects and assemble them into time series.

Figure 82-1, for instance, displays a model of the Colorado River as a directed graph of objects connected by information exchange links. This particular model is very simple, but more detailed models of this river (e.g., a model of the upper Colorado basin containing over 300 objects) have been constructed and are currently operational under the same object-oriented modeling scheme. This model shows the familiar branching of the Colorado River with the Flaming Gorge branch coming from the north, the Gunnison River branch coming from the east, and the main stem of the river running through Utah and Arizona until in ends below Lake Havasu near the Mexican border. Notice how the hydraulics of the model are built by simple links connecting Lake Mead to Lake Mohave, Lake Mohave to Lake Havasu, and so on. Also notice the intermediate power plant nodes. These take the water from the reservoirs, process it by generating power from it, and then pass it on to, for example, the downstream reservoir.

RIVER BASIN MODELING AND GIS TOPOLOGY

Building a river basin model in RSS is relatively simple; basically, there are two ways: (1) a user may query an attached georelational database for those river basin elements they want to include in the model, or (2) a user may graphically "draw" the network of objects representing the basin. For our discussion, the former method involving the database is the more interesting one. RSS provides a georelational database architecture or database "shell" that can be populated with the data on any river system. The system also contains facilities for rapid declaration of the requested contents of the simulation model in terms of which reservoirs, power plants, diversions, river reaches, and so on, need to be represented in the model. Upon completion of these declarations, RSS searches geographical network topology of the river basin represented in the database and generates the object-oriented representation of that network as a new simulation model. Next, for each node (water object) in the network, the relational part of the database is inventoried for historical and initial condition information pertaining to the node. The result is a network simulation model of a river basin, complete with initial conditions.

By this process, RSS allows users to request not only a specific model configuration, but also to equip the components of that configuration with data appropriate for a particular historical situation. For instance, RSS users can request the construction of a model containing a particular set of water class objects, as it was in, say, July 1967, November 1956, or January 1989. Thus, although the configuration of river basins may remain relatively stable over time, the initial conditions for the model run may vary greatly from month to month or from year to year. The data-centered modeling approach presented here accommodates these variations.

Note that all the data for automatically generating models for the Colorado River were available from the sponsoring organization (USBR). Both configuration and object-specific data have been maintained by organizations such as USBR over long periods of time.

To automatically generate simulation models, it was only a matter of populating the georelational database with these data.

POWER AND RESERVOIR SYSTEM MODEL (PRSYM): OPTIMIZATION/SIMULATION

Where RSS provides GIS-topology-based simulation modeling, a system called PRSYM (Power and Reservoir System Model) applies the same object-oriented technology for both simulation and optimization (CADSWES 1993). It is well known that object orientation offers attractive means for simulating discrete event types of system behavior. In cases where a system's behavior can be modeled as the comprehensive result of the behavior of its individual parts, object orientation offers attractive opportunities for the modeling of such a system. However, for the simulation of continuous phenomena or situations where the behavior of individual parts of a system are governed by the trajectory of the system as a whole, object orientation is much less attractive. A good example is the optimization of a river basin. Assume, for instance, that one would like to determine an operational strategy for a river basin such as the Colorado, such that the total amount of hydropower generated in the basin is maximized. In this scenario, the problem is that the behavior of the individual objects must be subject to measures that relate to the basin as a whole rather than to the object itself. Therefore, in this case, traditional linear and nonlinear optimization techniques better lend themselves to solving such a problem than object orientation can.

The conceptual "input deck" for the optimization of a river basin's operation consists of the following three components:

1. Objective function.
2. Continuity equations and physical constraints.
3. Additional operating constraints.

The objective function informs the optimization algorithm which variable to maximize or minimize (e.g., hydropower or the deviations for a reservoir storage guide curve). The additional operating constraints are all those constraints and forcing functions that need to be taken into account during optimization (expected local inflows, minimum and maximum allowed reservoir elevations, temperature ranges, etc.). The continuity equations, however, represent the physical characteristics of the river basin. Examples might include: "The outflow of the upstream reservoir must be equal to the downstream reservoir's inflow," or "A reservoir's storage cannot be larger than its maximum capacity at any time."

Now, although neither the objective function nor the additional operating constraints, nor historial or initial conditions are part of a river basin's topology, the physical constraints and continuity equations are. Put differently, the continuity equations that govern a river basin's mass balance characteristics are implicit in an object-oriented network model of the basin; the inflow and outflow variables of adjacent reservoirs are mapped through links between the reservoirs. Therefore, the same object-oriented model that is used to conduct (nonlinear) simulation can be used to generate a substantial amount of the input deck for the river basin optimization.

This is exactly the technique applied in PRSYM, RSS's successor system, sponsored by the Tennessee Valley Authority (TVA) and the Electric Power Research Institute (EPRI). PRSYM is currently under development and will provide simulation capabilities similar to RSS with the addition of an optimization component applying the technique mentioned previously (CADSWES 1993). Equipped with both nonlinear simulation and linear optimization, PRSYM will be used in an iterative manner. After first solving linearly for an optimal scenario, the simulator will accept release schedules set by the optimization and simulate the results of these schedules, thus taking into account the nonlinearities unaccounted for by the optimization. This most likely will result in a situation that further needs to be tuned toward either optimality or even acceptability in another series of optimizations/simulations, or just simulations alone.

CONCLUSION

A bright future can be expected for data-centered, object-oriented approaches to river basin modeling to support decision making in water resources and environmental management. Object orientation provides the means to take advantage of two unrelated factors that are increasing the demand for data-centered models and DSS: (1) organizations want to see their carefully collected and maintained data applied in models, and (2) object orientation allows for a very natural way to implement both state and process information for simulation purposes.

It has been shown that by adopting an object-oriented approach toward river basin modeling, a rather versatile and efficient set of simulation and optimization approaches can be acquired. The network character of the object-oriented river basin model lends itself elegantly to "autogeneration" based on GIS river network topology. Once in an object-oriented representation, that same topology can be used to generate the various continuity equations needed for, say, linear optimization of the basin's operations. Moreover, this chain of representations and transformations of representations is easily maintained and conducted within the object-oriented framework.

The object-oriented nature of both RSS and PRSYM has not yet been completely exploited. In the case of RSS, for example, we expect that it also will be possible to generate forcing hydrologies from a central, organization-specific database.

Recently, object-oriented databases that allow functions to be stored in the database have become available. At the same time, object-oriented dialects and extensions of traditional programming languages (C++, Object-Pascal, etc.) allow developers to explore the merits of object orientation. Perhaps these developments indicate that the days of writing large, generic simulation models are over. At any rate, data-centered, object-oriented approaches to modeling seem to be here to stay, at least for a while. It will be interesting to see how developers apply these techniques in next-generation models and DSS.

Acknowledgments

This work was partly sponsored by the U.S. Bureau of Reclamation (contract #8FC-81-12480) and by TVA-EPRI (contract #TV-86442V). Opinions expressed are those of the author and do not necessarily represent the opinions or positions of USBR, TVA, or EPRI.

References

Aho, A. V., J. E. Hopcroft, and J. D. Ullman. 1983. *Data Structures and Algorithms.* Reading: Addison-Wesley.

CADSWES (Center for Advanced Decision Support for Water and Environmental Systems). 1993. *Power and Reservoir System Model (PRSYM) Design Document for 1993 Implementation.* University of Colorado, Boulder.

Camara, A. S. et al. 1990. Decision support systems for estuarine water-quality management. *Journal of Water Resources Planning* 116:417–32.

Coplien, J. 1992. *Advanced C++: Programming Styles and Idioms.* Reading: Addison-Wesley.

DOI. 1966. *Delaware Estuary Comprehensive Study: Preliminary Report and Findings.* Philadelphia, Pennsylvania, Federal Water Pollution Control Administration.

Ellis, M. A., and B. Stroustrup. 1990. *The Annotated C++ Reference Manual.* Reading: Addison-Wesley.

Fedra, K. 1990. *Interactive Environmental Software: Integration, Simulation and Visualization.* IIASA RR-90-10, Laxenburg: IIASA.

———. 1991. *A Computer-Based Approach to Environmental Impact Assessment.* IIASA RR-91-13, Laxenburg: IIASA.

Hetling, L. J. 1966. *A Mathematical Model for the Potomac—What it Has Done and What it Can Do.* CB-SRPB Technical Paper 8. Philadelphia, Federal Water Pollution Control Administration.

Labadie, J. W., D. A. Bode, and A. M. Pineda. 1986. Network model for decision support in municipal raw water supply. *Water Resources Bulletin.*

Lippman, S. B. 1991. *C++ Primer.* Reading: Addison-Wesley.

Loucks, D. P., J. Kindler, and K. Fedra. 1985. Interactive water resources modeling and model use: An overview. *Water Resources Research* 21:95–102.

Reitsma, R. F. 1990. *Functional Classification of Space: Aspects of Site Suitability Assessment in a Decision Support Environment.* IIASARR-90-2, Laxenburg: IIASA.

Schuster, R. J. 1987. *Colorado River Simulation System Documentation: System Overview.* Denver, Colorado, U.S. Dept. of the Interior, Bureau of Reclamation.

Simonovic, S. P., and D. Savic. 1989. Intelligent decision support and reservoir management and operations. *Journal of Computing in Civil Engineering* 3:3.

Sutter, R. J., R. D. Carlson, and D. Lutes. 1983. Data automation for water supply management. *Journal of Water Resources Planning and Management* 109:237–52.

U.S. Army Corps of Engineers. 1982. *HEC-5 Simulation of Flood Control and Conservation Systems: User's Manual.* Davis, California, Hydrologic Engineering Center.

René Reitsma is a research associate at CADSWES and an assistant professor in the Department of Geography at the University of Colorado, Boulder. Reitsma is a graduate of the School of Policy Science at the University of Nijmegen, the Netherlands, where he worked on spatial decision-making modeling, multicriteria evaluation techniques, and multiattribute preference models. At the International Institute for Applied Systems Analysis (IIASA) in Laxenburg, Austria, Dr. Reitsma worked on the integration of such techniques in large-scale decision support systems for, among other things, the reorganization of a regional economy in China. Reitsma joined CADSWES in April 1990, where his main responsibility is the integration of decision-making tools and models in CADSWES' decision support systems. He is one of the creators of the RSS/MODM system.

Center for Advanced Decision Support for
Water and Environmental Systems
Department of Civil, Environmental, and Architectural Engineering
University at Colorado, Boulder
Phone: (303) 492-4828
Fax: (303) 492-1347

83

Scientific Visualization for Environmental Modeling:
Interactive and Proactive Graphics

Barbara P. Buttenfield

Poorly designed maps convey false ideas, and bias both analysis and interpretation. Many GIS users are not trained in graphical design, nor should they be. Their goal is to focus on an application (i.e., forestry, transportation, or geology), not on the mechanics of generating graphical displays. Current GIS do not include defaults and controls to ensure appropriate graphic design. Provision of effective defaults requires knowledge of cartographic formalisms. Examples of basic formalisms can be found in Bertin's system of visual variables, which can be applied readily to isolated icons and simple images. Extensions to Bertin's system will be presented for mapping data and data quality. More complex displays can be generated for thematic and environmental mapping using animation, multimedia, and hypermedia. Environmental issues often incorporate a temporal as well as a spatial component that lend themselves well to dynamic display. Additionally, it is sometimes more convenient to interact directly with a map for query and modeling, as opposed to interacting with a menu or command interface. This empowers users to attend to the application and transforms graphics in the GIS working environment from interactive to proactive.

CURRENT STATE OF GIS GRAPHICS

Integration of information systems technology with environmental modeling provides efficient storage of, and access to, large volumes of information. Given the synoptic nature of environmental data and of spatial models in general, this is mandatory. Users need access to information over a broad area at a point in time, over a single area over a period of time, and every combination in between. Information is not requested in items or in chunks so much as it is requested in streams. The metaphor of hydrologic flows is often applied in information science to describe how information is expressed and understood.

GIS technology relies heavily on users' visualization skills. Over half the neurons in the human brain are devoted to processing optical information (McCormick et al. 1987), meaning that people have a natural acuity for recognizing and interpreting visual patterns. One reason for the quick adoption of GIS technology is that visualization comes naturally to users. It is so natural, in fact, that one may be as easily informed as misled by a graphical display. This is hard to accept when one creates the displays personally, as is often the case in GIS, but it is common knowledge among cartographers and graphic designers alike (Weibel and Buttenfield 1992).

Poorly designed visual displays convey false ideas, and they bias analysis and interpretation. Many GIS users are not trained in graphical design, nor should they be. Their goal is to focus on applications in, for example, forestry or geology, and not on the mechanics of generating graphical displays. One would hope that a graphic is not designed to intentionally trick the viewer. However, it is easy to unintentionally fool the eye, and many principles of cartographic design compensate for this and protect against it. With few exceptions, current GIS do not include defaults and controls to ensure appropriate graphic design.

This chapter summarizes elements of map design that ensure creation of GIS displays that are visually credible and logically sound. Visualization methods for presentation of both static and animated displays are discussed and proposed as a next step for GIS visualization functions. The role of visualization is presented in the context of removing the barrier between the user and the system, that is, allowing the map itself to be treated as the GIS interface. Removing the barrier in this way transforms the GIS working environment from an interactive, command-driven environment into a proactive situation where users can focus on applications, not on system mechanics.

FORMALIZING VISUALIZATION GUIDELINES

Buttenfield and Ganter (1990) argue for a coordinated effort to formalize the study of visualization with GIS. Three roles are cited in their paper, including illustration, analysis, and decision making. Guidelines for each role vary somewhat.

Illustrative graphics are intended for atlas reference, to show output from models and computations, and to realistically depict the landscape. Illustration forms the traditional role for visualization, and many think that the creation of static illustrative displays constitutes the core of cartography. Principles for design of illustrative graphics underlie much of cartographic design theory. To date, the products most often generated by mapping systems and GIS software include static displays in hard-copy form (paper and film) and electronic "throw-away" graphics on CRT and video.

An emerging role of visualization is analytical. Design principles for analytical tools focus on efficiency, simplicity, and abstraction. Early proposals for graphics that analyze data were presented by Tukey (1977) and others (Becker et al. 1988; Cleveland 1985), whose interests centered on exploration and identification of patterns in statistical charts and graphs. One of the earliest analytical applications of visualization was inferential; an example can be found in the work of John Snow, who created a dot map of cholera outbreaks in London to locate and replace a contaminated water pump during an epidemic (Tufte 1990). Analytical graphics have been proposed for hypotheses testing (Bachi 1968), although this capability has not been implemented in GIS.

A third role of visualization relates to spatial decision support. Decision support and requisite visual tools may be required throughout all stages of GIS operation. Customarily, environmental decision support refers to assessing environmental impacts, monitoring hazards, planning evacuation routes, and other types of decisions that are products of analysis and modeling. However, decisions are made in the course of building and running a GIS model that also may benefit from visualization tools and support. Buttenfield (1993a) prototypes a location allocation model in which facility allocations are modified by the user changing points on the map display. McCormick et al. (1987) comment on the effectiveness of iterative displays to steer supercomputer-driven simulations. This role for visualization forms an area of great challenge and research opportunity, as it forms a foundation for implementing proactive graphics environments. This will be expanded upon later in the chapter.

Systematic Guidelines

Certain design principles relate to all three visualization roles (illustration, analysis, and decision support). Many are based on results of empirical research in psychology and statistics (Stevens 1951; Miller 1956; Kosslyn 1985; Cleveland and McGill 1984). A comprehensive itemization is impossible in the space of a single book chapter. However, Buttenfield and Mackaness (1991) provide a review, and readers interested in full details are encouraged to peruse either a text on cartographic design (Dent 1993; Robinson et al. 1984) or a reference with cataloged examples of good versus bad design solutions (Monkhouse and Wilkinson 1971; Tufte 1990). These guidelines apply whether one is viewing a GIS display made by another user or creating a map for one's own use. The user must be careful not to be misled by another's design, nor by his or her own.

It is true that only a finite number of elements can be manipulated in most graphical design problems. Data can be represented as points, lines, areal patches, or (less often) as solid volumes. Six geometric characteristics may be manipulated for each representation, with varying effects. The characteristics, called visual variables (Bertin 1983), include shape (also called form or pattern), orientation, color, texture (graininess or resolution of a shade or tone), value (or darkness), and size. Cross-tabulating the *Points–Lines–Areas* scheme with visual variables, is illustrated in Figure 83-1, which shows eighteen graphical possibilities.

This matrix is useful only within the constraints of a viewer's acuity for specific visual variables. When mapping for any purpose, it must be kept in mind that the map's purpose is to provide good contrast so that items easily may be distinguished. Visual contrast is required for pattern recognition (Stevens 1951) and may be categorized. Distinctions may be dichotomous or even categorical, as in: "two items are the same, the third is different." Distinctions may imply a progression, as in ordinal rankings of a phenomenon, as "small, medium, large" or "minor, major, severe." The most powerful distinctions can have metric associations, as shown in empirical testing (Stevens 1946).

Three regions are outlined in Figure 83-1 to identify which visual variables provide each of the visual distinction types. Next to each variable name is a symbol to indicate the distinction the variable can provide (Bertin 1983). The symbol next to the variables *Shape* and *Orientation* indicates equivalence. Basically, neither variable is good when used by itself for creating visual distinctions. Visual acuity for shape detection is not very strong, and thus maps whose symbols vary only in shape will not be seen as different by most viewers. This is true for point, line, and areal symbols. Likewise, symbols that differ only in their orientation will probably be hard to distinguish. At best, dichotomous distinctions are possible, but for the most part differences will not be seen. One exception can be found in geologic maps of linear faults, where the orientation of the line symbol is matched precisely with the orientation of the fault line. Geologists expect to see fault lines displayed in proper orientation. Thus, training and specific map purpose override the design principle.

The middle region in Figure 83-1 (*Color* and *Texture*) is characterized with a crossed-line symbol ($\not-$) to indicate ordinal distinctions. Displays containing symbols that differ in color or texture can be readily distinguished, although the differences are categorical (i.e.,

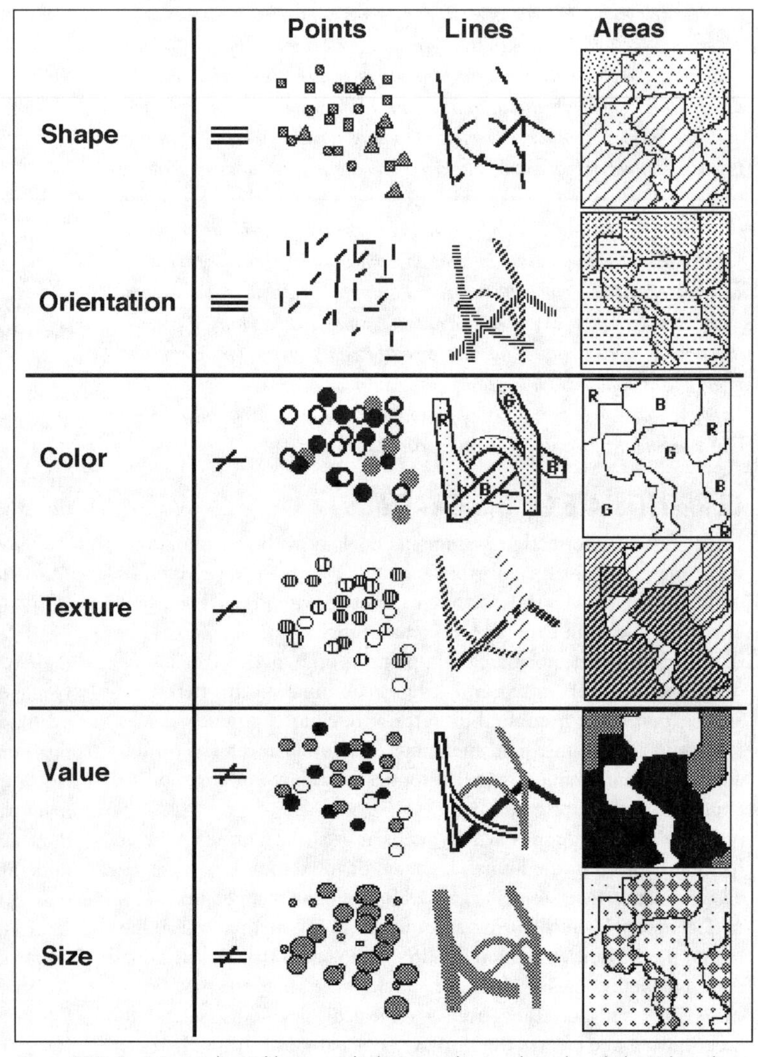

Figure 83-1. Bertin's visual variables as applied to point, line, and areal symbols. (Adapted from Bertin 1983.)

"this type of land cover differs from that one"). It is unlikely that map viewers will associate metric differences, regardless of what is shown in the map legend. For the most part, differences in hue (red, orange, yellow, green, blue) will indicate categorical differences but no clear progression. For example, yellow is not perceived to be "more" than red, or "less" than blue; yellow is simply seen as different. On the other hand, dark red is understood to represent "more" of a variable than pale red (but notice here that color is not being manipulated, since both items are the same color). Viewers may associate rankings with differences in symbol *Texture* (particularly areal patches using cross-hatched line screens). But the rankings are ordinal, not metric.

The use of color in particular has become frequent as the cost of color CRTs and color hard-copy devices continues to drop. The visual logic of color follows common-sense visual expectations. For example, more prominent colors will be associated with higher data values. Red and blue imply thermal distinctions, green implies vegetation, blue sinuous line features imply hydrography, blue areal patches imply wet areas or water bodies, and so on. A study by Patton and Crawford (1977) shows that in the absence of good legend information, viewers make such color associations even if the data do not refer to these specific environmental variables. This is particularly true of children and of viewers with little prior experience in GIS.

The bottom region in Figure 83-1 includes the strongest visual variables, *Value* and *Size*. The not equal symbol (≠) labeling this region indicates that viewers attach metric differences (visual progressions) to symbols varying in value or size. Value variation is accomplished with shades of gray for black and white graphics, and with darker and lighter shades of a color for monochromatic and multichromatic graphics. The visual progressions associate darker value with higher numbers. The visual progression of dark red and pale red described in the paragraph above has more to do with distinctions of value than of color.

In terms of size, the visual progression associates larger symbols with higher numbers. A large body of empirical research in cartography has established perceptually logical scaling factors for symbol size (e.g., graduated circles) and symbol value (percent black); however, few GIS packages include these empirical results in their graphical defaults. The general principle is that viewers will underestimate larger symbol sizes in relation to smaller ones. Thus, to indicate that one symbol is twice as large as an anchor symbol, the larger symbol needs to be scaled more than twice as large for viewers to get the correct impression of size. (Exact numeric values for overcompensation are available in most cartography texts.) (Dent 1993; Robinson et al. 1984).

The system of visual variables is elegant in its simplicity. Guidelines for choosing a particular variable for mapping a specific type of data (e.g., choosing point symbols and size to map incidence of nesting birds in a woodland) can be automated in rudimentary expert systems (Wang 1992). The more difficult problem involves combining visual variables to emphasize and to extend the range of visual contrast, and this has not been successfully automated. For example, to map the nesting populations of five different bird species, one can use graduated symbols of differing colors, one color for each species, to distinguish between species without implying a ranked progression. The graduated symbol *Size* would illustrate the number of nests observed at each site. Size differences and color differences are readily identifiable to viewers, creating a clear and logical visual display. Other combinations of visual variables are often more intuitive, however, and expert systems for automated map design remain a challenge for future research.

Extensions to the System

To represent data: Bertin adds one further variable to his system, stating that the position, or layout, of objects around the map display also can be manipulated to clarify a graphic display. Positioning obviously affects the layout of the legend, title, etc. It also provides the impression of solid volumes in a flat image. In a rendering of digital terrain, hills and surfaces in the foreground obscure those in the background, giving the viewer an impression of depth, solidity, and objects in the distance. The impression of the third dimension is provided in GIS using interpolation, shading, and rendering. Optimal defaults for viewing terrain range from 35°–40° elevation above the surface. Azimuth and viewing distance, of course, will depend upon the terrain image itself. Again this is difficult to automate.

Suggestions for extending Bertin's system have been proposed by several authors. Olson (1986) argues that color is not an unidimensional visual variable, but is in fact composed of hue as well as saturation. Color manipulation for terrain representation has been implemented by several authors, with varying success.

To represent data quality: One very powerful effect resulting from combined visual variables allows displays of data to be embedded with data quality (a current focus in cartographic research) (Beard et al. 1991; Buttenfield 1993b). Data quality may refer to accuracy, uncertainty, reliability of model results, logical consistency of the database, or a multitude of other metadata attributes. MacEachren (1992) demonstrates that defocusing an image gives an impression of fog, implying uncertainty. He adds that other authors have made similar proposals, citing David Woodward as the first to propose the technique. McGranaghan (1993) cites edge crispness, fill pattern, and fog as possible variables to use for mapping data quality. All of these proposed techniques require empirical testing to confirm that map viewers can understand the embedded information in a map display, mandating an agenda for research.

MOVING FROM INTERACTIVE TO PROACTIVE GRAPHICS

Recent advances in technology have reduced hardware costs and increased hardware speed to the point where substantial improvements in graphical realism and dynamics have taken place. With these improvements, environmental modelers should be able to ask more complex and realistic research questions. Specific advances in technology allow for dynamic visualization in the form of map animation. A growing body of research on graphical design focuses on dynamic mapping, as it becomes apparent that the system of visual variables cannot by itself accommodate the perceptual and cognitive complexities of understanding patterns from moving imagery (DiBiase et al. 1992; Gersmehl 1990; Buttenfield 1993a; Weber and Buttenfield 1993).

Of particular interest for spatial modeling is the research animating exploratory statistical techniques, such as data brushing (MacDougall 1992; Monmonier 1992). Many current products will run stand-alone on personal computers. None is currently integrated with GIS packages. Modelers are thus constrained to transfer data into and out of GIS software to utilize the technology, which of course limits adoption to a great extent.

Another constraint that frustrates the modeling process relates to the ability to directly interact with data in their visual form. It is great to have animation capabilities, but the "look-but-don't-touch" character impedes rather than facilitates data exploration and analyses (Buttenfield and Weber 1993). Consider the following generic modeling scenario: Parameters for a model are set by keyboard command,

and the model output is displayed as map or tabular output. The output indicates that parameter adjustment will refine the model results. Keyboard commands are applied to adjust the model, and the process iterates to subsequent solutions.

Two things are wrong with this scenario. First, limiting the modeler to keyboard or even menu input requires verbal responses to visual stimuli. It is indirect and requires learning whatever command syntax the system mandates. Furthermore, the user only can perform those operations anticipated by system designers. In the case of database query, this means users only can browse data along paths of access designed into the system. Second, many GIS users don't even recognize this as a limitation to efficient access and use of spatial data, simply because things have "always" been done that way. From a human factors perspective, it seems likely that using keyboard commands to modify models and screen displays reduces user efficiency, increases fatigue, and may contribute to what Beard (1989) calls "use error."

The development of multimedia and hypermedia and their integration into all types of information systems likely will resolve these impediments and provide users with direct access to data through direct manipulation of visual (map) displays. Multimedia documents contain a collation of disparate media forms (text, graphics, video, sound, numerical output) available on a single presentation device, typically a computer. Multimedia are intended to expand the channels of information available to users. Hypermedia extend multimedia by linking the multiple channels of information transparently. Users can select the mode of information presentation they prefer. It is possible to pursue threads of database query that may not have been anticipated by system designers (Barker and Tucker 1990).

What is needed to fully integrate the user into the GIS process is provision of visualization tools enabling proactive involvement, as opposed to interactive involvement.

> The term proactive refers to taking action before it is mandated by a system request. Interactive computing provides capabilities to take actions anticipated by system designers (such as opening and closing files in pre-determined formats, and generating displays using system defaults). Proactive computing simulates a system responsive to system commands and queries that may not have been anticipated by system designers. A fully operational proactive computing environment incorporates a scripting language that allows users to implement their own commands, and to design command structures beyond the syntax resident in the system (Buttenfield 1993a).

In GIS environments, where so much of the information stream is visual, proactive computing becomes synonymous with visualization. Users should be able to manipulate images to enable system commands or database queries, in order to steer modeling and computation.

CONCLUSION

Visualization tools are needed by users whose knowledge of particular data sets is deep, and whose interest in developing cartographic or systems expertise is overshadowed by their interest in immediate applications or domains of inquiry. These users include environmental planners, natural resource managers, and service providers. In many modeling situations (i.e., in emergency planning), the time available to become facile with system use is quite short, in contrast to the level of training required to learn most GIS command languages. Integration of visual interface tools that can provide proactive access to data and to modeling tools can only improve our abilities to understand the environmental data we archive and illustrate.

Having said this much, it is important to remember that proactive visualization places responsibility on the user to be aware when using and generating GIS graphical displays. MacEachren and Ganter (1990) cite two types of visualization errors, akin to Type 1 and Type 2 errors in inferential statistics. Type 1 visualization errors occur when a viewer sees incorrectly, such as misinterpreting pattern or attributing a map color to a wrong legend class. Type 2 errors amount to "not seeing" or failing to notice patterns contained in an image, that is, missing them due to a lack of contrast or distinction. The latter is the more serious error, since missing information can bias or distort interpretations and subsequent decision making. Given the present lack of attention to logical defaults in many GIS, responsible creation and use of GIS graphics fall on the shoulders of the user.

Acknowledgments

This research is part of NCGIA's Research Initiative 8, "Formalizing Cartographic Knowledge," funded by the National Center for Geographic Information and Analysis (NCGIA). Support by the National Science Foundation (grant #SBR 88-10917) is gratefully acknowledged. This chapter was completed while the author was a visiting research scholar in residence at the U.S. Geological Survey, Reston, Virginia. Sabbatical support by USGS National Mapping Division also is acknowledged.

References

Bachi. R. 1968. *Graphical Rational Patterns: A New Approach to Graphical Presentation of Statistics*. New York: Israel Universities Press.

Barker, J., and R. N. Tucker. 1990. *The Interactive Learning Revolution*. New York: Kogan Page; London: Nichols Publishing.

Beard, M. K. 1989. Use error: The neglected error component. In *Proceedings, AUTO-CARTO 9*, 808–17.

Beard, M., B. P. Buttenfield, and S. Clapham. 1991. *Visualizing the Quality of Spatial Information*. Scientific report on the specialist meeting. NCGIA Technical Report 91-26. Santa Barbara: NCGIA.

Becker R., W. S. Cleveland, and A. R. Wilks. 1988. Dynamic graphics for data analysis. In *Dynamic Graphics for Statistics*, edited by W. S. Cleveland and R. McGill, 1–72. Pacific Grove: Wadsworth and Brooks.

Bertin, J. 1983. *Semiology of Graphics: Diagrams, Networks, Maps*, translated by William J. Berg. Madison: University of Wisconsin Press.

Buttenfield, B. P. 1993a. Proactive graphics for GIS: Prototype tools for query, modeling and display. In *Proceedings, AUTO-CARTO 11*, 377–85.

———. 1993b. Representing data quality. *Cartographica* 30:1–7

Buttenfield, B. P., and J. H. Ganter. 1990. Visualization and GIS: What should we see? What might we miss? In *Proceedings, 4th International Symposium on Spatial Data Handling*, 307–16.

Buttenfield, B. P., and W. A. Mackaness. 1991. Visualization. In *GIS: Principles and Applications*, vol. 1, edited by D. MacGuire, M. F. Goodchild, and D. Rhind, 427–43. London: Longman Publishers Ltd.

Buttenfield, B. P., and C. R. Weber. 1993. Visualization and hypermedia in GIS. In *Human Factors in Geographic Information Systems*, edited by D. Medyckyj-Scott and H. Hearnshaw, 136–47. London: Belhaven Press.

Cleveland, W. S. 1985. *The Elements of Graphing Data*. Monterey: Wadsworth Books.

Cleveland, W. S., and R. McGill. 1984. Graphical perception: Theory, experimentation, and application to the development of graphical methods. *Journal of the American Statistical Association* 79(387):531–54.

Dent, B. D. 1993. *Cartography: Thematic Map Design*, 3rd ed. Dubuque: Wm. C. Brown.

DiBiase, D., A. M. MacEachren, J. B. Krygier, and C. Reeves. 1992. Animation and the role of map design in scientific visualization. *Cartography and GIS* 19(4):201–14, 265–66.

Gersmehl, P. J. 1990. Choosing tools: Nine metaphors of four-dimensional cartography. *Cartographic Perspectives* 5:3–17.

Kosslyn, S. M. 1985. Graphics and human information processing. *Journal of the American Statistical Association* 80:499–512.

MacDougall, E. B. 1992. Exploratory analysis, dynamic statistical visualization and geographic information systems. *Cartography and GIS* 19(4):237–46.

MacEachren, A. E. 1992. Visualizing uncertain information. *Cartographic Perspectives* 13:10–19.

MacEachren, A. E., and J. H. Ganter. 1990. A pattern identification approach to cartographic visualization. *Cartographica* 27(2):64–81.

McCormick B. H., T. A. Defanti, and M. D. Brown. 1987. Visualization in scientific computing. In *SIGGRAPH Computer Graphics Newsletter* 21(6).

McGranaghan, M. 1993. A cartographic view of spatial data quality. *Cartographica* 30:8–19.

Miller, G. A. 1956. The magical number seven plus or minus two: Some limits on our capacity for processing information. *Psychological Review* 63:81–97.

Monkhouse, F. J., and H. R. Wilkinson. 1971. *Maps and Diagrams: Compilation and Construction*. London: Methuen.

Monmonier, M. S. 1992. Authoring graphic scripts: Experience and principles. *Cartography and GIS* 19(4):247–60.

Olson, J. M. 1986. Color and the computer in cartography. In *Color and the Computer*, edited by H. J. Durrett, 205–21. Boston: Academic Press.

Patton, J., and P. V. Crawford. 1977. The perception of hypsometric colors. *Cartographic Journal* 14(20):115–27.

Robinson, A. H., R. D. Sale, J. L. Morrison, and P. C. Muehrcke. 1984. *Elements of Cartography*. New York: John Wiley and Sons.

Stevens, S. S. 1946. On the theory of scales of measurement. *Science* 103:677–80.

———. 1951. Mathematics, measurement and psychophysics. In *Handbook of Experimental Psychology*, edited by S. S. Stevens. New York: John Wiley and Sons.

Tufte, E. R. 1990. *Visualizing Information*. Cheshire: Graphics Press.

Tukey, J. W. 1977. *Exploratory Data Analysis*. Reading: Addison-Wesley.

Wang, Z. 1992. An intelligent interface for application of graphic elements. In *Proceedings, Fifth International Symposium on Spatial Data Handling*, vol. 2, 391–400.

Weber, C. R., and B. P. Buttenfield. 1993. A cartographic animation of average yearly surface temperatures for the forty-eight contiguous United States: 1897–1986. *Cartography and GIS* 20(3):141–50.

Weibel, W. R., and B. P. Buttenfield. 1992. Improvement of GIS graphics for analysis and decision making. *International Journal of Geographical Information Systems* 6(3):223–45.

Barbara P. Buttenfield
National Center for Geographic Information and Analysis (NCGIA)
Department of Geography
SUNY-Buffalo
Buffalo, New York 14261
E-mail: *geobabs@ubvms.cc.buffalo.edu*

84

A High-Resolution Moisture Index:
Gunnison River Basin, Colorado

Loey Knapp, Gregory J. McCabe, Jr., Lauren E. Hay, and Randolph S. Parker

The climate of Colorado is arid, with potential evapotranspiration exceeding precipitation on an annual basis. Monitoring moisture conditions in Colorado is important to detect the emergence of drought, to anticipate possible effects of drought, and perhaps, in particular locations, to take action to minimize these effects. Current indices used to monitor moisture conditions in Colorado, such as the Palmer Drought Severity Index, are inadequate due to low spatial resolution. This study presents a high-resolution moisture index (10-km grid cells) based on climate variables (i.e., precipitation and potential evapotranspiration) that describes spatial and temporal distributions of moisture conditions. Scientific visualization techniques are used to analyze the index and communicate the characteristics of the spatial and temporal distributions of moisture in the Gunnison River basin.

INTRODUCTION

After a series of droughts in Colorado, the Colorado Drought Response Plan was formulated in 1981 (Lamm 1981). The plan provides for various levels of decision making and actions to mitigate adverse effects of below-average hydrologic conditions. Key to this plan is the Water Availability Task Force (WATF), which is an intergovernmental group of state and federal agencies with access to meteorological, climatological, and hydrologic information used to assess moisture conditions. WATF is chaired by the director of the Colorado Office of Emergency Management. To evaluate various categories of drought in Colorado, WATF uses statistical summaries of climatic and hydrologic data (precipitation, temperature, reservoir storage, and streamflow) and a series of indices to assess surface water supply and climatic drought.

An example of an index used by WATF to assess moisture conditions is the Palmer Drought Severity Index (PDSI). The index uses a water balance to define a long-term "normal" for a given area and then identifies departures from this normal to determine wet and dry periods. Index values are reported monthly and provide valuable information for WATF. Alley (1984) identified a number of shortcomings with the PDSI. In the mountainous areas of the United States, one of the primary inadequacies of PDSI is its low spatial resolution. The National Climatic Data Center calculates PDSI values for climatic divisions, which in the western United States are large areas. For example, climate division three in Colorado includes approximately the western third of the state (approximately 100,000 km^2). Because topographic and climatic variability are large, PDSI does not reflect the variability of moisture conditions in this area. To provide a more reliable representation of the variability of moisture conditions in Colorado, the Office of State Climatology computes PDSI values for twenty-five regions in Colorado (Doesken et al. 1983), but even these areas are much larger than most watersheds (e.g., 1,000 km^2). Given these inadequacies, WATF seeks improvements in methods to assess moisture conditions in the state.

The objectives of this study are: (1) to develop a moisture index with a high spatial resolution (i.e., 100 km^2 grid cells) that can be interpreted in a historical perspective, and (2) to utilize recent developments in scientific visualization to analyze and communicate the characteristics of the spatial and temporal distributions of the index. In this chapter, the moisture index is tested in the Gunnison River basin, located in southwestern Colorado. The visualization of the data is performed by using IBM's Data Explorer software. (The U.S. Geological Survey does not endorse this or any other particular software package.)

THE STUDY AREA

The Gunnison River basin was chosen for this study because its attributes are similar to other basins in the western United States, and because it is a major tributary of the Colorado River. The Gunnison River basin has a drainage area of approximately 20,530 km^2. Topography in the basin is extremely variable with elevations ranging from 1,390–4,360 m. In addition, the Gunnison River contributes approximately 42% of the flow of the Colorado River at the Utah/Colorado state line (Ugland et al. 1990).

THE MOISTURE INDEX

Several moisture indices have been developed and are used for various purposes, such as the development of climate classifications (Thornthwaite 1948; Budyko 1956; Mather 1978). Moisture indices typically are comparisons of time-averaged precipitation to time-averaged moisture demand by the atmosphere (Willmott and Feddema 1992). The moisture index used in this study is based on precipitation (P) and potential evapotranspiration (PET). Precipitation represents the moisture input to an area, and PET represents the climatic demand for water.

Potential Evapotranspiration

Potential evapotranspiration is calculated using the Hamon model (Hamon 1961). The Hamon PET model is a temperature-based model and is calculated by:

$$PET_{Hamon} = 13.97D^2W_t \quad (84\text{-}1)$$

where PET_{Hamon} is Hamon PET in millimeters per month, D is the monthly mean hours of daylight in units of twelve hours, and W_t is a saturated water vapor density term calculated by:

$$W_t = \frac{4.95e^{(0.062TC)}}{100} \quad (84\text{-}2)$$

where TC is the mean monthly temperature in degrees Celsius.

The moisture index for a location is determined by subtracting potential evapotranspiration from precipitation for a given time period. Thus, positive values of the moisture index indicate a surplus of moisture in a given time period, and negative values indicate a deficit of moisture in a given time period. To account for the effects of stored soil moisture and accumulation of snow during winter months, five-month sums of P and PET were used (Carter and Mather 1966; Mather 1978). Thus, the moisture index for a given month represents the cumulative effect of P and PET during the given month and the four previous months. The moisture index is calculated in the following manner:

$$M_{ij} = P_{ij} - PET_{ij} \quad (84\text{-}3)$$

where M_{ij} is the moisture index for location i and month j, P_{ij} is the cumulative precipitation at location i for month j and the four preceding months, and PET_{ij} is the cumulative potential evapotranspiration at location i for month j and the four preceding months. The five-month period used to accumulate precipitation and potential evapotranspiration was chosen based on comparisons of various accumulation periods with: (1) April 1 snowpack accumulations in the Gunnison River basin, and (2) monthly Palmer Drought Severity Index values for the climate division in which the Gunnison River basin is located. For example, a comparison between basin mean April 1 snowpack in the Gunnison River basin, calculated by averaging April 1 snowpack accumulations at twenty-one snow course stations within and near the Gunnison River basin for the years 1982–1990, and mean moisture index values calculated using March moisture index values for grids with mean elevations greater than 2,300 m (2,300 m is the approximate elevation of the winter snow line), produced a correlation of 0.93 (significant at alpha equal to 0.01).

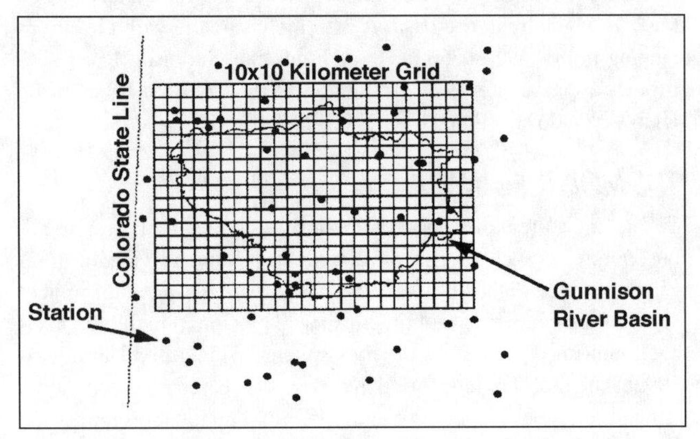

Figure 84-1. Location of meteorological stations providing data for the study and 10-km grid cells.

Moisture index values were calculated for 100-km^2 grid cells covering the Gunnison River basin (Figure 84-1). This spatial resolution provides estimates of moisture conditions for areas that are much smaller than subbasins of the Gunnison River basin.

Because there is a range of climatic conditions in the Gunnison River basin, the moisture index values were standardized to: (1) remove spatial biases in the values for each grid cell in order to compare different grid cells, and (2) permit the evaluation of moisture at each grid cell in a statistical and historical perspective. The standardization of the moisture index values was performed in the following manner:

$$M'_{ijk} = \frac{(M_{ijk} - \bar{M}_{ij})}{S_{ij}} \quad (84\text{-}4)$$

where M'_{ijk} is the standardized moisture index for location i, month j, and year k, M_{ijk} is the raw moisture index value for location i, month j, and year k, $\bar{M}_{ij}$ is the long-term mean moisture index value for location i and month j, and S_{ij} is the long-term standard deviation of moisture index values at location i and month j. The standardization of moisture index values produced temporal distributions of values for each grid cell with a mean of 0 and a variance of 1.

The development of moisture index values as described in this study offers several advantages. One advantage is the minimal climatic data required to determine the moisture index. The only climatic data needed to calculate the moisture index are monthly values of temperature and precipitation. Another advantage is that the moisture index used in this study is in the units of the original variables. In addition, because the moisture index is developed for numerous grid cells within and near the Gunnison River basin, details of the spatial distribution and variability of moisture conditions in the basin can be evaluated.

Temperature and Precipitation Interpolation

The precipitation and temperature data needed to calculate PET and the moisture index for 100-km^2 grid cells were estimated from temperature and precipitation data measured at meteorological stations located within and near the Gunnison River basin (Figure 84-1) by using empirical orthogonal function (EOF) analysis. The technique is based upon the work by Rao and Hsieh (1991), who use EOF analysis to estimate hydrologic variables at ungaged locations. Figure 84-1 shows the 10-km grid cells and the sixty-five stations used to produce estimates of precipitation and temperature on a monthly basis.

Empirical orthogonal function analysis was used to decompose monthly precipitation and temperature measurements at the sixty-five stations into their temporal and spatial components as shown in (84-5). The temporal component is $T_i(t)$ with $t = 1,2,\ldots,N$, where N is the number of measurements. The spatial component is $S_i(x)$ with $x=1,2,\ldots,m$, where m is the number of measurement stations. The method estimates $S_i(x)$ values at grid points and calculates the standardized series of precipitation and temperature ($f_{x,t}$) using (84-5):

$$f_{x,t} = \sum_{i=1}^{m} T_i(t)\, S_i(x) \quad (84\text{-}5)$$

The following steps outline the method used in this study, based on the method presented in Rao and Hsieh (1991):

1. Compute means ($\bar{y}_x$) and standard deviations (s_x) of the observed meteorological data for each station ($x=1,2,\ldots m$) for the time period $t=1,2,\ldots,N$.

2. Standardize the station measurements by:

$$f_{x,t} = \frac{y_{x,t} - \bar{y}_x}{s_x} \quad (84\text{-}6)$$

These standardized observations can be represented in an m x N matrix $\boldsymbol{F}$:

$$\boldsymbol{F} = [f_1 f_2 \ldots f_N], \quad (84\text{-}7)$$

where $[f_1 f_2 \ldots f_N]$ contain the m set of column values measured at time n.

3. Use $\boldsymbol{F}$ in EOF analysis to find the number of significant EOFs (**M**) and the corresponding spatial and temporal EOFs.
4. Use multiple regression (developed regressing the spatial EOFs against latitude, longitude, and elevation) to determine the value of the spatial EOFs for 100-km^2 grid cells.
5. Estimate the standardized series $\hat{h}_{x,t}$ for each grid cell (x) by:

$$\hat{h}_{x,t} = \sum_{i=1}^{m} S_i(x)\, T_i(t). \quad (84\text{-}8)$$

In this case $S_i(x)$ is the ith significant spatial component of EOFs at grid cell x and is determined from the multiple regression of the ith spatial EOF in step 4.

6. The estimated $\hat{y}_{x,t}$ series at each grid cell x is calculated using

$$\hat{y}_{x,t} = \bar{y}_x + \hat{S}_x \hat{h}_{x,t} \quad (84\text{-}9)$$

The mean and standard deviation values at each grid cell are calculated using linear multiple regressions developed regressing the spatial EOFs against latitude, longitude, and elevation.

DISCUSSION

The Moisture Index

Because the PDSI and the moisture index developed in this study are representations of moisture conditions, these indices are significantly correlated (Table 84-1); however, there are differences in the magnitudes of the moisture index values for each grid cell as illustrated in Figures 84-2 and 84-3. These spatial differences in the relative magnitudes of moisture conditions in the basin are important for assessing the water resources of the Gunnison River basin. The high-resolution moisture index values provide an improved means of analyzing the temporal and spatial distributions of moisture conditions in the basin. However, these high-resolution values create large data sets, and such large amounts of data are difficult to analyze. Thus, scientific visualization was used to aid the analysis process.

Table 84-1. Correlations between monthly moisture index values for grids 12, 146, and 432, and Palmer Drought Severity Index (PDSI) values for climate division two in Colorado. (All correlations are significant at alpha equal to 0.01.)

	Grid 146	Grid 432	PDSI
Grid 12	0.95	0.59	0.76
Grid 146		0.59	0.81
Grid 432			0.56

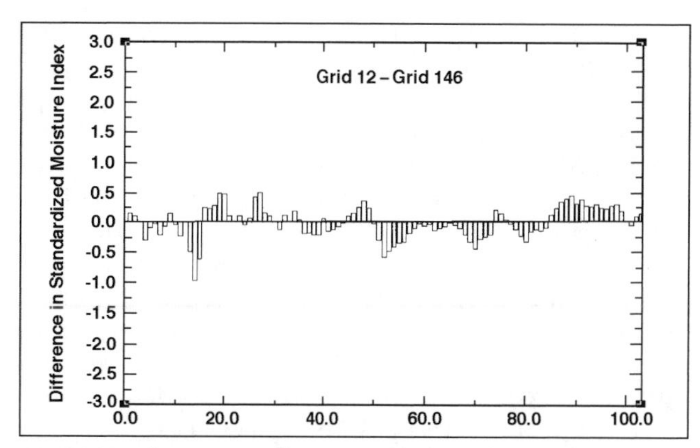

Figure 84-2. Difference in monthly moisture index values for moisture index values at grid cell 12, minus moisture index values at grid cell 146.

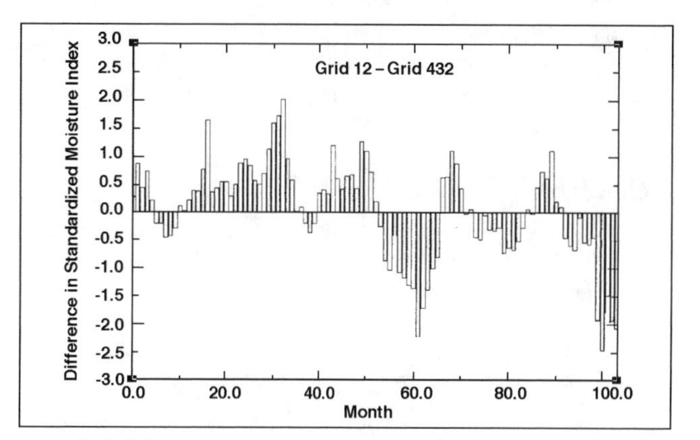

Figure 84-3. Difference in monthly moisture index values for moisture index values at grid cell 12, minus moisture index values at grid cell 432.

Scientific Visualization

Scientific visualization technology includes the capability to: (1) animate data time series, (2) interactively manipulate display variables, (3) select from multiple data representations, and (4) simultaneously display one-dimensional data to n-dimensional data. Scientific visualization techniques can be beneficial in data exploration, data analysis, communication of research results, and decision-making processes (Buttenfield and Ganter 1990). However, use of visual techniques without a clear understanding of the tasks they are to support does not necessarily result in improved understanding of the data (Casner and Larkin 1989). A task analysis approach to the determination of appropriate visualization techniques has been developed by Knapp (1993) and was used in this study. The objective of this approach is to develop designs that explicitly link the task and goal of the scientist with visualization techniques. Scientific visualization software was used to implement several research designs resulting from the task analysis.

To develop the visualization approach for this study, several members of WATF were interviewed to determine their perspectives on drought, and designs were created based on the resulting task analyses. Interviews with scientists also were conducted. These interviews outlined two primary sets of tasks, the first relating to the development and verification of precipitation and temperature fields that form the basis for the moisture index, and the second dealing with the verification of the index itself. These high-level tasks were further defined by the goals of the user, the activities required to complete the task, the data sets involved, and the visual judgments that the user must make based on the displays.

Because the data were time series of two-dimensional maps, the animation capabilities of the software were used extensively. One to four images were displayed simultaneously with various interactive display controls for user selection of time lags, sequencing, and data sets. In some cases the data were draped over a digital elevation model, and in other cases statistical relations were presented together with the spatial representation. The capabilities of the visualization software were clearly useful extensions to the modeling and geographic information systems (GIS) tools already in use at the U.S. Geological Survey. Results of this study indicate that full implementation of the designs based on the task analysis would require several extensions to the software, including integration with a GIS for ease of data import and analysis, addition of functions for time-series manipulation, and addition of advanced statistical capabilities. The task analysis approach proved useful in two ways: (1) the design of visualization techniques based on a user rather than a data perspective, and (2) the development of requirements for further software enhancement.

CONCLUSION

New approaches to moisture conditions monitoring are required in Colorado where topographic variability presents significant modeling challenges. This chapter describes development of a moisture index that can be implemented at any desired spatial or temporal resolution. Use of visualization techniques, including animation, simultaneous spatial and statistical display, and interactive display controls substantially enhanced the understanding of the characteristics of the index. Approaching visualization from a user perspective as opposed to a data perspective provided an important link between the scientists and this new technology.

References

Alley, W. M. 1984. The Palmer drought severity index—Limitations and assumptions. *Journal of Climate and Applied Meteorology* 23:1,100–109.

Budyko, M. I. 1956. *The Heat Balance of the Earth's Surface*, translated by N. A. Stepanova. Leningrad: Gidrometeoizdat.

Buttenfield, B. P., and J. H. Ganter. 1990. Visualization and GIS: What should we see? What might we miss? In *Proceedings of the 4th International Symposium on Data Handling, 307–16.*

Carter, D. B., and J. R. Mather. 1966. Climatic classification for environmental biology. *Publications in Climatology* 19:305–95.

Casner, S., and J. Larkin. 1989. *Cognitive Efficiency Considerations for Good Graphic Design.* Carnegie Mellon University Technical Report AIP-81.

Doesken, N. J., J. D. Kleist, and T. B. McKee. 1983. *Use of the Palmer Index and Other Water Supply Indices for Drought Monitoring in Colorado.* Climatology Report N-83-3, Department of Atmospheric Science, Colorado State University, Fort Collins.

Hamon, W. R. 1961. Estimating potential evapotranspiration. *Journal of the Hydraulics Division, Proceedings of the American Society of Civil Engineers* 97:107–20.

Knapp, L. 1993. Task analysis and geographic visualization. In *Proceedings of ASPRS 93,* New Orleans, Louisiana.

Lamm, R. D. 1981. *The Colorado Drought Response Plan: Executive Chambers.* State of Colorado.

Mather, J. R. 1978. *The Climatic Water Budget in Environmental Analysis.* Lexington: D. C. Heath and Company.

Rao, A. R., and C. H. Hsieh. 1991. Estimation of variables at ungaged locations by empirical orthogonal functions. *Journal of Hydrology* 123:51–67.

Thornthwaite, C. W. 1948. An approach toward a rational classification of climate. *Geographical Review* 38:55–94.

Ugland, R. C., B. J. Cochran, M. M. Hiner, R. G. Kretschman, E. A. Wilson, and J. D. Bennett. 1990. *Water Resources Data for Colorado, Water Year 1990,* vol. 2. Water-Data Report CO-90-2. Colorado River Basin, U.S. Geological Survey.

Willmott, C. J., and J. J. Feddema. 1992. A more rational climatic moisture index. *Professional Geographer* 44:84–87.

Loey Knapp
IBM Corporation and University of Colorado
Boulder, Colorado 80301
E-mail: *knappl@bldfvm9.vnet.ibm.com*

Gregory J. McCabe, Jr., Lauren E. Hay and *Randolph S. Parker*
U.S. Geological Survey
Water Resources Division
Denver Federal Center, MS 412
Lakewood, Colorado 80225
E-mail: *gmccabe@climate1.cr.usgs.gov*
E-mail: *lhay@lhay.cr.usgs.gov*
E-mail: *rsparker@rspdcolka.cr.usgs.gov*

85

Tools for Visualizing Landscape Pattern in Large Geographic Areas

Sidey P. Timmins and Carolyn T. Hunsaker

Landscape pattern can be modeled on a grid with polygons constructed from cells that share edges. Although this model only allows connections in four directions, programming is convenient because both coordinates and attributes take discrete integer values. A typical raster land cover data set is a multimegabyte matrix of byte values derived by classification of images or gridding of maps. Each matrix may have thousands of raster polygons (patches), many of them islands inside other larger patches. These data sets have complex topology that can overwhelm vector geographic information systems (GIS). Our goal is to develop tools to quantify change in the landscape structure in terms of the shape and spatial distribution of patches. Three milestones toward this goal are: (1) creating polygon topology on a grid, (2) visualizing patches, and (3) analyzing shape and pattern. An efficient algorithm has been developed to locate patches, measure area and perimeter, and establish patch topology. A powerful visualization system with an extensible programming language is used to write procedures to display images and perform analysis.

INTRODUCTION

Natural land cover patterns are created by plant and animal life occupying and modifying a landscape and its particular terrain, climate, soil, light, and water bodies (Krummel et al. 1987). These patterns vary greatly from one region to another and document the interaction between earth-shaping forces and biota over millions of years. Land cover patterns are visible from airplanes, on satellite images, and on land cover maps such as LUDA (Fegeas et al. 1983). For example, the Appalachian Mountains have different patterns than do the southeastern coastal plains (Figure 85-1). Natural patterns may be as large as mountain ranges or as small as anthills—and they are forever changing. Human influence is primarily that of management: agriculture, transportation, urbanization, forestry, and wilderness preservation. This management imposes a regularity on landscapes that vary from city block development to furrowed fields. To varying degrees, this ongoing management removes or modifies existing habitat in accordance with human design. In order to study landscape pattern and change, we need a model that is amenable to variable scales and metrics. Because most metrics are based on discrete measures of adjacency or entropy, calculation is simplified using discrete spatial data. Using a grid-based model, landscape pattern has been studied extensively with regard to metrics and their application to small regions or regions with rectangular map boundaries (O'Neill et al. 1988; Turner 1990; Graham et al. 1991; Baker and Cai 1992). Examples of landscape metrics are:

1. Dominance: As this measure approaches 1.0, the landscape tends to be dominated by a single cover type.
2. Contagion: As this measure approaches 1.0, the landscape tends toward larger, more contiguous patches.
3. Shape complexity: A measure of perimeter-to-area ratio, a value of 2.0 being the most complex (Figure 85-1) (Hunsaker et al. 1994).

Raster data for such a model are limited by available images that have cell sizes in the range of 30 m–1 km. Our emphasis on landscape pattern for large geographic areas is prompted by the need to assess landscape change for regional and national policy. The U.S. Environmental Protection Agency's Environmental Monitoring and Assessment Program plans to assess status and monitor trends in landscape pattern for the entire United States. The combination of large assessment regions (hundreds of km wide) and suitable grid modeling scales (sub-km cells) creates very large data sets. This chapter describes our raster patch model and efficient methods to prepare, analyze, and visualize large grids of landscape data.

CONCEPTS

For the calculation of landscape pattern metrics that depend on edge relationships, it is preferable for spatial data to be stored in a discrete form. Discrete representations code continuous information into integers and create counts. For example, area and perimeter become multiples of the cell units, cells and edges, respectively. Gridding spatial data also creates regular, though not equivalent, relationships between adjacent cells (Figure 85-2). Although connections may be assumed to exist between all adjacent cells, we use a model that builds polygons only from cells of the same type that also share edges. Each cell may share from zero to four edges with neighbor cells of the same attribute.

Raster Patch Model

The raster patch model describes polygons made from grid cells connected by their edges. A single cell patch has an area of 1 unit and a perimeter of 4 units and shares no edges with cells of the same attribute. Multicell patches always have perimeter cells but need not have interior cells. Cells on the perimeter share one to three edges with adjacent patch cells, whereas interior cells share all four edges. A two-cell patch has an area of 2 and a perimeter of 6, and its two perimeter cells share one edge with each other. By induction, we derive a perimeter formula for any cell or patch:

perimeter = (area × 4) - count of neighbor cells with same attribute (85-1)

In this equation, values are measured in counts: area is in cells, and perimeter is the edge count of both the external boundary and any inner perimeter(s) (like a doughnut hole). Many patches are actually islands in a larger patch; thus, their external boundaries form internal perimeters in the surrounding patch. For inner perimeters, travel in a clockwise direction orients the patch on the left, whereas for the outer perimeter, the patch lies on the right. This is the basis for a left-seeking rule that can be used to circumnavigate an outer perimeter, provided we begin with a clockwise orientation (see Left-Strutting Turtle Algorithm section).

As with all models, the raster patch model has both advantages and disadvantages. A grid tiling of two-dimensional space is simple and creates discrete cells and edges suitable for modeling landscape maps. Each cell in a grid can be referenced with a column and row address. Because typical land cover data sets have fewer than 256 categories, their attributes may be represented by a byte matrix of values. Gridding also permits many model calculations to be performed with integers because spatial coordinates are multiples of the area and perimeter. Disadvantages of gridding include classification error and anisotropic connectivity. Our model can only simulate connections in diagonal directions by step-shaped polygons. Also, when each cell is classified, a choice has to be made when there is more than one land cover within each cell. This may be ambiguous if two land covers, each covering 50% of the cell area, are present, or discriminatory for narrow land types such as roads, which may be consistently narrower than the cell width. Thus, for landscape pattern analysis, there are limitations in modeling reality using a grid.

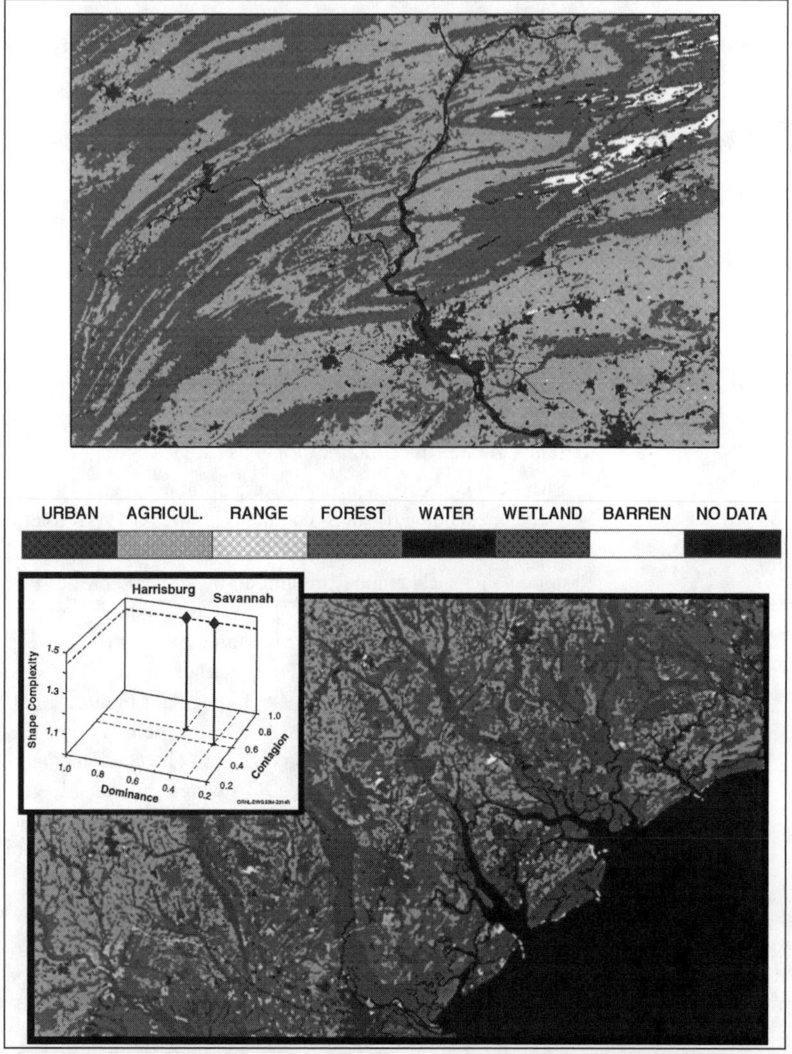

Polygon Topology

Although a cell model is sufficient to determine some metrics, landscape structure is better described by polygon topology. Topology consists of locating each polygon, finding its extent, and then describing its arrangement in the spatial data set. Vector data typically have a single link from each spatial entity (point, line, or polygon) to an attribute database. With a vector data structure, the cost of topology creation is largely dependent on rapid calculation of line intersections (Franklin et al. 1989). Vector polygon topology is determined by building an entity table to describe how each pair of lines intersects at a node, and then tracing around each polygon boundary. Vector GIS typically limit the number of polygons and line segments they can handle, and these limits may be exceeded by LUDA data sets (~1 Mb). Editing must always be followed by recreation of topology before spatial processing can proceed. Spatial operations for areas in a vector GIS are performed on a polygonal basis by using coordinates, attributes, and rules to interpret the description for each polygon in the entity table.

In the raster world, topology is not necessary for most spatial data operations, which can be performed on a cell-by-cell basis. Clipping, for example, can be done by using a mask to select data cells. Each polygon's area may be determined by a simple but inefficient algorithm that connects both perimeter and inner cells to determine shape. Because large patches usually have more interior cells with more connected edges than perimeter cells do, this process has a nonlinear cost with respect to data set size. This cost is significantly less for a more efficient algorithm, such

Figure 85-1. Land cover patterns: Appalachian Mountains near Harrisburg (above); southeastern coastal plain near Savannah (below). Inset shows the separation of these landscapes in three-dimensional space by three pattern metrics.

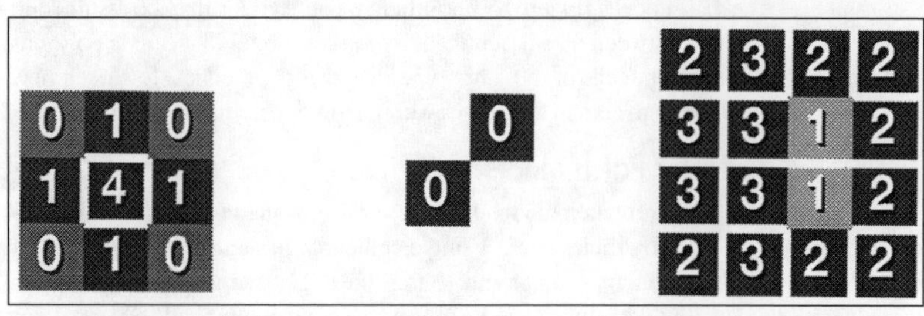

Figure 85-2. Raster model showing shared edges in white; cross-shaped patch with four diagonal neighboring patches (left), two patches (center), and island patch within another patch (right). Digits indicate the number of shared edges. All cells shown are perimeter cells except the interior cell at left with four white edges.

as the "left-strutting turtle," which determines shape from perimeter cells only. With raster format, polygons are constructed from cells that are easily modified, and thus cell-based operations may be used to process the data on a polygonal basis. However, processing patches on a grid is not a simple task because a complete entity table may take more space than the data set itself. Hence, patch processing programs must actually recreate each patch shape by using the data set (a perimeter-following rule) and a simple entity table that describes just the location of each patch.

Left-Strutting Turtle Algorithm

The procedure for creating patch topology has three components: (1) locate each patch, (2) measure areas and perimeters, and (3) create an entity table that describes interpatch relationships. Algorithms to measure both area and perimeter on a raster grid are not common. A method to locate patches is described by Gonzalez and Wintz (1987), and this method has been extended to find all islands. The left-strutting turtle algorithm was developed to measure area and perimeter efficiently by using a left-seeking perimeter-following rule to determine patch shape. First, for each cell, it counts like-attribute neighbor cells to calculate perimeter using equation (85-1). Then, for each patch, the algorithm locates an initial cell and travels through the external perimeter cells following a rule commonly used to exit a maze. Although the turtle doesn't actually walk on the perimeter edges (Figure 85-3), perimeter edges can be counted by using equation (85-1). Starting from the initial cell and with an initial orientation to the east, the left-strutting turtle attempts to turn left into the next perimeter cell. If the turn fails (because the new cell is not the same attribute), the turtle walks straight ahead; if that move fails, it turns right; if that move fails, it tries to back up. When a move succeeds, the rule is reapplied relative to the new orientation in the new cell, and the walk continues in a clockwise direction until the turtle returns to the initial cell. In this way, the external perimeter and the area it encloses (including any islands) are found for each patch.

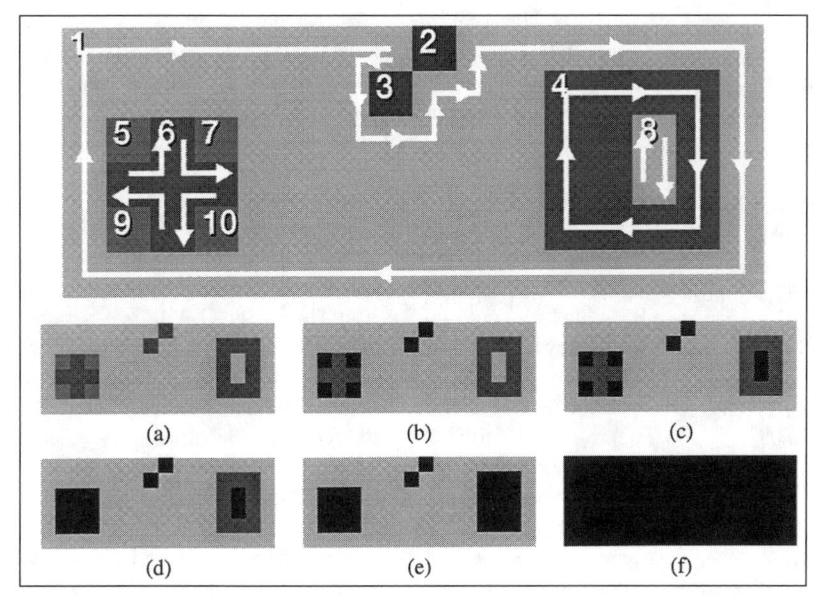

Figure 85-3. The turtle walks through the outer perimeter cells of each patch, keeping the boundary on the left. Each walk starts and ends in a numbered cell. These patch numbers are in the order patches are located (top to bottom, left to right). Area calculation is done in order of increasing enclosed area (below). As each patch is processed (a–f) its cells are marked null (black), so they are not counted again. For example, the cells from patch 8, which have the same attribute as patch 1, are not counted with patch 1.

To complete the algorithm, the values of the enclosed area are sorted from small to large. The area of each patch then can be measured by scanning the region enclosed by its outer perimeter and counting the patch cells only. This can be accomplished by proceeding from the inner smaller patches to the outer larger patches and marking counted cells with a null flag. In this way, all non-null data cells within the outer perimeter can be counted to measure patch area. The total perimeter count is completed simultaneously using equation (85-1). Internal perimeter for the patch (if any) can be calculated by difference between the total perimeter and the outer perimeter as measured by the turtle walk. The algorithm is very fast: Processing seven million 1-km cells for the United States is completed in two minutes on a DEC 3000 Model 300 ALPHA workstation (integer SPECmark is 66). Besides computing some simple metrics, the program writes a table describing the location of the initial cell, attribute value, area, and perimeter of each patch.

DATA PREPARATION TOOLS

Need for Efficiency

Preparing and storing raster data for landscape analysis is usually costly in terms of central processing unit (CPU) time, input/output time, and disk space because of high data volume. For example, with 1-km cells, the continental United States is about 14 Mb, whereas at 30 m, Florida alone is over 500 Mb. Consequently, efficient tools must be employed with a low per-cell cost. Also, import/export and input/output operations should be minimized by using a single GIS and combining tasks where possible. For, example the turtle program optionally performs reclassification and clipping prior to patch analysis. Storage space also can be reduced by compressing the data on disk and then restoring the original grid format in memory. (Compression trades CPU time for compactness, but the convenience of having more data on-line can be worthwhile.) When a second GIS or a user application program is required, data often must be reformatted. Most formats are proprietary, and neither descriptions nor software drivers are in the public domain. The use of a single public domain format, such as the hierarchical data format (NCSA 1989), could reduce reformatting costs. Similarly, the cost of transferring data is a concern. Because most GIS do not support a standard grid object (with a specific orientation and description of location and size), grids cannot be piped from one system to another via a UNIX pipe. Instead, GIS usually work in a disk-to-disk mode, with duplicate storage required for input and output data. Most current GIS also do not support dynamic linking to user-written software to add new capabilities. To avoid many of these limitations, we have prepared most of our data with a visualization system.

Language as a Tool

The tools we have written work in an X-Windows environment within a commercial visualization system (PV~WAVE™). The main advantage of this system is that it combines graphics with a high-level interpreted programming language. This structured language enables tools to be designed and tested interactively with visual or numeric feedback. The system has a distinct "hands-on" approach because data are processed immediately, in memory, using simple atomic commands. Because visualization is part of the language, results can be displayed with ease. Thus, after import, an image can be scaled by 2 and displayed by typing: *tv,image*2*. Errors are handled gracefully with a complete walk-back and description of the problem. The system can be extended by writing procedures or by dynamic linking to FORTRAN or C subprograms. Because the kernel of the system is not modified by this extension, PV~WAVE has high reliability.

A major advantage in using PV~WAVE for visualization is the speed at which it processes arrays and matrices. The language treats each variable as an array of a certain length of a particular data type (byte, 2-byte integer, etc.). Mathematical commands such as *result=image*2* work for any numeric data type and are implemented using a virtual machine. After setup, to interpret and initiate each command, the virtual machine processes arrays quickly because instructions are implemented in a tight loop, like compiled code. Spatial operations on grid cells can be performed with a "where" command that extracts cell addresses that meet a criterion. For example, a clip operation takes three commands:

1. output_grid = input_grid
2. unselected_cells = where(mask not equal to desired mask value)
3. output_grid(unselected_cells) = null

The ability to perform multiple functions in memory before output, coupled with the array processing capability, makes this language faster than many GIS. Human productivity is excellent because the language provides significant capabilities yet can be dynamically changed to perform new functions on large data sets.

LANDSCAPE PATTERN VISUALIZATION

Visualization converts data into a picture, which provides the viewer with an opportunity to make observations or draw conclusions that could not be made easily without the picture. To use visualization for landscape pattern would seem to be straightforward. The human eye is a very effective image processing system for storing and analyzing patterns. Its main defect is that it can be overwhelmed by data volume, especially if the images are unfamiliar. During the recognition process, information is grouped into regions and patterns rather than pixel by pixel. Entire images or regions within an image are assigned a significance based on their familiarity, which is determined by comparison with an image database in the brain. New (unfamiliar) information is of more interest than are objects that have been seen before and, consequently, it takes longer to recognize and store. Pattern perception is both size- and contrast-dependent. Thus, finding unclassified pixels in a color image on a monitor may be virtually impossible when the pixels are small, but may be simplified by increasing pixel size or by altering the colors so that the unclassified pixels appear white against a black background. Familiarity, color, and size are important discriminatory tools in the recognition process.

The development of tools crafted for analyzing landscape pattern has just begun. The turtle program is expected to be a valuable tool because landscape patterns are made from patches and not cells. However, significant progress is necessary to provide even rudimentary pattern matching to compare with the capabilities of the human eye. A visualization system with its own programming language creates an environment in which tools can be crafted for particular needs. The procedures we have developed are based on conventional cell-based imagery and are not yet adapted for patches. One tool imports a grid effortlessly into the visualization system and then, with the aid of a second tool, displays an overall view. As the mouse cursor is placed at a menu entry, small colored maps aid in the selection process. Once a selection is made, the data are imported, oriented correctly, and displayed at a suitable scale. This display tool, which fits the grid to the viewing screen, is important because it defines a graphics state for the image. From this view, a telescope tool can be used to pan and zoom the grid at an arbitrary scale. Vector lines and boundaries can be superimposed for reference. Images are displayed with the current color table, which can be modified using a palette. However, whereas changing image size to accommodate the human eye is straightforward, color perception is highly subjective. Interpolation between colors often produces muddy tints with low saturation, and few people, other than artists, know how to mix colors. Hence much work is required on color. However, the combination of simultaneous views, a feature of the X-Windows system, and the ability of the eye to integrate these views creates the potential to develop powerful tools. Ideally, the user should be able to switch rapidly from one tool or view to another to achieve visual feedback and continuity. Although the PV~WAVE visualization system supports multiple windows, only one window may be active at any particular instant and only a single procedure can be working in the virtual machine. Allowing multiple process threads, each with its own active window(s), would be an improvement.

The analysis of pattern can be aided by the image analysis capabilities of the human brain, which is self-programming. Assessment of landscape pattern is most effectively accomplished by a hierarchical approach: cell-based landscape metrics; robotic vision, which attempts to determine edges and patches within an image; and the human eye, which analyzes and detects patterns in entire images. Obviously, there is room for significant advances, particularly at the second level in this hierarchy—patches.

CONCLUSION

Efficiency with regard to data storage and algorithms is required to successfully analyze landscape pattern on large grids. Efficiency is increased by: (1) combining functions into a single step to reduce input/output, (2) dynamically linking user application programs, and (3) using a capable visualization programming language. The development of tools to analyze landscape pattern at the patch level may lead to better metrics to describe the differences in landscape pattern.

Acknowledgments

We thank Kurt Riitters of the Tennessee Valley Authority, and Raymond McCord and Michael Huston of Oak Ridge National Laboratory for reviewing the manuscript. Research was supported by the Environmental Monitoring and Assessment Program, U.S. Environmental Protection Agency (EPA) (under interagency agreement #DW89936104-01); with the U.S. Department of Energy (under contract #DE-AC05-84OR21400); and with Martin Marietta Energy Systems, Inc. (Environmental Sciences Division publication #4288). This research has not been subjected to EPA review and therefore does not necessarily reflect the views of the EPA, and no official endorsement should be inferred. The submitted manuscript has been authored by a contractor of the U.S. Government under contract number #DE-AC05-84OR21400. Accordingly, the U.S. Government retains a nonexclusive, royalty-free license to publish or reproduce the published form of this contribution, or allow others to do so, for U.S. Government purposes.

References

Baker, W. L., and Y. Cai. 1992. The *r.le* programs for multiscale analysis of landscape structure using the GRASS geographical information system. *Landscape Ecology* 7(4):291–302.

Fegeas, R. G., R. W. Claire, S. C. Guptill, K. E. Anderson, and C. A. Hallam. 1983. *Land Use and Land Cover Digital Data.* Geological Survey Circular 895-E. Washington, D.C.: U.S. Geological Survey.

Franklin, W. R., N. Chandrasekhar, M. Kankanhalli, D. Sun, M. Zhou, and P. Wu. 1989. Uniform grids: Technique for intersection detection on serial and parallel machines. *Proceedings Auto-Carto 9,* 100–09.

Gonzalez, R. C., and P. Wintz. 1987. *Digital Image Processing.* Reading: Addison-Wesley.

Graham, R. L., C. T. Hunsaker, R. V. O'Neill, and B. L. Jackson. 1991. Ecological risk assessment at the regional scale. *Ecological Applications* 1(2):196–206.

Hunsaker, C. T., R. V. O'Neill, B. L. Jackson, S. P. Timmins, D. A. Levine, and D. J. Norton. 1994. Sampling to characterize landscape pattern. *Landscape Ecology* 9(3):207–26.

Krummel, J. R., R. H. Gardner, G. Sugihara, R. V. O'Neill, and P. R. Coleman. 1987. Landscape patterns in a disturbed environment. *OIKOS* 48:321–24.

NCSA. 1989. *Hierarchical Data Format Specifications Manual.* Champaign: National Center for Supercomputing Applications.

O'Neill, R. V., B. T. Milne, M. G. Turner, and R. H. Gardner. 1988. Resource utilization scales and landscape pattern. *Landscape Ecology* 2(1):63–69.

Turner, M. G. 1990. Landscape changes in nine rural counties in Georgia. *Photogrammetric Engineering and Remote Sensing* 56(3):379–86.

Visual Numerics Inc. 1993. *PV~WAVE* (PV~WAVE Command Language Reference). Boulder: Visual Numerics, Inc.

Sidey Timmins is a GIS programmer and analyst, whose interests lie in spatial data handling, visualization, data compression, and algorithm optimization. His programming experience includes working as a geophysicist in the oil industry, and as a subcontractor to Oak Ridge National Laboratory analyzing and visualizing watersheds and landscapes.
Analysas Corporation
151 Lafayette Dr.
Oak Ridge, Tennessee 37830
E-mail: *stq@ornl.gov*

Carolyn Hunsaker is an environmental scientist and research staff member in the Environmental Sciences Division of Oak Ridge National Laboratory with research interests in landscape ecology, ecological risk assessment, spatially distributed water-quality modeling, and regional assessment methods.
Environmental Sciences Division
Oak Ridge National Laboratory
PO Box 2008
Oak Ridge, Tennessee 37831-6038
E-mail: *cth@ornl.gov*

86

Intelligent Browsing of Earth Science Data in the Sequoia 2000 Project

Len Wanger, Zahid Ahmed, and Peter Kochevar

The ability to directly access, analyze, and visualize Earth science data on the desktop currently eludes most end users. Although computer networks and powerful workstations provide the necessary infrastructure to accomplish these tasks, nondatabase and nonvisualization experts find the associated software too difficult to use. To address this problem, a research effort has been established under the auspices of the Sequoia 2000 project. Within Sequoia, a visualization management system is being developed that brings together database management, scientific visualization, and graphical user-interface building into one system. The heart of this system is an intelligent subsystem that makes use of a knowledge base to help construct interactive data visualizations. Using the system, end users specify a visualization task, and the system figures out how it is to be performed. With this functionality, end users will be able to concentrate more on their work rather than on the process of forming a visualization.

INTRODUCTION

The Sequoia 2000 project is a unique blend of Earth and computer scientists, who together are creating a computing infrastructure that will foster a better understanding of global environmental change (Dozier 1992). The entire field of global change is characterized by massive amounts of information pertaining to Earth that now are available or will be available soon, through such endeavors as NASA's EOS project (Dozier 1990). For this reason, Sequoia Earth scientists have expressed a need for comprehensive data management to augment scientific visualization (Olson et al. 1992).

A primary focus of the visualization research efforts within Sequoia 2000 is to put sophisticated database access and data-visualization capabilities into the hands of the scientists themselves. But in doing so, scientists should not be overburdened with the mechanics of accessing and rendering data. Scientists want only to concentrate on their science, *not* on the process of performing it. Therefore, it is important to make visualization and database access easy and useful for nonexpert data management users.

In the Earth sciences, the search for, and ready access to, appropriate data sets are as important to scientists as the actual visualization of data. The browsing of databases for data with which to conduct a study is a frequently performed task. Consequently, an Earth science information system that is to prove useful for studying global change must supply tools to make browsing efficient and effective.

The kind of data encountered in the study of planet Earth is quite varied in its structure and format. Within Sequoia, the goal is to visualize all kinds of data, not just raw data from satellite imagery or output from global climate computer models, for example. Metadata, ancillary information that describe data collection and storage, need to be visualized as well. Furthermore, the results of queries to a database are yet another kind of "data" that must be visualized in order to aid in browsing databases.

Visualizations of data must go beyond mere presentations of information. Systems that allow scientists to do their own interactive, exploratory visualizations are desirable. For this to occur, a visualization system needs to build virtual environments in which objects that represent the data being explored have "behavior." Events, such as "mouse picks," directed toward these objects may result in actions that convey additional information to scientists. Some of these actions may guide scientists as they "fly through" visual representations of data sets, or they can result in additional database queries and hence further visualizations of data. Such mechanisms provide a way to navigate through databases and link screen visuals to actual numbers that reside in databases.

In the study of global change, most data cover the globe and are sequenced in time. Consequently, most queries are for data that are associated with particular regions of the planet during particular periods of time. For this reason, a geographical information system (GIS) should form the basis for an Earth science information management system. Within such systems, domain-specific query tools should be created to help scientists locate data. For instance, a scientist would much prefer to specify regions of interest on a world map or globe rather than keying in textual queries in an arcane database query language.

The primary purpose of this chapter is to describe how these goals are being addressed in project Sequoia. The first section of the chapter describes a very simple visualization management system that links database management, scientific visualization, and graphical user-interface building into one package. This system, called the Visualization Executive, was created as a test bed to understand the interplay between database and visualization software within an Earth science setting.

The next section discusses the limitations of the Visualization Executive and some of the lessons learned during its design, implementation, and use. These lessons have been incorporated into a new visualization management system—Tioga—currently under development and described in the final section of this chapter. The heart of this new system is an intelligent server that makes use of a knowledge base and a knowledge-based management system to aid in the construction of automatic interactive data visualizations. With Tioga, scientists specify what to visualize and leave the actual visualization process to the system.

A SIMPLE VISUALIZATION MANAGEMENT SYSTEM

An early attempt to develop a visualization management system within the Sequoia project is described in Kochevar et al. (1993a). This system, dubbed the Visualization Executive, links the POSTGRES database management system (Stonebraker and Kemnitz 1991) with most any visualization package. The design and operation of the Visualization Executive are intended to be somewhat analogous to the operation of the Apple Macintosh folder browser. On the Macintosh, a user traverses the folder tree until a document of interest eventually is located. The browsing of folders proceeds visually by pointing and clicking on iconic representations of folders and documents. When a desired document is located, the user double-clicks on the representative icon and the document is then "visualized" within the application that created the document.

In a similar fashion, a user of the database browser in the Visualization Executive navigates through the database by pointing and clicking on iconic representations of database contents. Once a data set of interest is located, a user double clicks on the data set's icon and the corresponding data set is then visualized. The key to this functionality is that information about the visualization of each data set is stored in the database along with the data set itself.

In general, the Macintosh analogy cannot be pushed too far. For instance, in the Visualization Executive, an "application" is not the program that *creates* a data set but rather is the method for its *visualization.* In addition, the same data set can be visualized in many different ways using a variety of different visualization packages, whereas on the Macintosh, a single application is associated with only one document. In practice, information about each of the various ways a data set is to be visualized could be stored with the data set, and a way of choosing from among the possibilities could be presented to a user. But the prototype system made the assumption that only one visualization method was associated with a given data set.

Query Formation

The database browser in the current incarnation of the Visualization Executive comes with two interaction tools, both of which were implemented using the X-Windows-based Tcl/Tk graphical user-interface tool kit (Ousterhout 1990, 1991). The simplest yet most powerful of these is the text tool, which allows any legal database command to be sent to the database server for execution. The text tool is a visually oriented text editor with a built-in history mechanism. All commands submitted for execution with the tool are saved in a buffer, and each later can be retrieved, edited, and then resubmitted. At system start-up, the history buffer is loaded with commands from a file so that frequently used database commands do not have to be entered by hand each time the system is initialized.

Although the text tool offers the utmost flexibility in interacting with the database, it is provided as a last resort. The tool is difficult to use in that textual commands must be entered from the keyboard and a knowledge of the database query language is mandatory. The preferred method of interacting with the database is through a point-and-click interface, whose design is partially determined by the kind of information that is to be visualized. An example of such a domain-specific database interaction tool is provided within the Visualization Executive by the map tool.

The map tool is a simple GIS interface that allows limited forms of queries to be submitted to a database server. The intent of this tool is to allow nondatabase experts to make complex queries of the database in a way that is more natural and intuitive given the kind of data being sought. The tool is built around a two-dimensional world map, which forms a backdrop for graphically specifying latitude and longitude ranges. Aside from specifying such constraints, the map tool is used to enter date and time ranges as well. Once latitude/longitude and date/time ranges have been entered, a query for all Earth science data sets in the database whose positional and temporal attribute values fall within the specified constraints is formulated. All Earth science data sets whose latitude/longitude and date/time ranges nontrivially intersect those given in the query are retrieved from the database. The database system does not clip data sets to the requested ranges, although such a subsetting capability easily could have been added. A further shortcoming is that areas on the map cannot be zoomed in upon, thereby revealing more detailed geographical information.

Visual Representation of Information

Upon receipt of query results, the Visualization Executive invokes an icon generator to assign a graphic representation to each item that is returned. Each Earth science data set in the database has associated with it an icon stored as a pix map. If an element returned by the query processor corresponds to one of these data sets, then the icon generator assigns that pix map to the element as its icon. In all other cases, a generic "catch-all" pix map is assigned to an element by the icon generator.

When icons are generated, they are arrayed in groups on a scrollable canvas as seen in the top middle portion of Figure 86-1. When an icon is selected with the middle mouse button, a separate window appears which displays metadata for the element represented by the icon. The metadata values can be edited within the window, and the changes will be written back into the database whenever the window is closed. If an icon corresponds to raw scientific data, then double-clicking on it causes the data to be visualized according to the visualization information stored with the data.

LESSONS LEARNED

The prototype visualization management system described was designed and built to test new ideas about linking scientific databases to visualization packages. Experience gained in implementing and using the system has provided a good understanding of what kind of methodology and functionality works and what can be improved upon. Although far from a production system, this system was successful because of insight gained from its creation and use. Following are some of the lessons learned that have bearing on the interplay of visualization and database management.

1. The results of general database queries are a kind of data that should be visualized just like any other scientific data. One appropriate visualization technique for data of this kind is the positioning of icons onto an "informational landscape." In this way, the visualization encodes additional information that can be utilized by a scientist. For example, queries for data sets pertaining to regions of Earth might be positioned above a world map or globe so that a "light cone" highlights the exact area of a data set's coverage. In addition, the visualization of

query results must be able to handle the screen clutter that can result when queries generate thousands of items. The hierarchical display of data, such as the case with folders on the Macintosh, is one visualization technique for dealing with the clutter.

2. Interactivity is important in data visualization as a means for both exploring the information content of a chosen data set and for extending links back into the database. Objects that appear in a visualization can be thought of as "widgets" when various behaviors are assigned to screen visuals. For instance, these behaviors might be elicited when pointed at with a picking device via a callback subroutine. Such behaviors can trigger additional database interactions and visualizations, thus affording a mechanism with which to browse. As an example, consider a query for all snow gauges in the Sierra Nevada of California that resulted in a collection of icons representing each gauge displayed on a map. When an icon is pointed at, a history of snow measurements for that site could be displayed as a two-dimensional graph, with time as the independent axis and snow depth as the dependent axis. If, in addition, icons are provided to allow further database queries, a scientist might then request climate data for the same area and period of time in order to visually explore the possible correlation of mean temperature and snow depth.

3. A data set can be visualized in a great many ways. When scientists visualize data they are making use of the large bandwidth of the human visual system to make a visual computation. Exactly how a scientist wishes to participate in the computation is information that can be used to structure an appropriate visualization for the task at hand. Scientists may wish to select one item from many possibilities, to correlate or compare data in different data sets, or to search for patterns or anomalies, etc. In general, a visualization should be structured to make the visual computation as efficient as possible. Having a list of predetermined visualization programs associated with a data set is too limiting in that it may hinder exploration and preclude the reusability and sharability of the data.

4. Scientists should not be required to write visualization programs themselves, or through visualization experts acting as intermediaries. A scientist only should have to identify what is to be viewed via browsing and then let the visualization management system figure out how to do the visualization. To have this capability, a visualization management system requires built-in intelligence and access to a knowledge base that contains rules and definitions pertaining to data visualization. Simple systems have been developed that address the automatic visualization of certain kinds of data (Beshers and Feiner 1992; Casner 1991; Mackinlay 1986; Senay and Ignatius 1992), and these systems need to be extended to fully serve general scientific applications.

5. Databases need to be self-describing; in other words, the schema design and a complete characterization of all data classes must be stored in the database along with the data. Having this information on-line allows a visualization management system to "learn" about the structure of a database, thereby giving it the flexibility to dynamically create graphical interfaces tailored to the information that a user seeks. Information about classes of data and their relationship to one another should go beyond the mere description of syntactic structures. The semantics associated with classes of data are important information that need to be stored in a data/knowledge base as well. Such information can be used to give a domain-specific look and feel to interactive visualizations. For instance, if the independent variables of a data set range in latitude and longitude, then the data set can be placed on a globe with the appropriate topography and political boundaries in place.

6. To help determine effective and efficient visualizations automatically, an appropriate data model is needed. Depending on how data are structured within the model, only certain classes of visualization techniques can be used to render the data. The model must be general enough to describe relational data, such as metadata, as well as all the forms of nonrelational data encountered in the Earth sciences. Such nonrelational data may consist of polyhedra, as is the case for topography represented as polygonal surfaces or river networks represented as 3-space graphs; or they may consist of multidimensional arrays, as is the case for output from global climate computer models. In general, the data model is the foundation from which database schemas and knowledge structures are constructed.

KNOWLEDGE-BASED VISUALIZATION

Many of the lessons learned in the development of the Visualization Executive are being incorporated in a new visualization management system called Tioga, whose architecture is described by Kochevar et al. (1993b). At the core of Tioga is an intelligent subsystem built upon a knowledge representation and inferencing system. The intelligent subsystem is a server that receives descriptions of data sets to be visualized—"hints" as to how a scientist wishes to view the data—and then produces a visualization program that renders the given data. The visualization program is in the form of a textual encoding of a data flow network similar to those found in scientific visualization systems such as AVS (Advanced Visual Systems 1992), Explorer (Silicon Graphics 1992), and Khoros (Khoros Group 1992).

Data Characterization

The intelligent subsystem assumes that all data are defined in a format based on the mathematical notion of fiber bundles (Butler and Pendley 1989; Butler and Bryson 1992; Haber et al. 1991). A fiber bundle is determined by the specification of two spaces, a base space and a fiber space, in which elements from the fiber space are assigned to each element in the base space. As such, fiber bundles generalize the notion of the graph of a function—the base space being analogous to the independent function variables, while the fiber space corresponds to the function's dependent variables.

In actuality, the definition of fiber bundles makes no assumptions about the structure of either type of space. In fact, base and fiber spaces can be arbitrary topological spaces that may or may not have coordinate systems defined on them. A fiber bundle, whose base space consists of an unordered set of elements and whose fiber space is the Cartesian product of the field domains of a record, is a relation—the theoretical underpinning of relational database management systems. By using more complex spaces, topological information can be encoded, thus allowing the definition of arbitrary polyhedra that can be used to represent such objects as the topography of Earth or streamlines corresponding to the winds in the atmosphere. Finally, fiber spaces themselves can be collections of fiber bundles, thus allowing the representation of hierarchical structures.

Fiber bundles are described to the intelligent subsystem using descriptions specified in a data definition language (defined by

Kochevar et al. 1993b). Within this language, the particular type of fiber bundle is indicated as well as the structure, extent, and number of base and fiber space domains. The data specifications also indicate the number of elements present in the base space of a fiber bundle and whether or not the base elements are indexable using multi-indices defined in a specified index space (Kochevar et al. 1993b).

Data flow modules that appear in visualization networks treat fiber bundles as abstract data types. For each abstract type, a number of methods are defined that are used to operate on the components of fiber bundles by implementations of the data flow modules. These methods are used to iterate over elements of a fiber bundle's base space, to query for the cardinality and dimensionality of data spaces, and to return the fiber space element associated with a given element of the base space.

Task Specification

A task is the hint the scientist gives the intelligent subsystem as to what is desired from a visualization of a specified data set. Tasks are described in a high-level task-specification language that uses terminology from a scientist's particular field of study. Each specification written in the language consists of an operator and a list of fiber bundle domains upon which to operate.

For any given data set, a number of techniques may be employed to visualize the indicated domains of interest. The task operator is used to distinguish one visualization technique as being the most appropriate for the desired visual computation. For example, a *select* operator indicates that the scientist wishes to distinguish one data item from a collection of many. This task would lead to a visualization that allows data items to remain autonomous and easily distinguishable. Alternatively, a *correlate* operator indicates that the scientist is interested in determining if a general relationship exists between two or more data domains without requiring the autonomy of individual data items. Of the many task operators mentioned in the literature (Casner 1991; Wehrend and Lewis 1990), the Tioga system supports the operators *select, search,* and *correlate.*

In addition to the task operator, the scientist also needs to provide a list of the fiber bundle domains that are of interest at the current time. A given data set may be comprised of many variables, of which only a few are relevant to a particular study. Domains of interest are listed by the scientist in priority order to help guide the determination of an effective visualization. For instance, since the visual system is more accurate in resolving differences in position than differences in hue, glyph position would be used to encode the first, higher-priority domain, while glyph color would be used to encode the second in a task specification with two domains.

The Knowledge Base

The actual creation of visualization programs takes place within a component of the intelligent subsystem called the visualization planner. The planner utilizes an inferencing engine that has access to a knowledge base containing the concepts, facts, and rules needed to produce visualization networks from data and task specifications. Knowledge encoded in the knowledge base is divided into the following categories:

Data model: Knowledge regarding the fiber bundle data model. This includes the methods, type, structure, and extent of the base and fiber spaces of fiber bundles.

Visualization techniques: A taxonomy of general techniques appropriate for visualizing various types of fiber bundles. Examples of visualization techniques include scatter plots, graphs, isosurfaces, and so on, which are selected depending on the character and dimensionality of a fiber bundle's base space.

Task specification: Knowledge encoding the interpretation of the task language and the semantics of the task operators.

Data flow networks: Knowledge regarding the functional specification of the modules with which data flow visualization networks are constructed.

Domain knowledge: The semantics of fundamental variable domains such as temperature, latitude, longitude, and time. Also included are knowledge about conventions and symbology used within the application domain.

Visual perception: Perceptual considerations to be taken into account in forming visualizations. Examples include the spatial, temporal, and color acuities of the human visual system.

Output medium: Knowledge of the characteristics of various output media such as display monitors, printers, and film recorders.

User models: Information about the preferences and capabilities of the participants engaged in the visualization.

Visualization Planning

Data flow networks produced by the visualization planner are comprised of four sections, each of which includes one or more modules of a particular class. The first section is composed of reader modules that draw data out of a database and create appropriate fiber bundle objects. The second section is made up of data-manipulation modules that perform transformations on the fiber bundle objects. Such operations include the conversion of fiber space domains into base space domains, imposing an order on unordered sets, gridification of scattered data, and projections of fiber spaces. The third section of a network consists of modules that convert the fiber bundle objects into renderable forms composing a three-dimensional virtual environment. These forms then are displayed by the rendering modules that make up the fourth and final section of a data flow network. Renderable forms include geometric information, appearance attributes such as color and texture, and behavior specifications invoked when certain events occur. Renderable forms are used to define interactive objects that represent both data and scene "decorations" such as bounding volumes, axes, labels, and legends.

The creation of data flow visualization networks within the intelligent subsystem takes place according to the following six steps:

1. Determine a visualization technique—Based on rules in the knowledge base, an appropriate visualization technique is selected using information primarily about the type, structure, and extent of a fiber bundle's base space. Each technique determines a skeletal network that must be fleshed out with the determination of suitable parameters and appropriate module connections.
2. Add global manipulation modules—Determine the needed data-manipulation modules and their parameters. Depending on the visualization technique that is chosen, fiber bundles of one type may have to be transformed into bundles of another type.
3. Add renderable generation modules—Look at the fiber bundle structure and the semantics of the task to determine the renderable object(s) to use in the visualization. This process includes determining which domains of interest are to be mapped to the geometric, appearance, and behavior attributes of the chosen renderable forms. Additional data manipulation also may be specified depending on the input types required by individual renderable generation modules.

4. Add decoration modules—Add scene decorations such as axes, labels, and legends in order to clarify the visualization.
5. Add connections—Connect the modules selected in the previous steps together to make a single data flow network.
6. Create a network script—Translate the intelligent subsystem's internal representation of a network into a visualization system-specific script format.

RESULTS

A prototype server for the intelligent subsystem has been developed as a front end to the AVS visualization system. In this prototype, the visualization planner uses the CLASSIC knowledge representation system (Resnick et al. 1993) to manage the knowledge base and Common LISP (Steele 1990) to implement the planning algorithm. Currently, the knowledge base contains limited amounts of knowledge about the fiber bundle data model, a taxonomy of visualization techniques, the task language, the construction of data flow networks, and the problem domain of Earth science.

Figure 86-1 shows a visualization created by the prototype Tioga system for a relational data set representing California-based hydrological measurement stations. The intelligent subsystem used the task specification *select (longitude, latitude, elevation, stationType).* Although relational data do not have an intrinsic coordinate space associated with them, the planner chose to represent the data set using a scatter plot by restructuring the data set as scattered data embedded in a three-dimensional coordinate space. This visualization technique required manipulating the data set by "promoting" to base space domains some of the fiber domains listed in the task specification. Based on the rules in the knowledge base, the longitude, latitude, and elevation domains were promoted to be the X, Y, and Z axes of a virtual environment, respectively. In addition, the planner decided to use spheres to represent the location of individual stations (retaining autonomy for selection) and to map the station type to the color of the spheres.

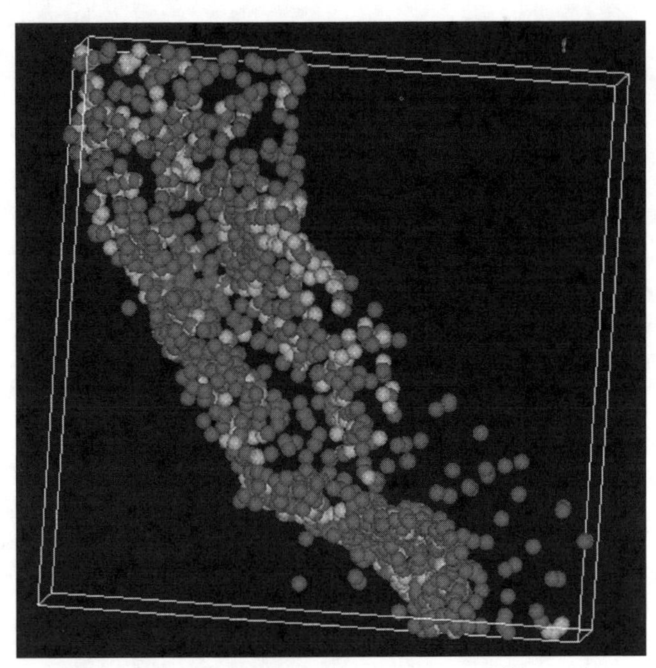

Figure 86-1. Visualization of hydrological survey stations in California, color coded by station type. This image was produced by the Tioga system when given the task *select (latitude, longitude, elevation, stationType).*

Figure 86-2 shows another visualization created by the Tioga system for a data set output from a regional climate model program. The data set is a three-dimensional array indexed by latitude, longitude, and elevation in which each array element is a record containing wind vector, cloud density, water content, and temperature values. Given the task specification *search (water-content),* the intelligent subsystem took advantage of the three-dimensional structure of the base space to create a coordinate space plot. In the case of base spaces, which inherently contain a coordinate space, it is assumed that the base space is to be used for the spatial component of the visualization. With the given base space, the given task, and the large number of data points, the planner chose to represent the water content data as an isosurface.

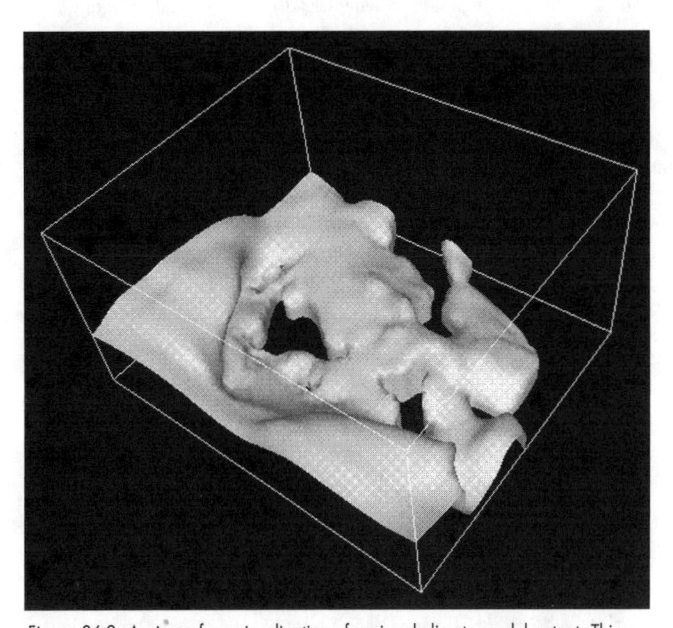

Figure 86-2. An isosurface visualization of regional climate model output. This image was produced by the Tioga system given the task *search (water-content).*

CONCLUSION

The intelligent subsystem within the Tioga visualization management system is still in the very early stages of its development. The prototype that currently is implemented is a fully functional system but only in a very limited sense. While its results are quite preliminary, there is guarded optimism about the potential of the system. With a more functional version of the system, it is believed that scientists will be able to perform exploratory visualizations of data in a manner similar to how one currently performs "what-if?" analyses of data using a spreadsheet package, and they will be able to do it quickly.

Even so, there is no expectation that effective visualizations will be automatically generated in all cases. A more realistic goal for the time being is to try to satisfy the "90/90" rule, that is, to provide scientists 90% of their visualization management needs 90% of the time. Because of the fallibility of automated systems, the intelligent subsystem must provide adequate feedback to scientists as to how well their visualization task requests have been satisfied. If requests are not adequately handled, scientists always have the option to edit by hand whatever data flow networks are produced by the intelligent subsystem.

One important lesson that has been learned in the initial development of the intelligent subsystem within Tioga is the difficulty in performing knowledge engineering for interactive data visualization. Unfortunately there are no firmly established rules for creating data visualizations, and even if a set of *ad hoc* rules is determined, its implementation within a knowledge representation system is difficult because of the great many options available for encoding knowledge. Currently, knowledge representation systems have trouble integrating different knowledge paradigms, such as those based on rules, forward and backward inferencing, fuzzy logic, and constraints, into a coherent whole. While each of these paradigms is useful for creating and accessing certain types of knowledge, none appears to be robust enough when used alone to fully handle interactive visualizations of Earth science data. At best, knowledge engineering is still a black art.

Nonetheless, the intent of the Sequoia visualization research group is to continue to expand the prototype visualization server into the full Tioga system. As the rate at which Earth science data are generated increases, the need for systems like Tioga will become essential if the data are to be put to use to improve global environmental conditions. There seems to be no other recourse than to build in expertise so that scientists can rapidly understand the processes that shape planet Earth.

Acknowledgments

This work was supported by the Digital Equipment Corporation, the University of California, and the San Diego Supercomputer Center. Special thanks to Mike Bailey, director of the SDSC visualization group, for his tremendous support. Thanks also to Jonathan Shade and Colin Sharp for their hard work in helping turn designs into reality. Finally, our appreciation goes out to John Roads and Larry Riddle of Scripps Institute of Oceanography for providing the data sets used in the examples.

References

Advanced Visual Systems, Inc. 1992. *AVS User's Guide.*

Beshers, C., and S. Feiner. 1992. Automated design of virtual world for visualizing multivariate relations. In *Proceedings, Visualization '92 Conference*, 283–90.

Butler, D. M., and S. Bryson. 1992. Vector bundle classes form powerful tools for scientific visualization. *Computers in Physics* 6(6):576–84.

Butler, D. M., and M. H. Pendley. 1989. A visualization model based on the mathematics of fiber bundles. *Computers in Physics* 45–51.

Casner, S. M. 1991. A task-analytic approach to the automated design of graphic presentations. *ACM Transactions on Graphics* 10(2):111–51.

Dozier, J. 1990. Looking ahead to EOS: The Earth observing system. *Computers in Physics* 4(3):248–59.

———. 1992. *How Sequoia 2000 Addresses Issues in Data and Information Systems for Global Change.* Sequoia 2000 Technical Report 92/14. Berkeley: University of California.

Haber, R. B., B. Lucas, and N. Collins. 1991. A data model for scientific visualization with provisions for regular and irregular grids. In *Proceedings, Visualization '91 Conference.*

Kochevar, P. et al. 1993a. Bridging the gap between visualization and data management: A simple visualization management system. In *Proceedings, Visualization '93 Conference.*

———. 1993b. *A Visualization Architecture for the Sequoia 2000 Project.* Sequoia 2000 Technical Report 93/XX. Berkeley: University of California.

Mackinlay, J. D. 1986. Automating the design of graphical presentations of relational information. *ACM Transactions on Graphics* 5(2):110–41.

Olson, M. et al. 1992. Visualization Benchmark Specification for the Sequoia 2000 Project. Unpublished report.

Ousterhout, J. 1990. Tcl: An embeddable command language. In *Proceedings of the 1990 Winter USENIX Conference.*

———. 1991. An X11 tool kit based on the Tcl language. In *Proceedings of the 1991 Winter USENIX Conference.*

Resnick, L. et al. 1993. *CLASSIC Description and Reference Manual For the Common LISP Implementation.* Murray Hill.

Senay, H., and E. Ignatius. 1992. *Vista: A Knowledge-Based System for Scientific Data Visualization.* Technical Report GWU-IIST-92-10. George Washington University.

Silicon Graphics, Inc. 1992. *Explorer User's Guide.*

Steele, G. L., Jr. 1990. *Common LISP: The Language*, 2nd ed. Digital Press.

Stonebraker, M., and G. Kemnitz. 1991. The POSTGRES next-generation database management system. *Communications of the ACM* 34(10):78–92.

The Khoros Group. 1992. *Khoros User's Manual.* Department of Electrical and Computer Engineering, University of New Mexico.

Wehrend, S., and C. Lewis. 1990. A problem-oriented classification of visualization techniques. In *Proceedings, Visualization '90 Conference.*

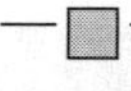

Len Wanger and Zahid Ahmed
San Diego Supercomputer Center
Peter Kochevar
Digital Equipment Corporation

EPILOG

One of the objectives of the 1993 Second International Conference/Workshop on Integrating GIS and Environmental Modeling, from which these chapters come, was to review recent progress in integrating GIS and environmental modeling. Another was to identify directions where progress might be made in the near future. Are GIS and environmental models more integrated than they were in 1991 at the conference in Boulder? Clearly they are, as the large number of chapters reporting experiences with coupled implementations of GIS and environmental modeling software demonstrate. The number and sophistication of such examples clearly increased many times over during this two-year period. The chapters in Part III are a clear indication of the wealth of new ideas that are just beginning to influence the field. On the other hand, it is difficult to see concrete measures of progress or to know how to answer the all-important question, "How will we know when we are done?" Will the process end with the development of a fully integrated modeling package, or will software integration always be in a state of flux—weighing the convenience of a single package against the disadvantages of less than optimal performance, or perhaps because of the "not invented here" syndrome.

On the final day of the 1993 Breckenridge conference, the organizers scheduled an open discussion to review the conference and to develop ideas for future activities. A series of preliminary meetings were held, and a questionnaire was completed by conference participants to help us identify issues of major concern. We would like to focus here on that discussion and the resulting suggestions in regards to GIS in the future.

The strongest consensus among participants was the general issue of data access. Standards, and particularly standards for metadata, reflect the concern of the spatial database community with the lack of generally accepted ways of describing the characteristics of data sets to others, particularly when producer and user of data are connected by such a tenuous link as the Internet. Some of the most striking examples of this problem occur in environmental modeling, such as the atmospheric scientist's need for access to data created by soil scientists, possibly for very different purposes. The development of the Federal Geographic Data Committee's proposed metadata standard is a major step forward, but it needs to be coupled with a greater willingness on the part of scientists to see that their data are archived, documented, and made accessible to others. Perhaps funding agencies could encourage this through attaching explicit data policies to grants.

The participants ranked metadata, standards, and accessibility highest among the six data-related issues on the questionnaire. Data quality was fourth, followed by the need for information on data lineage. In addition, participants were concerned about problems caused by lack of access to proprietary data formats, lack of generally accepted rules for aggregation and disaggregation of data, and lack of good techniques for conflation of data from different sources of varying quality.

Unlike many smaller countries, the United States has no generally accepted and simple grid system that can be used as a framework for environmental science. Environmental data for the United Kingdom can be georeferenced to a standard rectangular grid system, but the size of the United States and the importance of Earth curvature make this impossible, even for the conterminous states. UTM (Universal Transverse Mercator coordinate system) is widely used but problematic when zone boundaries must be crossed, and latitude/longitude are similarly inconvenient. Participants argued for research leading to the development of a national grid for the United States, perhaps along the lines of the sampling scheme devised for EPA's EMAP project.

Other participants stressed the need for a concerted national effort to improve the availability of data on widely used environmental variables, particularly climatic means, which are critical for environmental characterization, land use evaluation, and simulation modeling. They urged relevant government agencies to support development of a standard spatiotemporal data set of climatic variables, using a gridded format at a spatial resolution of 1 km, and focusing initially on mean monthly minimum and maximum temperatures and precipitations. Such climatic databases already have been developed for substantial portions of the earth's land surface, including Mexico, Africa, China, and Australia, and are playing a vital role in environmental analysis and policy formulation. The conference as a whole endorsed this proposal.

A related concern surfaced about the availability of social and economic data, which are vital for understanding the anthropogenic sources of environmental change, and the need for research on the effects of change on human populations. Social and economic data also could be made available on a gridded basis rather than on the basis of existing and arbitrarily shaped reporting zones that vary enormously in size and have little to do with environmental processes.

Finally, participants were strongly supportive of the notion that the means of processing data be encapsulated as far as possible with the data themselves. For example, climate data collected at sample points are almost invariably interpolated spatially before being used for analysis or modeling. The method of interpolation is traditionally selected by the user rather than the producer, and increasingly the user is likely to be from another discipline with little understanding of climatic processes. It would be far more effective if the means of interpolation were determined by the data producer and encapsulated with the data for dissemination. This concept, which originated in the object-oriented paradigm of computer science, seems to have great promise for the informed use of environmental data.

In an attempt to identify the functions that most need to be added to GIS to improve support for environmental modeling, David Maidment asked each willing respondent to imagine they were investing in software development and to allocate $100 among various categories. Forty-two people responded, and again the issues of standards emerged as the most important. On average, respondents allocated $20 to support open interoperability between systems, using open data structures that could be addressed externally. They allocated $18 to the development of a standard subroutine library of modules that could be linked into environmental models, and $16 to standardized linkages between packages. The remaining twelve topics fell well below these, although they included macrolanguages ($6), functions to handle time ($6), simplicity of user interfaces and command languages ($6), support for metadata ($5), and support for three-dimensional representations ($5). At the bottom of the list were support for remote operation ($2), implicit environmental functions ($2), and handling of discrete events ($2).

One widely supported suggestion was that a group be formed to monitor the progress of the GIS software industry in developing better support for environmental modeling. This might be done by a committee, perhaps under the auspices of the National Academy of Sciences, or by an academic research organization such as the NCGIA; such a group could report their findings at the next conference. Participants thought that better methods of communication were needed among the environmental modeling and GIS research communities, and they proposed the publication of new journals, special issues of existing journals, and new electronic media.

In particular, this discussion drew attention to the interdisciplinary nature of the group and the impediments that exist for fields that inherently must cross disciplinary lines. Young scholars are often discouraged from interdisciplinary activity because of their need to meet the expectations of tenure committees structured along traditional disciplinary lines; funding agency review panels tend to focus support on the central agendas of traditional disciplines. Therefore, the following resolution was passed by conference attendees:

A new research discipline involving fundamental scientific issues is being created by the synthesis of geographic information systems and environmental modeling. National and international research agencies, such as NSF and NIST in the United States, need to examine what programs are required to support this field.

Whether a lasting research discipline, such as environmental modeling, will evolve from the intersection of GIS with other disciplines, remains to be seen. It will depend in part on substantial progress being made in the next few years, both in finding new solutions to the technical problem of integrating software and in resolving outstanding research issues. If nothing else, this book surely demonstrates the significance of such progress in our ultimate goal of understanding how the systems function that create the environment of this planet.

The editors look forward to the next conference in January 1996 in Santa Fe, New Mexico, and the progress we hope will be reported there.